THE BIRDS OF TURKEY

The Ornithological Society of the Middle East was formed in 1978 as a successor to the Ornithological Society of Turkey and was expanded in 2001 to cover the Caucasus and Central Asia. The Society aims to focus attention on the birds of the Middle East, the Caucasus and Central Asia, partly through the publication of *Sandgrouse*, which contains papers and news on birds and bird conservation throughout the region. The Society also encourages interest in the ornithology of the Middle East through its financial support of surveys and research by individuals and organisations. A number of Turkish projects have received assistance, from the provision of GPS tracking units for atlas mapping to financial support for raptor counting projects on the Bosphorus. Help is also given to a number of birdwatchers in the region through a supported membership scheme, bringing the Society's literature to a wider audience. A number of young Turkish birdwatchers are supported in this way. OSME's website carries articles and news about birds and birding within the region, including trip reports and papers from *Sandgrouse*. There are plans to translate a number of key pages into several languages, including Turkish.

Membership costs £15 for those in the United Kingdom, £20 for those in Europe (including Turkey, Cyprus and Central Asia), and £25 for other countries. For more details visit www.osme.org or email secretary@osme.org.

OSME

The Birds of TURKEY

The Distribution, Taxonomy and Breeding of Turkish Birds

Guy M. Kirwan, Kerem A. Boyla, Peter Castell, Barbaros Demirci, Metehan Özen, Hilary Welch and Tim Marlow

with additional contributions from

Güneşin Aydemir, Sancar Barış, Bahtiyar Kurt, Gernant Magnin, Richard F. Porter and Geoff Welch

CHRISTOPHER HELM
LONDON

First published in 2008 by Christopher Helm, an imprint of Bloomsbury Publishing,
50 Bedford Square, London WC1B 3DP

Reprinted 2011, 2012, 2015

www.bloomsbury.com

Bloomsbury Publishing: London, New Delhi, New York and Sydney

A CIP catalogue record for this book is available from the British Library

Commissioning Editor: Nigel Redman
Project Editor: Jim Martin

Design by J&L Composition, Filey, North Yorkshire

IISBN (print) 978–1–4081–0475–0
ISBN (epub) 978–1–4081–0894–9
ISBN (e-pdf) 978–1–4081–3314–9

Printed and bound in Great Britain

This book is produced using paper that is made from wood grown in managed sustainable forests. It is natural, renewable and recyclable. The logging and manufacturing processes conform to the environmental regulation of the country of origin.

MIX
Paper from
responsible sources
FSC® C013604

Cover illustrations by John Gale
Front: Black-bellied Sandgrouse; spine: Krüper's Nuthatch; back: Caucasian Grouse

CONTENTS

DEDICATIONS

For Richard Porter, who pioneered so much of modern field ornithology in Turkey; Simon Albrecht, passionate conservationist, who dreamt of a Turkish bird atlas; Veronica Nogueira, for her treasured companionship; and, above all, for my dearest love, Marco Alejandro Kirwan Chávez, who does not yet (and perhaps never will) understand why his father spends all his time on walks gazing into the trees in search of birds.

GMK

PREFACE

by Guy M. Kirwan

Turkey (or Asia Minor as it was known to earlier visitors) has been subject to prolonged ornithological exploration, described in great detail by Hans Kumerloeve in a long paper in the 1970–73 Turkey Bird Report, and brought up to date by Richard Porter *et al.* in one of the introductory chapters to the present work. In contrast, my own first visit to Turkey was as recent as 1990, immediately following the completion of my university days. It was suitably eventful, as we lost a hire car to floods that claimed the lives of over 30 people in the mountains south of Trabzon, watched Green Warblers in the garden of an old bear-hunter whilst waiting for the waters to subside, and yet still drove many thousands of kilometres in search of some of the most exciting birds in the Western Palearctic. In a sense my heart never returned from that trip. This book's genesis came just a couple of years later (by which time I had already contrived to spend many more months in Turkey in a successful effort to escape regular employment), through a conversation with Rodney Martins, who had long harboured the notion of producing a Turkish avifauna.

This book might be fairly stated to possess several predecessors, all of which laid important groundwork. Unarguably chief amongst these is Kumerloeve's 1961 *Zur Kenntnis der Avifauna Kleinasiens* which, until the present work, is the only one to have attempted a complete review of the status, distribution and taxonomy of all Turkish birds. Of enormous importance and usefulness for recent researchers, as well as birdwatchers, has been Kees Roselaar's *Songbirds of Turkey* (1995), a work that provided, for the first time, detailed maps of the breeding distributions of all passerines known (or thought) to nest in the country. Even more crucially it offered the only in-depth study of the taxonomy of Turkish passerines, apart from that of Kumerloeve, and thus became a massive building brick in the development of the current project. Finally, we come to *Die Vögel der Türkei* (1992) by Max Kasparek, another German worker with almost as prodigious an output on Turkish avifauna as his predecessor Kumerloeve. Although his work offers only brief text and nothing on taxonomy, it covers all species and brought much of Kumerloeve's work up to date for the modern era. The ranges and temporal distributions of many species have been subject to only comparatively minor 'tweaks' in the intervening decade. I must also mention the pivotal role of the Ornithological Society of Turkey (OST), subsequently the Ornithological Society of the Middle East (OSME), in the development of modern Turkish ornithology. Principally through publication of the bird report, both OST and OSME have brought regular updates and refinements to our knowledge of Turkish bird status and distribution to a wider audience and, by acting as a repository for huge quantities of data generated by birdwatching visitors, have enabled a significant part of the present work.

Like all such long-term projects, started with near-boundless enthusiasm and energy before the realisation of the long hard slog ahead seeps inevitably in, *The Birds of Turkey* has enjoyed innumerable ups and downs, sometimes seeming almost to move only backwards. That it has been completed at all is testimony to the fantastic input that the project received at a comparatively late stage from a number of Turkish colleagues, some of whom joined as co-authors. Their role, and the assistance they have received from their fellow Turks, has been of keynote importance. It is a particular pleasure to witness the vast sea-change in the number and skills of home-grown observers. Recent years have seen the rise of Turkish-language periodicals on birds, the publication of a new Important Bird Areas inventory and national Red Data List authored solely by Turks, and co-editorship and much closer involvement with the production of the bird report. Turkish ornithology is now, at last, largely in the hands of Turks themselves.

Outside mainland Europe, some of the most interesting countries in the Western Palearctic, namely Armenia, Israel, Morocco and Egypt, have been the subjects in recent years of major overviews of their avifauna. For me, Turkey, long recognised by travelling birdwatchers as one of the most interesting and dramatic countries within this zoogeographical region and truly a place where east meets west (both avifaunally and culturally), needed such a book too. My hope is that the present work may be seen to nestle on a par with the likes of *The Birds of Morocco* and *The Birds of Egypt.*

ACKNOWLEDGEMENTS

This work has suffered (or enjoyed) an unusually long gestation period. My fieldwork in Turkey commenced in 1990 and continued on a near-annual basis until 1999, with more occasional visits since then. During this period, principally the first few years, I benefited immeasurably from a close cooperation with the former BirdLife International partner in the country, the now-defunct Doğal Hayatı Koruma Derneği, and in particular with several of its Bird Section members, especially Güven Eken (who also drafted four of the gamebird accounts), Gernant Magnin and Murat Yarar.

The following joined me in fieldwork in Turkey, sharing in many fascinating discoveries, a few tribulations, much drinking of çay, numerous scrapes with Turkish sheepdogs on the Central Plateau and many long days, sometimes for little tangible reward. They are: the late Louise Augustine, Sancar Barış, Vincent van den Berk, Nick Blackwood, Kerem Ali Boyla, Chris Bradshaw, Duncan Brooks, Mick Davies, Paul Doherty, Güven Eken, Mark Elwonger, Mike Flieg, Andrew Grieve, Burak Güneş, Suhendan Karauz, Cem Kiraç, Andrew Lassey, Karin Linders, Gernant Magnin, Tim Marlow, Rodney Martins, Uygar Özesmi, Raymond Pahlplatz, Firdous Raja, Don Rathbun, Dawn Ross, Malcolm Roxby, Naomi Stuiver, Lars Svensson, Graham Tucker, René Vos, Geoff & Hilary Welch, Jan van der Winden and Murat Yarar, with apologies for any unintentional omissions. In addition to occasional bird tours, the following have helped finance my fieldwork: the Bird Exploration Fund, Doğal Hayatı Koruma Derneği, Paul Doherty (Bird Images), the Field Museum of Natural History (Chicago) and the Ornithological Society of the Middle East, which organisation also sponsored some of my museum research at a time when personal funds were running low. Special gratitude is due to Julian Francis for sponsoring the book's colour section.

An early start on the text was made in 1992. The following have offered all manner of information, responded to queries and generally assisted my endeavours: Alexander Abuladze, Aydin Akin, Vasil Ananian, Mark Bakker, Özge Balkız, Dawn Balmer, Matthias Bechtolsheim, Soner Bekir, Arnoud van den Berg, Rob Bijlsma, Z. Boev (National Museum of Natural History, Sofia), Leo Boon, Colin Bradshaw, Mehmet Çetinkoç, Paul Collin, Pete Combridge, José Luis Copete, Andrea Corso, Dr René Dekker (National Museum of Natural History, Leiden), Edward Dickinson, Lieuwe Dijksen, Hugues Dufourny, Dr Renate van den Elzen (Alexander Koenig Research Institute & Museum of Zoology, Bonn), Gunnar Engblom, Mustafa Erturhan, Dr Clem Fisher and Tony Parker (Liverpool Museum, National Museums & Galleries on Merseyside), Peter Flint, Anita Gamauf (Natural History Museum, Wien), Steve Gantlett, Ian Green, Steven Gregory, Kari Haataja, Moira & Roy Hargreaves, Tom van der Have, Peter Hayman, Peter Heathcote, Mary Hennen, John Bates and David Willard (Field Museum of Natural History, Chicago), Mogens Henriksen, Chris Hewson, Janet Hinshaw (University of Michigan Museum of Zoology, Ann Arbor), Remco Hofland, Paul Hope, Stan Howe, Jocelyn Hudon (Provincial Museum of Alberta, Edmonton), Fatih Izler, Philip Jackson, Antoine Joris, Igor Karyakin, Max Kasparek, Shannon Kenney (American Museum of Natural History, New York), Okan Koçyiğit, Ahmet Kılıç, Andrew Knowles-Brown, Miriam Langeveld, Vladimir Loskot (Zoological Institute, Russian Academy of Sciences, St. Petersburg), Klaus Malling Olsen, Henry McGhie (The Manchester Museum, University of Manchester), Krister Mild, John Muddeman, E. N. Panov, Nik Peet, Göran Pettersson, Stefan Pfützke, Richard Porter, Richard Ranft and Nadine Svoboda (British Library National Sound Archive, London), Pamela Rasmussen, Mike Reed, Magnus Robb, the late Mike Rogers, the late Dave Rushforth, Konrad Sachanowicz, Napier Shelton, Jevgeni Shergalin, Hadoram Shirihai, Roy Slaterus, Gunter de Smet, Frank Steinheimer, Peter Stewart, Lars Svensson, Ray Symonds and Michael Brooke (Cambridge University Museum of Zoology), José Tavares, Rolf Uhlig, Magnus Ullman, Edwin Vaassen, Michael Warren and Emin Yoğurtçuoğlu. I also must thank all of the observers, both foreigners and Turks, who have submitted bird records to the Turkey Bird Report since its inception; without their contributions the present task would probably have proved impossible. Rodney Martins drafted a number of the species accounts and several of the taxonomic remarks. The following commented knowledgeably on the developing texts: Duncan Brooks, Phil Chantler, David Christie, Mark Cocker, John Faldborg, Andy Green, the late Gunter Groh, Süreyya Isfendiyaroğlu, Urban Olsson, Denis Summers-Smith and Geoff Welch. Rosamund Taylor translated some material from French, as did Ruth Maier and David Christie from the German language. Effie Warr, as always, provided absolutely sterling assistance in the library of the British Museum (Natural History), now The Natural History Museum, at Tring. Mark Adams, Robert Prŷs-Jones, Douglas Russell and Michael Walters generously offered their help with the collection of the same institution and Effie's successor as librarian, Alison Harding, was also most helpful during my visits there. At A&C Black, I am grateful to Nigel Redman, for keeping the faith in a relatively old-fashioned ornithological work, Jim Martin, for steering the project to a successful conclusion and, in particular, for revising the maps to incorporate many new data and for his assistance in organising the colour section, and Ernest Garcia for copy-editing the work. The typesetters, J&L Composition Ltd, transformed the original text and maps into a 'real' book and coped admirably with the many requested changes in proof.

Finally, I turn to my most personal debts of gratitude. In the early years, I was frequently made to feel a welcome visitor in the homes of Sancar Barış and Uygar Özesmi, in Samsun and Ankara, respectively. Both provided much welcome conversation, of birds and all manner of things Turkish, and helped me to overcome many logistical difficulties. The irrepressible Gernant Magnin was an always stimulating if suitably irreverent companion in Istanbul and all over western and central Turkey, and provided many useful comments on both the maps and species accounts in the later stages. Vincent van den Berk and family, and Karin Linders, made me a welcome part of the Göksu Delta project team in 1991. Ruth Bimson was always understanding of my long absences from the UK (not only in Turkey) throughout much of the 1990s. The entire Chávez family again made me a welcome guest in their home, in the Ciénaga de Zapata of Cuba, as did Arturo Kirkconnell, Rosita Posada and family in La Habana, during March–May 2003, in which time I made my first comprehensive revisions of the text for a number of years, in order to finally prepare the book for publication. My mother and late father have always supported my lack of conformity to society's norms and my decision to become an itinerant 'bird person'. My current partner, Veronica Nogueira Gama, has been ever understanding, patient and supportive of my need to spend yet more time at the computer, and stoically forbore several long absences in the museum and field, in my quest for still more data to make the book a little more complete. Some day, I will repay you for all the lost hours!

Guy M. Kirwan

The map data were based on information supplied by Guy Kirwan, Hilary Welch (the WWF Mediterranean Priority Forest Areas project, 2000–03, and GAP data), Gernant Magnin, Heike Tholl-Schmidt, Emin Yoğurtçuoğlu, Franziska Arıcı, Derya Kahraman, Barbaros Demirci, Sancar Barış, Nuri Özbağdatlı (Great Bustard), and Güven Eken (Konya Basin). I also wish to thank *Atlas* magazine for providing Turkey vegetation maps.

Kerem Ali Boyla

The 'breeding' sections have been compiled from published references, including birdwatchers' trip reports, but a considerable amount of information is from previously unpublished sources. Particularly useful sources of published material have been the Turkey Bird Reports (published by OST and more recently OSME), the 12 booklets in the Birds of Turkey series (published by Kasparek Verlag, Heidelberg), the books on Sultan marshes (1985) and Bafa Gölü (1988) by M. Kasparek, and the reports on surveys of the Kızılırmak Delta (Hustings & van Dijk 1994), the Çukurova region (van der Have *et al.* 1988), Uluabat Gölü (Welch & Welch 1998), and the Çamlihemşin-Ardeşen area of northeast Turkey (Faldborg 1994). Papers in *Ibis* by Danford (1877–78, 1880) and Selous (1900) provide a fascinating comparison with modern times. Contributions by K. Warncke (particularly 1964–65) contain important breeding records for several species, and papers by Kiziroğlu (1982, 1984, 1987) contain a wealth of data on species breeding in nest boxes in the Ankara region.

The meticulous personal diaries of Captain J. H. McNeile, held in the Edinburgh Museum, have been invaluable. He searched for nests for about seven weeks in the Aegean region in both 1950 and 1951 and then made shorter trips to various parts of Turkey in 1967, 1968, 1970, 1972 and 1973; he was 81 years old on the last of these. He made important contributions on the breeding biology of several species, notably Crimson-winged Finch, Woodchat Shrike and Rüppell's Warbler.

Richard Porter has painstakingly supplied extensive breeding data from his notebooks for the years 1966, 1967, 1969, 1970, 1973, 1975 and 1999. Many of these records are credited to him in the texts, but most were included in the published reports of the time, particularly the earlier Turkey Bird Reports, in which he played a considerable role; the start of modern ornithology in Turkey.

Geoff & Hilary Welch kindly searched the records of the GAP Biodiversity Project, the WWF Mediterranean Important Forest Areas Project, KuşBank, and their own field notes, and supplied important details for several under-recorded species. Henry McGhie (Manchester Museum), Bob McGowan (Edinburgh Museum) and Douglas Russell (NHM, Tring) allowed access to data and material as required.

Neil Burke, Malcolm Cowlard, Michael Demain, Jim Hodson, John Hollinrake, Tom Quittenden and Jim Whitaker have supplied details of their various nest-finding trips. My own nest-finding trips to Turkey have totalled 35 weeks between 1992 and 2007. Some of these (especially the longer ones) have provided some unique breeding records for Turkey and also much new information on breeding biology in general; with combinations of various friends in 2004, data were obtained on over half of the regular breeders. Special thanks are due to the most regular participants: Bernard Pleasance, Steven Williams and Lynda Castell; and many thanks also to Lawrence Baxter, Richard Castell, Grant Demar, Andrew Dixon, Mark Farmer, David Gregson, Barbara Hudson, Graham Jones, John Mason, Tim Maunsell, Terry Park, Tom Quittenden, Keith Robson, Alan Rustell, Brian Standley, Kenny Taylor, Roger Thomas and Nick Williams.

Pertti Koskimies and Roger Burch translated papers from the German language. Richard Castell assisted in the selection of photographs. Lynda Castell typed numerous drafts and revisions.

Peter Castell

My thanks go to Bahtiyar Kurt and Özge Balkız for their support, and Berat for her motivation.

Barbaros Demirci

I particularly wish to thank my lovely wife and daughter for their indispensable support.

Metehan Özen

I thank Geoff Welch and Süreyya Isfendiyaroğlu; all those who worked with me on the GAP Biodiversity Research Project, as well as the South East Anatolia Regional Development Administration (who commissioned the project), the United Nations Development Programme (who funded it), Doğal Hayatı Koruma Derneği (DHKD) and WWF Turkey (for their assistance with the project's implementation), the many Turkish birdwatchers involved in the fieldwork who helped to collect the data, and the close collaboration we enjoyed with the Turkish military stationed in the south-east. For cooperation on the Priority Forest Areas of the Mediterranean project, I thank again DHKD/ WWF Turkey, the British Consulate, FCO (who funded the project), as well as Geoff Welch, Jim Scott and Rick Vonk (all from the Royal Society for the Protection of Birds), who between them collected all of the data. Many thanks to the growing number of Turkish birdwatchers for submitting records, which have contributed immeasurably to the knowledge presented in this book. We hope the book deepens and encourages your interest in birds so that future editions will be even more complete!

Hilary Welch

I particularly thank Sarah Ward and Mark Baynes for their support and encouragement.

Tim Marlow

THE BREEDING SEASON

by Peter Castell

The breeding season in Turkey is essentially similar to that in other countries in southern and eastern parts of the Western Palearctic, i.e. nesting is mainly in spring and early summer.

The earliest eggs are laid in January (White-tailed Eagle, Lammergeier and Eurasian Griffon Vulture). A few species (seven) lay their earliest eggs in February (including Great Cormorant, Dalmatian Pelican, Grey Heron and Bonelli's Eagle). A further 34 species lay their earliest eggs in March, including European Shag, Eurasian Spoonbill, Greylag Goose, Eurasian Black Vulture, Eastern Imperial and Golden Eagles, Kentish and Greater Sand Plovers, Caspian Gull, Eurasian Eagle, Tawny and Long-eared Owls, Mistle Thrush, Great Tit and Rook. Thereafter the peak of egg-laying is in April, especially in the second half of the month (and in early May); by the end of April 187 species (62%) will have laid their earliest eggs; this total includes most ducks, some eagles, harriers and falcons, waders, gulls, some terns, owls, woodpeckers, larks, thrushes, tits, crows and finches. A further 99 species (33%) lay their first eggs in May, including swifts, bee-eaters, martins, wheatears (some), warblers (most), flycatchers, and shrikes. A few species (*c.*12) lay their first eggs in June, including Velvet Scoter, Eurasian Hobby, Corncrake, Alpine Accentor, Common Rosefinch and Grey-necked Bunting; and European Honey Buzzard and Lesser Spotted Eagle probably also do so. Eleonora's Falcon probably lays eggs in July. Thus, the overwhelming majority of species (over 80%) lay their first eggs in April and May, but these are the earliest records, and many species continue to lay eggs throughout May and June and even July. The peak of the breeding season is May and June and this is when activity is most evident.

Many species are double- (or even treble-) brooded and have a long breeding season. Little Grebes lay eggs from early April but have also been observed incubating in mid August. Graceful Prinias lay their first eggs in March, with pairs nesting throughout the spring and summer, and young in one nest as late as 11 September. Common House and Eurasian Crag Martins have been recorded at nests in September. Little is known concerning the interval between broods in a season, and this may be confused by replacements following failure. There may be several weeks between broods, but a male Isabelline Wheatear was observed feeding recently fledged young in Inner Anatolia while his mate was building a nest nearby for a second brood.

The breeding season is also protracted for certain single-brooded species. Black-winged Stilts and Pied Avocets lay their earliest eggs in late April (mainly in May), but there are records of eggs throughout June and even in mid July, the latter probably replacement clutches following earlier failure.

Many general factors are known to influence the timing of the breeding season: hours of daylight, latitude, height above sea level, temperature, growth of vegetation and the availability of food for nestlings, and many of these are inter-related. Where species have a general (or more widespread) distribution in Turkey, the earliest breeding records are usually in warmer coastal and lowland areas in the west and south of the country, with the latest records in the more mountainous areas of the east and northeast; there are many examples of this in the species texts.

The breeding season for European Stonechats on Uludağ (Marmara) is about one month earlier at 800m than at the summit (Jetz 1995). One would expect ground-nesting species at high altitude to commence breeding later, and many do so (e.g. Alpine Accentor), but Horned Larks lay eggs from mid April, and Caspian Snowcocks from April.

For certain summer visitors (*e.g.* Black-headed Bunting) there is evidence that nesting commences almost immediately on arrival on the breeding grounds. Other species (e.g. European Reed and Great Reed Warblers) may have to wait for growing vegetation in which to conceal their nests. Greater Flamingos have been recorded laying eggs from early April to mid June and timing may be dependent on water levels.

Considerably more is known about the start of breeding than the end. Birdwatchers tend to visit Turkey during April to June, with generally less observer activity in the hot months of July and August. There are significant gaps in this latter period, with for example no observations of fledged young of Corn Bunting. Richard Porter has heard many species singing in September, including European Goldfinch, Cirl Bunting and Eastern Black-eared Wheatear, and October singing by Finsch's Wheatear and Woodlark; some species will, of course, sing when not breeding, but few people would consider looking for nests at this time.

There is a danger that the breeding (and other) records will be directly correlated to the movements of birdwatchers. There is a wealth of data from well-watched areas, such as the Göksu Delta, but a distinct absence of records from large parts of the east. There is, for instance, a shortage of records for April and May in eastern Turkey (most birdwatchers visit in June). Wherever possible, records have been backdated to laying of first eggs, in order to provide a level baseline for comparison between the different regions.

Challenges. Despite the efforts of so many people, much breeding information is still required. Turkey is a huge country and habitats can be harsh – vast steppe and marsh in the burning heat of summer; reeds towering 3–4 m above waist-deep, leech-infested water; steep hillsides clothed in endless mature forests in the Black Sea Coastlands; and craggy peaks in the Taurus, snow-covered even in June. No surprise that the only detailed records of nests of Caspian Snowcock are from 1876; there is much to be learned about this species in Turkey, but this is a job for the young and very fit! There are many other challenges. Little is known about the breeding biology of Brown Fish Owl but it may nest in steep, rocky, forested river valleys in the Mediterranean region. Is River Warbler merely a (singing) migrant in Turkey or does it breed occasionally? Spotted Crake may breed, but there is no proof. Does White-tailed Lapwing breed regularly on wet steppe near lakes in Inner Anatolia? Some former breeding areas have been drained, but many potential sites remain. Only five species, all dry-land birds, have been recorded as host to the Common Cuckoo, but surely there are many others. Mediterranean Shearwater is assumed to breed on offshore islands or on mainland cliffs in Aegean and Mediterranean, but colonies are undiscovered. These are merely a few examples but we hope that this book will inspire others to seek and discover.

Breeding species. A total of 316 species has bred in Turkey; 300 are considered to be regular, and breeding has been proven for 291 of these; a further 11 species are considered to be possible breeding species, but there is no real evidence.

Table 1. The 300 regular breeders (breeding not proven for nine of these*). Scientific names are presented in the species accounts.

Chukar	Black-necked Grebe	Eurasian Griffon Vulture
See-see Partridge	Mediterranean (Yelkouan) Shearwater*	Eurasian Black Vulture
Black Francolin	Great Cormorant	Short-toed Snake Eagle
Grey Partridge	European Shag	Western Marsh Harrier
Common Quail	Pygmy Cormorant	Pallid Harrier
Common Pheasant*	Great White Pelican	Montagu's Harrier
Caucasian Grouse	Dalmatian Pelican	Northern Goshawk
Caspian Snowcock	Great Bittern	Eurasian Sparrowhawk
Mute Swan	Little Bittern	Levant Sparrowhawk
Greylag Goose	Black-crowned Night Heron	Common Buzzard
Ruddy Shelduck	Cattle Egret	Long-legged Buzzard
Common Shelduck	Squacco Heron	Lesser Spotted Eagle*
Gadwall	Little Egret	Eastern Imperial Eagle
Eurasian (Common) Teal	Great White Egret	Golden Eagle
Mallard	Grey Heron	Booted Eagle
Northern Pintail	Purple Heron	Bonelli's Eagle
Garganey	Black Stork	Lesser Kestrel
Northern Shoveler	White Stork	Common Kestrel
Marbled Duck	Glossy Ibis	Eurasian Hobby
Red-crested Pochard	Northern Bald Ibis	Eleonora's Falcon
Common Pochard	Eurasian Spoonbill	Lanner Falcon
Ferruginous Duck	Greater Flamingo	Saker Falcon
Tufted Duck	European Honey Buzzard*	Peregrine Falcon
Velvet Scoter	Black Kite	Water Rail
White-headed Duck	White-tailed Eagle	Little Crake
Little Grebe	Lammergeier	Corn Crake
Great Crested Grebe	Egyptian Vulture	Moorhen
Red-necked Grebe		Purple Swamphen

Eurasian Coot
Common Crane
Demoiselle Crane
Little Bustard
Great Bustard
Eurasian Oystercatcher
Black-winged Stilt
Pied Avocet
Stone-curlew
Collared Pratincole
Little Ringed Plover
Kentish Plover
Greater Sand Plover
Spur-winged Lapwing
Red-wattled Lapwing
Northern Lapwing
Common Redshank
Common Sandpiper
Mediterranean Gull
Black-headed Gull
Slender-billed Gull
Audouin's Gull
Yellow-legged/Caspian Gull
Armenian Gull
Gull-billed Tern
Caspian Tern
Sandwich Tern
Common Tern
Little Tern
Whiskered Tern
Black Tern
Black-bellied Sandgrouse
Pin-tailed Sandgrouse
Rock Dove
Stock Dove
Common Wood Pigeon
Eurasian Collared Dove
European Turtle Dove
Laughing Dove
Rose-ringed Parakeet
Great Spotted Cuckoo
Common Cuckoo
Barn Owl
Pallid (Striated) Scops Owl
European Scops Owl
Eurasian Eagle Owl
Little Owl
Tawny Owl
Long-eared Owl
Short-eared Owl
Tengmalm's Owl*
European Nightjar
Common Swift
Pallid Swift
Alpine Swift
Little Swift
White-throated Kingfisher
Common Kingfisher
Pied Kingfisher
Blue-cheeked Bee-eater
European Bee-eater
European Roller
Eurasian Hoopoe
Eurasian Wryneck
Grey-headed Woodpecker
European Green Woodpecker
Black Woodpecker*
Great Spotted Woodpecker
Syrian Woodpecker
Middle Spotted Woodpecker
White-backed Woodpecker*
Lesser Spotted Woodpecker
Desert Lark
Calandra Lark
Bimaculated Lark
Greater Short-toed Lark
Lesser Short-toed Lark
Crested Lark
Wood Lark
Eurasian Skylark
Horned Lark
Sand Martin
Eurasian Crag Martin
Barn Swallow
Red-rumped Swallow
Common House Martin
Tawny Pipit
Tree Pipit
Water Pipit
Yellow Wagtail
Citrine Wagtail
Grey Wagtail
White Wagtail
Yellow-vented (White-spectacled) Bulbul
White-throated Dipper
Winter Wren
Dunnock
Radde's Accentor
Alpine Accentor
Rufous-tailed Scrub (Rufous Bush) Robin
European Robin
Common Nightingale
Bluethroat
White-throated Robin
Black Redstart
Common Redstart
Whinchat
European Stonechat
Eastern (Siberian) Stonechat
Isabelline Wheatear
Northern Wheatear
Eastern Black-eared Wheatear
Finsch's Wheatear
Red-tailed Wheatear
Rufous-tailed Rock Thrush
Blue Rock Thrush
Ring Ouzel
Common Blackbird
Song Thrush
Mistle Thrush
Cetti's Warbler
Zitting Cisticola
Graceful Prinia
Savi's Warbler
Moustached Warbler
Sedge Warbler
Paddyfield Warbler
Marsh Warbler
European Reed Warbler
Great Reed Warbler
Eastern Olivaceous Warbler
Upcher's Warbler
Olive-tree Warbler
Blackcap
Garden Warbler
Barred Warbler
Lesser Whitethroat
Eastern Orphean Warbler
Common Whitethroat
Spectacled Warbler
Subalpine Warbler
Sardinian Warbler
Ménétriés's Warbler
Rüppell's Warbler
Green Warbler
Eastern Bonelli's Warbler
Common Chiffchaff
Caucasian Chiffchaff
Goldcrest
Firecrest
Spotted Flycatcher
Red-breasted Flycatcher
Semi-collared Flycatcher
Bearded Reedling
Long-tailed Tit
Marsh Tit
Sombre Tit

Coal Tit
Blue Tit
Great Tit
Krüper's Nuthatch
Eurasian Nuthatch
Eastern Rock Nuthatch
Western Rock Nuthatch
Wallcreeper
Eurasian Treecreeper*
Short-toed Treecreeper
Eurasian Penduline Tit
Eurasian Golden Oriole
Red-backed Shrike
Lesser Grey Shrike
Woodchat Shrike
Masked Shrike
Eurasian Jay
Common Magpie
Alpine Chough
Red-billed Chough
Western Jackdaw
Rook
Hooded Crow
Common Raven
Common Starling
Rose-coloured Starling
House Sparrow
Spanish Sparrow
Dead Sea Sparrow
Eurasian Tree Sparrow
Pale Rockfinch
Chestnut-shouldered Sparrow
Rock Sparrow
White-winged Snowfinch
Common Chaffinch
Red-fronted Serin
European Serin
European Greenfinch
European Goldfinch
Eurasian Siskin
Common Linnet
Twite
Common Crossbill
(Asian) Crimson-winged Finch
Desert Finch
Mongolian Finch
Trumpeter Finch
Common Rosefinch
Eurasian Bullfinch
Hawfinch
Yellowhammer*
Cirl Bunting
Rock Bunting
Cinereous Bunting
Ortolan Bunting
Grey-necked Bunting
Cretzschmar's Bunting
Common Reed Bunting
Black-headed Bunting
Corn Bunting

Table 2. Recent (R), irregular (I) or former (F) breeding species (16 species) (all have bred).

Black Grouse	(F)
African Darter	(F)
Hen Harrier	(I)
Steppe Eagle	(I)
Osprey	(I)
Baillon's Crake	(I)?
Macqueen's Bustard	(F)
Black-winged Pratincole	(I)
White-tailed Lapwing	(I)
Alexandrine Parakeet	(R)
Pied Wheatear	(I)
Desert Wheatear	(I)
Common Grasshopper Warbler	(I)
Icterine Warbler	(I)
Iraq Babbler	(R)
Common Myna	(R)

Table 3. Possible breeders (11 species).

Cory's Shearwater
Barbary Falcon
Spotted Crake
Cream-coloured Courser
Common Snipe
Eurasian Woodcock
White-winged Tern
Brown Fish Owl
River Warbler
Wood Warbler
Plain Leaf Warbler

MODERN ORNITHOLOGY IN TURKEY: REFLECTIONS AND FUTURE DIRECTIONS

by Richard F. Porter, Güneşin Aydemir, Bahtiyar Kurt, Guy M. Kirwan and Gernant Magnin

In 1966, when the first-named author made his inaugural visit to Turkey, the number of Turkish birdwatchers could be counted on the fingers of a single hand: Tansu Gurpınar, of the National Parks Department, and Ali Kizilay, warden extraordinaire of Turkey's first bird reserve – the Kuş Cenneti (bird paradise) at Manyas Gölü. Tansu, with his formal university education, was an ornithologist and conservationist who did a great deal to spearhead Turkey's conservation movement under the National Park's director, Dr Zekai Bayer. Ali, by contrast, was a self-taught birdwatcher and practical warden in the old-fashioned sense. He knew the scientific names, identification and habits of all the species that the Kuş Cenneti hosted during the year; and worked tirelessly to ensure their protection. In a mixture of Turkish, pidgin German and the odd word of English, he would talk enthusiastically to all visitors about his beloved reserve.

At that time, the two other key figures in the Turkish birding scene were German. Professor Curt Kosswig – who was behind the creation of the Kuş Cenneti as a national park – and the doyen of 20th century Turkish ornithology, Dr Hans Kumerloeve. Porter's first visit to Turkey, with three friends for eight months, relied heavily for reference on Kumerloeve's magnum opus of 1961, *Zur Kenntnis der Avifauna Kleinasiens* – Turkey's first avifauna, yet still frequently cited by the present work. From the 1930s until at least the 1970s, Kumerloeve was the major force behind research into Turkish birds, and hardly a month seemed to pass without a new paper appearing. In his review of 'The history of ornithology in Turkey', published in the Turkey Bird Report of 1970–73, Kumerloeve listed over 270 major papers that had been published on Turkey's birds, over 60 of them his own.

Other strong influences in developing Porter's interest were the writings of Alléon and Vian who, in the 1880s, first attempted to catalogue the Bosphorus soaring-bird migration. More recently, in the 1950s and early 1960s, Nisbet and Smout, and Ballance and Lee, also described their observations of migrating raptors in the pages of *Ibis*. So was the beginning of a love affair with the country that still remains as strong today.

Although, as mentioned above, the work of chronicling the keynote events in Turkish ornithology prior to the creation of the Ornithological Society of Turkey (OST), has already been tackled by Kumerloeve, we suspect that many readers will not have access to the latter work. There follows a by no means comprehensive and somewhat personal timeline, which endeavours to give a flavour of the events of the last five centuries, but more particularly concentrates on the last four decades.

Pre-1800

In 1548, Pierre Belon crossed Asia Minor, from Syria to Istanbul, and subsequently wrote *L'histoire de la nature des Oyseaux* (1555), the first book covering the birds of Turkey. Rather few wide-ranging species were first described from the country, but in 1720, W. Sherard, the British Consul in Smyna (Izmir), collected a kingfisher – which was named by Linné *Alcedo smyrnensis* (the White-breasted or Smyrna Kingfisher), and the first type specimen to be collected in Turkey. Between 1792 and 1796, the expeditions of G. A. Oliver were the first to report on the presence of numerous shearwaters in the Bosphorus, later to be named 'Yelkovans' in Turkish.

The 1800s

Based on observations in 1835–6, Hugh Strickland lists 129 species for the Smyrna area, including Cinereous Bunting (previously undescribed) and a peculiar chiffchaff, *Sylvia* (=*Phylloscopus*) *brevirostris*, whose taxonomic placement continues to intrigue even modern ornithologists. Strickland's specimens, including both the chiffchaff and the bunting, are housed in Cambridge (UK). Further east, Keith Abbott, another Englishman, was collecting around Trebizond (Trabzon) and Erzeroom (Erzurum) between 1835 and 1837. Amongst his most significant finds were the lark *Eremophila alpestris penicillata* and an attractive new species of finch, *Rhodopechys sanguineus*, both of which were described by 'The Bird Man', John Gould. A little later, in the 1850s, the Marchese Orazio Antinori was collecting and trading in bird skins, and in 1856 described the woodpecker now known as *Dendrocopus syriacus*, thus named because at the time Syria was part of the Turkish Empire. Antinori also described Rose-coloured Starlings nesting around present-day Izmir. During the period 1840–60, L. Rigler, a physician living in the Constantinople (Istanbul) area recorded 164 species (although details of these were not published until 1904, by O. Reiser). Indeed, a great deal of important ornithological work during the course of the second half of the 19th century was undertaken within the environs of modern-day Istanbul, although this period also witnessed pioneering exploration of hitherto ornithologically unknown parts of the country. In 1865, A. Günther, writing in *Ibis* on a new subspecies of Long-tailed Tit from around Constantinople, acknowledged the work of Thomas Robson who was then resident and collecting in this region, and had been responsible for discovering the new race (then, of course, considered a species). Robson's contribution to Turkish ornithology is oddly significant for someone who never published a single article under his own name. In addition to Günther, his material was extensively worked by Henry Eeles Dresser for his monumental *History of the Birds of Europe*, published between 1871 and 1890. Many of his specimens, with their distinctive heavily creased blue-paper labels, are still available to modern researchers, principally in the Natural History Museum (Tring), and number an impressive number of 'rarities'. In 1880, it was the turn of the Comte Amede Alléon (and his assistant Vian) to inform the world of the spectacular raptor and stork migration over the Bosphorus. Alléon, like Robson, assembled a reasonable collection of birds from this region, again including a remarkable number of novelties. His material is housed in Sofia. To the south, around the present-day city of Izmir (as well as in Greece), Theodor Krüper, an ornithologist and trader in skins and eggs, was active, erratically, between 1863 and 1894. In addition to the nuthatch that bears his name (one of the few species almost endemic to Turkey), he made valuable contributions on a number of species including Masked Shrike, Isabelline Wheatear, White-throated Robin and Levant Sparrowhawk. Until now, southern and central Turkey had been largely ignored by ornithologists. Another Englishman, Charles Danford, commenced the exploration of these regions during the late 1870s, during which period he discovered the Bald Ibis colony on the Euphrates at Birecik. Once again, it was left to Dresser to work on Danford's material (although the collector did publish two of his own papers) from which, amongst other novelties from the Taurus Mountains, he described a new subspecies of the Caspian Snowcock *Tetraogallus caspius tauricus*. Danford's specimens, held in Tring and Manchester, remain invaluable for understanding geographic variation in Turkish birds. Now lost as a breeder both from the Turkish and entire Western Palearctic avifauna, the African Darter colony at the, now sadly drained, Lake of Antioch (Amik Gölü) was discovered about the same time by the Frenchman, Chantre. Finally, at the close of the century, we must mention the commencement, by Russian ornithologists, of bird research in the Caucasus including north-easternmost Anatolia, and chiefly the leading contributions of Radde (1884) and Vilkonskij (1897).

1900–45

The new century opened with Braun (1901–11) reporting on birds from Istanbul and north-west Anatolia, but it is Woosnam's 1905 travels through north-east and eastern Turkey (and subsequently Persia), which produced novelties such as the first Grey-necked Bunting for Turkey, and Ramsay's work in the montane isolates of the southern Anatolian Plateau, which most catch the eye with hindsight. Mention must also be given of the great Hubert Lynes' study of White Stork migration in the Iskenderun area, work that eventually led to the recognition of the second of Turkey's great soaring-bird bottlenecks, the Belen Pass. In the north-east, the Russians were still active with important contributions coming from Nesterov, Bobrinkij and Dombrovskij in the first two decades. At Erzurum, McGregor continued the earlier work of English ornithologists in this region, and Weigold was the first person for a number of years to visit the Bald Ibis colony at Birecik, going on to make other important observations in this region and Mesopotamia. Even the First World War did not put paid to ornithological observations, with German, English and Russian birdwatchers and skin collectors recording on the front line in north-east Anatolia, around Ankara and during the Dardanelles campaign. See, for example, papers in the *Zoologist* by A. W. Boyd and D. A. J. Buxton.

During the 1920s the Turkish biologist, Ali Wahby, started to collect bird records for Istanbul and became the first person to start ringing birds (White Storks) in the country. His paper 'Les oiseaux de la region de Stamboul et de ses environs', which appeared in 1930 and 1934, was the first ornithological publication by a Turk. At the same time, 1931, Otto Steinfatt studied the spectacular migration at the Bosphorus between September and November and, six years later, Lutz Mauve undertook the first comprehensive spring survey of migrating raptors and storks at the same watchpoint. The 1930s also marked the first visits to Turkey by the German Hans Kumerloeve who, together with Gunther Niethammer, concentrated on areas hitherto ignored

Professor Kosswig, Manyas Gölü, April 1970 (Richard Porter).

Stanley Cramp, Central Anatolia, June 1973 (Richard Porter).

Salih & Belkıs Acar and Richard Porter, Bebek, Istanbul, June 1973 (Stanley Cramp).

by ornithologists. In 1933, the legendary Colonel Richard Meinertzhagen made his only visit to Turkey, to Amik Gölü. His observations represent one of the very few detailed reports on the state of the darter colony. The following year, 1934, Hugo Rossner and Gabriele Neuhauser independently collected specimens in northern Anatolia, including Grey-headed Woodpecker which, despite such evidence, was not unequivocally recognised as a part of the breeding avifauna until almost 50 years later. A spring and summer collecting trip to the eastern Taurus and Anti-Taurus, in 1935, by C. G. Bird and E. K. Balls, led to the discovery of Pale Rockfinch and Upcher's Warbler, and the publication of a detailed report in *Ibis* concerning the results of their trip. In 1937, Miklos Vasvari and Imre Patkai discovered a Little Bustard breeding area around Karacabey, near Apolyont (Uluabat) Gölü, long since abandoned. N. J. P. Wadley, one of the few people to watch and collect birds in Turkey during the Second World War, recorded a total of 205 species. His 'Notes on the birds of Central Anatolia' appeared in *Ibis* in 1951, and his specimens, once part of the Meinertzhagen collection, are now held in the Natural History Museum at Tring. At the close of the war, the first illustrated book on Turkish birds (*Turkiye Kuşları*) was published, by a Turk, Saadet Ergene; the pictures are in black and white, but the text provides brief descriptions and the distribution of every species.

The 1950s and 1960s

Curt Kosswig has already been mentioned earlier. At this time head of the Zoological Institute of Istanbul University, he promoted the importance of the large breeding colony of herons, Spoonbills and cormorants at Manyas Gölü, which led to the lake becoming a national park in 1958. Kosswig was also instrumental in encouraging Saadet Ergene to author his illustrated review of Turkish birds. However, much of the rest of this period was dominated by visitors from Western Europe. In 1951 Phil Hollom made the first of his several birding trips to Turkey, visiting the Taurus and south coast and recording Audouin's Gull and Red-fronted Serin amongst others. From 1953, Hans Kumerloeve recommenced his exploration of Turkey (initiated in the early 1930s) and from then until at least 1969 he comprehensively covered nearly the whole of Asia Minor. This resulted in innumerable papers and the classic works *Zur Kenntnis der Avifauna Kleinasiens* (1961) and 'Liste systématique revisée des espèces d'oiseaux de Turquie' (1966), which have served as a baseline almost until the present. One of his final contributions to our knowledge of the country's birds came in 1986, a bibliographic compilation of published literature on birds and mammals of Turkey and related regions. In 1956 I. C. T. Nisbet and T. C. Smout once again awakened interest in the Bosphorus, and Dardanelles, as an important bottleneck for migrant soaring birds; the results of their observations were later published in *Ibis* and similar studies were undertaken at the Bosphorus in 1959, by D. K. Ballance & S. L. B. Lee. Migration was also the focus of the 1960s studies of Hartmut Heckenroth and his colleagues, who focused their attention on the White Stork.

Following in Kumerloeve's footsteps, his fellow Germans in particular hastened to explore the interior of Anatolia. Special mention must be made of Gottfried Vauk and a team of ringers from Helgoland's spring 1964 ringing study in the Beyşehir Gölü area, and Heinz Lehmann and his colleagues, especially Klaus Warncke, who in 1964 started their ornithological expeditions to Turkey. Stretching over several years, they discovered a host of new breeding birds, including Dalmatian Pelican, Greylag Goose, Mute Swan, Saker Falcon, White-tailed Lapwing, Gull-billed Tern, Caspian Tern and Greater Flamingo, as well as Crimson-winged Finch on the Central Plateau. Also significant are Willy Bauer's and Günther Müller's studies of the birds of the Meriç Delta, on the Greek border, at a time when it was seriously threatened with drainage, as a result of which they published a management plan in 1968.

However, where the Germans led, the Britons were not slow to follow, and in 1965 R. A. D. Cameron, Lin Cornwallis and colleagues, picking up the mantle of Hubert Lynes from 60 years before, studied soaring-bird migration in south-east Turkey (the results being published in *Ibis* in 1967). Hot on their heels, 1966 saw a team of British birdwatchers make the first complete autumn count of soaring birds across the Bosphorus (also published in *Ibis*). Until now, wintering birds had been to a large extent ignored but in 1967 the Annual European and Asian midwinter wildfowl counts were extended to Turkey, following the first International Waterfowl Research Bureau (IWRB) visit to the country, by J. Szijj and Hayo Hoekstra. A second IWRB mission, in July–November 1968, by Jacques Viellard and his colleagues, made extensive observations in eastern Turkey discovering breeding Velvet Scoter. Thereafter the mantle of counting extraordinarily large numbers of birds at still-enormous natural wetlands passed to the Dutch, particularly the indefatigable Lieuwe Dijksen and, in the early years, Fred Koning.

A new order

The year 1968 saw the placement of Turkish ornithology on a more formal footing with the creation of the Ornithological Society of Turkey (OST). The organisation was proposed the previous year, at a milestone conference on Middle Eastern wetlands held in October, in Ankara, Bursa and Istanbul. Sponsored by the International Union for the Conservation of Nature, IWRB (now Wetlands International) and International Council for Bird Preservation (now BirdLife International), this gathering of world conservationists was to have a major influence on the future of Turkey's wildlife programme, and was a forerunner to the Ramsar Conference, which has played such a defining role in the protection of the world's wetlands.

The OST had both national and foreign sections and, during its ten years of life from 1968 to 1978, published four bird reports and 15 bulletins. Scanning through these historic documents one quickly gains a feel for the many contributions that were being made to Turkish ornithology. Their influence is, indeed, still palpable and it was through casually perusing the bird reports that Kirwan's interest in the country was first pricked. Birding in the final third of the 20th century was moving apace, as in so many other countries around the world. For example, the *Check List of the Birds of Turkey* published by the OST in 1971 lists 394 species, of which 251 were known to be regular breeders. By 2005 the Turkish list stood at 455 and the number of known regular breeders had reached 300 – clearly as a result of intensive effort and systematic surveys, especially in the more remote regions of the country.

It would be impossible to catalogue all of the ornithological happenings of the OST era, and certainly invidious (as well as nigh on impossible) to mention the names of all of the key players. A read through the bird reports and bulletins will reveal much of significance. But we must pay tribute to the late William Wilkinson, the first chairman of OST and a great Turcophile. William's brain and hard work were behind so many of the ornithological happenings of the early days, yet his name rarely appears in the literature. Secondly, it is important to record the publication of *Les Oiseaux du Proche et du Moyen Orient* by F. Hüe and R. D. Etchécopar which was, for many years, the standard reference work on the birds of the Middle East, but whose significance is perhaps overlooked in some quarters due to the fact that it was never translated into English.

Amongst ornithological field activities, the winter wildfowl counts (mentioned above), raptor migration watches and the start of bird ecotourism stand out. The raptor watches at the Bosphorus, Belen Pass and in the eastern Pontics – which reached their zenith in the OST era – probably did more than anything to put Turkey on the birding map. And birdwatching tours, which started in earnest in the late 1960s with most major tour operators making annual visits, clearly had a profound effect in opening up a country that to many birdwatchers was hitherto just a dream beyond Europe's doors.

During the late 1960s and early 1970s a parallel approach to birds and conservation was being made by that remarkable couple, Salih and Belkıs Acar. Their vision and tireless energy provided the inspiration for the Turkish voluntary conservation movement, which manifested itself in the creation of Doğal Hayatı Koruma Derneği (DHKD) – The Society for the Protection of Wildlife – in 1975. Belkıs was its first secretary. But, at the start, it was just the Acar's books and exhibitions that were the focus of public attention. In 1971 they opened their first exhibition, Goçmen Kuşlar (Migratory Birds), at the largest gallery in Istanbul. This led to the publication of a book for children by the same name and then, following its success, *Kuşlarımiz (Our Birds)* – a small guide to the birds of Turkey written in conjunction with Udo Hirsch. The Acars always acknowledged their gratitude to the OST for help in those formative years and referred to the society as their 'more scientific cousin'.

However, the days of the OST were coming to an end. Turkish ornithology was starting to move away from the grasp of the foreign visitor and into that of a developing Turkish birding movement. It was time too for the OST to move on. In the autumn of 1976, whilst watching the impressive raptor migration over the eastern Pontics, the idea of expanding into a society to embrace the entire Middle East was hatched. And so, in 1978, the Ornithological Society of the Middle East (OSME) was formed – the first of the world's regional bird clubs.

Conservation to the fore

With the dissolution of the OST, Turkish ornithology drifted freely for sometime. Foreign birdwatchers continued to make their input via the bird report (which OSME has remained committed to publishing as part of its journal *Sandgrouse*), but their contribution was mainly to find new sites for Turkish specialities such as Cinereous Bunting and Caucasian Grouse, although such visitors were also responsible for adding a varied *potpourri* of new breeding species to the Turkish avifauna, for instance: Red-wattled Lapwing, Pallid (Striated)

Scops Owl, Desert Lark, Paddyfield Warbler and Mongolian Finch. During the 1980s a new German force in Turkish ornithology arrived on the scene, Max Kasparek. Like his forbear, Kumerloeve, he has published widely on Turkish avifauna, producing innumerable papers and several books, most notably an annotated checklist of the country's birds, which served as the first detailed update of the Turkish bird fauna since 1961 and, in partnership with his wife and Asaf Ertan, the first inventory of Important Bird Areas (IBAs) for Turkey.

DHKD, which was responsible for publishing the IBA book, was by now an established force, with a modern office in Istanbul and some committed staff. In 1988, the International Council for Bird Preservation (now BirdLife International) recognising the importance of Turkey for birds decided to assist this institutional development by dispatching a Dutchman, Gernant Magnin, to the country. Together with one of a new league of Turkish birdwatchers, Murat Yarar, who had started working at DHKD in 1988, DHKD embarked on an active programme to bring Turkish birders together through the DHKD Bird Section, where many birders active today started their 'careers'. In addition to assuming responsibility for organising the midwinter counts, the DHKD Bird Section mounted major projects in some of the country's most important wetlands, aiming to survey these areas more thoroughly than had been attempted previously and, in some cases, to produce integrated management plans for their conservation and future development. Nonetheless, it was often keen to work with outside experts and organisations in the furtherance of such objectives, and amongst those bodies from outside Turkey involved in such partnerships were Tour du Valat (France), Wetlands International (UK), Foundation Working Group International Wader and Waterfowl Research (WIWO, the Netherlands) and OSME. Special mention must be made of the role of WIWO, whose work not only was instrumental in providing baseline data upon which IBA status could be established but also garnered much of our knowledge of waterbird (and to some extent passerine) migration across the country, all of which data were disseminated, usually commendably swiftly, through a series of well-produced reports. Various individuals, amongst them some of the authors of this book, were also involved in this research which chiefly aimed to identify new IBAs and monitor existing sites. Other important projects that were pursued during this period were intensive surveys of breeding White Storks, Caucasian Grouse, Eurasian Black Vultures and Lesser Kestrels in central Turkey, by Steve Parr and colleagues; a study of the wintering habits of White-headed Duck at Lake Burdur, which sought to build on the international symposium held in Burdur, in 1991, and several surveys of the same species at its summer haunts in Anatolia. Magnin and Yarar, as well as other members of the DHKD Bird Section, covered nearly half a million kilometres between them during 1992 to 1997 in order to collect bird and site information. The culmination of this intense period of activity came in 1997 with the publication of a revised, and much-improved, IBA inventory for Turkey, again published by DHKD and compiled by Gernant Magnin and Murat Yarar. For the purpose of this book, all Turkish birders were consulted in order to arrive at the first-ever fully accepted list of Turkish bird names, published as an annex to the IBA inventory.

The 1990s also witnessed the development of other local birdwatching groups. The majority of them were led by young people, who had volunteered in the DHKD Bird Section, and had thereafter moved to Ankara to study. The first of these, the Ankara Bird Watching Club (AKGT), was founded in 1991 and became the first birders' group and made possible local bird surveys. The following year saw the formation of Hacettepe University Bird Watching Club (Beykuş), the first university club. In 1995 the Middle East Technical University Bird Watching Club (OKGT) was established by Kerem Ali Boyla and became the most active birdwatching club during the period. Eventually, the creation and cooperation between the different clubs would permit their involvement in the IBA Monitoring Programme, organised by DHKD, which started in the year 2000. DHKD helped beginners by providing books, binoculars, telescopes and other assistance, and by these means encouraged swift growth in birds and birdwatching. However, due to the lack of a modern field guide in Turkish, this growth was limited, until, in 2002, the Collins field guide by Heinzel, Fitter and Parslow was translated to Turkish and adapted for a Turkish audience by Kerem Ali Boyla, and published by DHKD. Its publication encouraged both university-based bird clubs and individuals outside them, thereby enabling in a short time the language and terminology of birdwatching to switch from English to Turkish, thus greatly strengthening communication amongst birdwatchers.

By this time, DHKD was producing the *Kuşçu Bülteni* (Birder Bulletin) thrice per year, further enhancing the development of a birding network in the country. *Kuşçu Bülteni* has since 2004 metamorphosed into *Kuşçu Postası* (Birder Mail). All of this activity was also manifest in the development of new national bodies in Turkey designed to motivate and facilitate contact within the steadily growing birdwatching community. In 1997 the Turkish Ornithology Council (TOK) was briefly formed. TOK organised two birdwatching courses and also

Apolyont Köy, Apolyont (now Uluabat) Gölü, May 1966 (Richard Porter).

Hoyran Delta, January 1970 (Richard Porter).

Phil Hollom and Stanley Cramp, Amik Gölü (since drained), May 1970 (Richard Porter).

Richard Porter (left) with Tansu Gürpinar at the seventh Turkish Birdwatchers' Conference, Izmir, October 2004 (Richard Porter).

published a single issue of its ornithological journal *Turna*. The following year, the Bird Research Society (KAD) was founded by birdwatchers and conservationists, with the aim of promoting the study and conservation of birds, and to raise public awareness in conservation issues. It too launched a publication, *Ibibik*, which has become the first popular birdwatching and ornithology magazine, of which five issues have been produced to date. The arrival of a flourishing birding community in Turkey has also led to Turks becoming closely involved in the production of the bird report, for so long the preserve of few British birdwatchers with a deep-rooted interest in the country's birds.

Mention must also be made of an important benchmark publication that appeared during this period, produced in the Netherlands. Kees Roselaar, one of the mainstays of *The Birds of the Western Palearctic* project and one of Europe's foremost taxonomists, produced the *Songbirds of Turkey*. Only its subtitle, *An Atlas of the Biodiversity of Turkish Passerine Birds*, gave some clue as to the importance and scope of this book. It stands alone as the most important contribution to the study of geographical variation in Turkish birds, despite 'only' covering the passerines.

As the 20th century drew to a close, there was a new confidence to the Turkish birdwatching and conservation movement. And, with an increase in both manpower and relevant experience, this found expression in the launch of several distinctly more ambitious and far-reaching field surveys than had been previously attempted. The first of these, in 1998, was a survey of the Konya Basin, in the heart of Central Anatolia. This was followed by detailed yet wide-ranging studies of the Mediterranean forests and south-east Anatolia, although these surveys continued to rely heavily for their motivation and design on Geoff & Hilary Welch, from the UK, who have provided a consistent backbone to the development of a Turkish birdwatching and conservation community for over 15 years, inspiring many with their relentless determination to get things done. Such surveys were successful in pooling huge quantities of data, filling-in many gaps in our knowledge of Turkish bird distributions, and revealing how much there is to be done still. They have paved the way for the organisation of a Turkish atlas project, a dream that has existed in the minds of various people for as long as 30 years, but now appears tangible, provided sufficient funding can be found for such an undertaking.

There were losses too, however, and not just in birds and wetlands. However, from the ashes of DHKD has come a new organisation, Doğa Derneği (DD, Nature Society), which has swiftly assumed the mantle of BirdLife partner in Turkey. DD aims to protect Turkey's threatened species, especially birds, Important Bird Areas, Key Biodiversity Areas and priority habitats, through a national grassroots network. To this end, it has already published a new, updated listing of Turkish IBAs, compiled by Dicle Tuba Kılıç and Güven Eken, and incorporating a proposed list of Red Data birds in Turkey.

Symptomatic of the sea-change in the number of people interested in birds within Turkey has been the development of a national conference. Initially such efforts, spearheaded by the former DHKD, attracted only very small numbers, first in Istanbul in 1991 and secondly in Ankara, in 1992. Since then, as a result of birdwatching courses and through the efforts of many individuals, birders' numbers have increased and six additional meetings have taken place. With such interest now being maintained, it seems that Turkish ornithology has a promising future.

THE GAPS IN OUR KNOWLEDGE

by Geoff Welch, Hilary Welch and Sancar Barış

This book seeks to provide a new baseline in the study of Turkish avifauna and it is undoubtedly the case that, in quite large part due to the development of a homegrown birdwatching and conservation community, our knowledge of the country's birds has developed quite remarkably within the last 15–20 years. Nonetheless, there are still many opportunities for determined national and foreign-based observers to contribute to our knowledge store. This chapter seeks to identify some of the main problems restraining our knowledge of Turkey's birds, where those gaps lie in both general and some quite specific terms, and to look to the future and the next milestone in publications on Turkish avifauna.

SPECIES

(1) Distributions

Distribution of birds or of birdwatcher activity?

Until recently Turkey's pool of bird data has largely been a legacy of visiting birdwatchers on holiday from Western Europe. Although it is an enormously valuable resource, the data should be considered cautiously because, as a result of its origins, it contains errors and omissions. In the main, visiting birdwatchers want to see as many new species and types of habitat as possible in their short holiday. Travelling is no object and is sometimes done at night to save valuable birdwatching time. To arrive at the best ratio of good value for money against time spent, most of these birdwatchers also visit the same sites, following in the footsteps of those fortunate few who first located a regional speciality. All this results in a distinct bias; birdwatchers visit at the same time of year, go to the same sites and observe the same birds, which does not help in the understanding of a country's ornithology. On the other hand, native birdwatching activity is now increasingly significant, but information gathered still possesses the same drawbacks. Turkish birdwatchers have a different preferred site list than foreign visitors, but instead it tends to reflect where they live and understandably does not include many generally unknown sites. Understanding that birdwatchers will continue to act in similar fashion in the future is important because this means that other means of covering neglected parts of the country must be found. One such approach was taken at the initial stages for the atlas project and establishment of at least one active birdwatching group for every 150 × 150 km grid was planned.

Standardised data – species' presence and absence

Between 1998 and 2003 three major bird surveys of large regions of Turkey were conducted, which for the first time collected breeding bird data following standard methodologies – namely in the Konya Basin (Eken & Magnin 1999), Mediterranean forests (Zeydanli *et al.* 2005) and South-east Anatolia (Welch 2004). In the Konya Basin and South-east Anatolia, selection of the survey squares or points was by stratified random sampling (i.e. semi-randomly in that it was ensured each vegetation type was sampled in proportion to its occurrence in the study area). So, for the first time, survey locations were not biased by prior knowledge, the presence of concentrations of birds or regional specialities, and for the first time we possess datasets which inform us not only where birds definitely are but also where they almost certainly are not. We are also beginning to gather data on widespread and common species. Such surveys, whilst difficult and relatively expensive to organise, and demanding comparatively significant manpower, are needed too in other parts of Turkey, especially the central-east of the country, traditionally under-watched by birdwatchers in general. A brief glance at the maps for passerine birds in Roselaar (1995) reveals large gaps in the range of many species which might reasonably be expected to occur in this region.

What standardised data can reveal about 'overlooked' species

There are certain species which European birdwatchers take for granted because they are common in Europe, for example, European Robin. Without being conscious of the European Robin's distribution it is difficult to be excited about seeing one in Turkey. Yet here this species is on the edge of its breeding range and defining the limit of its distribution is of great interest to zoogeography. Prior to 2000 its breeding distribution was thought limited to northern Turkey, principally along the Black Sea coast, but survey work in selected Mediterranean forests over three breeding seasons in 2000–03 (Zeydanli *et al.* 2005) found the species to breed at scattered locations in the eastern Mediterranean, in a wide range of wooded habitats – deciduous, coniferous and sclerophyllous. It is interesting that of these somewhat isolated records the greatest concentration is in the Amanos Mountains, where pockets of unique climatic conditions have enabled a suite of more eastern and northern taxa (not only birds) to survive both the local Mediterranean climate and the effects of various Ice Ages down the millennia. Conversely, in South-east Anatolia in 2001/2, (Welch *et al.* in prep.) the European Robin was not recorded at all as a breeding bird. However, in this region the forested mountains of Şırnak were poorly covered and from here eastwards the likelihood of the European Robin occurring is greater due to the proximity of breeding populations in northern Iran.

Systematic surveys of defined areas have thus supported the European Robin's predicted distribution in central and eastern areas through revealing no new records nor areas of hitherto unknown suitable habitat in the Konya Basin and South-east Anatolia, but at the same time found small populations further south than was previously considered likely, in pockets of suitable habitat and climate in the eastern Mediterranean.

What do we really know about the distribution of 'common' species?

Staying with 'common' birds, all birdwatchers are guilty of under-recording common and widespread species on the assumption (usually unconscious) that they occur everywhere – crows, larks, House Sparrows, Common Starlings, Feral Pigeons – yet if recorded systematically we may discover intriguing gaps in their distribution. Eurasian Collared Dove is generally considered resident with only local movements and slow expansions of range through 'leap-frogging'. Yet, on two occasions, significant movements of this species have been recorded in southern Turkey. On 16 January 1996 'large numbers' were recorded on the Çukurova plain adjacent to the Demirkazık–Adana road (GW & HW pers. obs.), and on 14 April 2002, during fieldwork in Southeast Anatolia, equally large numbers were seen along the road between Birecik and Şanlıurfa – 332 in 10 km was a sample counted. Were these true migrations or local movements stimulated by adverse weather conditions?

Eurasian Collared Dove is an easy bird to survey from a moving vehicle; if none is seen in any settlements during a drive through a (10 × 10 km) square, despite concerted effort and scanning of rooftops and wires in villages and towns, the species is either absent or at best exceedingly rare. Exactly this was done in June 2005 during two weeks' fieldwork in northeast Turkey. During a journey from Ankara to Kars, and thence to the Black Sea at Hopa, Eurasian Collared Doves were found to 'disappear' north of Erzurum, seemed to be completely absent from Kars province, and only reappeared on the Black Sea coast at the Yeşilırmak plain. This and several other examples – such as the absence of Common Magpies from most of the Black Sea coast – await substantiation and, even more importantly, explanation.

Distribution changes in scarce species

At the other end of the scale, and of more interest to all birdwatchers, is the recent upsurge in records of Purple Swamphen. Previously restricted to the Mediterranean – at the Göksu and Çukurova deltas (and historically also at Amik Gölü in Hatay/Antakya), in 2002 it was found on the Euphrates, at Karkamış on the Syrian border. This is a logical extension of its range eastwards and may possibly be explained by birds moving north along the Euphrates from Syria. This theory is supported by records from five sites along the Euphrates by the OSME Syrian Wetland Expedition in January–February 2004 (Murdoch *et al.* 2004), and the species was considered possibly to be breeding at Mheimideh in June 2003 (Murdoch 2005).

But how can we explain the Purple Swamphens recorded at the Kızılırmak Delta on the Black Sea? The first observation by birdwatchers was made here in 1998, but claims by locals and hunters date from as early as 1993. By mid 2005 the population was thought to number several hundreds, with one report of over 1,800 individuals in winter, and rice farmers in the delta now consider the birds to be serious pests! Where have these birds come from? Is it possible they originated from a previously unknown population further inland, from which area the birds have been displaced by dam developments upstream on the Kızılırmak, or were the founders genuine long-range stragglers from the migratory North Caspian population? Are new immigrants still arriving or is the extraordinarily rapid increase purely the result of a series of highly successful breeding seasons? As the Kızılırmak was extensively surveyed by a WIWO project in spring 1992 (Hustings & van Dijk 1994), when no Purple Swamphens were recorded, it seems improbable that an undiscovered population in the delta defied detection until a local population explosion led to its discovery. A study of this population to identify its racial affinities and to understand its movements and colonisations is much-needed.

(2) Status

Developing local knowledge

Despite the continuing growth in the number of birdwatchers in Turkey, runs of data for all but a handful of sites, especially collected in a consistent manner, are still lacking. Additionally, observation of bird behaviour is comparatively scarce. Together, these omissions mean that the precise status of many species, especially with respect to breeding, is still unclear. A good illustration of this is Red-backed Shrike, a species which is very common and widespread on migration and is shown in the majority of field guides and handbooks as breeding throughout most of Turkey. Is this really the case? Red-backed Shrike appears to have a long migration period (with poorly known start and end dates), so birds may still be passing on migration at northern sites on dates when they are feeding young further south. To solve this riddle, repeat visits need to be made to sites where Red-backed Shrikes are observed in spring to confirm that the species is still present later in the breeding season, and to gather information on its behaviour. Currently this information is unavailable. The KuşBank bird database (at August 2005) contained 313 records of Red-backed Shrike spanning the period 11 May 1986 to 17 July 2005. Of these, 128 (41%) have a breeding code assigned but only seven (2%) provide unequivocal evidence of confirmed breeding. Of the remaining 121 records, all but 35 could easily relate to migrants. But for conservation, we need to know whether we are conserving migration corridors or breeding habitat. And, once we know where Red-backed Shrikes breed, we need to understand what the breeding population is doing – is it stable? Experience in Western Europe demonstrates that changes in farming, and potentially climate, will affect Red-backed Shrike, and the former changes are already occurring in Turkey. We need baseline information now on breeding distribution, preferred habitat and (at least) relative abundance so we know which areas are most important.

Local birdwatchers need to find a way to birdwatch regularly in the same areas – to develop their own 'local patches' and to discover, over a season's visits, which species they see regularly in each location and, most importantly, what the birds are *doing*. They should be encouraged to spend more time observing and reporting behaviour, a few minutes observation can change a 'possible' breeding record into a 'probable' or even 'confirmed', and enormously increases the value of the observation.

Collecting data consistently over a period of time

There is just one long-term monitoring programme ongoing in Turkey, since at least 1986, the Midwinter Waterbird Counts organised by IWRB/Wetlands International. However, although counts have been carried out in ten years during this period, many aspects of the way the counts are undertaken have varied markedly between years: this has included the number of sites visited, the coverage of each site, and the experience and number of observers involved. Thus, any use of the figures to identify changes in wintering bird numbers needs to consider the effect of the data-collection variables. Is a species really increasing or declining, or is it just that more or fewer sites were counted, or that the species is difficult to identify, or that sites were covered more or less thoroughly in a particular year? Additionally, there will always be variables which cannot be controlled: disturbance from hunting, weather conditions on the day/in Turkey/in Eastern Europe–Russia–Central Asia.

Perhaps of greatest concern, such inconsistently collected data can, potentially, lead to errors in the identification of Important Bird Areas (IBAs). Numbers of birds using a site will naturally vary from season to season and between years. A site deserving to qualify as an IBA could be entirely overlooked in a poor year and conversely, a site experiencing an exceptionally good year – perhaps due to unfavourable water levels at a neighbouring site – could be flagged as a high priority for protection when (in this example) its importance lies in its function as part of a network of sites. At the time of writing, all IBAs are being reassessed by Doğa Derneği to determine their current status. This process will enable a more targeted approach to developing conservation strategies and make it possible to prioritise work with more certainty according to the most important species, sites and greatest threats. However, to be reliable this process requires comparably collected, up-to-date datasets from a series of years; this is an ongoing process which needs the support of all birdwatchers.

(3) Overlooked species

Some birds are more conspicuous than others, and some you do not even need to see. At the opposite end of the spectrum are species which are hard to see, reach or identify. Nocturnal birds will always be under-recorded unless special techniques are used, and secretive species may not be noted even though they abound. For instance, the status of Barn Owl has remained controversial for many years simply because it is nocturnal and because few Turkish birdwatchers live in the countryside. Tengmalm's Owl offers a potentially even more striking example of a species which may be reasonably widespread in the country, which completely escaped detection until less than 25 years ago, and for which fewer than a handful of records were available until about ten years later. Species inhabiting closed-canopy forest or impenetrable thorn thickets and bushes are also easily overlooked, and one must rely more on hearing; these may be reasons for the poor understanding of the distribution and abundance of White-backed Woodpecker and for the delayed discovery of Grey-headed Woodpecker.

Colonial species are unevenly distributed even in an optimal habitat and some, like vultures, inhabit remote and inaccessible places. Once a colony is known, it is relatively easy to monitor the population but finding a new one may prove difficult. Consider the Eurasian Griffon Vulture colony at Ispir, for instance, which is known because it is located on the cliffs beside the road, but what is present in the remote hills and valleys nearby is unknown. Alpine species almost always are patchily distributed and occur at low densities, and their occurrence can be temporary. As a result, distribution of most highland species is incompletely known as they require much greater investment in terms of time, effort and preparation.

Other species are difficult to identify, especially those species most easily identified by their vocalisations. On the one hand there is a relative lack of expertise amongst the young Turkish birdwatching community. On the other there is deficient knowledge of the calls and songs of several species and local races. A peculiar call of the Turkish Common Redstarts sounds just like a Bullfinch, even to expert European ears. In some areas, singing Cinereous Buntings can sound exactly like Ortolan Buntings and so be completely overlooked in places where the latter is common. Similar reasons are at the root of our collective lack of knowledge concerning the status and distribution of all of the *Porzana* crakes: they are secretive and one needs to know their calls to establish their presence in most instances.

Some species, such as large raptors, occur at low densities and possess huge territories. Thus, an isolated observation is not very informative and estimating populations, breeding distributions and threats requires detailed studies spanning several years. Some improvement in the quality of records will definitely occur in the future, when the birdwatching community gains more experience, but most of the problems concerning low-density species demand carefully planned long-term studies with a sound methodology still to be resolved.

Another type of 'overlooking' occurs when we assume residency for a species recorded year-round. Turkey not only covers large distances latitudinally and longitudinally, but it is also a mountainous country of varied topography, which further contributes to the enormous habitat variety. We poorly understand altitudinal and latitudinal migrations of most bird species and, for instance, records of Eurasian Sparrowhawks probably relate to different wintering, breeding and migratory populations. This is also true for many other species and groups, *e.g.* birds of prey, waterbirds, herons, cranes, thrushes, chats and finches. Understanding the 'true' nature of occurrence is clearly a must for adequate conservation.

POPULATIONS AND HABITATS

Once we start to possess and can analyse all of the above types of data it will be possible to begin to develop a much better understanding of populations – changes, estimates and phenology.

(1) Population estimates

In Western Europe, detailed data on bird populations of all species are taken for granted and form the basis from which major conservation organisations develop their policies and programmes. Reliable and sometimes very precise numbers are available for each species' population. Runs of data indicate whether a species' population is stable, increasing or declining. The ecology of certain species has been studied in great detail, so that the factors affecting it are understood and effective conservation action can be taken to stabilise and restore populations. Most significantly, this type of scientifically sound data contributes to the formulation of EU policy which can have far-reaching benefits for conservation.

In Turkey reliable quantitative information is simply lacking or historical, making it extremely difficult to develop a clear, objectively sound and scientifically supported national plan of prioritised action. In contrast with Turkey's force of visiting birdwatchers from abroad, local birdwatchers are handicapped by the very large size of the country combined with the small number of birdwatchers, and especially the paucity of those with the ability – experience, time, money, mobility, training – to undertake systematic survey work. Furthermore, academic interest in ornithology is relatively recent, and only a handful of PhDs and Masters theses have yet been completed. Time is required for adequate experience and expertise to increase. Facilitating and accelerating this process should be a priority for all parties (Barış 2004).

Several single-species surveys have or are in the process of being carried out, on high-profile/conservation-value species such as Great Bustard, White Stork, Eurasian Black Vulture and Caucasian Grouse, and for these a variety of techniques for determining numbers are being used – from requests for information circulated to birdwatchers and local communities (White Stork) and radio telemetry (Eurasian Black Vulture), to computer-modeling of distribution and population size (Caucasian Grouse). However, for the majority of species nothing but the most rudimentary 'best guess' information is available.

At present, apart from the semi-captive Northern Bald Ibis at Birecik, there is probably no species in Turkey for which there is a reliable population estimate, the current very broad estimates we possess were agreed by pooling the ornithological knowledge of a handful of experienced individuals.

To improve understanding of any species' breeding population, reliable data on its national distribution (where it occurs and where it doesn't), its relative abundance where it occurs, and (ideally) its broad habitat preferences are required. Collecting all this information would require a national survey and significant funding – it is clearly a major task which the birdwatchers cannot achieve on their own. A breeding bird atlas was first identified as a need in the 1960s but, despite several valiant attempts (the most recent running in parallel with the South-east Anatolia project field work, in 2001–03) it has still to be undertaken.

To arrive at more finite population figures, particularly for common and widespread species, counts of birds are needed at a range of locations across Turkey, using simple but scientifically robust monitoring programmes which can be followed by both the growing number of 'ordinary' Turkish birdwatchers and by academicians. An ongoing annual monitoring programme could provide actual figures from the same locations each year, and could be used together with the species' distribution and habitat information to derive baseline population information for the entire country. These data could also be used to detect population changes.

Habitat

In the absence of a national bird atlas with extensive fieldwork, an understanding of Turkey's habitats and the suite of species each supports would be a great aid to predicting species distributions more precisely. However, developing a simple and meaningful means for birdwatchers to collect habitat data remains problematic. In forests, for example, the problem is not just to record the principal tree species, but also the forest structure – age classes, understorey, presence of dead wood, open areas, amount of grazing. This problem was underlined during fieldwork in Mediterranean forests in 2000, where surveys in maquis revealed varying bird communities which seemed to be related more to habitat structure (affected largely by grazing, cutting and burning activities) than to the principal woody species present. Most of the habitat classifications used are usually of Western European origin and, understandably, serve their original purposes best. For example, steppe might prove a useful definition elsewhere, but encompasses several different types of vegetation and structural levels in Turkey. Considering the wealth of the Turkish flora and the varied landscape, habitat classifications of different 'resolutions' or approaches, as stated above, are required to better collect and use field data.

Achieving a more even coverage of all habitat types would improve our overall understanding of Turkey's birds and provide a more representative dataset. This will require organised, systematic surveys, as it is unlikely birdwatchers will choose to spend much of their birdwatching time away from wetlands – the spectacular birdwatching these habitats provide has always made them natural magnets and most bird records continue to come from them. Over the years, this wetland bias to records has resulted in a depressing catalogue of the species and important areas lost as the State Drainage Authority has dutifully and efficiently gone about its work, draining and damming wetlands in accordance with the national master plan drawn up in the 1930s.

Since 1998, systematic and large-scale bird and biodiversity studies have started to fill the gaps in our knowledge of the birds of steppe, forest and montane habitats. In particular, projects implemented by WWF-Turkey (in South-east Anatolia and the Mediterranean regions) and the Middle East Technical University/TEMA (in the Lesser Caucasus), between 2000 and 2005, have helped establish a standardised and robust methodology for the large-scale studies Turkey needs. It is to be hoped they will continue and ultimately cover the entire country.

MIGRATION STUDIES AND MONITORING

Visible migration of raptors and other soaring birds

Despite Turkey's long-recognised importance for migrating birds, there has been little in the way of sustained, coordinated monitoring of migration. Observations of raptors and soaring birds at the Bosphorus, in north-east Turkey and at the Gulf of Iskenderun in the 1960s and 1970s (Porter & Willis 1968, Beaman & Porter 1977, Beaman 1977, Sutherland & Brooks 1981) awakened Western birdwatchers to the spectacle of migration in Turkey and resulted in a flurry of visits. However, with increasing mobility and accessibility of more exotic destinations and, importantly, new species to be seen further east, Turkey is now a less popular destination for visiting birdwatchers, and those who do visit tend to concentrate on seeking the specialities rather than observing migration. There is also the feeling – not entirely justified – that 'everything is now known', so there is no longer scope for ground-breaking research or valuable study.

The recognised Turkish migration bottleneck sites are gazetted as IBAs in the same way as any others, so as IBAs there is a need for their bird populations to be monitored to ensure the sites continue to meet the criteria and to monitor threats. To be of value, observations must be made systematically and form a long-term monitoring programme which results in comparable, annual datasets, in the same way that the Breeding Bird Survey in the UK relies on constant and systematic effort over a period of years. We cannot know a site is important without having runs of systematically collected data to demonstrate it.

Counting visible migration requires great discipline and long periods of concentration from a few committed counters, and this is a major reason why regular sustained raptor migration studies are undertaken at very few sites anywhere in the world; obtaining funding for counting raptors is also difficult, as the perceived conservation value of the data has still to be demonstrated, but the spectacle of migration is a powerful tool for raising awareness and engaging the general public with their natural environment and the need for its conservation; this tool should not be underestimated.

Another aspect of bird movements lies in understanding bird populations: are the Kentish Plovers present on a wetland in winter the same birds that breed there? This type of study, which usually involves ringing of migrating waders and passerines, started with the two studies by WIWO in the Çukurova (1990 – Kivit *et al.* 1992) and Kızılırmak deltas (1992 – Hustings & van Dijk 1994). The establishment of a national ringing scheme, in 2002, produced trained ringers and a coordinated programme involving spring and autumn ringing at several sites – Ankara, Manyas, Manavgat, Kars, Diyarbakır and the Kızılırmak Delta; the Kızılırmak being one of the longest established and the most consistently manned location (Barış 2005). In conservation terms, the long-term value of this type of study lies in establishing the numbers of birds using stopover sites, their turnover rates and changes in populations, and thus the real importance of each site for birds.

But these constant-effort sites are contributing far more than data to the long-term good of Turkish ornithology. These seasonal observatories are up to three-month-long camps, manned by volunteers (experienced and beginner birdwatchers), which provide unique opportunities for Turks to become more deeply connected and involved with birds and the natural environment, to share their experiences, ideas and passion for birds with like-minded individuals, and to work together to learn new skills which they can use in scientific and academic studies. There can be no better way to establish a solid foundation for the growth of birdwatching and ornithology in Turkey.

DATA

When all the above information is collected for Turkey, it is important that all the records are stored in a place and format accessible to anyone, particularly those who can use the information for bird conservation. The KuşBank web-based database – on which work commenced in 2001 and which was finally launched in 2004 – has been developed with precisely that aim. KuşBank has proved extremely successful and popular with birdwatchers – by the end of November 2005 the database contained over 97,000 records of 400 species, submitted by 350 observers.

A recent example of the value of the KuşBank data is its use for the confirmation and identification of IBAs for wintering waterbirds. This involved querying all of the counts from the 2004/05 national Midwinter Waterbird Count to identify those records where counts equalled or surpassed the IBA-qualifying thresholds.

This exercise identified two sites previously unsuspected as qualifying as IBAs and which will now be more closely monitored to determine their true importance. As the data in KuşBank increase, the number of sites and species where this type of exercise can be carried out will be that much greater and the dataset will enable trends for sites or species to be determined.

However, to be of real value, the quality of records in KuşBank must be assured and therefore a Records' Committee has been established to ensure such standards are met. At present the Committee only considers those records submitted to KuşBank, but it is hoped that it will eventually form the basis of a national Rarities' Committee.

KuşBank is very much an ongoing project with refinements being made to the system on a regular basis. Proposed improvements to the current system include adding a GIS-based data-input facility so that observers can click on a map to indicate the precise location of their observations, and the facility to enter data both offline and by importing spreadsheets, both of which will make the inputting of historical data that much easier.

Filling the gaps

Nationwide data collection and long-term monitoring are unachievable goals for the current generation of local individual birdwatchers. The problems are many.

- Turkey is a very large country and most birdwatchers are located in a few cities.
- Most birdwatchers are university students whose interest in birds is new, whose free time in spring is greatly restricted by exams, and whose interest in birds all too frequently ends once the course finishes and the need to earn money becomes paramount.
- There are too few (less than 20?) experienced birdwatchers sufficiently capable and motivated to follow standard methodologies without guidance and supervision.
- There are too few birdwatchers able to drive, and virtually none with access to the type of all-terrain vehicle needed for fieldwork.
- Professional ornithologists are few and academic interest is very recent.
- There are very few birdwatchers who can undertake surveys, even at weekends, without some financial support.

Despite all this, if a national data-collection programme worked systematically from region to region, year by year, following a rolling programme and provided training, regular support and encouragement to birdwatchers, it would be possible to obtain a detailed dataset providing species distributions and relative abundance for the entire country. More importantly, experiences during and since birdwatcher training in 2001–02 have demonstrated this exercise would also be an effective way of speeding-up the process of obtaining experienced and capable local birdwatchers. By the end of all fieldwork there would be a strong, capable and enthusiastic force ready to carry out long-term monitoring of species populations and contribute effectively to their conservation.

We believe it is important not to let another 40 years slip by but to make the Turkish Bird Atlas dream a reality in our lifetimes. To achieve it will require long-term funding, long-term commitment and vision. What are we waiting for?!

TURKEY'S ECOREGIONS: THEIR BIODIVERSITY AND CONSERVATION

by Hilary Welch and Guy M. Kirwan

The following serves as an introduction to the main ecoregions of Turkey and thus provides a background to the country's zoogeography. We describe the main physical features of each of Turkey's ecoregions, along with data on their climate, habitats and floral communities. Interesting data concerning typical mammals of each region and, especially, avifauna are given in many cases. The aim is to present an overview of the country's general ecology and the mechanisms shaping its biodiversity, with notes on particular conservation problems operating in each region. It should be noted that we give almost equal attention to two ecoregions which are primarily Iranian and Caucasian with only relatively small extensions into Turkey.

Some general notes on conservation are worth making as a background to the information which follows. Protected Areas, whilst better than nothing, are generally ineffective due to a lack of understanding of their individual importance for biodiversity and of how to actively conserve and promote them. All protected areas are under-staffed and forest rangers lack training and the skills to monitor wildlife and deal with visitors. First and foremost, protected area managers and staff tend to be foresters. Additionally most forested protected areas are also managed for timber as a means of income generation for the Ministry of Environment and Forestry, without regard for the needs of the area's biodiversity. Wetland protected areas come under the Directorate of National Parks but in times of national need the Directorate has no power and wetland management is, in truth, in the hands of DSI, the State Water Authority which has no nature conservation mandate. Turkey is also continuing to work to a national development plan initiated in the 1930s by Kemal Atatürk, and is following policies based on a world economy, technology and knowledge base which has changed considerably in the intervening years.

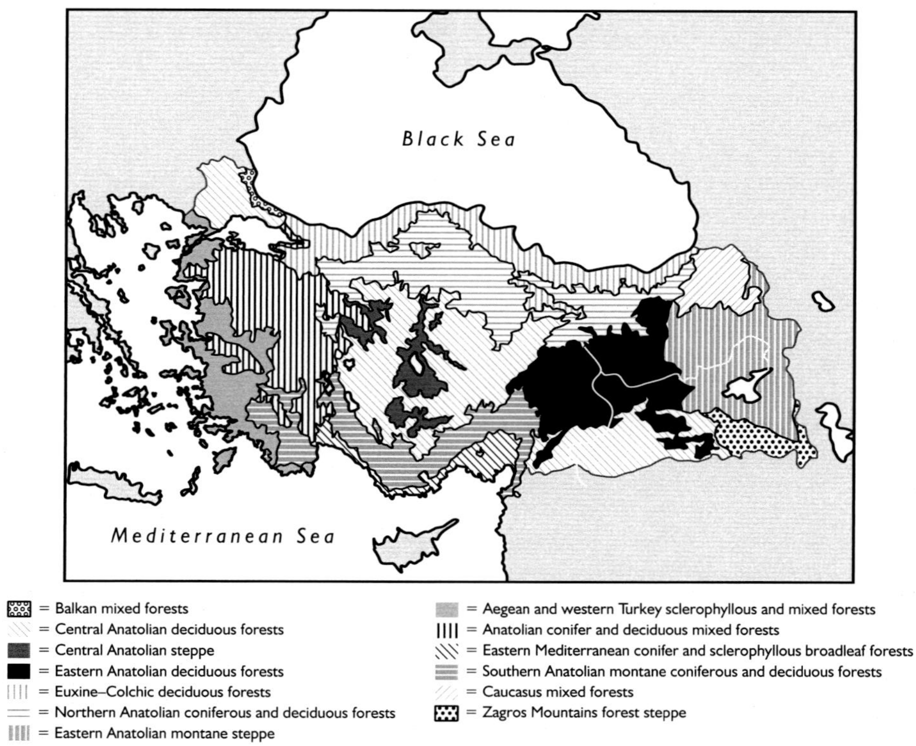

Figure 1. Map showing ecoregions of Turkey.

THE ECOREGIONS OF TURKEY

Balkan Mixed Forests (Thrace)

The vegetation of this ecoregion, especially its forests and grasslands, is Central European in character and also covers much of neighbouring Bulgaria. Floral and faunal diversity is relatively high compared to the rest of Europe, though bird species are generally those typical of such forests elsewhere in the Balkans, and with many endemic plant species. Mixed oak forests are characteristic, with *Quercus frainetto* dominant. Oak forests are interspersed with pine, silver fir *Abies alba* and Norway spruce *Picea abies* forests, wooded pastures, woodlands dominated by drought-resistant species and grasslands. High valleys and sheltered slopes are dominated by beech *Fagus sylvatica* and hornbeam *Carpinus orientalis* and *C. betulus* forests. The region's herpetofauna is among the most diverse in Europe.

Central Anatolian Deciduous Forests

Lying within the boundaries of the Inner Anatolian basin, surrounded by the mountains of the Black Sea and the Taurus, and contained within it is a second ecoregion (Central Anatolian Steppe), comprising salt steppes, a plain, and two river basins all with a distinctive treeless steppe character. The region's main community is steppe. Depending on topography, climate and soil, the vegetation is dominated either by woody communities with small trees, principally oaks, or by herbaceous vegetation. In montane areas and those with high precipitation the tree layer is denser, but elsewhere it is mainly herbaceous steppe with small patches of woodland or isolated trees. Climate is continental with cold wet winters and hot dry summers, whilst annual precipitation averages 400–500 mm.

The mountains to west and south act as barriers to Mediterranean flora and fauna, whilst in the north they mark the limit of Euro-Siberian flora, and the deep valleys of the Yeşilırmak, Kızılırmak, Sakarya and Göksu rivers permit the penetration of other floristic elements. The area is formed by a series of hilly plateaux and the lower slopes of the surrounding mountains. There are three main steppe formations: low mountain steppe, deciduous steppe woodlands (*Quercus pubescens*) and Anatolian black pine (*Pinus nigra pallasiana*) steppe forests. The most widespread of these is the first, occurring at 900–1,250m though with a variable distribution from south to north and west to east. The dominant communities are: *Thymus–Astragalus* steppe (woody plants with deep roots, and small leaves and thorns dominate as they are well adapted to drought and overgrazing); Gramineae steppe (a rapidly disappearing habitat, dominated by grasses, which is commonest at lower altitudes and in less disturbed areas); and *Artemisia santonicum* steppe (commonest in the south, especially at the boundary with the Central Anatolian steppe; due to agriculture and overgrazing this is the most degraded community in the ecoregion).

Low mountain steppe is generally considered to be man-made, its origin attributed to the over-exploitation and destruction of forests. Archaeological and historic records support this theory and indicate that the destruction started as early as 4,000 BC. There are reports of people in Cappadocia gathering wood from Erciyes mountain's no longer extant forests, and during the Roman occupation of Anatolia most forest cover was burnt. Palaeontological studies show that during these eras the climate was dry and it seems that this, together with continual human interference, prevented natural vegetation from regenerating.

Of deciduous steppe woodlands and Anatolian black pine steppe forests, it is widely accepted that the latter is the climax vegetation. Beynam Forest, south of Ankara, and the Ayaş Mountains are considered two of the best examples of the successional process between these two communities, and support both species. In the north of the ecoregion, where a more humid climate prevails, the successional stages create a mosaic of Anatolian black pine, *Quercus pubescens*, wild pears and hawthorns. *Q. pubescens* is the most common wooded habitat in the ecoregion, usually in the form of coppiced shrubland, but in protected areas it can grow to heights of 10–12m. In areas of deeper soils and more humid climate, *Q. robur* or *Q. cerris* are the dominant species with *Q. pubescens* forming a shrub layer. But for human use and disturbance, there is no doubt that the region would be covered with *Q. pubescens* woodland. Due to the harsh environmental conditions, it seems likely that the extent of Anatolian black pine forest would always have been limited, at least in recent times, as the region's extreme summer droughts cannot support tall forest. Wild orchards occur in the south of the ecoregion and comprise many small fruiting trees and shrubs including hawthorn, plum, pear, rowan, almond, oak, juniper and dog rose. Their fruits are an important source of food for many frugivorous birds and small mammals.

The area is rich in plant species due to the varied geology and topography over very small areas, and most taxa have a large number of subspecies and varieties. Erciyes Mountain, the Ayaş and Emir ranges, and

Beypazarı, Karaşar and Nallıhan regions are important centres of biodiversity with diverse and representative forest and steppe communities. Among these, Erciyes Mountain is particularly distinctive.

The 'Anatolian Diagonal' determines the eastern boundary of the ecoregion. This biogeographic line starts around Bayburt, crosses the eastern Anatolia mountains and reaches the Mediterranean at the Amanos Mountains and Mersin. East and west of the Anatolian Diagonal are different floristic and faunistic communities.

Together with the wetlands that occupy this region, the habitat diversity provided by the steppe, oak woodland and pine forests have created an ecoregion rich in birds. The representative bird species of the region's steppe are Chukar, Long-legged Buzzard, Saker and Lanner (both now very rare), Lesser Kestrel, Common Quail, Stone-curlew, Greater Sand Plover, Gull-billed Tern, Black-bellied Sandgrouse, Calandra Lark, Bimaculated Lark, Greater Short-toed Lark, Isabelline Wheatear, Upcher's Warbler, Rook, Rock Sparrow and Black-headed and Corn Buntings.

Conservation Lowlands have been converted to agriculture, foothills destroyed by overgrazing and the last patches of oak woodland and wild orchards converted to conifer plantations. Following agricultural modernisation in the 1950s, large areas were cultivated, destroying most of the *Artemisia fragrans* steppe in particular. In contrast the undulating and hilly areas less suitable for agriculture tend, instead, to be over-grazed and this too has largely destroyed the steppe cover. Wind and soil erosion are now major problems in many areas. Forestation, important in combating erosion, can also have negative effects. In eroded areas there tends to be less soil cover and a disturbed water regime, making forestation difficult. Wanting nonetheless to increase tree cover as per their remit, foresters are thus inclined to plant trees in less disturbed sites. As these areas often support natural plant cover, the result is a monoculture conifer plantation replacing the rich steppe community.

Central Anatolian Steppe

The main feature of this ecoregion is the lack of tree communities and it has thus been referred to as 'true steppe' although it is most likely that the area lost its tree cover due to human exploitation. The region includes the Konya closed basin, originally an inland sea, and an area rich in biodiversity associated with its inland wetlands, both saline and freshwater. Though many of the region's wetlands are now drained or much degraded, the region's characteristic steppic wetland species can still be seen. The ecoregion is composed of plains and river basins in three areas. Tuz Lake lies in the centre with the Kızılırmak (Anatolia's longest river) describing a wide arc from Cankiri in the north to Nevşehir in the east and forming the northern and eastern boundaries. In the south, separated by the Obruk Plateau, is the Karapınar Plain and to the west lie the Porsuk and Sakarya basins.

The region's mean altitude is *c.*1,000m and its main geological formations are young (Oligocene) alluvial deposits of gypsum and salt deposits. Climate is continental with cold wet winters and hot dry summers. Annual precipitation averages 400–500mm.

Salt steppes and lakes are the main formations and are represented by Tuz Lake, Tersakan Lake, Eşmekaya Marsh, Çumra Plain, Ereğli Plain, Çalıkdüzü Plain and Karapınar Plain. However, all of the waterbodies are under great stress from pollution, water extraction and/or drainage to the extent that those at Eşmekaya and Ereğli now barely exist. Due to the high salinity there are no woody formations and only halophytic plants grow.

Tuz Lake is the second-largest lake in Anatolia covering 1,900km^2. In summer much of it dries up and reveals the 1–30cm salt layer below (salt pans here provide more than half of Turkey's salt). Salinity, affected by distance from the lake and micro-topography, determines the plant communities. In the most saline areas plant communities are almost monospecific, with variety increasing as the salinity reduces. From most to least saline the plant communities around the lake are: *Salicornietum europaeae* (dominates nearest the water); *Halocnemum strobilaceum* (another largely monospecific association); *Atropis distans–Limonium gmelinii* (less salt tolerant and patchily distributed); *Petrosimonia birandii* (on muddy sandy soil close to the lake); *Frankenia hirsuta–Limonium icomium* (on gently sloping areas); and *Salsola inermis* (growing adjacent to the surrounding *Artemisia* steppe). The lifecycle of the vegetation is completely different from that of the surrounding steppe. *Tamarix* can also be common, and in areas with freshwater and brackish marshes, *Phragmites australis* and *Juncus heldreichianus orientalis* are dominant.

Amongst the more typical birds of the steppe habitats are larks, with six species, and the region is of particular conservation importance for the small populations of Great Bustard and Little Bustard it supports. Other characteristic birds of conservation importance are: Pygmy Cormorant, Great White Pelican, Dalmatian Pelican, various herons, Glossy Ibis, Eurasian Spoonbill, Greater Flamingo, Marbled Duck, Red-crested Pochard,

Ferruginous Duck, White-headed Duck, Pallid Harrier, Lesser Kestrel, Common Crane, Collared Pratincole, Kentish Plover, Greater Sand Plover, Spur-winged Lapwing, Gull-billed Tern and Little Tern. Unsurprisingly, the region's ten Important Bird Areas are all wetlands. They are important for their populations of breeding birds, and also for migrating and wintering waterbirds. In winter the region supports significant numbers of Greater White-fronted Geese and Ruddy Shelducks.

Conservation Of the region's few protected areas, Sultan Marshes is more often dry than wet due to the demands on its water, and Gölbaşı Lake, though close to Turkey's capital and the statutory bodies responsible for environmental and nature conservation issues, is under great pressure from extensive urbanisation, development of recreation facilities, road construction and pollution. All remaining areas of steppe are over-grazed, particularly the *Artemisia fragrans* steppe, and threatened by conversion to agriculture. Though the region is dry and its soils salty, irrigation schemes using the fresh water feeding the salt lakes have succeeded in converting salt steppe to cultivatable land. This has led to a loss of salt steppe habitats and has diverted fresh water from the wetlands, increasing their salinity and lowering water tables. In 2001 Tuz Lake was declared a Specially Protected Area (SPA) covering the whole lake, surrounding marshes and some important areas of steppe. Eşmekaya Marshes has been protected as a Wildlife Reserve since 1994, which prohibits hunting. This protection has not been sufficient to prevent an irrigation scheme by the State Water Authority (DSI) extracting most of the water from the marshland and the area's value for birds has now been lost. The principal threat to wetlands is mismanagement, particularly through changes in the hydrology and over-consumption of underground water resources.

Eastern Anatolian Deciduous Forests

At the western end of the East Anatolian mountains, at the boundaries of the ecoregion, is a rich zone where four biogeographical realms meet: the Dry Central Anatolian Plateau, Arabian Plateau, Black Sea and East Anatolian Plateau. The main geographical features characterising the region are the Upper Euphrates, Anti Taurus Mountains and the western East Anatolian mountains. The Tahtalı Mountains define the western boundary, whilst the arc of the Anti Taurus between the Euphrates and Tigris determines the southern. The eastern and northern boundaries of the ecoregion are less apparent due to the area's generally montane nature. In the north the gradually increasing influence of the Black Sea climate provides an indistinct border, whilst in the east it is the increasing effect of the continental climate and the greater height of the plateaux. The highest points are in the Anti Taurus, which rise to 3,462m in the Munzur range. The upper catchments of the Ceyhan, Tigris and Euphrates are the important rivers, whilst Hazar Lake and Nemrut Crater Lake are the main natural waterbodies. There is also a network of recently constructed reservoirs. Climate is continental with extreme winter temperatures. Annual rainfall varies between 600 and 1,000mm and snowfall can be heavy.

The Anti Taurus Mountains include many Mediterranean elements. Inland the Mediterranean forests and shrublands are replaced by steppe forests dominated by oaks, with *Quercus brantii*, *Q. libani*, *Q. boissieri* and *Q. ithaburensis macrolepis* particularly obvious. Due to the dry continental climate, these forests are resistant to cold and drought, and are referred to as dry or xeric forests. The main forested communities, characterised by their open canopy, well-developed herb layer and patchy distribution, due to human use, are: *Quercus robur pedunculiflora* forest; *Pinus sylvestris* montane forest; wild orchards; *Rosa pimpinellifolia* / *R. canina* shrublands; *Platanus orientalis* valley forest; humid relict Euxinic forests of *Alnus* / *Castanea sativa*, *Acer*, *Lonicera caucasica*; and *Juniperus communis nana–Convolvulus calvertii* woodlands.

The extent of steppe in the ecoregion has increased due to widespread forest loss. The main steppe communities are: *Artemisia fragrans*; mixed *Astragalus* sp., *Gundelia tournefortii*, *Noaea mucronata*, *Thymus* sp. and *Salvia cryptantha*; various grass formations; and high mountain steppes of *Achillea vermicularis*, *Ajuga chia*, *Helianthemum nummularium*, *Malcolmia africana* and *Marribum parviflorum*.

The Anatolian Diagonal (see previous ecoregion) runs from the southern foothills of the Black Sea Mountains around Bayburt, through the Munzur and Anti Taurus and reaches the Mediterranean at the Amanos and Bolkar Mountains. A high proportion of Turkey's plant species occurs either west or east of this line indicating its great biogeographic significance. The ecoregion is considered one of the most important in Turkey for plant biodiversity with 3,200 vascular plant species, 725 of them endemic to Turkey.

Conservation Over the millennia, forests have been extensively cut for construction and fuel. In 1974–94 more than half of the forests around Bingöl were cleared. There is only one protected area in the ecoregion, Munzur Valley National Park (48,000ha).

Euxine–Colchic Deciduous Forests

This ecoregion of humid and old-growth forests extends from the Istranca Mountains (Turkey) to Abkhazia (Georgia) and is mainly restricted to the southern coasts of the Black Sea. Floristically, it is also known as the Colchic region. The Melet River, which empties into the Black Sea near Ordu, constitutes the boundary between the Colchic and Euxine subprovinces. Here the high ridges of the eastern Black Sea Mountains end and the low ridges of the western Black Sea Mountains begin. The high mountains act as isolating barriers for the distinctive Caucasus region, determining its climate and vegetation. The ecoregion has a moist, mild climate and is the most humid in the Caucasus. In the Colchic region mean annual precipitation is 1,500–2,500mm, rising to 4,000mm, whilst in the Euxinic region mean precipitation is 1,000–1,500mm.

Geologically the region is young, its formation having started in the last 5,000–6,000 years. Most soils are alluvial. The larger rivers carry away snow, rain, ground and glacial waters, and alluvial deposits from the mountains to the Black Sea's mouth; here hardwood and softwood bottomland forests grow. Bottomland forest is an extremely rare habitat throughout Europe. On the Black Sea coast there are three examples (all protected as Strict Nature Reserves), at İğneada (Kırklareli), Sarıkum (Sinop) and Hacıosman (Samsun). Bird species associated with these forests include Black Stork, White Stork, Pygmy Cormorant, Lesser Spotted Eagle and a variety of woodpeckers.

At the mouths of the Kızılırmak and Yeşilırmak large and important deltas have formed. Here important species include Dalmatian Pelican, Purple Heron, Eurasian Spoonbill, Gadwall, Red-crested Pochard, Ferruginous Duck, Common Crane, Purple Swamphen, Stone-curlew and Collared Pratincole. Several passage and wintering birds also occur in significant numbers, and Cattle Egret is regularly seen in the Yeşilırmak Delta.

With such high precipitation, the main vegetation is broadleaf mixed deciduous forest, dominated by *Fagus orientalis* above 1,000m. This grows together with *Castanea sativa*, *Quercus petrea iberica*, *Acer leatum*, *A. cappadocicum*, *Zelkova carpinifolia*, *Tilia* sp. and *Pterocarya fraxinifolia*. There is also a diverse understorey including *Ilex aquifolium*, *Hedera helix*, many rhododendrons, *Laurocerasus officinalis*, *Buxus colchica*, *B. sempervirens* and *Ruscus colchicus*. The diverse forests are unique for their humidity, intactness and scenic beauty. In the west they are characterised by their species richness (12–15 tree species per 500 m^2 in the Bolu, Zonguldak and Bartın regions) and in the east by their maturity. They support one of the largest populations of Brown Bear *Ursus arctos* in Europe. In the Çoruh Basin, on the fringe of the Lesser Caucasus, is a Mediterranean enclave where mesic forest, referred to as pseudo maquis, dominates.

Also in this ecoregion are some rare and threatened peatlands, of which less than 2% (<2,000ha) now remain, though this is still the largest remaining fragment of peatland in the eastern Mediterranean. On the coast there are also 23 dune systems of particular importance for their rare plants.

Migratory bottlenecks are a feature of both the western and eastern ends of the region and involve birds from breeding populations in Eastern Europe, Fenno-Scandinavia and Russia. The migration includes storks and raptors, waterbirds following the Black Sea coast and an even larger movement of passerines, although these last are much less well studied.

Conservation Very little of the original Colchic forest remains mainly due to uncontrolled logging and urban and agricultural development. These high-productivity lowland forests are heavily logged following forestry management strategies, rather than biodiversity-oriented ones. Around 60% of the wetlands have been drained for agriculture (mainly for tea and citrus plantations) and rural development. In eastern Turkey, the Black Sea motorway continues to reduce the remaining coastal habitats to small patches, despite strong objections by environmentalists. Here, as elsewhere in Turkey, despite the extensive areas of suitable habitat, hunting is a major threat to wildlife. On the coast there is extensive use of nets in autumn to catch birds that have crossed the Black Sea.

Northern Anatolian Coniferous and Deciduous Forests

This ecoregion extends through north Anatolia, south of the coastal zone. The landscape is mountainous, forming a barrier between the continental climate of Inner Anatolia and the oceanic Black Sea. Thus, on the north-facing slopes grow humid broadleaf deciduous forests, and on the southern slopes drier needle-leaf conifers. The east is characterised by the high ridges of the eastern Black Sea Mountains, rising to more than 3,000m, and in the west by the north-west and central-north Anatolian mountains, which rarely exceed 2,000m. Due to the shelter provided by the high coastal mountains and the continental climate to the south, the region has a transitory climate with high precipitation (500–1,000mm) its main feature. In west and central areas the wettest period is the winter, and in the east it is the spring.

The region lies within the Euxinic realm. Forest composition and structure varies from west to east and north to south. In the west *Abies bornmulleriana* is the dominant species, forming mixed forests with *Fagus orientalis* and/or *Pinus sylvestris*. In the east *Picea orientalis* dominates. In central areas small patches of *Cedrus libani* are found, though their naturalness is questioned. In the south, where conditions are more steppic, *Pinus* spp. are widespread with the dominant species *Pinus nigra pallasiana* in the west and *P. sylvestris* in the east. Many variations occur, depending on climate and topography, but other common species are *Quercus* spp., *Acer* spp., *Sorbus* spp. and, particularly at high altitude in the east, *Juniperus oxycedrus, J. communis* and *Betula* sp. form patchy scrub. As with the previous ecoregion, an important feature is the intact forest cover, with high woody species diversity in the west, and maturity (old-growth features) in the east. The region is notably rich in centres of plant endemism and the intact forests are thought still to support particularly strong populations of large mammals, including the Brown Bear *Ursus arctos*.

Breeding raptors are an important feature, with Eurasian Black Vulture, Lammergeier, Eurasian Griffon Vulture and Booted Eagle. In the north-east, the Black Sea Mountains are a major migration route used by many species of raptor on their journeys between breeding grounds in Russia and wintering areas in Africa and Arabia.

Conservation As elsewhere, it is likely that the intact forest cover is threatened by the Ministry of Forestry's management plans, which are oriented towards timber production and threaten the quality rather than the total area of the forests. Dams on the principal rivers are rapidly destroying Turkey's remaining wild rivers in the name of electricity generation. Since the year 2000 three dams have been built on the Çoruh River, destroying its outstanding natural beauty, limiting the region's considerable tourism potential and severely reducing its value for wildlife: at Muratlı, on the Georgian border, after less than 12 months of operation, the dam is reported to be silting up unexpectedly rapidly, reducing the dam's useful life by *c.*40%. Illegal hunting threatens a number of the region's larger mammals.

EASTERN ANATOLIAN MONTANE STEPPE

A mountainous ecoregion dominated by cold winters and strong winds. The mosaic of steppe and patches of woodland, and the remoteness and intactness of the landscape, make the region rich in wildlife. It is located at the junction of the Lesser Caucasus and the Iranian and Mediterranean biogeographic zones. There is a great range in altitude (375–5,165m) and an attendant diversity of climatic zones that have produced a diversity of landscapes and distinctive ecological communities, including many regionally endemic, relict and rare species. The dominant vegetation community is montane steppe, but there are also significant areas of desert and semi-desert, forest and woodland, and alpine and subalpine meadow.

At 1,500–2,500m the landscape is dominated by wide undulating plateaux; above this tower the high peaks of Ağrı (5,165 m) and Alagöa mountains (4,095 m). The plateaux are swept by strong cold winds which prevent growth of trees and shrubs. Only in lower areas and protected pockets supporting their own microclimate can patches of woodland be found. Precipitation is more or less evenly distributed with no summer drought, unlike adjacent ecoregions which have a Mediterranean-type climate. Annual rainfall varies between 400 and 600mm. However, in the rainshadow of the high mountains rainfall is just 200–300mm, but in the surroundings of large lakes reaches 800–1,000mm.

Van Lake (3,173 km^2) dominates the Turkish portion of this ecoregion. Typical emergent wetland vegetation is *Phragmites australis, Typha* spp., *Carex acuta, C. diluta, Scirpus tabernaemontani* and *Bolbolshoenus maritimus*. Desert habitats are characterised by a scarce vegetation cover composed mostly of drought-loving and turf-forming species. In contrast, the vegetation of semi-deserts consists mostly of ephemeral plants, such as *Artemisia fragrans, Capparis spinosa, Kochia prostrata* and *Poa bulbosa*. The ecoregion is recognised as a centre of endemism within the Irano-Turanian phytogeographic region.

Dry climatic conditions, a high percentage of vegetation cover and a rich flora characterise the mountain steppes (1,500–2,200m), which include formations of cushion-forming plants (e.g. *Artemisia* and *Astragalus*) and of grasses (e.g. *Festuca* and *Poa*). In the high mountains (2,200–2,700m) umbellifers are a dominant group. Some of the most characteristic species of the alpine vegetation on the high peaks are *Draba* spp., *Dracocephalum* spp., *Oxyria* spp., *Polygonum* spp., *Veronica* spp. and bulbous species like *Gentiana verna* and *Primula* sp. *Juniper–Amygdalus* steppe woodlands are the most characteristic woody formation of the ecoregion, with species composition varying. The open crown encourages a dense shrub layer (including *Pistacia*, *Berberis* and *Rosa*) and herbs dominated by *Astragalus* or *Artemisia* spp. Oak woodlands and shrublands also occur.

Characteristic bird species of the ecoregion include: Egyptian Vulture, Long-legged Buzzard, Golden Eagle, Lesser Kestrel, Peregrine Falcon, Great Bustard, Caspian Snowcock, Wallcreeper, Alpine Chough and White-winged Snowfinch.

Conservation Habitat loss and modification, mainly by agriculture, unsustainable use of biological resources, and the impact of introduced and non-native species have resulted in declines of animals and plants, and degradation of ecosystems. Protected areas appear insufficient for effective conservation. However, in mountainous areas the inhospitable climate and remoteness make the region unattractive for large-scale development, and the only threat comes from the nomadic lifestyle of the local people, herding their livestock to the high mountains in summer. In some areas, due to overgrazing, steppe compositions have changed and *Euphorbia* spp., *Astragalus* spp., *Acanthalimon* spp. and *Gundelia tournefortii* are now dominant.

Aegean and Western Turkey Sclerophyllous and Mixed Forests

This region covers the westernmost part of Turkey (and also most of Greece). Large numbers of small islands in the Aegean are also part of this ecoregion. Three seas, the Ionian, Aegean and Mediterranean surround the ecoregion. The eastern part consists mainly of coastal lowlands and plains. In the north this is reduced to a thin belt, with the northern boundary delineated by the Rhodope Mountains. The volcanic history of the region is apparent in the mountains which extend from east to west in long deep furrows caused by the ancient subsidence of land between two natural fault lines. These valleys carry the principal rivers: the Meriç, Edremit, Bakırçay, Gediz, and Küçük and Büyük Menderes.

Though the Mediterranean climate prevails in the area it shows considerable variation from south to north. In the north annual precipitation is more evenly spread with summers less dry and winters less wet. Overall, winter has the most and summer the least rainfall, with annual precipitation ranging from 600 to 1,350mm. Eastern Mediterranean vegetation is centred on the Aegean region, with the addition of a number of elements from surrounding floristic regions. During the Pleiocene, Greece and western Turkey were connected, accounting for their similar vegetation and flora today. The most common communities are maquis and *Pinus brutia*. From south to north the dominant maquis species change and the annual growth rate of *P. brutia* decreases: in the south, the upper altitudinal limit is considerable higher than in the north. These pine forests support a limited range of typical European woodland bird species: Winter Wren, Common Blackbird, Common Chiffchaff, Coal Tit, Blue Tit, Great Tit, Short-toed Treecreeper, Eurasian Jay and Common Chaffinch. Krüper's Nuthatch is common and is probably the habitat's most important bird species as the vast majority of its population occurs in Turkey.

Principal maquis species are *Arbutus andrachne, A. unedo, Spartium junceum* and *Laurus nobilis*, but in areas where *Pinus brutia* has been removed *Quercus coccifera, Calicotome villosa, Paliurus spinacristii* and *Erica arborea* are the main species. Characteristic birds of maquis include *Sylvia* warblers (e.g. Subalpine Warbler), buntings (Cirl, Rock and Black-headed), and Chukar. Species considered important on a global basis are Eastern Olivaceous Warbler, Olive-tree Warbler, Rüppell's Warbler, Masked Shrike, Cinereous Bunting and Cretzschmar's Bunting.

In the south, the Aydın and Samsun Mountains possess rich and varied maquis communities featuring local species such as *Quercus aucheri, Cupressus sempervirens* and *Juniperus phoenicea*. Although such communities are extremely disturbed everywhere these mountains contain a well-protected example of this habitat.

Liquidambar orientalis is endemic to southwest Turkey and is of highly restricted distribution, forming small patches from the Çine River in the northwest to Fethiye in the southeast. In 1949, these forests covered 6,312ha; now just 1,215ha remain. The main stronghold is around Köyceğiz where at least 50% of all *Liquidambar orientalis* occurs. The trees grow best in low-lying damp areas, along streams or river valleys, and form a dense forest canopy with hanging creepers and *Rubus* and *Smilax* forming dense thickets. A distinct bird community is associated with these forests. Principal species are Common (Rufous) Nightingale, Eastern Olivaceous Warbler and Cetti's Warbler. Levant Sparrowhawk occurs in larger and more mature patches of habitat.

Conservation As in all Mediterranean coastal areas, the dense human population and agricultural activities have destroyed or fragmented natural habitats. Pine forests and maquis have been lost to urbanisation and agriculture, or degraded by overgrazing and illegal logging, so there are very few patches of undisturbed habitat remaining. The Gelibolu Peninsula in the north and the Datça Peninsula in the south, both with low human population densities, are relatively undisturbed areas of habitat. The remaining coastal wetlands are under increasingly heavy pressure from the spread of tourism, both for their land and for the precious freshwater resource they can provide.

Anatolian Conifer and Deciduous Mixed Forests

This region lies in the Inner Aegean. The biogeographical features of the Anatolian Plateau vary in two directions, from east to west and north to south. The south–north shift is from Mediterranean maquis and scrub to deciduous broadleaf Euro-Siberian formations, whilst the west–east change is to steppic Irano-Turanian formations. Since Euro-Siberian formations dominate the Black Sea region and Irano-Turanian formations the central Anatolia plateau, this ecoregion acts as a bridge. Mountains delineate it, with the highest Uludağ (2,543m) in the north. In the north-east are the Sündiken Mountains (1,786m), in the east the Afyon Mountains (1,992m), and in the west the Aydın Mountains (1,831m), Boz Mountains (2,159m) and Kaz Mountains (1,774m).

A Mediterranean climate dominates, but its character changes from west to east, becoming increasingly affected by the continental climate of the central Anatolian basin. In the north, winters are colder and precipitation is less. Annual precipitation is 400–600mm. Dominant vegetation is *Pinus nigra pallasiana* and *P. brutia* with mixed oak woods and shrublands. *P. brutia* forests occur mainly in the western foothills and low-lying areas, but their distribution is also influenced by the sea. As continentality increases, this species is replaced by *P. n. pallasiana*. *P. brutia* forms communities with oaks, *Quercus cerris*, *Q. ithaburensis macrolepis* and *Q. coccifera* being commonest. *P. n. pallasiana*, one of the most dominant formations of the region, is found mainly above 1,000m in the eastern mountains and continental areas. It can be found in pure stands or mixed with *P. brutia* (in the west) or with various oaks (in the east). At higher altitudes it occurs with junipers and a fir endemic to the Kaz Mountains. Oaks are the region's other important formation. They can dominate in both humid and dry areas, mix with pine forests and cover degraded areas. *Quercus cerris* is the most widespread and commonest forest species whilst *Q. pubescens* is the principal form in steppic and degraded areas. More localised species forming forests are *Carpinus orientalis*, *Castanea sativa* and *Juglans regia*, the latter two mainly along streams. On Uludağ there are communities of the endemic *Abies bornmulleriana* and *Fagus orientalis*.

Although the region has Mediterranean characteristics it does not support as wide a range of endemics as southern Anatolia, due to the lack of high mountains and deep valleys forming refuges for warm-climate species during glaciation. The northeastern mountains are forested with a mixture of *P. n. pallasiana* and oaks and are rich in steppic plants, with a high number of endemic species.

Conservation The region's *P. n. pallasiana* forests are still heavily managed for timber production. Although 12,762ha of the mountain is protected as a National Park, Uludağ is also a winter tourism centre, which creates conservation problems due to the pressure for further construction.

Eastern Mediterranean Conifer and Sclerophyllous Broadleaf Forests

This region includes part of the northern Mesopotamian basin where agro-pastoral activities and many important civilisations started. Thus, it has long been influenced by human activities. In the west the ecoregion starts in the central Mediterranean coast and extends to Southeast Anatolia (and over much of Syria, Lebanon, Palestine, Israel and Jordan). The Turkish part of the ecoregion occupies two parts: the south Mediterranean coast from Kemer to Iskenderun Bay, with the northern border formed by the southern slopes of the Taurus Mountains, penetrating inland via the Aksu, Göksu, Seyhan and Ceyhan valleys; and Southeast Anatolia and the north Syrian basin, which is bordered to the north by the Anti Taurus and to the south by the Syrian Desert. The latter part is heavily influenced by the desert so the principal rivers, the Euphrates and part of the Tigris, have a significant effect on ecology.

Warm, wet winters and hot, dry summers characterise the climate. Precipitation decreases from west to east, with 1,250–1,500mm around Antalya, 600–800mm in the Adana area, and just 400mm in the east (decreasing still further to the south). The high temperatures, low humidity and poor soil conditions limit the ability of the vegetation to recover following human disturbance, and the botanical record provides evidence of such deleterious effects from as long ago as 3,000 BC.

Vegetation falls into three main groups: broadleaved sclerophyllous vegetation (maquis), coniferous forests of *Pinus halepensis* and *P. brutia*, and dry oak woodlands and steppe formations. The boundary of the ecoregion delineated by the Amanos and Anti-Taurus Mountains is a humid-weather barrier which affects species composition. Maquis is dominant in areas closest to and most affected by the Mediterranean. Here the main species are *Olea europea*, *Quercus coccifera*, *Pistacia palestina*, *P. lenticus*, *Ceratonia siliqua* and *Arbutus andrachne*. However, more than 40 sclerophyllous species contribute to the vegetation, e.g. *Philyrea latifolia*, *Pistacia terebinthus*, *Calicotome villosa*, *Genista acantha*, *Rhamnus oleoides*, *Myrtus communis*, *Laurus nobilis*, *Styrax officinalis* and *Spartium junceum*. In the coastal belt, the *Olea–Ceratonia* alliance is the climax community though this is much degraded everywhere, *Ceratonia siliqua* in particular is now represented by very few individuals. *Quercus coccifera* is one of the commonest and widespread communities because it has a broad ecological tolerance and a good capacity for recovery, including in areas of degraded *P. brutia*. The latter tree is the main forest-forming species, occurring from sea level to 1,800m, but it does not now grow naturally in the Mesopotamian part of the region. Of note are the few remnant patches of *Pinus pinea*, which harbour the southernmost distribution of *Rhododendron ponticum*, a characteristic of Euxinic vegetation in the Black Sea region.

Due to the low rainfall, further east is poorer in terms of woody vegetation, and Viranşehir delineates the boundary between desert and Mediterranean formations. This dry region stretches as far as the Tigris and the Cizre area; it is considered part of the same ecoregion due to the occurrence of *Olea europaea* and other shrub communities with Mediterranean characteristics. The east Anatolian part has drought-tolerant vegetation formations dominated by plants such as *Artemisia* sp. and many others. At higher elevations (*c.*700m) *Quercus ithaburensis*, *Q. cerris*, *Q. brantii* and *Q. infectoria boissieri* form woodland communities, whilst along rivers *Populus euphratica* and *Salix triandra* form gallery forests.

This ecoregion lies on a major bird migration route due to its geographic location, at the junction of three major landmasses, Europe, Asia and the Middle East, and at the north-east corner of the Mediterranean. In Turkey, the Upper Rift Valley ends, with the Belen Pass and the wetland of Gavur Gölü, south of Kahramanmaraş, providing a conduit for species migrating from and to Africa. Two of Turkey's most ornithologically exciting wetlands are found in this region; the Göksu and Çukurova deltas (the former a Specially Protected Area). Both support globally threatened breeders and important wintering waterbirds. Species include breeding Marbled Duck, Ferruginous Duck, Black Francolin, Purple Swamphen, Stone-curlew, Kentish Plover, Spur-winged Lapwing, Little Tern and White-breasted Kingfisher, whilst wintering birds include Greater Spotted Eagle and Eastern Imperial Eagle. Arguably, the most important species is the small, now feral, colony of the eastern population of the Critically Endangered Northern Bald Ibis.

Conservation This region is one of the most degraded due to long and increasingly intensive human settlement. Most coastal areas are either developed for tourism or agriculture. Whilst agricultural activities destroy the natural habitats of the lowland plains, tourism destroys coastal dunes, one of the most fragile and rarest of habitats. Maquis has been over-grazed for centuries, threatened by fire, disappearing under tourism developments and second housing, and not valued by foresters, yet with a rich biodiversity which is still little known and quite different structurally from forest. From most setbacks, the ecosystem is capable of recovering but usually human pressure prevents it. In the east agriculture is extensive and becoming more so with the irrigation and other projects along the Euphrates and Tigris; generally only hilly areas remain uncultivated. The extensive plains and broad river valleys are now principally devoted to cereals; the broad plains of the Syrian Desert, extending from Harran to Cizre, and at their most impressive when seen from the advantage of Mardin, are the wheat-basket of Turkey. In most areas the only remaining natural communities are on the hills and, as these represent the only remaining pasture, they are highly degraded due to overgrazing. Dam construction on the Euphrates and Tigris is a source of major change in the ecoregion. The Euphrates is now dammed regularly along its length and the infamous Ilusu Dam, on the Tigris, has once again been approved for construction.

Southern Anatolian Montane Coniferous and Deciduous Forests

This is one of the most diverse ecoregions in the Mediterranean basin and one of the main centres of endemism in the Western Palearctic. The reasons for this are geology: as the deep valleys of the karst system have created isolated areas with peculiar local conditions, and floristic mix (namely the combination of Mediterranean flora and Irano-Turanian elements), providing unique circumstances in which new species can evolve. The region is delineated by the high mountains of southern Anatolia which, in places, penetrate to the central Anatolian basin. Due to the Irano-Turanian influence the ecoregion is rich in herbaceous plants; >30% are endemic, the most of any Mediterranean region.

The climate is Mediterranean with summer drought and high winter precipitation (800–2,000mm). The main features determining the vegetation are the high mountains rising steeply from the sea and the orientation of these mountains, particularly to the prevailing southwest winds which carry most rain. The high mountains are the western Taurus (3,024m), central (Lycian) Taurus (3,585m), eastern (Cilician) Taurus (3,734m) and the Amanos (2,262m). The entire region is extremely mountainous and the most common parent rock is limestone. Coniferous forests, deciduous forests and alpine meadows are the main vegetation formations. Most forests are coniferous formations of *Pinus nigra*, *Cedrus libani*, *Abies cilicia*, *Juniperus foetidissima* and *J. excelsa*, the latter forming the treeline. *Cedrus libani* occurs in areas with a Mediterranean climate, whilst *Pinus nigra* prefers more continental sites. On southern slopes, the treeline reaches 2,000m, and 2,200–2,400m further inland. Due to higher temperatures and lower precipitation, the further east in the region one goes the higher the treeline.

Deciduous forests are concentrated in the Geyik and Amanos Mountains in the east. This bias is mainly due to the locally high precipitation (1,500–2,000mm). Dominant species include: *Carpinus orientalis*, *Ostrya carpinifolia*, *Quercus cerris*, *Q. libani*, *Q. trojana*, *Q. petrea pinnatilob*, *Acer hyrcanum*, *A. platonoides*, *A. campestre* and *A. monspessulanum*. Another formation is dry alpine meadows where the main associations are *Tanacetion praeteriti*, *Agropyron–Stochyon*, *Alyssum propinquum*, *A. masmenaeum* and many *Astragalus* spp. The region is especially rich in bulbous plants and the Taurus and southwest Anatolia support *c*.950 plant species. The Sandras Mountains support 670 species, 76 of which are Anatolian endemics and 11 are known only from this range. Besides Mediterranean vegetation, it is possible to find relicts of the Euro-Siberian flora on the south slopes of Dedegöl and the west slopes of the Amanos range, which experience high rainfall even in summer, an extremely rare event in this region. Thus, one finds *Fagus orientalis* forests with *Tilia argentea*, *Sorbus graeca*, *Acer* spp., *Rhododendron ponticum* and *Ilex colchica*. These enclaves are apparently relicts of vegetation that spread south during cool wet periods of the Pleistocene and which survived because of the microclimate with high rainfall. In addition the Amanos Mountains mark the southern end of the Anatolian Diagonal and provide a 'bridge', facilitating the penetration of northern flora to Lebanon and Syria. This area is designated an Important Bird Area due to its importance as a migration route for birds rounding the northeast corner of the Mediterranean. Another important feature is the world's largest and most intact *Cedrus libani* forest covering 89,810ha on Akdağı.

Conservation Intensive destruction of natural habitats dates from 2,000 BC when the human population increased rapidly following the successful establishment of agro-pastoral systems. The Anatolian and Mesopotamian basins are one of the principal regions where agriculture and subsequent human settlement began. Despite this, the ecoregion's naturalness is more intact than that of the Southeastern Anatolian and Middle East Dry Conifer and Broadleaf Forests Ecoregion simply due to the rugged terrain. In high montane areas the main threat is overgrazing. As well as this, the species concerned (feral goat) increases destruction through its liking for young seedlings and shoots. Inappropriate forestry activities which favour timber production are another major threat. Collection and over-harvesting of bulbous plants threaten wild populations despite regulations and controls. Tourism, which is one of the most destructive activities in the lowlands and coastal areas of Turkey, has now also started to affect montane areas.

Caucasus Mixed Forests

These are located at a biogeographical crossroads where flora and fauna of at least three biogeographic provinces converge. Speciation levels are high and, accordingly, the Caucasus forests have one of the highest levels of endemism in the temperate world. Unsustainable forest use, including poor management and illegal cutting in combination with uncontrolled timber export, are the chief threats. The climate is temperate but fluctuates with elevation and region, whilst annual precipitation averages 1,500–2,000mm (max. 4,500mm). Temperate mixed forests are concentrated at mid elevations (400–2,200m) and cover most of the region. Forests are predominantly broadleaf with *Quercus iberica* and *Carpinus caucasica* dominating up to 1,000–1,200m; *Castanea sativa* in the humid west at 500–1,000m; *Fagus orientalis* at 1,000–1,800m, replaced in drier areas by *Quercus macranthera*. The treeline is mainly composed of *Betula litwinowii*, *B. raddeana* in humid areas, and *Pinus kochiana* in drier areas. Coniferous forests are mainly composed of *Abies nordmanniana*, *Picea orientalis* and *Carpinus caucasica*. Between the treeline and snowline grasslands occur composed of tall herbaceous vegetation, alternating with thickets of *Rhododendron caucasicum* and rock scree vegetation.

This is a unique ecoregion due to its evolutionary phenomena. The eight ecoregions bordering it differ significantly in climatic conditions and species composition. Apart from this, the area partly coincides with the Colchic refuge of Tertiary Flora, the most important refuge and relict area of the arctotertiary forests in West Eurasia. These are the only places in this part of the world where warm-temperate deciduous forests have existed without interruption since the Tertiary. Even now, many relict forms still appear as dominant or co-dominant, e.g. *Quercus hatwissiana*, *Q. pontica*, *Zelkova carpinifolia*, *Betula medwedewii*, *B. megrelica*, and rhododendrons. Because of such conditions, it has one of the highest levels of endemism in the temperate world with 23% of vascular plants and 10% of vertebrate animals being endemic to the region. In addition, landscape and habitat diversity favours high species richness: *c.*5,000 vascular and 7,000 lower plants, and 700 vertebrate animals are found in this ecoregion.

Faunal richness is high; characteristic mammals include Chamois *Rupicapra rupicapra*, Caucasian Red Deer *Cervus elaphus maral*, Wild Goat *Capra aegagrus*, Mouflon *Ovis orientalis gmelini*, Brown Bear *Ursus arctos*, Grey Wolf *Canis lupus*, Lynx *Lynx lynx*, European Otter *Lutra lutra*, and the Critically Endangered subspecies of Leopard *Panthera pardus ciscaucasica*. Many species of bird are also found here including such rare species as Golden Eagle and Lammergeier and restricted-range species such as Caucasian Grouse. Other characteristic species include Gadwall, Common Pochard and Dalmatian Pelican. The East Caucasian Tur *Capra cylindricornis*, West Caucasian Tur *C. caucasica*, Caucasian Snowcock *Tetraogallus caucasicus*, Great Rosefinch *Carpodacus rubicilla* and Güldenstädt's Redstart *Phoenicurus erythrogastrus* are also characteristic of the Greater Caucasus but unknown in the Turkish sector.

Conservation Urban and rural development have converted most lowland forests to agricultural and unsustainable forest use. Poor management and illegal cutting, in combination with uncontrolled timber export, are the main threats for forest ecosystems. Overgrazing is a major cause of habitat degradation at the upper limits of forests and grasslands, and over-exploitation or poaching of game and economically valuable species is another very significant threat to biodiversity here. Georgia and Armenia have developed national Biodiversity Strategy and Action Plans, wherein conservation of these habitats is listed as one of the priority actions.

Zagros Mountains Forest Steppe

This ecoregion supports oak-dominated deciduous forests and pistachio–almond forests, amidst a diversified steppe flora. This ecoregion is located primarily in Iran, ranging northwest to southeast and roughly paralleling the country's western border. It occurs along the Zagros–Makran mountain arch, starting in the north around the Diyala River, near the Iraq–Turkey border, passing through southern Azerbaijan, Iranian Kurdistan and Faristan, and extending to Makran in southeast Iran. Further to the southeast the mountains continue into the districts of Fars, Laristan and Makran. Here the terrain mostly comprises lowlands and a plateau, interrupted by isolated mountains such as the Shiraz range, and with large flats and saltmarshes. The forest and steppe-forest areas of the Zagros possess a semi-arid temperate climate, with annual precipitation ranging from 400mm to 800mm, falling mostly in winter and spring. Winters are severe, with winter minima often below –25°C, and summer aridity is extreme.

The Kurdo-Zagrosian steppe-forest consists mainly of deciduous broadleaf trees or shrubs with a dense ground cover of steppe vegetation. Dominant species are *Quercus* spp., *Pistacia* spp. and a few others. In the northern reaches, lower altitudes (400–500m) host communities dominated by *Astragalus* spp., *Salvia* spp., or others, whilst at higher elevations (700–800m) forests or forest remnants of *Quercus brantii* and/or *Q. boissieri* occur up to *c.*1,700m. Above the timber line (1,900–2,000m) appears a relatively wide zone of subalpine vegetation. Further south, the forest becomes more impoverished and a richer steppe flora develops among the trees.

The Zagros Mountains have traditionally supported a rich variety of animal life, including Brown Bear *Ursus arctos*, Asiatic Black Bear *U. thibetanus*, Wild Goat *Capra aegagrus*, Grey Wolf *Canis lupus*, Leopard *Panthera pardus* and other wild cats. Five taxa of lizards are endemic to the Zagros and the contiguous mountains of Anatolia. The Zagros form a corridor for the southward distribution of northern (European) avifaunal elements (Martins & Hirschfeld 1998), although the range also appears to be the source of some flora that are also now found in the eastern Mediterranean mountains, e.g. *Quercus libani* and *Q. boissieri*. Some typically Iranian avifauna find their westernmost limits in extreme southeast Turkey, e.g. Plain Leaf Warbler *Phylloscopus neglectus*.

Conservation In areas of the Zagros where rain-fed cultivation and grazing occur, the natural vegetation has been reduced to remnants. Overgrazing has also led to changes in vegetation composition, the retreat of the original woodlands and a wider distribution of thorny shrubs and thorny dwarf-shrubs. Advanced degradation is visible even in densely forested areas and species that are better able to withstand grazing and pruning have crowded out original species, such as *Carpinus* spp. and *Quercus* spp., in many areas. No conservation units are currently designated within the tiny Turkish range of this ecoregion.

INTRODUCTION TO THE SPECIES ACCOUNTS

by Guy M. Kirwan

Each of the 463 species that we deem to have been reliably recorded in Turkey is discussed under two or three different headings: Subspecies and taxonomy; Status and distribution; and Breeding (for those species proven or thought to nest in Turkey). The first section is often merely abbreviated to Subspecies where there are no broader issues of taxonomy (e.g. species limits, generic assignment etc.) meritorious of discussion. Each account commences with the species' English name, and Dutch (Du), French (Fr), German (Ger), Swedish (Swe) and Turkish names. We have relied on *BWP* for most of the European-language names, with occasional reference to more recent sources (e.g. Dutch names are taken from the recently published list of *Dutch Birding* names for Western Palearctic birds), whilst Turkish names are taken, in their entirety, from the Collins field guide (Heinzel, Fitter & Parslow, translated into Turkish by Boyla), which is the only generally accurate such book currently available in the Turkish language. Some species recorded or claimed in Turkey are not treated in the latter book, and thus presently lack widely accepted Turkish names. *BWP* has also served as our principal source for the scientific citation of each species, but we have endeavoured to remain alert to necessary changes, particularly those proposed by David & Gosselin (2002a,b) and Dickinson (2003, and addenda distributed privately).

Order and nomenclature largely follow Beaman's (1994) list of Palearctic birds, in line with our earlier checklist (Kirwan *et al.* 1999), with a handful of modifications to names as a result of more recent and widely accepted work (*e.g.* Gill & Wright 2006). The most obvious departure from Beaman is our following the order, nomenclature and species limits for the genus *Sylvia* of Shirihai *et al.* (2001). Nonetheless, despite the existence of other recently prepared lists covering either just the Western Palearctic or much larger areas, broadly similar to that treated by Beaman, we still consider the latter to be the most formidably defended and admirably chosen series of English names for Palearctic birds, as a whole, currently available. We are cognisant of the fact that English names and the systematic order of taxa appear to have become increasingly emotive issues in recent years, at least amongst birdwatchers. Because the present work commenced prior to the widespread testing of the sequence suggested by Sibley & Monroe (1990), which garnered some subsequent support in scientific circles and seemingly widespread disapproval amongst laymen, and our volume was far advanced by the time the sequence suggested by Cracraft *et al. in* Dickinson (2003) became available, we have 'played safe' with a sequence that should be broadly familiar to most readers. Our sole diversion from what will otherwise be recognisable as a Voous sequence is our adoption of the proposition that the Anseriformes and Galliformes (the Galloanseres) should be moved to a position between the Tinamiformes (tinamous) and Gaviiformes (loons), which has accumulated much evidence and support (see Banks *et al.* 2003). Notwithstanding our just-stated conservatism, it is plain that the massive amount of phylogenetic work devoted to birds in recent decades has thrown up, and continues to throw up, much new (and surprising) information concerning avian interrelationships and the evolution of birds. Lists of the birds of the world or a small part of it must increasingly embrace the results of such studies, whilst acknowledging man's ultimate inability to divine an absolute truth through such analyses.

Those species whose occurrence in Turkey we deem to be unproven receive an abbreviated account, usually with no discussion of subspecies or taxonomy, and are placed within square brackets. Some are plainly very unlikely vagrants to Turkey, but others could occur or may have done so in the past. All have been included within the main body of species accounts, rather than relegating them to an appendix. The list of such species is principally based on that in Kirwan *et al.* (1999), with a handful of recent additions, although five species (Lesser Yellowlegs *Tringa flavipes*, Bar-tailed Lark *Ammomanes cinctura*, Arctic Warbler *Phylloscopus borealis*, Iraq Babbler *Turdoides altirostris*, and Lapland Longspur *Calcarius lapponicus*) cited in the earlier listing of rejected species have now been accepted onto the main list as a result of new information or documented sightings. Our cut-off date for the inclusion of new data in the species accounts has, generally, been 2006, although we have done our utmost to incorporate both published and unpublished information that has come our way since.

Subspecies and taxonomy. These sections were almost entirely prepared by GMK, the only exceptions being the small number of taxonomic remarks originally drafted by Rodney Martins. Our starting point for the discussion of subspecific limits in Turkish birds has been Roselaar (1995) and, to a lesser extent, the extensive commentaries of Kumerloeve and Vaurie. The writings of these authors were then checked against specimens and the evidence of our own field observations, as well as the remarks of other authorities. GMK worked on Turkish, Iranian and related material held at the following institutions during the course of the project: the Natural History Museum (Tring, UK), The Manchester Museum (Manchester, UK), the University Museum of Zoology (Cambridge, UK), the Muséum national d'Histoire Naturelle (Paris, France), the Field Museum of Natural History (Chicago, USA), the Robert's College (Istanbul, Turkey), the Zoological Institute, Russian Academy of Sciences (St. Petersburg), the National Museum of Natural History, Smithsonian Institution (Washington DC, USA) and the National Museum of Natural History (Sofia, Bulgaria). Furthermore, the following institutions, in particular, offered invaluable assistance with certain material held in their collections: the Liverpool Museum, National Museums & Galleries on Merseyside (Liverpool, UK), Alexander Koenig Research Institute & Museum of Zoology (Bonn, Germany), and the American Museum of Natural History (New York, USA). Unfortunately, it proved impossible to visit the latter collection during the course of the project.

In general a thorough re-evaluation of subspecific limits in Turkish birds has been attempted in this work, using the guidelines for recognising such taxa in Barrowclough (1982) and Haffer (2003). However, due to limited time and resources, we have occasionally left unexamined the validity of certain subspecies in cases where only one was known to occur in Turkey. Furthermore, our efforts have been spearheaded towards those species that breed in the country. The validity of subspecies known (or suspected) to occur only as vagrants in Turkey has generally not received detailed attention.

None of us is a trained taxonomist and we do not seek to espouse any one of the multifarious species concepts currently available. Instead, we have been content to note some of the principal issues in species and higher-level taxonomy affecting those forms recorded from Turkey, only occasionally (where we feel our particular knowledge can be brought to bear most advantageously) commenting directly on these issues. In fewer than a handful of additional cases, e.g. Cinereous Bunting *Emberiza cineracea*, we would have wished to comment further on, for instance, whether two phylogenetic species were involved, but have been defeated by lack of time and absence of key data. *Thus, our overriding desire has been to evaluate, as best we can, the diagnosability of described taxa.* In a great many instances we find our thoughts at some variance with those of Roselaar (and hence also *BWP*), principally because we adhere to a concept of subspecies that does not recognise the necessity of naming minor step-like variation within long clines, or where such taxa almost certainly hybridise. Thus, many taxa treated as valid by these latter sources are not recognised here. Below, we list those taxa originally described from (and sometimes endemic to) Turkey that we consider to be mere synonyms. The list is based on that provided by Kumerloeve (1984), with additions from Roselaar (1995).

In our opinion, whatever the relative merits of other species concepts, it seems imperative if the Biological Species Concept (BSC) is to survive as a meaningful tool for taxonomists that the BSC concentrates on the recognition of real-world diversity rather than becoming bogged down in the needless naming of incredibly minor, and sometimes only doubtfully valid, geographic variation.

Taxa described from Turkey considered synonyms by this work. Those marked * were considered endemic to Turkey by Roselaar (1995). Those denoted † were not treated as valid by Roselaar (1995). *Tetraogallus caspius tauricus Dresser* 1876 (=*T. c. caspius*); †*Tetraogallus caspius challayei* Oustalet 1877 (=*T. c. caspius*); **Francolinus francolinus billypayni* Meinertzhagen, 1933 (*F. f. francolinus*); *Anhinga rufa chantrei* (Oustalet 1882) (=*A. r. rufa*); *Falco cherrug gurneyi* Menzbier, 1888 (=*F. c. cherrug*?); *Ceryle rudis syriaca* Roselaar, 1995 (=*C. r. rudis*); **Dendrocopus major paphlagoniae* Kummerlöwe & Niethammer, 1935 (=*D. m. pinetorum*); **Dendrocopus medius anatoliae* Hartert, 1912 (=*D. m. medius*); †*Melanocorypha calandra hollomi* Kumerloeve, 1969 (=*M. c. calandra*); †*Melanocorypha calandra dathei* Kumerloeve, 1970 (=*M. c. gaza*?); **Calandrella brachydactyla woltersi* Kumerloeve, 1969 (=*C. b. hermonensis*?); **Calandrella rufescens niethammeri* Kumerloeve, 1963 (=*C. r. aharonii*); †*Galerida cristata weigoldi* (Kollibay, 1912) (=*G. c. subtaurica*); †*Galerida cristata ioniae* (Kollibay, 1912) (=*G. c. caucasica*); †*Galerida cristata ankarae* Kummerlöwe & Niethammer, 1934 (=*G. c. subtaurica*); **Eremophila alpestris kumerloevei* Roselaar, 1995 (=*E. a. penicillata*); †*Cinclus cinclus amphitryon* Neumann & Paludan, 1937 (=*C. c. caucasicus*); **Prunella modularis euxina* Watson, 1961 (=*P. m. modularis*?: more research required); †*Erithacus rubecula balcanicus* Watson, 1961 (=*E. r. rubecula*); †*Saxicola rubetra senguni* Kumerloeve, 1969 (=*S. r. rubetra*); †*Saxicola torquata gabrielae* Neumann & Paludan, 1937 (=*S. rubicola*); **Monticola saxatilis coloratus* Stepanyan, 1964 (=*M. s. saxatilis*); †*Turdus viscivorus bithynicus* Keve, 1943 (=*T. v. viscivorus*); †*Sylvia communis traudeli* Kumerloeve, 1969 (=*S. c. icterops*); **Panurus biarmicus* *kosswigi* Kumerloeve, 1958 (=*P. b. biarmicus*?: more research required); **Aegithalos caudatus tephronotus* (Gunther, 1865) (=*A. c. alpinus*); *Parus lugubris anatoliae* Hartert, 1905 (=*P. l. lugubris*); **Sitta europaea levantina* Hartert, 1905 (=*S. e. caesia)*; **Sitta neumayer zarudnyi* Buturlin, 1908 (=*S. n. neumayer*); †*Certhia brachydactyla harterti Hellmayr*, 1901 (=*C. b. brachydactyla*); **Certhia brachydactyla stresemanni* Kummerlöwe & Niethammer, 1934 (=*C. b. brachydactyla*); †*Remiz pendulinus persimilis* Hartert, 1918 (=*R. p. pendulinus*); *Garrulus glandarius anatoliae* Seebohm, 1883 (5G. g. krynicki); †G. g. nigrifrons Buturlin, 1906 (5G. g. krynicki); †G. g. lendlii Madaràsz, 1907 (=*G. g. krynicki*); **G. g. hansguentheri* Keve, 1967 (*G. g. glandarius*?: more research required); †*Passer domesticus colchicus* Portenko, 1962 (=*P. d. domesticus*); **Passer domesticus mayaudi* Kumerloeve, 1969 (=*P. d. biblicus*); **Montifringilla nivalis leucura* Bonaparte, 1855 (=*M. n. alpicola*?: more research required); †*Montifringilla nivalis fahrettini* Watson, 1961 (=*M. n. alpicola*); *Carduelis carduelis niedieckì* Reichenow, 1907 (=*C. c. carduelis*); †*Loxia curvirostris vasvarii* Keve, 1943 (=*L. c. guillemardi*); **Pyrrhula pyrrhula paphlagoniae* Roselaar, 1995 (=*P. p. pyrrhula*).

Status and distribution. Responsibility for these sections was divided between the different authors according to the list below. Each account commences with the species' conservation status, for all those species listed as globally endangered or included within the recent Turkish Red Data Book. We also mention the apparent population trend within Turkey, but have elected not to repeat the population estimates themselves as published in the Red Data Book, for the simple reason that we find these too much based on guesswork to be meaningful. Producing reasonably accurate population estimates for all Turkish breeding species is an important future objective. Immediately thereafter, as far as possible, the text attempts to summarise a given species' overall status in the country, before going on to provide a more detailed summation of our current knowledge. Each account should be read in conjunction with the species map and the baseline map of Turkey which shows the different subregions of Turkey employed in this work. Unlike in the Turkey Bird Report, we have adopted the same regional subdivisions of Turkey as employed in Kasparek (1992), a system based on Erol (1982) and which more closely echoes zoogeographical factors impacting bird distributions (see Fig. 2).

Within each species account we have always located a given locality by region or, in the case of smaller and less easily traced sites, also by province or a nearby major town. Our principal guide for the spelling of place names has been the gazetteer of Burgett *et al.* (1984), although we have occasionally deviated from this where we are aware of a more modern alternative in widespread local use. For most species, references are rarely provided except for the most interesting and important facts, as most information has usually been culled from one of the principal data sources, e.g. Kumerloeve (1961), Kasparek (1992a), Roselaar (1995) or the Turkey Bird Reports (OST 1969, 1972, 1975, 1978, Beaman 1986, Martins 1989, Kirwan & Martins 1994, 2000a, Kirwan *et al.* 2003, in press). However, for those species that we consider to be particularly interesting for readers, e.g. near-endemics or those with particularly important Western Palearctic populations in Turkey, and especially

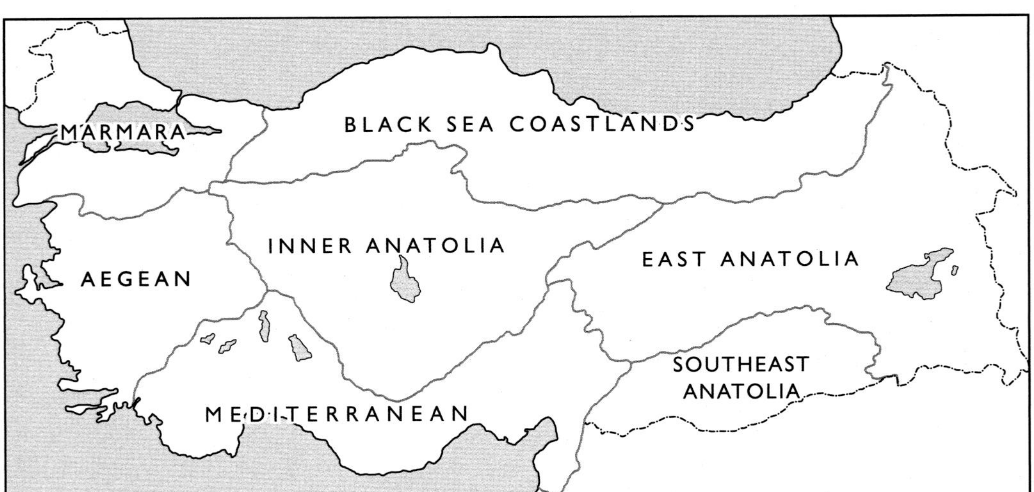

Figure 2. Map to show the regions of Turkey as used in this book, based on Kasparek (1992a).

those for which important historical data are available, we have been more expansive in our use of references. Unpublished personal observations are mentioned where appropriate, and indicated by the relevant authors' initials in parentheses.

Vagrants are usually dealt with rather summarily, but where we have considered it particularly interesting to do so, we have provided information concerning status elsewhere in the Middle East and/or other neighbouring states, e.g. Armenia, Bulgaria, Georgia and Greece.

Breeding. This section includes nesting data, where known, for each of the breeding species. There is a description of the habitat, nest site and nest, and details of the number of eggs, young and broods raised; for certain species there is information on breeding densities, success rates, and comparison with neighbouring countries; detailed records attempt to identify the breeding season (see earlier) for each region of the country. All data are from Turkey, unless otherwise stated. Species that breed, or have bred, in Turkey (316), or may possibly do so (11), are detailed in the separate chapter on the Breeding Season.

The maps. The first basis for the maps were those prepared, by GMK, for the *Field Guide to the Birds of the Middle East* (Porter *et al.*; originally published in 1996). Subsequently, KAB assumed sole responsibility for creating the maps, using as an initial base for the passerines Roselaar's *Songbirds of Turkey* (1995). The maps were further refined with reference to, and in conjunction with, the developing text, with the basis for revision with respect to those maps in the above-mentioned works being personal observations; personal communication with other observers, principally in Turkey; published literature; and dedicated field surveys, usually organised by DHKD, in various regions of Turkey, e.g. Southeast Anatolia, the Mediterranean forests and the Konya Basin. It is worth emphasising that the maps attempt to provide a rather greater level of distributional detail than would ever be possible in the species texts, but, it is also important to be aware that: (a) the maps do not always show every outlying or vagrant record, especially those which we consider doubtful; for these the texts need to be consulted; and (b) occasionally the maps over- or underestimate known ranges, especially in winter, based on our perception of (i) what might be the true distribution or (ii) what is likely to be the normal range. It is the intention of the compiler (KAB) to continue to update the maps with future unpublished and published data, and in this respect observers with information that supersedes that presented here are requested to contact him via the publishers.

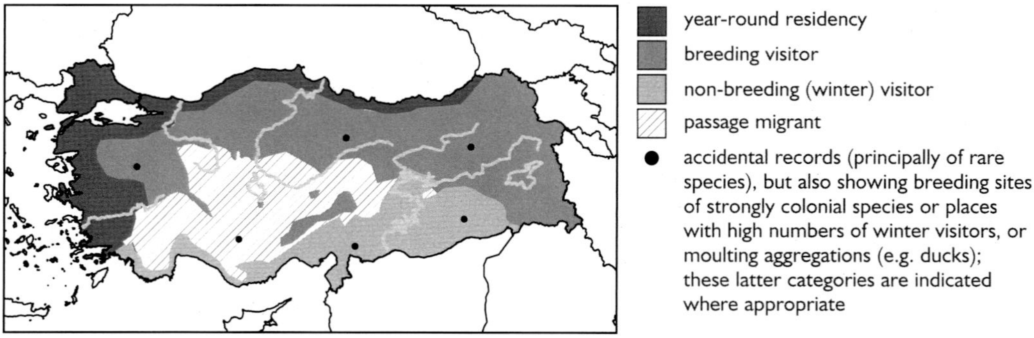

Figure 3. Key to the species distribution maps.

Species accounts. Unless otherwise stated below the Status and distribution accounts were prepared by GMK.

BD: Eurasian Oystercatcher *Haematopus ostralegus*, Black-winged Stilt *Himantopus himantopus*, Pied Avocet *Recurvirostra avosetta*, Stone-curlew *Burhinus oedicnemus*, Little Ringed Plover *Charadrius dubius*, Common Ringed Plover *C. hiaticula*, Kentish Plover *C. alexandrinus*, Greater Sand Plover *C. leschenaultii*, European Golden Plover *Pluvialis apricaria*, Grey Plover *P. squatarola*, Spur-winged Lapwing *Vanellus spinosus*, Northern Lapwing *V. vanellus*, Sanderling *Calidris alba*, Little Stint *C. minuta*, Temminck's Stint *C. temminckii*, Curlew Sandpiper *C. ferruginea*, Dunlin *C. alpina*, Ruff *Philomachus pugnax*, Jack Snipe *Lymnocryptes minimus*, Common Snipe *Gallinago gallinago*, Eurasian Woodcock *Scolopax rusticola*, Black-tailed Godwit *Limosa limosa*,

Eurasian Curlew *Numenius arquata*, Spotted Redshank *Tringa erythropus*, Common Redshank *T. totanus*, Marsh Sandpiper *T. stagnatilis*, Common Greenshank *T. nebularia*, Green Sandpiper *T. ochropus*, Wood Sandpiper *T. glareola*, Pallid (Striated) Scops Owl *Otus brucei*, European Scops Owl *O. scops*, Eurasian Eagle Owl *Bubo bubo*, Little Owl *Athene noctua*, Tawny Owl *Strix aluco*, Long-eared Owl *Asio otus*, and Short-eared Owl *A. flammeus*.

MÖ: Rock Dove *Columba livia*, Stock Dove *C. oenas*, Common Wood Pigeon *C. palumbus*, Eurasian Collared Dove *Streptopelia decaocto*, European Turtle Dove *S. turtur*, Laughing Dove *S. senegalensis*, Great Spotted Cuckoo *Clamator glandarius*, Common Cuckoo *Cuculus canorus*, European Nightjar *Caprimulgus europaeus*, Winter Wren *Troglodytes troglodytes*, Dunnock *Prunella modularis*, Alpine Accentor *P. collaris*, Rufous-tailed Scrub Robin *Cercotrichas galactotes*, European Robin *Erithacus rubecula*, Thrush Nightingale *Luscinia luscinia*, Common (Rufous) Nightingale *L. megarhynchos*, Bluethroat *L. svecica*, Black Redstart *Phoenicurus ochruros*, Common Redstart *P. phoenicurus*, Whinchat *Saxicola rubetra*, European Stonechat *S. rubicola*, Eastern (Caucasian) Stonechat *S.* (*maurus*) *armenicus*, Northern Wheatear *Oenanthe oenanthe*, Pied Wheatear *O. pleschanka*, Cyprus Wheatear *O. cypriaca*, Black-eared Wheatear *O. hispanica*, Red-tailed Wheatear *O. xanthoprymna*, Rufous-tailed Rock Thrush *Monticola saxatilis*, Blue Rock Thrush *M. solitarius*, Ring Ouzel *Turdus torquatus*, Common Blackbird *T. merula*, Fieldfare *T. pilaris*, Song Thrush *T. philomelos*, Redwing *T. iliacus*, Mistle Thrush *T. viscivorus*, Spotted Flycatcher *Muscicapa striata*, Marsh Tit *Parus palustris*, Sombre Tit *P. lugubris*, Coal Tit *P. ater*, Blue Tit *P. caeruleus*, Great Tit *P. major*, Eurasian Treecreeper *Certhia familiaris*, Short-toed Treecreeper *C. brachydactyla*, Penduline Tit *Remiz pendulinus*, and Golden Oriole *Oriolus oriolus*.

TM: Spotted Crake *Porzana porzana*, Little Crake *P. parva*, Baillon's Crake *P. pusilla*, Corn Crake *Crex crex*, Common Moorhen *Gallinula chloropus*, Eurasian Coot *Fulica atra*, Common Crane *Grus grus*, Pallas's Gull *Larus ichthyaetus*, Mediterranean Gull *L. melanocephalus*, Little Gull *L. minutus*, Black-headed Gull *L. ridibundus*, Slender-billed Gull *L. genei*, Common Gull *L. canus*, Lesser Black-backed Gull *L. fuscus*, Gull-billed Tern *Gelochelidon nilotica*, Caspian Tern *Hydroprogne caspia*, Sandwich Tern *Sterna sandvicensis*, Common Tern *S. hirundo*, Little Tern *Sternula albifrons*, Whiskered Tern *Chlidonias hybrida*, Black Tern *C. niger*, Black-bellied Sandgrouse *Pterocles orientalis*, Pin-tailed Sandgrouse *P. alchata*, White-throated Dipper *Cinclus cinclus*, Radde's Accentor *Prunella ocularis*, White-throated Robin *Irania gutturalis*, Isabelline Wheatear *Oenanthe isabellina*, Finsch's Wheatear *O. finschii*, Red-breasted Flycatcher *Ficedula parva*, Semi-collared Flycatcher *F. semitorquata*, Collared Flycatcher *F. collaris*, Pied Flycatcher *F. hypoleuca*, Bearded Reedling *Panurus biarmicus*, Long-tailed Tit *Aegithalos caudatus*, Krüper's Nuthatch *Sitta krueperi*, Eurasian Nuthatch *S. europaea*, Eastern Rock Nuthatch *S. tephronota*, Western Rock Nuthatch *S. neumayer*, and Wallcreeper *Tichodroma muraria*.

Güven Eken: Chukar *Alectoris chukar*, Grey Partridge *Perdix perdix*, Common Quail *Coturnix coturnix*, and Water Rail *Rallus aquaticus*.

BLACK GROUSE *Tetrao tetrix*
Orman Tavuğu

(Du: Korhoen; Fr: Tétras lyre; Ger: Birkhuhn; Swe: Orre)

Subspecies and taxonomy. Presumably nominate *tetrix*. Has also been placed in the genus *Lyrurus* (see next species).

Status and distribution. Reputedly extirpated, although a relict population apparently persisted in the vicinity of Istanbul (Marmara) until around the end of the 19th century (Kasparek 1990a), with the species first being mentioned for Turkey by Rigler (1852), then by Tchihatchef (1864) from the middle years of that century, and finally by Reiser (1904) and Mathey-Dupraz (1923), who specifically reported that it could be found around the Alem Dağı, on the Asian side of the Bosphorus, and noted the presence of dead males for sale in the city, albeit of uncertain provenance. During the same period, Elwes & Buckley (1870) noted a marked decline in the Bulgarian population, apparently due to hunting, and the species is also long-since extinct there, as its entire range has contracted north and westwards (Hagemeijer & Blair 1996, Madge & McGowan 2002). Handrinos & Akriotis (1997) mentioned four records for Greece, all since 1935 and all in the vicinity of Thessaloniki or the Rodopi Mountains.

CAUCASIAN GROUSE *Tetrao mlokosiewiczi*
Huş Tavuğu

(Du: Kaukasisch Korhoen; Fr: Tétras du Caucase; Ger: Kaukasusbirkhuhn; Swe: Kaukasisk orre)

Subspecies and taxonomy. Monotypic. Has also been placed in the genus *Lyrurus*, as it usually is in the Russian-language literature (for a résumé of different treatments, see Gokhelashvili *et al.* 2003). Short (1967) arguably provided the most detailed discussion of the relationships between these grouse, arriving at the conclusion that two superspecies, within a single genus, should be recognised. Nonetheless, there is some basis for assuming an expanded *Tetrao* (*i.e.* including both capercaillies and black grouse) to represent a non-monophyletic lineage, thus suggesting that removing the two black grouse to *Lyrurus* is an available solution, but equally we have not seen any well-supported, published evidence that unequivocally demonstrates this to be necessary to preserve monophyly in *Tetrao*. As suggested by Short (*op. cit.*), the peculiarly shaped tails of *Lyrurus*, and the slightly metallic plumage and the elongated neck feathers of males of *Tetrao* are not necessarily taxonomically informative characters. Indeed, a recent mtDNA study of grouse phylogeny suggested a very close relationship between the two black grouse and the two capercaillies (Dimcheff *et al.* 2002).

Status and distribution. Conservation status: Data Deficient (BirdLife International 2000, 2004); national conservation status: Vulnerable (population apparently decreasing). Turkish population currently estimated at 1,508–2,675 individuals and the species is known from 45 localities (Isfendiyaroğlu *et al.* 2007), of which 29 were only discovered recently, in four relatively discrete subpopulations. Very poorly known in Turkey until most recent decades (G. Neuhäuser collected a pair in the mountains between Rize and Erzurum, in September 1943, which was probably the first record in Turkey: Kumerloeve 1961), with an extreme paucity of records prior to the 1980s, since when it has proved to be a locally scarce or not uncommon resident at subalpine elevations (1,500–2,900m) in the Pontics (eastern Black Sea Coastlands), where apparently most frequently encountered on the northern slopes of the north ridge of these mountains. Reports from above 3,000m (e.g. Başkaya 2003) require confirmation, but almost certainly descends (perhaps up to 15km) closer to habitation and slightly below the treeline in autumn/winter, perhaps lower in particularly severe weather. Distribution is centred on slopes with substantial dwarf *Rhododendron* and birch *Betula* scrub below the alpine herb zone (Atkinson *et al.* 1995), but the western limits of the species' range are unclear, despite increased fieldwork in recent years. Gottschalk *et al.* (2007) used computer modelling in an attempt to more accurately delimit the species' range, and to gauge its population, in Turkey, but did not substantially expand the known distribution as mapped here. However, their estimate of almost 4,900 birds in Turkey is much higher than previous best guesses. Principally known from the Kaçkar Dağları, where probably 90% of the Turkish population occurs, especially from the area around Sivrikaya, where a survey in 1993 located 134 lekking males, and the total population at this time was speculated to be in excess of 2,000 individuals (Magnin & Yarar 1997). Also occurs within a narrow band in the Yalnızçam Dağları, from southeast of Artvin to Posof on the Georgian border. Reports from local people suggest that the species might also occur further west in Gümüşhane and Giresun provinces, and perhaps even south to the Bingöl area, in the Cilo Dağları, but such claims require confirmation.

The Turkish population is affected and potentially threatened by habitat dereliction and fragmentation, due to greatly increased road construction, and by disturbance, as summer villages become increasingly used as holiday homes, as well as, to a much lesser extent, by hunting (which is principally a problem in autumn). Over-grazing, which has been listed as a source of concern for the species, is presently having a negligible effect, although the situation requires monitoring. There is insufficient historical information concerning population and range to determine whether the population is stable or declining, although the latter is suspected to be the case.

Breeding. On high-altitude, steep, grassy slopes in areas of dwarf *Rhododendron* scrub, below the alpine herb zone. Males display communally, mainly at dawn and dusk, at leks, where the sexes meet only for copulation. Only two nests have been found in Turkey, both at Sivrikaya (**Black Sea Coastlands**), the well-known locality for the species. The first was at 2,800–3,000m, in an area of *c.*3ha of dwarf *Rhododendron* scrub; the nest was in denser, taller (c.1m) and more tangled scrub; well concealed, a shallow scrape in the ground between the branches of a stump, lined with dry grass and several dry *Rhododendron* leaves, and it had five eggs on 6 July 1991 (Temple-Lang & Cocker 1991). The other records involved a female flushed from a nest with four eggs on 12 July 1993 (A. Hogg), and an eggshell found on 11 June 1997 (GMK, P. Doherty); two half-grown young, with an adult female, seen at Posof, near Ardahan (East Anatolia) on 12 June 2003 (A. Ün), indicate egg laying in early May. Elsewhere, usual clutch 5–6 (2–10). In Armenia, a nest had eight eggs on 30 May 1984, in which it was estimated that the first egg was laid 21–23 May 1984, another had eggs on 20 May 1985 (first egg estimated 13–16 May 1985), and another had eggs on 31 July 1994; a female was seen with three young (*c.*25% of adult size) on 5 June 1980, and one with five full-grown young on 26 July 1980 (Adamian & Klem 1999). The somewhat earlier dates in Armenia raise some doubts as to whether the Turkish records represent the normal breeding season.

[WESTERN CAPERCAILLIE *Tetrao urogallus*]
Büyük Orman Tavuğu

(Du: Auerhoen; Fr: Grand Tétras; Ger: Auerhuhn; Swe: Tjäder)

Subspecies. Unknown, although *T. u. major* reaches the Balkans.

Mentioned, presumably in error, by Huş (1974) for northeast Turkey, around Kars, the Zigana Dağları, south of Trabzon, and by Artvin (eastern Black Sea Coastlands), all of which are known areas for Caucasian Grouse *Tetrao mlokosiewiczi*, which was presumably the source of confusion (Kasparek 1990a). The species is unknown from the Caucasus or southern Russia (Hagemeijer & Blair 1996, Snow & Perrins 1998).

[CAUCASIAN SNOWCOCK *Tetraogallus caucasicus*]
Kafkas Urkekliği

(Du: Kaukasisch Berghoen; Fr: Tétraogalle du Caucase; Ger: Kaukasuskönigshuhn; Swe: Kaukasisk snöhöna)

Subspecies. Monotypic.

Species not accepted onto the Turkish list (Kirwan *et al.* 1999). Listed as occurring in Turkey by Serez (1992 and *in litt.* 1995) and del Hoyo *et al.* (1994), on the basis of unsubstantiated and undated 'records' from Artvin province (Black Sea Coastlands). Both Ergene (1945) and Turan (1990) listed the species as hypothetical in Turkey. In the absence of any adequately documented evidence, such statements must be regarded as unproven.

CASPIAN SNOWCOCK *Tetraogallus caspius*
Urkeklik

(Du: Kaspisch Berghoen; Fr: Tétraogalle de Perse; Ger: Kaspikönigshuhn; Swe: Kaspisk snöhöna)

Subspecies. *T. c. tauricus* (including the synonym *challayei* described from Erzurum) throughout Turkey, although Roselaar (1995) speculated that *semenowtianschanskii* might be the form occurring in the Hakkari range (East Anatolia). However, as noted by Dement'ev & Gladkov (1967) the latter form was described on the basis of the holotype alone, and is almost certainly a synonym of nominate *caspius*. The same authors further postulated that the taxonomic validity of *tauricus* might prove to be questionable and, under the terms of reference adopted here, we consider that *T. caspius* is best treated as a monotypic species. Kumerloeve (1961, 1967a) had also questioned the validity of *tauricus*, in part following Vaurie (1965), given that those from the northeast appear inseparable from nominate *caspius*, despite having previously been considered within *tauricus*. Given that various authors have noted that those in Armenia are inseparable from birds taken at Erzurum and are definitely referable to nominate *caspius* (see Dement'ev & Gladkov *op. cit.*), our conclusion to return the species to monotypy appears to be the only defensible course.

Status and distribution. National conservation status: Vulnerable (population apparently decreasing). Not uncommon but generally a rather local resident of montane areas in the eastern third of the country, with an apparently isolated population in the eastern Taurus recorded west as far as the Bolkar and Melendiz mountains (Mediterranean/Inner Anatolia), with isolated older records west to the Geyik Dağ, north of Alanya, and even in the mountains around Antalya (Kumerloeve 1961). Further west still, there appear to be suitably high, permanently snow-covered peaks in the Akdağlar and Beydağları (R. P. Martins). Three principal populations: in the eastern Pontics (Black Sea Coastlands); montane East Anatolia, south as far as Yüksekova and Hakkari, and east to the borders with Iran; and the aforementioned westernmost population in the Taurus. The westernmost limits of those breeding in the Black Sea Coastlands and East Anatolia are presently unclear, although the species has been noted west to at least the Munzur Dağları and the Zigana Dağları, south of Trabzon. Principally recorded above 2,400m in summer, but noted down to *c.*2,000m at this season, e.g. in the mountains north of Mersin (Mediterranean). Occasionally gathers into large groups of up to 60 in autumn, with perhaps some descending from the highest areas in midwinter.

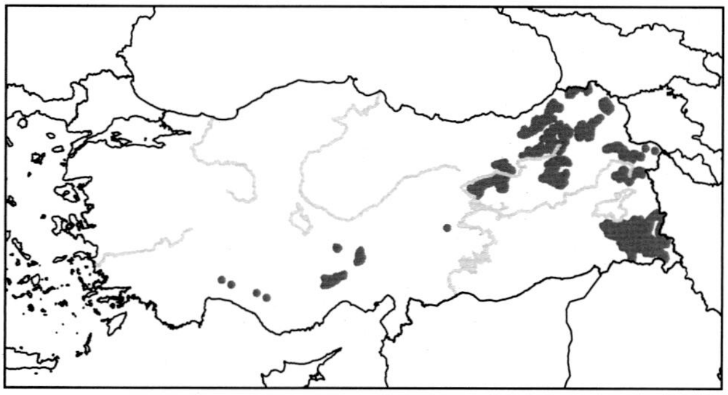

Breeding. Mainly above 2,400m, on precipitous mountains with alpine meadows, steep crags and gullies, and year-round snowfields. Danford (1878) searched for nests in the Aladağ area of the Taurus (**Mediterranean**) in April 1876. At 2,100m on Karanfil Dağ, on 23 April 1876, a female flew from a small ledge in a steep narrow gully, where the nest was sheltered by an overhanging rock and a gnarled juniper stump; it was a deep round hollow scraped in the stony earth, scantily lined with dry grass and a few of the bird's feathers; it had six eggs. Two nests in the Bolkar Dağları on 25 April 1876 were similar, but one was lined with green fir needles; they had six and four eggs. Two eggs, taken from two of the three nests above, are in Manchester Museum, and six eggs, from two of these nests, are in the Natural History Museum, Tring, which also has an egg taken in 'Asia Minor' on 1 June 1894 and three (probably an incomplete clutch) from Aladağ on 5 April 1901. More recently, also at Aladağ: an adult was with 5 young, 1–2 weeks old, on 8 July 1986 (A. Harrop: first egg would have been laid about 21 May); a female was with 5–6 young, the size of Grey Partridges, on 3/5 August 1967 (Vielliard 1968); there were at least three family groups, each with 3–4 young, some noticeably smaller than the adults, on 19 August 2000 (A. & K. Gancz); and two adults and two young were at 3,000m on Karanfil Dağ on 6 August 1966 (Sutton & Gray 1972). At Sivrikaya (**Black Sea Coastlands**), an adult with five young, was crossing a small snowfield on 12 June 1989 (D. Page: first egg would have been laid no later than 2 May, even if the young were small). In Armenia, nests with eggs in May–June, typically backdated to egg laying on 10–15 May, and incubation 20–30 May, with records of young from 13 June (Adamian & Klem 1999); some earlier records of young indicate laying of first eggs as early as 23 April and, in two cases, late April/early May, roughly coinciding with the dates of Danford's nests in Turkey.

CHUKAR *Alectoris chukar*
Kinali Keklik

(Du: Aziatische Steenpatrijs; Fr: Perdrix choukar; Ger: Chukarhuhn; Swe: Berghöna)

Subspecies. The race *cypriotes* occurs throughout the south of the country, with *kleini* in the north and *kurdestanica* in southeastern Turkey, where some show influence of the Levantine form *sinaica*. We have not addressed the issue of geographical variation in this species but, as noted by Madge & McGowan (2002), much of it appears to be markedly clinal and with strong intergradation. The latter authors admit 16 forms, but they acknowledge that at least four are probably synonyms, whilst Dickinson (2003) admitted 14 names.

Status and distribution. National conservation status: Vulnerable (population apparently decreasing). Widespread and still fairly common resident over much of Turkey, although in obvious decline and apparently always absent from much of Thrace and the western Black Sea Coastlands (where principally recorded at 700–2,000m in the hinterland). Principally found in dry, rocky hills and montane regions up to at least *c*.2,800m. Fairly large coveys, often mainly comprising juveniles and numbering up to almost 40 individuals, have been recorded. Formerly more common in areas such as the Hatay (Mediterranean) and eastern Black Sea Coastlands, in particular.

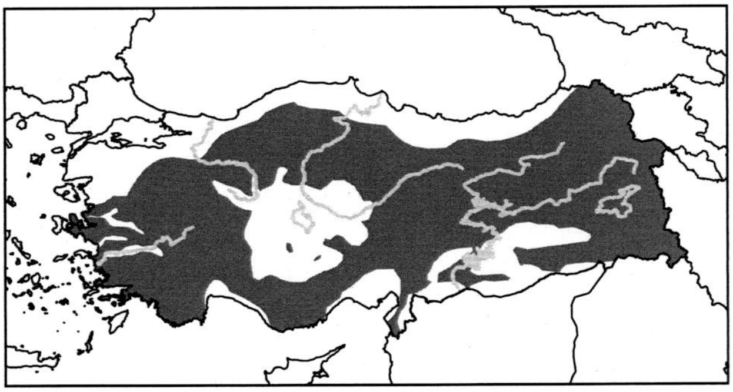

Breeding. In rocky and stony hills, often with scrub, in dry cultivation, and very arid barren areas. Nests on the ground, under a bush or concealed at the base of vegetation; one nest (with 22 eggs) was on a rock ledge, screened by a bush, on the inner wall of a crater near Ereğli (**Inner Anatolia**) on 24 May 1998 (T. Quittenden). Nine to 20 eggs are laid from late April (Banoğlu & Burr 1953). An incubating adult was flushed from a nest with 14 eggs near Izmir (**Aegean**) on 7 May 1951; the nest was a grass-lined scrape, with many of the bird's own feathers, sheltered by a prickly oak; the eggs varied from (mostly) ivory-white to reddish, minutely spotted reddish-brown (McNeile). In **Mediterranean**, a nest had five eggs (probably an incomplete clutch) at Acıgöl on 11 May 1899 (Selous 1900). At Karadağ (Inner Anatolia), two nests had fresh eggs in late May 1907, and fresh eggs were found in June (Ramsay 1914). An adult was accompanying 5–6 well-grown young at the pass between Burdur and Bucak, on 13 July 1968 (Rokitansky & Schifter 1971). At Nemrut Dağ (**East Anatolia**), a nest had 15 cold eggs in the morning of 9 June 2004 but, later in the day, an adult was sitting, possibly on its first day of incubation; the nest was a deep scrape in the ground, with a scant lining of grass stems and a few of the adult's feathers, partially concealed under a low bush within sparse deciduous woodland (PC). Records of broods include small young at Işıklı (**Southeast Anatolia**) on 2 June 2001 (PC), and broods of six, seven and eight at Menemen, near Izmir (Aegean), on 22 June 1966; in Mediterranean, there was a brood at Dalyan on 22 May 1989, and young, a few days old, at Aladağ on 30 June 1966. An unusually large brood of *c*.40, with a single adult, at Kızılcahamam (Inner Anatolia), on 13 August 1967 (Barış *et al.* 1984), is considered to be the product of several broods merging.

[ROCK PARTRIDGE *Alectoris graeca*]
Kaya Kekliği

(Du: Steenpatrijs; Fr: Perdrix bartavelle; Ger: Steinhuhn; Swe: Stenhöna)

Subspecies. Nominate *graeca* is the only form that could plausibly occur.

Species not accepted onto the Turkish list (Kirwan *et al.* 1999). Considered to occur in Turkey by many authors, including Kumerloeve (1961), prior to the publication of Watson (1962), who demonstrated that the apparent zone of contact between this species and Chukar *A. chukar* lies entirely within Thracian Greece and Bulgaria. There appears to be no justification for subsequent statements in the literature that this species has occurred in Turkey, e.g. by Kiziroğlu (1989), Turan (1990) and Serez (1992), the latter repeated in del Hoyo *et al.* (1994).

SEE-SEE PARTRIDGE *Ammoperdix griseogularis*
Kum Kekliği

(Du: Perzische Woestijnpatrijs; Fr: Perdrix si-si; Ger: Persisches Wüstenhuhn; Swe: Visselhöna)

Subspecies. Monotypic.

Status and distribution. National conservation status: Vulnerable (population apparently decreasing). Confined to Southeast Anatolia, where it is resident and restricted to principally lowland arid rocky hillsides and wadis, often with light cover, from just west of Gaziantep east to at least Cizre, and north to about Diyarbakır and Elbistan. Somewhat local but apparently not uncommon, until recently being known from just *c.*15 localities, most of these within 50km of the Syrian border, and rarely recorded in more than low single figures, except in autumn when loose groups of up to 40 recorded. However, recent surveys of the southeast of the country revealed the species to be present at almost 50 sites and present in most areas of suitable habitat that were surveyed (Welch 2004).

Breeding. In arid rocky country with sparse vegetation, in **Southeast Anatolia**. Nests and eggs undescribed in Turkey but, elsewhere, nest a shallow scrape in the ground, with a scant lining of grass or other nearby plant material, usually sheltered by a rock or a plant tuft; eggs usually C/8–12, occasionally up to 16. In Turkey, seen in pairs from April. An adult with three young, aged three weeks, at Halfeti on 5–7 June 1973, indicates laying of the first egg in mid April, which is normal elsewhere. There are several records of young from the Birecik area in 2004–05 (H. Göktekin *et al.*). A family party (no details of size of young) was at Cizre on 11 August 2001 (A. Comas).

BLACK FRANCOLIN *Francolinus francolinus*
Turaç

(Du: Zwarte Frankolijn; Fr: Francolin noir; Ger: Halsbandfrankolin; Swe: Svart frankolin)

Subspecies. Nominate *francolinus* throughout, but Roselaar (1995), unlike virtually all other recent commentators (Kumerloeve 1984 being one notable exception), chose to recognise the apparently endemic and now extinct *billypayni*, which was described by Meinertzhagen from the region of the former Amik Gölü (Mediterranean). Like other authorities, we do not acknowledge the validity of *billypayni*.

Status and distribution. National conservation status: Near Threatened (population apparently increasing). Status and ecology recently subject to a detailed review by van den Berk (1988), although range has apparently since expanded slightly in the east, due to intensification of agriculture in this region, which has provided some additional habitat for the species. Generally local, but not uncommon, resident in Mediterranean, west to about Antalya, and Southeast Anatolia, being usually found in moist scrub or agricultural areas, principally in southern border regions and the coastal littoral. Also noted north almost to the Turkish Lake District and the Anti-Taurus, where recorded very locally up to 600m (Kumerloeve 1961). In Southeast Anatolia recently recorded from *c.*15 sites, all relatively close to the Syrian border and recorded east to around Cizre (Welch 2004). The species was formerly resident in southeast Marmara and parts of coastal southern Aegean (Kumerloeve 1963c), but was extirpated in both regions in the 19th century, with an 1850s record from the Bosphorus, although there is a recent record of three at Tavas, Denizli province (Aegean), in October 2003 (D. Kahraman), which presumably relates to escapees. More recently seems to have become extirpated in western Mediterranean, e.g. around Köyceğiz Golu, where recorded in 1960 but not since. Current status in the Hatay unknown; however, special protective measures were planned for the species in some areas elsewhere in southern Turkey. In Mediterranean, best known from the largest deltas: in Çukurova, current breeding population in excess of 85 pairs, with most (75 pairs) around Akyatan Gölü (Magnin & Yarar 1997). At Göksu Delta, there are 50 breeding pairs (Magnin & Yarar 1997), with many of these in the dunes.

Breeding. Breeds mainly in Göksu and Çukurova deltas and their immediate environs (**Mediterranean**), and along the Euphrates and Tigris near the Syrian border (**Southeast Anatolia**). Inhabits dunes covered with grass and other herbs, and depressions between them with scrub, bushes and low plantations; damp (not wet) areas, with luxuriant undergrowth, bordering rivers; and grazed tussocky grassland with *Tamarix* thickets and cornfields; highest densities in sand dunes (one male per 2–5ha) and agriculture (one male per 15–30ha) (van den Berk 1988). Adults call from extensive cornfields near the Syrian border, south of Birecik, and must surely breed in such habitat, as they do in Cyprus (PC). In Göksu Delta, a nest in the dunes on 17 June 1992 contained hatched eggshells, sun-bleached and faded, possibly from the previous year; it was an unlined, shallow hollow in the sand, with merely a few plant fragments, at the base of a clump of grasses (PC). Eggs: no data; elsewhere, usual clutch 8–12 (7–18). In Göksu, a female with at least seven young, 1–2 days old, on 5 May 2004, indicating date of first egg as 9 April (PC); an adult and a young bird were observed on 20 July 1986 (R. Webb). In Çukurova, local people caught chicks on 10 May 1986 (reportedly common), indicating laying about mid April; peak calling is in the second half of April/first half of May (van den Berk 1988). The following account is from Banoğlu & Burr (1953): 'hen birds are busy with their eggs and nests at the end of March and in April. The eggs are the size of the partridge (Chukar) and of a light green colour. They lay on the ground among the thickets on the shores of the Ceyhan and Seyhan (Çukurova) and their feeders. Around Dörtyol and Alik they nest at the edges of the cliffs of the steep and deep ravines, or among the reed beds between the islands, in thickets and clumps of bush. In the old days they used to be numerous in the flat country around Adana.' Three eggs in Manchester Museum (an incomplete clutch) were taken near Izmir (Aegean) on 10 May 1899; of four eggs in the Natural History Museum, Tring, two were taken at Mersin (Mediterranean) on 7 May 1884, and two in 'Asia Minor' on 15 May 1899.

GREY PARTRIDGE *Perdix perdix*
Çilkeklik

(Du: Patrijs; Fr: Perdrix grise; Ger: Rebhuhn; Swe: Rapphöna)

Subspecies. The form *canescens* occurs throughout Turkey, except Thrace, where apparently replaced by nominate *perdix*. Geographical variation throughout the range of this species is rather complex and complicated by the existence of introduced birds that have interbred with local populations (Madge & McGowan 2002). We have not studied the situation in Turkey.

Status and distribution. National conservation status: Vulnerable (population apparently decreasing). Scarce resident in agricultural areas and dry areas of Inner Anatolia, the Turkish Lake District, East Anatolia and very locally in Southeast Anatolia. Probably extinct in Thrace and is certainly very local in Black Sea Coastlands. The population has dramatically decreased across Turkey over the last 20 years, probably due to agricultural intensification, overgrazing and hunting. Breeds mainly in agricultural areas in lowlands, foothills with tall-grass communities and primary semi-steppe to at least 2,250m. The species no longer occurs over large areas of the Konya basin, whilst the lower Sakarya and Kızılırmak basins, and especially the Sivas area are the main regions where it continues to occur in reasonable numbers. The northernmost Konya Basin, around Çöl Gölü and Çalıkdüzü is another area where the species is not uncommon. In the east, it is confined to alluvial plains and their surroundings, particularly in Muş province along the Murat River. Recently found breeding in southeast Turkey. Although hunting has been prohibited since 1997, illegal hunting continues in various places, and many hunting organisations across the country have reported rapid population declines due to over-use of chemicals for agriculture.

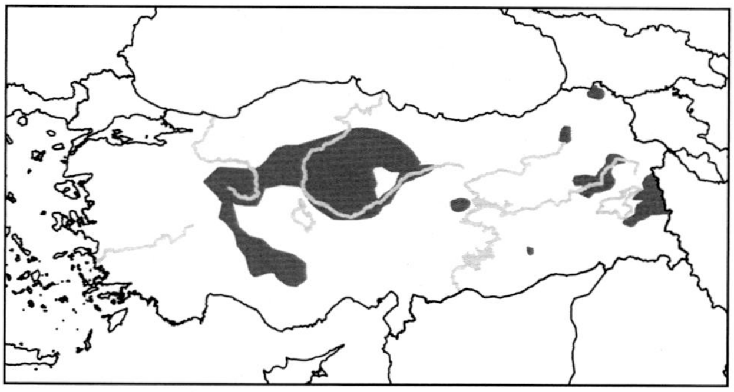

Breeding. Generally in open agricultural country. No description of nest from Turkey but elsewhere is a shallow hollow in the ground, lined with grass and dead leaves, usually well concealed in vegetation. In **Inner Anatolia**, an adult was with a juvenile at Mogan Gölü on 11 June 1977, eight juveniles were at Emir Gölü on 16 July 1977, and a family party of seven were at Çavuşçu Gölü on 28 August 1984. At Çınar-Göksu Reservoir, near Diyarbakır (**Southeast Anatolia**), a nest on a slope with thorny bushes (*Astralegus* sp.) held ten eggs on 16 May 1999 and 19 eggs by 27 May (Karakaş & Kılıç 2002). Elsewhere, the usual clutch is 9–20 (23), with larger broods possibly from two females.

COMMON QUAIL *Coturnix coturnix*
Bıldırcın

(Du: Kwartel; Fr: Caille des blés; Ger: Wachtel; Swe: Vaktel)

Subspecies. Nominate *coturnix* occurs throughout.

Status and distribution. National conservation status: Vulnerable (population apparently decreasing). Widespread summer visitor across Turkey, being a particularly common breeder in Inner Anatolia, but much rarer in coastal regions, where grain crops are relatively scarce, and breeds up to at least 2,300m in the east of the country. In Southeast Anatolia principally confined to the west of the region. Otherwise, common on migration, across the entire country, with spring passage peaking late March–April, whilst autumn movements peak in late August–September and even earlier in the north. Some overwinter in Aegean and on the coastal littoral of Mediterranean, making it difficult to establish extreme dates of passage, but recorded in Black Sea Coastlands as late as early November (Albrecht 1986) and probably rarely arrives in the extreme north prior to late March. There is a very late record from Inner Anatolia, at Seyfe Gölü, from 26 November 2003 (Balmer & Betton 2004a), which surely refers to a migrant.

Largely confined to agricultural land with grain crops, both during breeding and on migration, and in Inner Anatolia is abundant in dry wheat fields. Also breeds in semi-steppe, meadows and other suitable habitats with tall grass. Like many other agricultural-land birds, the species has declined across the country as a result of agricultural intensification, expansion of non-grain crops and hunting. Migrants are heavily hunted and the species is still illegally trapped during migration, especially in the north and particularly in Black Sea Coastlands.

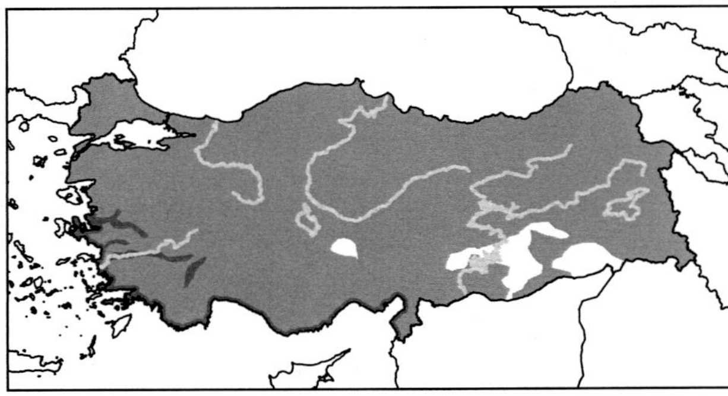

Breeding. Mainly in dry cornfields and grassy steppe; occasionally heard in other habitats, such as montane areas and sand dunes with denser vegetation. Breeding exceptionally difficult to prove, and no records of nests in Turkey. Elsewhere, a shallow hollow in the ground, lined with grass and nearby plant material, and usual clutch 7–12 (6–18). A record of about nine fully feathered young, some three weeks old, hiding under piles of cut corn at harvest time, and caught by locals for the pot, southeast of Birecik (**Southeast Anatolia**), on 4 June 1993, and at least another six on 5 June 1993, indicates laying *c.*20 April (PC). Young observed near Uluabat Gölü (**Marmara**) on 22 May 1966. Earliest arrivals in most regions in mid April, and the records of young suggest that breeding commences almost immediately; calling birds heard in August, perhaps suggesting a prolonged breeding season. Pairs with young observed in Syria in July (Baumgart *et al.* 1995).

COMMON PHEASANT *Phasianus colchicus*
Sülün

(Du: Fazant; Fr: Faisan de Colchide; Ger: Fasan; Swe: Fasan)

Subspecies. Apparently nominate *colchicus* (Dement'ev & Gladkov 1967, Roselaar 1995), although ascribed to *septentrionalis* by Kumerloeve (1961). Madge & McGowan (2002) considered that the Turkish population was only doubtfully natural, but Kasparek (1988c) demonstrated that given the availability of records from around Istanbul from as long ago as 1792, it seems likely that the original population was, in fact, native. However, the current stock may contain very few elements of the original population, which has been swamped by the release of birds for hunting.

Status and distribution. National conservation status: Vulnerable (population apparently decreasing). Sparsely and uncommonly distributed through Black Sea Coastlands and Marmara where resident, but almost the entire native population has disappeared as a result of the introduction of game pheasants originating from various other subspecies, as well as hunting. Kasparek (1988c) mapped all known localities, both historical and modern: most are in close proximity to the Bosphorus, in southern Marmara and the central Black Sea Coastlands, although the species has been reported as far east as Trabzon. Currently, remnants of the native population perhaps persist in the Sinop region and at some other localities in Black Sea Coastlands (e.g. Kızılırmak Delta). Kasparek (*op. cit.*) reported that game pheasants were being introduced in other parts of Turkey, with individual sightings from various parts of Inner Anatolia, Aegean and Mediterranean undoubtedly referring to such introduction attempts. The native population in Turkey was most probably confined to coastal forests, pseudo-maquis and heathlands of Black Sea Coastlands and Marmara.

Breeding. No records.

MUTE SWAN *Cygnus olor*
Sessiz Kuğu

(Du: Knobbelzwaan; Fr: Cygne tuberculé; Ger: Höckerschwan; Swe: Knölsvan)

Subspecies. Monotypic.

Status and distribution. National conservation status: Vulnerable (population decreasing). A scarce and local resident, principally at several marshy wetlands in Black Sea Coastlands, Mediterranean and Inner Anatolia, and less frequently Marmara, including Thrace. Now largely only a winter visitor to Aegean, although formerly resident there (e.g. bred at Marmara Gölü and Bafa Gölü). Recently recorded in summer on Van Gölü and in midwinter on Pınarbaşı Gölü, Malatya province (East Anatolia), with one recent winter record from Southeast Anatolia. Currently, the breeding population is probably less than 20 pairs, with very few recent records of confirmed breeding, mostly from Ereğli marshes (Inner Anatolia) and even this population has apparently become extinct in the last few years. Many summer records, especially from Black Sea Coastlands and Mediterranean, apparently refer to non-breeding birds, although plausibly bred in Kızılırmak Delta (Black Sea Coastlands) in 1968 (Dijksen & Kasparek 1985). Formerly bred in considerably greater numbers: a reasonably well-documented decline at the Ereğli marshes has been noted (Kılıç & Kasparek 1990), whilst Kumerloeve (1961) recorded that the populations at Marmara Gölü and Akşehir Gölü (Inner Anatolia) were *c.*50 and 100 pairs, respectively. Neither site is presently utilised and several other former breeding localities have now been largely or entirely drained.

More widespread in winter, generally in small, more occasionally moderate or large (see below), numbers scattered throughout coastal wetlands in the western two-thirds of Turkey. At this season Inner Anatolia hosts a regular but small population, generally at large reservoirs, but more widespread in harsh periods. However, former records indicate existence of a larger wintering population there. Now most regularly recorded in Marmara including Thrace, though occasional large concentrations in Black Sea Coastlands are probably weather-related. The total wintering population has frequently been in excess of 1,000 individuals in recent years, formerly it was usually not more than 200. Most winter in the west of the country, in 1993, 1,244 were counted in the Meriç Delta (Marmara) on 24 January (DHKD 1993), an incredible 8,900 (of a total midwinter population of 9,088) were present there in early 1999 (DHKD 1999, Magnin *et al.* in Heath & Tucker 2000), and 2,000 were present at Lake Gala, on the Greek border, alone on 2 January 2003.

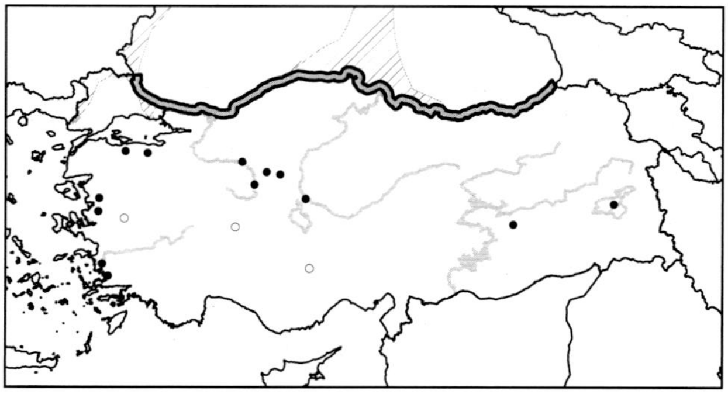

Breeding. At large lakes and marshes with extensive reeds and open water. Site, nest and clutch size undescribed in Turkey but, elsewhere, nests are at water's edge, on land or on a small island, or built up in shallow water in reeds; a large heap of reeds and aquatic vegetation, with a central hollow in the top; usual clutch 5–7. At Ereğli marshes (**Inner Anatolia**), Lenz found 11 occupied nests in 1968, one in 1969 and three in 1970, a pair with 4–5 young was observed on 6 July 1976, a pair with four young on 17 July 1982 and, on 16 May 1987, a nest was located from which young had hatched (M. Kasparek), indicating egg laying no later than early April; three pairs, with one juvenile, were observed on 17 May 1987, plus a nest with hatched young. At Karataş Gölü (**Mediterranean**), one adult and three well-grown young were observed from15–17 August 1971. In **Aegean**, on 13 May 1899, gypsies informed Selous (1900) that a pair had nested in a reedbed, and that the young had already hatched.

BEWICK'S SWAN *Cygnus columbianus*
Küçük Kuğu

(Du: Kleine Zwaan; Fr: Cygne de Bewick; Ger: Zwergschwan; Swe: Mindre sångsvan)

Subspecies and taxonomy. *C. c. bewickii.* New World *C. c. columbianus* (Whistling or Tundra Swan) and Old World *C. c. bewickii* are readily distinguished, given a reasonable view, by the former's black bill and essentially black facial skin. It is also slightly larger than *bewickii*, though measurements overlap. The two groups are sometimes considered at species level (e.g. Johnsgard 1978, Rasmussen & Anderton 2005), which is relevant, considering that *columbianus* could potentially wander to Turkey (there being several European records). However, traditional taxonomy has been maintained by the Taxonomic Advisory Committee of the Association of European Records and Rarities Committees. For further details see discussion in AOU (1998).

Status and distribution. Rare winter visitor in irregular numbers, which has been proven to be annual in recent years, especially at Nallıhan (Inner Anatolia) and other reservoirs, between mid November and late February, including regular juveniles. About 45 records, all but one since 1991, with most from the western two-thirds of the country (two from further east, but with the majority in Inner Anatolia, the Turkish Lake District and immediate surroundings) between November and mid April, most involving small groups of fewer than ten, but occasionally up to 50+, as follows: two adults and four immatures at the Göksu Delta (Mediterranean) on 3 January 1983 (Kasparek 1990a), one at Burdur Gölü (Mediterranean) on 3 December 1991 (KAB), five adults at Süleymanhacı Gölü (Inner Anatolia) on 22 January 1992; five adults (the same?) at Burdur Gölü on 26 January 1992 (DHKD 1992), 51 at Burdur Gölü on 28 January 1993, 18 at Manyas Gölü (Marmara) on 6 February 1993 (DHKD 1993), three at Kabaklı Reservoir, Diyarbakır (South-east Anatolia) in March 1993 (Biricik 1996), one at Kulu Gölü (Inner Anatolia) on 16 April 1993, one at Kızılırmak Delta (Black Sea Coastlands) on 26 December 1993, one at Burdur Gölü on 24 December 1994, 17, in two groups, at Eğirdir Gölü (Mediterranean) on 26 January 1996 (Kirwan & Martins 2000a), 50 at the Meriç Delta (Marmara) on 25 January 1997 (Boyla & Eken 1998), six at the Göksu Delta on 19 February 1998, one at Burdur Gölü on 9 January 1999 (KAB *et al.*), six at Cavuşçu Gölü (Inner Anatolia) on 19 January 1999, two at Eber Gölü on 19 January 1999, one at Balıkdamı (Inner Anatolia) on 20 January 1999, 25 at the Meriç Delta on 27 January 1999, two at Kulu Gölü in November 1999, one at Kızılırmak Delta on 12 December 1999, ten at Burdur Gölü on 30 January 2000, two at the Göksu Delta on 17 February 2000 (all Kirwan *et al.* 2003), one at Burdur Gölü on 13 January 2002 (Demirci 2002), one at Lake Gala on 20 January 2002 (Balmer & Betton 2003a), 52 at Sarıyar Reservoir (Inner Anatolia) on 27 January 2002 (Demirci 2002), two at Marmara Gölü (Aegean) on 3 January 2003 (Kirwan *et al.* in press), 34 again at Sarıyar on 18 January 2004 (J. Tavares *et al.*), six at Balıkdamı on 7 March 2004 (G. Welch & H. Welch), up to 31 between 13 November 2004 and 11 January 2005 at Sarıyar Reservoir, five in the same place from 2005 until at least 21 January 2006 (BD, KAB & S. Bekir), 22 in Kocaeli Bay (Marmara) on 25 January 2006 (B. Bilgen *et al.*), three at Gala Gölü on 4 February 2006, nine at Eğirdir Gölü on 11 February 2006, three at Büyükçekmece, Istanbul, on 3–9 December 2006 (Kirwan *et al.* in press), 21 at the Göksu Delta on 20 January 2008 (Balmer & Betton 2008), 25 at Enez, Edirne (Marmara) on 26 January 2008 (E. Bucak *et al.*), and 140 at Hirfanlı (Inner Anatolia) on 9 February 2008 (BD *et al.*), the latter constituting the largest-ever flock in Turkey. Midwinter counts in 2005 (Çağlayan *et al.* 2005) revealed a dispersed total of 144 birds: four at Sarıyar on 11 January, 68 at Meriç Delta on 15 January (with three still present on 4 February), 36 at Hirfanlı Reservoir on 16 January (Inner Anatolia), up to nine at Göksu Delta from 4 January–16 February, 18 at northern Van Gölü (Eastern Anatolia) and nine near Trabzon (Black Sea Coastlands) on 1 February. Small numbers (up to ten) were also present at Hirfanli Reservoir in

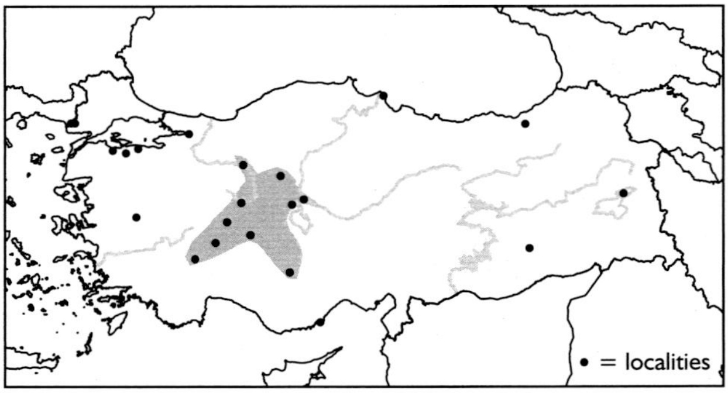

late 2005 (Kirwan *et al.* in press). Furthermore, there is a record of a herd of 192 individuals, either this or the next species, at Burdur Gölü, on 27 January 1996 (KAB *et al.* pers. obs.). We have not seen details of a record of one together with the two other swan species, photographed at Mogan Gölü (Inner Anatolia), in February during the mid 1990s (A. Aykurt).

Listed as a winter visitor to northern Iran and as a vagrant to Iraq, Israel, Oman, north-west Saudi Arabia and the United Arab Emirates (Porter *et al.* 1996), as well as Armenia (Adamian & Klem 1999) and Greece (Handrinos & Akriotis 1997).

WHOOPER SWAN — *Cygnus cygnus*
Ötücü Kuğu

(Du: Wilde Zwaan; Fr: Cygne chanteur; Ger: Singschwan; Swe: Sångsvan)

Subspecies. Currently considered monotypic, although Brazil (2003) discussed apparently geographically related variation across the species' range.

Status and distribution. Local winter visitor, recorded in small to moderate numbers, apparently becoming increasingly regular, presumably in part due to increased observer coverage of wetlands at this season (see Table 1). The majority of records are from coastal wetlands throughout the western two-thirds of Turkey, although wintering is also regularly reported from Inner and Eastern Anatolia (Adızel 1998) and the species has occurred once in Southeast Anatolia. Most observations are from November to February, although birds were recorded in late October and remained into the first half of April (*e.g.* 14 April 2007 at Göksu Delta), exceptionally late April in northern areas, while one individual at Kızılırmak Delta, Black Sea Coastlands (perhaps injured) was recorded as late as 6 June 1993 (Kirwan & Martins 2000a). Whooper Swans may reach western Turkey across the Black Sea from the Ukraine or via the west coast of the Black Sea (Brazil 2003).

In recent years, the species has occurred in Black Sea Coastlands, at Kızılırmak Delta, Yeniçağa Gölü, Sarıkum Gölü and small deltas of the streams between Kastamonu and Rize, from 19 November (2005), with a max. 44 at Akliman, Sinop on 19 December 1993 (S. Barış). In Thrace, birds have occurred mainly at Meriç Delta, Iğneada, and Büyükçekmece in January–February, with a maximum of 1,200 on 25 January 1998 at Meriç Delta (Boyla & Eken 1998) (this site on the Greek border is clearly an important, and presumably regular, wintering area for the species in the Balkans region). In Aegean, the species has been recorded at Bafa Gölü, Marmara Gölü, Büyük and Küçük Menderes Delta and Gediz Delta in January–February, with a maximum of 32 at Marmara Gölü on 27 January 1993. In Mediterranean, records are in December–April at Göksu Delta, Burdur Gölü and in Çukurova deltas, with a maximum of 67 at Göksu Delta on 24 January 1993. In Inner Anatolia, it has occurred at Kulu Gölü, Beyşehir Gölü, Ereğli marshes, Mogan Gölü, Hirfanlı and Sarıyar reservoirs, and Tuz Gölü in December–April with the largest count being 158 at Kulu Gölü on 8 December 1993. The Van Gölü basin, especially Sodalı Gölü, holds a significant population with a local maximum of 164 on 30 October 1995 (Adızel 1998). The only record from Southeast Anatolia was at Kabakli Reservoir, Diyarbakır, in February 1993. A total of 172 was at seven sites in January/February 1993 (DHKD 1993), whilst 133 were recorded in early 1992 (DHKD 1992) when numbers were also exceptionally high in the south Caspian region (Iran) (D. A. Scott *in litt.* 1992). However, these totals were eclipsed by the 259 reported in winter 1993/94, 446 in winter 1995/96 and the exceptional numbers recorded in 1998 (see Table 1).

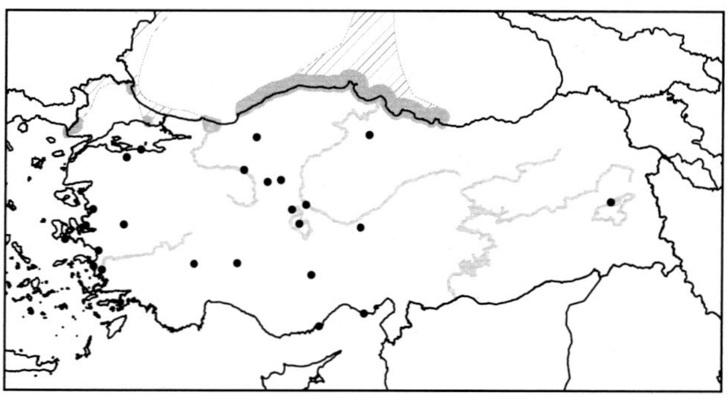

Table 1. Numbers of Whooper Swan *Cygnus cygnus* in Turkey by year/season according to reports submitted to the Turkey Bird Report and the authors.

Year	No. of birds	No. of sites present
1967	2	1
1968	5	1
1968/69	21	6
1969/70	47	1
1972	15	2
1973	9	1
1974	26	3
1976	2	1
1978	5	1
1986	16	1
1987	16	3
1988	64	3
1988/89	29	3
1990	29	1
1991	7	1
1992	143	9
1993	182	8
1993/94	259	4
1994/95	92	6
1995/96	446	7
1998	1,200	1
1999	69	5
2000	42	4
2005	210	9
2006	67	1

[SNOW GOOSE *Anser caerulescens*]
Kar Kazı

(Du: Sneeuwgans; Fr: Oie des neiges; Ger: Schneegans; Swe: Snögås)

Subspecies. Two recognised, but the identity of the Turkish claims was not specified.

Species not admitted to the Turkish list (Kirwan *et al.* 1999). Krüper (1875), Kasparyan (1956) and Kiziroğlu (1989) listed the species for Turkey, the last-named claiming it occurred as a winter visitor to the eastern Mediterranean and southern Aegean regions of the country. Adequate details were not made available and the occurrence of the species appears highly unlikely, although Vaurie (1965) mentioned two or three records from the southern Caspian region.

BEAN GOOSE *Anser fabalis*
Tarla Kazı

(Du: Taigarietgans / Toendrarietgans; Fr: Oie des moissons; Ger: Saatgans; Swe: Sädgås)

Subspecies and taxonomy. None of the Turkish records has been specifically assigned, although recent records from Kulu Gölü (Inner Anatolia) and Büyükçekmece (Marmara) were considered likely by the observers in question to have referred to *rossicus* from northwest Siberia (Richardson 2003). It is unclear, therefore, which form/s occur in the country, and given the lack of previous interest in the identification of the different taxa within the Bean Goose complex their winter distribution within Europe is unclear. Sangster & Oreel (1996) proposed specific status, under the Biological or Phylogenetic Species Concepts, for Taiga Bean Goose *A.* (*f.*) *fabalis* (including *fabalis*, *johanseni* and *middendorffii*) and Tundra Bean Goose *A.* (*f.*) *serrirostris* (consisting of *serrirostris* and *rossicus*), based on a range of criteria. This proposal has not been adopted by most authorities as yet, e.g. Snow & Perrins (1998) continued to treat these taxa within a single polytypic species.

Status and distribution. Vagrant or perhaps an irregular and scarce winter visitor, principally to wetlands in Aegean, Mediterranean and Inner Anatolia, occasionally in large numbers, e.g. *c.*90 between Ereğli and Karapınar (Inner Anatolia) on 2 March 1965 (Kumerloeve 1970a), 61 at Marmara Gölü (Aegean) on 30 January 1992 (DHKD 1992) and, most exceptionally, 376 at Büyükçekmece, Istanbul, on 9 January 1993 (KAB *et al.* pers. obs.). Formerly recorded from the western Black Sea Coastlands (see Kumerloeve *op. cit.*) and the eastern part of the Mediterranean coast, where collected by Schrader (1891) in the 1870s or 1880s at Mersin, with one at the Göksu Delta (Mediterranean) on 24 January 1993 (DHKD 1993).

Sixteen modern (post-1966) reports, most within the period January–March, with one old (1933) and four recent April records. Earliest were 13 at Karamık marshes (Mediterranean) on 15–16 October 1969 (OST 1972), with one at Seyfe Gölü (Inner Anatolia) on 30 April 1988 the latest. The species may plausibly be occasionally overlooked among the large flocks of Greater White-fronted Geese *A. albifrons* that winter in some parts of the country, particularly in Inner Anatolia.

There are two unsubstantiated records from Cyprus, in 1904 and 1908 (Flint & Stewart 1992), fewer than 20 records in Greece since the mid-19th century (Handrinos & Akriotis 1997), and the species is occasional in Iran (Porter *et al.* 1996) and a vagrant to Israel, Jordan and Lebanon (Shirihai 1996, Balmer & Betton 2007b, Ramadan-Jaradi *et al.* 2008).

[PINK-FOOTED GOOSE *Anser brachyrynchus*]
Küçük Tarla Kazı

(Du: Kleine Rietgans; Fr: Oie à bec court; Ger: Kurzschnabelgans; Swe: Spetsbergsgås)

Subspecies. Monotypic.

Species not admitted to the Turkish list. Kasparek (1985, 1990a) mentioned one with Greylag Geese *A.anser* at the Sultan marshes (Inner Anatolia) on 17 April 1984, supported by a very basic description. The normal winter range of the species lies entirely in northwest Europe, principally the British Isles and the Netherlands. It breeds in two discrete breeding areas; eastern Greenland and Iceland, and Spitzsbergen and probably Franz Josef Land and the Karla Peninsula, making it an extremely unlikely, although potential, vagrant. There are no other records from the Middle East, although there is a specimen record from Bulgaria (Nankinov 2004). In the absence of a detailed description we do not admit this to the Turkish list. Given the relative similarity of Pink-footed Goose to the *rossicus* form of Bean Goose *A. fabalis* (see e.g. Svensson *et al.* 1999), to which one recent record of the latter species in Turkey has been plausibly ascribed (Richardson 2003), the possibility that the Sultan marshes record involved an individual of the latter cannot be eliminated.

GREATER WHITE-FRONTED GOOSE *Anser albifrons*
Sakarca

(Du: Kolgans; Fr: Oie rieuse; Ger: Bläßgans; Swe: Bläsgås)

Subspecies. *A. a. albifrons.*

Status and distribution. Widespread and common passage migrant and winter visitor to wetlands almost entirely in the western two-thirds of Turkey, with the majority concentrated around Tuz Gölü and the Konya Ovası (Inner Anatolia). Significant numbers congregate at other wetlands in this region and also in Aegean (especially at the Büyük Menderes Delta) and Mediterranean. Much less common further east, where occurs chiefly in small flocks on passage, though creation of new dam lakes in Southeast Anatolia may lead to small numbers wintering regularly in parts of this region. Maximum count during the midwinter waterfowl censuses conducted by the International Waterfowl Research Bureau (now Wetlands International) was 98,600 in 1968/69. More recently, there were 84,000 in 1987. The mean winter count is approximately 50,000, although most recent winters have seen smaller numbers, usually 20,000–30,000, with just 11,822 in January/February 1993 (DHKD 1993), 3,956 in early 1999 (DHKD 1999) and 3,891 in early 2005 (Çağlayan *et al.* 2005), but more than 15,000 at Büyükçekemce (Marmara) alone on 11 February 2006 (MÖ). The majority arrive in mid to late October (earliest 6th) remaining until late March or even late April (29th). One mid-May record from Black Sea Coastlands. Small numbers of non-breeders, occasionally in flocks of 30–80 birds, regularly oversummer in Turkey, particularly in Inner, but also in Southeast and East Anatolia.

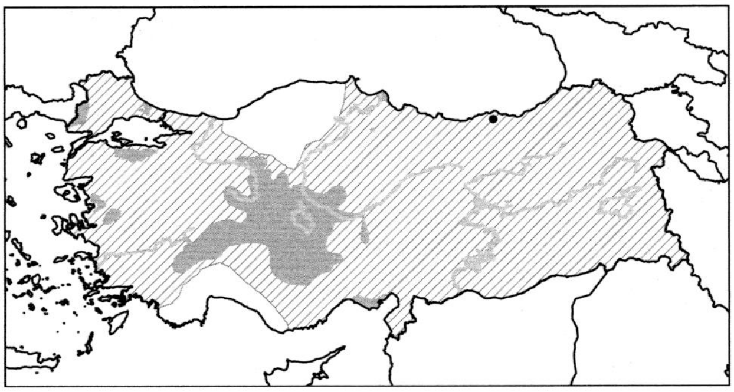

LESSER WHITE-FRONTED GOOSE *Anser erythropus*
Küçük Sakarca

(Du: Dwerggans; Fr: Oie naine; Ger: Zwerggans; Swe: Fjällgås)

Subspecies. Monotypic.

Status and distribution. Conservation status: Vulnerable (BirdLife International 2000, 2004). Vagrant or very rare winter visitor. Eight records since 1966: two at the mouth of the Kocabaş River, Canakkale province (Marmara) on 16 February 1967 (OST 1969), an adult and two immatures at Bafa Gölü (Aegean) on 24 December 1986 (Kasparek 1988a), 12 at Seyfe Gölü (Inner Anatolia) on 6 April 1990 (Kirwan & Martins 1994), one at Büyükçekmece (Istanbul) on 23 January 1993 one at the Göksu Delta (Mediterranean) on 24 January 1993 (DHKD 1993) with another at the latter locality on 29 December 1997 (Kirwan *et al.* 2003), one tracked via satellite was at Haçlı Gölü (Eastern Anatolia) on 20 November 2004 (Morozov & Aarvak 2004), and another, also satellite-tracked, moved from the environs of the Turkey/Iran border to Al-Hassakh, northeast Syria, between 30 November and 3 December 2006 (Kirwan *et al.* in press). In addition, Kumerloeve (1970a) noted 12 records, all but one (dated

24 October 1952) from December and January, between 1945 and 1965. All were from either the Bosphorus or Thrace, with the majority from the Büyük/Kücükçekmece (Istanbul) area. However, no documentation is available for any of these reports.

In neighbouring states, Lesser White-fronted Goose is also very rare: there are fewer than 30 documented records in Greece since the mid-19th century (Handrinos & Akriotis 1997) and *c.*12 in Armenia since the 1980s (Adamian & Klem 1999), and it must be now very rare in both the Iraq marshes and northern and western Iran, with only vagrants reported in Cyprus (Balmer & Betton 2004a), Israel, Kuwait, Oman and Syria (Porter *et al.* 1996).

GREYLAG GOOSE *Anser anser*
Boz Kaz

(Du: Grauwe Gans; Fr: Oie cendrée; Ger: Graugans; Swe: Grågås)

Subspecies. *A. a. rubrirostris.*

Status and distribution. National conservation status: Vulnerable (population apparently decreasing). Rather local resident, or perhaps migrant breeder in some areas, in marshy wetlands in Inner and East Anatolia, scarcer in Mediterranean and perhaps only formerly present in Marmara and Black Sea Coastlands. There is evidence to suggest a decline, in part due to wetland drainage in its core range, in all currently occupied regions from the number of records submitted to the Turkey Bird Report. Usually occurs at relatively low densities but occasionally fairly common at favoured breeding localities, particularly in Inner Anatolia (e.g. 120 nests and 300 pairs estimated at the now largely drained Ereğli marshes in April 1970, and, despite much-reduced wetland area, 160 individuals were present in late July 1996, although current status is unknown). The total number of breeding pairs in this region is now perhaps fewer than 250. Post-breeding flocks congregate at certain favoured wetlands, e.g. 800 at Kulu Gölü (Inner Anatolia) in July 1984.

More widespread and common on passage in all regions and as a winter visitor from October (late September) when substantial numbers immigrate: exceptionally 12,000 in Sultan marshes, Inner Anatolia and 10,000 at Kuyucuk Gölü, East Anatolia, in September 2004 (C. Şekercioğlu, C. Özsemir & K. Erciyas unpubl.). The majority have departed by the end of March, but some winter residents may delay departure until at least mid April. Principally winters in coastal areas but also found inland across the western two-thirds of Turkey. The majority winter in the Black Sea Coastlands (principally at Kızılırmak Delta). Typically in flocks of under 300 birds, especially during recent winters when the total population has rarely exceeded 1,000 (although 1,897 in January/February 1993; DHKD 1993). Formerly more common with an average midwinter population of around 5,000 and a maximum of 11,200 in 1967, when there were 5,000 at the Kızılırmak Delta, 4,500 in the Meriç Delta (Marmara) and 1,500 at Hotamış marshes (Inner Anatolia).

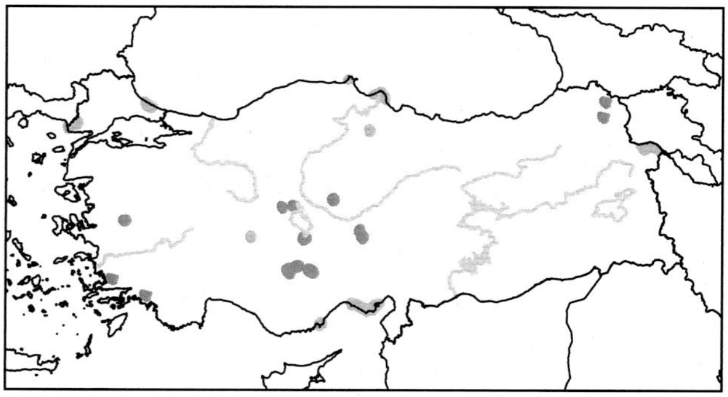

Breeding. Usually in small groups; on islands in lakes, e.g. Kulu Gölü (Inner Anatolia), where makes a shallow scrape in dry ground with a scant lining of nearby vegetation, down and small feathers; also at wet sites, in reedbeds, e.g. Ereğli marshes (Inner Anatolia), where constructs a mound of reeds and aquatic vegetation, built up from the water level at the base of reeds. No reliable data on clutch size in Turkey (4–6 is usual elsewhere), but broods of up to five goslings observed. Most records are from **Inner Anatolia**, where earliest goslings were at Kulu Gölü on 23 April 1988, Sultan marshes on 27 April 1988, Mogan Gölü on 30 April 1968, and Ereğli marshes on 30 April 1973 (where there were 120 nests on 28 April 1970); laying commences, therefore, in late March. Goslings seen in **Marmara** on 20 April 1996 and 14 May 1969, in **Black Sea Coastlands** on 16 May 1970; and **East Anatolia** on 24 June 1983 (23 small young), suggesting a later breeding season.

BARNACLE GOOSE *Branta leucopsis*
Ak Yanakli Kaz

(Du: Brandgans; Fr: Bernache nonnette; Ger: Weißwangengans; Swe: Vitkindad gås)

Subspecies. Monotypic.

Status and distribution. Vagrant; perhaps even only hypothetical in Turkey, given the lack of available documentation associated with either of the two records listed by Kumerloeve (1970a): one in the Sakarya Delta, north of Adapazarı (Black Sea Coastlands) in the winter of 1946/47, and another shot in the same area in autumn or early winter 1961 that turned up in Ankara in September 1964, but was unavailable for sale (Kumerloeve 1966d). One modern occurrence: one observed at Büyükçekmece, Istanbul, on 5 January 2003 (E. Çağlayan, B. Akyıldırım). No other reports from Turkey.

The species is listed as a vagrant to Egypt (Goodman & Meininger 1989) but there are no other records from the Middle East (Hollom *et al.* 1988) or other neighbouring states, except Bulgaria, where one was present on the Black Sea coast in February–March 1997, with further records in January 2000 and one (photographed) in November 2005 (Nankinov 2004, Nikolov *et al.* 2006), and also recently photographed in Iran (Winkel & de Weerd 2007). Most winter in north-west Europe, occasionally south to the northern Mediterranean coast (Cramp & Simmons 1977).

BRENT GOOSE *Branta bernicla*
Yosun Kazı

(Du: Rotgans; Fr: Bernache cravant; Ger: Ringelgans; Swe: Prutgås)

Subspecies and taxonomy. Few data are available concerning the form/s occurring in Turkey; the only specific identification involved a record of the dark-bellied race, *B. b. bernicla* (see below). All other records are presumably attributable to this form, which winters solely in northwest Europe (Cramp & Simmons 1977). The only documented record in Greece also involved this form (Handrinos & Akriotis 1997). Shields (1990) suggested that the dark-bellied, pale-bellied (*B. b. hrota*) and so-called black brant (*B. b. nigricans*) forms be recognised as species, which proposal was supported by Sangster *et al.* (1997), although some apparent hybrid *nigricans* × *bernicla* have been observed in Western Europe in recent winters (Wynn 2003). Attendant interest in these forms' field identification has mushroomed in recent years, but we have not studied the issue and offer no opinion here.

Status and distribution. Vagrant. Mathey-Dupraz (1923) recorded that the species was regular at Maltepe, Istanbul during a cold spell in January 1889, and in February 1891 a large flock was at Kadiköy, on the Asian shore of the Bosphorus. Zeynab shot one individual, again in the vicinity of Istanbul, at Büyükçekmece on 7 February 1945, although the specimen has not been preserved

(Kumerloeve 1970a). More recently, two dark-bellied individuals (*B. b. bernicla*) were offshore at Ardeşen, Rize province (Black Sea Coastlands) on 3–4 September 1973 (OST 1975) and two were in the Küçük Menderes Delta (Aegean) on 6 April 1981 (Beaman 1986). A published record from Acıgöl (Mediterranean) in 1969 is generally considered to be unacceptable (Dijksen & Kasparek 1987).

The species is known as a vagrant to Egypt (Goodman & Meininger 1989) and Greece (Handrinos & Akriotis 1997), but a record from Cyprus (shot at Kyrenia in December 1908) is now rejected as the specimen was not retained (Flint & Stewart 1992).

RED-BREASTED GOOSE *Branta ruficollis*
Sibirya Kazı

(Du: Roodhalsgans; Fr: Bernache à cou roux; Ger: Rothalsgans; Swe: Rödhalsad gås)

Subspecies. Monotypic.

Status and distribution. Conservation status: Endangered (BirdLife International 2008). A rare winter visitor to the western two-thirds of Turkey. Pre-1966 records all come from Thrace: Kosswig (*in* Kumerloeve 1970a) reported this species at Küçükcekmece in winter 1946/47, whilst Kumerloeve (*op. cit.*) recorded a group of nine between Büyükçekmece and the Meriç Valley from 27 November to 6 March (year uncertain 1947/54) with two at Beylik Mandra at the same time. In 1959 Ishakoğlu shot one from a party of eight at an unspecified locality, while Makatsch (1950) also listed the species for Thrace. Finally, Kumerloeve (1966d) mentioned one at Kuyucuk Gölü (East Anatolia) on 12 November 1964 and that three were seen by E. Hirzel at Çekmece (Istanbul) on 17 January 1965.

Between 1966 and early 2008 there were 63 records in Turkey, with 15 of these in Marmara, eight in Black Sea Coastlands, four in Aegean, six in Mediterranean and 12 in Inner Anatolia (see Table 2). They span the period 30 October to 18 March, with the vast majority in late December–February. Most records concerned single birds, but several comprised small groups of 3–9 birds, and five large flocks, of 38–94 individuals, were involved. Winter 2001/2002 appears to have been an exceptional season for the species in Turkey, with at least 192 birds reported. The incidence of records at Büyükçekmece (eight) is presumably due to the frequency of visits by birdwatchers from nearby Istanbul, rather than a true reflection of its attractiveness to the species. Several records involve birds among Greater White-fronted Geese *Anser albifrons*, whilst judging from the number of stuffed birds seen in hunting shops in Turkey and reports from hunters (see e.g. Dijksen & Kasparek 1985), the species is rather more regular than birdwatchers' records indicate. Indeed, virtually all of the Inner Anatolian records were made during systematic counts of geese flocks in that region, perhaps indicating that at least some go unnoticed in the still-vast Greater White-fronted Geese herds there each winter.

Elsewhere the species winters in the northern and western Black Sea (largest numbers in Romania), the Caspian and former Aral Seas, being occasional in Iran (D. A. Scott) and, at least formerly, a rare winter visitor to Iraq (Allouse 1953). Also known as a vagrant to Israel (Shirihai 1996), Egypt (in the last century; Goodman & Meininger 1989) and recently Cyprus (February 1991 and December 1993; Flint & Stewart 1992, J. Sadler pers. comm.), where a dead individual in Larnaca market in 1928 may also be assumed to relate to a bird obtained locally.

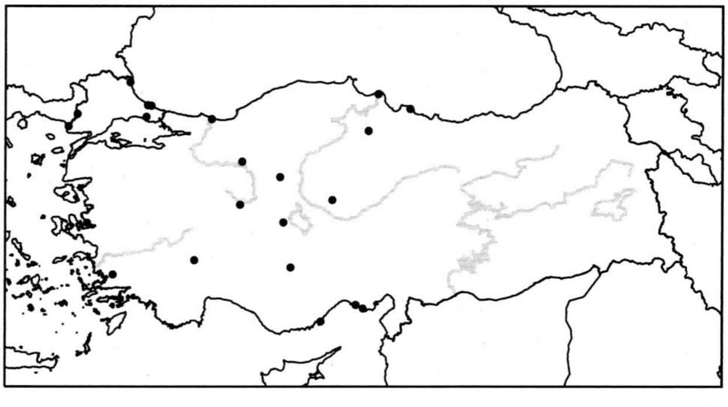

There is one further startling record from Turkey, which lacks explanation or precedent; 14 adults and eight young at Erçek Gölü (East Anatolia) on 5 August 1982 (Kasparek & van der Ven 1983). No description was published and, as additional details have not subsequently emerged, this record is best regarded as erroneous and ignored.

Table 2. Post-1966 records of Red-breasted Goose *Branta ruficollis* in Turkey. Not included here is a published claim (Kiziroğlu & Kiziroğlu 1987) of an unknown number of individuals at Manyas Gölü (Marmara) in January (year unknown).

Locality	Region	Date	No. of individuals
Mogan Gölü	Inner Anatolia	27 November 1966	1
Meriç Delta	Marmara	11 January 1967	24
Yeşilırmak Delta	Black Sea Coastlands	22–24 January 1967	1
Büyükçekmece	Marmara	15 February 1968	1
Balıkdamı	Inner Anatolia	30 October 1968	1
Near Cihanbeyli	Inner Anatolia	19 December 1970	9
Hotamış marshes	Inner Anatolia	8–9 January 1973	1
Mogan Gölü	Inner Anatolia	1 November 1975	7
Seyfe Gölü	Inner Anatolia	3 January 1989	4
Büyükçekmece	Marmara	14 December 1991	1
Kızılırmak Delta	Black Sea Coastlands	29 December 1991	7
Kızılırmak Delta	Black Sea Coastlands	December/January 1991/92	3 shot by hunters
Büyükçekmece	Marmara	11 January 1992	6
Göksu Delta	Mediterranean	31 January 1992	3
Büyükçekmece	Marmara	23 February 1994	2
Between Küplü and Ipsala	Marmara	January 1996	2
Büyükçekmece	Marmara	11 February 1996	75
Büyükçekmece	Marmara	16 February 1996	1
Terkos Gölü	Marmara	1 January 1997	1 dead
Yedikır Reservoir	Black Sea Coastlands	15 January 1997	8
Karaburun	Marmara	5 February 1997	6
Mert Gölü, Iğneada	Marmara	20 January 2001	2
Gediz Delta	Aegean	30 December 2001	94
Burdur Gölü	Mediterranean	30 December 2001	5 (3 until 13 January 2002)
Göksu Delta	Mediterranean	5 January 2002	39 (15 on 7th)
Tuzla Gölü	Mediterranean	5 January 2002	9
Akyatan Gölü	Mediterranean	3 January 2002	6 (1 still on 8th)
Büyükçekmece	Marmara	19 January 2002	80
Gediz Delta	Aegean	26 January 2002	119
Nallıhan	Inner Anatolia	27 January 2002	57
Aliağa, Izmir	Aegean	4 February 2002	1
Nallıhan	Inner Anatolia	18 January–18 February 2004	1
Kızılırmak Delta	Black Sea Coastlands	26 December 2004	1
Gölbaba, Edirne	Marmara	28 January 2006	1
Balıkdamı	Inner Anatolia	29 January 2005	1
Kızılırmak Delta	Black Sea Coastlands	2 February 2006	1
Gölkent Gölü	Black Sea Coastlands	5 February 2006	38
Tuzla Gölü	Mediterranean	10 February 2006	1
Sarıyar Reservoir	Inner Anatolia	16 February 2006	21
Sarıyar Reservoir	Inner Anatolia	21 January 2006	5
Sakarya Delta	Black Sea Coastlands	5 February 2006	38
Gulluk Delta, Muğla	Aegean	25 February 2006	1
Büyükçekmece	Marmara	27 January–18 February 2006	up to 320
Sarıyar Reservoir	Inner Anatolia	18 March 2006	1
Büyükçekmece	Marmara	10 January 2008	2

EGYPTIAN GOOSE *Alopochen aegyptiaca*
Nil Kazı

(Du: Nijlgans; Fr: Ouette d'Egypte; Ger: Nilgans; Swe: Nilgås)

Subspecies. Monotypic. Note the correct spelling for the specific name (David & Gosselin 2000b).

Status and distribution. Vagrant. Reported three times: a doubtful record (not admitted by Kasparek 1992a) is mentioned by Weigold (1912/13): two, south of Urfa (South-east Anatolia) on 11 April 1911, which the observer was unwilling to claim as definite. A recent record, one at Kulu Gölü (Inner Anatolia) on 28 April 1986 (K. Wilson) is acceptable. Furthermore, on 6–13 July 2002, a pair was present and photographed at a park in Ankara, but was not seen again during regular visits to the area over a four-month period (E. Yoğurtcuoğlu). The species is known from a single zoo in Turkey (in Istanbul), but the Ankara birds did not escape from this source, whilst the nearest-known feral populations are in the Netherlands (Hagemeijer & Blair 1997), making the provenance of the Ankara birds unknown.

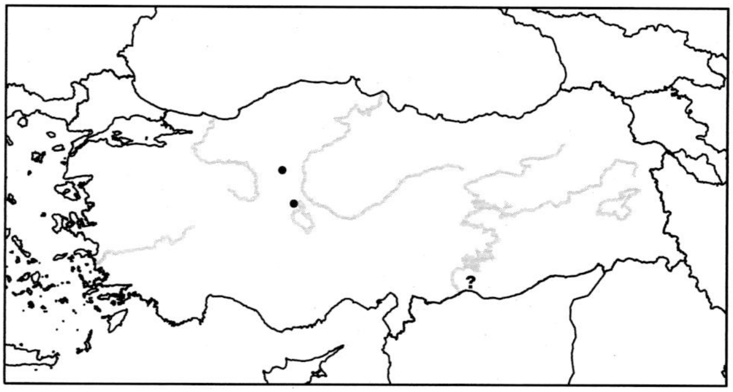

Vaurie (1965) reported that the species once bred in both Syria and Palestine, but neither Kumerloeve (1967a) nor Baumgart *et al.* (1995) listed any acceptable records in the former. Scarce in Cyprus, where regarded as a rare winter visitor between 1906 and 1928, but only three records since: in 1958, 1962 and 1989.

RUDDY SHELDUCK *Tadorna ferruginea*
Angıt

(Du: Casarca; Fr: Tadorne casarca; Ger: Rostgans; Swe: Rostand)

Subspecies. Monotypic.

Status and distribution. National conservation status: Least Concern (population apparently decreasing). Widespread and reasonably common resident, or partial migrant, in wetlands over much of the country, but especially numerous in Inner and East Anatolia, as well as the northern fringes of Mediterranean, in the Taurus. Less common in the interior of Marmara (particularly scarce in Thrace), Aegean and Southeast Anatolia, and less widespread in Black Sea Coastlands, where the most important site is Yedikır Reservoir. Prefers smaller pools, reservoirs, damp grassland areas and streams, often in upland areas, for breeding, although moderate numbers oversummer in larger wetlands apparently without breeding. Large flocks assemble to moult at certain wetlands following the breeding season from July to September, with the largest concentrations usually recorded in the former month. Max. counts include: at Erçek Gölü, East Anatolia (20,000), Sultan marshes, Inner Anatolia (11,000), Kulu Gölü, Inner Anatolia (10,000), Emir Gölü (now drained), Inner Anatolia (10,000–15,000 in September 1936), Sarıyar Reservoir, Inner Anatolia (8,000 in November 2004) and Kuyucuk Gölü, East Anatolia (6,000). More widespread in winter, sometimes in large numbers, although it is evident that some emigrate, as the winter maximum is only 10,155 (Çağlayan *et al.* 2005) and numbers are usually between 4,000–4,500 (only 711 in January/February 1993, but 5,636 at Sarıyar Reservoir alone on 18 January 2004 and 7,641 there on 18 February 2006), principally in the western two-thirds of Turkey.

Tucker & Heath (1994) suggested a population of between 4,000 and 8,000 breeding pairs, but it is considered to have declined rapidly in recent years, principally on the basis of midwinter counts, and Emirogullari *et al.* (in prep.) estimated 1,200–5,100 pairs, with an absolute minimum of 1,000 pairs and max. 9,300 pairs, but considered that it was not possible to make definite statements concerning trends at any season due to variable survey effort. Four important wintering areas – Tuz Gölü, Tuzla Gölü, Karapınar Ovası and Ereğli marshes (all in Inner Anatolia) – held just 65 birds in late 1990, although this is apparently a severe underestimate (BD). Emirogullari *et al.* (in prep.) considered the winter population to be 4,100–13,300 individuals, with a maximum of 28,500 and minimum 2,600 individuals There is some evidence to suggest that birds are wintering in a more dispersed fashion, with small increases in numbers from traditionally less-favoured localities evident in recent winters. A decline in post-breeding flocks, especially in the interior of the country has unquestionably occurred but, due to the increasing number of dam lakes, there is still much available habitat for nesting and the species appears to be one of the least-affected wildfowl by the large-scale wetland drainage that has occurred in recent decades.

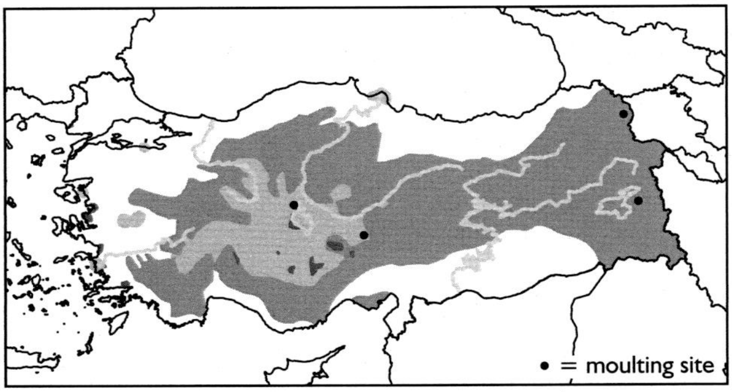

Breeding. In open country, generally in cliffs and hills around lakes, in holes and crevices in hillsides; on an old nest on a cliff (below); amongst rocks and ruins on an island in Beyşehir Gölü (Mediterranean) (A. Kullberg). In Aegean, a female was observed entering a burrow (probably rabbit) on a steep hillside near Akköy on 30 April 2003, but the nest was deep and could not be examined (PC). A female was incubating on an old stick nest (probably of Long-legged Buzzard *Buteo rufinus*) on a cliff near Ereğli (Inner Anatolia) on 22/24 May 1998 (T. Quittenden). The nest (undescribed in Turkey) is a hollow, lined with plant material, down and some feathers. Eggs: a clutch of eight, in the collection of R. Nicoll, was taken by Revd. W. Freeman at Ankara on 25 May 1895 (A. Dixon pers comm.), and a clutch of 12 eggs in the Natural History Museum, Tring, was collected from a nest in a burrow in the ground (no exact location) on 27 April 1887; elsewhere the usual clutch is 8–12 (16). Young: B/5–16 (31), mean 8.7; if broods of 5–6 young are excluded (because numbers possibly reduced by predation etc.), then the average of B/7–16 (23) is 9.9, of which 11 broods comprised 9–10. There is a record of a pair with an amazing 18 young (BD). Earliest broods, with estimated date of first egg in brackets, as follows: **Mediterranean:** 23 April (18 March); **Aegean:** 4 May (29 March); **Inner Anatolia:** 9 May (3 April); **Black Sea Coastlands:** 12 May (6 April); **Marmara:** 20 May (14 April), but only five broods noted, and these data are considered unrepresentative, with earlier dates expected; **East Anatolia:** 24 May (18 April); **Southeast Anatolia:** no data. Many other records do not specify the number of young observed, or their size or age, but may assist in providing a better picture of overall times of breeding. The following show the total number of broods reported from each region, with the number of broods reported before 31 May in brackets : Mediterranean: 14 (10); Aegean: 9 (9); Inner Anatolia: 13 (5); Black Sea Coastlands: 11 (6); Marmara: 5 (2); East Anatolia: 10 (4).

COMMON SHELDUCK *Tadorna tadorna*
Suna

(Du: Bergeend; Fr: Tadorne de Belon; Ger: Brandgans; Swe: Gravand)

Subspecies. Monotypic.

Status and distribution. National conservation status: Vulnerable (population apparently increasing). Generally a rather local resident (or partial migrant in some areas) in moderate numbers in Inner and East Anatolia, but usually scarcer at inland wetlands in Aegean, and at both inland and coastal wetlands in Mediterranean. In former areas breeds in larger, saline wetlands or at smaller, upland pools. Post-breeding, moulting birds assemble in significant numbers (up to 2,500) at Erçek Gölü (East Anatolia) in August and September. Elsewhere up to 700 have been recorded at Kulu Gölü (Inner Anatolia) in late October.

Generally less localised and more common in winter, largely in southwest Aegean, Mediterranean and Inner Anatolia, scarcer elsewhere, in Aegean, Marmara (including Thrace) and Black Sea Coastlands. Normally recorded in relatively small flocks, the total wintering population is usually 1,000–1,500, rarely more, especially in recent years (although there were 2,928 in 2005 and 1,802 were present at Acıgöl, Mediterranean, alone on 18 January 2004; D. Kahraman), exceptionally 4,900 in 1968/69 and 3,450 in 1969/70. Most remarkable is the recent count of 5,390 at Yumurtalık, Çukurova (Mediterranean) on 16 February 2006 (S. Eksioğlu), the highest-ever winter total in Turkey.

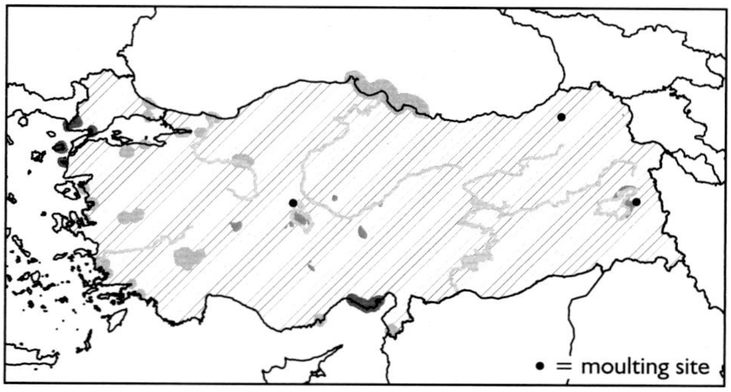

Breeding. At saline lakes, mainly Bolluk, Kulu and Tuz in **Inner Anatolia**, at Gediz Delta in **Aegean**, and Van Gölü in **East Anatolia**. In most parts of range, nests usually 1–2m inside rabbit burrows, but only nests in Turkey, 12 found on an island in Bolluk Gölü on 24 June 1992, were on the ground, some completely open, some partially or completely underneath bushes, and only one slightly inside a natural hole; they had C/1, C/4 (incomplete clutches), C/6 (2), C/8 (2), C/9 + 1 newly hatched young, C/10, C/14, C/15, C/16, C/18 (PC); the largest clutches may partly have resulted from egg-dumping by other females. On the same island the following year, the birds had not yet nested by 7 May. Earliest broods were seen at Kulu Gölü on 5 June 1975 (size unknown), but a brood of ten small ducklings on 6 June 1996 (GMK) indicates date of first egg as *c.*28 April. There are many records of broods at Kulu Gölü in July, including eight and 11 small ducklings on 27 July 1971; brood sizes were 1–15, plus a large brood of 20 (Kasparek 1987a), perhaps the result of more than one female. Small ducklings also seen elsewhere in Inner Anatolia until 28 July. At Gediz Delta (Aegean), population estimated at eight pairs in 1996, and a female with nine ducklings was observed in the first half of June (Eken 1997a). In East Anatolia, half-grown young were noted on 18 August 1972, indicating date of first egg as mid June; and other broods have been seen, mainly in August.

[COTTON PYGMY-GOOSE *Nettapus coromandelianus*]

(Du: Coromandeleend; Fr: Anserelle de Coromandel; Ger: Koromandelzwergente)

Subspecies. Presumably only the nominate would be capable of reaching our region.

Vagrant, with one record of a male, at Birecik (Southeast Anatolia), on 23 May 2002 (I. Weiss). The record has been submitted to the German rarities committee and, for now, we prefer to await the result of their deliberations, before adding such an exceptional rarity in the Western Palearctic to the Turkish list. Away from its main Asian range, the species has wandered west to Bahrain, Iran, Iraq, Jordan, Oman, the United Arab Emirates and Yemen (Porter *et al.* 1996, Bashford 1997, Snow & Perrins 1998, Balmer & Betton 2004b).

EURASIAN WIGEON *Anas penelope*
Fiyu

(Du: Smient; Fr: Canard siffleur; Ger: Pfeifente; Swe: Bläsand)

Subspecies and taxonomy. Monotypic. MtDNA and morphological data subject to phylogenetic analyses initially suggested that two clades exist within ducks currently treated in the genus *Anas*. Eurasian Wigeon, along with its American counterpart, *A. americana*, and Falcated Duck *A. falcata* and Gadwall *A. strepera* form one clade, and were therefore removed to the genus *Mareca* on this basis (Livezey 1991). Other than Sangster *et al.* (1997), a publication whose decision was subsequently reversed (Sangster *et al.* 2003), we have noted little support for this generic rearrangement and it is not followed here (e.g. AOU 1998).

Status and distribution. Very common and widespread passage migrant and winter visitor to all wetlands in the western two-thirds of Turkey, only on passage in the eastern third. At well-watched sites in Inner Anatolia, return passage appears to peak in late March–early April. Largest wintering concentrations in Aegean, Mediterranean and Inner Anatolia with several thousand regularly counted at many wetlands in these regions. There is evidence of a large decline in numbers since the late 1960s and early 1970s, when there were regularly over 150,000 wintering in Turkey, exceptionally 208,600 in early 1968 and 458,800 in the mild winter of 1968/69. Between 1986 and 2005 there were only four counts in excess of 40,000. The majority arrives in September and October (earliest 29 August), very small numbers remaining into late April and even May (latest 23rd). Passage through Inner Anatolia is evident in March to early April. Occasionally oversummers, without breeding, in East Anatolia, from where there are several recent records in late May, and possibly Inner Anatolia. Also recorded twice each in late May in Black Sea Coastlands and Marmara, once in late July in Thrace, and once in early August in Inner Anatolia.

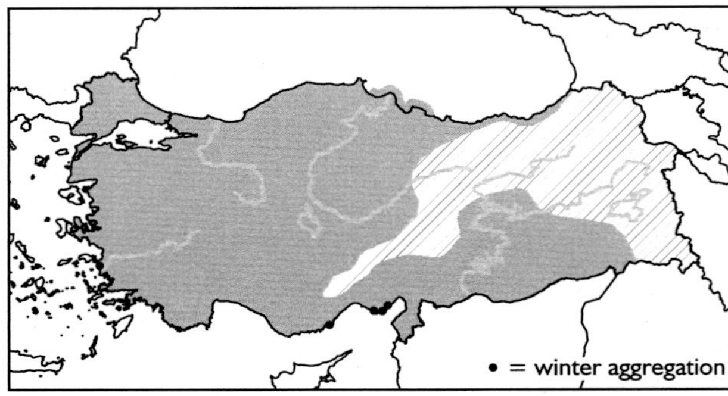

[FALCATED DUCK *Anas falcata*]
Büyük Çamurcun

(Du: Bronskopeend; Fr: Canard à faucilles; Ger: Sichelente; Swe: Praktand)

Subspecies. Monotypic. See taxonomic note under Eurasian Wigeon *A. penelope*.

Conservation status: Near Threatened (BirdLife International 2008). Species not admitted to the Turkish list. Hypothetical vagrant. One record; a male, on Van Gölü (East Anatolia) in November 1968 (J. Vielliard *in* Cramp & Simmons 1977) was considered likely to be a wild bird, but no details have ever been made available. Other records from the Middle East are a female at Azraq, Jordan, in January 1969 (Savage 1969) and two records from Iraq: a drake in March 1916 (Allouse 1953) and one (unsexed) in winter 1968/69 (Savage 1969). The species has also been listed for Iran without details (Porter *et al.* 1996) and has recently (November 2006) been claimed in Oman (Balmer & Betton 2007a).

GADWALL *Anas strepera*
Boz Ördek

(Du: Krakeend; Fr: Canard chipeau; Ger: Schnatterente; Swe: Snatterand)

Subspecies. Monotypic. See taxonomic note under Eurasian Wigeon *A. penelope*.

Status and distribution. National conservation status: Vulnerable (population apparently decreasing). Relatively widespread and common resident breeder, or partial migrant, in marshy wetlands in Inner and East Anatolia, as well as adjacent parts of Aegean, Mediterranean, Black Sea Coastlands (the 200 pairs at Kızılırmak Delta make this a particularly important breeding area) and Southeast Anatolia, but more local elsewhere and perhaps absent from Thrace. Previously, Tucker & Heath (1994) considered the Turkish breeding population to be stable at between 500 and 5,000 pairs, but since then a decline has become evident.

More widespread and common on passage and in winter, when especially numerous, across the country, particularly in Aegean and Inner Anatolia. Spring passage through the latter region is evident in March to early April. Less usual in large numbers (up to 1,000) in coastal wetlands in Mediterranean. Rarely more than 200 in

Black Sea Coastlands. Formerly more common in midwinter, typically more than 3,000 during the International Waterfowl Research Bureau (now Wetlands International) counts in 1967–73, including exceptional single-locality counts of 5,000 at Manyas Gölü (Marmara) in January 1967, 7,500 at Akşehir Gölü (Inner Anatolia) in January 1969, and 2,490 at Hotamış marshes (Inner Anatolia) in January 1971. However, in recent winters (1986–2005) always fewer than 1,500 birds in total and generally less than 1,000 (e.g. 765 in 2005: Çağlayan *et al.* 2005).

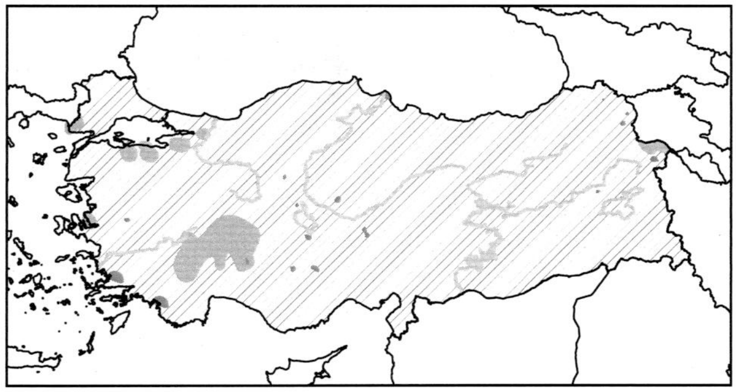

Breeding. Beside lakes, or on islands in lakes, in dense waterside vegetation; also in marshes and flooded areas, usually with dense cover. Nest a hollow in the ground, lined with plant material and the female's own down. Eggs: 6–15 (13), a number confused by seven nests at various locations in Inner Anatolia (7–15 eggs), which also contained 1–6 eggs of other ducks; the remaining records are C/6 (1), C/7 (2), C/8 (1),C/10 (1), C/11 (1), with a record of three eggs considered to be an incomplete clutch. Young: B/4 (1), B/5 (2), B/12 (1). In **Inner Anatolia**, at Kulu Gölü, three nests had three, ten and 11 eggs on 6 May 1972, and another seven eggs on 14 July 1971 (Kasparek 1987a); this latter is considered a replacement; there was a nest with eight eggs at Bolluk Gölü on 17 May 2004 (L. Baxter); and ducklings were observed in 1969 in late April–July at Mogan Gölü. In **Mediterranean**, an adult with five young was at Karamik Gölü on 18 July 1972. At Kızılırmak Delta (**Black Sea Coastlands**), a nest had seven eggs on 5 May 1992, another had six eggs on 4 June 1992, and a female with 12 ducklings, 2–3 days old, was seen on 11 May 1992 (Hustings & van Dijk 1994); the latter indicates first eggs in early April. In **East Anatolia**, broods were observed on 26 June 1991 (four young at Ahlat), 22 July 1992 (two families with young at Haçlı Gölü) and 6 September 1973 (five young at Van Gölü).

EURASIAN (COMMON) TEAL *Anas crecca*
Çamurcun

(Du: Wintertaling; Fr: Sarcelle d'hiver; Ger: Krickente; Swe: Kricka)

Subspecies and taxonomy. Nominate, which is increasingly recognised as a species distinct from New World *carolinensis*, Green-winged Teal (see Sangster *et al.* 2001, Collinson 2006). *A. crecca* is best considered monotypic.

Status and distribution. National conservation status: Near Threatened (population apparently decreasing). Local and rather scarce breeding resident, and probably partial migrant in East Anatolia, generally in very small numbers in marshy wetlands in Black Sea Coastlands (e.g. just 15–20 pairs in Kızılırmak Delta in 1992), Inner and East Anatolia, possibly elsewhere. Few confirmed breeding records from the latter region.

Widespread and common on passage (September/October, occasionally from mid August, and March/April) and in winter, principally throughout the western two-thirds of Turkey, with the largest concentrations in Inner Anatolia and in coastal wetlands in Aegean and Mediterranean. Most have departed central regions by early April. Less

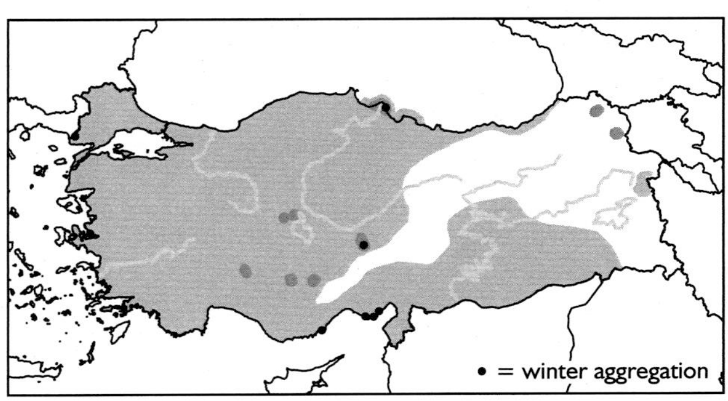

regularly in large numbers in Marmara and Black Sea Coastlands. As with other species of dabbling ducks, a substantial decline has occurred since the instigation of the International Waterfowl Research Bureau (now Wetlands International) midwinter waterfowl counts. Formerly numbers regularly exceeded 150,000 birds, e.g. 270,400 in 1968/69 and 326,700 in 1969/70 (including 200,000 at Sultan marshes, Inner Anatolia alone; of a further 400,000 unidentified ducks a large proportion were thought to be this species). More recently numbers have consistently been less than 100,000, with the lowest counts being 21,000 and 13,400 in 1988 and 1989.

Breeding. In marshes and at lakes. Nest and clutch size undescribed in Turkey but, elsewhere, nest is usually a hollow in the ground, lined with leaves, plant material, down and some feathers, and usual clutch is 8–12 (16); nests also on tussocks in wet areas, occasionally at considerable distance from water. In Kızılırmak Delta (**Black Sea Coastlands**), a nest had eggs on 29 May 1979, a female was observed with nine young on 28 July 1971 and another with five young on 6 August 1971 (Dijksen & Kasparek 1985) and, in 1992, the population was 15–20 pairs, with distraction behaviour observed on 5 May, but no nest was found (Hustings & van Dijk 1994). In **Inner Anatolia**, several adults with young were at Hotamış marshes on 14 May 1991 (Kirwan 1993a), indicating laying no later than mid April; and two females, with broods of seven and ten, were seen at Kulu Gölü, 5–6 August 1972 (Kasparek 1987a). In **East Anatolia**, a female with a single duckling was at Haçlı Gölü on 24 June 1983.

MALLARD *Anas platyrhynchos*
Yeşilbaş

(Du: Wilde Eend; Fr: Canard colvert; Ger: Stockente; Swe: Gräsand)

Subspecies. Monotypic.

Status and distribution. National conservation status: Least Concern (population apparently decreasing). Widespread resident breeder, usually in moderate numbers in all types of wetlands in Inner Anatolia, but perhaps less widely distributed over much of the rest of the country, and potentially absent as a breeder from Southeast Anatolia, where there is a paucity of suitable habitat.

Numerous in autumn and winter (until the first week of April), being more widespread and rather more evenly distributed across the western two-thirds of Turkey, with the largest numbers in Marmara, southwest Aegean and Black Sea Coastlands (400–600 pairs in Kızılırmak Delta in 1992). Fewer in Mediterranean and Inner Anatolia but occasionally recorded in significant numbers in Southeast and East Anatolia at this season. In recent winters numbers have averaged approximately 20,000, although rather larger numbers were apparently present in January 2005 (see below). Formerly more common, particularly in the winters of 1967–70 inclusive, when peak counts included; 52,000 in the Kızılırmak and Yeşilırmak Deltas (Black Sea Coastlands) and 42,000 in the Büyük Menderes Delta (Aegean) in early 1967, 42,000 at Manyas and Uluabat lakes (Marmara) in early 1968; an estimated 80,000 in the Büyük Menderes Delta, 40,000 at Akyatan lagoon, Çukurova and 30,000 at Amik Gölü (Mediterranean) in 1968/69, and 34,500 at the Meriç Delta (Marmara) and 30,000 at the Sultan marshes (Inner Anatolia) in 1969/70. The midwinter 2005 total of 106,140 represented a significant increase in wintering numbers, with the 50,000 present at Kızılırmak Delta being the largest single-site total in recent years.

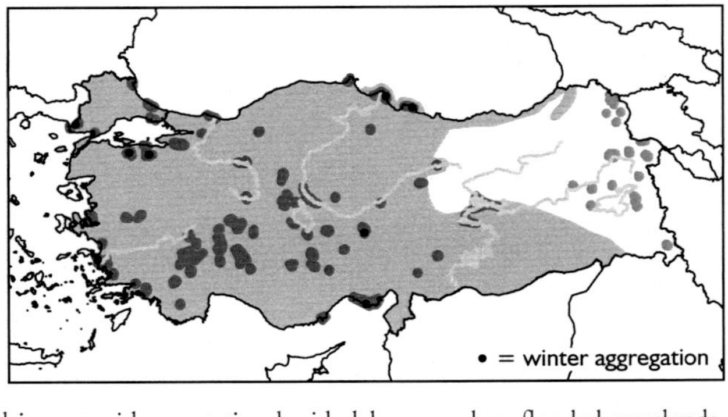

Breeding. On islands in lakes, in reeds, and in waterside vegetation beside lakes, marshes, flooded grasslands. Nest a hollow in the ground, usually concealed by vegetation. There are no records in Turkey of nests in tree cavities, or in old nests of other bird species, such as corvids. Eggs: C/2–14, usually 5–9, with smaller numbers probably incomplete clutches. Many nests at Yarma (**Inner Anatolia**) in 1971, with up to 23 eggs (often mixed with eggs of other species) – one held 17 of Mallard, three of Gadwall *Anas strepera* and three of Red-crested Pochard *Netta rufina*. Nests with more than *c.*14 eggs are probably the result of laying by more than one female. Earliest breeding record a female with young at Koçaçay Delta (**Marmara**) on 18 April 1993 (Ertan 1996). In **Inner Anatolia**, most nests with eggs are in May–June, with eight nests at Kulu Gölü (2–12 eggs) on 13–15 July 1971 (Kasparek 1987a); ducklings observed from mid May. In **Black Sea Coastlands**, two nests had eggs, and ducklings were present, at Yeniçağa Gölü on 19–20 May 1992 (PC); first ducklings were seen at Kızılırmak Delta on 5 May 1992 (Hustings & van Dijk 1994). A nest had nine eggs at Manyas Gölü (Marmara) on 16 May 1967 (R. F. Porter). In Thrace, a female with six young was seen on 20 June 1973. In **East Anatolia**, the earliest record is of ducklings at Erçek Gölü on 14 June 1968, and two nests had five and eight eggs there on 28 June 1968 (Kasparek & van der Ven 1983); two nests had eggs at Balık Gölü on 9 June 2001 (PC).

NORTHERN PINTAIL *Anas acuta*
Kilkuyruk

(Du: Pijlstaart; Fr: Canard pilet; Ger: Spießente; Swe: Stjärtand)

Subspecies. Nominate.

Status and distribution. National conservation status: Vulnerable (population apparently decreasing). Very local and uncommon breeding resident, or perhaps in some areas partial migrant, in marshy wetlands in Black Sea Coastlands and Inner and East Anatolia, and perhaps also elsewhere in the eastern two-thirds of Turkey. Bred in 1998 and 1999 at Girdev Gölü (Mediterranean). Recently discovered as a breeding bird and still very few confirmed instances reported. Tucker & Heath (1994) tentatively placed the Turkish breeding population at between 500 and 1,000 pairs, but the true total is now probably much lower due to wetland drainage. Some presumed records may relate to non-breeders.

On passage common and widespread at wetlands across the entire country, except Southeast Anatolia. Winters, from mid September to late March/early April, predominantly in the western two-thirds of Turkey. Most winter in the south, southwest and Inner Anatolia, with fewer in Marmara and Black Sea Coastlands. A dramatic decline has been documented. Total numbers have rarely exceeded 10,000 in recent winters (25,700 in 1986, 11,070 in 1992 and 13,573 in 1999), although formerly regularly recorded in numbers exceeding 100,000. However, some older counts were based on crude estimates. Nevertheless, some more specific and accurate single-locality counts indicate the scale of the decrease: e.g. 60,000 at the Büyük Menderes Delta (Aegean) and 70,000 at Emer Gölü (Mediterranean) in 1967; 100,000 at Akyatan Gölü, Çukurova and 50,000 at Gâvur Gölü (Mediterranean) in 1968/69 (a particularly mild winter when most wildfowl were present in unusually large numbers). Despite the possibility that some earlier counts were based on exceptional over-estimates, e.g. 160,000 at Sultan marshes (Inner Anatolia) in 1969/70, which unduly distorted the annual totals, large numbers, regularly up to several thousand, used to winter at wetlands which now only support hundreds. Several wetlands important for this and other species of ducks have been partially or wholly drained during the period covered by the International Waterfowl Research Bureau (now Wetlands International) counts. This species, in common with other dabbling ducks, is apparently starting to utilise dam lakes in winter.

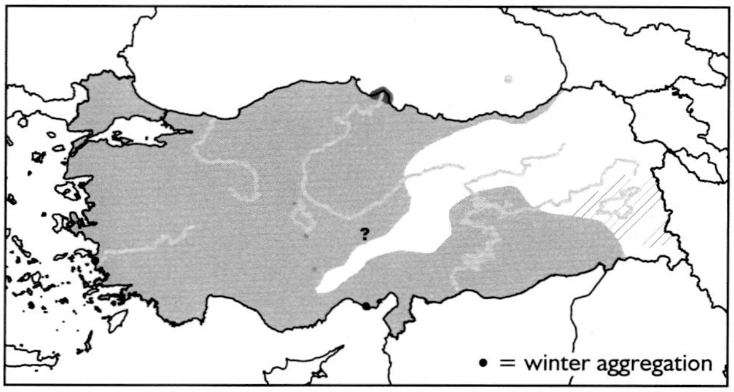

Breeding. There are many records of adults in spring and summer in **Inner Anatolia**, but the only proven breeding records are two nests (seven and ten eggs) on 22 May 1992, and five nests (6–9 eggs) on 19 June 1992, all in vegetation at the edges of the largest island in Kulu Gölü, and a nest with 11 eggs, sheltered by a bush, at Bolluk Gölü, on 24 June 1992 (PC). Young were observed in June at Girdev Gölü, near Fethiye

(**Mediterranean**) in both 1998 and 1999. In Kızılırmak Delta (**Black Sea Coastlands**), behaviour possibly indicative of breeding has been observed, but no definite proof (Hustings & van Dijk 1994). The nest is a hollow in the ground, lined with plant material, down and some feathers.

GARGANEY *Anas querquedula*
Çıkrıkçın

(Du: Zomertaling; Fr: Sarcelle d'été; Ger: Knäkente; Swe: Årta)

Subspecies. Monotypic.

Status and distribution. National conservation status: Near Threatened (population apparently decreasing). Summer visitor usually from mid February (11th) to mid September or late October (29th), apparently breeding only locally and in small numbers in marshy wetlands in Marmara, Aegean, Mediterranean and Black Sea Coastlands. More common in Inner and East Anatolia. Two potential breeding localities identified in Southeast Anatolia, but principally a migrant in this region. Breeds from mid April but passage continues until at least the end of May. Widespread and often common passage migrant in all regions, often in large flocks numbering up to several hundred or even several thousand birds in both spring and autumn. Considerable movements regularly witnessed along Black Sea coast in late August–early September. Only occasionally recorded in winter, usually in very small numbers, in Marmara, Aegean and Mediterranean, e.g. exceptionally 3,000 at the Göksu Delta and 5,000 at Gâvur Gölü (Mediterranean) in the unusually mild winter of 1968/69. More recently, 65 in Güllük Delta in January 2002, 58 at Bafa Gölü in February 2002 and 89 in Çukurova in December 2002.

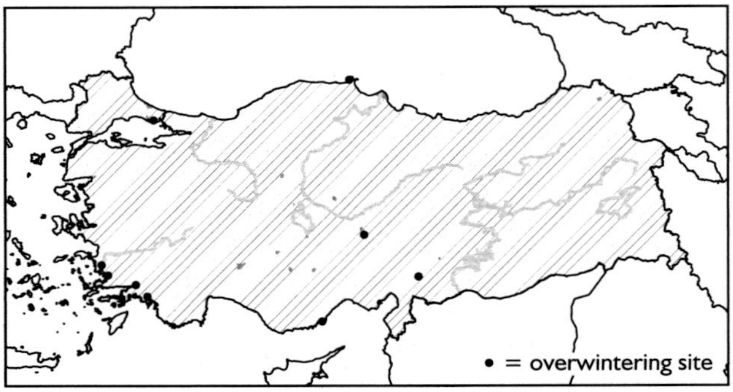

Breeding. In damp grassland near lakes and in marshes, reeds and lakeside vegetation. There was a nest with recently destroyed but obviously fresh eggs on 6 May 1993 in damp grassland near Yeniçağa Gölü (**Black Sea Coastlands**), where two nests similarly destroyed by predators were found on 19 May 1992 (PC). No data in Turkey but, elsewhere, usual clutch is 8–11. In **Inner Anatolia**, several pairs with large broods were observed at Hotamış marshes on 10–15 May 1991 (Kirwan 1993a), six pairs with large young were seen at Kulu Gölü on 27 July 1971; 1–2 broods were at Mogan Gölü in June–July 1968, and at least four pairs with large young were at Yarma on 27 July 1971. In Black Sea Coastlands, a female with 17 young at Abant Gölü on 13 May 1986 (M. Ullman), indicates egg laying in mid–late April; a pair and seven young were observed at Kızılırmak Delta on 2 August 1971.

NORTHERN SHOVELER *Anas clypeata*
Kaşıkgaga

(Du: Slobeend; Fr: Canard souchet; Ger: Löffelente; Swe: Skedand)

Subspecies. Monotypic.

Status and distribution. National conservation status: Endangered (population apparently decreasing). Distinctly local resident breeder and partial migrant in very small numbers in marshy wetlands in Black Sea Coastlands, Inner and East Anatolia, and perhaps adjacent parts of Aegean and Mediterranean. Two potential breeding sites identified in Southeast Anatolia during recent fieldwork.

Widespread and common passage migrant, occasionally in large numbers, e.g. 7,000 at Kulu Gölü and 9,000 at Sultan marshes (Inner Anatolia) in September, and 4,500 at Kızılırmak Delta (Black Sea Coastlands) in late March, in all regions from early March to late April, and mid September to early November. Winters throughout the western two-thirds of Turkey. Total numbers have rarely exceeded 5,000 in recent years (7,898 in January/February 1993, 13,114 in early 1999 and 13,576 in early 2005), but it was formerly more common, with site maxima of 23,000 at Büyük Menderes Delta (Aegean) and 8,000 in Kızılırmak Delta (Black Sea Coastlands) in January 1967 (4,564 at the latter in January 1993), and regularly up to 3,000 at various localities in Inner Anatolia in 1967–73.

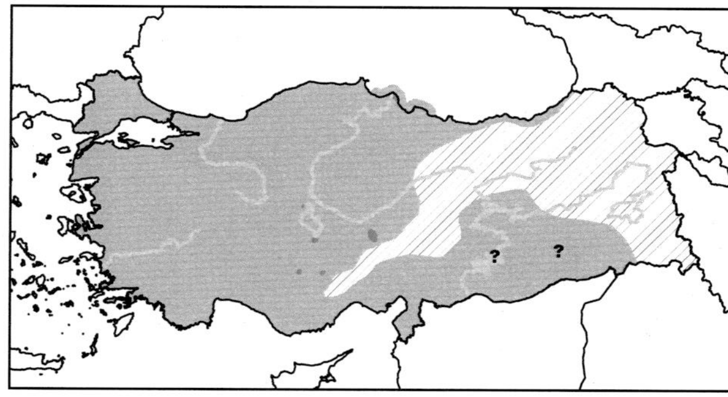

Breeding. Most records are from Kulu Gölü (**Inner Anatolia**), where two nests had eight and ten eggs on an island on 14/15 July 1971, broods of two and four were seen on 5/6 August 1972, and most birds seen in the breeding season were males, suggesting that females were on nests (Kasparek 1987a). Young were seen at Kulu on 31 May 1987, and a nest with nine eggs was found in scant vegetation on an island on 19 June 1992; a shallow hollow in bare ground, lined with down, mixed with grass and plant stems (PC). A female and five ducklings were at Eşmekaya (Inner Anatolia) in June 1977 (Schubert 1979). At Kızılırmak Delta (**Black Sea Coastlands**), two females, with four and five young, were seen 6/7 July 1972 (Dijksen & Kasparek 1985) and, in 1992, the population was estimated at 0–1 pairs, with breeding unproven (Hustings & van Dijk 1994). Pairs also seen in spring/summer elsewhere in Turkey, including display at Van Gölü (**East Anatolia**) on 29 May 1969 (R. F. Porter). Insufficient data to comment reliably on season in Turkey; elsewhere, commencement variable, from April to late May. Records of nests with eggs in mid-July 1971 (above) are considered replacements, following earlier failure.

MARBLED DUCK *Marmaronetta angustirostris*
Yaz Ördeği

(Du: Marmereend; Fr: Sarcelle marbrée; Ger: Marmelente; Swe: Marmorand)

Subspecies. Monotypic.

Status and distribution. Conservation status: Vulnerable (BirdLife International 2000, 2004); national conservation status: Endangered (population apparently decreasing). Local breeder in moderate numbers in lowland, marshy wetlands (typically shallow and eutrophic) in Mediterranean and in smaller numbers in several, usually large wetlands in Inner Anatolia and very locally in small marshes in East Anatolia, where the population has possibly declined in recent years. In Mediterranean, breeds in Çukurova deltas, typically favouring areas of fresher water, as opposed to brackish or saline, for nesting (32 pairs in 1987, with a total of 50–100 pairs suspected) and at the Göksu Delta (*c.*50 pairs in both 1989 and 1991). Formerly bred at Amik Gölü, perhaps in significant numbers before this site was drained (Kumerloeve 1963a). Confirmed breeding at the following wetlands in Inner Anatolia: Ereğli marshes (probably fewer than five pairs), Hotamış marshes (10–15 pairs), Sultan marshes (probably fewer than five pairs) and formerly at the Yarma marshes, at Gogenç Gölü and possibly several other wetlands. However, virtually all of these populations are currently in significant decline and some are now extinct. Karapınar Ovası, listed by some authors (e.g. Grimmett & Jones 1989) as a potential breeding site, is no longer suitable. Several other areas, e.g. Mogan Gölü and Eber Gölü, were probably former nesting localities, but are almost certainly no longer suitable, in part due to habitat destruction and water pollution. Small numbers breed in East Anatolia, where more limited areas of suitable habitat exist, principally on the fringes of Van Gölü. Breeding has been confirmed at the Erciş and Van marshes and is presumed or thought possible at an additional four sites; near Ağrı, Ahlat marshes, Bendimahi Delta and Kuyucuk Gölü. Also observed during the breeding season in Marmara, Aegean, Black Sea Coastlands (only one record) and Southeast Anatolia (at Birecik in late June 2000) but no evidence of breeding.

Principally a summer visitor from early March to early October. Formerly wintered in moderate numbers in Mediterranean and Inner Anatolia, but few now remain in the Mediterranean deltas at this season (e.g. just four recorded in early 1993, although 35 were present in Çukurova in February 1997). Turkey still holds an internationally important breeding population of this species despite evidence of significant decline due to habitat degradation in some areas, and possibly hunting, during the past two decades. Post-breeding flocks in both the Çukurova and Göksu deltas now number 100–200, while in August 1967 over 2,000 and 450 were counted at these two sites repectively. Remains occasional in very small numbers at other wetlands on passage in both spring and autumn.

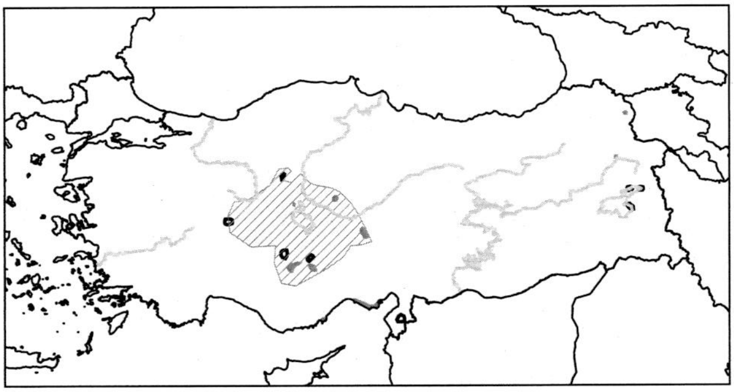

Breeding. Around shallow lakes, often with reed islands, and adjacent pools and marshes with dense submerged vegetation; prefers larger wetlands. Most important breeding sites are Göksu and Çukurova deltas (**Mediterranean**). In 1991, at Göksu Delta, *c.*50 pairs bred and at least 31 broods were seen; laying began late April, peaked second half of May and early June; the first brood was seen on 7 June and small young continued to be seen until late July, with fully fledged broods from early July; brood size 2–12/mean 6.5 (Green 1993). A nest had six eggs, and another 13 eggs (the latter possibly laid by two females, and included one egg of a Mallard *Anas platyrhynchos*), at Yarma marshes (**Inner Anatolia**) on 4/5 June 1971. A nest with an incomplete clutch of two at Göksu on 9 June 1993 was in a clump of *Juncus*, *c.*1m in diameter, in an area of shallow pools in marshland dominated by *Juncus*, with *Phragmites* nearby; 0.7m above water, well concealed, and a substantial pad of reeds and stems, with a deep cup, obviously intended for many more eggs, and lined with finer plant material, but no down (PC). Typical clutch size, elsewhere, is 9–13 (5–18). Young: of 17 broods with numbers recorded, 11 had 8–13 young, five had 4–6, and one was of 15; brood amalgamation observed in Göksu in 1991, when two females with broods met, and a female was seen with 32 ducklings on 15–18 July 1992 (Green 1993). In Mediterranean, the earliest record, a brood of six in Çukurova on 22 May 1971, indicates egg laying in second half of April; in Göksu, broods observed from early June, but mainly July, including both large and small young on 10 July 1967 (Vielliard 1968). In Inner Anatolia, an adult with one young was at Kulu Gölü on 12 June 1998, and young seen at three other sites in July. In **East Anatolia**, two pairs with very small young were at Van marshes on 22 July 1987, indicating egg laying in mid June; and several records of young there in late July–August.

RED-CRESTED POCHARD *Netta rufina*
Macar Ördeği

(Du: Krooneend; Fr: Nette rousse; Ger: Kolbenente; Swe: Rödhuvad dykand)

Subspecies: Monotypic.

Status and distribution. National conservation status: Vulnerable (population apparently decreasing). Fairly widespread and usually fairly common resident breeder (partial migrant in some areas) in large brackish or freshwater marshy wetlands in Inner Anatolia (particularly common at Ereğli and Sultan marshes; up to 500 pairs at the former in the early 1990s, though just 20 pairs were found in 1998), adjacent parts of Mediterranean and also in much smaller numbers in East Anatolia, Black Sea Coastlands (breeding proven only in Kızılırmak Delta, where 50–75 pairs estimated in 1992; Hustings & van Dijk 1994) and perhaps elsewhere. Recently recorded from Thrace during the breeding season. Formerly bred more widely in Mediterranean. The Turkish breeding population was earlier tentatively placed at 1,000–5,000 pairs by Tucker & Heath (1994), but recent declines at several sites in Inner Anatolia mean that the current total is considerably lower.

More widespread on passage throughout Turkey, although less common in the eastern third. Winters in the western two-thirds of Turkey in irregular numbers, principally in the south and in Inner Anatolia, often favouring the older dam lakes which are less prone to surface freezing. Max. total count was of 12,000 in January 1967, which included 7,000 at Amik Gölü, Mediterranean (now drained), and more recently 5,249 in January/February 1992, 6,522 in January 1996, and 6,228 in early 1999. In February 2003 and January 2005, there were exceptional single-locality totals of 10,000 and 20,000, respectively, at Beyşehir Gölü (Mediterranean).

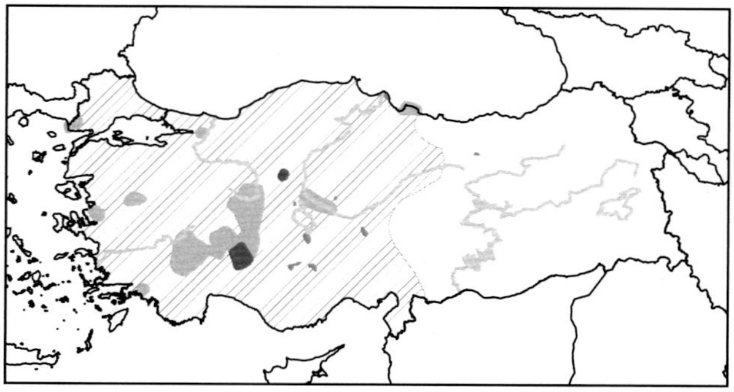

Breeding. At freshwater and brackish lakes with dense reeds and waterside vegetation, and in marshes with open water. Nest a hollow in the ground, lined with plant material, down and feathers, usually in dense vegetation, occasionally in the open (e.g. on islands) or, at wet sites, built up just above water level, in tussocks of reeds and other aquatic vegetation, usually well hidden. Eggs: C/4–12, mean 8.3 (18); plus a nest with 24 eggs, presumably involving more than one female. Young: B/2–12; mean 6.2 (16) is significantly reduced by six broods of only 2–4, perhaps due to predation or other causes. In **Inner Anatolia** (many records at many sites); eggs from 25 April, most in May, and a late nest with 12 eggs on 19 June 1992 (PC); young from 11 May, most in June, but small young seen 8 July 1967 (Vielliard 1968) and 5 August 1972. In **Marmara**, a nest had eggs at Koçaçay Delta on 1 May 1993 (Ertan 1996). In **East Anatolia**, three broods of 7–8 were at Van Gölü on 21–22 July 1986. In **Mediterranean**, a brood of small young was at Karamık Gölü on 18 July 1992. In **Black Sea Coastlands** (Kızılırmak Delta), a nest held five eggs on 27 May 1992, and the first downy young, *c.*1 week old, were seen 4 June 1992 (Hustings & van Dijk 1994), and a brood of eight was seen on 27 May 1979 (Dijksen & Kasparek 1985).

COMMON POCHARD — *Aythya ferina*
Elmabaş Patka

(Du: Tafeleend; Fr: Fuligule milouin; Ger: Tafelente; Swe: Brunand)

Subspecies. Monotypic.

Status and distribution. National conservation status: Least Concern (population apparently stable). Generally a rather local resident, breeding in fairly small numbers in wetlands in Inner and East Anatolia (although generally more common in the eastern third of Turkey and in Kızılırmak Delta, Black Sea Coastlands, where 300–350 pairs estimated in spring 1992; Hustings & van Dijk 1994). Even more local elsewhere in Black Sea Coastlands and in Southeast Anatolia, due to a lack of suitable habitat, and perhaps other regions. Considerable numbers of non-breeders oversummer in Inner and East Anatolia, more occasionally elsewhere throughout Turkey, potentially distorting true breeding status.

Widespread and common across the country on passage and in winter, with an average in excess of 67,000 in recent years (although only *c.*40,000 were present in 2005) and maxima of 248,334 in 1996 (including over 47,000 on Beyşehir Gölü, Mediterranean, and 42,000 on Uluabat Gölü, Marmara) and 136,510 in 1999 (over 40,000 on Eğirdir Gölü, Mediterreanean). The overall mean taken from a compilation of all International Waterfowl Research Bureau (now Wetlands International) counts (18) is 93,200, with an exceptional 355,000 (based largely on estimates) recorded in the 1968/69 winter. Large numbers arrive from mid

October, e.g. 40,000 at the Göksu Delta in 1970, and remain until March. A concentration of between 100,000 and 130,000 on Sodalı Gölü (East Anatolia), on 10 October 2002, was exceptional and presumably included many migrants. Additionally, approximately 45,000 were at Kulu Gölü (Inner Anatolia) in November 1970 (with 28,000 there the following November): both are exceptional counts for this region.

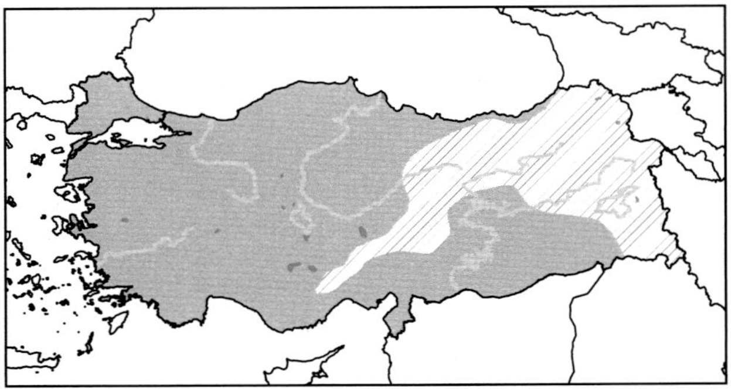

Breeding. On lakes with reeds and in extensive marshes with open water. A nest at a small lake near Erçek Gölü (**East Anatolia**) on 19 June 1904 was attached to the base of reeds in thick cover, and built up from the water, like a Eurasian Coot *Fulica atra* nest; a substantial structure of dead reed stems and other plant material, interwoven to form a deep neat cup, lined with plentiful down and some feathers; it had one unhatched (cold) egg and several hatched and trampled eggshells; a female and seven small ducklings (less than a week old) were nearby (PC). Elsewhere, nests are usually in similar sites, occasionally on dry ground, in dense cover at the water's edge. Clutch size unrecorded in Turkey; elsewhere usually 6–9. In **Inner Anatolia**, a nest had two eggs (incomplete clutch) in early June, and two, four and five eggs were laid in nests of Gadwall *Anas strepera* in early-June 1971, and a nest had eggs at Hotamış on 13 May 1991 (Kirwan 1993a). Young: B/2–11: 2–5 young: (4); 6–11 young: (11), mean 6.6. At Kızılırmak Delta (**Black Sea Coastlands**), the earliest record, downy young, only a few days old, on 11 May 1992 (Hustings & van Dijk 1994), indicates laying in the first week of April; a brood of young, *c.*5 days old, and two broods, about three weeks old, were seen on 14 June 1984 (Dijksen & Kasparek 1985). In Inner Anatolia, five broods of small young were at Eşmekaya in late-May 1970, six broods were at Sultan marshes on 1 June 1969 (R. F Porter), and there were broods at other sites in June–July. In East Anatolia, records include eight females with ducklings at Van marshes on 19 June 1983 (R. P. Martins).

FERRUGINOUS DUCK *Aythya nyroca*
Pasbaş Patka

(Du: Witoogeend; Fr: Fuligule nyroca; Ger: Moorente; Swe: Vitögd dykand)

Subspecies. Monotypic.

Status and distribution. Conservation status: Near Threatened (BirdLife International 2000, 2004); national conservation status: Endangered (population apparently decreasing). Rather local and uncommon summer migrant, breeding in wetlands in all regions except Southeast Anatolia, where three possible breeding localities were identified during recent fieldwork. Most numerous in Inner and East Anatolia. Also breeds in significant numbers in Kızılırmak Delta, Black Sea Coastlands (150–200 pairs estimated in spring 1992; Hustings and van Dijk 1993), in Marmara, at Kocaçay Delta (70 pairs estimated in 1993; Ertan 1996) and Uluabat Gölü (32 pairs in 1988; Welch & Welch 1998b), and at Göksu Delta, Mediterranean (*c.*30 pairs). The Turkish population is potentially one of the most important in the world; Tucker & Heath (1994) postulated that between 1,000 and 3,000 pairs may breed, but Kirwan (1997b) more conservatively estimated 500–600 pairs based on a detailed

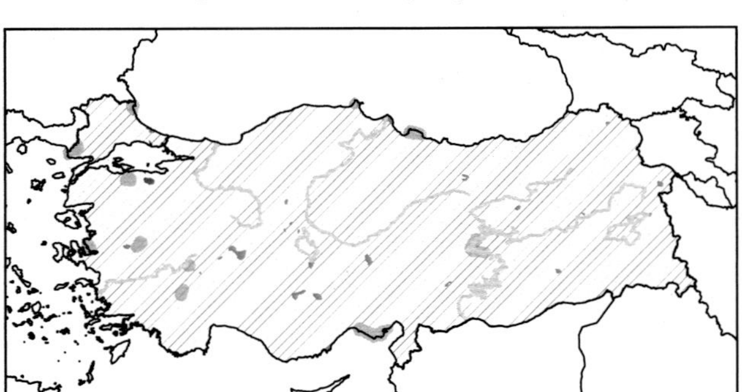

review of status and distribution. Given the drastic decline in numbers in many other parts of its European range, a baseline survey of the species' Turkish population is urgently required. Declines were noted at several localities in Inner Anatolia in the late 1990s. Small numbers are resident. Summer migrants are present mid March–late September.

Slightly more widespread across the whole country in small to moderate numbers on passage. Presently over-winters only in very small numbers: maximum of 105 in 1992 but more usually fewer than 50, with none 1986–88 and 1993. There is some evidence for a resurgence of winter records since the mid 1990s, perhaps due to improved coverage at this season, although numbers are still rather small. Formerly an irregular, but more numerous wintering bird throughout the western two-thirds of Turkey, depending upon weather conditions. Usually between 50 and 450 in most years from 1968–74, but a max. 860 at Marmara Gölü (Aegean). Recent winter records from East and Southeast Anatolia, which peaked in 2005 (44), are of some note.

Breeding. Freshwater lakes with fringes of reeds and dense emergent vegetation; often with more extensive reeds and marsh. Nests in dense vegetation near the water's edge: a nest, on an island in Kulu Gölü (Inner Anatolia), was on bare ground, between low shrubs; a slight hollow lined with grass and down (Pforr & Limbrunner 1982; A. Limbrunner pers. comm.). Eggs: C/6 (2), C/8 (1); elsewhere, usual clutch 7–11 (14). Young: B/8–9 (2), B/6–7 (9), B/3–5 (6), the latter possibly already reduced by predation etc. At Kızılırmak Delta (**Black Sea Coastlands**), most pairs were in marshy areas, a nest had six eggs on 5 May 1992, and two broods of three and four were seen 1 June 1992 (Hustings & van Dijk 1994), indicating laying no later than late April; a brood of seven ducklings was observed there 6 August 1971. In **Marmara**, there were several broods of ducklings (some small) at Uluabat Gölü on 19 June 1999 (PC), and young also at Manyas Gölü in 1966. In **Mediterranean**, a nest had eight eggs in Çukurova on 15 May 1962 (Kirwan 1997b); at Amik Gölü, a nest had eggs on 8 May 1953 (Kirwan 1997b), and a female, shot on 27 May 1933, had an egg in the oviduct (Meinertzhagen 1935); broods observed at seven sites, the earliest on 17 June, at Göksu Delta. In **Inner Anatolia**, a nest at Sultan marshes had eggs on 28 April 1982 (Kirwan 1997b), and another had six eggs at Kulu Gölü in May 1973 (A. Limbrunner); broods observed at Cöl Gölü, Çogenc Gölü, Sultan marshes, Mogan Gölü, Kulu Gölü and Eber Gölü (earliest 20 June at Eber). In **East Anatolia**, broods observed at Van Gölü in late-June 1985 and 1987, and Edremit marshes on 29 June 1987 (Kirwan 1997b), indicating egg laying by late May.

TUFTED DUCK *Aythya fuligula*
Tepeli Patka

(Du: Kuifeend; Fr: Fuligule morillon; Ger: Reiherente; Swe: Vigg)

Subspecies: Monotypic.

Status and distribution. Very scarce and local breeder (proven on four occasions), but regularly oversummers, largely in Inner Anatolia, more occasionally in Black Sea Coastlands (where breeding proven in Kızılırmak Delta in 1968 and 1992), East Anatolia (breeding proven in August 1967 and August 1981, and a population of up to 20 pairs suspected at Çalı Gölü, Kars province) and occasionally elsewhere (e.g. at least four August and two July records from Mediterranean, and three July records and one August record from a locality in Thrace), usually without evidence of breeding. Recently also observed in summer in southern Marmara, but all these latter records may merely indicate dispersing or wandering non-breeders. There is a somewhat anomalous record, described as a moulting concentration, of 700 at Kulu Gölü (Inner Anatolia) in July 1982 (Kasparek 1987a).

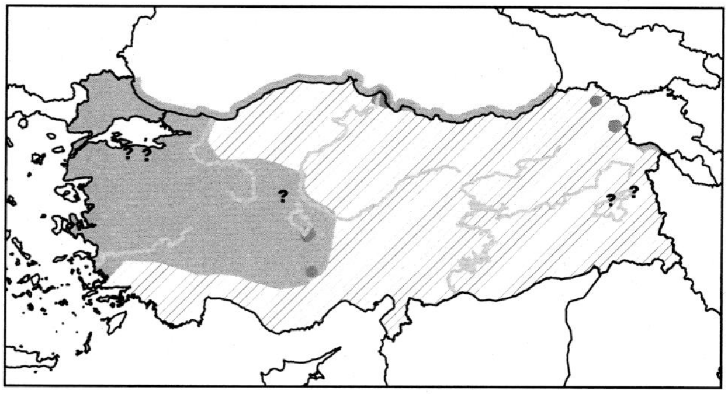

A widespread and common passage migrant and winter visitor from early (4th), more usually late, September to at least the second half of April (20th; some late-April records may pertain to migrants but are from known summering/breeding sites), principally to the western two-thirds of Turkey, but small numbers also recorded in the eastern third (where 1,200, mostly males, at Sodalı Gölü in September 1967). Frequently winters at sea, especially in Black Sea Coastlands. Until recently, the winter maxima were 20,500 and 20,800 in early 1968 and the winter of 1968/69 respectively, but with the exception of 1996, when 58,271 were present, more recent counts have typically been lower, usually 5,000–10,000, although nearly 13,000 were recorded in 1992, 16,965 (including 10,478 on Eğirdir Gölü, Mediterranean) in 1993 and 18,512 in 1999.

Breeding. Freshwater lakes with emergent vegetation. In Kızılırmak Delta (**Black Sea Coastlands**), a nest with eight eggs was at the base of a clump of *Juncus acutus* in marshland on 5 May 1992 (Hustings & van Dijk 1994), and breeding also confirmed 28 May 1968 (Dijksen & Kasparek 1985). In **East Anatolia**, two females with young were seen 19 July 1992 at Çalı Gölü, where 20 breeding pairs have been reported (Magnin & Yarar 1997). No evidence of breeding in other regions, but pairs present in suitable habitat in the breeding season, e.g. at Uluabat Gölü (**Marmara**) and Uyuz Gölü (**Inner Anatolia**).

GREATER SCAUP *Aythya marila*
Karabaş Patka

(Du: Topper; Fr: Fuligule milouinan; Ger: Bergente; Swe: Bergand)

Subspecies. Nominate.

Status and distribution. Unclear: recent records suggest this is an uncommon and perhaps irregular winter visitor, in small numbers, from November to late March/early April. Historical records, largely from the Istanbul area, are as follows: Alléon (1880) noted a number, mainly females or immatures, rarely males, and collected a male at Çekmece, just west of Istanbul, in February 1893 (housed in the National Museum of Natural History, Sofia), Mathey-Dupraz (1924) reported a female held in the Robert's College, Istanbul collection (apparently no longer extant; Kirwan 1997a), Ogilvie (1954) saw an unspecified number offshore at Çatalağzi, Zonguldak province (Black Sea Coastlands) in the winters of 1946/7 and 1947/8, while six were collected at an unspecified locality on 15 January 1950 (Kumerloeve 1970a). At Büyükçekmece, Istanbul, there was a male in January 1963 and a female in late February 1964 (Kumerloeve *op. cit.*). Records mentioned by Braun (1908) of an individual below Aydin (Aegean) on the Büyük Menderes River in mid-March 1906 and a bird offered for sale in the Istanbul poultry market in 1907, as well as a specimen collected at Haydarpaşa on the Bosphorus by Wahby in February 1927 were all questioned by Kasparek (1990a) and Kumerloeve (1961).

The advent of modern recording witnessed records of large numbers in early 1967 and 1968 at several localities in Marmara and Mediterranean which were later rejected (OST 1975), but subsequent records have established the species' unequivocal occurrence. Accepted (by OST) records were: seven at the Sakarya Delta (Black Sea Coastlands) and four at Manyas or Uluabat Gölü (Marmara) in January/February 1969. Up to 38 wintered on Liman Gölü, Kızılırmak Delta (Black Sea Coastlands) in the early 1990s suggesting that unsubstantiated records from the same area in the 1970s may have been valid (Dijksen & Kasparek 1985). There are additional winter records from five other sites in Black Sea Coastlands, seven in Marmara (most in Thrace), one in Aegean, one in Inner Anatolia, one in Mediterranean and one in Southeast Anatolia (Kirwan & Martins 2000a, Kirwan *et al.* 2003, Demirci 2003; KAB, V. Konrad). Twelve were present at four sites in early 2006 (Kirwan *et al.* in press). In 1992 two males at Sodalı Gölü (East Anatolia) on 30 May (Kirwan & Martins 1994) were the first summer record. However, two at Kulu Gölü (Inner Anatolia) on 19 April 1981 (A. Goodwin *et al.*), *c.*20

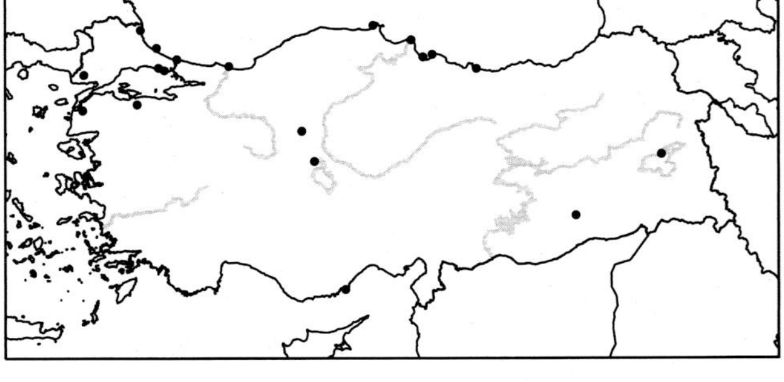

at Göksu Delta (Mediterranean) on 12 April 1990 (Kirwan & Martins 2000a) and six at Çınar-Göksu Reservoir, Diyarbakır (Southeast Anatolia) on 22 April 1999 (Karakaş & Kılıç 2002) were all unusually late.

The species is a rare and irregular winter visitor to Greece, with a maximum count of 700 in January 1984 (Handrinos & Akriotis 1997), a scarce winter visitor to northern Iran and a vagrant to Iraq and Israel (Hollom *et al.* 1988). There are also four records, all November–January from Cyprus, prior to 1973 (Flint & Stewart 1992).

COMMON EIDER *Somateria mollissima*
Pufla

(Du: Eider; Fr: Eider à duvet; Ger: Eiderente; Swe: Ejder)

Subspecies. Unknown, although the race involved must be either *borealis* (from the Arctic North Atlantic, in which males have an orange-yellow bill and females the body more rufous-brown), or nominate *mollissima* (from north-west Europe).

Status and distribution. Vagrant, or increasingly regular visitor. Twelve records, apparently the only reports from the Middle East: a male on Çernek Gölü, Kızılırmak Delta (Black Sea Coastlands) on 20 September 1983 (Dijksen & Kasparek 1985), a female dead at the Göksu Delta (Mediterranean) on 3 January 1984 (Kasparek 1990), a well-described first-winter male and two females off Kefken, west of the Sakarya River delta (Marmara) on 1 February 1997 (Welch & Welch 1998a), an eclipse male and two immatures off Riva, at the north-east mouth of the Bosphorus (Marmara), on 8–14 September 2002 (KAB, BD), one on Terkos Gölü (Marmara) on 8 December 2002 (M. Bozdoğan, S. Isfendiyaroğlu), two off Riva on 3 January 2004, with one still present on 22 March 2004 at least (B. Bilgen, M. Bozdoğan, S. Isfendiyaroğlu & S. Armağan), one again near Riva on 25–26 September 2005 (E. Yoğurtçuoğlu & B. Bilgen), three near Cernek Gölü on 16–17 December 2005 (S. Bekir & N. Yavuz), singles in Kızılırmak Delta on 21 April and 18–19 August 2007 (the latter, a male, photographed; Balmer & Betton 2008), one in Izmit Bay (Marmara) on 7 October 2007 (*per* A. Atahan) and three at Karasu, Sakarya (Marmara) on 20 January 2008 (S. Bekir). Another, mentioned by Kumerloeve (1970a), is considered unacceptable here due to the lack of description; an adult male between Tarabya and Beykoz, Bosphorus (Marmara) in the first half of February 1929. All reliable reports have occurred since the discovery of a breeding site on the Ukranian Black Sea coast (in 1975); this population had reached 1,000 pairs by the mid 1990s and has continued to increase until the present (*Dutch Birding* 23: 354).

There has been a total of at least 20 records in Greece (Handrinos & Akriotis 1997), some of the more recent occurrences involving small flocks of 20–30 individuals, and at least one on the Turkish border (at the Evros Delta), suggesting that the species should also be searched for in Aegean Turkey and the Sea of Marmara.

[KING EIDER *Somateria spectabilis*]
Süslü Pufla

(Du: Koningseider; Fr: Eider à tête grise; Ger: Prachteiderente; Swe: Praktejder)

Subspecies. Monotypic.

Species not admitted to the Turkish list (Kirwan *et al.* 1999). Although occurrence of King Eider appears extremely unlikely it was doubtfully listed by Kumerloeve (1970a), based upon a male claimed by Abbas Celaloğlu (an often unreliable correspondent). No locality or date for this unprecedented observation was presented. Vaurie (1965) refers to a record from Italy and Snow & Perrins (1998) mention records from Spain and Hungary.

LONG-TAILED DUCK *Clangula hyemalis*
Telkuyruk

(Du: Ijseend; Fr: Harelde boréale; Ger: Eisente; Swe: Alfågel)

Subspecies. Monotypic.

Status and distribution. Unclear, perhaps only a vagrant (12 records). The first record for Turkey was of a juvenile female collected by Alléon at Çekmece, just west of Istanbul (Marmara) in February 1893 (housed in the National Museum of Natural History, Sofia), followed by an immature at Izmit (Marmara) on 13 November 1968 (OST 1975). In addition, four recent summer records (all in May and June between 1977 and 1987) from Van Gölü (East Anatolia): two at Horkum, west of Gevaş and three between Tatvan and Ahlat on 10 June 1977 (Beaman 1986), a male, southwest of Van on 22 May 1985 (Martins 1989), a male at Van marshes on 9 June 1987 and an unsexed individual 10km south of Van on 22 June 1987 (Kirwan & Martins 1994). There are four other recent winter records, from Paradeniz Gölü, Göksu Delta (Mediterranean); one on 1–2 January 1986 and a female on 5 January 1989 (Kasparek 1990a); off the Sakarya River mouth (Black Sea Coastlands) on 18 February 2004 (Balmer & Betton 2004b); one photographed off the Fırtına River mouth, Rize province (Black Sea Coastlands) on 26 February 2006 (N. Yavuz & S. Bekir *per* E. Yoğurtcuoğlu *in litt.* 2006); with, most recently, an adult female photographed at Iğneada, Kırklareli province (Marmara) on 17–19 January 2008 at least (*per* E. Yoğurtcuoğlu), and a male photographed at Kıyıköy, Kırklareli province (Marmara) on 13 February 2008 (Ö. Necipoğlu).

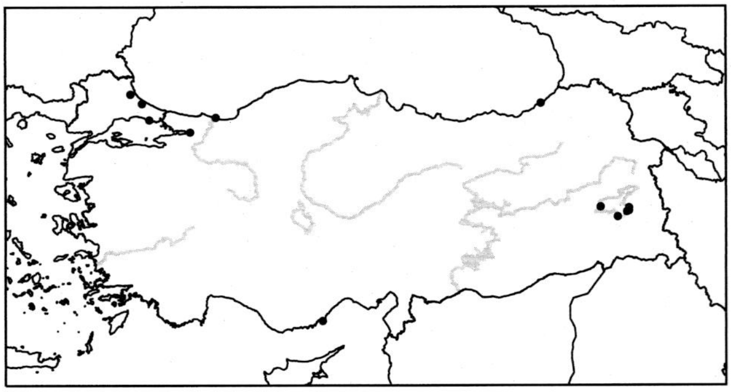

There are at least 16 records from Greece, with most in the north, particularly Thrace, including six at the Evros Delta, on the Turkish border (Handrinos & Akriotis 1997). Elsewhere in the Middle East, it is a vagrant to northern Iran, the Gulf of Aqaba (Israel/Jordan) and Kuwait (Porter *et al.* 1996). There is also a recent summer record from Armenia, from the Armash fishponds (on the Turkish border) in June 2002 (Ananian & de Rouw 2003).

COMMON SCOTER *Melanitta nigra*
Kara Ördek

(Du: Zwarte Zee-eend; Fr: Macreuse noire; Ger: Trauerente; Swe: Sjöorre)

Subspecies. Nominate.

Status and distribution. Irregular winter visitor (24 records) in very small numbers to Black Sea Coastlands, chiefly between September and March, but recorded into early May. Very scarce in the Sea of Marmara and only recently certainly reported from the Mediterranean coast. Kumerloeve (1961) repeated doubtful and unsubstantiated records and comments by Krüper (1875) and other authors which suggested that the species was a rare winter visitor to the Aegean region, and also cited dubious records from the Bosphorus (Marmara) and the Ceyhan Delta, Çukurova (Mediterranean).

One at the Kocaçay Delta (Marmara) on 9 April 1967 is the first acceptable record for Turkey (OST 1969). Subsequently ten were at the Bosphorus (Marmara) on 20 August 1968 (OST 1972) and two offshore at Trabzon (Black Sea Coastlands) on 14 September 1970. Other records, all from the same region (except where stated), involve two females at the mouth of the Gülüç River, Ereğli, between 30 November and 24 December 1977 (Beaman 1986), two off Samsun on 6 January 1987 (Dijksen & Wolf 1987), three between Hopa and Of on 6 September 1988 (Kirwan & Martins 1994), seven flying south at the Bosphorus (Marmara) on 22 July 1991 (KAB, J. Klinker), 17 near Liman Gölü, Kızılırmak Delta on 25 September 1993 (S.Barış), two at Sarıkum Gölü on 19 December 1993 (S. Barış), 12 off Yeşilırmak Delta on 29 January 1995, two there on 9 May 1995 and 12 again on 29 December 1995, six off Eynesil on 11 March 1995 (Kirwan & Martins 2000a), two females off Limanköy (Marmara) on 6 February 1997 (Welch & Welch 1998a), one off Rumelifeneri, north of Istanbul (Marmara) on 27 September 1999 (Kirwan *et al.* 2003), 15 off Fethiye (Mediterranean) on 23 December 1997 (P. Hope) and two at Gediz Delta (Aegean) on 10 February 2008 (O. Gul & O. Bilge). The most regular locality is the Kızılırmak Delta (Black Sea Coastlands) where there was one on 11 September 1973 (OST 1975), two males on 18 March 1992, a pair on 30 March 1992, a pair on 4 April 1992, two on 20 April 1992 (Hustings & van Dijk 1994), three females on 18 October 1992 (Kirwan & Martins 2000a), and a male and two females on 23 January 1997 (Kirwan *et al.* 2003). Kiziroğlu & Kiziroğlu (1987) mentioned a January record (year unknown) from Manyas Gölü (Marmara), but no details are available, and this claim is best disregarded.

Elsewhere in the Middle East regarded as a vagrant or occasional visitor to Cyprus (Flint & Stewart 1992) and a vagrant to Iran (D. A. Scott). There have been fewer than 20 records in Greece (Handrinos & Akriotis 1997).

VELVET SCOTER *Melanitta fusca*
Kadife Ördek

(Du: Grote Zee-eend; Fr: Macreuse brune; Ger: Samtente: Swe: Svärta)

Subspecies and taxonomy. Nominate. Collinson *et al.* (2006) recommend that the three taxa within the *M. fusca* species-group all be considered species on the basis of genetic (Livesey 1995) and morphological evidence.

Status and distribution. National conservation status: Endangered (population apparently decreasing). The first record for Turkey was of two off Zeytinburnu, near Istanbul, on 4 February 1917. Now known to be a very local summer visitor, only very occasionally in moderate numbers, in East Anatolia. Breeds in small numbers on high-altitude lakes (up to at least 3,000m). Only four sites are known; the crater lake at Nemrut Dagı, Tatvan province (up to 20 pairs); Balık Gölü, Ağrı province (numbers never quantified, but apparently small and none present in June 2001); Çıldır Gölü (breeding strongly suspected but unconfirmed); and Aktaş Gölü, Ardahan province. The last-mentioned colony is poorly known but appears to be the largest: 100 birds were present on 3 October 1980 (van der Ven & Gheyselinck 1982) and 725 including many immatures on 14–15 July (Yarar 1995). Regularly oversummers, apparently without breeding, at other localities in the same region (e.g. Aygır Gölü, Kars province, where up to 32 individuals noted, and Nazik Gölü, Muş province), as well as in the eastern half of Black Sea Coastlands.

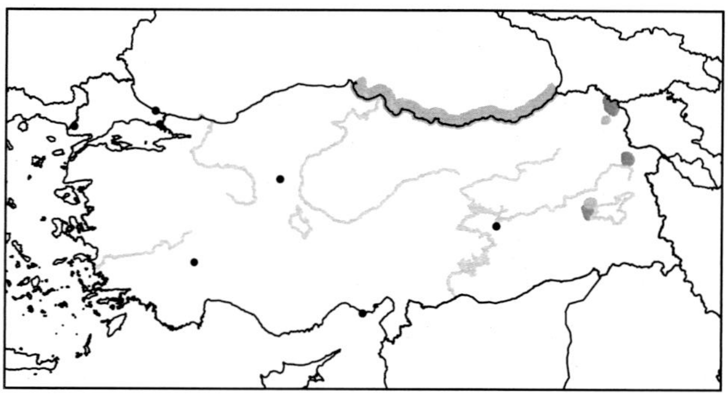

A relatively local winter visitor usually in small numbers along the Black

Sea coast (max. 870 at Yeşilırmak Delta in late December 1995), more occasionally west to the Sea of Marmara and Thrace (e.g. at Gala Gölü in February 2002) and south to the Mediterranean coast (e.g. off Çukurova and Fethiye). Also recorded inland in winter in Mediterranean (at Burdur Gölü, where exceptionally 27 in January 1970), Inner Anatolia (at Mogan Gölü, in February 1966) and even East Anatolia (Hazar Gölü, in January 2005).

Breeding. At two or three high-altitude lakes in **East Anatolia.** Nest not found in Turkey but, elsewhere, is on the ground, concealed in dense vegetation, and often on islands in lakes; usual clutch 7–10. At Nemrut Dağı crater lake, three females had broods of two, seven and nine small downy ducklings on 10 July 1967 and, at Balık Gölü, three females were seen with broods of four, five and six ducklings on 20 August 1967 (Vielliard 1968). Assuming small ducklings were one week old on average, then first eggs would have been laid in the early days of June. A female was seen with seven young (about half grown) at Nemrut Dağı on 23 August 1972; eggs would have been laid in the last week of June. Five pairs and two young were also observed at Nemrut Dağı on 9 July 1985. No recent breeding records, and a thorough search of the island in Balık Gölü was made on 9 June 2001, when no nest was found, nor adults seen (PC).

COMMON GOLDENEYE — *Bucephala clangula*
Altıngöz

(Du: Brilduiker; Fr: Garrot à oeil d'or; Ger: Schellente; Swe: Knipa)

Subspecies. Presumably the nominate, though the species may be best regarded as monotypic.

Status and distribution. Uncommon winter visitor from September (earliest 15th, but more usually October) to late April (27th), to coastal, less frequently inland wetlands, in the western two-thirds of Turkey. Most regular in Black Sea Coastlands and perhaps most infrequent in Mediterranean. Numbers rarely exceed 100 birds each winter, but 200 were off Kızılırmak Delta (Black Sea Coastlands) on 3 February 1992 (L. J. Dijksen & M. Klenman), and it is also clear that unusually large numbers reached Turkey in winter 2005/06, with several records from Aegean in February, amongst them a flock of 72 at Gediz Delta. Otherwise, there is another unusually large, single-site count of 60 on Gala Gölü (Marmara) on 3 February 2002 (Demirci 2002). Recently recorded in eastern Black Sea Coastlands in late spring. Between 1977 and 1993 there was an apparently regular series of summer records from East Anatolia, the majority on Van Gölü, some involving multiple individuals. These referred to non-breeders (most were immatures or adults in heavy moult), although Kasparek (1992a) questioned whether an as yet undiscovered breeding population might exist in this part of Turkey or nearby.

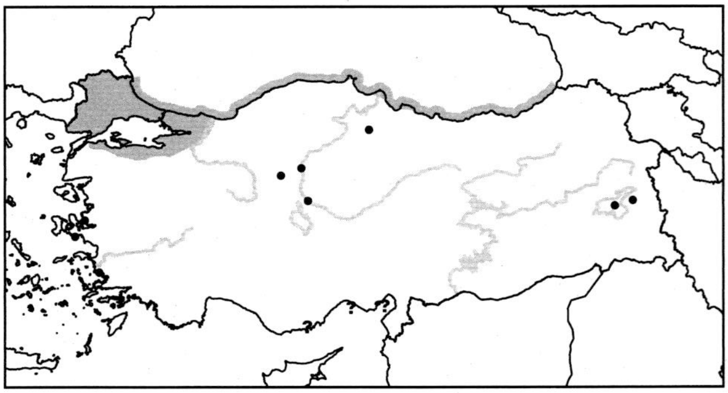

SMEW — *Mergellus albellus*
Sütlabi

(Du: Nonnetje; Fr: Harle piette; Ger: Zwergsäger; Swe: Salskrake)

Subspecies and taxonomy. Monotypic. Livezey (1995) demonstrated that this species is more closely related to goldeneyes (*Bucephala*) than other *Mergus*, thus justifying its removal to the genus *Mergellus*. This move has acquired support from, e.g., Sangster *et al.* (1997) and the AOU (1998) but, given that Hooded Merganser *Mergus* (*Lophodytes*) *cucullatus* also required separate generic treatment on the basis of results from the same study, it does mean that a genus comprising just six extant species must now be viewed as comprising three genera in order to avoid polyphyly. Furthermore, subsequent mt-DNA-sequencing work (Donne-Gousse *et al.* 2002), attempting to reconstruct a phylogeny for the entire Anseriformes, did not support such an arrangement. Species described from Turkey.

Status and distribution. Typically a scarce and possibly irregular winter visitor from November to mid April in wetlands and coastal areas across the western two-thirds of Turkey. Most frequently recorded in Marmara (including Thrace, where regular fieldwork in the Istanbul area revealed the species to be apparently not infrequent at two sites, Büyükçekmece and Rumelifeneri, in the 1990s), Black Sea Coastlands and Inner Anatolia, occasionally in East Anatolia but very scarce elsewhere. Ironically, the earliest records of the species in Turkey, appear to be two specimens taken by Strickland at Izmir (Aegean), a male on 6 January 1836 and a female in February of the same year (Univ. Mus. Zool., Cambridge collection). With the creation of large new dam lakes in eastern and southeastern Turkey in recent winters, the species clearly may be expected to occasionally occur further south and east in the future; the observation of over 1,000 at Lake Razazah, south of Baghdad (Iraq) in January 1979 (Scott & Carp 1982), demonstrates the extent to which large numbers may sometimes move well to the south of their normal, expected range.

Usually fewer than 100 birds each winter but sometimes recorded in moderate to large numbers, e.g. Ogilvie (1954) found it to be fairly common at sea off Çatalağzi, Zonguldak province (Black Sea Coastlands) in 1946–48, 300 were at Uluabat Gölü in January 1967, exceptionally 1,300 were there in 1969/70, 555 in January 1973 and 111 also in the same place in January 1989. Elsewhere, 102 were at Manyas Gölü (Marmara) in January 1992 and 79 at Büyükçekmece on 23 January 1993. The winters of 1995 (when 248 were present at Uluabat Gölü and 52 at two other sites in mid January), 1996 (at least 191 at 11 sites throughout the western half of the country), 1999 (400 at seven sites, including two counts of over 100 individuals) and 2006 (several hundreds in the north of the country and significant numbers in Aegean) also witnessed significant numbers reach Turkey. Recently recorded in late April in Southeast Anatolia (at Diyarbakır in 1987). Three summer records; a male at Eymir Gölü, Ankara province (Inner Anatolia) on 11 June 1969 (OST 1972), a female at Göründü, Van Gölü (East Anatolia) on 27 June 1987 (Kirwan & Martins 1994) and a male and two females at Uluabat Gölü on 27 May 1995 (Kirwan & Martins 2000a).

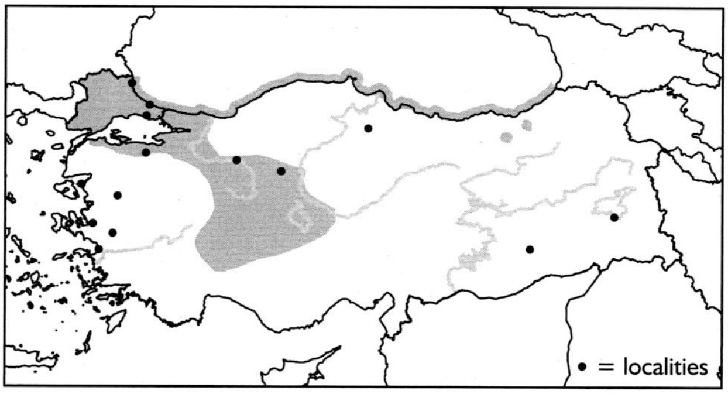

RED-BREASTED MERGANSER *Mergus serrator*
Tarakdiş

(Du: Middelste Zaagbek; Fr: Harle huppé; Ger: Mittelsäger; Swe: Småskrake)

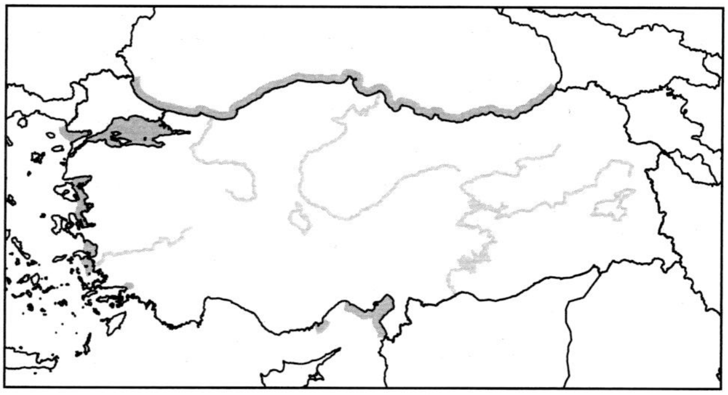

Subspecies. Monotypic.

Status and distribution. Has been viewed as a scarce winter visitor (27 October–27 April), occasionally recorded in considerable numbers, to the western two-thirds of Turkey, usually in coastal areas. Eken (1997d) considered it to be regular in some numbers at coastal deltas in Aegean and, to a lesser extent, Mediterranean.

The mid-1990s produced several exceptional counts from three areas, two in the former and one in the latter region: Gediz Delta (up to 397, in February 1996, and regularly in excess of 100 individuals); Büyük Menderes Delta (up to 67, in February 1993) and Yumurtalık, Çukurova (up to 44, in January 1996). Most frequently recorded in Marmara (including Thrace), Aegean and Black Sea Coastlands (where reported as far east as the Georgian border). Despite the evidence of Eken (*op. cit.*), it is rarely reported from the Mediterranean coast (although the species is reasonably regular in Cyprus) and Red-breasted Merganser is very scarce in Inner Anatolia. One summer record of seven or eight at Amik Gölü, Antakya province (Mediterranean) on 11 June 1964 (Kumerloeve 1966–67), with a late individual at the Göksu Delta (Mediterranean) on 20–21 May 1994 (*Birdquest Newsletter* 23: 59).

GOOSANDER — *Mergus merganser*
Büyük Tarakdiş

(Du: Grote Zaagbek; Fr: Harle bièvre; Ger: Gänsesäger; Swe: Storskrake)

Subspecies. Nominate.

Status and distribution. Very rare and highly irregular winter visitor, in very small numbers, to the western two-thirds of Turkey, e.g. only ten records since 1997, despite the advent of much-increased observer activity in winter during this period (Kirwan *et al.* 2003). Usually found in coastal wetlands, most regularly in the two large Black Sea Coastlands deltas (where noted as late as 20 May). Recorded twice inland in East Anatolia in late winter or early spring (February/March) and three times in the same region during summer: a pair on the Aras River between Pasinler and Horasan, Erzurum province, on 11 June 1970, one at Van Gölü on 7 June 1986, and a female or immature at Bendimahi Delta on 29 June 1988. No evidence of breeding in any of these cases, although the species formerly bred in the Lake Sevan basin of Armenia, prior to drainage (Adamian & Klem 1999).

[RUDDY DUCK — *Oxyura jamaicensis*]
Kara Başlı Dikkuyruk

(Du: Rosse Stekelstaart; Fr: Erismature rousse; Ger: Schwarzkopf-Ruderente; Swe: Amerikansk kopparand)

Subspecies. Monotypic.

Hypothetical. One record: at Büyükçekmece, Istanbul (Marmara) in April 1988 (exact date uncertain) (J. Alexander pers. comm.) relates either to this species or potentially to a hybrid between this and White-headed Duck *O. leucocephala*. Dark-headed examples of the latter have been recorded, e.g. in Spain (see Madge 1984), but a pure-bred individual of this species was eliminated. There are no other records from the Middle East but Ruddy Ducks have now been reported in at least 19 European countries, including the Ukraine (Kear 2005).

WHITE-HEADED DUCK *Oxyura leucocephala*
Di kkuyruk

(Du: Witkopeend; Fr: Erismature à tête blanche; Ger: Weißkopf-Ruderente; Swe: Kopparand)

Subspecies. Monotypic.

Status and distribution. Conservation status: Endangered (BirdLife International 2000, 2004); national conservation status: Endangered (population apparently decreasing). Local resident, although more usually a partial migrant, in small to moderate numbers in dense, marshy wetlands (usually shallow, eutrophic lakes with either fresh or occasionally brackish water) in Inner Anatolia, and perhaps in adjacent areas of Mediterranean and East Anatolia. May also nest in suitable areas in Black Sea Coastlands from where there are several recent records well into the breeding season. Summer records from the south coast and the immediate hinterland almost certainly relate to non-breeding birds. There is recent evidence that the species is able to utilise older dam lakes for breeding (e.g. at Akkaya Reservoir, Inner Anatolia; Kirwan 1994b). Breeding status in East Anatolia is unclear as many potential localities are underwatched and some even unvisited for many years. Nesting has been confirmed from a number of areas, principally around Van Gölü and at a small wetland in Kars province, but it seems likely that there are other localities in the extreme east of the country where the species breeds (many of which are not marked on small-scale maps). The breeding population in Inner Anatolia probably exceeded 150 pairs, mostly at four key sites – Ereğli, Hotamış and Sultan marshes, and Kulu Gölü – in the late 1980s and early 1990s, but has since declined dramatically, with both the two first-named sites being largely drained in the mid to late 1990s, and the importance of the latter being steadily reduced for unknown reasons (Richardson 2003).

Migrants moving through Turkey en route to breeding areas elsewhere distort perceptions of the species' breeding status between March and early May. Substantial concentrations are regularly recorded from both known breeding and non-breeding sites at this time, e.g. 1,246 at Kızılırmak Delta (Black Sea Coastlands) on 23 March 1992 and 508 at Ereğli marshes (Inner Anatolia) in March 1990. However such flocking behaviour, especially after the first week of May, probably presages dispersal to breeding grounds within Turkey. In August and September, occasionally in early July, large numbers of post-breeding birds, many possibly from outside Turkey (as such large numbers apparently do not breed in eastern Turkey), assemble at certain localities, e.g. maxima of 500 at Kulu Gölü in July and 600–1,000 at Sodalı Gölü (East Anatolia) in August.

In winter large numbers occur at several wetlands in Mediterranean, generally in much smaller numbers in Inner Anatolia and more occasionally elsewhere in the western two-thirds of the country, principally during harsh weather. Large numbers irregularly winter in Black Sea Coastlands. Until very recently, typically in excess of 50% of the world population wintered at Burdur Gölü (Mediterranean) with recent maxima from this locality as follows: 6,400 in late January 1987, 9,230 in late January 1988, 6,700 in mid-January 1989, 10,927 in early February 1991 (Green & Anstey 1992) and 2,054 in early April 1991 (a late large concentration). Totals were significantly lower in 1992 (3,264), 1993 (3,010) and 1994 (3,337). The 1993 count at Burdur Gölü was considered to be highly accurate, as it was performed by a large team, counting in ideal weather conditions (A. Green *in litt.*). It appears that in years when numbers wintering on Burdur Gölü are high most of the Turkish population is concentrated there. When lower numbers are recorded at this site this may in part reflect wider dispersal, perhaps including areas outside Turkey, e.g. Russia, Azerbaijan and Syria, and, more recently, Israel.

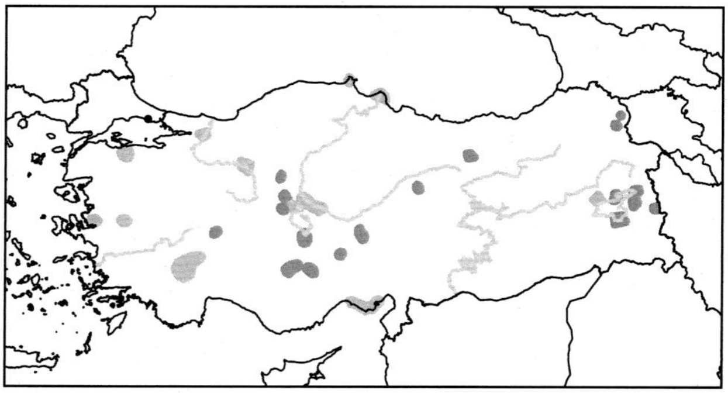

There is evidence of a decline in the Turkish breeding population since the advent of modern recording (1966), due principally of course to wetland drainage, although some of the losses have been 'offset' by the discovery of new breeding sites in recent years. However, as more systematic monitoring of wintering areas, e.g. Burdur Gölü, continues to advance understanding of the species' status and conservation requirements, there is evidence to suggest that a substantial decline has taken place in the wintering population since 1991. Hunting has certainly been a contributory factor (perhaps in excess of 1,000 were hunted at Burdur Gölü in winter 1992/93; A. Green *in litt.*), although fewer young birds were counted at Burdur Gölü in 1993 than in 1991, suggesting that lower breeding success (in Kazakhstan and elsewhere) is also important in this decline.

Breeding. Mainly at freshwater or brackish lakes, usually small (often less than 10ha), and shallow (less than 2m), with abundant submerged vegetation, extensive reeds with open water, and often near larger wetlands (Anstey 1989). Several pairs may breed at the same site. A nest at a small lake near Erçek Gölü (East Anatolia) was built up from the water in a dense stand of vertical *Phragmites*, forming an island near the centre of the lake; it was constructed of aquatic plant material, built at the base of the reeds, and contained an incomplete clutch of two eggs on 11 June 2001; eight males and seven females were present but, despite a fairly thorough search, no other nests were found, indicating that breeding had not yet properly commenced (PC). A nest, at the same lake, had a full clutch of ten eggs on 19 June 2004; it was well concealed in a dense reedbed, in knee-deep water, attached to the base of upright reeds, and built up from the water; a substantial structure of dead reed stems and other aquatic vegetation, interwoven to form a neat cup, with a few tufts of pale grey down (B. Pleasance); incubation was still ongoing on 26 June 2004 (PC). A nest in reeds at Kulu Gölü (**Inner Anatolia**) had seven eggs on 13 July 1987 (I. A. Green); the nest and eggs were partially submerged, possibly as an aid to intermittent incubation by the female (Anstey 1989); however, White-headed Ducks lay exceptionally large (and heavy) eggs, relative to body size, and it is more likely that the nest had merely sunk slightly, due to the weight, a phenomenon also observed in nests of Ruddy Duck *O. jamaicensis* (PC). Elsewhere, the usual clutch is 5–12. Most records in Turkey are of broods of young: B/3–10. Juveniles seen in **East Anatolia** in June–September; the earliest records, at Van Gölü, on 9 June 1987 and 14 June 1990, indicate egg laying about mid May, but most records are much later, in July–August, suggesting laying from mid June. In Inner Anatolia, earliest records are of three large young and three downy young at Kulu Gölü on 5 June 1975; this record indicates egg laying in early May, but one of large young in April (possibly early April) appears exceptionally early, especially as display is observed mainly from May. At Çogenc Gölü, two females were observed with broods of five young (both 20 days old) and one with six young (four days old) on 6 August 1972; these records indicate egg laying in mid June and early July. There are many records of broods in Inner Anatolia in July–August. In neighbouring Armenia, a nest at Shorzha had three eggs on 22 June 1928 and two nests at Armash had five and six eggs respectively on 14 June 1991 and 30 June 1991 (Adamian & Klem 1999); the main breeding season in Turkey (or at least in East Anatolia) is probably similar to that in Armenia.

RED-THROATED LOON (RED-THROATED DIVER) *Gavia stellata*
Kızılgerdanlı Dalgıç

(Du: Roodkeelduiker; Fr: Plongeon catmarin; Ger: Sterntaucher; Swe: Smålon)

Subspecies. Monotypic (see Collinson 2006).

Status and distribution. Apparently a genuinely scarce (rather than merely overlooked), but widespread, winter visitor from the end of September (occasionally mid August) to at least early June, to the Black Sea, Marmara, Aegean and Mediterranean coasts. Most records are from the former region, although this perhaps reflects more extensive coverage of this area during the winter. Exceptionally, 20 off the Yeşilırmak Delta (Black Sea

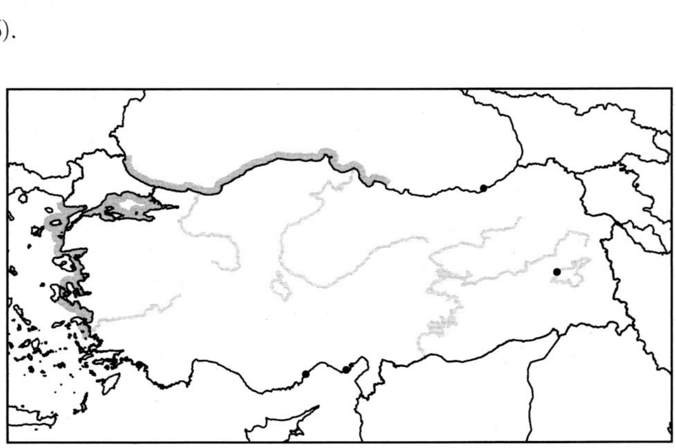

Coastlands) on 9 January 1969. Non-breeding individuals are occasionally recorded in late spring and summer, usually on the Black Sea coast, but there is also one inland record from this period; one at Ahlat, Van Gölü (East Anatolia) on 14 June 1977 (Beaman 1986). Two specimens, presumably taken in the vicinity of the Bosphorus, are held in the Robert's College, Istanbul collection (Kirwan 1997a).

BLACK-THROATED LOON (BLACK-THROATED DIVER)
Karagerdanlı Dalgıç *Gavia arctica*

(Du: Parelduiker; Fr: Plongeon arctique; Ger: Prachttaucher; Swe: Storlom)

Subspecies. Nominate.

Status and distribution. Widespread and common winter visitor to Black Sea Coastlands: where less common in western half, and Marmara coasts, from early September (occasionally mid August) to April. Towards the end of the winter period numbers in the Black Sea are augmented by returning migrants from further south, e.g. numbers off Kızılırmak Delta probably reach a peak in late February/early April (max. 517 on 27 March 1992), and the highest-ever single-day total in Turkey is from this period, 2,100 off Artvin (Black Sea Coastlands) on 26 February 2006 (S. Bekir & N. Yavuz). However, 1,500 were present there on 25–27 January 1967, and 50 per hour moved west off the Yeşilırmak Delta on 9–10 January 1969. Distinctly scarce off Aegean and Mediterranean coasts during the same period, e.g. listed as perhaps only accidental off well-watched Göksu and Çukurova deltas, with one record from Antakya (Mediterranean). Rather scarce inland in the western two-thirds of Turkey in winter, with apparently no records since 1997. Non-breeders, occasionally in significant numbers, remain until late June (when display has been noted), and perhaps later, especially in the central and eastern Black Sea Coastlands, more rarely inland (four records from Van Gölü, East Anatolia, in June 1978, June 1983, July 1987 and a bleached bird in July 1995, one photographed at Bayındır Reservoir, Ankara, in 1996, one from Burdur Gölü, Mediterranean, in June 1965, and one at Aygir Gölü, East Anatolia, in June 2005), and exceptionally on the Mediterranean coast.

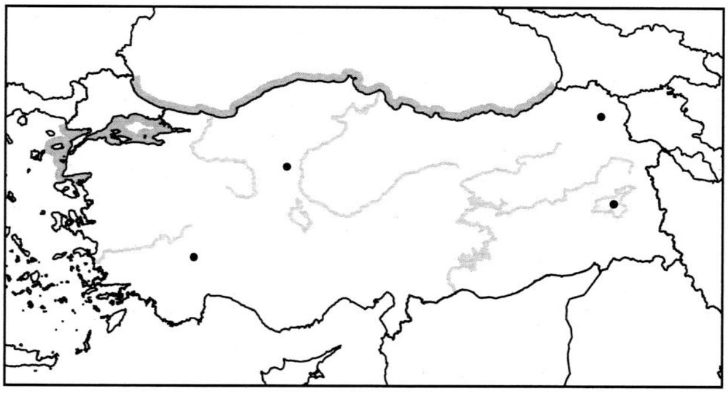

GREAT NORTHERN LOON (GREAT NORTHERN DIVER)
Buz Dalgıcı *Gavia immer*

(Du: Ijsduiker; Fr: Plongeon imbrin; Ger: Eistaucher; Swe: Svartnäbbad islom)

Subspecies. Monotypic.

Status and distribution. Vagrant. A specimen, apparently held in the Robert's College, Istanbul collection, listed by Mathey-Dupraz (1923) and Kasparek (1990a) could not be located in early 1996 (Kirwan 1997a), although many specimens once housed there have been lost or destroyed since 1992. M. Kasparek (*in litt.* 1996) was able to photograph the skin during a visit to the museum in 1986. Four other records from Turkey: three from the Sea of Marmara; one in summer plumage, on the sea off Büyükçekmece, Istanbul on 13 May 1964 (Warncke 1965); eight, including some in partial summer plumage, at the same place, on 29 April 1968 (Groh 1968); one off the mouth of the Serefli River on 27 March 1981 (Goriup & Parr 1983), and a dead individual on the beach adjacent to Akgöl, Göksu Delta (Mediterranean) on 13 May 1989 (Kirwan & Martins 1994). Other old and more recent records from Black Sea Coastlands have since been proven to be erroneous, although the species might be expected to occur very rarely in this region.

LITTLE GREBE — *Tachybaptus ruficollis*
Küçük Batağan

(Du: Dodaars; Fr: Grèbe castagneux; Ger: Zwergtaucher; Swe: Smådopping)

Subspecies and taxonomy. Nominate *ruficollis* in western Turkey and (perhaps) *capensis* in eastern Turkey (Roselaar 1995), but their precise ranges are unknown. Van Swelm (2001) postulated that apparent eye-colour differences between predominantly European *ruficollis* and Afro-Asian *capensis* were significant, readily diagnosable and merited recognition at species level. Unfortunately, the author chose to base his research solely on sight records and a limited number of photographs, and failed to add weight to his views through an extensive study of specimen material and their labels. Nonetheless, the issue requires additional research and, should the two forms prove worthy of specific treatment, the question of whether *capensis* is present in eastern Turkey will acquire heightened interest. Turkey may prove an ideal location to study the potential relationships between the two forms.

Status and distribution. National conservation status: Least Concern (population apparently decreasing). Common resident breeder, being widespread in relatively small numbers in marshy wetlands across Turkey, although scarcer in Southeast Anatolia due to a general lack of suitable breeding habitat. The 350–500 pairs estimated at Kızılırmak Delta (Black Sea Coastlands) in 1992 make this perhaps the single-most important breeding site in Turkey. Those in Inner Anatolia (and probably East Anatolia) principally move south to winter on Aegean and Mediterranean coasts, returning from March. Will utilise even very small pools, provided there is sufficient cover for breeding, though more usually found in larger wetlands.

Post-breeding concentrations assemble in all regions from July, although numbers rarely exceed 300–500 birds, more exceptionally up to 1,000, e.g. on Manyas Gölü (Marmara), Erçek Gölü and Van Gölü (East Anatolia), with a highest count of 2,625 at Sarıyar Reservoir (Inner Anatolia) in September 2000 (Perktaş & Ayaş 2005). Winters largely in the western two-thirds of Turkey in moderate to large numbers, in freshwater lakes, marshes, reservoirs, coastal deltas and, less usually, on the sea. At Ereğli (Black Sea Coastlands) in the late 1970s, birds arrived early in the second half of October and remained until late March. The largest concentrations (up to 1,000) at this season are usually on the south and west coasts, typically at Bafa Gölü/Büyük Menderes Delta (Aegean), Köyceğiz Gölü and Göksu Delta (Mediterranean), although significant numbers (up to 250) will remain on the wetlands of Inner Anatolia (e.g. 538 at Sarıyar Reservoir on 11 February 2005), and well inland in the Mediterranean region if the weather is relatively mild, e.g. 1,155 at Eğirdir Gölü, on 20 January 2005. Also common in Thrace at this season, with peak count of 1,120 at Küçükçekmece on 3 February 1991 (J. Klinker).

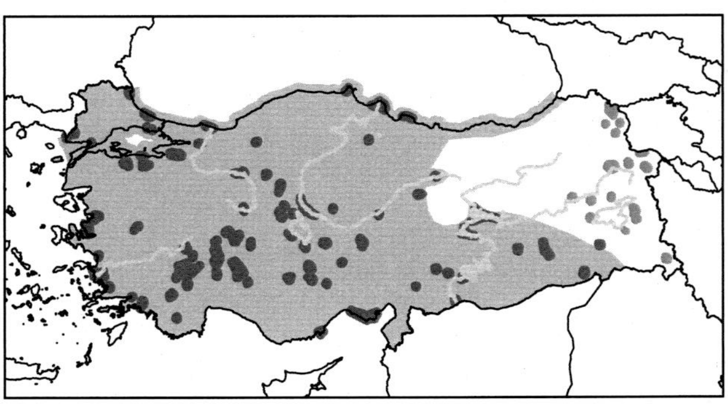

Breeding. On freshwater and brackish lakes with luxuriant vegetation, marshes and reedbeds with open water, and old river arms. In Kızılırmak Delta (Black Sea Coastlands) in 1992 the breeding population was 350–500 pairs, semi-colonial in some areas, with 25–45 pairs/100 ha (Hustings & van Dijk 1994). Nest usually floating, anchored to live vegetation; a heap of aquatic vegetation, with a shallow hollow on top. Eggs: C/4 (2), C/5 (2), in other parts of the range, usual clutch 4–6 (2–7) eggs. Young: B/1 (2), B/2 (4), B/3 (2), B/4 (2), B/5 (1); the lower numbers may have been reduced by losses. In **Inner Anatolia**, a nest had four eggs at Uyuz Gölü on 18 May 1998, and another four eggs at Eşmekaya on 19 May 1998 (T. Quittenden); at Sultan marshes, adults were incubating on a roadside canal, and one nest had five eggs, 14 May 2004 (PC); some adults were still incubating at Akşehir Gölü, 8 August 1971; there are four records of young, the earliest at Sultan marshes, 7 June 1982 (Kasparek 1985). In **Mediterranean**, breeding commenced in Çukurova in early April 1987, and downy young were observed from 4 May (van der Have *et al.* 1988); well-grown young were seen near Antalya, 18 May 1970. In **Aegean**, there are five records of young, including four from Bafa Gölü, where four young on 5 May 1995 indicate egg laying in early April. In **Marmara**, there are five records of young, all in June, the earliest at Forest of Belgrade on 4 June 1996 (GMK); a nest had five eggs at Uluabat Gölü on 2 June 2006 (R. Castell). In Kızılırmak Delta (**Black Sea Coastlands**), two full-grown young were seen on 10 June 1995 and four young on 15 July 1971 (Dijksen & Kasparek 1985), and the first young on 26 May 1992 (Hustings & van Dijk 1994). Recently hatched young were seen at Birecik (**Southeast Anatolia**) on 7 June 2006 (M. Cowlard). In **East Anatolia**, near Erçek Gölü on 19 June 2004, a nest had an incomplete clutch of two fresh eggs, and another two warm hatching eggs, with hatched eggshells in the nest, and an adult close by, presumably with swimming young (B. Pleasance); an adult was still incubating at Sodalı Gölü on 18 August 1972, and there are seven records of young (earliest 25 June). Later records are almost certainly second broods.

GREAT CRESTED GREBE — *Podiceps cristatus*
Bahri

(Du: Fuut; Fr: Grèbe huppé; Ger: Haubentaucher; Swe: Skäggdopping)

Subspecies. Nominate.

Status and distribution. National conservation status: Least Concern (population apparently decreasing). A common breeding resident (although as yet no confirmed breeding records from Southeast Anatolia) which is particularly common in the large marshy wetlands of Inner Anatolia, and to a lesser extent East Anatolia. Large populations of more than 400 pairs at Uluabat Gölü (Marmara) and 250–300 pairs at Kızılırmak Delta (Black Sea Coastlands). This species is less tolerant of smaller pools and marshes for breeding than the previous species but will utilise barrage lakes.

More widespread in the post-breeding and winter periods, although still usually shuns small wetlands. In the coldest periods of the year it also moves into suitable areas in Thrace and East and Southeast Anatolia. In winter the population largely moves into coastal areas and lakes in the south and west, e.g. the Marmara lakes, with generally only moderate numbers remaining in central and eastern parts during particularly harsh winters: however some of the large reservoirs such as Sarıyar in Inner Anatolia (5,500), Karakaya (12,000) and Keban (10,000) in East Anatolia support large populations. Particularly favoured wintering areas such as the Kızılırmak and Yeşilırmak deltas (Black Sea Coastlands), Küçükçekmece (Marmara) and Büyük Menderes Delta (Aegean) may regularly support several thousand, especially in harsh winters. Exceptionally up to 20,000 have been recorded in the former region, where immigrants apparently arrive in mid October and depart in late April (see Albrecht 1986). With more extensive coverage in recent years, winter totals peaked at 30,853 in 2005.

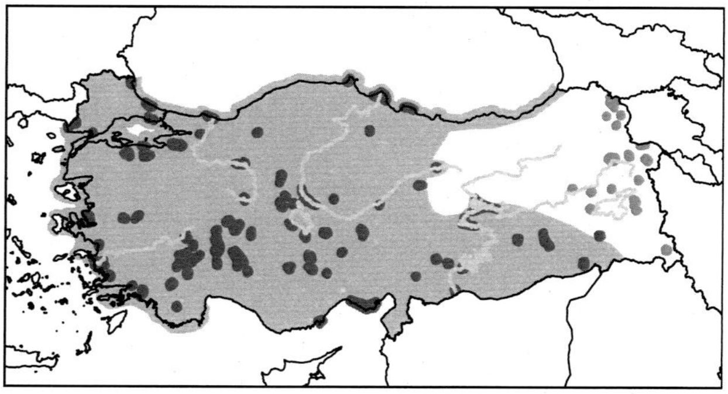

Breeding. On lakes with fringing reeds: at Uluabat Gölü (Marmara) the highest breeding densities in 1998 were in broken reed fringe, where many pools and channels provided sheltered feeding and roosting areas (Welch & Welch 1998b). Nests in reeds, in water lilies or partly submerged branches; nest built-up from ground in shallow water, or floating and attached to vegetation in deeper water; of accumulated aquatic vegetation with shallow hollow on top. Eggs: C/4 (1), C/5 (4). Young: B/1–5, mean brood 2.6 (n=231) at Uluabat Gölü (Welch & Welch 1998b). In **Marmara**, several pairs were incubating on 25 April 2003, and there were many pairs with mainly large young at Uluabat Gölü on 19 June 1999, where large young, *c.*3 weeks old, on 13 May 2007 indicate egg laying in late March (PC); nest building was observed on 25 April 1970 and two broods (four and five) were at Uluabat Gölü on 2 June 1967 (R. F. Porter); three nests were half built on 23 April 1966, and earliest young were on 24 May 1966 at Manyas Gölü (R. F. Porter); and four pairs with very small young were at Iznik Gölü on 6 July 1966 (R. F. Porter). In **Aegean**, Selous (1900) found *c.*20 nests 13 May 1899, and there are many records of young 21 May–21 September. At Kızılırmak Delta (**Black Sea Coastlands**) in 1992 nests had eggs from 27 April, and first young were seen on 20 May (Hustings & van Dijk 1994). In **Inner Anatolia**, adults were incubating from late April at Hotamış (Kirwan 1993a), and there are many records of young at other sites (earliest, well grown 6 June). In **Mediterranean**, nest building observed, Beyşehir Gölü, 1 May 1967 (R. F. Porter), young from 10 June, with eight unfledged young at Göksu Delta as late as 14 September 1972. In **East Anatolia**, three adults incubating in mid July; earliest young, Van Gölü, 8 June 1975, indicating egg laying early May.

RED-NECKED GREBE *Podiceps grisegena*
Kızıl Boyunlu Batağan

(Du: Roodhalsfuut; Fr: Grèbe jougris; Ger: Rothalstaucher; Swe: Gråhakedopping)

Subspecies. Nominate.

Status and distribution. National conservation status: Vulnerable (population apparently decreasing). Seemingly an uncommon and rather local breeding summer visitor, largely to Inner and East Anatolia, but also to immediately adjacent areas of other regions, including the extreme northeast Black Sea Coastlands and perhaps the central part of the same region, at Kızılırmak Delta. Probably scarcer in the western part of its range. In eastern areas utilises small, marshy wetlands to at least 2,250m for nesting and thus, because many such wetlands are poorly or wholly unknown ornithologically, may be more common than records suggest. However, in Inner Anatolia, the species is virtually wholly dependant upon larger wetlands, such as Ereğli or Sultan marshes, and is only rarely recorded on the small, eutrophic wetlands it favours further east. In this region its range shows some evidence of having contracted in most-recent decades due to drainage of several of its favoured sites, although this requires confirmation. Certainly more common in the first half of the 20th century: e.g., Wadley (1951) noted *c.*20 pairs breeding at Mogan Gölü, south of Ankara, in 1945–6, where the species' current status is uncertain. Earliest arrival is in the third week of March (17th) and most have departed by late October (29th). Non-breeders or perhaps failed nesters typically over-summer on the Black Sea coast or Van Gölü (East Anatolia). An unusual concentration of 40, on Sodalı Gölü (East Anatolia) in mid-July 2001, presumably involved such birds. There was also an exceptional concentration of 73 at Eşmekaya, Aksaray province (Inner Anatolia) on 9 June 1998 (Eken & Magnin 1999).

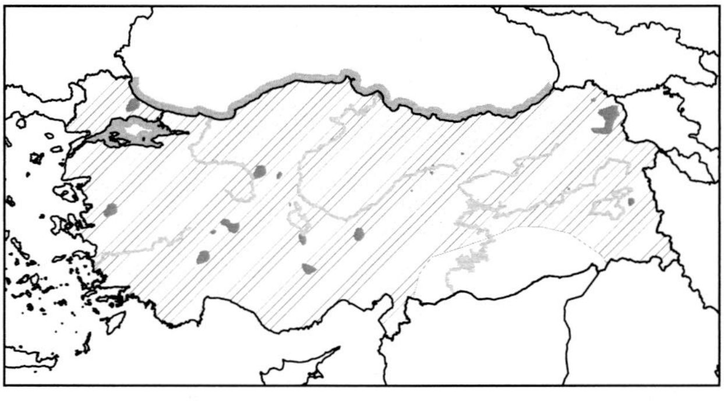

On passage distinctly more widespread, being recorded in all regions except Southeast Anatolia, but typically in coastal waters in Marmara and Black Sea Coastlands; also in inland areas across the country, with an exceptional concentration of 150 at Burdur

Gölü on 1–3 September 1980. Rarely recorded in winter, most frequently in Mediterranean (exceptionally ten at Burdur Gölü in mid-January 1970) but also in Marmara (including Thrace) and more occasionally in Black Sea Coastlands (up to seven).

Breeding. In fringing reeds on lakes, and wet marshes and reedbeds with open water. Nest usually floating and anchored to growing vegetation: a shallow hollow on a low mound of rotting aquatic vegetation. Eggs: C/3 (1); elsewhere usual clutch is 4–5 (3–5). Young: max. brood 3 (2), and several records of only one. In **Inner Anatolia**, display observed late April–early May; two nests at Eşmekaya had incomplete clutches of one and two eggs, 27 May 1993 (PC) and 20 May 1998 (T. Quittenden), and a well-grown young was seen there on 21 June 1998 (GMK); a nest at Ereğli had a (probable) full clutch of three eggs on 19 May 1971 (M. Kasparek); young were at Akşehir Gölü on 13 July 1977, at Kulu Gölü on 17 July 1986 (R. Webb), and Sultan marshes in late-August 1982 (Kasparek 1985). In **Mediterranean**, courtship was observed at Karamık Gölü on 12 April 1973. In **East Anatolia**: nest building was seen near Van on 29 May 1969 (R. F. Porter), and Çaldıran Gölü on 1 June 1990; a nest was near Kars on 18 July 1992, and there are six records of young (earliest 27 June).

HORNED (SLAVONIAN) GREBE — *Podiceps auritus*
Kulaklı Batağan

(Du: Kuifduiker; Fr: Grèbe esclavon; Ger: Ohrentaucher; Swe: Svarthakedopping)

Subspecies. Monotypic or, if treated as polytypic, the nominate.

Status and distribution. A very scarce winter visitor (21 records), from late September (25th) to early May (9th), typically to coastal waters in Marmara and Black Sea Coastlands (largely in the western half), but also inland in Inner, Southeast and East Anatolia. (We have ignored published records from September and November, at Manyas Gölü, Marmara, as being almost certainly unreliable; Kiziroğlu & Kiziroğlu 1987.) Usually only single individuals recorded, once a group of six (off Küçükçekmece, Istanbul on 2 October 1972). Ogilvie (1954) found the species offshore at Çatalağzi, Zonguldak province (Black Sea Coastlands) between late February and 11 March 1948, in company with Black-necked Grebe *P. nigricollis*. Old records of summering individuals in Inner Anatolia (Mogan Gölü) and East Anatolia (Kars province) are unconvincing, and a claim of six at Nazik Gölü, Bitlis province (East Anatolia) in late-September 1979 (van der Ven & Gheyselinck 1980) lacks a description and is strongly suggestive of confusion with Black-necked Grebe. Fifteen recent records: one in summer plumage at Çaldıran, Van province (East Anatolia) on 1 July 1985 (C. Husband *et al.*), one at the Göksu Delta (Mediterranean) in February 1989 (Kirwan *et al.* 2003), two at Büyükçekmece, Istanbul, on 3 December 1989 (J. Klinker), one off Kurna, Istanbul on 27 January 1991 (Kirwan & Martins 2000a), one in summer plumage at Kızılırmak Delta (Black Sea Coastlands) on 20 April 1992 (Hustings & van Dijk 1994) with another in winter plumage there on 6 March 1993 (Kirwan & Martins 2000a), four at Ören, Burhaniye (Aegean) on 29 January 1997 (Kirwan *et al.* 2003), one (videotaped) at Mimarsinan, west of Istanbul (Marmara) on 1 January 1998 (M. Freedman), one at Çınar-Göksu Reservoir (Southeast Anatolia) on 22 April 1999 (Karakaş & Kılıç 2002), one photographed off Riva, Istanbul, on 12 December 2004 (Balmer & Betton 2005), one at Meriç Delta (Marmara), on 16 January 2005 (B. Bilgen), one at Büyükçekmece, Istanbul, on 28 January 2006 (E. Yoğurtocuoğlu), one at Sariyer, Istanbul, on 30 January 2006 (Balmer & Betton 2006b), one at Fındıklı (Black Sea Coastlands), on 3 February 2006 (A. Abuladze) and one at Sarıyar Reservoir (Inner Anatolia), on 27 January 2008 (BD *et al.*).

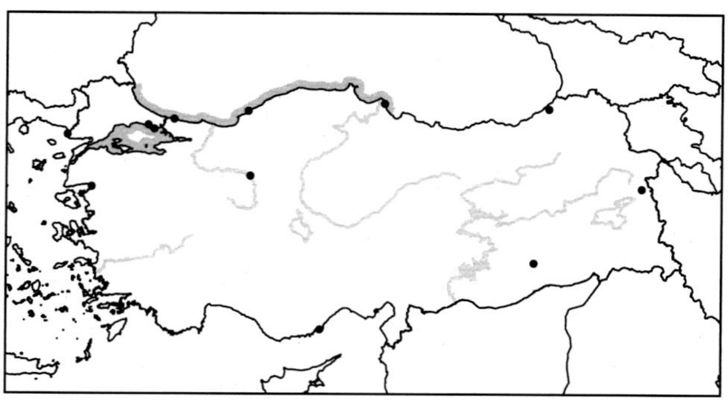

BLACK-NECKED (EARED) GREBE *Podiceps nigricollis*
Kara Boyunlu Batağan

(Du: Geoorde Fuut; Fr: Grèbe à cou noir; Ger: Schwarzhalstaucher; Swe: Svarthalsad dopping)

Subspecies. Nominate.

Status and distribution. National conservation status: Near Threatened (population apparently decreasing). Resident breeder or, more usually, partial migrant. Colonial nester, in groups of up to 200 pairs, usually on small, eutrophic, marshy wetlands to 2,500m. Local, but not uncommon in such wetlands in Inner and East Anatolia (e.g. 330 pairs at Kuyucuk Gölü, Kars province) between mid April and August, but a distinctly scarce breeding bird in adjacent areas of all other regions, except perhaps Southeast Anatolia, where very few records. Severe declines have been noted in some areas recently, e.g. at Kulu Gölü, Inner Anatolia (Richardson 2003). Failed breeders begin to congregate in July at favoured wetlands and are joined by post-breeding individuals, as well as by migrants from outside Turkey, over the following two months. Such concentrations can number several thousands, e.g. at Acıgöl, Mediterranean (max. 1,800), Kulu Gölü (max. 2,000), Erçek Gölü, East Anatolia (max. 4,430, although 10,000 were reported on 9 September 2000; *Birding World* 13: 406) and Sodalı Gölü, East Anatolia (max. 4,000–5,000).

Considerable numbers winter in Turkey, although in recent years the national average has been fewer than 2,000. Only rarely recorded on the sea, except in Black Sea Coastlands, where immigrants apparently arrive in mid October and depart in mid April (see Albrecht 1986). Formerly wintered in much larger numbers, particularly in the Lake District (Mediterranean/Inner Anatolia), with the largest counts at Burdur Gölü (regularly up to 5,000, exceptionally 18,662 in December 1970), in coastal areas of Aegean and, irregularly, Black Sea Coastlands and Marmara. May remain in wintering areas until at least early May. The few records available indicate that birds may remain at their breeding lakes and moulting grounds in East Anatolia until early December. Regularly recorded in small numbers at breeding and non-breeding localities in Inner Anatolia throughout the winter.

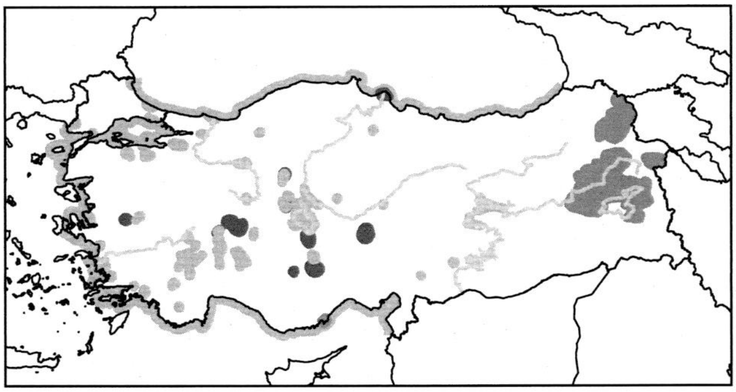

Breeding. On shallow, freshwater and brackish lakes with reeds and submerged aquatic vegetation. Usually in small colonies: e.g. two colonies totalled 20–25 pairs at Hotamış (Kirwan 1993a), occasionally solitary pairs, but 330 pairs Kuyucuk Gölü (East Anatolia). Nest usually floating, anchored to aquatic vegetation, in reeds and other vegetation; a low mound of aquatic and waterside vegetation, with a shallow hollow in the top. At Kulu Gölü (Inner Anatolia) a colony was at the margin of an islet of floating *Ruppia* sp., *Scirpus* sp. and *Phragmites communis*: on 13–15 July 1971 there were 120 nests and most young had already left, but 42 nests had eggs: C/1 (16), C/2 (11), C/3 (8), C/4 (5), C/5 (1), C/6 (1); on 22 July 1971 13 nests had C/1 (2), C/2 (4), C/3 (5), C/4 (2), with many young on the water, but some adults nest building (Kasparek 1987a). In **Inner Anatolia**, a summer resident at Mogan Gölü from early April, and paired and displaying by mid April (Wadley 1951); at Çöl and Uyuz lakes, five pairs were displaying on 1 June 1991, and a pair was with two chicks 5/6 July 1991 (S. Karauz); there are three other records of young, earliest 31 July. In **Mediterranean**, in Çukurova display was observed 14/15 April 1987 (van der Have *et al.* 1988); two pairs with young were at Karamık Gölü 18/19 July 1972, and an adult was with three young at Seyhan Reservoir on 29 July 1972. In **East Anatolia**, there were 196 nests at Kuyucuk Gölü on 18 July 1992, and there are three records of young at other sites (earliest 19 July). At Kızılırmak Delta (**Black Sea Coastlands**), in 1992, display was observed in May, and distraction display on 11 May, when a bird was flushed from a probable nest (Hustings & van Dijk 1994). A few pairs with young, 1–2 weeks old, were at Uluabat Gölü (**Marmara**) on 20 June 1999, indicating egg laying around 21 May (PC).

CORY'S (SCOPOLI'S) SHEARWATER *Calonectris diomedea*
Boz Yelkovan

(Du: Scopoli's Pijlstormvogel; Fr: Puffin cendré; Ger: Gelbschnabel-Sturmtaucher; Swe: Gulnäbbad lira)

Subspecies and taxonomy. Nominate (Scopoli's Shearwater). Recent work on Cory's Shearwater has tended to favour the recognition of two, if not three species. *C.* (*d.*) *edwardsii*, which breeds in the Cape Verdes, is now widely accorded specific status (see Hazevoet 1995, Snow & Perrins 1998, Gill & Wright 2006), whilst nominate *diomedea* (which breeds on Mediterranean islands) and *borealis* (subtropical northeast Atlantic islands) are also increasingly regarded as species (e.g. Sangster *et al.* 1998) on the basis of their differing mtDNA profiles (Heidrich *et al.* 1996), vocalisations (Bretagnolle & Lequette 1990) and a phenetic analysis of their morphologies (Granadeiro 1993). However, other authors prefer to maintain *diomedea* and *borealis* as one species and consider the specific separation of *edwardsii* as premature (Thibault *et al.* 1997). At-sea identification of *diomedea* and *borealis* was ably covered by Gutiérrez (1998) and Howell & Patteson (2008).

Status and distribution. National conservation status: Data Deficient (no information on trends). Uncertain: possibly nests along the Mediterranean and Aegean coasts, but no confirmed breeding records. Summer visitor to west and south coasts from early March (10th) to at least mid October, which (as various authors since Krüper have speculated) may breed in Turkey. As yet, however, there are no strong indications of this. Fairly widespread and not uncommon off Aegean and Mediterranean coasts, with concentrations of up to 50 being noted south of Izmir in late September, and from the Bodrum peninsula, Muğla province (Aegean) in June and July, and 65 moving west in one hour off Mavi Ege, between Assos (Marmara) and the Greek island of Lesbos, in late August. Perhaps less frequent in Sea of Marmara (except at Dardanelles where regular; max. 100 on 10 March 2001, the highest-ever count in the country) and Bosphorus, and has been recorded at least twice in autumn in Black Sea Coastlands, as far east as Rize with Yelkouan Shearwaters *Puffinus yelkouan*. No recent winter records, though it is mentioned in the literature as being rarely seen at this season. The species is probably significantly under-recorded.

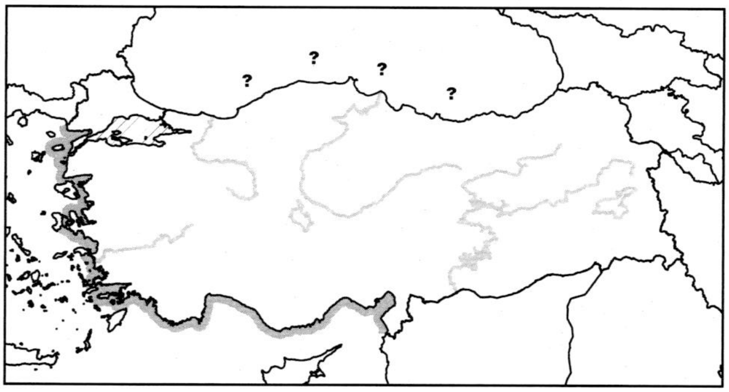

[SOOTY SHEARWATER *Puffinus griseus*]
Kül Rengi Yelkovan

(Du: Grauwe Pijlstormvogel; Fr: Puffin fuligineux; Ger: Dunkler Sturmtaucher; Swe: Grålira)

Subspecies. Monotypic.

Species not accepted onto the Turkish list. An unpublished claim of one in Iskenderun Bay (Mediterranean) among small numbers of Yelkouan Shearwater *P. yelkouan* on 8 January 1986 lacks sufficient descriptive details (especially of the underwing pattern) to substantiate the claim adequately (Kirwan *et al.* 1999).

MEDITERRANEAN (YELKOUAN) SHEARWATER *Puffinus yelkouan*
Yelkovan

(Du: Yelkouanpijlstormvogel; Fr: Puffin yelkouan; Ger: Mittelmeer-Sturmtaucher; Swe: Medelhavslira)

Subspecies and taxonomy. Monotypic, but until recently *P. yelkouan* was generally regarded as conspecific (under the name Mediterranean Shearwater) with the allopatric *P. mauretanicus* (Balearic Shearwater) which breeds in the western Mediterranean. The two are sister taxa, with relatively low DNA-sequence divergence (Austin 1996, Heidrich *et al.* 1998) and the majority of individuals are not diagnosable in the field (Yésou *et al.* 1990, Yésou & Paterson 1999), nor do they respond differentially to one another's voices (Bretagnolle & Zotier 1998). However, as discussed by Sangster *et al.* (2002), there are marked osteological differences between the two taxa (Walker *et al.* 1990), their DNA are almost as different as those of *yelkouan* and *P. puffinus* (Manx Shearwater) and they are clearly following separate evolutionary trajectories. As all three forms, *yelkouan*, *mauretanicus* and *puffinus*, are reciprocally monophyletic to treat *puffinus* as one species and *yelkouan* and *mauretanicus* as another makes little taxonomic sense. We follow the recent decision of several of the principal West European rarities committees to treat all three at the specific level (e.g. Sangster *et al.* 1997, Collinson 2006).

Status and distribution. National conservation status: Data Deficient (no information on trends). Presumed breeding resident, though no colonies have been discovered. Common in all months in Aegean, Marmara and Black Seas, although slightly less regular in the latter between November and February, and numbers apparently lower in Aegean during winter (Eken 1997d). Sometimes recorded in very large numbers (up to 10,000) especially off the Black Sea coast of Thrace, and in the Bosphorus (up to 6,800 per hour moving in both directions on 18–22 April 1966 and in late June 1973) and Dardanelles (up to 8,200 per hour), but also east as far as Samsun (Black Sea Coastlands). Reported irregularly from the Mediterranean, east as far as Çukurova, in March–October, perhaps more occasionally in the winter months.

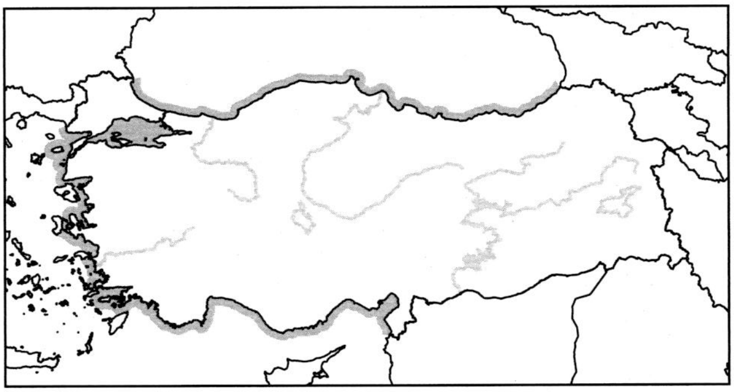

Breeding. Assumed on offshore islands or mainland cliffs in **Aegean** and **Mediterranean**, but no confirmed records. Elsewhere in range, breeds in colonies, in burrows, usually *c.*1m long, or in natural crevices within fallen rocks; a slight but variable amount of plant material is used to form a nest pad, and a single egg is laid in April–May.

[LITTLE SHEARWATER *Puffinus assimilis*]
Küçük Yelkovan

(Du: Kleine Pijlstormvogel; Fr: Puffin semblable; Ger: Kleiner Sturmtaucher; Swe: Dvärglira)

Subspecies and taxonomy. The subspecies most likely to occur in Turkey is presumably the north-temperate Madeiran *baroli*, which has been proven to wander as far eastern North America and Western Europe, and is presumably the form that has been recorded in the western Mediterranean east as far as Italy (Bonaccorsi 2003). *P. assimilis* represents a significant problem to taxonomists, it traditionally being considered with *P. lherminieri* (Audubon's Shearwater) to form a large group of 18 allopatric taxa, of which at least three could be regarded as species. No comprehensive review of the issues presented by this grouping appears to be available, but Hazevoet (1995) took the unilateral step of recognising Cape Verdean *boydi* (long considered a 'puzzling race') at species level (under the Phylogenetic Species Concept), and it is conceivable that this form does not even belong within the *assimilis* superspecies (Shirihai 2002). The latter author also provides a résumé of potential future developments in the taxonomy of the seven Southern Ocean forms within the *assimilis* complex. The British Ornithologists' Union Records Committee has recently taken the decision to split *baroli* and *boydi* from other forms of Little Shearwater, under the name North Atlantic Little Shearwater (see Collinson 2006). Though breeding almost as close to Turkey as *baroli*, the form *boydi* appears an unlikely candidate for vagrancy, although it may disperse reasonably widely in the tropical Atlantic and one resembling this taxon was caught on St Helena (Hazevoet 1995).

Species not admitted to the Turkish list. An unpublished claim of one with 20 Yelkouan Shearwaters *P. yelkouan* from the Eceabat (Ecebat) to Çanakkale ferry in the Dardanelles on 31 May 1988 was not accompanied by a description (Kirwan *et al.* 1999). Subsequently, *c.*10 were observed in the Sea of Marmara on 7 September 2003 (M. Warren). Though the observer noted a number of features generally associated with this species, the description appears insufficient to establish a first country record.

EUROPEAN STORM-PETREL *Hydrobates pelagicus*
Fırtınakırlangıcı

(Du: Stormvogeltje / Mediterraan Stormvogeltje; Fr: Pétrel tempête; Ger: Sturmschwalbe; Swe: Stormsvala)

Subspecies. Usually considered monotypic (e.g. Cramp & Simmons 1977), but Lalanne *et al.* (2001) recently confirmed the distinctiveness of the Mediterranean populations, for which the name *melitensis* is available, on the basis of larger bill and wing measurements and weight, and the Dutch Birding Association has recently elected to treat *melitensis* specifically (*Dutch Birding* 30:21). However, the same study revealed a degree of variation in these characters between insular Mediterranean populations, which was in some cases more marked than between Mediterranean versus Atlantic colonies, demonstrating the relative genetic isolation of each colony studied by Lalanne *et al.* (*op. cit.*) but not, in our view, proving the need to recognise *melitensis* or indeed any variation, thus far studied, within this species at nomenclatural level.

Status and distribution. Three on a seawatch 10km west of Kaş, Antalya province (Mediterranean) on 29 April 1988 (Haaß 1990) were published as the first record for Turkey, but were pre-dated by six off Karaada (Aegean) on 15 March 1972 and one off Kaş (Aegean) on 3 March 1972 (T. Gürpinar), and followed by seven off the former locality on 17 March 1992 (Eken 1997c). There are two confirmed records from Cyprus, in March and July (Flint & Stewart 1992), three from Lebanon, all recent (Ramadan-Jaradi *et al.* 2008) and several in summer from the Greek Islands, where it bred at least twice in the 1980s (Handrinos & Akriotis 1997), but overall the species is genuinely scarce in the eastern Mediterranean. Hollom *et al.* (1988) also mention records from the Black Sea, implying that some occasionally move through the Bosphorus.

[LEACH'S STORM-PETREL *Oceanodroma leucorhoa*]
Çatal Kuyruklu Fırtınakırlangıcı

(Du: Vaal Stormvogeltje; Fr: Océanite culblanc; Ger: Wellenläufer; Swe: Klykstjärtad stormsvala)

Subspecies. Monotypic.

Species not admitted to the Turkish list. Listed by Ergene (1945) without supporting details. Nonetheless, this species is a potential visitor to Turkish waters, having been recorded in Mediterranean Israel (Shirihai 1996, including significant numbers in early 1998 (Kirwan 1998b), Egypt (Goodman & Meininger 1989) and Lebanon (Ramadan-Jaradi & Ramadan-Jaradi 1999), as well as in the Gulf of Aqaba and off the United Arab Emirates (Porter *et al.* 1996).

NORTHERN GANNET *Morus bassanus*
Sümsük

(Du: Jan-van-gent; Fr: Fou de Bassan; Ger: Baßtölpel; Swe; Havssula)

Subspecies and taxonomy. Monotypic, but has frequently been placed in the genus *Sula*, whereby the specific name becomes *bassana*. Recent osteological studies of boobies and gannets have suggested that the differences between these groups warrants their generic-level separation (van Tets *et al.* 1988), as noted by Sangster *et al.* (1997), but others have preferred to view all of the Sulidae as more closely related (Friesen *et al.* 2002).

Status and distribution. Probably under-recorded. Sixteen winter (late December–February) and three autumn (August, October and November) records, with most other reports (at least 34) in March and mainly April. Almost all observations are from two well-watched localities on the Mediterranean coast, usually single immatures, but occasionally involve small numbers (max. eight, in April and 11 in late December). Recently recorded in late December in extreme eastern Black Sea Coastlands (Kirwan *et al.* 2003). The first record for Turkey was a first-year dead in the Gulf of Iskenderun, Antakya province, on about 5 March 1965, which had been ringed as a pullus on the Bass Rock, Scotland, the previous July (*OST Bull.* 7:1).

GREAT CORMORANT *Phalacrocorax carbo*
Karabatak

(Du: Aalscholver; Fr: Grand Cormoran; Ger: Kormoran; Swe: Storskarv)

Subspecies. *P. c. sinensis.*

Status and distribution. National conservation status: Least Concern (population apparently increasing). Local breeder, and in some areas almost certainly declining, in Marmara, Black Sea Coastlands and Inner Anatolia. Formerly bred in East Anatolia (at Van Gölü and Hazar Gölü) but no recent records. There is evidence to suggest a significant decline in the breeding population in Inner Anatolia: colonies at Akşehir Gölü, Beyşehir Gölü and Ereğli marshes have either become extinct or decreased to one or two pairs, and other small colonies have been extirpated due to wetland drainage. Following an increase, in part due to official protection, at Manyas Gölü there are now perhaps as many as 2,650 pairs there (Magnin *et al.* in Heath & Evans 2000; maximum in the 1960s was 544 pairs). The current status of the breeding population at Demirköprü Reservoir (Aegean), where there were *c.*300 pairs in 1966, is unknown. There are still a number of (and perhaps many) small colonies along the Black Sea coast. In Southeast Anatolia, known only from the Euphrates south of Birecik.

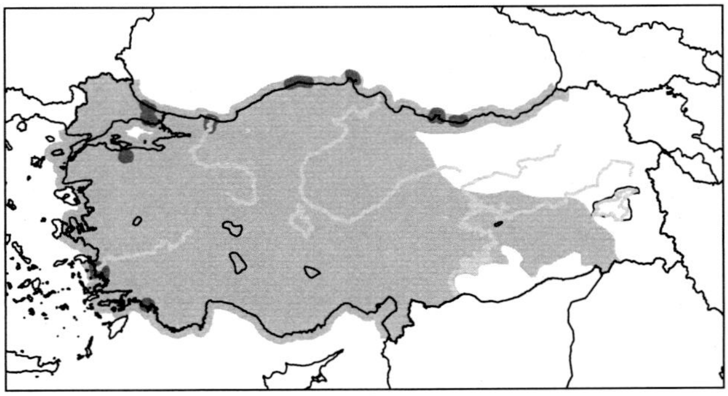

Much more widespread in autumn and winter, sometimes in considerable numbers, especially in coastal areas and inland waters of Marmara (up to 10,200 at Küçükçekmece, just west of Istanbul, in 1993, 9,531 at the Bosphorus on 19 January 2002 and 9,300 at Meriç Delta, in 1999), Aegean (where 7,651 at Marmara

Gölü at 26 January 2005, 3,112 at Köyceğiz Gölü on 2 February 2002 and up to 3,005 at Büyük Menderes Delta in November 1988) and, to a lesser extent, Mediterranean, Inner Anatolia (where up to 1,000 recorded at a single site and, exceptionally 3,645 at Sarıyar Reservoir, in February 2005) and Eastern Anatolia (up to 283). Non-breeders regularly oversummer in both these regions.

Breeding. Colonies range from 20 individuals, including immatures, at Trabzon (Black Sea Coastlands) to more than 2,000 pairs at Manyas Gölü (Marmara). Sometimes single-species only, or in mixed colonies with other species, such as Grey Heron *Ardea cinerea* and Little Egret *Egretta garzetta*, e.g. Bafa Gölü (Aegean). A solitary pair nested on the ground with Dalmatian Pelicans *Pelecanus crispus* and Grey Herons on an island in Karine Gölü (Aegean) on 26 May 2004 (M. Demain). Larger colonies in trees, sometimes partially flooded, close to, or on, islands in inland lakes; also nests on coastal cliffs and offshore islands, and in reeds. Nests in trees are large structures of sticks, lined with grass, leaves and aquatic vegetation. Coastal and reed nests apparently undescribed in Turkey but, elsewhere, coastal nests are usually mainly seaweed and sticks, and those in reeds mainly of reeds. At Bafa Gölü (**Aegean**), on 1 May 2003 the colony comprised *c*.200 pairs, breeding in trees on an island in the lake; nest height was 1–5m above ground, and nests had 1–5 eggs (mainly 3) and 3–5 young (mainly 3); most had recently hatched (oldest *c*.1 week), indicating earliest egg laying late March (PC). At Manyas Gölü (**Marmara**), some nests had young on 6 April 1967 (R. F. Porter), and almost-fledged young on 26 April 1970, indicating egg laying late February; breeding season protracted, and this colony recorded as active in mid June, with first young in nests not seen until 5 June in 1966 (R. F. Porter). In a mixed colony at Uluabat Gölü, nests of Great Cormorant contained large young on 3 June 2006, and were higher in the trees than six smaller species also present (PC). At Ereğli marshes (**Inner Anatolia**), there were 13 nests with fledged young nearby on 8 June 1971. In **Black Sea Coastlands**, a colony of 80–100 birds was at Perşembe 2/14 May 1970, with 79 (including young) there 10 June 1975, and young also observed at other coastal colonies in June–July.

EUROPEAN SHAG — *Phalacrocorax aristotelis*
Tepeli Karabatak

(Du: Mediterrane Kuifaalscholver; Fr: Cormoran huppé; Ger: Krähenscharbe; Swe: Mellanskarv)

Subspecies and taxonomy. *P. a. desmarestii*. The taxonomic status of this rather well-marked form demands further investigation. Has been placed in the genus *Stictocarbo*, for which monophyly has been demonstrated (Siegel-Causey 1988) and refuted (Kennedy *et al.* 2000, or, more recently, *Leucocarbo* (Gill & Wright 2006).

Status and distribution. National conservation status: Least Concern (population apparently stable). Resident breeder, often in mixed colonies with Great Cormorant *P. carbo* and Yellow-legged Gull *Larus cachinnans* in Black Sea Coastlands, Aegean and Mediterranean, favouring rocky islets, sea-caves and cliffs. Perhaps more common in the eastern half of Black Sea Coastlands than currently known. Less common and rather local resident in Marmara and Aegean, often without breeding. The most important breeding sites of the 12 confirmed localities (three additional sites are unconfirmed nesting areas) include the Şile Islands, in the Black Sea just east of Istanbul (175 pairs); the Ayvalik Islands, Aegean (100–150 pairs); the Foça Islands, off Izmir Bay, Aegean (59 pairs); the Ildır Islands, Aegean (84 pairs); and Akkuş Island, Ordu province, Black Sea Coastlands (90 pairs), leading Eken (1997c) to place the Turkish breeding population at 600–2,500 pairs, considerably higher than the 50–350 pairs suggested by Kasparek & Bilgin (1996). Slightly more widespread outside the breeding season in coastal waters, with, for example, an unusual concentration of 97 at Aliağa Kuşcenneti (Aegean) on 29 December 2001, up to 300 at Kadiköy,

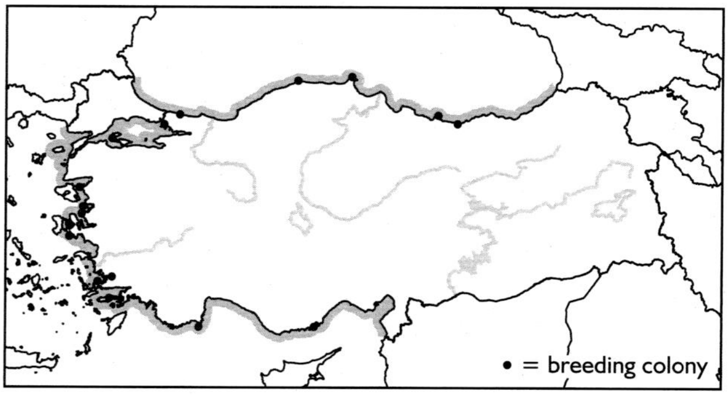

Istanbul, in winter 1979/80 and 600+ off Riva, Istanbul, in autumn 2003, but dispersal is usually limited, and thus far there are no certain inland records (those made by Vielliard in northeast Turkey, in 1967, appear suspect, due to the numbers involved and because they have never been repeated).

Breeding. On small, barren or sparsely vegetated coastal islands and stacks, usually close to shore, and mainland cliffs. Mainly colonial: 90 pairs breed on Akkuş, less than 100m from land, west of Ordu (Black Sea Coastlands), nesting mainly in crevices; and 175 pairs, mainly on two of four small islands, only 50m offshore, at Şile, near Istanbul (Marmara), nesting in crevices or on barren ground (Magnin & Yarar 1997). Small colonies also noted: e.g. *c.*20 pairs on an offshore stack near Sinop (Black Sea Coastlands), 7 June 1996 (A. Whitehouse), and a few scattered pairs on small islands between Taşucu and Aydıncık (Mediterranean), 1992–93 (PC). Nest undescribed in Turkey; elsewhere an accumulation of seaweed and plant stems, with a hollow in centre, lined with finer material. Eggs: C/3 (1); usual clutch elsewhere 3–4 (2–5). Witt (1976) visited several **Mediterranean** islands in 1973–74: on one, a nest had two nestlings on 5 May 1973, and on another, three nests had young on 20 May 1973; three nests had a total of six young, and another three eggs, on 1 June 1974; a young bird, nine days old on 13 April 1974, indicates egg laying early March. Recently fledged young were at the base of cliffs on an island off Aydıncık (Mediterranean) on 13 June 1992, indicating egg laying mid March (PC). Two adults were incubating eggs, or brooding small young, and another nest had 2–3 small young, on mainland cliffs, near Perşembe (**Black Sea Coastlands**), on 10 June 1975.

PYGMY CORMORANT — *Phalacrocorax pygmeus*
Küçük Karabatak

(Du: Dwergaalscholver; Fr: Cormoran pygmée; Ger: Zwergscharbe; Swe: Dvärgskarv)

Subspecies and taxonomy. Monotypic. Has been placed in the genus *Microcarbo*, for which monophyly has been suggested (Siegel-Causey 1988), but Kennedy *et al.* (2000) recommended that all shags and cormorants be retained within a non-monophyletic *Phalacrocorax* until such time as robust divisions between different genera in this grouping become known. However, all latter-day commentators are agreed that neither shags nor marine cormorants represent monophyletic lineages.

Status and distribution. Conservation status: Near Threatened (BirdLife International 2000, 2004); national conservation status: Least Concern (population apparently decreasing). Very local breeding resident, generally in small to moderate numbers, in marshy wetlands in Marmara, Aegean, Mediterranean (at Köyceğiz Gölü and formerly at Amik Gölü prior to drainage) and Inner Anatolia, occasionally in large colonies within the latter region. The most important nesting areas were: Akşehir/Eber lakes (*c.*150 pairs), Ereğli marshes (*c.*600 pairs) and Sultan marshes (200–250 pairs) but all have almost certainly substantially declined in recent years, e.g. the Ereğli population was most recently estimated at just 20 pairs in the 1990s (Eken & Magnin 1999); thus, the *c.*823 nests counted at Uluabat Gölü (Marmara) in 1998 make this currently the most important breeding colony in Turkey. The once-important colony at Hotamış marshes (Inner Anatolia) is now extirpated due to habitat destruction. Recently discovered breeding at one locality in Southeast Anatolia, at Karkamış Reservoir, south of Birecik, where a max. 550 on 6 July 2001, with large numbers rather frequent and birds apparently commuting along the Euphrates between Birecik and Syria. May also breed in smaller numbers at localities in Thrace, Mediterranean and East Anatolia. Breeding in latter long suspected, particularly in the Bulanık Ovası (where finally proven in spring 2002; *Sandgrouse* 25: 6) and the Bendimahi Delta; more occasionally in other areas, but remains unproven in most instances. Has recently been recorded in significant numbers at Çıldır Gölü, Kars province (East Anatolia) during the breeding season, and huge numbers occur in the Aras Valley, near Iğdır, in the same region, principally on the Armenian side of the border. Several small colonies in the western two-thirds of the country have declined or are now extirpated. Regular presence in at least one wetland in Black Sea Coastlands has recently been confirmed but no evidence of breeding.

More widespread in autumn and winter when more common in Marmara (including Thrace), Aegean, Black Sea Coastlands, Mediterranean and Southeast Anatolia. Regular in small numbers, up to 250, in many wetlands in all but latter region, as well as in Inner Anatolia. Regular non-breeding localities include Büyük Menderes Delta/Bafa Gölü (Aegean; up to 490), Beyşehir Gölü (Mediterranean; up to 64), Gediz Delta (Aegean; up to

1,000), Göksu Delta (Mediterranean; up to 180), Işıklı Gölü and Marmara Gölü (Aegean; up to 130 and 300 respectively), Meriç Delta, including Gala Gölü (Marmara; up to 20,000) and Uluabat Gölü (up to 2,023). Significant numbers remain throughout the winter in Inner Anatolia, while numbers in other nesting areas are augmented by limited dispersal of and immigration by birds from outside Turkey. Regular in small numbers at several sites in Thrace at this season, including Büyükçekmece (Istanbul), where max. 209 in December 2003 (E. Çağlayan).

Breeding. Forms colonies at inland lakes with extensive reedbeds; largely or wholly single-species, but usually mixed with herons, egrets and spoonbills. Some small, e.g. 15+ pairs at Yarma (Inner Anatolia) in 1971, more usually much larger, e.g. Manyas Gölü (**Marmara**). Nests in trees or reeds: in 1998, the colony at Uluabat Gölü was in widely spaced clumps of willows within a vast reedbed (a mixed colony of six species). Most colonies are in dense reedbeds, the nests up to 3m above water level. Those in a colony of *c.*200 pairs at Uluabat, on 26 April 2003, were in flooded willows on a large reed island, 1.5–5.0m above water level, and were shallow saucer-shaped structures of willow twigs, with most having only one or two eggs, and breeding just starting; *c.*100 nests were visible in this colony on 3 June 2006 and all contained almost full-grown young (youngest *c.*2 weeks) (PC). On 7 June 1998, nests in part of colony at Uluabat had almost-fledged young, whereas nests in another part had small young, and many still had eggs (Welch & Welch 1998b). Adults were incubating at Manyas on 3 April 1967, and most nests (of 20 examined) had young on 15 May 1967 (R. F. Porter). In **Inner Anatolia**, 20 nests had eggs in a mixed colony at Ereğli marshes on 25 May 1998 (T. Quittenden); of 500–600 pairs, in a remote reedbed at Ereğli, all examined nests had eggs on 16 May 1987, but some had young on 18 May 1987 (M. Kasparek); nests examined on 28 April 1982 at Sultan marshes (where eggs seen 13 April 1982) had four eggs (3), five eggs (9), or six eggs (10) (Kasparek 1985); fledglings observed at Yarma as early as 9 June 1971. In **Aegean**, many nests had eggs, but most other species had not yet laid, in a mixed colony at Marmara Gölü, 30 April 1951 (McNeile); a nest at Bafa Gölü had three young on 13 May 1980 (Kasparek 1988a). In **Mediterranean**, Meinertzhagen (1935) found two colonies at Amik Gölü (now drained) on 26 May 1933; the nests were attached to reeds, constructed of dead reeds, without lining, and contained mainly 4 (–6) eggs, with young up to one week old in many.

AFRICAN DARTER *Anhinga rufa*
Yılanboyun

(Du: Slangehalsvogel; Fr: Anhinga roux; Ger: Schlangenhalsvogel; Swe: Afrikansk ormhalsfågel)

Subspecies and taxonomy. *A. r. chantrei* (Oustalet 1882), which reportedly has greyer upperwing-coverts and a paler foreneck, is no longer regarded as valid (see Cramp & Simmons 1977), thus *A. rufa* is considered monotypic here. Vaurie (1965) upheld *chantrei* as valid, but having examined specimens from Iraq and Meinertzhagen skins from Amik Gölü, held at the Natural History Museum (Tring), we can see no rationale for such a view. We follow most latter-day authorities (e.g. Sibley & Monroe 1990, Rasmussen & Anderton 2005) in considering *A. melanogaster sensu stricto* to represent three species, one in Africa (*A. rufa*), one in Asia (*A. melanogaster*) and the other in Australasia (*A. novaehollandiae*), or two species, one in Africa and the other in Asia and Australasia (Gill & Wright 2006). Oddly, Sibley & Monroe (*op. cit.*) considered Turkish and Iraqi populations to represent *melanogaster*. Presumably this was merely an unintentional error: no other commentator has considered the Middle Eastern populations to be representative of *melanogaster*, and that they are, indeed, *rufa* can be confirmed by examination of skins (GMK pers. obs.).

Status and distribution. Extinct. First described as a Turkish bird almost coincidentally by Tristram (1882) and Chantre (1883), from observations and specimens taken at Amik Gölü, Antakya province (Mediterranean), formerly the only known breeding area in the Western Palearctic, away from the marshlands of southeast Iraq. Further observations were made at Amik Gölü in 1910, when Aharoni collected a further 21 individuals for two museums in Western Europe, but the colony was not counted again until 1933 (although several collectors visited the area, some providing quite detailed observations) when Meinertzhagen discovered a total of 55 pairs nesting in a tightly packed, mixed colony with Pygmy Cormorant *Phalacrocorax pygmeus*. Nesting apparently commenced in late March (27th) and was completed by the end of June (Aharoni 1930), a small percentage of the population being apparently resident, with up to 50 in February 1948 and 1950, although Tristram, on the evidence of local people, had reported that eggs did not hatch until early June and that the colony disbanded as soon as the young were capable of flight, not reappearing until late April. The majority, however, wintered in Israel, from the Hula marshes south through the Jordan Valley to the mouth of the River Yarmuk, from mid September to March, more occasionally departing for the breeding grounds in April. None was observed there after 1957 (Shirihai 1996), until 2004 when one was reported at Lake Kinnaret on 31 May (Ottens 2006). During the 1950s a decline in the Turkish population became evident, with a coincident decrease in the numbers wintering in Israel being noted (Kumerloeve 1966–67). By 1956 mechanised draining of the lake had already begun. In the 1960s only 40–50 km^2 remained, and by 1975 drainage was complete. This situation served to compound a decline, in part instigated by over-intensive collecting for collections in Europe, until its eventual extinction in Turkey in the early 1960s. The final, documented observation was of a single bird on 9 May 1962 (Kumerloeve 1963a).

Breeding. No longer does so (see above). Nests constructed of reeds, mainly at water level, with some up to 1.5m up in the reeds; on 26 May 1933 many nests contained eggs (C/3–5), but most had young at all stages, from newly hatched to those as large as Western Jackdaws *Corvus monedula*, which were able to slip off the nests into the water (Meinertzhagen 1935). Back-dating from these larger young, the first eggs in earliest nests would have been laid in early April, with nest building in March.

GREAT WHITE PELICAN *Pelecanus onocrotalus*
Ak Pelikan

(Du: Roze Pelikaan; Fr: Pélican blanc; Ger: Rosapelikan; Swe: Pelikan)

Subspecies. Monotypic.

Status and distribution. National conservation status: Endangered (population apparently decreasing). Very localised, apparently highly irregular breeding bird, with no reports of confirmed nesting between 1986 and 1991. All recent breeding observations are from Inner and East Anatolia, although formerly bred at Amik Gölü (Mediterranean) until drainage; Ereğli marshes (1,500–2,000 pairs in 1968–71, with much smaller numbers the following year, 23 pairs in 1993 and perhaps ten pairs in 1998, but this site is almost certainly now defunct due to wetland drainage), Karapınar Ovasi (30–50 pairs in 1985), Seyfe Gölü (max. 80 pairs in 1992 and 65 occupied nests in 1993), Aktaş Gölü, Ardahan province, East Anatolia (*c.*50 pairs in 1995 and either this species or Dalmatian Pelican *P. crispus* was mentioned as breeding by the late Nihat Turan *in* Environmental Foundation of Turkey, 1993) and on Van Gölü (40 pairs in 1967 and probably even greater numbers the following year). Two pairs bred unsuccessfully at Tuz Gölü (Inner Anatolia) in 1998. Nesting was suspected in the early 1970s at Hotamış (where an egg was found in 1971), the Sultan (small colony in 1970) and Yarma marshes (Inner Anatolia), and at the Akşehir/Eber wetland complex in the late 1980s. Groups of up to several hundred were regular at Ereğli marshes during the summer months but breeding was not suspected between 1973 and 1992. Smaller flocks have often been seen in northeastern Turkey, especially in the vicinity of Çıldır Gölü, Kars province (East Anatolia) during this season in recent years, presumably from the newly discovered colony at Aktaş Gölü (Yarar 1995). Some may also be wanderers from the Iranian breeding population. Occasionally recorded in summer in Black Sea Coastlands but no evidence of breeding.

More widespread on passage at extensive wetlands across the whole country at both seasons, typically in flocks of fewer than 300 birds, especially in Inner Anatolia. Regularly recorded in larger numbers (1,500 or more, max. 10,000 at Manyas Gölü, Marmara, in late September) at certain, favoured stopover sites, e.g. Akşehir Gölü, Bendimahi Delta and elsewhere around Van Gölü (East Anatolia), Çukurova and Göksu deltas (Mediterranean) and the Marmara lakes. Spring migration occurs from late March (20th) to late May (31st), or early June (11th) in East Anatolia, although most passage is in April, only stragglers occurring after the second week of May. A south-coast record from Adana, on 8 March 2001, perhaps refers to one that had overwintered. Autumn passage commences as early as mid July (12th), continuing until late November (25th), although numbers peak in the Belen/Çukurova/Gulf of Iskenderun area (the best watchpoint) from late September to mid October. Sutherland & Brooks (1981a) recorded 8,000 over Belen between 2 August and 23 September 1976, the only year with relatively complete coverage at this site. Scarce in winter, being only occasionally recorded in the western two-thirds of Turkey, normally at coastal wetlands, with maxima of 14 at the Göksu Delta in January 1974 and December 1975, and 16 at Gediz Delta in January 1995. Only six inland records at this season.

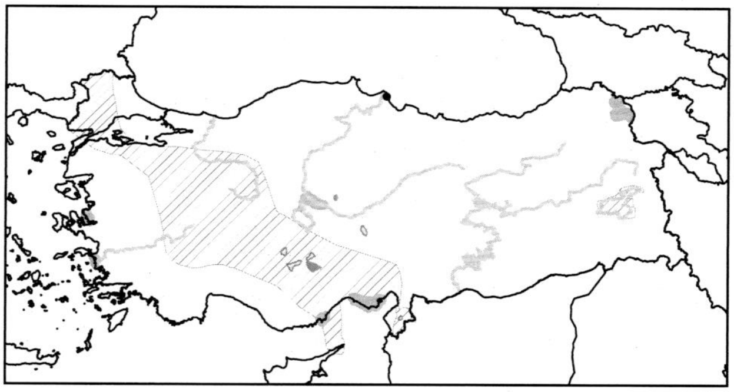

Breeding. Three sites: in **Inner Anatolia**, at Seyfe Gölü, a large shallow brackish lake, and at Ereğli marshes, an extensive complex of shallow marshes, reedbeds and freshwater lakes and, in **Black Sea Coastlands**, at Aktaş Gölü, a shallow lake at 1,800m. At Seyfe in 1993 *c.*70–80 pairs bred in a dense colony on a low flat mud island, and most nests were atop old nests of Greater Flamingo *Phoenicopterus ruber*; they were mainly of thin reed stems, grasses and aquatic vegetation; on 13 May 1993, of *c.*50 nests, many had two eggs, some only one egg, and one had three eggs; on 28 May 1993, there were 65 nests, still mainly with eggs, but some had an egg and a recently hatched chick, and others chicks up to 14 days old; C/1 (26); C/2 (21), C/1 + B/1 (9), B/1 (4), B/2 (5), the oldest young indicating earliest egg laying *c.*11 April; on 14 June 1993, two nests still had two eggs (possibly replacements), but most young were 15–31 days old (PC). There were 55 young in this colony on 18 June 1992 (Kirwan 1992c) and 155 young on 26 June 1992 (Magnin & Yarar 1994). At Ereğli, 2,000 pairs were present in May 1970 and on 8 June 1971 a colony of 420 pairs was found on an island in the vast reedbed; most nests had two eggs and a few had young, up to 18 days old, indicating earliest egg laying 18 April; on 16 May 1987, two nests had two eggs, and the third a newly hatched chick (M. Kasparek), indicating egg laying 13 April; on 16/17 June 1993, there were 250 adults, but only 23 nests, *c.*40 cm high, made of stems of *Phragmites*, and each with two eggs, on an island (Magnin & Yarar 1994); this record is much later than other breeding dates. In **East Anatolia**, there were 85 chicks, ranging from small to already flying, on three islands in Lake Van, in mid-August 1967 (Vielliard 1968), but no recent breeding records there. Breeding was proven in 1995 at Aktaş Gölü, in northeast Black Sea Coastlands, on the Turkish–Georgian border, where there were 50 pairs (Magnin & Yarar 1997).

DALMATIAN PELICAN *Pelecanus crispus*
Tepeli Pelikan

(Du: Kroeskoppelikaan; Fr: Pélican frisé; Ger: Krauskopfpelikan; Swe: Krushuvad pelikan)

Subspecies. Monotypic.

Status and distribution. Conservation status: Vulnerable (BirdLife International 2008); national conservation status: Vulnerable (population apparently increasing). A very local and scarce resident breeder, and (in some areas) partial migrant. Formerly bred in moderate numbers but subsequently declined, both in Turkey and elsewhere throughout its range. Despite these losses, principally due to wetland drainage, the population has most recently shown a small increase. At present five active colonies are known and one further colony requires confirmation, whilst a few other areas may also support small breeding populations. Recognised colonies are; Büyük Menderes Delta, Aegean (usually 35–42 pairs), Gediz Delta, Aegean (10–35 pairs), Ereğli marshes, Inner Anatolia (1–5 pairs, formerly 50 pairs in 1968–71), Kızılırmak Delta, Black Sea Coastlands (up to six pairs in recent years, formerly 20–25 pairs with 60–70 pairs in 1970–73), Manyas Gölü, Marmara (30 pairs in recent years, formerly 50) and Aktaş Gölü, Ardahan province, East Anatolia (*c.*20 pairs nesting in mixed colony with White Pelican *P. onocrotalus* in July 1995; Yarar 1995). Also probably at Akşehir Gölü (Inner Anatolia) where in 1992 3–5 pairs were thought to be breeding and birds were again present in May 1995. Has also been suspected to breed at Marmara Gölü (Aegean) in recent years (Magnin & Yarar 1997). One pair possibly bred at Hotamış marshes (Inner Anatolia) in 1998. Other now defunct breeding sites were at Beyşehir Gölü (Mediterranean; 83 pairs in 1964); Yarma marshes (Inner Anatolia), where ten deserted nests were found in 1970 and adults with fledged young, probably from Ereğli, were present in 1971; and at Hazar Gölü (East Anatolia). In addition birds may have bred at Ağyatan Gölü, Çukurova in the mid 1950s and almost certainly did at Amik Gölü (Mediterranean), where 50–100 where seen in a colony of over 1,000 White Pelicans in mid-June 1966, and Meinertzhagen had considered it possibly breeding in 1933 and Aharoni certainly so in 1910. There were no confirmed records from the latter locality before its drainage. Tiny numbers of non-breeders are reasonably regular at Göksu Delta (Mediterranean).

Some, presumably largely non-breeders, are rarely recorded at other wetlands in Mediterranean, Inner and East Anatolia; records from some sites may indicate breeding, particularly in the latter region. Recorded on passage in East Anatolia, where regular at Çıldır Gölü, Kars province. In winter more widespread, in moderate numbers, throughout the western two-thirds of Turkey, with most of the population currently wintering in coastal wetlands (Crivelli *et al.* 1991). Sarıgül (2001) demonstrated that numbers wintering increased in 1970–2000 and recent winter maxima involved 1,162 individuals at just three localities in Aegean in early February 2002 (Balmer & Betton 2003a), but the 2005 survey, which did not cover Bafa Gölü / Büyük Menderes Delta, produced 1,097 birds, suggesting the presence of an even larger population (Çağlayan *et al.* 2005). Formerly more regular in Inner Anatolia at this season. In recent years up to 75% of the wintering population has been present in the Bafa Gölü/Büyük Menderes Delta area with over 430 in early 1996 (exceptionally *c.*1,000 in late January–early February 1972, when the Turkish breeding population was significantly larger). With the exception of Gediz Delta (regularly up to 341 in recent years, exceptionally 615 in December 2001), Marmara Gölü (Aegean; 490 on 26 January 2005) and Meriç Delta (Marmara; up to 610 in 1998), few other localities support more than 20–50 birds. Formerly up to 90 in winter in Kızılırmak Delta and up to 375 at Göksu Delta. The majority of wintering Dalmatian Pelicans are resident, augmented perhaps in varying numbers by immigrants from Bulgaria, Greece and Romania, principally at western localities. Hunting remains a major threat to this species in the wintering areas.

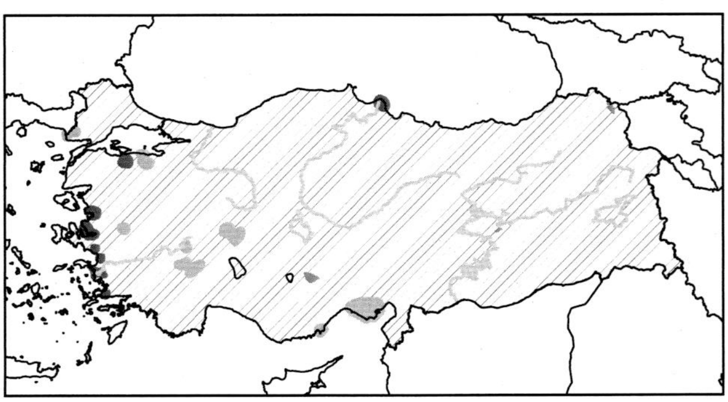

Breeding. Seven breeding sites (at two probably recently extirpated), involving colonies of 6–42 pairs: three colonies at coastal river deltas and four at inland lakes and marshes. Nests are on the ground on low bare islands (in coastal lagoons), on mud islands or of floating reeds, in dense reedbeds; in trees, and on artificial platforms erected especially for this species in flooded willows at Manyas Gölü (**Marmara**). At Karine Gölü, in Büyük Menderes Delta (Aegean), pelicans nest on low islands of mud and shells. Nests may be solitary, but most are in groups of 10–12, consolidated into one mass, with larger and higher nests (up to 0.8m) in the centre, and lower ones (typically 0.3m) at the edge; the nest is mainly of cut bamboo canes (used by local fisherman for their nets, and washed up on the islands), reeds and other plant material, all consolidated by excreta as the breeding cycle proceeds, with a shallow hollow on top, lined with grass, string and feathers (PC). Nests in reeds are large platforms of dead reeds, with a slight hollow in the middle (McNeile). Eggs and young: usually 2 (1–3). A colony at Kızılırmak Delta (**Black Sea Coastlands**), in 1992, was on a mud island in an open area of 40m^2 in an extensive reedbed; an adult was observed carrying nest material on 18 March 1992, six nests had a total of seven eggs on 5 April 1992, and there were six eggs and four chicks, 2–3 weeks old, on 10 May 1992; egg laying would have started 20–30 March (Hustings & van Dijk 1994). Other records are: 25 pairs in July 1966; 12 adults and 33 unfledged young at two locations on 6 August 1971 (Dijksen & Kasparek 1985); 34 young, *c.*4 weeks old, on 7 July 1972; 30–50 breeding pairs in 1984. At Manyas Gölü, there were 13 nests in two colonies in 1966, including five nests in one tree, and adults were incubating at least as early as 29 April; and 25 pairs were nesting on the new artificial platforms on 15 May 1969 and 30 pairs on 26 April 1970 (R. F. Porter); incubation commenced about 6 April 1970, and large young were seen on 29 June 1970; five pairs with young, 2–3 weeks old, were observed on 20 April 1996 (L. & L. Vallotton), indicating egg laying early March; there were 51 adults and 47 swimming young on 3 June 1991 (GMK); an adult and two chicks (age unspecified) were observed on a nest on 22 August 1980, possibly a replacement following earlier failure. At Gediz Delta (**Aegean**), there were 35 pairs in three small colonies in 1995, with colony formation in the second half of January and in the second week of April, with overall breeding success of 0.83 young per pair; 31 pairs bred in 1996, with the first sign of colony formation in late March (Eken 1997a); ten adults were sitting next to nests on distant islands on 30 April 1995 (T. Bakker). At Karine Gölü, on 23 April 2003, there were 60–80 pairs, at least 20 nests had two eggs, a few only one egg, and one had three eggs, and 30–40 young, ranging from newly hatched to *c.*5 weeks old; and some eggs were recently laid (white, rather than stained brown); on 3 May 2001, young ranged from newly hatched to *c.*5–6 weeks old, and six nests had eggs (PC); on 3 June 1991, there were 56 adults and 38 large young (GMK); these records indicate earliest egg laying late February, but some eggs were laid in late April. At Ereğli marshes (**Inner Anatolia**), an irregular breeder; there were 50 pairs, including 7–8 pairs with large young, on 8 June 1971; three pairs bred in 1987, two adults with nesting material were observed on 20 April 1988, and 19 adults and two young (in a nest on a sandy island) on 13 June 1991. Also breeds at Aktaş Gölü (**Black Sea Coastlands**), a shallow, high-altitude lake on the Georgian border, and a few pairs possibly breed on islands of reeds at Akşehir Gölü (Inner Anatolia). Large numbers formerly bred at Marmara Gölü (Aegean), in a dense reedbed, where the reeds were up to 6m high; on 13 May 1950, in a narrow channel, 'there was a string of nests for over 40 yards [37m] or more', some almost 1m high, and most contained young, ranging from newly hatched to half grown, and some still contained eggs, some on the point of hatching (McNeile). A few pairs bred near Yarma (Inner Anatolia) in 1971. Dalmatian Pelicans formerly bred at Beyşehir Gölü (**Mediterranean**), where there were 83 pairs in 1964.

GREAT BITTERN — *Botaurus stellaris*
Balaban

(Du: Roerdomp; Fr: Butor étoilé; Ger: Grosse Rohrdommel; Swe: Rördrom)

Subspecies. Nominate.

Status and distribution. National conservation status: Vulnerable (population apparently decreasing). Local but perhaps not uncommon resident in certain large wetlands, more usually in smaller numbers in suitable areas in Marmara, Mediterranean, Black Sea Coastlands, Inner and East Anatolia, perhaps also in other regions (Kasparek 1986a). Recently proven to breed in large numbers, possibly up to 240 pairs (Hustings & van Dijk

1994), in Kızılırmak Delta, and probably not uncommon in remaining areas of marshland in the Yeşilırmak Delta (Black Sea Coastlands). Elsewhere less common, though frequently recorded in wetlands debatably large enough to support the species, with probably the majority in Inner Anatolia and immediately adjacent areas, and also in East Anatolia, where recently recorded from several sites in the Van Gölü basin and to at least 2,000m. The Turkish breeding population is clearly more substantial than previously realised, both in a European and global context.

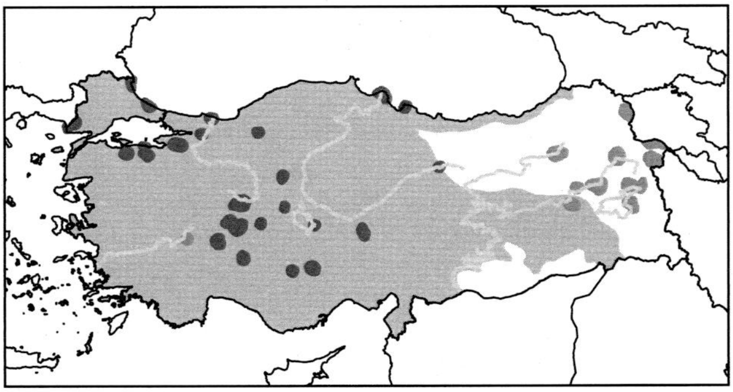

Also a scarce passage migrant and winter visitor, mostly in the western two-thirds of Turkey, the resident population being augmented by immigrants, even in Inner Anatolia where conditions can be especially harsh. Very few records at any season from Southeast Anatolia.

Breeding. Main breeding locality is at Kızılırmak Delta (**Black Sea Coastlands**) but males heard booming in late April–June (and July), the presumed breeding season, suggest small populations in **Aegean**, **Inner Anatolia**, **Mediterranean** and **East Anatolia**. An adult was observed carrying nesting material into a large reedbed at Kızılırmak Delta on 7 April 2005 (A. Gursoy) and, at Hafik Gölü (Inner Anatolia), a nest contained three ready-to-fly juveniles on 22 May 2005 (S.Bekir). No further details. Elsewhere, the nest is at water level, built up from the ground in shallow water, or attached to growing reeds in deeper water; a low accumulation of reeds and aquatic vegetation, lined with finer material; the usual clutch is 4–6 (3–7).

LITTLE BITTERN *Ixobrychus minutus*
Küçük Balaban

(Du: Woudaap; Fr: Blongios nain; Ger: Zwergdommel; Swe: Dvärgrördrom)

Subspecies. Nominate.

Status and distribution. National conservation status: Near Threatened (population apparently decreasing). Fairly widespread and reasonably common summer visitor and passage migrant. Breeds in all regions, though apparently scarcer in Black Sea Coastlands (15–30 breeding pairs estimated at Kızılırmak Delta in spring 1992; Hustings & van Dijk 1994) and perhaps Thrace. In Southeast Anatolia principally restricted to localities along the Euphrates and Tigris rivers. Arrives in small numbers from late March (24th) with the majority arriving or passing through in the first half of May and passage continuing until at least the end of that month (24th). Recorded to at least 2,100m on passage. Consistently arrives 10–14 days later in Inner and East Anatolia. Return movement begins in mid August, perhaps peaking in early to mid September (at least in the north of the country), continuing through October, exceptionally into the first week of November and once in late November (26th). Breeds in marshy wetlands, utilising even small areas of suitable habitat. Significant sites for the species include the Meriç Delta (Marmara: 200 pairs) and Uluabat Gölü (Marmara: *c.*150 pairs). The large population at Sultan marshes (Inner Anatolia) has almost certainly declined from the reported estimate of 300 pairs in the 1980s by Kasparek (1985).

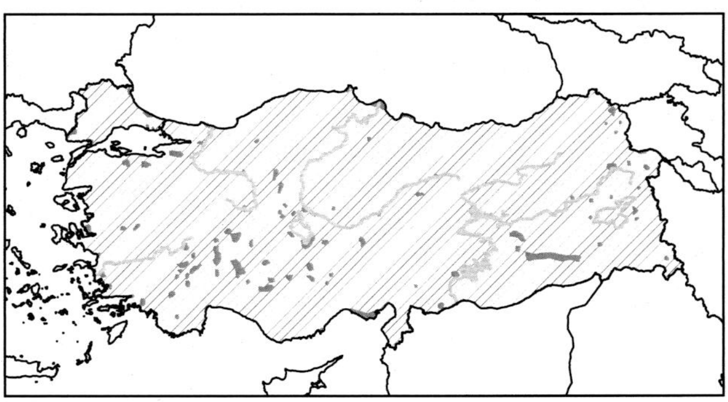

Breeding. In reedbeds (large and relatively small) and reedy fringes to lakes, canals and river meanders. Solitary breeder but, in optimum habitat, several pairs may nest fairly close together in loose association. Small numbers only at most breeding sites in Turkey, but at Uluabat Gölü (Marmara) numerous in dense *Phragmites*, with 122 territorial males in 1998 (Welch & Welch 1998b). Main arrival at breeding sites is in early May with some records in late April. Nest building observed in Antakya area (**Mediterranean**) on 8 May 1962 (Kumerloeve 1963a). A nest (no details) was found at Manyas Gölü (**Marmara**) in 1968. In **Inner Anatolia**, juveniles were observed at Mogan Gölü in late July 1968. At Kızılırmak Delta (**Black Sea Coastlands**), a nest, *c.*10–20cm above water level, containing three eggs, was found in an extensive reedbed, near a colony of Purple Herons *Ardea purpurea*, on 10 June 1992 (Hustings & van Dijk 1994). In other parts of the range, there are records of nests in low bushes and trees, as well as at the base of reeds, and usual clutch size is 5–6 (4–10).

BLACK-CROWNED NIGHT HERON *Nycticorax nycticorax*
Gece Balıkçılı

(Du: Kwak; Fr: Bihoreau gris; Ger: Nachtreiher; Swe: Natthäger)

Subspecies. Nominate.

Status and distribution. National conservation status: Least Concern (population apparently increasing). A rather local summer migrant breeder, generally in small to moderate numbers (but *c.*500 pairs at Manyas Gölü in 1967, which had declined to *c.*150 pairs by the mid 1990s), to marshy wetlands throughout most of the country, although most common in Marmara (200 pairs at the Meriç Delta and 250 pairs at Iznik Gölü), Aegean, Inner and East Anatolia. Generally scarcer in Mediterranean (the most important colony in this region is that of 150 pairs at the Göksu Delta) and Black Sea Coastlands, and largely restricted to the Euphrates and Tigris rivers in Southeast Anatolia. There is recent evidence to suggest a decline in Inner Anatolia, where several colonies are now apparently extirpated (Eken & Magnin 1999). Principally nests in trees, often in poplars (at Nallıhan, Inner Anatolia, in a mixed colony with Little Egret *Egretta garzetta* and Grey Heron *Ardea cinerea*), but small colonies also nest in reedbeds. Twice recorded nesting in the same Rook *Corvus frugilegus* colony in East Anatolia (OST 1972/1975).

Rather more common and widespread on passage: from early March (2nd) to at least mid May (18th) in spring, and in autumn from mid August to late October (29th), perhaps peaking in September but juveniles dispersing from mid July. Occasionally recorded in winter, usually at coastal or hinterland localities, with apparently exceptional concentrations (for this season) of 113 in the Iğdır Ovası (East Anatolia) on 3 January 1999 (DHKD 1999), 83 at Manyas Gölü (Marmara) on 3 February 2002 (Demirci 2002) and 41 at Yeşilırmak Delta (Black Sea Coastlands) on 20 January 1997 (Welch & Welch 1998a). A recently discovered colony at Karakaya Reservoir (Southeast Anatolia) is apparently resident and hosts the largest known winter population in the country, with at least 210 birds in January 2005 and January 2006 (M.Özen).

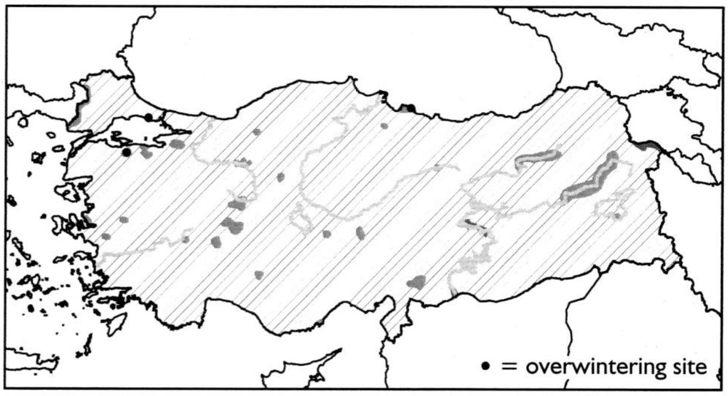

Breeding. In dense colonies, often with other herons and cormorants, mainly at large inland lakes, coastal river deltas and near rivers, but numbers in such colonies range from a single pair in a colony of *c.*100 pairs of Rooks, near Pasinler (East Anatolia), in June 1970, to 500 pairs (nesting with herons, egrets, spoonbills and cormorants) at Manyas Gölü (Marmara) in 1967. Colonies known at ten IBAs, ranging from nine pairs at Yeşilırmak Delta (**Black Sea Coastlands**) to 250 pairs at Iznik Gölü (Marmara). Nests in trees (sometimes in flooded areas) or extensive reedbeds. The nest is a shallow structure, mainly of twigs if built in trees, or of reed stems if in reeds. Eggs: usually 3–4 (5). Young: see below. In **Aegean**, Selous (1900) found a dense colony in reeds which also included Glossy Ibis *Plegadis falcinellus*, Squacco Heron *Ardeola ralloides*, Little *Egretta garzetta* and Great White Egrets *E.alba* and Pygmy Cormorant *Phalacrocorax pygmeus*; nests of Black-crowned Night Heron were slightly above water level, and most were empty, but some had incomplete clutches of up to two eggs, on 13 May 1899. At Manyas, adults were already on nests in flooded willows by 5 April 1967 (R. F. Porter); photographs in Pforr & Limbrunner (1982) show a nest with three young, 2–3 weeks old, in May 1971, and two young, 1–2 weeks old, on 30 June 1970, the latter indicating egg laying in late May. At Uluabat Gölü (**Marmara**), there were 105 nests in a mixed colony of six species, in widely spaced clumps of willows in a reedbed, in 1998; on 7 June 1998 some still had eggs, but most young were close to fledging (Welch & Welch 1998b), indicating earliest egg laying about mid April. In a mixed colony of seven species at Uluabat Gölü, *c.*40 nests in willows contained young at all stages, and some with 3–5 eggs, on 3 June 2006 (PC). In **Inner Anatolia**, *c.*20 pairs nested in reeds, with *c.*20 pairs of Pygmy Cormorants, and a few pairs of Great White Egrets and Grey Herons *Ardea cinerea*, at Ereğli marshes, on 25 May 1998; most nests had 3–4 eggs, but several were hatching, indicating egg laying late April (T. Quittenden); 50 adults were seen nesting between Grey Herons, near Ankara, on 7 May 1981 (H. Hols); fledged young were observed at Eber Gölü in late-June 1994 (M. Langeveld) and near Konya on 4 July 1986 (R. Webb), both records indicating egg laying no later than late April. In **East Anatolia**, ten nests had 1–4 eggs, in a colony of *c.*20 pairs, on Akdamar Island in Van Gölü, on 3 June 1972, and juveniles were seen there on 8 August 1974 (T. A. Walsh); a total of 12 pairs bred in two small colonies in willows near Ağrı in 1969.

CATTLE EGRET *Bubulcus ibis*
Sığır Balıkçılı

(Du: Koereiger; Fr: Héron garde-boeufs; Ger: Kuhreiher; Swe: Kohäger)

Subspecies and taxonomy. The nominate subspecies is involved. Ridgely & Greenfield (2001) suggested that more than one species might be involved and Rasmussen & Anderton (2005) split nominate *ibis* (Western Cattle Egret) from the eastern *coromandus*. The monospecific genus *Bubulcus* is sometimes subsumed within *Egretta*, but DNA-DNA hybridisation work suggested that it is equally close to *Ardea* (Sheldon 1987). We elect to maintain *Bubulcus* hesitantly.

Status and distribution. National conservation status: Vulnerable (population apparently increasing). Uncertain. Breeding proven only on a small number of occasions. Before drainage nested at Amik Gölü, Antakya province (Mediterranean); Tristram (1882) obtained eggs there in 1881 and Meinertzhagen (1935) found the species nesting in good numbers in a mixed colony with African Darter *Anhinga melanogaster* and Pygmy Cormorant *Phalacrocorax pygmeus* in mid-May 1933. There were no subsequent confirmed breeding records at this locality. However, 150 were present at the end of September 1965, 94 in September 1967, 130 in October 1968 and 30 in May 1970, suggesting that a population may have survived until comparatively recently. A colony of 30 pairs was found at Ereğli marshes (Inner Anatolia) in 1968, but only one pair in 1969, and none was recorded thereafter until 1993, when a group of 60 were seen in July, although breeding was not confirmed. No subsequent records are available from this locality that suggest nesting. Recently the species has been found to be regular in moderate numbers at the Göksu Delta (Mediterranean), where it is resident, and nesting was confirmed in 1992. Although unconfirmed, the presence of groups of up to 150, often in full breeding plumage, in the Aras Valley near Iğdır, Kars province (East Anatolia), and smaller numbers in the Yeşilırmak Delta (Black Sea Coastlands), suggests the existence of further breeding sites. Also recorded less regularly in small to moderate numbers in the Çukurova region (Mediterranean) and Sultan marshes (Inner Anatolia), which are also potential breeding sites.

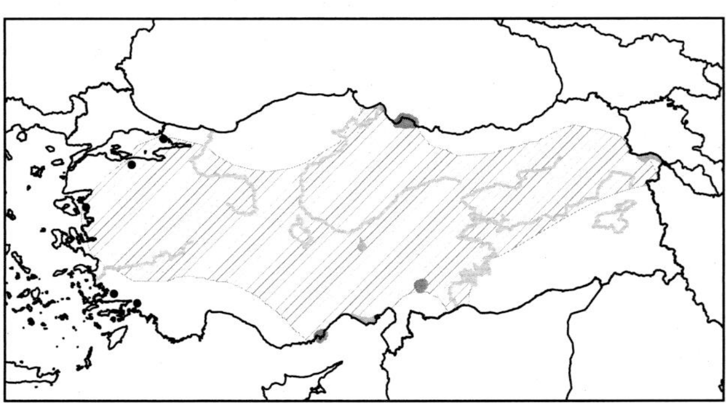

Elsewhere an irregular visitor, perhaps a local resident or partial migrant, to wetland areas in Black Sea Coastlands (e.g. a few records at Kızılırmak Delta, all since 1977), Aegean, Mediterranean, Inner and East Anatolia, usually singly or in small groups. Occasionally recorded in small numbers in Southeast Anatolia (noted at nine sites during recent surveys), and recorded once in autumn on the Princes Islands, Istanbul. Most records are in spring/early summer (March–late May) and may thus relate to passage birds; there are very few winter records, all from the south coast (e.g. totals of nine in January 1996, 13 in January 1999 and ten at the former Amik Gölü, in December 1964). Low winter temperatures throughout much of the country must be a significant constraint to the species becoming better established.

Breeding. 'A good many pairs' were breeding at Amik Gölü (**Mediterranean**), on 26 May 1933; they were breeding with darters and cormorants (in reeds), 'their nests occupying the top story of the slum, and they were beginning to lay' (Meinertzhagen 1935). More recently, there are breeding records at Ereğli marshes (**Inner Anatolia**) in 1968 and 1969; at Amik Gölü in 1981, and 25 pairs currently breed in Göksu Delta (Magnin & Yarar 1997). No description of nest or details of clutch size in Turkey, except a report of one on a nest-like structure atop flattened reeds as if brooding eggs (BD). Elsewhere the nest is a shallow structure, mainly of dead reeds if built in reedbeds, or mainly of sticks if in trees or bushes, and usual clutch size is 4–6.

SQUACCO HERON — *Ardeola ralloides*
Alaca Balıkçıl

(Du: Ralreiger; Fr: Crabier chevelu; Ger: Rallenreiher; Swe: Rallhäger)

Subspecies. Monotypic.

Status and distribution. National conservation status: Vulnerable (population apparently decreasing). Apparently a rather local breeding bird and summer migrant, usually in moderate numbers, to marshy wetlands throughout the country. However, few colonies, mostly in trees, have been recorded although, in view of the number of suitable localities from which breeding is strongly suspected, many colonies probably remain undiscovered. The incidence of reedbed nesting in Turkey appears to be higher than in many other parts of the species' range, especially in the virtually treeless areas of Inner and East Anatolia. Few records from Southeast Anatolia, where largely restricted as a breeder to the Euphrates and Tigris rivers. Significant sites for the species include: Meriç Delta, 300 pairs; Uluabat Gölü, *c.*110 pairs; Manyas Gölü, *c.*100 pairs (all Marmara); Marmara Gölü, 200 pairs (Aegean); Akşehir and Eber lakes, 70 pairs; Sultan marshes, 70 pairs (all Inner Anatolia); and Göksu Delta, 70 pairs (Mediterranean).

More common and widespread across the entire country on passage. In spring birds arrive from mid March (17th), with the bulk of passage occurring between late April and mid May. Arrival in Inner Anatolia is generally from mid April and further north not until the third week. A few stragglers continue to pass through Turkey until the end of May and possibly early June (8th). Departs from breeding areas between late July/early August and late September/early October with a few remaining as late as mid November (12th). Occasionally recorded in exceptional concentrations of up to 150 individuals in non-breeding areas at this season, particularly on the Black Sea coast, indicative of immigrants from outside Turkey. There are no winter records.

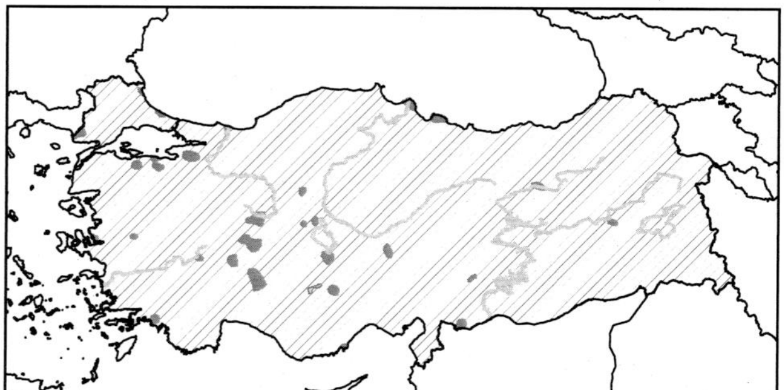

Breeding. In dense, usually mixed colonies, mainly at large inland lakes and at coastal river deltas, but numbers of Squacco Herons are usually low, typically 10–30 pairs, and usually represent a small proportion of the total colony. Nests in trees or bushes, and in reeds, usually in extensive reedbeds. Nest is a shallow structure of thin twigs if built in a tree or bush (and noticeably smaller than those of Black-crowned Night Heron *Nycticorax nycticorax* and Little Egret *Egretta garzetta*), and of reeds if in reedbeds. Eggs: 3–5; in other parts of range, usual clutch is 4–6. Young: B/4 (1). At Manyas Gölü (**Marmara**), where nests in trees, a nest had four eggs, and another four young, 1–2 weeks old on 4 June 1970 (Pforr & Limbrunner 1982), indicating egg laying early May. At Uluabat Gölü, 109 nests were counted in a colony, in clumps of willows, totalling 1,180 pairs, of six species, mostly Pygmy Cormorants *Phalacrocorax pygmeus*; on 7 June 1998, nests of Squacco Heron had almost fledged young and some had eggs (Welch & Welch 1998b); on 3 June 2006, *c.*50 nests examined contained mainly 3–5 eggs but some had young, ranging from newly hatched to half-grown, indicating earliest laying in late April, but mainly in May (PC). In Kızılırmak Delta (**Black Sea Coastlands**), seven adults were carrying nesting material on 5 May 1992 (Hustings & van Dijk 1994); ten pairs in a tree colony (with Grey Herons *Ardea cinerea* and Little Egrets) on 27 August 1984, when several dead young were seen below the colony (Dijksen & Kasparek 1985). At Sultan marshes (**Inner Anatolia**), adults were carrying nest material on 15 May 1979 and breeding was in June (Kasparek 1985). In **Aegean**, Selous (1900) found two nests with eggs at a lake in Izmir district on 13 May 1899: the nests were built up from the water at the base of reeds; other species in the colony were Great White *Casmerodius albus* and Little Egrets, Black-crowned Night Heron, Glossy Ibis *Plegadis falcinellus* and Pygmy Cormorant. In **Mediterranean**, Squacco Herons were starting to lay on 26 May 1933, in colonies of darters and Pygmy Cormorants at Amik Gölü (Meinertzhagen 1935).

LITTLE EGRET — *Egretta garzetta*
Küçük Ak Balıkçıl

(Du: Kleine Zilverreiger; Fr: Aigrette garzette; Ger: Seidenreiher; Swe: Silkeshäger)

Subspecies: Nominate.

Status and distribution. National conservation status: Near Threatened (population apparently decreasing). Local but not uncommon resident breeder or partial migrant, often in considerable numbers in Marmara (e.g. up to 470 pairs in the Meriç Delta), Aegean, Mediterranean and Inner Anatolia, where the species has declined at a number of localities in recent years, though perhaps coincident increases have been reported at other sites in the central part of the region. Scarcer due to a lack of suitable habitat in Black Sea Coastlands (though a minimum of 230 pairs estimated at Kızılırmak Delta) and East Anatolia. No recorded instances of breeding in Southeast Anatolia.

Widespread and common on passage, in spring from mid March (19th) to late May (30th); perhaps peaking in mid to late April/early May over much of the country (slightly earlier in the south), and in autumn from late July to at least late September/early October, with large numbers occasionally recorded in eastern Black Sea Coastlands and Thrace (e.g. up to 1,700 at Meriç Delta). Moderate numbers overwinter, principally in the western two-thirds of Turkey, with the majority in Aegean and Mediterranean. Only occasionally recorded in Marmara (including Thrace) and Black Sea Coastlands at this season. Small to moderate numbers remain at wetlands in Inner Anatolia in milder winters. A dark-phase individual was photographed at Mogan Gölü (Inner Anatolia) on 3 May 1997, the first record for Turkey of this colour morph (Boyla & Eken 1998).

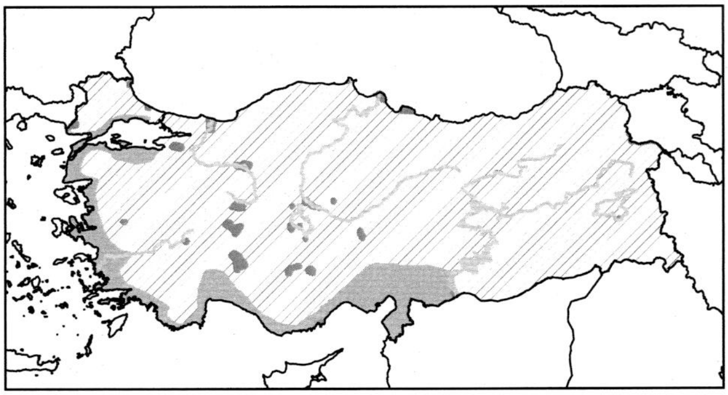

Breeding. In dense, usually mixed colonies, mainly at large inland lakes and at coastal river deltas; in East Anatolia also on a river island on the border with Armenia (G. Sandve). Mixed colonies may include hundreds, but numbers of Little Egrets range from one pair among (mainly) Black-crowned Night Herons *Nycticorax nyc-*

ticorax at Akdamar Island, Van Gölü (**East Anatolia**), to 400 pairs in Meriç Delta (Marmara); typically there are 30–230 pairs. Nests in trees or bushes, sometimes in flooded areas, and in reeds; *c.*10 nests were on bare ground on an island in Bolluk Gölü (**Inner Anatolia**) in 2004 (R. Thomas). Constructs a shallow structure of thin twigs (in a tree or bush) or of reeds. Eggs: C/4–6. Young: no data. In **Aegean**, *c.*100 pairs bred in trees on an island in Bafa Gölü, with Great Cormorants *Phalacrocorax carbo* and Grey Herons *Ardea cinerea*: on 1 May 2003, *c.*20 nests were examined, and most had four eggs, some 5–6 eggs, all apparently laid recently (PC). At Marmara Gölü, McNeile found a few nests, with fresh eggs, amongst Glossy Ibis *Plegadis falcinellus*, Pygmy Cormorant *Phalacrocorax pygmeus* and spoonbills, in dense reeds on 30 April 1951. At Uluabat Gölü (**Marmara**), there were 76 nests, in a mixed colony within clumps of willows, in a reedbed; on 7 June 1998 young were mostly close to fledging but some nests still had eggs (Welch & Welch 1998b), indicating earliest laying in late April, and those with eggs may have been replacements, following earlier failure, or merely later breeding. At Manyas Gölü, two nests were 6m up in mature willows, and had four and five eggs on 2 May 1966; adults returned to the colony on 29 March 1967 and commenced building/repairing nests on 12 April 1967 (R. F. Porter). At Bolluk Gölü, two nests on an island had three and four eggs, whilst others were ready but empty, on 23 April 2004; on 16 May 2004, *c.*10 nests had 4–5 eggs, and one had four eggs and a newly hatched chick, indicating laying *c.*19 April (D. Gregson). At Kızılırmak Delta (**Black Sea Coastlands**), where the species breeds in trees, most young had already fledged but many nests still contained young on 27 August 1984 (Dijksen & Kasparek 1985); in 1992, many were nest building on 2 and 5 May, and breeding was in progress, but many were still building on 3 June (Hustings & van Dijk 1994). Thus breeding dates vary in different areas, and nesting in a colony may not commence synchronously.

GREAT WHITE EGRET — *Casmerodius albus*
Büyük Ak Balıkçıl

(Du: Grote Zilverreiger; Fr: Grand Aigrette; Ger: Silberreiher; Swe: Ägretthäger)

Subspecies and taxonomy. Nominate. The species was traditionally placed within *Egretta*, but Sibley & Monroe (1990) reported that DNA-hybridization data suggested that a more appropriate treatment would be to include it within *Ardea*. Separate molecular studies confirmed that Great White Egret is closer in its relationships to *Bubulcus* and *Ardea*, but did not provide a clear result that would justify inclusion within the latter genus. The monotypic genus *Casmerodius* is available for the species and others have already used this stopgap solution until further evidence becomes available (see Sangster *et al.* 1997)

Status and distribution. National conservation status: Endangered (population apparently decreasing). A very local resident breeder or partial migrant in some areas, usually in small to moderate numbers, in marshy wetlands across Inner and East Anatolia. Also reported breeding even more locally in Aegean, Mediterranean and Black Sea Coastlands (e.g. 11–15 pairs at Kızılırmak Delta in 1992).

Widespread and regularly recorded on passage except from eastern areas, often in considerable numbers; in spring from early March (5th) to mid May, with the bulk passing through Inner Anatolia in early April, and in autumn from August to early October. Also recorded in larger numbers in winter (presumably as a result of immigration from more northerly areas), being particularly common in coastal areas in the western two-thirds of Turkey (e.g. up to 416 at Akyatan Gölü, Mediterranean; 428 at Marmara Gölü and 312 at Büyük Menderes Delta, Aegean; 260 at Meriç Delta, Marmara; and 210 in Kızılırmak Delta, Black Sea Coastlands). Less common inland, including East and Inner Anatolia, where remains in flocks of up to 100 birds in milder winters, and 217 were at Sarıyar Reservoir on 18 January 2004.

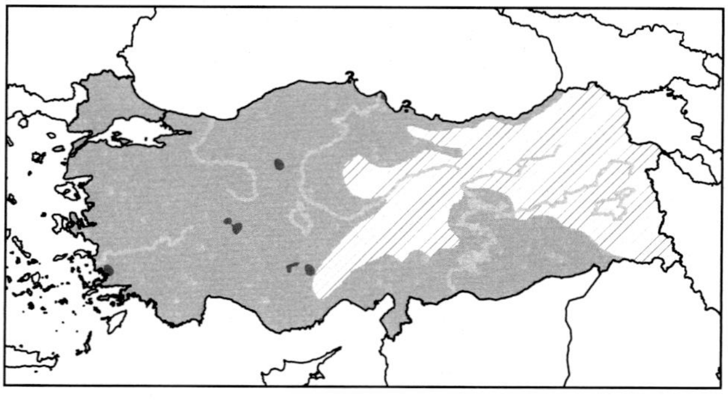

Breeding. In extensive reedbeds, with a few pairs loosely associated rather than in colonies. Nests are scattered over a wide area, with nests of other large herons often close by. Numbers range from 1–2 pairs in Kocaçay Delta (**Marmara**) to up to 50 pairs at Hotamış (Inner Anatolia) in 1991 (Kirwan 1993a). Tall dense mature reeds are required for breeding. They are broken over by the birds to provide a platform for the bulky but shallow nest of dead reeds, usually *c.*1m above water level. Eggs: C/2–4, average 3.2 (*n*=9). Young: B/1–3, mainly 3. At Ereğli marshes (**Inner Anatolia**), on 23 May 1998, two separate groups of two and three pairs, associated with similar numbers of breeding Grey Herons *Ardea cinerea*; the egrets' nests each had about three young *c.*2 weeks old (T. Quittenden), indicating that laying was initiated in the first week of April. At Kızılırmak Delta (**Black Sea Coastlands**), in 1992, the population bred in reeds amidst much larger numbers of Purple Herons *Ardea purpurea*; eight nests examined on 2–10 June 1992 had mainly eggs, but also young ranging from newly hatched to *c.*2–3 weeks old: for three nests, where reliable back-dating was possible, first eggs were laid in the last week of April, first week of May, and not earlier than 17 May (Hustings & van Dijk 1994), somewhat later than those at Ereğli. Old breeding records include two nests, with two and four eggs, in **Aegean** on 13 May 1899 (Selous 1900), and a clutch of four taken in 'Asia Minor' (location unknown) on 28 May 1895 (A. Dixon pers. comm.).

GREY HERON — *Ardea cinerea*
Gri Balıkçıl

(Du: Blauwe Reiger; Fr: Héron cendré; Ger: Graureiher; Swe: Häger)

Subspecies. Nominate.

Status and distribution. National conservation status: Least Concern (population apparently stable). Relatively local resident breeder or partial migrant, usually in moderate numbers (e.g. 190–200 pairs bred in Kızılırmak Delta, Black Sea Coastlands, in 1992), in all types of wetlands across the country, although scarcer in the eastern third, where absent as a breeder from Southeast Anatolia, and in Inner Anatolia.

Widespread and relatively common on passage throughout the whole country (especially noticeable along the Black Sea coast, from at least 13 March to 30 May in spring; Albrecht 1986), but most pronounced in autumn, from August to mid October, perhaps with majority in September (e.g. 550 at the Bosphorus in September 1971). Considerable numbers also winter in the western two-thirds of Turkey, principally in coastal areas, more occasionally inland.

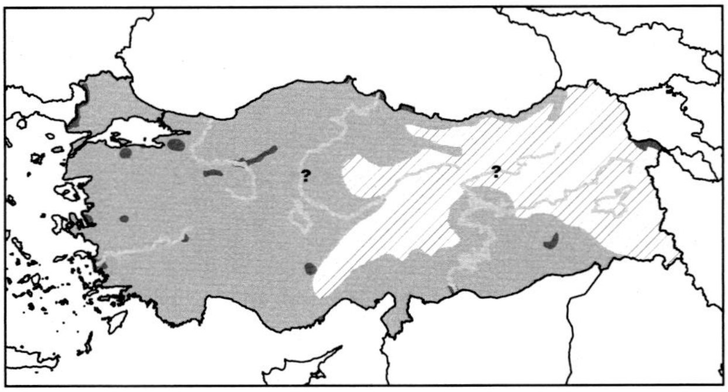

Breeding. Near water, usually in trees or reeds, in colonies of a few to several hundred pairs, either single-species or mixed with others, including White Storks *Ciconia ciconia*, cormorants, egrets and other herons. Also nests on the ground on islands in lakes. Nests in trees are of sticks, those in reedbeds of reeds. Eggs C/3–5, records of smaller numbers probably incomplete clutches. Season protracted, commencing mostly March, and late records are probably replacement clutches. Earliest records, in **Aegean**, of pairs with full-grown young at Karine Gölü on 23 April 2003, and large young (some flying) at Bafa Gölü on 1 May 2003, indicate laying mid February (PC). Also in Aegean, 13 nests with large young at Bafa Gölü on 13 May 1980 (Kasparek 1988a), young at all stages, but some nests still with eggs, on 3 May 2001, and nests with eggs and young at all stages on 27 June 1999 (probably replacements) at Karine Gölü (PC). In **Inner Anatolia**, nests had large young at Ereğli marshes on 23 May 1998 (T. Quittenden); 19 nests were occupied at Kızılcahamam on 18 March 1984 (Baris *et al.* 1984); 20 nests were in one tree at Eskişehir, in which young mainly hatched on 19 May 1907 (Ramsay 1914). At Kızılırmak Delta (**Black Sea Coastlands**), 30 nests had young close to fledging, 14–21 June

1984, and 48 and 25 nests were occupied in July 1971 and July 1972 (Dijksen & Kasparek 1985); there were 111 nests (in trees), with some young at fledging stage on 3 June 1992, indicating that breeding started on 20 March; and 47 nests (in reeds), with eggs and young at all stages, on 7 June 1992 (Hustings & van Dijk 1994). In Thrace, there are records of occupied nests from March, with 30 nests, mainly with young, on 25 April 1970.

PURPLE HERON *Ardea purpurea*
Erguvani Balıkçıl

(Du: Purperreiger; Fr: Héron pourpré; Ger: Purpurreiher; Swe: Purpurhäger)

Subspecies: Nominate.

Status and distribution. National conservation status: Vulnerable (population apparently decreasing). Generally a rather local summer migrant and breeding bird in marshy wetlands in Marmara (including Thrace), Aegean, Mediterranean and Inner Anatolia, where there is some evidence for a recent decline at several wetlands. Generally scarcer in Black Sea Coastlands (although 475–500 breeding pairs estimated at Kızılırmak Delta in spring 1992, making it the most important nesting site in the country; Hustings & van Dijk 1994) and East Anatolia, due to a relative lack of suitable breeding habitat. In spring recorded in southern and coastal localities from mid March (16th), arriving on average approximately 2–3 weeks later at inland, northern and eastern sites. Passage persists until at least mid May in Black Sea Coastlands, almost certainly later, as peak passage in Inner Anatolia probably occurs in early May, and apparent migrants have been observed as late as 18 May at Birecik (Southeast Anatolia).

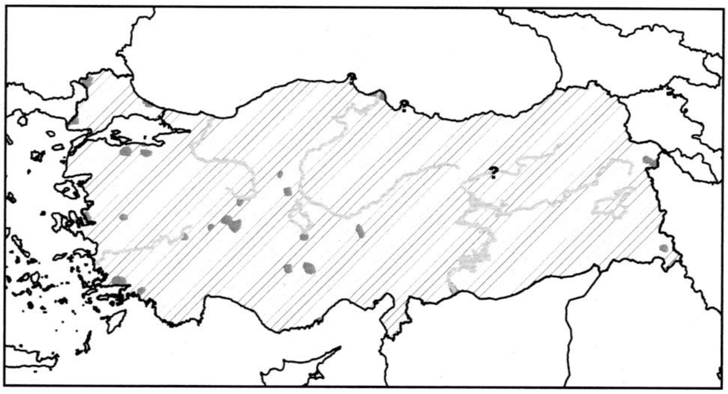

Widespread and slightly more common on passage across the whole country (though generally scarce in Southeast Anatolia, where small numbers follow the major river valleys), principally in autumn (when occasionally recorded in large numbers along the Black Sea coast; e.g. over 1,400 in Kızılırmak Delta on 30 August 1982), from mid August, peaking between late August and late September. Exceptionally recorded until late October or even mid November (13th).

Breeding. Mainly in extensive beds of mature *Phragmites*, in groups or loose associations of 3–5 pairs, e.g. at Kocaçay Delta, Marmara (Ertan 1996), to four groups or colonies totalling 475–500 pairs in Kızılırmak Delta (see above); also breeds in trees, e.g. at Manyas Gölü (Marmara), in mixed colonies with cormorants, other herons and spoonbills. Nests in reeds are usually widely spaced and built above water level on old reeds, which are broken to form a platform, on which a substantial, but shallow, nest of dead reeds is constructed; nests in trees or bushes are of sticks. At Uluabat Gölü (**Marmara**), several nests, each with two eggs, were found *c.*2m up in tall reeds on 9 May 1970; these were probably incomplete clutches. At Manyas, in 1966, six nests were 1–3m above ground in willows within a reedbed; one nest was solely of willow twigs, another a combination of reeds and willow branches; one was empty on 27 April 1966, the first egg was laid on 28 April 1966, and there were three eggs on 3 May 1966; another had one egg on 28 April 1966 and three eggs on 3 May 1966; the others contained C/5 on 29 April 1966, and C/4, C/4, C/4 on 3 May 1966 (R. F. Porter). At Durusa Gölü, near Istanbul (Thrace), 15 adults were observed nest building in small trees on islands, on 24 April 1981 (H. Hols). The following information is from the detailed survey of Kızılırmak Delta (**Black Sea Coastlands**) in 1992 (Hustings & van Dijk 1994): four colonies were all in extensive reedbeds with uncut reed, the nests 0.5–0.8m above water. Eggs: C/4 (158), C/3 (153). Egg laying was from 21 April in one colony, about one week later in others. Hatching from 17 May, with many hatching late May.

YELLOW-BILLED STORK *Mycteria ibis*
Sarı Gagalı Leylek

(Du: Afrikaanse Nimmerzat; Fr: Tantale ibis; Ger: Nimmersatt; Swe: Afrikansk ibisstork)

Subspecies. Monotypic.

Status and distribution. Vagrant with two records; an immature at Amik Gölü, Antakya province (Mediterranean) on 7–20 May 1962 (Kumerloeve 1963a, 1989), and one at the Göksu Delta (Mediterranean) on 28 and 30 May 1986 (Martins 1989).

Elsewhere in the Middle East, there was a total of 18 records from Israel prior to 1996 (Shirihai 1996). The species is a regular non-breeding visitor to Abu Simbel, Lake Nasser, southern Egypt (Goodman & Meininger 1989) and an immature was photographed at Sharm el Sheikh, south Sinai, in August–September 1995 (*Birding World* 8: 292, 335). Recently, recorded in Bulgaria in June–July 2002 (Ragyov *et al.* 2003) and Qatar in November 1999 (Hellyer 2000).

BLACK STORK *Ciconia nigra*
Kara Leylek

(Du: Zwarte Ooievaar; Fr: Cigogne noire; Ger: Schwarzstorch; Swe: Svart stork)

Subspecies. Monotypic.

Status and distribution. National conservation status: Least Concern (population apparently stable). A relatively widespread and not uncommon, but rather localised, summer visitor, breeding in drier, open hilly and more occasionally mountainous areas, and in forested lowlands, usually in the immediate vicinity of wetlands, e.g. at Kızılırmak Delta (Black Sea Coastlands, where up to 50 pairs in recent years), or forest streams. Nests from mid March in Marmara (probably including Thrace), Aegean, Black Sea Coastlands and Inner Anatolia, as well as in the northern fringe of Mediterranean (in the central and western Taurus), and in northern parts of East Anatolia, locally as far south as Malatya.

Even more widespread on passage, although active migration only rarely reported in Southeast and East Anatolia. Appears to be commoner in the western two-thirds of Turkey being recorded in considerable numbers at the Bosphorus, particularly in autumn, although sometimes recorded in significant numbers there in spring, e.g., 1,118 between mid March and late May 2006 (KAB *et al.*). In spring recorded from mid March (16th) to early June (1st) on migration, and in autumn from early August (2nd) to early November (13th), the bulk occurring between early September and early October. Birds well away from known breeding areas occasionally reported as early as mid July presumably reflective of post-breeding dispersal. The maximum count at the Bosphorus at the latter season was 8,318 in 1973, including 2,556 on 25 September, although the highest single-day count at this watchpoint was 5,333 on 18 September 1978. Few detailed or complete counts at this site subsequently

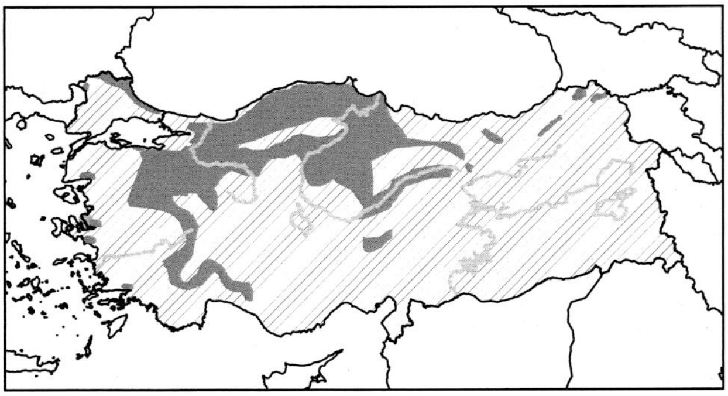

but 2,588 noted on 20 September 1995 (KAB *et al.*). Relatively uncommon at both Borçka and the Belen Pass, but movements involving several hundreds recorded in Çukurova (Mediterranean) in early October. Only occasionally recorded in winter (sometimes in small flocks of 20+), principally in Aegean (where apparently regular at Gediz Delta) and Mediterranean, very rarely in Marmara (including Thrace), Black Sea Coastlands and Inner Anatolia (Kirwan 1993d, Eken 1997d, Demirci 2002, 2003).

Breeding. In wooded and dry rocky hills (nests mainly on cliffs, but also in trees), and in lowland deciduous forest, usually near wetlands, e.g. at Kocaçay Delta (**Marmara**) and Kızılırmak Delta (**Black Sea Coastlands**), where nests are always in mature trees, both deciduous and coniferous, 2.5–6.0m above ground (n=6). Elsewhere, on ledges and in recesses and caves on cliffs, usually >10m above ground. A substantial structure of sticks and branches with a shallow cup on top, lined with moss and grass. New nests may be relatively small but many are re-used annually and may become enormous. Eggs: C/3–4 (2). Young B/2–4, mean 2.9 (10), but this relies heavily on observations of broods of larger young, seen from the ground, which would not take account of possible infertile eggs or of young that may have died earlier, and most pairs probably lay at least three, perhaps usually four, eggs. In Kızılırmak Delta, two broods, observed in the nest on 13 July 1972 fledged before 23 July (Dijksen & Kasparek 1985), indicating first eggs in the third week of April; in 1992, there were 30–35 pairs, with occupied nests from 17 March and young in nests from 26 May (Hustings & van Dijk 1994), indicating first eggs in late April. In **Inner Anatolia**, a nest had three young, *c.*10 days old, at Ürgüp on 3 June 1971, indicating first egg *c.*22 April; the first of three eggs in a nest at Kızılcahamam was laid on 6 May 1993 (PC). Two nests on cliffs near Aksaray held eggs and small young on 29 April 2007, the latter indicating egg laying in late March (A. Dixon).

WHITE STORK *Ciconia ciconia*
Leylek

(Du: Ooievaar; Fr: Cigogne blanche; Ger: Weißtorch; Swe: Vit stork)

Subspecies. Nominate.

Status and distribution. National conservation status: Least Concern (population apparently decreasing). A widespread and relatively common summer migrant breeder across the whole country, to at least 2,200m, although generally found at rather lower densities in Southeast Anatolia, and only very locally in the Black Sea Coastlands, being largely absent from the eastern half. Kılıç & Kasparek (1986) remarked that a significant decline (*c.*60%) in the breeding population had become evident since the late 1960s, although this was not based on any systematic survey and van den Berk (1994) pointed to limited evidence from three coastal areas which suggested that the population has in fact remained stable or even increased locally. The current breeding population was estimated as between 7,000 and 30,000 pairs (Kasparek 1992a) and by Tucker & Heath (1994) as being between 15,000 and 35,000 pairs, but is almost certainly nearer the upper limit of this range. Parr *et al.* (1996) surveyed a substantial area of Inner Anatolia in spring 1993, estimating the population in this part of Turkey to be between 1,000 and 3,000 pairs.

Recorded in very large numbers on passage; in spring from late February (20th) on the Mediterranean coast, although generally not becoming widespread until mid to late March and perhaps largely absent from East Anatolia until early April. Few are recorded on migration beyond the end of May (30th), though exceptionally recorded until mid June (18th). Peak passage is in the second half of April and early May; a total of 52,458 passed the Bosphorus between mid March and late May 2006. In autumn, the return movement begins occasionally from late July (with post-breeding aggregations forming from mid month, perhaps earlier), strengthening from mid August and continues until late October or even November (25th); the latter record a flock of over 100 individuals. Peak movements occur between mid August and early September, with peak days at the Bosphorus usually between 13 and 31 August. The autumn maximum at the Bosphorus was 338,353 in 1972 (including three counts of over 35,000, the highest being 52,954 on 29 August). No long-term observations are available from this watchpoint for recent seasons. At both passage periods birds follow a southeast to northwest axis, from the Göksu Delta (Mediterranean), where flocks of up to 25,000 have been recorded in autumn, and to a lesser extent the Çukurova region (Mediterranean), to the Bosphorus. Most birds migrate on a path close to, or to the west of, this flight line, with comparatively few east of it. Very few are recorded in the Borçka–Hopa area during spring or autumn, e.g. just 50 on 21 March–14 May 1994 at the latter (M. Henriksen). At the Belen Pass 103,576 were recorded in 1976 (Sutherland & Brooks 1981a), the only year for which full details are available. More casual observations made by Kumerloeve (1967a) in the vicinity of Amik Gölü, also hint at very large numbers passing through Antakya province. Occasionally recorded in winter mainly in the south and west (north as far as Thrace), but also irregularly in Inner and even Southeast Anatolia.

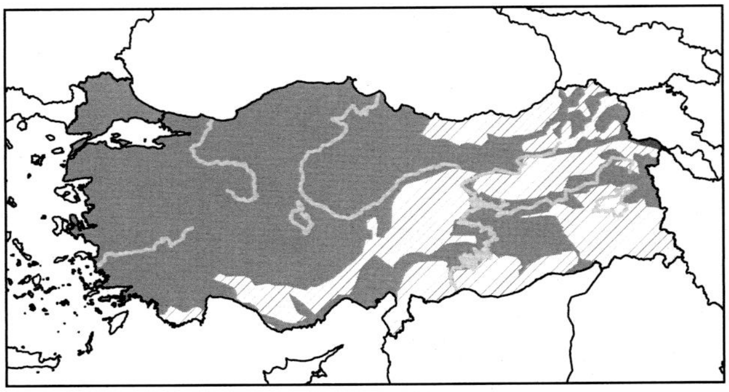

Breeding. Near habitation, mainly villages and towns, but also in nearby woodland or at isolated farms or buildings; rarely more than a few kilometres from people. Has a preference, in central Turkey, for settlements adjacent to low-lying, flat areas with wet grassland and marsh (Parr *et al.* 1996); and, in 1966, most of *c.*170 nests in Aegean/Marmara were in villages. Nests on buildings, ruins, chimneys, telegraph poles, electricity pylons, trees and stumps; a pair nested on a cliff at Kızılcahamam (Inner Anatolia) in 1983–84 (Barış *et al.* 1984), and another on a rock between Amasya and Osmancık (Black Sea Coastlands) in 1992 (G. Magnin). Commonly uses platforms, especially erected for nesting, on roofs and poles in villages. In a survey of central Turkey in 1993 (Parr *et al.* 1996), most nests were on buildings and most pairs were solitary, but some settlements held a few pairs, and the highest number was 12 at Eşmekaya. At Kızılırmak Delta (Black Sea Coastlands) in 1992 (Hustings & van Dijk 1994), the population was *c.*125–130 pairs; most nests were in trees, often near houses, in loose colonies in small woods, with 11 nests in one tree, and only one nest on a house roof. Trees used include poplar, willow, pine, olive and juniper, with 18 nests on a single oleaster tree at Incesu (Inner Anatolia) on 1 June 1975. Most nests are higher than 5m, but one was only 3m above ground in a pollarded tree at Kızılcahamam in 1992 (PC). Odd pairs may also nest in trees in which herons and egrets are nesting. The nest is a large structure of sticks and branches mixed with grass and earth, with a shallow cup in the centre top, lined with grass, feathers and rubbish. Old nests, used for many years and added to annually, may become huge. House Sparrows *Passer domesticus* and Spanish Sparrows *P. hispaniolensis* often construct their nests in the base of White Stork nests, with up to 50 nests in a single stork nest. Eggs: C/5 (1). Young: B/2 (5), B/3 (7), B/4 (4), B/5 (1); 22 nests contained 55 young in Kızılırmak Delta on 26 July 1971, mean 2.5 per nest. In **Mediterranean**, adults return to nests in late March (earliest 16th); several adults were incubating at Ceyhan as early as 27 March 2000 (L. Dijksen); a nest had five eggs in Çukurova on 7 May 1987 (van der Have *et al.* 1988); most records are of adults incubating (or brooding small young) in May, and of large young in nests in late June–early July; there were at least 21 pairs nesting in the village of Yeşildağ, near Beyşehir Gölü, in 1998 (A. Kullberg), and 10–19 pairs nesting in a colony in pines near Dalyan (Kiliç & Kasparek 1989). In **Aegean**, of 30 pairs breeding, many had small young at Milet on 19 May 1970, indicating egg laying mid April; more than 50 pairs were nesting in a village near Söke, and some still had eggs, but most held young, 13 May 1899 (Selous 1900), indicating egg laying no later than 11 April; there are many records of adults at nests in early April, and incubating from 21 April. In **Marmara**, adults have been recorded at nests from 26 March and as late as 9 August 1968; several adults were incubating at Uluabat Gölü on 25 April 2003 (PC); large young have been observed in nests in late June. In **Inner Anatolia**, adults are at nests from March (earliest 15 March) until 4 August, with many records of adults incubating in April–May (earliest 21 April), and an adult was observed feeding small young in a nest at Sultan marshes on 14 May 2004 (PC), indicating egg laying in mid April; young have been observed in other nests from 25 May, and some nests still contained young at Kızılcahamam on 16 July 1986 (R. Webb), the latter indicating egg laying no earlier than late April. In **Black Sea Coastlands**, the earliest incubating adult was observed on 3 May and the first young in a nest in 1975 were at Ilgaz on 7 June (R. F. Porter), indicating egg laying no later than early May; young were seen in nests at Kızılırmak Delta from 21 May in 1992 (Hustings & van Dijk 1994), indicating egg laying no later than 20 April; young, aged *c.*4 weeks and 3 weeks, in nests on electricity pylons, near Ispir, on 12 June 2004, can be back-dated to egg laying about 15 and 22 April (PC); 22 nests still contained young at Kızılırmak Delta on 26 July 1971. In **Southeast Anatolia**, five nests had young, between Birecik and Cizre, on 17 May 1989, indicating egg laying no later than mid April, but young were still in several nests, between Şanlıurfa and Diyarbakır, on 1 August 1992, indicating egg laying not before early May. In **East Anatolia**, the earliest young were seen in a nest near Van Gölü on 30 May 1969 (R. F. Porter); there are several records of young in nests in July–early August, latest 4 August 1974.

GLOSSY IBIS *Plegadis falcinellus*
Çeltikçi

(Du: Zwarte Ibis; Fr: Ibis falcinelle; Ger: Sichler; Swe: Bronsibis)

Subspecies. Monotypic.

Status and distribution. National conservation status: Endangered (population apparently decreasing). A rather local, but not uncommon, summer migrant breeder. Sometimes recorded in considerable numbers, though most known colonies number fewer than 50 pairs; a notable exception are the 100 pairs that nest in the Meriç Delta, Marmara. Principally found in dense marshy wetlands in Marmara (including Thrace), Inner Anatolia and perhaps adjacent areas of Aegean and East Anatolia. Particularly local and generally recorded in very small numbers in the latter region and some evidence of a recent decline in numbers from Inner Anatolia. Formerly bred at Amik Gölü, Antakya province (Mediterranean) in the first half of the 20th century, and in the Büyük Menderes Delta (Aegean) in the 1800s, but not since. Small groups still regularly oversummer in wetland areas in other regions including Black Sea Coastlands. The extended spring passage period obscures breeding status.

More widespread on passage, although less common in the eastern third of Turkey and few records from Southeast Anatolia. Locally recorded in large numbers, e.g. maxima of 1,200 at Meriç Delta (Marmara) on 26 July 2004 and 590 at Kızılırmak Delta (Black Sea Coastlands) on 27 April 1992. Passage in spring from mid March (14th) to late May/early June (2nd), with the majority in the second half of April and perhaps a second peak in early May, and in autumn from early August (occasionally mid July; 16th) to early October (9th), although most have departed by mid September. Very rarely reported in winter (maximum up to 25), in southern and western areas from Thrace to Mediterranean, even more occasionally in inland areas.

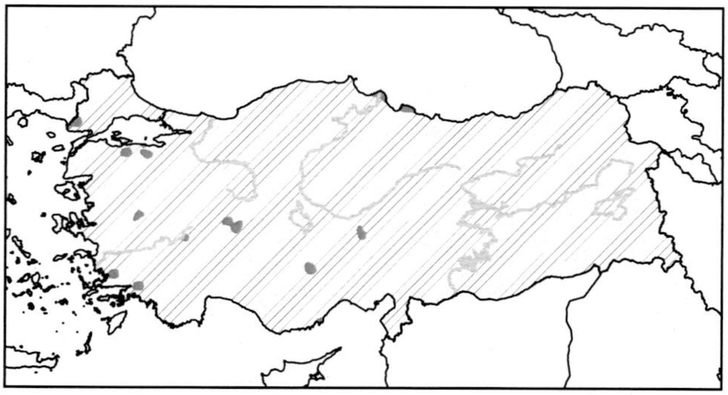

Breeding. In dense, mixed colonies, mainly at large inland lakes and marshes, but also on coasts. Colonies sometimes include hundreds of birds but numbers of Glossy Ibis are usually low (typically 10–30 pairs) and represent only a small proportion of a colony, although 200–300 bred at Manyas Gölü (Marmara) in 1968. Nests in trees and bushes, sometimes in flooded areas, or in extensive reedbeds. The nest is a shallow structure of twigs, lined with grass and leaves, if in trees or bushes, or of reeds. Eggs: C/3–5, mainly C/4. Young: B/3–4. In **Inner Anatolia**, 12 nests in a mixed heron/Pygmy Cormorant *Phalacrocorax pygmeus* colony at Ereğli marshes on 18 May 1987, of which most had four eggs (one had five and one had three), and the others incomplete clutches (M. Kasparek). Many fledged young were observed at Yarma on 22–25 July 1971. In **Aegean**, at a mixed colony at Marmara Gölü, on 30 April 1951, some nests of spoonbills, Pygmy Cormorant and Little Egret *Egretta garzetta* had eggs, but all nests of Glossy Ibis were empty, i.e. laying had not commenced (McNeile). Selous (1900) found a mixed colony near Izmir on 13 May 1899, in which egg laying was starting for the ibis. At Amik Gölü (**Mediterranean**), Meinertzhagen (1935) found a few pairs in the colonies of darters and other species in the reeds; the nests of ibis were at the base of the reeds and most had fresh eggs on 26 May 1933. In **Marmara**, a tree nest at Manyas Gölü had newly hatched young on 3 June 1970 and another three young, *c.*2 weeks old, on 4–6 June 1970 (Pforr & Limbrunner 1982), indicating egg laying *c.*10 May and 1 May respectively; at Uluabat Gölü, there were 14 nests of ibis in the mixed colony of six species in trees; on 7 June 1998 some of the nests had young, almost fledged, and some still had eggs (Welch & Welch 1998b). There were at least 40 nests of Glossy Ibis in the same mixed colony on 3 June 2006; most were in the lower branches of flooded willows but some were built up from the water in reeds, and most contained small or newly hatched young, indicating peak egg laying in early May; but there were also a few large young and a few nests still with 3–5 eggs (PC).

NORTHERN BALD IBIS — *Geronticus eremita*
Kelaynak

(Du: Heremietibis; Fr: Ibis chauve; Ger: Waldrapp; Swe: Eremitibis)

Subspecies and taxonomy. Monotypic, but the results of recent genetic research indicate that there are consistent genetic differences between the Turkish (and presumably Syrian) and Moroccan populations of Bald Ibis.

Status and distribution. Conservation status: Critical (BirdLife International 2000, 2004); national conservation status: Critical (population apparently stable). The truly wild population is extinct in Turkey, although there is a record of uncertain provenance concerning an immature on the north side of Uludağ, Bursa province on 22 August 1995 (Jetz 1995). Whether this was a wandering individual from the feral population at Birecik or a wild bird is unknown. Until 1989 was a summer migrant to a small area of the Euphrates Valley, breeding at Birecik and at a number of other smaller colonies between there and Halfeti (Southeast Anatolia). It was present from mid to late February to late June or early July. The exact age of the former colony at Birecik, first observed by Ainsworth in June 1839, is unknown. At this time he also collected a bird at Yaylak, Şanlıurfa province, 70km northeast of Birecik. No other travellers documented the colony until Danford in 1879, who recorded a large colony at the former locality. Tristram also mentions the presence of an impressive colony there two years later. Weigold counted over 1,000 birds at Birecik in 1911. In the 1950s the colony there numbered 400–500 pairs, Kumerloeve counting as many as 1,300 individuals, presumably including young, in June 1953.

However, between 1956 and 1959 DDT and dieldrin was sprayed throughout the Euphrates Valley to combat malaria. Simultaneously, a scheme designed to save newly reclaimed agricultural areas from locust swarms was instigated using prophylactic pesticide sprays. The effect upon the population was catastrophic: 70% (600–700 birds) were poisoned while reproductive success declined to very low levels, at which they remained for the next 10–12 years. In 1965 70–75 pairs were present. In 1970 36 occupied nests were counted, and by 1972–73 the number of occupied nests had fallen to 26. In the latter year a conservation programme, under the auspices of the World Wildlife Fund, was initiated. It was designed to protect the original colony site above the town from hunting, disturbance and the tipping of human waste, but also to attract the population away to a new site to the north of the town, and to establish a captive breeding and associated reintroduction programme. Although the now protected colony was eventually relocated successfully (in 1982) the aim of sustaining and rebuilding the population using captive breeding and release failed due to inadequate knowledge of the birds feeding and other biological requirements. Indeed, the breeding success of the wild birds remained marginally higher than that of the captive population. In 1977–83 only 23–34 healthy immatures were released and most, if not all, failed to migrate with the wild birds, remaining instead at Birecik where many perished during the cold winters. Despite improved breeding success the wild population declined to six pairs in 1979–82. In 1989 only three birds returned to the colony, two of which were later killed or disappeared during a storm, effectively marking the extinction of the Turkish wild population. In recent years the Birecik birds have been the subject of

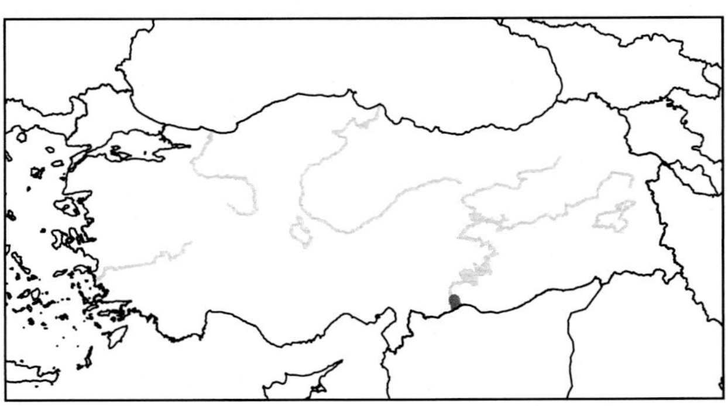

renewed conservation efforts, given the imperiled state of the Moroccan and recently rediscovered Syrian colonies. Thus, although the population had declined to as few as 42 birds, by 2000, additional efforts to monitor the colony and improve its breeding success appear to have been successful; 17 young successfully fledged in 2002 (*Sandgrouse* 25: 7). At present, approximately 50–60 free-flying, but largely dependent birds are present at Birecik. The decline and extinction of the wild Turkish population is fully documented in Peter (1990).

Elsewhere in Turkey, Chantre (1883) recorded the species and collected two individuals at the former Amik Gölü, Antakya province (Mediterranean). These were presumably migrants from the Birecik population or from Syria.

Breeding. On ledges, and in large purpose-made nest boxes serving as artificial crevices and ledges, on cliffs alongside the Euphrates at Birecik (**Southeast Anatolia**). Most pairs breed at the WWF centre, but some in an adjacent wadi, and odd pairs still nest on cliffs on the outskirts of town. The nest is a relatively shallow accumulation of plant stems and roots, and the usual clutch is 3–4. Prior to 1990, when the birds were truly wild, they returned to the breeding ledges in February; in 1879 the first pair arrived on 16 February and the first large party on 18 February (Danford 1880); egg laying was late March to mid April in 1969 (Warncke 1972). There were 36 nests on 5 May 1970 and, of 11 nests examined, four had three young, four had two young, and three had one young; in 1971, adults were incubating at 12 nests by 2 April, 33 nests by 5 April and 35 nests by 10 April, *i.e.* peak egg laying was in late March–early April (Warncke 1972). In 1973, 80 eggs were laid in 23 nests (mean 3.5), and 21 young fledged. In 1974, first arrivals were on 21 February, and there were *c.*45 adults present on 29–30 March, but only four adults and four young remained by 1 July; there were 54 adults and 36 young on 30 May 1975.Judging from the size of young observed in nests in 1993, 2001 and 2004, the breeding season in recent years was broadly similar to the pre-1990 period (PC).

EURASIAN SPOONBILL *Platalea leucorodia*
Kaşıkçı

(Du: Lepelaar; Fr: Spatule blanche; Ger: Löffler; Swe: Skedstork)

Subspecies. Nominate.

Status and distribution. National conservation status: Endangered (population apparently decreasing). Local summer visitor or partial migrant. Apparently a rare (probably declining) breeding bird in Marmara, Inner Anatolia (where several colonies have been extirpated due to wetland drainage, and remaining colonies generally number fewer than 30 pairs) and at single localities in Black Sea Coastlands (Kızılırmak Delta; up to 76 pairs in recent years) and Aegean (Bafa Gölü). Relatively few regularly occupied colonies are known. Formerly bred at Amik Gölü, Antakya province (Mediterranean) until the late 1960s (Vielliard 1968), perhaps later, and along the Karasu River, near Erzurum (East Anatolia) in 1839 (Dickson & Ross 1839), and possibly near Izmir (Aegean) in the 19th century (von Gonzenbach 1852). The Turkish breeding population may be declining, e.g. the well-known colony at Manyas Gölü (Marmara), where part of the population appears to be resident at least in some years, has undergone a partially documented, long-term decline (e.g. from up to 835 pairs in 1966 to *c.*200 pairs in 1995). No recent confirmed breeding records from East Anatolia, although regularly recorded from the Murat River around Balatos and Bulanık and at nearby Haçlı Gölü, Muş province, in summer during the 1980s. May breed elsewhere, e.g. at wetlands on the northern fringe of the Mediterranean region, but confirmation required. Nesting commences in March.

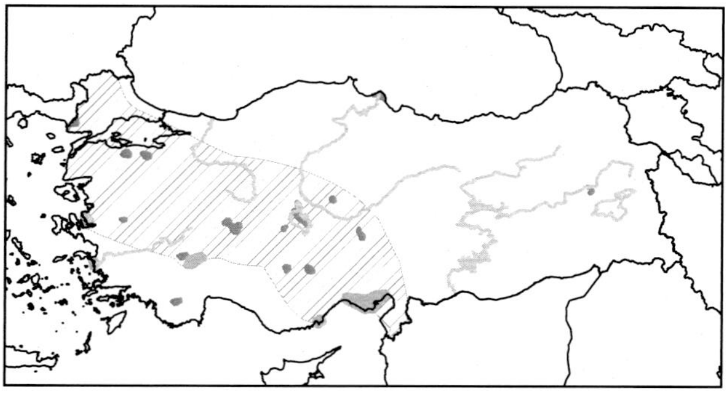

Widespread, typically in small numbers, throughout the western two-thirds of Turkey on migration. Spring passage occasionally commences from mid March (13th), more usually early April, and persists to late May (24th) but presumably continues later. Autumn movement occurs from early August (5th) to early October (8th). Scarce in both seasons at the Bosphorus, but recorded in moderate numbers at the Belen Pass, e.g. 816 between 6 August and 21 September 1976 (Sutherland & Brooks 1981a), and at several Inner Anatolian wetlands. Winters in very small numbers (rarely in groups of up to 60+) in coastal areas, principally in Aegean and Mediterranean, more occasionally inland (e.g. at Burdur Gölü and in Inner Anatolia) and elsewhere in the western two-thirds of Turkey (including Thrace).

Breeding. In usually dense colonies, the nests often touching, from a few pairs to several hundred, often mixed with, or close to, herons, egrets and cormorants. Nests are in dense reedbeds built up in the water, e.g. at Kızılırmak Delta (**Black Sea Coastlands**), in trees (widespread clumps of willows) in lakes, e.g. Uluabat Gölü (Marmara), and on bare mud on islands in lakes, where tall nests (up to 0.6m) of twigs and debris are built on the ground, e.g. Bolluk Gölü (**Inner Anatolia**). Eggs: mainly C/4 (49), but also C/3 (31); smaller clutches of two probably incomplete. Young: mainly B/4. Season protracted, with late nests having eggs in late June (probably replacement clutches) when other pairs in same colony have fledged young. Colonies begin laying on different dates. In Kızılırmak Delta, in 1992, laying commenced at one colony in late March and at another on 17–27 April (Hustings & van Dijk 1994). At Bolluk Gölü, nests had eggs and newly hatched young on 7 May 1993, indicating laying in early April, and some still had eggs and young on 24 June 1992, when most had already fledged (PC); on 23 April 2004, 25 nests had mainly 3–4 eggs, some incomplete clutches of two eggs, and one had newly hatched young, indicating laying in late March (R. Thomas). At Marmara Gölü (**Aegean**), a few nests had eggs but most were still empty on 13 May 1950, and a few nests had slightly incubated eggs, but most were still empty on 30 April 1951 (McNeile). At Manyas Gölü (**Marmara**), four nests had C/3–4, and two nests C/1–2, on 2 May 1966, all in mature willows, with three nests over 6m above ground (R. F. Porter); young hatched as early as 15 April 1970 (indicating laying in mid March), with well-grown young on 9 May 1970, and most young had fledged but some were still in the nest as late as 28 June 1972; there were 500 pairs at this colony in May 1969, and at least ten young were able to fly on 14 May 1969. A few pairs nested in the mixed colony of seven species at Uluabat Gölü in 2006; on 3 June one nest contained four eggs and another two large young: both nests were 4m up in flooded willows (PC).

GREATER FLAMINGO — *Phoenicopterus ruber*
Flamingo

(Du: Flamingo; Fr: Flamant rose; Ger: Rosaflamingo; Swe: Flamingo)

Subspecies and taxonomy. *P. r. roseus*. Greater Flamingo has traditionally been viewed as a polytypic species comprising two subspecies, Old World *roseus* and New World *ruber* (Caribbean or American Flamingo), but they are now regarded as separate species by the British Ornithologists' Union Records Committee (Knox *et al.* 2002) and others (e.g. Gill & Wright 2006). The two exhibit differences in plumage, bill, behaviour and vocalisations, and may be best treated at species level, as was suggested by Morony *et al.* (1975) and Sibley (1996); useful overviews of the problem were recently presented by Sangster (1997) and Collinson (2006).

Status and distribution. National conservation status: Endangered (population apparently stable). A very local breeding resident and/or partial migrant in large numbers at saline or semi-saline wetlands at two localities in Aegean and in several areas in Inner Anatolia. Kumerloeve (1962a) remarked upon the relative scarcity of observations of this species prior to the 1960s. These principally came from what are now well-known wintering areas in Aegean and Mediterranean, but also from Inner Anatolia in summer and in Southeast and East Anatolia, around Diyarbakır and Van, in the 19th century.

In 1969 suspicions that the Greater Flamingo might breed in Inner Anatolia were confirmed when eggshells were discovered at Tuz Gölü, an apparent colony of 1,500–2,000 birds was found on a small island, and at least 14,000 individuals were present at the lake in late May. In 1970, a colony containing 5,000 pairs was found there and further observations at this huge, largely inaccessible salt lake involved 83 nests in 1971 (a year of high water levels), 3,500 pairs in 1972, 3,700 pairs in 1973, *c.*1,000 young with 2,000 attendant adults in 1974, and 5,000 chicks estimated in 1978. Birds probably also bred here in 1968. Further information was unavailable until 1991 when an aerial survey of the lake revealed a colony of approximately 11,000 nests and 4,100 young. In 1992 a similar survey counted 14,000 young. Further counts of the numbers of young were made in 1997 (4,000), 1998 (11,400), 1999 (1,200), 2000 (8,000–10,000), 2002 (4,750), 2003 (3,059) and 2004 (7,312). Although erratic occupancy of nesting sites is a well-known aspect of the species' biology, its perceived status as an irregular breeding bird at Tuz Gölü, as well as elsewhere in Turkey, is due more to the inaccessibility of its known breeding areas, fluctuations in observer coverage and other related factors, than to genuine scarcity as a breeder.

Sporadic breeding is also recorded from three or four other localities in the same general area as Tuz Gölü. At Sultan marshes 1,500–2,000 pairs were found in 1970 and of 8,000 birds in June 1974, 200 pairs were starting to nest; at Seyfe Gölü nests were seen in 1970, while 2,000 pairs bred at this site in 1992, several hundred fresh nests were located in late May 1993 and 240 young in summer 1994; at Karapınar Ovası observations suggest that a colony may have existed in 1976 and 1977 but this could not be confirmed; and at Ereğli marshes 35–40 unoccupied nests were located in 1986. Subsequently 217 nests, of which 68 had eggs, were found at Ereğli marshes in 1991, while in 1993, 300 pairs, many with young, were found there and perhaps 20 pairs bred in 1998. No visit was made to the colony during the intervening years. The increased salinity of Ereğli has perhaps enhanced the attractiveness of the area to breeding flamingos in recent years.

At Çamaltı Tuzlası (Aegean) a colony of 100–150 pairs became established in 1982 and this colony has been active annually since, with a max. 1,450 pairs in 1995 (Magnin & Yarar 1997). However, in recent years, due to conservation efforts, the breeding population has increased, with nearly 3,000 chicks in 2003 and 3,000–3,500 chicks in 2004 (Balkız *et al.* in prep.). In 1993 another colony of up to 100 pairs was located at Acıgöl (Aegean), where the species may also have bred in 1964 and 1968 (Dijksen & Kasparek 1988). May have formerly bred, in small numbers, at Ceyhan Delta, Çukurova (Mediterranean) in the 1960s (see Warncke 1965). The total Turkish breeding population is currently estimated at 17,000 pairs (Magnin & Yarar 1994). At Erçek Gölü (East Anatolia) significant numbers (up to 1,000), predominantly adults, are regular in late summer and early autumn (until mid September). These, and smaller numbers at Sodalı Gölü and other wetlands around Van Gölü (East Anatolia), almost certainly wholly originate from the large colony at Lake Uromiyeh (Lake Rezaiyeh), in northwest Iran. Individuals hatched and ringed in Iranian colonies have been sighted/recovered in Turkey on at least 22 occasions (Johnson 1989, Behrouzi-Rad 1992).

A fairly widespread winter and non-breeding visitor in large numbers to suitable wetlands in Aegean, Mediterranean and East Anatolia, largely between late September and late April. Few (if any) remain in the latter region during severe winters. Very rarely recorded on passage in Southeast Anatolia. In late autumn the majority of the Turkish population formerly concentrated at the Sultan marshes; in October 1980 60,000–80,000 were present, although flocks of 20,000–30,000 were more usual. Most move into coastal areas in Aegean and Mediterranean for the winter, principally to Akyatan Gölü, Çukurova (Mediterranean), where a maximum of 19,000 was recorded in January 1972. But large numbers may occasionally remain in Inner Anatolia during mild winters, e.g. 1,700 at Tuz Gölü on 25 January 1969 and 2,000 at Sultan marshes earlier in the same month. More usually, 5,000–10,000 are present in Çukurova (but see below), and significant numbers may remain until early May. Until recently, the maximum midwinter count in Turkey (1972) was 25,900, with 22,210 recorded in early 1993, but 51,755 were counted in January 1999 (including 18,930 at Akyatan Gölü, 14,889 at Büyük Menderes Delta and 15,413 at the Gediz Delta; DHKD 1999) and in early February 2002, well over 29,000 were present at just two Aegean-coast sites: 9,524 at Büyük Menderes Delta and 19,656 at Gediz Delta (Balmer & Betton 2003a). Totals of 31,694, 26,418 and 38,743 were recorded in 2003–05 (Balkız *et al.* in prep.). An unknown proportion of the population disperses beyond Turkey at this season, although the majority is probably resident and may be augmented by an unknown number of immigrants. Iranian birds (mainly immatures) largely move south and west to Europe, north and northeastern Africa and the Arabian Peninsula, although ringing recoveries

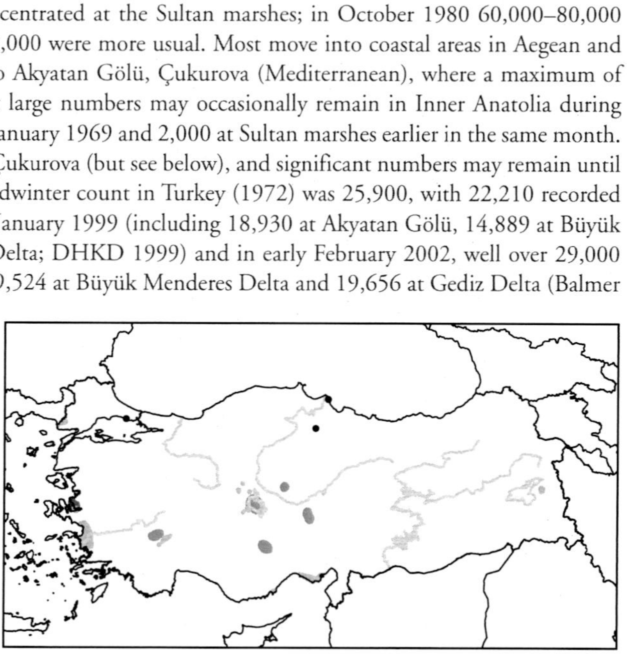

also indicate that smaller numbers move east to the Indian subcontinent (Behrouzi-Rad 1992). Flamingo chicks ringed at Gediz Delta in 2003–04 have been observed in nine countries in the west Mediterranean, North Africa and east Mediterranean, and there have been up to 200 sightings of French, Spanish and Italian-bred flamingos in Turkey (Johnson 1989, Balkız *et al.* in prep.).

Breeding. At seven sites since 1969, all with shallow water and mud, one at sea level, the others at 836–1,110m; six are saline (or brackish) and one fresh water. Breeds in dense colonies, ranging from 100 pairs to 14,000 pairs. Very sensitive to disturbance and the entire colony may desert; lost eggs are not replaced. Nests in shallow water or on low-lying islands of drying-out mud. The nest is a conical mound of mud, drying to form a hard structure, with a shallow hollow in the top; feathers may also be incorporated; usually 25–40cm high, but variable, and occasionally just 10cm. A single egg is laid early April–mid June, perhaps according to water levels. The largest, most regular and most inaccessible colony is at Tuz Gölü (**Inner Anatolia**), the second largest lake in Turkey, which is extremely saline and mostly dries out in summer; courtship and copulation were observed on 31 March 1969 (Warncke 1971), and eggshells and a presumed breeding colony were found in May 1969; on 18 May 1970, there were 5,000 nests (25–30cm high), of which 70% had an egg, 20% young (up to 10 days old), and 10% still empty, with breeding starting early April (Warncke 1971); only 83 nests were found in 1971, 3,500 pairs were nesting on 24 May 1972, with almost all chicks one week old; 3,700 pairs were nesting on 15 June 1973, with all chicks 1–3 weeks old, the latter records indicating laying mid–late April; 1,000 young, aged 30–40 days, were present 11 June 1974 (Kahl 1975), indicating laying early April; there were 11,000 nests and 4,500 chicks in 1991, and 14,000 chicks in 1992 (Magnin & Yarar 1994); a group of 1,000–2,000 chicks, *c.*4 weeks old, was seen on 6 June 1992, indicating laying in early April (PC). An aerial survey the same year located the colony on islands in the south of the lake (Magnin & Yarar 1997), indicating that the chicks seen on 6 June 1992 had walked several kilometres; 40 young of the year were observed at nearby Kulu Gölü on 25/26 August 1974 (T. A. Walsh), and presumably these had flown from Tuz Gölü. See also Status and Distribution for other counts of young at this site, principally based on aerial surveys. At Seyfe Gölü (Inner Anatolia), a shallow brackish lake, flamingos bred on two small islands in 1970 (Husband & Kasparek 1984); 2,000 pairs bred in 1992–93 (Magnin & Yarar 1994), and chicks up to 15 days old were seen on 18–22 June 1992 (G. Magnin *et al.*), indicating laying in early May; all flamingo nests were old on 13 May 1993 but, on 14 June 1993, several hundred nests (some 40cm high) were new and fresh, the mud still wet in many, but only *c.*20 had an egg, and breeding was only beginning; in 1993 a colony of Great White Pelicans *Pelecanus onocrotalus* built 70–80 nests atop old nests of Greater Flamingo here (PC). At Ereğli marshes (Inner Anatolia), there were 35–40 nests, perhaps old, on 10 May 1987; at a freshwater area, on 1 June 1991, there were 1,100 adults and 227 nests on three small islands and most nests were new but still empty, although 68 had an egg (Kirwan 1992d); on 16–17 June 1993, 300 pairs were breeding on four islands and a count was made on two of these, finding 17 empty nests, 54 with one egg, and 33 with young, all no older than one week (Magnin & Yarar 1994), the latter indicating earliest laying *c.*10 May. At Sultan marshes (Inner Anatolia), 1,500–2,000 pairs bred in Yay Gölü (a saline lake) in 1970 (Warncke 1971) and 200 pairs were commencing nesting on 8–10 June 1974 (Kahl 1975). There is a claim of a colony of 500 pairs at Karapınar Ovası (Inner Anatolia) in 1977 (Kılıç 1988, Kirwan 1992d). There were 150 nests in a colony at Acıgöl (**Mediterranean**), the second-most saline lake in Turkey, in 1993 (Magnin & Yarar 1997). At Gediz Delta (**Aegean**), an extensive coastal wetland, there was regular breeding by 100–200 pairs during 1987–91, and 600 pairs in 1993 (Magnin & Yarar 1994); in 1995, 1,450 pairs nested in the salt pans, with first signs of colony formation in the second half of March, but the colony was abandoned in late May (Eken 1997a).

LESSER FLAMINGO — *Phoenicopterus minor*
Küçük Flamingo

(Du: Kleine Flamingo; Fr: Flamant rose; Ger: Rosaflamingo; Swe: Flamingo)

Subspecies and taxonomy. Monotypic. The species is also frequently placed in the monospecific genus *Phoeniconaias* (e.g. Gill & Wright 2006).

Status and distribution. Vagrant. An adult was photographed with Greater Flamingos *P. ruber* at Ereğli marshes (Inner Anatolia) on 10 April 2006 (N. Sullu) and was still present on 16 April at least (E. Yoğurtcuoğlu, KAB).

Another Lesser Flamingo was present at Eilat, Israel between 19 March and 20 May 2006 at least (Y. Perlman *in litt.* 2006), supporting the notion that the Turkish bird was of wild origin. There are a small but increasing number of other records in the Western Palearctic, and the species has bred at least once in southwest Mauritania. Up to 2003 there have been 12 accepted records in Spain (de Juana 2006) and several more since, including attempted breeding in 2007 at Fuentedepiedra (Málaga). There are also at least 21 from France. Some of these European records are known to have involved birds of captive origin but others are almost certainly wild birds. In addition there are at least 11 records of single birds from Morocco (Thèvenot *et al.* 2003), with another in southern Egypt, in November 1992 (Lassey 1994) and one photographed in Kuwait, in December 2007 (K. Al-Nasrallah & R. Al-Hajji).

EUROPEAN HONEY BUZZARD *Pernis apivorus*
Arı Şahini

(Du: Wespendief; Fr: Bondrée apivore; Ger: Wespenbussard; Swe: Bivråk)

Subspecies. Monotypic.

Status and distribution. National conservation status: Near Threatened (population apparently stable or perhaps declining). A local and uncommon summer visitor, breeding in forested areas of Marmara (especially in Thrace), Black Sea Coastlands, parts of Aegean and almost certainly elsewhere (e.g. possibly breeds in Southeast Anatolia). Breeding status and distribution are perhaps partially obscured by late passage birds.

Common and widespread on passage in spring and autumn throughout the country with large numbers recorded at the three major migration watch sites: Bosphorus, Borçka and the Belen Pass/Gulf of Iskenderun area. Spring passage occurs between mid April (doubtfully late March) and late May or early June (exceptionally as late as 18th) in virtually all areas, with peak passage in early to mid May. Spring passage rarely exceeds a few hundreds at the Bosphorus, with a published max. of 1,104 between 23 March and 6 April 1965 (Collman & Croxall 1967), although 8,981 passed between mid March 2006 and late May (KAB *et al.*). Further systematic counts covering the entire spring passage at this site are urgently required. However the 1965 observation, along with other records before mid April, is highly suspect. At well-watched Eilat (Israel) the first migrants usually arrive in mid April, with an earliest date (during six spring counts between 1977 and 1988) of 2 April (Shirihai & Christie 1992). At the Fırtına River mouth, Rize province (Black Sea Coastlands) 29,237 were counted between 27 April and 10 June 1993 (including 10,471 on 4 May), and the majority of 6,855 unidentified raptors were probably also this species (Faldborg 1994), whilst 25,183 passed over Hopa, Artvin province (Black Sea Coastlands) between 21 March and 14 May 1994 (M. Henriksen), with 1,964 there between 21 April and 5 May 1995 (Peške 1995). A total of 1,868 passed through Kızılırmak Delta (Black Sea Coastlands) between 16 April and early June 1992 (Hustings & van Dijk 1994).

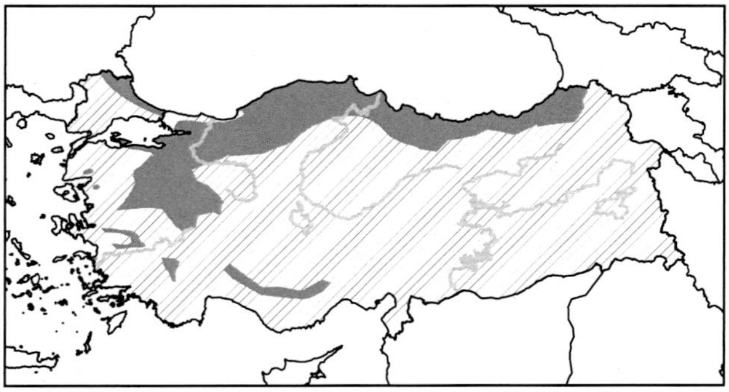

Autumn passage is also protracted, but clearly peaks in early September. Early arrivals appear in the first half of August (8th), while stragglers persist until mid October (13th). At the Bosphorus the autumn maximum was 25,751 in 1971, with peak numbers in late August and early September. The single-day maximum is 6,655 on 29 August 1968. At Borçka autumn passage peaks in late August, also extending into the first half of September, with a maximum of 138,000 in 1976 and a single-day maximum of 37,000 on 4 September 1976 (the only year for which published data based upon extended observations are available; Andrews *et al.* 1977). A total of 20,000 passed Hopa (Black Sea Coastlands) on 5 September 1988 (M. Raes). The Belen pass area has received little systematic coverage, especially in recent years; only in autumn 1976 was it extensively watched, when 15,971 were recorded. Peak numbers apparently usually occur in mid September at this site. The single-day maximum was 5,170 (Sutherland & Brooks 1981a). One report from winter: in the Hatay (Mediterranean) on 31 December 2000.

Breeding. Present in suitable forest and woodland, May–August (the presumed breeding season) but there are no confirmed nesting records. At Kızılırmak Delta (**Black Sea Coastlands**), in 1992, breeding population estimated at 12–20 pairs, mainly in Yürükler Forest; adults were observed displaying, and flying just below and above the canopy, indicative of breeding but no nest had been found when the survey ended on 10 June, and most pairs would only have started to lay by that date (Hustings & van Dijk 1994). Elsewhere, the nest is a large structure of sticks, lined with leaves, built high in a tree; two eggs laid, usually in June, and young fledge mainly in the second half of August.

CRESTED HONEY BUZZARD *Pernis ptilorhyncus*
Tepeli Arı Şahini

(Du: Aziatische Wespendief; Fr: Bondrée orientale; Ger: Schopfwespenbussard; Swe: Tofsbivråk)

Subspecies. No specimen material is available, but the subspecies involved must be *orientalis* on distributional grounds.

Status and distribution. Vagrant. Two records: an adult female photographed at Borçka (Black Sea Coastlands) on 27 September 1979 (Laine 1996) and an adult male there on 25 September 1996 (Kirwan & Martins 2000a).

The discovery that this species is an apparently regular migrant through the Middle East has been among the more dramatic revelations in the region's ornithology in recent years. Shirihai *et al.* (2000) summarised all records known prior to the 21st century: one in Egypt (suggesting that most winter in Africa), 17 in Israel, one in Oman, two in Saudi Arabia and at least 12, most of them in winter, in the United Arab Emirates. There are also two recent records from Lebanon (Ramadan-Jaradi *et al.* 2008). Given steadily increasing observer awareness of the species, there have been many additional records since, with a maximum spring passage of 98 in Israel in 2002, the first records in Jordan, and *c.*18 new reports from the United Arab Emirates (see e.g. Balmer & Betton 2002b).

BLACK-SHOULDERED KITE *Elanus caeruleus*
Ak Çaylak

(Du: Grijze Wouw; Fr: Elanion blanc; Ger: Gleitaar; Swe: Svartvingad glada)

Subspecies. Nominate.

Status and distribution. Vagrant. Ten records as follows. In January 1876, Danford (1878) saw two, possibly a pair at Zebil (now Sebil), Içel province (Mediterranean). Bird (1937) noted two together, at Adana (Mediterranean) on 11 April 1935. Single birds were at the Bosphorus on 14 and 21 October 1931, in March and on 6 April 1933, with two at the same locality on 17 April 1953 (Kumerloeve 1958b). One was well watched at Uluabat Gölü (Marmara) on 17 April 1968 (OST 1972). The most recent records were of an adult, *c.*20 km east of Niğde (Inner Anatolia), on 1 August 1984 (Martins 1989), one at Diyarbakır (Southeast Anatolia) on 6 May 1999 (Kirwan *et al.* 2003), and a second-calendar year individual photographed at Göksu Delta (Mediterranean) on 24–25 April 2006 (*Dutch Birding* 28: 170, 174).

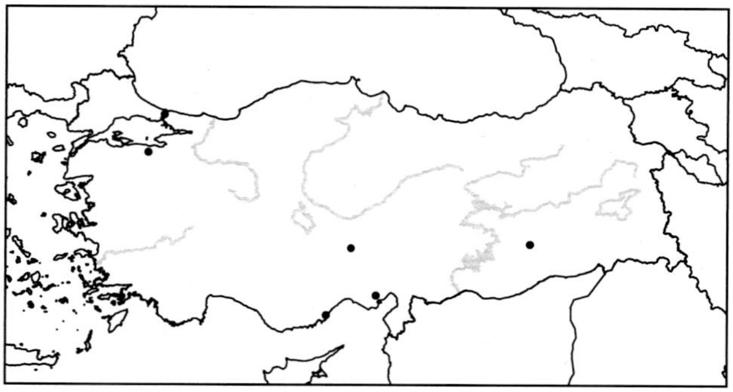

There is recent evidence to suggest the species is expanding its range in the Middle East, with an apparent increase reported in Egypt, and several recent records in southern Iran and four widespread records in Iraq (where it has been discovered breeding: see Salim 2002 and references therein).

BLACK KITE *Milvus migrans*
Kara Çaylak

(Du: Zwarte Wouw; Fr: Milan noir; Ger: Schwarzmilan; Swe: Brunglada)

Subspecies. Nominate breeds throughout Turkey. Whether the form *lineatus*, which has been afforded species status by some recent authors (e.g. Sibley & Monroe 1990), has definitely occurred in Turkey as a migrant/vagrant is unknown. It has been claimed on a number of occasions, principally in the northeast.

Status and distribution. National conservation status: Endangered (population apparently decreasing). Formerly an abundant summer migrant breeder and passage migrant, based on observations at the Bosphorus and Istanbul prior to the advent of modern recording (in 1966). Still widespread, but probably declining summer visitor, breeding in all regions and to at least 3,000m, although now generally local and uncommon. Usually more common in Inner and East Anatolia but apparently virtually absent from much of Mediterranean and Black Sea Coastlands (although still breeds regularly in Istanbul). Nests in or around habitation, rarely away from buildings, though two pairs were recorded within a mixed heron colony at Manyas Gölü (Marmara) in April 1967.

Rather more widespread and locally numerous on passage, particularly during autumn. Spring migration begins in early March (8th), although few are recorded until late in the month, peaking in the first half of May, and substantial numbers continue to pass through northern Turkey until early June. Stragglers continue until mid June. At the Fırtına River mouth, Rize province (Black Sea Coastlands) 2,664 were counted between 8 April and 10 June 1993 (with 727 on 1 May) (Faldborg 1994), and 9,069 passed over nearby Hopa, Artvin province between 21 March and 14 May 1994 (M. Henriksen). At Kızılırmak Delta (Black Sea Coastlands) 698 were recorded between 16 March and 10 June 1992 (Hustings & van Dijk 1994). Autumn passage is from mid August (occasionally late July) to the end of October. At the Bosphorus the autumn maximum was 2,707 in 1971 (with 440 on 31 August) and at Borçka 5,775 in 1976 (Andrews *et al.* 1977), but just 506 at the Belen Pass in 1976 (Sutherland & Brooks 1981a). Sporadically reported in winter, principally from the western two-thirds of the country, although there are records from all regions. Reported twice from Southeast Anatolia at this season,

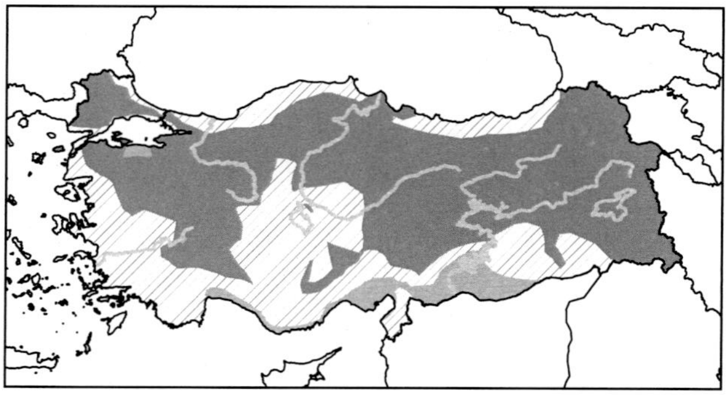

exceptionally 400 at Gaziantep on 23 January 2001 and at least 600 going to roost 25 km southeast of Gaziantep on 11 December 2001, suggesting that large numbers perhaps winter in this region of the country. A group of 100 roosting beside the Seyhan River, Çukurova (Mediterranean) on 11 February 1996 were either very early migrants or a significant wintering concentration.

Breeding. Frequents open country with trees, and woodland, often near habitation. Nests mainly in trees, but uses buildings in Istanbul (Marmara), including a mosque. The nest is of sticks, with a shallow hollow on top, lined with grass and various debris, including paper, rags and fur. Eggs: C/2 (5), C/3 (7). Young: B/3 (2). At Sarayonu (**Inner Anatolia**), 10–15 pairs nesting in woodland on 23 April 1964, most still displaying and nest building, but two nests had two eggs and one had three eggs; also there, on 10 May 1964, 15–20 pairs, and five nests had three eggs and three nests had two eggs, but some eyries in a small colony east of Akşehir Gölü were still unoccupied 23 April 1964 (Warncke 1964–65). At Pozantı (**Mediterranean**), two adults, defending three newly fledged young, observed 24 July 1971, indicating egg laying first half of May. In Inner Anatolia, common near towns and villages on the plateau in 1907; a nest at Karadağ was empty on 26 May, a nest at Eskişehir had fresh eggs on 19 May and, on 22 May, a nest in a garden at Konya held two eggs and a chick, and the chicks were fully fledged, but had not yet flown, by 30 June (Ramsay 1914); this indicates laying in late April. In **Marmara**, very common in Istanbul in 1907 and egg laying on 11–12 May (Ramsay 1914); numbers declined in Istanbul from *c.*500 pairs in the 1930s to fewer than ten pairs in the mid 1970s (Açar *et al.* 1977); adults were incubating at two nests in trees in Istanbul on 17–22 April 1966 and a pair copulated on a school roof on 21 April 1970 (R. F. Porter); pairs bred in a mixed heronry in 1966–67 (at Manyas Gölü), taking up territory in mid April, and a nest in a willow tree had three eggs on 16 May 1967 (R. F. Porter). In **East Anatolia**, 40–50 observed regularly in Van and Erzurum in summer, and a pair was at a nest east of Van on 2 May 1983. Adults were carrying nesting material near Erçek Gölü on 27 May 1969 (Kasparek & van der Ven 1983) and in coniferous forest at Selim (near Kars) on 4 June 2002 (G. Sandve): these dates are rather late for original nest building, and Black Kites, like many other raptors, will add to and freshen nests with additional material as the season progresses.

RED KITE — *Milvus milvus*
Kızıl Çaylak

(Du: Rode Wouw; Fr: Milan royal; Ger: Rotmilan; Swe: Glada)

Subspecies. Monotypic (see Hazevoet 1995, Ferguson-Lees & Christie 2001, Sangster *et al.* 2003).

Status and distribution. Conservation status: Near Threatened (Birdlife International 2008); national conservation status: Data Deficient (population trend unknown). Apparently a very scarce summer visitor to the eastern Black Sea Coastlands, East Anatolia and perhaps adjacent areas of other regions, but no recent records at this season. Occurs principally in forested areas, without evidence of breeding. Otherwise a rare passage migrant and irregular winter visitor, principally in the western two-thirds of the country, although there are very few recent winter records, e.g. just two in the last decades, from Black Sea Coastlands (Kirwan & Martins 2000, Kirwan *et al.* 2003, in press). In spring recorded from early April (earliest 8th), usually in very small numbers, although at least 77 were claimed on passage through the Kızılırmak Delta (Black Sea Coastlands) between 21 April and 31 May 1992 (Hustings & van Dijk 1994), but this total was probably a significant over-estimate and almost certainly included a number of misidentifications (Kirwan & Martins 2000). Latest 9 June (East Anatolia).

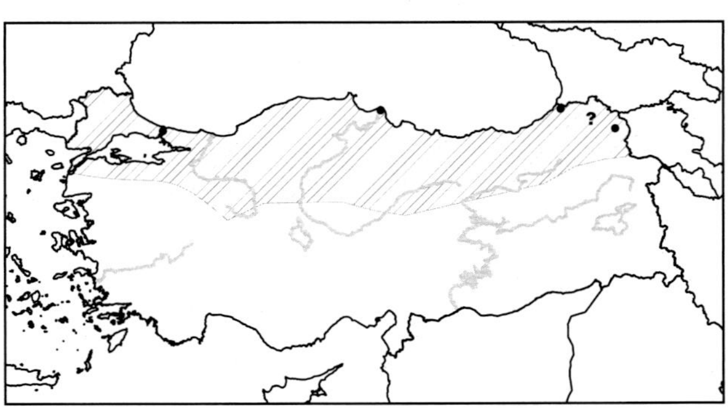

More regular on passage in autumn with single-figure numbers formerly annual at the Bosphorus between late

August and late September, more occasionally early October (26 August–19 October; one East Anatolia record from 9 August and noted in Inner Anatolia from 11 August to 27 October). Rare at other regular migration watch sites, and only one spring record at the Bosphorus (Collman & Croxall 1967). An exceptional *c.*150 birds were claimed between Sero and Van (East Anatolia) on 21 September 1968, the highest single-day total. Confusion has often occurred with pale, especially immature, Black Kites *M. migrans*, thus many records including this one are almost certainly erroneous. Kumerloeve's (1967) account of status and distribution in northeast and east Turkey undoubtedly reflects such confusion.

[PALLAS'S FISH EAGLE *Haliaeetus leucoryphus*]
Moğol Kartalı

(Du: Witbandzeearend; Fr: Pygargue de Pallas; Ger: Bindenseeadler; Swe: Bandhavsörn)

Subspecies. Monotypic.

Status and distribution. Species not accepted onto the Turkish list (Kirwan *et al.* 1999). Records in January/February 1876 and March 1879 (Danford 1878, 1880) are now regarded as referring to immature Golden Eagle *Aquila chrysaetos* (Kasparek 1990a). At Sebil, Çamliyayla, Içel province (Mediterranean), Danford claimed one in January or February 1876 and in March 1879 he described the species as 'by no means rare' around Pınarbaşı, east of Kayseri (Inner Anatolia). Ergene (1945) listed the species as occurring in Anatolia, without details, but presumably based on the above records, and Kiziroğlu (1989) considered the species a rare winter visitor to eastern Turkey, based on records from the Van area. Whilst vagrancy to Turkey is conceivable, there are no adequately documented reports, although Beme (Bohme) noted records from Armenia in winter 1910 and 1911, and the species bred in Crimea in the mid-19th century (Elwes & Buckley 1870). Allouse (1953) recorded it as a winter visitor in small numbers to Iraq and Vaurie (1965) noted winter occurrences in 'Iran (on the Caspian, in Seistan, and on the coast of the Gulf west to the marshes of Iraq and east to Baluchistan)'. Current status in Iraq is unknown, but winter records in Iran have declined dramatically in the last 50 years (Khaleghizadeh 2004). In Europe, there are vagrant records from Finland, Norway and Poland (Cramp & Simmons 1980), and Farman (1869) noted its possible occurrence, and even nesting, in Bulgaria in the mid 1860s. In view of this history we feel that occurrence in Turkey awaits formal documentation.

WHITE-TAILED EAGLE *Haliaeetus albicilla*
Ak Kuyruklu Kartal

(Du: Zeearend; Fr: Pygargue à queue blanche; Ger: Seeadler; Swe: Havsörn)

Subspecies. Monotypic.

Status and distribution. National conservation status: Critical (population apparently decreasing). Until recently a local resident breeder in small numbers, in wetland areas in Marmara (including Thrace, where at least 15 pairs bred in 1966), Aegean, Mediterranean (at Amik Gölü, now drained, and perhaps elsewhere), Black Sea Coastlands and the northern fringes of Inner and East Anatolia. However, now seriously endangered with exact breeding status unclear. No recent breeding season records from Black Sea Coastlands, very few recent breeding records from Marmara, and just three recent breeding season records from East Anatolia. Presumed extinct in many former breeding areas. It has, however, been recorded in small numbers at Çıldır Gölü, Kars province (East Anatolia) in early autumn and it is unknown whether these records reflect passage or perhaps local breeding. A few pairs, almost certainly fewer than 20 and perhaps no more than ten, still breed at a small number of localities in Aegean, parts of Mediterranean and adjacent areas of Inner Anatolia, all areas where the species is known to have bred since the 19th century or the first half of the 20th century. Tucker & Heath (1994) considered the breeding population to be between ten and 30 pairs, but now even smaller (see above). Perhaps also breeds in inland areas in the western half of the Black Sea Coastlands.

More widespread on passage and in winter when the breeding population is augmented by small numbers of immigrants, principally in the western two-thirds of Turkey. Very small numbers, usually fewer than five, are reported from the Bosphorus and the Borçka/Hopa area in September (more exceptionally from early August). More occasional and even scarcer on spring passage, in late March to May. Eight birds, probably all migrants, were recorded in Kızılırmak Delta (Black Sea Coastlands) between 21 March and 23 May 1993 (Hustings & van Dijk 1994). Less common than formerly in winter, with few recent records from Inner Anatolia or Black Sea Coastlands; and none from Southeast Anatolia, where it occurred until the first half of the 20th century. Extent of decline is obvious when one reads accounts such as that of Ogilvie (1954), who recorded the species as 'fairly common on the [Black Sea] coast' in the mid 1940s. Still regular, in very small numbers in Aegean and Mediterranean, particularly in the Bafa Gölü/Büyük Menderes Delta area in the former region (usually 1–2 birds) and at the Göksu Delta in the latter (3–4). There are very few recent records from the Kızılırmak Delta (Black Sea Coastlands) where, until the mid 1970s the species bred (Dijksen & Kasparek 1985) and was regular in winter (up to five birds).

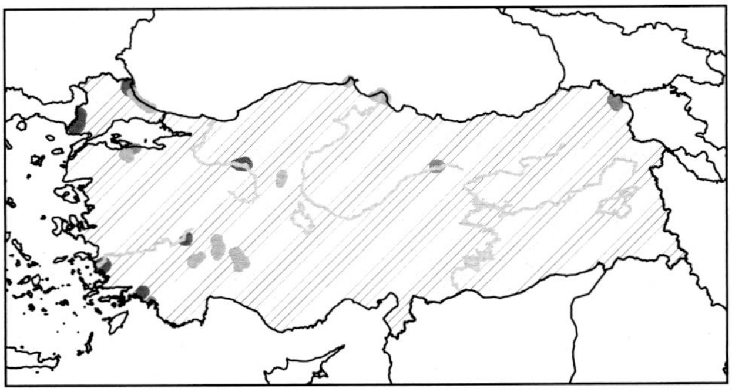

Breeding. Most current, past or suspected breeding sites are at sea level, on nearby hillsides or at low elevation near lakes, but two current sites are well inland, in montane areas near lakes. Nests in a tree, usually mature and tall, on flat or sloping ground; mainly in forest, but occasionally in solitary trees. Most commonly on pine, but also poplar, oak and ash; has also nested on inland cliffs (Danford 1877–78, 1880). At least two nests per territory, used in alternate years. The nest is a massive structure, growing over many years of use; of branches, with a shallow hollow in the centre at the top, lined with fresh leafy twigs which are renewed regularly throughout the season. Eggs: C/1 (3), C/2 (5), C/3 (2). Young: no data. Danford (1877–78) stated that nesting was in 'low cliffs in the interior'; he visited an empty nest, in the Aladağ (**Mediterranean**) on 30 April 1876; in May 1876, he took the single egg from a nest on the topmost branches of a tall, slender tree. In Birecik (**Southeast Anatolia**), Danford took clutches of two and three eggs, all in an advanced stage of incubation, on 14 February 1879, from two nests on cliff ledges with bushes, as well as eggs from a nest, in a solitary fir, near Malatya (**East Anatolia**), on 9 March 1879 (Danford 1880). Selous (1900) visited a large lake near Izmir (**Aegean**) on 14 May 1899, finding two pairs and a nest (a huge pile of sticks), in a large fir in hills above the lake; eight eggs (4 × C/2) in the Natural History Museum, Tring, were taken by Selous at Bafa Gölü (?) (Aegean), on 20 January 1902, 1 February 1902 and 2 February 1902; and he took two further eggs (2 × C/1), also in Tring, at Beşparmak Dağı, northwest of Acıgöl (Mediterranean), on 8 February 1902 and 16 February 1902. Kumerloeve (1963) mentioned a nest with three eggs in the Antakya area (Mediterranean), on 5 April 1930. At Kızılırmak Delta (**Black Sea Coastlands**), an adult flew from an occupied nest, 10m high in a tree in woodland at Balık Gölü, on 9 June 1969; there were five disused nests in July–August 1971: four in ash trees and one in an oak, and two adults and two young were present on 22 July 1971 (R. F. Porter; Dijksen & Kasparek 1985), but no subsequent breeding records there (Hustings & van Dijk 1994). An adult carrying nesting material at Düden Gölü, south of Elmali (Mediterranean), on 14 May 1972, was more likely adding to a nest, rather than starting nest building, given the late date. An immature was at Yeniçağa Gölü (Black Sea Coastlands) on 19 May 1992 (PC). One or two pairs breed in Kocaçay Delta (Marmara), an abandoned eyrie was found there in 1993 (Ertan 1996), and another unoccupied nest on 24 June 1999 (PC). An occupied nest was found on a tall pine, in a long inaccessible valley at Akdağ, near Işıklı Gölü (**Aegean**), on 4 May 1996 (Heredia *et al.* 1997), A pair of adults was at Köyceğiz Gölü (Mediterranean) on 26 May 2001 (C. Cameron & J. Dawson), and breeding is presumed in the surroundings (Magnin & Yarar 1997). At a well-known breeding area, a huge unoccupied nest, with grass growing on the rim and 20m above ground in a pine, was found on a wooded hillside, with impenetrable thickets below the trees, near Bafa Gölü (Aegean) on 30 April 2003; a local fisherman claimed to have found an occupied nest, on which the bird was incubating, in March the same year in a tall pine on the opposite side of the lake (PC).

LAMMERGEIER *Gypaetus barbatus*
Sakalli Akbaba

(Du: Lammergeier; Fr: Gypaète barbu; Ger: Bartgeier; Swe: Lammgam)

Subspecies and taxonomy. Nominate. Roselaar (1995) listed the weakly defined subspecies *aureus* for Turkey, but like Orta *in* del Hoyo *et al.* (1994) and Ferguson-Lees & Christie (2001) we prefer not to recognise this form. Genetic evidence, recently published (Lerner & Mindell 2005, Griffiths *et al.* 2007), supports recognition of two families of Old-World vultures, the Gypaetinae (including the present genus and *Neophron* amongst the four constituent genera) and the Aegypiinae (all other genera).

Status and distribution. National conservation status: Endangered (population apparently decreasing). A fairly widespread resident breeder in most mountainous areas, usually at 1,500–4,000m, but occurring erratically at lower altitudes, most frequently immature non-breeders or outside the breeding season. Occurs in all regions, although there is only one record from Thrace and it is more local in regions lacking extensive suitable habitat, e.g. Marmara, Aegean and much of Southeast Anatolia. Perhaps formerly more common, although the evidence for this supposition is largely circumstantial: e.g. Danford (1878) remarked that the species was 'so common throughout the Taurus that hardly a day passed without our seeing some of them', while other authors including Kumerloeve (1961) referred to occupied breeding localities from which there are no recent records.

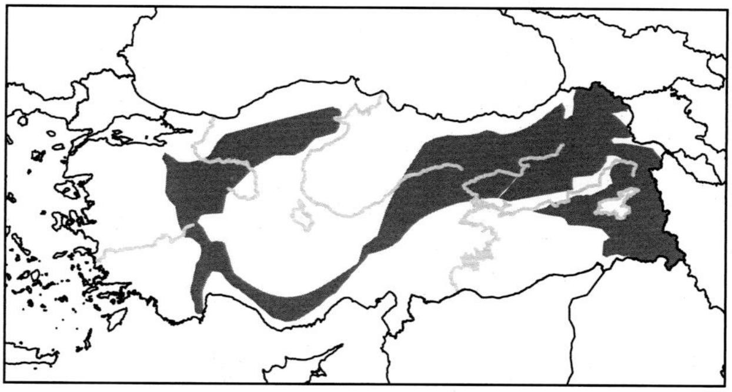

Breeding. On huge cliffs in high mountains. Most suitable cliffs are tenanted in the Köroğlu Dağları and Sündiken Dağları, in **Inner Anatolia** (Heredia *et al.* 1997); 1–3 pairs probably breed in the eastern summit region, above 2,000m, at Uludağ (**Marmara**), which is probably the most important breeding ground in the west (Jetz 1995); and 20 pairs breed in the Doğu Karadeniz Dağları in the northeast Black Sea Coastlands (Magnin & Yarar 1997). Nests in a hollow on a high cliff (Danford 1878). Observations elsewhere are of a large and relatively shallow nest, of branches and twigs, lined with animal bones, the site heavily marked by droppings (Harrison & Castell 2002); a nest near Pozantı (Mediterranean) was lined with rags and a lock of woman's hair (Danford 1878). Eggs: C/1 (2), C/2 (1). Young B/1 (1). In **Mediterranean**, breeding commences in the Taurus about late January (Danford 1878); a nest had two eggs at Pozantı on 2 February 1876, and another a single fresh egg in late February 1876 (Danford 1878); an egg in the Natural History Museum, Tring, was collected by F. C. Selous at Beşparmak Dağ, northwest of Acıgöl, on 8 February 1902; an immature, at Demirkazık on 15 April 1996 (L. & L. Vallotton), must have been from a previous year, as young could not fly until mid June, even if eggs were laid on 1 January (the entire breeding period takes 5.5 months). In **Black Sea Coastlands**, a nest contained one young, at Çiçekli (south of Rize), on 6 May 1970, and an adult and juvenile were at Alıcık, southwest of Merzifon, on 25 July 1974. Eggs in the Natural History Museum, Tring, collected by Krüper, include three (3 × C/1) from neighbouring Armenia (one taken on 3 January 1881), and two (2 × C/1) from Greece, on 14 January 1872 and in February 1873.

EGYPTIAN VULTURE *Neophron percnopterus*
Küçük Akbaba

(Du: Aasgier; Fr: Vautour percnoptère; Ger: Schmutzgeier; Swe: Smutsgam)

Subspecies and taxonomy. Nominate. See remarks under Lammergeier.

Status and distribution. Conservation status: Endangered (BirdLife International 2008); national conservation status: Vulnerable (population apparently decreasing). Widespread and generally fairly common summer visitor, breeding in most if not all regions, but apparently very local or absent from coastal areas in Marmara, Aegean, Mediterranean and the western half of the Black Sea Coastlands. In Southeast Anatolia commonest in the east of the region. Found in a wide variety of montane and other open habitats, usually breeding at between 1,000 and 2,200m. Occasionally wanders to adjacent lowlands, such as the major river deltas, in summer. Arrives in Turkey between mid and late March (earliest 7th). Most have departed by the end of September but stragglers remain until mid October (12th).

More widespread on passage across the country, being commoner in autumn from mid August to early October, peaking September. Spring passage may persist until the end of the first week of June. Rather scarce and irregular at Borçka but common at both other major migration watch sites: largest seasonal autumn totals include 544 at the Bosphorus in 1971 and 874 at the Belen Pass in 1976 (Sutherland & Brooks 1981a), with single-day maxima of 152 (on 13 September 1974) and 124 (on 12 September 1976) respectively. One winter record: at Kocaçay Delta (Marmara) on 26 January 1993 (B. Güneş *et al.*).

Apparently, formerly much more common. The species at one time even nested in Istanbul: Alléon & Vian (1869) estimated that over 1,000 young birds were raised in and around the city each year, and in the 20th century Bird (1937) and Hollom (1955) described it as common around towns in southern Anatolia. Kasparek (1992a) speculated that the population entered decline around the beginning of the 20th century, stabilising at its present diminished level in the early 1960s.

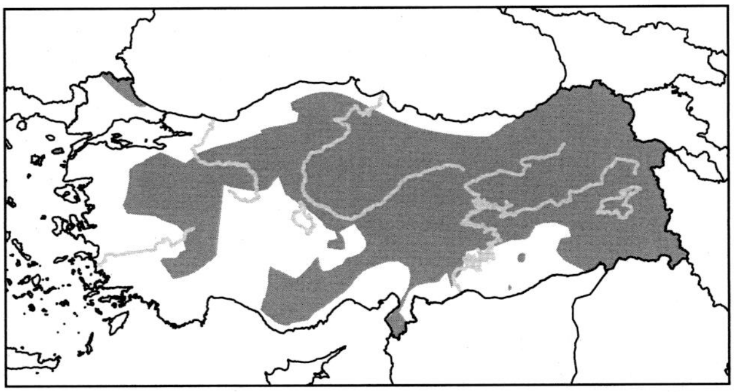

Breeding. Frequents hilly and montane areas with cliffs and gorges, but also open country with suitable nesting sites, namely small caves and recesses in cliffs; also recorded nest building on Van castle (East Anatolia). Sites often re-used annually, becoming plastered with excreta. A nest at Karadağ (Inner Anatolia) was constructed of sticks and various items, including wool, and the floor of the cave was thickly strewn with shells of small tortoises (Ramsay 1914). Eggs: C/2 (5). Young: B/1–2 (2). In **Aegean**, Selous (1900) found six occupied nests near Aydin on 7–8 May 1899: two had two slightly incubated eggs (first egg presumably laid late April), one was ready for eggs, and adults were incubating on the remaining three nests, but they were unreachable. In **Inner Anatolia**, nest building observed early April, an adult was already incubating near Ereğli marshes on 21 April 2004 (A. Dixon), and six nests had eggs from 6 May. In **Mediterranean**, two nests had eggs on 10 May and 26 May and two pairs were at nests, presumably containing young, in June. In **East Anatolia**, an adult was incubating eggs or brooding small young, near Van, on 2 June 2001 (T. Quittenden); two pairs were at nests, presumably with young, in late June; and a large young was fed at a nest at Kağızman on 16 August 1972; the latter indicates egg laying in early May.

EURASIAN GRIFFON VULTURE *Gyps fulvus*
Kızıl Akbaba

(Du: Vale Gier; Fr: Vautour fauve; Ger: Gänsegeier; Swe; Gåsgam)

Subspecies and taxonomy. Nominate. See remarks under Lammergeier.

Status and distribution. National conservation status: Endangered (population apparently decreasing). Resident breeder or partial migrant, most common and widespread (in both summer and winter) in Mediterranean, adjacent areas of Inner Anatolia (now very scarce), eastern areas of Black Sea Coastlands and East Anatolia. Very local as a breeding bird in western areas of the country, in Aegean (only a handful of pairs) and Marmara (especially Thrace), but also relatively uncommon in much of Southeast Anatolia: where largely present in the east of the region. Generally recorded in high mountainous, more occasionally well-forested areas, principally above 1,500m, locally as high as 4,500m.

Formerly more common, significant concentrations being noted in the past in several regions where the species is now rare or no longer occurs (e.g. at Gaziantep (Southeast Anatolia) and Bürücek (Mediterranean); see Kumerloeve 1961). In the second half of the 19th century the species' breeding range was considerably more extensive, it occupying even lowland areas, e.g. around Izmir and the Büyük Menderes Delta (Aegean; von Gonzenbach 1860, Kasparek 1988), and even as late as the 1950s Hollom (1955) was able to note the species over central Adana and Kumerloeve (1957c) as a breeder around Bergama (Aegean). The population was presumed to have stabilised in the 1960s (Kasparek 1992a) at between 100 and 1,000 pairs (Tucker & Heath 1994), but further declines have been evident since, leading Vaassen (2001) to postulate that the current population may number no more than 150 to 500 pairs, with 35–50 of these in the central Taurus.

More widespread on (principally autumn) passage throughout the country, although generally uncommon as a migrant at Borçka, where 30 were recorded in mid-October 1977 (Beaman 1977). Autumn maxima greater at other well-watched sites, e.g. at the Bosphorus 165 in 1931 (Steinfatt 1932) and, at the Belen Pass, 125 in just six days in 1976 (in Sutherland & Brooks 1981a), but numbers at both sites much lower in recent years, e.g. just 13 at the former between 19 September and 2 October 2002 (*Birding World* 15: 423). Autumn passage is usually mid September (occasionally late August) to early October. At the Bosphorus in spring, only recorded by Mauve (1938). Overwinters generally in small to moderate numbers, principally in the western two-thirds of Turkey, the majority descending to lower altitudes at this season, e.g. it is still occasionally recorded in the environs of Izmir.

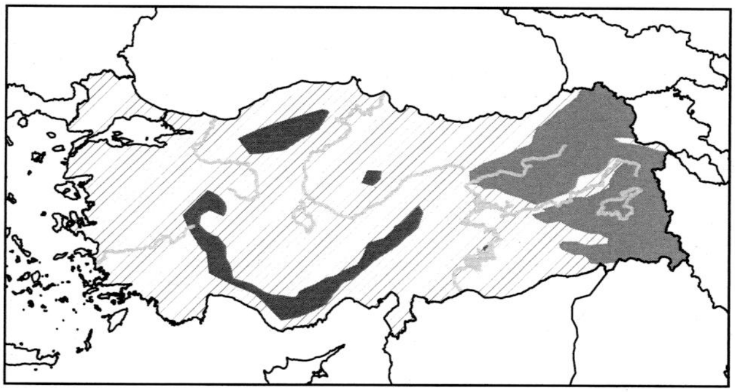

Breeding. In mountainous areas with high cliffs, gorges and rocky outcrops; usually in small colonies of 2–8 pairs, but at least 15 nests were in a colony near Ispir (Black Sea Coastlands) in 1993 (Faldborg 1994), and a solitary pair was at Nemrut Dağ in East Anatolia (Magnin & Yarar 1997). The breeding population in the central Taurus (Mediterranean), estimated at 35–50 pairs, is mainly in the Sütcüler area and the canyon of the River Göksu, but there are also several other small colonies of 2–3 pairs (Vaassen 2000). Nests in a cavity, or on a ledge, of a cliff; the nest is a small flat structure of sticks, lined with leafy twigs and debris. Eggs: C/1 (6). Young: B/1 (17). In **Mediterranean**, adults nest building late February–early March in the Bolkar Dağları (Danford 1878); on steep cliffs facing the sea, between Alanya and Silifke one nest had an egg, another probably a young bird and another a chick *c.*5 days old, on 13 April 1971. In the same region, on 4 June 1971, there were eight nests with young almost ready to fly (Warncke 1972); the chick, aged five days, indicates laying in about mid February, but we consider that the young in the eight nests were probably not ready to fly on 4 June 1971 in view of the early date (egg laying would have been in December, and this is highly unlikely); young are feathered at *c.*10 weeks old (but can fly after *c.*16 weeks), so these birds had perhaps laid in early February. At Karanfil Dağ there were at least five nests, each with a single juvenile, on 12 September 1972, and presumably these were young birds still attending nests, but already fledged. Two eggs (C/1 × 2) in the Natural History Museum, Tring, were collected by F. C. Selous at Beşparmak Dağ on 25 March 1897; and three further clutches (C/1 × 3) were also collected by Selous in Turkey on 22 March 1897, 30 January 1902 and 20 February 1902. In **East Anatolia**, there were at least five nests in a colony, of which at least three had a single chick, 30km west of Tuzluca, on 3 September 1992 (G. Sandve). A colony of five pairs, with juveniles, was at Arastal (west of Kağızman) on 3 September 1973; 20 adults were nesting at Iğdır on 4 July 2001; 11 adults were at a colony near Hakkari on 17 May 1975. At Ispir (**Black Sea Coastlands**) there were 35 adults and 17 nests on 10–11 May 1991, 13 adults and eight nests in June 1992, and 18 adults and 15 nests on 10 May 1993 (J. Faldborg), and probably more than 15 nests in 1992 (Yarar & Magnin 1993). There are also small colonies in other regions, including three pairs at Akdağ, near Denizli (**Aegean**), two pairs at Kızılcahamam (**Inner Anatolia**), and breeding is suspected on Uludağ (**Marmara**) (Magnin & Yarar 1997).

EURASIAN BLACK VULTURE *Aegypius monachus*
Kara Akbaba

(Du: Monniksgier; Fr: Vautour moine; Ger: Mönschgeier; Swe: Grågam)

Subspecies and taxonomy. Monotypic. See remarks under Lammergeier.

Status and distribution. Conservation status: Near Threatened (BirdLife International 2000, 2004); national conservation status: Endangered (population apparently decreasing). A fairly widespread but generally uncommon resident breeder and perhaps partial migrant in most regions. Probably very local in Thrace (no recent records and also absent from adjacent Strandja mountains of Bulgaria; Milchev 1994) and absent from Southeast Anatolia. Principally breeds in montane forests at moderate altitudes (usually above 900m) but occasionally also recorded in arid, mountainous areas and plains, sometimes at lower altitudes in western and, locally, northern areas. The core area for the Turkish breeding population is apparently the northern and western fringe of Inner Anatolia (e.g. around Kızılcahamam and Bolu, where up to 60 birds were reported in July 2007) although the species is also reported regularly from the central and eastern Black Sea Coastlands and much of East Anatolia. A smaller population survives in the central and eastern Taurus, but that in the Konya Basin is almost extirpated (Eken & Magnin 1999) except in the extreme west of this region (M. Yarar). Occasionally reported from other areas, usually outside the breeding season, including the Bosphorus and Dardanelles (Marmara) where irregularly recorded, apparently in declining numbers on autumn passage late August to end of September, with only one recent spring record (7 April 2006). Also recorded at Borçka at the same season, where also very scarce. Formerly regarded as common on passage in Marmara (Alléon & Vian 1869).

The Turkish breeding population has apparently diminished substantially during the 20th century and this trend probably continues. Based largely on circumstantial evidence, the species is now apparently absent from several old localities, principally in the Taurus, but also Beynam Forest, near Ankara (Inner Anatolia), which in the early 1970s supported a small population that has been lost, seemingly through poaching and poisoning. Kasparek (1992a) suggested that perhaps fewer than 50 pairs breed in Turkey but this figure is almost certainly a significant underestimate, especially in the light of a recent record of 40 between Beypazarı and Kıbrısçık, in northernmost Inner Anatolia, on 9 October 2005 (G. Eken & Y. Lise). In recent years fuller understanding of the breeding distribution in eastern Turkey has been achieved. In addition, several formerly important areas, e.g. the Ilgaz Dağları, Çankırı/Kastamonu provinces (Black Sea Coastlands), have received little or no recent ornithological attention. Some such areas may retain healthy populations of the species.

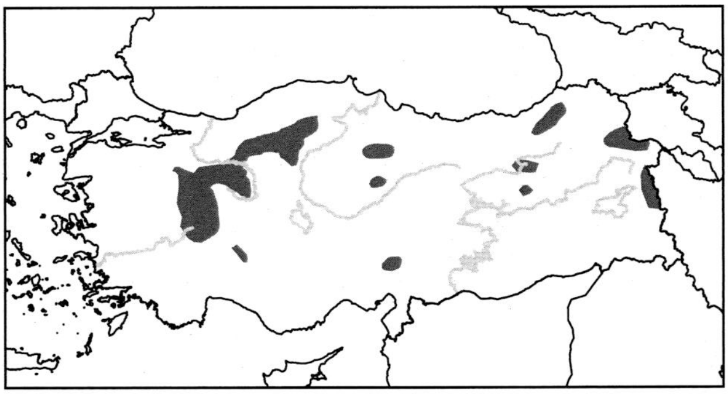

Breeding. Favours coniferous forest, mainly pine and preferably old, in steep-sided and usually remote valleys and on slopes in montane areas. A solitary breeder, although a few pairs may nest in the same valley in loose association, but with the nests well spaced. Nest built high in a mature pine, usually at the top; of large sticks, lined with leafy branches and bark, and debris, including skin. New nests are large but old nests, re-used annually, become enormous. Nesting period extremely long: incubation lasts *c.*54 days, and young remain in the nest for *c.*100 days; thus, a young bird would fledge in late July from an egg laid in early March. Eggs: C/1 (3). Young: B/1 (5). Three eggs (3 × C/1) in the Natural History Museum, Tring, were collected by F. C. Selous at Beşparmak Dağ, northwest of Acıgöl (**Mediterranean**), on 8 March 1897, 10 March 1897 and 25 March 1897. At Kızılcahamam (**Inner Anatolia**), a nest in a pine tree contained a young bird 'the size of a duck' on 15 June 1983 (Barış *et al.* 1984). All details below are from Heredia *et al.* (1997): at Soğuksu National Park, four nests each contained a large fully feathered young bird in July 1994. The same nests were occupied in 1995 (7–21 May 1995), and two additional ones were found: five of these were in one valley, near a tributary stream, at 1,500m. At Haman Dağı (Inner Anatolia), two nests had young, *c.*14 days old, in pine trees in a small valley at 1,300m on 11 May 1995, indicating egg laying about 5 March. At Murat Dağı (**Aegean**), two occupied nests were located on 1–2 May 1996 at 1,400–1,600m, in pines within a long valley intensively managed for forestry.

SHORT-TOED SNAKE EAGLE *Circaetus gallicus*
Yılan Kartalı

(Du: Slangenarend; Fr: Circaète Jean-le-Blanc; Ger: Schlangenadler; Swe: Ormörn)

Subspecies. Monotypic.

Status and distribution. National conservation status: Vulnerable (population apparently decreasing). A reasonably widespread and fairly common summer visitor, breeding in all regions although perhaps more local in the eastern third of the country. Principally occurs in dry, open and/or hilly country, typically at low to moderate altitudes: from near sea level to over 2,000m. Breeding commences in April (see below).

More widespread and relatively common on passage throughout the country. Formerly regarded as the commonest migratory eagle (see Nisbet & Smout 1957). Passage in spring from mid March (more occasionally early March) to the end of May (and probably early June). Passage in autumn from mid August (perhaps regularly late July in northern areas) to late October (29th), peaking in the second half of September and early October. In spring, significant numbers may move through Turkey as early as late March, with a total of 204 at the Bosphorus in 1993 (1 March–9 April). Particularly common at the Bosphorus, where the autumn maximum is 2,342 in 1971 (including 850 on 26 September) and the Belen Pass, where the maximum is 727 in 1976 (including 114 on 18 September; Sutherland & Brooks 1981a). A total of 837 passed over Çukurova from 18 September–13 October 1994 (R. Thorpe & J. Young). Rather less common at Borçka where the maximum count is 243 in autumn 1976 (Andrews *et al.* 1977) and 395 at nearby Hopa in spring 1994 (M. Henriksen), although 1,156 were counted at nearby Batumi (Georgia) in autumn 2002 (*Sandgrouse* 25: 89). Winters, apparently irregularly and in very small numbers, in southern coastal areas, especially Mediterranean.

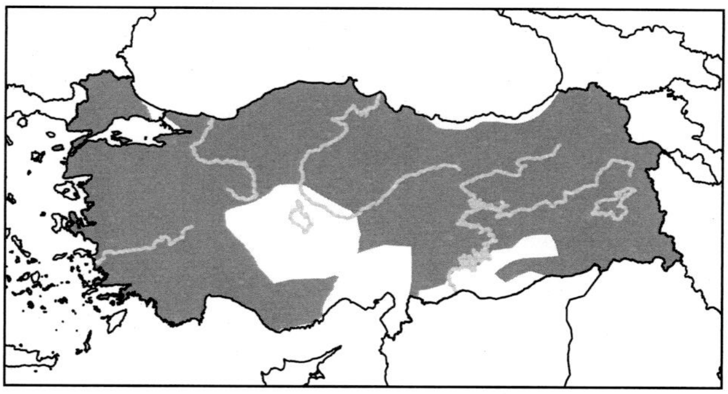

Breeding. In dry open or sparsely wooded country, mainly in hills. Nests in a tree or, occasionally, on a cliff. The nest is a broad but rather thin structure of small branches and twigs, with a hollow in the centre of the top lined with green leafy twigs. A new nest may be built each year but, more usually, there are 2–3 nests in a territory, used in succession, but the same nest may be used in consecutive years. Eggs: C/1 (4). Young: B/1 (2). In **Inner Anatolia**, an eyrie in an almond tree at Avanos, near Ürgüp, had one egg on 3 June 1971, one young, 25 days old, on 22 July 1971, and one egg, in the same nest, on 30 May 1972: the record in 1971 can be back-dated to egg laying about 11 May. An eyrie at Delice was on a cliff ledge and had one young, *c.*18 days old, on 25 June 1977 (Pforr & Limbrunner 1982), indicating egg laying about 21 April. Display was observed at Kızılcahamam on 13 May 1986 (M. Ullman). Two eggs (C/1 × 2), in the Natural History Museum, Tring, were collected by Krüper, in the Izmir area (**Aegean**), on 18 April 1871 and 26 May 1872. In **Marmara**, display was observed in Forest of Belgrade on 23 April 1970 and west of Terkos on 21 June 1973 (R. F. Porter); in 1993, at least two pairs bred in the hills around Kocaçay Delta, and an adult was seen carrying food (Ertan 1996). At Gaziantep (**Southeast Anatolia**), an adult was observed carrying a snake on 7 May 1992, and a female was incubating at a nest on a crag on 8 June (M. Eldridge). In **Mediterranean**, display was observed by a pair in the hills north of Taşucu on 23 March 1993 (J. van der Winden), by at least one pair at Suğla Gölü on 11 April 1991 (M. Bakker), and by three birds in the Dalyan area, on 3 May 1996 (A. & S. Dijksen). An adult was seen carrying a snake at Darim Dağı on 9 May 1995, and another, also carrying a snake, at Termessos on 22 May 1995 (I. A. Green). A pair was located on a large rocky island, covered by junipers and pines, in Beyşehir Gölü, 20 May–2 June 1998 (A. Kullberg). A nest was being protected by the local forestry department, near Çandır (near Sütcüler), on 6 April 1991 (M. Bakker).

WESTERN MARSH HARRIER — *Circus aeruginosus*
Saz Delicesi

(Du: Bruine Kiekendief; Fr: Busard des roseaux; Ger: Rohrweihe; Swe: Brun kärrhök)

Subspecies. Nominate.

Status and distribution. National conservation status: Near Threatened (population apparently decreasing). Resident breeder and partial migrant throughout most of Turkey, usually in small to moderate numbers, principally in marshy wetlands with extensive reed cover. An estimated 250–275 pairs in Kızılırmak Delta (Black Sea Coastlands) in spring 1992 (Hustings & van Dijk 1994). Apparently absent from much of Southeast Anatolia and the region south and southeast of Van Gölü during the breeding season.

Widespread and relatively common on passage throughout Turkey: in spring from mid March (with most adults passing through south-coast localities by the end of March) to at least mid May, in autumn between late August (22nd) and mid October. Extreme dates of passage strongly obscured by presence of wintering birds. A count of 385 at Borçka in autumn 1976 (Andrews *et al.* 1977) is by far the largest total at any migration watch site in Turkey and one of the highest from any area in the Western Palearctic. Similarly 251 at nearby Hopa in spring 1994 was also a significant total (M. Henriksen), but is dwarfed by the 1,205 that passed Batumi, in neighbouring Georgia in autumn 2002 (*Sandgrouse* 25: 89). Fairly common in winter throughout the western two-thirds of Turkey, when particularly common in the major river deltas, and also recorded wintering in small numbers around Malatya (East Anatolia).

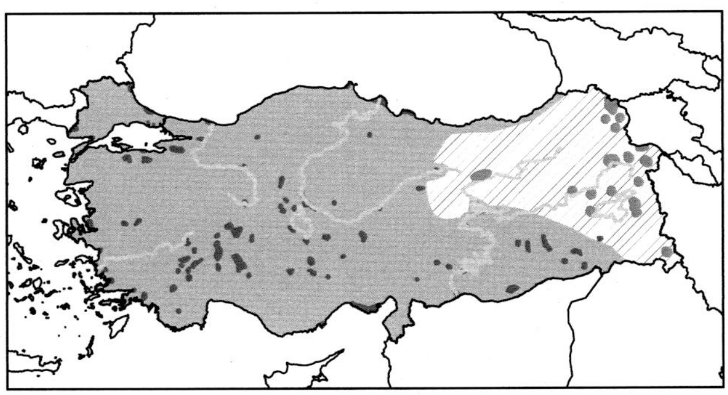

Breeding. Usually in extensive and dense cover of tall *Phragmites* but sometimes in quite small areas of mature reed, e.g. in 400m^2 at Bolluk Gölü (Inner Anatolia) in 1993 (PC); also in *Carex* sp. and other lower marsh vegetation, e.g. at Eşmekaya marshes (Inner Anatolia). Nest built up from the ground in shallow water, or raised on a platform of broken-down reeds; it is a large mass of sticks, reeds etc. lined with grass. Eggs: C/3 (2), C/4 (10), C/5 (1), C/6 (2). Young: B/3 (3). In **Inner Anatolia**, nest building at Sultan marshes on 15 April 1996 and at Mogan Gölü on 8 May 1946; 11 nests had full clutches in May (earliest 23 April) and incomplete clutches of two eggs and one egg on 20 May 1998 and 16 May 2004; three nests had eggs/newly hatched young 24–27 May, indicating laying of first eggs *c.*20 April. One nest had three newly hatched young and an egg as early as 15 May 2004 (D. Gregson), indicating laying of first egg *c.*5 April. A nest had three eggs at Tödürge Gölü on 13 June 2006 (M. Cowlard). In **Black Sea Coastlands**, a nest had four eggs at Yeniçağa Gölü on 19 May 1992 (PC); in Kızılırmak Delta, a nest had five eggs on 26 April 1992, and many were seen carrying food in the first ten days of June (Hustings & van Dijk 1994). In **Marmara**, nests found on 2 May 1993 and 13 June 1993 in Kocaçay Delta (Ertan 1996), and display observed at Durusu Gölü (Thrace) on 26 April 1981 (H. Hols). In **East Anatolia**, two pairs were nest building at Çenge Gölü on 27–28 May 1969, but, in view of the late dates, this is considered to be nest repair rather than original nest building; many juveniles were at Ardahan on 29 July 1999.

HEN HARRIER *Circus cyaneus*
Gökçe Delice

(Du: Blauwe Kiekendief; Fr: Busard Saint-Martin; Ger: Kornweihe; Swe: Blå kärrhök)

Subspecies. Nominate; New World *hudsonicus* may warrant species status.

Status and distribution. National conservation status: Data Deficient (population trend unknown). Principally a winter visitor and passage migrant, usually to wetland areas below 1,500m in the western two-thirds of Turkey. Its status as a breeding bird is unclear. Kumerloeve (1969a) recorded a pair with two fledged young in the floodplain of the Altındere, north of Erciş, Van province (East Anatolia) on 3 August 1968, with a male in suitable breeding habitat at Ödland (untraced locality in East Anatolia) on 10 June 1968, and an observation of a nest on an island at Aktaş Gölü (East Anatolia) (G. Sarıgül *in* Bilgin & Demirci 2002). There are about five recent records of individuals apparently summering in northeastern, eastern and southern Turkey, and the species breeds in nearby Georgia (Massa & Fontana 2004) but not in Armenia (Adamian & Klem 1999). Recorded once in July in Inner Anatolia. Otherwise, arrives from the beginning of September (occasionally late August; earliest 29th) and remains until the end of April (less often the second week of May; latest 15th) over much of Turkey. Usually recorded singly, although small roosts may form at some sites. Peak return passage is probably completed by early April. Very few records are available from the major migration watchpoints, although 17 moved through the Bosphorus in early spring 1993 (P. Denes *in* Zalles & Bildstein 2000), but large numbers clearly must move through the northeast, as 1,386 moved through the Batumi area of neighbouring Georgia in autumn 2002 (*Sandgrouse* 25: 89).

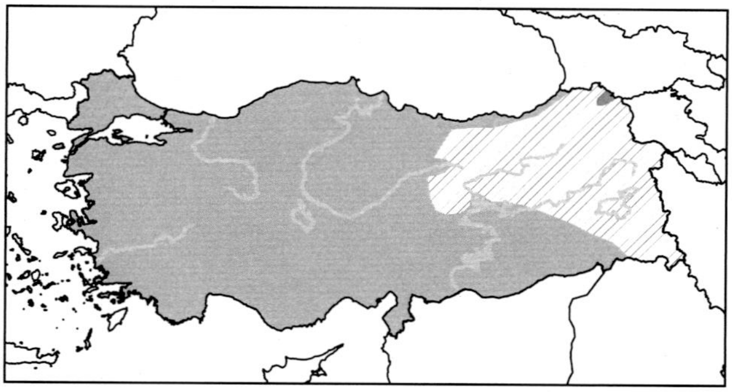

PALLID HARRIER *Circus macrourus*
Bozkır Delicesi

(Du: Steppekiekendief; Fr: Busard pâle; Ger: Steppenweihe; Swe: Stäpphök)

Subspecies. Monotypic.

Status and distribution. Conservation status: Near Threatened (BirdLife International 2000, 2004); national conservation status: Critical (population apparently decreasing). Fairly widespread and relatively common passage migrant, principally through the eastern two-thirds of Turkey; in spring from the third week of March (15th) to late May (29th), peaking in mid April (with females generally slightly later), and in autumn from late August (22nd) to mid October (latest, 29th), with the majority in September. Usually recorded in small numbers, although exceptional observations include description as 'abundant' around Erzurum (East Anatolia) in spring 1838 by Dickson & Ross (1839) and 133 (but probably many more) through Borçka in autumn 1976 (Andrews *et al.* 1977). In addition, Alléon & Vian (1869, 1870) claimed thousands on passage at the Bosphorus, although probably over a period of several years (see Nisbet & Smout 1957). Confusion with other harrier species seems likely. Nonetheless, at Batumi, in neighbouring Georgia, 1,314 were counted moving south in autumn 2002 (*Sandgrouse* 25: 89).

Recorded several times in recent decades in late May–June in Inner and East Anatolia, usually without evidence of breeding (e.g. Groh 1971, Kirwan *et al.* 2003). For evidence of breeding at Eşmekaya (Inner Anatolia), see below, but note that Warncke's (1970) claim to have discovered a breeding area with 10–13 pairs in the vicinity of Tuz Gölü (Inner Anatolia) in late May 1970, is surely indicative of confusion with Montagu's Harrier. A late spring bird in Southeast Anatolia, at Siverek on 2 June 2002, was perhaps still on migration. Also recorded in winter in the western two-thirds of Turkey, with records from Marmara (two), Aegean (two), Black Sea Coastlands (one), Mediterranean (probably regular, including five together in Göksu Delta in December 2001) and Inner Anatolia (two).

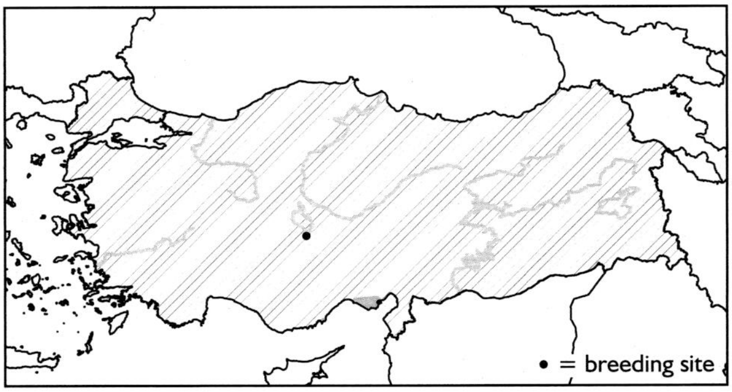

Breeding. Has bred only near Eşmekaya (Inner Anatolia), in a formerly extensive shallow marsh, dominated by *Carex* sp. and surrounded by dry cultivation, orchards and rough grazing. A nest had three fairly large young in late June 1977 (P. Zeininger pers. comm.). Two nests were found in 1993 (see Pleasance 1997): they each comprised a pad of dried grasses and sedges (one *c.*100 mm deep, and thicker than the other), with a pronounced central depression, on the ground on a flattened tussock, and indistinguishable from nearby nests of Montagu's Harrier *C. pygargus*. One nest had four eggs on 11 May 1993, three eggs and a newly hatched chick on 27 May 1993, three chicks up to three days old and an infertile egg on 30 May 1993, and three chicks (oldest 20 days) on 15 June 1993. The first egg in this nest would have been laid on 25–26 April 1993. The second nest had an incomplete clutch of three on 12 May 1993, and four eggs on 16 May 1993, and had been added to considerably in the intervening days; it was empty on 27 May 1993, presumably destroyed by a predator; the first egg in this nest would have been laid on 7–8 May 1993, which is consistent with two nearby nests of Montagu's Harrier that contained one and two eggs on 11 May 1993. At the same marsh an adult male Pallid Harrier was observed 'food-passing' with a female Montagu's Harrier on 2 June 1996 (Davidson & Kirwan 1997) and the male had been previously observed 'displaying vigorously' to the female on 12 May 1996 (I. A. Green). Hybridisation events involving these two species are apparently extremely rare (Forsman 1993).

MONTAGU'S HARRIER *Circus pygargus*

Çayır Delicesi

(Du: Grauwe Kiekendief; Fr: Busard cendré; Ger: Weisenweihe; Swe: Ängshök)

Subspecies. Monotypic.

Status and distribution. National conservation status: Endangered (population apparently decreasing). A rather local and uncommon summer visitor, breeding in moderate numbers in Inner and East Anatolia, and very locally in adjacent areas of Black Sea Coastlands (perhaps even in Kızılırmak Delta) and Southeast Anatolia, where widespread suitable habitat now exists. Plausibly breeds in Thrace (from where there are recent mid-June records), as a small but significant population breeds in the Strandja Mountains of adjacent Bulgaria (Milchev 1994), but is generally scarcer in the west of range. Breeds principally in marshy wetlands, damp grassland and agricultural fields in steppe areas. The total breeding population is probably not more than 500 pairs, with the largest concentrations in East Anatolia.

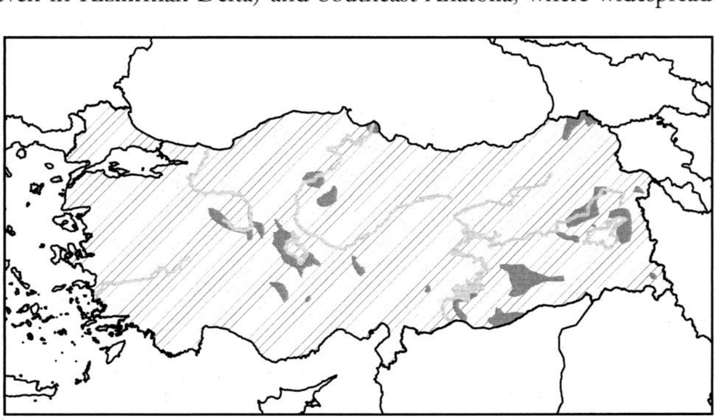

More common and widespread on passage throughout Turkey, including Southeast Anatolia, in spring from the end of March (23rd) to at least late May (28th), peaking in late April and early May; and in autumn from early August (2nd) to mid October (17th), peaking in September. Few recorded at the three major watchpoints at either season. In contrast, 115 passed through the Çıldır area (East Anatolia) on 9–10 September 1989 (Huber & Barbalat 1990) and 63–64 were at Fırtına River mouth, Rize province (Black Sea Coastlands) on just six days between 19 April and 27 May, with a peak of 41 on 27 April (Faldborg 1994). In neighbouring Georgia, 1,281 passed through the Batumi area in autumn 2002 (*Sandgrouse* 25: 89). One winter report, one at the Göksu Delta (Mediterranean) in December 2001 (*Kuşçu Bulteni* 10: 7).

Breeding. Frequents marshes, reeds and flooded grassland. Nests on the ground, usually in tall vegetation, making a thick pad of sedges and dried grasses with a central hollow. Eggs: C/4 (5), C/5 (2). Young: B/4 (1). Most data are from Eşmekaya (**Inner Anatolia**): at least 21 breeding pairs in June 1977, 18 of which were carrying prey and, on 20–21 June 1977, five nests had respectively C/4, C/4, C/5; two young + two eggs, and four young, 5–10 days old; two nests were just 50m apart (Schubert 1979). A nest held small young on 11 June 1971. In 1993, two nests had incomplete clutches of one and two eggs on 11 May 1993, and another four eggs on 16 May 1993 (PC). A female carried material to a nest with one egg on 19 May 1998 (T. Quittenden). A nest had C/5 on 15 May 2004 (D. Gregson) and another four eggs on 16 May 2004 (M. Farmer). A total of 35 adults, most males, were present 12 May 1996 (I. A. Green). Adults were observed carrying food to a suspected nest at Yarma (Inner Anatolia) on 13–14 May 1970: this date is too early for chicks to be in a nest and possibly relates to courtship feeding. At Kızılırmak Delta (**Black Sea Coastlands**), there were four pairs and five young in July 1971, and one pair was still feeding young (Dijksen & Kasparek 1985); and adults were carrying nest material in the first half of May 1992 (Hustings & van Dijk 1994). In **Southeast Anatolia**, a pair nested in a ditch at the edge of a small conifer plantation on an island in the Euphrates at Birecik on 19 June 1988 (M. Telfer). In **East Anatolia**, children were playing with very young birds near Bulanık on 23 July 1987, and two juveniles were at Bendimahi on 30 July 2000 (K. Haataja).

NORTHERN GOSHAWK *Accipiter gentilis*
Çakır

(Du: Havik; Fr: Autour des palombes; Ger: Habicht; Swe: Duvhök)

Subspecies. Presumably *marginatus* (see Roselaar 1995).

Status and distribution. National conservation status: Near Threatened (population apparently stable). A relatively local resident breeder in small numbers in all regions, except Southeast Anatolia, and almost certainly very rare in East Anatolia. Almost certainly commoner than observations suggest; during a breeding bird census of the Strandja Mountains of Bulgaria in 1988–90, Milchev (1994) recorded birds in 40 (29%) of 137 5-km squares surveyed. Breeds in forested or wooded areas from near sea level to at least 2,000m. Breeding commences in March and display persists until late May.

Perhaps more widespread in autumn and winter, with some immigration occurring, especially through northeast Turkey, from mid August to late October. Very small numbers have been recorded in autumn at the Bosphorus (mainly September, but from mid August); very occasionally recorded in spring, March–April, in this region, although 19 were noted in early spring 1993. Also reported in spring and autumn (mainly late September) from the Borçka–Hopa area (max. 52 in spring 1995). The true scale of this passage is perhaps evident from counts

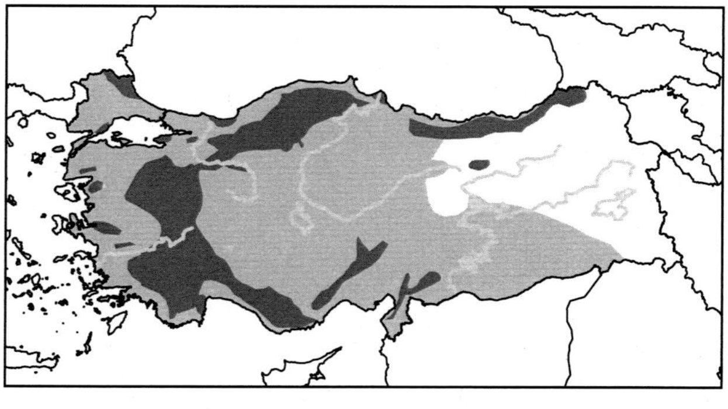

on the Georgian side of the border, where 795 were registered during autumn 2002 (*Sandgrouse* 25: 89). Recorded in more open habitats, including coastal wetlands, in winter.

Breeding. In mature woodland, both deciduous and coniferous; in flat areas at sea level and on slopes in montane areas. Nests in a tree, usually high; a large shallow structure of sticks. Nest-lining and clutch size undescribed in Turkey; elsewhere, a slight hollow in the top of nest, lined with leafy twigs or conifer needles, or pieces of bark, is constantly renewed throughout the season; usual clutch 3–4 (6), laid mainly April in southern Europe. In **Black Sea Coastlands**, an adult was carrying food in evergreen forest near Ikizdere on 11 June 1975, and a male was holding territory in pine forest near Çiftlik on 17 June 1975; at Kızılırmak Delta, in 1992, four breeding pairs, all with occupied nests: one in an isolated wood on 31 March 1992 and three in Yürükler Forest in the second half of April, with an adult carrying food to one of these three on 4 June 1992 (Hustings & van Dijk 1994). In **Marmara**, a breeding resident in deciduous woodland adjacent to Kocaçay Delta and there was a nest near Ekmekçi on 3 May 1993 (Ertan 1996). Display observed over coniferous forest at Uludağ in May 1966, and breeding suspected. In **Mediterranean**, a male was displaying at Bolkar Dağ on 6 June 1999 (R. F. Porter), and an adult was displaying at Akseki on 11 May 1988 (R. Beddard). An adult was at a nest, in dry hills and woods inland of Taşucu, on 13 May 1984 (Cygnus tour report). In **Inner Anatolia**, a male was displaying over poplar woods, near Nevşehir, on 22 May 1972, and there was a nest near Ayrancı, on 23 July 1994 (R. van der Vliet). At Soğuksu National Park, an adult mobbed a Common Raven *Corvus corax* on 21 May 1996 (I. A. Green), and two fledged young were heard food-begging, and subsequently seen, on 15 July 1993 (A. Hogg), the latter indicating egg laying no later than mid April.

EURASIAN SPARROWHAWK — *Accipiter nisus*
Atmaca

(Du: Sperwer; Fr: Epervier d'Europe; Ger: Sperber; Swe: Sparvhök)

Subspecies. Nominate.

Status and distribution. National conservation status: Near Threatened (population apparently decreasing). Local resident breeder or partial migrant in all regions, although distinctly uncommon in Southeast Anatolia. Occurs in small numbers in wooded or forested areas. Recorded in just ten (or 7%) of the 5-km squares surveyed by Milchev (1994) in adjacent Bulgaria (see Northern Goshawk *A. gentilis*). Breeding range, distribution and habits in Turkey are poorly known, although densities seem to be very low in some apparently suitable areas of Inner Anatolia and central Black Sea Coastlands.

Common and widespread on passage throughout the entire country, including Southeast Anatolia, occurring in rather smaller numbers at the Belen Pass during both spring and autumn. In spring, 3,966 passed Hopa, Artvin province (Black Sea Coastlands) between 21 March and 14 May 1994 (M. Henriksen), 1,730 moved through the same area on 21 April–5 May 1995 (Peške 1995), 695 moved over the Fırtına river mouth, Rize province (Black Sea Coastlands) between 8 April and 25 May 1993 (Faldborg 1994) and 186 passed through Kızılırmak Delta (Black Sea Coastlands) between 14 April and 21 May 1992 (Hustings & van Dijk 1994). Relatively common at the Bosphorus in both spring (mid March to late May; peak perhaps late March, but numbers apparently fairly constant throughout much of this period elsewhere in northern Turkey), and in autumn (late August to mid October, with most apparently in September), with maxima of 994 in spring 1993 (195 were counted in just two days in April 2005), 1,701 in spring 2006, and 428 recorded there in autumn 1969. Also common on passage at Borçka, occurring from mid August to late October. In 1976 688 were counted

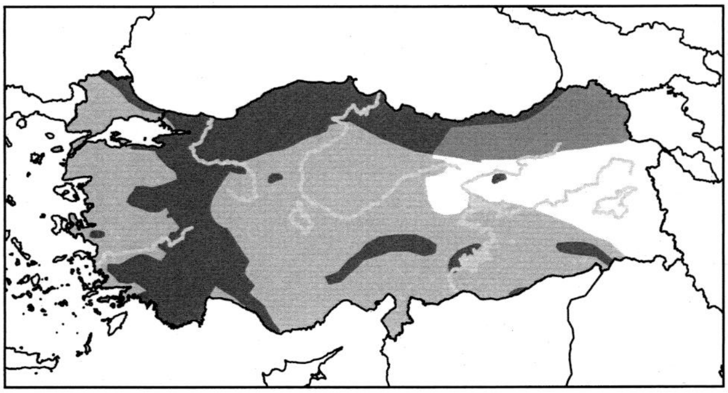

between 17 August and 10 October (Andrews *et al.* 1977); however, in 1977, 1,057 moved through on 11–25 October (Beaman 1977), suggesting that a substantial proportion of migrants may pass through late in the autumn, and 339 were counted there on 20–24 September 1998 (KAB). In this region, falconers using Red-backed Shrikes *Lanius collurio* as decoys commonly trap Sparrowhawks. Magnin (1989) studied the falconers trade and practices in autumn 1987. The results of his investigation suggest that the real total of Sparrowhawks migrating through northeastern Turkey is much higher than the 1977 figure, and, indeed, 3,250 moved through Batumi (Georgia) in autumn 2002 (*Sandgrouse* 25: 89). Although only small numbers have ever been recorded at Belen (probably largely due to periods of coverage at this watchpoint), opportunistic observations in four years between 1982 and 1990 suggest that considerable numbers may pass through the Çukurova region (Mediterranean), with a total of 142 noted on 15–16 October 1988 alone at Akyatan Gölü (van den Berk 1991a). The species is widespread in winter virtually throughout the country, except in particularly arid and mountainous areas.

Breeding. In deciduous and coniferous woodland, in flat lowlands as well as in hilly and montane areas. Nest site, nest and clutch size undescribed in Turkey; elsewhere, nest usually positioned higher than 5m, in a tree; a shallow structure of twigs, lined with leafy twigs and bark; usual clutch 4–5. In **Mediterranean**, a family party was near Akseki on 23 July 1986 (R. Webb), an adult was carrying prey at Dalyan on 11 May 1989, and a young bird was near Demirkazık on 11 August 1998 (A. Barfull). There was probably a resident pair in woodland on the northwest side of Karanfil Dağ in 1966 (Sutton & Gray 1972). In **Black Sea Coastlands**, an adult was mobbing a Golden Eagle *Aquila chrysaetos* in mountains between Ispir and Ikizdere on 12 June 1975. At Kızılırmak Delta, there were an estimated 1–3 pairs in 1992; breeding unproven, but a male twice visited an isolated wood, perhaps to a nest, on 5 June 1992 (Hustings & van Dijk 1994). In **Marmara**, 1–2 pairs probably bred in alluvial forest at Kocaçay Delta in 1993 (Ertan 1996); and recorded year-round, apparently breeding in semi-open land below 1,300m at Uludağ (Jetz 1995). In **East Anatolia**, nested in tall poplars in the cemetery and in gardens at Erzurum in 1910–11 (McGregor 1917), and an adult was observed carrying food in the Ardahan area on 10 June 2004 (G. Welch).

LEVANT SPARROWHAWK *Accipiter brevipes*
Yaz Atmacası

(Du: Balkansperwer; Fr: Epervier à pieds courts; Ger: Kurzfangsperber; Swe: Balkanhök)

Subspecies. Monotypic.

Status and distribution. National conservation status: Vulnerable (population trend unknown). Recently established to be rather local summer visitor, in small (perhaps very small) numbers to many areas of Turkey. Breeding recently proven in Marmara (including Thrace), Mediterranean, Black Sea Coastlands and Inner Anatolia (Hasan Dağ, near Aksaray), and strongly suspected from Aegean and Southeast Anatolia, usually in areas of deciduous woodland. Recorded on at least five occasions recently in East Anatolia in suitable habitat during the breeding season (Demirci & Mülayim in prep.; M. Bechtolsheim; H. Dufourny; G. Magnin & L. Nauta; Kirwan *et al.* in press). In adjacent Bulgaria, Milchev (1994) found evidence of breeding in only 2% of the 5-km squares censused in 1988–90.

Relatively uncommon on passage throughout but principally through the western two-thirds of Turkey, perhaps from early April (exceptionally March) to mid May (latest 4 June) in spring, and in autumn from late July (20th), more usually late August, to early October (latest, 29th), peaking in the second half of September. Some early records almost certainly reflect confusion with *A. nisus.* Locally recorded in considerable numbers at the Bosphorus (autumn maxima 5,707 in 1971, including 2,243 on 15 September, and 6,516 in 1978, with the vast majority on 18 September) and in the Belen Pass area, where 2,951 passed in autumn 1976 (Sutherland & Brooks 1981a). Also at the Belen Pass, 1,914 passed on 16–18 September 1991 alone, including the single-day maximum of 1,062 on the former date (A. Forsten & T. Numminen). Kasparek (1992a) mentioned an autumn total of 7,750 (including a single-day count of 6,625) at the Bosphorus although no year or date is specified. In autumn reported in much smaller numbers from northeast Turkey: a total of 290 was noted in 1976 at Borçka (Black Sea Coastlands), the only year for which detailed information is available (Andrews *et al.* 1977). In spring

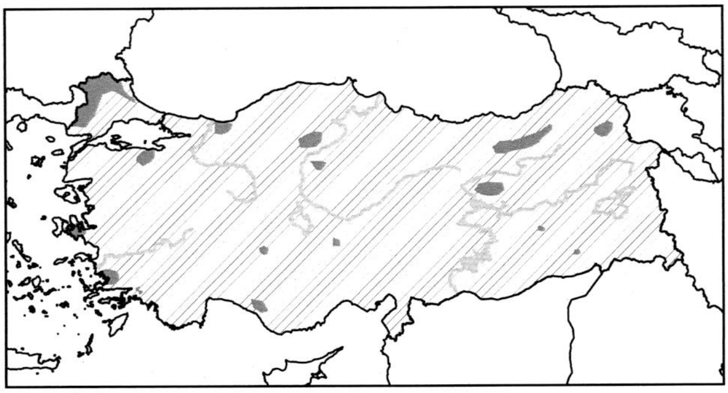

1993, Faldborg (1994) counted 1,945 through the Fırtına river mouth area, Rize province (Black Sea Coastlands), between 29 April and 25 May, with 1,286 on 30 April (M. Henriksen *in litt.*). There were 1,092 at Hopa, Artvin province, in spring 1994 (including 737 on 2 May) and Peške (1995) recorded 1,163 through the same area in just two weeks in spring 1995, demonstrating that much larger numbers may pass through this part of Turkey than previously supposed.

Breeding. In woodland, both deciduous and coniferous. McNeile was shown a nest at the top of a large oak near Izmir (**Aegean**); a pair had been seen earlier at this nest, but it was empty on 25 May 1950. A pair was observed at a nest in dense pinewoods, below a ridge of cliffs, at Taşağıl (Mediterranean) on 15 June 1988 (Kirwan & Martins 1994). Nest undescribed in Turkey; photographs in Pforr & Limbrunner (1982), taken in Greece, show a nest with five eggs on 23 May 1978, and nestlings, aged *c.*2 weeks, on 6 July 1977, the latter indicating egg laying in mid May: the nests were relatively small, of sticks, lined with green leaves, and both were at *c.*5m in slender trees (A. Limbrunner pers. comm.). Four eggs in the Natural History Museum, Tring, were collected by Krüper from three nests inland of Izmir (Aegean): one egg on 28 May 1863, one egg (of a clutch of three) taken from a nest in a pine on 1 June 1871, and two eggs (of four) on 18 May 1872. Three further clutches (C/4, C/3, C/3) were collected in Turkey on 25 May 1870, 28 May 1901 and 19 May 1907. In **Marmara**, a pair, with four juveniles, was at Çamlıca on 20 August 1979 and may have bred locally; the young were 'playing' together in tumbling flight, with much calling (Beaman 1986; R. F. Porter). Adults were observed in suitable breeding habitat on 3 August 1978 and 11 July 1980 (Beaman 1986) and a pair with fledged young was photographed at Edirne on 27 June 2000 (Kirwan *et al.* 2003). A pair was 'probably breeding' in orchards near Ispir (**Black Sea Coastlands**) on 2 August 1990 (R. Thewlis). Apparent breeding behaviour was observed in 2004 (*per* J. L. Copete) and a nest with a female brooding eggs or small chicks was discovered there on 20 May 2005 (J. L. Copete). Display was observed at Ayvacık (**Marmara**) on 7 June 1984 (Martins 1989) and near Uludere (**East Anatolia**) on 14 July 1990. In **Mediterranean**, two occupied nests were found in clumps of pines surrounded by agricultural land near Manavgat on 23–26 May 2007; one nest was 15m from the ground, and an adult could be seen incubating; an adult flew from the other nest (13m), and it contained three eggs, all slightly stained from the green leaves lining the nest (J. Hollinrake); display was observed at Köyceğiz on 15 May 2001 (Kirwan *et al.* 2003).

COMMON BUZZARD *Buteo buteo*
Şahin

(Du: Buizerd / Steppebuizerd; Fr: Buse variable; Ger: Mäusebussard; Swe: Ormvråk)

Subspecies. *B. b. vulpinus* ('Steppe Buzzard') and perhaps *menetriesi*. *B. b. buteo* almost certainly also occurs (A. Vittery *in* Cramp & Simmons 1980, Ferguson-Lees & Christie 2001). The form *vulpinus* has long been considered a rather dubious and poorly distinguished taxon (see Voous *et al.* 1948), with much evidence for intergradation between it and nominate *buteo* (see Ferguson-Lees & Christie 2001), and recent genetic work by Screiber *et al.* (2001) has demonstrated a lack of significant differentiation between any of the various races of *B. buteo*, with the exception of east Asian *japonicus*, a potentially surprising result given the levels of morphological variation exhibited within this species. There is no evidence that *japonicus* (which is closer in morphology to pale nominate *buteo*) and *vulpinus* are sympatric (Ferguson-Lees & Christie 2001). Rasmussen & Anderton (2005) also pointed to the possibility that *japonicus* is vocally distinct, and therefore meritorious of specific status, but also postulated that *vulpinus* is vocally different, which is not our experience. Characters exhibited within continental populations of *buteo* and the three forms placed within the *vulpinus* group (*vulpinus*, *menetriesi* and *refectus*) are neither constant, nor confined to region (see Ulfstrand 1977, who found much clinal variation, including smaller and redder birds, somewhat close to *vulpinus*, as far west as southern Sweden), and it may prove to be the case that most named variation within *B. buteo* is unworthy of taxonomic recognition, although Himalayan *burmanicus* (and synonym *refectus*) appears specifically distinct (Riesing *et al.* 2003, Kruckenhauser *et al.* 2004, Rasmussen & Anderton 2005). Certainly, in our view and field experience, *menetriesi* is unworthy of recognition at the nomenclatural level, notwithstanding that at least some birds within its purported range show the distinctive characters of this taxon.

Status and distribution. National conservation status: Least Concern (population apparently stable). Fairly widespread and generally not uncommon resident breeder or partial migrant usually in hilly or montane-forested areas (from near sea level to at least 2,200m) in Marmara (including Thrace), parts of Aegean and throughout the Black Sea Coastlands. Generally very local elsewhere in adjacent regions, although possibly breeds in most areas with suitable habitat in Southeast Anatolia. Dispersal from breeding areas apparently commences as early as the second half of July (OST 1975).

Widespread and relatively common passage migrant, locally recorded in very large numbers, in autumn. Spring passage is from mid March to early June, with birds occurring regularly in very small numbers into mid May. However 136,327 passed Hopa, Artvin province (Black Sea Coastlands) between 21 March and 14 May (with 16,333 on 6 April; M. Henriksen) and at Fırtına river mouth, Rize province (Black Sea Coastlands) 17,928 were recorded between 31 March and 10 June 1993, including 3,879 on 30 April (Faldborg 1994) indicating that large numbers pass through northeastern Turkey at this season. At the Bosphorus, 16,348 were counted in spring (mid March to late May) 2006 (KAB *et al.*). In autumn, passage recorded from early August to late October and peak numbers occur in the latter half of September and early October. There have been no protracted observations at the Belen Pass during the main part of the passage period. In autumn 1969, 32,895 were recorded at the Bosphorus, where the highest single-day total is 6,142 on 25 September 1972. Few counts are available from northeast Turkey but huge numbers appear to be regular at Borçka (where 205,000 were counted in autumn 1976, with 135,000 on 28 September alone; Andrews *et al.* 1977). Other impressive single-day counts at Borçka include 73,000 on 19 September 1978 (Magnin 1989), 70,000 (of a total of 165,871 between 20 September and 2 October) on 22 September 1980 (P. Heathcote *et al.*) and 152,000–170,000 on 1 October 1990 (Kasparek 1992a). Winters in small numbers at low or moderate altitudes, principally in the north and west of the country, but also recorded more locally elsewhere throughout the western two-thirds of Turkey.

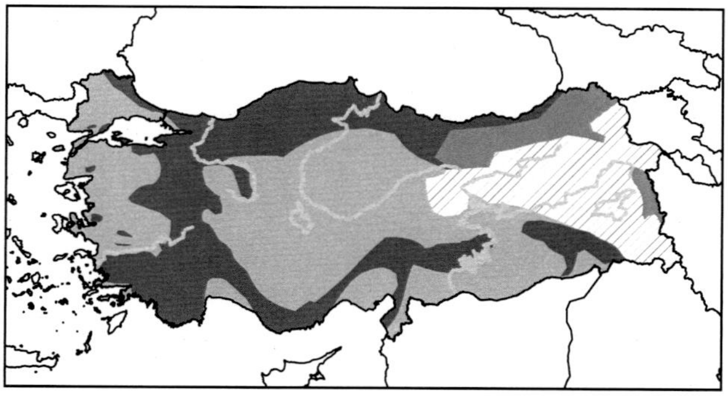

Breeding. Mainly in hilly or montane forested country. Nests in mature trees or on cliffs. Nest is a large structure of sticks with a slight hollow in the central top, lined with leafy twigs. Eggs: no data in Turkey; usual clutch elsewhere, 3 (2–4). Young: B/3 (1). In **Marmara**, regular sightings in 1993 indicated 2–3 pairs breeding in the hills south of Kocaçay Delta (Ertan 1996); a pair was at a nest between Akyazi and Mudurnu on 30 July 1992 (G. Magnin). In **Black Sea Coastlands**, an adult was carrying food near Ikizdere on 13 June 1975 (R. F. Porter). Breeding population in Kızılırmak Delta estimated at 16–20 pairs in 1992, of which 12 were in Yürükler Forest; two occupied nests, and one of these had young on 23 May 1992 (Hustings & van Dijk 1994), indicating egg laying by mid April, possibly earlier, depending on the size of the young. In **Inner Anatolia**, four adults were in suitable breeding habitat at Kalecık on 7 June 1975; they included one carrying nest material presumably for freshening the nest, rather than original nest building, in view of the late date. A cliff nest had three large, feathered young, exercising their wings and close to fledging, near Kızılcahamam on 26 June 1992, indicating egg laying in mid April (PC). In Lesbos (Greece), a flimsy nest, *c.*4m up in a tree, had two eggs on 4 May 1995, and two young, about half grown, on 28 May 1995 (T. Quittenden), indicating egg laying in early April.

LONG-LEGGED BUZZARD *Buteo rufinus*
Kızıl Şahin

(Du: Arendbuizerd; Fr: Buse féroce; Ger: Adlerbussard; Swe: Örnvråk)

Subspecies. Nominate.

Status and distribution. National conservation status: Near Threatened (population apparently decreasing). A reasonably widespread and fairly common resident breeder throughout much of interior Turkey, but relatively local in coastal areas of Marmara (even more so in Thrace), Black Sea Coastlands and Mediterranean. Slightly more widespread in autumn and winter, when generally more common in coastal areas, although only very small numbers have been recorded at both the Bosphorus and Borçka in autumn. Most frequently recorded in open habitats, particularly in dry cultivated areas, steppe and montane regions up to 4,000m.

Vittery (2000) proffered the view that most buzzards breeding in Inner and East Anatolia were *B. b. menetriesi* not *B. rufinus*, a hypothesis we reject on the basis of ample field experience to the contrary. Furthermore, some of the evidence brought forward in support of his idea, e.g. the relative lack of records of migrants from bottleneck sites (which merely proves the point that *rufinus* is largely resident), can easily be dismissed. To tout the long- (and best) forgotten supposition that *menetriesi* may even warrant specific status is categorically refuted by the overall findings of Screiber *et al.* (2001; see Common Buzzard *B. buteo*).

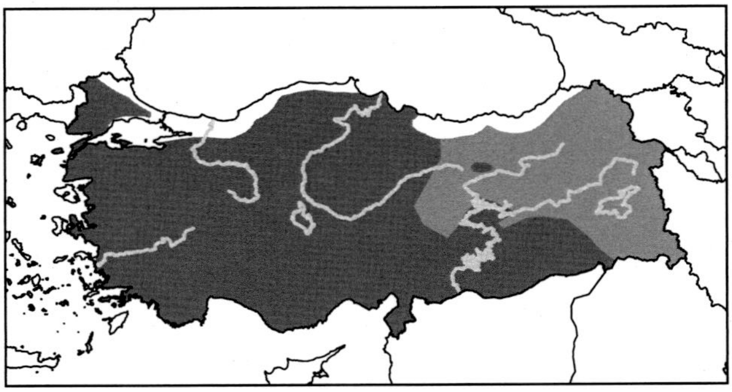

Breeding. In arid open country, in hills with cliffs and in steppe with rock outcrops. Nests usually on a cliff ledge, but also in trees and on pylons: the nest is a large structure of sticks, with a shallow hollow in the top, lined with grass, finer twigs, plant material and wool; a nest near Eskişehir (Inner Anatolia), in mid-April 1879, was lined with camel hair (Danford 1880). The same nest may be used for several years, or two or more nests, often on the same cliff, may be used alternately. A nest on an electricity pylon near Elmali (Mediterranean) on 14 May 2004 was a thin flat structure of sticks with a blue polythene bag incorporated and several pairs of Spanish Sparrows *Passer hispaniolensis* nesting in the base. Another pylon nest, near Horasan (East Anatolia), also incorporated polythene bags (PC). Old nests may be used in subsequent years by Saker Falcon *Falco cherrug*. Eggs: C/5 (1), C/4 (3), C/3 (2), C/2 (2), C/1 (1). Young: B/1 (1), B/2 (2), B/3 (4) B/4 (4). In **Inner Anatolia**, display seen in late March; nest building on 21 March 1969. Ten nests had eggs, mainly in April (earliest 3 April) and early May, with a late record on 28 May. Seven nests had young (*c*.4 weeks old) in the last ten days of May, indicating laying in late March–early April. Recently fledged young seen on 10 June 1973. In **Mediterranean**, a displaying pair was north of Taşucu on 23 March 1993 (J. van der Winden), and copulation was seen at Akseki on 19 April 2004 (A. Dixon). A nest had four well-incubated eggs in the Aladağ on 30 April 1876 (Danford 1878), indicating egg laying early April. In **Southeast Anatolia**, a nest had four young, 2–3 weeks old, at Durnalık on 9 May 2004 (J. Mason), indicating egg laying in the third week of March, and this nest was also used in 2006, with a recently fledged young bird close by on 12 June 2006 (PC). A pair with four young was at Birecik on 29 May 1987 (Eames 1989). In **East Anatolia**, an adult was feeding two chicks, 3–4 weeks old, near Horasan on 25 June 2004, indicating egg laying in late April (PC); two nests had young in early June; and four family parties were observed in early August 1992 (J. Frost & J. Hornbuckle).

ROUGH-LEGGED BUZZARD *Buteo lagopus*
Paçalı Şahin

(Du: Ruigpootbuizerd; Fr: Buse pattue; Ger: Rauhfußbussard; Swe: Fjällvråk)

Subspecies. Nominate.

Status and distribution. A rare, usually local winter visitor in very small numbers, principally to Marmara and Black Sea Coastlands (where comparatively regularly reported from Kızılırmak Delta). Less frequently recorded in Aegean, Mediterranean (noted as far south as the former Amik Gölü, Antakya province; Kumerloeve 1967), Inner and East Anatolia. Perhaps formerly more frequent (see Kumerloeve 1957d). Records are principally from late December to February, but the period of occurrence extends from mid October to late April, and the species is regularly recorded until late March. Usually in or around wetlands but also occasionally found in steppe or even montane areas. Some individuals remain in the same area for many weeks, especially during periods of severe weather. Recorded on spring passage at the Bosphorus on a small number of occasions (Kumerloeve 1957a,d, Collman & Croxall 1967, P. Denes *in* Zalles & Bildstein 2000); at Hopa, Artvin province (Black Sea Coastlands) annually in 1993–95, and at Kızılırmak Delta (12 between 22 March and 25 April 1992, Hustings & van Dijk 1994); however there are no recent autumn records from any of the major watch sites, though Kumerloeve (1967d) noted small numbers moving through northeast Turkey in the early 1960s. In contrast, in neighbouring Georgia, the extraordinary total of 2,145 was counted moving through Batumi in autumn 2002 (*Sandgrouse* 25: 89).

LESSER SPOTTED EAGLE *Aquila pomarina*
Küçük Orman Kartalı

(Du: Schreeuwarend; Fr: Aigle pomarin; Ger: Schreiadler; Swe: Mindre skrikörn)

Subspecies and taxonomy. Nominate, though Lesser Spotted Eagle may be better treated as a monotypic species, as it has recently been suggested that the form *hastata* from the Indian subcontinent merits specific status, which it was frequently afforded in the late-19th and early-20th centuries (Parry *et al.* 2002, Väli 2006). Recent genetic data suggest that *pomarina* and *hastata*, and the next species, are more closely related to the African genus *Lophaetus* than to other *Aquila* (Helbig *et al.* 2005, Lerner & Mindell 2005).

Status and distribution. National conservation status: Endangered (population apparently decreasing). Breeding status and distribution unclear; a local summer visitor, probably in very small numbers (though see below) to lowland and montane forested areas, often near water, with records of possible breeding from all regions except Southeast Anatolia. Perhaps formerly more common, as Alléon (1886) mentioned breeding in the Forest of Belgrade (Marmara). Probably absent from much of Inner Anatolia, with most records from northern and western areas. Milchev (1994) found evidence of breeding in 42% of 5-km squares that he surveyed in adjacent Bulgaria in 1988–90, suggesting that the species may be commoner than is realised.

Widespread and relatively common on passage throughout Turkey, although most numerous in the western two-thirds. Numbers are smaller on spring passage from mid March (18th) to late May (28th), e.g. just 277 at Fırtına River mouth, Rize province (Black Sea Coastlands) between 20 April and 25 May 1993 (Faldborg 1994), and 600 at Alahan (Mediterranean) on 3 April 2002 (H. Schmaljohann *et al.*), with the exception of 17,326 at the Bosphorus between 1 March and 9 April 1993 (max. 6,370 on 27 March: P. Denes *in* Zalles & Bildstein 2000) and 15,232 in spring 2006 (KAB *et al.*). Occasionally numerous on autumn migration, from mid August (17th) to early October (8th), rarely to late October in central and southern areas, with stragglers recorded until mid November (13th). Commonest at the Bosphorus and Belen Pass. At the former, max. 18,984 in autumn 1971 and day totals in the order of 5,000 are virtually annual: 15,035 at Büyük Çamlıca on 27 September 2005 (İKGT) is by far the largest single-day total, whilst on 26 September 1990 separate groups of observers recorded 11,379 at Sariyer, on the European shore (Robel & Bräuning 1992) and 11,703 at Küçük Çamlıca, on the Asian side (H. Nijmeijer *et al.*). Additionally, 9,732 passed Büyük Çamlıca on 20 September 1997 (Boyla & Eken 1998). Studies have revealed that passage over the Bosphorus occurs on a broad front and that relatively little duplication is involved when counting from both watch sites on the same day (Bijlsma 1987). At the Belen Pass 1,299 were recorded in autumn 1976 (including 513 on 13 September; Sutherland & Brooks 1981a), but 3,424 were noted there on 28 September 2001 alone, demonstrating that true numbers moving through this region are probably considerably higher. A total of 10,584 moved northeast through Çukurova between 18 September and 13 October 1994, including 6,168 on 2 October (R. Thorpe & J. Young). In contrast, at Borçka in 1976, the autumn total was much lower: 290 birds (Andrews *et al.* 1977). The max. there was 729 in 1980 (P. Heathcote *et al.*). Occasionally recorded in winter, usually in Thrace; e.g. one at Meriç Delta on 24 January 1993 and one or two at Iğneada in January–February 2001.

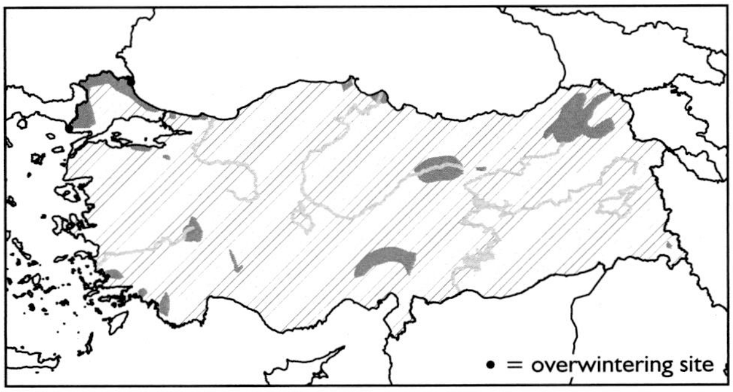

Breeding. In lowland or hilly wooded country, mainly near water. Adults (including pairs) are regularly observed, mainly in Black Sea Coastlands and Marmara, and breeding is assumed to occur, but there are apparently no definite records. Elsewhere, the nest is 6–25m up in a tree, usually near forest edge or a clearing; a large structure of sticks, with a hollow in the central top, lined with grass and leaves; usual clutch 2 (1–3). In Forest of Belgrade (**Marmara**), display by three pairs observed on 24 April 1970; on 29 June 1973, a pair was displaying, and a pair probably changed-over at the nest, but the nest was not found (R. F. Porter). At Kızılırmak Delta (**Black Sea Coastlands**), two were observed carrying food to an area of forest daily in late July 1969, but no nest was found, whilst display was observed on 27 May 1979 (Dijksen & Kasparek 1985); in 1992, the population (in Yürükler Forest) was estimated at 4–5 pairs, but, again, no nests were found, and no adult was observed carrying food (by 10 June, when the survey ended), but eggs would probably not have hatched until mid June (Hustings & van Dijk 1994). An adult was carrying food between Çıldır and Ardahan (**East Anatolia**) on 12 June 2005 (G. & H. Welch).

GREATER SPOTTED EAGLE — *Aquila clanga*
Büyük Orman Kartalı

(Du: Bastaardarend; Fr: Aigle criard; Ger: Schelladler; Swe: Större skrikörn)

Subspecies and taxonomy. Monotypic. Recent genetic data suggest that this, and the previous species, are more closely related to the previously exclusively Afrotropical genus *Lophaetus* than to other *Aquila* (Helbig *et al.* 2005, Lerner & Mindell 2005).

Status and distribution. Conservation status: Vulnerable (BirdLife International 2000, 2004). A rather local winter visitor, from late October to late March (more occasionally April) in small numbers to wetlands throughout the western two-thirds of Turkey. Formerly more common, e.g. numbers on passage at the Bosphorus declined between 1931 (Steinfatt 1932) and 1956 (Nisbet & Smout 1957), and had probably decreased still further by the advent of modern recording (1966). Regularly recorded from the Black Sea Coastlands (tiny numbers at Kızılırmak Delta), Aegean and Mediterranean (5–15 regular at Göksu Delta in recent years, and formerly regular at Amik Gölü, Antakya province, in similar numbers; Kumerloeve 1967). Less common in Inner Anatolia than in the late 1960s, while current status in Thrace and elsewhere in Marmara is unclear, although there are sporadic reports from the Marmara Lakes region. The species was formerly (1960s and early 1970s) more regular at the Meriç Delta.

Regular in very small numbers on passage throughout Turkey, in spring from the end of March to early May (exceptionally late May) at the Bosphorus and Belen Pass. Also in autumn from mid August to early November, with most in late September and early October, at the three major migration watch sites. However, around Ankara (Inner Anatolia), most 'winter' sightings are from the second week of February to the first week of March, suggesting even earlier movement within the country (BD). In spring, D. Ristow (see Kumerloeve 1967) recorded a small but significant movement (nine birds) through the northeast near Rize (Black Sea Coastlands) in late March and early April 1963, and M. Henriksen (*in litt.*) saw 24 during the same period in 1994 at Hopa, where seven were recorded in two weeks in 1995 (Peške 1995), whilst seven were observed migrating north over Ağrı (East Anatolia) in mid-May 1970. At Batumi (Georgia) 42 were counted in autumn 2002, perhaps indicating the true scale of the passage (*Sandgrouse* 25: 89). There were several summer records during the 1970s in extreme north and northeast Turkey although there was no evidence of breeding (see e.g. Groh 1971), and the species was claimed recently (June 1999) from Southeast Anatolia at this season (Karakaş & Kılıç 2002). The pale morph, so-called '*fulvescens*', is very rare in Turkey, but there is a recent record from Terkos Gölü (Marmara), on 2 February 2008 (S. Bekir).

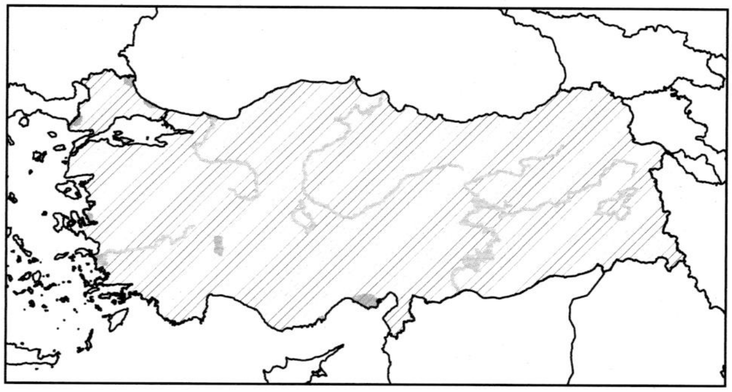

STEPPE EAGLE *Aquila nipalensis*
Bozkır Kartalı

(Du: Steppearend; Fr: Aigle des steppes; Ger: Steppenadler; Swe: Stäppörn)

Subspecies and taxonomy. Nominate, although Roselaar (1995) suggested that the form *orientalis* may also occur; we have been unable to examine any specimen material from Turkey. However, Clark (2005) recommended that Steppe Eagle be considered monotypic, a proposal which we deem to be well founded. The results of one recent and wide-ranging genetic study (Lerner & Mindell 2005) of modern raptor species suggest that *A. nipalensis*, along with *A. heliaca* and Tawny Eagle *A. rapax*, form a separate clade from other eagles currently placed in the genus *Aquila*, perhaps necessitating recognition of a separate genus for these three. Griffiths *et al.* (2007) also recovered support for the proposition that *Aquila*, as traditionally constituted, is paraphyletic.

Status and distribution. National conservation status: Critical (population apparently decreasing). Breeding status and distribution unclear (see below); perhaps, formerly at least, bred regularly at one site in Inner Anatolia (until the 1980s), with two additional confirmed breeding records at different sites in the same region since then. Most recent breeding-season records are of immatures, though intensive fieldwork in the Konya Basin in summer 1998

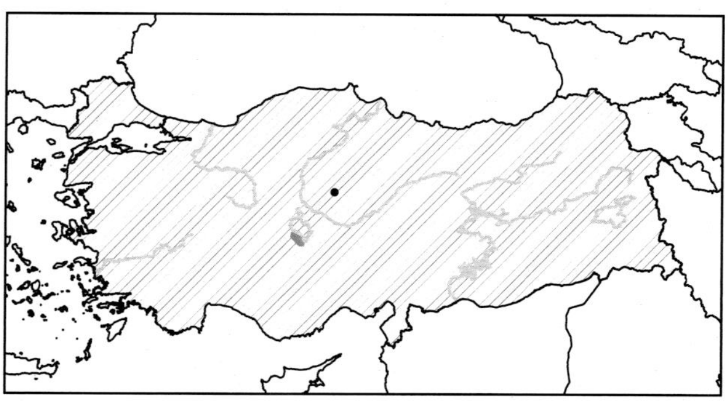

produced several records of adults (Eken & Magnin 1999, Kirwan *et al.* 2003). There was also an adult at Malazgirt, Ağrı province (East Anatolia) in early July 1999 (J. Olah), one near Hafik (East Anatolia) in July 2001 (M. Panayotopoulou *in* Bilgin & Demirci 2002) and another adult at Tendürek Dağı (East Anatolia) in July 2006 (E. Yoğortcuoğlu *in litt.* 2006).

Until the latter half of the 1960s virtually all records of this species were regarded with some scepticism. Since then it has been recognised as a regular passage migrant, usually in small numbers, during both spring and autumn, principally through the eastern half of Turkey. In spring, recorded from early April (9th) to mid May (19th) and, in autumn, from August to late October (25th). Very rarely (in some instances perhaps doubtfully) recorded from the Bosphorus in spring and autumn. At Borçka 271 were recorded in autumn 1976 (Andrews *et al.* 1977), but 434 were counted on 11–25 October 1977 (Beaman 1977), perhaps indicating that the majority are late migrants. Five winter records; from Thrace, Mediterranean and Inner Anatolia.

Breeding. Details of two nests found in Inner Anatolia were presented by Lehmann (1977). Photographs in Pforr & Limbrunner (1982), of two nests in Turkey, show nest linings of grass, fur and debris. These latter, like several other Turkish records of breeding Steppe Eagles, were originally assigned to *A. heliaca*. However, recent re-analysis of photographic evidence reveals that in all cases the birds were *A. nipalensis*, based on the plumage of the young, nest site and use of debris in the construction materials (*per* A. Dixon, A. Knowles-Brown & I. Karyakin). At Tuz Gölü (**Inner Anatolia**), a pair nested on the ground on a slope on a remote island. The eyrie was occupied in 1969, but not in 1970 (Warncke 1971); it had two chicks, *c.*10 days old, on 31 May 1971, indicating egg laying *c.*10 April, and a young bird, *c.*4 weeks old, on 10 June 1973, indicating laying in late March/early April. This eyrie was still intact, but old and unused, on 9 May 1993 (PC). Lehmann (1977) presented details and photographs, presumably appertaining to the same eyrie, based on visits in 1972 and early May 1975, as well as data supporting the identification of these birds as Steppe Eagles. A nest at Bolluk Gölü (Inner Anatolia) was on the ground, on a low, flat island; it had two young, *c.*1 week old, and an egg, on 6 May 1975 (Pforr & Limbrunner 1982), indicating laying in mid March, and was also occupied in 1980 (P. Zeininger pers. comm.), but there was no trace of it in early May 1993 (PC). The most recent nesting event (also on the ground and documented with photographs) was at Seyfe Gölü (Inner Anatolia), on 11 June 2003 (L. Haraszthy); the nest contained three downy young.

EASTERN IMPERIAL EAGLE — *Aquila heliaca*
Şah Kartal

(Du: Keizerarend; Fr: Aigle impérial; Ger: Kaiseradler; Swe: Kejsarörn)

Subspecies and taxonomy. Monotypic. See comments under the previous species.

Status and distribution. Conservation status: Vulnerable (BirdLife International 2000, 2004); national conservation status: Endangered (population apparently decreasing). A rather local resident or partial migrant, breeding in all regions except perhaps Southeast Anatolia (where formerly presumed to breed but now only recorded, very rarely, in winter, with one recent record in possible breeding habitat). Probably very local in East Anatolia and now on the brink of extinction as a breeder in Thrace, although formerly bred in the Meriç Delta and elsewhere. The available evidence strongly suggests that the species has declined over the past two or three decades, especially in Aegean, Mediterranean (in the Taurus) and in central and southern parts of Inner Anatolia (the Beynam

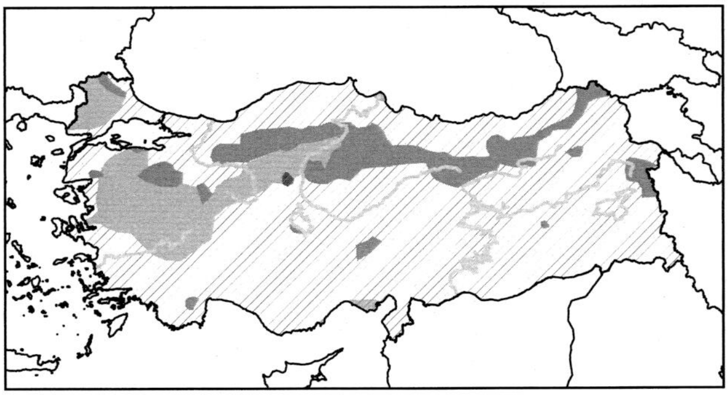

Forest, close to Ankara, is a regular breeding locality). Kumerloeve (1961, 1967, 1970) produced a reasonably comprehensive list of known and suspected breeding areas, many of which are apparently no longer occupied. De Roeck (1993) proposed a population of 50–150 pairs for Turkey, whilst Acar *et al.* (1977) suggested 50–99 pairs. Despite the marked decline it seems feasible that the lower figure is too conservative an estimate. Principally breeds in wooded lowlands (both deciduous and coniferous) or montane areas (to approximately 2,000m) from late March. More rarely nests in virtually treeless areas.

Rather uncommon on passage throughout the country, although the lack of spring records is in part due to low observer coverage at this season. It is recorded at the major watch sites from early March. A maximum of 20 recorded in spring 1994 at Hopa, Artvin province (Black Sea Coastlands), with a further 69 birds belonging to either this species or Steppe Eagle *A. nipalensis* (M. Henriksen *in litt.*). In autumn, maxima of 18 at the Bosphorus in 1966 and 29 at Borçka in 1977 (Beaman 1977), and recorded in both areas from mid August to late November, peaking in late September to mid October. Slightly more widespread in winter, largely in the west and south and Inner Anatolia, often frequenting coastal and wetland areas, e.g. up to five were at the Göksu Delta (Mediterranean) in winter 1991/2 (van den Berk & Guder 1992).

Breeding. In open woodland: isolated patches, scattered or solitary trees, and in treeless areas; e.g. in a small isolated forest of black pine and oak, at Beynam, near Ankara (Inner Anatolia), where two pairs breed; in an isolated patch of Black Pines, at 1,000m, on a plateau, near a swamp, in near-treeless country near Bolu, Black Sea Coastlands (Wirth 1995); and in isolated trees on steppe in Inner Anatolia (see Danford below). Nests mainly in mature, tall trees, including white poplar, willow and pine, but also on cliffs; the nest is a large structure of branches and twigs, with a shallow hollow in the central top, lined with grass, twigs and live foliage, which is renewed throughout the season. Usually 2–3 nests in a territory, used in different years; additional material is added, and old nests become huge. Eggs: C/1 (4, but at least two of these perhaps incomplete), C/2 (2), C/3 (2). Young: B/1 (5), B/2 (5), B/3 (1); 2–3 eggs may hatch, but often only the largest chick survives, and nests with larger young are, therefore, more likely to contain one. Danford (1880) found several nests in central Turkey in 1876 and 1879: he 'took a few nests' from willows and other trees between the Aladağ (**Mediterranean**) and Kayseri (**Inner Anatolia**); the breeding season was variable, because fresh eggs, incubated eggs, and young were all obtained in the second week of May 1876; an adult was shot off a nest, which had two young, on a cliff, near Kayseri, on about 8 May 1876 (Danford 1877–78). Three eggs from the above nests, C/1 and C/2, collected on 9 May 1876, are now in the Natural History Museum, Tring. Three eggs were taken from a nest near Kayseri, in the second half of March 1879, and two 'richly marked' eggs in late March 1879 from a nest, near Köprüköy, near Ankara (Inner Anatolia); a pair of Common Kestrels *Falco tinnunculus*, and both House *Passer domesticus* and Spanish Sparrows *P. hispaniolensis*, were nesting in the base of the latter nest. In the Ankara region 'the nests [of Imperial Eagle] burden most of the few and far between trees of this dreary region' (Danford 1880). At Beynam, south of Ankara, three immatures were seen August 1968; a nest had one 'well-grown young', in a pine, at 1,400m, on 12 July 1971; two pairs reared one and two young at nests in trees in 1995 (Wirth 1996); a nest had one young, *c.*3 weeks old, on 5 June 2002 (H. & J. Eriksen), indicating egg laying in early April. Elsewhere in Inner Anatolia, a pair reared 2–3 young, near Ankara, in 1966, and a pair reared two young in 1968 at Yozgat, where an eyrie was also found on 5 March 1971. In **Black Sea Coastlands**, an eyrie had two young, on a cliff face near Havza, on 8 June 1970, and two adults were at a tree eyrie in Kızılırmak Delta on 24 July 1971. A nest at Gümüşhane had at least one chick in June 1986. A nest had a fully feathered chick, in a large nest, 20m above ground in a Black Pine, in an isolated woodland ($0.5km^2$), east of Bolu, on 1 July 1989 (Wirth 1996). An occupied nest on a high pylon along the main road between Bayburt and Gümüshane had two downy chicks on 6 June 2005. In **Marmara**, a female was incubating one egg in a nest, 10m above ground, in a white poplar, near Meriç Delta, on 22 April 1965 (Warncke 1964–65); a juvenile was near Uluabat Gölü on 29 June 1973; an adult was carrying food in Kocaçay Delta on 29 May 1993 and was considered to be breeding in nearby hills (Ertan 1996). Five eggs in the Natural History Museum, Tring, were collected by Krüper in the Izmir area (**Aegean**) on 15 April 1872 (C/3), April 1873 (1), and April 1874 (1), the latter two eggs possibly form incomplete clutches.

GOLDEN EAGLE *Aquila chrysaetos*
Kaya Kartalı

(Du: Steenarend; Fr: Aigle royal; Ger: Steinadler; Swe: Kungsörn)

Subspecies. *A. c. homeyeri* in Anatolia, whilst those that appear in Thrace in winter are apparently referable to the nominate (Roselaar 1995). However, DNA evidence suggests that *homeyeri* and *chrysaetos* are only very weakly distinguished taxa (Wink *et al.* 2004), and thus *homeyeri* might be better regarded as a synonym of the nominate.

Status and distribution. National conservation status: Least Concern (population apparently stable). Widespread and not uncommon resident, breeding in mountainous country (generally above 1,500m) throughout the country, although very local in Marmara, Aegean and Southeast Anatolia (where principally restricted to the east of the region). Occasionally recorded from other areas at lower altitudes outside the breeding season, when reported from Thrace (recorded in this part of the country as late as early May, but no direct evidence of breeding). Irregularly recorded, in very small numbers, from both the Bosphorus (Alléon & Vian 1869, 1870, Reiser 1904, OST 1975) and Borçka (Andrews *et al.* 1977, Beaman 1977) from mid September to late October, with one early August record at the former watchpoint (Ballance & Lee 1961), and in spring from late February.

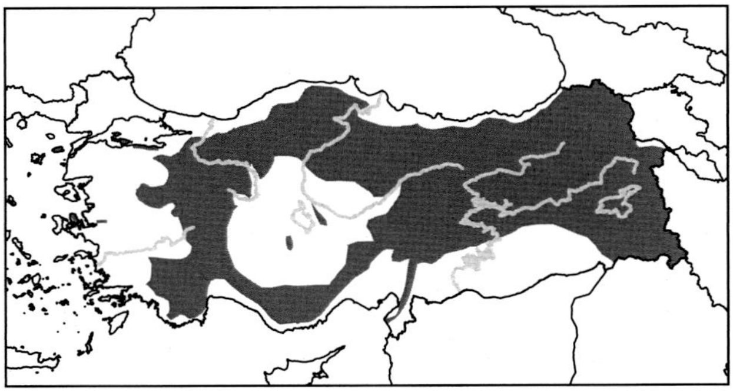

Breeding. In mountains, with cliffs, gorges and rock outcrops, or large trees. Nests on a cliff ledge, usually in a sheltered site in a slight recess or beneath an overhang, or in a large tree. Several nests, used in alternate years, may be built on the same stretch of cliff. The nest is a large stick structure, becoming huge if consistently reused; a nest in Inner Anatolia was lined with pine needles (Pforr & Limbrunner 1982), and another in Mediterranean with green fir twigs (Danford 1877–78). Eggs: C/2 (5). Young: B/1 (3), B/2 (1); elsewhere, the younger chick usually dies or is killed by the larger chick in the first few weeks, and so usually only one survives. In **Inner Anatolia**, a nest at Kızılcahamam had two eggs on 27 May 1975, and another two young, *c.*2 weeks old, on 31 May 1973 (Pforr & Limbrunner 1982), indicating egg laying on 1 April. A nest 30m up in a tree, near Tuzla Gölü, had two eggs on 8 May 1965 (Warncke 1964–65); two adults and an immature were near Kızılcahamam on 16 July 1986 (R. Webb); and an adult was feeding two small young in a nest near Şereflikoçhisar on 13 June 1998 (GMK). At Demirkazık (**Mediterranean**), two were displaying on 15 April 1996 (L. & L. Vallotton); one young was visible at a nest in a gorge on 27–28 May 2000 (P. & M. McGough); both pair members flew to a nest in a gorge in early June 1993 (T. Park); both adults brought fresh green material, and one a snake, to a nest which had at least one downy young, on 3 June 1999 (R. F. Porter). Danford found many nests in 1876 and 1879 in this region: he visited several nests in the Taurus on 9 March 1879, but eggs had not yet been laid (Danford 1880); a nest in Aladağ had one very small young and a yolkless egg on 30 April 1876 (Danford 1877–78), indicating laying in mid March. In Bolkar Dağları, he took two richly marked eggs from an eyrie on 30 March 1876, and complained that 'all other eggs found were poor in colour, some near white'; he subsequently collected 'fine coloured eggs' in Aladağ in the first half of April 1879. A nest, south of Mut (Mediterranean), was in a recess on a huge cliff in a river gorge; it had one chick, 3–4 weeks old, on 15 May 2004, indicating laying in early March (PC). At Ispir (**Black Sea Coastlands**), display noted in April–May (J. Langhorn), an adult was carrying prey on 12–14 June 1975, and two adults and a juvenile were on crags on 4/5 July 1988 (M. Telfer). A pair was displaying at Kocaçay Delta (**Marmara**) on 1 May 1993 (Ertan 1996). In **East Anatolia**, an eyrie had one young at Horasan on 14 June 1973, and another one young east of Pasinler on 15 August 1972, the latter a very late date for young in a nest, especially given that fledged young have been observed on 21 and 22 June 1994 (J. Francis); a pair alarm-calling, in mountains west of Görentaş on 31 May 1969 (R. F. Porter), is indicative of breeding. Golden Eagles, and many other raptors, replenish nests with green material, or add to nests continually during the nesting period; adults carrying nesting material at Ispir on 12/14 June 1975 and at Demirkazık on 31 May 1991 (D. J. Bates) are considered to be engaged in such behaviour, rather than original nest building, in view of the late dates.

BOOTED EAGLE *Hieraaetus pennatus*
Küçük Kartal

(Du: Dwergarend; Fr: Aigle botté; Ger: Zwergadler; Swe: Dvärgörn)

Subspecies and taxonomy. Monotypic. It has been suggested that *Hieraaetus* should be subsumed within *Aquila* on the basis of genetic evidence (Wink *et al.* 2004, Collinson 2006), but another study found that *pennatus* formed a separate clade (apart from *Aquila*) along with most other small *Hieraaetus*, but not *fasciatus*, thus supporting the maintenance of a separate genus for these taxa (Helbig *et al.* 2005), within which the African species, Wahlberg's Eagle *Aquila wahlbergi* might also be best placed (Lerner & Mindell 2005). The most recent genetic study, by Griffiths *et al.* (2007), also found *Hieraaetus*, as traditionally constituted, to be paraphyletic.

Status and distribution. National conservation status: Vulnerable (population apparently decreasing). Widespread and not uncommon summer visitor principally to northern Turkey, breeding in Marmara, Black Sea Coastlands, the northern fringes of Inner and East Anatolia and much more locally in other regions. Apparently absent from much of Southeast Anatolia, except the extreme east of the region. Breeds principally in wooded montane areas (although few records above 2,500m), but also occurs in more open, drier regions.

Generally more widespread and more numerous on passage throughout the country; occurring in spring from late March (27th) to mid May, peaking in late April (latest records attributed to migrants, 13 June), and in autumn from mid August (11th) to mid October, peaking in late September (latest 4 November). Recorded in highest numbers in autumn. Max. at the Bosphorus 525 in 1971 (131 on 26 September) and at Borçka 473 in 1976 (Andrews *et al.* 1977). None was recorded at the latter in late autumn (11–25 October) 1977 (Beaman 1977), suggesting that passage rapidly declines following its peak. At the Belen Pass 588 were counted in 1976, with the highest single-day total being 126 on 17 September (Sutherland & Brooks 1981a). The majority (often placed at *c.*70%) of migrants are pale-phase birds; among breeders a higher proportion, though still less than half, are dark phase.

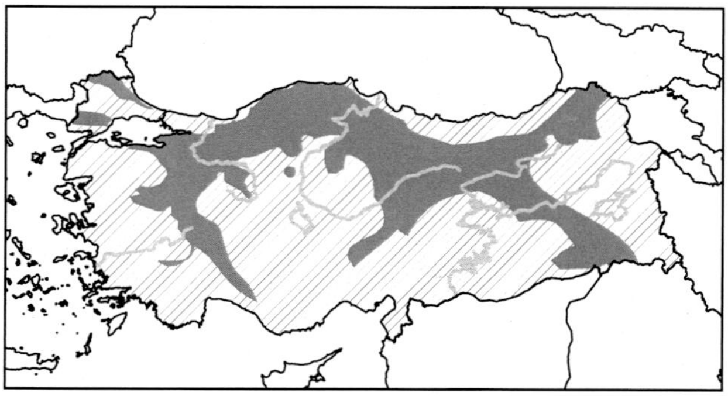

Breeding. In forest in hilly or montane areas, sometimes in quite arid regions. A nest near Çatalağzi (**Black Sea Coastlands**) was 6.2m above ground in very dense forest; it was *c.*1.3m in diameter, of sticks, and had two eggs, laid early May (Ogilvie 1954). A nest at Aladağ (**Mediterranean**) was in a tall Aspen and lined with fresh leaves (usual elsewhere), although no eggs were present by 5 May 1876 (Danford 1878). A nest at Kızılcahamam (**Inner Anatolia**) on 21 June 1983 was atop a ridge (Barış *et al.* 1984), presumably in a tree. Copulation, by two dark-morph birds, was seen at a pass near Yalvaç (Mediterranean) on 29 May 1999 (R. F. Porter). Display by two adults (pale and dark) was seen at Kemerburgaz, near Istanbul (**Marmara**), on 21 April 1996 (L. & L. Vallotton), and ten were displaying at Akdağ Dağı (**Aegean**) on 4 May 1996 (Heredia *et al.* 1997). Two adults and two juveniles were together near Pozantı (Mediterranean) on 24 July 1971. A clutch of two eggs in the Natural History Museum, Tring, was taken by Krüper at Parnassus, in Greece, on 16 April 1896.

BONELLI'S EAGLE *Hieraaetus (Aquila) fasciatus*
Tavşancil

(Du: Havikarend; Fr: Aigle de Bonelli; Ger: Habichtsadler; Swe: Hökörn)

Subspecies and taxonomy. Monotypic. It has been suggested that *Hieraaetus* should be subsumed within *Aquila* on the basis of genetic evidence (Wink *et al.* 2004), but the results of a separate study (Helbig *et al.* 2005) suggested that only *Hieraaetus fasciatus* and *H. spilogaster* should be removed to *Aquila*, maintaining all other, smaller species within *Hieraaetus*.

Status and distribution. National conservation status: Endangered (population apparently decreasing). Breeding status and distribution only partially understood. Kasparek (1992a) calculated that the Turkish breeding population numbered approximately 50 pairs, whilst Tucker & Heath (1994) suggested that up to 100 pairs could breed. It is undoubtedly a rare and local resident throughout Aegean, Mediterranean (where apparently particularly scarce), Southeast Anatolia and probably adjacent areas of Inner Anatolia. Recent records in the breeding season in the eastern half of the Black Sea Coastlands, Inner Anatolia and East Anatolia almost certainly all relate to wandering birds, usually immatures. Generally recorded below 1,500m and some may move short distances to coastal deltas in Aegean. Much confusion appears to have occurred before the advent of modern recording (in 1966) between this species and immature European Honey Buzzards *Pernis apivorus*. At least some published records are almost certainly based upon erroneous identifications, e.g. both Mathey-Dupraz (1921) and Braun (1904) described this species as fairly common at the Bosphorus, where it is nowadays only extremely rarely noted. There are ten records, all since 1980, including eight in 1994, apparently on passage through the Borçka–Hopa area between late September and early October (Kok & Onengae 1995, Mrlík *et al.* 1995).

Breeding. On tall cliffs and rock outcrops, often in gorges and riverine cliffs. Nests on a cliff ledge or in a recess. The nest is a large structure of sticks, lined with leafy twigs. May be re-used in subsequent years, becoming massive; several alternative nests may be used in different years, and there may be 3–4 nests, all fairly close, on the same cliff. All records from **Southeast Anatolia.** Danford (1880) found nests in the Birecik area, and all eggs were well incubated on 14 February 1879. Another nest at Birecik was on a cliff in a dry wadi, and was mainly of sticks, but had a blue plastic bag in the lining; it held two young, *c.*2 weeks old, on 10 May 2004 (D. Marshall), indicating egg laying in mid March, somewhat later than other records. The following data are from Kılıç *et al.* (2003): a nest in a recess on a cliff immediately above the River Tigris near Hasankeyf, *c.*10m from the clifftop and *c.*42m above the river, was of sticks and measured more than 1m in diameter; four unused nests were nearby, presumably from previous years. On 25 April 2003 it held one chick, 30–35 days old, which was almost adult-sized by 16 May 2003, and fledged by 23 May 2003; egg laying was estimated as 10 February. There are many breeding records from the well-known site at Halfeti, with young in the nest from 21 April to early June. A nest had two young, *c.*10 days old, west of Kilis on 7 April 1971, indicating egg laying in mid February. A family of two adults and two juveniles was between Cizre and Eruh on 16 June 1983. An egg in the Natural History Museum, Tring, was taken in Greece on 21 February 1861.

OSPREY *Pandion haliaetus*
Balık Kartalı

(Du: Visarend; Fr: Balbuzard pêcheur; Ger: Fischadler; Swe: Fiskgjuse)

Subspecies and taxonomy. Nominate. Wink & Saner-Gurth (2004) found high levels of genetic distinctiveness between American (*P.h. carolinensis*), European (nominate) and Australian (*P.h. melvillensis*) populations of Osprey.

Status and distribution. National conservation status: Data Deficient (population apparently decreasing). A very local summer visitor in very small numbers. Has bred at two localities in northern Turkey: at Terkos Gölü/Forest of Belgrade (Marmara) in 1860 and 1964 (Alléon 1880, Kumerloeve 1964) and at Kızılırmak Delta (Black Sea Coastlands) in 1966 (Dijksen & Kasparek 1985), but there has been no confirmed breeding at either site, or elsewhere, since (despite an extensive breeding bird survey at Kızılırmak in 1992). Kasparek (1989b) speculated that fewer than 20 pairs breed in Turkey and identified a further four localities which have held birds during the breeding season; Alıcık, near Merzifon, Amasya province (Black Sea Coastlands) in 1962, at Giresun (Black Sea Coastlands) in 1966, at Tatvan, Bitlis province (East Anatolia) in July 1966 (where one was described to be 'calling as though near a nest') and at Yeniçağa Gölü (Black Sea Coastlands) in 1983. Subsequently, recorded during the 1995 breeding season on the Greek/Turkey border at Küplü, and at Iğneada (Thrace) in May–June 2004 (*per* G. Magnin), as well as during the breeding season near Sinop (Black Sea Coastlands) in the 1990s (C. Kiraç). See also Breeding (below).

More widespread as a scarce passage migrant throughout Turkey. A total of 240 records was analysed by Kasparek (1989b). In spring, recorded from mid March (12th) to the end of May/first week of June, peaking in mid to late April (median date 22 April), and in autumn from mid August to early November (3rd) peaking from mid September to mid October (median date 25 September). Commoner in autumn. Relatively few recorded at the major raptor watch sites, although 24 in autumn 1976 at Borçka (Andrews *et al.* 1977) was a significant total, albeit dwarfed by the 199 that moved through Batumi, in neighbouring Georgia at the same season in 2002 (*Sandgrouse* 25: 89). There are a number of winter records in the last decade, when the species has been regular at the Göksu Delta (Mediterranean) in tiny numbers, and can be expected to occur elsewhere in the south and west at this season. There are winter records from Alpullu, Kırklareli province (Marmara) on 6 February 1917 (Gengler 1920), two in Çukurova (Mediterranean) in January 1996 (DHKD 1996), at Tuzla Gölü, Çukurova, on 5 January 2002 (Balmer & Betton 2003a) and 15 January 2006 (KAB *et al.*), and in Gediz Delta (Aegean) in November 2004, December 2005 and January 2006 (Ö. Döndüren), with other undated annual records from Fethiye (Mediterranean) in December and February in recent years (P. Hope).

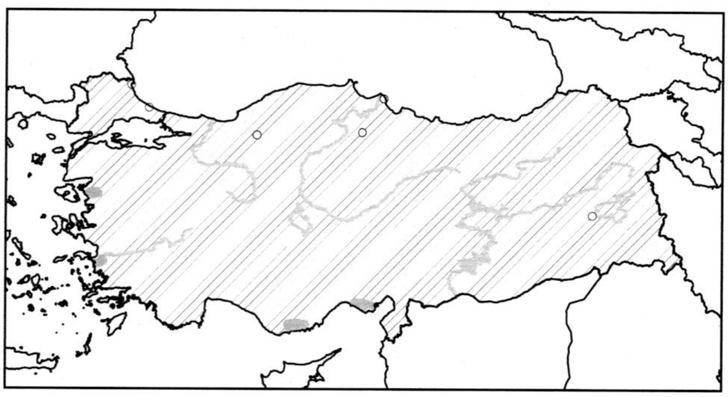

Breeding. An occupied nest was found at Kızılırmak Delta (**Black Sea Coastlands**) in July 1966, where the species was present early May to mid September. Has bred in recent years in Thrace (Açar *et al.* 1977). No other data.

LESSER KESTREL — *Falco naumanni*
Küçük Kerkenez

(Du: Kleine Torenvalk; Fr: Faucon crécerellette; Ger: Rötelfalke; Swe: Rödfalk)

Subspecies. Monotypic.

Status and distribution. Conservation status: Vulnerable (BirdLife International 2000, 2004); national conservation status: Vulnerable (population apparently decreasing). A widespread and generally common summer visitor over much of the country. Recorded from mid March (10th) although does not become widespread until mid April (exceptionally 100 at Kozanlı Saz Gölü, Inner Anatolia, on 3 April 2004), when breeding usually commences. Passage possibly continues until at least mid May (15th). Apparently absent as a breeding bird from the Black Sea coastal fringe. Almost certainly declining in western and southern areas, in Marmara, Aegean and Mediterranean, and perhaps also in Southeast Anatolia, with evidence from all these regions that colonies have disappeared during recent decades. Remains common and widespread (although these populations are probably also declining) in both Inner and East Anatolia, generally breeding in small colonies (of 2–5, sometimes over

20 pairs) in towns, villages and ruins, as well as in dry, treeless agricultural areas, usually close to habitation. Apparently most numerous in flat areas, where the dominant habitats are cereal agriculture, dry and wet grasslands and marshes. Also utilises cliff-faces for nesting in some areas, occasionally some distance from settlements. Breeds up to at least 2,600m (perhaps 3,000m) in several areas. Parr *et al.* (1995) surveyed Lesser Kestrels in 60 randomly selected 10-km squares in central Turkey between 11 April and 15 May 1993, visiting 231 settlements. Extrapolation of their results suggested that the Inner Anatolia population could be conservatively estimated at between 1,500 and 3,500 pairs. This is the second largest population in the world, after that in Spain (Biber 1990).

Post-breeding concentrations of up to 100 birds congregate in some areas (particularly Inner Anatolia), usually in late July and August, prior to migration. The majority depart by late September, although some remain until late October (27th). There is evidence (13 together on the south coast on 27 November 1970, with up to 18 at three localities in this region in late November/early December 1969) to suggest that some occasionally winter in parts of southern and western Turkey (OST 1972, 1975), although the only recent records are of small numbers at Gediz Delta (Aegean). Small numbers are recorded on passage at all three major migration watch sites, principally in autumn, between late July and early October, but passage is often much more obvious well inland with large numbers often seen at sites in Inner Anatolia, e.g. Mogan Gölü, where up to 300 are regular between late August and early October.

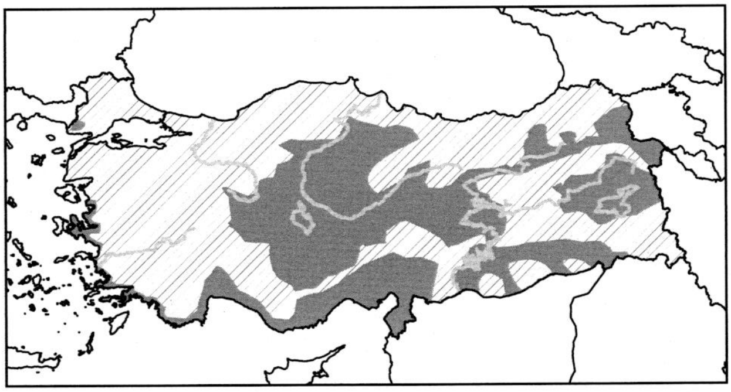

Breeding. In villages or isolated buildings surrounded by arable fields, grasslands or marshes. Colonies usually 2–30 pairs. Commonly nests under roof tiles, especially ridge tiles, but also in crevices in walls, ruins, stones and holes in cliffs. Up to six pairs in a single roof (Wadley 1951). The nest is a slight hollow or scrape, with no material. Eggs: C/4 (1), C/5 (1). Young: B/2 (1), B/3 (1), B/4 (2). B/5 (2). Adults return to breeding sites mainly late March/early April. In **Aegean**, pairs and females were at nest-holes at Akköy on 23–29 April 2003, but laying had not commenced (PC); a male copulated with two females at Izmir on 3 May 1967; many eggs were found at two colonies on 13–15 May 1899, but laying had not started at two others on 8–11 May 1899 (Selous 1900); a nest had one egg (incomplete clutch) at Bafa Gölü on 12 May 1980 (Kasparek 1988). Nests had two young, *c.*2 days old, and five young, up to five days old, at Milet on 19 May 1970, indicating laying in mid April, which appears early. In **Marmara**, the first young fledged at Susurluk on 14 June 1966, also indicating laying in mid April. In **Mediterranean**, copulation was seen on 4 May 1951 (Hollom 1955), and adults were displaying in Çukurova on 5 May 1987 (van der Have *et al.* 1988). In **Southeast Anatolia**, a nest had four young near Gaziantep on 5 June 1935 (Bird 1937), and fledged young were near Birecik on 29 June 1982. In **Inner Anatolia**, at Sultan marshes, two nests had four and five eggs on 25 May 1975, and two had three and five young on 19 June 1977 (Kasparek 1985). A nest had four young, 1–2 weeks old, on 17 June 1977 (Pforr & Limbrunner 1982), indicating laying in early May. Several nests near Konya had young on 7 July 1907 (Ramsay 1914); fledglings recorded mainly in July (earliest 19 June 1996). In **East Anatolia**, adults were at colonies near Van and in villages near Nemrut Dağ, between Malatya and Adıyaman, in the breeding season, but there are no details of nest contents.

COMMON KESTREL *Falco tinnunculus*

Kerkenez

(Du: Torenvalk; Fr: Faucon crécerelle; Ger: Turmfalke; Swe: Tornfalk)

Subspecies. Nominate.

Status and distribution. National conservation status: Least Concern (population apparently decreasing). A fairly widespread and relatively common resident breeder throughout Turkey in a wide variety of habitats from

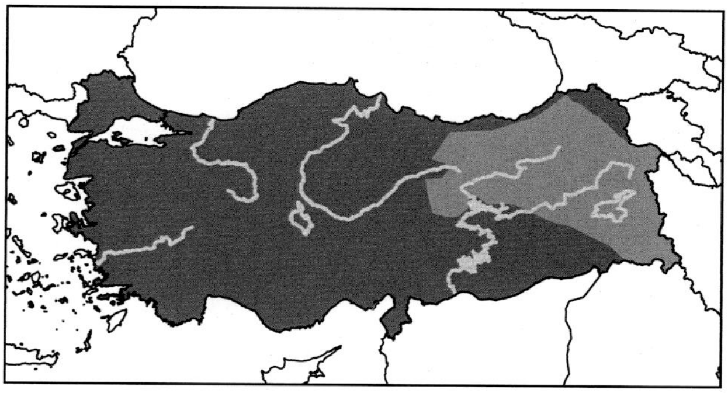

open agricultural areas, steppe and hilly and mountainous country (to at least 4,000m) to woodland.

Widespread on passage in small numbers throughout the country, and in winter, when usually found in coastal localities. Passage is perhaps heaviest through the eastern half of Turkey; only small numbers are recorded at the Bosphorus and the Belen Pass between mid August (12th) and early October (4th). However, 450 passed Borçka on 11–25 October 1977 (Beaman 1977). Only 30 were recorded there in 1976, between 17 August and 10 October (Andrews *et al.* 1977), suggesting that the bulk of the passage occurs in late autumn.

Breeding. Frequents open agricultural areas with scattered trees, woodland and edge, treeless hilly or montane areas with cliffs and gorges, old buildings and ruins, and quarries. Nests in holes in cliffs and on ledges, in old nests of other species, especially Common Magpie *Pica pica* and Hooded Crow *Corvus corone cornix*; and in buildings, e.g. on a mosque at Silifke (Mediterranean). Danford (1880) found a pair nesting, with House Sparrows *Passer domesticus* and Spanish Sparrows *P. hispaniolensis*, in the base of an occupied nest of Eastern Imperial Eagle *Aquila heliaca*, in Inner Anatolia in 1879. There are no records of tree-hole nests in Turkey, which are quite common elsewhere. The nest is a slight hollow, with no material. Eggs: C/5 (3). Young: B/2 (1), B/3 (1), B/4 (1), B/5 (2). At Karapınar (**Inner Anatolia**), two nests had incomplete clutches of two eggs on 24 April 1964, and one of these had five eggs on 9 May 1964 (Warncke 1964–65); a nest at Karapınar had five eggs on 8 May 1990 (J. Whitaker). Five young, *c.*1 week old, at Eşmekaya on 30 May 1993, indicate laying *c.*17 April (PC). In **Southeast Anatolia**, there were seven clutches near Birecik on 30 April 1983 (S. Barış), and three almost full-grown young in a cliff hole there on 7 June 2006 (M. Cowlard). Four recently fledged young near Gaziantep, on 15 May 2004 (GMK), indicate egg laying mid March, much earlier than other records. A nest at Akköy (**Aegean**) had four small young and a hatching egg on 20 May 1999 (J. Hollinrake), indicating laying in mid April. Several nests were in Kızılırmak Delta (**Black Sea Coastlands**) on 17 May 1992 (Hustings & van Dijk 1994). A nest had two fully feathered young at Gelinkaya on 23 June 2004, indicating laying in late April (PC). In **Marmara**, a pair and four fledged young were in a quarry at Büyükçekmece, on 19 June 1973, and display was observed at Manyas Gölü on 27 April 1966 (R. F. Porter). In **Mediterranean**, courtship was observed at Köyceğiz/Dalyan on 29 March 1987 (Kiliç & Kasparek 1989); several records of adults flying into holes in cliffs in May–June, and a nest had young at Demirkazık on 7 June 1998 (T. Quittenden).

RED-FOOTED FALCON *Falco vespertinus*
Aladoğan

(Du: Roodpootvalk; Fr: Faucon kobez; Ger: Rotfußfalke; Swe: Aftonfalk)

Subspecies. Monotypic.

Status and distribution. Conservation status: Near Threatened (BirdLife International 2008). Generally widespread and fairly common, occasionally numerous passage migrant, largely through the western two-thirds of Turkey. In spring recorded from early April (10th) in small numbers, increasing through mid to late April, and usually peaking in mid to late May, with numbers swiftly declining thereafter. Nonetheless, stragglers occur until at least mid June. There are a number of summer (late June to mid July) observations, some in suitable nesting habitat, although without specific evidence of breeding, in all regions except Southeast Anatolia, with the majority from the western and central Pontics (Black Sea Coastlands). Generally scarce in the eastern third of the country, even on passage. Return passage is from early August (9th) to mid/late October (26th; stragglers sometimes occurring into early November on the south coast), with few before late August and the peak in late September–early October. Relatively uncommon at Borçka but usually commoner through the Bosphorus, where there was a maximum of 391 in autumn 1972, including 150 on 29 September. In other years, e.g. 1971, fewer than 20 recorded. There is one late July record from the latter watchpoint. Apparently passed through this area in much larger numbers in the 19th century (see Nisbet & Smout 1957). Quite frequently reported in flocks numbering up to 100 individuals, with exceptional numbers on 22–23 May 1993 when several thousand were at various localities in northern and western Turkey, from the Kızılırmak Delta (Black Sea Coastlands) in the east (where there were 500) to the Marmara Lakes in the west (S. Barış, G. Magnin, R. P. Martins).

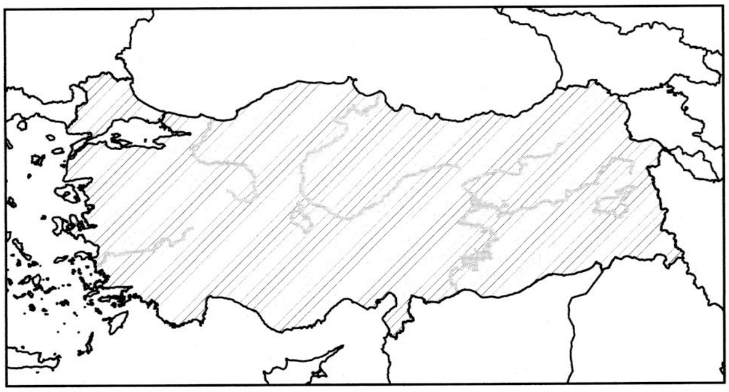

Breeding. According to Danford (1878) seen in flocks near Kayseri (**Inner Anatolia**), and found breeding at a village further north, on 8–14 May 1876. This appears highly doubtful, not only due to the location, but because the species only usually commences breeding *c.*2 weeks later. Five eggs at Manchester Museum, taken in Turkey (location unknown) on 29 April 1911, are eggs of Common Kestrel *Falco tinnunculus* or Lesser Kestrel *F. naumanni*, given the date. No recent evidence of breeding. Elsewhere, 3–5 (2–6) eggs laid, late May–early June, in old nests of other birds, especially Rook *Corvus frugilegus* and Common Magpie *Pica pica*.

[AMUR FALCON *Falco amurensis*]
Amur Doğanı

(Du: Amoerroodpootvalk; Fr: Faucon de l'Amour; Ger: Amurfalke; Swe: Amurfalk)

Subspecies. Monotypic. Often previously considered conspecific with Red-footed Falcon *F. vespertinus*.

Hypothetical, given six records in the Western Palearctic, all in southern Italy, of which three have been accepted by the Italian rarities committee (Corso & Catley 2002), and from several Middle Eastern countries, namely Oman, Saudi Arabia, Yemen and the United Arab Emirates (Clement & Holman 2001). Several unconfirmed claims from Turkey are apparently available, from the Bosphorus (Marmara) and Borçka (Black Sea Coastlands), though we have not seen details of any of these (*per* R. Garavaglia *in* Corso & Catley 2003). A. Corso *in litt.* (2003) advised that the Bosphorus report related to a male, but neither a specific date or named observer is available, whilst the Borçka information is even more unsatisfactory; one or two have apparently been seen there, but no dates, observers or other details are available. Corso & Catley (*op. cit.*) emphasised that some recent claims of *amurensis*, both in Europe and the Middle East, have involved transitional-plumaged second-calendar-year *vespertinus* and that particular care in the two species' separation should thus be exercised.

MERLIN *Falco columbarius*
Boz Doğan

(Du: Smelleken; Fr: Faucon émerillon; Ger: Merlin; Swe: Stenfalk)

Subspecies. Most appear to belong to the form *aesalon*, but there is one Natural History Museum (Tring) specimen (85.8.19.2483), a female, from the Hume Collection that is labelled as being from Izmir (Smyrna), Aegean, which is attributable to *insignis* (Bird 1937; GMK pers. obs.). In addition, Vaurie (1965) mentioned that a specimen of *pallidus* had been taken at Trabzon, in the eastern Black Sea Coastlands, and records from around Malatya (East Anatolia) in winter are also attributed to this subspecies (MÖ).

Status and distribution. Fairly widespread and not uncommon passage migrant and winter visitor throughout Turkey, although principally wintering in the western two-thirds of the country. Only occasionally reported from the major raptor watch sites from late September, although 21 were recorded at Borçka in mid to late October 1977 (Beaman 1977). However, the astonishing total of 972 was recorded at Batumi, in neighbouring Georgia, in autumn 2002, indicating the true scale of passage along the eastern Black Sea route (*Sandgrouse* 25: 89). Generally recorded from mid September to mid April, although occasionally into early May (latest 10th, in Thrace) and exceptionally recorded as early as 20 July at the Bosphorus. Appears to favour wetland localities, particularly in Black Sea Coastlands and Inner Anatolia. Relatively uncommon south of the Taurus (Mediterranean).

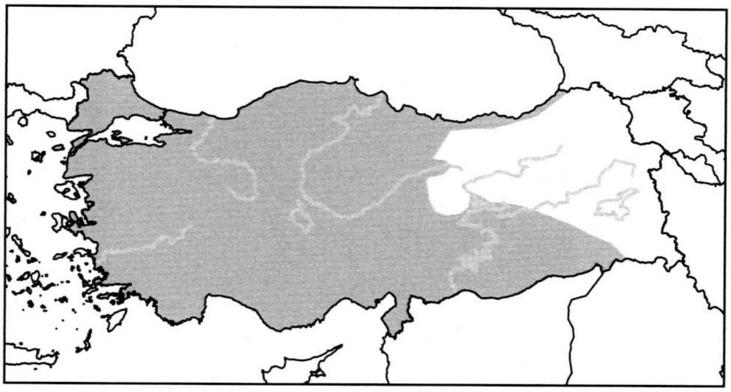

EURASIAN HOBBY — *Falco subbuteo*
Delice Doğan

(Du: Boomvalk; Fr: Faucon hobereau; Ger: Baumfalke; Swe: Lärkfalk)

Subspecies. Nominate.

Status and distribution. National conservation status: Least Concern (population apparently stable). Widespread and fairly common summer visitor over much of the country, although perhaps somewhat more local in East Anatolia, but apparently a relatively widespread breeder in Southeast Anatolia. Generally recorded in open areas with scattered woodland in the lowlands, steppe and in montane areas up to at least 2,800m (twice to 4,000m). Principally recorded from mid April to early October but has been reported from mid March (12th) to late October (30th). Ogilvie (1954) claimed a December record at Catalağzi, Zonguldak province (Black Sea Coastlands) in the mid 1940s, but this would appear erroneous.

Rather more common and widespread on passage throughout the country, although still scarcer in the eastern third of Turkey. Spring passage dates are from mid March (10th) to mid May, with a few still passing through until late May, and in autumn from mid August to end of October (though most passage is usually completed by mid October). Usually recorded in small numbers at the major watch sites, e.g. 189 at Borçka in autumn 1976 (Andrews *et al.* 1977); 910 were counted at Batumi (Georgia) in autumn 2002, again indicating how the scale of passage along the eastern Black Sea route may have been significantly underestimated for some species on the basis of the Turkish counts.

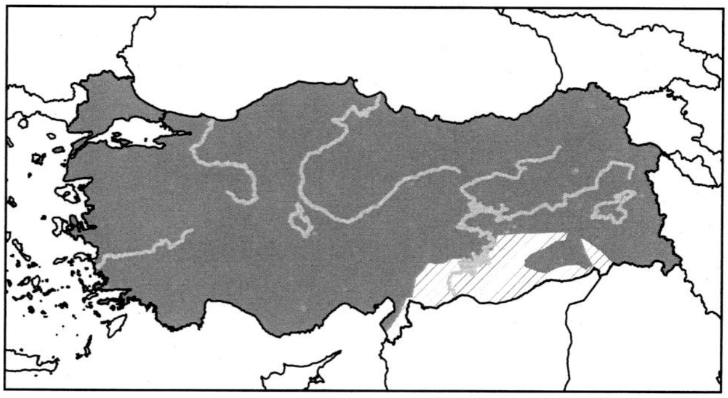

Breeding. In open country with trees, copses, patches of woodland and small plantations, and in large areas of open woodland, both deciduous and coniferous; in lowlands and in hilly or montane regions, with several records of territories in poplars, typically small plantations, beside rivers, in agricultural areas; regularly in gardens in Ankara, where two pairs in 1945 (Wadley 1951). Old nests of other birds, in trees, are used for nesting, and no material is added; in Turkey, only Hooded Crow *Corvus corone cornix* has been recorded, but must surely also use nests of Rook *C. frugilegus* and Common Magpie *Pica pica*, and possibly tree nests of egrets and herons, and Black Kite *Milvus migrans*. Eggs: no data in Turkey; elsewhere, usual clutch three (occasionally two). At Kızılırmak Delta (**Black Sea Coastlands**), there were 12 occupied nests in the period 7 July–6 August 1971, and one still had eggs on 27 July 1971; an adult passed prey to a juvenile on 29 August 1984 (Dijksen & Kasparek 1985); in 1992, the breeding population was 45–50 pairs, of which 23 were in Yürükler Forest; at least five occupied nests were found in trees, the first on 17 May 1992, and calling had not ceased by 10 June, when the survey ended (Hustings & van Dijk 1994), indicating that incubation had probably not commenced. Elsewhere in this region, adults were at a nest at Ispir on 11 May 1986 (M. Ullman), 'breeding was confirmed' on 15 June 1987 (J. Eames), and an adult was carrying prey at Gelinkaya in early September 1992 (Gosney 1993). In **Marmara**, a pair was at a nest at Beykoz on 6 May 1997 (S. Christmas); a food pass was observed near Kesan, on 24 June 1973 (R. F. Porter); at Çamlıca, a nest had four small young, still in down, on 14 August 1966, in a conifer on a hillside with pines and oaks, and scrub; the first young fledged on 2 September 1966, and there were three young out of the nest by 4 September (R. F. Porter pers. comm.), indicating egg laying in late June/early July. A juvenile was at Kuşadası (**Aegean**) on 20 August 1996. In **Mediterranean**, display observed in Çukurova on 20 May 1987 (van der Have *et al.* 1988), and there was a nest near Demirkazık, on 8 May 1992 (N. McMahon). In **East Anatolia**, a pair was displaying on 30 May 1969 at Edremit, near Van Gölü (R. F. Porter), where two pairs were nesting 26 June–2 July 1987 (I. A. Green); 'breeding was confirmed' at Çatak in early June 1987 (J. Eames); a nest near Malatya on 19 May 1935 (Bird 1937); a calling pair was recorded, one of which flew from an old nest of Hooded Crow, *c.*12m up, in a small poplar plantation, near Güzelsu, on 18 June 2004; this nest was not examined, but incubation had clearly not commenced, but it probably contained an incomplete clutch, or egg laying was imminent (PC). Elsewhere, Eurasian Hobby is a late breeder, typically laying in mid June, hatching in mid July, with young fledging in mid August; pairs occupy nests some weeks before laying eggs and, at this stage, are often noisy and conspicuous, but relatively silent during incubation; given this, it is considered that nests in May (above) were probably still empty.

ELEONORA'S FALCON — *Falco eleonorae*
Ada Doğanı

(Du: Eleonora's Valk; Fr: Faucon d'Eléonore; Ger: Eleonorenfalke; Swe: Eleonorafalk)

Subspecies. Monotypic.

Status and distribution. National conservation status: Endangered (population apparently decreasing). A very local breeder on at least one island in the Sea of Marmara (where a colony of 18–21 pairs was discovered in 1978; Kaymas 1980b), and eight, considered to be 'probably breeding' were noted at Çanakkale, on 17 July 2000 (Kirwan *et al.* 2003); also recently confirmed to breed on islands off the south-west coast (*per* KAB). However, an intensive three-day search of Marmara Island, in early October 1994, failed to find the species. Warncke (1972) noted an island, apparently ideal for breeding, in early June 1971, near Gelindere, Mersin (untraced locality, but perhaps Kelenderis, near Aydıncık), and eight pairs were reported at Karaburun, Izmir (Aegean) on 15 May 2004. Nests July–October elsewhere in range. Eken (1997c) spotlighted further areas, principally in Aegean, with records from the relevant period that may prove worthy of additional searching for the species. Widespread prior to breeding.

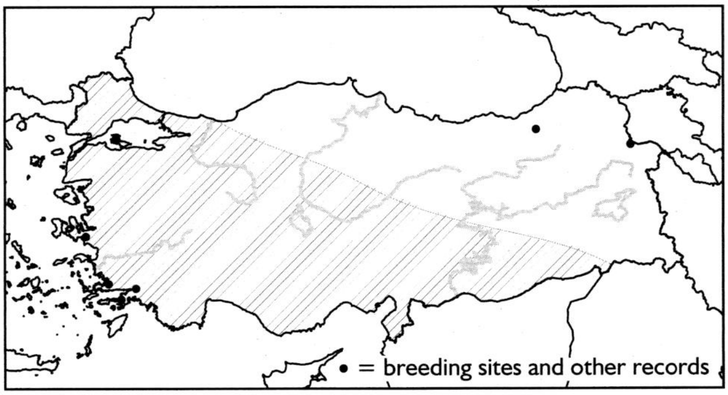

Regularly recorded, usually in small groups of not more than ten, along the entire Aegean and Mediterranean coastlines, presumably as both a passage and summer visitor prior to breeding. Exceptional concentrations include 22 at Muğla, Aegean, on 25 June 1994 (KAB pers. obs.), 29 between Uluabat Gölü and Yeniköy, Bursa province in late May 1990, 30 at Manyas Gölü in late June 1973 and a max. 112 in the

Bodrum area (Aegean) in early June 1981 (Ristow & Wink 1995). Reported from early April (earliest, 9th) to late October (latest, 29th), mainly between late April and mid June. Principally recorded at coastal wetlands and their environs, e.g. the Göksu Delta (Mediterranean) and riverine habitats and cliffs, but also, less commonly, inland (from where there are two ringing recoveries, see Ristow & Wink 1995) and from Inner Anatolia as far north as the environs of Ankara (usually in spring and early summer, but occasionally until late October), as well as elsewhere in Aegean, Marmara and Mediterranean. Usually in lowland areas and foothills from sea level to 1,000m, but occasionally at higher altitude, e.g. at over 2,000m on Demirkazık (Mediterranean), in addition to other records from Sivrikaya and East Anatolia (mentioned below).

It is difficult to interpret these widely scattered records, as it seems possible that undiscovered colonies may influence their spatial distribution. Scarce at the Bosphorus in spring and early summer but regular there from late August to at least early October. These birds presumably originate from the breeding population in the Sea of Marmara. Infrequently reported from Black Sea Coastlands (four records, and as far east as Sivrikaya, Rize province), three recent records in East Anatolia (including a group of four at Tuzluca, Kars province on 2 October 1980) and Southeast Anatolia (ten records, principally from Birecik, but also from Cizre, Mardin province, Diyarbakır and Durnalık, Gaziantep province).

Breeding. No data; see above. Elsewhere, lays 2–3 eggs on a bare ledge, or crevice, of a cliff, usually on an island. Breeding late, timed to coincide with the autumn migration of avian prey and thus a good food supply for the young; egg laying mainly late July, hatching late August, fledging in late September/early October.

SOOTY FALCON — *Falco concolor*
Gri Doğan

(Du: Woestijnvalk; Fr: Faucon concolore; Ger: Schieferfalke; Swe: Sotfalk)

Subspecies. Monotypic.

Status and distribution. Vagrant. One record of two individuals, dark and light morphs, at Birecik (Southeast Anatolia) on 24 May 1973 is remarkable but well described (Mertens 1974). Two other, more recent, records from the same locality, one on 15 June 1999 and another on 6 July 2001 (Kirwan *et al.* 2003). In addition, a dark-morph falcon of either this species or Eleonora's Falcon *F. eleonorae* was also seen near Birecik, hunting bats in the evening of 4 July 1976 (Beaman 1986, Kasparek & Ristow 1986). Recorded once from Cyprus in September 1962 (Flint & Stewart 1992) and three times in Malta (Sultana *et al.* 1975). The nearest known breeding localities are in southern Israel (Cramp & Simmons 1980).

LANNER FALCON — *Falco biarmicus*
Bıyıklı Doğan

(Du: Lannervalk; Fr: Faucon lanier; Ger: Lannerfalke; Swe: Slagfalk)

Subspecies. *F. b. feldeggii.*

Status and distribution. National conservation status: Vulnerable (population apparently decreasing). First confirmed as a Turkish breeding species in 1933 (Kumerloeve & Niethammer 1935), in the Ilgaz Dağları, Kastamonu/Çankırı province (Black Sea Coastlands), but breeding status and distribution remain poorly understood and badly need further clarification. This is largely due to past confusion with other large and medium-sized falcons, especially Saker *F. cherrug*, Peregrine *F. peregrinus* and even pale-morph Eleonora's Falcon *F. eleonorae*. A significant percentage of claimed records may be erroneous. Almost certainly a rare resident in all regions except Thrace where perhaps absent. Probably very local in Black Sea Coastlands. Largely recorded in arid steppe regions and treeless, montane zones, occasionally in wooded areas up to at least 3,000m. There are few confirmed breeding records, which are limited to only a few regions. Breeding distribution can only be inferred from the very small number of relevant observations, e.g. breeding proven in Aegean, Black Sea Coastlands (see above), Inner Anatolia and East Anatolia. There are no confirmed breeding records from Southeast Anatolia, although it perhaps breeds in the extreme east of this region.

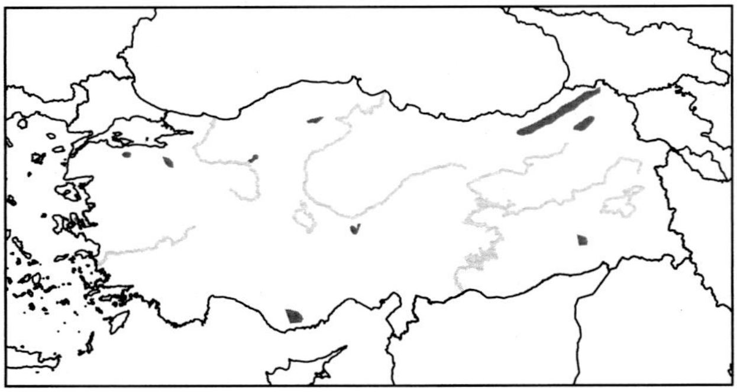

More widespread but relatively uncommon on passage (principally in autumn) and in winter, throughout the country, although apparently scarcer in the eastern third of Turkey. Recorded from mid September to mid October at least at the major watch sites. In winter found most frequently in coastal areas, often around wetlands throughout much of the country.

Breeding. In open country with cliffs and rocky outcrops. In **Inner Anatolia** a pair breeds on cliffs at the Sariyar Reservoir and another pair breeds on cliffs, *c.*100m in height, within a narrow gorge *c.*1km long, north of Beypazari (Magnin & Yarar 1997). Nest and eggs: no data in Turkey; elsewhere, nest merely a shallow scrape, with no material, and usual clutch 3–4 (5). Two adults, presumably a pair, were on a cliff at Kızılcahamam (Inner Anatolia) on 8 April 1984, and one was there on 12 May 1984 (Barış *et al.* 1984). Warncke (1964–65) found a nest, containing an infertile egg and three young, *c.*10 days old, on a steep cliff near Karapınar (Inner Anatolia) on 8 May 1964; when approached, the young lay on their sides, with talons outstretched, calling continuously, but the adults were secretive and quiet; this record indicates egg laying about 3 April.

SAKER FALCON *Falco cherrug*
Uludoğan / Sungur

(Du: Sakervalk; Fr: Faucon sacre; Ger: Würgfalke; Swe: Tatarfalk)

Subspecies. Nominate formerly occurred in Thrace and perhaps even throughout Turkey and into Iran (Ferguson-Lees & Christie 2001), but those in Anatolia have been referred to *milvipes* (see Roselaar 1995).

Status and distribution. Conservation status: Endangered (BirdLife International 2008); national conservation status: Critical (population apparently decreasing, due to young being taken from their nnsts, apparently to supply the demand in Arabia for falconry). A fairly widespread, but rare resident breeder in Inner and East Anatolia and in adjacent areas of Black Sea Coastlands, Mediterranean and Southeast Anatolia. Formerly (perhaps not since 1966) nested in both Marmara and Aegean, even nesting on the outskirts of Istanbul in 1875 (Kumerloeve 1961) and regular around the city until the mid 1950s (P. J. Hayman pers. comm.). During the breeding season generally found in high, arid plains and montane areas near suitable breeding sites. Breeds from mid March. Nesting sites are typically on cliff-faces.

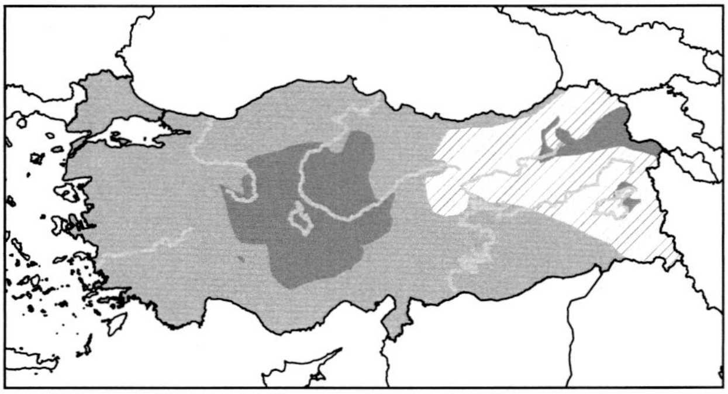

Slightly more widespread on passage, largely in autumn, and in winter throughout the country. Recorded in small numbers at all three major migration watch sites, between mid August and at least late October.

Breeding. Frequents areas with rock outcrops and cliffs, adjacent to dry, open steppe and broad valleys, in central Turkey, and on cliffs and gorges in high valleys in the east. Nests on ledges and in recesses in cliffs, usually high, but one old site, on a remote island in Tuz Gölü (Inner Anatolia), was only 3m above ground (P. Zeininger); also nests on old stick nests of Long-legged Buzzard *Buteo rufinus* on cliffs; and elsewhere also breeds in trees (on old nests of other species), e.g. in Hungary, but no records of this in Turkey. The nest is merely a slight scrape, with no material. Nest sites sometimes used in consecutive years. Eggs: C/3 (1), C/4 (1). Young: B/1 (4, but in two of these records, just one chick was visible, and there may have been more), B/2 (2), B/3 (2), B/4 (1), B/5 (1). At Tuz Gölü (**Inner Anatolia**), a pair reared at least one young in 1966 and two in 1967, at the same site; adults were present at a breeding site near Tuz Gölü in 1968–69. A nest on an island in Tuz Gölü had an unhatched egg and three young, *c.*25 days weeks old on 31 May 1971, indicating laying of the first egg *c.*7 April. A nest north of Şereflikoçhisar had one young, ten days old, on 9 May 1975, indicating laying in late March. Elsewhere in this region, a nest in a hole in a crater cliff at Hasan Dağ had at least one young, *c.*3 weeks old, on 1 June 1972, indicating laying *c.*11 April. Two nests east of Ankara had two and four young, 7–30 June 1977 (Schubert 1979). A nest at Karadağ, a shallow scrape on a broad earth ledge near the top of a 70m cliff, contained five young, *c.*10 days old, on 26 April 2007; another, near Yeşilhisar, in a cavity 3m from the top of a 50m cliff in a canyon, contained three young, *c.*21 days old, on 6 May 2007 (A. Dixon); both of these records indicate egg laying in mid March. Adults were observed landing on a nest several times, near Konya, on 25–26 April 1980 (M. Kasparek). A pair were at a basalt cliff in a crater east of Karapınar on 27 April 1968. An adult was defending a probable nest site, near Ürgüp, on 22 July 1971; and an adult with two juveniles was at Selima, on 21 June 1990. In **Southeast Anatolia**, a nest near the Euphrates and close to the Syrian border had three eggs on 8 April 1971; the incubating female left the nest, and caught food in the air, delivered by the male; four pairs of Spanish Sparrows *Passer hispaniolensis* were nesting in the base of the nest (Warncke 1972). In **East Anatolia**, there was a nest between Tutak and Namur on 10 May 1970, and a probable nest near Ercis, where two adults pursued an intruding Long-legged Buzzard, on 1 June 1971. Adults were carrying food to nestlings on a cliff at Akdamar Island, Van Gölü, on 9 June 1975. An immature and two juveniles were at a nest between Doğubeyazıt and Karabulak on 6 June 2002 (H. Dufourny). In **Black Sea Coastlands**, a juvenile was at Merzifon on 26 June 1978, and two adults and one young were at Ispir on 18 August 1989. A clutch of four eggs, in the Natural History Museum, Tring, was collected in 'Turkey' on 22 April 1872.

PEREGRINE FALCON *Falco peregrinus*
Gökdoğan

(Du: Slechtvalk; Fr: Faucon pèlerin; Ger: Wanderfalke; Swe: Pilgrimsfalk)

Subspecies. *F. p. brookei* breeds throughout; whether *peregrinus* and *calidus*, both of which are long-distance migrants, have occurred on passage and/or in winter is unknown (see Shirihai *et al.* 2000).

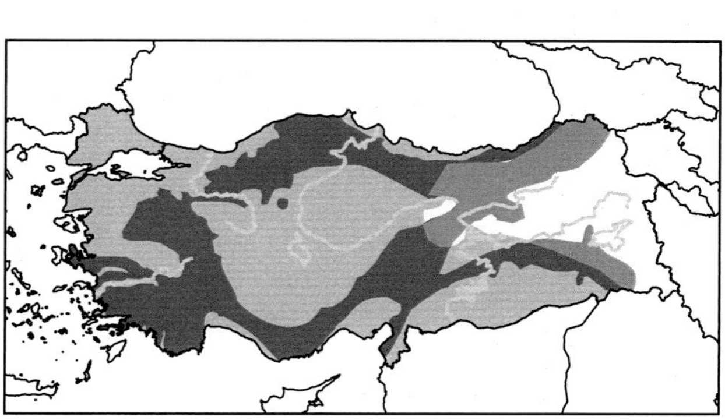

Status and distribution. National conservation status: Vulnerable (population apparently decreasing). Widespread but local resident (or partial migrant in some areas), breeding in all regions, although most common in Aegean, Black Sea Coastlands and Mediterranean. Considered a very rare breeder in Southeast Anatolia. Breeds in hilly and mountainous country, usually above 800m, but also utilises sea cliffs in some areas.

Disperses outside the breeding season, becoming more widespread and often occurring in coastal lowlands, principally in the western two-thirds of Turkey. There is perhaps some limited winter immigration by birds from further north, although there are few observations from the major migration watch sites (22 August–21 October at least).

Breeding. In hilly or mountainous areas, from near sea level on sea-cliffs at Ardreşan in Mediterranean (Eames 1989), to 2,800m in the Aladağ, also in Mediterranean. Nests on a ledge, or in a recess, on rock buttresses, in gorges, and on cliffs; a pair nested on the Upper Citadel at Trabzon (Black Sea Coastlands) in 1958 (Smith 1960). The nest is a shallow scrape, with no material added. Eggs: no reliable data in Turkey; elsewhere, usual clutch 3–4 (2–6). Young: B/2 (1, but only two young visible and there may have been more), B/3 (1). In **Mediterranean**, a pair was displaying at a nesting cliff, in hills north of Taşucu, on 23 March 1993 (J. van der Winden). A pair was defending cliffs, east of Aydıncık (west of Taşucu), on 15 April 1993 (R. & C. Smith). An eyrie had three young, *c.*14 days old, on a coastal rock face west of Gelindere on 13 April 1971, indicating laying of the first egg late February. Two pairs were nesting on sea-cliffs at Adreşan on 16 May 1987, and a fisherman said he could obtain a nestling for a high price (Eames 1989). There were pairs breeding in at least two gorges near Beyşehir Gölü, 20 May–2 June 1998 (A. Kullberg). In **Southeast Anatolia**, there was a nest at Birecik on 3 July 1970, and a nest at Karkamis had one egg on 26 March 1973, the latter presumably an incomplete clutch. An adult was carrying prey at Işıklı on 15 June 1996 (GMK). Eleven adults were in the Van area (**East Anatolia**), 25–30 May 1969, of which two pairs were displaying over suitable breeding territory. In **Black Sea Coastlands**, two adults were aggressive towards a pair of Golden Eagles *Aquila chrysaetos*, at Çat on 21 April 1993 (J. Faldborg). In **Inner Anatolia**, a nest in a recess on a high cliff at Hasan Dağ had at least two young, aged 3–4 weeks on 25 May 1992, indicating laying in late March (PC). At Ürgüp, a female was circling a rock face on 3 June 1970, and there were fledged young in the same place on 22 July 1970. A pair probably bred in 1970 at Uludağ (**Marmara**), where four birds, probably a family group, were seen mid August–late September. On Lesbos (Greece), a noisy adult and recently fledged young were on a nesting cliff on 14 May 1995 (T. Quittenden), thus indicating laying in mid March. The nesting period, from laying of the first egg to first flight of the young is *c.*2 months, but young are dependent for a further 2+ months. Whilst data in Turkey are sparse, appears to lay eggs in late March–April, the young fledging in late May–June.

[BARBARY FALCON *Falco pelegrinoides*]
Kızıl Enseli Doğan

(Du: Barbarijse Valk; Fr; Faucon de Barbarie; Ger: Wüstenfalke; Swe: Berberfalk)

Subspecies. Monotypic, or polytypic, if disjunct race, *babylonicus*, is also included within present species.

Best treated as hypothetical in Turkey. First mentioned for the country by Gosney (1996) from an observation at Birecik (Southeast Anatolia) in August 1990, and subsequently recorded in the same area by H. Shirihai and J. Francis in June 1994. The observers concerned are known to us and, in the case of HS, have very strong familiarity with the species. Nonetheless, details have never been published and other observers (including ourselves) have failed to record it, despite active searches and greatly increased ornithological coverage of all parts of Southeast Anatolia in recent summers. Given these factors, we deem it best to regard the species as hypothetical here, although we acknowledge that the species has probably occurred in Turkey and is even plausibly resident in, presumably, tiny numbers there. Baumgart *et al.* (1995) listed the species as a regular visitor to Syria that perhaps breeds, and there are three records in Lebanon (Ramadan-Jaradi *et al.* 2008). The same may prove to be true in extreme southeast Turkey. Most recently, one, apparently closest to the form *babylonicus* claimed at the Göksu Delta (Mediterranean), on 15 July 2006, but no details supplied, and this taxon, which occurs from Iran eastwards, seems highly unlikely to occur in Turkey.

WATER RAIL *Rallus aquaticus*
Sukılavuzu

(Du: Waterral; Fr: Râle d'eau; Ger: Wasserralle; Swe: Vattenrall)

Subspecies and taxonomy. Nominate. Only four subspecies are generally recognised (Dickinson 2003), of which one, *indicus*, of southern and eastern Asia, has been recommended for species status (Rasmussen & Anderton 2005).

Status and distribution. National conservation status: Least Concern (population apparently decreasing). We suspect that the breeding population is probably being severely affected by ongoing wetland drainage. Widespread breeder throughout the country where suitable habitat exists. Resident in those regions with a milder climate, namely Black Sea Coastlands, Marmara, Aegean, Mediterranean and Southeast Anatolia, whilst breeders in Inner Anatolia and the east move to coastal areas in winter, where more common both at this season and on migration. Some winterers may originate from outside the country. The breeding population is thought to have decreased over the past 15 years, due to the gradual loss of freshwater habitats.

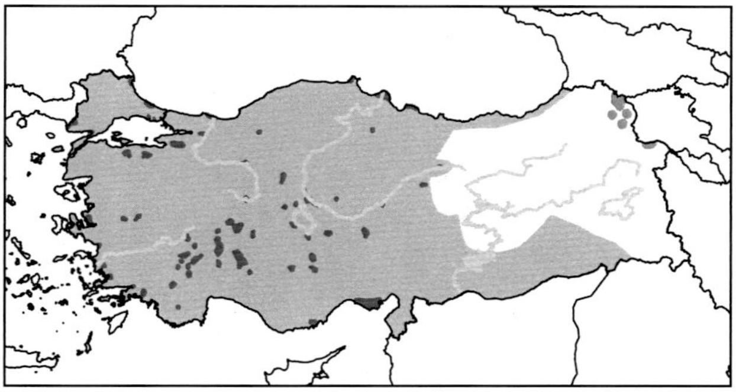

Breeding. In reedbeds, marshes, partially flooded grassland, and waterside vegetation around lakes and pools. The nest is a shallow cup of leaves, reeds and waterside plants, usually low in dense cover, slightly above water level. Eggs: no reliable data; elsewhere usual clutch 6–10 (5–16). The species is also double-brooded, and calling birds, from late March to July, suggest that this is the case in Turkey. At Hotamış (**Inner Anatolia**), a nest had one egg, and another one freshly destroyed eggshell, both at the base of tall *Phragmites* beside a narrow channel of deep water, on 25 May 1993 (PC). At Eşmekaya, a nest had small young on 25 June 1977, and young were heard near an alarm-calling adult, on 20 June 1977 (Schubert 1979). In **Black Sea Coastlands**, a nest, ready for eggs and with an adult calling nearby, was in dense vegetation by a pool, near Yeniçağa Gölü, on 20 May 1992 (PC). At Kızılırmak Delta, adults and juveniles were observed in July 1971, and two juveniles were caught by locals on 28–29 July 1971 (Dijksen & Kasparek 1985); the breeding population at this site was estimated at 500–750 pairs in 1992, and an adult and downy young were observed on 28 May 1992 (Hustings & van Dijk 1994), indicating egg laying in early May. An adult and two young were at Iznik Gölü (**Marmara**) on 7 July 1966 (R. F. Porter).

SPOTTED CRAKE *Porzana porzana*
Benekli Suyelvesi

(Du: Porseleinhoen; Fr: Marouette ponctuée; Ger: Tüpelsumpfhuhn; Swe: Småfläckig sumphöna)

Subspecies. Monotypic.

Status and distribution. National conservation status: Data Deficient (population apparently decreasing). Status uncertain. Recorded in summer in Marmara, Inner Anatolia, Black Sea Coastlands (see below) and East Anatolia, but principally known as a widespread and reasonably common passage migrant, although seems scarcer in the eastern third of the country. The Spotted Crake's true status is difficult to ascertain, as passage almost certainly obscures breeding in many suitable areas, and birds may call on northbound passage. Arrives mid–late March, the earliest date being 17th (Mediterranean), but most spring records are from April, with one early record of a singing bird on 9 April. Records from May seem equally likely to be breeders, as Spotted Crake arrives early on its breeding grounds and egg laying commences early April in central Europe and by mid May in northern Europe (Cramp & Simmons 1980).

Spring passage generally involves small numbers (the species tending to be generally less conspicuous at this season, unless calling), although 33 were noted between 14 April and 10 June at Kızılırmak Delta (Black Sea Coastlands) in 1992 (Hustings & van Dijk 1994). Post-breeding dispersal commences mid July, with the earliest autumn record being dated 19 July. Most records at this season are from August–September, but passage continues well into October, with the latest records dated 27 October (Aegean), exceptionally 28 November (Mediterranean). The majority of autumn records refer to single birds, but the species is certainly commoner than in spring and than records suggest, with several single-day counts in excess of ten; e.g. 20 at Balıkdamı (Inner Anatolia) on 18 September 1995 (Kirwan & Martins 2000).

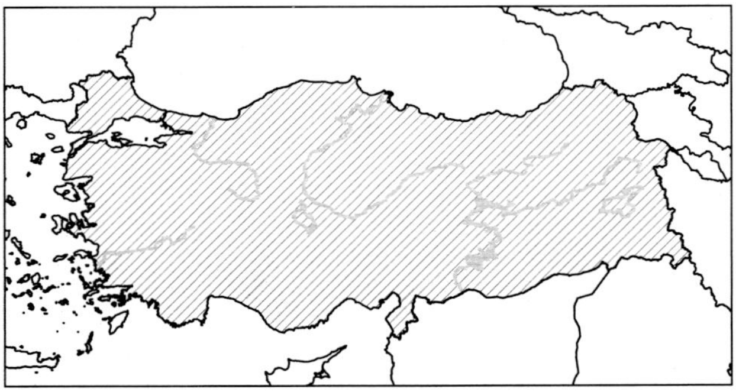

Breeding. Calling males in marshes and wet grassland may indicate breeding in several regions, but no proof. In Kızılırmak Delta (**Black Sea Coastlands**) a male heard calling on 7 June 1992 was in the same place as one heard on 6 May 1992 (Hustings & van Dijk 1994).

LITTLE CRAKE — *Porzana parva*
Bataklık Suyelvesi

(Du: Klein Waterhoen; Fr: Marouette poussin; Ger: Kleines Sumpfhuhn; Swe: Mindre sumphöna)

Subspecies. Monotypic.

Status and distribution. National conservation status: Data Deficient (population apparently decreasing). Status uncertain. Recently proven to breed in Mediterranean and almost certainly also does so in Inner Anatolia, Black Sea Coastlands and East Anatolia (see below). In fact, in our experience, this species is probably the commonest and most widespread of the three *Porzana* spp. as a breeder, and it potentially breeds in all regions, including Southeast Anatolia.

Otherwise, a widespread passage migrant that is almost certainly commoner than evidence suggests, and (compared to Spotted Crake) migration seems to be on a broader front across Turkey. Passage obscures breeding in many areas of the country, especially as birds do call whilst on migration. Arrives in mid March, the earliest record being dated 18th (Inner Anatolia), and occurs throughout April and May. Given that Little Crakes lay eggs in early May in central Europe (Cramp & Simmons 1980), late spring records must, in many cases, refer to prospective breeders, although it is possible that even birds recorded as late as the first week of June could be en route to breeding areas further north and east, where the breeding season is probably later. Numbers in spring are generally small, although 42 were at Kozanlı Saz Gölü (Inner Anatolia) on 19 April 1996, 39 at one site in Southeast Anatolia on 22 April 1996 (Kirwan & Martins 2000), 71 were seen in Çukurova (Mediterranean) between

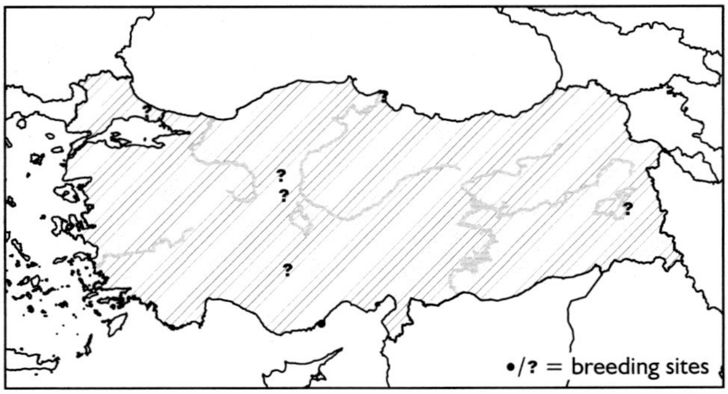

25 March and 15 May 1990 (Kivit *et al.* (1994), 32 were recorded at Kızılırmak Delta (Black Sea Coastlands) in spring 1992 (Hustings & van Dijk 1994) and Balmer & Betton (2003a) noted the presence of 'dozens' at Mogan Gölü (Inner Anatolia) in March–April 2002, hinting at a particularly heavy passage. The onset of autumn migration is unclear but it is certainly underway by August. Two early July records in Mediterranean are both from Göksu Delta, where the species has since been proven to breed. Most autumn records are from September and October. Numbers are usually low, especially compared to spring, and the latest recorded date is 4 November (Inner Anatolia), although passage more usually appears to terminate in mid October.

Breeding. The only nesting record is of three new, but empty, nests on 23 May 1993 in wet reeds and sedges at Göksu Delta (**Mediterranean**), in the immediate vicinity of where two adults, probably a pair and possibly with unseen young, were flushed on the previous two days; all were shallow cups of rushes, much smaller than the nest of Water Rail *Rallus aquaticus*, and well hidden in dense vegetation, only slightly above water level (PC). There are many other records of adults in Göksu, including males and females together, possibly pairs, and males calling in late March–April. In **Inner Anatolia**, a pair was at Mogan Gölü on 25 April 1974 and 1 May 1974, and two females and two males were at Çöl Gölü on 2 May 1992 (Karauz Kiraç 1993). At Kızılırmak Delta (**Black Sea Coastlands**), several were calling in 1992, and considered a possible breeder (0–5 pairs) (Hustings & van Dijk 1994). At Çenge Gölü (**East Anatolia**), there were two singing males on 20 May 1993 (Kirwan & Martins 2000a) and two singing males on 8/9 June 1997 (Kirwan *et al.* 2003). Recorded regularly, including singing birds, at Büyükçekmece, Istanbul (Marmora), in May–June 1997 and 1998, suggesting the presence of a breeding population (M. Freedman, GMK)

BAILLON'S CRAKE *Porzana pusilla*
Küçük Suyelvesi

(Du: Kleinst Waterhoen; Fr: Marouette de Baillon; Ger: Zwergsumpfhuhn; Swe: Dvärgsumphöna)

Subspecies and taxonomy. Presumably all *intermedia* but, as noted by Roselaar *in* Cramp & Simmons (1980), the boundaries between this subspecies and the nominate are somewhat diffuse and difficult to draw, thus perhaps invalidating *intermedia*, at least under our terms of reference. In contrast, Rasmussen & Anderton (2005) speculated that the Asian nominate race might prove sufficiently distinct to warrant specific status.

Status and distribution. National conservation status: Data Deficient (population apparently decreasing). Status uncertain. Just one proven breeding record, but birds noted at a number of additional potential breeding localities in summer (see below). Apparently a widespread passage migrant that is almost certainly commoner than evidence suggests, although it is probably the rarest of the three *Porzana* crakes in Turkey, and it appears to principally move through the eastern two-thirds of the country. True status is difficult to ascertain as passage almost certainly obscures breeding, and birds may sing on northbound passage. Arrives in March, the earliest recorded dates being 8th (Inner Anatolia) and 13th (Mediterranean), and a singing bird was noted in the former region on 19th–28th. Occurs throughout April and May. In central Europe egg laying commences May, thus late-spring records could refer to prospective breeders. Numbers are generally low at this season with a maximum of five on 20 April. Summer records are available from late June (Southeast Anatolia) and early July (Mediterranean). In autumn, recorded from August to October, principally in September with a latest date of 1 October and no records involving more than one individual. Two winter records: singles at Manyas Gölü (Marmara) on 15 December 1969 (OST 1975) and Kızılırmak Delta (Black Sea Coastlands) on 13 January 2001 (A. Gürsoy & S. Barış).

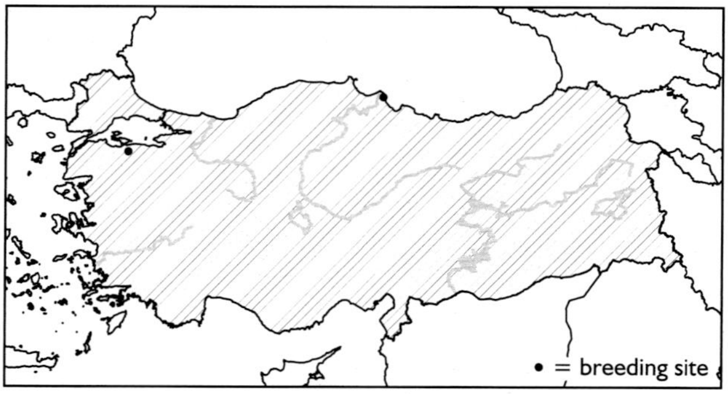

Breeding. Beaudoin (1967) was shown an active nest in a rice field near Manyas Gölü (**Marmara**) on 9 August 1965, and this is the only confirmed breeding record in Turkey; it was a rounded cup of sedge stems, slightly above water level, in a clump of sedge (probably *Scirpus*) *c.*50cm high. Three well-incubated eggs (olive background with fine brown markings) were still in the nest, and four tiny black chicks and both adults (alarm-calling) were in the immediate vicinity; egg laying would have commenced *c.*18–19 July and, in view of the late date, it was considered a second brood. Local people informed Beaudoin that these nests were often found in the rice fields. Elsewhere, calling males in marshes and wet grassland may indicate breeding in several regions, but no proof of nesting. An adult, flushed from a marsh with shallow water at Eşmekaya (**Inner Anatolia**) on 23 May 1992, was in suitable breeding habitat (PC).

CORN CRAKE *Crex crex*
Bıldırcınkılavuzu

(Du: Kwartelkoning; Fr: Râle des genêts; Ger: Wachtelkönig; Swe: Kornknarr)

Subspecies. Monotypic.

Status and distribution. Conservation status: Vulnerable (BirdLife International 2000), subsequently downgraded to Near Threatened (BirdLife International 2004); national conservation status: Endangered (population trend unknown). Breeding proven twice in Black Sea Coastlands (see below) and there are several other records of calling birds in northern Turkey in June–July (Maas Geesteranus 1959, Kirwan & Martins 1994). Von Gonzenbach (1852) considered it a summer bird, as well as a passage migrant, in western Anatolia, based on his experiences from around Izmir (Aegean) in the mid 19th century. Pasquali (1991) regarded the species as a breeder, but the justification for his statement is unknown.

Otherwise, apparently a widespread but generally rather uncommon passage migrant across the whole country. Spring migration in southern Europe commences in March, peaking in late April, with birds arriving on the breeding grounds from mid April (Cramp & Simmons 1980). In Turkey, spring records are from the period March–May, with the majority in late April/May. The earliest spring record is dated 8 March (Black Sea Coastlands). Numbers are small with a maximum of three in late May (East Anatolia). In autumn, the species' migration through Europe commences in August and continues into November (Cramp & Simmons 1980). In Turkey, autumn records are all dated August–October (earliest 9 August) with most in September and the latest records are dated 5 October (Mediterranean), 6 October (Marmara) and 22 October (Black Sea Coastlands), and, again, numbers are low. Additionally, Kasparek (1992a) mentioned a February record (locality unknown), which suggests that the species might occasionally overwinter in Turkey.

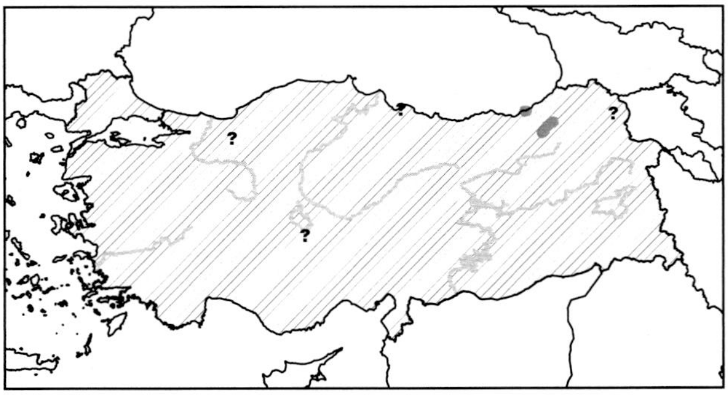

Breeding. First proven in hay meadows near Cam Geçidi (**Black Sea Coastlands**) on 3 August 1996, when an adult was flushed and two black downy young, the size of Common Quail *Coturnix coturnix*, ran off through the sedges; the dominant plants were sedges *Carex* sp., rush and grass, uncut, *c.*30–40cm tall. The meadows were cut by hand-scything in late July/early August (Green 1997). Assuming the usual clutch of 8–12 eggs, the first eggs would have been laid in the last ten days of June. Two adults were observed carrying food at Rize (Black Sea Coastlands) on 10 July 2001 (Kirwan *et al.* 2003); adults feed young only for the first few days, and this record can be back-dated to laying of the first egg *c.*7 June. A single bird near Kars (**East Anatolia**) on 21–22 June 2005 is also suggestive of breeding (Balmer & Betton 2005b), as is one calling in fields near Abant Gölü

(Black Sea Coastlands) on 31 May 2004 (T. Brush *et al.*). Elsewhere in its range, the nest is a hollow in the ground, lined with dry grasses and nearby plant material, and well hidden.

COMMON MOORHEN *Gallinula chloropus*
Sutavuğu

(Du: Waterhoen; Fr: Gallinule poule-d'eau; Ger: Teichhuhn; Swe: Rörhöna)

Subspecies. Nominate.

Status and distribution. National conservation status: Least Concern (population apparently decreasing). Widespread and common resident, and possibly a partial migrant, although rare across much of eastern Turkey. In winter, more numerous in western and central Turkey, particularly in coastal areas, suggesting that some are immigrants from outside the country. Occurs in a variety of wetland habitats, including very small areas of marshland and pools. Numbers generally small in Marmara, with a maximum of 40 at Meriç Delta, and in Aegean and Mediterranean, with a maximum of 62 at Corak. Widespread in Inner Anatolia, but again generally in small numbers; exceptionally 300 at Mogan Gölü in April 2002. In Black Sea Coastlands principally recorded at Kızılırmak Delta, with possibly slightly smaller numbers in winter. In East and Southeast Anatolia generally recorded only in very small numbers, although there was an exceptional *c.*400 on Karkamis Reservoir, south of Birecik (Southeast Anatolia) in September 2004, and this site appears to support quite large numbers regularly.

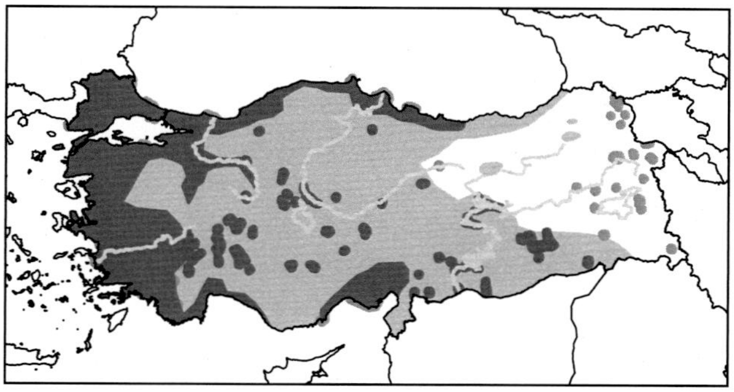

Breeding. In dense vegetation at lakes and pools, reedy waterways and in reedbeds and marshes with open water. In Çukurova (Mediterranean), breeds in freshwater areas, including canals, riversides and lake edges (van der Have *et al.* 1988). In Kocaçay Delta (Marmara), breeding population estimated at 400–450 pairs in 1993 (Ertan 1996). In Kızılırmak Delta (Black Sea Coastlands), population estimated at 200–250 pairs in 1992 (Hustings & van Dijk 1994). Nest usually built over water or close to water, well hidden in reeds and other aquatic vegetation. A nest at a gravel pond at Birecik (Southeast Anatolia), was in waist-deep water, well concealed in tall *Phragmites*, and built up from the water, attached to the base of reeds; a substantial structure of flat dead leaves of reeds, interwoven to form a neat cup. Eggs: C/7 (2), C/11 (1). In **Aegean**, a nest had three eggs, and several newly hatched young, at Büyük Menderes Delta on 26 May 2004 (M. Demain), indicating laying in late April/early May. In **Marmara**, most adults did not appear to have commenced nesting at Uluabat Gölü on 25 April 2003 (PC), and a nest contained 11 eggs there on 2 June 2006 (R. Castell); the first juvenile noted at Manyas Gölü in 1966 was on 30 June (R. F. Porter). In **Black Sea Coastlands**, an adult with at least five young, *c.*10 days old, was at Abant Gölü on 6–13 July 1975, indicating egg laying in early June. At Kızılırmak Delta, there were many young in July 1971 (Dijksen & Kasparek 1985), and a nest on 26 May 1992 (Hustings & van Dijk 1994). In **East Anatolia**, an adult and two immatures were at Bulanık on 29 July 1991 (N. Littlewood). In **Southeast Anatolia**, two nests each had seven eggs at Birecik on 11 May 2004 (K. Taylor), and recently hatched young were seen there on 7 June 2006 (M. Cowlard). In **Mediterranean**, young were seen in Çukurova from 10 May in 1987 (van der Have *et al.* 1988), indicating earliest egg laying in mid April. A pair and two young were at Aydıncık on 20 May 1999 (R. F. Porter). In **Inner Anatolia**, there was a family party at Hotamış on 5 July 1986 and *c.*30 pairs nesting there in 1991 (Kirwan 1993a); young were seen at Mogan Gölü on 15 August 1968, at Akşehir Gölü on 3 August 1971, and at Seyfe Gölü on 3 August 1983 (Husband & Kasparek 1984). Records indicate at least two broods a year, as elsewhere in the range.

PURPLE SWAMPHEN *Porphyrio porphyrio*
Sazhorozu

(Du: Purperkoet / Grijskoppurperkoet; Fr: Talève sultane; Ger: Purpurhuhn; Swe: Purpurhöna)

Subspecies and taxonomy. Sangster (1998) proposed species status for all six subspecies-groups in the *P. porphyrio* complex on the basis of evidence for paraphyly in the species (Trewick 1997) and consistent morphological differences (in a near-reversion to the arrangement adopted by Peters 1934). The English name Grey-headed Swamphen is proposed for the *poliocephalus* group – including the potentially dubious taxon *caspius* in southern Turkey (admitted by Cramp & Simmons 1980) – which occurs from southern Asia west to Turkey. Based on present evidence, Sangster (*op. cit.*) considered all other subspecies described within this grouping, *seistanicus*, *caspius* and *bemmeleni*, as synonyms of *poliocephalus*, a treatment we would echo, regardless of the ranking one applies to the latter. Aharoni (1930) considered those that formerly bred at Amik Gölü (Mediterranean) to be *seistanicus* and specimens from this area, taken by Kumerloeve (1963a), accord closely with that form. Additionally, irrespective of whether one considers *seistanicus* a synonym of *caspius*, as noted by Kasparek *et al.* (1989) morphometrics of these Turkish birds suggest a much closer relationship with other *seistanicus* populations than with those of the Caspian (*caspius*). Given such low levels of variation, we advocate that only the name *poliocephalus* warrants continued recognition, and that other names within this grouping (listed above) are hereafter considered junior synonyms. Mathey-Dupraz (1923) described the observation of a bird of one of the west Mediterranean races – Western Swamphen *P. porphyrio* (Sangster *op. cit.*) – near Istanbul on 15 July 1893.

Status and distribution. National conservation status: Vulnerable (population apparently increasing). Known as a breeder (resident) only from the Göksu Delta, the Çukurova region and the former Amik Gölü (Mediterranean), but new populations were discovered in the late 1990s in Kızılırmak Delta (Black Sea Coastlands) and, in 2003, at Birecik (Southeast Anatolia) (Balmer & Betton 2004b). Kasparek (1992a) estimated the Göksu population at *c.*70 pairs and that in Çukurova at *c.*20 pairs, where it became centred on Akyatan Gölü (Ovaa & Vos 1990), following the drainage of Aynaz Gölü, in the Tarsus Delta, in 1972 (although there was a record from this area as late as 1982). The Amik Gölü population has, of course, been lost, with the last record being in 1962 (Kasparek *et al.* 1989). However, more recent and intensive fieldwork placed the Göksu population at 300 pairs (Heath & Evans 2000), but drainage of the relevant area has again placed the Çukurova population at risk. The Birecik population is currently very small, but that in Kızılırmak Delta, where the species was first noted in the late 1990s (although it had been reported about five years previously by local people) has grown near-exponentially; in late December 2005 and October 2007, *c.*1,800 were counted in rice fields in the area (G. Serra; Balmer & Betton 2008).

Apart from the Istanbul record mentioned above (and the species was also listed for the city by Rigler [1852]), there are several other records from elsewhere in Turkey, some of which might even have indicated occasional breeding events: at Akşehir Gölü (Beaman 1986), Sultan marshes (Kasparek 1985, Kirwan & Martins 1994) and Golbaşı Gölü, Adıyaman province (Southeast Anatolia) (Kasparek *et al.* 1989), an adult (considered to be race *caspius*) at Ahlat marsh (East Anatolia) on 1 June 2002 (H. Dufourny *et al.*), one at Karakaya Reservoir, Malatya province (East Anatolia) on 27 November 2005 (MÖ) and one at Iğdır (East Anatolia) on 8 September 2006 (*Dutch Birding* 28: 376). The latter observation is the only one available from this regularly watched locality. However, the claim that the species at least formerly occurred at Manyas Gölü, Marmara (Kiziroğlu & Kiziroğlu 1987) seems completely lacking in foundation.

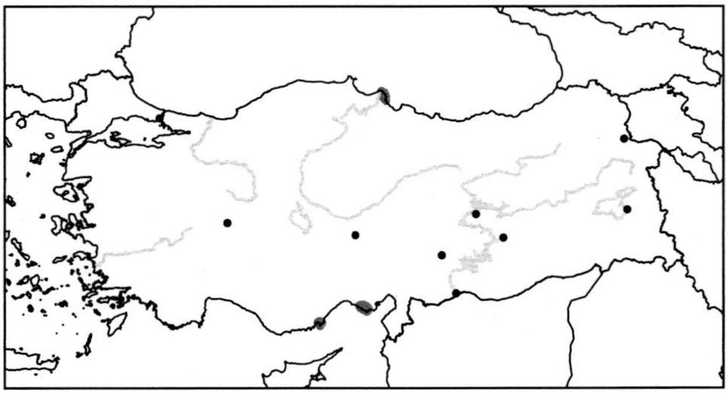

Breeding. In Göksu and Çukurova deltas (**Mediterranean**), and Kızılırmak Delta (**Black Sea Coastlands**); also recorded in Inner Anatolia, where status uncertain; formerly bred at Amik Gölü (Mediterranean). In Göksu, nests in rushes and reeds growing in shallow pools beside Akgöl. The nest is a large platform of dead stems and leaves of aquatic plants, built up from the water to form a shallow cup; taller rushes around the nest are bent over by the incubating bird to conceal it from above. Nests in use for a few days become rather conspicuous, especially where the adults clamber out of the muddy water to the nest. Most are in tussocks growing in water, but one was on dry ground on a small island of grasses, surrounded by reeds and rushes, with a muddy enclosed tunnel leading 2m through dense vegetation to the nest, which was 1.3m above water level and completely screened from above. Eggs: C/3 (2), C/4 (2), C/5 (4), C/6 (1); smaller numbers considered incomplete clutches. Most mobile young are seen as B/1 or B/2 (15), but are probably incomplete broods, due to losses after hatching or because observers see only some of the young; also B/3 (5), B/4 (3). Of eight nests at Göksu Delta, 21–23 May 1993, four had incomplete clutches, in which additional eggs were subsequently laid, and one hatched on 24 May 1993, indicating laying of the first egg on 25 April (PC); a nest there on 26 May 1998 had four eggs (T. Quittenden); a late nest had three eggs, probably well incubated, on 17 June 1992 (PC). Young mostly observed mid May and June; the earliest date 3 May 1989 (Kasparek *et al.* 1989), indicating laying early April. Late records include newly fledged young in late August. In Çukurova there was an adult with a single chick on 8 May 1990, and there were seven territories, with at least nine adults and ten chicks, on 9 May 1990 (Ovaa & Vos 1990). In Kızılırmak Delta, there was an adult with two well-grown young on 27 June 1999 (GMK). At Amik, on 27 May 1933, a nest with five eggs was found in a dense reed brake, and a newly hatched chick was caught (Meinertzhagen 1933), and five eggs were collected on 6 June 1929 (Kasparek *et al.* 1989).

EURASIAN COOT *Fulica atra*
Sakarmeke

(Du: Meerkoet; Fr: Foulque macroule; Ger: Bläßhuhn; Swe: Sothöna)

Subspecies. Nominate.

Status and distribution. National conservation status: Least Concern (population apparently stable). Widespread and common resident and passage migrant/winter visitor, although very local in Southeast Anatolia where only recently demonstrated to be a summer resident (Welch 2004). Particularly numerous in winter, especially in western and central Turkey, at which season numbers since the early 1970s seem to be increasing, with a current winter population of 400,000–1,000,000 birds, e.g. 1,012,910 in early 1999. Occurs in wetlands with bodies of open water.

In Marmara, fairly common in the breeding season in Meriç Delta, but far commoner in winter, with e.g. 81,400 in November 1970 and huge numbers regular at this season on the southern Marmara lakes, particularly Uluabat, where there were 273,000 on 24 January 1999 and a maximum count of over 320,000 was made at this site. In Aegean common at several localities, with largest numbers at Büyük Menderes Delta, where up to 80 present in the breeding season and winter totals regularly reach over 12,000 and a maximum of 32,535 (January 1999), and up to 68,500 (in 1996) on nearby Bafa Gölü. Widespread in Mediterranean with large concentrations at several sites and counts in excess of 20,000 reported from a number, e.g. 80,000 (January 1970) and 112,640 (January 1971) at Akyatan Gölü; 95,000 at Karamik marshes, and 54,350 (January 1971) at Göksu Delta, although recent numbers at all of these sites have been considerably lower (Magnin *et al. in* Heath & Evans 2000). The maximum winter count in this region was 138,925 (January 1995) at Burdur Gölü, but up to 300,000 have been recorded at the same site on passage. In Inner Anatolia, common during the breeding season at many localities with high counts of 600+ at Kulu Gölü, 300+ near Yarma, 250 pairs at Ereğli and 200 pairs at Gogenç Gölü, although the species must unquestionably have declined as a breeder in this region as a result of wetland drainage. Winter numbers are considerably higher and probably more stable, typically in excess of 5,000 at many sites, with high counts of 28,464 at the former Hotamış marshes (January 1970) and a regional maximum of 134,100 at Seyfe Gölü (January 1970), where numbers are now greatly reduced. However, larger numbers are now recorded on the barrage lakes, such as Hirfanlı, in this region. In Black Sea Coastlands fairly common in the breeding season at Kızılırmak Delta, but even commoner in winter, with up to 57,186 there in January 2005. Numbers are rather lower at the few other wetlands in this region, but some winter on the sea itself, e.g. nearly 15,000 were counted along the entire length of this coast in early 1997 (Welch & Welch 1998a). Much scarcer in East Anatolia, with numbers generally only in the low hundreds and, with the exception of autumn flocks, e.g. 10,000 at Sodalı Göl (September 2004) and 1,580 at Çıldır Golu (August 1972), the largest concentrations have been observed at Karakaya Reservoir, Malatya, with a maximum of 50,000 on 20 January 2006. Numbers in Southeast Anatolia are also generally small, except in winter when new reservoirs may support populations of global significance; 110,131 were counted at Karkamis Reservoir, south of Birecik, on 16 January 2005 (Çağlayan *et al.* 2005).

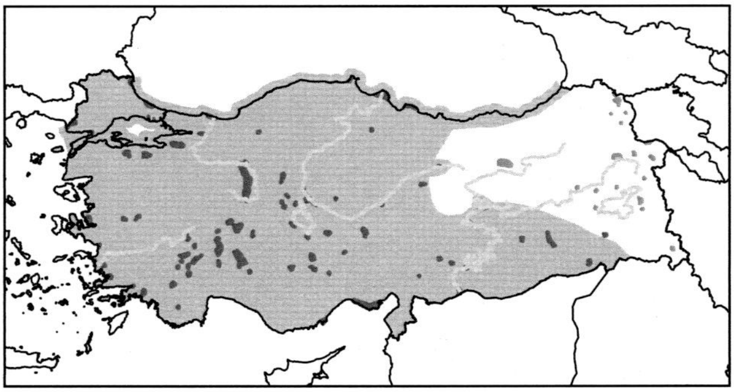

Breeding. In lakes with reeds and waterside vegetation, and marshes with open water. Nest built up from water level in reeds; a substantial cup of dead leaves and stems of aquatic plants. Eggs: C/8–9 (2), C/7 (1); but probably also fewer eggs than these limited data suggest (6–9 elsewhere). Young: broods of up to nine observed, but mainly 2–5. In Kızılırmak Delta (**Black Sea Coastlands**), in 1992, flock sizes decreased in April, with pair formation during the last ten days of April, first nest on 1 May, the second on 20 May, many in late May, and pairs with downy young from 27 May; population 500–1,000 pairs, with most nests in *Scirpus* (Hustings & van Dijk 1994). In **Aegean**, 20 nests had eggs on 13 May 1899, including one C/9 (Selous 1900), and another 13 May 1950 (McNeile); adults with young on 4 June 1972 indicate laying in early May. In **Marmara**, a brood of two, three-quarters grown, on 19 June 1973, indicates laying early May. A nest had seven eggs at Manyas Gölü on 7 May 1966 (R. F. Porter). In **Inner Anatolia**, eight nests had eggs (1 C/8) at various locations from 18–25 May. There are many records of young, mainly from 1 June, but there was a brood, three days old, at Mogan Gölü on 11 May 1970 and newly hatched young at Sultan marshes on 14 May 2004 (PC), indicating egg laying in mid April. In **Mediterranean**, many large young were at Göksu Delta on 18–19 July 1972. In **East Anatolia**, earliest young were at Erçek Gölü on 27 May 1969 (Kasparek & van der Ven 1983), and small young were at Tatvan as late as 18 August 1972, a probable case of double-brooding, which is usual elsewhere.

COMMON CRANE — *Grus grus*
Turna

(Du: Kraanvogel; Fr: Grue cendrée; Ger: Kranich; Swe: Trana)

Subspecies. Considered monotypic, following del Hoyo *et al.* (1996) and other commentators, although Roselaar (1995) continued to acknowledge the validity of *lilfordi* for those populations in Turkey, Transcaucasia and northern Asia.

Status and distribution. National conservation status: Endangered (population apparently decreasing). Fairly widespread but now rather uncommon breeder across eastern and central Turkey, as well as being a partial migrant and winter visitor over much of the country. Overall, winter numbers appear to have declined strongly in recent years (e.g. just 202 were present during the 1999 midwinter count); previously thousands were often present at this season, the population being swollen by immigrants from outside the country. An extensive survey of passage Common Cranes at wetlands in west and central Turkey, in autumn 1982, revealed a total of 12,960 birds (van den Berk *et al.* 1986). In Marmara, principally occurs on migration in autumn, with much smaller numbers in spring. Away from the Bosphorus, high counts are of 200 between Istanbul and Edirne (18 September 1973) and 1,078 moving southwest over Edirne in 1.5 hours (6 October 1973). Regularly recorded at the Bosphorus in early spring (from mid March), and in autumn (from mid August). High counts include 175 moving north on 15 March, 451 moving west between 19 and 24 March and 163 moving east between 21 and 30 September, but current numbers are generally much smaller. The historical scale of this passage is not well known. In Black Sea Coastlands breeds in Kızılırmak Delta where there were an estimated 50 pairs recently (Magnin *et al. in* Heath & Evans 2000), and reasonable numbers are also present in winter. Also in this region, at least formerly bred at Yeniçağa Gölü, a long-established site (Wadley 1945), where up to 40 birds have been found in the breeding season (see also below), but this area seems unlikely to support a population now. The species has long been known principally as a passage and winter visitor in Aegean and Mediterranean (see von Gonzenbach 1852), although it is occasionally recorded during the breeding season in both regions. Generally recorded in small numbers with maxima of 800 at Acıgöl (Mediterranean) on 12 September 1961 (Kumerloeve 1961) and 556 there in late November 1971. Breeds at several localities in Inner Anatolia, but population certainly in significant decline due to wetland drainage, with a former maximum of 50 pairs at Kurbağa Gölü in May 1970 and 38 pairs there in May 1972. In autumn, the largest concentrations were around Tuz Gölü, with maximum counts of 4,186 (November 1970) and 4,023 (November 1971), but numbers are again currently much lower, although up to 1,200 were recorded on passage at Sultan marshes in autumn 1994 and 1,000 in winter at Ereğli marshes in 1993. Fairly widespread in East Anatolia, with birds present in small numbers throughout the breeding season at several localities, although again there is quite some evidence of recent declines at local level due to habitat degradation and destruction. The maximum count in the breeding season is 40+ between Eleşkirt and Ağrı (May 1970). In autumn higher counts include: 80 near Ağrı (September 1970), 165 near Horasan (November 1970) and 224 at Aşvan (August 1971). Apparently only a rare straggler to Southeast Anatolia, although recently recorded in June, in suitable breeding habitat at Gölbaşı in the extreme northwest of this region (Welch 2004).

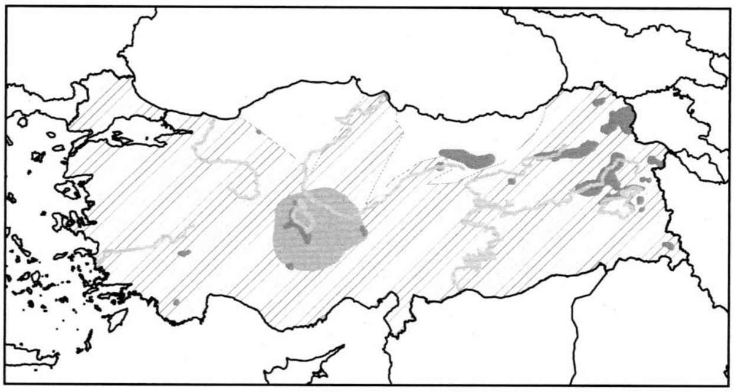

Breeding. In less-disturbed extensive shallow marshes, often adjacent to lakes, usually with sedges and grasses up to *c.*60cm tall, or in wet meadows with scattered *Phragmites* in lower parts (e.g. Yeniçağa Gölü, Black Sea Coastlands). The nest is merely an area of flattened vegetation, with scant additional plant material, built on a slight, but dry, eminence, in vegetation sufficiently tall to conceal the incubating adult. In deeper water, the nest may be built up to form a large heap of vegetation, but no records of this in Turkey. Eggs: C/1–3, mean 1.9; (C/2 (5), C/1 (2), C/3 (1). Young: B/1 (24), B/2 (8). Earliest record, a pair with two young, *c.*3 weeks old, at Eşmekaya (**Inner Anatolia**) on 11 May 1993, indicates egg laying *c.*21 March (PC). In **Mediterranean**, five nests had eggs and a recently hatched young was observed at Acıgöl on 10 May 1899 (Selous 1900), where display was noted on 12 May 1974, nests were found on 21 April 1981, five young were seen on 27 June 1964 and one young on 16 July 1986 (Dijksen & Kasparek 1988). In Inner Anatolia, eggs hatched on 23 May 1992 and 11 May 1993 at Eşmekaya (PC), where a late nest still had eggs on 14 June 1973 (Pforr & Limbrunner 1982); records of young at various sites, late May–July, including 38 pairs with 29 small young at Sultan marshes on 22–27 May 1972 (Kasparek 1985). In **East Anatolia**, an adult was incubating at Bendimahi on 29 May 1969 (R. F. Porter), six adults were incubating at Fahril Gölü on 14 May 1970, and a nest had three eggs at Sarıkamıs on 24 May 1975. There were two young, *c.*3 weeks old, near the Murat River on 9 June 1984 (A. Schubert). A late record, of small young at Bulanık on 27 June 1989, indicates egg laying in late May, later than elsewhere in Turkey. In **Black Sea Coastlands**, at Kızılırmak Delta, 11 broods were seen in July (Dijksen & Kasparek 1985); a downy young, 7–10 days old, on 26 May 1992, indicates egg laying commenced 15–20 April, and two young were seen on 5 June (Hustings & van Dijk 1994). At Yeniçağa Gölü, pairs were displaying and copulating on 2 May 1978; four nests, some with eggs, were recorded on 25 April 1962, and four broods were observed in June (earliest 2 June) (Kılıç & Kasparek 1987).

SIBERIAN WHITE CRANE *Grus leucogeranus*
Ak Turna

(Du: Siberische Witte Kraanvogel; Fr: Grue de Sibérie; Ger: Schneekranich; Swe: Snötrana)

Subspecies and taxonomy. Monotypic. Has been considered congeneric with Wattled Crane *Bugeranus carunculatus* of Africa and even to warrant its own genus, *Sarcogeranus*.

Status and distribution. Conservation status: Critically Endangered (BirdLife International 2000, 2004). Status uncertain, perhaps formerly a vagrant or even scarce but regular passage migrant, but given the relative lack of definitive evidence perhaps best considered as hypothetical. The basis for this species' inclusion on a list of Turkish birds is the following. Kasparek (1987b) outlined the basis for accepting the validity of reports by Danford (1880), who reported over 100 in the Ankara region (Inner Anatolia) in April 1879, commenting on their stork-like appearance and that he and his fellows were able to approach the birds closely, and Sandwith (1856), who obtained one bird from two parties of 4–5 and 20 individuals in the Erzurum Ovası (East Anatolia) in September 1854. Unfortunately, the specimen was not retained and although Sandwith distinguished the birds from Demoiselle Crane *G. virgo* his remarks probably do not conclusively eliminate *G. grus*. Harrap (1987) provided a swift rejoinder to Kasparek's note advocating acceptance of these records, which ably sets on record reasons for non-acceptance on the Turkish list, though advocacy of modern standards for judging historical records were in turn criticised by Walters (1988).

In the present-day, Davidson (1985) described the strange and remarkable observation of an apparent hybrid Common Crane *G. grus* × Siberian Crane *G. leucogeranus* west of Bulanık, in the Murat Valley (East Anatolia), on 21 May 1985. Martins (1989) commented that such an event was more likely the result of a conservation-management programme in Russia than through a natural pairing. Most recently a Siberian White Crane was discovered in captivity (in a chicken coop!) at Gerze, near Sinop (Black Sea Coastlands) in early June 1999. The bird had originally appeared in the area two or three months prior to this and was taken into captivity by a local man. Subsequently, the bird was proven to have escaped from Oka Biosphere Reserve, in Russia, in August 1996, whereupon it had joined a party of local Common Cranes *G. grus*, which migrated south, presumably explaining its appearance in Turkey (see Bradshaw & Kirwan 2000).

In defence of the 19th-century records is the fact that the western population of this species was formerly considerably more numerous, e.g. up to 300 were recorded in the Volga Delta (Russia) on spring migration (March–April, exceptionally May) during the same century, and the dates of the Turkish records accord well with the known migration periods (see BirdLife International 2001). Elwes & Buckley (1870) noted a 19th-century record from Macedonia, although a similar lack of documentation exists for that report, and the species has allegedly occurred in Bulgaria (Snow & Perrins 1998). There is even a mediaeval record from the Nile Delta of Egypt (Provençal & Sørensen 1998). What is certain is the current perilous state of the western population. Just 7–11 individuals have been recorded in Iran, on the shores of the Caspian, in recent winters, and two in northern India, with a maximum of 13 noted in the Volga Delta in 1996 and a recent estimate placed numbers on the breeding grounds at eight pairs (Snow & Perrins 1998, BirdLife International 2001). Satellite telemetry has recently confirmed those breeding grounds to lie in the Ob River basin, 650km south of the Kunovat River basin, and that the species probably usually migrates along the western border of the Caspian (Kanai *et al.* 2002). The potential for future wandering to Turkey would appear to be remote but a record of three birds at Qa' Khanna, in the eastern desert of Jordan, in February 2001, has recently been accepted (Hamidan 2003).

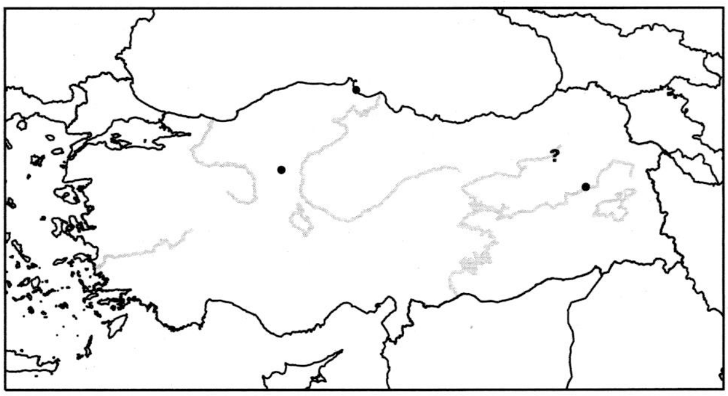

DEMOISELLE CRANE *Grus virgo*
Telli Turna

(Du: Jufferkranvogel; Fr: Grue demoiselle; Ger: Jungfernkranich; Swe: Jungfrutrana)

Subspecies and taxonomy. Monotypic. The species has traditionally been placed in the genus *Anthropoides*, but recent DNA hybridisation data and those from cytochrome-*b* gene sequences reveal that *Anthropoides* and *Grus* do not form reciprocally monophyletic clades, and these independent datasets confirm that a more appropriate placement of *Anthropoides* is within *Grus* (Krajewski 1989, Krajewski & Fetzner 1994, Krajewski & King 1996).

Status and distribution. National conservation status: Critical (population apparently decreasing). Very rare and localised summer visitor to northeast Turkey, where it currently breeds at perhaps only three localities. Kasparek (1988b) estimated the population at just 25–30 pairs whereas, more recently, Eken *et al.* (2003) considered there to be even fewer, just 10–20 pairs. Due to dam construction on the Murat River and agricultural expansion in the Murat basin the species may continue to decline in the near future. Thus, it is certainly among the country's rarest breeders, along with Little Bustard *Tetrax tetrax*, White-tailed Eagle *Haliaeetus albicilla*, Mute Swan *Cygnus olor* and others. First noted as a Turkish breeding bird by Sandwith (1856) and thereafter by Dresser (1891) from observations around Erzurum (East Anatolia), where it no longer breeds. There were no further breeding records until 1967. Since then there have been regular records from a relatively small area of eastern Turkey bounded to the south by Van Gölü and to the north by the Upper Aras Valley, with recent records from further north almost certainly referring to migrants. Three or four nesting sites were known until recently, with an additional five localities, such as Saz Gölü, Doğubeyazıt, either presumed to hold breeders or having done so formerly. The majority of records, especially in recent years, comes from the Upper Murat Valley, between Malazgirt in the east and Bulanık and Yoncalı in the west (i.e. within the confines of the Bulanık Ovası IBA), with a maximum of 22 individuals present in late summer. However other records involving flocks, of 15 in April 1986 and 21 birds in April 1981 (Kasparek *op. cit.*) from the same general area of the country surely concerned migrants. Other breeding sites may await discovery as the species can be surprisingly hard to locate when nesting, and several apparently suitable areas have yet to receive intensive ornithological attention. Such areas may lie on the northern course of the Murat River, between Patnos and Malazgirt, as well as in the Muş Plain, southwest of Yoncalı. The breeding range appears to be occupied from mid April (17th) until at least mid August (10th).

More widespread, but very scarce on passage across Turkey, occasionally west as far as the Sea of Marmara and, principally in the 19th century, Thrace. Other than vague statements in the literature, e.g. that by Alléon, referring to the species occasionally being noted in the vicinity of Istanbul, there is a total of 36 records of passage migrants, usually involving small flocks of fewer than ten individuals; exceptionally 38 moved south at the Bosphorus in late September 1977. Records are available from all regions as follows: Marmara (five, the most recent that cited above from the Bosphorus), Aegean (two: the first in either April 1863 or 1864, and the other in the Gediz Delta in September 1995), Black Sea Coastlands (three), Mediterranean (one, that cited by Martins 1989 was actually at Mogan Gölü, near Ankara), Inner Anatolia (23), Southeast Anatolia (one) and East Anatolia (three). Not all are specifically dated (some just referring to month), but spring passage appears to span the period 24 March to 23 May and autumn passage from 2 September to 14 October. There is also a recent winter record: two with Common Cranes *G. grus* at Göksu Delta (Mediterranean) on 9 February 1997 (Vaassen 1998).

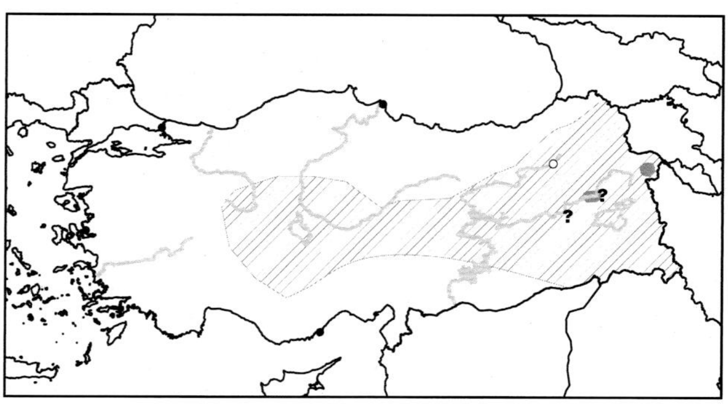

Breeding. In extensive open country near water, in grasslands, marshes and on shingle islands in rivers, now principally in the valley of the Murat River and in adjacent East Anatolia. The nest is a shallow scrape in the ground, in a grassy or bare site. Eggs: C/1 (1), C/2 (5). Young: B/2 (2), B/1 (1). A pair was copulating west of Ağrı on 30 April 1981 (P. Flint). An adult was incubating on a nest near Balatos 11/13 June 1987 and another was between Ağrı and Horasan on 1 June 1988. A nest on a small island in a marsh near Doğubeyazıt had two eggs on 4 June 1971, and a nest had two eggs on a gravel island in a river near Soylemez on 20 May 1975. There are several records of young, mainly from the Bulanık area in June–July; the earliest, 1 June 1988 and 1 June 1989, indicating egg laying early May. Seven eggs in the Natural History Museum, Tring, were collected on 20 May 1864 (C/1), 22 May 1864 (C/2), 10 May 1865 (C/2) and 15 May 1865 (C/2).

LITTLE BUSTARD *Tetrax tetrax*
Mezgeldek

(Du: Kleine Trap; Fr: Outarde canepetière; Ger: Zwergtrappe; Swe: Snåtrapp)

Subspecies. Monotypic. Formerly placed in genus *Otis* (see Collar *in* del Hoyo *et al.* 1996).

Status and distribution. Conservation Status: Near Threatened (BirdLife International 2000, 2004); national conservation status: Critical (population apparently decreasing). Like Great Bustard *Otis tarda*, population is in long-term decline as a result of changes in agricultural practice and associated habitat change, as well as hunting. Its historical distribution and status were comprehensively reviewed by Kasparek (1989a), who documented the species' former range as encompassing parts of southern Marmara, very locally in coastal Aegean, as well as in Inner Anatolia, Mediterranean, East Anatolia and Southeast Anatolia, where the species was still common at Ceylanpinar in the 1960s (T. Gürpınar), with the overwhelming majority of records from the three latter regions being outside the breeding season. Kasparek (*op. cit.*) located just fewer than 40 localities for Little Bustard, of which ten were known to have been occupied during the breeding season, although the vast majority of literature concerning the species in Turkey was pre-1950.

Current status somewhat uncertain, although it appears to be a very rare and local resident in Inner Anatolia, where two small breeding populations were discovered in 1998 (Eken & Magnin 1999) and display and prolonged tenure was observed at a third site in 2003 (I. Richardson), and perhaps adjacent parts of Mediterranean (no recent records in breeding season), and recorded several times recently in East Anatolia, in late autumn and summer, where almost certainly also breeding (*e.g.* near Muş). Spring records are also available from Southeast Anatolia, where at least formerly the species also wintered, but intensive recent fieldwork failed to locate any Little Bustards in this region of the country (Welch 2004). In winter the species is also known from Black Sea Coastlands, from where there a handful of recent records (e.g. in Kızılırmak Delta, most recently in December 2005), as well as Marmara and Mediterranean (several recent winter records from Göksu Delta), but records at this season are, as at all times of year, very rare and infrequent.

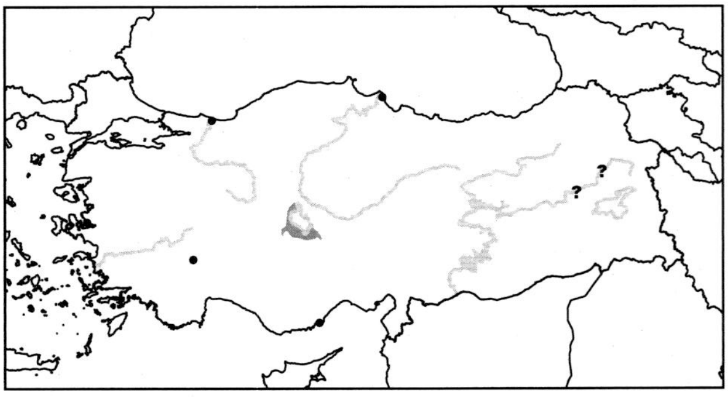

Breeding. No data in Turkey but elsewhere breeds in open grasslands and cornfields. The nest is a shallow unlined hollow in the ground, usually screened by vegetation, and the usual clutch 3–4 (2–5). A dead bird, near Karamık marshes (**Mediterranean**), on 12 July 1969, was considered to be a pullus almost fledged. Seven displaying males were shot within an area of *c.*60 ha in the Karacabey–Bursa area (**Marmara**) in May 1937, and breeding was confirmed (Kasparek 1989a). A displaying male was observed at Ceylanpınar (**Southeast Anatolia**) on 8 April 1981 (Parr 1981), and another in the Konya Basin (Inner Anatolia) on 23 June 1998 (Kirwan *et al.* 2003). Two displaying birds, a male calling and chasing a female, were observed near Kulu Gölü (**Inner Anatolia**) on 27 March 2004 (I. Richardson).

MACQUEEN'S BUSTARD *Chlamydotis (undulata) macqueenii*
Yakalı Toy

(Du: Oostelijke Kraagtrap; Fr: Outarde houbara; Ger: Kragentrappe; Swe: Kragtrapp)

Subspecies and taxonomy. Conservation status: Vulnerable (*C. undulata sensu lato*; BirdLife International 2008). Gaucher *et al.* (1996), through molecular analysis, study of the sequence of postures adopted and associated vocalisations during display among the three allopatric populations of *Chlamydotis undulata* recognised a 'need for revision of the taxonomy of genus *Chlamydotis* which should separate the *macqueenii* and *undulata* populations at species level'. A more general study of bustard phylogeny, based on mtDNA analysis has recently supported the separation of *undulata/fuertaventurae* and *macqueenii* at species level (Broders *et al.* 2003), and this has been followed by several West European records committees (see Sangster *et al.* 2004, Collinson 2004, 2006). Nonetheless, this decision has been criticised (Cowan 2004) and Collar (*in* del Hoyo *et al.* 1996) suggested that pre-copulatory displays might be more important in isolating (*i.e.* preventing breeding) different species of bustards, in which ceremony the two proposed species of houbara appear largely identical (Gaucher *et al.* 1996). The separation of *undulata* and *macqueenii* apparently occurred *c.*430,000 years ago (Idaghour *et al.* 2004), whereas Iberian and more easterly populations of Great Bustard *Otis tarda*, which are not accorded separate taxonomic status, have not been in contact for about 200,000 years (Pitra *et al.* 2000). The patchy distribution of *macqueenii* extends from Sinai (Egypt) east through Arabia to Mongolia; northern breeders are annual long-distance migrants. Records from Turkey clearly referred to *macqueenii*. English names used for this taxon in recent literature include both Macqueen's Bustard (though it seems to be more appropriately spelt MacQueen's in light of the name of its collector: see Beolens & Watkins 2003) and Asiatic Houbara Bustard (Sangster 1996, van Heezik & Seddon 1999). Houbara Bustard *sensu lato* is sometimes considered congeneric with Great Bustard (Dement'ev & Gladkov 1968).

Status and distribution. Conservation status: Vulnerable (BirdLife International 2004); national conservation status: Extinct. Apparently bred formerly. No recent records in Turkey, although Goriup & Parr (1985) considered there to be still suitable habitat in remote parts of eastern Turkey in 1981 and that the species might occasionally occur in winter. The only mentions of the species for Turkey are that it was recorded around in Kars in 1912 or earlier (Satunin 1912), and that a tiny population existed in the Aras Valley on the Turkish/Armenian border prior to 1917 (Glutz von Blotzheim *et al.* 1973). As mentioned by Kumerloeve (1963a), it is unfortunate that Aharoni did not preserve the young bird that he apparently collected in the vicinity of the former Amik Gölü (Mediterranean) in 1910, when the species probably bred as close to Turkey as western Syria.

GREAT BUSTARD *Otis tarda*
Toy

(Du: Grote Trap; Fr: Outarde barbue; Ger: Großtrappe; Swe: Stortrapp)

Subspecies. Nominate.

Status and distribution. Conservation status: Vulnerable (BirdLife International 2000, 2004); national conservation status: Endangered (population apparently decreasing; 764–1,250 individuals [Doğa Derneği 2005]). Population in long-term decline due to agricultural intensification, hunting and habitat degradation and conversion, and the species only appears to persist in extensive treeless areas with mosaic of agriculture and semi-steppe, and little human disturbance, occurring to 1,800m in the east of the country. Kasparek (1989a) provided a detailed résumé of the species' former range in Turkey, which encompassed all regions of the country (including Thrace), albeit with the same population strongholds as currently exist, and identified a total of 83 major bustard sites (some of them in fact large areas) and including many former breeding areas, most of which are not currently known to harbour bustards.

The Great Bustard is now a very local resident and partial migrant (although movements are only partially understood) in Inner Anatolia and adjacent parts of Aegean and Mediterranean, apparently in very small numbers, and in Southeast and East Anatolia, where, at least formerly, locally more common. Recorded more widely in Aegean, Marmara and on the south coast from autumn to spring. Most records from Black Sea Coastlands (both recent and old) are from the winter period, and the species has apparently always been largely absent as a breeder from this region. Considerable numbers still wintered in border regions of Southeast Anatolia in the early 1980s (Goriup & Parr 1985), but there appears to be just one recent record from this region at the appropriate season (in December 2004), and recent and intensive fieldwork has found rather little evidence of the species' continued presence anywhere in Southeast Anatolia during the breeding season (Kirwan *et al.* 2003, Welch 2004, Kılıç & Karakaş 2005b). Such concentrations appeared too large to originate entirely from breeders in Turkey and may have also comprised birds from outside the country (given some evidence of immigration and passage in northern coastal regions). Despite an early 1980s survey by Goriup & Parr (1985) that indicated that the short-term prospects for the species' survival might be less dire than previously believed (with the population being estimated at 800–3,000 pairs as recently as 1997), and that Turkey might be one of the more important centres of distribution, recent data (e.g. Eken & Magnin 1999, Heunks *et al.* 2001, 2002) suggest that Great Bustard has declined significantly during the last two or three decades, even in former strongholds such as the Murat River valley, around Bulanık (East Anatolia), although 145 were counted in this region during spring 2002. The first survey of populations in the eastern third of the country for over 20 years, undertaken in 2002, revealed just 251 birds, with the most important areas being Bulanık and Muş (see Balmer & Kirwan 2003).

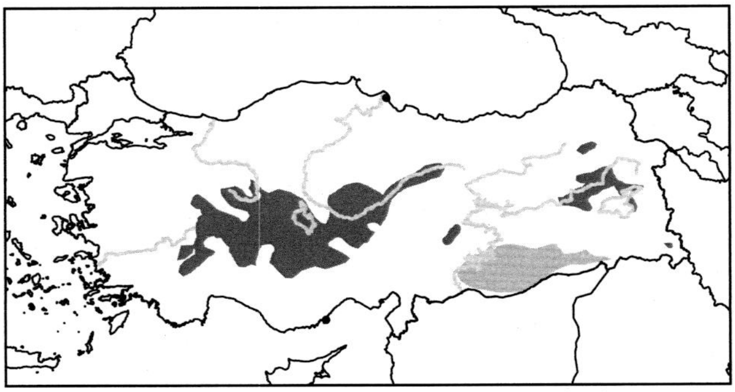

Breeding. Frequents extensive open, treeless areas, preferably undisturbed; in large agricultural fields, fallow ground and grasslands. At Aliken, southwest of Sivrihisar (Inner Anatolia), an important area with up to 45 adults, 50% of the land is wheat and barley fields, 40% fallow, and 10% uncultivated rocky steppe. Almost half of the crop fields are left fallow each year at Altintaş Ovası near Kütahya (Inner Anatolia), another important area, with 40 birds, and where breeding is confirmed annually (Magnin & Yarar 1997). In a survey of four areas around Tuz Gölü (Inner Anatolia) in April 2000, three display grounds were located, all relatively far from human settlements, in treeless areas of large crop fields, bare ground (old cornfields), and in natural primary steppe; males usually displayed on low hills, in groups of 2–6 or solitarily, and copulation was observed once (Heunks *et al.* 2002). Females usually nest fairly close to where they copulated, and incubate eggs and rear young alone, remaining with the young until the next spring. Nest and clutch size undescribed in Turkey; elsewhere, the nest is an unlined shallow hollow, in growing corn or in grass, but also in short sparse vegetation with virtually no cover; usual clutch 2–3 (4). Selous (1900) stayed near Acıgöl (**Mediterranean**) from 9–11 May 1899; he did not personally find a nest, but a fresh egg was brought to him, and a small population almost certainly still survives there (Magnin & Yarar 1997). In **Inner Anatolia**, Danford (1877–78) 'plundered a bustard's nest' on his journey between Kayseri and Çorum on 8–14 May 1876. A Great Bustard's egg was found in a colony of gulls on an island in Tuz Gölü in spring 1972. A female, with one young, was at Çubuk Ovası, north of Ankara, on 27 June 1951, and display was observed at Ereğli on 25 April 1965 and near Tuz Gölü in May 1969 (Kasparek 1989a). Display was seen on the east side of Tuz Gölü on 15 April 1996 (H. Schekkerman). In **Southeast Anatolia**, an adult female behaved as if it had young, west of Viranşehir on 14 May 1975, and a female with a part-grown chick was at Cöltepe, near Batman, on 14 June 1983. In **East Anatolia**, *c.*30 birds, including six displaying males and an immature male were at Bulanık on 30 May 1992 (I. A. Green). Young were seen at Hazar Gölü, near Sivrice, and a female behaved as if she had young, near Van Gölü, on 7 June 1987 (Kasparek 1989a). Up to 32 birds have been seen in the breeding season east of Sodalı Göl, where breeding has been confirmed (Magnin & Yarar 1997).

[GREATER PAINTED-SNIPE *Rostratula benghalensis*]
Boyaliçulluk

(Du: Goudsnip; Fr: Rhynchée peinte; Ger: Goldschnepfe; Swe: Rallbeckasin)

Subspecies and taxonomy. If this species were to wander to Turkey, the nominate form would clearly be involved. The subspecies *australis* (from Australia) may warrant specific status (Lane & Rogers 2000).

Species not admitted to the Turkish list (Kirwan *et al.* 1999). Hartert (1910–22) treated the species as an occasional visitor to Asia Minor, and Dement'ev & Gladkov (1951–54) stated that it had even bred in southeast Turkey, but neither statement was substantiated by the authors.

EURASIAN OYSTERCATCHER *Haematopus ostralegus*
Poyrazkuşu

(Du: Scholekster; Fr: Huîtrier pie; Ger: Austernfischer; Swe: Strandskata)

Subspecies. Nominate according to del Hoyo *et al.* (1996) but considered to be *longipes* by Roselaar (1995). We have not examined any specimens from Turkey, but would not, in any case, recognise *longipes* as a valid subspecies.

Status and distribution. National conservation status: Near Threatened (population apparently stable). Widespread but generally rather uncommon summer visitor to wetlands in Aegean, Mediterranean, Inner and East Anatolia, where it principally utilises river valleys for breeding, and to one site in Black Sea Coastlands (see below). Recorded nesting up to 2,100m in East Anatolia. The most regularly occupied breeding areas are Büyük Menderes Delta (Aegean), Kulu Gölü (Inner Anatolia), Meriç Delta (Marmara), and the river valleys and small lakes from Erzincan eastwards (East Anatolia).

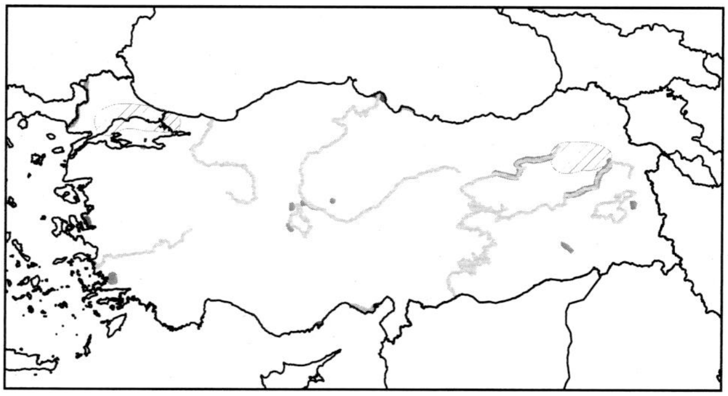

Slightly more common on passage, but still distinctly uncommon at most inland wetlands and, except small family groups observed in July, the species is generally solitary. Passage is principally noted in March–May and July–October. In winter, small numbers occur at coastal wetlands throughout the country (e.g. 73 counted during the 1999 midwinter count), including most of the relevant breeding areas, where the species is mainly resident.

Breeding. On open areas along rivers and shingle islands in rivers (especially East Anatolia), on low islands and unvegetated edges of inland lakes, and on beaches, dunes, shingle spits, lagoons and offshore islets. Breeding population at Gediz Delta (Aegean) estimated at 27 pairs in 1995 and 25 pairs in 1996 (Eken 1997a). Nest a shallow hollow in the ground, unlined, or with small plant material/small stones. Eggs: C/3 (5); nests with single eggs on 7 May 1993 and 14 May 2007 are considered incomplete clutches. Young: B/1 (2), B/2 (4). At Göksu Delta (**Mediterranean**), display seen on 30 April 1970 and two nests had three eggs, 50m and 100m from the sea, on 27 May 1973 (Witt 1976). In **East Anatolia**, there was a nest at Erçek Gölü on 29 May 1969 (Kasparek & van der Ven 1983). A pair with one young was near Selim on 17 May 1970, indicating egg laying no later than 20 April. A pair and two young were at the River Murat on 27 May 1994, and another pair and two young were there on 9 June 1984 (A. Schubert). A pair and a chick, 3–4 weeks old, were on a shingle river island river near Ağrı on 25 June 2004, indicating egg laying early in May (PC). Ten pairs were at the River Murat on 13 May 1992, and several pairs on shingle islands were giving alarm-calls on 6 June 2001 (PC). In Kızılırmak Delta (**Black Sea Coastlands**), breeding pairs were only found by the Kızılırmak River, adults with downy young on 31 May 1992, and two chicks on 18 July 1992 (Hustings & van Dijk 1994). In **Inner Anatolia**, an incubating adult was on an island in Kulu Gölü on 22 May 1992, and two nests, on islands in Bolluk Gölü, had one egg (an incomplete clutch) on 7 May 1993 (PC) and three eggs on 15 May 2004 (J. Mason). At Kulu, a nest had three eggs on 31 May 1972, and at least two fledged young were present on 14–15 July 1971 (Kasparek 1987a). In **Aegean**, a nest had three eggs at a coastal lagoon, near Izmir, on 17 May 1899 (Selous 1900) and on an offshore islet near Aliağa a pair were 'alarming hysterically' on 23 May 1950, suggesting they must have had young (McNeile); a nest with one egg (incomplete clutch) was on an island in Karine Gölü on 14 May 2007 (PC).

BLACK-WINGED STILT *Himantopus himantopus*
Uzunbacak

(Du: Steltkluut; Fr: Echasse blanche; Ger: Stelzenläufer; Swe: Styltlöpare)

Subspecies and taxonomy. Nominate *himantopus*. However, recent global reviews (*e.g.* Sibley & Monroe 1990, Gill & Wright 2006) have tended to consider *H. himantopus sensu lato* as four allospecies: *himantopus* in Eurasia and Africa, *leucocephalus* in the Philippines, Indonesia and Australasia, *mexicanus* in North America south to northern South America, and *melanurus* throughout the rest of South America.

Status and distribution. National conservation status: Least Concern (population apparently decreasing). Very common and widespread summer visitor to shallow lakes, ponds, lagoons, estuaries and wet meadows across the country, to 1,800m, but main strongholds are the environs of Tuz Gölü (*c.*500 pairs), Akşehir Gölü (150 pairs), Seyfe Gölü (500 pairs), Sultan marshes (200 pairs) and Ereğli marshes (100 pairs, all Inner Anatolia); the Gediz Delta and Marmara Gölü (100 pairs, both Aegean); Burdur Gölü and Göksu Delta (Mediterranean); and around Van Gölü (East Anatolia). Relatively scarce and local in Marmara and Black Sea Coastlands, with few breeding records away from the Kızılırmak (250 pairs) and Meriç (300 pairs) deltas. Apparently absent as a breeder from most of Southeast Anatolia, although there are recent records of probable breeding from a handful of sites.

More common and widespread on passage, which commences early March in Aegean and Mediterranean (4th), mid March in Inner Anatolia (15th) and late March/early April elsewhere. Spring passage continues until late May (21st). Post-breeding dispersal recorded in most regions, with maxima of 4,000 at Kulu Gölü (Inner Anatolia) and 1,000 at Çöl Gölü (Inner Anatolia), on 17 July 1986, and 1,073 at Gediz Delta on 6 October 2000, and most records from mid July. Significant movement continues through August, with many fewer in September and very few later records, e.g. at Van Gölü, on 6 October, and Erzurum Ovası (East Anatolia), on 25 October 2000. In November recorded in coastal regions, where occasionally overwinters in Aegean and Mediterranean, as well as in the Black Sea coastal deltas, should weather conditions be sufficiently mild. However, winter records generally involve very small numbers: e.g. one at Kızılırmak Delta on 10 January 1969, four at Göksu Delta on 31 January 1992 (MÖ), two at Tuzla, Çukurova (Mediterranean) on 22 January 1993 (MÖ), two at Acıgöl (Mediterranean) on 27 January 1996 and eight in the same place on

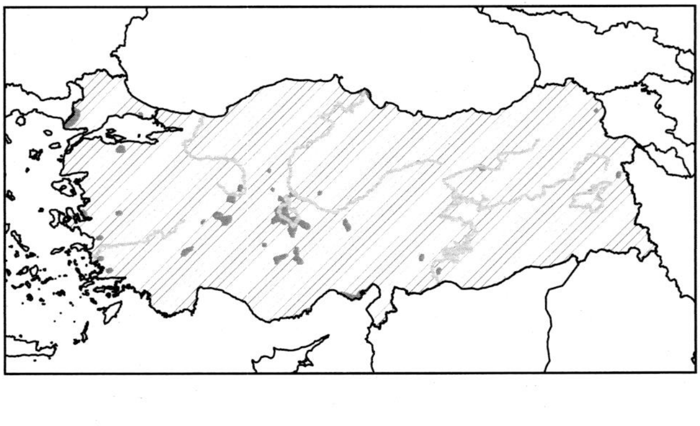

12 January 2003 (D. Kahraman), and up five annually in winter at Gediz Delta since 2002/2003 (O. Onmuş), with one at Karakaya Reservoir (East Anatolia) on 18 December 2005 (MÖ *et al.*).

Breeding. In open areas by lakes and marshes, both coastal and inland. Usually in colonies, typically of 10–50 pairs, but up to 500 pairs, e.g. at Seyfe Gölü in Inner Anatolia (Magnin & Yarar 1997), and solitary pairs also recorded. Nests on flat islands in lakes, along shores of lakes, and in flooded marshland and grassland, typically wet in spring but drying out in May. Builds on a low tussock in shallow water or up from the ground in very shallow water, or on mud near the water's edge. The nest is a shallow hollow with a variable amount of nest material, merely a few fragments of plants in dry sites or a thick pad in wet sites, and further material may be added if the water level rises. Eggs: C/4 (95), C/3 (5); smaller numbers considered incomplete clutches. Young: all broods are of four or fewer. In **Aegean**, adults were on territory, but not incubating, at Miletos on 23 April 2003, first eggs were laid in the last week of April, and five adults were incubating by 2 May 2003 (PC). Nests with eggs and several newly hatched young were at Bafa Gölü on 29 May 1999 (J. Hollinrake), indicating earliest eggs in late April/early May. An adult with three small young and a nest with four eggs were at Marmara Gölü on 1 July 1966 (R. F. Porter). In **Mediterranean**, 200–250 breeding pairs, some displaying in late April, in Çukurova (van der Have *et al.* 1988), and several pairs were there with small young (and nests with eggs) on 8–9 May 1964 (Warncke 1964–65), indicating egg laying about 12 April. At Göksu Delta, several adults were incubating on 21 May 1993 and 15 June 1992, and a young bird, *c.*2 weeks old, was seen on 17 June 1992, indicating egg laying in early May (PC); more than 50 adults were incubating on sandbanks there on 19 May 1999 (R. F. Porter). Unfledged young were at Karamık Gölü on 18 July 1972 and Acıgöl on 27 June 1964 (Dijksen & Kasparek 1988). In **Inner Anatolia**, adults on islands in Bolluk Gölü had not nested by 23 April 2004 (R. Thomas), but many nests had full clutches on 7 May 1993 (PC). A nest had newly hatched young near Eşmekaya on 5 June 1992, indicating egg laying early in May (PC). At Sultan marshes, there were young as early as 21 May 1979 (Kasparek 1985), indicating egg laying about 23 April. At Kulu Gölü, there were *c.*50 nests with eggs on 22 May 1992 (PC), several pairs with small young on 6 June 1996 (GMK) and seven nests still with eggs on 13–15 July 1971 (Kasparek 1987a), these last probably replacements following earlier failures. At Seyfe Gölü there were *c.*500 pairs, some with young, on 18–22 June 1992 (G. Magnin), and many nests had eggs on 14 June 1993 (PC). There are several records of young at other sites in Inner Anatolia in June–July. In **Marmara**, adults were incubating in June 1966 at Manyas Gölü, and a nest there had eggs on 1 July 1966. Copulation was observed at Uluabat Gölü on 10 June 1998 (Welch & Welch 1998b). At Kızılırmak Delta (**Black Sea Coastlands**), the population in 1992 was *c.*250–300 breeding pairs, present from mid March, increasing in April, with territorial behaviour on 24 April 1992, copulation on 30 April 1992, first nests with eggs on 10 May 1992, but no young birds seen by 10 June 1992, when the survey ended (Hustings & van Dijk 1994). In **East Anatolia**, egg laying commenced at Sodalı Gölü on 4 June 2001, when several nests had 1–2 eggs (T. Quittenden). There were young at Yoncalı on 2 August 1992 and at Bendimahi on 21 July 1992, but there is no further data for East Anatolia, and breeding may commence earlier than these records suggest. The breeding season in Turkey is clearly protracted, but later records are probably replacement clutches.

PIED AVOCET *Recurvirostra avosetta*
Kılıçgaga

(Du: Kluut; Fr: Avocette élégante; Ger: Säbelschnäbler; Swe: Skärfläcka)

Subspecies. Monotypic.

Status and distribution. National conservation status: Vulnerable (population apparently decreasing). Common and widespread summer visitor to the saline wetlands of Inner Anatolia and Aegean, and a local breeder in Mediterranean, East Anatolia and the southern fringes of Marmara. In the east recorded to 1,800m. Main breeding areas are around Tuz Gölü, Seyfe and Kulu lakes, Sultan marshes (all Inner Anatolia); Gediz Delta and Acıgöl (Aegean); and Göksu Delta (Mediterranean).

More widespread on passage, when recorded from all regions but more common in the western half of Turkey. Spring passage commences mid March (14th) and continues until late May (21st). In the first half of July, local breeders are augmented by passage migrants in many areas, e.g. 3,000+ at Tuz Gölü on 17 August 1969, 2,000+ at Acıgöl in mid-September 1968, 1,500–2,000 at Erçek Gölü (East Anatolia) in September and 6,000+ at Gediz Delta in October. Birds finally vacate the breeding areas by late November (27th). Considerable numbers overwinter in the western two-thirds of the country, where the climate is milder, e.g. 650 at Acıgöl on 5 January 1969 and 142 on 18 January 2003 (D. Kahraman), *c.*100 at Tuz Gölü in winter 1968/69, 160 at Bafa Gölü (Aegean) on 20 January 1974, and 125 at Büyük Menderes Delta (Aegean) on 28 February 1974.

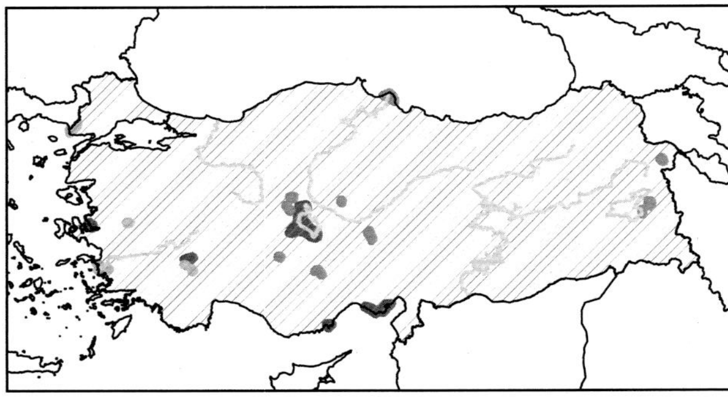

Breeding. By open saline or brackish lakes and saltpans, often on low islands and small sandbanks, but also in drying marsh, sparse grassland and mud. Mainly colonial but solitary pairs occur. Colonies have sometimes only a few pairs, but there were up to 750 pairs at Seyfe Gölü (Inner Anatolia) in 1992, mainly on islands, but also on the shore (G. Magnin). Nests usually well spaced at the edge of lakes, but may be close (only 2–3m apart) on small islands, where available space is limited. On larger islands, nests are often at the edge, together with those of Black-winged Stilt *Himantopus himantopus*, with nests of more strongly colonial species, e.g. Slender-billed Gull *Larus genei* and Gull-billed Tern *Gelochelidon nilotica*, in the centre. The nest is a shallow hollow in bare ground, unlined or with a tiny amount of plant material. Eggs: almost invariably C/4, rarely 3, and one record of C/5 (M. Farmer). The season is protracted, possibly due to replacements following earlier failure. At Bolluk Gölü (**Inner Anatolia**) there were 12 nests on 23 April 2004, of which 11 had incomplete clutches (1–3 eggs, and only one a full clutch of four) (R. Thomas); on 14–15 May 2004, on the same islands, there were 50–100 nests, almost all with four eggs, a couple with only three, one with an unusually large clutch of five, and newly hatched young in one nest (M. Farmer), indicating laying of the first egg on 18 April; most nests had four eggs on 7 May 1993 and, on 23/24 June 1992, there were young at all stages but also 14 nests with eggs (PC). At Kulu Gölü, there were 156 nests and probably more on another island, on 6 May 1972; 36 nests had eggs on 6 May 1974; and there were several pairs with young, up to ten days old, on 20 May 1974 (indicating earliest eggs 14 April). There were 105 breeding pairs on 5 June 1975, including six with small young; both small and very large young on 14–15 July 1971 (indicating latest egg laying mid June) (Kasparek 1987a); *c.*100 nests had eggs, but no young, on 22 May 1992, and several nests still had eggs on 19 June 1992 (PC), whilst at least one pair had a brood of five small young on 12 June 1998 (GMK). At Seyfe Gölü, there were 16 pairs with young on 22 May 1983 (indicating earliest eggs no later than 24 April); most of 300 breeding pairs already had young on 2 June 1971, but 75 nests still had eggs (Husband & Kasparek 1984); most nests had eggs, but some small young on 28 May 1993, and many nests still had eggs on 14 June 1993 (PC). Other breeding sites in Inner Anatolia include Sultan marshes (150 pairs), Ereğli marshes (75 pairs), Çöl Gölü, Tersakan Gölü and Tuz Gölü (Magnin & Yarar 1997). In **Mediterranean**, at Göksu Delta, Hollom (1955) found a young bird, no more than one week old, on 16 May 1951, indicating egg laying in mid April; and four nests each had four eggs on 6 May 1973 (Witt 1976); two nests each had four eggs in Çukurova on 23 May 1971. Four pairs were at Kızılırmak Delta (**Black Sea Coastlands**) in 1992, and a nest had four eggs on 10 May 1992 (Hustings & van Dijk 1994). At Van marshes (**East Anatolia**), many adults were alarm-calling on 10 June 2001 (PC), and juveniles were present on 6/7 July 1974. At Gediz Delta (**Aegean**), there was a breeding population of 55 pairs (Magnin & Yarar 1997), nesting mainly on islands in the saltpans (Eken 1997a).

CRAB-PLOVER — *Dromas ardeola*
Yengeç Yağmurcunu

(Du: Krabplevier; Fr: Drome ardéole; Ger: Reiherläufer; Swe: Hägerpipare)

Subspecies. Monotypic.

Status and distribution. Vagrant. One record: a group of five at the Göksu Delta (Mediterranean) on 15 July 1986 was photographed (Bouwman 1987). This species apparently seldom wanders (the Turkish record is seemingly the northernmost ever), although there is a record, also involving a group of five, from the Syrian/Lebanese border, in 1885 (Hüe & Etchécopar 1970, Ramadan-Jaradi & Ramadan-Jaradi 1999).

STONE-CURLEW — *Burhinus oedicnemus*
Kocagöz

(Du: Griel; Fr: Oedicnème criard; Ger: Triel; Swe: Tjockfot)

Subspecies. Nominate in north-east Turkey, *harterti* in the extreme south-east and *saharae* over the rest of the country (see Vaurie 1963). Geographic variation in this species is clouded by considerable individual variation and the probability that populations are locally responsive to soil and humidity conditions at the micro-scale. Given such factors, *B. oedicnemus* is probably much better considered monotypic, with the plausible exception of South Asian *indians*, which Rasmussen & Anderton (2005) chose to recognise at specfic level.

Status and distribution. National conservation status: Vulnerable (population apparently decreasing). A widespread and not uncommon summer visitor across the country, except to most of Black Sea Coastlands. Mainly recorded in steppe-like areas, on traditionally managed and arable farms, sand dunes and riverbanks. Thus, Inner Anatolia is arguably most important breeding area. Inland parts of Aegean, Marmara, riverbanks in East Anatolia and sand dunes in coastal deltas are also utilised for breeding, which is suspected up to 2,300m in Southeast Anatolia, and even higher in East Anatolia. Earliest arrivals are in late February in Mediterranean (28th, Göksu Delta), but most arrive in early April. In August–September small migrant flocks form, mainly at wetlands (max. 75, at Kulu Gölü, on 6 September 1968). Latest record is from the former Aynaz marshes (Mediterranean), on 26 November 1969, with three winter occurrences, in Çukurova (Mediterranean), on 6 December 1984 (G. Pettersson) and 3 January 2002 (M. Ullman), and at Samandağ (Mediterranean) on 2 January 2008 (A. Atahan).

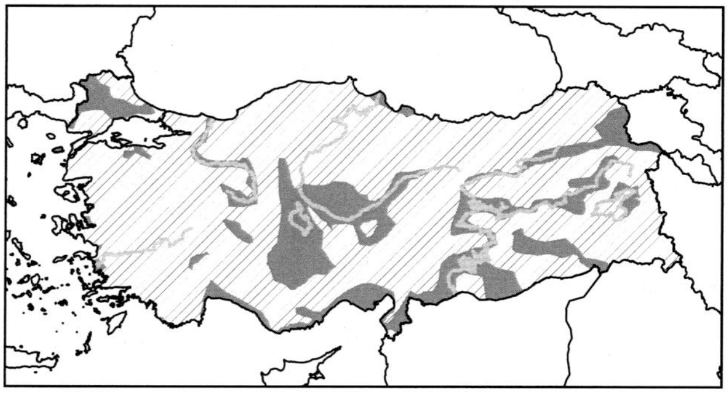

Breeding. In open steppe, agricultural fields and dunes, usually on bare, often stony ground or in short vegetation. The nest is a shallow unlined scrape, usually with a few fragments of plant debris, small stones and grit, mainly around the edge. Eggs: C/2 (15), C/1 (3); a record of four eggs, at Sultan marshes (Inner Anatolia) on 9 June 1982 (Kasparek 1985), is considered a misidentification. Records from **Southeast Anatolia** considerably earlier than those from other regions: an adult with one unfledged young at Ceylanpınar on 7 May 1970, and a pair with large young at Cizre on 6 May 1996 (M. Ullman), both indicate egg laying on *c.*18 March. Records from **Mediterranean** are also somewhat earlier, with first eggs from 25 April; a nest in Göksu Delta had one egg on 5 May 2004, and the second egg was laid early on 7 May 2004 (T. Maunsell); in the same area on 9 June 2006, two young *c.*2 weeks old indicate laying in late April (PC). Eggs from 10 May in **Marmara,** 11 May in **Inner Anatolia,** 12 May in **Aegean** and 15 May in **Black Sea Coastlands,** where there were also eggs on 8–9 June. A well-grown but unfledged young bird at Kızılırmak Delta (Black Sea Coastlands) on 30 July 1971 (Dijksen & Kasparek 1985) indicates laying *c.*11 June (possibly a replacement). Elsewhere, occasionally double-brooded, but no evidence in Turkey.

CREAM-COLOURED COURSER *Cursorius cursor*
Çölkoşarı

(Du: Renvogel; Fr: Courvite isabelle; Ger: Rennvogel; Swe: Ökenlöpare)

Subspecies. *C. c. bogolubovi* (Dickinson 2003) or perhaps nominate *cursor* (Roselaar 1995). No specimen material is available from Turkey, so it is impossible to be sure, but given that *cursor* is the form present in the Syrian and Iraqi deserts, those in adjacent Turkey are more likely to be of this form, rather than *bogolubovi* (which otherwise reaches west to southern Transcaspia). In any case, variation between these forms is slight and clinal, and *bogolubovi* is not diagnosable under the terms of reference adopted here.

Status and distribution. National conservation status: Data Deficient (population apparently decreasing). Apparently a very local breeder in Southeast Anatolia, where comparatively regularly recorded in the Birecik region, especially in recent years, and also known from the environs of Suruç, Urfa, Harran and Akçakale, all in Urfa province (Kasparek 1992a). Whether the species is certainly a summer visitor (as is frequently assumed, given the lack of winter records and that Cream-coloured Courser is a partial migrant elsewhere in the Near East) or a resident is unknown. A group of ten at Birecik in late September (1973) may indicate a pre-migratory gathering, and the earliest record is dated 3 April, but observations in this region outside the period April–August are still extremely few. Records from the Hatay (Mediterranean), at Iskenderun and Kırıkhan, also apparently relate to breeders (being in mid June and early July), but one at Tuzla Gölü, Çukurova (Mediterranean), in late May, surely relates to a migrant (Székely 1998). A late-autumn record from East Anatolia (one, perhaps an immature, at Doğubeyazıt, on 12 November 1970) presumably also concerns a migrant from further north; indeed, it was considered to be of the Caspian race, *bogolubovi,* which is rarely afforded taxonomic recognition (see above). There is also a potentially erroneous claim from Büyükçekmece, just west of Istanbul (Marmara), on 23 August 1946 (Kumerloeve 1970a). Although the observer concerned was the unreliable Abbas Celaloğlu, it should be remembered that westward vagrancy was much more frequent in the first half of the 20th century than it is currently (see, e.g., Snow & Perrins 1998).

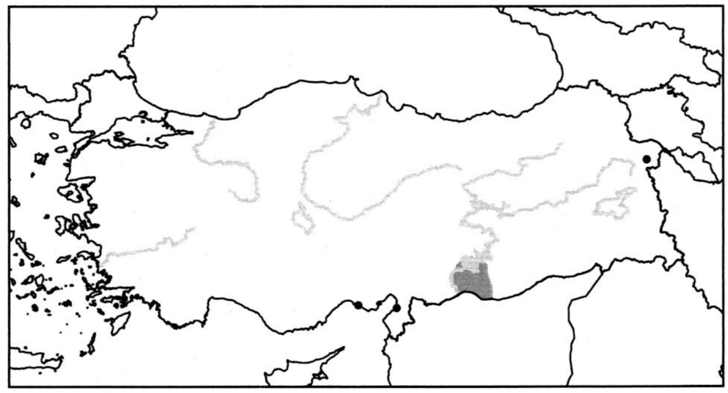

Breeding. A presumed breeder in **Southeast Anatolia** and adjacent parts of **Mediterranean.** Adults present in the breeding season, apparently paired, and display noted, mainly on a bare stony area with ridges, near Birecik, but no definite proof of nesting, although up to ten birds including a single chick were noted on 29 May 2005.

A pair, with two juveniles, plus a single bird were at Birecik on 7 September 1994 (Davidson & Kirwan 1995), and four there on 10 July 2004 (A. Demir & C. Türkman) may also indicate local breeding. In adjacent Syria, and elsewhere, the nest is an unlined shallow hollow in the ground on bare stony or sandy desert, and two eggs are laid, mainly in April.

COLLARED PRATINCOLE *Glareola pratincola*
Bataklıkkırlangıcı

(Du: Vorkstaartplevier; Fr: Glaréole à collier; Ger: Rotflügel-Brachschwalbe; Swe: Vadarsvala)

Subspecies. Nominate.

Status and distribution. National conservation status: Vulnerable (population apparently decreasing). Fairly common summer visitor, breeding in comparatively small numbers in parts of Marmara (including Thrace), principally coastal Aegean, Mediterranean (mainly in the Göksu and Çukurova deltas) and Inner Anatolia (where most widespread). The most important breeding localities are as follows: Meriç Delta (Marmara) 200 pairs; Kocaçay Delta (Marmara) 80 pairs; Büyük Menderes Delta (Aegean) 85 pairs; Kızılırmak Delta (Black Sea Coastlands) up to 115 pairs; Hotamış marshes (Inner Anatolia) 100 pairs; Tuz Gölü (Inner Anatolia) 200 pairs; Göksu Delta, 300 pairs; and Tuzla Gölü, Çukurova, 150 pairs (Magnin *et al. in* Heath & Evans 2000). Formerly bred in considerable numbers at Amik Gölü (Mediterranean) prior to drainage. Kasparek (1992a) reported that it bred throughout the country, being particularly local in Southeast Anatolia, East Anatolia and the Black Sea Coastlands. There is very little proof of breeding in either of the two first-named regions, with Kumerloeve's (1970a) claim that the species had nested at Van Gölü being based solely on an observation of two pairs present in suitable breeding habitat in late May. However, the species was recently (1999) observed during the second half of June in the Murat Valley (East Anatolia), from where several previous records of Black-winged Pratincole *G. nordmanni* are available (perhaps casting doubt on some of these). For Southeast Anatolia, there is only a June observation at Cizre (Martins 1989) and another from near Akçakale, in May 2001 (HW *et al.*), to suggest breeding, whilst Kızılırmak Delta appears to be the only nesting site in the Black Sea region.

Considerably more widespread and common on passage, although still rather uncommon in the eastern third of the country. Occasionally recorded in flocks of several hundred (e.g. 500 at Uluabat Gölü, Marmara, in September 1971, and exceptionally 1,500 at Amik Gölü, in September 1967), though usually groups are much smaller. Spring passage commences in late March (29th) and continues until at least late May/mid June (15th), perhaps peaking in the middle part of April, with autumn migration persisting from August (probably late July; 21st) until mid October (19th).

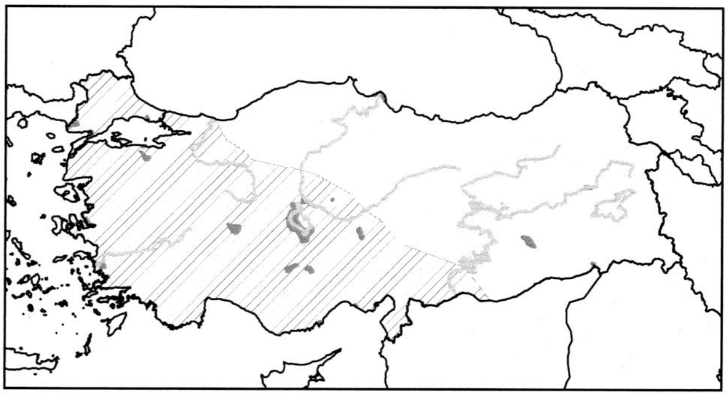

Breeding. Colonies occur on open flat ground, typically drying-out edges of lakes and marshes, with bare ground and low scattered shrubs, often at the transition between marsh and grassland; also on dry saltmarshes with *Salicornia*; in damp cultivation, including ploughed fields; on low flat coastal sand dunes; and occasionally on bare flat islands in inland lakes. Solitary pairs occur occasionally but usually in small colonies of 10–30 pairs, but 100 pairs were at at Kavak (Marmara) on 24 June 1973. Nest a shallow scrape in bare ground, unlined or with a few small pieces of plant material. Eggs: C/3 (13), C/2 (5), C/1 (5), but only one of the latter was definitely a complete clutch (adult incubating), and the others were probably incomplete. Young: B/2 (4), B/1 (1). In Çukurova (**Mediterranean**), in 1987, adults arrived second week of April, numbers increased in May (van der Have *et al.* 1988), and three nests had one, two, and two eggs on 6 May 1990, and several nests 1–2 eggs on 11 May 1990 (J. Whittaker). In Göksu Delta, the second egg in a full clutch of two was laid on 22 May 1993; two young, 5–7 days old, on 9 June 1993, indicate egg laying in mid May (PC). Recently fledged young were observed on 28 July 1971 and 19 August 1986 (J. McLoughlin). At Amik Gölü, nests with eggs were found on 18–20 June 1966. At Sultan marshes (**Inner Anatolia**), on 3 June 1982, three nests had three eggs, and another two newly hatched young and an egg (Kasparek 1985), indicating egg laying in mid May. At Hotamış, one nest was empty, but another had one egg on 15 May 1987; two young were observed on 4 June 1975 (Kirwan 1993a); the latter were just able to fly (R. F. Porter), indicating egg laying in late April, earlier than other records. At Bolluk Gölü, four nests on an island had C/3, C/3, C/2, C/1 on 15 May 2004 (M. Farmer). In **Aegean**, several nests had three eggs at coastal lagoons near Izmir, on 16 May 1899 (Selous 1900), and two full-fledged young (no precise location) were observed on 1 July 1966. In Kocaçay Delta (**Marmara**), three colonies, totalling 80–100 pairs, bred in 1993, and the earliest arrival was mid April (Ertan 1996). A small colony at Uluabat Gölü (Marmara) was displaced on 11 June 1998, when the breeding area was re-ploughed, but the birds joined another small colony nearby (Welch & Welch 1998b). At Kavak, two young, just able to fly, were present on 24 June 1973 (R. F. Porter). In Kızılırmak Delta (**Black Sea Coastlands**), in 1992, the breeding population was mainly in drying-out marshland on dry clay soil partly grown with *Rumex* (10–20cm high); display was observed 30 April 1992 and, on 10 May 1992, five nests had three eggs, and one nest had one egg (Hustings & van Dijk 1994).

[ORIENTAL PRATINCOLE *Glareola maldivarum*]
Kısa Kuyruklu Bataklıkkırlangıcı

(Du: Oosterse Vorkstaartplevier; Fr: Glaréole orientale; Ger: Orientbrachschwalbe; Swe: Orientalisk vadarsvala)

Subspecies. Monotypic.

Status and distribution. One recent claim, at Göksu Delta (Mediterranean), on 19 May 2004 (T. Clarke / Birdquest; *Dutch Birding* 26: 262), for which details are not yet available. Given records in Cyprus, in April 1993 (Rowlands 1994); Egypt, also in April 1993 (Baha el Din & Baha el Din 1996); United Arab Emirates, in November 1992 (Richardson & Aspinall 1998), January–March 1999 (Kirwan 1999), August 2000 (Kirwan 2001c), November 2002 (Balmer & Betton 2003a), April 2003 (Balmer & Betton 2003b), August 2003 (Balmer & Betton 2004a) and September 2004 (*Dutch Birding* 26: 396); and the recent claim of one in Iran, in January 2004 (van Diek *et al.* 2004), as well as records even further west, in Europe, occurrence in Turkey would be relatively unsurprising. Note, however, that Driessens & Svensson (2005) questioned the certain identification of the Cyprus and Egyptian individuals.

BLACK-WINGED PRATINCOLE *Glareola nordmanni*
Kara Kanatlı Bataklıkkırlangıcı

(Du: Steppevorkstaartplevier; Fr: Glaréole à ailes noires; Ger: Schwarzflügel-Brachschwalbe; Swe: Svartvingad vadarsvala)

Subspecies. Monotypic.

Status and distribution. Conservation status: Near Threatened (www.birdlife.org), with the breeding population recently demonstrated to be somewhat lower than thought (Belik & Lebedeva 2002). National conservation status: Data Deficient (population trend unknown). Status in Turkey not precisely known: until recently, regularly recorded in tiny numbers in summer in East Anatolia, principally in the Murat Valley, where probably has bred, but there are insufficient records to understand whether these are isolated and perhaps infrequent events or whether they represent a tiny, but regular breeding population. At present, the former hypothesis appears more likely, especially given the lack of recent records. Additionally, the recent (1999) discovery of Collared Pratincole *G. pratincola* apparently breeding at Bulanık (East Anatolia) may discredit at least some earlier claims of summer occurrence in this region (perhaps, most particularly, the claim of a mixed pairing in 1987). It should be borne in mind that the white trailing edge present on the wing of *pratincola* in fresh plumage is frequently lost by midsummer, and identification of *nordmanni* should be based on a range of features. Several July records from Inner Anatolia in the mid 1970s and one in the mid 1980s appear also strongly suggestive of such confusion and are therefore not included below.

Black-winged Pratincole is better known as a generally uncommon and local passage migrant across the eastern two-thirds of the country, principally in autumn, when it is occasionally recorded in considerable numbers, especially in East Anatolia. There are only seven records (none of them recent and including two Alléon specimens) from Marmara, none from Aegean, and only a handful (virtually all recent) from Southeast Anatolia. Maxima include: 230 at Mogan Gölü (Inner Anatolia) on 15 September 1968, 160 at Ereğli marshes (Inner Anatolia) on 7 September 1980, 160 southeast of Muş (East Anatolia) on 18 September 1989, *c.*1,000 at Bulanık on 20 September 1989, and 189 photographed at Van Gölü (East Anatolia) on 15 September 2000. Regular, usually in very small numbers (maximum 60) through the south coast and Kızılırmak (Black Sea Coastlands) deltas, but the vast majority of passage records are from Inner and East Anatolia: extreme dates are 4 April to 4 June in spring, and 1 August to 24 October in autumn; peak numbers occur in the first half of May and the latter half of September.

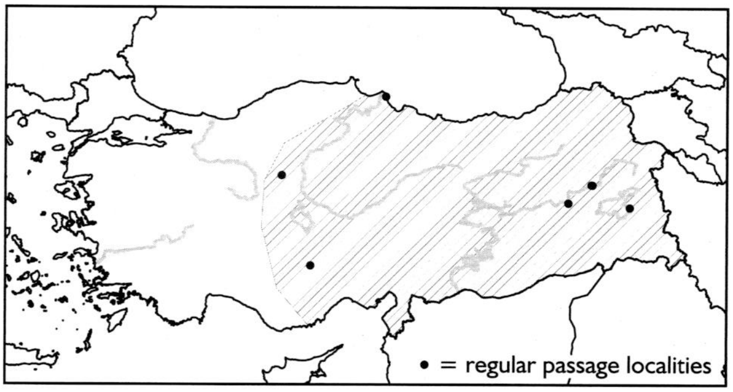

Breeding. All records from **East Anatolia.** A pair displayed at a marsh near Van on 30 May 1969. A pair, holding territory, was at Bulanık on 10 June 1987; one bird performed a broken-wing distraction display, and returned regularly to the same small area of riverbank but, despite extensive searching, no nest was found (Eames 1989); such behaviour is more likely to indicate the presence of young, probably very small. An adult was apparently paired with a Collared Pratincole at Balatos on 10 June 1987; at Yoncalı, on the River Murat, 20 adults and juveniles were observed on 23 July 1987, and two adults and two unfledged juveniles on 13 August 1989. The recent discovery of Collared Pratincoles apparently breeding at Bulanık may cast doubt on some previous claims of Black-winged Pratincole at this locality (see above).

LITTLE RINGED PLOVER *Charadrius dubius*
Halkalı Küçük Cılıbıt

(Du: Kleine Plevier; Fr: Petit Gravelot; Ger: Flußregenpfeifer; Swe: Mindre strandpipare)

Subspecies. *C. d. curonicus.*

Status and distribution. National conservation status: Near Threatened (population apparently decreasing). The most widespread breeding wader, scattered across the entire country, but local in much of Black Sea Coastlands and montane East Anatolia. Even more widespread on passage, especially in autumn, but generally in small numbers; exceptionally there were 300 at Kulu Gölü (Inner Anatolia), on 1 September 2002 (I. Richardson). Due to winter records, earliest spring arrivals are difficult to determine, but records from Karamık (Mediterranean) on 5 March and Uluabat Gölü (Marmara) on 10 March certainly involve migrants. Thus, spring passage lasts from late February (24th) or early March to late May (21st), peaking in late April, whereas autumn passage is from mid July (17th) to late October (27th), peaking in mid August. Occasionally winters in wetlands on the Mediterranean, Aegean and Black Sea coasts, in small numbers. More recently there have been reports of larger concentrations, e.g. ten at Gediz Delta (Aegean) on 20 February 2002 (S. Bekir) and 95 at Tuzla, Muğla province (Mediterranean) on 5 January 2000 (H. Tholl-Schmidt).

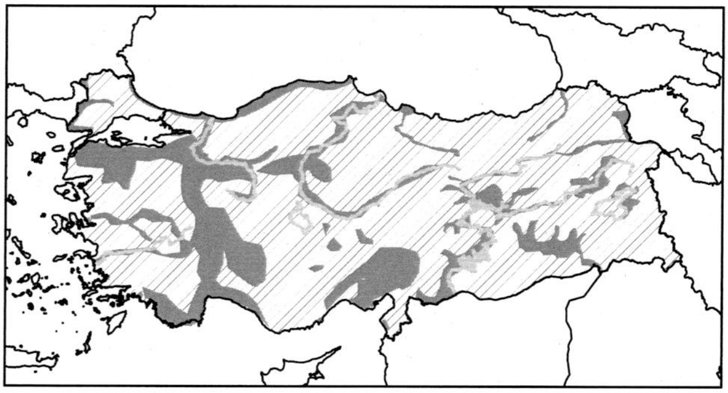

Breeding. Nests on open level areas of sand, shingle and dried mud in riverbeds, coastal flats and beaches, edges of lakes, gravel-diggings, and sandbanks in lakes and rivers, to 2,000m. Nest a shallow scrape in the ground, with a lining of small stones, grit and debris. Eggs: C/4 (5), C/3 (2), C/1 (1). Young: B/1–4 (18); broods of 1–2 frequent, because of predation etc. In **Mediterranean**, display observed on 20 April; a nest had three eggs at Dalyan on 11 May 2003; at Göksu Delta, there were broods of newly hatched young on 22 May 1993, and well-feathered young on 15 June 1992: both indicate laying of first eggs on *c.*21 April (PC). In **Aegean**, four nests had eggs in May (earliest 14 May) and early June. In **Marmara**, newly hatched young, at Kocaçay Delta on 3 June 2006 and 24 June 1999, indicate laying of first egg on 30 April and 21 May (PC). Display was observed near Istanbul on 3 May, and distraction display at Gelibolu on 8 May 1970. In **Inner Anatolia**, display observed early May, a nest had eggs at Ereğli on 22 May 1998 (T. Quittenden), another had eggs, near Karapınar, on 8 June 2006 (R. Castell), and three pairs with young were near Yarma on 26 July 1971. At Kızılırmak Delta (**Black Sea Coastlands**), a nest had eggs on 5 July 1983, three pairs, apparently with young, were seen on 14 June 1984, and there were three pairs, each with four young, on 24 July 1971 (Dijksen & Kasparek 1985). In **East Anatolia**, display was observed on 13 May, a nest had eggs on 1 June 2001 (T. Quittenden), and another three eggs at Van on 8 June 2004 (PC). The earliest record of young at Erçek Gölü, on 27 May 1969 (Kasparek & van der Ven 1983), indicates egg laying in late April. At Sarıcan, 57 km northwest of Ağrı, on 21 June 2004, an adult of a pair was flushed with three fully feathered flying young only 30m from where the other adult was incubating a single egg (PC). Season protracted, and records are suggestive of further cases of double-brooding.

COMMON RINGED PLOVER *Charadrius hiaticula*
Halkalı Cılıbit

(Du: Bontbekplevier; Fr: Grand Gravelot; Ger: Sandregenpfeifer; Swe: Större strandpipare)

Subspecies. Many appear to show the characteristics of *tundrae* (see van der Have *et al.* 1988), and there is a Wadley (1951) specimen taken at Mogan Gölü (Inner Anatolia) in October 1945, as proof of the subspecies' occurrence in Turkey, although GMK was unable to locate this individual at the Natural History Museum (Tring).

Status and distribution. Quite common passage migrant in generally small numbers throughout the country. Main migration periods are late April–late May and late August–early November, with extreme dates in spring of 19 March and 4 June, and the latest birds remain inland until late November or even mid December (13th). Modest numbers may gather at major stopover sites, e.g. 75 at Meriç Delta (Marmara), 60 at Kızılırmak Delta (Black Sea Coastlands), 50 at Ereğli marshes, and 40 at Tuz Gölü and Çavuşçu Gölü (Inner Anatolia).

A local but not uncommon winter visitor to coastal areas. Principal wintering areas are on the Mediterranean coast, with the lakes of Çukurova holding the majority of the population. Generally, the largest numbers are noted in mid January/early February, with 180 at Akyatan Gölü, on 29 January 1992, the highest number ever recorded (IWC 1992). Other important wintering areas are: Gediz, Güllük, Büyük and Küçük Menderes deltas (Aegean), Göksu Delta (Mediterranean), Büyükçekmece (Marmara) and the Yeşilırmak and Kızılırmak deltas (Black Sea Coastlands). Away from the coast, there is an inland record, from Akgöl, near Çeltikçi (Inner Anatolia) on 23 January.

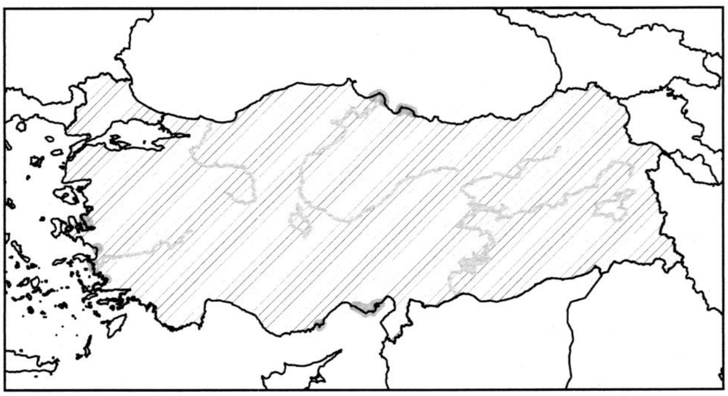

KENTISH PLOVER *Charadrius alexandrinus*
Akça Cılıbıt

(Du: Strandplevier; Fr: Gravelot à collier interrompu; Ger: Seeregenpfeifer; Swe: Svartbent strandpipare)

Subspecies and taxonomy: Nominate *alexandrinus*. Species limits in Kentish Plover probably require reappraisal, as the two subspecies-groups, one each in the New and Old Worlds, may be better considered as species (see del Hoyo *et al.* 1996); their calls and morphology appear to differ (Ridgely & Greenfield 2001, Rasmussen & Anderton 2005). However, the extraordinary publication of van Swelm (2002), proposing a new subspecies of *C. alexandrinus* of unknown range (beyond 'Oman'), based solely on the photograph of a lone juvenile, does not represent a constructive contribution to any such debate and is, furthermore, scientifically ridiculous.

Status and distribution. National conservation status: Vulnerable (population apparently decreasing). Common summer visitor to all regions except Southeast Anatolia, but relatively more widespread in Inner Anatolia, Aegean and Mediterranean. Most important breeding areas include Yumurtalık lagoon (Mediterranean), with up to 1,190 pairs (Székely 1998), the Gediz Delta, Acıgöl (Aegean) and Tuz Gölü (Inner Anatolia). More local in Black Sea Coastlands (Kızılırmak Delta), East Anatolia (around Van Gölü) and Marmara, where only small numbers breed.

Much more widespread and common on passage across the country. Determining onset of spring migration is difficult, as most wetlands in Aegean and Mediterranean support wintering populations, but records elsewhere in Anatolia commence in the second week of March (12th). Spring passage continues until the first week of June (7th), especially in the east, where largest numbers noted (e.g. 150 at Erçek Gölü on 6 June). Autumn migration commences in July (1st), but most pass in late August/mid September, and movement remains obvious until November (27th).

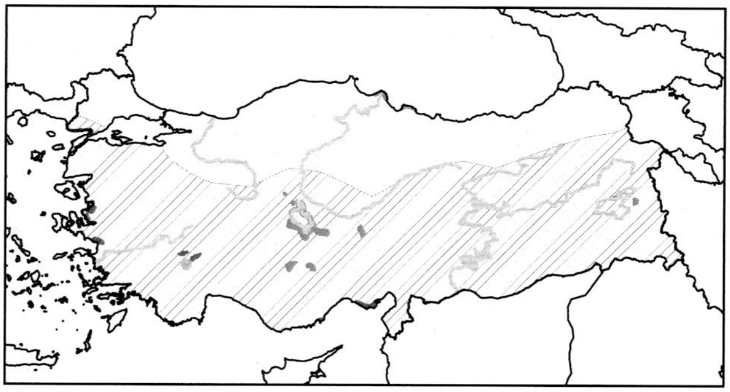

Winter records involve considerable numbers in Aegean and Mediterranean, with occasional records from Black Sea Coastlands. Gediz, Güllük and Büyük Menderes Delta (Aegean), and Çukurova (Mediterranean) host an overall wintering population of *c.*2,000 birds.

Breeding. Nests on open mud and sand, with scant vegetation, at the edges of inland lakes (especially saline or brackish), and on saltmarsh, mudflats, beaches, shingle and edges of lagoons in coastal areas. Nest a shallow hollow in the ground, unlined or with a few fragments of plant material; usually in an open site, but a nest at Kocaçay Delta (Marmara) was well concealed in a clump of rushes (R. Castell). Eggs: C/3 (29) C/2 (1); four records of C/1–2 are considered incomplete clutches; there are no records of clutches of four eggs in Turkey, which is occasional elsewhere. At Tuzla Gölü, in Çukurova (Mediterranean), in late May/June 2000, it was found that Kentish Plovers actively regulate the amount of nest material: they cover the eggs with nest material during the day and uncover them at night. The purpose is unclear, but may be intended to shade the eggs from the sun, when the adults leave the nest for short periods in the day, or to partially conceal the eggs from predators in daylight, or possibly as a regulator in relation to ambient temperature (Szentirmai & Székely 2004). There were 2–6 nests/ha at Tuzla Gölü in 1996–97, but breeding success was low; in 1997, young hatched in only 11 of 82 nests located (*c.*80% of nests were destroyed by domestic dogs and other mammals), and *c.*62% of these young reached the fledging stage (Székely 1998). Young: all broods seen are of three or fewer. In **Mediterranean**, most records are from Göksu and Çukurova deltas, where the population was estimated at more than 2,000 breeding pairs in 1987 (van der Have & van den Berk 1988): the earliest records, newly hatched chicks on 19 April 1973 and 21 April 1974, indicate egg laying on 22–24 March; there are records of 24 nests with eggs in April–May, and 17 records of chicks, mainly in May–June. In **Inner Anatolia**, 17 records of nests with eggs (earliest at Karapınar on 10 April 1970, latest at Kulu Gölü on 19 June 1992), and 11 records of young (earliest 21 April 1967). In **Aegean**, the breeding population at Gediz Delta was estimated at 900–1,000 pairs in 1995–96 (Eken 1997a); the earliest breeding record, newly hatched young, near Izmir on 6 May 1899 (Selous 1900), indicates egg laying about 10 April. In **Marmara**, there were many small chicks along Yeniköy beach in Kocaçay Delta on 26 June 1999 (PC), a nest with three eggs there on 4 June 2006 (R. Castell), and a pair with three young were at Gulpınar on 26 June 1973 (R. F. Porter). At Kızılırmak Delta (**Black Sea Coastlands**), there was a chick on 7 May 1992, indicating egg laying about 11 April (Hustings & van Dijk 1994). In **East Anatolia**, a nest had three eggs, and three broods of chicks were seen, near Van, on 28 May 1969, indicating egg laying early May. Season protracted and double-brooding is likely.

[LESSER SAND PLOVER *Charadrius mongolus*]
Moğol Cılıbıtı

(Du: Mongoolse Plevier; Fr: Gravelot mongol; Ger: Mongolenregenpfeifer; Swe: Mongolpipare)

Subspecies. Unknown.

Species not admitted to the Turkish list (Flint *et al.* 1997, Kirwan *et al.* 1999). Flint *et al.* (1997) provided rationale for removing the species from the Turkish list, and its non-acceptance on the Cyprus list. Specimen records from 1856 (Kumerloeve 1966a) and 1865 (Hollom 1971) lack documented provenance, and the latter is almost certainly the result of a misprint. A specimen taken on 11 May 1876, probably between Kayseri and Çorum, Inner Anatolia (Danford 1878) has not proved traceable and, given chronic confusion between Lesser Sand Plover and Greater Sand Plover *C. leschenaultii* of the race *columbinus* (see, e.g., Hirschfeld *et al.* 2000), is not substantiable. A sight record from the Göksu Delta (Mediterranean) on 21 April 1980 (Bezzel 1986) does not convincingly eliminate Greater Sand Plover, whilst another there on 24 August 1989 that was originally accepted (Kirwan & Martins 1994) has since been rejected as the bird showed the typical structure of *columbinus* (E. Hirschfeld *in litt.* 1996). Additional sight records from the 1980s, early 1990s (one listed by Kirwan & Martins 2000a) and 2003, at least two of which were photographed, are all either clearly *columbinus* or lack substantiation, as does the comment by Turan (1990) that the species is a rare visitor to the east and southeast of the country. The most recent claim, from Sodalı Gölü (East Anatolia) on 18 September 2004, was rejected by the Swiss rarities committee for lack of sufficient evidence (Kirwan *et al.* in press).

GREATER SAND PLOVER *Charadrius leschenaultii*
Büyük Cılıbıt

(Du: Woestijnplevier; Fr: Gravelot de Leschenault; Ger: Wüstenregenpfeifer; Swe: Ökenpipare)

Subspecies. All *columbinus*, a generally rather small, short-legged and small-billed race, which is frequently confused with the previous species (see above, and Hirschfeld *et al.* 2000).

Status and distribution. National conservation status: Endangered (population apparently decreasing). Local and uncommon summer visitor to saline steppes of Inner Anatolia, being almost wholly confined to the environs of wetlands, although breeding has recently been proven on montane steppe, at Bozdağ reserve, Konya province (Inner Anatolia), and on undisturbed farmland (Özen & Kurt 2001). Marked decline evident in the last decade due to agricultural reclamation and habitat destruction, and the species has disappeared from many former breeding sites, e.g. Mogan Gölü and Temelli Gölü. Breeding is principally confined to Inner Anatolia, from Balıkdamı and Akşehir Gölü in the west to Sultan marshes in the east, with strongholds in the east of this region, at Karapınar and Ereğli, Zengen near Nığde, and Sultan marshes. The largest concentration are from Sultan marshes, 494 on 4 June 1994, but local maxima are usually at most 120. Other possible nesting areas include the Euphrates near Birecik (Southeast Anatolia), where two adults and a juvenile were seen on 5–7 June 1973. Given records from Aegean, at Bafa Gölü on 26 June 1966 and at least five records at Büyük Menderes Delta since July, the species might also breed in this region.

Generally a very local passage migrant throughout the country, except Marmara. Regularly noted in the south-coast deltas in spring, between late February (27th) and mid May (12th). In autumn remains in the same region until late November (26th) and overwintering might be expected to occur, at least occasionally, as perhaps evidenced by one at Gediz Delta (Aegean) on 5 February 2006. Records from Gâvur Gölü (Mediterranean) and around Van and Erçek lakes (East Anatolia) may principally involve birds en route to breeding areas in Azerbaijan and Armenia, but the possibility of breeding should not be ignored. However, spring and autumn records from Black Sea Coastlands apparently relate entirely to overshoots.

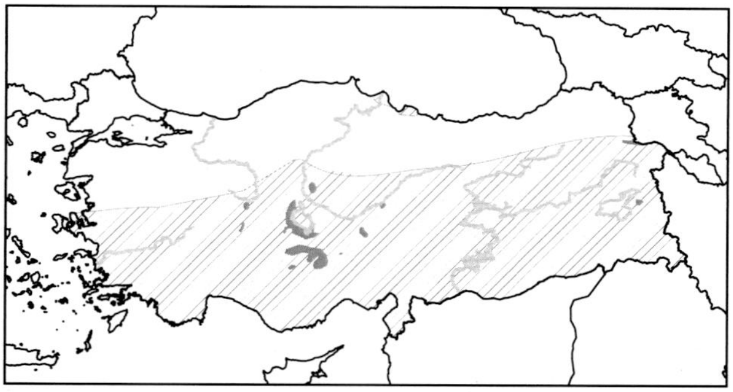

Breeding. On vast open, level mud and steppe near water; in wetter areas of mud, devoid of vegetation, with depressions of shallow water and a crust of salt as it dries out, but also in adjacent drier mud and steppe, with short, sparse plants and tufts of vegetation. Nests on the ground, in the open or against a tuft of vegetation; the nest is a shallow scrape lined with variable plant debris (Lehman 1969) or with small stones and a few stalks (OST 1975). Eggs: C/3 (36), C/2 (two, plus three other nests with two eggs, of which two were incomplete clutches). Young: 25 records of unfledged young; numbers, where recorded, B/1 (3), B/2 (3), B/3 (5). Many records from Karapınar Ovası (**Inner Anatolia**): 50 pairs in early June 1971; during 3–8 May 1968 and 3–4 May 1969, Lehman (1969) found seven nests with eggs, saw three broods of young on 8 May 1968, and many broods on 3 May 1969; from the age of young observed, and from the extent of incubation in the eggs collected, he calculated that first eggs were laid in late March, main laying dates 1–20 April, the first young hatching *c.*20 April, and clutches in May were considered second attempts. In three visits, 1970, 1972 and 1973, all on 14–27 April, McNeile found 12 nests with eggs, at all stages of incubation, including a clutch of two and another of three, which were close to hatching on 16–17 April 1970. J. Whitaker found two nests with eggs on 26–27 April 1988, a nest with one egg on 5 May 1989 (which had three eggs on 8 May), and a nest with eggs on 10 May 1990. A nest contained three newly hatched young on 20 April 2004, another contained two eggs (incomplete clutch) on 21 April 2004, and young, *c.*5 days old, were seen on 21 April 2004 (R. Thomas), the latter indicating date of first egg in the third week of March. Three nests each had three eggs 12–15 May 2004 (K. Robson); three nests each had three eggs on 8 May 2006 (N. Burke); four nests each had three eggs and another two eggs (complete clutch) on 9 May 2005 (J. Hodson). Elsewhere in Inner Anatolia, four birds were in display flight near Bor on 4 April 1981 (Parr 1981). At Bezirhane, a nest had three eggs on 17 June 1987, the latest record of a nest with eggs, and three juveniles were observed the same day (B. Anderson), and at least 13 juveniles were seen there 17 July 1986. Around Tuz Gölü, nests had eggs on 28 May 1967, on 16 May 1969 (Lehman 1969) and on 20 May 1970, and a brood of young was seen on 21 May 1969 (R. F. Porter); at Bolluk Gölü, a nest had three eggs on 26 May 1993 (B. Pleasance), a young bird, *c.*2 weeks old, was seen 8 May 1993, and large feathered young on 26 May 1993 (PC). At Çöl Gölü, a nest had three eggs on 23 April 1988 (H. Schekkerman), and 16 birds, mostly juveniles, were seen on 17 July 1986 (Karauz Kiraç 1993). A nest had three eggs near Hortu on 5 April 1971, the earliest recorded nest and eggs in Turkey. At Sultan marshes, two broods of three young were seen on 21 May 1989 (D. Richardson), a brood of two young on 26 May 1982 (Kasparek 1985), and three broods of young in June; broods (two and one) were at Ayranci on 11 May 1967 and 6 May 1968 (Lehman 1969). Two broods (small and 14 days old), were north of Konya, on 14/20 May 1970, and a brood (two) was near Inçesu on 1 June 1975. Fledged young were at Ereğli marshes on 13 June 1975 and at Hotamış on 5 July 1979. In **Mediterranean**, two pairs with small young were at Göksu Delta on 20 June 1989 (D. Richardson). Two adults and a recently fledged juvenile were north of Birecik (**Southeast Anatolia**), on 5–7 June 1973, possibly indicating local breeding. In **East Anatolia**, a female behaved as if nesting, on high, flat pebbly steppe between Iğdır and Tuzluca, on 7 June 1994 (Kirwan 1994d). Reported as a breeder at Erçek Gölü (Magnin & Yarar 1997). It is unclear whether later records are second broods or replacements following earlier failure.

CASPIAN PLOVER *Charadrius asiaticus*
Doğu Cılıbıtı

(Du: Kaspische Plevier; Fr: Pluvier asiatique; Ger: Wermutregenpfeifer; Swe: Kaspisk pipare)

Subspecies. Monotypic.

Status and distribution. An extremely rare, occasional passage migrant, recorded from Black Sea Coastlands, Mediterranean, Inner Anatolia and East Anatolia, in spring in March, April and June, and in autumn during August to October. A total of ten records: a male and female at Amik Gölü (Mediterranean) on 19 March 1965 (Kumerloeve 1966d), one at Mogan Gölü (Inner Anatolia) on 25–26 October 1966 (OST 1969), 14 at Sodalı Gölü on 12 August 1967 (Vielliard 1968), one east of Alanya (Mediterranean) on 14 April 1971 (Warncke 1972, OST 1975), one at Erçek Gölü (East Anatolia) on 5 September 1971 (OST 1975), 15 in dry fields by Tuzla (Palas) Gölü (Inner Anatolia) on 23 August 1988 (Kirwan & Martins 1994), two photographed at Bulanık (East Anatolia) on 3 August 1991 (Kirwan & Martins 1994; see *Birding World* 5: 252), one at Kulu Gölü (Inner Anatolia) on 29 July 1998, one photographed at Tuzla Gölü, Çukurova (Mediterranean) on 8 June 1999 (see *Sandgrouse* 22: 76), and one at Kızılırmak Delta (Black Sea Coastlands) on 17 October 1999 (Kirwan *et al.* 2003). Other claims in Turkey lack sufficient detail to be acceptable given the species' undoubted rarity in the country (see Kirwan 1995a).

EURASIAN DOTTEREL *Charadrius morinellus*
Dağ Cılıbıtı

(Du: Morinelplevier; Fr: Pluvier guignard; Ger: Mornellregenpfeifer; Swe: Fjällpipare)

Subspecies and taxonomy. Monotypic. Is occasionally placed in the genus *Eudromias.*

Status and distribution. Generally scarce migrant at both seasons, but occasionally recorded in large 'trips' of up to 800 individuals, principally in Inner Anatolia, especially in the vicinity of Tuz Gölü (where regular in the early 1970s in late autumn and more occasionally recorded in exceptional numbers in spring). Recorded throughout the country, but away from central regions usually in groups of fewer than ten birds, with the vast majority apparently moving through the western two-thirds of Turkey. No modern records are available from East Anatolia (but there is an extreme paucity of observations at the relevant periods), and none at all from Southeast Anatolia. Spring passage dates span the period 26 March to 7 May (one record from 14 June), and the species has been recorded in autumn between 10 September and 28 November. Two winter records (December and February) are available, suggesting that the species occasionally over-winters in the country. All records available to us are presented in Table 3.

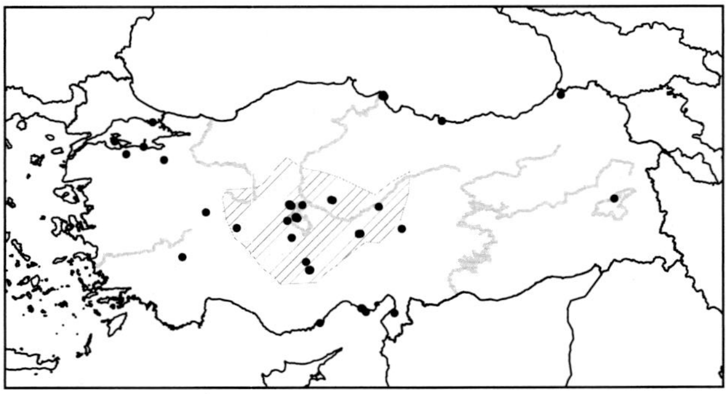

Table 3. Dated records of Eurasian Dotterel *Charadrius morinellus* from Turkey. More vague mentions in the literature include those of Dickson & Ross (1839) and Calvert, *in* Curzon (1854) from Erzurum; Alléon and Mathey-Dupraz (1923) at the Bosphorus in 1860, 1880 and 1890/94; Krüper (1875) at Bornova (Aegean) in early April, and by Ereğli (Inner Anatolia) in November (mentioned by Kumerloeve 1961). Five specimens were formerly held in the Robert's College (Bebek, Istanbul) collection (Mathey-Dupraz (1923).

Locality	Region	Date	No. of birds	Source
"Maslak", Istanbul	Marmara	4 April 1869	male	T. Robson specimen, The Manchester Museum, accession no. B09278
"Haskerey"	Marmara	3 November 1869	1	T. Robson specimen, The Manchester Museum, accession no. B09277
Küçükcekmece	Marmara	28 September 1914	2	Mathey-Dupraz (1923)
Seyfe Gölü	Inner Anatolia	3 November 1951	1	Kumerloeve (1970a)
Ordu	Black Sea Coastlands	28 November 1964	male	Kumerloeve (1967a)
Karapınar	Inner Anatolia	26 March 1967	5	Kumerloeve (1970a)
Tuz Gölü/Hirfanlı	Inner Anatolia	2 April 1967	80	OST (1969)
Kocaçay Delta	Marmara	9–10 April 1967	1	OST (1969)
Near Acıgöl	Mediterranean	10 September 1968	1	OST (1972), Dijksen & Kasparek (1988)
Tuz Gölü	Inner Anatolia	14 June 1969	1	OST (1972)
Tuz Gölü	Inner Anatolia	20 November 1969	46	OST (1975)
Ereğli marshes	Inner Anatolia	23 November 1969	7	OST (1972)
Akyatan Gölü	Mediterranean	28 November 1969	39	OST (1972)
Siverek	Southeast Anatolia	3 December 1969	5	OST (1975)
Seyfe Gölü	Inner Anatolia	12 December 1969	9	OST (1975), Husband & Kasparek (1984)
Tuz Gölü	Inner Anatolia	21 November 1970	558	OST (1975)
Ereğli marshes	Inner Anatolia	24 November 1970	34	OST (1975)
Hortu	Inner Anatolia	5 April 1971	300+	OST (1975)
Kurbaga Gölü	Inner Anatolia	20 November 1971	20	OST (1975)
Tuz Gölü	Inner Anatolia	22–23 November 1971	369	OST (1975)
Kulu Gölü	Inner Anatolia	23 November 1971	1	OST (1975)
Akşehir Gölü	Inner Anatolia	28 November 1971	21	OST (1975)
Tuz Gölü	Inner Anatolia	3 October 1973	455	OST (1975)
Uludağ	Marmara	16 September 1975	3	OST (1978)
Akyatan Gölü	Mediterranean	18 February 1974	4	OST (1978)
Karataş	Mediterranean	11 September 1976	2	Beaman (1986)
Borçka	Black Sea Coastlands	22 October 1977	1	Beaman (1986)
Sultan marshes	Inner Anatolia	28 October 1977	20+	Beaman (1986)
Kulu Gölü	Inner Anatolia	3 May 1978	5	Kasparek (1987a)
Paşalimanı Island	Marmara	20 September 1979	1	Beaman (1986)
Sodalı Gölü	East Anatolia	9 October 1980	2	van der Ven & Gheyselinck (1982)
Tuz Gölü	Inner Anatolia	2 April 1981	800+	Goriup & Parr (1983)
SW of Tuz Gölü	Inner Anatolia	3 April 1981	200+	Goriup & Parr (1983)
Palas Gölü	Inner Anatolia	21 October 1985	3	van den Berk *et al.* (1986)
Sultan marshes	Inner Anatolia	22–25 October 1985	40	van den Berk *et al.* (1986)
Akyatan Gölü	Mediterranean	27 October 1985	20	van den Berk *et al.* (1986)
Göksu Delta	Mediterranean	30 October 1985	21	van den Berk *et al.* (1986)
Sultan marshes	Inner Anatolia	26 April 1987	1–2	Kirwan & Martins (1994)

Manyas Gölü	Aegean	23 September 1987	7	Kirwan & Martins (1994)
Sultan marshes	Inner Anatolia	15 November 1987	1	Kirwan & Martins (2000a)
Tuz Gölü	Inner Anatolia	13 April 1988	8	Schekkerman & van Roomen (1993)
Belen Pass	Mediterranean	16 September 1988	1	M. Huhta-Koivisto *et al.* (unpubl.)
Kızılırmak Delta	Black Sea Coastlands	26 March 1992	1	Hustings & van Dijk (1994)
Kızılırmak Delta	Black Sea Coastlands	7 April 1992	2	Hustings & van Dijk (1994)
Kızılırmak Delta	Black Sea Coastlands	29 April 1992	2	Hustings & van Dijk (1994)
NE of Cihanbeyli	Inner Anatolia	17 April 1993	400	Kirwan & Martins (2000a)
Kızılırmak Delta	Black Sea Coastlands	25 September 1995	1	Kirwan & Martins (2000a)
North of Afyon	Inner Anatolia	3 April 1996	28	Kirwan & Martins (2000a)
Ereğli marshes	Inner Anatolia	6 May 1997	50	Kirwan *et al.* (2003)
Ereğli marshes	Inner Anatolia	7 May 1997	103	Kirwan *et al.* (2003)
NW of Tuz Gölü	Inner Anatolia	April 2000	40	Heunks *et al.* (2002)
S of Tuz Gölü	Inner Anatolia	April 2000	275–300	Heunks *et al.* (2002)
Kulu Gölü	Inner Anatolia	25 September 1997	3	Kirwan *et al.* (2003)
Sultan marshes	Inner Anatolia	12 November 2000	9	Kirwan *et al.* (2003)
Kulu Gölü	Inner Anatolia	15 September 2001	1	Kirwan *et al.* (2003)
Kulu Gölü	Inner Anatolia	20 October 2001	1	Kirwan *et al.* (2003)
Göksu Delta	Mediterranean	4 April 2002	2	H. Schmaljohann *in litt.* (2005)
Altıntaş Ovası	Aegean	April 2002	20+	S. Bekir *et al.*
Güllük Delta	Aegean	28 October 2002	5	I. Richardson & B. W. Stoneman
Tuz Gölü	Inner Anatolia	30 April 2003	300+	Demirci (2003)
Kızılırmak Delta	Black Sea Coastlands	9 April 2005	4	Balmer & Betton (2005b)
Göksu Delta	Mediterranean	4–9 April 2005	2–5	T. van Gastel *in litt.* (2005)
Kulu Gölü	Inner Anatolia	18 May 2005	7	L. Veyrunes *in litt.* (2005)
Kulu Gölü	Inner Anatolia	27 October 2005	6	(R. Gül)
Ereğli marshes	Inner Anatolia	13 November 2005	1	Balmer & Betton (2006a)
Kulu Gölü	Inner Anatolia	14 October 2006	1	Balmer & Betton (2007a)
Kulu Gölü	Inner Anatolia	14 October 2007	2	(Balmer & Betton 2008)
Cihanbeyli	Inner Anatolia	20–29 October 2007	up to 9	(Balmer & Betton 2008)

AMERICAN GOLDEN PLOVER *Pluvialis dominica*
Amerikan Altın Yağmurcunu

(Du: Amerikaanse Goudplevier; Fr: Pluvier bronzé; Ger: Amerikanischer Goldregenpfeifer; Swe: Amerikansk tundrapipare)

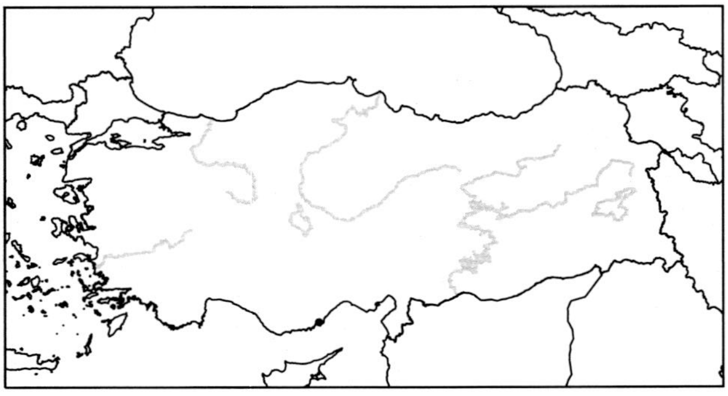

Subspecies. Monotypic, but formerly considered conspecific with *P. fulva*. See Sangster *et al.* (2003) for a summary of the reasons to use *dominica* rather than *dominicus* for this taxon.

Status and distribution. Vagrant. A first-summer or female on the northwest shore of Akgöl, Göksu Delta, on 15 May 1993, was comprehensively documented and remains the only unequivocal record of *dominica* in Turkey (Kirwan 1994c). See also Pacific Golden Plover *P. fulva* for additional records of *dominica/fulva*.

There is only one other record from the Middle East: at Sur, Oman, on 23 November–26 December 2003 (Grieve *et al.* 2005).

PACIFIC GOLDEN PLOVER *Pluvialis fulva*
Küçük Altın Yağmurcun

(Du: Aziatische Goudplevier; Fr: Pluvier fauve; Ger: Pazifischer Goldregenpfeifer; Swe: Sibirisk tundrapipare)

Subspecies. Monotypic, but formerly considered conspecific with *P. dominica.*

Status and distribution. Vagrant. One, seen by E. Thieme & H. Schramm, at Büyükçekmece, near Istanbul (Marmara) has been formally accepted by the German rarities committee (Kasparek 1990a, 1992a) and is the only definite record.

Additionally, there are two observations of 'Lesser Golden Plover' which pre-date the emergence of widespread understanding of field criteria for distinguishing between this species, to which they seem most likely to refer, and American Golden Plover *P. dominica* (Kirwan 1994c): three, seen well with European Golden Plover *P. apricaria* at Karataş, Çukurova (Mediterranean) on 30 March 1981 (Parr 1981, Goriup & Parr 1983), and an adult in summer plumage at the Göksu Delta (Mediterranean) on 6 May 1990 (Kirwan 1994c).

EUROPEAN GOLDEN PLOVER *Pluvialis apricaria*
Altın Yağmurcun

(Du: Goudplevier; Fr: Pluvier doré; Ger: Goldregenpfeifer; Swe: Ljungpipare)

Subspecies. Monotypic.

Status and distribution. Fairly local passage migrant and winter visitor across the country, but very rarely recorded in the eastern third. Spring passage is from early March (late February?) to early May, with peak numbers passing through Inner Anatolian wetlands in the first half of March and, by May, generally none. Latest spring records are from Meriç Delta (Marmara) and Aliağa (Aegean), both on 11 May. Two summer records, a single at Mogan Gölü (Inner Anatolia) on 16 June, and even more exceptionally, 50 at Çavuşçu Gölü (Inner Anatolia) on 21 June. However, the main autumn passage is from late August to late November, with peak numbers in late October.

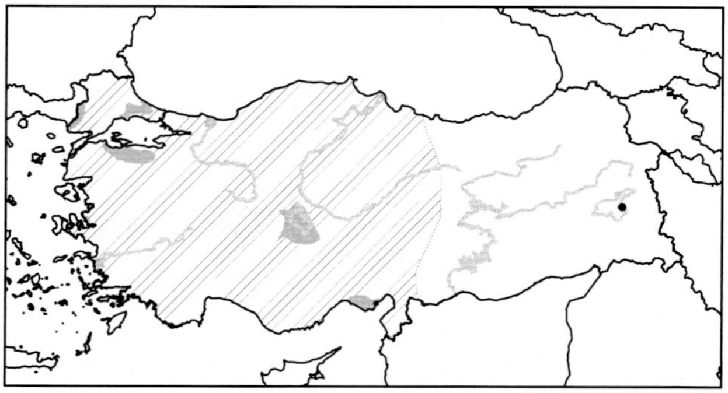

The main wintering grounds in Turkey are wetlands on northern, western and southern coasts, but winter records from Inner and East Anatolia are available, e.g. on 3–5 December 1969, at Van Gölü. Birds form flocks in

November which remain together until February, when they start to disperse. Largest flocks are in the north: 1,000 at Büyükçekmece (Marmara) on 18 December 2001 and 1,389 at Kızılırmak Delta (Black Sea Coastlands) on 14 January 1996. The species is probably under-recorded given the availability of apparently suitable but under-watched areas, such as lowland pastures, in the north and north-east of the country.

GREY PLOVER *Pluvialis squatarola*
Gümüş Yağmurcun

(Du: Zilverplevier; Fr: Pluvier argenté Ger: Kiebitzregenpfeifer; Swe: Kustpipare)

Subspecies. Monotypic. Is very occasionally placed in the monotypic genus *Squatarola.*

Status and distribution. Local passage migrant in small numbers at wetlands and coastal areas throughout the country. Spring passage commences in March on south and west coasts, peaks in mid April and continues steadily until late May (29th). Autumn passage persists from late July (24th) until late November (28th). Compared to spring, autumn passage is more confined to coastal areas but involves larger numbers, e.g. 200–300 in Saros Gulf (Aegean). Kasparek (1992) mentioned a single summer record, but this should not be taken to imply breeding.

In winter, relatively more common in the western two-thirds of Turkey, generally in coastal areas from Iğneada (Marmara) in the northwest, to Yumurtalık lagoon in the south, with single inland records from Inner Anatolia and Aegean. Coastal wetlands in the west support the largest numbers. Gediz Delta, both Menderes deltas and Bafa Gölü (Aegean) combined hold more than 50% of the Turkish winter population. Recent surveys at Gediz produced 374 on 12 January 2003, the largest number ever recorded in Turkey. In the south, important wetlands include Akyatan Gölü, Yumurtalık lagoon and Göksu Delta (Mediterranean).

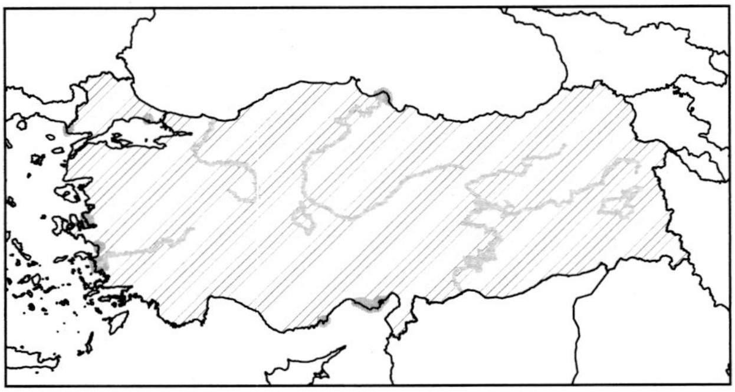

SPUR-WINGED LAPWING *Vanellus spinosus*
Mahmuzlu Kızkuşu

(Du: Sporenkievit; Fr: Vanneau éperonné; Ger: Spornkiebitz; Swe: Sporrvipa)

Subspecies. Monotypic. Often placed in the genus *Hoplopterus.*

Status and distribution. National conservation status: Vulnerable (population apparently decreasing). Local summer visitor to wetlands in all regions, except Black Sea Coastlands, and in Southeast Anatolia is principally restricted to the banks of the Euphrates and Tigris rivers. Prefers open marshes and environs of lakes, riversides and deltas. The main strongholds are in Aegean and Inner Anatolia, e.g. Gediz Delta and Marmara Gölü (Aegean), and Ereğli marshes (Inner Anatolia). Elsewhere, the species is regularly found in Marmara, at Uluabat Gölü, Manyas Gölü and Meriç Delta (the only breeding area in Thrace), and in Mediterranean at Göksu Delta (the single-most important breeding area) and Burdur Gölü. There is evidence to suggest that the species has only recently (late 1990s) colonised the Van Gölü basin (East Anatolia), where the species now appears to be present at virtually all suitable sites (formerly only a casual visitor to this region) north to the Murat River valley, with a single record from Ekşisu marshes (near Erzincan) during the breeding season. A marked decline throughout its main range in Turkey has occurred in recent decades, due to the destruction of suitable habitat, particularly in Inner Anatolia, where there are currently only a handful of occupied sites, e.g. Sultan marshes and Seyfe Gölü.

More common and widespread on migration throughout the country. Earliest arrival is 15 March (in Mediterranean) and as of mid April most breeding grounds are occupied. However, passage continues, even in the south, until late May, during which period it is most frequently observed in Southeast Anatolia, usually at Birecik. From late July, post-breeding flocks form in Aegean and Mediterranean wetlands, e.g. 64 at Salihli (Aegean), 55 at Burdur Gölü, 36 at Acıgöl (Aegean) and an exceptional 200 at Göksu Delta, on 27 August 1988. The latest observation is surprisingly from the northernmost breeding site, Meriç Delta, on 7 October 1977, but there is also a recent (2007) record from Izmir (Aegean) in November (A. Atahan).

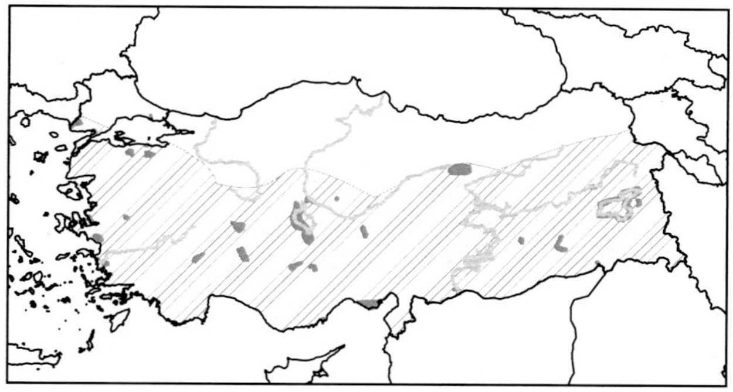

Breeding. On open areas near water, on mudflats, bare ground and grassy areas with sparse vegetation, especially ground drying out around lakes and marshes. Solitary, but several pairs may breed in loose association, and certain areas hold many pairs, e.g. Çukurova (Mediterranean), where there were *c.*80–100 pairs in spring 1987 (van der Have *et al.* 1988). Nest is a shallow hollow in open ground, lined with plant material. Eggs: usually 4 (27), occasionally 3 (7) or 2 (1); other records of two eggs or fewer considered incomplete clutches; five eggs in one nest were probably laid by the same female (PC). In **Mediterranean**, eight nests had eggs, four of which were full clutches, near Silifke and Karataş, on 3–7 May 1990 (J. Whitaker), and there are many other records of eggs in May, and one on 17 June 1992; small young mainly in June, with a record on 5 May (N. Williams), which indicates laying of the first egg *c.*9 April. In **Inner Anatolia**, the earliest record of a full clutch is 7 May 1970, with several others in May, and a late record of an adult incubating at Seyfe Gölü on 19 June 1975 (Husband & Kasparek 1984); small young were at Eşmekaya on 5 June 1992. In **Aegean**, the earliest record is an incomplete clutch of two eggs at Bafa Gölü on 30 April 1975; there are five other records of nests with eggs in the second half of May, and the latest record, a nest, which contained five eggs, was near Akköy on 26 June 1999 (PC). In **Marmara**, a nest had two eggs at Manyas Gölü on 31 May 1967, and territories were established at Uluabat Gölü on 11 April 1967 (R. F. Porter); three pairs were displaying there on 17–24 May 1998, but not subsequently, and they perhaps failed (Welch & Welch 1998b).

RED-WATTLED LAPWING — *Vanellus indicus*
Büyük Kızkuşu

(Du: Indische Kievit; Fr: Vanneau indien; Ger: Rotlappenkiebitz; Swe: Rödflikvipa)

Subspecies and taxonomy. *V. i. aigneri*, but variation within this species is principally clinal (del Hoyo *et al.* 1996). Has also been placed in *Hoplopterus* or *Lobivanellus.*

Status and distribution. National conservation status: Endangered (population apparently stable). A highly local resident or summer visitor to an apparently tiny area of extreme Southeast Anatolia, breeding in areas of gravel deposits with associated pools along the River Tigris close to the Syrian border at Cizre. First recorded in June 1983 (Murphy 1984, Martins & Robson 1988) and subsequently discovered at several additional localities north of the

town towards Siirt, west towards Nusaybin and Ceylanpınar, and southeast as far as Silopi (Balmer & Betton 2002b). Recently recorded using dam lakes and farmland, and seems able to breed in such areas, as well as on riverine gravel islands. Kasparek (1992a) placed the Turkish population at just ten pairs, which is clearly an underestimate, but the total number of breeding pairs is unlikely to be large.

Breeding. Until recently only recorded at gravel deposits, shingle islands and pools, along the Tigris, near the town of Cizre (**Southeast Anatolia**), but see above. The nest (undescribed in Turkey) is a shallow scrape in the ground, unlined or with a sparse lining of plant stalks and nearby debris. Eggs: no data but, elsewhere, usual clutch four. First recorded on 15 June 1983, when adults were behaving in an agitated manner; subsequent records include two chicks on 31 May 1984, a pair with three chicks on 18 May 1985, an adult incubating on 6 May 1986 (M. Ullman), and three chicks on 7 June 1987 (B. Anderson).

SOCIABLE LAPWING *Vanellus gregarius*
Sürmeli Kızkuşu

(Du: Steppekievit; Fr: Vanneau sociable; Ger: Steppenkiebitz; Swe: Stäppvipa)

Subspecies and taxonomy. Monotypic. Is frequently placed in genus *Chettusia* (but see BOURC 1997 for rationale for the latter genus' non-recognition).

Status and distribution. Conservation status: Vulnerable (BirdLife International 2000), but subsequently upgraded to Critically Endangered (BirdLife International 2004, and see below). A rare passage migrant recorded from all regions, but only once from Marmara (Ertan 1996), with no observations anywhere in the country between 1995 and 2002, but, in March 2007, 1,017 were discovered at Ceylanpınar (Southeast Anatolia) and in late October 2007, over 3,200 were found in the same area, as a result of a Kazakhstan-bred juvenile having been radio-tagged (Balmer & Betton 2008). The October flock was the largest number recorded anywhere for 100 years. Most frequent in the eastern half of Turkey. Kasparek (1992b) noted 21 records in Turkey; there have been a further 16 since. Occurs mainly late April and mid September to late October; earliest date 7 March and latest date 17 November. Most observations are of up to five individuals, the largest counts involved three flocks of 115, 125 and 189 in East Anatolia in October 2002, 2003 and 2006 (Demirci 2003, Murdoch & Serra 2006, Balmer & Betton 2008, Kirwan *et al.* in press) and a flock of 30 and several others, one of which was collected, on plains at Ceylanpınar and Harran (Southeast Anatolia) in late March 1969 (OST 1972), although two flocks, totalling 38 birds, which passed over Nemrut Dağı, near Tatvan (East Anatolia) on 27 September 1986, and were recorded by the observers as White-tailed Plovers *V. leucurus* (Martins 1989) almost certainly were of the present species (Kasparek 1992b). In general, numbers reflect patterns shown by records from other Middle Eastern countries. Kasparek (rather presciently) noted that 'wintering numbers in the Negev (Israel) suggest that many more must pass through Turkey on migration than are actually seen', preferring dry sandy plains, cultivated fields and areas of fallow land, which are extensive in Turkey, especially in the eastern half. Observers rarely visit the latter part of the country during the main passage periods.

Former breeding in the Ukraine may explain the apparently more frequent 19th-century occurrences. When first recorded in Turkey, flocks were observed repeatedly in the Erzurum area (East Anatolia), in mid September to mid October (Dickson & Ross 1839) and apparent further occurrence there was supported by Curzon (1854). However, a general long-term decline throughout the species' range is well documented (BirdLife International 2000, Eichhorn & Khrokov 2002, Tomkovich & Lebedeva 2002) and is now apparently continuing unabated

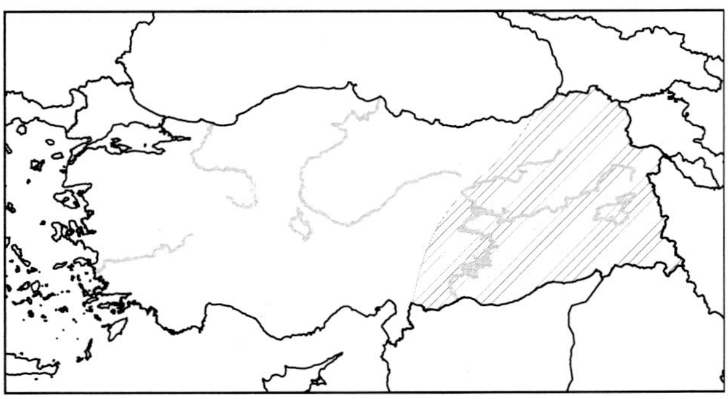

(Moseikin *in litt.* to Oriental Birding March 2003), and it would seem inevitable that such decline would be expressed in the history of the species' occurrence in Turkey, as elsewhere.

WHITE-TAILED LAPWING *Vanellus leucurus*
Ak Kuyruklu Kızkuşu

(Du: Witstaartkievit; Fr: Vanneau à queue blanche; Ger: Weißschwanzkiebitz; Swe: Sumpvipa)

Subspecies and taxonomy. Monotypic. Is frequently placed in genus *Chettusia* (but see BOURC 1997 for rationale for the latter genus' non-recognition).

Status and distribution. Status uncertain, but has bred in extremely small numbers (most recently, apparently in 1997 and 2000; Brugière 2002, Kirwan *et al.* 2003). No evidence strongly suggestive of breeding between 1986 and 1997 and possibly now principally an occasional, erratic wanderer to Turkey from core of breeding range to the south and east. All proven breeding areas are within the southern part of the Konya Basin (Inner Anatolia) or in the south-coast deltas. Kasparek (1992b) assessed all previous records from Turkey, interpreting them with regard to breeding status and distribution in the Middle East and, as a winterer, in Africa and Arabia. There is no subsequent evidence requiring adaptation of any conclusions reached therein, from which this account is largely derived.

Most records are post-1970 but the species is apparently not necessarily a recent, albeit intermittent colonist: the first record, a pair in the Dalman Collection, Jerusalem, was shot by Aharoni at Amik Gölü on 17 June 1910 (Kumerloeve 1966d, 1969d). Over the last 30 years, recorded as a very scarce, semi-colonial breeder between 5 April and 26 August (maximum at any one site nine pairs) at seven locations as follows. Göksu (maximum 15 individuals and most recently in 1997) and Çukurova deltas (Mediterranean), the former Yarma wetlands – now wholly destroyed – (never more than an exceptional nine pairs, usually fewer than five), Karatepe (at least two pairs in 2000), Hotamış (last recorded 1986, maximum 13 individuals in 1985) and Sultan marshes (Inner Anatolia), where most recently noted in 1996. Also present in circumstances suggesting possible breeding at Ereğli marshes (Inner Anatolia). Additionally, recorded during the breeding season at six locations, at least some of which may suggest occasional wider breeding distribution, as follows: Kızılırmak Delta (Black Sea Coastlands), Amik Gölü (Mediterranean), Van Gölü and Iğdır Ovası (East Anatolia), Birecik (Southeast Anatolia), where display flight was observed, and in the Aralık area of the Aras Valley (East Anatolia). Several pairs recently (1999) bred in the Armash area of the same valley, in neighbouring Armenia, with one bird remaining into late December at least (Ananian *et al.* 2002). Records in late April and October from Iğneada and Manyas Gölü (Marmara), and April from Büyük Menderes Delta (Aegean) almost all certainly refer to migrants. Recorded from 2 April, though most arrive in May. Latest record 20 October.

The irregular nature of recorded breeding is certainly not a function of variability in observer coverage. For example, clearly absent from Göksu Delta in most years, e.g. in 1988–91 when the area received extensive systematic coverage, and not located at what remained of the most established breeding site of Hotamış marshes during late April and mid-May 1991 (Kirwan 1993a). There has been no increase in records in Turkey commensurate with the radical increase in observer activity over the last 25 years. Drainage and destruction of apparently favoured shallow, steppe (playa) wetlands in the Konya Basin – Yarma is certainly a defunct area and Hotamış almost equally so – must certainly have contributed to the striking variability in patterns of occurrence and breeding. Nonetheless, it seems clear that available evidence is most suggestive of random fluctuations such as are typical in species with very small and widely separated subpopulations at the range periphery. Winter climate may be a factor constraining any long-standing colonisation and, given the low latitudes at which the species typically breeds and

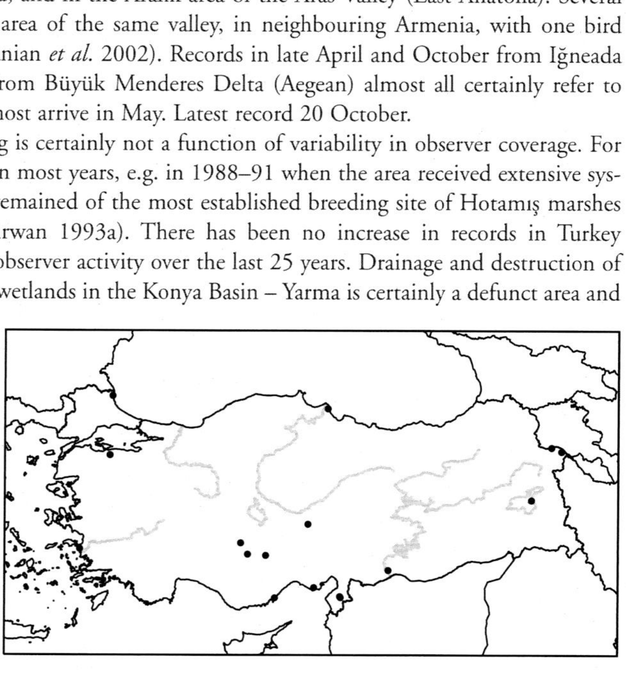

the irregularity and general scarcity of its occurrence in the winter range, opportunistic breeding at sites where environmental conditions are temporarily suitable could be a significant adaptive aspect of the breeding ecology of this poorly studied species. Such factors should be borne in mind when evaluating the suggested trend of recent westward expansion; see Dean *et al.* (1977). This has not been sustained in Turkey, although recent breeding in consecutive seasons in coastal Romania (Kiss & Szabó 2000, *Birding World* 15: 20) and the first records in Bulgaria (Nankinov 2004) do support such interpretation.

Breeding. Bred regularly until 1985 and may still do so, but recent records sporadic; in open marshy areas by fresh or saltwater, and, in Inner Anatolia, typically on wet and partially flooded steppe beside lakes. Nests on open ground, using a shallow hollow, lined with plant material. Eggs: C/3 (2, but unclear if these were complete clutches), C/4 (1); elsewhere, usual clutch four, and breeds from late April. Near Yarma (**Inner Anatolia**), Warncke saw three pairs, and copulation by one of these, on 3 June 1971; on 9 June 1971 he saw 5–6 pairs and found two nests (two and four eggs); on 11 June 1971 there were at least nine pairs and a third egg had been laid in one of the nests; the site had dried out by 23 July 1971, and adults were scattered (Kasparek 1992b). At Hotamış, there were three adults on 26/27 August 1974, a pair and a single bird on 3–4 June 1975, and one bird was very agitated, suggesting breeding; subsequent records of adults were on 10 May 1977 (1), 21 May 1984 (1), 14 June 1985 (13) and 24/25 June 1985 (2); three adults were alarm-calling on 6 July 1985, and breeding was assumed (Kasparek 1992b). Two adults, perhaps breeding, were at Sultan marshes on 9 July 1996. At Karatepe, west of Tuz Gölü, there were six adults, two of which were sitting irregularly on nests in an area of *Carex* sp. on 19 May 2000; next morning there were at least nine adults, and two were incubating steadily, but the nests were not examined (Brugière 2002). In Göksu Delta (**Mediterranean**), Lehman found a nest with three eggs on 27 May 1971, another nest on 29 May 1971, and a recently fledged young was seen 27 July 1971 (Kasparek 1992b). Up to six, in courtship and showing territorial behaviour, were seen from 12 May–11 June 1997. In Çukurova (Mediterranean), a pair and three single birds, which behaved as if breeding, were in flooded areas of the Berdan River on 1 June 1971 (Kasparek 1992b). Display was observed at Birecik (**Southeast Anatolia**) on 18 May 1992. Whilst some former breeding sites have been drained, there are many excellent potential breeding sites in Inner Anatolia, and pairs may easily be overlooked in the vastness of the marsh and steppe.

NORTHERN LAPWING — *Vanellus vanellus*
Kızkuşu

(Du: Kievit; Fr: Vanneau huppé; Ger: Kiebitz; Swe: Tofsvipa)

Subspecies and taxonomy. Monotypic. The genus *Vanellus* has, in the past, frequently been restricted to this species, but most modern commentators have preferred to expand it to incorporate the vast majority of the other long-legged plovers within the subfamily Vanellinae.

Status and distribution. National conservation status: Least Concern (population apparently decreasing). A widespread and very common resident or partial migrant in Inner Anatolia, East Anatolia, locally in smaller numbers in Aegean, Black Sea Coastlands, Marmara, inland parts of Mediterranean and perhaps the eastern part of Southeast Anatolia. Prefers fields, meadows and wetland margins. The largest numbers inhabit Inner Anatolia's wetlands and arable fields near water supplies. More widespread and numerous on passage, when observed in all regions. Flocks form from late July with largest numbers in mid October–late December, prior to the coldest winter weather. Largest known concentrations are of 4,500 at Kızılırmak Delta (Black Sea Coastlands) on 26 December 2004, and 4,350 at Gogenç Gölü (Inner Anatolia) on 22 November 1971. In December–

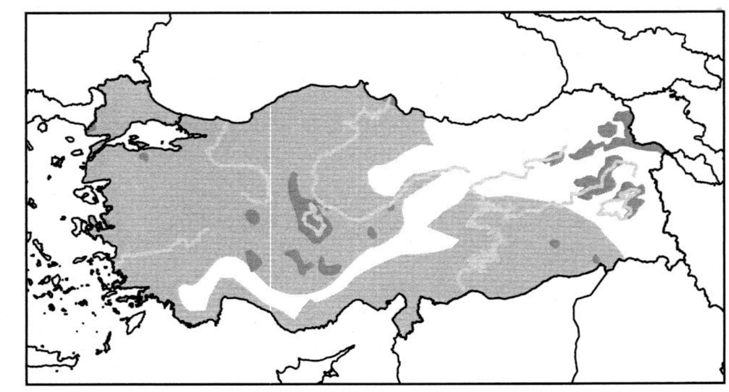

February, recorded in all regions, but most gather at wetlands along the Black Sea, Aegean and Mediterranean coasts, with 4,500 at Kızılırmak Delta (Black Sea Coastland) on 26 December 2004 (OMÜKUŞ) and 4,000 at Gediz Delta (Aegean) on 11 January 2003 (EKGT) being the largest counts during this period, but Çukurova (Mediterranean), Çivril Gölü (Aegean) and Yeşilırmak Delta (Black Sea Coastlands) also support up to 1,000 birds.

Breeding. Mainly in wet or damp grasslands, and beside marshes, lakes and rivers. The nest is a shallow hollow in the ground, lined with plant material. Eggs C/3 (1), C/4 (3); smaller numbers considered incomplete clutches. The earliest record, two broods of young at Çöl Gölü (**Inner Anatolia**) on 11 April 1992, indicates laying of the first egg in mid March, and is much earlier than other records. Other records in this region include seven nests with eggs from 14 April–9 June, and 12 broods of young in May–July. At Manyas Gölü (**Marmara**), territories established on 13 April 1967 (R. F. Porter). In Kızılırmak Delta (**Black Sea Coastlands**), in 1992, most pairs appeared to lay between 20 April and 10 May, with first young observed on 11 May (Hustings & van Dijk 1994). In **East Anatolia**, two nests had eggs on 6 June and 28 June, and young were seen on 30 May 1969, 30 May 1986 and 28 June; two young, at Patnos on 9 May 1986, indicate egg laying about 10 April. Season evidently protracted in Turkey, possibly due to replacement clutches.

RED KNOT *Calidris canutus*
Büyük Kumkuşu

(Du: Kanoet; Fr: Bécasseau maubèche; Ger: Knutt; Swe: Kustsnäppa)

Subspecies. Nominate *canutus* (which winters in western and southern Africa and breeds in northern Siberia and possibly in Yakutia).

Status and distribution. A regular passage migrant that overwinters at suitable coastal wetlands (principally in Aegean and Mediterranean) in small numbers (usually <100) and recorded from all regions. Mathey-Dupraz (1923) considered the species to be a rare passage migrant in the Istanbul area (Marmara) and Kasparek (1990a, 1992a) treated it as a very scarce passage migrant, listing 18 records for the period 1914–1989. There were an additional five acceptable records between 1987 and 1991 (Kirwan & Martins 1994, M. Kasparek *in litt.* 1995) prompting a comprehensive review, with special reference to Turkey, presenting the pattern of records, either published or reliably reported to the author, from 11 countries or islands extending from Egypt and Cyprus east through the Middle East and the Arabian Peninsula to Iran (Kirwan 1996a). The species' status as a very long-distance migrant throughout this region is clearly the key to interpreting the likely significance of the series of records from Turkey. All 41 records from Turkey reported to the author or 'listed in the available literature' were presented. This analysis showed that occurrence in spring is largely during 20 March to 4 June, perhaps peaking in the first half of May, and in late summer/autumn from 21 July to 30 October. Recent systematic counts of most wetlands in the western two-thirds of Turkey, organised by DHKD, confirm wintering in small numbers, typically three to 100, but up to 200 (on 2 February 2002; see Demirci 2002, Balmer & Betton 2003a), mainly at Gediz Delta (Aegean) and in Çukurova (Mediterranean), but 204 were at Kızılırmak Delta (Black Sea Coastlands) on 9 March 2005. Single birds are occasionally reported elsewhere in the southwest in winter. The nominate race, *canutus*, the only one likely to be recorded in Turkey, principally undertakes a loop migration (described as the Great Circle route) following the flyway along the northwest Atlantic maritime fringe to the principal wintering grounds in West Africa and returning through Europe and the East Mediterranean and West Asian flyway west, to southwest Asia in spring. Given this route it is unsurprising that only small numbers are typically recorded in Turkey, with most in spring, and

elsewhere in the eastern Mediterranean. Perhaps under-recorded, as coverage of suitable habitats (primarily coastal estuaries and brackish inland wetlands) at key migration periods remains weak.

SANDERLING *Calidris alba*
Ak Kumkuşu

(Du: Drieteenstrandloper; Fr: Bécasseau sanderling; Ger: Sanderling; Swe: Sandlöpare)

Subspecies and taxonomy: Monotypic. Occasionally accorded a monospecific genus, *Crocerthia.*

Status and distribution. Local passage migrant across the country, staging at coastal wetlands and, more occasionally, at inland lakes. Generally in small numbers (fewer than five), but during peak migration, flocks exceeding 100 are encountered: e.g. 150 at Karataş (Mediterranean), on 25 August 1967, 288 in Çukurova (Mediterranean), on 24 April 1990 (Kivit *et al.* 1994) and 300 there, on 29 April 1997 (M. Hemprich *et al.*). Peak migration in spring is between early May (10th) and early June (6th), with the first definite migrants (*i.e.* at inland localities) noted in the first part of April. In autumn passage occurs from mid August (13th) to early November, with one record from Inner Anatolia as early as 25 July.

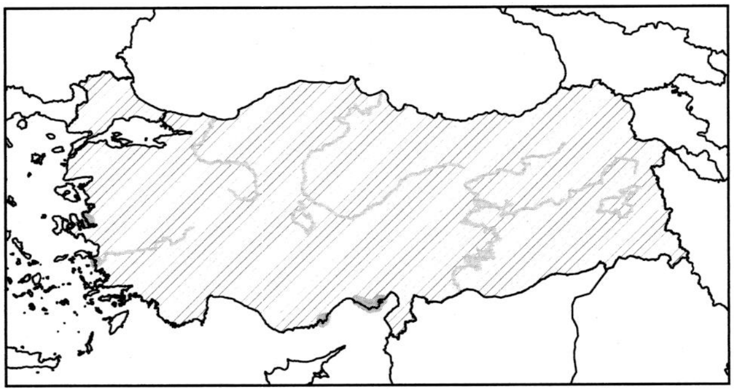

From the latter month until late May, wintering birds gather locally on Mediterranean, Aegean, Marmara and Black Sea coasts. Büyükçekmece (Marmara) and the Tarsus Delta (Mediterranean), with maxima of 130 and *c.*160, respectively, are among the most important wintering areas, with small groups elsewhere, from Iğneada (Marmara) in the northwest to the Göksu Delta (Mediterranean) in the south.

LITTLE STINT *Calidris minuta*
Küçük Kumkuşu

(Du: Kleine Strandloper; Fr: Bécasseau minute; Ger: Zwergstrandläufer; Swe: Småsnäppa)

Subspecies. Monotypic.

Status and distribution. One of the most regular and widespread passage waders, although rather scarce in Southeast Anatolia due to a lack of suitable habitat. Generally found in all types of wetland with muddy fields. Recorded in all months, even in late June in Mediterranean, where autumn migration for most species usually commences one month later than elsewhere. Main spring passage is from early April (8th) to early June, with an exceptional 8,000–10,000 recorded at Ereğli marshes (Inner Anatolia) in mid May (Kasparek 1992a), but most counts merely involve low double-figures. The scale of the passage through large wetland complexes is well illustrated by the *c.*35,000 estimated to have passed through Çukurova (Mediterranean) in spring 1990 (Kivit *et al.* 1994). In autumn, birds arrive from late July and up to 8,000 birds have apparently been recorded at Ereğli marshes in this period (Kasparek 1992a), although figures usually do not exceed low hundreds then. In many other areas, migration peaks in mid

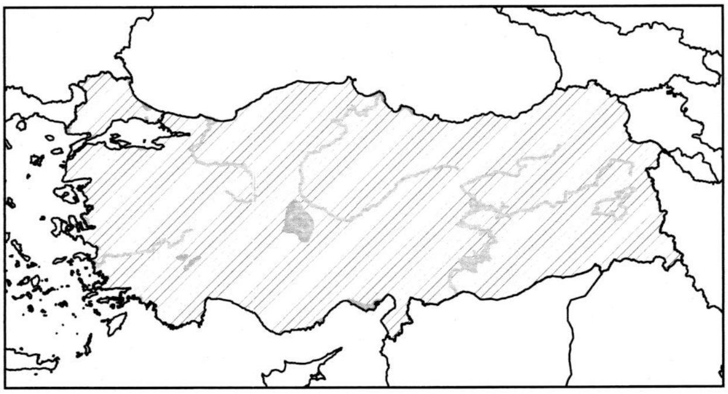

September and continues until early November. Acıgöl (Aegean) and Tuz Gölü (Inner Anatolia) are among those areas hosting more than 1,000 birds in this period.

A regular but local wintering bird throughout the country, except in the east, where the winter status of most waders is unknown. Largest numbers winter in the Mediterranean coastlands, where two-thirds of the total winter population was present during January 1996 (3,149 of a national total of 4,728). Most were at Yumurtalık, where 2,200 were present. Other important wintering areas include Ağyatan Gölü (max. 1,560) and Göksu Delta (max. 1,550), in Mediterranean, and Gediz Delta (max. 1,550) and Acıgöl (Aegean). In the north, important sites are Büyükçekmece and Kocaçay Delta (Marmara), Kızılırmak Delta (max. 881), in Black Sea Coastlands, and Tuz Gölü (max. 1,000), in Inner Anatolia.

TEMMINCK'S STINT *Calidris temminckii*
Sarı Bacaklı Kumkuşu

(Du: Temmincks Strandloper; Fr: Bécasseau de Temminck; Ger: Temminckstrandläufer; Swe: Mosnäppa)

Subspecies. Monotypic.

Status and distribution. A regular but local passage migrant to wetlands throughout the country, generally in small numbers but locally more common. Despite Anatolia being on a main migration route, the number of records is small (e.g. just 103 observed in Çukurova, Mediterranean, in spring 1990), and the species is apparently relatively overlooked, in part perhaps because of its distinct preference for the more vegetated parts of wetlands. Spring passage commences in late March and continues until late May (25th). The earliest records are from the Mediterranean coast and southern Inner Anatolia, with the largest concentration being 100 at Göksu Delta (Mediterranean), on 21–25 April 1973, although there is unsubstantiated report of 300 at Kulu Golu (Inner Anatolia), on an unknown date (Ertan *et al.* 1989). Mainly recorded in Inner Anatolia at marshy wetlands and generally in flocks smaller than ten, with a record of 50 at Eşmekaya (22 May) being noteworthy. Autumn passage is from late July (21st) until late September, with relatively few in October (latest 20th).

Occasionally recorded in winter in Mediterranean, Aegean and Southeast Anatolia, always in small numbers. Four at Karamık marshes (Mediterranean) on 14 December 1970 were the furthest inland. The most important wintering site is Göksu Delta, where up to 50 have been recorded. Other wintering sites include Birecik (Southeast Anatolia) and Gediz Delta (Aegean), where the species is much less regular and generally occurs only in small numbers.

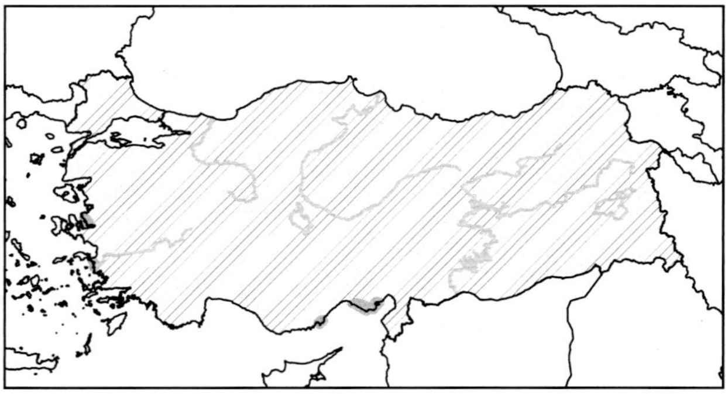

WHITE-RUMPED SANDPIPER *Calidris fuscicollis*
Ak Sokumlu Kumkuşu

(Du: Bonapartes Strandloper; Fr: Bécasseau de Bonaparte; Ger: Weißbürzel-Strandläufer; Swe: Vitgumpsnäppa)

Subspecies. Monotypic.

Status and distribution. Vagrant with one well-documented record: one at the Göksu Delta (Mediterranean) on 17 May 1996 (Browne 1997), the first record in the Middle East.

PECTORAL SANDPIPER *Calidris melanotos*
Çizgili Kumkuşu

(Du: Gestreepte Strandloper; Fr: Bécasseau tacheté; Ger: Graubrust-Strandläufer; Swe: Tuvsnäppa)

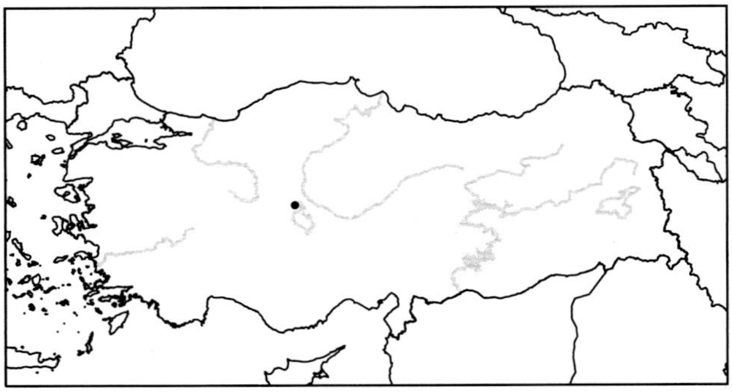

Subspecies. Monotypic.

Status and distribution. Vagrant with a single, documented record: a male photographed at Kulu Gölü (Inner Anatolia) on 9–10 May 2005 (W. Remmelzwaal), which is one of the few available records from the Middle East.

CURLEW SANDPIPER *Calidris ferruginea*
Kızıl Kumkuşu

(Du: Krombekstrandloper; Fr: Bécasseau cocorli; Ger: Sichelstrandläufer; Swe: Spovsnäppa)

Subspecies. Monotypic. Has been placed in its own genus, *Eriola* (see Voous 1992).

Status and distribution. Regular passage migrant in relatively small numbers across the entire country. Generally found in small flocks or as a few individuals, but much larger numbers are recorded at coastal wetlands, with 1,000 at Camalti Tuzlasi on 18 May 1993 and 478 at Gediz Delta (Aegean) on 13 May 2001. Almost 3,000 moved through Çukurova (Mediterranean) in spring 1987, mostly in the second week of May (van der Have *et al.* 1988) and 6,029, including 2,808 on 8 May, in spring 1990 (Kivit *et al.* 1994). Peak passage at Turkish wetlands occurs in late April–late May (extreme dates: 26 March and 17 June). Southbound passage occurs from mid July (16th), and mid September, with the greatest concentrations (adults) usually in the second half of August, e.g. 200 at Tuzla Gölü, Çukurova, on 23 August 1988, with a second peak, involving first-year birds, in mid–late September. Latest migrants in Mediterranean, on 10 October, and Inner Anatolia, 6 November, but recent midwinter counts have revealed the presence of small flocks at coastal wetlands in the extreme southwest of the country, e.g. Büyük Menderes Delta (Aegean), and, even less regularly, in the south, e.g. Göksu Delta (Mediterranean), where the species may prove to be regular at this season.

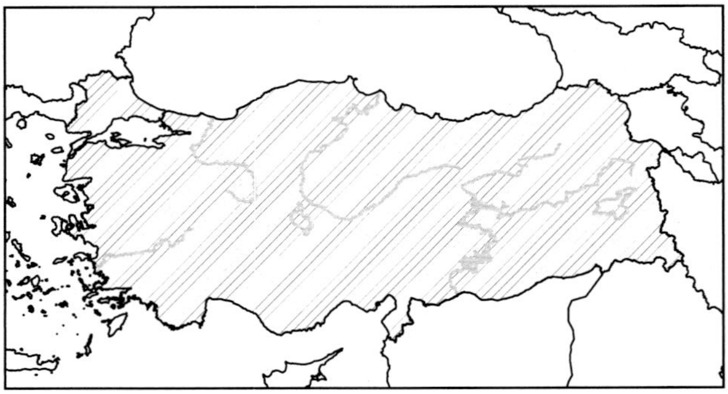

[PURPLE SANDPIPER *Calidris maritima*]
Mor Kumkuşu

(Du: Paarse Strandloper; Fr: Bécasseau violet; Ger: Meerstrandläufer; Swe: Skärsnäppa)

Subspecies and taxonomy. Monotypic. Has been placed in its own genus, *Arquatella* (see Voous 1992).

Species not admitted to the Turkish list (Kirwan *et al.* 1999). Listed by Ergene (1945) without supporting details. There are no Middle Eastern records (Porter *et al.* 1996) and, despite records in Greece (most recently in 1966 and 1980; Handrinos & Akriotis 1997), Bulgaria (in September 1993; Nankinov *et al.* 1988), Malta, Morocco and even as far south as Mauritania (Snow & Perrins 1998), the species appears to be a reasonably unlikely vagrant to Turkey.

DUNLIN — *Calidris alpina*
Kara Karınlı Kumkuşu

(Du: Bonte Strandloper; Fr: Bécasseau variable; Ger: Alpenstrandläufer; Swe: Kärrsnäppa)

Subspecies and taxonomy. Presumably the majority relate to the nominate form, with much smaller numbers of *schinzii* (according to ringing studies). Even the eastern form, *sakhalina*, which is considered to be possibly accidental in the Western Palearctic (Cramp & Simmons 1983), may have occurred in Turkey (for further discussion, see van der Have *et al.* 1988, Kivit *et al.* 1994). However, geographical variation among all three above-mentioned (and several other named) forms of Dunlin is principally clinal and obscured by individual variation. Has been placed in its own genus, *Pelidna* (see Voous 1992).

Status and distribution. A widespread but only locally common passage migrant to wetlands across the country. Peak migration periods are early April–late May and late July–mid November, but migrants appear as early as March (11th) at freshwater lakes of Inner Anatolia, where the species is unknown to winter, and from mid July in Black Sea Coastlands. Generally forms loose flocks, but Kasparek (1992a) mentioned a May record of 1,500 birds from Acıgöl (Aegean), whilst van der Have *et al.* (1988) noted 10,539 between 4 to 15 April 1987 in Çukurova (Mediterranean) and Kivit *et al.* (1990) a max. 3,688 on 10 April 1990 during a survey of the same region. The largest autumn total appears to the *c.*1,000 in Çukurova on 24 September 2004 (H. Schmaljohann *et al.*).

Locally numerous winter visitor over the western two-thirds of the country, mainly at coastal wetlands in Mediterranean and Aegean, and more occasionally, sometimes in 100s, at brackish lakes inland, e.g. Tuz Gölü (Inner Anatolia) and Burdur Gölü (Aegean), with small numbers even found along rivers in Southeast Anatolia at this season and an early December record from East Anatolia (Van Gölü, in 1969). Large flocks form in late November and are maintained until early February, with currently the most important wintering area being Gediz Delta (Aegean), where 3,649 were present on 12 January 2003 and more than 5,000 on 2 February 2003. Büyük Menderes Delta (Aegean) and Çukurova (Mediterranean) are also important with totals exceeding 2,500, exceptionally up to 5,000 in the past. Numbers inland are generally insignificant, and the overall wintering population is probably between 10,000 and 16,000.

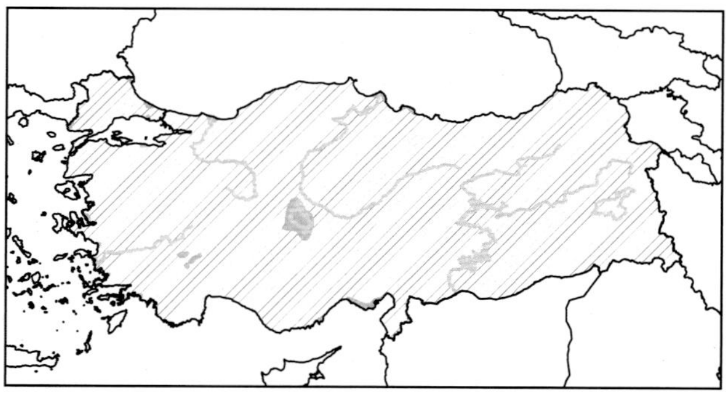

BROAD-BILLED SANDPIPER — *Limicola falcinellus*
Sürmeli Kumkuşu

(Du: Breedbekstrandloper; Fr: Bécasseau falcinelle; Ger: Sumpfläufer; Swe: Myrsnäppa)

Subspecies. Nominate.

Status and distribution. A regular, uncommon to fairly common, passage migrant to wetlands and coastal areas throughout the country, which is generally more frequent in central and eastern parts of Turkey, particularly the Van Gölü basin (East Anatolia) in autumn, e.g. a total of up to 206 at Van marshes between 3 August and 11 September 1986 (Martins 1989), 80+ at Bendimahi Delta on 16 August 1989, and over 100 at several sites in this area in late August 1990 (Kirwan & Martins 1994). Passage in the extreme west of the country typically involves very small numbers, and there are very few available records from Southeast Anatolia. Maxima from central and southern areas include: a total of 123 in Çukurova deltas (Mediterranean) between 24 April and 20 May 1987 (van der Have *et al.* 1988), 134 in this area between late April and the end of May 1990 (Kivit *et al.* 1994), 73 at Ağyatan Gölü in the same area on 30 April 1997 (M. Hemprich *et al.*), 110 at Tuz Gölü in the same area on 8 May 2004 (G. Pettersson *et al.*), 150 at Yumurtalık on 9 May 2005 (Balmer & Betton

2005b), and a total of up to 218 in Kızılırmak Delta (Black Sea Coastlands) during 10 May to 7 June 1992 (Hustings & van Dijk 1994). In spring noted from 17 April (once 11 March) to 29 June (with the peak in very late April and May, especially the middle two weeks), and in autumn recorded between 10 July and 27 September (with the majority in late August/early September). Passage almost certainly continues into early October, but there are no available records from this month. A group of six was seen at the Gediz Delta (Aegean) on 27 January 2008 (O. Döndüren *et al.*), the only winter record from Turkey. Groups rarely consist of more than ten, even in Inner Anatolia and the major south-coast deltas in autumn. Comparatively few records were available prior to the advent of modern, systematic recording.

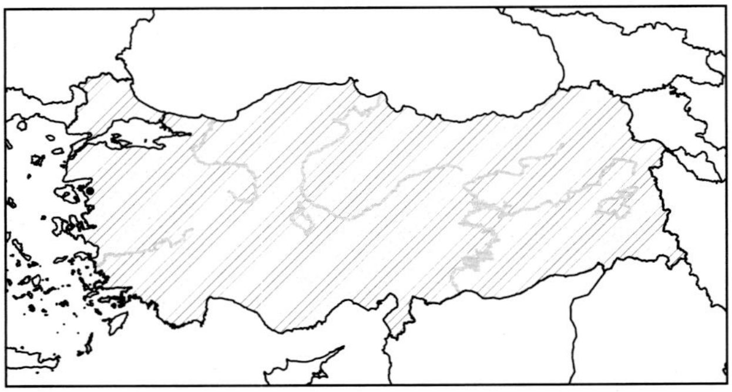

• = winter record

[BUFF-BREASTED SANDPIPER *Tryngites subruficollis*]
Çayırkoşarı

(Du: Blonde Ruiter; Fr: Bécasseau rousset; Ger: Grasläufer; Swe: Prärielöpare)

Subspecies. Monotypic.

Species not admitted to the Turkish list (Kirwan *et al.* 1999). Kumerloeve (1975a) regarded the report (Wahby 1930) of one collected at Küçükçekmece lagoon, just west of Istanbul, in January 1927, and retained in the University of Istanbul museum collection as probably erroneous. Wahby claimed some experience of American wader species, knowledge of the appropriate literature and experience through examination of skins. However, as we have not seen the specimen and other claims made by Wahby appear fanciful, it seems prudent to list the species as unconfirmed. There are no other Middle Eastern claims of the species (Porter *et al.* 1996).

RUFF *Philomachus pugnax*
Döğüşkenkuş

(Du: Kemphaan; Fr: Combattant varié; Ger: Kampfläufer; Swe: Brushane)

Subspecies. Monotypic.

Status and distribution. The commonest migrant wader, occurring in all wetlands and on farmland throughout the country. Migration commences in February and continues until early June, with most between the second half of March and first three weeks of April. Numbers are generally considerable and totals over 1,000 are not uncommon at many sites, e.g. Ereğli, Sultan and Balıkdamı marshes (Inner Anatolia), Kızılırmak Delta (Black Sea Coastlands), and Göksu Delta (Mediterranean). Perhaps the largest numbers are recorded in Inner Anatolia, where there were 10,000 at Sultan marshes on 19 April 1994, and 6,856 at Hotamış marshes on 20 February 1974,

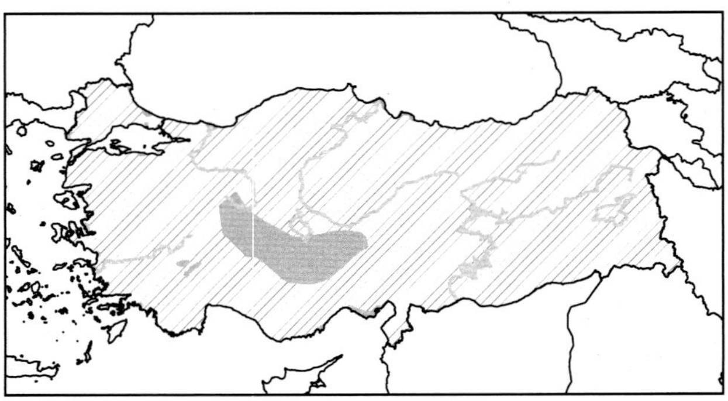

but Kivit *et al.* (1994) recorded a total of over 34,000 passing through Çukurova (Mediterranean) in spring 1990. As Kasparek (1992a) noted, some of the records from late June or early July suggest over-summering.

Also, a not uncommon but irregular winter visitor to wetlands in Black Sea Coastlands, Aegean, Mediterranean, Marmara and Inner Anatolia. The wintering population is quite small in comparison to migrant totals, with most in the southern Konya Basin (max. 320 at Hotamış marshes), on the Mediterranean coast, the lakes of inner and coastal Aegean, Büyükçekmece (Marmara) and Kızılırmak Delta. Larger numbers occur in November and Februrary, but these probably represent pre- or post-migration flocks.

JACK SNIPE — *Lymnocryptes minimus*
Küçük Suçulluğu

(Du: Bokje; Fr: Bécassine sourde; Ger: Zwergschnepfe; Swe: Dvärgbeckasin)

Subspecies. Monotypic.

Status and distribution. Very local and uncommon passage migrant, and winter visitor, to wetlands across the western half of the country, with migrants also recorded in the eastern Black Sea Coastlands (Kumerloeve 1967a, Faldborg 1994). The relatively small number of observations suggests that it is either severely under-recorded or genuinely scarce, although the former seems more likely the case, given that intensive surveys of suitable areas have occasionally yielded more frequent sightings, e.g. 31 were noted in Çukurova (Mediterranean) in March 1990 (Kivit *et al.* 1994) and 15 in Kızılırmak Delta (Black Sea Coastlands) in early spring 1992 (Hustings & van Dijk 1994), and the species was considered to be a fairly regular passage migrant at Kulu Golu (Inner Anatolia) in 2001 and 2002 (Richardson 2003); at both of the two latter sites, it had not been recorded previously despite regular previous ornithological coverage. Occurs in coastal areas, including saltmarshes, as well as inland, in all manner of wetlands and other moist areas, between September (earliest 20th) and late April (last 30th). Virtually always found solitarily, but occasionally in loose groups of a few individuals (max. 12).

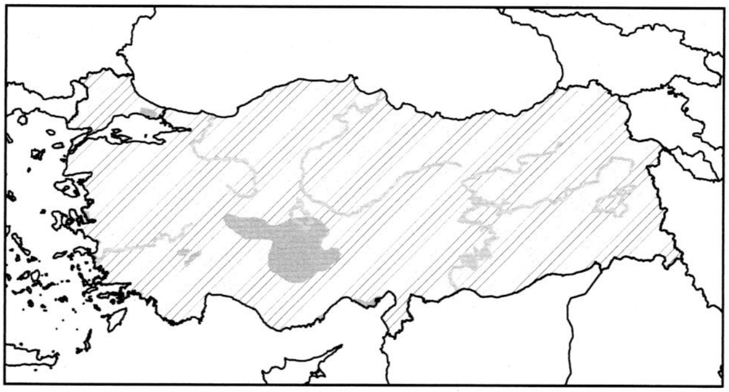

COMMON SNIPE — *Gallinago gallinago*
Suçulluğu

(Du: Watersnip; Fr: Bécassine des marais; Ger: Bekassine; Swe: Enkelbeckasin)

Subspecies and taxonomy. Nominate *gallinago*. New World *delicata* is increasingly recognised as being specifically distinct, under the name Wilson's Snipe (see, e.g., Sangster *et al.* 1999, Banks *et al.* 2002), especially as the two apparently breed sympathetically in the same marshes in western Alaska (K. J. Zimmer pers. comm.).

Status and distribution. National conservation status: Critical (population apparently decreasing). Status imprecisely known. There are several records indicative of breeding from Inner and East Anatolia and Black Sea Coastlands (see Kasparek 1992a, Marlow *et al.* 2001), with the most precise data involving drumming individuals between Eskil and Cihanbeyli (Inner Anatolia) on 22–23 June 1998. Given further records of drummers at Kozanlı Gölü, Inner Anatolia (I. Richardson) and considering the poor coverage of wetlands in the northeast, it seems probable that the species is part of the Turkish breeding avifauna, but direct proof is still lacking.

A very common passage migrant and a regular but less numerous winter visitor over most of the country, albeit rather local in Southeast Anatolia. It is difficult to determine arrival and departure dates in spring, but peak numbers are apparently in March and the first half of April, with a maximum of over 1,000 at Kızılırmak Delta (Black Sea Coastlands). Numbers thereafter decline rapidly in most regions, except East Anatolia, where birds do not depart until late May. Returning migrants appear in the first half of July (10th) and passage continues until the first half of November, after which the species is only recorded on its wintering grounds. The largest number at this season was at the former Amik Gölü (Mediterranean), where 300+ were observed and perhaps several thousand were present on 9–10 October 1968. Coastal wetlands hold the majority of winterers, although there are many fewer than during passage, with 200 at Tarsus (Mediterranean) and 100 at Gönen (Marmara) among the most significant totals.

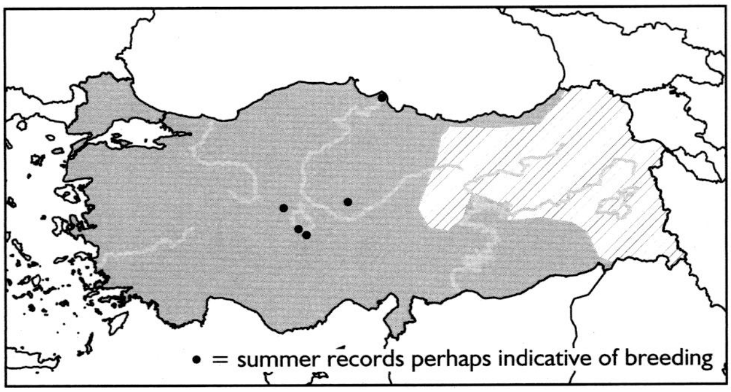

Breeding. Frequently seen in the breeding season in several regions; displaying birds were seen in May 1970 in **East Anatolia**, and even in June, in Kızılırmak Delta (**Black Sea Coastlands**) in 1992 (Hustings & van Dijk 1994), as well as in the Konya Basin (**Inner Anatolia**) in 1998 (Marlow *et al.* 2001) and Palas Gölü (Inner Anatolia) in 2002 (P. Hope), but there is no further evidence of nesting from anywhere in Turkey.

GREAT SNIPE *Gallinago media*
Büyük Suçulluğu

(Du: Poelsnip; Fr: Bécassine double; Ger: Doppelschnepfe; Swe: Dubbelbeckasin)

Subspecies. Monotypic.

Status and distribution. Conservation status: Near Threatened (BirdLife International 2000, 2004). A scarce passage migrant to all regions of the country, but scarcest in the extreme west. Uses farmland as well as wetland margins. Also recorded in winter in Inner Anatolia (one at Akşehir Gölü in mid-January 1969) and on the south coast (two at the Tarsus Delta in late January 1970). The numbers recorded on migration through Turkey have almost certainly declined over the course of the past century, in line with the species' overall reduction in numbers and range, as evidenced by the remarks of Mathey-Dupraz (1923), who reported that it was a regular passage migrant at the Bosphorus; Schrader (1891), who noted the species as a winter visitor around Aydin (Aegean) and Mersin (Mediterranean), which occurred in virtually all suitable areas in periods of harsh weather (thus perhaps suggesting confusion with Common Snipe *G. gallinago*), and Sandwith (1856) at Erzurum, where he remarked that it appeared in large flocks in May and small numbers in autumn.

Kirwan (1992a) listed all available records from Turkey prior to 1990, including full details for all those recorded post-1966. Of the 73 records (involving 140 individuals) since the advent of modern recording, almost 50% (and over 60% of individuals) were at just four localities: the Çukurova and Göksu deltas (Mediterranean) and the South Van marshes and Bulanık/Balatos region of the Murat Valley (East Anatolia). Taking into account subsequently published records (of which we have seen *c.*60), spring migration covers the period 2 March to 8 June, with the bulk in May, and in autumn Great Snipe has been noted between 2 August and 8 October (perhaps 24th), with peak

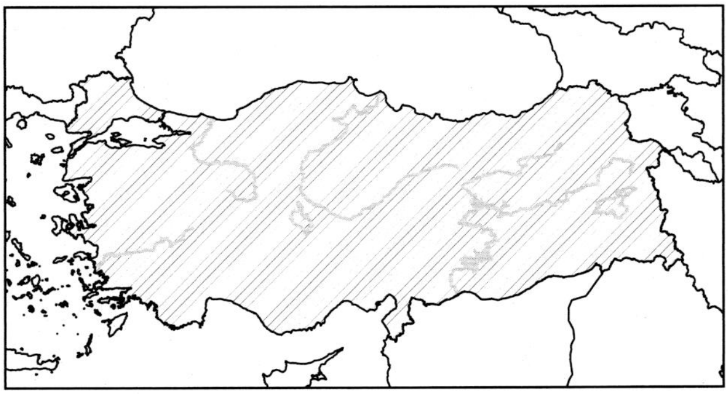

numbers in late September/October. Occasionally recorded in small groups of up to nine individuals, and once a total of 30 (at Van marshes in late September 1982).

EURASIAN WOODCOCK *Scolopax rusticola*
Çulluk

(Du: Houtsnip; Fr: Bécasse des bois; Ger: Waldschnepfe; Swe: Morkulla)

Subspecies. Monotypic.

Status and distribution. National conservation status: Data Deficient. May occasionally breed in extreme northern Turkey, see below. Otherwise, a widespread but rather uncommon passage migrant and winter visitor across much of Turkey, with most in the north and very few reaching the southeast. Most frequently recorded between early September (5th) and late March (29th), but few are present on the wintering grounds prior to mid November. Occasionally recorded at passage sites in montane valleys of the northeast, on Uludağ (Marmara) and in the Taurus (Mediterranean), and spring migration may persist until the end of the first week of April.

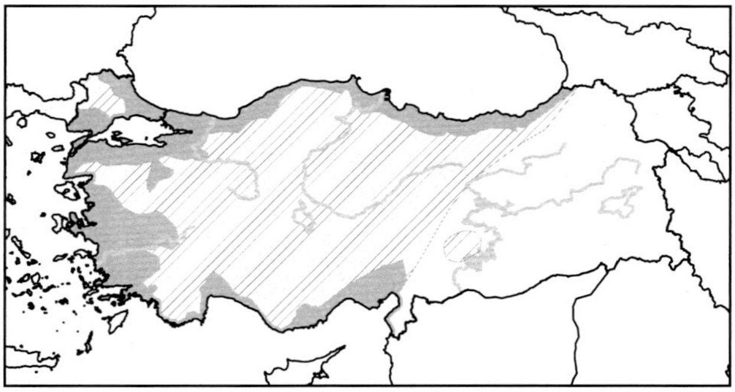

The most important wintering areas are in Black Sea Coastlands and Marmara, where the species is hunted heavily in wet grasslands and along streams. Also occurs at this season in Inner Anatolia, Aegean and Mediterranean in smaller numbers, and recorded once near Van Gölü (East Anatolia) in December.

Breeding. Four records from forested areas of the Kızılırmak Delta (**Black Sea Coastlands**) in April 1992, including one of roding on 17 April (GMK; Hustings & van Dijk 1994). Although the latter authors dismissed the possibility of breeding, due to the lack of subsequent observations during the same season, roding flights are strongly indicative of attempted breeding.

BLACK-TAILED GODWIT *Limosa limosa*
Çamurçulluğu

(Du: Grutto; Fr: Barge à queue noire; Ger: Uferschnepfe; Swe: Rödspov)

Subspecies. Nominate *limosa*.

Status and distribution. Conservation status: Near Threatened (BirdLife International 2008). Common passage migrant throughout wetland areas of Turkey, with the largest spring flocks recorded at Tuz Gölü (Inner Anatolia) in March 1970 (1,500) and Çukurova (Mediterranean) in late March 1990 (3,725). Spring passage commences in February and most have passed through by late May, but small numbers frequently oversummer in Inner Anatolia and in the north, and there is a report of 2–3 pairs considered to be almost certainly nesting, from the Ağrı

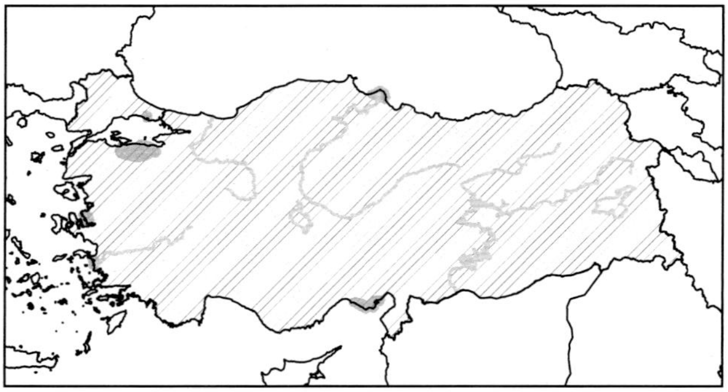

area (East Anatolia), in April–May 1968 (OST 1969). From July, flocks begin to build up again, with 1,030 at Meriç Delta (Marmara) on 12 July 1997, 741 at Kızılırmak Delta (Black Sea Coastlands) on 3 September 2005 (OMÜKUŞ), and 1,200 at Ereğli marshes (Inner Anatolia) on 23 September 1974, being particularly noteworthy. Generally moves through the eastern third of the country in smaller numbers at both seasons.

In winter, considerable numbers may remain in coastal areas, principally in the south and southwest (e.g. 800 at several sites in January 1999) and in mild winters surprisingly large numbers even congregate at inland wetlands, e.g. 1,080 at Çivril Gölü (Aegean) on 20 December 1969.

BAR-TAILED GODWIT *Limosa lapponica*
Kıyı Çamurçulluğu

(Du: Rosse Grutto; Fr: Barge rousse; Ger: Pfuhlschnepfe; Swe: Myrspov)

Subspecies. Nominate, although van der Have *et al.* (1988) speculated that some might belong to one of the longer-billed eastern races, presumably *menzbieri* (though this form is only weakly differentiated from *baueri* of even further east and North America; see del Hoyo *et al.* 1996).

Status and distribution. A scarce passage migrant through wetlands and coastal areas at both seasons across the whole country, with fewest recorded in the eastern third due to a relative lack of suitable habitat. Most records are from the major deltas, with many fewer from the large Inner Anatolian wetlands. No records from Southeast Anatolia. In spring noted from 11 March to 8 June, with most in the second to fourth weeks of April, and in autumn between 1 August and 15 October, with the majority in late August and the first half of September. Maxima: 20 at Büyükçekmece, Istanbul, in April 1976, 20+ at Kızılırmak Delta (Black Sea Coastlands) on 29 April 1998, and totals of 14 and 13 in the Çukurova deltas (Mediterranean) in May 1987 and May 1991; other records all involve single birds or small flocks of up to ten individuals.

Also recorded in winter on the Aegean coast, with single records at the Büyük Menderes Delta in early February in the 1880s (Schrader 1891), in January 1970 and January 1973 (OST 1975), an exceptional flock of 28 at Akyatan Gölü, Çukurova (Mediterranean) on 1 February 2007 (Balmer & Betton 2008), and ten at the Gediz Delta on 20 December 2003 and two there on 18 December 2004 (Balmer & Betton 2005a). As with several other of the rarer shorebirds, there were very few records of this species prior to the advent of modern recording in 1966. It is interesting to note the strong degree of similarity in passage dates recorded in Bulgaria: 1 March to 18 May and 7 August to 22 October, but there are many more records in the latter, though the overwhelming majority (97%) of observations are from one locality, Atanassovsko Lake, where flocks of up to 150 birds have been recorded (Nankinov 1999b).

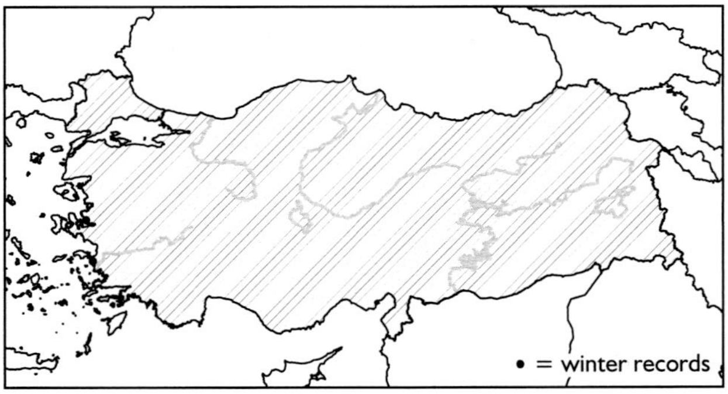

WHIMBREL *Numenius phaeopus*
Sürmeli Kervançulluğu

(Du: Regenwulp; Fr: Courlis corlieu; Ger: Regenbrachvogel; Swe: Småspov)

Subspecies. Nominate.

Status and distribution. Regular passage migrant in relatively small numbers across the entire country, although no records appear to be available from Southeast Anatolia. The majority appears to move through the central third of the country, using the important south- and north-coast deltas and the large playa lakes of Inner Anatolia (Kasparek 1990b, 1992a), although to what extent this spatial pattern may be equally reflective of observer activity is unclear. Spring passage occupies the period 20 March to 7 June, with most in April (Kasparek 1990b identified the median date as 23 April). Autumn migration is prolonged, lasting from 7 July to 29 October (10 November; see below), with the peak in August/September (median date 29 August). One mid-June record (16th) from the south coast. Most records involve single birds or small groups of fewer than five individuals, although there are both historical and modern-day reports of groups of 20–30 in the eastern Black Sea Coastlands, Aegean and Mediterranean regions. In the present-day, a total of 54 moved through the Çukurova (Mediterranean) between mid March and 27 April 1990 (Kivit *et al.* 1994) and 70 through Kızılırmak Delta (Black Sea Coastlands) from 29 March to 7 June 1992 (Hustings & van Dijk 1994), whilst the following single-day totals are significant: 20 in the Göksu Delta (Mediterranean) on 25 September 1981 and 40 at the Büyük Menderes Delta (Aegean) on 7–10 November 1983 (Kasparek 1990b). As much passage is nocturnal, the species is almost certainly under-recorded in Turkey. Two winter records: single birds in the first week of January (year and locality unstated; Kasparek 1990b) and at Tuzla (Aegean) on 26 January 2002 (Demirci 2002), although Krüper & Hartlaub (1875) noted the species' occurrence at Izmir (Aegean) at this season; Kasparek (*op. cit.*) questioned whether this statement may have reflected confusion with Slender-billed Curlew *N. tenuirostris.*

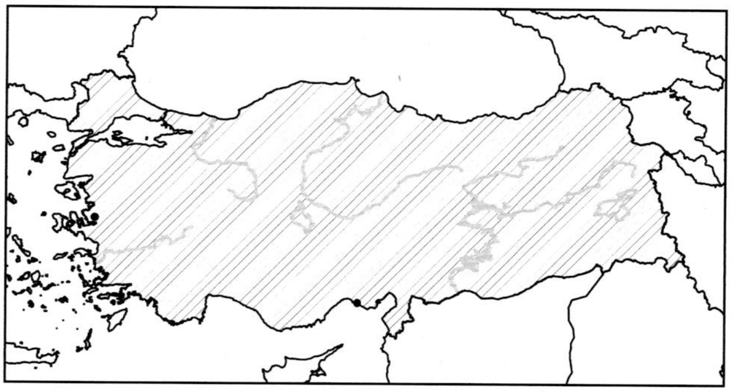

SLENDER-BILLED CURLEW *Numenius tenuirostris*
İnce Gagalı Kervançulluğu

(Du: Dunbekwulp; Fr: Courlis à bec grêle; Ger: Dünnschnabel-Brachvogel; Swe: Smalnäbbad spov)

Subspecies. Monotypic.

Status and distribution. Conservation status: Critically Endangered (BirdLife International 2000, 2004). An extremely rare passage migrant, last certainly recorded in 1986. Records span the period 10 July–12 May, with all but one in the western two-thirds of the country and the majority in the well-watched Mediterranean coast deltas or playa lakes of Inner Anatolia. Most reports are pre-1975. The species is seemingly heading inexorably for extinction (Gretton *et al.* 2002). The following records are available from Turkey (all from Gretton 1991 or the BirdLife International database, unless otherwise stated): two specimens presumably taken in the environs of Istanbul in the late 19th or early 20th century (Kirwan 1997a), an adult female taken at Istanbul in December 1892 and another female taken at Makriköy, 10km west of Istanbul, on 11 September 1895 (both from the Alléon collection, and currently held in National Museum of Natural History, in Sofia, Bulgaria; Boev 2003), one at Hoyran Gölü (Mediterranean) on 14 January 1967, two at Mogan Gölü (Inner Anatolia) on the same date, two there on 19 March 1967 and another in the same place on 23 March 1967, two at Balıkdamı (Inner Anatolia) on 26 March 1967, one at Sodalı Gölü (East Anatolia) on 12 August 1967, one at Amik Gölü (Mediterranean) on 10 October 1968, three near Tekirdağ (Marmara) on 5 April 1969, two moving west past Side in Antalya province (Mediterranean) on 15 April 1969, one at Seyfe Gölü (Inner Anatolia) on 12 December 1969, one near Yarma (Inner Anatolia) on 28 April 1970, one at Eber Gölü, on 12 May 1970, one at Çivril Gölü (Mediterranean) on

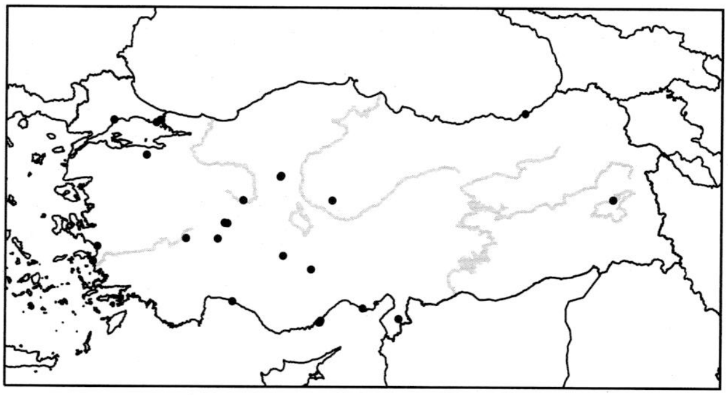

17 January 1971, one near Rize (Black Sea Coastlands) on 25 August 1971, one at Göksu Delta (Mediterranean) on 28 August 1973, two at Uluabat Gölü (Marmara) on 30 September 1973, one at Küçük Menderes Delta (Aegean) on 23 September 1979, single birds at Akyatan Gölü, Çukurova (Mediterranean) on 5 January 1982 and 4 March 1985, four at Göksu Delta on 9 March 1985, one at Ereğli marshes (Inner Anatolia) on 6 April 1986, one at Göksu Delta on 10 July 1986 and one at Büyük Menderes Delta (Aegean) on 26 December 1986. Some documentation exists for all of these reports. Gretton (1991) further lists three records, involving four individuals, from Büyükçekmece, just west of Istanbul, in 1946, 1947 and 1951, and one from the Meriç Delta (Thrace) involving two on 27 February 1959 (Kumerloeve 1970a), which are unacceptable (Kumerloeve 1961, Kirwan 1997a). Additionally, records of single birds in Çukurova on 30 April 1986, at the Dalyan Delta (Aegean) on 22 August 1990 and at Kulu Gölü (Inner Anatolia) on 10 May 1993 cannot be regarded as certain, while one at Tuzla, just southeast of Istanbul (Marmara) on 15 September 1985 was proven to be an *orientalis* Curlew *N. arquata* by Konrad (1994).

EURASIAN CURLEW — *Numenius arquata*
Kervançulluğu

(Du: Wulp / Oostelijke Wulp; Fr: Courlis cendré; Ger: Großer Brachvogel; Swe: Storspov)

Subspecies. Most are presumably nominate *arquata*, but *orientalis* has been recorded on a number of occasions (see e.g. Konrad 1994).

Status and distribution. Widespread and fairly common passage migrant to wetlands throughout the country, and recorded locally in considerable numbers during winter in the southwest and south. Passage commences mid March with peak numbers observed in the first half of April, although even in this period most groups are rather small, *e.g.* in Çukurova (Mediterranean), van der Have *et al.* (1988) recorded a maximum of 27 in spring 1987, and Kivit *et al.* (1994) a maximum of 42 in spring 1990. Passage persists reasonably strongly until at least early May, and there are irregular summer records from several regions in the western two-thirds of the country. In early July, autumn migration commences and, compared to that in spring, is relatively rapid, although passage persists until at least mid November in some years. Meriç Delta (Marmara) is among the most important stopover sites at this period, with regular flocks of 250 recorded.

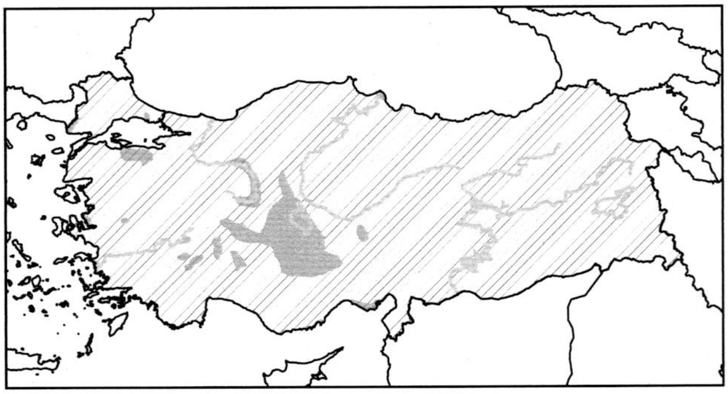

Numbers in winter are concentrated in the southwest and south, although moderate numbers are noted elsewhere in coastal areas in the western half of the country, with largest numbers at Büyük Menderes Delta (e.g. 1,254 in January 1971). More occasionally recorded inland at this season. The midwinter counts of 1999 and 2005 produced totals of 1,362 and 1,688 birds, respectively.

SPOTTED REDSHANK — *Tringa erythropus*
Kara Kızılbacak

(Du: Zwarte Ruiter; Fr: Chevalier arlequin; Ger: Dunkler Wasserläufer; Swe: Svartsnäppa)

Subspecies. Monotypic.

Status and distribution. Common and widespread passage migrant throughout the country, and winters in much smaller numbers, principally in coastal areas through the western two-thirds of the country. Main migration periods are mid March–early June and early July–early November, with peaks in the second week of March and third week of September. Generally in small groups of fewer than ten birds, but exceptional flocks involving

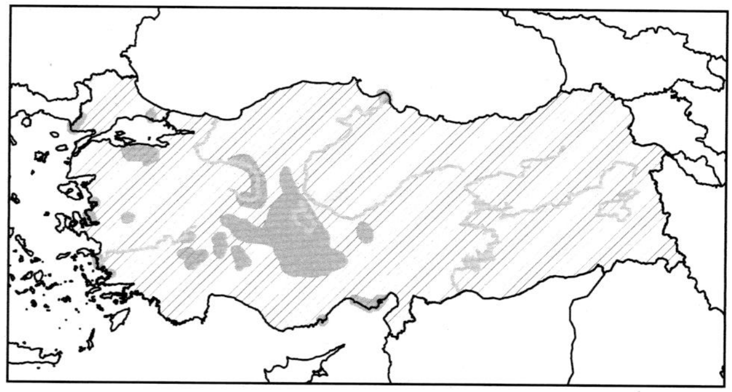

several hundreds recorded in late autumn in Inner Anatolia, e.g. 1,300 at Sultan marshes on 21 November 1969 and 1,000 at Hotamış on 23 September 1974, and up to 605 on 10 April 1990 at Çukurova (Mediterranean) (Kivit *et al.* 1994).

Winters generally in relatively small numbers, occasionally up to *c.*300 at a single site, but in 1999 just 53 were recorded during the midwinter count. Even recorded inland in East Anatolia as late as mid December. Individuals occasionally oversummer inland in Anatolia.

COMMON REDSHANK *Tringa totanus*
Kızılbacak

(Du: Tureluur; Fr: Chevalier gambette; Ger: Rotschenkel; Swe: Rödbena)

Subspecies. Nominate *totanus* and presumably also the weakly differentiated *ussuriensis* (see Roselaar *in* Cramp & Simmons 1983, Kivit *et al.* 1994). Subsequently, Roselaar (1995) elected to restrict *totanus* to northern Europe, and considered that those breeding in Turkey were possibly either *britannica* (considered a synonym of the nominate in *BWP*) or an unknown form. Wadley (1951) tentatively claimed to have collected a male of the race *eurhina* (which breeds in the Pamirs, in Central Asia, and winters in the Indian subcontinent) near Ankara, in November 1945, on the basis of its size, and Kumerloeve (1967a, 1969a) also collected specimens, in eastern Turkey, that he attributed to *eurhinus*, but also admitted the difficulties inherent to diagnosing this form. GMK was unable to locate the Wadley specimen at the Natural History Museum (Tring) in July 2003, but two other specimens from Turkey: one taken on 5 February 1869 (locality and collector not noted) and the other (a male) taken by Danford on 13 May 1876, accord with the rather weakly defined *eurhina*.

Status and distribution. National conservation status: Near Threatened (population apparently decreasing). A regular and common summer visitor to wetlands of East and Inner Anatolia, with smaller numbers in Black Sea Coastlands and Marmara (e.g. Kocaçay Delta, Büyükçekmece) and Aegean. Tersakan and Bolluk lakes, Akşehir Gölü and Sultan marshes (Inner Anatolia), and the wetlands around Van Gölü and Erzurum are the most important breeding areas (e.g. 400+ at Erçek Gölü in mid June). Once important, most of the wet and marshy areas near Yüksekova (East Anatolia) and Ereğli marshes (Inner Anatolia) are currently much less so, due to changes in farming practice, drainage and grazing pressure.

A regular and common passage migrant over much of the country, from late March (16th) to early June (6th) and from late June (17th) to late November. Largest numbers are recorded in late August, e.g. *c.*1,000 at Meriç Delta (Marmara) on 26 August 1971, and at least 2,500 at Tuz Gölü (Inner Anatolia) on 17 August 1969. In winter, flocks occur at coastal wetlands and more locally elsewhere, with most in Aegean and Mediterranean, including 1,260 at Tuzla-Milas on 5 January 2002 (H. Tholl-Schmidt) and *c.*1,500 in Gediz Delta in December 2004 (E. Yoğurtçuoğlu *et al.*), the largest winter concentrations.

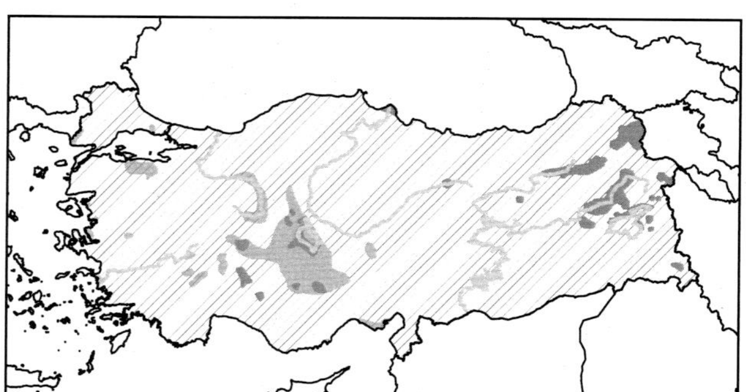

Breeding. In marshes, lakesides, waterside meadows and larger islands in lakes. Nest a hollow in the ground, lined with plant material and usually well hidden in grass or tall herbage. Eggs: C/3 (4), C/4 (10); records of

1–2 eggs considered incomplete clutches. In **Inner Anatolia**, five nests had full clutches in mid May; three nests (C/4, C/2, C/1) at Eşmekaya on 5 June 1992 may have been replacements; a large chick, at Bolluk Gölü on 24 June 1992, indicates egg laying in early May (PC); three nests with four eggs at Bolluk Gölü on 15 May 2004, one of these hatching on 16 May 2004 (L. Baxter), indicating egg laying *c.*20 April. An adult and two young were at Gediz Delta (**Aegean**) in June 1995 (Eken 1997a). In **Black Sea Coastlands**, four newly hatched chicks at Kızılırmak Delta on 4 June 1992 (Hustings & van Dijk 1994), indicate egg laying in the second week of May. In **East Anatolia**, the earliest record is a nest with eggs on 31 May 1990 at Erçek Gölü. Three other nests with full clutches were found there, one on 3 June 2001 and two on 28 June 1968. A nest had four eggs, and several other adults were alarm-calling and probably with young, near Tatvan, on 8 June 2004 (B. Pleasance); a nest with four newly hatched young at Van on 1 July 1985 indicates egg laying in early June.

MARSH SANDPIPER *Tringa stagnatilis*
Bataklık Düdükçünü

(Du: Poelruiter; Fr: Chevalier stagnatile; Ger: Teichwasserläufer; Swe: Dammsnäppa)

Subspecies. Monotypic.

Status and distribution. Scarce but widespread passage migrant to lagoons, mudflats, ponds and marshes across the country, although appears distinctly scarcer in the eastern third. Forms very small flocks during spring passage, which is noted from mid March (19th) to early June (10th), with the bulk in the first half of April, e.g. 167 in Çukurova (Mediterranean) on 5 April 1990, of a total spring passage involving 498 birds (Kivit *et al.* 1994). Autumn passage again generally involves small flocks, commencing 7 July, although considerable numbers may be recorded locally, especially beyond the first half of August, e.g. 135+ were at Hotamış marshes (Inner Anatolia) on 22–23 September 1974. Passage continues until late October (29th) and very small numbers are noted even later in the western half of the country. Very small numbers are occasionally reported in winter, particularly at Gediz Delta (Aegean), Acıgöl and the Göksu and Çukurova deltas (Mediterranean) and, even more rarely, in Inner Anatolia.

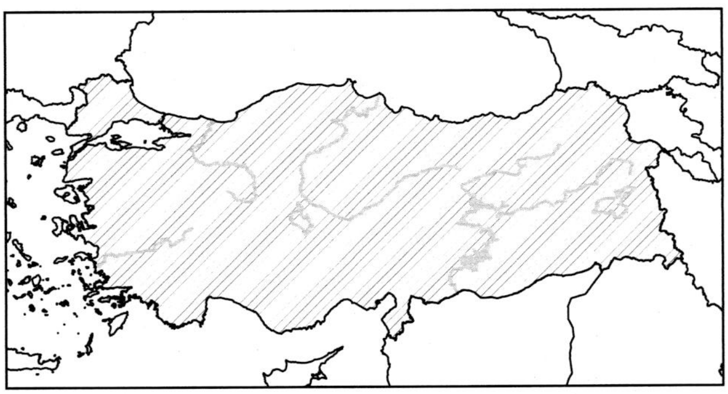

COMMON GREENSHANK *Tringa nebularia*
Yeşilbacak

(Du: Groenpootruiter; Fr: Chevalier aboyeur; Ger: Grünschenkel; Swe: Gluttsnäppa)

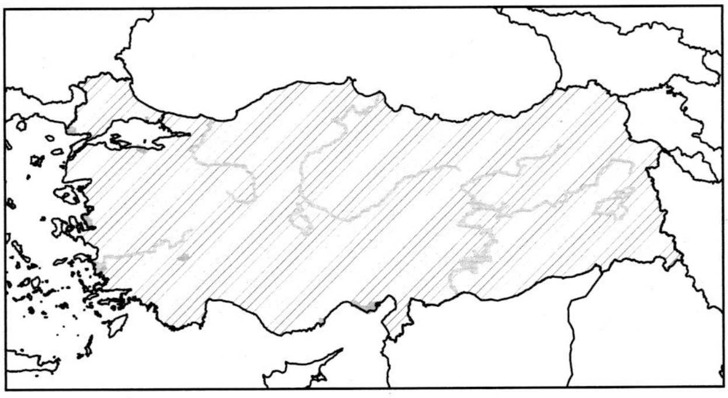

Subspecies. Monotypic.

Status and distribution. Common and widespread passage migrant to wetlands throughout the country. Passage commences mid March and continues until at least mid May, with the bulk in April. Thereafter occasional individuals are observed, some of which oversummer in Inner Anatolia without breeding (e.g. see OST 1978). From mid July, return migration commences and continues until late November (24th). Flocks are generally rather small, comprising generally fewer than five, but larger groups of more than 50 are infrequently observed. The scale of passage is evidenced by the total of just 242 in Çukurova (Mediterranean) in spring 1990 (Kivit *et al.* 1994). Winters in the western two-thirds of the country in small numbers mainly in coastal areas, but unfrozen wetlands inland, such as Sarıyar Reservoir (Inner Anatolia) may also support regular wintering attempts.

LESSER YELLOWLEGS *Tringa flavipes*
Küçük Sarıbacak

(Du: Kleine Geelpootruiter; Fr: Chevalier à pattes jaunes; Ger: Kleiner Gelbschenkel; Swe: Mindre gulbena)

Subspecies. Monotypic.

Status and distribution. Vagrant. An individual claimed near Ardahan, Kars province (East Anatolia) on 25 August 1988 lacked substantiating details (Kasparek 1990a), but subsequently a first-winter was trapped and photographed at Kızılırmak Delta (Black Sea Coastlands) on 27 November 2006 (Erciyas *et al.* 2008). There is one record from Greece, in August 1986 (Handrinos & Akriotis 1997) and single occurrences in Israel and Oman.

GREEN SANDPIPER *Tringa ochropus*
Yeşil Düdükçün

(Du: Witgat; Fr: Chevalier culblanc; Ger: Waldwasserläufer; Swe: Skogssnäppa)

Subspecies. Monotypic.

Status and distribution. Common passage migrant to wetlands throughout the country, generally in groups or aggregations of fewer than 50, but flocks in excess of 200 have been recorded. Only very small numbers generally move through the large south-coast deltas, and most passage is rather dispersed, with birds utilising most small pockets of suitable habitat. Peak passage periods are early April–mid May and late June–mid October, but significant numbers may remain in Inner Anatolia until at least late June (GMK). Recorded above 3,000m in the Pontic Alps (Black Sea Coastlands) on several occasions during migration (BD), and there is a record of a displaying individual in a wet grassy area at 3,000m in the mountains between Ikizdere and Ispir (Black Sea Coastlands) on 12 June 1975 (OST 1978), but breeding seems distinctly unlikely (given that display is known to occur on passage and that the species is not known to nest south of the Ukraine: Hagemeijer & Blair 1997). Winters regularly in the western two-thirds of Turkey, at which season it is also found even more frequently than on passage at small ponds, estuaries and along concrete irrigation channels, and it is the commonest wader inland during this period.

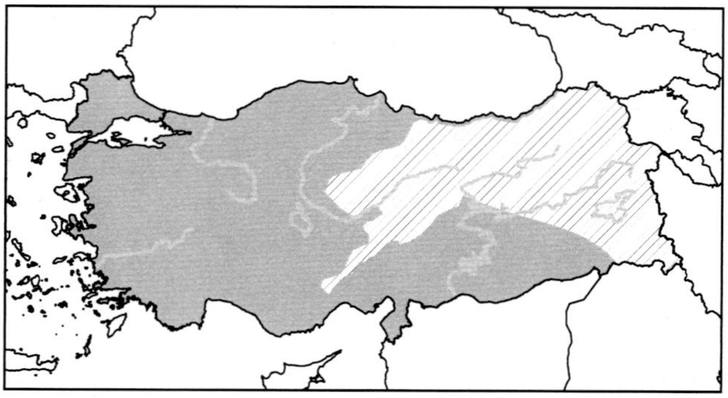

WOOD SANDPIPER *Tringa glareola*
Orman Düdükçünü

(Du: Bosruiter; Fr: Chevalier sylvain; Ger: Bruchwasserläufer; Swe: Grönbena)

Subspecies. Monotypic.

Status and distribution. Very common passage migrant to wetlands across the entire country, generally in small groups of less than a few dozen, but with larger numbers sometimes present at wetlands in Inner Anatolia and elsewhere, e.g. at least 1,000 at Manyas Gölü (Marmara) on 17 September 1996 (KAB), is the maximum count. Passage commences mid March (18th), peaks in the second quarter of April and continues until late May, whilst small numbers may oversummer. Return flocks already appear from late June and passage continues until late November. Very rarely overwinters, always in very small numbers, in the south and west.

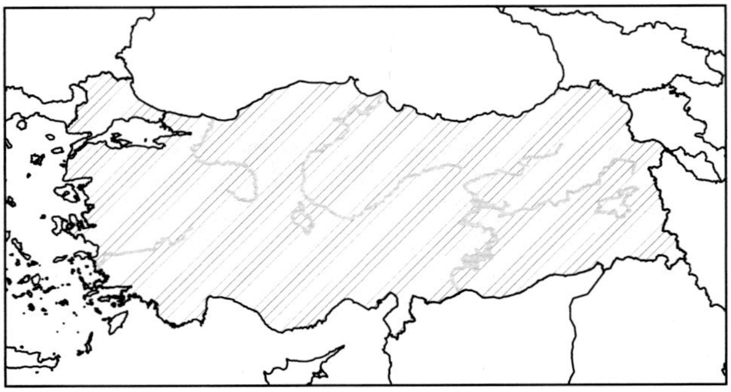

TEREK SANDPIPER *Xenus cinereus*
Terek Düdükçünü

(Du: Terekruiter; Fr: Bargette de Terek; Ger: Terekwasserläufer; Swe: Tereksnäppa)

Subspecies and taxonomy. Monotypic. Occasionally placed in the genus *Tringa*.

Status and distribution. A regular but very uncommon and local passage migrant to wetlands and coastal areas throughout central and eastern Turkey. It is probably most frequently recorded in the Van Gölü basin (East Anatolia) at either season (e.g. maxima of 16 at Van marshes in May 1989, 15 at Haçlı Gölü in July 1986 and 11 at the Bendimahi Delta in May 1985). Passage across the rest of the country typically involves very small numbers, usually in very loose parties of fewer than five, exceptionally ten in Çukurova (Mediterranean) in May 1990. Kasparek (1992a) was aware of just over 50 records. There have been many more since, although the overall total probably does not much exceed 150. In spring noted from 20 March to 27 June (with the peak in May, especially the first half), and in autumn recorded between 16 July and 26 October (with the bulk in mid August; it is distinctly rare beyond the first week of September). One recent winter record: at the Göksu Delta (Mediterranean) on 5 January 2002 (Demirci 2002); the species' nearest regular wintering grounds are in the Persian Gulf. Apart from vague reference to a pair shot in 'summer' 1875 in the vicinity of Istanbul, which were added to the Robert's College, Bebek (Istanbul) collection (Sclater & Taylor 1876), there were no records in Turkey prior to that by Hollom & Jamieson at Erçek Gölü (East Anatolia) in May 1966 (Kumerloeve 1966d, 1967a).

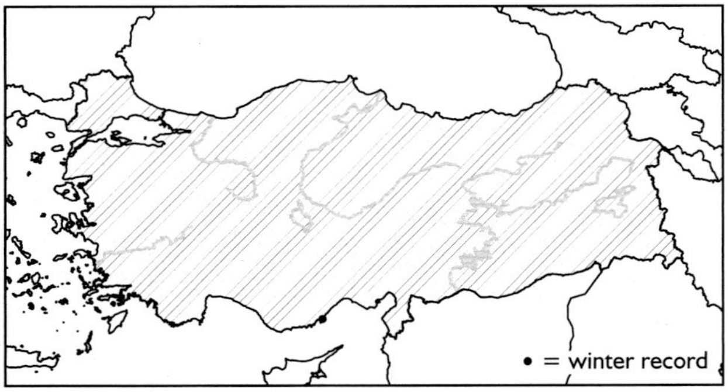

COMMON SANDPIPER *Actitis hypoleucos*
Dere Düdükçünü

(Du: Oeverloper; Fr: Chevalier guignette; Ger: Flußuferläufer; Swe: Drillsnäppa)

Subspecies and taxonomy. Monotypic. Sometimes placed in the genus *Tringa* (e.g. Sibley & Monroe 1990).

Status and distribution. National conservation status: Vulnerable (population apparently decreasing). A widespread and not uncommon summer visitor to upland areas of Black Sea Coastlands, East Anatolia (e.g. in the Upper Aras and Murat valleys, around Erzurum and elsewhere; Kumerloeve 1967a), the eastern Taurus (Mediterranean) and perhaps in other regions (e.g. there are several breeding-season records from the Dalyan area, Aegean; Kılıç & Kasparek 1989, KAB pers. obs.; Kralkızı dam, Southeast Anatolia; Karakaş & Bırıcık 2005b; and the Kocaçay Delta, Marmara; Ertan 1996). However, there are relatively few confirmed records of breeding (see below), although the species must surely do so more regularly in East Anatolia and Black Sea Coastlands. It may even breed at lower elevations at least occasionally (including once recently in Southeast Anatolia), and even at sea level in Black Sea Coastlands (Hustings & van Dijk 1994). Nonetheless, many records, considered by their observers perhaps to indicate breeding, probably involved passage birds (see, e.g., Warncke 1966) and at least one published record of breeding (Brush 2006) is erroneous; the nest concerned was that of a lark, presumably Horned Lark *Eremophila alpestris*.

Even more widespread and locally common on passage, at wetlands and on coasts throughout the country, from mid March (19th) to at least mid May in spring (19th), with the peak, at least in central and southern areas being in the second half of April. Occurs from late July (with failed breeders, perhaps from within Turkey, reappearing at wetlands from late June) to late October (23rd) in autumn, and one individual was as late as mid November in Inner Anatolia, which perhaps intended to overwinter. Winters regularly, albeit in tiny numbers, principally in suitable coastal areas of Aegean and Mediterranean, but there are a handful of inland records from the same regions at this season, with one recent mid-December record and one from mid January in Inner Anatolia, a January record from Black Sea Coastlands and a late January record from Southeast Anatolia.

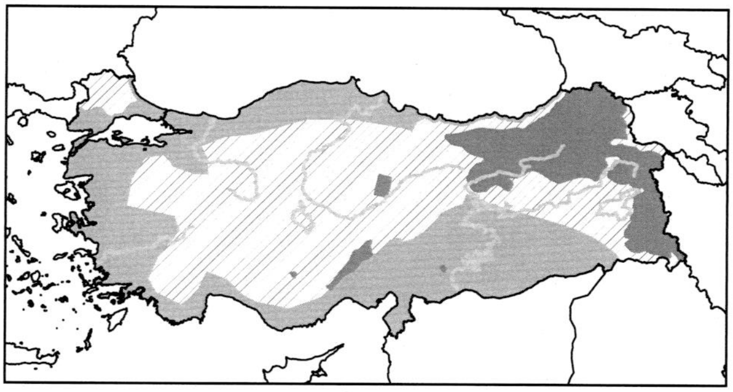

Breeding. Along fast-running rivers and at lakes, mainly in uplands (to at least 2,600m), especially in areas with islands and banks of sand and shingle, usually with a mix of open sand with some rank vegetation and scrub, e.g. willow. Pairs usually solitary but there may be two or more pairs fairly close together in optimum habitat. A nest at Sarıcan, 57km northwest of Ağrı (**East Anatolia**), was concealed in sparse, long grass in a flat sandy area near a river; it was a shallow hollow in the sand, lined with plant stems, grass and leaves, and had four eggs on 21 June 2004, which hatched on 25 June 2004, indicating the first egg was laid on 1 June (PC). There are two observations of juveniles by Kumerloeve (1969a), at Yüksekova and at Çatak (both East Anatolia) in June 1968. A pair and one chick were seen east of Gürpınar (East Anatolia), in early June 1969; this chick was *c.*1 week old on 2 June 1969 (R. F. Porter pers. comm.), indicating that the first egg was laid in early May. A pair with young were observed on a river, north of Demirkazık (**Mediterranean**), on 3 June 1999 (R. F. Porter). At Kralkızı Reservoir, near Diyarbakır (**Southeast Anatolia**), pairs formed in late March with display noted on 28 April 2000 and 22 April 2001 (Karakaş & Kılıç 2005b). A record of a nest at Giresun (**Black Sea Coastlands**) on 11 April 1975 is considered doubtful, in view of the early date. There are several records of adults displaying in May, and alarm-calling, and so presumably with young, in June–July.

SPOTTED SANDPIPER — *Actitis macularius*
Benekli Düdükçün

(Du: Amerikaanse Oeverloper; Fr: Chevalier grivelé; Ger Drosseluferläufer:; Swe: Fläckdrillsnäppa)

Subspecies and taxonomy. Generally considered monotypic; sometimes placed in the genus *Tringa* (e.g. Sibley & Monroe 1990). Note the correct spelling of the specific name (David & Gosselin 2002b).

Status and distribution. Vagrant with one well-documented record: an adult in summer plumage photographed at the Göksu Delta (Mediterranean) on 2 August 1988 (de Bruin 1989). Other claims involve three specimens, apparently formerly held in the Robert's College, Istanbul, collection (Mathey-Dupraz 1923) which apparently are no longer extant (Kirwan 1997a), and the notoriously unreliable Wahby (1930) mentioned one at Fikirtepe,

Istanbul, in October 1920. Also recorded in Greece, on Crete in May 1992 (Handrinos & Akriotis 1997) and in Bulgaria, e.g. in April 1973 (Nankinov *et al.* 1988).

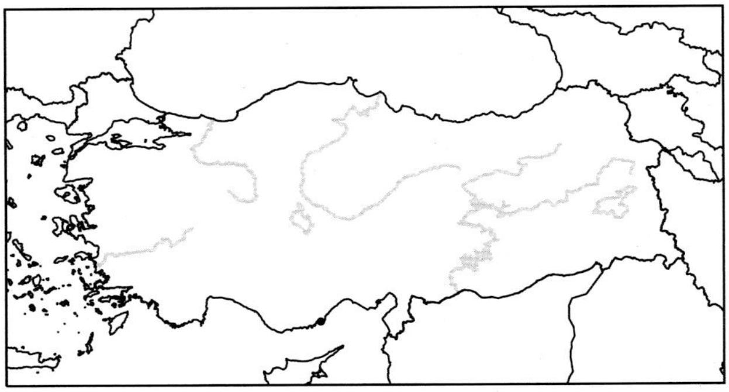

RUDDY TURNSTONE *Arenaria interpres*
Taşçeviren

(Du: Steenloper; Fr: Tournepierre à collier; Ger: Steinwälzer; Swe: Roskarl)

Subspecies. Nominate.

Status and distribution. A regular and uncommon to fairly common passage migrant, principally through the western two-thirds of the country, with many fewer records from the eastern third of Turkey. Most records are from coastal wetlands, but it is much less scarce at inland (playa) lakes than in Bulgaria (Nankinov *et al.* 1988). In spring recorded from 9 March to 9 June (with most in late April and May, median date, 10 May; Kasparek 1992c) and autumn passage is noted from 18 July to 30 October (with most in late August and September, median 14 September; Kasparek 1992c). Flocks typically number fewer than ten individuals, but more exceptional records include a total of 154 in Çukurova (Mediterranean) between 6 and 15 April 1987, with an additional 73 during 2–22 May (van der Have *et al.* 1988), at least 50 at Göksu Delta (Mediterranean) on 14 September (1972), a group of 43 in October (Kasparek *op. cit.*), 22 in Çukurova on 13 April (1990), and a total of 41 passing through Kızılırmak Delta (Black Sea Coastlands) between 1 May and 9 June 1992. Rather uncommon and irregular in winter (December–February), on the Aegean and Mediterranean coasts, when occasionally recorded in larger flocks, *e.g.* up to 40 at Büyük Menderes Delta (Aegean). Kasparek (*op. cit.*) listed other February records involving flocks of 30 and 50 individuals.

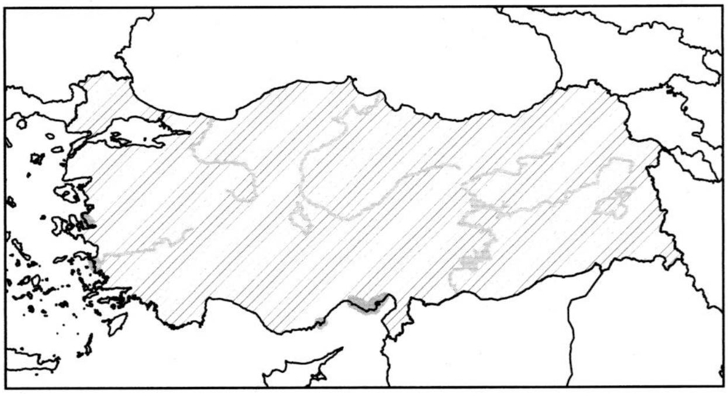

WILSON'S PHALAROPE *Phalaropus tricolor*
Büyük Denizdüdükçünü

(Du: Grote Franjepoot; Fr: Phalarope de Wilson; Ger: Wilsonwassertreter; Swe: Wilsonsimsnäppa)

Subspecies and taxonomy. Monotypic, but debate centres over whether the monospecific genus, *Steganopus*, in which it was originally described, should be resurrected for this rather different phalarope, which appears closer to *Tringa* than its congeners (see Sangster *et al.* 1999).

Status and distribution. Vagrant with two records: one, published without description, in winter plumage at Balık Gölü, Kızılırmak Delta (Black Sea Coastlands) on 19 September 1983 (Bräuning 1984, Dijksen & Kasparek 1985), and an adult in summer plumage at Sultan marshes (Inner Anatolia) on 5 May 1984 (Kasparek

1990a), for which no description has been published, although the sighting was apparently well described. Recently (1992) recorded in Bulgaria (Nankinov 1998), and in the Middle East this Western Hemisphere species has also reached Oman (Porter *et al.* 1996).

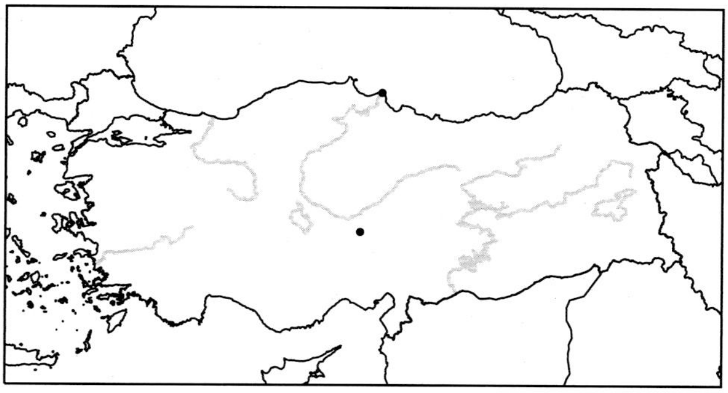

RED-NECKED PHALAROPE *Phalaropus lobatus*
Denizdüdükçünü

(Du: Grauwe Franjepoot; Fr: Phalarope à bec étroit; Ger: Odinshühnchen; Swe: Smalnäbbad simsnäppa)

Subspecies. Monotypic.

Status and distribution. A regular but uncommon and local passage migrant to wetlands and coastal areas throughout Turkey, which is generally more frequent and has been recorded in substantial flocks in the eastern third of the country, especially in the Van Gölü basin (East Anatolia) in spring. Passage across much of the rest of the country typically involves groups of fewer than 50 individuals, although 98 were recorded at Çöl Gölü (Inner Anatolia) on 31 May 1989, 95 in Çukurova (Mediterranean) in mid May (1987) and *c.*70 off Ardeşen (Black Sea Coastlands) in early September 1973, and there are very few available records from Southeast Anatolia. Maxima from the Van/Erçek Gölü area include: *c.*900 on 20–21 May 1966 (Kumerloeve & Hollom 1967), an exceptional 1,800 individuals on 12–13 May 1973 (Groh 1986), 380 on 7 May 1986 (Martins 1989), a minimum of 855 at three sites in the region on 1–11 May 1989 (van den Berk *et al.* 1993: to date the only attempt to survey wader passage on the eastern Turkish flyway systematically) and *c.*500, at Sodalı Gölü, on 2 May 1990 (M. R. K. Plaxton). Several other records involving flocks of up to 280 are available, but there are comparatively few records from this area in autumn (in part due to a relative lack of observer activity) and none of large flocks. In spring noted from 26 March to 16 June (with the peak in May, especially the first half), and in autumn recorded between 5 July and 5 October (with the bulk in August and early September). Groups rarely consist of more than 30 at this season. Comparatively few records are available prior to the advent of modern, systematic recording.

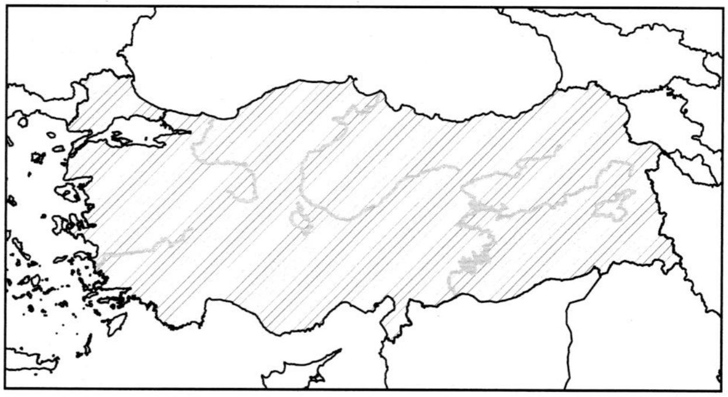

RED PHALAROPE *Phalaropus fulicarius*
Kızıl Denizdüdükçünü

(Du: Rosse Franjepoot; Fr: Phalarope à bec large; Ger: Thorshühnchen; Swe: Brednäbbad simsnäppa)

Subspecies. Monotypic. Note (following David & Gosselin 2002a) the spelling of the specific name is *fulicarius* not, as occasionally listed, *fulicaria*. Like Beaman (1994), we elect to use the North American name for this species, rather than the English Grey Phalarope, for reasons succinctly elucidated by Parkes (1982).

Status and distribution. Vagrant. Three recent sight records and one old specimen record: a breeding-plumaged female at Kulu Gölü (Inner Anatolia) on 1 June 2002 (S. Holmstedt *et al.* in Demirci 2003), one, which was reportedly well watched, at Birecik (Southeast Anatolia) on 3 May 2003 (H. Huhtinen, P. Laaksonen, R. Mäki & W. Velmala) and one photographed at Sodalı Gölü (East Anatolia) on 25 May 2005 (Balmer & Betton 2005b). Details concerning the first observation are very scant, although it was made by observers familiar with the species. *P. fulicarius* is mentioned for Turkey by both Ergene (1945) and Kasparyan (1956), two notoriously unreliable sources; both authors probably based their reports on that of Mathey-Dupraz (1923), who mentioned a female taken at Istanbul in the late 19th century, but we have been unable to examine the skin. Note that this record does not appertain to the female taken by Robson at the Bosphorus, as erroneously suggested by Kirwan (1997a), because the original remark (by Elwes & Buckley 1870) refers to *P. lobatus*, as a closer reading of the latter demonstrates. Brinkmann *et al.* (1990) mentioned one at the Büyük Menderes Delta (Aegean) in April–May 1989, but no substantiating details were presented. In Greece there are three records (Handrinos & Akriotis 1997), including two at the Evros Delta on the border with Turkey, the first one that was photographed on 22 December 1963 (Bauer & Müller 1969), and there are several from other countries in the Middle East (Porter *et al.* 1996).

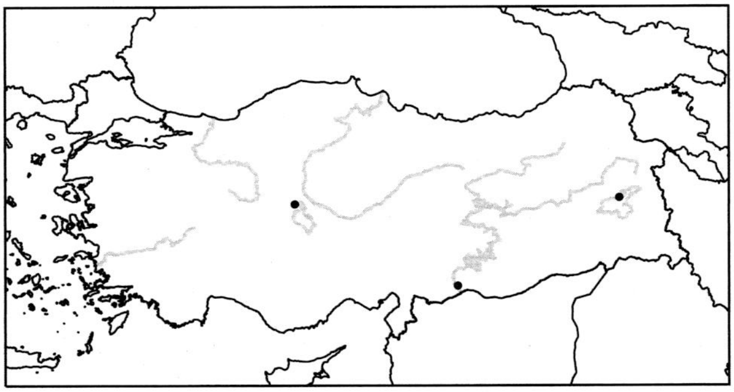

POMARINE JAEGER (POMARINE SKUA) *Stercorarius pomarinus*
Küt Kuyruklu Korsanmartı

(Du: Middelste Jager; Fr: Labbe pomarin; Ger: Spatelraubmöwe; Swe: Bredstjärtad labb)

Subspecies and taxonomy. Monotypic. MtDNA sequencing data suggest that this species is very closely related to the *Catharacta* skuas, particularly Great Skua *C. skua*. It may even form a link between the two genera. However, rather than transferring Pomarine Jaeger to *Catharacta*, some authors have preferred to subsume the entire species-group within *Stercorarius* (e.g. Andersson 1999, Cohen *et al.* 1997). See also Sangster *et al.* (1999) and the references cited therein, and Collinson (2006). Nonetheless, it should be noted that alternative views of relationships among jaegers have recently also been espoused, and some authors (e.g. Braun & Brumfield 1998) prefer to regard *Catharacta* as a perfectly valid, monophyletic grouping.

Status and distribution. Probably a rare passage migrant. There are at least 25 published and four previously unpublished records as follows. Alléon (1880) regarded the species as regular in the Bosphorus during November and December and there are two specimens of his, labelled 'Marmara' and dated spring 1888 and May 1892, in the National Museum of Natural History, Sofia, while Mathey-Dupraz (1924) reported the species six times (30 October 1888, 17 November and 15 December 1890, end of January 1891, 3 December 1892, and 15 December 1893) in the Bosphorus, off Kadiköy and between Üsküdar and Haydarpaşa (with hindsight, it is questionable whether sufficient was known of skua identification at this period to regard these sight records as anything but unconfirmed). Thereafter, three were off Antalya (Mediterranean) on 24 April 1959 (Vader 1965), and in 1967 single birds were at the Göksu Delta (Mediterranean) on 11 July, Tuzla Gölü, Çukurova (Mediterranean) on 12 July, off Ünye (Black Sea Coastlands) on 26 September, and two offshore between Terme and Ünye on 30 September (Vielliard 1968), with a pale-phase adult off Trabzon (Black Sea Coastlands) on 26 August 1969, one at Tuzla Gölü, Çukurova on 5 January 1971 (OST 1975), an immature off Tarsus (Mediterranean) on 28 August 1977 (Schmidtke & Utschik 1980), a pale-phase adult off Antalya on 6 April 1978 (Beaman 1986), an adult off Trabzon on 23 September 1981, two adults off Izmit (Marmara) on 24 May 1987, an immature at Bendimahi (East Anatolia) on 26 July 1989 (Kasparek 1990a), three at the Küçük

Menderes Delta (Aegean) on 18 June 1993, one in the Black Sea between Demirciköy and Kilyos (Marmara) on 20 August 1995 (KAB), three at Kulak beach, Çukurova, on 22 March 1996 (Kirwan & Martins 2000a), two at Kızılırmak Delta (Black Sea Coastlands) on 14 May 1997, two off Yumurtalık, Çukurova, on 27 March 1999, six at Trabzon harbour (Black Sea Coastlands) on 9 June 2000 (Kirwan *et al.* 2003), six off Ormanlı, west of Terkos Gölü (Marmara) on 13 October 2002 (KAB *et al.*), six off Saros Bay (Marmara) on 17 April 2006 (O. Dochy *et al.*), six off Meriç Delta (Marmara) on 22 April 2006, one off Muğla (Aegean) on 25 April 2006 (*per* E. Yoğurtcuoğlu), one off Kuşadasi (Aegean) on 31 May 2007, one off Boğazi, in the Bosphorus, on 15 September 2007 (Balmer & Betton 2008) and an exhausted bird at Mersin (Mediterranean) in April 2008. There was also a probable off the Göksu Delta on 27 May 1996 (Kirwan *et al.* 2003).

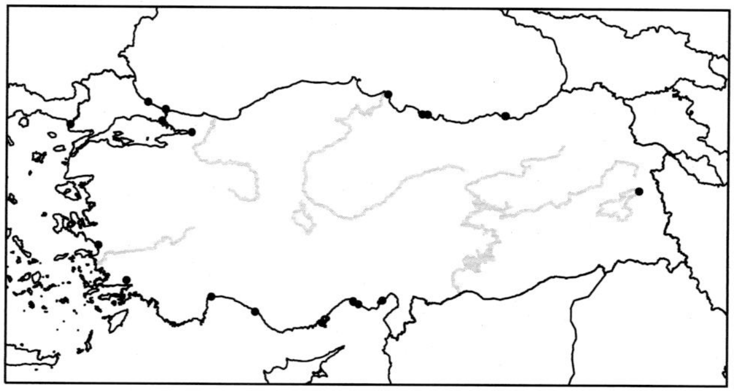

Most of the records listed above have lacked descriptive details. Nonetheless, Pomarine Jaeger is plausibly even more regular in Turkey than records suggest: for Lebanon, Ramadan-Jaradi *et al.* (2008) mentioned 20 records, and there are five records in Armenia (Adamian & Klem 1999) and six for Greece (Handrinos & Akriotis 1997).

PARASITIC JAEGER (ARCTIC SKUA) *Stercorarius parasiticus*
Korsanmartı

(Du: Kleine Jager; Fr: Labbe parasite; Ger: Schmarotzerraubmöwe; Swe: Labb)

Subspecies. Monotypic.

Status and distribution. A scarce but almost certainly regular passage migrant along the Black Sea coast, and through the Bosphorus and Sea of Marmara. It is also apparently regular on Aegean and Mediterranean coasts, as well as occasionally migrating overland (like all jaegers), being recorded inland in Mediterranean (e.g. at Burdur Gölü), Inner Anatolia (three times at Kulu Gölü and once at Ereğli marshes) and East Anatolia (Bendimahi, Van Gölü and near Van town). In spring, noted between 10 March and 22 June (the majority in May) and in autumn from 11 July to 10 October (with most in September). There are a number of winter records (January–February) from the entire length of the Black Sea Coastlands (e.g. Welch & Welch 1998a recorded a total of seven, at six localities, during January 1997), at least four from the Gulf of Iskenderun (Mediterranean) and one, involving 13 birds, off Fethiye (Mediterranean) in December 1997 (P. Hope). There is also a possible January 2004 record from Inner Anatolia at Kulu Gölü (I. A. Richardson). Kasparek (1992a) reported nearly 50 available records; not including undetermined skua species, we have located a total of 114 records (some involving as many as 15 individuals), as well as four specimens, all collected by Alléon, in 'Marmara'. Not recorded from Southeast Anatolia.

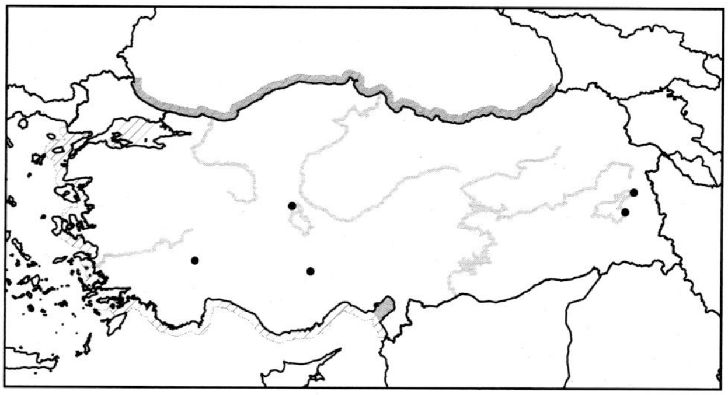

LONG-TAILED JAEGER (LONG-TAILED SKUA)

Uzun Kuyruklu Korsanmartı *Stercorarius longicaudus*

(Du: Kleinste Jager; Fr: Labbe à longue queue; Ger: Falkenraubmöwe; Swe: Fjällabb)

Subspecies. Presumably the nominate race (which breeds in Arctic and subarctic Scandinavia and Russia east to the Lena River delta) is involved, but of the Turkish records none has been subspecifically identified.

Status and distribution. Perhaps only a vagrant. Eight records as follows: one obviously of this species, given the description by Mathey-Dupraz (1924) at Terkos Gölü (Thrace) in late September 1888, two adults or subadults in the Bosphorus (Marmara) on 23 September 1979 (Martins 1989), an adult photographed at Kulu Gölü (Inner Anatolia) on 3 June 1991, two immatures off Taşucu (Mediterranean) on 11 June 1991 (Kirwan & Martins 1994), one off Karaburun, near Istanbul, on 17 September 1991 (Robel 1992), an adult off Kale, west of Finike (Mediterranean) on 24 May 1992, a pale-morph adult at Sultan marshes (Inner Anatolia) on 10 June 1994 (Kirwan 1995a), and a second-summer off Yumurtalık, Çukurova (Mediterranean) on 8 May 1999 (Kirwan *et al.* 2003). Not all records are supported by a description. Recorded once in Armenia (Adamian & Klem 1999), but never in Greece (Handrinos & Akriotis 1997) or Lebanon (Ramadan-Jaradi *et al.* 2008).

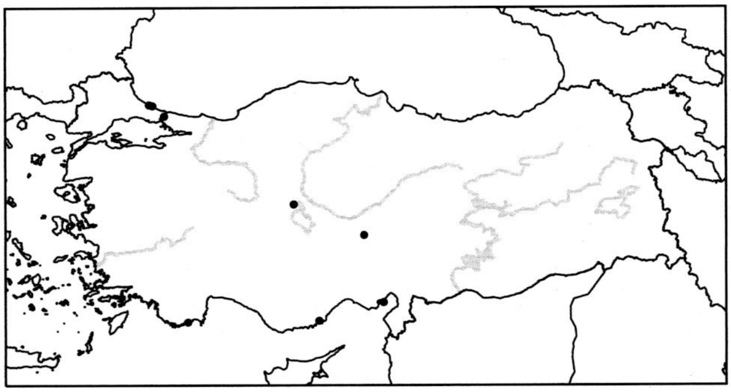

GREAT SKUA *Catharacta skua*

Büyük Korsanmartı

(Du: Grote Jager; Fr: Grand Labbe; Ger: Skua; Swe: Storlabb)

Subspecies and taxonomy. Generally considered monotypic by most modern authorities. Although some recent commentators have preferred to subsume the *Catharacta* group of skuas within *Stercorarius* (see under Pomarine Skua), we prefer to highlight the relative distinctiveness and differing uniformity of the former group by retaining separate generic status for Great Skua and its closest relatives (although see Sangster *et al.* 1999 and Collinson 2006 for the opposing view).

Status and distribution. Vagrant. Six sight records (all but the most recent two apparently undescribed or published without description): an immature near Ahtamar Island, Van Gölü (East Anatolia) on 15 September 1967 (Vielliard 1968), and single birds at Tuzla Gölü, Çukurova (Mediterranean), on 30 December 1983, in Iskenderun Bay (Mediterranean) on 7–8 January 1986 (D. Simpson in *Sea Swallow* 36: 41; very brief descriptive notes supplied), off the Ceyhan Delta, Çukurova, on 9 January 1990 (Kasparek 1990a), off the Göksu Delta (Mediterranean) on 15 April 1996 (Kirwan & Martins 2000a) and from a ferry in the Sea of Marmara (Marmara) on 5 April 2003 (G. Magnin).

Rare elsewhere in the eastern Mediterranean, with five records between 1958 and 1993 in Greece (Handrinos & Akriotis 1997), none from Cyprus (Flint & Stewart 1992) or

Lebanon (Ramadan-Jaradi *et al.* 2008), four (including one inland) in Israel (Shirihai 1996) and at least five records of indeterminate *Catharacta* skuas from Egypt (Goodman & Meininger 1989).

WHITE-EYED GULL *Larus leucophthalmus*
Kızıldeniz Martısı

(Du: Witgoogmeeuw; Fr: Goéland à iris blanc; Ger: Weißaugenmöwe; Swe: Rödahavssotmås)

Subspecies. Monotypic.

Status and distribution. Vagrant. One record: a first-winter, which was well described, at Marmaris (Aegean), on 28 March 1988 (Kirwan & Martins 1994). There is an old (mid-19th century) record for Greece (Handrinos & Akriotis 1997) but none from Cyprus (Flint & Stewart 1992) or Lebanon (Ramadan-Jaradi *et al.* 2008).

[RELICT GULL *Larus relictus*]
İçdeniz Martısı

(Du: Relictmeeuw; Fr: Mouette relique; Ger: Reliktmöwe; Swe: Reliktmås)

Subspecies. Monotypic.

Species not admitted to the Turkish list (Kirwan *et al.* 2003). Two extraordinary records have recently come to light, both involving ring recoveries, but neither of the dead birds was retained. The first was in Bulgaria, in March 1978, and the second, at Adana (Mediterranean), on 30 March 1990. The ring recovered in Turkey belonged to a bird that had been ringed at Lake Alakol, Kazakhstan, on 22 June 1979 (Gavrilov & Gavrilov 2000). Given that these are the only Western Palearctic reports of this rare species, other than a recent sight record, of 14 birds, in European Russia (see Nankinov 2003), it seems safest to await further evidence of Relict Gull's occurrence in the region, but observers should clearly be aware of the possibility of the species appearing again, *e.g.* in the south-coast deltas in winter.

PALLAS'S (GREAT BLACK-HEADED) GULL *Larus ichthyaetus*
Büyük Karabaş Martı

(Du: Reuzenzwartkopmeeuw; Fr: Goéland ichthyaète; Ger: Fischmöwe; Swe: Svarthuvad trut)

Subspecies. Monotypic. Has been placed in its own genus, *Ichthyaetus* (see Wolters 1975–82).

Status and distribution. An uncommon winter visitor (mean winter population is placed at 200–400 individuals) and apparently a passage migrant, principally across the eastern two-thirds of the country. It was formerly considered very rare, but there has been a substantial increase in winter records recently. This increase may well be indicative of a genuine change in the species' winter status, rather than solely reflecting improved observer coverage. Recorded in coastal areas and wetlands in all regions, but particularly scarce in the extreme west (e.g. in Thrace) and far east of the country. Wintering birds usually present from mid December (earliest 26 November) to April (latest 13 May). Two extremely early records – an immature at the Bosphorus (Marmara) on 25 September 1995 (Ecsedi 1996) and one at Araklı, Trabzon province (Black Sea Coastlands) on 15 September 2005 (S. İsfendiyaroğlu & E. Gem) – are without precedent, whilst a record from Göksu Delta (Mediterranean) on 25 May 1993 (Kirwan & Martins 2000) was either a very late winterer or a migrant. Most records are from Black Sea Coastlands, Inner Anatolia (where it regularly uses inland reservoirs) and Mediterranean (where most regularly found in the coastal deltas), in all of which regions the species has been recorded at several localities. Historically, recorded in very low numbers, but considerably more numerous in recent years, with records in the above-mentioned regions typically involving 5–20 individuals, but there are larger counts of 71 (January 1996), 168 (December 1995), 219 (January 1997) and 273 (January 1999) at Yedikır Reservoir (Black Sea Coastlands).

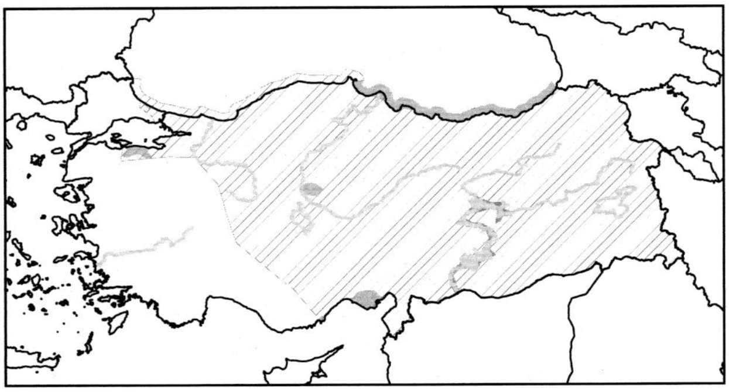

MEDITERRANEAN GULL *Larus melanocephalus*
Akdeniz Martısı

(Du: Zwartkopmeeuw; Fr: Mouette mélanocéphale; Ger: Schwarzkopfmöwe; Swe: Svarthuvad mås)

Subspecies. Monotypic.

Status and distribution. National conservation status: Vulnerable (population apparently stable). Local summer visitor to wetlands in Inner Anatolia and Aegean. Also recorded in summer in coastal Marmara, but no evidence of breeding there. Also occurs in East Anatolia in summer. Widespread and common on passage on coasts of west and central Turkey, although wintering numbers are generally small, e.g. 357 were recorded on the Black Sea coast in February 1997 and the usual wintering population is 2,000–3,000 individuals in the entire country. The largest breeding colonies (400–2,250 pairs) are in Inner Anatolia and Aegean. In Marmara recorded from early March to early June and from mid July to mid November, with large numbers noted at the Bosphorus and on the Marmara coast as far west as the Dardanelles. High counts include 600 (July), 800 (August) and 1,200 (October) at Tekirdağ, but even larger numbers have been recorded between Küçükçekmece and Silivri, where there were 6,000+ (October) and 7,300 (September), and from Büyükçekmece, where up to 10,000 have been recorded on passage, but inland it is rare.

Spring migration seems to peak during the last week in March and first half of April, with a maximum of 500+ seen during a one-hour boat trip along the Bosphorus on 4 April 2003 (G. Magnin). In Aegean it occurs on passage in moderate numbers and in winter, with counts of up to 250 in Küçük Menderes Delta and 430 near Çanakkale (Eken 1997d). In Mediterranean it is widespread along the coast from mid March to early May and from early August to late November, although numbers are generally small and in winter it is generally uncommon and only seen in low numbers. In Inner Anatolia, the species is a scarce migrant away from breeding localities, and in the eastern third of the country the species is generally rare.

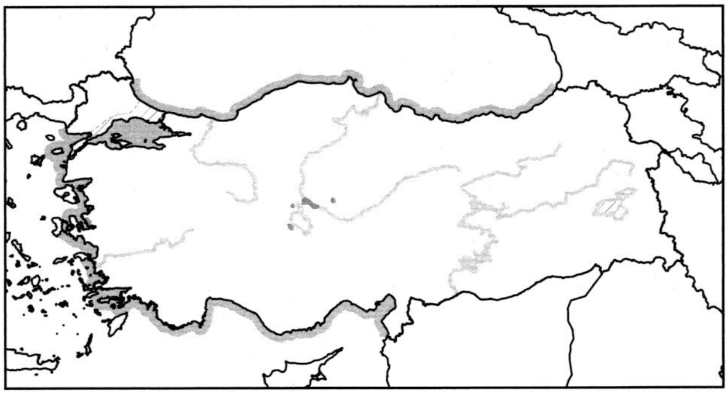

Breeding. In colonies, now mainly large, of 400–2,250 pairs (Magnin & Yarar 1997) on low flat islands of bare earth and scant vegetation, mainly at three saline or brackish shallow lakes (Kulu, Seyfe, and Bolluk) in **Inner Anatolia**, and at coastal lagoons in Büyük Menderes and Gediz deltas (**Aegean**). Sometimes breeds in loose association with other species, e.g. Gull-billed Tern *Gelochelidon nilotica* and Black-headed Gull *Larus ridibundus.*

Nest is a shallow hollow in the ground, lined with plant stems, grass and some feathers, or a larger accumulation of material, but much smaller and lower than nests of Black-headed Gull where they breed together. Eggs: mainly three, sometimes two, occasionally one. First eggs laid in late April and early May, with earliest young hatching in late May–early June: a fairly consistent pattern between colonies. Season protracted, probably due to replacement clutches; at Kulu Gölü on 19 June 1992 most young were already more than half grown, but some nests still had eggs and small young; at Büyük Menderes Delta on 26 June 1999, most young were large but there were also newly hatched young, a difference in age of *c.*4 weeks (PC).

LITTLE GULL *Larus minutus*
Küçük Martı

(Du: Dwergmeeuw; Fr: Mouette pygmée; Ger: Zwergmöwe; Swe: Dvärgmås)

Subspecies and taxonomy. Monotypic. A recent review of gull systematics based on a phylogenetic methodology (Chu 1998) suggested that this species is most closely related to Ross's Gull *Rhodostethia rosea*, and that they might both be best transferred to the genus *Hydrocoloeus*, given that the gulls are clearly a polyphyletic group. However, as suggested by the paper's author, we prefer the option of expanding the genus *Larus* to include all gulls, rather than continuing to recognise any subgroups that are poorly supported by currently available data. For further discussion see Sangster (1999), Crochet *et al.* (2000) and Sangster *et al.* (2007).

Status and distribution. Widespread and common passage migrant to all coasts, often in large numbers, wintering in smaller numbers (typically 1,500–2,500 individuals), but large concentrations are recorded occasionally. There are a few summer records. Inland, the species is much scarcer on passage and in winter. In Marmara widespread on all coasts, generally in moderate numbers, with maxima of 410 between Istanbul and Sariyer (September), at Meriç Delta (up to 2,000) and Büyükçekmece (max. 1,740). In Aegean, generally present from mid August to mid April, with occasional summer records. Numbers are usually small but larger concentrations do occur with a maximum of *c.*320 at Büyük Menderes Delta (January). In Mediterranean generally present late August to early May, with occasional summer records, in moderate numbers but with locally larger concentrations in spring. Widespread on the coast, but scarce inland in this region. In Inner Anatolia occurs throughout the year but is scarce and only ever recorded in small numbers. In Black Sea Coastlands widespread on the coast, generally in moderate numbers, but with high counts of 530 at Trabzon (August and September), 400+ between Samsun and Trabzon (October) and 1,138 along the entire coast in February 1997 (Welch & Welch 1998a). In this region, most frequent and recorded in largest numbers at Kızılırmak Delta, with a maximum of 41,000 in early spring 1992 (Hustings & van Dijk 1994). Generally very rare in the eastern third of the country.

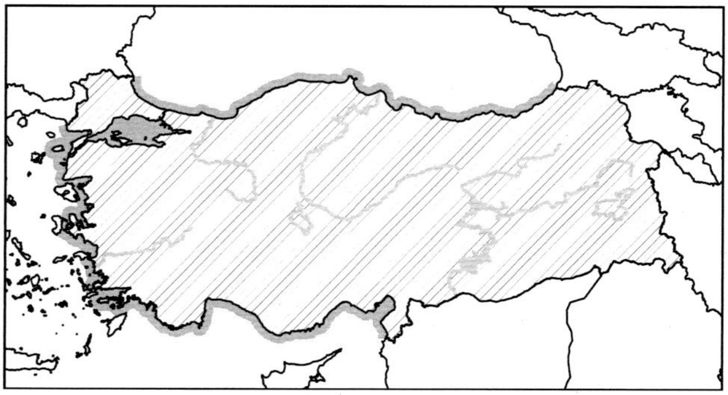

BLACK-HEADED GULL *Larus ridibundus*
Karabaş Martı

(Du: Kokmeeuw; Fr: Mouette rieuse; Ger: Lachmöwe; Swe: Skrattmås)

Subspecies and taxonomy. Monotypic. Under the evolutionary tree recognised by Chu (1998) this species could be transferred to the genus *Chroicocephalus*, which action was indeed further promulgated by Sangster *et al.* (2007). As noted under the previous species, for now, we prefer to consider all gulls within *Larus*, rather than attempt to resolve any ambiguities inherent to their relationships by endeavouring to recognise poorly supported, smaller genera within the family. See above.

Status and distribution. National conservation status: Least Concern (population apparently increasing). Resident throughout the country, although there is some contraction of range into western and central Turkey in winter, when totals for the entire country are usually between 90,000 and 130,000 birds. Breeds in Inner and East Anatolia and Black Sea Coastlands. In regions where it does not breed generally present from early August to late May. In Marmara widespread in winter on all coasts and inland, with the largest concentration at Uluabat Gölü (6,200 in January 1996). In Aegean widespread in winter in considerable numbers, both on the coast and inland. Maximum counts are of 3,000 at Acıgöl (January 1970 and September 1972), with summer records from this region but no confirmed breeding records. In Mediterranean, widespread on the coast and at wetlands throughout the region, being recorded in considerable numbers in winter, e.g. 2,520 at Çorak Gölü (November 1971). In Inner Anatolia a widespread breeder at several wetlands, but has undoubtedly declined due to habitat conversion and drainage. Important breeding sites (many of them now destroyed or largely so) include: Seyfe Gölü (max. 120 pairs), Ereğli marshes (max. 361 pairs), wetlands near Yarma (max. 215 pairs) and Gogenç Gölü (max. 100 pairs). In winter widespread at wetlands across the region, e.g. 1,070 at Seyfe Gölü (November 1971), but it is passage periods when largest concentrations are observed, e.g. 20,000 at Sarıyar Reservoir on 24 September 2004 (I. Richardson & E. Yoğurtçuoğlu) and regularly up to 10,000 at Mogan Gölü in March (BD). In Black Sea Coastlands widespread in winter on the coast, with 17,986 recorded in this region in 1996. In East Anatolia it is widespread at wetlands throughout the region on passage, generally in moderate numbers, but with several counts of up to 300. Rather rare in Southeast Anatolia.

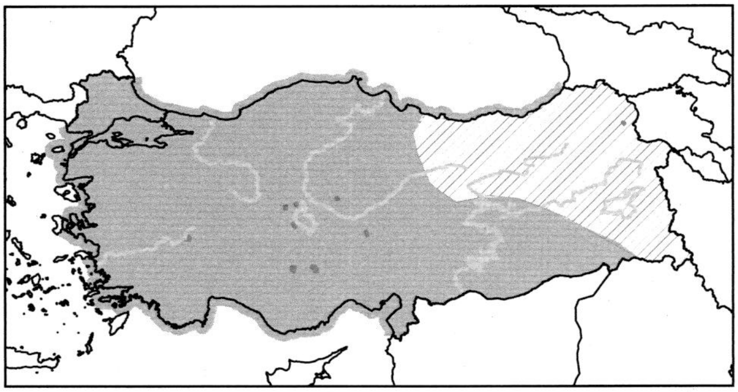

Breeding. In colonies at a few sites in Inner and East Anatolia, on islands in lakes and marshes, sometimes in mixed colonies or in close association with other species, including Mediterranean Gull *Larus melanocephalus* and Gull-billed Tern *Gelochelidon nilotica*. Nests are on the ground and are mainly of plant stalks; variable in size and up to 0.3m in height, much higher than those of Mediterranean Gull, where they nest together. Eggs: 2–3 (mainly three, occasionally one). At Kulu Gölü (**Inner Anatolia**), on 22 May 1992, *c.*350 pairs were breeding on an island; nests had mainly three eggs, with some newly hatched young, indicating date of first eggs as *c.*26 April. Latest records of nests, still with eggs, at Seyfe Gölü (Inner Anatolia) on 14 June 1993, and Kulu Gölü on 19 June 1992, when most young were already large, and some had nearly fledged (PC).

SLENDER-BILLED GULL *Larus genei*
İnce Gagalı Martı

(Du: Dunbekmeeuw; Fr: Goéland railleur; Ger: Dünnschnabelmöwe; Swe: Långnäbbad mås)

Subspecies and taxonomy. Monotypic. Could be considered within the genus *Chroicocephalus* (see previous species for discussion), but Sangster *et al.* (2007) did not pursue such a course for this species.

Status and distribution. National conservation status: Vulnerable (population apparently increasing). Fairly widespread and common summer visitor to wetlands in Inner Anatolia, where breeds at a number of sites in colonies of up to *c.*2,000 pairs (e.g. 2,100 pairs at Seyfe Golu in 1971 and 1,900–2,000 pairs at Bolluk Gölü in 1984, although some smaller colonies, such as at Ereğli and Sultan marshes, have certainly declined in recent years). Also a widespread and fairly common passage migrant through coastal areas and inland wetlands throughout the country, often in large numbers. Winters in moderate numbers, mainly in coastal areas in the extreme south and west, where the current winter population is placed at 800–1,600 individuals. In Marmara recorded on spring passage from early March and in autumn until mid November. Away from the Bosphorus, Slender-billed Gull has been noted at several localities, usually in small numbers, and is very scarce in winter. In Aegean recorded late August to February at a number of localities, with high counts of 1,284 at Büyük Menderes

Delta in January 1989 and 413 in Gediz Delta in late January 1992. In Mediterranean, present late July to late April, generally in moderate numbers, with high counts of *c*.600 in Çukurova (January 1992) and 1,000+ at Acıgöl (October 1970). In Inner Anatolia, away from the breeding areas, recorded in moderate numbers between late July and mid May. In Black Sea Coastlands recorded in coastal areas between early March and late November, generally in small numbers but with maxima of 1,500 at Ereğli in late autumn 1977 and up to 85 at Ardeşen in autumn 1973, with a handful of (mainly) recent winter records from this region (see Welch & Welch 1998a). In East Anatolia recorded in small numbers from early August to early October in the Van / Erçek Gölü area. Elsewhere in this region it is occasional, with some summer records, but always in small numbers. It is also rather rare and always recorded in small numbers in Southeast Anatolia.

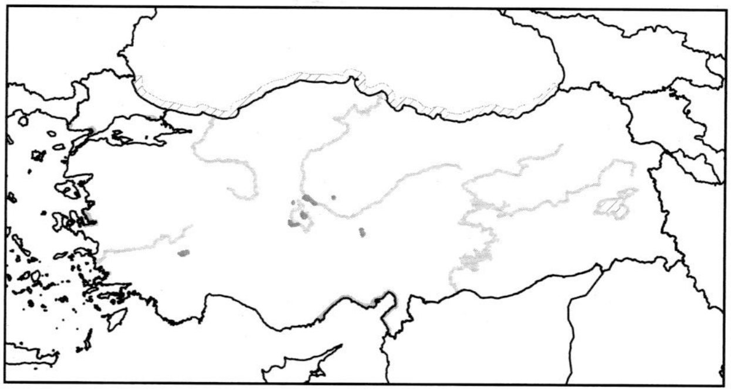

Breeding. In compact colonies, of 50–2,100 pairs, sometimes in loose association with other species, e.g. Gull-billed Tern *Gelochelidon nilotica* and Mediterranean Gull *Larus melanocephalus*, on low flat islands of bare earth and scant vegetation, mainly at three saline or brackish shallow lakes (Kulu, Seyfe and Bolluk) in **Inner Anatolia**. Nest is a shallow hollow in the ground, sparsely lined with plant debris, grass and a few feathers, slightly raised at the rim, which becomes well defined, whitewashed and solidified by excreta, as incubation proceeds. Eggs: C 2/3, occasionally one. First eggs are laid in late April–early May, with earliest hatchings in late May/early June, and this is fairly consistent between colonies. Season protracted, probably due to replacement clutches; at Bolluk Gölü, on 24 June 1992, most young were already large and well feathered but a few nests still had eggs and small young (PC).

AUDOUIN'S GULL *Larus audouinii*
Ada Martısı

(Du: Audouins Meeuw; Fr: Goéland d'Audouin; Ger: Korallenmöwe; Swe: Rödnäbbad trut)

Subspecies. Monotypic.

Status and distribution. Conservation status: Near Threatened (BirdLife International 2004); national conservation status: Vulnerable (population apparently increasing). Population estimate of *c*.500 pairs in Turkey, made by Burger & Gochfeld (*in* del Hoyo *et al.* 1996) appears grossly exaggerated and its basis is unknown. Scarce and rather local resident on southern and southwestern coasts, in Mediterranean and southern Aegean, with most records from the period early April to early October. Occasionally recorded from the south shore of the Sea of Marmara (although Kasparek 1992a suggested that it was reasonably regular there), with an exceptional record from the eastern Black Sea Coastlands: a second-summer at Trabzon on 23 May 1995, and another record from Batumi, in adjacent Georgia, in October 1984, suggesting that further Black Sea occurrences are possible (King & Shirihai 1996). Claimed once far inland, on Eğirdir Gölü (Mediterranean) on 9 September 1961 (Kumerloeve 1961), but such an occurrence seems very unlikely. Numbers observed usually very small, with most counts involving fewer than ten, but up to 15 seen in February–May at Fethiye (Mediterranean) (P. Hope),

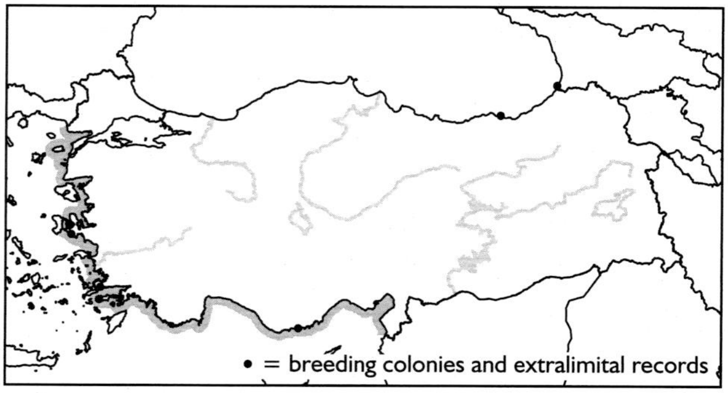

and also regular at the Göksu Delta (Mediterranean), from where the first certain Turkish record was made as recently as 1951 (Hollom 1959), and non-breeders gather at this and other sites from July. First discovered breeding in Turkey in 1973 (see below), but the population on the Aydıncık Islands (Mediterranean) has declined and may no longer be extant. Recently discovered breeding in very small numbers in Aegean, at Karaburun (see below), and there are other observations perhaps suggestive of breeding from small and uninhabited islands off the coast of Mediterranean in the 1970s (OST 1975). Eken (1997c) postulated that the Alaçatı (possible), Güllük Bay (probable), Bodrum Yarımadası (probable) and Datça islands (possible), all in Aegean, might also hold small colonies, although this remains unproven.

Breeding. Until recently, bred only on a small, low well-vegetated island, one of two islands off the south coast near Aydıncık (**Mediterranean**). There were 25 pairs in 1973 and 28 pairs in 1974, with eggs in late April and newly hatched young by 28 May (Witt 1976); *c.*20 pairs on 16 April 1985 and 30 pairs in 1987, but only six pairs in 1996, the decline presumably due to an increase in the breeding population of Yellow-legged Gulls *Larus cachinnans* on the island from none in 1974 to 40 pairs in 1996 (Magnin & Yarar 1997); on 6 May 2004, Audouin's Gull was not present, but there were several nests of Yellow-legged Gull (M. Farmer). A second breeding population is at Karaburun Island, west of Izmir (**Aegean**), where at least one pair bred successfully in 1996 (Boyla & Eken 1998) and 2–4 birds were present in late May 1997 (Eken 1997c). Nest and clutch size undescribed in Turkey; elsewhere, the nest is on the ground, a shallow scrape lined with plant stems, seaweed and other plant material; usual clutch 2–3.

MEW (COMMON) GULL — *Larus canus*
Küçük Gümüş Martı

(Du: Stormmeeuw; Fr: Goéland cendré; Ger: Sturmmöwe; Swe: Fiskmås)

Subspecies and taxonomy. Nominate and probably also *L. c. heinei* (see Malling Olsen & Larsson 2003), which intergrade. Recent phylogenetic work by Chu (1998), mtDNA sequencing by Zink *et al.* (1995) and comparisons of the morphology of immature plumages (Tove 1993) suggested that Mew Gull may be best considered as two or three species, *L. canus*, *L. kamtschatschensis* (Kamchatka Gull) and *L. brachyrhynchus* (Short-billed Gull), with the two latter forms more closely related to one another than either is to nominate *canus*. Malling Olsen & Larsson (2003) separated *L. brachyrhynchus*, under the English name Mew Gull, and placed all of the remaining taxa within *L. canus* (Common Gull). All of the principal taxa involved have a propensity to wander long distances, even to opposite continents.

Status and distribution. A very common winter visitor to the Black Sea coast, and local and not uncommon at wetlands across Turkey, but with most records in the west and north (earliest 17 August and latest 16 May). The current wintering population is placed at 8,000–12,000 individuals, with the bulk along the Black Sea coast. Rather regular in Marmara, where common around Istanbul and the Bosphorus in winter, especially after the arrival of the first cold front in November or December, but perhaps less regular elsewhere. More occasional in Aegean, where usually recorded in very low numbers, but *c.*200 were at Büyük Menderes Delta in February 1972. Generally rather scarce and uncommon in Mediterranean and Inner Anatolia, although occasionally noted in larger numbers of up to *c.*150 even in these regions, and is regular at some of the larger reservoirs in the latter region between November and February. Most regularly recorded and in largest numbers in Black Sea Coastlands, principally on the coast, where 5,279 were counted in February 1997 (Welch & Welch 1998a). We have seen very few records from either East or Southeast Anatolia.

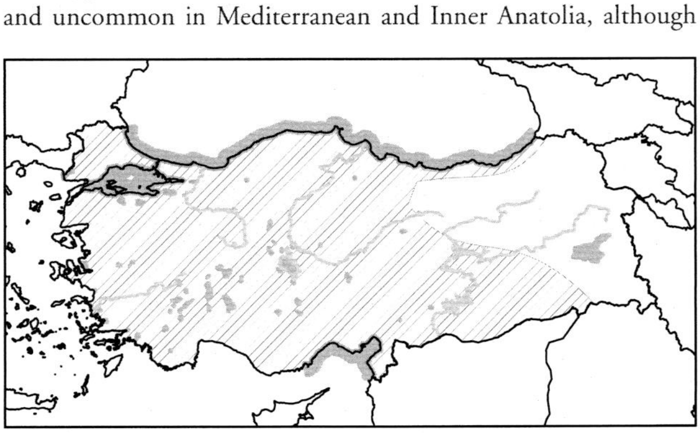

LESSER BLACK-BACKED GULL *Larus fuscus*
Kara Sırtlı Martı

(Du: Kleine Mantelmeeuw; Fr: Goéland brun; Ger: Heringsmöwe; Swe: Stilltrut)

Subspecies and taxonomy. Only nominate *fuscus* is regularly mentioned for Turkey, which form breeds around the Baltic and in northwest European Russia, and principally winters in northeast and East Africa (Malling Olsen & Larsson 2003). If included within the *fuscus* group, *heuglini* (Heuglin's or Siberian Gull) is a rather less regular (but probably overlooked) passage migrant throughout the country (including inland), and somewhat rarer winter resident, principally to coasts in the south and west. Other forms seem unlikely to occur, although given that *intermedius* is recorded rarely on passage through Israel (Shirihai 1996), it should clearly be searched for in Turkey.

Sangster *et al.* (1998) elected to split Lesser Black-backed Gull into two species: Baltic Gull *L. fuscus* and Lesser Black-backed Gull *L. graellsii*, but this proposal appears to have acquired few proponents. Thus most debate surrounds the form *heuglini*, which until recently was considered by most authors to belong within the Herring Gull *L. argentatus* complex, although a few (e.g. Cramp & Simmons 1980) treated it within Lesser Black-backed Gull, whilst Beaman (1994) drew early mainstream attention to the potential that this taxon might even deserve specific status. The current consensus appears to favour the recognition of the following taxa within the *fuscus* group: *L. f. graellsii, intermedius, fuscus, heuglini* and *barabensis* (with *taimyrensis* if recognised also part of this grouping). At least some authors (e.g. Yésou 2002) appear to favour subdividing the group into two species: Lesser Black-backed Gull *L. fuscus* (including *graellsii, intermedius* and *fuscus*) and Heuglin's Gull *L. heuglini* (including *heuglini, taimyrensis* and *barabensis*), which course was followed by Malling Olsen & Larsson (2003), with the exception that they treated *barabensis* within *Larus cachinnans* (Caspian Gull). The validity of *taimyrensis* and the position of *barabensis* are in need of additional research. Although *barabensis* (Steppe Gull) is genetically very closely related to *heuglini* (Liebers *et al.* 2001), it clearly differs in biometrics and morphology, and the two breed in quite different environments. In one of the widest-ranging studies, Liebers *et al.* (2001), who sampled mtDNA from nine taxa of large white-headed gulls, found evidence for two separate evolutionary lineages, which despite, at least some, modern-day contact, remain genetically quite distinct, with the *fuscus* group clustering within a northwest Palearctic/Central Asian clade that also includes *cachinnans, barabensis* and *mongolicus*.

Status and distribution. Widespread and not uncommon in coastal areas of Turkey on passage, and recorded in all months, occasionally in large numbers, but generally rather scarce and only ever noted in small numbers inland. Wintering population presently estimated at just 30–90 individuals, and just one bird was recorded during the January 1999 midwinter waterfowl count, although quite a number of suitable areas for the species were not surveyed at this time. In Marmara, apparently scarce and generally in small numbers away from the Bosphorus / Istanbul area, where it seems reasonably common. In Aegean there are relatively few records and numbers always seem to be very low. In Mediterranean present in moderate numbers late February to late May, with several counts involving more than 200 individuals. In Inner Anatolia very scarce and always recorded in small numbers. In Black Sea Coastlands widespread along the coast from mid August to mid May, but usually only noted in rather small numbers (e.g. just eight in a count of the entire Black Sea coastline in February 1997), although much larger numbers (up to 1,750) are regularly recorded at the Bosphorus, especially on autumn passage, but larger counts on passage seem to be rather scarcer than expected in this region (max. 600 in early May). In the inland eastern third of the country Lesser Black-backed Gull is very scarce and only ever recorded in small numbers, principally on large lakes and reservoirs.

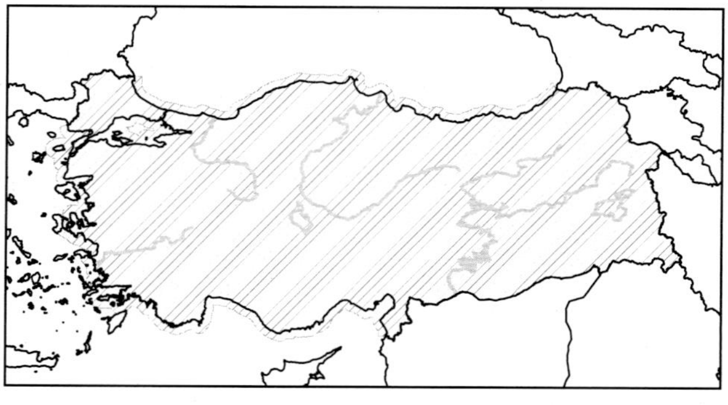

YELLOW-LEGGED GULL — *Larus michahellis*
CASPIAN GULL — *Larus cachinnans*
Gümüş Martı

(Du: Geelpootmeeuw; Fr: Goéland leucophée; Ger: Weißkopfmöwe; Swe: Gulfotad trut)

Subspecies and taxonomy. Two forms (increasingly recognised as species) occur: *michahellis* and *cachinnans*. Their respective ranges and status in Turkey are somewhat debated. Roselaar (1995) considered that *michahellis* breeds on Mediterranean coasts and those on the Black Sea coast to be *cachinnans*, whereas Malling Olsen & Larsson (2003) stated *michahellis* to be resident on the Black Sea and north Aegean coasts, and *cachinnans* to be merely a migrant across Turkey. In fact, as ably demonstrated by Liebers & Helbig (1999) *michahellis* also breeds relatively far inland, in a mixed colony with Armenian Gull *L. armenicus* (which see) on Beyşehir Gölü (Mediterranean). As yet, with the meritorious exception of the last-cited work, relatively little taxonomic work on this complex of gulls has been undertaken in Turkey. Nonetheless, it is clear that most of the large colonies of 'yellow-legged' gulls in Turkey are, in fact, *michahellis*, and that the breeding range of this form in the country spans the Mediterranean coast (and inland), as well as the Aegean, Marmara and Black Sea coasts. However, we also consider that some colonies are of *cachinnans*, in Black Sea Coastlands and Marmara (as long ago stated in OST 1975, 1978), making Turkey an excellent region in which to undertake further studies of these gulls and their interactions.

The 'Herring Gull' complex is increasingly being considered to comprise six species: *Larus argentatus* (Herring Gull), *L. smithsonianus* (American Herring or Smithsonian Gull), *L. vegae* (Vega or East Siberian Gull), *L. cachinnans* (Caspian Gull), *L. michahellis* (Yellow-legged Gull) and *L. armenicus* (Armenian Gull), based in particular on the important contributions of Liebers *et al.* (2001), Crochet *et al.* (2002) and Yésou (2002) (see *Brit. Birds* 98: 109). Malling Olsen & Larsson (2003) also recognised all of these six taxa at species level. We have not studied the complicated issue of whether to acknowledge both *michahellis* as well as *cachinnans* specifically, but consider it likely that two species are involved, especially given that the two nest in close proximity but in different habitats in the western Black Sea, in Romania (Klein & Buchheim 1997).

Status and distribution. National conservation status: Least Concern (population apparently increasing). Two forms – *cachinnans* and *michahellis* – which are probably best considered as species, breed in Turkey (see above). Eken (1997c), without discriminating between these forms, identified 22 confirmed coastal nesting sites and a further 15 possible or potential breeding sites on coasts. All these colonies are distributed between Giresun Island, on the eastern Black Sea coast, and islands off the Göksu Delta, on the eastern Mediterranean coast, with most definite or potential breeding sites in Aegean and, to a lesser extent, Mediterranean. Most colonies are rather small, mostly fewer than 500 pairs, but Eken (*op. cit.*) suggested a population of 3,000–5,000 pairs in Istanbul and that up to 3,000 pairs breed on islands in the Sea of Marmara. In addition, one inland colony is known, on a small island in Beyşehir Gölü (Mediterranean) where a total 40–50 pairs (including some *L. armenicus* and others showing mixed phenotypes) was present in May 1999, and *c.*40 pairs were present in 1964 (Vauk 1973). All other inland colonies appear to be pure *L. armenicus* (pers. obs.; Suter 1990).

In the non-breeding season, the relative status of the two forms is even more confused, with very few observers attempting to distinguish them. Considered together, they are widespread and common throughout most of the country in winter, especially in coastal areas, with smaller numbers remaining at inland wetlands and rivers, including Inner, Southeast and East Anatolia. The mean wintering population was placed at 100,000–120,000 individuals (Red Data Book), with large numbers regular at a number of sites, particularly in Marmara and Aegean (e.g. up to 40,000 in Gediz Delta), but with large numbers also reported from the Black Sea coast (e.g. almost 34,000

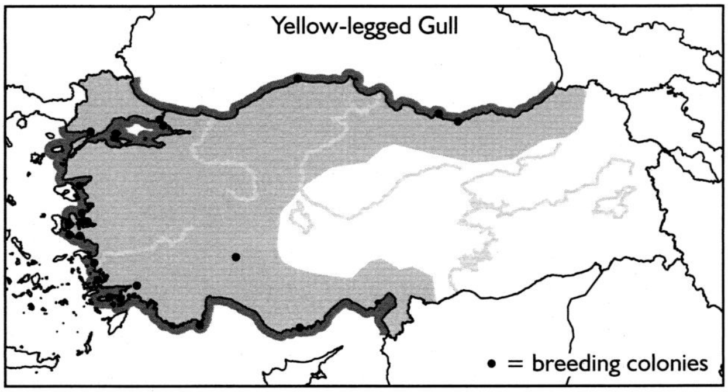

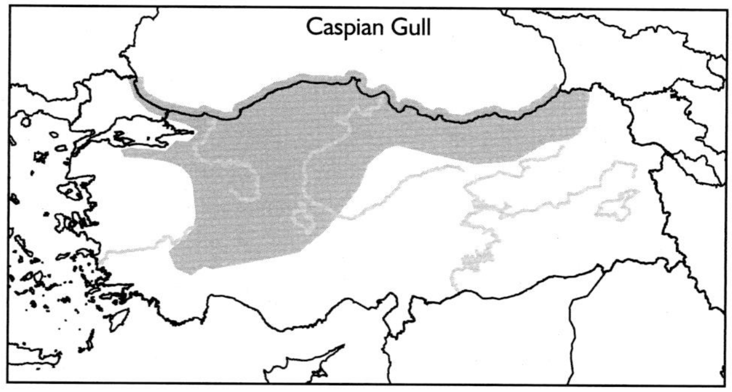

in February 1997). Most, but probably not all, wintering birds are *michahellis* (*cachinnans* reportedly winters chiefly in Arabia: Malling Olsen & Larsson 2003). On passage, both forms are rather widespread, principally across the western two-thirds of the country, but also in smaller numbers in the eastern third (where the only breeding birds are *L. armenicus*), including inland, where passage continues in spring until at least late May and commences in autumn in August.

Breeding. No attempt has been made to differentiate between *michahellis* and *cachinnans* in the following account, although most observations almost certainly relate to the former taxon. Nests on offshore rock stacks and islands; flat and low-lying as well as steep and rocky, on islands in coastal lakes, and on roofs of buildings, notably in Istanbul (**Marmara**). Nest is a large accumulation of plant material, seaweed and debris, with a central hollow lined with finer material. Eggs: mainly 2–3; C/3 (15), C/2 (11), C/1 (3). Earliest record, a young bird, *c.*1 week old, on an island near Anamur (**Mediterranean**) on 28 April 1972, indicates laying date for first egg as 20 March.Other records in this region include young, mainly fledged, at Gelindere on 5 July 1971 and on an island off Aydıncık on 13 June 1992 (PC), indicating laying in late March/early April. At Beyşehir Gölü, in May 1964, two nests had one egg, five nests had two eggs and 32 nests had three eggs (Vauk 1973). In **Aegean**, nests on an offshore island near Çeşme had eggs on 8 May 1950 (McNeile), and there were nests with eggs and newly hatched young on an island in Karine Gölü on 23 April 2003 (PC), indicating laying on 22 March. In **Black Sea Coastlands**, a juvenile, probably just fledged, was at Giresun on 11 June 1975, a similar date to those in other regions, but adults, some on nests, at colonies in Perşembe, Ordu and Giresun on 10–11 June 1975 are much later. Two young on a rooftop in Istanbul were able to fly in mid July, indicating egg laying in late April.

ARMENIAN GULL — *Larus armenicus*
Van Gölü Martısı

(Du: Armeense Meeuw; Fr: Goéland d'Arménie; Ger: Armenienmöwe; Swe: Armenisk trut)

Subspecies and taxonomy. Monotypic. Originally described as a subspecies of *Larus taimyrensis* and thereafter frequently treated as a subspecies of Herring Gull *L. argentatus* or, if considered specifically, Yellow-legged Gull *L. cachinnans* (see Beaman 1994). Liebers *et al.* (2001), in one of the most complete efforts to date to understand the systematics of the large white-headed gull complex, using mtDNA-sequencing techniques, considered *armenicus* to cluster within the Atlantic/Mediterranean clade (also including *atlantis* and *michahellis*). Only recently has it become broadly accepted that *L. armenicus* represents a valid, diagnosable species, or even taxon (Devillers & Potvliege 1981, Haffer *in* Glutz von Blotzheim & Bauer 1982, Snow & Perrins 1998, Sangster *et al.* 1998, Collinson 2006), although at least some of these authors noted that the principal (if not only) basis for accepting Armenian Gull at species level seemed to be its geographical isolation, which is demonstrably only partial (Liebers & Helbig 1999). Bourne (1989) and Filchagov (1993) pointed to phenotypic similarities between *armenicus* and *barabensis* (and thus *heuglini*), but the mtDNA profile of *armenicus* is apparently closest to *michahellis* (Liebers & Helbig 1999) and the most divergent within any of the large white-headed gulls sampled thus far (Liebers *et al.* 2001); it appears to be the relict of an early colonisation event from the Atlantic via the Mediterranean basin. Whilst as noted by several earlier authors, no single character can be used to diagnose *armenicus*, in combination wingtip pattern, bill marking (and mtDNA) can be used to successfully identify the species within its core breeding range. In areas of contact, e.g. with *michahellis*, the situation is much less clear-cut and intermediates are not uncommon (Liebers & Helbig 1999). The work of Chu (1998) offered further evidence of its species-level validity, but suggested (rather aberrantly) that it is more closely related to the

Mew *L. canus* and Ring-billed Gull *L. delawarensis* complex, although as noted by Sangster (1999) this proposition requires much additional support given the findings of others (see above). Nonetheless, species status for *armenicus* appears to have gained near-universal support (see *Brit. Birds* 98: 109).

Status and distribution. National conservation status: Least Concern (population apparently increasing). Liebers & Helbig (1999) estimated a Turkish population of *c.*2,400 pairs based on visits to three colonies (those in Van Gölü, Tuz Gölü and Beyşehir Gölü) in the summer of 1999, and suggested that the total population was in decline. However, it seems certain that the population is considerably larger than this. The Turkish population is extremely important in global terms, although large numbers also breed at Lake Sevan and Lake Arpi in Armenia, and Lake Uromiyeh in northwest Iran, with just 100–250 pairs in Georgia (A. Abuladze *in litt.* 2004). Currently breeds at six colonies in East Anatolia (Aktaş Gölü, Balık Gölü, Çıldır Gölü, Nemrut Gölü, Sodalı Gölü and Van Gölü), with one in Inner Anatolia (Tuz Gölü: 450–500 pairs), and one (a small outlier where very small numbers breed mixed with *L. michahellis* and birds of hybrid parentage are capable of reproducing) at Beyşehir Gölü (Mediterranean). Largest numbers are present in the East Anatolian colonies, with for instance that on Lake Van numbering 3,450 pairs in 1995 (Magnin & Yarar 1997), and only the colony recently discovered at Nemrut containing fewer than 300 pairs. Large numbers of non-breeders are also present in East Anatolia, especially in the vicinity of Van, but also west at least as far as Diyarbakır (Bertault *et al.* 1988) and Elazığ, where a colony of 70–100 pairs on Hazar Gölü was present in 1965 (Kumerloeve 1967a). We suspect that other small colonies await discovery.

The mean wintering population is placed at 8,000–12,000 individuals, with the 2005 midwinter count revealing 11,585 birds, although lower counts (e.g. 1,450 in midwinter 1999) are not unusual. Most seem to winter at two dam lakes, Karakaya and Keban, on the Euphrates, with e.g. over 4,300 at the former in midwinter 2006. At this season, some move to the Mediterranean coast (e.g. Çukurova and Göksu deltas) but numbers vary and significant numbers are also occasionally recorded at wetlands and even rivers in Inner Anatolia, as well as suitable (*i.e.* milder) areas of the east (e.g. the Iğdır Ovası) and perhaps also Southeast Anatolia, although this region is not well surveyed at this time of year. Only very scarcely reaches as far west as Aegean or Marmara coasts. Only small numbers appear to be present on the Black Sea coast (Welch & Welch 1998a) in winter, although reasonably large numbers appear to be present in autumn (Kumerloeve 1967a) and possibly even summer (pers. obs.), the latter presumably being non-breeders. Return to and departure from the colonies is poorly documented.

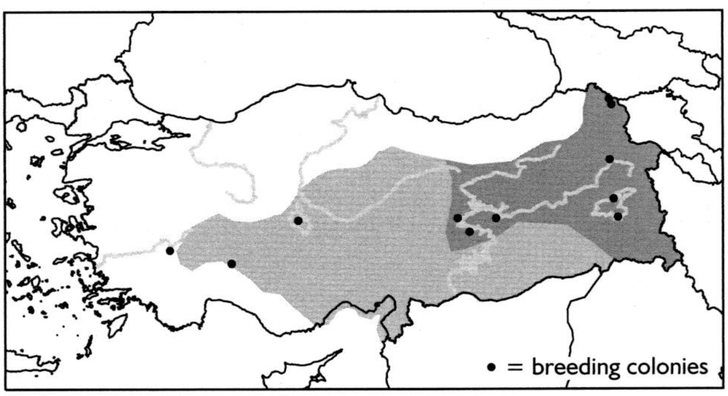

Breeding. Colonially, mainly on the ground, on both bare and well-vegetated islands in high-altitude lakes (to at least 2,000m) in **East Anatolia**, particularly at Van Gölü, Balık Gölü and Çıldır Gölü, and at Tuz Gölü (**Inner Anatolia**) and Beyşehir Gölü (**Mediterranean**). Colonies usually large, e.g. 400–500 pairs at Balık Gölü in 2001 (PC) and 1,100 pairs at Çıldır Gölü (Magnin & Yarar 1997), but solitary pairs breed on small islets, e.g. at Nemrut Gölü, in 2001 (T. Quittenden), where 12 pairs were nesting on a small island in June 2004 (PC). Nest size varies, probably according to available material; usually an accumulation of grass and plant material, with a central hollow lined with finer material. Eggs: C/2–3 (mainly three). Most nests had eggs, but also many young, up to *c.*1 week old, at Balık Gölü, on 9 June 2001 (PC). At Ahtamar Island (Van Gölü), there were 500 pairs and most nests had three eggs, on 30 May 1971, and *c.*1,500 pairs, with some chicks and eggs seen, on 8 June 1975, but just 15 pairs and *c.*50 empty nests (most, if not all, apparently destroyed by predators) there on 1 June 1999, at which time most of this colony were on Çarpanak Island, where c.2,000 pairs were present on 2 June 1999, with most having eggs that had either just hatched or were ready to do so (Liebers & Helbig 1999). At Çıldır Gölü, 900 pairs bred on three islands, with 291 young counted, most of them large but few able to fly, on 17–18 June 1992 (G. Magnin & L. Nauta). These records indicate that first eggs are laid early in May, consistent with colonies in Armenia, where the main hatching period is in late May (Adamian & Klem 1999). At Beyşehir Gölü

Armenian Gulls nest in a mixed colony with Yellow-legged Gulls *L. michahellis* (of which a total 40–50 pairs were present on 23–25 May 1999, when most had young *c.*2 weeks old and three pairs had recently laid replacement clutches). At Tuz Gölü, most nests had three eggs but some had newly hatched young on 9 May 1993, somewhat earlier than at colonies in East Anatolia (PC), and on 28–29 May 1999 there were 350–370 pairs in two colonies on islands 7km from the northwest shore, of which 95% of pairs had chicks from 1 day–3 weeks old, whilst the rest were still incubating: most clutches had three eggs (Liebers & Helbig 1999).

GLAUCOUS GULL *Larus hyperboreus*
Kutup Martısı

(Du: Grote Burgemeester; Fr: Goéland bourgmestre; Ger: Eismöwe; Swe: Vittrut)

Subspecies. We have not been able to examine the Turkish specimen, which may well no longer be extant, but on distributional grounds it is most probable that the bird referred to the nominate.

Status and distribution. Vagrant. Known in Turkey solely from an old specimen record. A young female was collected in the Bosphorus, Istanbul, on 23 February 1874 (Mathey-Dupraz 1924), who further noted that Alléon had claimed two adults of this species in the Bosphorus in autumn. There is a single (1914) record from Israel/Jordan (Andrews 1995, Shirihai 1996) but none from any of the immediately surrounding countries.

GREAT BLACK-BACKED GULL *Larus marinus*
Büyük Kara Sırtlı Martı

(Du: Grote Mantelmeeuw; Fr: Goéland marin; Ger: Mantelmöwe; Swe: Havstrut)

Subspecies. Monotypic.

Status and distribution. Vagrant or very scarce winter visitor to the western two-thirds of the country, which occasionally lingers into late spring (May and June). Recorded once inland in Inner Anatolia. Kumerloeve (1961) mentioned that the species was noted by von Gonzenbach at Izmir (Aegean) and by Schrader at Mersin (Mediterranean) on 23 April (in the early 1880s), the latter date and locality coinciding unusually closely with subsequent reports. An adult male was reportedly collected at the Bosphorus in the late 19th or early 20th century (Mathey-Dupraz (1924) and the same author noted a total of ten records from this region between 1888 and 1894, some involving multiple individuals and including at least three in July. An immature was reported, without details, at Gemlik, Bursa province (Marmara) on 23 July 1966 (Ganso & Spitzer 1967) and the species was mentioned for Manyas Gölü (Marmara) in April (year unknown; Kiziroğlu & Kiziroğlu 1987), but this record is best regarded as dubious. The vast majority of the 21 modern records (involving 29 individuals) have been in two coastal deltas, and 1997 was clearly an unusual year for the species in Turkey with records from three localities. Since the advent of modern recording the following records are available: one at Keban Reservoir, Elazığ province (East Anatolia) in late September 1979 (van der Ven & Gheyselinck 1980); a first-year at the Göksu Delta (Mediterranean) on 5 December 1984 (Kirwan *et al.* 2003), with another first-year on 6 May 1985, one on 10 April–1 May 1986 (K. Wilson), and one at sea nearby on the exceptionally late date of 22 June 1986 (Martins 1989), and additional records in the delta on 31 March 1995 (two), 11 February 1997, 2 November 1997 and 16 March 2000 (Kirwan & Martins 1994, Kirwan *et al.* 2003); one in Istanbul (Marmara) on 26 April 1986 (K. Wilson); an adult at the Kızılırmak Delta (Black Sea Coastlands) on

1–2 February 1992 (DHKD 1992), four on 3 February 1992, three on 22 March 1992, an adult on 20 May 1992 (Hustings & van Dijk 1994), one on 29 January 1995 (Kirwan & Martins 2000a) and one on 2 October 2005 (Balmer & Betton 2006); an adult at Üsküdar, Istanbul (Marmara) on 13 February 1996 (Kirwan & Martins 1994); four at Tuzla Gölü, Çukurova (Mediterranean) on 22 April 1997; one at Hirfanlı Reservoir (Inner Anatolia) on 14 May 1997 (Kirwan *et al.* 2003); one near Cernek, Kızılırmak Delta on 2 October 2005 (G. Welch); a long-staying individual near Haydarpaşa, Istanbul, between at least 20 September 2005 and 20 March 2006 and again present on 27 September 2006, with an adult at Kadiköy, Istanbul, on 1 March 2008 (E Cappe). Of interest, there are three records from the Greek part of the Evros (Meriç) Delta, the first in January 1965 (Bauer & Müller 1969) and one a long-staying adult from 1982 to 1994 (Handrinos & Akriotis 1997).

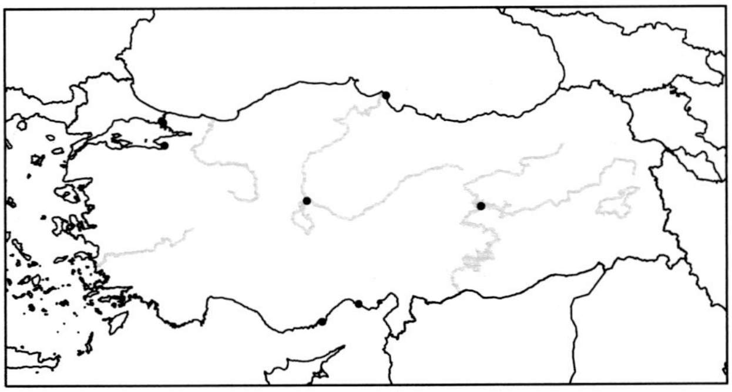

BLACK-LEGGED KITTIWAKE *Larus tridactylus*
Kara Ayaklı Martı

(Du: Drieteenmeeuw; Fr: Mouette tridactyla; Ger: Dreizehenmöwe; Swe: Tretåig mås)

Subspecies and taxonomy. Nominate. We place this species in *Larus* (a masculine genus, hence the specific name becomes *tridactylus*), although the genus *Rissa* has enjoyed considerable and prolonged currency for this species and its red-legged counterpart (*L. brevirostris*). The phylogenetic study of Chu (1998) upheld the validity of *Rissa*, but as noted under Little Gull *L. minutus*, a proposal that endeavours to recognise a number of smaller (and several poorly supported) genera within the gull family is less endorsable than that of expanding *Larus* to encompass the entire group. Like several other available genera for some of the smaller gulls, *Rissa* may prove to be a valid monophyletic grouping but for now we prefer to subsume it. A recent study (Chardine 2002) of geographic variation in wingtip patterns among the species found evidence for consistent and non-clinal differentiation between Pacific and Atlantic populations, thus upholding recognition of the form *pollicaris* in the former ocean (see also Collinson 2006). Chu's (1998) evolutionary tree suggested that *pollicaris* might even warrant specific recognition, although those on islands in northeast Siberia are reportedly intermediate between *pollicaris* and *tridactyla* (Malling Olsen & Larsson 2003).

Status and distribution. Formerly considered a vagrant, but recent records demonstrate it to be a reasonably regular, if scarce, winter visitor and almost certainly a passage migrant, based on the incidence of records in the south and inland in spring (and once in autumn). Mathey-Dupraz (1924) noted the presence of two immatures in the Robert's College (Bebek, Istanbul) collection, and that the species had been observed in winter at the Bosphorus. Records are available from all regions, with the exception of East Anatolia, although the vast majority is from the north, and extreme dates are 23 September and 5 May. The following are all records that we have noted: two immatures collected during the 19th century at the Bosphorus (Marmara), and thereafter an adult and seven immatures at the Euphrates by Birecik (Southeast Anatolia) on 23 September 1982 (Kinzelbach 1985), two between Antalya and Silifke (Mediterranean) on 5 January 1984 (Kasparek 1992a), an adult off Iskenderun (Mediterranean) on 6 January 1986 (*Sea Swallow* 36: 41), an immature at Taşucu (Mediterranean) on 31 December 1991 (Kasparek 1992a), an adult at Kulu Gölü (Inner Anatolia) on 27 April 1992, an immature at Kızkalesi, Içel province (Mediterranean) on 5 May 1992, single birds off Rumelifeneri, near Istanbul (Marmara) on 27 December 1992 and 3 January 1993, three first-summers off Karataş, Çukurova (Mediterranean) on 14 April 1994, three immatures in Karaburun harbour (Marmara) on 28 December 1994, a first-winter off Karabiga, Çanakkale province (Marmara) on 21 February 1996 (Kirwan & Martins 2000a), one in Izmit Bay (Marmara) on 11 January 1997 (Kirwan *et al.* 2003), a total of 20 (all first-winters) along the entire length of the Black Sea coast between 16 January and 7 February 1997, including 11 between Sarp and

Kemalpaşa, Artvin province (Black Sea Coastlands) on the first-named date (Welch & Welch 1998a), one in Foça harbour, Izmir province (Aegean) on 10 February 1997, one in the Bosphorus (Marmara) on 26 September 1998, a first-summer off Manavgat, Antalya province (Mediterranean) on 10–11 May 1999, a first-winter off Iğneada (Marmara) on 3 February 2001, a first-winter off Karataş, Çukurova (Mediterranean) on 14 April 2001 (Kirwan *et al.* 2003), with two more in the same place on 3–6 January 2002 (M. Ullman; Demirci 2002), five off Gökova, Muğla province (Aegean), on 14 February 2003 (Demirici 2003), one off Şile (Marmara), on 22 January 2005 (Balmer & Betton 2005b), three in Rize harbour (Black Sea Coastlands) on 27 March 2006 (S. Eksioğlu), one off Riva (Marmara) on 1 November 2006 (Balmer & Betton 2007a), one in the Gulf of Kocaeli (Marmara) on 5 November 2007 (S. Bekir), and four photographed off Foça (Aegean) on 4 February 2008 (F. Bengil *et al.*) and one at Gediz Delta (Aegean) on 3 May 2008 (O. Gül).

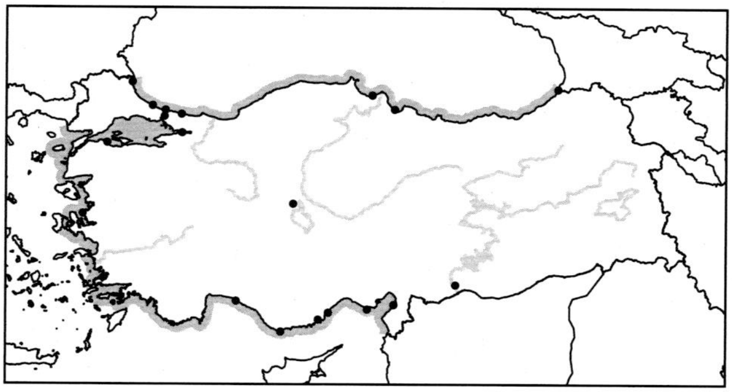

GULL-BILLED TERN *Gelochelidon nilotica*
Gülen Sumru

Du: Lachstern; Fr: Sterne hansel; Ger: Lachseeschwalbe; Swe: Sandtärna)

Subspecies and taxonomy. Nominate. Has also been placed in the genus *Sterna*. Indeed, the British Ornithologists' Union Records Committee (BOURC) quite recently (1996) elected to transfer this species to *Sterna*, thus subsuming *Gelochelidon* therein, on the basis that no cogent rationale for maintaining the latter genus had been published. Equally, there did not seem to be powerful reasons for the BOURC decision to treat *nilotica* within *Sterna*, which bucked the concept of stability within avian taxonomy and systematics without strong support for a change. Beaman (1994) also preferred to place the species in *Sterna*, but did not proffer any rationale for such treatment. Gochfeld & Burger (*in* del Hoyo *et al.* 1996) drew attention to this species' 'behaviour intermediate between typical terns and crested terns, and unique combination of characters [supporting] separate generic treatment'. More recently, an in-depth attempt to reconstruct a phylogeny of Sternini has once again provided support for the recognition of *Gelochelidon*, and has led the BOURC to overturn its previous decision (see Collinson 2006). The situation among gull relationships, where it is possible to recognise several small genera within a large family possessing a number of derived characters, might seem parallel, but evidence for the recognition of smaller genera is currently not consistently well supported (Chu 1998), making it arguably best to treat all gulls within a single genus, *Larus*, as done here.

Status and distribution. National conservation status: Vulnerable (population apparently decreasing). Widespread and common summer visitor to Inner Anatolia and in smaller numbers to Aegean (where several small colonies are known) and East Anatolia (where the largest colony, of up to 300 pairs, is at Bulanık, with much smaller numbers around Van Gölü). The largest colonies are in Inner Anatolia: at Kulu Gölü (max. 1,000 pairs, but more recently usually fewer than 450 pairs), Seyfe Gölü (max. 1,000 pairs, but more recently only up to 500 pairs), Ereğli marshes (max. 200 pairs, but probably no longer breeds at this site) and Karapınar Ovası (max. 80 pairs), but numbers are certainly declining in this region due to wetland drainage and degradation. Smaller

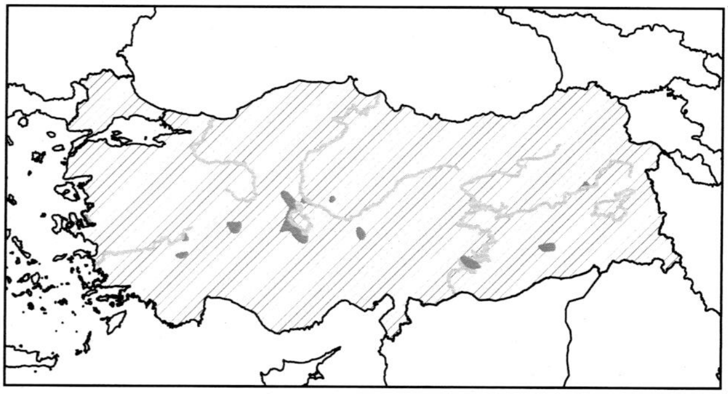

numbers (fewer than 50 pairs) breed or have bred at several other sites. Occurs in a variety of wetlands and sometimes encountered far from water.

Widespread throughout the country on passage, but generally in moderate numbers. In Marmara principally recorded in small numbers from Meriç Delta between early May and late August. In Mediterranean, Gull-billed Tern is recorded from early March, with an earliest date of 14th, and in reasonable numbers, and in autumn from mid August to October, with a late date of 19th. In Black Sea Coastlands it is a rare migrant in spring and autumn, as in Southeast Anatolia. Occasionally overwinters, usually singly, in Aegean, Mediterranean and even on the Black Sea coast.

Breeding. In colonies, on islands in saline or brackish lakes and on sandbanks and shingle islands in rivers. Colonies are usually of 80–300 pairs, but 1,000 nests were at Kulu Gölü (Inner Anatolia) on 2 June 1971 (Kasparek 1987a). Nests sometimes in loose association with other species, including Mediterranean Gulls *Larus melanocephalus*, Slender-billed Gulls *L. genei* and Black-headed Gulls *L. ridibundus*, and Common Terns *Sterna hirundo* and Little Terns *Sternula albifrons*. The nest is a shallow hollow in bare ground, sparsely lined with grasses or nearby vegetation. Eggs: 2–3, mainly three, occasionally one. In **Inner Anatolia**, first eggs are laid in late April/early May, earliest young hatching mid May, and large or recently fledged young appear in late June; dates consistent in the main colonies. There are records of nests still with eggs as late as mid June (Seyfe Gölü), late June (Bolluk Gölü) and even mid July (Kulu Gölü), when most young had already fledged, but these were probably replacements. In **East Anatolia** and **Aegean**, breeding dates are also consistent, with adults incubating in mid May and feeding recently fledged young in late June/early July.

CASPIAN TERN — *Hydroprogne caspia*
Hazar Sumrusu

(Du: Reuzenstern; Fr: Sterne caspienne; Ger: Raubseeschwalbe; Swe: Skräntärna)

Subspecies and taxonomy. Monotypic. Debate has centred on whether or not to recognise the monospecific genus *Hydroprogne* (see, *e.g.*, Gochfeld & Burger *in* del Hoyo *et al.* 1996, *Ibis* 147: 821–826, and Collinson 2006). Whilst ethological evidence supports the merger with *Sterna* and morphological recognition of *Hydroprogne* appears to be solely based on the very large bill, recently published genetic data certainly support the maintenance of *Hydroprogne*. See also the discussion under *Gelochelidon nilotica*.

Status and distribution. National conservation status: Vulnerable (population apparently increasing). Local summer visitor to saline wetlands in Aegean (principally at Gediz Delta and Büyük Menderes Delta) and Inner Anatolia (mainly at Tuz Golu, where the colony has declined from a maximum of 25 pairs to just a handful of pairs recently), in which regions it always occurs in small numbers. Occasionally recorded in summer at sites around Van Gölü and at Bulanık (East Anatolia), in both of which areas it breeds, but only in tiny numbers and apparently infrequently. Perhaps infrequently breeds at the Göksu Delta (Mediterranean), as speculated by Kasparek (1992a), although evidence for this is lacking (van den Berk & Guder 1992).

Otherwise, a more widespread but still very uncommon passage migrant throughout the country, principally on coasts, much rarer inland, and recorded in very small numbers in winter on Aegean and Mediterranean coasts, most frequently in Çukurova (Eken 1997d). One bearing a Finnish ring was recovered in Aegean in winter (Kasparek 1992a). Particularly rare at any season in Black Sea Coastlands. Passage through Mediterranean is typical of elsewhere in the country, the species being very rare in spring in this region, and scarce in autumn (August–October), and occasionally recorded in winter (most records at this season are from January), with the highest count being 23 at Tuzla lagoon, Adana (Mediterranean), on 20 January 2005.

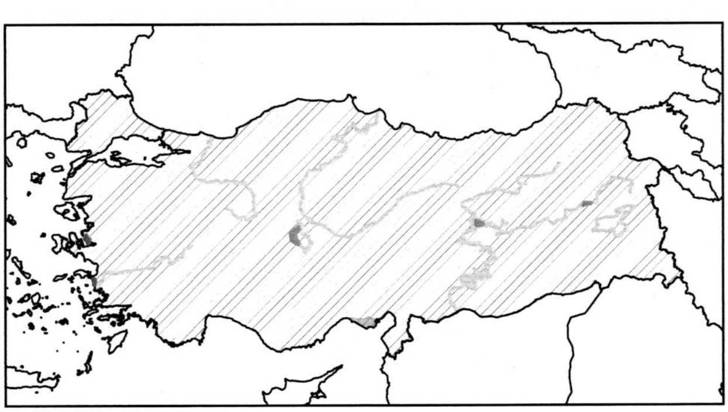

Breeding. In small colonies on low, shell-covered or bare islands in saline lakes. In Gediz Delta (Aegean), 103 pairs bred in four colonies in 1995 (Eken 1997a). Colonies may be compact and dense, with several pairs breeding in close proximity, or widely spaced, with several metres between pairs. Nest is a shallow hollow, unlined or with a sparse lining of plant debris. Of 15 pairs at Tuz Gölü (**Inner Anatolia**) in May 1969, 13 nests had two eggs and two nests had three eggs; on 31 May 1971, 19 nests had 2–3 eggs or both eggs and young totalling 2–3, and the oldest of 12 young was *c.*8 days, indicating first laying on *c.*2 May; adults were also incubating on fairly widely spaced nests at this colony on 28 May 1974, when one nest had a newly hatched young (Pforr & Limbrunner 1982). At Karine Gölü, Büyük Menderes Delta (**Aegean**), in 2003, 28 pairs nested in two separate compact groups on a small island; on 23 April three nests had three eggs, ten nests 1–2 eggs, and there were ten new, but empty, scrapes; on 9 May, nine nests had three eggs, 17 had two eggs, and two nests had one egg: a mean of 2.2 eggs in 28 nests. At the same location, a late nest had two eggs on 27 June 1999, but was probably a replacement (PC); all nests in 2004 still had eggs on 26 May (M. Demain).

LESSER CRESTED TERN — *Sterna bengalensis*
Tepeli Sumru

(Du: Bengaalse Stern; Fr: Sterne voyageuse; Ger: Rüppellseeschwalbe; Swe: Mindre iltärna)

Subspecies and taxonomy. The Turkish records were not subspecifically identified. Has been placed in the genus *Thalasseus* (e.g. Gochfeld & Burger *in* del Hoyo *et al.* 1996) and a recent genetic study provided weak support for this option.

Status and distribution. Vagrant. Three records: a first-year apparently well seen (but only briefly described) at the Göksu Delta (Mediterranean) on 28 May 1973 (Witt 1976). Another was reported flying past at sea at the same site on 5 June 2003 (S. Hartill & G. Beck), but no details are available. One was at Küçük Menderes Delta (Aegean) on 13 March 2004 (Balmer & Betton 2004b), for which a brief description is available. The species is a regular but rare non-breeding spring and summer visitor to Eilat, Israel (Shirihai 1996) with two records in neighbouring Jordan (Andrews 1995). It formerly bred on one island off the Lebanese coast (Ramadan-Jaradi & Ramadan-Jaradi 1999) and there are three recent records in Greece, all since 1987 (Handrinos & Akriotis 1997).

SANDWICH TERN — *Sterna sandvicensis*
Kara Gagalı Sumru

(Du: Grote Stern; Fr: Sterne caugek; Ger: Brandseeschwalbe; Swe: Kentsk tärna)

Subspecies. Nominate. Has been placed in the genus *Thalasseus* (e.g. Gochfeld & Burger *in* del Hoyo *et al.* 1996) and a recent genetic study provided weak support for this option.

Status and distribution. National conservation status: Endangered (population apparently stable). Recently (mid 1980s) discovered to breed at Gediz Delta (Aegean) in small numbers (see below). Otherwise a fairly widespread and not uncommon passage migrant in coastal areas, wintering in very small numbers in west and central Turkey. In Marmara recorded from late September to early April, with highest counts *c.*450 between the Büyük Islands and Yalova in January 1972, 231 at Büyükçekmece in March 1974 and 150 there in November 1970, and 170 at the Kavak Delta in January 1987. In Black Sea Coastlands, recorded from late September to early April, usually in small numbers but regular in Kızılırmak Delta and around Samsun, usually in numbers less than 100. In Aegean generally recorded between late October and late May, mostly in small numbers but with a maximum of *c.*100 in Büyük Menderes Delta in 1972. In Mediterranean recorded from early November to late March, usually in small numbers but with highest counts of *c.*50.

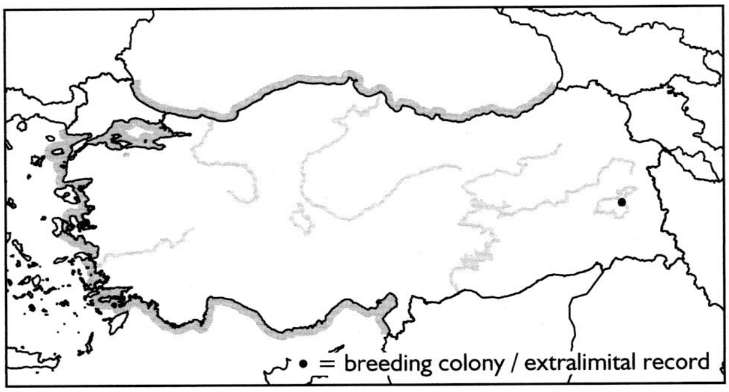

Breeding. Has bred only in Gediz Delta (**Aegean**), where confirmed in 1984–88 on islands in saline lagoons, on which Mediterranean Gulls *Larus melanocephalus* were also breeding. In early May 1996, the colony comprised 53 pairs of Sandwich Terns and 97 pairs of Mediterranean Gulls (Eken 1997a). No data in Turkey but elsewhere colonies are dense, with nests usually less than 0.5m apart; nests are shallow hollows in the ground, typically unlined; usual clutch two, occasionally one, laid mainly May.

[ROSEATE TERN *Sterna dougallii*]
Pembe Sumru

(Du: Dougalls Stern; Fr: Sterne de Dougall; Ger: Rosenseeschwalbe; Swe: Rosentärna)

Subspecies. The only race likely to occur is nominate *dougallii.*

Species not admitted to the Turkish list (Kirwan *et al.* 1999). Tristram (1889) mentioned a male specimen taken at Smyrna (Izmir) in 1851 that was held in his collection. We have not examined this specimen and its whereabouts appear to be unknown. Listed without details by Pasquali (1991) and treated as hypothetical by Ergene (1945); Renkhoff (1972) reported a pair feeding a juvenile daily at the Kızılırmak Delta during July 1971. This claim has not been fully documented and was rejected by Dijksen & Kasparek (1985). On distributional considerations, occurrence appears highly unlikely.

COMMON TERN *Sterna hirundo*
Sumru

(Du: Visdief; Fr: Sterne pierregarin; Ger: Flußseeschwalbe; Swe: Fisktärna)

Subspecies. Probably all are nominate, although birds recalling *tibetana* of Middle Asia have been claimed in the Middle East.

Status and distribution. National conservation status: Least Concern (population apparently stable). Widespread and common summer visitor to coastal areas and a variety of inland wetlands, principally in Marmara (e.g. 500 pairs at Manyas Gölü), Aegean (up to 2,150 pairs at Gediz Delta), Mediterranean (smaller numbers, but up to 100 pairs at Göksu Delta), Inner Anatolia (most colonies are of 50–100 pairs) and East Anatolia, but distinctly rarer and recorded at many fewer localities in the breeding season in Black Sea Coastlands (nesting confirmed at just one site) and Southeast Anatolia.

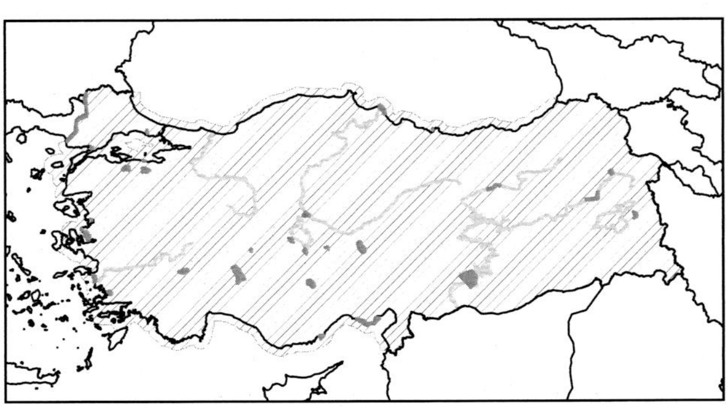

Commoner and even more widespread on passage, principally in autumn between at least early August and mid October, peaking in September when large numbers sometimes recorded in eastern Black Sea Coastlands, and more rarely noted to mid November. Less common in spring from mid March (14th), with occasional winter records, mainly from the Black Sea Coast, but a late December and early to mid February records from Mediterranean may also refer to overwintering birds rather than very late and early migrants respectively.

Breeding. In colonies, usually of fewer than 80 pairs, but up to 500 pairs, e.g. at Manyas Gölü (Marmara) in 1966. The population in Gediz Delta (Aegean) estimated at 2,150 pairs in 1996, is the most important in Turkey (Eken 1997a). Occasionally breeds solitarily and sometimes in close association with other species, e.g. Gull-billed Terns *Gelochelidon nilotica* and Little Terns *Sternula albifrons*, mainly on low islands of sand, mud or gravel in lakes (both coastal and inland), in flooded areas and along rivers, and on offshore coastal islands; also nests on floating vegetation at Uluabat Gölü, Marmara (Welch & Welch 1998b). The nest is usually a hollow in the ground, unlined or variably lined with nearby plant material. Eggs: C/3, occasionally two. In **Inner Anatolia**, in a colony of 60 pairs at Seyfe Gölü, 48 nests had eggs and the remaining pairs young on 2 June 1971 (Husband & Kasparek 1984), indicating egg laying about 10 May; some were still incubating eggs at Yarma on 26 July 1971; 100 pairs were nesting on an island in Çavuşçu Gölü on 29 May 1999 (R. F. Porter). In **Marmara**, four fledged young at Büyükçekmece (Thrace) on 19 June 1973, indicate egg laying in early May; four nests with eggs were in a colony of *c.*30 pairs at Uluabat Gölü on 2 June 1966 (R. F. Porter); other records in Marmara include nests with eggs in May (earliest 14 May) and June, and fledged young at Kocaçay Delta on 11 July 1966 (Ertan 1996). At Karine Gölü (**Aegean**), most pairs had young, ranging from newly hatched to almost full grown, but several nests still had eggs, on 27 June 1999, and breeding was still in its early stages on 14 May 2007, when two nests had incomplete clutches of one egg. (PC); McNeile collected two well-incubated eggs from a small coastal islet near Aliağa on 23 May 1950. In **Mediterranean**, adults were incubating at Göksu Delta from 15 May, with *c.*100 incubating on nests on sandbars in Akgöl on 19 May 1999 (R. F. Porter), and fledged young were seen from early July. In **East Anatolia**, adults were incubating on nests at Erçek Gölü in late May, and on shingle islands in the Murat River and at Karakaya Reservoir, Malatya, in early June. In **Black Sea Coastlands**, 30–50 pairs were nesting on a small offshore islet at Fatsa on 10 June 1975. Records of late nests with eggs are probably replacements.

ARCTIC TERN — *Sterna paradisaea*
Kutup Sumrusu

(Du: Noordse Stern; Fr: Sterne arctique; Ger: Küstenseeschwalbe; Swe: Silvertärna)

Subspecies. Monotypic.

Status and distribution. Vagrant. Six records as follows: two adults collected in the Sea of Marmara (Marmara) on 27 May 1915 (Mathey-Dupraz 1924), one in the Bosphorus (Marmara) on 4 September 1959 and two there the following day (Ballance & Lee 1961), four adults (photographed) 2km west of Ahlat, Van Gölü (East Anatolia) on 25 May 1969 (OST 1972), two between Taşlıçay and Yoncalı, Ağrı province (East Anatolia) on 1 May 1970 (OST 1975) and, most recently, several off Iztuzu beach, Dalyan (Aegean) on 27 April 1996 (Kirwan & Martins 2000a) and up to four, at Tuzla Milas, near Bodrum (Aegean) between 12 and 27 May 2005 (B. W. Stoneman *et al.*). Records by Renkhoff (1972) at Kızılırmak Delta (Black Sea Coastlands) in late July 1971 were regarded as unproven by Dijksen & Kasparek (1985). The species is a vagrant to Cyprus and Israel (Porter *et al.* 1996) and there are at least eight records from Greece (Handrinos & Akriotis 1997).

LITTLE TERN *Sternula albifrons*
Küçük Sumru

(Du: Dwergstern; Fr: Sterne naine; Ger: Zwergseeschwalbe; Swe: Småtärna)

Subspecies and taxonomy. Nominate. Until very recently usually placed in the genus *Sterna*, but recent genetic data support the removal of this species, Least Tern *S. antillarum* and Saunders's Little Tern *S. saundersi* to their own genus (see *Ibis* 147: 821–826 and Collinson 2006).

Status and distribution. National conservation status: Near Threatened (population apparently decreasing). Local summer visitor, principally to coastal sites and inland playa lakes, generally between mid April (earliest 6th) and late September (latest 29th), to Marmara (largest colony 200 pairs in Meriç Delta), Aegean (largest colonies 300 pairs in Büyük Menderes Delta and 205 pairs in Gediz Delta), Inner Anatolia (largest colony 500 pairs at Seyfe Gölü), and Mediterranean (largest colony in Çukurova, of *c.*300 pairs), with just single known colonies in Black Sea Coastlands and East Anatolia. In Southeast Anatolia there were 34 pairs at Birecik in 1973, but breeding was not noted subsequently until the late 1990s when the species was discovered breeding at several riverine sites, mainly on the Tigris. More widespread on passage, especially in May and September, and generally in small numbers, although significant numbers have been recorded in eastern Black Sea Coastlands and Mediterranean in autumn.

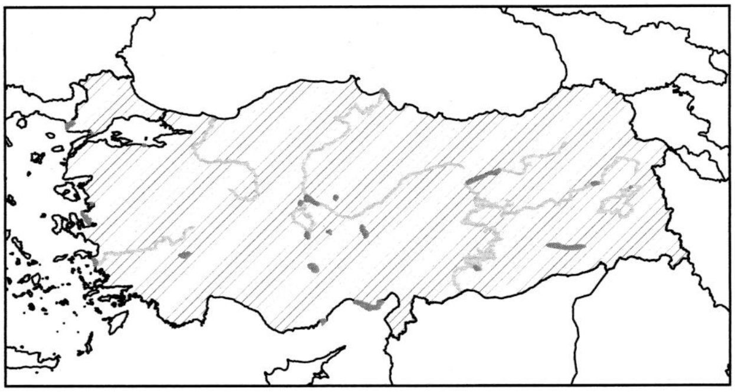

Breeding. On islands and sandbanks, on coasts and at inland lakes and rivers. Colonies range from 2–3 pairs to 67 pairs, in Sultan marshes (Inner Anatolia) (Kasparek 1985), and there were 270 pairs, evenly distributed over all coastal lagoons, in Çukurova (Mediterranean) in 1987 (van der Have *et al.* 1988). See also above. Nest is a shallow hollow in bare sand, shingle, gravel and dried mud. Eggs: usually three (91) or two (16), occasionally one (4). Breeding commences mainly in the second half of May and June. The earliest records are of many nests, with full clutches of three eggs, in Çukurova on 9 May 1989 and 11 May 1990 (J. Whittaker). At Karine Gölü (**Aegean**), adults were on territory but not yet nesting on 9 May 2003; on 14 May 2007, there were *c.*20 new but empty nest scrapes, 16 nests with incomplete clutches of one egg, and one nest with two eggs, and several of these were unusually close, some only 1m apart, on a small shingle island (PC); elsewhere in Büyük Menderes Delta, two nests had eggs and one a newly hatched chick on 26 June 1999 (PC), and several adults were incubating on nests in coastal lagoons on 25 May 2004 (M. Demain). In **Marmara**, there were four nests with eggs in a colony of *c.*15 pairs at Uluabat Gölü on 2 June 1966 (R. F. Porter), and a pair feeding young at Kocaçay Delta on 16 September 1995 (Ertan 1996). In **Inner Anatolia**, 50 nests had 2–3 eggs, one had one egg, and two nests had newly hatched young on 3 June 1982 in Sultan marshes (Kasparek 1985); 48 nests had 2–3 eggs at Seyfe Gölü on 2 June 1971 (Husband & Kasparek 1984); a nest had two eggs on an island at Kulu Gölü on 19 June 1992 (PC). In **Mediterranean**, pre-nuptial courtship was seen near Silifke on 30 April 1970 (R. F. Porter); at Göksu Delta adults were incubating on 26 May 1998 (T. Quittenden) and 11 June 1992 (PC), four nests each had three eggs on 9 June 1974, and six nests had newly hatched young on 16 June 1974 (Witt 1976); two fledged young were seen on 8 June 1999 (R. F. Porter), indicating egg laying in early May. At Kızılırmak Delta (**Black Sea Coastlands**) there were 40–45 pairs in 1992, mostly along the river, and they had just started to breed by 9 June 1992 (Hustings & van Dijk 1994). In **East Anatolia** 5–10 pairs were feeding juveniles at Karakaya Reservoir, Malatya, on 2 July 2005 (MÖ).

WHISKERED TERN *Chlidonias hybrida*
Bıyıklı Sumru

(Du: Witwangstern; Fr: Guifette moustac; Ger: Weißbart-Seeschwalbe; Swe: Skäggtärna)

Subspecies. Nominate. Note that the correct spelling of the specific name is *hybrida*, not *hybridus* (David & Gosselin 2002a).

Status and distribution. National conservation status: Near Threatened (population apparently decreasing). Local summer visitor (generally from late April to late August/September) to wetlands in west and central Turkey, where it often occurs in large numbers; *e.g.* in colonies of 500 pairs at Meriç Delta and 660 pairs at Uluabat Golu (Marmara), and widespread in Inner Anatolia (largest colony of up to 400 pairs at Sultan marshes), with much smaller numbers in Aegean and Mediterranean. Apparently largely absent as a breeder from Black Sea Coastlands, Southeast and East Anatolia, where only a handful of potential breeding localities have been identified.

Frequent on passage across virtually the whole country, although largest numbers are again noted in the western two-thirds of Turkey. Spring passage continues from at least late April (earliest 30 March in Mediterranean) to early June, and in autumn migration is obvious from mid to late August until late September (more rarely noted until mid October). Recorded in very small numbers in winter from Aegean, Inner Anatolia and Mediterranean, usually at south-coast wetlands but also inland, from late November to late February.

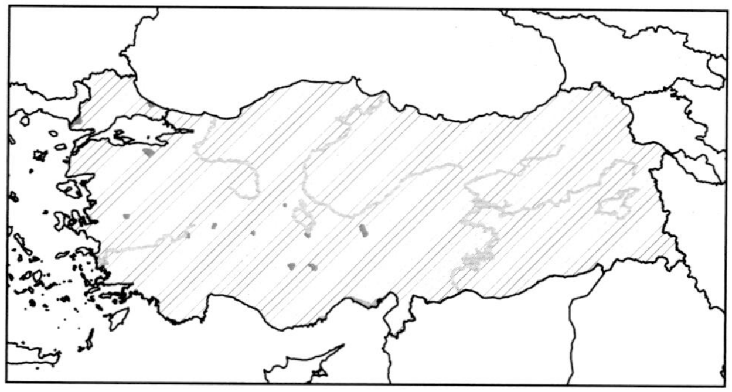

Breeding. In colonies, in sheltered areas of lakes with emergent vegetation, e.g. water lilies. These may be large, with 500 pairs in Meriç Delta (**Marmara**) and 400 pairs in Sultan marshes (**Inner Anatolia**); at least 660 nests at Uluabat Gölü (Marmara) in 1998, all on floating lily pads, of which 392 pairs were breeding together in dense water lilies in one bay (Welch & Welch 1998b); this is a regular and long-established breeding site. Smaller colonies also recorded in Inner Anatolia, e.g. ten pairs at Cogenç Gölü on 2 June 1972, 30 pairs at Hotamış on 27 May 1975 and 100 pairs at Çavuşçu Gölü (Magnin & Yarar 1997). The nest is a slight accumulation of reeds and other aquatic vegetation, interwoven into the floating lilies. Eggs: two nests, with one and three eggs, at Sultan marshes on 2 June 1975 (Kasparek 1985), and four nests with two, two, three, and three eggs, at Uluabat Gölü on 2 June 2006 (M. Farmer), are the only recorded contents; elsewhere, usual clutch three (occasionally 2–4). Colonies are active from mid May; at Uluabat Gölü, many adults were carrying nest material on 2/3 June 2006 (PC) and adults were carrying food to young in nests on 28 June 1973 (R. F. Porter); in Inner Anatolia, there were newly fledged young near Yarma on 25 July 1971, at Akşehir Gölü on 3 August 1971 and at Sultan marshes on 22 July 1981 (Kasparek 1985), indicating peak egg laying in early June.

BLACK TERN *Chlidonias niger*
Kara Sumru

(Du: Zwarte Stern; Fr: Guifette noire; Ger: Trauerseeschwalbe; Swe: Svarttärna)

Subspecies. Nominate.

Status and distribution. National conservation status: Vulnerable (population apparently decreasing). Local summer visitor to parts of Marmara and over much of Inner Anatolia, and speculated also to breed in East Anatolia (Kasparek 1992a), although there seems to be no proof of this. Most colonies in Marmara are of 25+ pairs and the species is rather widespread in Inner Anatolia, often in significant numbers.

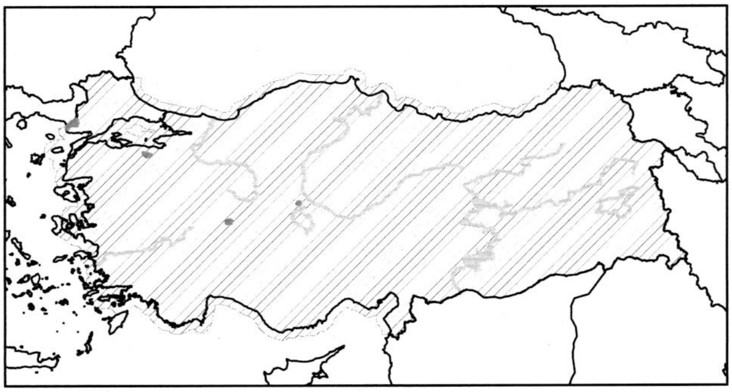

Even more widespread and rather commoner on migration (albeit in generally moderate numbers and still distinctly less frequent in the eastern third of the country). Passage is principally in April–May and August–October (peaking September), although migration occasionally commences by early July in Mediterranean and a concentration of over 2,000 birds has been noted in this region (at Burdur Gölü). Principally recorded between mid April and mid October. There is a recent winter record from Kızılırmak Delta (Black Sea Coastlands), on 18 December 2005 (Balmer & Betton 2006a).

Breeding. In small colonies in sheltered areas of lakes, with emergent or floating vegetation, e.g. water lilies. The nest is a heap of reeds and water plants, attached to or resting on growing vegetation, and with a hollow in the top lined with finer material. At Uluabat Gölü (**Marmara**): 15–20 pairs nested on floating vegetation on 2 June 1966 (three nests examined by R. F. Porter had C/2, C/1, empty, and were probably incomplete clutches); there were four colonies, totalling *c.*100 pairs, in May 1967; up to 40 birds were present in May 1969 and nesting activity was observed; and at least three pairs were nesting within a colony of Whiskered Terns *Chlidonias hybrida* on 15 May–20 June 1998 (Welch & Welch 1998b). At Eber Gölü (**Inner Anatolia**), there were two colonies totalling 15 pairs in 1967, with nest building observed on 8 May (R. F. Porter). No detailed records of clutch size in Turkey but, elsewhere, usual clutch three.

WHITE-WINGED TERN *Chlidonias leucopterus*
Ak Kanatlı Sumru

(Du: Witvleugelstern; Fr: Guifette leucoptère; Ger: Weißflügel-Seeschwalbe; Swe: Vitvingad tärna)

Subspecies. Monotypic.

Status and distribution. National conservation status: Near Threatened (population apparently decreasing). A fairly widespread and common passage migrant to wetlands and coastal areas, especially across the eastern two-thirds of the country, which is regularly recorded in summer and often assumed to breed. Indeed, Kasparek (1992a) considered breeding events to be not rare in Inner and East Anatolia, as well as in Marmara and elsewhere in the north of the country. However, our detailed dissection of the literature, both published and unpublished, has revealed very few records that can be considered even strongly suggestive of breeding (see below), and recommend that any definite future nesting records be thoroughly documented. Nonetheless, it does seem likely that breeding occurs in Turkey at least occasionally.

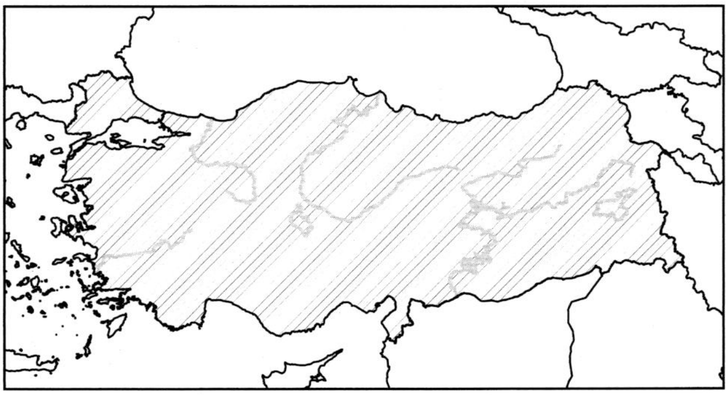

Common on passage at both seasons, especially spring, from mid April (earliest 8th) until early June, with most in late April to mid May (especially in the south and east). Migrants are generally noted in northern localities up to two weeks later than in the south. Non-adults tend to move through later than adults. In autumn noted from mid July (earliest 10th) to mid October, with the bulk in late September and the latest records, from Inner Anatolia, on 25 October, and Mediterranean, on 21 October. Generally recorded in much smaller numbers in the western third of country, where flocks are rarely more than 100 strong, but huge numbers are occasionally noted in eastern Turkey, where counts in the low thousands even prove to be regular (see van den Berk *et al.* 1993); e.g. 15,000–20,000 at Erzurum Ovası (East Anatolia) on 13 May 1993 (Faldborg 1994), and 11,500 at Sodalı Gölü (East Anatolia), on 6 May 2004 (GMK *et al.*).

Breeding. Adults observed over marshland during the breeding season, and nesting occasionally assumed. In **Inner Anatolia**, a fully fledged juvenile at Mogan Gölü on 5 July 1969 and three juveniles at Temelli on 31 July 1969, may have been reared locally; a pair, two other adults and three young were at Seyfe Gölü on 19 June 1975 (Husband & Kasparek 1984). In **Black Sea Coastlands**, two adults and a fledged young were at Yeniçağa Gölü on 16 June 1983 (Kılıç & Kasparek 1987), and adults were observed feeding juveniles in Kızılırmak Delta on 17 July 1992 (Hustings & Van Dijk 1994). In **East Anatolia**, nine birds, including three juveniles, were at Haçlı Gölü on 17 July 1986, but there is no proof that the juveniles were bred in Turkey; at Saz Gölü, near Doğubeyazıt, there were 150 birds, presumed to be in small colonies, on 4 June 2002, and *c.*100 adults were seen there on 28 May 2003, but no conclusive proof of breeding was obtained on either occasion (H. Dufourny); at Karakaya Reservoir, near Malatya, where there is suitable breeding habitat, *c.*20 adults were present on 10 June 2005 and more than 35 on 18 July 2005, of which 10–15 were juveniles, some being fed by adults (MÖ). There is no real evidence of breeding in Turkey and birds observed are considered to be on passage, remaining at sites for short periods only. At breeding colonies, birds are present for 2–3 months and are conspicuous and noisy, being unlikely to be overlooked.

PALLAS'S SANDGROUSE *Syrrhaptes paradoxus*
Paçalı Bağırtlak

(Du: Steppehoen; Fr: Syrrhapte paradoxal; Ger: Steppenflughuhn; Swe: Stäpphöna)

Subspecies. Monotypic.

Status and distribution. Vagrant, but perhaps best regarded as hypothetical. During one of the species' large invasions into western and central Europe, in 1888, four Pallas's Sandgrouse, apparently taken at Rumeli (presumably modern-day Rumelifeneri), were noted in the Istanbul poultry market in mid November (Mathey-Dupraz 1923). In addition to the 1888 date, Ergene (1945) mentioned that the species reached Turkey during the irruptions of 1859 and 1863, but provided no further details. In 1962, Tansu Gürpınar (pers.comm.) saw a stuffed individual at the Çumra hunting club, near Konya (Inner Anatolia), but no further details are available. The 1863 and 1888 movements into Europe were among the largest ever recorded (Madge & McGowan 2002). There are no modern records in Turkey and, elsewhere in the Middle East, the species is listed only as a 'probably irregular visitor to Iranian steppes east of Caspian Sea' (Porter *et al.* 1996).

[CROWNED SANDGROUSE *Pterocles coronatus*]
Taçlı Bağırtlak

(Du: Kroonzandhoen; Fr: Ganga couronné; Ger: Kronenflughuhn; Swe: Kronflyghöna)

Subspecies. The form *vastitas* is the only one likely to occur.

Species not admitted to the Turkish list (Kirwan *et al.* 1999). Listed by Ergene (1945) without supporting details. Although the species is known to be nomadic, occurrence in Turkey appears unlikely.

SPOTTED SANDGROUSE *Pterocles senegallus*
Benekli Bağırtlak

(Du: Sahelzandhoen; Fr: Ganga tacheté; Ger: Tropfenflughuhn; Swe: Ökenflyghöna)

Subspecies. Monotypic.

Status and distribution. Vagrant, though the possibility that the species is more regular than records suggest was discussed by Kirwan (2000). Two records, both at Birecik (Southeast Anatolia): a female on 18 July 1986 (Martins 1989) and a pair on 20 June 1999 (Kirwan *op. cit.*). Both occurrences were well described by the observers concerned. Additionally, there is at least one other observation of two, possibly of this species, also at Birecik, on 17 July 1987 (C. Beel & F. Verbelen), for which some descriptive details are available. Spotted Sandgrouse formerly bred in central and southern Syria, but there is only one record from the north of the country (in September 1945) and just one recent occurrence in central Syria, in April 1994 (Baumgart *et al.* 1995). The nearest currently occupied breeding grounds to Turkey appear to be southern Iraq and the Levant.

[CHESTNUT-BELLIED SANDGROUSE *Pterocles exustus*]
Kahverengi Bağırtlak

(Du: Roodbuikzandhoen; Fr: Ganga à ventre brun; Ger: Braunbauch-Flughuhn; Swe: Brunbukig flyghöna)

Subspecies. Only *erlangeri* (from Arabia) could conceivably reach Turkey.

Species not admitted to the Turkish list (Kirwan *et al.* 1999). Listed by Ergene (1945) without supporting details. As with other species of sandgrouse, known to be nomadic. Within the Middle East, breeds only in Arabia and, although it has reached Israel and Jordan as a vagrant (Porter *et al.* 1996) occurrence in Turkey appears unlikely.

BLACK-BELLIED SANDGROUSE *Pterocles orientalis*
Bağırtlak

(Du: Zwartbuikzandhoen; Fr: Ganga unibande; Ger: Sandflughuhn; Swe: Svartbukig flyghöna)

Subspecies. Nominate. Species described from Turkey.

Status and distribution. National conservation status: Vulnerable (population apparently decreasing). A fairly common resident and/or partial migrant over much of Inner and East Anatolia and, at least formerly, in Southeast Anatolia, reaching at least 2,300m (albeit perhaps only locally). Particularly widespread in Inner Anatolia, where largely absent only from the extreme north of the region, and (at least formerly) often recorded in large flocks (several counts in excess of 100 and a maximum of 500+ in September 1974). Small numbers also appear to breed in neighbouring parts of extreme southern Black Sea Coastlands, extreme eastern Aegean and extreme northern Mediterranean, in all of which regions the species is principally a winter visitor (see below). In East Anatolia, where almost certainly more widespread than records suggest, it is regularly encountered from the environs of Van Gölü north through Horasan and Ağrı to Kağızman, Iğdır and east to the Doğubeyazıt area, with records west as far as Hafik, but is generally only recorded in moderate numbers (rarely more than *c*.50). This sandgrouse is now generally very scarce and local in Southeast

Anatolia, even at Birecik, where it was formerly numerous with hundreds regularly being recorded there (Welch 2004). In all three of these regions, the species seems to be declining as a result of the modification of dry steppe for agriculture.

Black-bellied Sandgrouse is a partial migrant in many areas, moving to milder regions in winter, e.g. being absent from many parts of Inner Anatolia between late October and early March. It is still generally rather scarce in Aegean and parts of Mediterranean at this season, with most records usually coming from the extreme southeast of the former region (although formerly recorded as far west, at least occasionally, as Gediz; von Gonzenbach 1852), but reaching as far south as the Hatay in the latter. Numbers are usually small, although there was a maximum of 120 at Acıgöl in January 1970 and the species is regular at Göksu Delta (Mediterranean), where it occasionally remains as late as the end of April (Davidson & Kirwan 1997).

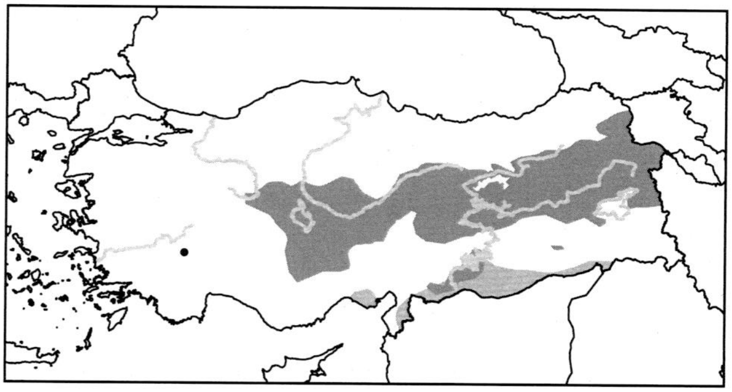

Breeding. On dry flat steppe on bare ground or with low scattered bushes, and on arid, gently sloping hillsides. Usually breeds solitarily, but there may be several widely spaced pairs in optimum habitat. Nest is a shallow hollow in the ground, usually unlined. Eggs: C/3 (11); three nests with one egg considered incomplete clutches. All records are from **Inner Anatolia**. Four nests each had three fresh eggs, north of Kayseri, 8–14 May 1876 (Danford 1878). At Karapınar, three nests had three eggs, and one had one egg on 3 May 1989, and another had three eggs on 10 May 1990 (J. Whitaker); a nest had an incomplete clutch of one egg near Bolluk Gölü on 14 May 1993 (PC); a nest had three eggs at Sultan marshes on 29 May 1983 (Kasparek 1985) and, nearby, a nest had three eggs on 27 May 1972; photographs of a nest with three eggs on 31 May 1974 and of a newly hatched chick on 12 June 1973, both near Cihanbeyli, are in Pforr & Limbrunner (1982).

PIN-TAILED SANDGROUSE *Pterocles alchata*
Kılkuyruk Bağırtlak

(Du: Aziatisch Witbuikzandhoen; Fr: Ganga cata; Ger: Spiebflughuhn; Swe: Vitbukig flyghöna)

Subspecies. *P. a. caudacutus.*

Status and distribution. National conservation status: Critical (population apparently decreasing). There is just one confirmed breeding record in Turkey (see below). Principally recorded in Southeast Anatolia, where it formerly occurred in large numbers and was presumably a locally common breeder. Most counts were formerly in the range 50–500, with up to 2,000 recorded near Akçakale (May 1970) and at Birecik in the 1980s. However, a large decline was evidenced by recent survey work in this region, which revealed a single-site maximum of just 31 birds and registered the species in just four 10-km squares (Welch 2004). As with Black-bellied Sandgrouse *P. orientalis*, numbers visiting the Euphrates at Birecik to drink recently have declined seriously, perhaps in part due to recent changes in water levels there (as a result of dam projects upstream of the Syrian border), but also due to conversion of

dry steppe for agriculture, which is presumably the principal reason for the species' dramatic reduction in numbers in Turkey. Occasionally recorded in very small numbers from Inner Anatolia, where, e.g., recorded on the east side of Tuz Gölü in May 1971 and at two localities in the south of the Konya Basin in May 1986. Recently recorded, once but in large numbers, at Van (East Anatolia) in July 1998 (Kirwan 1999), and 19th-century records are available from Mersin (Mediterranean) in mid March, and by Izmir (Aegean) (see Kumerloeve 1961).

Breeding. Photographs of a nest with two eggs, and of an adult incubating eggs or brooding small young (probably the same nest), were taken by K. Warncke in Turkey in June 1977 (*in* Pforr & Limbrunner 1982), but there are no further details. No recent breeding records, but probably still nests in the southeast of the country.

ROCK DOVE *Columba livia*
Kaya Güvercini

(Du: Rotsduif; Fr: Pigeon biset; Ger: Felsentaube; Swe: Klippduva)

Subspecies. Nominate *livia*, though Roselaar (in Cramp 1985) speculated that separate subspecific status might be warranted for populations in Turkey and the Caucasus, and Kumerloeve (1961) ascribed Turkish birds to *gaddi* in the north and *palaestinae* in the south.

Status and distribution. National conservation status: Least Concern (population apparently stable). Widespread and abundant as a feral pigeon throughout the country, particularly in Istanbul, Inner, Southeast and East Anatolia. Pure (or more or less pure) populations are present but much less numerous, being restricted to rocky uplands (up to at least 3,000m in the south and 4,000m in the north and east) and on sea-cliffs. Observers are encouraged to clarify the presence of wild-type birds to more closely identify the distribution of such populations, although at least in Southeast Anatolia it may no longer be possible to differentiate between feral and wild populations, especially as the latter may, at least occasionally, breed in close proximity to humans.

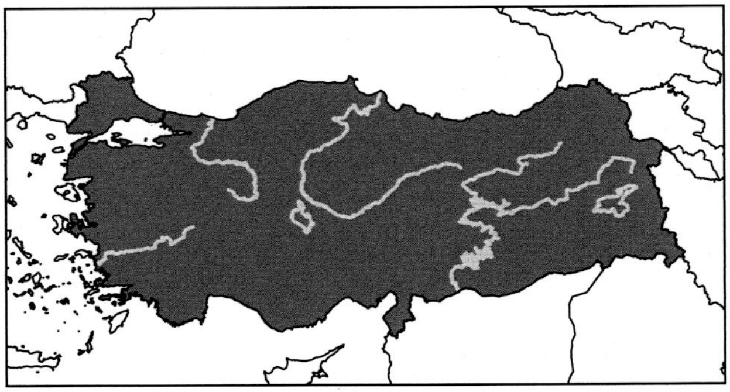

Breeding. Mainly in small colonies, usually fewer than 20 pairs, and sometimes solitarily. Nests in crevices and caves in cliffs (coastal, offshore and inland), in piles of boulders, and in buildings, ruins and below ground in wells; also on the ground, under bushes. Sites sometimes shared with other bird species, including Western Jackdaw *Corvus monedula*. Nest is a thin platform of twigs, fine stems and roots. Eggs: C/2 (15), C/1 (2), possibly incomplete clutches. Young: B/2 (2). Two eggs and two well-grown young, in the same nest in **Inner Anatolia** on 22 May 2007 (Ramsay 1914), are considered to be the product of two pairs, probably because the young had wandered from their own nest. In Inner Anatolia, recently fledged young on 7 May indicate egg laying in mid March; a colony on an island in Bolluk Gölü was nesting on the ground under bushes, intermixed with nests of Common Shelduck *Tadorna tadorna*, on 24 June 1992 and 7 May 1993; there were *c.*5 nests with two eggs on both dates, clearly indicative of double-brooding, which is usual elsewhere. In **Aegean**, a nest had two well-incubated eggs, on flat ground under a bush on an islet off Ilıca, near Çeşme, on 8 May 1950 (McNeile). Selous (1900) found a nest with two eggs in a ruin near Izmir on 14 May 1899. At Balık Gölü (**East Anatolia**), several pairs were breeding (with Western Jackdaws) in piles of boulders on an island in the lake on 9 June 2001 (PC).

STOCK DOVE — *Columba oenas*
Gökçe Güvercin

(Du: Holenduif; Fr: Pigeon colombin; Ger: Hohltaube; Swe: Skogsduva)

Subspecies. Nominate.

Status and distribution. National conservation status: Vulnerable (population apparently decreasing). A rather local resident or partial migrant in fairly small numbers in well-wooded uplands over much of the country, reaching at least 2,100m, but apparently absent from Thrace, Southeast Anatolia and much of Inner Anatolia. A relatively common breeder in parts of the northeast below 400m (Faldborg 1994).

More widespread on passage, when recorded at the Bosphorus from mid September to late October, with largest numbers in mid October (Porter 1983, Beaman 1986), peak numbers through northeast Turkey in mid–late October (e.g. Beaman 1986), and spring passage at Kızılırmak Delta (Black Sea Coastlands) perhaps peaking in late March and continuing until mid April (Hustings & van Dijk 1994), but in the latter case the early part of the migration was probably missed. Autumn passage, in particular, also occasionally manifest on the south coast, and perhaps regular through the eastern third of the country. Winters across the western two-thirds of the country, principally in the west, southwest and at the southern fringes of Inner Anatolia, close to the Taurus; and locally appears in considerable numbers. Small numbers may persist in non-breeding areas until as late as mid May.

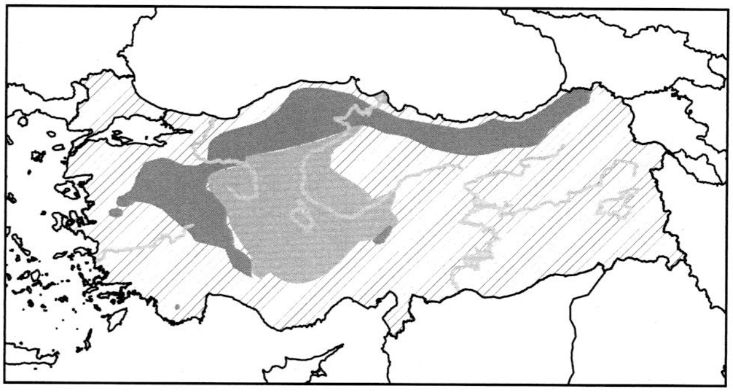

Breeding. Two nests contained young in the Antalya area (**Mediterranean**) on 21 October 2004 and 6 November 2004 (V. Kilinc *et al.*). These records are late, and late nesting is fairly frequent elsewhere, but they are unlikely to be typical of the main breeding season in Turkey. No further details.

COMMON WOOD PIGEON — *Columba palumbus*
Tahtalı

(Du: Houtduif; Fr: Pigeon ramier; Ger: Ringeltaube; Swe: Ringduva)

Subspecies. Nominate, although Roselaar (1995) noted that some intergradation with *iranica* was noticeable in southeast Turkey. Gibbs *et al.* (2001) remarked that, further east, *iranica* also graded into *casiotis*, and suggested that *iranica* might represent a synonym of *casiotis*. We are unaware of the sample size available to Roselaar (*op. cit.*), but we note that *C. palumbus* is not especially common in southeast Turkey, and therefore are unsure of the basis of the latter's remarks. Nonetheless, the probability that variation between the three forms is clinal strongly negates the validity of either *casiotis* or *iranica* under our terms of reference.

Status and distribution. National conservation status: Near Threatened (population apparently decreasing). A fairly widespread and reasonably common resident or partial migrant in Mediterranean, and rather more locally over the much of the rest of the country, e.g. in Inner and Southeast Anatolia, where principally breeds in more montane areas of the north and east in the latter region. Apparently absent from Thrace. Generally breeds in drier coniferous forest, usually in uplands, but rather thinly distributed. A scarce breeder down to sea level, but principally recorded at this season above 900m (and 1,200m in the north) and reaching to at least 2,200m (and probably somewhat higher).

More widespread across the country on passage, including East and Southeast Anatolia, with movements at the Bosphorus noted early September to early October (thus earlier than *Columba oenas*): 647 were counted there on 2–4 October 1974, and 11,780 between 24 September and 8 October 1973 (maximum over 3,000 on 2 October). Passage occasionally evident on the south coast but it is generally very scarce in the Mediterranean coastal deltas. Even inland, large numbers (*i.e.* several hundreds) may be noted on migration, e.g. in Marmara and Inner Anatolia. Winters principally in western and southern parts of the country, occasionally in Inner Anatolia, and found locally in considerable numbers. Immigration principally accounts for winter flocks in the west and south, e.g. 950 at Korkuteli, Antalya province, and 740 at Burdur Gölü (Mediterranean), in December 1969. Spring migration is concentrated between mid March and mid April, with exceptionally large counts of *c.*10,000 at Kuşcenneti (Marmara) on 24–31 March 1972, at least 2,000 between Sandıklı and Afyon (Aegean) on 19 November 2005 (O. Koçyiğit) and thousands in Meriç Delta (Marmara) in December 2004 (Ü. Başaran). May remain in winter areas as late as the first week of May.

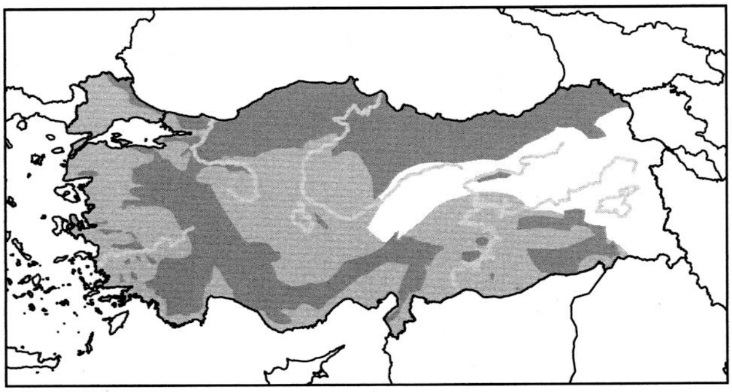

Breeding. Although a thinly, but well-distributed breeder in woodland areas in many regions, there is no detailed record of breeding. At Yürükler Forest in Kızılırmak Delta (**Black Sea Coastlands**), some were displaying in April 1992 and a bird was seen near a nest on 3 June 1992 (Hustings & van Dijk 1994). Juveniles were seen at Zigana Dağ, near Torul (Black Sea Coastlands) on 16 June 1975 (R. F. Porter). Display flight was observed at Karkamis, south of Birecik (**Southeast Anatolia**) on 3 May 2004 (GMK).

EURASIAN COLLARED DOVE *Streptopelia decaocto*
Kumru

(Du: Turkse Tortel; Fr: Tourterelle turque; Ger: Türkentaube; Swe: Turkduva)

Subspecies. Nominate. Species described from Turkey.

Status and distribution. National conservation status: Least Concern (population apparently stable). Kasparek (1996a, 1998) examined the change in modern status and range of this species in the Balkans and Middle East, documenting this dove's range expansion in considerable detail, including that in Turkey. Widespread and common resident, generally at lower altitudes, in Marmara, Aegean and Southeast Anatolia, and parts of Mediterranean, locally in smaller numbers in Inner and East Anatolia (where absent from large parts of the extreme northeast), and in the western and central Black Sea Coastlands. Much less regularly recorded in the eastern Black Sea Coastlands (where seems to be very local or even largely absent) and in the extreme east, and such records may even be the result of erratic movements. The breeding range is still expanding in the Middle East and Caucasus. Partial migrations may occur in some regions, with birds moving away from those areas in the interior of the country subject to most extreme winter climate, and large flocks of several hundreds are sometimes recorded in winter and spring (e.g. in Southeast Anatolia).

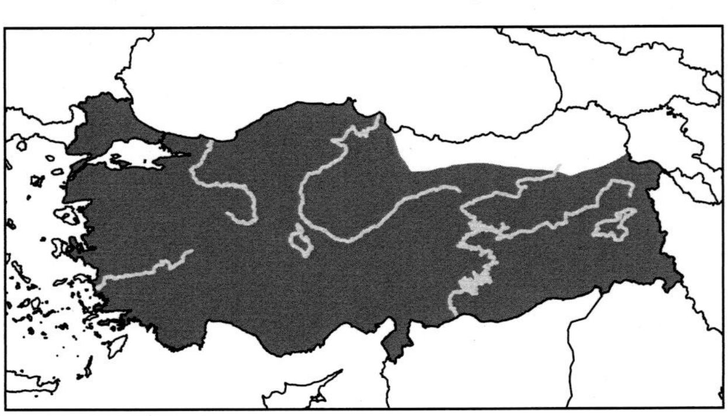

Breeding. In cultivated and urban areas, especially towns and villages with mature gardens and parks, ideally with a mix of deciduous and coniferous trees. Nests in a tree or tall bush, or on buildings/structures, e.g. a nest on a telegraph pole at Mersin, Mediterranean (Hollom 1955). Nest is a thin platform of fine twigs and plant stems, with a scant lining of finer plant and other material. Eggs: C/2 (3). Young: B/2 (2). In **Aegean**, adults incubating eggs at Akköy on 23 April 2003 (PC) and Izmir on 9 May 1950 (McNeile); a nest at Izmir had two fresh eggs on 2 June 1954 (McNeile), and an adult was incubating on a nest in a wicker basket 5m up in a lilac tree at Akhisar on 29 June 1966 (R. F. Porter). In **Inner Anatolia**, a nest was found at Hotamış on 24 April 1991 (Kirwan 1993a), a nest in a Black Pine near Incesu had two eggs in the first week of June 1991 (BD), an adult was incubating on a nest in a roadside tree at Şereflikoçhisar on 23 April 2004 (A. Dixon), and a nest had two newly hatched young at Cihanbeyli on 15 May 2004 (M. Farmer). In **Mediterranean**, an adult was lining a nest with hair at Mersin on 18 May 1951 (Hollom 1955), an adult was building a nest on a pylon at Iskenderun on 3 May 1970 (R. F. Porter), an adult was incubating on a nest in a line of conifers at Dalaman airport on 5 May 2003 (PC), and adults were incubating on two nests in Göksu Delta on 6 May 2004 (R. Thomas). In **Southeast Anatolia**, nest building was observed near Halfeti on 11 May 2004 (PC), two nests at Birecik had two eggs on 3 June 1998 (T. Quittenden) and two eggs on 11 May 2004 (A. Rustell), and two large nestlings were being consumed by a snake in a nest at Birecik on 25 June 2001 (I. Merrill). The species is at least double-brooded in Turkey, as elsewhere.

EUROPEAN TURTLE DOVE — *Streptopelia turtur*
Üveyik

(Du: Zomertortel; Fr: Tourterelle des bois; Ger: Turteltaube; Swe: Turturduva)

Subspecies. Nominate.

Status and distribution. National conservation status: Least Concern (population apparently decreasing). A widespread and generally fairly common summer visitor to wooded and agricultural areas throughout most of the country, but more common in the west and apparently rather local in East Anatolia. Breeds at 0–1,500m in western Black Sea Coastlands and Marmara, 200–1,300m in the Taurus, and up to *c.*2,000m in the northeast.

Even more widespread on passage, often in very large numbers, principally arriving in early April and departing early October. No evidence of large-scale passage at the Bosphorus, but generally abundant on passage through the eastern third of the country, particularly the extreme northeast. Hundreds observed between Silifke and Antalya (Mediterranean) in late September 1968, *c.*1,000 between Zigana pass and Trabzon (Black Sea Coastlands) on 16 June 1975, *c.*3,000–4,000 between Erzurum and Gümüşhane (East Anatolia/Black Sea Coastlands) on same date, and possibly as many as 2,000 roosting south of Birecik (Southeast Anatolia) on 2 May 2004. Spring passage may continue in some numbers until early and even mid June, especially in the north. Earliest records from Marmara on 30 March, and Mediterranean on 3 April, with return passage commencing in the last week of July and generally continuing until early October (peaking in September). There is an exceptionally late record from Inner Anatolia, on 28 October.

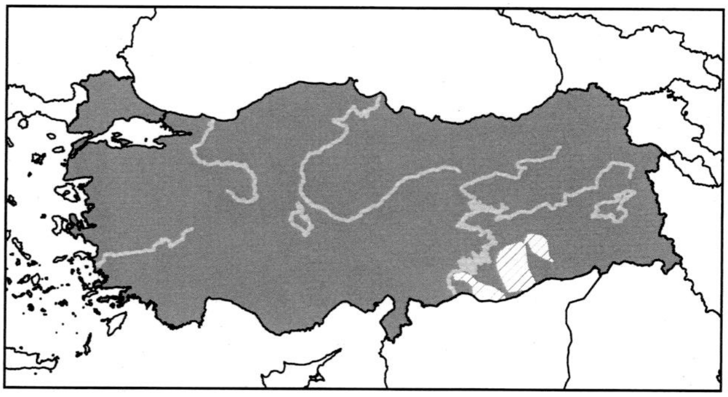

Breeding. In areas with trees, hedges and taller bushes, in both cultivated and natural areas, in orchards and olive groves, in woodland and at woodland edges. Nests in a bush or low tree, usually 2–4m above ground. The nest is a shallow platform of thin twigs, with a slight hollow lined with grass and fine rootlets. Eggs: usually C/2 (13), occasionally C/1 (1, but this perhaps an incomplete clutch). Young: B/2 (3). At Kızılırmak Delta (**Black Sea Coastlands**), there were eight occupied nests in 1ha of woodland in July 1971 (Dijksen & Kasparek 1985), and a population of 600–800 pairs was estimated in 1992 (Hustings & van Dijk 1994). In **Aegean**, the earliest record of a nest with (fresh) eggs is 10 May 1950, with several others in the second half of May/early June in the Izmir (McNeile) and Akköy areas (J. Hollinrake); a nest had two eggs at Bafa Gölü on 24 June 1966 (R. F. Porter). In **Southeast Anatolia**, a nest had two eggs at Birecik on 18 May 1993 (PC), two nests each had two eggs, 1.7m up in small orchard trees near Halfeti on 11 May 2004 (R. Thomas), and two young were in a nest near Gaziantep on 10 June 1990. Adults nest building near Gaziantep on 8 June 1993 (PC) and 14 June 1996 (GMK) were possibly constructing their second nests of the year. In **Inner Anatolia**, four adults were displaying at Göreme on 30 May 1999 (R. F. Porter), nests with one and two eggs at Hasan Dağ on 25 June 1992 were probably second broods (PC), but a nest with two young a few days old at Karadağ on 13 June 1907 (Ramsay 1914) was a first brood. Elsewhere, the species is double-brooded, sometimes treble-brooded, with breeding continuing well into August, but there are no such records in Turkey.

[ORIENTAL TURTLE DOVE *Streptopelia orientalis*]
Büyük Üveyik

(Du: Oosterse Tortel; Fr: Tourterelle orientale; Ger: Orientturteltaube; Swe: Större turturduva)

Subspecies and taxonomy. Unknown, but *meena* has recently been confirmed to breed at the extreme eastern edge of the Western Palearctic, in the Ural Mountains (see Wilson & Korovin 2003), and has been reported as a vagrant on several occasions in the Middle East and Europe. Nominate *orientalis* has also reached Western Europe as a vagrant. L. Svensson (*in* Wilson & Korovin *op. cit.*) has recently raised the intriguing possibility that *meena* and *orientalis* may warrant separation at the specific level, based on the dissimilarity in their plumages and vocalisations.

Species not admitted to the Turkish list (Kirwan *et al.* 1999). Listed as a rare visitor to the southeast and east of the country by Turan (1990). There are no published details to support this statement, although the species is clearly a potential vagrant and, subsequently, BD has received a claim, without supporting details, from Artvin (Black Sea Coastlands), and another, of the race *orientalis*, from near Terkos Gölü (Marmara) on 21 July 2004 (B. Bilgen). Most of the main features separating from the species from *S. turtur* were described by the observer of the July 2004 record. The species is scarce on passage through eastern Iran and rare in Oman, and is a vagrant elsewhere in the Arabian Gulf and in Israel (Porter *et al.* 1996). The vast majority of Omani records are in autumn, from mid September (Eriksen & Sargeant 2000), whilst Israeli records are also principally from the same, late autumn period (Shirihai 1996, Balmer & Betton 2003a), and European records also conform to this pattern (Alström *et al.* 1991), with history of overwintering. Thus, the most recent and only credible Turkish claim must be considered rather unusual temporally, and may perhaps refer to an escapee.

LAUGHING DOVE *Streptopelia senegalensis*
Küçük Kumru

(Du: Palmtortel; Fr: Tourterelle maillée; Ger: Palmtaube; Swe: Palmduva)

Subspecies. *S. s. phoenicophila*, although Kasparek (1991) considered that nominate *senegalensis* also occurs, perhaps as an introduction, e.g. in Istanbul. We have not examined specimens of this species from Turkey.

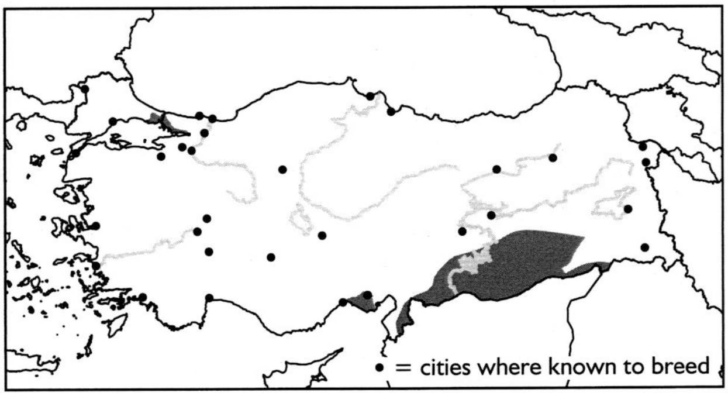

Status and distribution. National conservation status: Least Concern (population apparently increasing). Its range and status in Turkey were subject to a detailed study by Kasparek (1991). A local resident, often in considerable numbers, in cities and towns in Mediterranean and Southeast Anatolia, and Istanbul, as well as very locally in Thrace, Inner Anatolia and the west, and patchily in Black Sea Coastlands and East Anatolia (almost to the Iranian border at Doğubeyazıt). In some areas may out-compete Eurasian Collared Dove *S. decaocto*, but in others the latter is dominant. Regularly recorded in larger numbers from, e.g., Istanbul, Adana, Mersin, Adıyaman, Gaziantep, Urfa, Mardin and Diyarbakır, with rather smaller populations in, e.g., Antalya, Afyon, Ankara, Samsun, Erzincan and Malatya, and scarce records from, e.g., Izmir, Çanakkale, Edirne, Tekirdağ, Konya, Hakkari and Van.

Breeding. Breeds in Istanbul, where perhaps introduced, and in other towns, and villages and cultivation, mainly in the south and southeast of the country. Nests in a bush, tree or on the ledge of a building; makes a flimsy, shallow platform of thin twigs, lined with grass and fine plant stems. The same nest may be used for subsequent broods in the same year, which is evident from the different layers of nest material, and usually from the extent and condition of excrement. Eggs: C/2 (1). Young: B/2 (1). In Istanbul, a pair was observed mating on 28 March 1967 and a pair were at a nest on a window ledge, 10m above ground, on 22 April 1970 (R. F. Porter); and a pair had young in late July 1968. In **Mediterranean**, a pair was nesting in a palm at Adana airport on 28 March 2000 (L. Dijksen), and recently fledged young were seen in Adana city centre on 18 May 2004 (PC). In **Southeast Anatolia**, most pairs were feeding young, but a nest had two eggs at Birecik on 3 May 1964 (Warncke 1964–65), a recently fledged young, still with traces of down and *c.*1 week out of the nest, was near Halfeti on 11 May 2004, a nearby nest was 4m up on a ledge of an occupied house and had two young, *c.*7 days old, indicating egg laying about 20 April: however, it was evident that this nest had already been used for an earlier brood, probably in February or March (PC).

NAMAQUA DOVE *Oena capensis*
Kap Kumrusu

(Du: Maskerduif; Fr: Tourterelle masquée; Ger: Kaptäubchen; Swe: Kapduva)

Subspecies. Nominate.

Status and distribution. Vagrant. One record: a female photographed just north of Birecik (South-east Anatolia), on 23–24 May 2005 (F., S. & L. Veyrunes; *Sandgrouse* 27: 164). An increasing population has been resident in Israel's Arava Valley since 1961 (Shirihai 1996), but vagrants have only rarely reached further north with an uncertain record in Syria (Baumgart *et al.* 1996), a recent claim in Lebanon (Haraldsson 2008), a specimen in south-east Iraq (Salim 2008) and an accepted record from Cyprus (Lamsdell & Lamsdell 2000).

ROSE-RINGED PARAKEET *Psittacula krameri*
Yeşil Papağan

(Du: Halsbandparkiet; Fr: Perruche à collier; Ger: Halsbandsittich; Swe: Halsbandparakit)

Subspecies. Unknown: no specimen material from Turkey is apparently available, although most Middle Eastern populations are apparently thought to be of the race *borealis*, which is native to the Indian subcontinent (Juniper & Parr 1998).

Status and distribution. First noted in Turkey, at Göksu Delta (Mediterranean), in July 1990, and in Istanbul (Marmara), in September 1991, by Kasparek (1992a), although subsequently records from Ankara (Inner

Anatolia) in 1975 and 1976 came to light (Kasparek & Bilgin 1996). Thereafter, Boyla *et al.* (1998), in a comprehensive review of this escapee's status in Turkey, noted that the species had been present in Istanbul since at least March 1990 and that this parakeet may even have been present in Izmir (Aegean) since the 1950s. The same authors in particular, and Kirwan *et al.* (2003), documented the spread and dramatic increase of Rose-ringed Parakeet in the country, where it is now known from many of the major cities in the western two-thirds of the country, with particularly large concentrations in Istanbul (flocks of up to 450), Izmir (up to 250) and, to a much lesser extent, Ankara (up to 35). There are few records from rural localities but the species has been recorded in all regions, except East Anatolia, and only very sparsely in Black Sea Coastlands and Inner and Southeast Anatolia, from where it is known as far east as Diyarbakır and Cizre.

Breeding. Details are from Boyla *et al.* (1988) unless otherwise stated. Breeds mainly around larger cities (Istanbul, Ankara, Izmir), suggesting that the population originated in accidental escapes from captivity. A pair probably bred in a hole in the minaret of a mosque in Istanbul (**Marmara**) in May 1990, and a pair was confirmed to breed in a hole of an old building in Bornova, Izmir (**Aegean**) on 16 May 1995. A pair probably bred in woods at Eymir Gölü, Ankara (**Inner Anatolia**) in 1997. At Gulhane Park, near Topkapi Palace in Istanbul, an occupied nest was found on 31 May 2001, and there were several pairs present, described as a colony (C. Cameron). No data from Turkey: elsewhere the nest is an unlined cavity, usually in a tree hole, but also in a hole in a rock outcrop or a building; usual clutch 4 (3–6).

ALEXANDRINE PARAKEET *Psittacula eupatria*
İskender Papağani

(Du: Grote Alexanderparkiet; Fr: Perruche alexandre; Ger: Grober Alexandersittich; Swe: Alexanderparakit)

Subspecies. Unknown, as no specimen material is available and the only photographs appear to be of a female, which lacks the collar markings helpful in racial identification. Five races are generally accepted.

Status and distribution. An escapee that appears to be becoming established in Turkey. The species was first mentioned for Turkey by Boyla *et al.* (1998), based on an observation in Ankara (Inner Anatolia), on 20 February 1998, of a bird that was roosting with a small group of Rose-ringed Parakeets *P. krameri*. Subsequently, the species was found breeding in Istanbul (Marmara) in both 2003 and 2004 (see below), as well as at two localities in the southwest, and has also been observed as far east as Diyarbakır (East Anatolia), in May 2004 (GMK *et al.*). Observers should beware of falsely concluding that any sighting of a *Psittacula* in Turkey relates to *krameri*, and the true range of this species may be even wider than currently believed.

Breeding. The species bred at Topkapı Palace, Istanbul (**Marmara**) in 2003, with the nest and birds being photographed (S. Oksay: *Ibibik* 4:30). A pair was present at Gülhane Park, also in Istanbul, on 15 October 2004; the female flew into a nest hole in a tree, and the male was observed aggressively driving a Hooded Crow *Corvus corone cornix* from the area (G. Langley). No further details.

GREAT SPOTTED CUCKOO *Clamator glandarius*
Tepeli Guguk

(Du: Kuifkoekoek; Fr: Coucou geai; Ger: Häherkuckuck; Swe: Skatgök)

Subspecies. Monotypic.

Status and distribution. National conservation status: Least Concern (population apparently stable). Fairly widespread and not uncommon summer visitor to Aegean, more local in Thrace, Mediterranean and Inner and western parts of East Anatolia, and perhaps Black Sea Coastlands. Only a handful of records from Southeast Anatolia suggestive of breeding. Recorded regularly in the extreme west, in small numbers, west of a line from Uluabat Gölü (Marmara) to Çivril Gölü and Acıgöl (Mediterranean), between early April and early September. Breeds up to at least *c.*1,200m. Juveniles are recorded in autumn (August and September), almost certainly on passage, in East Anatolia and there are also recent spring records (also of birds probably on passage) from this region. The earliest record is from Karataş (Mediterranean) on 10 March, and the latest is from Göksu Delta (Mediterranean) on 22 September, but the temporal extent of passage in spring and autumn is difficult to determine, as are its peaks, although spring passage on the south coast apparently persists until at least the end of April (Kivit *et al.* 1994).

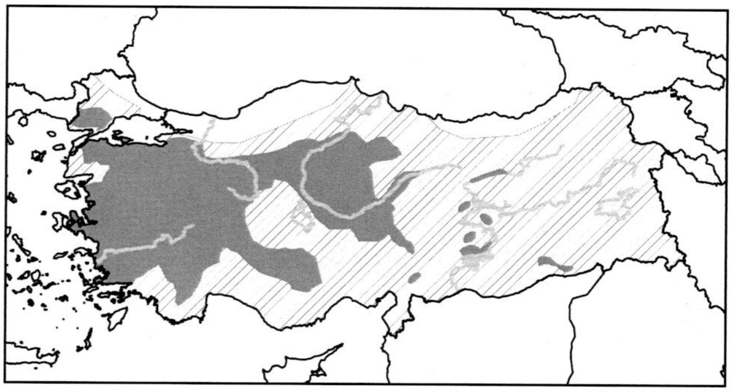

Breeding. Elsewhere, parasitises mainly corvids, especially Common Magpie *Pica pica*, which in Turkey is the only recorded host, and the breeding season is, therefore, synchronous with that species. In **Inner Anatolia**, a nest had five eggs of the host, and one of the cuckoo, on 8 May 1945 (Wadley 1951); a single young cuckoo (no host young) was found in a nest at Ankara on 21 May 1972; at Kirşehir, a nest had two eggs of the cuckoo and a broken egg of the host on 14 June 1977, and a nearby nest, on the same day, had a fresh cuckoo's egg but none of the host (Schubert 1979); two immatures accompanied a flock of Common Magpies on 7 August 1967; an immature was seen at Gordium on 11 July 1969; two juveniles (together) were at Ereğli marshes on 4 July 1993 (A. Hogg). In **Aegean**, fledged young were seen on 25 June 1966 at Bafa Gölü (R. F. Porter) and as early as 4 June 1971 at Gulluk Gölü; a juvenile was mobbed by Crested Larks *Galerida cristata* in farmland at Akköy on 23 May 1993 (G. Harper). In **Marmara**, display was observed, with much calling, by paired birds near Manyas Gölü on 25 May 1967 (R. F. Porter). At Kızılırmak Delta (**Black Sea Coastlands**) a juvenile was seen on 10–11 August 1972. Elsewhere, several eggs (up to 13) of the cuckoo may be laid in one nest. In Bulgaria, Common Magpie is also possibly the only host, and eggs are laid mainly in April (March to mid May) (Nankinov 1999a).

COMMON CUCKOO *Cuculus canorus*
Guguk

(Du: Koekoek; Fr: Coucou gris; Ger: Kuckuck; Swe: Gök)

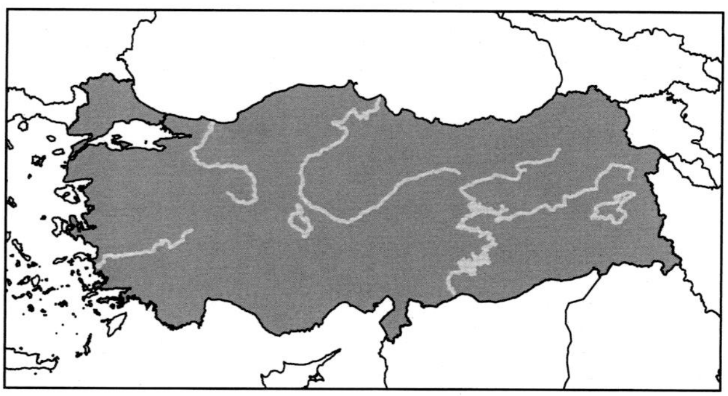

Subspecies. Nominate.

Status and distribution. National conservation status: Data Deficient (population trend unknown). A widespread and common summer visitor throughout most of the country, but apparently rather more local (but still quite widespread) in

Southeast Anatolia. Breeds from sea level in a wide variety of habitats, in forested and agricultural areas, and even in quite dry, open areas. Also inhabits mountains, up to at least 2,000m.

Recorded widely throughout the country on passage, being regular in all regions between early April (late March in the south and southwest) and late September, although still more local in Southeast Anatolia. The earliest record is from Tuzla Gölü, near Milas (Aegean) on 12 February 2005, with the latest record from Uluabat Gölü (Marmara) on 3 October 1993.

Breeding. Frequents a wide range of habitats: forest, agriculture, scrubland, marshland and even quite dry open areas, presumably depending on host species. Arrives early to mid April. In Kızılırmak Delta (Black Sea Coastlands), population estimated at 70–90 territories, mainly in Yürükler Forest, and much scarcer in marshlands, despite the large number of potential hosts, such as European Reed Warbler *Acrocephalus scirpaceus* (Hustings & van Dijk 1994). Elsewhere, females usually specialise on one host species, possibly that by which they were reared; it is unknown if this is true in Turkey, but a resident of Izmir (**Aegean**) told McNeile in 1950 that Orphean Warbler *Sylvia hortensis* was the usual host (in the scrublands) of the district. In the same area, McNeile found an Orphean Warbler's nest with a cuckoo's egg and three eggs of the host on 10 May 1950 (all four were fresh), and another with a cuckoo's egg and two of the host on 26 May 1951 (deserted); he found a cuckoo's egg in the nest of a Rufous-tailed Scrub Robin *Cercotrichas galactotes*, which also included four eggs of the host, on 2 June 1951 (all five were well incubated). He also located a fledged young cuckoo (host unspecified) on 28 May 1951, which indicates egg laying in late April. This apparently rather early date fits quite well with observations of two females at Altınkum on 22 April 2002, which were watching, and being attacked by, pairs of Rüppell's Warblers *Sylvia rueppelli* which were on the point of building; the cuckoos were clearly searching for nests (PC). In **East Anatolia**, two juveniles were begging food from Tawny Pipits *Anthus campestris* at Kağızman on 16 August 1972, and another juvenile, apparently reared by a White Wagtail *Motacilla alba*, was observed there on 17 August 1972; a juvenile was at Bulanık on 1 August 1986. In **Mediterranean**, a juvenile was caught at Pozantı on 4 August 1967. In **Southeast Anatolia**, a juvenile was at Işıklı on 13 August 1986. A dead juvenile was found at Kızılırmak Delta (**Black Sea Coastlands**) in July 1971 (Dijksen & Kasparek 1985). These records identify only five host species in Turkey, all dry-land birds; one can only speculate that the true total must be considerably higher than the rather limited data suggest, and must surely also include reedbed species such as European Reed and Great Reed Warblers *Acrocephalus arundinaceus*.

BARN OWL *Tyto alba*
Peçeli Baykuş

(Du: Kerkuil; Fr: Effraie des clochers; Ger: Schleiereule; Swe: Tornuggla)

Subspecies. Kasparek (1986b) reviewed all available information on the respective distributions of white-breasted (*alba/erlangeri*) and brown-breasted (*guttata*) forms or races, although many more data are required. Other authors (e.g. Bruce *in* del Hoyo *et al.* 1999, Roselaar 1995, König *et al.* 1999) do not consider *guttata* to occur in Turkey, and *erlangeri* has been treated as only a very weakly defined form of marginal validity (König *et al. op. cit.*). Kasparek (*op. cit.*) tentatively concluded that a 'hybrid zone' exists along a line from the surroundings of Izmir to Manyas Gölü and onward to Thrace, with white-breasted birds to the northwest of this, and also that brown-breasted *guttata* extends from southeast Europe (e.g. Romania and Bulgaria) through the Marmara and Aegean regions to the south coast, with white-breasted birds reappearing along the eastern south coast and in the southeast. However, breeding-season records from Inner Anatolia have also been white-breasted (GMK pers. obs.) and the sole observation from East Anatolia (see below) was photographed and ascribed to the subspecies *alba*, whilst those from Kızılırmak Delta (Black Sea Coastlands) have involved both white-breasted and dark-breasted individuals (the latter ascribed to *guttata*; e.g. M. Reed photographs). Supporting Kasparek (*op. cit.*), those in the southeast and east Mediterranean coast observed by us have been white-breasted.

Status and distribution. National conservation status: Near Threatened (population apparently decreasing). A widespread and uncommon resident in almost any lowland habitat, recorded from all regions, but apparently only once seen in East Anatolia: at Karakaya Reservoir, Malatya, on 15 January 2006 (MÖ, M. Erturhan). Apparently most common within the maritime fringe of peninsular Thrace and southern Marmara, and south through the Aegean region to the Hatay (to Syria), and the southeast to at least Ceylanpınar (Urfa province) and Diyarbakır. Fifteen road casualties noted within a 55-km stretch of highway in Mediterranean Turkey, in January 1992, may be perhaps the best indication of its true abundance in its core range, but were perhaps also reflective of winter immigration by starving individuals (van den Berk 1994). May be largely restricted to this area, as Kasparek (1986b) noted only two records from deeply inland locations: one at Sivas (Black Sea Coastlands) on 16 March 1967 (Erard & Etchécopar 1968) and a road mortality, found by L. J. Dijksen & F. J. Koning, near Ilgın, northwest of Konya (Inner Anatolia) on 18 January 1986), neither of which are on dates suggestive of local breeders. These may reflect wandering or dispersing individuals, as perhaps was one at Sultan marshes (Inner Anatolia) on 3 November 1990 (I. Kiziroğlu *et al.*), one in Ankara on 31 January 2000 (E. Uçar) and another there in autumn 2004 (E. Yoğurtçuoğlu). However, two other records, both on 31 May 1990, at Eşmekaya and Sultanhanı (Inner Anatolia), hint that the species may be commoner than previously considered in this region, and is probably breeding (Kirwan & Martins 1994). Apparently very local and rather rare along the fringe of the Black Sea with strong evidence of absence from many regions, e.g. not recorded during 325 days of field observations in the environs of Ereğli (Albrecht 1986). In this region of Turkey known only from the environs of Istanbul (Kasparek 1992a) and the Kızılırmak Delta (Hustings & van Dijk 1994). Probably under-recorded in Thrace, traditionally a region of low observer activity, as Handrinos & Akriotis (1997) considered it widespread and fairly common throughout the Greek mainland, including Grecian Thrace.

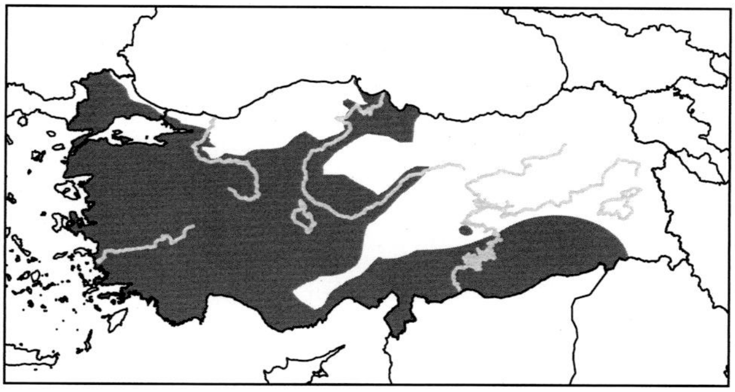

Breeding. Pellets and feathers are found in caves in cliffs, in ruins, buildings, farm buildings, and in holes in trees (Kasparek 1986b) and earth banks (GMK). In Çukurova (**Mediterranean**), on 14 May 1987, a breeding pair was found in the loft of an abandoned house, and five eggs had been laid on pellets (van der Have *et al.* 1988). Two adults were observed entering buildings at Gökçeli, in Çukurova, on 16 April 1988 (C. & C. Bradshaw). An adult was observed carrying food to a nest in a hotel roof at Side (Mediterranean) on 20 May 1996 (A. Bloomfield). Seven Barn Owls, brought to Ankara Zoo in August 1986, were found in the roof of a house at Serik, near Antalya (Mediterranean), as juveniles of a single brood (Kasparek 1986b). In Europe, egg laying usually commences in April and the breeding cycle, from laying to fledging, occupies slightly over three months, with second broods quite common; 4–7 (2–12) eggs are laid on the flat base of a cavity, with no material, but a shallow scrape may be made in existing debris, which usually includes many discarded food pellets.

PALLID (STRIATED) SCOPS OWL *Otus brucei*
Çizgili İshakkuşu

(Du: Gestreepte Dwergooruil; Fr: Petit-duc de Bruce; Ger: Streifenohreule; Swe: Blek dvärguv)

Subspecies. Apparently *obsoletus*, on evidence of photographs, as no specimen material is available. It seems dubious whether any geographical variation within this species demands recognition at nomenclatural level.

Status and distribution. National conservation status: Critical (population apparently decreasing). A marginal migrant breeder to at least three localities between Birecik and Halfeti (Southeast Anatolia), and where it was first reported as recently as April 1982 (van den Berg *et al.* 1988) from arguably the most famous site for the species in the world: the tea garden in Birecik town. The northernmost site is Savaşan, near Rumkale, Halfeti, and further east it has very recently been recorded at Ceylanpınar, on the Syrian border. A recent record of a bird ringed at Diyarbakir, in autumn 2003 (M. Biricik), hints either at a wider distribution than heretofore expected or to the possibility that Central Asian birds might occasionally pass through the country en route to Near Eastern wintering areas. The earliest record is 1 April and there are no published observations between November and March; thus, the species is presumed to be only a summer visitor. The nearest known wintering area is in Israel.

Breeding. In arid woodland; all breeding records are from **Southeast Anatolia**, mainly from the well-known café at Birecik, but has also bred elsewhere in the Euphrates Valley, in an orchard north of Birecik, and near Halfeti. Nests are mainly in holes in trees, 7–10m above ground (n=4), but uses lower holes elsewhere, and it is probable that low holes are also used around Birecik, where breeding must surely be overlooked. The same tree hole was used in two consecutive years. It also nested in the old nest of a Dead Sea Sparrow *Passer moabiticus*, 8m above ground in a poplar at Halfeti (van den Berg *et al.* 1988); and assumed (but not proven) to be nesting in an old nest of Hooded Crow *Corvus cornix* at Birecik on 29 May 1985 (S. M. Andrews & A. J. Eardley). No nest material is added. Eggs: no data in Turkey; usual clutch, elsewhere, 4–6. Young in nest: no data. Broods of fledged young as follows: B/2 (2), B/3 (4), B/4 (3), B/5 (2); these figures include the largest number of young observed in a given year, where differing numbers were seen on different dates; observations of single young (and probably broods of two) considered incomplete. Most records prior to 1988 are summarised by van den Berg *et al.* (1988): earliest date 1 April; copulation 11 April 1982, 12 April 1984 and 3–5 May 1987, when an adult brought food to the other, and one entered the nest hole. Incubation probably commenced on 28 April 1987 at a nest at Halfeti found on 15 April 1987. Both adults carried food to a tree hole on 1–2 June 1988. Fledged young were seen on 2 July 1971 (two), 2 July 1985 (four), 9–11 June 1986 (four), 4 July 1987 (five), early July 1988 (four, also seen 19 June 1988: M. Telfer), indicating egg laying no later than 25 April.Two adults and three juveniles were seen on 30 June 1984, and there are three observations in 1986 (max. 5 young), perhaps all of the same brood. A brood of four young was reported in 1986 by van den Berg *et al.* In 1989, an adult entered a nest hole on 7 May, both adults were seen there on 22 May and two young were seen on 24 June (many observers). A chick, examined in mid-June 1996, one of a brood in a nest, was more than half grown (GMK *et al.*), indicating egg laying in early May. Fledged young observed 29 July 1992 (one), 25 June 1994 (up to four), 2 July 1999 (three), 26 June 2001 (three) and 3 July 2004 (three) (many observers). Records indicate a single brood each year, which is usual elsewhere.

EUROPEAN SCOPS OWL — *Otus scops*
İshakkuşu

(Du: Dwergooruil; Fr: Petit-duc scops; Ger: Zwergohreule; Swe: Dvärguv)

Subspecies. Nominate in northern Turkey and *cycladum* in the south, with *turanicus* perhaps penetrating the extreme southeast. However, the vocalisations of all three are very similar and it is doubtful whether any geographical variation within this species should be recognised, with the exception of the vocally distinct *cyprius* from Cyprus.

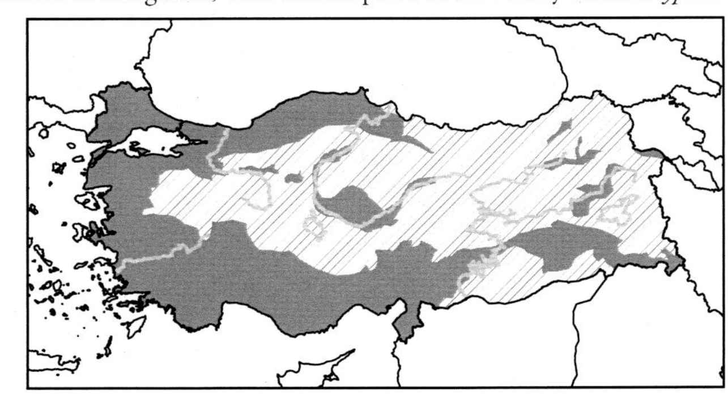

Status and distribution. National conservation status: Least Concern (population apparently decreasing). Widespread and common summer visitor across much of the country, especially to southern and western regions. Elsewhere, much more common on passage and nesting is relatively local. Prefers open montane and wooded areas, usually rather close to habitation, and breeds up to at least 2,300m in the Anti-Taurus (Mediterranean). Earliest migrants arrive on southern coasts in late March (21st) and by late April most migrants have moved through, e.g. recorded in the extreme northeast by mid April (Faldborg 1994). In autumn, migrants are noted from the second half of August around Istanbul (Marmara) and the latest birds usually depart by mid September from southern coasts, but has been recorded as late as 14 October in Mediterranean. The species has recently been found to be wintering in small numbers on the central Aegean coast, near Burhaniye (Eken 1997b), and further such discoveries can be expected from neighbouring areas.

Breeding. In open woodland, cultivation with trees, rocky montane areas: including gorges, and often near habitation, in villages and even towns. Nests in holes in cliffs and walls, in old nests of other birds and in nest boxes; it must also nest in holes in trees and buildings (particularly under roof tiles) in Turkey, but no records. The nest is an unlined cavity, with no material. In a study using nest boxes, near Ankara in 1976–85, 2–4 pairs bred each year, at a mean density of 0.27 breeding pairs/ha, and the mean clutch (of seven nests) was 3.7 eggs, with 3.3 young fledged per pair (Kiziroğlu *et al.* 1987). Elsewhere, in **Inner Anatolia**, three fledged young at Sultan marshes on 22 June 1983 (Kasparek 1985), indicated egg laying by late April. In **Mediterranean**, an adult was perched on a nest (probably an old nest of a Red-billed Chough *Pyrrhocorax pyrrhocorax*) in a rock face in the Aladağ on 28–29 April 1970, and a pair was feeding two chicks in a nest in a wall at Maden, at 1,700m, on 17–23 July 1971. In **Aegean**, a pair was nesting in a crevice near Marmaris on 1 June 1997. Eggs in the Natural History Museum, Tring, were collected near Izmir: C/4 (12 May 1870), C/2, probably incomplete (12 May 1872), C/4 (1 June 1872), C/1, probably incomplete (20 June 1878). In Armenia, breeds in tree cavities, in roofs, in nest boxes, and there are two records of nesting in old nests of Common Magpie *Pica pica*; season protracted, with the final egg laid in two nests on 30 May 1994 and 11 June 1980, and nests with eggs as late as 27 July 1969, and young in one nest, which were close to fledging, in late August (Adamian & Klem 1999).

EURASIAN EAGLE OWL — *Bubo bubo*
Puhu

(Du: Oehoe; Fr: Grand-duc d'Europe; Ger: Uhu; Swe: Berguv)

Subspecies and taxonomy. Most, if not all, are apparently *interpositus*. König *et al.* (1999) remarked that DNA evidence suggested that *interpositus* is sufficiently distinct to be recognised at species level, and considered the forms *ascalaphus* and *desertorum*, which have traditionally been placed in *B. bubo*, as a separate species, Pharaoh Eagle Owl *B. ascalaphus*. Vaurie (1960b) discussed seven specimens, from Palestine and Syria, which appear to represent hybrids between *interpositus* and *ascalaphus*, and for which the name *aharonii* is available. He considered this evidence to be the most compelling for continuing to recognise *ascalaphus* as conspecific with *bubo* (their separate treatment was first proposed in 1920). Given that Roselaar (1995) opined that *interpositus* might grade into *nikolskii* in southeast Turkey and that H. Shirihai (in *Birding World* 9: 218) suggested that those at Birecik appeared to represent intergrades between *interpositus* and *desertorum* (though photographs from this area examined by us do not appear to support this proposition), we prefer to await additional evidence before making any judgement over the taxonomic position of Turkish populations. The relationships between *ascalaphus* and *desertorum* also clearly merit further research (the latter name was considered a mere synonym of *ascalaphus* by Vaurie 1960b). The DNA results for *interpositus* are interesting, but thus far do not appear to be supported by other data, e.g. vocalisations, unlike the case for *B. ascalaphus*.

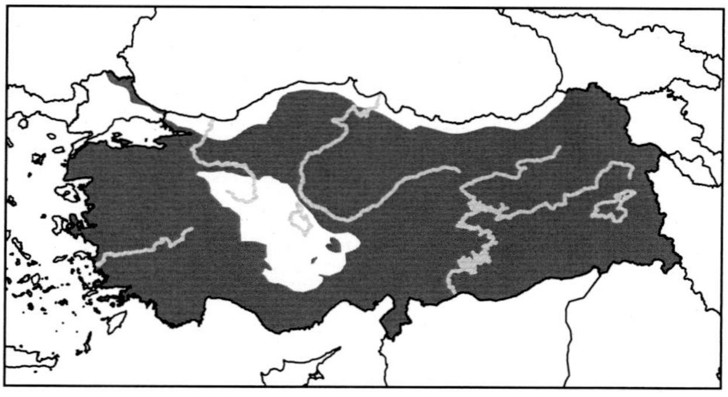

Status and distribution. National conservation status: Least Concern (population apparently stable). A sparsely distributed resident in rocky areas over much of the country, except the Black Sea Coastlands littoral and Thrace, from where there are only four records: single birds at Çamlıca Hills on 4 October 1970, near Tuzla on 24 January 1998

(KAB) and at Terkos Gölü on 3 November 2003 (O. Mülayim & C. Gezgin), with a further record from the latter on 4 October 2004 (M. Bozdoğan & S. Bekir). Occurs at all altitudes, from sea level (Yumurtalık, Mediterranean) to high mountains (e.g. Demirkazık, Mediterranean, and Nemrut Dağı, East Anatolia). The relative abundance of winter records, principally around wetlands of Inner Anatolia, perhaps reflects altitudinal movements and increasing activity prior to breeding. Despite reports of high densities at some localities, e.g. gorges near Göksu Delta (Mediterranean) (van der Winden & van den Berk 1997), and increasing observer coverage in many areas, numbers of records have decreased at many localities, especially in Inner Anatolia, where the species is not rare. If such a decline is real, it may be related to persecution and traffic-based mortality which are both well documented.

Breeding. In areas with cliffs and rock buttresses; in valleys, gorges and wadis, in both dry, virtually treeless areas, as well as forested ones. Danford (1880) found a nest in a rocky gorge, and another in a cave, in a valley 'clothed in pine and oak scrub'. Nests in a cave or crevice on a cliff, or on a cliff-ledge. There are no records in Turkey of nests in trees, either in natural cavities, or on old nests of other birds. The nest is a shallow scrape, with no material. Eggs: C/2 (1), C/3 (1). Young (in nest): B/2 (4), B/3 (1); fledged young, at well-known site at Birecik: B/1 (3), B/2 (1), B/3 (1); fledged young elsewhere in Turkey: adult and juvenile at Işıklı (**Southeast Anatolia**) on 10 July 1986, two adults and four young near Van (East Anatolia) on 30 May 1989, and two young on a steep slope behind Kocaçay Delta (**Marmara**) on 24 May 1993 (Ertan 1996). Many breeding records from Birecik 1989–99, with an adult incubating or brooding small young in early May 1992 and 1997 and early April 1999, young in the nest until early June, and earliest fledged young on 22 May (many observers); young leave the nest at 6–10 weeks old, before they can fly properly, indicating egg laying by early March and possibly up to four weeks earlier. In **Inner Anatolia**, Danford (1880) found a nest with eggs ('left in peace') near Ankara in early April 1879, and another near Eskişehir in mid-April 1879; a clutch of two in the Natural History Museum, Tring, was collected by Danford near Kayseri, apparently on 28 March 1879, possibly from one of the nests above; and Eagle Owl was also 'found nesting', near Kayseri on 7 May 1876 (Danford 1877–78). A nest on a cliff in a crater near Karapınar had two young, 10–14 days old, on 8 May 1964 (Warncke 1964–65), indicating egg laying in mid March. A nest had three young, near Sarıyar Reservoir, on 15 June 1997 (Boyla & Eken 1998). An adult flew three times (once carrying food) into a small valley, near Kızılcahamam, on 15 May 1980 (Barış *et al.* 1984). Young were heard at a nest site at Ereğli on 30–31 July 1981. An adult flew from an empty, but probable, nest in a crevice, with many disgorged pellets, on a cliff near Yeşilhisar, on 30 May 1998 (T. Quittenden). An adult flew from a ledge on a cliff, a likely nest site, overlooking Hotamış, on 24 May 1993 (PC). In **East Anatolia**, a pair and two recently hatched young were in a gorge near Erçek, on 21 May 1985. In **Mediterranean**, two large young were still in the nest at Girdev Gölü, near Fethiye, in August 1999 (P. Hope). A clutch of three eggs, in the Natural History Museum, Tring, was taken in 'western Turkey' on 17 March 1902. A single brood is reared in a year.

BROWN FISH OWL *Bubo zeylonensis*
Balık Baykuşu

(Du: Bruine Visuil; Fr: Kétoupa brun; Ger: Fischuhu; Swe: Brun fiskuv)

Subspecies. *B. z. semenowi*, which is considered paler, more buffish and less boldly marked than any of the other described forms, and occurs from Turkey east to Pakistan, but, together with *orientalis*, is often treated as a colour morph of *K. z. leschenault*. *Ketupa* is best subsumed within *Bubo* (see König *et al.* 1999) and this proposal has recently been endorsed by the Taxonomic Advisory Committee of the Association of European Records and Rarities Committees (see Collinson 2006).

Status and distribution. National conservation status: Critical (population apparently decreasing). On the brink of extinction. For obvious reasons, the precise whereabouts of a pair photographed in October 2004 are not presented here (see Yöntem 2007), but the species may also persist in another deeply incised river valley in the *Pinus* forest zone north of Adana, in the southern foothills of the Taurus (Mediterranean). The only other modern proof of its occurrence in Turkey is one, which had swallowed a fish caught on a line of baited hooks, and was caught and photographed at an undivulged locality in late April 1990 (Magnin 1991, Ebels 2002). The bird was subsequently released, one week later, during which period it did not feed, approximately 20km from the site of capture along the same river. Magnin visited the area on 22–23 June 1990 to assess and search the site and habitat, which consisted of natural pine forest with little riparian vegetation, at 170m altitude, and to interview the fisherman, who claimed to have shot 'a bird of the same species' about 20 years previously, less than 1km downstream of the capture site. The topography of the area presents a significant challenge to ready censusing or observer activity, and it is probably now flooded by a dam lake. Modern proof that the species may persist in Turkey focused renewed attention on the trail of conjecture, uncertainty and guesswork, which had previously characterised reports or suspicions of the species' occurrence. G. Schrader, a collector working in southern Turkey received four specimens, three from wooded gorges near Mersin and the other from a deep gorge near Aydin, in the late 19th century (Kumerloeve 1961), and Kumerloeve himself (*op. cit.*) found pellets in natural forest between Haruniye and Dumanli Dağı, in the valley of the River Ceyhan (Mediterranean), in June 1953, which he considered may have been of this species (certainly the remains of the freshwater snail *Potamon fluviatile* that Kumerloeve identified would have been unusual, though not unrecorded prey for *Bubo bubo*). Hollom *et al.* (1988) stated that there are old records from the bird market in Adana (Mediterranean) but this reportedly may be in error (Hollom *in* Magnin 1991). Kumerloeve (1970a) refers to sightings by Rothmann *et al.* in *Tamarix* and other bushes beside the lagoon (presumably Akgöl) at Silifke (Mediterranean) on 3 May 1965 and 4km north of the same locality in *Salicornia* on 3 May 1967, but *B. bubo* cannot be eliminated and the latter is indeed now known to be rather common in this area of the south coast (van der Winden and van den Berk 1997).

Breeding. No data from Turkey. Elsewhere, breeds in wooded or rocky habitats near water, with 1–2 eggs laid on a platform or hollow in a tree, on a ledge or hole in a cliff, or rarely on a large stick nest of another species; most aspects of breeding are unknown. In Iran, an adult and a juvenile were seen in a wadi at Sirik, Hormozgan province, on 8 April 2004 and 16 April 2004, where a pair had been seen on 18 January 2004 (Balmer & Betton 2004b).

LITTLE OWL — *Athene noctua*
Kukumav

(Du: Steenuil / Woestijnsteenuil; Fr: Chevêche d'Athéna; Ger: Steinkauz; Swe: Minervauggla)

Subspecies and taxonomy. *A. n. indigena* across much of the country, except the southeast, where replaced by the very pale sandy-brown race, *lilith*, which also has less-streaked underparts and more distinct occipital face. The latter has been considered, on the basis of DNA evidence and differences in its song, possibly to represent a separate species (König *et al.* 1999). The latter authors further considered the combined forms *spilogastra* and *somaliensis* from northeast Africa also to be potentially worthy of specific status. Both König *et al.* (*op. cit.*) and Roselaar (1995) noted that *lilith* overlapped with the form *bactriana* in Iraq and western Iran, but only the latter author suggested that the two intergraded. Given that König *et al.* (*op. cit.*) retained *bactriana* within *noctua*, the possibility that it intergrades with *lilith* clearly requires elucidation and may impact the proposition that *lilith* is specifically distinct. Thus, we prefer to await additional evidence for the latter, but agree that the issue demands further research.

Our own field research suggests that *lilith* or *lilith*-type birds occur over much of southeast Turkey, north to Diyarbakır and around Bitlis, and even north of Van Gölü, to at least Bulanık, but are replaced by *indigena* in the Pontics and their foothills, as well as in the Mediterranean region. Previous discussions of the range of these taxa in Turkey (and elsewhere) have been very vague (e.g. König *et al.* 1999).

Status and distribution. National conservation status: Least Concern (population apparently decreasing). A fairly common resident over much of the country, except the Black Sea littoral, where the species is very sparsely represented. With the exception of dense forested areas, occurs in virtually all habitats, including treeless steppes, and is thus the most commonly encountered owl in Turkey. Recorded from sea level to at least 3,300m in East Anatolia. In winter, dozens of individuals may be noted atop poles and roofs near highways in Inner and Southeast Anatolia: e.g. at least 33 in 300km between Diyarbakır and Bitlis on 3 December 1969.

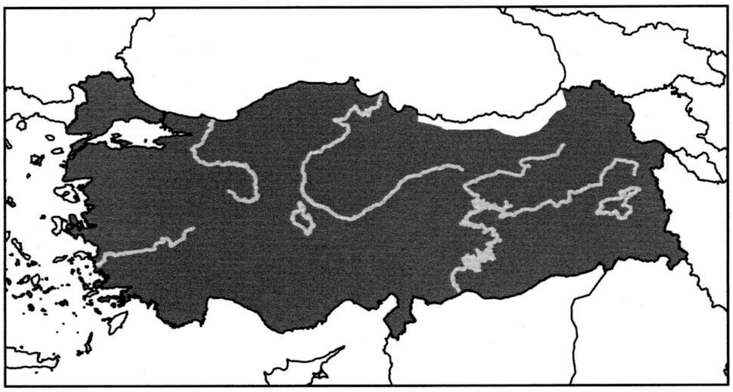

Breeding. Inhabits open areas, often rocky, in cultivation with trees, and around villages, but not in dense forest. Nest sites include holes and crevices in trees and cliffs, piles of stones, walls, wells, buildings and ruins, and nests of other birds. Eggs are laid at the base of the cavity, with no material. Eggs: C/5 (2). Young: B/3–4 (2), but up to five fledged young observed. In **Aegean**, around Izmir, five fresh eggs were in a Common Magpie *Pica pica* nest on 9 May 1950, and five eggs (with chicks calling inside the shells) were under a wooden trough, overturned on the ground, on 25 May 1950 (McNeile). Two fledged young, at Pamucak, on 20 June 1993, indicate egg laying in the last week of April. In **Marmara**, a pair was observed displaying, mutual preening and calling, near Manyas Gölü on 5 April 1967 (R. F. Porter). In **Inner Anatolia**, nests had full-grown young at Sultan marshes on 30 May 1998 (T. Quittenden), and three young, three weeks old, were at Şereflikoçhisar on 20 May 1970, indicating egg laying in late March/early April. Five recently fledged young were near Ereğli marshes on 21 June 1990 (A. Forsten), and an adult was feeding two fledged young near Alacatı on 18 June 1996 (GMK). An adult appeared briefly at the entrance to a nest in a well at Ereğli on 21 April 2004 (PC). In **Southeast Anatolia**, a nest had four young near Gaziantep on 21 May 1935 (Bird 1937), and fledged young have been observed near Halfeti in early June. On the outskirts of Van (**East Anatolia**), a calling adult attacked a dog on 20 June 2004 and a cat on 26 June 2004, presumably in defence of a nest or (unseen) young in a large wall nearby (PC). In **Black Sea Coastlands**, a pair was observed mating at Ereğli on 14 February 1977 (Albrecht 1977), and there are three records of pairs at nests in early May.

TAWNY OWL *Strix aluco*
Alaca Baykuş

(Du: Bosuil; Fr: Chouette hulotte; Ger: Waldkauz; Swe: Kattuggla)

Subspecies. The form *sylvatica* is found in west and central Turkey with *willkonskii* in the northeast. *S. a. clanceyi* has been mentioned for Asia Minor, but this is clearly in error and, in any case, the form is best regarded as a synonym of *sylvatica*. Roselaar (1995) considered that *sanctinicolai* might occur in the extreme southeast.

Status and distribution. National conservation status: Least Concern (population apparently stable). A local resident across much of the country, except East and Southeast Anatolia, where very rare. Found mainly in wooded areas, but records from areas such as Nemrut Dağı (Southeast Anatolia) and Tuz Gölü (Inner

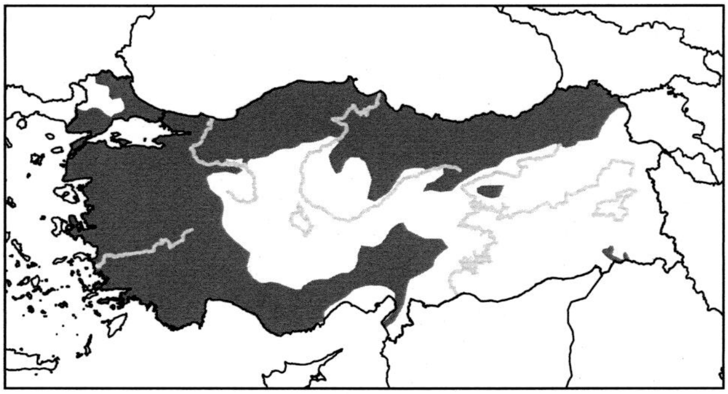

Anatolia) indicate that it is not wholly dependent on such habitat in arid regions. Recorded from sea level (e.g. at Fethiye and Çukurova, Mediterranean) to the treeline (Uludağ, Marmara). True numbers may be even higher over much of the country, given that most areas of suitable habitat are rarely visited by birdwatchers. Significant numbers appear to be concentrated in coastal regions of Mediterranean, Aegean, Black Sea Coastlands and Marmara. The observation of 11 individuals at eight different localities in Mediterranean, between Fethiye and Çakırlar (10km west of Antalya), is perhaps indicative of the species' real abundance in suitable habitat (M. Trzoska). Around Van and Kemaliye (East Anatolia), Işıklı, Birecik, Nemrut Dağı and Güreniz (Southeast Anatolia) are the only known localities in these regions (not mapped).

Breeding. Inhabits deciduous and coniferous woodland, and found in cliffs and rocks in more open areas. The population at Kızılırmak Delta (Black Sea Coastlands) was estimated at 30–40 pairs in 1992 (Hustings & van Dijk 1994), and there were 15–20 pairs in alluvial forest in Kocaçay Delta (Marmara) in 1993 (Ertan 1996). Nests in a hole in a tree or in a rock face; the nest is a scrape with no material. Eggs: no data in Turkey but, elsewhere, usual clutch three (2–4). Young (in nest): no data, but broods of up to three fledged young seen. In **Mediterranean**, two adults flew from a hole in the roof of a cave, in which young were heard calling, near Uzuncaburç on 12 June 1992 (PC). An adult and a fledged young were in the Tarsus area on 20 May 1991 (D.J. Bates), indicating egg laying by mid March. Fledged young were observed west of Taşucu on 8 June 1993 (J. Berg), at Taşağıl (two young) on 11 June 1989 and near Akseki on 4 July 1986 (R. Webb). In Kızılırmak Delta (**Black Sea Coastlands**) there was a pair near a nest on 13 May 1992, a juvenile on 8 June 1992 (Hustings & van Dijk 1994) and three young on 12 May 1991, indicating egg laying no later than 10 March. In Kocaçay Delta (**Marmara**), a family party on 21 May 1999 (K. Haataja), indicated egg laying in mid March. At Bafa Gölü (**Aegean**), there was a juvenile on 10 June 1981 (H. Hols). At Birecik (**Southeast Anatolia**), a family party of at least four was present on 28 May 1991 (D.J. Bates).

LONG-EARED OWL *Asio otus*
Kulaklı Orman Baykuşu

(Du: Ransuil; Fr: Hibou moyen-duc; Ger: Waldohreule; Swe: Hornuggla)

Subspecies. Nominate.

Status and distribution. National conservation status: Least Concern (population apparently increasing). A common resident and winter visitor throughout most the country, but rather more local in East Anatolia. Has a scattered distribution, being found in all types of habitat with trees: forests, wooded areas, plantations, parks and gardens. The principal breeding areas are in Inner Anatolia, southern Marmara and Aegean, and more locally in Black Sea Coastlands: e.g. very common in Rize (S. Ekşioğlu), Mediterranean, Southeast Anatolia and northern East Anatolia south to the Lake Van basin. Wintering groups are widely recorded, suggesting immigration from further north, but the numbers of birds in these rarely exceed 20: larger counts include 39 at Eymir Gölü (Inner Anatolia) on 27 February 2000, and 30 at Erzurum Atatürk University in late October 2003.

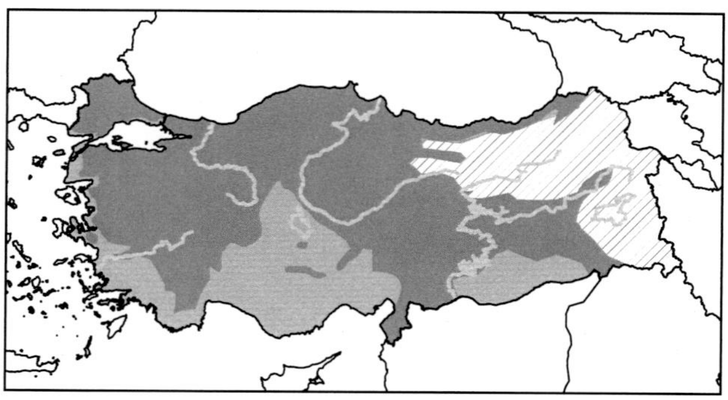

Breeding. Mainly in woodland, including plantations and orchards. Nests in old nests of other species, particularly Common Magpie *Pica pica* (n=5) at 3–5m above ground (n=5), but probably also in higher nests of Hooded Crow *Corvus cornix* and Rook *C. frugilegus*; also recorded nesting on the ground, under a low bush, on an island in Uluabat Gölü (Marmara) (Welch & Welch 1998b). The nest is merely a scrape, with no material. Eggs: C/2 (1), C/3 (1), C/5 (1). Young (in nest): B/2 (2), B/3 (3), B/4 (2). Records of fledged young are mainly of one (probably part of larger brood) (eight) or two (four), but there are also records of broods of five (one) and six (two). Usual clutch appears to be 2–6, compared with 4–5 (3–8) elsewhere. In **Mediterranean**, a nest had three small young at Ceyhan on 10 May 1964 (Warncke 1964–65), indicating the date of the first egg as early April. A fledged young in Çukurova on 18 May 1988 indicates laying in the last week of March. In **Inner Anatolia**, nests with eggs were recorded in late March 1977 and 21 May 1970, and a late nest, near Kulu Gölü, had two eggs on 15 May 1993, which had not hatched by 30 May 1993 (PC). Two nests had three and five eggs at Eşmekaya on 22 April 2004 (R. Thomas). A nest had three young, *c.*2 weeks old, at Şereflikoçhisar on 22 April 2004 (A. Dixon), indicating the first egg was laid in early March. Young in two nests at Eşmekaya fledged on 30 May 1993 and 25 June 1992, indicating first eggs in early April and early May (PC). Fledged young were observed at the Sultan marshes on 22 June 1982 and 25 May 1983 (Kasparek 1985). In **Marmara**, a nest had four small young at Uluabat Gölü on 2 June 1998 (Welch & Welch 1998b), indicating a first egg in late April. Two young were out of the nest, and another dead in the water, at Manyas Gölü, on 14 May 1969 (R. F. Porter). In **East Anatolia**, a nest at Van had two young almost ready to fly on 7 June 1984, indicating a first egg in mid April. In **Southeast Anatolia**, all records are from Birecik: nests with young on 4 May 1986 and 20 May 1986, and 15 records of fledged young from mid May.

SHORT-EARED OWL *Asio flammeus*
Kır Baykuşu

(Du: Velduil; Fr: Hibou des marais; Ger: Sumpfohreule; Swe: Jorduggla)

Subspecies. Nominate.

Status and distribution. National conservation status: Endangered (population apparently decreasing). Scarce passage migrant throughout the country, an uncommon winter visitor to the western two-thirds of Anatolia (with a max. 20 recorded at a single site in Inner Anatolia at this season), and a very local summer visitor or resident in the Konya Basin (Inner Anatolia) and parts of East Anatolia.

Spring passage commences late March in Mediterranean and early April elsewhere, with no clearly defined end point to the passage, as most of the May records are from East Anatolia, where the species certainly oversummers and possibly breeds. In Inner Anatolia, despite regular observations from the Konya Basin during the breeding season, the status of this population is unknown. The other known breeding area is situated *c.*1,000km to the east, within the area bounded by Yüksekova, Erzurum, Ağrı and Muş (East Anatolia). In 2003, there were still some numbers in this region, but the future of the species might be partially threatened by ongoing dam projects and irrigation schemes.

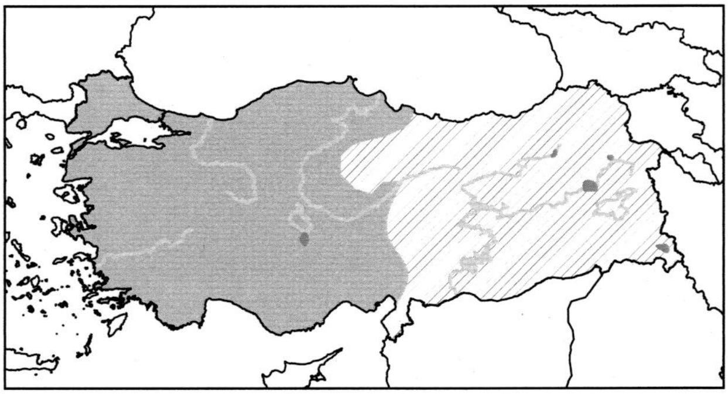

Autumn migration commences early September, with all records of passage birds at Çamlıca Hills (Istanbul) from the period 17 September to 1 October, but there are later records, e.g. from Kızılırmak Delta (Black Sea Coastlands) on 14 October, of obvious or presumed migrants. Unsurprisingly, observations such as those from Ares Island and Sinop (Black Sea Coastlands) indicate that at least some migrate over the sea.

Breeding. There are only two definite breeding records, both in **Inner Anatolia**. A pair was observed carrying food to at least two well-feathered young in a nest in *Juncus* northeast of Konya on 14 May 1970, and an incubating adult was flushed from a nest which had three eggs in rough grass in a dry part of Eşmekaya marsh on 23 May 1992; the nest was an unlined hollow and contained young *c.*1 week old on 5 June 1992 (B. Pleasance). These two records indicate laying of the first egg in early April and early May; breeding commonly commences

▲ The mountain of Uludağ, above Bursa (Marmara) in northwest Turkey, is home to a predominantly European avifauna, with a smattering of species typical of more easterly regions, such as Common Rosefinch *Carpodacus erythrinus*, as well as the near-endemic Krüper's Nuthatch *Sitta krueperi* (Gernant Magnin).

▲ Krüper's Nuthatch *Sitta krueperi*. This species has recently been classified as Near Threatened by BirdLife International. Kızılcahamam, Ankara (Soner Bekir).

▼ Common Rosefinch *Carpodacus erythrinus*. Sivrikaya (Emin Yoğurtçuoğlu).

▲ Coniferous forest at Akseki (Mediterranean) harbours many woodpeckers including Grey-headed *Picus canus* and White-backed *Dendrocopus leucotos*, whilst at the edges one may find more typically Mediterranean species such as Cretzschmar's Bunting *Emberiza caesia* (Bernard Pleasance).

▼ Cretzschmar's Bunting *Emberiza caesia*. Denizli (Okan Koçyiğit).

▲ Lily pads at Uluabat Gölü (Marmara). One of the most wildlife-rich lakes in northwest Turkey, Uluabat Gölü boasts huge numbers of Pygmy Cormorants *Phalacrocorax pygmeus* throughout the year (Peter Castell).

▼ Pygmy Cormorant *Phalacrocorax pygmeus*. Izmit Bay, Kocaeli (Soner Bekir).

▲ The Turkish Lake District, in the centre-west of the country, is home to almost innumerable waterfowl in winter; counting them, however, is beset by problems such as low temperatures and the huge size of some of these waterbodies. This photograph is of Eğirdir Gölü (Gernant Magnin).

▼ Dalmatian Pelican *Pelecanus crispus* colony. Büyük Menderes Delta, on the Aegean (Peter Castell).

▲ As recently as the mid-1990s the Turkish central plateau was still home to many birds and fantastic wetlands, but those days are now gone and most of the largest marshes have been drained. Nonetheless, globally threatened species such as this White-headed Duck *Oxyura leucocephala* cling on in small pockets of habitat. Sodalı Gölü (Mustafa Erturhan).

▲ Hotamış was one of the largest of the shallow Inner Anatolian playa lakes; this photograph was taken in 1990 when the site was still in reasonable condition, but as a result of a long drought and unsustainable water use (mainly for agriculture) the marshes are now all but gone (Gernant Magnin).

▼ Egg and newly hatched Common Crane *Grus grus*. Eşmekaya, Inner Anatolia (Peter Castell).

▲ The great salt lake in the heart of Turkey, Tuz Gölü (Inner Anatolia), is home to an enormous Greater Flamingo *Phoenicopterus ruber* colony, and Armenian Gulls *Larus armenicus* are regular here in summer. Unfortunately, Saker Falcon *Falco cherrug* and Steppe Eagle *Aquila nipalensis* may no longer breed (Gernant Magnin).

▼ Greater Sand Plover *Charadrius leschenaultii columbinus* on the nest. Karapınar Ovası (Lawrence Baxter).

▼ Huge numbers of Ruff *Philomachus pugnax* move through central Turkey on passage, especially in spring (Fatih Izler).

▲ Slender-billed Gull *Larus genei* breeds in sometimes quite large colonies in the Turkish interior. Bolluk Gölü, Inner Anatolia (Nick Williams).

▼ Armenian Gull *Larus armenicus* (Kars, Soner Bekir) with (inset) two newly hatched chicks (Balık Gölü, Peter Castell).

▲ The high slopes of Demirkazık (Mediterranean), in the Taurus Mountains, are among the most famous sites for Caspian Snowcock *Tetraogallus caspius*, whilst in summer the slopes below the cliffs harbour abundant Horned Larks *Eremophila alpestris*, Finsch's Wheatears *Oenanthe finschii* and Red-fronted Serins *Serinus pusillus* (Peter Castell).

▼ Caspian Snowcock *Tetraogallus caspius*. Yedigöller, Ispir (Soner Bekir).

▲ White-winged Snowfinch *Montifringilla nivalis*. Aladağlar, Niğde (Emin Yoğurtçuoğlu).

▼ Male Finsch's Wheatear *Oenanthe finschii*. Şanlıurfa (Soner Bekir).

▼ Red-fronted Serin *Serinus pusillus*. Denizli (Okan Koçyiğit).

▲ Kızılcahamam, Inner Anatolia, in May. A bastion of Pontic forest on Turkey's central plateau, this area is home to species such as Eurasian Black Vulture *Aegypius monachus*, Krüper's Nuthatch *Sitta krueperi*, Eastern Bonelli's Warbler *Phylloscopus orientalis*, abundant Hawfinches *Coccothraustes coccothraustes* and even Green Warbler *Phylloscopus nitidus*, as well as many typical European species (Guy M. Kirwan).

▼ Hawfinch *Coccothraustes coccothraustes*. Denizli (Okan Koçyiğit).

▼ Eastern Bonelli's Warbler *Phylloscopus orientalis*. Akseki, Antalya Province (Soner Bekir).

▲ The hills around Gaziantep (Southeast Anatolia) harbour a great number of birds of interest to the Western Palearctic birder, amongst them Pale Rockfinch *Carpospiza brachydactyla*, Eastern Rock Nuthatch *Sitta tephronota* and Upcher's Warbler *Hippolais languida*; the orchards below these slopes harbour Cinereous Bunting *Emberiza cineracea* and mark the easternmost limit of the breeding range of Olive-tree Warbler *Hippolais olivetorum* (Peter Castell).

▼ Eastern Rock Nuthatch *Sitta tephronota*. Gaziantep (Soner Bekir).

▲ Upcher's Warbler *Hippolais languida*. Gaziantep (Soner Bekir).

▼ Cinereous Bunting *Emberiza cineracea semenowi*. Gaziantep (Soner Bekir).

▼ Male Desert Finch *Rhodospiza obsoleta*. Birecik (Fatih Izler).

▲ Pale Rockfinch *Carpospiza brachydactyla*. Yesılce (Chris Bradshaw).

▼ Adult Long-legged Buzzard *Buteo rufinus* at a nest with three young. Near Gaziantep (Peter Castell).

▼ Isabelline Wheatear *Oenanthe isabellina,* a familiar breeding bird of the semi-steppe of central and eastern Turkey (Keith Robson).

▲ A view of the Birecik area (Southeast Anatolia). This region has long been famous amongst birdwatchers for its once-large population of Bald Ibis *Geronticus eremita*, which is now present in the area only as a feral bird, as well as a reliable place to find the localised See-see Partridge *Ammoperdix griseogularis* (Peter Castell).

▼ A nesting pair of Bald Ibis *Geronticus eremita*. Birecik (Soner Bekir).

▲ See-see Partridge *Ammoperdix griseogularis*. Şanlıurfa (Soner Bekir).

▼ Iraq Babbler *Turdoides altirostris* has only recently been discovered in Turkey, on the Euphrates at Birecik, following its apparent expansion through Syria. Birecik (Soner Bekir).

▲ Many of Turkey's cultural sites also provide good birdwatching opportunities. Nemrut Dağı in Southeast Anatolia is one such place; the Commagene detached-head sculptures provide a suitable home for Red-tailed Wheatears *Oenanthe xanthoprymna* (Guy M. Kirwan).

▼ Adult female Red-tailed Wheatear *Oenanthe xanthoprymna*. Gaziantep (Soner Bekir).

▼ Adult male Ménétriés's Warbler *Sylvia mystacea rubescens*. Birecik (David Cottridge).

▲ Ispir (Black Sea Coastlands) in May, home to one of the last Eurasian Griffon Vulture *Gyps fulvus* colonies in Turkey, along with Wallcreeper *Tichodroma muraria* and Ring Ouzel *Turdus torquatus amicorum* (Guy M. Kirwan).

▲ Eurasian Griffon Vulture *Gyps fulvus*. Çamlıca Hills, east of Istanbul (Soner Bekir).

▼ Ring Ouzel *Turdus torquatus amicorum*. Uzundere, Erzurum Province (Soner Bekir).

▲ Wallcreeper *Tichodroma muraria*. Demirkazık (Soner Bekir).

▼ Gelinkaya, northwest of Erzurum (Black Sea Coastlands) in May (Guy M. Kirwan). On the south-facing slopes of the Pontic Alps, one enters the heartland of Caucasian Chiffchaff *Phylloscopus lorenzii* and the rather less ubiquitous Semi-collared Flycatcher *Ficedula semitorquata*. Other common birds in this habitat include Eurasian Penduline Tit *Remiz pendulinus* and Common Rosefinch *Carpodacus erythrinus*, as well as typical European species including a variety of tits.

▲ Male Semi-collared Flycatcher *Ficedula semitorquata*. Ispir, Erzurum (Soner Bekir).

▼ Caucasian Chiffchaff *Phylloscopus lorenzii*. Ispir, Erzurum (Soner Bekir).

▲ The valley of the River Murat between Bulanık and Malazgirt, East Anatolia. One of the Western Palearctic's rarest breeding birds is found here, Demoiselle Crane *Grus virgo*. Many larks, including Bimaculated *Melanocorypha bimaculata* and Calandra *M. calandra* are common, and a declining population of Great Bustards *Otis tarda* also occurs here, in one of its last strongholds in Turkey (Guy M. Kirwan).

▼ Great Bustards *Otis tarda*. Bulanık Plain, Muş (Soner Bekir).

▲ Calandra Lark *Melanocorypha calandra*. Gaziantep (Soner Bekir).

▼ Demoiselle Cranes *Grus virgo*. Bulanık Plain, Muş (Soner Bekir).

▲ The Çatak Valley, East Anatolia. Plain Leaf Warbler *Phylloscopus neglectus* was discovered here in the early years of the 21st century. More regular fare on these slopes includes Crimson-winged Finch *Rhodopechys sanguineus* (Guy M. Kirwan).

▼ Crimson-winged Finch *Rhodopechys sanguineus*. Uzundere, Erzurum (Soner Bekir).

▲ Eurasian Black Vulture *Aegypius monachus*. Erzurum Plain (Soner Bekir).

▼ Grey-necked Bunting *Emberiza buchanani* (Guy M. Kirwan), with (inset) a nest (Peter Castell). This bunting, which gradually replaces Ortolan Bunting *E. hortulana* in the east of the latter's range, is common in the Van region of eastern Turkey.

▲ Van castle, East Anatolia. The castle itself is home to breeding Alpine Swifts *Apus melba* and Lesser Kestrels *Falco naumanni*, whilst the marshes from where this photograph was taken harbour a range of waterbirds, including Paddyfield Warbler *Acrocephalus agricola*; the latter was only discovered in Turkey in the mid-1980s, at this site (Guy M. Kirwan).

▼ Black-headed Wagtail *Motacilla flava feldegg* (top) and Reed Bunting *Emberiza schoeniclus* (bottom), both from the Van marshes (Guy M. Kirwan).

▼ Male Eurasian Penduline Tit *Remiz pendulinus*. Penduline Tits occur over much of Turkey, except the northern coastal strip and parts of the southeast. Denizli (Okan Koçyiğit).

▲ A view of the raptor watchpoint at Borçka, northeast Turkey (Black Sea Coastlands). The Çoruh River valley functions as a funnel for hundreds of thousands of migrant raptors each spring and autumn, although the significance of this route for soaring birds was not realised until the mid-1970s (Gernant Magnin).

▲ European Honey Buzzard *Pernis apivorus* is a very common passage migrant through Turkey. Sarıyer, Istanbul (Mehmet Çetinkoç).

▼ Common Buzzards *Buteo buteo* soaring above the Borçka Pass (Soner Bekir).

▲ A rhododendron-covered hillside at Sivrikaya (Black Sea Coastlands). This is probably the world's most accessible site for Caucasian Grouse *Tetrao mlokosiewiczi* (Guy M. Kirwan).

▲ Green Warbler *Phylloscopus nitidus* is a common constituent of the avifauna at mid-altitudes in the eastern Pontics (Ruedi Aeschlimann).

▼ Radde's Accentor *Prunella ocularis* is principally restricted as a breeding species to the eastern Taurus and East Anatolia, but the birds move to lower elevations in winter. Birecik (Fatih Izler).

▲ Male Caucasian Grouse *Tetrao mlokosiewiczi*. Kazbegi (Georgia) (Ruedi Aeschliman).

▼ Female Caucasian Grouse *Tetrao mlokosiewiczi*. Yusufeli, Artvin (Soner Bekir).

▲ Another view of the pass at Sivrikaya (Black Sea Coastlands) in June. A good habitat for Alpine *Prunella collaris* and Radde's *P. ocularis* Accentors, and Water Pipits *Anthus spinoletta coutellii* are in abundance (Peter Castell).

▼ Water Pipit *Anthus spinoletta* is a common breeder at high altitudes throughout the country, moving to lower areas in winter. Gaziantep (Fatih Izler).

▼ Alpine Accentor *Prunella collaris*. Three species of accentor breed in Turkey, of which this is the most widespread (a fourth has been recorded as a vagrant). Demirkazık (Soner Bekir).

▲ One of the rarest breeding raptors in Turkey, White-tailed Eagle *Haliaeetus albicilla* is markedly more widespread and numerous in winter, when the population is augmented by substantial number of immigrants. Beypazarı, Ankara (Mehmet Çetinkoç).

▼ Male Lesser Kestrel *Falco naumanni*. This globally threatened raptor is still reasonably widespread in Turkey, being especially numerous in the central part of the country, where it remains one of the commonest and most frequently seen birds of prey. Mogan, Ankara (Mehmet Çetinkoç).

▲ Pool near Erçek Gölü (East Anatolia) in June. Such areas of marshland provide valuable breeding habitat for Common Pochard *Aythya ferina*, White-headed Duck *Oxyura leucocephala*, Moustached Warbler *Acrocephalus melanopogon mimicus* and Citrine Wagtail *Motacilla citreola* (Peter Castell).

▼ Citrine Wagtail *Motacilla citreola* (Soner Bekir) with (inset) a nest (Peter Castell). Despite being a recent addition to the Turkish breeding avifauna, Citrine Wagtail is already quite widespread.

▼ Breeding site of Velvet Scoter *Melanitta fusca*, the crater lake at Nemrut Dağ, near Tatvan (East Anatolia). That this seaduck should breed so far south is a biogeographical anomaly. The sparsely vegetated slopes of the inactive volcano harbour White-throated Robins *Irania gutturalis* (Guy M. Kirwan).

▲ Moustached Warbler *Acrocephalus melanopogon*. Akyatan, Çukurova (Mehmet Çetinkoç).

▼ Purple Swamphen *Porphyrio porphyrio*, Göksu Delta, Mersin. This species has been recorded several times in eastern Turkey, but it is principally restricted to the major coastal deltas (Soner Bekir).

▲ European Bee-eater *Merops apiaster*. Çatalca, Istanbul (Mehmet Çetinkoç).

▼ Male White-throated Robin *Irania gutturalis*. Afyon (Okan Koçyiğit).

▲ White Stork *Ciconia ciconia* is a common breeder almost throughout Turkey, and its nests frequently provide a breeding site for House *Passer domesticus* and Spanish *P. hispaniolensis* Sparrows (Guy M. Kirwan).

elsewhere in late April, but sometimes a few weeks earlier. In **East Anatolia**, adults have been observed in display flight at Bulanık on 19 May 1975, at Malazgirt on 9 May 1986 and at Balatos on 10 May 1986, and the species was suspected to be nesting at the first of these sites on 24 May 2005.

TENGMALM'S OWL *Aegolius funereus*
Paçali Baykuş

(Du: Ruigpootuil; Fr: Nyctale de Tengmalm; Ger: Rauhfußkauz; Swe: Pärluggla)

Subspecies. The form occurring in Turkey is unknown (no specimen material is available and most records have involved only vocalising birds). The nominate occurs in Bulgaria, whilst the race *caucasicus*, which is smaller and darker (more chocolate-brown), with a shorter wing (no overlap in measurements; König *et al.* 1999), smaller white spots, browner facial disk and larger dark throat patch (Adamian & Klem 1999), is known from the Caucasus including Armenia (Ananian *et al.* 2002), where it occurs at two localities close to the border with Turkey. On the basis of photographs of one found dead in northeast Turkey in 2004, which had very dark upperparts, it seems that *caucasicus* penetrates the easternmost Pontics at least, but nominate *funereus* could occur in western and southern Turkey, as already noted by Roselaar (1995). Further research on this issue is clearly desirable. A similar biogeographical pattern, with Caucasian forms in northeast Turkey and European forms in the west and centre of the country, is witnessed in many other species.

Status and distribution. National conservation status: Vulnerable (population apparently stable). Initially discovered in Turkey in 1979 and there have now been at least ten records, from the Pontics and Taurus, as well as montane isolates south of the Sea of Marmara and in Inner Anatolia. Given the recent discovery that the species is widespread and even reasonably common in Bulgaria (Nankinov 1996b, Shurulinkov *et al.* 2003), it is plausible to assume that Tengmalm's Owl is significantly more numerous and geographically widespread than currently known, and must almost certainly be present in highland northeast Thrace. All records in Turkey have been opportunistic. Specific surveys for the species have not been conducted in the country and fieldwork in the relevant habitats during the principal seasons when the species is most easily detected (March–May and October) is almost non-existent in Turkey. Recorded as follows: one singing on Uludağ (Marmara) on 15–16 June 1979 (Mertens 1981), one singing at Kızılcahamam (Inner Anatolia) on 20–27 June 1981 (Barış *et al.* 1984), one at Sivrikaya (Black Sea Coastlands) on 15 September 1986 (Martins 1989), one between Sumela and Maçka (Black Sea Coastlands) on 21 June 1990 (Kirwan & Martins 1994), two singing at Güzeller in the Yugluk Dağı (Inner Anatolia) on 15 May 1998, one at Kocaçay Delta (Marmara) on 13 March 2000 (Kirwan *et al.* 2003), one singing throughout much of the night (tape-recorded) at Sivrikaya on 23–27 May and 13–14 June 2002 (*Dutch Birding* 24: 241 and M. Robb *in litt.* May 2003), one tape-recorded in the Aktaş Valley in Bolu province (Inner Anatolia), on 7 April 2003 (Demirci 2003), one found dead 1km north of Elevıt in the upper Fırtına Valley at *c.*1,800m (Black Sea Coastlands) on 8 August 2004 (G. Magnin *in litt.* 2005; photographed), and one seen and tape-recorded at Kızılcahamam on 18 September 2004 (Balmer & Betton 2005a, Şekercioğlu 2006). Details of another record, from the western Taurus (Mediterranean) in February 2001 (J. Blanchemain *in litt.* March 2001), are currently awaited.

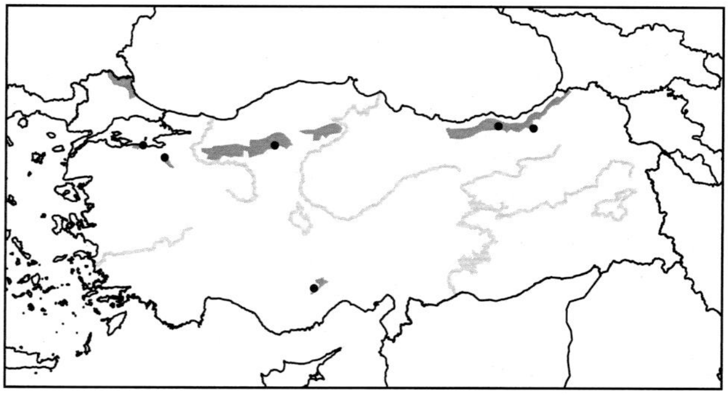

Breeding. Presumably a rare breeder in montane coniferous forest, but there are no definite nesting records. In other southerly parts of the range, breeding commences in mid April, and nests are in tree holes, including those of Black Woodpecker *Dryocopus martius*. No nest material is added to the base of the hole: 3–6 eggs are laid.

EUROPEAN NIGHTJAR *Caprimulgus europaeus*
Çobanaldatan

(Du: Nachtzwaluw; Fr: Engoulevent d'Europe; Ger: Ziegenmelker; Swe: Nattskärra)

Subspecies. *C. e. meridionalis*, but variation in this species is almost entirely clinal. The rationale for maintaining *meridionalis* was discussed by Vaurie (1960a), but we consider the case for its recognition to be rather weak.

Status and distribution. National conservation status: Least Concern (population apparently decreasing). Fairly widespread and not uncommon summer visitor to the west and south, and locally over much of the rest of the country, except (apparently) Southeast Anatolia, where seems only to be present in southern areas, although is probably more widespread than records suggest. Generally recorded in areas of dry scrub or open woodland during the breeding season, up to at least 2,300m. Somewhat more widespread on passage. Present from at least mid May to late September in the north and until late October in the south, with an exceptionally early individual found exhausted at Alanya (Mediterranean) on 6 March 1970. The latest record is also from this region, at Erdemli, on 10 November 1970. In both spring and autumn, groups of up to 40–100 have been reported, from Marmara, at Uluabat and Manyas lakes, and in Mediterranean, at Göksu Delta (see also van den Berk & van der Winden 1992). One claimed winter occurrence: at Sultan marshes (Inner Anatolia), on 26 December 1996 (Kirwan 1997d).

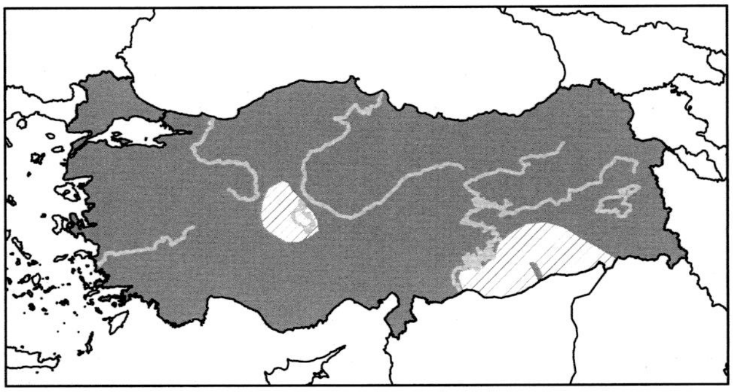

Breeding. Requires areas of dry bare ground, which is essential for nesting, in dry scrub with or without scattered trees, in open deciduous and coniferous woodland, especially woodland edges and open areas near trees, on coastal dunes with scrub, and in rocky gullies with scattered bushes; up to 2,300m in the Taurus. An incubating adult was flushed from an egg (probably an incomplete clutch) on a hillside, with scrub *c.*1m tall, and bare patches of ground at Akköy (**Aegean**) on 25 May 2004 (M. Demain). A nest (contents unknown) was on a scrub-covered hillside at Çay (Inner Anatolia) on 12 May 1970. A probable pair was flushed from an area of thin deciduous woodland, with rocky outcrops and small bare clearings, at 2,250m, close to the largest crater lake at Nemrut Dağ (East Anatolia), on 9 June 2004: the female was flushed from a bare patch, under a tree, ideal for nesting, and the male was subsequently flushed, *c.*80m away, where he had been roosting (but churred briefly) atop a large rock (B. Pleasance). Nightjars are present in Turkey early May–September; churring is heard and wing-flapping displays observed during these months, and it is assumed that two broods are reared, as elsewhere; the usual clutch is two eggs, laid on bare ground, with no nest material added.

[EGYPTIAN NIGHTJAR *Caprimulgus aegyptius*]
Çöl Çobanaldatanı

(Du: Egyptische Nachtzwaluw; Fr: Engoulevent du desert; Ger: Pharaonenziegenmelker; Swe: Ökennattskärra)

Subspecies. Unknown, although the nominate is presumably the more likely, of the two generally recognised forms, to occur.

Species not admitted to the Turkish list (Kirwan *et al.* 1999). Listed as hypothetical in Turkey by Ergene (1945) and Turan (1990), who stated that the species possibly occurred near the Syrian border, and claimed as a vagrant at Nemrut Dağı, Tatvan (East Anatolia), by Green & Moorhouse (1995). No details concerning this record have been published and occurrence in Turkey seems unlikely, although there are records from six European countries as far west as Britain, and the species breeds rarely in Iraq (Snow & Perrins 1998). Baumgart *et al.* (1995) mentioned a specimen record from central Syria in 1904, when the species was apparently more common in adjacent Iraq (Snow & Perrins 1998).

COMMON SWIFT *Apus apus*
Ebabil

(Du: Gierzwaluw; Fr: Martinet noir; Ger: Mauersegler; Swe: Tornseglare)

Subspecies. Nominate.

Status and distribution. National conservation status: Least Concern (population apparently stable). Widespread and common summer migrant, breeding in towns and villages, more occasionally in caves and on cliffs, throughout Turkey, although it appears to be very local and even absent from many towns in the Black Sea region. Breeds from sea level to at least 2,300m. Arrives from mid (13th, exceptionally early) March in most southern areas, although more frequently recorded from early April (4th). At Kızılırmak Delta (Black Sea Coastlands) in 1992, arrival noted from mid April and passage peaked in the final two weeks of May (Hustings & van Dijk 1994), whilst in the south passage probably peaks between late April and the end of the first week of May (see van der Have *et al.* 1988), and in the centre and east of the country during the first two weeks of May.

Passage may involve huge numbers passing in a constant stream and on a broad front, e.g. 3,000 were counted in just 15 minutes at Mardin (Southeast Anatolia) on 7 May 2002. Departs from early August (an exceptional concentration of *c.*10,000 was noted at Aşkale, East Anatolia, on 1 August 1974). Many have left by the end of this month: peak passage at the Belen Pass in 1976 was 19–29 August (Sutherland & Brooks 1981b) May remain relatively widespread, if uncommon, into late September (29th). At the Bosphorus in the mid 1960s Porter (1983) found that the majority had left by 10 September, although birds have been recorded in Istanbul (Marmara) until late October (latest 25th) and Inner Anatolia until mid November (18th; both KAB pers. obs.).

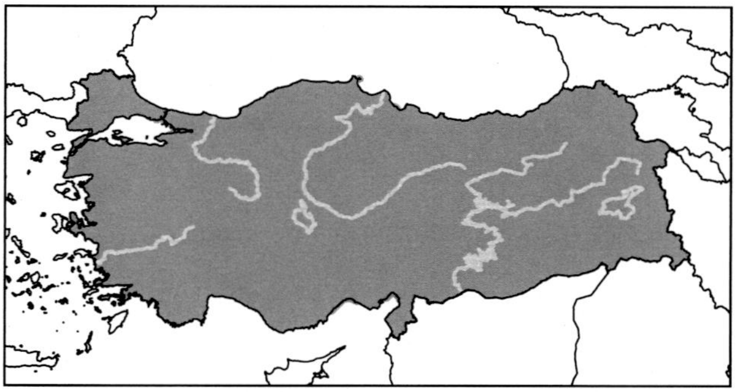

Breeding. Mainly in towns and cities, with large concentrations in Istanbul and Izmir, e.g. thousands over the latter on 15 June 1966. Also in more remote areas with cliffs, both coastal (including offshore islands) and inland (including gorges). Colonies vary from a few pairs to 200 birds breeding in hotel roofs above 1,900m on Uludağ (Marmara) (Jetz 1995). Nests usually in buildings, often in roofs and under eaves of (mainly tall) buildings, but also in ruins, under bridges, and in holes and crevices in cliffs; an adult was observed entering a mud nest on a cliff near Ardahan (East Anatolia) on 13 May 1970, and this nest had, presumably, been built by Common House Martins *Delichon urbicum*. McNeile visited two small rocky islets off Ilıca, near Çesme (**Aegean**) on 8 May 1950; he found a nest with one egg (an incomplete clutch), and another with three eggs in a narrow horizontal fissure in the low cliffs. The nest is of grass, debris and feathers, collected in flight, and consolidated with saliva to form a shallow cup, placed on a ledge or flat surface. Adults observed copulating in flight in **Inner Anatolia** at Eber Gölü on 19 April 1967, at Akşehir Gölü on 24 April 1967 (R. F. Porter), and at Cihanbeyli on 15 May 2004 (J. H. Mason), where cliff-nesting birds were seen in the Inkuyu Valley on 24 June 1998 (GMK). An adult was carrying nesting material at Gulpınar (**Marmara**) on 26 June 1973 (R. F. Porter). Adults observed flying to nesting holes May–July; latest 5 August 1966 in Van (**East Anatolia**). The entire breeding cycle, involving a single brood, takes 8–10 weeks, exceptionally long for a small bird.

PALLID SWIFT *Apus pallidus*
Boz Ebabil

(Du: Vale Gierzwaluw; Fr: Martinet pâle; Ger: Fahlsegler; Swe: Blek tornseglare)

Subspecies. *A. p. brehmorum*, except on the Dalmatian coast, where apparently replaced by *illyricus*, which is differentiated by its slightly darker plumage (although considerable variation is known to exist within *brehmorum*), with the pale throat patch consequently more obvious, and perhaps overall larger size (Chantler & Driessens 1995, Chantler *in* del Hoyo *et al.* 1999). Vaurie (1965) considered *illyricus* to be poorly differentiated, an assessment with which we could concur. A specimen in the Natural History Museum, Tring (1951.13.740) from Cape Greco, Cyprus, taken in April, accords well with *illyricus*, clearly suggesting the possibility that this form could occur in Turkey. Nominate *pallidus* may occur in eastern Turkey, according to Roselaar (1995). In very worn plumage this form can be extremely pale on both surfaces, which accords well with our own sight observations of birds in midsummer in southeast Turkey, suggesting that *pallidus* may well occur there. As previously noted, differentation between *illyricus* and *brehmorum* is clearly very slight, but even the degree of variation between either of these forms and nominate *pallidus* is marginal, obscured by a degree of overlap and probably strongly influenced by wear. Under the terms adopted here, *brehmorum* and *illyricus* are best considered synonyms of nominate *pallidus*.

Status and distribution. National conservation status: Data Deficient (population trend unknown). Rather uncertain. First documented from Turkey by Kumerloeve (1966d, 1970a), based on observations in the Amik Gölü–Antakya area (Mediterranean) and from Birecik, Halfeti and Urfa (Southeast Anatolia). Observations were first made on Uludağ (Marmara), now known to hold a small breeding population, and from Istanbul and the Bosphorus in 1966–67 (OST 1969). A local summer visitor in rather small numbers to Uludağ: where recorded at 1,800–2,500m, Istanbul and at Işak Paşa Sarayı, Doğubeyazıt (East Anatolia). In Istanbul, it is now known to be a fairly common breeder between Kadıköy and Bostancı on the Asian side, and near Bakırköy and Ataköy on the European shore. More rarely reported elsewhere, most frequently in extreme east Mediterranean (OST 1975, 1978, Beaman 1986, Martins 1989) and adjacent parts of Southeast Anatolia, where it probably breeds (e.g. near Nemrut Dağı, Adıyman province). Also reported from the southwest coast and southwest Black Sea Coastlands in summer in potential breeding habitat.

A scarce passage migrant in other areas, principally in the west and south, although up to 150 were recorded at Göksu Delta (Mediterranean) in early April. In spring recorded from early April, with one record from 27 February in Mediterranean. In autumn on passage from mid August to late September/early October (with single-day maxima of 150 at Uludağ on 21 September 1987 and 85 at the Bosphorus on 2 October 1997), occasionally until mid October (18th) on the western Black Sea coast and late October (25th) in Istanbul.

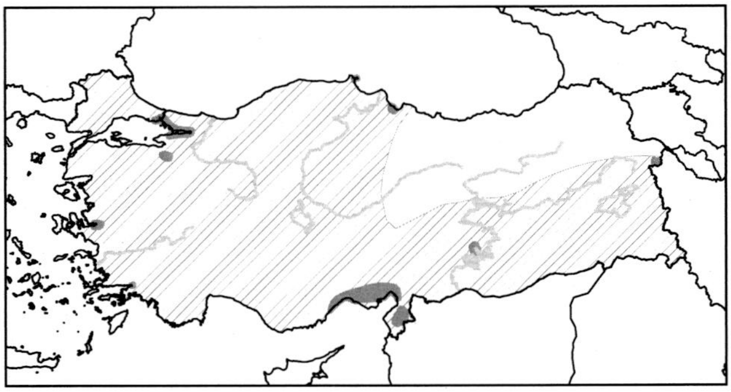

Breeding. Since 1967, 25–40 have been regularly observed entering roofs, with Common Swifts *A. apus*, in buildings at 1,850m on Uludağ (**Marmara**), with several records late May–August, the presumed breeding season. Several pairs found nesting at Işak Paşa Sarayı (**East Anatolia**) on 22 May 1985. Nest and clutch size undescribed in Turkey but, elsewhere, nests are built in crevices in cliffs and walls, as well as in roofs; of grass and feathers, collected in flight, and consolidated with saliva to form a shallow cup; usual clutch is 2–3. In Istanbul it prefers newer buildings of less than 30 years old, whereas *A. apus* breeds in older buildings, often in narrower streets, and in other man-made structures such as walls, generally in the older part of the city.

ALPINE SWIFT *Apus melba*
Ak Karınlı Ebabil

(Du: Alpengierzwaluw; Fr: Martinet à ventre blanc; Ger: Alpensegler; Swe: Alpseglare)

Subspecies and taxonomy. Nominate. Frequently placed in the genus *Tachymarptis* (e.g. Snow & Perrins 1998, Chantler *in* del Hoyo *et al.* 1999, Gill & Wright 2006) but see Sangster *et al.* (1999) for cogent rationale for its retention in *Apus*, at least until such time as further, and more robust, molecular data become available.

Status and distribution. National conservation status: Least Concern (population apparently increasing). Fairly widespread and common (or locally abundant) summer visitor breeding throughout Turkey, commonest in the south and west, particularly in the Taurus (Mediterranean), and least common in Southeast Anatolia, where restricted to the north and east. Several large breeding colonies are known, including one in Istanbul, with several other large-city breeding sites known, e.g. Elazığ (East Anatolia). Breeds from sea level to at least 2,700m, but principally recorded in upland and montane areas.

Widespread on passage, when locally recorded in substantial numbers. Regularly recorded from early or mid March (with three records in late February, earliest 21st), although seldom common and widespread until early April, especially in the north. In 1987, peak numbers moved through Çukurova in late March (van der Have *et al.* 1988), but large numbers noted passing through Göksu Delta in early–mid April 1971 and 1973 (OST 1975). Return passage occurs between mid August and mid October, peaking at the Bosphorus during the first ten days of September (Porter 1983) and at the Belen Pass (Mediterranean) in the middle two weeks of September (Sutherland & Brooks 1981b), although stragglers are occasionally recorded until late October or mid November in Thrace (latest, 17th, in Istanbul) and, more rarely, into late November (21st) in Mediterranean. Hundreds, even thousands, may remain in Istanbul until at least late October. Exceptionally, there are two December records (6th and 19th) from Istanbul, both in 1994 (KAB pers. obs.).

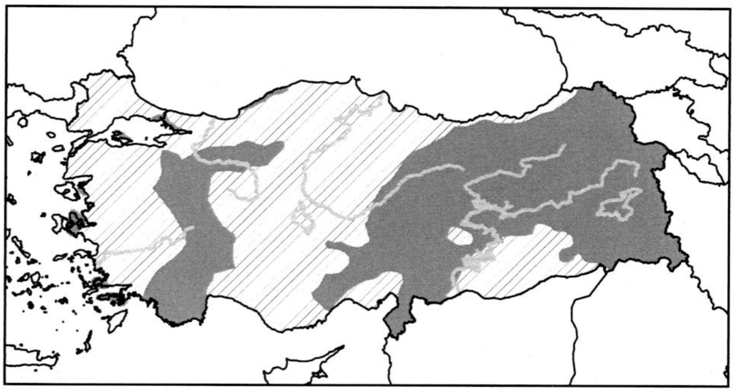

Breeding. Colonies are in cliffs (both coastal and inland), tall buildings, ruins, e.g. Van castle (East Anatolia), man-made structures, such as bridges, and within 'fairy chimneys' in Cappadocia (Inner Anatolia). Colonies vary from a few pairs, e.g. 12 in the attic of a 20-m building in Istanbul (PC) to hundreds, e.g. 300 pairs west of Hakkari (East Anatolia). Nests in crevices in the cliff face, often in narrow fissures; on a ledge or adhering directly to the crevice; the nest is a mixture of plant debris and feathers, collected in flight, and consolidated with saliva to form a shallow cup, as a bracket if attached to a vertical surface. Eggs: C/1 (2), C/2 (3), C/3 (6). Young: B/3 (4), B/2 (2). No data from Turkey but, elsewhere, young remain in the nest for 6–8 weeks. In Istanbul (**Marmara**), a pair was observed building a nest that was almost complete, between a wall and a wooden shutter of a school, on 21 April 1970 (R. F. Porter). Two nests had three eggs on 3 June 1992 (PC). In 2004, at least ten pairs nested in the attic of an old building, and two pairs nested just below, on ornamental wooden shelving brackets, forming narrow vertical crevices in which the nests were attached to one side. Adults were flying to and fro on 7 June 2004, and a pair copulated on the wing as they flew from a nest; on 27 June 2004, an adult was twice observed flying to a nest and, clinging to it with both wings open, fed at least one large young. There were eight occupied nests on this building on 13 June 2006, and two pairs were nest building and one adult incubating on 17 May 2007 (PC). At least 12 pairs were nesting amidst debris on the floor of an attic on 14 June 2006; three nests had three eggs, three newly hatched young and two young, *c.*4–5 days old, indicating egg laying in mid May (R. Castell). In **Inner Anatolia**, a colony of 15 pairs, with 11 nests visible, was in a dovecote at Göreme on 14–15 June 1977; six had well-incubated eggs and four had newly hatched young, with two nests only 30cm apart (Schubert 1979). In another fairy chimney at Göreme, one nest was on a vertical wall, empty, but ready for eggs, and the other had two eggs, at the base of a cavity for doves, on 14 June 1993 (PC). In **Aegean**, a nest had one egg, inside a narrow crack in a rock face at Ilıca, near Çesme, on 8 May 1950 (McNeile). There are several records of adults entering holes, mainly May–July (earliest mid April), in all regions of Turkey.

LITTLE SWIFT *Apus affinis*
Küçük Ebabil

(Du: Huisgierzwaluw; Fr: Martinet des maisons; Ger: Haussegler; Swe: Stubbstjärtseglare)

Subspecies. *A. a. galilejensis.*

Status and distribution. National conservation status: Endangered (population apparently decreasing). A very local and generally uncommon summer visitor, breeding in the extreme east and southeast Mediterranean, around the northern end of the Gulf of Iskenderun, in Çukurova and in Antakya province, and at a number of localities in Southeast Anatolia, apparently east to about Siirt (East Anatolia), from where there are several records during the breeding season. Breeding is also possible in adjacent areas of southwest East Anatolia, given its presence as far north as the Atatürk barrage (Southeast Anatolia). The largest colonies are in the Euphrates River cliffs, at Birecik and Halfeti, and away from there at Kilis (Southeast Anatolia). Slightly more widespread on passage, when reported from a number of neighbouring areas, e.g. Göksu and Çukurova deltas, Mersin and elsewhere (Mediterranean). Turan (2006) mapped occurrence in westernmost Thrace and around Izmir (Aegean), but we have not seen evidence of this and these surely involve passage birds, if correct. In spring recorded from mid March until at least late April on passage, with regular reports of up to 20 at passage-only sites; in autumn most have departed by mid September, rarely reported until late September (30th), though there is a report from Mersin as late as 29 September (2001), involving 85 birds (Turan 2006). Three recent winter claims perhaps pertain to the resident northern Israel poplation which is known to disperse in winter (Chantler *in* del Hoyo *et al.* 1999): two at Antakya (Mediterranean) on 6 January 2007, with further records from the same locality on 4 January (six) and 13 January 2008 (six) (A. Atahan).

The record of 8–10 near the Gulf of Iskenderun (Mediterranean) on 2 July 1970 was erroneously reported as the first record for Turkey (*OST Bull.* 7:1). The species was collected by Chantre (1883) at Antakya (Mediterranean) in summer 1881 and was also reported, perhaps doubtfully, by Radde (1884) in the vicinity of Büyük Ağrı Dağı (Mount Ararat), Kars province (East Anatolia) in August 1871. The first modern report was in April 1962 (Eggers & Lemke 1964, Kumerloeve 1966d) and by the late 1970s at least 11 reliable records had emerged at or near the Mediterranean coast (Kumerloeve 1970a).

Breeding. In dry areas with cliffs and gorges. At Birecik (**Southeast Anatolia**), the best-known site, small colonies are on high cliffs on the east bank of the Euphrates, and in nearby wadis with tall cliffs. Breeds in colonies, typically of 4–15 nests but 25 pairs were at a colony near Kilis (Southeast Anatolia) in 1971 (Warncke 1972), and 120 birds were in suitable breeding habitat near Kilis on 14 May 1989. There may be several small colonies along the cliffs at Birecik, and a total of 62 pairs bred there in summer 1973. Nests are on cliffs, usually attached to the underside of an overhang, or the roof of a recess or cave; elsewhere, e.g. Morocco and Israel, nests are also in sheltered sites on buildings and structures, but there are no such records in Turkey. The nest is a globular structure: there may be a single nest, or several may be attached, each with an individual entrance and chamber, but the whole forming a contiguous mass; of grass and feathers, collected in flight, bound with saliva, and lined with finer grass and feathers. Nests are also constructed inside old, and often partially damaged, mud nests of Common House Martins *Delichon urbicum*. The same nest may be repaired and used in successive years, but there are no details from Turkey. In a colony of *c.*15 pairs, at Birecik on 23 May 2004, three nests were individual, four were in old nests of Common House Martin, and *c.*8 nests were attached en masse (PC). Eggs: no data; usual clutch, elsewhere, is 2–3. The breeding cycle is exceptionally long for a small species, *c.*9 weeks from laying the first egg to young leaving the nest, and two broods are usually reared annually. The earliest breeding record is from Kilis (Southeast Anatolia), near the Syrian border, where Warncke (1972) found 25 pairs in a cave within a cliff on 7 April 1971, and 60 birds were seen there on 19 August 1972; *c.*15 pairs were nesting in a cave in a narrow gorge, near Iskenderun (Mediterranean), on 11 April 1971 (Warncke 1972). At Birecik, none had returned to their nests by 8–19 April 1988 (C. Bradshaw); there are many records of adults at nests there, May–August, and adults were entering nests as late as 7 September 1994 (J. Taylor). A colony (20+ birds) was found near Siirt (East Anatolia) on 8 July 1986 (R. Webb).

WHITE-THROATED KINGFISHER *Halycon smyrnensis*
İzmir Yalıçapkını

(Du: Smyrnaijsvogel; Fr: Martin-chasseur de Smyrne; Ger: Braunliest; Swe: Smyrnakungsfiskare)

Subspecies. Nominate, which was originally described from Izmir (Aegean).

Status and distribution. National conservation status: Endangered (population apparently decreasing). Described by Linnaeus from specimens taken at Smyrna (Izmir) by William Sherard in the early 18th century. Also recorded from this area by Krüper & Hartlaub (1875) who referred to breeding along the major rivers around Izmir (Aegean). Krüper collected a clutch of eggs in early May 1894 but there has been only one subsequent, confirmed record from this region, near Camaltı, in 2002/03. A specimen on display in the Aegean University at Izmir was probably collected from the neighbouring area within recent decades (van den Berk & Kasparek 1988), but this remains unproven. In the last century, both Danford and Selous knew this species from the lower Büyük Menderes Valley (Aegean) and it has recently been recorded in the Güllük Delta, just to the south. This region is apparently the most northerly contemporary breeding area: a pair in suitable breeding habitat at Troja, Çanakkale province (Marmara) in June 1990 (Deppe 1990) and one at the Gediz Delta (Aegean) in February 2003 are the only recent records further north (not mapped).

A local and generally uncommon resident breeder in coastal lowlands and along rivers in southern Aegean and Mediterranean. Rarely recorded in Southeast Anatolia (five records, all in the breeding season) in the Euphrates and Tigris rivers as far east as Cizre and Siirt (Mardin province). The most significant breeding areas are all in Mediterranean; around Köyceğiz Gölü (a few pairs), on the lower Göksu river (8–10 pairs in recent years, van den Berk & Güder 1992) and in Çukurova, with the largest population in the Tarsus and Ceyhan deltas (40–45 pairs in the former in 1987). A tentative recent population estimate suggested 100–150 pairs (van den Berk & Kasparek 1988). With the exception of the Çukurova population, where intensified agricultural practices appear to be beneficial, the national population is apparently undergoing a long-term decrease. Several former, known and suspected breeding localities, *e.g.* around Izmir, the Küçük Menderes Delta (Aegean) and Amik Gölü (now drained), Antakya province (Mediterranean) have been deserted, while others, e.g. at the Göksu Delta, appear to hold smaller numbers than in the past.

A few records from outside the known breeding areas suggest limited and perhaps irregular dispersal within the Aegean and Mediterranean regions outside the breeding season, principally between late August and late February. Exceptionally the species is listed by both Rigler (1852) and Wahby (1930), two often rather suspect correspondents, among the birds of Istanbul. Unfortunately, neither proffered more information. More certainly, one was present on the banks of the Kızılırmak River, at Avanos, Nevşehir province (Inner Anatolia) on 17 December 2001 (G. Güven).

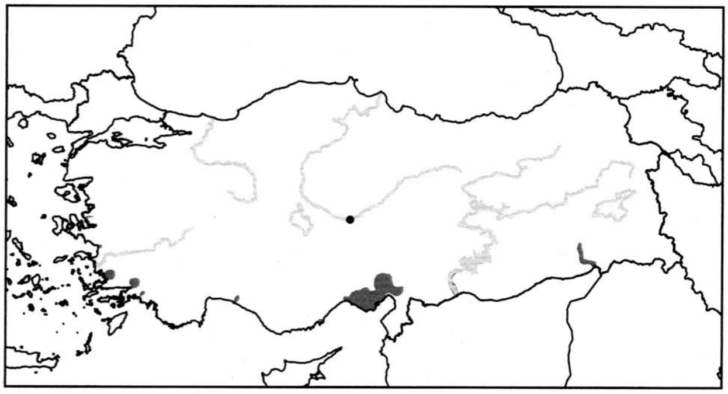

Breeding. Along rivers and adjacent areas (including wooded regions), in coastal lowlands, mainly in **Mediterranean** (particularly Çukurova and Göksu deltas), but also in **Aegean**. Most data are from van den Berk & Kasparek (1988), unless otherwise stated. The most important breeding area is the Berdan River, Çukurova, with at least 40–45 pairs; the banks are up to 4m high, and 19 occupied nests (200–800m apart) were found in the outer meander bends on 12–13 May 1987, with two other nests in roadside verges in forestry areas (van der Have *et al.* 1988). Display in front of breeding holes was observed early April. An incomplete clutch of three, and a complete clutch of five, were found on 27 April 1965; two complete clutches of six (one fresh) were found 6 May 1964, and the nesting tunnel in the bank was 60–80cm long (Warncke 1964–65). Adults were carrying small lizards and agamas to a nest with young, near Mersin, 21 May 1974. On the Göksu River, between Silifke and Kurtuluş, there were five nests in both 1989 and 1991, and three other territories were in farmland east of the river; nests have been found in river banks in April–May; an adult was observed by a nest hole on 12 June 2002 (H. & J. Eriksen), and an adult was entering a presumed nest in a bank on 10 July 1991 (N. Littlewood). Display, including fish carrying, was observed on the Aksu River, near Antalya (Mediterranean), on 24 April 1967 (R. F. Porter). In Aegean, Krüper collected a clutch near Izmir on 4 May 1894; the species bred at Bafa Gölü in 1980–81, and breeding has been confirmed several times near Söke since 1980. The nest is an unlined chamber at the end of a tunnel (Harrison & Castell 2002). Eggs: C/5 (1), C/6 (2). Young: no data.

COMMON KINGFISHER *Alcedo atthis*
Yalıçapkını

(Du: Ijsvogel; Fr: Martin-pêcheur d'Europe; Ger: Eisvogel; Swe: Kungsfiskare)

Subspecies. Presumably nominate during breeding season, with *ispida* from northwest Africa and Spain west to Russia, which winters south to Cyprus and Iraq, presumably occurring in large numbers at the latter season.

Status and distribution. National conservation status: Near Threatened (population apparently decreasing). An apparently rather local resident breeder in wetlands and along rivers through most of Turkey, typically in very small numbers. Much more local in the eastern third of the country and apparently absent as a breeder from most of Inner Anatolia. Breeding is rarely proven.

More widespread and often common on passage throughout the country, principally in the western two-thirds; in spring from March to late May (with peak numbers in Black Sea region perhaps in the middle two weeks of April), in autumn from at least late August (e.g. appeared from 28th at Sultan marshes; Kasparek 1985) to late September when more common in coastal areas. Albrecht (1986) noted passage at Ereğli (Black Sea Coastlands) until late December, with peak numbers from 20 October–18 November. Largely restricted to the western two-thirds of the country in winter, when often recorded in considerable numbers at certain coastal sites especially Göksu Delta, Çukurova (Mediterranean) and Kızılırmak Delta (Black Sea Coastlands): totals of up to 1,000–2,000 individuals have been recorded at these localities. Russian-ringed birds have been recovered at the latter site and an autumn migrant, presumably also from further north, moved between Kızılırmak and Çukurova deltas in just three days in 2003 (Gürsoy *et al.* 2004).

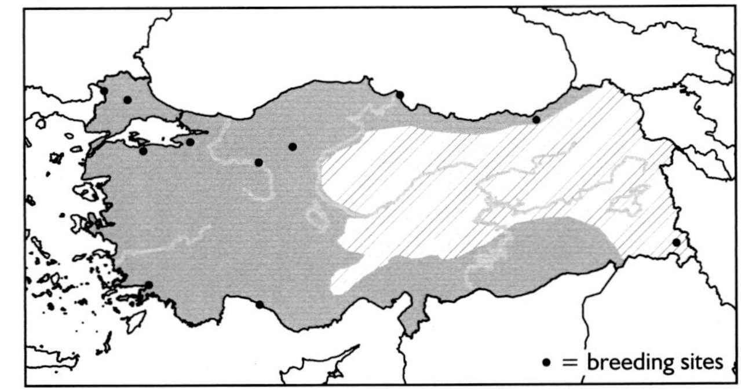

Breeding. Along rivers and streams, the adults digging a tunnel in a vertical bank; the tunnel slopes slightly upwards to an unlined nesting chamber, which becomes gradually lined with cast-up fish bones. Clutch size unrecorded in Turkey but elsewhere usually 6–7 (4–8). There were occupied nests at Yüksekova (**East Anatolia**) on 4 June 1969, near Kızılcahamam (**Inner Anatolia**) on 16 May 1970, at Yatagan (**Aegean**) on 5 June 1971, and at the Fırtına River in northeast Turkey (**Black Sea Coastlands**) in late May 1993 (Faldborg 1994). A pair was feeding young at Luleburgaz (**Marmara**) 30 May 1970. One bird carrying food, and another soliciting, were on a river north of Çaykara (Black Sea Coastlands) on 16 June 1975. Five, including juveniles, at the mouth of the River Karpuz (**Mediterranean**), 13–14 April 1993 (R. & C. Smith), indicate laying in late February. The records above suggest possible double-brooding, as in other parts of the range.

PIED KINGFISHER
Alaca Yalıçapkını

Ceryle rudis

(Du: Bonte Ijsvogel; Fr: Alcyon pie; Ger: Graufischer; Swe: Gråfiskare)

Subspecies. Nominate, although Roselaar (1995) described a new subspecies, *syriaca*, from Turkey (Izmir, Antalya and 'Harpara'; an untraced locality), Cyprus, the Levant, Iraq and Iran, based on its consistently smaller measurements (except bill length); this proposal was subsequently vigorously refuted by Kasparek (1996b) because Bergmann's rule permits for increases in such variables as wing length from south to north. We agree with Kasparek (*op. cit.*) and, in any case, consider that recognition of such inconsequential variation is unwarranted at a nomenclatural level. However, note that if *syriaca* is recognised its name should be spelt *syriacus* (David & Gosselin 2002b).

Status and distribution. National conservation status: Critical (population apparently decreasing). A rather local but not uncommon resident breeder, apparently with occasional limited winter dispersal, in wetlands and along rivers in coastal lowlands in south Aegean and Mediterranean. Formerly more numerous and widespread in both regions. Formerly recorded by many observers around Izmir and its environs (Aegean), but now recorded regularly only from the Büyük Menderes Delta, although breeding may persist also in the Gediz Delta, Izmir province, and the Güllük Delta, Muğla province. In Mediterranean previously bred at (the now drained) Amik Gölü, Antakya province and the Göksu Delta but apparently now absent from these sites. Still breeds in the Köyceğiz wetland complex and apparently in Çukurova, from which there are few recent records (mostly in winter), with one, unprecedented, recent report (September 2004) of a female at Akseki (Mediterranean) (H. Schmaljohann *et al.*). Frequents major rivers, such as the Tigris and Euphrates in Southeast Anatolia and immediately adjacent areas of East Anatolia. Northern limits of range appear to correspond to a line between Gölbaşı, Adıyaman province, Malatya, Diyarbakır and Bitlis. There is little evidence, as yet, to suggest a decline in the eastern half of range, although dam construction has flooded some former habitat.

Generally more widespread in winter within the same regions. Limited evidence of wider dispersal at this season is illustrated by records from Karacabey, Balıkşehir province (Marmara) in January 1969, west of Denizli (Marmara) in December 1970 (OST 1972, 1975), an undated record from Burdur Gölü (Mediterranean) and another from Tuz Gölü (Inner Anatolia) (Kasparek 1992a, Kumerloeve 1961).

Breeding. Along rivers and by pools adjacent to rivers, e.g. the Euphrates at Birecik (**Southeast Anatolia**). Nest an unlined chamber at the end of a narrow tunnel excavated by the birds in a vertical bank. Eggs: C/6 (1); elsewhere, the usual clutch is four (3–6). Young: no data from nests, but fledged broods have comprised 3–5 (three). At Birecik, display was observed on 12 April 1996 (L. & L. Vallotton), an adult was in the early stages of digging a tunnel on 20 May 1993 and, *c.*50m away, an adult was obviously incubating eggs in another tunnel on 5 June 1993 (PC); a family party of six was observed on 24 July 1998 (B. M. Zold & T. Emri). In **Aegean**, Selous (1900) saw an adult enter a tunnel near Izmir on 14 May 1899, and dug it out, caught the adult, and took six fresh eggs. Two pairs in and out of holes at the Büyük Menderes River, on 23–27 June 1966, presumably had young; an adult and three fledged young were observed there on 25–26 June 1966 (R. F. Porter). In **Mediterranean**, adults were visiting an active nest at Göksu on 12 May 1982. Two adults, being followed by two fledged young, at Dalyan, in late May 1983 (Kılıç & Kasparek 1989), indicate egg laying in late April. In **East Anatolia**, two adults and five juveniles (no details of development) were at Asvan on 7–18 August 1971. Elsewhere, the species is single-brooded, and the records above clearly demonstrate that the commencement of breeding in Turkey can vary by at least a month.

[GREEN BEE-EATER *Merops orientalis*]
Küçük Yeşil Arıkuşu

(Du: Kleine Groene Bijeneter; Fr: Guêpier d'Orient; Ger: Smaragdspint; Swe: Grön dvärgbiätare)

Subspecies. Should the species be confirmed to have occurred in Turkey, the form involved would presumably be that breeding in the Near East, *orientalis*.

Species not admitted to the Turkish list (Kirwan *et al.* 1999). Listed as occurring in Anatolia by Ergene (1945) without supporting details.

BLUE-CHEEKED BEE-EATER *Merops persicus*
Yeşil Arıkuşu

(Du: Groene Bijeneter; Fr: Guêpier de Perse; Ger: Blauwangenspint; Swe: Grön biätare)

Subspecies. Nominate.

Status and distribution. National conservation status: Endangered (population apparently decreasing). A local and generally very uncommon summer visitor principally to dry open agricultural country in Southeast Anatolia, breeding at Birecik and at various localities between Birecik and Karkamis, Gaziantep province, at Ceylanpınar (also not recorded recently), the Atatürk Reservoir, Silopli, near the River Karababa, and almost certainly elsewhere. Apparently, a small population also exists near Iğdır (Eastern Anatolia). Formerly a common breeder between Birecik and Karkamis (see Tristram 1882), but now very rare and almost certainly extirpated in this region. Also occurs in adjacent parts of Mediterranean, including Amik Gölü, Antakya province and in the Aras Valley near Iğdır, Kars province (East Anatolia).

Slightly more widespread on passage, especially in autumn from mid August (19th) to late October (22nd), when not infrequently recorded from the Çukurova and Göksu deltas, and East Anatolia, north and west to Erzurum and Anı (Kars province). Occasionally recorded in large numbers in East Anatolia at this season (Vielliard 1968, Balmer & Betton 2005a, 2008). Also reported in autumn from Inner Anatolia, at Mogan Gölü and perhaps even more exceptionally, by Krüper (1876) at Izmir (Aegean) and Russell at Sardis (Aegean; Kumerloeve 1961). Spring passage, from early April (2nd) to late May, may obscure breeding status and distribrution.

Breeding. In open arid areas, near water; also in light woodland and plantations, and near habitation at Birecik (**Southeast Anatolia**). Colonies usually small, 2–15 pairs, occasionally solitary pairs; the maximum was 50–100 birds in two colonies at Birecik on 27 May 1987. Nests in banks, ditches and sand-pits; the bank may be low, merely a slight rise in ground level. Nest a tunnel, excavated horizontally into a bank, leading 1–2m to an unlined nest chamber. Eggs: no data; elswhere, usual clutch 4–5; two eggs (probably an incomplete clutch) in the Natural History Museum, Tring, were collected in 'Asia Minor' on 15 May 1882, earlier than other records indicate. Young: no data. At Birecik, the early stages of nest excavation by three pairs was observed 18 May 1993 (PC), and by one pair during 28–31 May 1984 (A. Schubert); 12 birds were digging holes on 26 May 1991, and display and copulation was observed on 27 May 1991 (D. J. Bates); adults observed carrying food to young in holes on 3 July 1970 and 6 July 1973, and fledged young leaving the nest on 26–27 July 1994 (R. van der Vliet); adults still present 7–8 August 1988, but colonies deserted by 9 August 1998 and 10 August 1989. Whilst data are limited, the breeding cycle at Birecik appears to be: nest building second half of May, egg laying early June (*c.*5 days), incubation most of June (18–19 days), hatching late June, nestlings July (27–29 days), and young fledging late July. In **Mediterranean**, an adult was observed digging a tunnel in a sandy bank in agricultural land near Amik Gölü on 2 May 1970. Breeding was suspected east of Iğdır (**East Anatolia**) on 7 July 1990 (Eames 1991) and 26 June 2005 (S. İsfendiyaroğlu & E. Bacak).

EUROPEAN BEE-EATER *Merops apiaster*
Arıkuşu

(Du: Bijeneter; Fr: Guêpier d'Europe; Ger: Bienenfresser; Swe: Biätare)

Subspecies. Monotypic.

Status and distribution. National conservation status: Least Concern (population apparently decreasing). Widespread and generally common summer visitor, breeding in all regions, although apparently absent from much of the Black Sea Coastlands during the breeding season. The only known location along the Turkish Black Sea coast is in the Kızılırmak Delta, where a small colony was found in 1992 (Hustings & van Dijk 1994). Scarce and more local in parts of the south. Breeds from sea level to at least 2,400m, predominantly at lower altitudes, from late April in the south and west, later in the east and north.

Often recorded in substantial numbers on passage, when widespread and very common throughout Turkey. In spring recorded on passage from mid April (earliest 6th) to early June, although passage is usually completed in Inner Anatolia by mid May. However, in Black Sea rarely arrives before very late April and numbers peak in mid May (Albrecht 1986, Hustings & van Dijk 1994). In autumn, passage is noted from mid August to late September (e.g. 13 August to 24 September at the Bophorus in 1966), and regularly as late as the end of October, exceptionally mid November (18th) in Thrace and the south. A common passage migrant at all major watch-points, e.g. a total of 1,928 passed over the Belen Pass (Mediterranean) in late August and September 1976 (Sutherland & Brooks 1981b), and V. van den Berk *et al.* (unpubl.) recorded 'many thousands' passing through the Borçka area (Black Sea Coastlands) during 24 September to 3 October 1980. Spring observations, e.g. in May 2004, suggest that such impressive numbers may be virtually commonplace through many parts of the eastern half of the country.

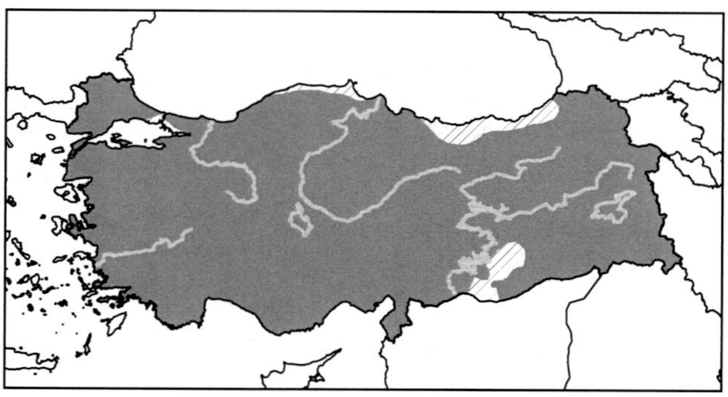

Breeding. In open country, especially near rivers, in valleys and plains, but also in cultivation, scrub and open woodland, and on the outskirts of villages. Colonies usually small, e.g. most in Thrace have fewer than ten pairs, typically 2–10 pairs in most regions, but up to 100 pairs in Aegean. Numbers of pairs in 19 colonies in Marmara (including Thrace), in 1966–73, were 2–6 (11 colonies), 8–12 (four), 20 (three) and 50 (one), and earliest nest excavations were on 19 May and 26 May (R. F. Porter). Nests in river banks, roadside cuttings, sand-pits, mounds of earth and sand dunes. The nest is a tunnel, excavated horizontally into a bank, 1–2m long, leading to a rounded, unlined nesting chamber, in which the castings of insect remains accumulate as breeding proceeds. A nesting tunnel in flat ground at Akköy (Aegean) on 15 May 2007 (PC) is the only such record for Turkey, but this is not unusual elsewhere in the range. Old nesting tunnels are used for nesting, in subsequent years, by Rock Sparrows *Petronia petronia*, House Sparrows *Passer domesticus* and Tree Sparrows *Passer montanus*, and (in enlarged holes) by European Rollers *Coracias garrulus*. Eggs: no recent data; elsewhere, usual clutch 4–7. A clutch of five in the Natural History Museum, Tring, was collected from a nesting tunnel of 2.5m, in 'Asia Minor' on 1 June 1887; a clutch of six was collected in neighbouring Georgia on 30 May 1907. Young: no data. In **Aegean**, daily visits were made to two small, long-established, colonies, with many old nest holes, near Akköy, on 23 April–10 May 2003; birds were present at both colonies, apparently for the first time that year, on 30 April and 1 May respectively and by 4 May one pair (in a colony of *c.*10 pairs) had already started a new nesting tunnel, *c.*2.5cm long, with both adults present at the hole; by 9 May many adults were digging tunnels

(PC). McNeile dug out a nest, which had one egg, an incomplete clutch, in the Izmir area, on 21 May 1951. In Çukurova (**Mediterranean**), in 1987, adults were seen daily from 10 April and were in small colonies from 6 May. Nest building was observed at Birecik (**Southeast Anatolia**) on 4 May 2003 and 11 May 2004 (N. Williams) and on 28 May 1991 (D. J. Bates), and at Van Gölü (**East Anatolia**) on 26 May 1969. Adults were observed carrying food to nest holes, at Enez and Kesan (**Marmara**), on 23 June 1973, and at Gulpınar on 26 June 1973 (R. F. Porter). Elsewhere, the breeding cycle, from first egg to fledging, usually lasts *c.*50 days; if 21 May is typical of the first egg (McNeile above), then young would hatch in mid June, and fledge in mid July; there appears to be no record of fledged young observed in Turkey.

EUROPEAN ROLLER *Coracias garrulus*
Gökkuzgun

(Du: Scharrelaar; Fr: Rollier d'Europe; Ger: Blauracke; Swe: Blåkråka)

Subspecies. Nominate.

Status and distribution. Conservation status: Near Threatened (BirdLife International 2008); national conservation status: Least Concern (population apparently decreasing). A widespread and fairly common summer visitor, breeding throughout most of Turkey. Extremely local on the Black Sea coast and perhaps entirely absent from some areas. Breeds in open, wooded country, or in virtually any open area with scattered trees or suitable alternative nest sites such as crags, barns or ruins, to at least 2,000m.

Common on passage throughout Turkey; in spring from late March (29th) on the south coast, peaking in late April and early May, but only arriving in the north during this period; in autumn from early August to late October, as late as the end of November (latest 29th) in Mediterranean and Inner Anatolia, although most have departed the north and west by mid September (latest 9 October in these regions).

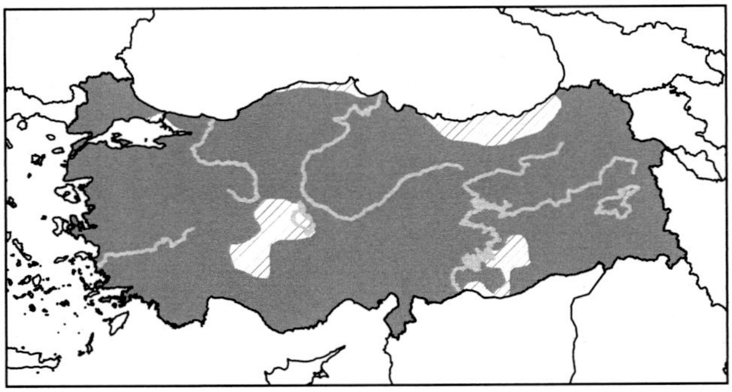

Breeding. Occurs in varied habitats, possibly related to the availability of nesting sites, including forest edge, light woodland, olive groves, open country with scattered trees and treeless steppe. The nest is an unlined cavity in a mature tree, especially in olives, cliffs, earthbanks, ruins, buildings, bridges, and even in a crane used for excavating the historic site of Ephesus (Aegean); also in holes in the ground, and in European Suslik *Spermophilus citellus* burrows in flat, treeless areas, particularly in Inner Anatolia. Eggs: C/3 (2), C/4 (1), C/5 (4). In **Aegean**, Selous (1900) flushed many adults from holes in cliffs near Aydin on 8 May 1899, but they had not yet laid eggs; three clutches (C/5, C/4, C/5) in the Natural History Museum, Tring, were collected by Selous at Izmir on 1 May 1887, at the ruins of Miletos on 19 May 1901, and at the Büyük Menderes River on 29 May 1901. Adults, at inaccessible nesting holes on 14 May 1991 (McNeile) and 19 May 1999 (J. Hollinrake), were almost certainly incubating. A nest in a hole in a stone bridge near Izmir had five well-incubated eggs on 28 May 1951 (McNeile), indicating date of first egg as *c.*8 May. In **Southeast Anatolia**, a nest, in a bank at Birecik, had four naked young on 5 June 1998 (T. Quittenden); adults were in nest holes and, presumably, incubating, there on 20 May 1993 (PC) and 4 June 1998 (T. Quittenden), and a nest had one egg (incomplete clutch) on 20 May 2007 (PC). In **Mediterranean**, a pair was displaying and courtship feeding at a nest in a rock face near Adana, on 9 May 1970 (R. F. Porter); a nest hole, 2.7m above ground, in a pine at Side, had at least two well-grown young on 2 July 1972 and a nest contained five eggs in a bridge near Manavgat on 24 May 2007 (J. Hollinrake). In **Inner Anatolia**, a pair was incubating in the same bank as a colony of 15 pairs of European Bee-eaters *Merops apiaster*; juveniles were seen in July–August. In **East Anatolia**, an adult (presumed to be incubating) flushed out of a large hole in a tall roadside bank, also used by Sand Martins *Riparia riparia*, near Van on 8 June 2004

(S. Williams). At Ispir (**Black Sea Coastlands**), on 24 June 2004, an adult flew into a large hole (probably an enlarged nesting tunnel of a bee-eater) in a tall roadside bank, in which a colony of bee-eaters was also nesting; thereafter, two adult rollers flew out and it was assumed that the nest contained young, probably large, as small young are continuously brooded. At Kızılırmak Delta, pairs were seen disputing tree holes with Western Jackdaws *Corvus monedula* on 28 May 1979, in display on 9–10 June 1975, four pairs were carrying food to holes in July 1971, two fledged young were seen on 29 July 1971, and fledged young daily from 5–12 July 1983 (Dijksen & Kasparek 1985); in 1992, display was seen mainly in the second half of May, with alarm-calling pairs near their nests in the first ten days of June (Hustings & Dijk 1994). At Manyas Gölü (**Marmara**), a nest in a tree hole 1m above water level in a partially submerged wood of willows, had one egg on 5 June 1966 and three eggs on 9 June 1966; another nest in a similar site had three eggs, 0.26m inside the hole on 6 June 1966; another nest one egg on 6 June 1966 (R. F. Porter); a late nest still had one young at Manyas in early August 1968; and five pairs were observed carrying food, in sites in Marmara, 21–27 June 1973 (R. F. Porter). Breeding is evidently variable in Turkey, with egg laying commencing early May to mid June.

INDIAN ROLLER *Coracias benghalensis*
Hint Gökkuzgunu

(Du: Indische Scharrelaar; Fr: Rollier indien; Ger: Hinduracke; Swe: Indisk blåkråka)

Subspecies. Presumably the nominate, which reaches west to extreme east Iraq and eastern Arabia.

Status and distribution. Sclator & Taylor (1876) reported that in 1875 an individual was collected between Haydarpaşa, Istanbul and Izmit, Koçaeli province (Marmara). This came into the possession of Pearce and was later transferred to the Robert's College (Istanbul) collection. Subsequently, Danford (1878) claimed one between Giaour Keui (Gavur Köy) and Bereketlu (Bereketli) in the central Taurus (Mediterranean) and another between Çamardı and Karanfil Dağ at the foot of Aladağ, Niğde province (Mediterranean), both in late April 1876.

Kumerloeve (1961) made no mention of the specimen record, although its identity had been confirmed by Dresser (1878), and questioned Danford's records. Given the latter's tendency to claim other extremely rare and unlikely vagrants without description, the two sight records seem best withdrawn from the Turkish list, the specimen record remaining.

EURASIAN HOOPOE *Upupa epops*
İbibik

(Du: Hop; Fr: Huppe fasciée; Ger: Wiedehopf; Swe: Härfågel)

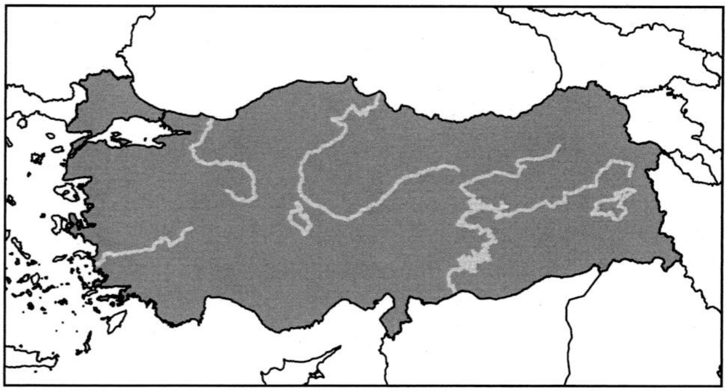

Subspecies. Nominate.

Status and distribution. National conservation status: Least Concern (population apparently decreasing). A widespread and reasonably common summer visitor, breeding throughout Turkey. Occurs in an extremely broad range of habitats, breeding to at least

3,000m, from late April. More common on passage. Spring arrival commences in early March (9th), more usually in late March, peaking in the first half of April, and passage persisting until mid May. Breeding is usually complete by late July, return passage commencing promptly (first migrants from further north observed arriving on Black Sea coast on 28 July). Most have departed by early (3rd) October in northern and central areas, or mid October (10th) in the south, although late individuals are occasionally recorded in the latter region into early November (3rd). At least two winter records: singles south of Ağyatan Gölü, Çukurova, on 31 January 1998 (E.Vaassen & O. Arihan) and at the River Seyhan, Çukurova (Mediterranean) on 2 January 2002 (M. Ullman).

Breeding. In a variety of habitats, provided suitable crevices are available as nest sites, particularly in open rocky areas, villages, farmland with trees and walls, but also in open woodland, including orchards and olive groves. Nests in holes in trees, including old woodpecker holes, in cliffs, under roof tiles and in buildings, in piles of stones and in walls, especially low, where the birds go down to soil level to lay eggs. No nest material. Droppings accumulate at the nestling stage, and become foul-smelling. Eggs: C/7 (2). Young: B/2 (1), B/4 (1), B/7 (2). In **Inner Anatolia**, a nest in a pile of rocks at Eşmekaya had three eggs on 11 May 1993, five eggs on 16 May 1993, six eggs and a newly hatched chick on 27 May 1993, and all had hatched by 30 May 1993 (PC). At Sultan marshes, a nest had large young on 17 May 1979, and flying young were observed 10 June 1982 (Kasparek 1985). A nest at Kızılcahamam had two young on 20–27 June 1981 (Barış *et al.* 1984). A nest near Bolluk Gölü had newly hatched young on 14 May 2004 (L. Baxter), indicating egg laying in late April. A nest near Kulu Gölü had six eggs and one newly hatched young on 8 May 2005 (seven young by 14 May 2005), indicating egg laying in mid April (J. Hodson). There are several records of adults carrying food to nest-holes, mainly in late May–mid June. In Ankara, a pair was feeding three fledged young on 22 July 1969, and a pair and a juvenile were seen on 2 July 1968. In **Southeast Anatolia**, a nest in an old Syrian Woodpecker *Dendrocopos syriacus* hole at Birecik had four large young on 4 June 2001 (PC). A pair carried food to a nest with young at Birecik on 26 April 1981 (P. Flint), indicating egg laying in early April. In **Mediterranean**, adults were feeding young at a cliff near Uzuncaburç on 7 May 2004 (N. Williams). In **Aegean**, an adult flew out of a tree hole at Akköy, on 20 May 1999 (J. Hollinrake), and a nest at Altınova had young on 12 May 1970. In **Marmara**, a pair and two fledged young were at Manyas Gölü on 13 June 1966 (R. F. Porter). At Kızılırmak Delta (**Black Sea Coastlands**), two nests had young in July 1971, two pairs were with fledged young on 8–9 June 1969 (Dijksen & Kasparek 1985), and three pairs were feeding fledged young on 19 July 1975. In 1992, there were 33 territories in Yürükler Forest, and 12 in the immediate surrounds, with several nests found, one with young on 10 June 1992 (Hustings & van Dijk 1994). In **East Anatolia**, a nest had young near Van on 11 June 2001 (PC), a pair was feeding young at Erçek Gölü on 28 May 1969 (R. F. Porter) and a family party was on the banks of the Euphrates near Erzincan on 5 July 2004 (BD & O. Mülayim). Most nest records are of adults carrying food to holes; if both adults are involved, the nest has young, but a single adult may be the male feeding his incubating female. There is wide variation in breeding dates, which may indicate double-brooding, an occasional feature elsewhere in the range.

EURASIAN WRYNECK *Jynx torquilla*
Boyunçeviren

(Du: Draaihals; Fr: Torcol fourmilier; Ger: Wendehals; Swe: Göktyta)

Subspecies. Presumably nominate (migrants through central Turkey in autumn all belonged to this form; van den Berk 1990), although *tschusii*, which breeds on several Mediterranean islands from Corsica east to the Adriatic coast, could occur, at least on passage, although Vaurie (1959c) described this form as being 'not highly migratory'.

Status and distribution. National conservation status: Near Threatened (population apparently stable). A very local, apparently uncommon, summer migrant, breeding in small numbers in forested areas to at least 1,700m in the western Black Sea Coastlands and almost certainly elsewhere, including the central Taurus (Mediterranean). Also recently recorded in East Anatolia and the eastern Black Sea Coastlands during the breeding season. May also breed elsewhere in northern and western Turkey; throughout Marmara, including Thrace, and northern and eastern parts of Aegean, although there are few recent reports from these areas. In the nearby Strandja Mountains of Bulgaria, Milchev (1994) found evidence of breeding in 47% of 5-km squares surveyed (with nesting confirmed in over 25%) in 1988–90. Usually found breeding in open woodland, plantations and orchards. Passage obscures breeding distribution.

Also a passage migrant in small numbers throughout Turkey, principally in the western two-thirds. Usually commoner in autumn, especially in Inner Anatolia (V. van den Berk *in* Schekkerman & van Roomen 1993). Kasparek (1989c) analysed 216 migrant records, establishing the median passage dates as 15 April and 6 September. In spring, recorded from late March, occasionally early March (3rd) to mid or late May (31st) and in autumn from early August to early October, occasionally late October (19th), peaking markedly in mid September. Apparently commoner on passage formerly; both von Gonzenbach (from observations around Izmir, Aegean, in the mid-19th century) and Braun (from records at Istanbul in autumn in the early years of the 20th century) describe the species as a common, even very common, migrant. Occasional in winter in coastal areas in Aegean and Mediterranean, although there are merely eight available records, all between December and February (from Izmir, Büyük Menderes Delta (two), Erdemli, Antalya (two) and Demircili), and three records from November, one from Side (Mediterranean) in mid month, which also behaved as if it had established a winter territory (A. Vittery), one at Samandağ, Antakya (Mediterranean) on 24th, and one at Birecik (Southeast Anatolia) on 25th (G. Bauer).

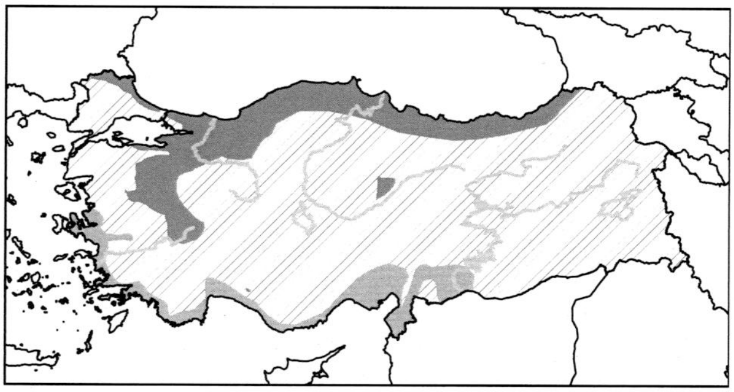

Breeding. In forests, open woodland, and agricultural areas with trees. Nests in a hole in a tree; an unlined cavity, with no material added. Eggs: no data, but 7–10 is the usual clutch elsewhere. Young: no data. In **Black Sea Coastlands**, three unbroken eggs (two slightly incubated and one infertile) were found on the ground below a nest hole, at 1,300–1,400m, in the Ilgaz range on 11 June 1977; the eggs were positively identified, and five birds were heard calling there on 11–12 June 1977 (Schubert 1979). An adult was observed investigating a potential nest hole at Ereğli on 16 April 1978, and two were heard calling there on 14 June 1978; breeding was suspected but not proven (Albrecht 1986). At Kızılırmak Delta, an occupied nest, where the adults were feeding, was found on 31 May 1979 (Dijksen & Kasparek 1985). In **Mediterranean**, a juvenile with traces of down (thus bred locally) was seen 60km north of Akseki on 17 July 1986 (Martins 1989), indicating egg laying in early June. Two adults were interacting and prospecting a hole at Akseki on 4 June 2000, and another was calling nearby the same day (P. & M. McGough). Adults heard singing at several other locations in the breeding season in Black Sea Coastlands, at Kızılcahamam (**Inner Anatolia**), and at Uludağ (**Marmara**). Breeding probably far commoner than the records suggest, as the species is quiet and unobtrusive during incubation and when feeding young, and is, therefore, easily overlooked.

GREY-HEADED WOODPECKER *Picus canus*
Küçük Yeşil Ağaçkakan

(Du: Grijskopspecht; Fr: Pic cendré; Ger: Grauspecht; Swe: Gråspett)

Subspecies. Nominate.

Status and distribution. National conservation status: Data Deficient (population trend unknown). Status recently summarised by van den Berg (1988) but extent of distribution remains unclear. Previous assessments of status now appear incorrect. Thus, specimens collected in Thrace (two from Istanbul in February 1910) and Black Sea Coastlands (three from Elemen, near Mengen, Bolu province in August 1934, one from Bolu in October 1934 and one from Büyükdere, northwest of Istanbul on 15 May 1949), apparently discounted by

Kumerloeve (1961) as stragglers from the Bulgarian population, merit reinterpretation. Other sight records: one in the Forest of Belgrade, Istanbul in April 1969; one at Demirköy, Kırkaleri province (Marmara) on 12 May 1967; a male near Kızılcahamam (Inner Anatolia) on 29 April 1972 (where confirmed to occur in the mid 1990s); and one, just outside modern Turkey, near Mandra on the Greek side of the River Meriç on 31 January 1917, were similarly dismissed, perhaps incorrectly. In 1988–90 Milchev (1994) found the species in 35% of 5-km squares during breeding-bird censuses in the adjacent Strandja mountains of Bulgaria, and late 1990s fieldwork has confirmed its presence in reasonable numbers on the Turkish side of the border, in northeast Thrace. Krüper & Hartlaub (1875), as well as Mathey-Dupraz (1921), referred to occurrences in the Bosphorus area between 1850–1900, and at least one specimen from this region is extant (Kirwan 1997a), whilst recent sight records from the Forest of Belgrade have confirmed its continued presence (Kirwan & Martins 1994). The lack of documentation to support additional records away from north-west Turkey: two at Pozantı, Adana province, on 18 June 1964; one at Gulek, Tarsus province (Mediterranean), on 13 April 1965; four at the Zigana pass, Trabzon province, on 30–31 August 1971 and one or two at nearby Istavri (Black Sea Coastlands), was noted by van den Berg (1988). However, present knowledge appears to strengthen the credibility of many of these older records.

Elsewhere in Turkey, the species appears to be a widespread resident breeder in the Pontic forests (Black Sea Coastlands), based on the available records from widely scattered areas, although there is only one documented record, from near Çamlıhemşin in 1993 (Faldborg 1994). It is perhaps scarcer and more local in western areas. Both Heiser (1984) and van den Berg (1988) observed the species around Ikizdere, Rize province (Black Sea Coastlands), in July 1983 and May 1987 respectively and it has also been recorded at Ayder, Rize province, in late July 1991 (O. Thorup). In spring 1993 there were regular observations in the Çamlıhemşin area, Rize province (Black Sea Coastlands), where the species was considered fairly common at *c.*300–400m to the tree-line. Breeding was proven (Faldborg 1994). European Green Woodpecker *P. viridis* is apparently absent from this area. Additionally, one was between Alaçam and Kızlan, Samsun province, on 29 May 1993 (GMK pers. obs.) and in the same year several were recorded from Sülüklu Gölü, Bolu province (Black Sea Coastlands) in May (G. Magnin & M. Yarar), where the species appears well established and there is much suitable habitat (GMK pers. obs.). There are several other recent records from Black Sea Coastlands, in Yedigöller National Park, Bolu province (O. Mülayim), Yenice Forest, Karabük province (G. Gençoğlu & Y. Lise), and the Ilgaz Dağları, Çankırı/Kastamonu province. Also frequently recorded in small numbers, since 1989 until the present, from Akseki, Antalya province, in the western Taurus (Mediterranean), and also recently recorded in the foothills of the central-eastern Taurus, above Uzuncaburç (Kirwan & Martins 2000a), suggesting the true distribution may be more extensive than presently realised. Recorded from near sea level to above 2,300m. Most records are from coniferous woodland, although several (especially in Thrace) are from mixed forest. Further information is required to establish its range fully.

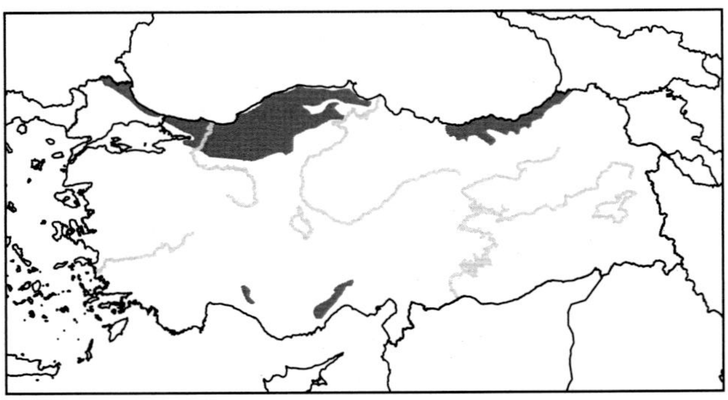

Breeding. Observed April–June, the presumed breeding season, in Thrace (**Marmara**) and in **Mediterranean** (especially at Akseki, in open pine forest), but no evidence of nesting. The only definite breeding record is from the Çamlıhemşin–Ardeşen area in eastern **Black Sea Coastlands**: a common breeder, heard almost daily; an occupied nest was found, *c.*27m above ground, in a *Fagus orientalis*, *c.*37–40m high, on 3 May 1993, when incubation was in progress, with both adults involved in shifts of 3–4 hours; on 8 June 1993 it was clear the eggs had hatched, with both adults frequently visiting the hole (Faldborg 1994). Eggs: no data; elsewhere, usual clutch 4–5.

EUROPEAN GREEN WOODPECKER *Picus viridis*
Yeşil Ağaçkakan

(Du: Groene Specht; Fr: Pic vert; Ger: Grünspecht; Swe: Gröngöling)

Subspecies. *P. v. karelini*, which is slightly smaller, duller and less yellow than the nominate, but with which it is known to intergrade in Europe.

Status and distribution. National conservation status: Least Concern (population apparently decreasing). A relatively local resident breeder in both deciduous woodland and, perhaps more frequently, in coniferous woodland throughout most of Turkey. Very local in much of the interior and absent from large parts of Inner and East Anatolia, *e.g.* there are just two records, both in autumn, from the Van Gölü basin (van der Van & Gheyselinck 1982; L. J. Dijksen & L. C. van Beckhoven). Recorded three times in Southeast Anatolia, east as far as Siirt, where the species' status is unclear and requires clarification. Occurs from sea level but most commonly between 700 and 1,900m, occasionally to 2,200m in the Pontics and Taurus mountains. In some areas may be more common than records suggest. Milchev (1994) during breeding-bird surveys of the Strandja mountains of adjacent Bulgaria found the species in 91% of 5-km squares censused.

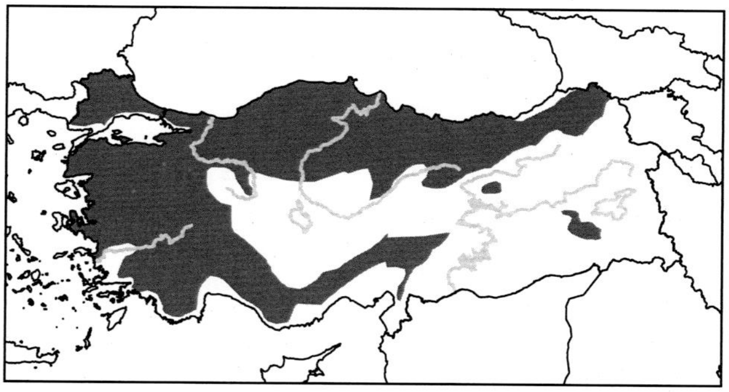

Breeding. In deciduous and coniferous woodland. Nest is a hole in a tree, excavated by the birds. Eggs: no data but, elsewhere, usual clutch 5–7. At Kocaçay Delta (**Marmara**) there were 0–29 pairs/ha in a 31ha plot in 1993, and a nest was found on 19 May 1967 (Ertan 1996); the young were being fed at the nest entrance, and were, therefore, large (R. F. Porter pers. comm.), indicating laying of first egg *c.*10 April. In **Aegean**, juveniles observed with adults in July 1966; and a juvenile, near Marmaris on 1 June 1995, indicates egg laying mid April. At Kızılırmak Delta (**Black Sea Coastlands**), a pair with two young, and two pairs with one young, were seen in July 1971 (Dijksen & Kasparek 1985). A family party, of which one juvenile was collected, was observed at Porsuk (**Inner Anatolia**) on 9 July 1968 (Rokitansky & Schifter 1971). Juveniles were observed at Akseki (**Mediterranean**) on 15 July 2004 (K. P. Kendricks).

BLACK WOODPECKER *Dryocopus martius*
Kara Ağaçkakan

(Du: Zwarte Specht; Fr: Pic noir; Ger: Schwarzspecht; Fr: Spillkråka)

Subspecies. Nominate.

Status and distribution. National conservation status: Near Threatened (population apparently stable). A local and uncommon resident in montane coniferous forest throughout the Black Sea Coastlands, southeastern Marmara (and reportedly extreme western Thrace) and the extreme north of Inner Anatolia. Status in the Taurus (Mediterranean/Inner Anatolia) is unclear. Danford (1878) described it as

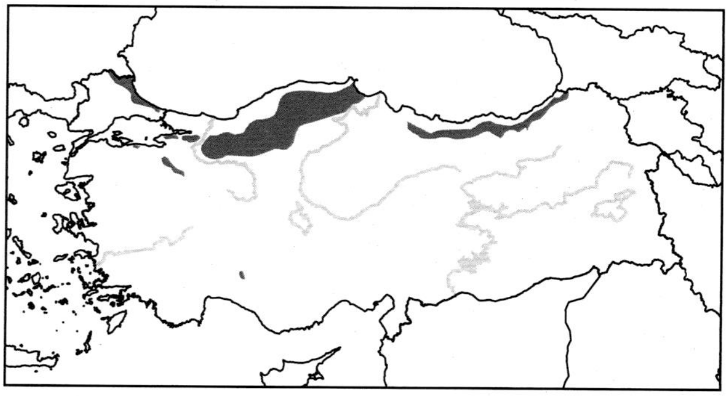

'rare, but occasionally seen' in the central area of this range, but few observers have recorded it since. The most recent records from this area were in 1993 and 1998 (in the Gidengelmez Dağları). Recorded mainly from 900m to at least 2,000m in northern Turkey, occasionally as low as 500m (e.g. recorded once in deciduous forest, at around this altitude, by Yeniköy, near Kocaçay Delta, Marmara; Ertan 1996), and even at sea level (at Iğneada, Thrace). Discovered in May 2008 in pine forest south of Osmaniye (Mediterranean) at 700m (J. L. Copete *et al.*) Most frequently reported in the eastern Black Sea Coastlands (e.g. Sumela and Borçka, Rize province) and regularly on Uludağ (Marmara). Probably occurs regularly elsewhere in northwestern areas including Termal, Yalova province, Yenice Forest, the Ilgaz mountains and adjacent areas (from where there are a small number of recent records): these have received little attention since Kumerloeve and Niethammer's travels in the 1930s, and also in northern Thrace. In adjacent Bulgaria it was recorded in 28% of 5-km squares surveyed during breeding-bird census in 1988–90 (Milchev 1994).

Breeding. No information.

GREAT SPOTTED WOODPECKER *Dendrocopos major*
Orman Ağaçkakanı

(Du: Grote Bonte Specht; Fr: Pic épeiche; Ger: Buntspecht; Swe: Större hackspett)

Subspecies. Debatably *paphlagoniae* (admitted by Roselaar in *BWP*), which is endemic to Turkey, but the latter is subsumed within *pinetorum*, which occurs from Britain east to the Caucasus, by Vaurie (1959b), and Winkler & Christie *in* del Hoyo *et al.* (2002).

Status and distribution. National conservation status: Least Concern (population apparently decreasing). A widespread but generally local resident breeder in small numbers, in montane coniferous forest throughout Black Sea Coastlands, Marmara, Mediterranean and the fringes of Inner and northern parts of East Anatolia (distribution particularly poorly known in the latter, but has been recorded south as far as the Bitlis Dağları; van den Berk *et al.* 1993). Absent from Southeast Anatolia. Commonest in the two former regions. Recorded from 300m in eastern Black Sea Coastlands, but commonest above 1,000m. Elsewhere typically occurs at 1,000–2,100m, perhaps lower in parts of Marmara (including Thrace) and the Taurus (Mediterranean), although records from these areas in late summer and autumn perhaps reflect post-breeding dispersal, when the species may occur in the same areas as Syrian Woodpecker *D. syriacus*. Several authors in the 19th and early 20th centuries confused the two species, thus distorting understanding of distribution.

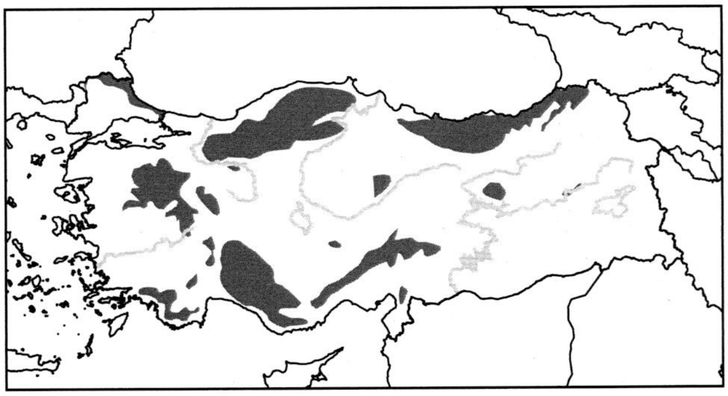

Breeding. Mainly in montane coniferous forest. Nest is a hole in a tree, excavated by the birds, and eggs are laid at the base of the hole, with no material. Eggs: no data but, elsewhere, usual clutch 4–7. Young: no data. In **Marmara**, adults were feeding young at a hole at Uludağ on 3 June 2002 (H. & J. Eriksen), and breeding was also confirmed there at 1,400m on 7 June 1988 (Jetz 1995). Fledged juveniles were observed in Forest of Belgrade on 4 June 1996 (GMK), indicating egg laying in late April. At Kızılcahamam (**Inner Anatolia**), more than 15 were calling on 17 March 1984 (Barış *et al.* 1984), a pair was nesting on 19 April 1996 (L. & L. Vallotton), a nest was found in a telegraph pole on 22 May 1996 and another occupied nest was nearby (I. A. Green), and a nest with several young was found 1.5m up in a conifer on 23 May 2006 (GMK). In **Black Sea Coastlands**, an adult was observed carrying food, near Torul, on 17 June 1975 (R. F. Porter), a juvenile was at Ispir on 21 July 1996, a family party was at Şavşat on 7 August 1992 (J. Frost & J. Hornbuckle), and a recently fledged young, barely able to fly, was at Sivrikaya on 14 June 2004, indicating egg laying about 8 May (PC).

SYRIAN WOODPECKER *Dendrocopos syriacus*
Alaca Ağaçkakan

(Du: Syrische Bonte Specht; Fr: Pic syriaque; Ger: Blutspecht; Swe: Balkanspett)

Subspecies. We follow Vaurie (1959b) and Winkler & Christie (*in* del Hoyo *et al.* 2002) in electing to recognise no geographical variation within this species.

Status and distribution. National conservation status: Least Concern (population apparently decreasing). A widespread and fairly common resident breeder at low to moderate altitudes over much of Turkey. In many areas of the country the most common woodpecker. More local but still common in most of Inner Anatolia, Southeast Anatolia (more common in the north and west of this region) and parts of the Black Sea Coastlands. Absent only from the eastern Black Sea Coastlands, e.g. from the Çamlıhemşin area, Rize province, where unrecorded during intensive fieldwork in April–May 1993 (Faldborg 1994), and much of the extreme east of East Anatolia (though present in the Aras Valley). Recorded from sea level to at least 2,100m in northern and western areas. In the Taurus, where overlaps more extensively than in other regions with Great Spotted Woodpecker *D. major*, rarely occurs above 900m. Breeds in all types of woodland, although only occurs in coniferous forest at lower elevations, being replaced by Great Spotted Woodpecker at 1,000–2,000m, and often at the lower edge of this range. Occurs more commonly in more open habitats including orchards, olive groves, agricultural areas with scattered trees and poplar plantations around human settlements. Apparently replaced by Middle Spotted Woodpecker *D. medius* in some agricultural areas in the southwest Aegean and perhaps also in parts of northern Thrace. In the adjacent Strandja mountains of Bulgaria, Syrian Woodpecker is almost wholly replaced by Great Spotted Woodpecker (Milchev 1994).

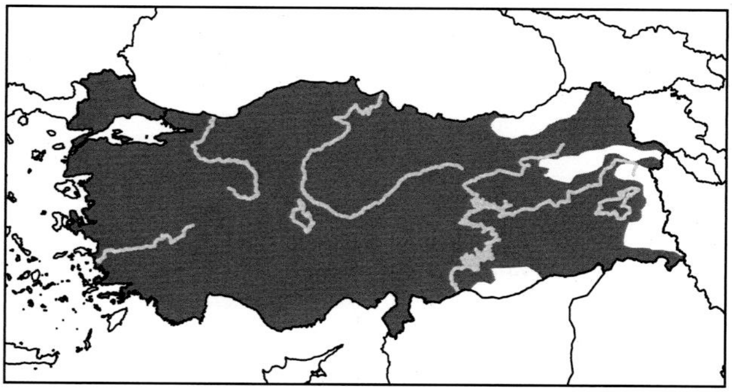

Breeding. Orchards, olive groves, open woodland, and agricultural and open country with copses or scattered trees. Nest is a hole in the trunk (or branch) of a tree, excavated by the adults, leading down to an unlined nest chamber at the base. In pistachio orchards at Birecik (Southeast Anatolia), prefers trees with thicker trunks; nests are usually low, less than 2m, with some less than 1m above ground. In other habitats, holes may be as much as 10m up. Eggs: C/2 (1), C/5 (1). Young: B/1 (1), B/2 (1), B/4 (1), B/5 (1). Few nests have been closely examined, and these limited data may not be representative; elsewhere the usual clutch is 5–7. In **Aegean**, adults reported at nest-holes from 7 May, and a nest had almost full-grown young at Kuşadası on 2 June 1974, indicating egg laying in late April. In **Southeast Anatolia** (mainly at Birecik), five records of large or full-grown young in nests, in late May/early June, indicate egg laying in late April; a nest at Durnalik had five eggs on 9 May 2004 and newly hatched chicks on 16 May 2004, indicating egg laying in late April. Another nest at Durnalık had eggs on 9 May 2004 and small young on 23 May 2004 (L. Baxter). In **Marmara**, an adult and a fledgling were in Forest of Belgrade on 18 June 1973, and an adult was carrying food to a nest hole near Kesan, on 23 June 1973 (R. F. Porter); a nest had large young at Troy on 31 May 1988, indicating egg laying in late April. In **Black Sea Coastlands**, the first nests with begging young were found at Kızılırmak Delta on 27 May 1992 (Hustings & van Dijk 1994). In **Inner Anatolia**, nest building was observed at Hasan Dağ on 27 May 1993, which seems late, but was perhaps a second attempt following earlier failure (PC). In **Mediterranean**, a nest had two eggs, near Taşucu, on 15 June 1992; large young in a nest at Uzuncaburç on 9 June 1993 indicate egg laying in early May (PC); three nests examined with an endoscope, were new and ready for eggs but empty, near Uzuncaburç on 7 May 2004 (PC). Fledged young have been observed in most regions, mainly in late June–July. In subsequent years, nest holes excavated by Syrian Woodpeckers are used by several other species, especially Common Starlings *Sturnus vulgaris* and sparrows, including Chestnut-shouldered Sparrow *Gymnoris xanthocollis*, which may be partially dependent on this woodpecker for the provision of nest sites.

MIDDLE SPOTTED WOODPECKER *Dendrocopos medius*
Ortanca Ağaçkakan

(Du: Middelste Bonte Specht; Fr: Pic mar; Ger: Mittelspecht; Swe: Mellanspett)

Subspecies. Variation across the range of Middle Spotted Woodpecker is apparently almost entirely clinal, both north to south and east to west (see Vaurie 1959b, 1965). Winkler & Christie *in* del Hoyo *et al.* (2002) acknowledged only *D. m. caucasicus* (in northern Turkey), which has brighter underparts than the nominate, with deeper but less extensive red undertail-coverts and golden-yellow belly, and more extensively streaked flanks, and *anatoliae* (endemic to western and southern Turkey), which is very similar to *caucasicus*, but slightly smaller, with a paler ventral region and slightly heavier underparts streaking. Roselaar (1995) admitted nominate *medius* in European Turkey and possibly Iranian *sanctijohannis* in the extreme southeast (some of Danford's specimens from the Taurus are labelled as being of this form, but are clearly not). Between the Balkans and Iran, there is apparently a general cline towards reduced and posteriorly more restricted red and yellow-buff on the underparts, reduced black streaking on the upper flanks, and overall whiter appearance to the underparts. *D. m. sanctijohannis* represents the extreme in the obvious paleness to the underparts. *D. m. caucasicus*, of which we have seen very few specimens and none from Turkey, is obviously larger than *D. m. anatoliae*, and in plumage is unsurprisingly closest to *sanctijohannis*, but our field observations from northern Turkey would suggest that *caucasicus* is variable in its morphology and can approach *anatoliae*/nominate *medius*. It is difficult to see much justification, under the terms adopted here, for recognising *anatoliae* or the very doubtful *splendidior* (for which quite a number of specimens from western Turkey are available), both of which apparently reflect weakly defined, clinal variation from nominate *medius*. The position regarding *caucasicus* and *sanctijohannis* is more interesting, but both clearly represent reasonably well-marked extremes within a cline, and *caucasicus* provides proof of Bergmann's rule. Vaurie (1965 and note in Natural History Museum, Tring) considered *anatoliae* to be a synonym of *caucasicus* (see also Kirwan 2005b).

Status and distribution. National conservation status: Least Concern (population apparently stable). A fairly widespread and not uncommon resident breeder in much of Aegean and Mediterranean, more local, although still fairly common, in Marmara (including Thrace) and the western half of the Black Sea Coastlands (e.g. 20–30 pairs in Kızılırmak Delta in spring 1992, which is one of the very few estimates of breeding density available for the species from Turkey). Kumerloeve (1961, 1970a) documented occurrence in western areas of East Anatolia and the eastern half of the Black Sea Coastlands. There are fewer than ten recent records from the former area, in part due to a lack of ornithological activity and a probable reduction in the amount of suitable habitat. Faldborg (1994) did not record this species in the eastern Black Sea Coastlands during spring 1993. It occurs principally in coastal areas and their immediate hinterland, in deciduous, occasionally mixed, woodland and olive groves. Rare in coniferous forest. Oak (*Quercus*) appears to be an important constituent of optimal habitat. Typically recorded from sea level to 800m in the south and west (including the Thracian highlands), occasionally higher (to 1,400m) in the northeastern Pontics and elsewhere. Kumerloeve & Niethammer doubtfully reported it from around 1,900m in the Ilgaz Dağları (Black Sea Coastlands).

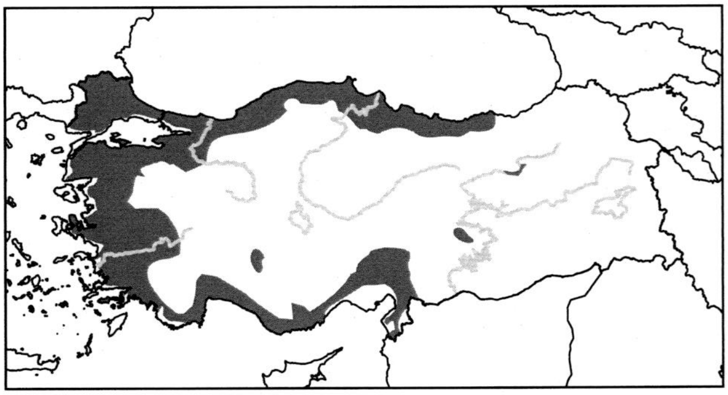

Breeding. In deciduous woodland, especially oak, in both dry and damp areas; in *Liquidamber* forest at Dalyan (Mediterranean); in olive groves, *e.g.* on the southern shore of Bafa Gölü (Aegean), where ten pairs were estimated on 24 June 1966; in alluvial forest at Kocaçay Delta (**Marmara**), where up to ten pairs may breed (Ertan 1996); in pine forest on lower montane slopes (OST 1972); and appears to prefer small oaks at the edge of open mature pine forest at Akseki (Mediterranean) (GMK). Nest is a hole in a tree, excavated by the birds, and leading to an unlined cavity at the base. Eggs: no data but, elsewhere, usual clutch 4–8. Young: no data. At Kızılırmak Delta (**Black Sea Coastlands**) a nest had young on 27 May 1992, and another had young on 10 June 1992 (Hustings & van Dijk 1994). At Bafa Gölü (**Aegean**), a nest had young on 21 May 1980 (Kasparek 1988a). In **Mediterranean**, two pairs were nesting at Ağla and two pairs at Dalyan in late April/May 1989 (P. J. Heath); a nest had young near Akseki on 5 June (M. I. Eldridge); a pair and at least two fledged young were at Akseki in late May 1990, indicating egg laying by 21 April. In Lesbos (Greece), two nests had young in holes in olive trees on 8 May 1995 (G. Demar).

WHITE-BACKED WOODPECKER *Dendrocopos leucotos*
Ak Sırtlı Ağaçkakan

(Du: Witrugspecht; Fr: Pic à dos blanc; Ger: Weißrückenspecht; Swe: Vitryggig hackspett)

Subspecies. *D. l. lilfordi* is plausibly sufficiently distinct to warrant specific status, given that it is separated by altitude and ecology from the parapatric nominate race (see Winkler & Christie *in* del Hoyo *et al.* 2002), specifically the apparent overlap at different altitudes in eastern Serbia with nominate *leucotos*, although variation appears largely clinal (see Cramp 1985, Winkler & Christie 1995) and with quite a degree of overlap in some of the major features used to distinguish *lilfordi*, namely the amount of streaking on the flanks and outertail feathers, and the extent of red on the posterior underparts (GMK pers. obs.). The *lilfordi* population in Turkey differs from the nominate in possessing a more distinct yellow-buff forehead, lores, ear-coverts and chin; distinct black line from rear of ear-coverts to nape (absent or nearly so in *leucotos*); broader black malar stripe; longer scapulars, and back barred more evenly black and white (rather than being largely white); rump black with white confined to feather tips only (not largely white); tertials, flight feathers and wing-coverts irregularly barred black and white, with bars of equal width (not largely white with narrow black bars); flanks, sides of breast and belly broadly streaked black, flanks partially barred black and ground colour of underparts pale yellow-buff (*leucotos* is paler cream in this respect). Nonetheless, the taxonomy of those forms currently included within *D. leucotos* requires more extensive re-examination than is possible here, as several of the other Asian forms appear highly morphologically distinct.

Status and distribution. National conservation status: Vulnerable (population apparently decreasing). Apparently a rare and very local resident breeder, occurring at low densities in Marmara (including northern Thrace), Aegean (where particularly rare and local), Mediterranean (throughout the Taurus, although records from the southern coastal foothills in winter appear doubtful) and throughout much of the Black Sea Coastlands. Several historical and modern records almost certainly reflect confusion with congeners; equally, the rather different appearance of the form *lilfordi* (which is rarely illustrated in field guides and other popular identification literature) may have led to its occasionally being overlooked. Most frequently recorded from mixed or coniferous forests in the eastern Black Sea Coastlands, but even here is extremely hard to locate, e.g. Faldborg (1994) recorded just one during spring 1993 in the Çamlihemsin area, Rize province. Also in parts of the north, west, and central Taurus, where generally recorded at 300–1,200m. Typically above 1,500m elsewhere. Secretive, scarce and often inconspicous: it is perhaps more widely distributed than available records suggest.

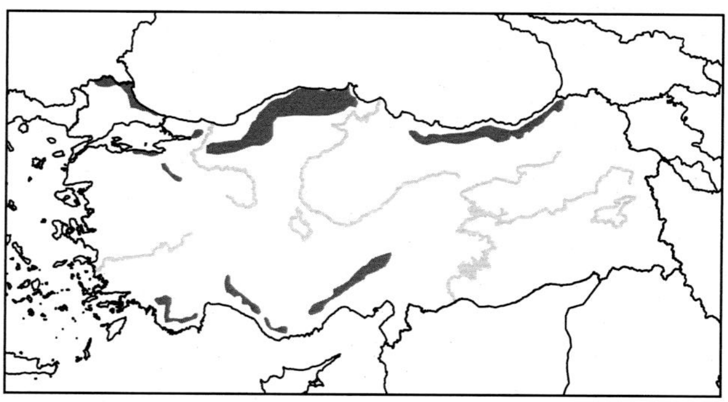

Breeding. Observed in pairs in May–June, the presumed breeding season, but no nesting records.

LESSER SPOTTED WOODPECKER *Dendrocopos minor*
Küçük Ağaçkakan

(Du: Kleine Bonte Specht; Fr: Pie épeichette; Ger: Kleinspecht; Swe: Mindre hackspett)

Subspecies. *D. m. danfordi*, which is dark below, being sometimes brown-tinged on the underparts and possesses a black band from the hindcrown to the rear ear-coverts.

Status and distribution. National conservation status: Least Concern (population apparently stable). A local resident, usually in small numbers (although locally common) in parts of Mediterranean, Marmara (e.g. Forest of Belgrade, Istanbul, and Kocaçay Delta, where over 200 pairs may be present; Ertan 1996) and Black Sea Coastlands, e.g. regarded as a common breeding species at lower altitudes around Çamlıhemşin, Rize province, in April–May 1993 (Faldborg 1994) and 75–125 pairs estimated in Kızılırmak Delta in 1992 (Hustings & van Dijk 1994). Principally occurs in deciduous, occasionally mixed, woodland and orchards in Marmara (including Thrace), Aegean, Mediterranean, Black Sea Coastlands (mainly in the western half, although see above) and very locally in Inner, Southeast (e.g. recently recorded near Diyarbakır; Karakaş & Kılıç 2002, and in the montane region south of Siirt; Goriup & Parr 1983) and East Anatolia. Altitudinal range unclear although certainly occurs throughout low and moderate altitudes. Only rarely recorded above 1,000m, but present to 2,000m in Armenia and the Caucasus (Kumerloeve 1961).

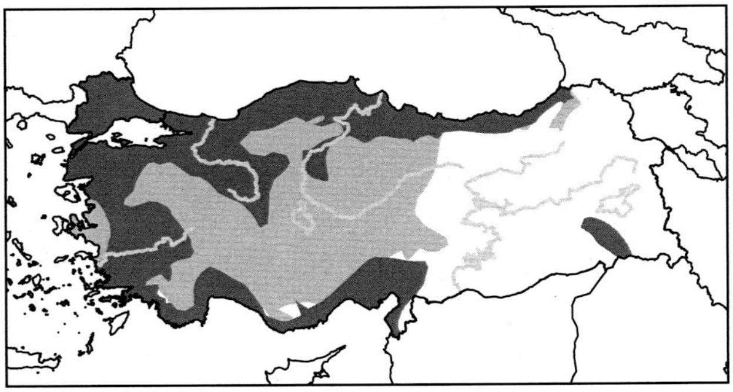

Breeding. In forests, smaller woodlands and rows of trees. At Kocaçay Delta (**Marmara**), prefers moister woodland, usually the ecotone between forest and canals or flooded areas, with six nests, all in *Fraxinus excelsior*, 0.29 pairs/ha in a 31ha plot, and newly fledged young in the second week of June (Ertan 1996), and an adult was observed at a nest hole on 16 May 1966. Also in Marmara, a juvenile (with an adult) was observed on Çamlıca Hills on 22 July 1966 (R. F. Porter), and an adult was at a hole in Forest of Belgrade on 4 June 1996 (GMK). In **Mediterranean**, two occupied nests were located in the Dalyan area on 20 April–25 May 1989 (P. J. Heath), a pair was at a nest at Akseki on 14 May 2001 (T. Harrop) and another nest attended by a female was found there on 26 May 2005 in a hole 50cm down from the top of 3.5m high *Quercus* stump: it had at least two young, judging from calls emanating from the nest (GMK). At Kızılırmak Delta (**Black Sea Coastlands**), an occupied nest hole was found on 9 May 1977, with several records of fledged young in July, and a family party observed on 10 June 1975 (Dijksen & Kasparek 1985); the latter indicates egg laying in early May. In 1992, the first food-begging calls were heard on 27 May and newly fledged young were seen in the second week of June (Hustings & van Dijk 1994). Eggs: no data; elsewhere, usual clutch 4–6.

DESERT LARK *Ammomanes deserti*
Çöl Toygarı

(Du: Woestijnleeuwerik; Fr: Ammomane isabelline; Ger: Steinlerche; Swe: Stenökenlärka)

Subspecies. Presumably *coxi*, which occurs in neighbouring Syria, although no specimens from Turkey are available.

Status and distribution. National conservation status: Critical (population apparently decreasing). A presumed resident and local breeder in small numbers at two sites. Murphy (1984) and Martins & Robson (1988) both reported the first record from Turkey when three were discovered on the plateau north of Birecik (Southeast Anatolia) in June 1983. Up to 15 individuals have subsequently been reported from this locality on many occasions, and breeding has been proven (Martins 1989, see below). Known from at least one other similar, suitable breeding locality, 15km further north since 1989 (A. Broome *et al.*), where up to five have also been regularly observed. Presumed resident, although not recorded between October and February. Additionally, one was claimed 25km west of Iğdır, Kars province (East Anatolia) on 6 August 1992 (J. Hornbuckle *et al.*), although Roselaar (1995) suggested that confusion with another lark species, perhaps immature Horned Lark *Eremophila alpestris*, may have occurred; certainly, the record is not easily explained, if correctly identified. In late May 1975 (an unspecified number of) individuals showing breeding behaviour, reported as Bar-tailed Lark *A. cincturus* (no descriptions published) were claimed at four unnamed localities, on low stony hillsides between Birecik and Diyarbakır (Southeast Anatolia) (*OST Bull.* 13:4). No further information concerning this record has emerged, but it is perhaps plausible to suggest the birds observed were Desert Larks. If so, the distribution of the species may be more extensive than records suggest. Only recently certainly recorded away from the Birecik region, one northeast of Nusaybin (Southeast Anatolia) on 2 June 2001 (Kirwan *et al.* 2003). There is a single record of this species in Cyprus, in April 1973 (Flint & Stewart 1992), making it also a potential vagrant to Turkey's south coast.

Breeding. Confirmed, on a few occasions, since 1984, at a dry (and mainly bare) stony plateau with gullies, at Birecik (**Southeast Anatolia**). Nest and clutch size undescribed in Turkey but, elsewhere, nests in a hollow in the ground, sheltered by a low rock or tuft of vegetation; builds a cup of grass and plant material, lined with finer plant material, and small stones or pieces of dried earth around the exposed side of the nest, forming a rampart; usual clutch 3–5 eggs. Two adults were feeding two recently fledged juveniles on 31 May 1984, indicating egg laying in early May. Adults, each with a single juvenile, were observed on 18 July 1986 and 17 August 1986, and seven, including two juveniles, were seen on 10 August 1986. Four adults and two young were seen on 6–7 May 1989 (N. Green), indicating egg laying no later than 9 April. Anxious birds were seen on 28 May 1991 (D. J. Bates). Nest building was observed on 14 April 1996 (L. & L. Vallotton). Elsewhere, the species is sometimes double-brooded.

BAR-TAILED LARK — *Ammomanes cinctura*
Küçük Çöl Toygarı

(Du: Rosse Woestijnleeuwerik; Fr: Ammomane élégante; Ger: Sandlerche; Swe: Sandökenlärka)

Subspecies. Presumably *arenicolor*. Note the correct spelling of the specific name (David & Gosselin 2002b).

Status and distribution. Vagrant. In addition to the reports from Southeast Anatolia mentioned under Desert Lark *A. deserti*, to which they are arguably referable, there are three recent claims from the Göksu Delta (Mediterranean): one, well described, on 12 May 1987 (Kirwan & Martins 2000a) has now been accepted by the Danish rarities committee (K. Pedersen *in litt.* to T. Andersen 1997); two, apparently without description, were claimed on 27 April 2000 (*Birding World* 13: 231, Kirwan *et al.* 2003); and one well-watched in dry fields north of Akgöl, on 21 May 2005 (GMK). Kiziroğlu (1989) asserted that *A. cinctura* was a summer visitor to southeast Turkey (presumably

on the basis of the *OST Bull.* report), a comment that can be discounted given the lack of any additional evidence for the species' occurrence in that part of the country, although it is notable that there have been three summer records in Syria: in May 1972, August 1976 and April 1994 (Snow & Perrins 1998) and a healthy breeding population has recently been discovered in northern Lebanon (Bradshaw & Kirwan 2000, Ramadan-Jaradi *et al.* 2008). Additionally, the species has recently been recorded in Cyprus: one in March 1999 (P. Flint *in* Kirwan *et al.* 1999), up to four (at least one of which was photographed) in April 2000 (Bradshaw & Kirwan 2000) and one in March 2006 (*Dutch Birding* 28: 174).

GREATER HOOPOE LARK *Alaemon alaudipes*
İbibik Toygarı

(Du: Witbandleeuwerik; Fr: Sirli du désert; Ger: Wüstenläuferlerche; Swe: Ökenlöplärka)

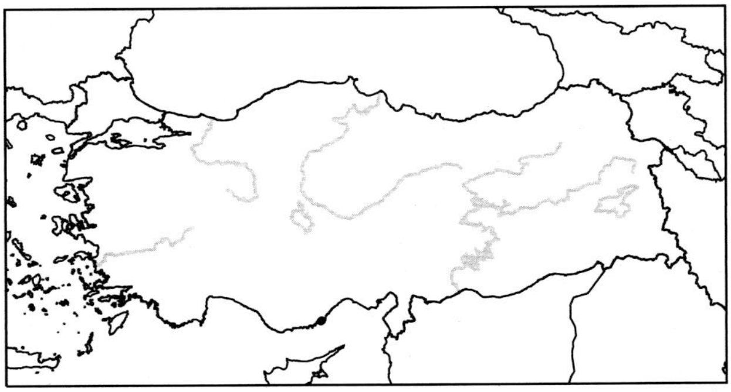

Subspecies. Presumably the nominate.

Status and distribution. Vagrant. The only record for Turkey involved one, well described, at the Göksu Delta (Mediterranean) on 2 May 1997 (Shelton 1997).

[THICK-BILLED LARK *Ramphocoris clotbey*]
Kalın Gagalı Toygar

(Du: Diksnavelleeuwerik; Fr: Alouette de Clotbey; Ger: Knackerlerche; Swe: Tjocknäbbad lärka)

Subspecies. Monotypic.

Species not admitted to the Turkish list (Kirwan *et al.* 1999). A published claim of one in flight in the vicinity of the former Amik Gölü, Antakya province (Mediterranean) on 31 May 1975 (*OST Bull.* 13:4) lacks substantiating details. However, the species clearly could occur in Turkey. Aharoni (1931) mentioned that the species bred in the 'Syrian desert' and, as detailed by Kumerloeve (1969c), in 1930 two pairs were found and in 1931 a pair with four eggs. Aharoni (*op. cit.*) provided a detailed account of the nest's discovery. The records were recently considered as not certainly being from within the present-day borders of the Syrian Arab Republic by Baumgart *et al.* (1995), although Cramp (1988) treated the record as confirmed, albeit commenting on the lack of detailed locality information. Subsequent investigation into the provenance of this record appears to place the record surely within modern Syrian territory (Kirwan 2004a).

CALANDRA LARK *Melanocorypha calandra*
Boğmaklı Toygar

(Du: Kalanderleeuwerik; Fr: Alouette calandra; Ger: Kalanderlerche; Swe: Kalanderlärka)

Subspecies. As acknowledged by Roselaar (*in* Cramp 1988) and Roselaar (1995), the situation is complex and very much influenced by local conditions. Nominate *calandra* (throughout Turkey except parts of Southeast Anatolia), *hebraica* (Birecik and much of the Hatay, although specimens from the former locality show some influence of *calandra*) and *gaza* (Ceylanpınar). *M. c. gaza* (spelt *gazae*, in error, by Roselaar 1995) was not admitted by Cramp (1988) and was considered by its finder, Meinertzhagen, to represent a new subspecies of

M. bimaculata (*Bull. Brit. Orn. Club* 39: 84–85), further demonstrating the complexity of the issue. It is unclear which subspecies occurs in other areas in Southeast Anatolia. The subspecies *hollomi*, known from nine specimens from Erzurum, Van and Yüksekova, and described by Kumerloeve (1969a), is considered a synonym of *calandra* by Roselaar (1995); it not being marked enough to warrant recognition as a separate taxon. Roselaar (*op. cit.*) speculated that the generally rather pale and grey-looking *psammochroa* which winters in Iraq, Jordan and Israel may also occur in Turkey at this season.

Though micro-scale soil (and climate) conditions clearly influence the appearance of this group, and individual variation is relatively pronounced, it is also readily feasible to identify the majority of individuals to a given form, *calandra*, *hebraica* or *gaza* (in this respect, we agree with Roselaar *op. cit.*, that recognition of this form appears as equally valid as *hebraica*). We have not seen any specimens of *dathei*, which was described on the basis of five specimens from Ceylanpınar (Southeast Anatolia), by Kumerloeve (1970c), but was treated as a synonym of *gaza* by Roselaar (*op. cit.*). Variation between *calandra*, *hebraica*, *gaza* and, probably, *psammochroa* appears clinal, perhaps invalidating either of the latter two forms under the terms of reference adopted here, but nonetheless we consider all four forms recognised by Roselaar (*op. cit.*) to be broadly diagnosable.

Status and distribution. National conservation status: Least Concern (population apparently stable). A widespread and locally very common (even abundant) resident breeder throughout Turkey except in Black Sea Coastlands, where largely absent in the breeding season. Occurs principally in lowland areas, from sea level to around 1,800m, occasionally higher in East Anatolia, but usually replaced by Bimaculated Lark *M. bimaculata* above 2,000m. One of the most characteristic breeding species of cultivated steppe.

Large flocks, occasionally numbering several thousand, gather in winter throughout Turkey, although remains less widespread within Black Sea Coastlands: e.g. 2,000 at Merzifon, Amasya province, Black Sea Coastlands, on 8 January 1969 (OST 1972) and 5,000 at Kulu Gölü, Inner Anatolia, on 1 December 2001 (*Kuşçu Bulteni* 10: 7). There is some evidence of immigration and passage through Turkey to more southerly wintering areas in late autumn (mid October to mid November) and return passage between mid March and late April (e.g. 17 March–25 April 1992 in Kızılırmak Delta, Black Sea Coastlands).

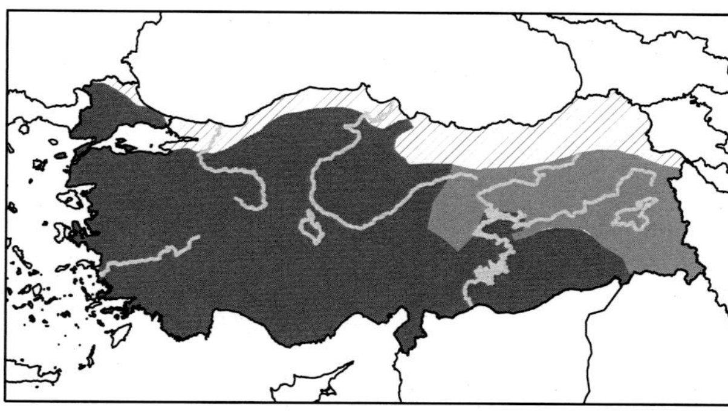

Breeding. In dry open country, in flat or undulating steppe and tussocky grasslands, and arable land, particularly cornfields. Nests on the ground, in a hollow; nest is a deep cup of dry grass and plant material, rootlets and hair. Eggs: C/3 (1), C/4 (4), C/5 (1), C/6 (1). Young: details below. In **Southeast Anatolia**, fledged young at Akçakale on 6 May 1970, indicate egg laying in early April. In **Mediterranean**, a nest at Acıgöl had four eggs on 11 May 1899 (Selous 1900). On the coast near Manisa (**Aegean**), on 12 May 1950, a nest had an incomplete clutch of four eggs, and another two large young and four infertile eggs (McNeile), the latter indicating laying in mid April. An adult was carrying food near Uluabat Gölü (**Marmara**) on 22 May 1966 (R. F. Porter). In **Inner Anatolia**, a nest had five eggs at Tuz Gölü on 26 May 1974, and a recently fledged young was photographed on 25 June 1977 (Pforr & Limbrunner 1982), indicating egg laying in late May; a nest had four eggs near Mogan Gölü on 13 May 1967 (R. F. Porter); at Bolluk Gölü, on 15 May 2004, a nest had three eggs and another three newly hatched young and one egg (M. Farmer), the latter indicating egg laying in late April; many juveniles were at Çöl Gölü on 17 July 1986 (Kiraç 1993). In **East Anatolia**, several adults were carrying food at Sodalı Gölü on 5 June 2001, one of which clearly had fledged (and dispersed) young, as it was delivering food to several different places; the latter indicates egg laying in early May (PC). Probably double-brooded in Turkey, as elsewhere, but the data are too sparse to draw firm conclusions.

BIMACULATED LARK *Melanocorypha bimaculata*
Küçük Boğmaklı Toygar

(Du: Bergkalanderleeuwerik; Fr: Alouette monticole; Ger: Bergkalanderlerche; Swe: Asiatisk kalanderlärka)

Subspecies. *M. b. rufescens* (southern, central and eastern Turkey, e.g. in the Taurus Mountains, Inner Anatolia and around Birecik and Ceylanpınar), but many populations indistinguishable from nominate *bimaculata* or are intermediate. Birds from extreme northeast Turkey, e.g. around Iğdır and Kars are apparently identical to *bimaculata* whilst populations around Van and Erçek are fully intermediate (Roselaar 1995). As with most larks, local soil conditions may play a far more important, micro-scale, role in determining geographical variation within this species. For instance, those specimens available from around Gaziantep (Southeast Anatolia), all in heavily worn plumage, are immediately identifiable among skins held at the Natural History Museum (Tring) due to their strongly pinkish-washed underparts. The other differences between nominate *bimaculata* and *rufescens* are, as evidenced above, subject to individual variation, apparent intermediates and wear and bleaching. Therefore, whether the naming of such variation is of real value in attempting to understand the systematics of these larks appears doubtful. Indeed, Alström (*in* del Hoyo *et al.* 2004) considered *M. bimaculata* to be a monotypic species.

Status and distribution. National conservation status: Least Concern (population apparently increasing). A fairly widespread and common summer migrant, breeding in East Anatolia, in smaller numbers in Inner and Southeast Anatolia and adjacent areas of north and west Mediterranean, e.g. around Beyşehir Gölü and the upper Çukurova. Also recorded on several occasions in suitable breeding habitat in Black Sea Coastlands. Nests principally on stony hillsides, steppe and semi-steppe, as well as in cultivated areas at moderate to higher altitudes. Rarely below 400m (in Southeast Anatolia) and otherwise usually between 1,500m and at least 3,000m during the breeding season (from mid May). In areas of sympatry with Calandra Lark *M. calandra* usually selects slightly different (often higher altitude) nesting sites.

Occasionally reported in large flocks on passage, usually in autumn, when also more widespread in Mediterranean, with occasional records from the extreme west of this region. Flocks of up to 250 appear common; exceptionally 2,800 were recorded at Hotamış (Inner Anatolia) in late September 1974 (OST 1978). Spring arrival commences in earnest in mid April (though the vanguard is noted from 14 March on the south coast) and passage probably continues until at least mid May (14th). The autumn departure is in early August, most having departed by mid to late October (21st), although recorded into the third week of November (21st). Recorded as a vagrant at Gokova (Aegean).

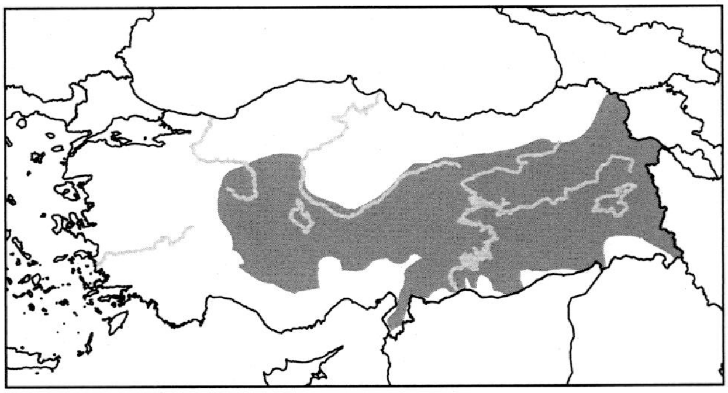

Breeding. In open, stony, sparsely vegetated areas, dry hills with stones, and roughly ploughed fields. Nest is a hollow in the ground, lined with grass and plant material, in the open or partially sheltered by a tuft of vegetation or a stone. Eggs: C/4 (seven). Young: B/4 (two). In **Inner Anatolia**, display and courtship feeding was seen at Hasan Dağ on 24 April 2004 (B. Pleasance); there are three records of nest building in late May and one record on 10 June; six nests had eggs in late May–June, including earliest eggs laid on 26 May (Ramsay 1917); a nest with four eggs, near Ankara on 5 July 1970, is more likely a second attempt than a second brood; a nest had young, *c.*2 days old, at Hasan Dağ on 18 June 1992, indicating the first egg was laid on 1 June (PC); near Karapınar, a nest had four eggs on 9 May 2006 (N. Burke) and another four young, *c.*3 days old, on 8 June 2006 (S. Williams). Recently fledged young are mainly seen in late June, but an early record on 21 May 1998 (T. Quittenden), indicates the first egg in late April. In **Southeast Anatolia**, an adult was carrying food at Işıklı on 28 June 2003. In **East Anatolia**, two nests had four eggs, at Erçek Gölü, on 14 June 1968 (Kasparek & van der Ven 1983).

WHITE-WINGED LARK *Melanocorypha leucoptera*
Ak Kanatlı Toygar

(Du: Witvleugelleeuwerik; Fr: Alouette leucoptère; Ger: Weißflügellerche; Swe: Vitvingad lärka)

Subspecies. Monotypic.

Status and distribution. Vagrant, formerly a sporadic (perhaps even regular) winter visitor. The modern status has most recently been reviewed by Kirwan (2001b). Hartert (1910/22) described the species as not uncommon in autumn and winter in Turkey, a view shared by Alléon (1880), who collected four; a male and female at the village of San Stephano [=Yeşilköy, Istanbul] (Marmara) on 5 and 6 January 1894, and single males at Makriköy [=Bakırköy, Istanbul] (Marmara), on 29 February 1896 and 26 December 1897 (accession numbers 1009–1012, National Museum of Natural History, Sofia). The Natural History Museum, Tring, holds a female (accession number 81.5.1.2283) taken by Robson, at Haskeuy (Hasköy), on the European side of the Bosphorus, on 1 January 1868. Sharpe (1890) mentioned a second specimen, a male, also taken by Robson at the same place on the same day, which can no longer be located in the collection. There is another Robson specimen, which he sent to Canon Tristram, a male (but apparently not the missing Tring specimen) held in Liverpool Museum (National Museums & Galleries on Merseyside), simply labelled as taken at Istanbul, probably on 23 January 1865 (accession number T17881). It should be noted that many Tristram specimens were apparently relabelled, not always accurately (T. Parker *in litt.* 2003), and that possible reasons for a specimen's 'disappearance' from the Tring collection were discussed by Knox & Walters (1992). There was an exceptional influx of the species in February 1871, when Robson secured a male at Hasköy on 17th, two females at Havanköy on 20th, a male at Pendik on the same day and another male at Izmit also on 20th, and then a further two males at Izmit on 21st. All are currently held in The Manchester Museum (BB0271–0277). Dresser (in his *A History of the Birds of Europe*, vol. 4, p. 374) mentioned two additional Robson specimens, now apparently lost, both males and also taken at Hasköy on 17 February 1871. Braun (1903) recorded the species near Istanbul (Marmara) and two specimens shot on 14 October 1914 at Küçükcekmece, Istanbul, were in the Robert's College, Istanbul collection (Kumerloeve 1961), although only a male remained in early 1996 (Kirwan 1997a). Much of the collection has either been lost or destroyed during the course of the last 20 years.

Four subsequent claims, none of which is documented, have been published: at Gelibolu, Çanakkale province (Marmara) on 18 April 1965 (Schweiger 1965), at least 12 around Van town (East Anatolia) in late September 1979 (van der Ven & Gheyselinck 1980), three, north of Çaldıran, Van province (East Anatolia) on 24 September 1981 (van der Ven & Gheyselinck 1982) and north of Arpaçay, Kars province (East Anatolia) on 20 September 1986 (Mycock 1987). The species undertakes (erratic?) winter movements, occurring irregularly in Turkestan, Transcaspia, northern Iran, the Caucasus, the Ukraine and to the north of the Black Sea (Cramp 1988). Formerly more common, the breeding range in the former USSR has contracted within the last century (G. G. Buzzard *in* Cramp 1988).

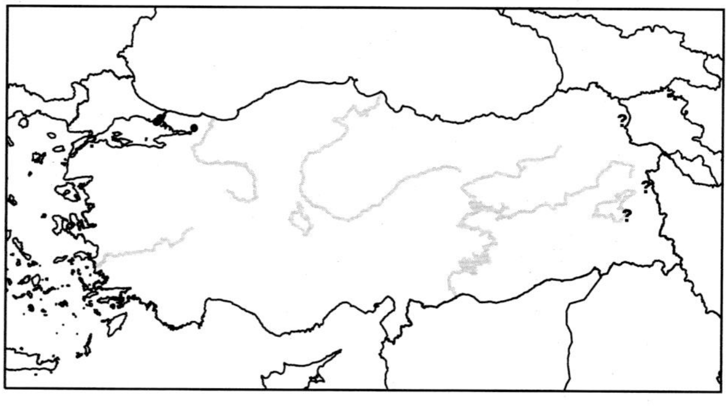

BLACK LARK *Melanocorypha yeltoniensis*
Kara Toygar

(Du: Zwarte Leeuwerik; Fr: Alouette nègre; Ger: Mohrenlerche; Swe: Svartlärka)

Subspecies. Monotypic.

Status and distribution. Kasparek (1990a) noted that Mathey-Dupraz (1922) had reported a female, shot at Küçükcekmece, Istanbul (Marmara), on 14 October 1914 (see also White-winged Lark *M. leucoptera*), which was held in the Robert's College, Istanbul collection. The existence of the specimen was not noted by Kumerloeve (1961, 1970a) and it could not be found in the mid 1980s (M. Kasparek *in litt.* 1996). Neither does Kumerloeve mention the female reported, without details, by Braun (1901). A possible Black Lark was observed at Rize (Black Sea Coastlands) in December 2001 (S. Ekşioğlu). The species undertakes irregular and nomadic winter dispersal, principally to the south and west, beyond the breeding grounds in the Central Asian steppes (from around the northern Caspian Sea east through Kazakhstan and Kyrgyzstan; Sibley & Monroe 1990). It has wandered west as far as Austria, the Czech Republic, Finland, Germany (two late 19th-century records), Greece (four 20th-century records), Malta, Lebanon (one collected at Baalbek, on 31 October 1958), Poland, Romania, Sweden (in spring 1993) and the UK (the most recent Western Palearctic records, in June 2003 and April 2008), and east to Xinjiang (China) and Mongolia (Cramp 1988, Snow & Perrins 1998, MacKinnon & Phillipps 2000, Lindroos & Tenovuo 2002), with a recent (1995) record from Bulgaria (Lindroos & Tenovuo 2002).

GREATER SHORT-TOED LARK *Calandrella brachydactyla*
Bozkır Toygarı

(Du: Kortteenleeuwerik; Fr: Alouette calandrelle; Ger: Kurzzehenlerche; Swe: Korttålärka)

Subspecies. Unclear in many areas as populations show characteristics of several races or are variable. The following is based on Roselaar (1995): nominate *brachydactyla* (Thrace and perhaps coastal Aegean, though all specimens examined from latter in spring were *longipennis*), *artemisiana* (eastern Turkey west through Inner Anatolia as far as Eskişehir and Burdur, Mediterranean), *woltersi* (endemic to the region around Amik Gölü) and *hermonensis* (border with central and eastern Syria, e.g. around Ceylanpınar and Urfa), with *longipennis* widespread in winter. Those from Birecik and Gaziantep are apparently intermediate between *hermonensis* and *woltersi* though nearer the latter, while those in Mediterranean from Silifke east to Karamanmaraş and in the foothills of the Taurus north as far as Kayseri and near Beyşehir are closest to *artemisiana*, but with more cinnamon upperparts and shorter bills. In Southeast Anatolia, specimens from Diyarbakır and Siirt and west to Siverek are intermediate between either *hermonensis* or *woltersi* and *artemisiana*. The level of geographic variation recognised by Roselaar (*op. cit.*) appears unsupportable to us, although he examined a greater number of specimens and broader range of described plumages. Material representative of some of the forms mentioned above, e.g. *woltersi*, was unavailable to us. Thus, we decline to comment on the worthiness of endeavouring to recognise such minor variation as was admitted by Roselaar, but suggest that a re-evaluation of the taxonomy of this, the next species and Asian Short-toed Lark *C. cheleensis* is clearly warranted.

Status and distribution. National conservation status: Least Concern (population apparently increasing). A widespread and common summer visitor, breeding in dry open agricultural habitats including steppe and semi-steppe in Marmara (including Thrace), Aegean, Mediterranean, Inner, Southeast and East Anatolia. Most common in central and south-eastern areas. Apparently largely absent from the Black Sea Coastlands

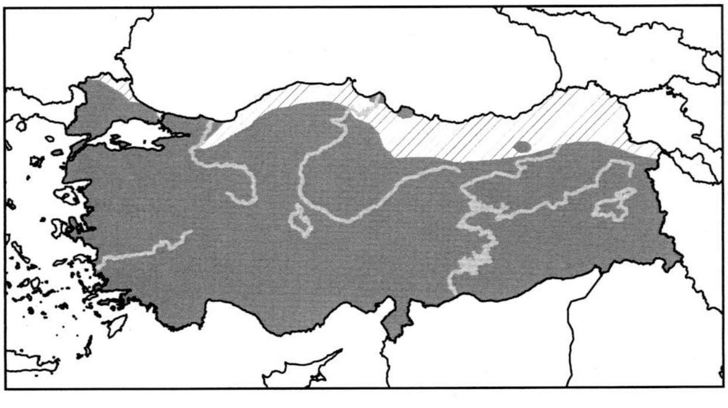

although has bred irregularly. Breeding occurs from near sea level up to at least 2,000m, commencing mid to late April.

Common on passage, when also regular in Black Sea Coastlands, locally in large numbers, e.g. *c.*10,000 moving southwest at Tuz Gölü (Inner Anatolia) in early September 1968 (OST 1972). Flocks numbering several hundred are regularly observed throughout, including Black Sea Coastlands. Arrives mid March (14th) on the south coast and in the west, and from late March (29th) or early April in central, eastern and northern areas, continuing until at least early May in the north. Return passage is from mid August, most having departed by mid October (latest on the Mediterranean coast, 12 November).

Breeding. In dry open steppe, cultivation, field margins and coastal saltmarshes. Nest a deep cup, in a hollow in the ground, usually sheltered by a low shrub or alongside a rock; of dried grasses and roots, lined with down, hair and wool. Several pairs breed in loose associations in optimum habitat, especially in Inner Anatolia (Tuz Gölü), in Mediterranean (Çukurova) and in parts of Aegean. Eggs: C/3 (10), C/4 (6). Earliest record, four pairs with fledged young at Bafa Gölü (**Aegean**) on 19 May 1970, indicates laying of the first egg on 22 April.Most records, in Aegean, **Inner Anatolia** and **Mediterranean**, are nests with eggs in late April and throughout May, with earliest nestlings on 20 May 1970; latest nests, with eggs on 19 June 1973 and newly hatched young on 25 June 1992, both in Inner Anatolia, were probably second attempts. In **Marmara**, an adult was carrying food at Kavak on 24 June 1973, and fledged young were at Uluabat Gölü on 2 June 1967 (R. F. Porter). In **Southeast Anatolia**, a nest had three eggs at Akçakale on 6 May 1970, and another full-grown nestlings, near Birecik, on 4 June 1993 (PC), with a fledged juvenile at Nemrut Dağı on 16 June 1996 (GMK).

LESSER SHORT-TOED LARK *Calandrella rufescens*
Çorak Toygarı

(Du: Kleine Kortteenleeuwerik; Fr: Alouette pispolette; Ger: Stummellerche; Swe: Dvärglärka)

Subspecies and taxonomy. Personal field and museum work reveals that birds indistinguishable from *minor* breed in southern and southeastern Turkey (from Gaziantep to around Nusyabin and along the border with Syria), and perhaps also penetrate further north towards the Van basin, whilst in eastern Turkey (northeast of a line between Aşkale to Tatvan and Yüksekova) at least some breeders appear indistinguishable from *heinei*, which had been speculated by Roselaar (1995) to occur only in winter, presumably following Mayr & Greenway (1960). However, populations in this region of the country are not entirely uniform in their characteristics, and some at least (e.g. in the Iğdır Ovası) conform more closely to the type description of '*pseudobaetica*' (which was considered to occur throughout the easternmost part of Turkey by Roselaar *op. cit.*), although this form is probably best considered intermediate (see, *e.g.*, Mayr & Greenway *op. cit.*) and is very poorly described (Stegmann 1932). In addition, Mayr & Greenway (*op. cit.*) listed *C. r. leucophaea* as a potential accidental visitor to Turkey.

Roselaar (1995), followed by Snow & Perrins (1998), considered all breeding birds in Inner Anatolia (e.g. near Eber and Akşehir Gölü, Tuz Gölü and around Konya and Ereğli) to belong to *niethammeri* (a subspecies endemic to Turkey; see Kumerloeve 1963b) which he placed within the Asian Short-toed Lark *C. cheleensis* complex, and which he considered to be characterised by their marginally longer, deeper based bills and relatively longer wings and tail, the latter tending to show more white in the outer feathers. In addition, the *cheleensis* group reportedly has the upperparts a paler ground colour, with narrower and darker streaks, although these characters are untrue of the nominate form. The subspecies *persica*, which breeds from central Iraq east through southern Iran to Afghanistan, was also placed in *cheleensis* by Roselaar (*op. cit.*); it has not certainly been recorded in Turkey but, as it winters widely in Syria and south to Kuwait, it may conceivably occur at this season, and Nesterov (1911) collected four examples (at three localities) and observed it at another, all within the modern boundaries of the country, in August 1910 (although the measurements he provided do not conclusively eliminate *C. r. pseudobaetica*). In contrast, Beaman (1994) treated *cheleensis* as a subspecies of *rufescens*, a proposal with some merit, albeit arguably too simplistic a solution to the obvious problems posed by this complex. Beaman & Madge (1998) quite perceptively hinted that the key to understanding the taxonomy of the *cheleensis*/*rufescens* complex might lie in study of the form *leucophaea*, which is usually considered part of the *cheleensis* group. Clearly, a discussion of this broader issue is beyond our scope here but, in our opinion on the basis of specimen examination and field observations from many parts of the range of these larks, albeit not including *leucophaea*, either one very variable species should be recognised, or additional species-level taxa, in addition to *rufescens* and *cheleensis*, should be considered.

In the extensive experience of Alström *et al.* (in prep.), the only constant characters for separating *cheleensis*-type and *rufescens*-complex birds are outer tail-feather pattern and wing formula. Independent specimen study suggests that the characters proposed by Roselaar (1995) for separating *cheleensis* from *rufescens* are uncertain and cannot be used in many cases. On the basis of tail pattern and wing formula it is certain that Inner Anatolian birds belong to *rufescens* and not *cheleensis*, *contra* Roselaar (*op. cit.*). Furthermore, field experience garnered elsewhere in Turkey reveals that these characters are quite probably constant and that no Turkish population can be ascribed to *cheleensis*. In addition to photographs of the type series (held in Bonn) and broad field experience, including of live birds in the hand, we have examined only two specimens from the range of *niethammeri*, both taken by Wadley (1951) and held in the Natural History Museum (Tring). These are labelled as *C. r. aharonii* and, like others (Svensson 1992, Beaman & Madge 1998), we prefer to consider *niethammeri* as a junior synonym of *aharonii*, *contra* Roselaar (1995) who suggested that *aharonii* must be closer to *persica* than to *niethammeri*. Some authors (Dickinson 2003, de Juana & Suárez *in* del Hoyo *et al.* 2004) have considered *aharonii* to be a mere synonym of *persica*. In practice, *aharonii* and *heinei* are not dissimilar, especially in size and overall coloration, based on our examination of specimens held at Tring, but are separable using tail-feather pattern and wing formula. Further field and museum studies are desirable and, given the small sample size, additional specimen material is required.

Status and distribution. National conservation status: Least Concern (population apparently decreasing). A summer migrant (also apparently locally resident) in considerable numbers, breeding in Inner, East and (slightly more locally and usually in smaller numbers) Southeast Anatolia, where many formerly suitable areas have been converted to agriculture in recent years, although it remains locally numerous. May also breed in coastal lowlands of eastern Mediterranean, although reports from Antalya, Göksu Delta, Çukurova and Amik Gölü all require confirmation and many suggest confusion with Greater Short-toed Lark *C. brachydactyla.* (It is worth reporting that, in our experience, *rufescens* and *brachydactyla* are actually rather rarely found syntopically in Turkey, and where they do one or other species usually far outnumbers the other.) Additional reports of breeding in coastal Aegean, e.g. at Bafa Gölü and around Menemen, are also unsubstantiated (Roselaar 1995). Breeds from mid April, more usually from mid May, mainly in dry salt steppe (with which the species is strongly associated), semi-steppe (unmodified or slightly modified grassland), or on stony plains and hillsides and cultivated areas, to at least 2,000m.

More widespread on passage when recorded from all regions (except, apparently, Thrace). Locally reported in large numbers in both spring and autumn, especially in Inner Anatolia (*e.g. c.*200 at the former Hotamış marshes, 22 July 1975, OST 1978). Recorded from late February on the south coast though not normally widespread inland until mid April (but see below). Departs between mid August and late September. Apparently winters in irregular numbers being resident in some areas of (southern) Inner Anatolia and highly locally in adjacent parts of Aegean and Mediterranean. Winter status and distribution is perhaps strongly influenced by weather conditions, as apparently commoner and more widespread during mild winters.

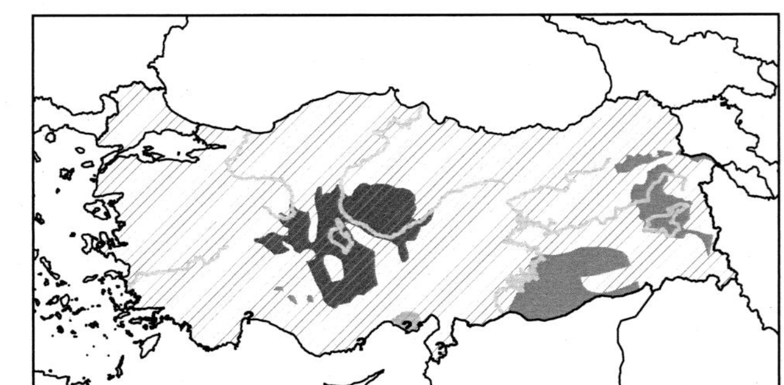

Breeding. Mainly found in flat open steppe with scant vegetation, and in open stony plains, dry saltmarsh and drying out edges of inland lakes; also in cultivation, and nesting has been recorded in ploughed fields. Nests in a hollow in the ground, usually sheltered by a low tuft of grass or a low shrub; a deep cup of dry grasses, plant stems and roots, lined with plant down, hair and fine rootlets. Eggs: C/3 (six), C/4 (four), C/2 (four, but these not necessarily complete clutches). Young: B/2 (1), B/3 (3). In **East Anatolia**, a nest in a ploughed field at Sodaligöl had three almost fully grown young on 5 June 2001, indicating egg laying in mid May (PC), and a juvenile, out of the nest, but not flying, was seen (no location) on 27 July 1966. At The Manchester Museum, there is a clutch of four eggs taken at Izmir (**Aegean**) on 15 May 1860, and two others from the same area (no date). All records below are from **Inner Anatolia**. At Karapınar, four nests had three eggs, one nest had one egg, and another had eggs and newly hatched young on 23–25 April 1988 and, on 6 May 1988, a nest had two eggs, and another small young (J. Whittaker); these records of young indicate egg laying on 11 April and 19 April. At Sultan marshes, a nest had two eggs on 27 May 1972, another had four eggs on 4 June 1982, and recently fledged young were seen on 22 May 1979 (Kasparek 1985), indicating laying by 30 April.In Konya Basin, four nests had 3–4 eggs during 14–27 May, and fledged young were observed on 21–22 May 1969, 13 June 1967, 3 July 1993 (Karauz Kiraç 1993), 13–15 July 1971 and 24 July 1973. Nest building was seen at Kulu Gölü on 13 May 1985 (A. J. Eardley). Many of these records indicate double-brooding, which is frequent and probably usual elsewhere.

CRESTED LARK *Galerida cristata*
Tepeli Toygar

(Du: Kuifleeuwerik; Fr: Cochevis huppé; Ger: Haubenlerche; Swe: Tofslärka)

Subspecies and taxonomy. Following Roselaar (1995) and, partially, Watson (1962c) we are presented with a complex situation, as follows. *G. c. caucasica* (northern Turkey from the Georgian border as far west as the Bosphorus), *subtaurica* (endemic to central Turkey west to Kütahya and Elmali, and east to Van Gölü and the Iranian border, but not in the Anadolu or Taurus mountains, and including *G. c. ankarae*), *zion* (southern areas around Ceylanpınar, though also claimed, erroneously, from around Adana; Bird 1937), and *meridionalis* (western Turkey – in Thrace and from Manyas Gölü south and west as far as Muğla and along the Mediterranean coast east to the Çukurova deltas and neighbouring parts of the Taurus). Birds showing characteristics apparently intermediate between *zion* and *subtaurica* occur in the southeast around Gaziantep, Birecik, Urfa and Elazığ (and have been ascribed the name *weigoldi*), whilst specimens examined from the adjacent Hatay and north to Haruniye and Osmaniye are best considered as intermediate between *meridionalis* and *cinnamomina* (which is restricted to coastal Lebanon and northwest Israel) or between *meridionalis* and *zion* (Roselaar *op. cit.*).

As admitted by previous commentators on the subject of geographical variation within *Galerida cristata* in the eastern Mediterranean, divining overall patterns is extremely difficult due to micro-scale variation according to soil type and therefore climate, wear and bleaching, individual variation and possibly even additional factors. Roselaar (*op. cit.*) remarked that if one wished to acknowledge every permutation of regional variation it would probably be necessary to name hundreds of subspecies. Clearly, such an idea is nonsensical. Variation in this species is chiefly in the ground colour of the upper- and underparts, overall size, bill shape, quantity of dorsal spotting and extent and character of the breast streaking. Though we broadly accept the descriptions and maps of subspecific variation proposed by Watson (*op. cit.*) and Roselaar (*op. cit.*), we have serious reservations as to the advisability and necessity (for the understanding of meaningful and significant evolutionary processes) of recognising such intricate and extremely minor variation, and suspect that analyses of vocalisations and DNA from birds in the eastern Mediterranean region would not produce surprises warranting reappraisal of such a view (see Guillaumet *et al.* 2006 for confirmation that those in Europe, western Asia, northeast Africa and northwest Morocco form a single, monophyletic clade, *cristata*).

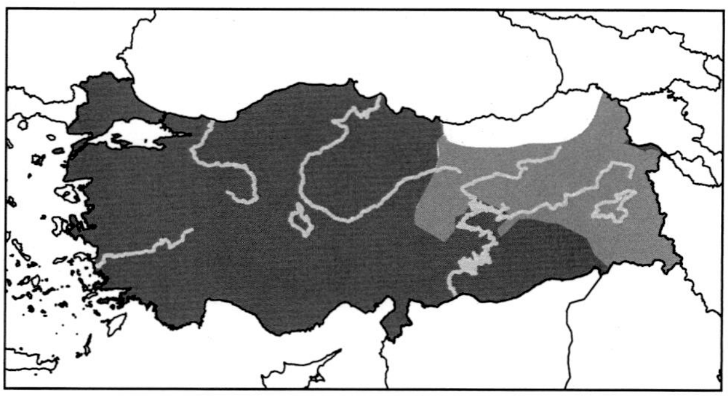

Status and distribution. National conservation status: Least Concern (population apparently stable). A widespread and common resident breeder throughout most of Turkey, e.g. 500–1,000 pairs were in Kızılırmak Delta, Black Sea Coastlands, in spring 1992, during one of the very few intensive breeding-bird surveys conducted in the country. However, apparently scarce (or even absent) from the eastern Black Sea Coastlands and rather sparsely distributed in extreme East Anatolia. Found in a broad range of habitats at low and moderate altitudes, up to *c.*2,000m. Less social in autumn and winter than other larks.

Breeding. Found in open areas, usually with sparse and low vegetation, including dry grassland, poor arable cultivation, dunes, drier saltmarshes, arid hillsides and steppe; also in scattered scrub on areas of bare ground, and roadside verges and wasteland beside crop fields; a pair was on territory on a low island, 10km from shore, in Tuz Gölü (Inner Anatolia) on 22 April 2004 (PC). Nests in a hollow in the ground, often dug by the birds, of dry grasses and other plant material, lined with fine grass and hair. Eggs: C/3 (7), C/4 (10), C/5 (11), C/6 (1); nests with 1–2 eggs considered incomplete. Young: B/3 (2), B/4 (5). Double-brooded in Turkey, as elsewhere, with nests containing eggs mainly late April–late June, and young in nests until July. Earliest records of nests with eggs are: **Mediterranean** (Çukurova) C/4 on 23 March 1987 (van der Have *et al.* 1988); **Aegean** (Altınkum area) four nests, C/4–6 on 22–24 April 2002 (PC); **Southeast Anatolia** (Birecik) a nest with three young on 5 May 1992 back-dated to egg laying about 23 April; **Inner Anatolia** (Karapınar) C/1 (incomplete clutch) on 10 April 1970 (McNeile); **Marmara** (Bulgarian border in Thrace) C/4 on 20 April 1964 (Warncke 1964–65); **Black Sea Coastlands** (Kızılırmak Delta) a nest with young in the last ten days of May can be back-dated to egg laying about early May (Hustings & van Dijk 1994); **East Anatolia** – no breeding records.

WOOD LARK — *Lullula arborea*
Orman Toygarı

(Du: Boomleeuwerik; Fr: Alouette lulu; Ger: Heidelerche; Swe: Trädlärka)

Subspecies. *L. a. pallida* (following Roselaar 1995) occurs throughout Turkey.

Status and distribution. National conservation status: Least Concern (population apparently stable). Widespread though usually uncommon or local resident breeder, or partial migrant, in Marmara (including Thrace), Aegean, Mediterranean and the eastern half of the Black Sea Coastlands and more locally throughout most of the remainder of Turkey, especially in Southeast Anatolia. During the breeding season found in open forest or scrub, typically with juniper or conifers in hilly or montane areas generally between 1,100 and 2,000m, although occasionally recorded up to 3,000m, or down to sea level in Thrace and Black Sea Coastlands.

Widespread in moderate numbers in autumn and winter, when social, forming small flocks (rarely up to several hundred, e.g. in Büyük Menderes Delta, Aegean, in January–February 1972; OST 1975). Winters principally in the south and southwest, more occasionally elsewhere in the western two-thirds of Turkey. Some evidence of immigration into the country at this season, e.g. 32 passed southeast at the Bosphorus between 24 September and 25 October 1966 (Porter 1983). May be regular at this season on the Black Sea coast.

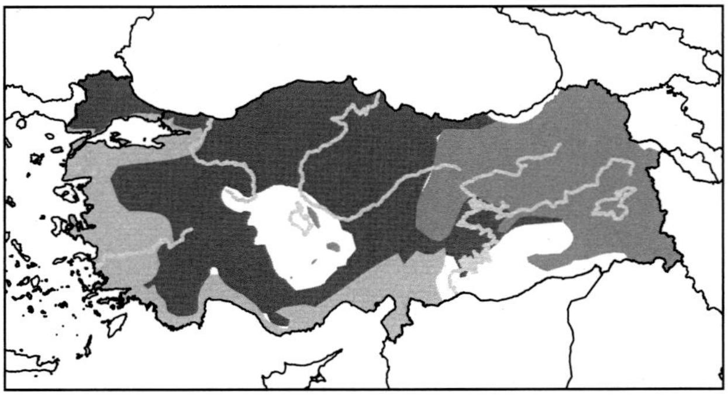

Breeding. In open woodland, both coniferous and deciduous, especially with open areas of short or scant vegetation, heaths with isolated trees, farmland with some trees, hillsides with scattered scrub, dunes with bushes and trees and occasionally on bare hillsides with short vegetation, devoid of trees. Mainly in hilly or montane areas, but also in lowlands and in flat areas at sea level, e.g. in dunes at Kocaçay Delta, Marmara (Ertan 1996), up to 3,000m in southeast and east Turkey (Roselaar 1995); at Uludağ (Marmara), in semi-open and open farmed areas, and on lower slopes to just above the treeline (Jetz 1995); in open woodland of small trees, with tiny clearings, scattered tufts of grass and isolated low bushes, at 2,200m, in the crater at Nemrut Dağ (East Anatolia) (PC). Nests on the ground, sheltered by a tuft of vegetation. Nest undescribed in Turkey; elsewhere, a cup of grass and moss, lined with fine grass, occasionally with some hair. Eggs: C/4 (1); the usual clutch, elsewhere, is 3–4 eggs. Young in nest: no data but see broods of fledged young (below). In **Marmara**, Warncke (1964–65) saw a female feed young in a nest near Osmaneli on 22 April 1964; a nest was in dunes at Kocaçay Delta on 14 April 1990 (Ertan 1996), an adult with nesting material was at 1,900m on Uludağ on 18 May 1969, and fledged young were near Kosedere on 25–26 June 1973 (R. F. Porter). In **Aegean**, a family party of nine on hills near Menemen on 22 June 1966 was an exceptionally large party, probably more than one family. An adult and three fledged young (a typical brood) were flushed (and flew well) between Manisa and Menemen, on 24 April 2003, indicating egg laying no later than 20 March (PC). In **Mediterranean**, two adults were feeding two juveniles at Mut on 10 July 1967, and there were many other family parties in July (Vielliard 1968); adults were feeding young in juniper forest in a valley near Pozantı on 18 May 1970, and a nest had four eggs at Akseki on 25 April 1973. A pair was feeding young in the nest, and an adult was seen with two juveniles, at Halfeti (**Southeast Anatolia**), on 4 May 1997. In **Black Sea Coastlands**, breeding proven at Abant Gölü on 5 June 1996 (GMK). No reliable data on breeding densities in Turkey, but R. F. Porter heard seven singing males from his camp at Yüksekova (**East Anatolia**) on 4 June 1969. Adults sing mainly March–July, and records indicate double-brooding in Turkey, which is usual elsewhere.

EURASIAN SKYLARK *Alauda arvensis*
Tarlakuşu

(Du: Veldleeuwerik; Fr: Alouette des champs; Ger: Feldlerche; Swe: Sånglärka)

Subspecies. *A. a. armenica* (extreme eastern Turkey) with birds from Ağrı, Erzurum and Kars most closely resembling *cantarella* (found in the Caucasus and eastern Europe) in plumage colour and bill size but with longer wings. Elsewhere in Turkey the subspecific situation is unclear. One specimen examined by Roselaar from Inner Anatolia was intermediate between *armenica* and *cantarella* but closer to the latter in plumage. The subspecies occurring in Marmara, Aegean, Mediterranean, much of the Black Sea Coastlands and Southeast Anatolia is unclear, although *cantarella* which occurs in Bulgaria and Greece might be expected to breed throughout much of western Turkey, whilst in winter both *arvensis* (from central and northern Europe) and *dulcivox* (from the Volga steppes and Kazakhstan) could be expected to occur (Roselaar 1995).

Status and distribution. National conservation status: Least Concern (population apparently decreasing). Fairly widespread resident breeder or partial migrant in some areas (principally Inner and East Anatolia), usually at relatively low densities. Rather local in all regions and apparently absent during the breeding season from Southeast Anatolia. Most common and widespread in Inner and East Anatolia. Found through a broad range of open habitats, from 1,000m to 3,000m, only occasionally at lower altitudes during breeding season, although breeding has been recorded at Kızılırmak Delta, Black Sea Coastlands.

More common and widespread, often in large numbers, on passage throughout Turkey: from March to mid April, in spring, and from mid September to early November, in autumn, when peak numbers apparently occur during the second half of October (e.g. 200 passed the Bosphorus between 5 October and 5 November 1966, with peak numbers on 25 October). Winters in small or moderate numbers, primarily in lowland and coastal areas, in the western two-thirds of Turkey. Exceptionally 5,000–10,000 were at the Meriç Delta (Marmara) on 3 January 1969 (OST 1972). Wintering populations in Inner Anatolia reflect the severity of weather conditions; largely absent in very severe winters. Also winters, generally in small numbers, in Southeast Anatolia, though rarely recorded in East Anatolia during this season.

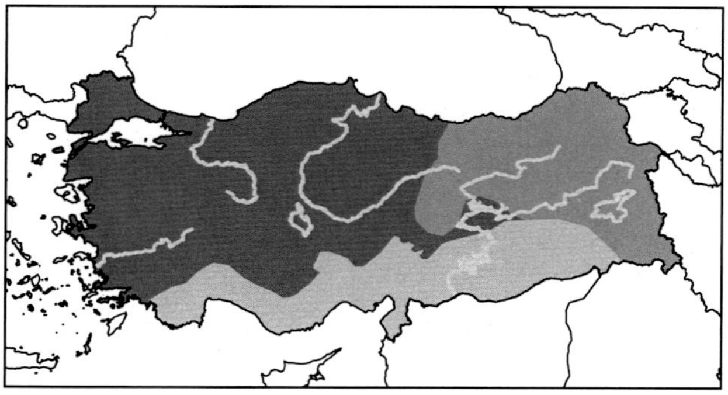

Breeding. Inhabits treeless open country, rolling grasslands, meadows, edges of marshes and lakes; also found in arable fields, including ploughed land, occasionally on rocky slopes, and high-altitude pastures in East Anatolia. A pair bred in 2004 on a low island in Bolluk Gölü (Inner Anatolia), and the nest had three eggs on 17 May (L. Baxter). Nests in a hollow in the ground, usually sheltered by a tuft of vegetation; the nest is a cup of grasses, lined with finer grasses. Eggs: C/3 (1), C/4 (1). Young: no data. In **Inner Anatolia**, a nest had four eggs at Seyfe Gölü on 13 May 1993 (PC); nest building was seen at Hasan Dağ on 24 April 2004 (S. Williams); at least 25 singing males and pairs, some carrying food and alarm-calling, were at Kurbağa Gölü on 14 June 1977, and an adult fed a fledged young at Eşmekaya on 21 June 1977 (Schubert 1979), indicating egg laying in late May. In **Marmara**, an adult was carrying food at Manyas Gölü on 5 June 1966 (R. F. Porter). In **Black Sea Coastlands**, adults carrying were food at Kızılırmak Delta on 8 June 1992 (Hustings & van Dijk 1994), near Ilgaz on 9 June 1977, and near Gerede on 29 June 1977 (Schubert 1979). In **Mediterranean**, an adult was carrying food at Sultandağlar on 7 May 1967 (R. F. Porter). In **East Anatolia**, there were three nests (no details) at Erçek Gölü on 23 June 1968 (Kasparek & van der Ven 1983). Probably double-brooded in Turkey, as elsewhere, but the data are too sparse to draw any firm conclusions.

HORNED LARK — *Eremophila alpestris*
Kulaklı Toygar

(Du: Strandleeuwerik; Fr: Alouette haussecol; Ger: Ohrenlerche; Swe: Berglärka)

Subspecies. *E. a. kumerloevei* (which is endemic to western and central Turkey) and *penicillata* (eastern Turkey) have black of cheek connecting with black of chest (Roselaar 1995). The former is a newly described subspecies (see Roselaar *op. cit.* for full details) and is known from Uludağ (Marmara), a number of localities in the Taurus (Mediterranean) and Elma Dağ near Ankara (Inner Anatolia). The type specimen was taken at Bereketli, near Çamardı, Nığde province (Mediterranean) by Danford in 1876. Birds from Inner Anatolia were previously assigned to *balcanica* by, e.g., Kumerloeve (1961). In our opinion, *kumerloevei* could easily be regarded as a synonym of *penicillata* but unfortunately we have been unable to examine many specimens of *balcanica* and we have no field experience of this form. Features for its recognition are poorly expressed and clearly subject to variation (Kirwan 2006), and furthermore the degree of overlap with *penicillata* and extreme subtlety of the differences elucidated by Roselaar (*op. cit.*) in their upperparts colorations and patterns, suggest that *kumerloevei* is only doubtfully meritorious of nomenclatural recognition. Though we have not conducted a detailed examination of specimens from across the Palearctic range of this species, it appears obvious from field and museum studies that at least some variation, e.g. in the face pattern and coloration, is clinal.

Status and distribution. National conservation status: Least Concern (population apparently stable). Widespread and common resident breeder and altitudinal migrant, principally in the eastern two-thirds of Turkey. Absent as a breeder from much of Aegean (where recorded only on Boz Dağı, Izmir province and in the extreme east) and Marmara, except on Uludağ. Its range in Southeast Anatolia is still very inadequately known, as most of the suitable habitat has not been surveyed. Generally found on rocky and grassy slopes from moderate to high altitude during the breeding season. Occurs above 700m in central and southern areas (Mediterranean and Inner Anatolia), at 1,500–3,150m in the Taurus (Mediterranean), and at 1,000–4,500m in East Anatolia.

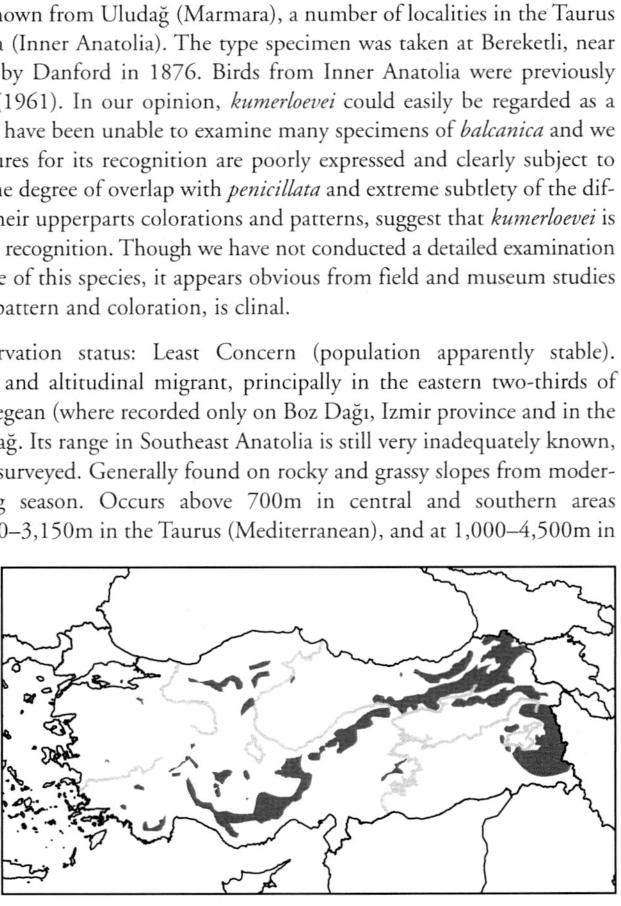

Descends in winter, when noticeably more widespread over Inner Anatolia, even occurring as far afield as the Marmara lakes and irregularly in the vicinity of Istanbul (Braun 1901), from which region there are several recent records at Kilyos. Frequently utilises cultivated steppe at lower altitudes at this season. However, even at this season some remain at high altitude, with birds recorded at 2,100m, above the snowline, in the southern Taurus in January 1995 (Uhlig & Uhlig 1999b). Forms substantial flocks numbering even up to several thousands in late autumn and winter. Also recorded on the south coast, apparently on passage, in autumn (OST 1975) and on the Black Sea Coast during midwinter cold-weather movements (Ogilvie 1954).

Breeding. On open rocky or grassy slopes, but often in bare stony areas with only sparse grass, in alpine vegetation above the treeline, generally above 2,000m, but also in very stony desert habitat with sparse vegetation, and in rocky abandoned fields, near Kursunlu (Black Sea Coastlands) at only 1,100m (Schubert 1979). Nests in a hollow in the ground, usually sheltered by a low shrub, tuft of grass or a low rock; the nest is of dry grass and plant stems, lined with hair and fine grass. Eggs: C/2 (2), C/3 (5), C/4 (3), C/5 (1). Young (in nest): B/2 (2), B/3 (3), B/4 (2). In **Mediterranean**, early records include fledged young in Aladağ on 15 May 1970, indicating egg laying in mid April, nest building at Karanfil Dağ on 9 May 1951 (Hollom 1955), and a nest with four eggs at Demirkazık on 4 May 1989. Later records are nests with three eggs on 7 June 1998, one egg (incomplete clutch) on 12 June 1993, four eggs on 19 June 1990, a juvenile on 18 July 1986, and two family parties on 10 August 1988. In **Inner Anatolia**, a nest at Hasan Dağ had two eggs on 24 April 2004 (S. Williams), a nest there had three eggs and one newly hatched young on 25 May 1992 (fledged on 4 June 1992), and a nearby pair was also feeding newly fledged young on 4 June 1992, indicating egg laying about 7 May (PC). Fledged young near Niğde on 23 May 1970, indicate egg laying in late April; a nest at Erciyes Dağ had three young on 10 June 1971; fledged young were at Karadağ on 19 June 1996 (GMK); 12 pairs were feeding young between Göreme and Yeşilhisar on 17 June 1977, and three pairs were feeding young on Erciyes Dağ on 16 June 1977 (Schubert 1979). In **Black Sea Coastlands**, a nest (no details) was near Abant on 23 April 1966; in the Kursunlu area, Schubert (1979) found several nests in June 1977: a nest with three young, six days old, on 9 June 1977, a nest with two young which fledged on 10 June 1977, two nests with one egg (incomplete clutch) and three eggs on 11 June 1977, and he observed nest building on 11 June 1977, and a pair with fledged young on 11 June 1977; he also found a nest near Gerede in which the first egg was laid on 30 June 1977. Several pairs were nesting on rocky and grassy slopes between snowfields at 2,800m above Sivrikaya, and three nests each had three eggs and one nest had two eggs on 13–14 June 2004; two of the nests were only 50m apart (PC). Two adults were feeding juveniles at Sivrikaya on 27 July 1986 (J. McLoughlin), and a pair was feeding fledged young at Kopdaği pass on 15 June 1975 (R. F. Porter). At Uludağ (**Marmara**), an adult was carrying food on 29 May 1988, a nest had four broken eggs on 20 June 1959, and there was a pair with four young on 1 July 1981. In **East Anatolia**, at least 50 pairs, some carrying food, were at 2,100–3,200m in the Kurdish Alps, west of Görentaş, on 31 May–4 June 1969, and fledged young were there on 31 May 1969 (R. F. Porter). A nest had five eggs at Başkale on 3 June 1984 (A. Schubert); a nest had four young, four days old, near Balık Gölü on 7 June 2001 (PC), and a nest near Doğubeyazıt had three feathered young on 27 June 2004 (S. Williams). There are five records of fledged young at Nemrut Dağ between 8–16 June, indicating egg laying about mid May. There were recently fledged young at Erçek Gölü on 7 May 2004 (GMK), indicating egg laying *c.*10 April. The wide range of breeding dates clearly indicates double-brooding, which is usual elsewhere.

SAND MARTIN — *Riparia riparia*
Kum Kırlangıcı

(Du: Oeverzwaluw; Fr: Hirondelle de rivage; Ger: Uferschwalbe; Swe: Backsvala)

Subspecies. Apparently all nominate *riparia*. Plumage characters and measurements of specimens taken in Ankara and Van areas accord with this race, although a bird taken in Kayseri province (Inner Anatolia) on 27 September 1974 was markedly darker than all other Turkish skins examined, more closely resembling *eilata* (recorded on migration in southern Israel) but the wing was too long for the latter subspecies, being nearer *ijimae* of eastern Asia. It may have been attributable to the poorly known *fuscocollaris*, a dark taxon recorded as a migrant on the former Yugoslav coast (Roselaar 1995). Fifty-one trapped in Inner Anatolia in spring 1988 had wing-lengths closely according with existing data for *riparia* (Schekkerman & van Roomen 1993). However, a reappraisal of the validity of some of these forms appears necessary.

Status and distribution. National conservation status: Vulnerable (population apparently decreasing). A locally common summer visitor breeding throughout Turkey. Breeds up to around 1,750m (although perhaps expectable at higher altitudes) from mid to late April.

More widespread on passage throughout the country, locally in enormous numbers in wetlands and coastal areas, e.g. 35,000 were at Van, East Anatolia, on 17 September 1973 (OST 1975) and 10,000–50,000 were there on 5–8 May 1989 (van den Berk *et al.* 1993), and possibly in excess of 100,000 were in Kızılırmak Delta, Black Sea Coastlands, in mid-May 1992, including a roost of 42,000 at Cernek Gölü alone on 16 May (Hustings & van Dijk 1994). Recorded on spring passage from mid March (11th) to early June (10th), most arriving in late April and the first half of May (reaches northern areas *c.*2 weeks later than the south). Autumn passage is from mid August (14th) until late October (28th), peaking during September. Late individuals are occasionally recorded until the end of November, especially at south-coast localities. The majority has departed by early October, but exceptionally recorded as late as 24 November (OST 1972). A record of two at the Büyük Menderes Delta (Aegean) on 28 February 1974 (OST 1978) almost certainly reflects early arrival (not overwintering).

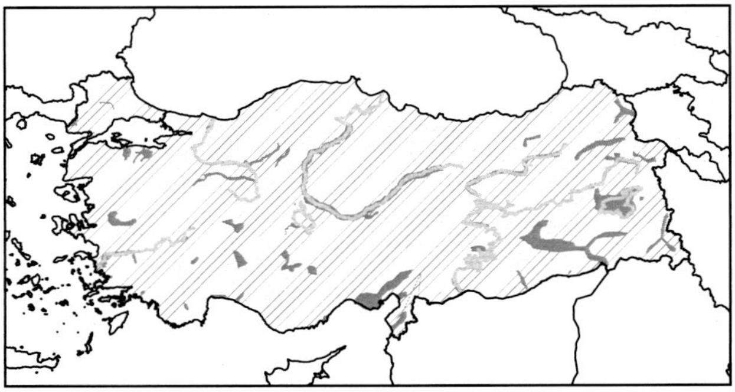

Breeding. In colonies near water in open country. Nests mainly in banks of rivers, lakes, ponds and irrigation channels, but also in sand-pits, roadside cuttings and mounds of earth; from sea level, e.g. Göksu Delta (Mediterranean), to 1,750m in the east, e.g. Van Gölü (East Anatolia). Colonies may comprise just 20–30 pairs, but more typically 50–100 pairs; there were 500 pairs in two adjacent colonies at Bulanık (East Anatolia) on 10 June 1987 (Eames 1989), many small colonies in Çukurova (Mediterranean) in 1987, the largest of *c.*500 nests (van der Have *et al.* 1988), and a colony at Birecik (Southeast Anatolia) comprised *c.*2,200 pairs on 10–11 June 1996 (GMK), whilst previous observers have estimated this colony at 3,000–5,000 nests. Colonies of this species and European Bee-eaters *Merops apiaster* may use the same bank for nesting. The nest is an untidy pad of grass stems and feathers, in a cavity at the end of a horizontal tunnel, 0.5–1.0m long, excavated by both adults in a vertical bank. Eggs: C4 (1); elsewhere, usual clutch 4–5. Young: no data. In **Mediterranean**, adults were flying near nesting tunnels in Göksu Delta on 31 March 2000 (L. J. Dijksen), and colonies in Çukurova were occupied from mid April in 1987 (van der Have *et al.* 1988). At Birecik (**Southeast Anatolia**), on 10 May 2004, hundreds of adults were observed taking grass and straw from the ground for nest building, but no adult was observed excavating a tunnel (PC); many adults flew out of nesting tunnels (presumably incubating) in part of the colony, and a nest had four eggs on 10 May 2004 (R. Thomas). Adults were excavating nesting tunnels at Istanos (**Inner Anatolia**) on 14 May 1944 (Wadley 1951). At Van (**East Anatolia**), two adults were still excavating nesting tunnels on 8 June 2004 (possibly replacements), and large young in three nests were being fed at the entrances to the tunnels, indicating laying of the first egg *c.*7 May (PC); elsewhere in this region, young were being fed in nests in a riverbank near Ağrı on 25 June 2004 (PC), and breeding colonies (no details) were still occupied on 9 August 1966, presumably for second broods, which is usual elsewhere.

[PALE CRAG MARTIN *Ptyonoprogne obsoleta*]
Küçük Kaya Kırlangıcı

(Du: Vale Rotszwaluw; Fr: Hirondelle isabelline; Ger: Steinschwalbe; Swe: Blek klippsvala)

Subspecies and taxonomy. Unknown, although if the species was to reach Turkey it would surely be nominate *obsoleta* that was involved. Some authorities (e.g. Keith *et al.* 1992, Snow & Perrins 1998) have subsumed *obsoleta* within *H. fuligula*, under either the name African Rock Martin or Pale Crag Martin. Forms a superspecies with the next species (Sibley & Monroe 1990).

Species not admitted to the Turkish list (Kirwan *et al.* 1999). Listed as a summer visitor to the Mediterranean region of the country and Cyprus by Kiziroğlu (1989). No documentation exists to substantiate this statement and, in the absence of other reports from this comparatively well-watched region of Turkey, it seems safest to disregard this claim.

EURASIAN CRAG MARTIN *Ptyonoprogne rupestris*
Kaya Kırlangıcı

(Du: Rotszwaluw; Fr: Hirondelle des rochers; Ger: Felsenschwalbe; Swe: Klippsvala)

Subspecies and taxonomy. Monotypic. Has also been placed in the genus *Hirundo*.

Status and distribution. National conservation status: Least Concern (population apparently stable). A widespread and generally fairly common summer visitor and/or winter visitor, according to region, principally to the eastern two-thirds of Turkey, although rather uncommon and local in Southeast Anatolia. Generally more local in the west and apparently absent from Thrace. Breeds mainly in upland and montane areas, from around 500–2,500m (occasionally lower at inland sites), locally as high as 3,900m, but also at low altitudes on sea cliffs. Inland breeding sites are occupied from mid or late April to early October (e.g. latest record 4 October on Uludağ).

The species winters, sometimes in considerable numbers in coastal areas in southwest Aegean and (more widely) in Mediterranean to at least 500m. Rarely recorded inland in winter, when observed once in Black Sea Coastlands: eight at Samsun on 31 December 1970 (OST 1975). There are comparatively few observations from western and central parts of this region, the majority of breeding records coming from the east, as well as most of those referring to passage birds, e.g. daily passage was observed at Borçka during 24 September to 3 October 1980, with a max. 150 on 1 October.

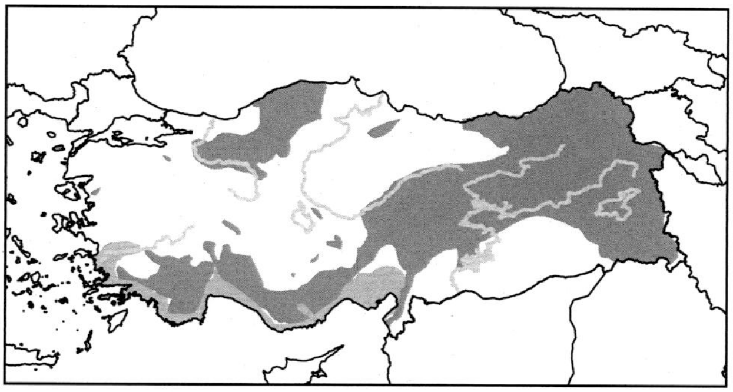

Breeding. Frequents areas with cliffs and rock outcrops, mainly in montane areas with open ground; also in similar habitat nearer sea level; on sea cliffs in Black Sea Coastlands (Ogilvie 1954). Breeds in small colonies, typically 2–5 pairs, sometimes solitarily, but 15 were nesting at one site in Göreme (Inner Anatolia) and there were 20 nests at Cimil Dağı (Black Sea Coastlands) (R. F. Porter). Nests mainly on cliffs, usually more than 6m above ground, and usually under and close up against a slight overhang, creating a sheltered niche; also in roofs of caves and recesses in cliffs; commonly under bridges over rivers, in quarries, and in road tunnels; also on walls of ruins and tall buildings; and a pair nested on a high concrete wall of a canal in central Adana on 5 May 1987 (van der Have *et al.* 1988). Nest a half cup, attached as a bracket to a vertical cliff; of mud pellets, thickly lined with feathers and some grass, and closely resembles that of Barn Swallow *Hirundo rustica*, but lacks the obvious binding material of grass stems which is mixed into the mud nest of the latter. Eggs: no data; usual clutch, elsewhere, 4–5. Young: no data. In **East Anatolia**, two nests at Nemrut Dağ were 8m and 12m above ground and *c.*30m apart; on 8 June 2004, an adult was incubating at a nest and an adult was lining the other with white feathers (PC); nests were on the walls of Van Castle on 9 May 1990; several adults were collecting mud in a valley near Görentaş on 31 May 1969, and there were two active nests in a road tunnel, near Hakkari, on 4 June 1969. In **Black Sea Coastlands**, there were occupied nests on 20 July 1994 at Sumela Monastery (I. A. Green); at Çatalağzi, a few pairs nested in sea cliffs, with Common House Martins *Delichon urbicum*, which vacated the site *c.*9 September, the Crag Martins remaining later (Ogilvie 1954); 20 active nests were at Cimil Dağı on 12 June 1975, and nest building was observed at Kopdağı pass on 15 June 1975 (R. F. Porter); at Ispir, on 12 June 2004, an adult fed chicks in a nest *c.*15m above ground; nearby *c.*5 pairs were incubating inside a short road tunnel, and several House Martins were nesting at the entrance, but not inside the tunnel, clearly differing in preferences for nest sites (PC). In **Inner Anatolia**, there was a colony on the walls of a hotel at Kızılcahamam, and occupied nests under a nearby bridge on 23 May 1982 (Baris *et al.* 1984), and adults were gathering mud there on 4 June 2002 (H. & J. Eriksen); a pair was nest building, and another adult incubating nearby, at Ereğli on 25 May 1998 (T. Quittenden); an occupied nest was at Karadağ on 11 June 1907 (Ramsay 1914). In **Mediterranean**, three nests (under a bridge, in a tunnel, and on a rock face), were being constructed at Çiftehan on 28 April 1970 (one complete, but presumably unlined, on 29 April 1970), and many pairs were building in the Aladağ on 29 April 1970 (R. F. Porter); there were occupied nests at Demirkazık on 1 May 2000 and on 28 May 1998 (T. Quittenden). In **Aegean**, some pairs were nesting near Pamukkale on 13 May 1974. About ten adults were collecting mud at 2,300m on Uludağ (**Marmara**) on 29–30 May 1966. Elsewhere double-brooded; the breeding period, from laying of the first egg to fledging, takes *c.*6 weeks. Adults return to breeding sites in Turkey early April–early May (later in some areas) and lay the first eggs mainly early May–mid June, with first-brood young fledging mid June to late July; some later records possibly relate to second broods. The habit of nesting on buildings is increasing in Bulgaria, with up to 25% doing so in recent years (Uhlig 1992).

BARN SWALLOW *Hirundo rustica*
Kır Kirlangıcı

(Du: Boerenzwaluw; Fr: Hirondelle rustique; Ger: Rauchschwalbe; Swe: Ladusvala)

Subspecies. All are regarded as nominate *rustica* by Roselaar (1995), though tending towards *transitiva* found in Lebanon, Israel and Jordan, presumably explaining records of single birds between Burdur and Afyon (Mediterranean/Inner Anatolia) in early May 1973 (OST 1975) and at the Göksu Delta (Mediterranean) in late September 1974 that apparently resembled the Levant race (OST 1978).

Status and distribution. National conservation status: Least Concern (population apparently stable). A widespread and common summer visitor, breeding throughout Turkey. Nests around habitation from sea level to at least 2,000m, from late April or early May.

More abundant on passage, when enormous numbers congregate at favoured reedbed roosting sites, especially in autumn, e.g. 100,000 at Emir Gölü, Ankara province (Inner Anatolia) on 21–25 August 1974 (OST 1978). Huge numbers also move on a broad front throughout the country at both seasons, e.g. at the Bosphorus/Dardanelles (Marmara) in autumn, Nisbet & Smout (1957) estimated that possibly half a million birds passed through this region in just two days in late September 1956. Spring arrival commences early March (3rd) in southern, western and central areas, but can apparently occur up to three weeks later in northern and eastern localities. There are a total of six February records (earliest 1st). Passage persists until early or mid June in northern areas (e.g., until 4 June at Ereğli, Black Sea Coastlands; Albrecht 1986). A particularly impressive movement of 500,000 birds was recorded at Alanya, Antalya province (Mediterranean) on 1–3 May 1975 (OST 1978). In autumn passage occurs from mid July (14th) to mid October (latest 22nd in Black Sea Coastlands), peaking in September and persists into late November (17th) or even early December (5th). There are six winter records: single birds at the Göksu Delta (Mediterranean) on 31 January 1995 and at Antakya (Mediterranean) on 31 December 2000 (A. Atahan), with three at the latter on 2 January 2006, singles there on 29 and 31 December 2007, and one on 2 January 2008 (A. Atahan).

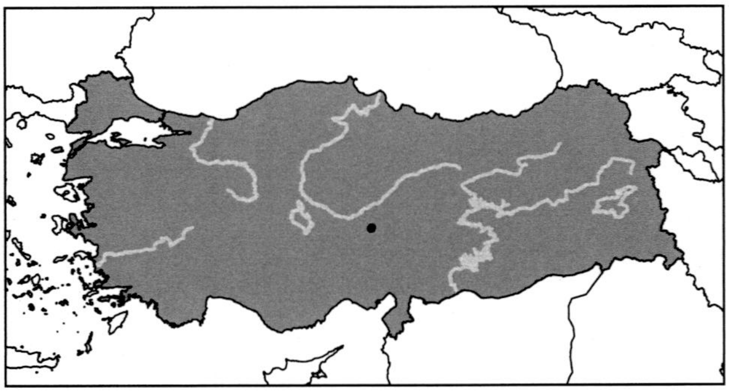

Breeding. Frequents open country near water, usually near habitation, where it nests in barns and buildings, under bridges and under verandas in small towns, e.g. Dalyan (Mediterranean). There are no records in Turkey of nests in natural sites such as caves in cliffs. The nest is usually attached to a vertical surface, but partially supported by a ledge or rafter; a cup of mud pellets, mixed with grass and lined with feathers. Eggs: C/2 (1), C/4 (1), C/5 (1). Young: B/5 (1). In **Mediterranean**, an adult was incubating just two eggs (a small clutch) in an underground bunker near Adana on 23 May 2004 (PC); at Dalyan, there was nest building on 28 March 1986, four occupied nests on 29–30 March 1986, and nests also occupied June 1987 (Kılıç & Kasparek 1989); there were many nests there in 1999, at all stages, with adults incubating and large young in nests on 29 June (PC); in Çukurova, breeding commenced in early April, in 1987, and nests had young on 7 May and 19 May (van der Have *et al.* 1988). In **Aegean**, a nest ready for eggs and three others with incomplete clutches were at Bafa Gölü on 1 May 2003 (PC); there were two almost complete nests at Salıhlı on 27 April 1970 (R. F. Porter); a nest had five fresh eggs near Izmir on 28 May 1951 (McNeile); there are several records of adults incubating from 3 April; large young, ready to fledge, at Marmaris on 10 May 2003, indicates egg laying early April (PC). In **Marmara**, first adults returned at Manyas Gölü on 14 April 1967, and a nest had four eggs on 13 June 1966 (R. F. Porter); there was an occupied nest at the border post with Bulgaria on 26 April 1969. In **Inner Anatolia**, there was nest building at Çavuşçu Gölü on 29 April 2005 (PC), breeding in several villages near Hotamış 10–15 May 1991 (Kirwan 1993) and adults feeding fledged young at Kulu town on 12 June 1998 (GMK). At Kızılırmak Delta (**Black Sea Coastlands**), a nest was occupied on 7 June 1992 (Hustings & van Dijk 1994). In **Southeast Anatolia**, adults were collecting mud near Viranşehir on 7 May 1970. In **East Anatolia**, two pairs were nesting near Hakkari on 2 June 1969 and a pair was nesting near Van on 30 May 1969 (R. F. Porter). Clearly double-brooded in Turkey, as elsewhere.

RED-RUMPED SWALLOW *Cecropis daurica*
Kızıl Kırlangıç

(Du: Roodstuitzwaluw; Fr: Hirondelle rousseline; Ger: Rötelschwalbe; Swe: Rostgumpsvala)

Subspecies and taxonomy. *C. d. rufula* occurs throughout (Roselaar 1995). Formerly placed in *Hirundo*, but Sheldon *et al.* (1999) suggested that this species should be removed to the genus *Cecropis* (see also *Ibis* 147: 821–826, Collinson 2006).

Status and distribution. National conservation status: Least Concern (population apparently stable). Moderately widespread and not uncommon summer visitor to Marmara (including Thrace), Aegean, Mediterranean and in the western half of Southeast Anatolia and the Mardin Mountains at least. Also breeds, very locally, in parts of the Black Sea Coastlands and Inner Anatolia. Locally outnumbers Barn Swallow *Hirundo rustica* as a breeder e.g. in Çukurova deltas (van den Berk *et al.* 1993). Often found in hilly country, locally up to *c.*1,050m (e.g. around Malatya) during the breeding season, although mainly nests from sea level to 500m.

Slightly more widespread on passage, but only very rarely recorded from western areas of East Anatolia. On spring passage recorded from late March (exceptionally 14th) to mid May, generally in small numbers, although *c.*1,000 moved northeast at the Göksu Delta (Mediterranean) on 14 May 1993 (GMK). In 1978, spring passage at Ereğli (Black Sea Coastlands) was apparently completed as early as 11 April (Albrecht 1986) though, in 1992, birds were observed until 24 April in Kızılırmak Delta (Hustings & van Dijk 1994). In autumn most have departed by mid September, stragglers remaining until early October (11th).

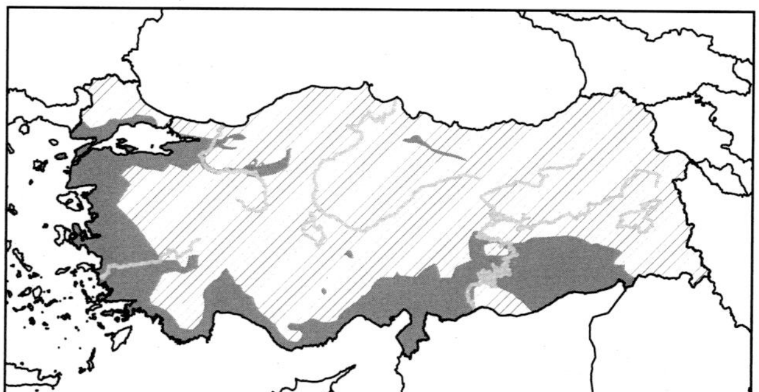

Breeding. Near water in open country, mainly in lowlands. The nest is attached to the underside of a horizontal surface, especially on bridges and culverts, buildings and ruins, but also in natural sites, e.g. caves and recesses in cliffs; it is an enclosed, rounded, bowl of mud pellets, with an entrance spout extended along the surface, and lined mainly with feathers. Eggs: McNeile took six fresh eggs from a nest near Ephesus (**Aegean**) on 27 April 1950, and the bird subsequently repaired the nest, and laid a further three eggs (possibly an incomplete clutch) by 11 May 1950; he also found a nest with three eggs near Izmir on 21 May 1951. Eggs at The Manchester Museum were collected by Krüper in the Izmir area on 10 June 1871 (two) and 2 June 1912 (one). Elsewhere in Aegean, there are four records of nest building, mainly in early May, with several occupied nests in May. In **Marmara**, nest building was seen at Gulpınar on 26 June 1973 and at Uvecık on 27 June 1973 (R. F. Porter), and Rokitansky & Schifter (1971) described a nest at Güngörmez on 29 July 1968. In **Southeast Anatolia**, there were adults at nests in June, and fledglings on 8 July 1986 (R. Webb) and 10 August 1998 (A. Barfull). In **Mediterranean**, nest building observed in May, with an early record at Dalyan on 7 April 2003 (B. Swann), and there are many records of occupied nests from early May. Almost certainly double-brooded in Turkey, as elsewhere.

COMMON HOUSE MARTIN — *Delichon urbicum*
Ev Kırlangıcı

(Du: Huiszwaluw; Fr: Hirondelle de fenêtre; Ger: Mehlschwalbe; Swe: Hussvala)

Subspecies. Indeterminate. All specimens (three) examined by Roselaar (1995) were intermediate between nominate *urbicum*, but marginally closer in wing length to *meridionalis* of the Mediterranean basin east to Iran. David & Gosselin (2002b) demonstrated that the correct spelling of the specific name is *urbicum*.

Status and distribution. National conservation status: Vulnerable (population apparently decreasing). A widespread and common summer visitor, breeding in all regions, often in large colonies, although generally more local and uncommon in much of Inner and East Anatolia. Nests from early or mid April, from sea level to at least 3,000m.

Common on passage, usually through coastal areas but occurs on a broad front: in spring from early to late March (earliest 2nd, in Black Sea Coastlands) to early, even late, May (passage complete at Ereğli, Black Sea Coastlands by 15 May in both 1977 and 1978; Albrecht 1986), and in autumn from late July (21st) to late October (27th), but very rare in the second half of October. Over 2,000 moved through Borçka (Black Sea Coastlands) on both 28 September and 2 October 1980. There are eight spring records, four single birds at coastal sites in Aegean and Mediterranean in the last week of February 1974 (OST 1978), one in Ankara (Inner Anatolia) on 25 February 1996 (KAB), four at Antalya (Mediterranean) on 23 February 2000 (T. Albayrak) and single birds at Fethiye (Mediterranean) on 10 February and 28 February 1998 (P. Hope), which were unusually early.

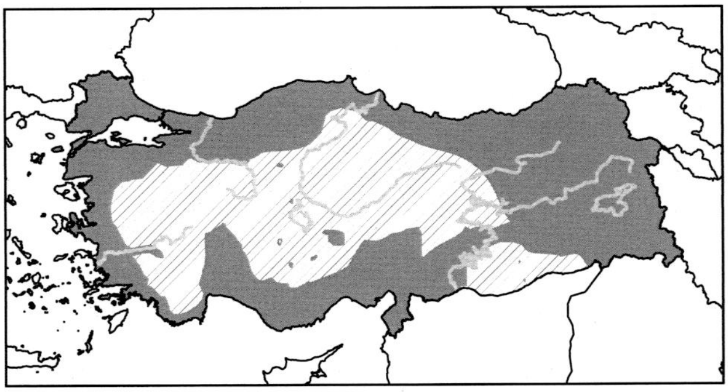

Breeding. In open country, from sea level in all regions with a coastline, to 2,000m on Uludağ (Marmara), 2,400m in the Aladağ (Mediterranean), and 3,000m in the east. Nests on buildings and structures, e.g. bridges, on rock outcrops, gorges and cliffs, both inland and coastal, in overhangs of cliffs, and in recesses or shallow caves, but not in deep caves; at Ispir (Black Sea Coastlands), Eurasian Crag Martins *Ptyonoprogne rupestris* were nesting inside a road tunnel (a deep artificial cave), whilst the House Martins were nesting at the entrance but not inside (PC). Usually in colonies, commonly of 20–50 pairs on large buildings, or smaller numbers (2–5 pairs), but 400 nests were in a colony on sea cliffs near Çatalağzi, Black Sea Coastlands (Ogilvie 1954). The nest is a rounded structure of mud pellets, strengthened with plant fibres, on a vertical surface and so close to an overhanging projection that only a narrow entrance slit is present at the top; the base of the interior is usually lined with dry grass and feathers, but adults were taking thin woolly string, the casing for a wire cable, to use in the nest lining, at Oludeniz (Mediterranean), on 6 May 2003 (PC), and moss was frequently picked from branches of *Fagus orientalis* in northeast Turkey (J. Faldborg). Nests usually separate, but may also be touching or joined together, typically a line or row under the eaves of a building, but not a contiguous mass, as in Little Swift *Apus affinis*. Nests remain in reasonable condition for some years, and are subsequently used by House Sparrows *Passer domesticus* and Little Swifts. In **Aegean**, early stages of nest building were noted at Altınkum on 24 April 2002, at Pamukkale on 7 May 2003 and at Marmaris on 20 April 2004, but a nest was virtually complete at Altınkum on 29 April 2003 (PC). In **Mediterranean**, nest building was seen as early as 15 April 1993 and early May (at the coast) and, in the Aladağ, on 8 June 1998 (T. Quittenden) and July/August 1966. In **Marmara**, there were ten occupied nests at the customs post with Bulgaria on 26 April 1969, and nest building was seen at 2,000m on Uludağ on 18 May 1969 (R. F. Porter). Most records are of active colonies in May–July. Late records include young being fed in nests, in **Black Sea Coastlands**, between Rize and Ispir on 14 August 1972 and at Yeniçağa on 18 August 1985 (Kılıç & Kasparek 1987), and active colonies at Horasan and Tahir (**East Anatolia**) on 3 August 1974 (T. A. Walsh), at Uludağ on 16 August 1987, and at Mut (Mediterranean) in early August 1957 (Ballance 1958). Clearly double-brooded in Turkey, as elsewhere.

RICHARD'S PIPIT *Anthus richardi*
Mahmuzlu İncirkuşu

(Du: Grote Pieper; Fr: Pipit de Richard; Ger: Australspornpieper; Swe: Större piplärka)

Subspecies and taxonomy. Presumably all records are referable to nominate *richardi* although no specimen material from Turkey has been examined. In recent years, despite some disagreement among different commentators, through the application of molecular techniques, the *Anthus novaeseelandiae* complex has increasingly been viewed as paraphyletic. Given that morphological and vocal characters provide a degree of support to this result, Richard's Pipit has recently been viewed as between four and six species (see e.g. Sibley 1996). For now, a more conservative line, such as that adopted by Alström & Mild (2003) appears sensible; these authors preliminarily recognise four species-level groups: the *cinnamomeus*-group in Africa and southwest Arabia (Grassland or African Pipit), the *rufulus*-group in southern Asia (Paddyfield Pipit), *richardi* in northern Asia and China (Richard's Pipit) and the *australis*/*novaeseelandiae*-group (Australian/New Zealand Pipits, which may represent two distinct groups with separate evolutionary lineages). In our view, this represents a potentially minimalist view of the number of taxa that may prove worthy of recognition at species level under either the Phylogenetic or Evolutionary Species Concepts. Indeed, some authors have considered the form *camaroonensis* from West Africa also to be specifically distinct, but this suggestion was vigorously refuted by Dowsett & Dowsett-Lemaire (1993). The latter authors also tackled the problems raised by other forms of large pipit in Africa that have been elevated to species rank in recent years, with apparently similarly little justification. However, it is clear that much work remains to be done with respect to the African, Indonesian and Australasian taxa within the complex. For an overview of some of the problems presented by the group in its widest sense, readers are referred to Alström & Mild (2003) and the references cited therein.

Status and distribution. A regular, but perhaps very scarce passage migrant throughout Turkey in very small numbers and an occasional winter visitor. The bias toward records in spring perhaps reflects increased observer coverage during this season. Described as a winter visitor to both Asia Minor and Persia by Hartert (1910/22) and Hartert & Steinbacher (1932/38). Mathey-Dupraz (1922) reported an unusual pipit at Kağıthane, Istanbul on 25 March 1894; however, both Kumerloeve (1961) and Kasparek (1990a) were unsure whether it was this bird's skin that was subsequently lodged in the Robert's College, Istanbul collection (as this species). The earliest occurrence of this species is therefore rather unclear. This specimen appears to have been lost in recent years: it could not be found in February 1996 (Kirwan 1997a).

Subsequently, there have been at least 29 published records (several involving more than one individual), suggesting that the species may be an occasional passage migrant in small numbers. Spring records (20 March–4 June) are more numerous but there are five autumn occurrences (12 September–12 November), discounting the possible seen by Gallner (1976) in late August and the south-coast record of 1 December 1969, which may be better treated as a winter record. The majority are from regularly watched coastal deltas, with seven from the Göksu Delta

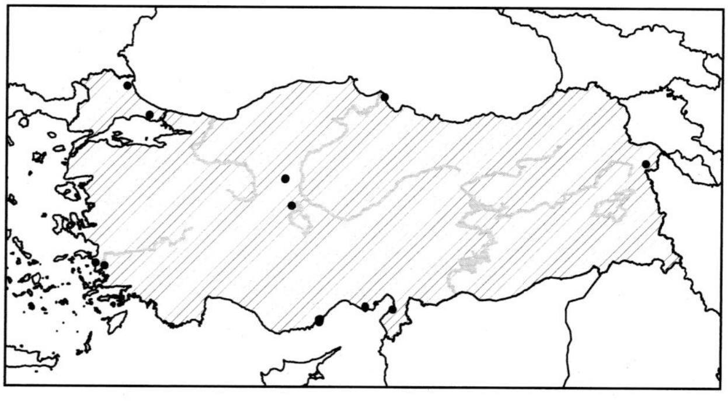

area (Mediterranean) and a remarkable ten (involving 13 individuals) from Kızılırmak Delta (Black Sea Coastlands), all but one in spring 1992. Systematic observations from this area in autumn are limited. Regarded as a scarce passage migrant and winter visitor to Cyprus, but its status in the wider Middle East is rather unclear (Hollom *et al.* 1988, Porter *et al.* 1996).

Table 4. Published records of Richard's Pipit *Anthus richardi* in Turkey since 1966.

Locality	Region	Date	No. of birds	Source
Silifke	Mediterranean	27 March 1967	2	OST (1969)
Beynam, Ankara	Inner Anatolia	9 April 1967	1	OST (1969)
Karakot Gölü (near Iskenderun)	Mediterranean	1 December 1969	2	OST (1972)
Doğubeyazıt	East Anatolia	12 November 1970	1	OST (1975)
Göksu Delta	Mediterranean	8 April 1971	2	OST (1975)
Akbulak	East	25 August 1973	Possible	Gallner (1976; not mapped)
Büyük Menderes Delta	Aegean	2 April 1986	2	Kasparek (1990a)
Göksu Delta	Mediterranean	8 April 1986	1	Kasparek (1990a)
Göksu Delta	Mediterranean	5 and 8 April 1988	1	Kasparek (1990a)
Doğubeyazıt	East Anatolia	31 May 1988	1, perhaps 3	Kasparek (1990a)
Kızılırmak Delta	Black Sea Coastlands	20 March 1992	1	Hustings & van Dijk (1994)
Kızılırmak Delta	Black Sea Coastlands	31 March 1992	3	Hustings & van Dijk (1994)
Kızılırmak Delta	Black Sea Coastlands	3 April 1992	2	Hustings & van Dijk (1994)
Kızılırmak Delta	Black Sea Coastlands	5 April 1992	1	Hustings & van Dijk (1994)
Kızılırmak Delta	Black Sea Coastlands	17 April 1992	1	Hustings & van Dijk (1994)
Kızılırmak Delta	Black Sea Coastlands	25 April 1992	1	Hustings & van Dijk (1994)
Kızılırmak Delta	Black Sea Coastlands	27 April 1992	1	Hustings & van Dijk (1994)
Kızılırmak Delta	Black Sea Coastlands	29 April 1992	1	Hustings & van Dijk (1994)
Kızılırmak Delta	Black Sea Coastlands	4 June 1992	1	Hustings & van Dijk (1994)
Kulu Gölü	Inner Anatolia	2 October 1994	1	Kirwan & Martins (2000a)
Çukurova	Mediterranean	12 October 1994	4	Kirwan & Martins (2000a)
Göksu Delta	Mediterranean	11 April 1995	2	Kirwan & Martins (2000a)
Near Iğneada	Marmara	12 September 1996	1	Kirwan & Martins (2000a)
Bucak–Heraklia, near Bafa Gölü	Aegean	30 September 1996	3	Kirwan & Martins (2000a)
Göksu Delta	Mediterranean	27 April 1997	1	Kirwan *et al.* (2003)
Kızılırmak Delta	Black Sea Coastlands	29 April 1998	1	Kirwan *et al.* (2003)
Göksu Delta	Mediterranean	21–22 April 2004	15	E. & S. Durand
Büyükçekmece	Marmara	19 September 2006	1	M. Ullman (*in litt.*)
Samandağ	Mediterranean	29 December 2007	1	(*per* A. Atahan; photographed)

TAWNY PIPIT — *Anthus campestris*
Kır İncirkuşu

(Du: Duinpieper; Fr: Pipit rousseline; Ger: Brachpieper; Swe: Fältpiplärka)

Subspecies. *A. c. boehmei* occurs throughout Turkey according to Roselaar (1995). However, like Alström & Mild (2003), we prefer to view geographical variation within this species as not sufficiently marked to warrant nomenclatural recognition, and therefore consider it monotypic.

Status and distribution. National conservation status: Least Concern (population apparently decreasing). A fairly widespread and not uncommon summer visitor breeding in all regions, but distinctly local in Black Sea

Coastlands (see below; with single birds in suitable breeding habitat at Kızılırmak Delta in early June 1992: Hustings & van Dijk 1994), and most commonly in Aegean, Thrace, Inner and East Anatolia. Breeds in steppe, semi-steppe and cultivated areas, principally in uplands to 3,000m, but also more locally down to sea level.

More widespread and locally numerous on passage, when also widely recorded in Black Sea Coastlands, from mid March (10th) to mid May (15th) in spring and from mid August (late July in north) to mid or late October (27th) in autumn, exceptionally to 12 November in Mediterranean and 28 November in Black Sea Coastlands (Ogilvie 1954). Occasionally observed in coastal localities in the western two-thirds of Turkey in winter, e.g. single birds at Miletos (Aegean) on 19 January 1969, at Alanya (Mediterranean) on 20 February 1972, at Kızılırmak Delta (Black Sea Coastlands) on 4 January 1973 and Göksu Delta (Mediterranean) on 21 December 2007.

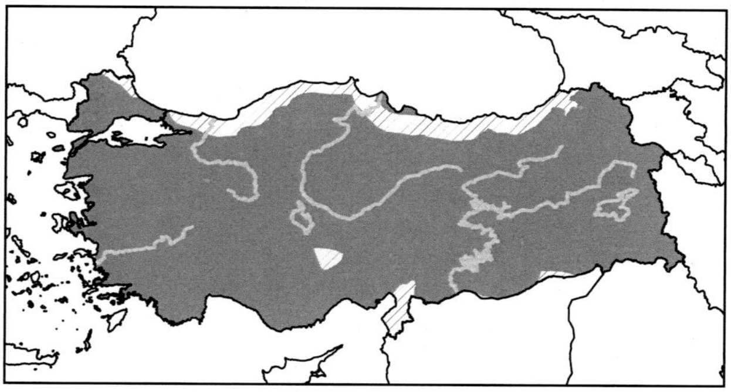

Breeding. In open dry areas with sparse vegetation and bare ground, typically arid steppe and dry cultivation; also on hillsides and saltmarshes, from sea level to 3,000m. There were 23 pairs in a 13-km stretch of the west side of Manyas Gölü (Marmara) in late May 1966. The nest is a hollow in the ground, usually sheltered by a tuft of vegetation or a low rock or bush; a cup of grass and roots, lined with plant fibres, fine grass and hair. Eggs: C/4 (1), C/5 (4). Young: no data. In the Bolu area (**Black Sea Coastlands**), a nest had five eggs on 21 June 1977 and young on 29 June 1977, and a fledged young was nearby in June 1977 (Pforr & Limbrunner 1982). In **Aegean**, a nest had five eggs at Büyük Menderes Delta on 29 May 2004 (M. Demain). In **Marmara**, a female collected nesting material, while the male sang nearby, near Manyas Gölü, on 25 May 1967, a pair was carrying nesting material, also near Manyas Gölü, on 15 May 1969, and a pair was carrying food in Forest of Belgrade on 18 June 1973 (R. F. Porter). In **Inner Anatolia**, an adult and a juvenile were at Kulu Gölü on 7 June, and two nests, each with five eggs, were at 1,500–1,600m on Karadağ on 31 May 1972 and 1 June 1972; a pair and three young were at Susuz on 27 July 1969, and two juveniles at Ankara on 2 August 1969; an adult feeding young was at Kızılcahamam on 9 July 1962 (Baris *et al.* 1984). In **East Anatolia**, a nest had four eggs at Erçek Gölü on 1 June 1968 (Kasparek & van der Ven 1983); two juvenile Common Cuckoos *Cuculus canorus* were observed begging food, apparently from a Tawny Pipit, at Kağızman on 16 August 1972. Males heard singing as late as July, and double-brooding possible (occasional elsewhere).

OLIVE-BACKED PIPIT — *Anthus hodgsoni*
Yeşil Sırtlı İncirkuşu

(Du: Siberische Boompieper; Fr: Pipit à dos olive; Ger: Waldpieper; Swe: Sibirisk piplärka)

Subspecies. The one Turkish record presumably involved the form *yunnanensis* (which breeds east from the White Sea to Japan and south into China, and is that recorded elsewhere in the Middle East and Europe) rather than the principally Himalayan *hodgsoni*.

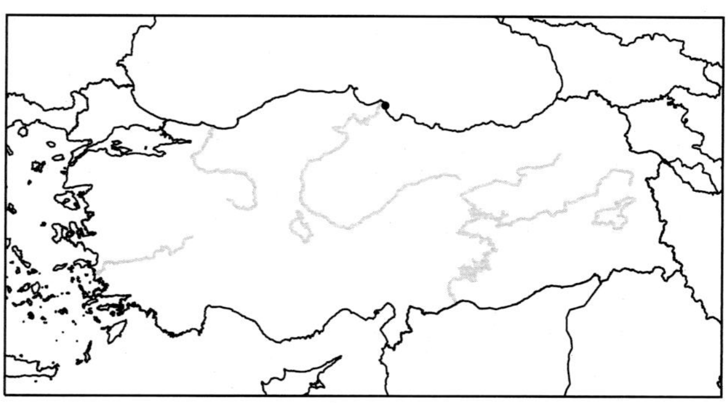

Status and distribution. Vagrant. One was seen well together with a Tree Pipit *A. trivialis* at Kızılırmak Delta on 13 April 1992 (Kirwan 1993b). There are single records from Cyprus in February

1989 (Flint & Stewart 1992) and in January 1992 from Iran (Scott 1994). Scarce elsewhere in the Middle East (Porter *et al.* 1996), with most records from Israel (Shirihai 1996) and the United Arab Emirates (Richardson 1990).

TREE PIPIT *Anthus trivialis*
Ağaç İncirkuşu

(Du: Boompieper; Fr: Pipit des arbres; Ger: Baumpieper; Swe: Trädpiplärka)

Subspecies. Nominate *trivialis* occurs both as a breeder and on passage (Roselaar 1995). Clancey (1987) considered Turkish breeding birds to belong to a previously undescribed subspecies, *differens*, based on specimens from Bolu Dağ, Rize, above Eleşkirt and at Sarıkamıs, but this taxon is regarded as a synonym of *trivialis* by Roselaar (1995) and Alström & Mild (2003), it being very similar to specimens from northern Europe also in heavily worn plumage.

Status and distribution. National conservation status: Data Deficient (population trend unknown). A rather local but not uncommon summer visitor, breeding in Marmara (including Thrace) and throughout the Black Sea Coastlands, and probably many immediately adjacent parts of Aegean, Inner and East Anatolia south to the Upper Aras Valley. Records in mid July and on 12 August from Tatvan, noted by Kumerloeve (1969a) and Roselaar (1995), are assumed here to refer to non-breeders, but the species may have bred in the southwest Konya Basin (Inner Anatolia) in 1998 (Eken & Magnin 1999). Found principally in deciduous or mixed woodland, and adjacent scrubby areas, in montane areas to 2,600m during the breeding season. Breeding areas occupied from early April.

Common and widespread on passage throughout Turkey, occasionally in very large numbers (e.g. up to 900 per hour moved southeast between Izmit and Çamlıca, Black Sea Coastlands, on 5 October 1973). In spring recorded from mid March (10th), more usually late March, to early June (7th), peaking in the second half of April, and passage probably largely completed through southern and central areas by mid May. Autumn passage commences from mid August (late July in the north), persisting until the second half of October (27th), peaking in the latter half of September and early October. Late birds are regularly reported through October (29th), occasionally until mid November (12th). Overwintering has been noted (Kumerloeve 1961, 1970a), but there is only one recent record at this season: one at the Büyük Menderes Delta (Aegean) on 17 January 1969 (OST 1972).

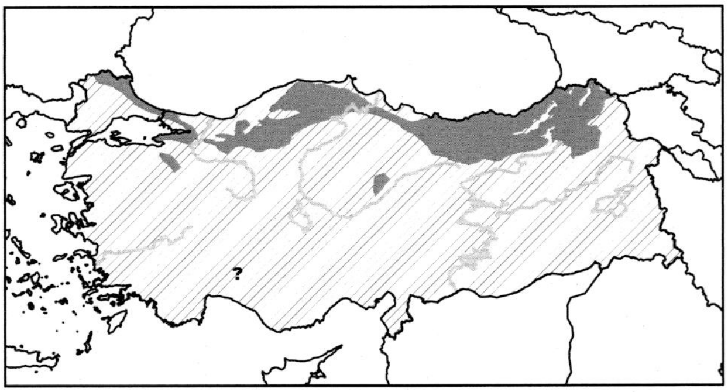

Breeding. Frequents areas with patches of bare ground, scattered low scrub and usually some trees; in open woodland (mixed or deciduous), edge and clearings, sometimes in virtually treeless areas at higher altitude; at Yeniçağa Gölü (Black Sea Coastlands) in poplar plantations and willows, but also in mountains south of the lake (Kılıç & Kasparek 1987); at Uludağ (Marmara) in the transition zone between alpine forest and open alpine scrub (Jetz 1995). Nest and clutch size unrecorded in Turkey but elsewhere the nest is on the ground, usually sheltered by a low shrub or grass tuft; a cup of grass with some moss, lined with grass, hair and plant fibres; usual clutch 4–6. In Turkey, adults are on territory mainly from the second half of April, with regular singing reported to mid July, and as late as 18 August 1985 at Yeniçağa Gölü (Kılıç & Kasparek 1987), which may indicate double-brooding. In **Black Sea Coastlands**, a pair was carrying food to a nest in oak scrub at 2,300m at the Kop Dağı pass on 15 June 1975; an adult was carrying food between Ispir and Erzurum on 13 June 2001; and adults were carrying food near Trabzon on 16 June 1975 and near Torul on 17 June 1975 (R. F. Porter); these records indicate egg laying in the second half of May. In **East Anatolia**, a pair and fledged young were at Tendürek Geçidi on 8 May 2004 (GMK), indicating laying of first egg about 10 April; an adult was alarm-calling and carrying food in a treeless area, with only willow scrub, c.1m high, at Sarıcan, on 25 June 2004 (PC).

MEADOW PIPIT — *Anthus pratensis*
Çayır İncirkuşu

(Du: Graspieper; Fr: Pipit farlouse; Ger: Wiesenpieper; Swe: Ängspiplärka)

Subspecies. Monotypic following Alström & Mild (2003).

Status and distribution. A widespread and common spring and autumn passage migrant throughout Turkey (although much less common in the eastern third) and a winter visitor, apparently restricted to the western two-thirds. Recorded on spring passage during March and April, late birds occurring until mid May (20th). In autumn from mid September (17th) to early November, usually in small or moderate numbers (e.g. 86 at the Bosphorus between 9 October and 5 November 1966), but a total of 1,134 moved through the Kızılırmak Delta (Black Sea Coastlands) between mid March and mid May in 1992. Gallner (1976) claimed one in East Anatolia as early as 24 August. Present in wintering areas from late September, more typically mid October, until mid April (latest 22nd). In Inner Anatolia and Black Sea Coastlands, passage involving birds from further south is evident in early April. An apparent preference for wetland and coastal areas in winter almost certainly reflects observer coverage. Occasional records, largely from northern Turkey in June and July, indicate that some non-breeders may irregularly oversummer (see, e.g., Jetz 1995). However Boyd's (1916) claim of breeding at Gallipoli (Marmara) is insufficiently substantiated.

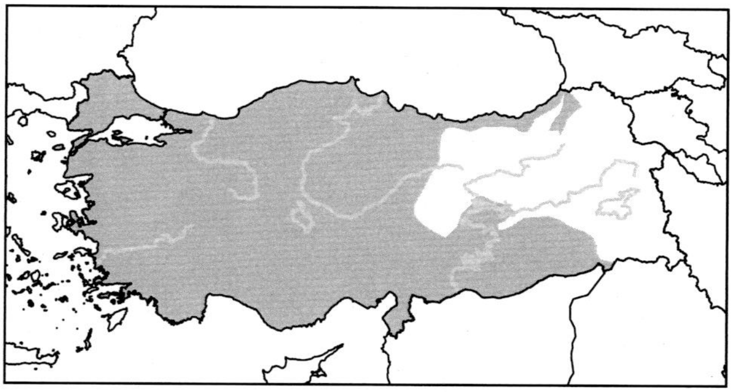

RED-THROATED PIPIT — *Anthus cervinus*
Kızıl Gerdanlı İncirkuşu

(Du: Roodkeelpieper; Fr: Pipit à gorge rousse; Ger: Rotkehlpieper; Swe: Rödstrupig piplärka)

Subspecies. Monotypic following Alström & Mild (2003).

Status and distribution. A widespread and common spring and autumn passage migrant, principally through the western two-thirds of Turkey, often in considerable numbers. Spring passage is from mid March to late May (26th), peaking strongly in late April and early May (e.g. over 1,000 at Palas Gölü, Inner Anatolia, in late April 1998), with comparatively few thereafter. Rare in Inner Anatolia and further north prior to early April. Autumn passage is from late August until late October, peaking strongly in late September and early October, with much smaller numbers occurring into the second half of November (18th). Regularly recorded in winter, generally in very small numbers, principally in coastal wetlands in Mediterranean and southern parts of Aegean. More rarely

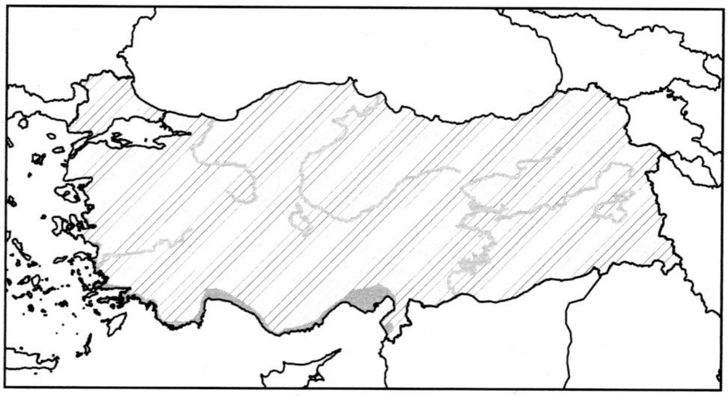

recorded in winter in southern areas of Inner Anatolia. Perhaps absent throughout in very harsh winters. There is one December report from Marmara, at Büyükçekmece, Istanbul, on 8 December 1990 (KAB *et al.*).

WATER PIPIT *Anthus spinoletta*
Dağ İncirkuşu

(Du: Kaukasische Waterpeiper; Fr: Pipit spioncelle; Ger: Bergpieper; Swe: Vattenpilärka)

Subspecies. *A. s. coutellii* apparently occurs throughout Turkey; specimens from northeast Turkey and Uludağ, Marmara being indistinguishable from this subspecies (Roselaar 1995). Birds taken in autumn or winter from three localities in western and central Turkey most closely resemble nominate *spinoletta*.

Status and distribution. National conservation status: Least Concern (population apparently stable). A fairly widespread, presumably resident breeder and partial/altitudinal migrant. Generally in small numbers. In the breeding season found in montane areas, generally from 1,500m to at least 3,200m, in all regions except Thrace and Southeast Anatolia. There is only one published record from the latter; at Birecik on 6 April 1967 (Kumerloeve 1970a). Commonest in the eastern Black Sea Coastlands and adjacent parts of East Anatolia.

Descends as low as sea level in autumn and winter, generally from mid October (though recorded at Yeniçağa Gölü as early as 1 September; Kılıç & Kasparek 1987) to late March. Rarely recorded in coastal areas later than the first week of April, although present in Çukurova, Mediterranean, until 3 May in 1990, but noted at wintering areas in Inner Anatolia as late as mid April. In winter more widespread in wetland and coastal areas throughout the western two-thirds of Turkey and more rarely further east. Numbers in winter are presumably augmented by immigrants from elsewhere (see discussion of subspecies above), although there is little evidence of passage from northern, coastal localities.

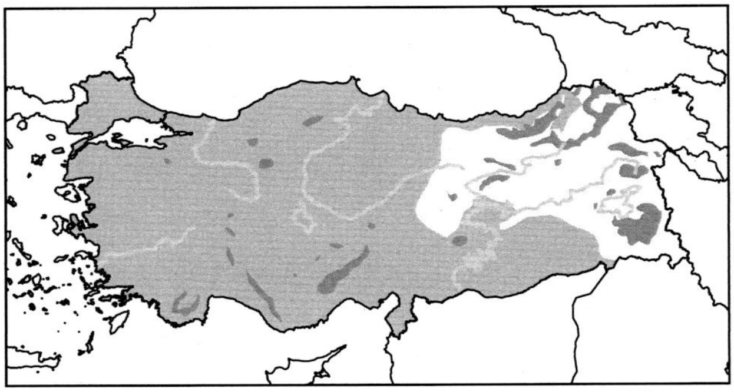

Breeding. Inhabits open alpine meadows, often with patchy low scrub, between the treeline and the snowline, above 1,800m (1,500–3,500m) in high mountains. At Uludağ (**Marmara**) it is the commonest breeder in open alpine scrub and meadows above 1,900m, breeding May–July, with 19 individuals in *c.*15ha on 18 May 1993 (Jetz 1995). An adult was carrying food here on 8 July 1966, and another carried food to a nest site on 17 May 1969 (R. F. Porter). A nest at *c.*2,700m above Sivrikaya (Black Sea Coastlands) was in a hollow in the side of a low overhanging bank and partially screened by a rock embedded in the bank; it was a cup of dead grass and plant stems, lined with finer grass and some hair, and had five eggs on 13 June 2004 (B. Pleasance); elsewhere, the outer cup often includes moss and the lining some leaves. Eggs: C/4 (2), C/5 (2). Young: no data. In **Black Sea Coastlands**, a nest at over 2,500m, near Çiçekli (50km south of Rize) on 6 May 1970 had five eggs, and two nests each had four eggs at Zigana pass on 14 June 1975. A pair was feeding two fledged young near the summit of Karadağ (**Inner Anatolia**) on 19 June 1996 (Kirwan 1998a), indicating egg laying in mid May. The species is double-brooded in other parts of the range.

(WESTERN) YELLOW WAGTAIL *Motacilla flava*
Sarı Kuyruksallayan

(Du: Gele Kwikstaart; Fr: Bergeronnette printanière; Ger: Schafstelze; Swe: Gulärla)

Subspecies and taxonomy. *M. f. feldegg* (breeding); *flava, lutea, cinereocapilla, beema, superciliaris, thunbergi* and, doubtfully, *dumbrowskii* (passage). The polytypic Yellow Wagtail offers an outstanding problem to taxonomists.

Whilst much has been written concerning this complex, in recent years two different groups of ornithologists have attempted to grasp the nettle. Firstly, Sangster *et al.* (1999) took the bold, but often-mooted move among birdwatchers, to recognise a large number of forms as species. Of those recorded (or at least claimed) in Turkey, the following were elevated to specific status: *feldegg* (Black-headed Wagtail), *flava* (Blue-headed Wagtail, including *beema*), *lutea* (Yellow-headed Wagtail), *cinereocapilla* (Ashy-headed Wagtail) and *thunbergi* (Grey-headed Wagtail). In contrast, *superciliaris* and *dumbrowskii* were both deemed to be the result of hybridisation. Though such a pronouncement offers a real 'field day' for listers, does it provide an accurate reflection of evolutionary history? Alström & Mild (2003) stated 'not quite'. These authors continued to recognise 13 subspecies of Yellow Wagtail, including all of the taxa reported from Turkey, with the exception of the always-dubious *dumbrowskii.* Their review, one of the most wide-ranging and in-depth yet performed, offers some interesting lines for future research. Genetic distances between all taxa are small, but the evidence of DNA sequencing (though somewhat contradictory in some respects to relationships among *Motacilla*, depending on the materials used) does support the hypothesis of a deep split between eastern forms (*taivana*, *macronyx* and *tschutschensis*) and the other, western taxa (note that *leucocephala* was not included in these analyses). (Rasmussen & Anderton 2005, on the basis of these data elected to treat *M. flava* as two species, Western *M. flava* and Eastern Yellow Wagtails *M. tschutschensis.*) Furthermore, morphological and vocal data support the conclusion that Yellow Wagtail is not monophyletic, which suggests that two or more species should be recognised. With appropriate caution, especially with regard to *thunbergi* versus *macronyx* and *lutea* with *flavissima*, Alström & Mild considered that *flava* (including *beema*), *lutea* (including *flavissima*), *thunbergi* (including *plexa*), *feldegg* (including *melanogrisea*), *cinereocapilla* (including *iberiae*, which the Dutch review had promoted to specific status, and *pygmaea*), *leucocephala*, *macronyx*, *taivana* and *tscutschensis* (including *simillima*, *angarensis* and *zaissanensis*) might all qualify as species, but at least for now they prefer subspecies treatment for all taxa within Yellow Wagtail.

Status and distribution. National conservation status: Least Concern (population apparently decreasing). Nonethless, it remains a widespread and common summer visitor, breeding in and around suitable wetland localities throughout the country, albeit scarcer and more local in the west and south and apparently absent from much of Southeast Anatolia. Commonest, occurring in high densities in freshwater marshes and associated wet grassland, in Inner and East Anatolia, although 1,500–2,000 pairs were estimated in Kızılırmak Delta (Black Sea Coastlands) in spring 1992. Breeds from mid April in the south and west, more usually from mid May in the eastern third of Turkey. Rarely recorded breeding above 2,000m, with no published records from above 3,000m.

Widespread and common on passage throughout the country. In addition to breeding *feldegg*, seven races have been reliably recorded in Turkey: *flava*, *thunbergi*, *lutea*, *cinereocapilla*, *beema* and *superciliaris*. All are relatively regular on passage, although the latter four are generally recorded in very small numbers. The race *dumbrowskii* has also been doubtfully reported (Kumerloeve 1961). The races *feldegg* and, to much lesser extent, *superciliaris* usually predominate in early spring (between mid March and mid April), with *flava* and *thunbergi* (scarcer) becoming more common into the second half of May, although both these latter forms are observed from the third week of March. Peak numbers are usually in the second half of April. Most records of *beema* are from mid or late April, although it has been recorded from late March (28th) and occurs well into May. Extreme spring passage dates for all forms are 10 March–25 May. Considerable attention needs to be paid to these forms to elucidate their spring phenologies, as current data are very few. Autumn passage commences in mid July to early August, peaking in mid September and diminishing through October (latest 19 November). Often recorded in very large numbers: e.g. 40,000 were at Aynaz swamp, Çukurova (Mediterranean) on 4 October 1968 (OST 1972) and up to 3,000 were moving east per hour at Mersin, Içel province (Mediterranean) between 30 September and 2 October

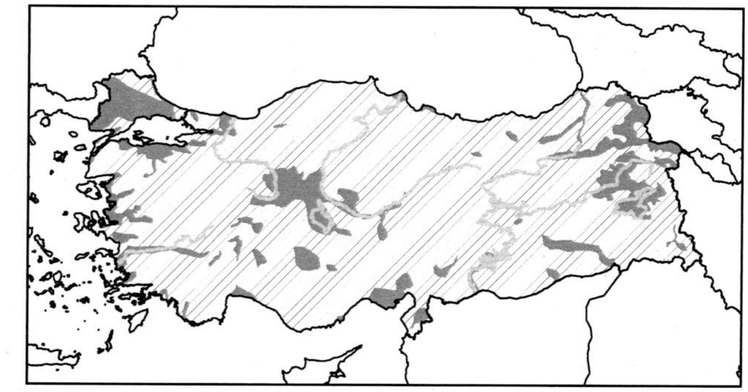

1974 (OST 1978). Occasionally recorded in winter (December–January) in lowlands in the western two-thirds of Turkey, although probably very irregular at this season.

Breeding. In damp pastures and meadows, freshwater marshes and coastal saltmarshes, and sometimes in damp areas with scrub, such as *Tamarix*. Nest a cup, in a hollow in the ground, sometimes in the hoof print of an animal, and usually well hidden in vegetation; of grasses, plant stems and roots, lined with hair. Eggs: C/6 (3), C/5 (1). Young: B/5 (2), B/6 (1). In **Black Sea Coastlands**, three nests had six, six and five eggs at Yeniçağa Gölü on 19–20 May 1992 (PC), and recently fledged young were there on 16 June 1978 (Kılıç & Kasparek 1987); at Kızılırmak Delta, in 1992, song commenced second half of April, nest building on 30 April, and many fledglings were seen in early June (Hustings & van Dijk 1994). At Göksu Delta (**Mediterranean**), a nest had six eggs on 5 May 2004 (L. Baxter), another had five young, about five days old, on 1 June 1993, and another six young, about six days old, on 22 May 2004 (PC). A nest had one egg (incomplete clutch) near Manyas Gölü (**Marmara**) on 17 May 1967 (R. F. Porter). At Van marshes (**East Anatolia**), a nest had five small young on 1 June 2001, and many adults were food carrying on 10 June 2001 (T. Quittenden). In **Inner Anatolia**, fledged young were observed at several sites from 19 June.

CITRINE WAGTAIL *Motacilla citreola*
Sarı Başlı Kuyruksallayan

(Du: Citroenkwikstaart; Fr: Bergeronnette citrine; Ger: Zitronenstelze; Swe: Citronärla)

Subspecies and taxonomy. Apparently all *werae* (GMK pers. obs.). A (presumed migrant) female taken at Beyşehir Gölü (see Vauk 1973 and below) is *werae* on measurements but no other specimen material from Turkey is apparently available (Roselaar 1995). H. Dufourny *in litt.* (2003) reported a pair, showing all of the features of nominate *citreola* at the Bendimahi Delta (East Anatolia) in late May 2003. However, Alström & Mild (2003) regard *werae* as a synonym of nominate *citreola*, the differences in plumage between them being insufficiently marked to warrant upholding *werae*. The same authors go on to remark that, under the Phylogenetic Species Concept, the form *M. c. calcarata* (which breeds from the Tien Shan Mountains eastwards) demands recognition at the specific level. In contrast, another phylogeographic study of 'yellow' wagtails, suggested that Citrine Wagtail comprises two species-level, western and eastern groups, *werae* on the one hand, and *citreola* and *quassatrix* on the other, which are apparently not even sisters (Pavlova *et al.* 2003). These authors did not include *calcarata* in their mtDNA analysis. Further, Alström & Mild (*op. cit.*) considered Citrine Wagtail *sensu lato* as being sister to the White Wagtail *M. alba* superspecies.

Status and distribution. National conservation status: Near Threatened (population apparently increasing). A relatively recent addition to the Turkish list, a female at Beyşehir Gölü (Mediterranean) on 12 May 1964 (Vauk 1973, Kasparek 1990a) was the first record, but now well established as a locally fairly common breeding summer visitor in both Inner and East Anatolia, and even more locally in Black Sea Coastlands, west as far as Abant Gölü, Bolu province. There were fewer than ten published records from Turkey prior to 1980, perhaps all of which related to passage migrants. Since 1981 breeding has been proven at several localities in East Anatolia (and suspected at others), especially around Van Gölü and Çıldır Gölü, as well as elsewhere in Kars province, e.g. at marshes around Ardahan and Doğubeyazıt, but also west as far as Erzurum. Subsequently, breeding was also proven or strongly suspected at many wetland areas in Inner Anatolia in the latter half of the 1980s and through the 1990s, e.g. Palas Gölü, Hotamış marshes, Ereğli marshes and Kulu Gölü. Observations from at least three other sites, in Southeast Anatolia, Mediterranean and even west as far as Aegean (Büyük Menderes Delta) during the breeding season indicate the possibility of further expansion. All breeding sites lie between 900m and 2,700m, but there are additional breeding-season records from altitudes down to sea level. Earlier observers in East Anatolia make no mention of this species, which was still regarded as a rare vagrant until the 1980s (OST 1975, 1978). It would appear that Turkey was swiftly colonised during the latter decade, reflecting the strong westward expansion observed in other areas on the western fringe of the species' range (Cramp 1988).

Also regular on passage, particularly in spring and principally through the eastern two-thirds of Turkey, from mid or late March (earliest published record 26th) to mid or late May (28th), peaking in mid April, but also

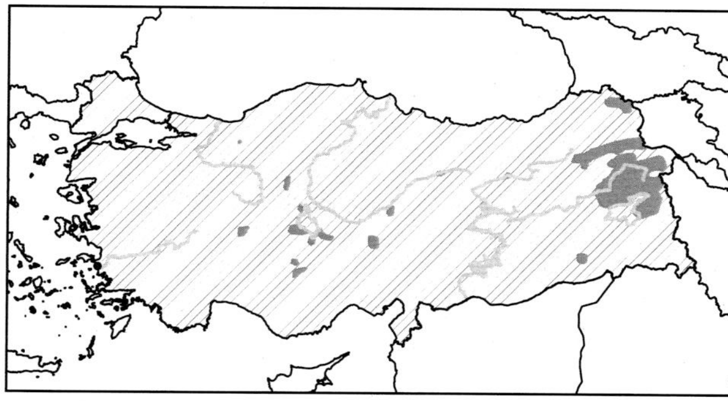

increasingly through the western half of the country. Kasparek (1992a) analysed all available passage records, identifying 19 April as the median spring date. The scale of passage though Turkey was demonstrated by Hustings & van Dijk (1994), who noted 127 observations in the Kızılırmak Delta (Black Sea Coastlands) in spring 1992. Apparently less common (although perhaps a reflection of observer coverage) in autumn, from early August (6th) to mid October (12th), with no clearly identifiable peak. Recorded to 3,500m (on Suphan Dağı, Muş province, East Anatolia) on passage. Three winter records, all recent: at Göksu Delta on 19 December 2004 (V. Konrad), at Büyük Menderes Delta on 24 December 2005 (A. Atahan *et al.*) and at Nizip, Gaziantep province (Southeast Anatolia) on 27 January 2008 (*per* A. Atahan).

Breeding. In marshes and damp meadows, usually near lakes, sometimes with scattered patches of drying-out *Phragmites*. Nests in a hollow in the ground, sheltered by a low shrub, or well hidden at the base or in the roots of a clump of *Phragmites*; builds a cup of plant stems and moss, lined mainly with hair. Eggs: C/4 (1); a clutch of two eggs on 21 May 1985 is considered incomplete. Young: B/3 (1), B/4 (2). In **East Anatolia**, nest building was reported at Saz Gölü on 8 June 2001. At Van marshes, on 11 June 2001, three nests were located: C/4, B/4 (newly hatched) and B/4 (virtually fully fledged); in the two nests containing young, egg laying would have commenced on 25 May and 11 May (PC); also there, a nest had three young, about ten days old, on 20 June 2004, and at least two recently fledged young, with adults, were seen on 8 June 2004, indicating egg laying about 9 May (PC). Adults were carrying food at various sites in East Anatolia in June (earliest 17 June), and there are several other records of fledged young (earliest 26 June). In **Inner Anatolia**, nest building at Palas Gölü on 17 May 1988, and agitated behaviour, as if breeding, was seen at Akşehir Gölü on 19 June–5 July 1994, food carrying at Eskil on 21 June 1998 and five fledged young seen nearby on 22 June 1998, whilst a male was feeding three fledged young at Eşmekaya marshes also on 21 June 1998 (GMK), and three broods of fledged young were at Tödürge Gölü on 13 June 2006 (M. Cowlard, G. Hinchon *et al.*). In **Black Sea Coastlands**, an immature (age unknown) was at Kaçkar on 31 July 1996.

GREY WAGTAIL — *Motacilla cinerea*
Dağ Kuyruksallayani

(Du: Grote Gele Kwikstaart; Fr; Bergeronnette des ruisseaux; Ger: Gebirgsstelze; Swe: Forsäla)

Subspecies. Nominate *cinerea* occurs throughout.

Status and distribution. National conservation status: Least Concern (population apparently stable). A widespread and fairly common resident and partial migrant, in upland areas throughout Turkey. More local in the western half (e.g. in the Taurus and the mountains of northern and western Turkey) and largely absent from much of Inner and Southeast Anatolia during the breeding season, although recently discovered to breed along many streams in montane areas of the former region. In summer found in montane areas, always in close proximity to running

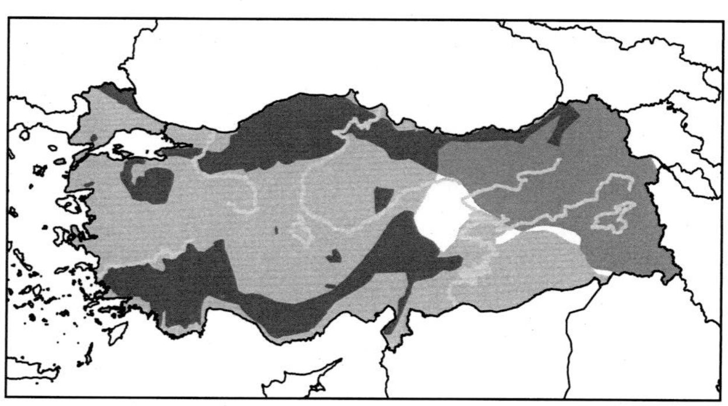

water, to at least 2,700m, occasionally from as low as 300m in the Taurus. May breed locally down to sea level, e.g. at Kızılırmak Delta, Black Sea Coastlands (Hustings & van Dijk 1994).

More widespread and recorded in larger numbers on passage, principally in autumn (from mid August to late October, peaking in late September and early October) and winter. Breeders descend in winter (between early October and late March, more rarely until late April), frequently occurring down to sea level. Winters principally in the north (on the Black Sea coast), west (Marmara and Aegean) and south (Mediterranean), and is a common wintering bird in many large cities, such as Istanbul and Ankara, where it even penetrates the most built-up areas.

Breeding. Along streams and rivers, with banks and rocky edges, in hilly and montane areas. Nest is in a recess or crevice of a bank, rock, wall or bridge, at the edge of, or near, running water; a cup of twigs, roots, grass and moss, lined mainly with hair. Eggs: no data but usual clutch 4–6 elsewhere. In **Black Sea Coastlands**, a nest at Sivrikaya was built atop a dilapidated nest of White-throated Dipper *Cinclus cinclus*, in a recess on a bridge, and another was hidden behind a plastic container on an upstairs window ledge of a house, *c.*100m from the river but adjacent to a small fish pond; one of these nests had four (or possibly five) large young on 14 June 2004, and the other one fully fledged young, with at least three others nearby, already out of the nest, on 13 June 2004, indicating egg laying in mid May (PC); in the Çamlıhemşin–Ardeşen area, in 1993, judging by adults carrying food, the first eggs hatched in late April, but the first fledged young were not seen until mid May (Faldborg 1994); most other records from this region are of fledged young in June–July, including a pair with five young, and four other pairs with young, near Ikizdere, on 10 July 1974. In **Mediterranean**, there are four records of fledged young, the earliest near Dalyan on 17 May 1989.

WHITE WAGTAIL *Motacilla alba*
Ak Kuyruksallayan

(Du: Witte Kwikstaart; Fr: Bergeronnette grise; Ger: Bachstelze; Swe: Sädesärla)

Subspecies. Traditionally, the populations in Turkey have been ascribed to *dukhunensis* (eastern Turkey as far west as Sumela in the north and Hakkari in the south) and nominate *alba* (western and central Turkey east to Bolu in the north and to Haruniye in the south). Roselaar (1995) identified a large area of intergradation: those from around Kastamonu and Zonguldak east to Tirebolu in the north, and from near Gaziantep to Cizre in the south he considered to be intergrades between *alba* and *personata*, rather than *alba* and *dukhunensis*, and perhaps warranted recognition as a third taxon in Turkey – *intermedia*. However, this proposal was rejected by Alström & Mild (2003), who considered that all populations in Turkey (and adjacent Iran) can be considered to refer to nominate *alba*, and relegated *dukhunensis* to a synonym of the nominate. We uphold their decision to not admit *dukhunensis*.

Status and distribution. National conservation status: Least Concern (population apparently stable). A common and widespread resident over much of Turkey, although only a summer migrant in some areas. Breeds in all regions, although apparently very local in Inner and Southeast Anatolia and generally occurring in smaller numbers in Aegean. Usually occurs to at least 2,900m during the breeding season. Even more widespread and locally very common on passage (e.g. 500 at Kızılırmak Delta, Black Sea Coastlands, on 4 April 1992; Hustings & van Dijk 1994): in spring between early March and late April, peaking in late March and mid April, in autumn from mid August (11th) to at least late October (27th) in the north, and early/mid November in the south, peaking in the last days of September and early October. Large numbers appear to move on a broad front

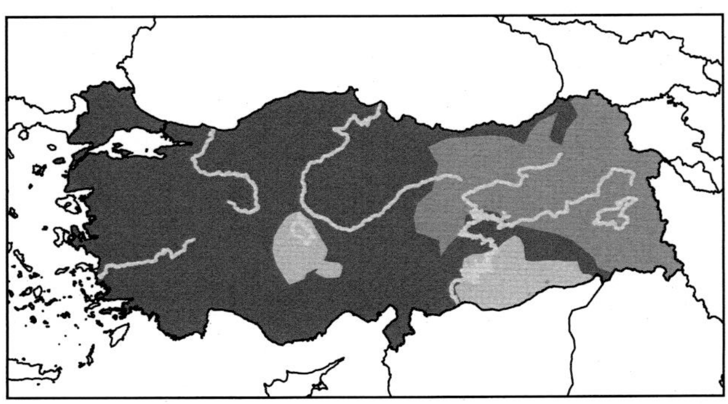

through the Bosphorus/Dardanelles region (Marmara) in both spring and autumn (Boyd 1916, Huxley 1949, Nisbet & Smout 1957). Winters mainly in coastal and wetland localities in Marmara, Aegean and Mediterranean: there are relatively few winter records from elsewhere in the western two-thirds of Turkey.

Breeding. Around settlements near water, typically in open areas and agriculture around farms and villages, and buildings near lakes, but also in more remote areas. Nests in a variety of sites: in walls, buildings, piles of rocks, under bridges; a nest near Karapınar (Inner Anatolia), was 1m below ground in the wall of a well (PC), and an unusual nest in a garden in the suburbs of Ankara (Inner Anatolia) was 1.6m above ground in a small conifer (Wadley 1951). The nest is a cup of plant stems, roots, leaves and moss, lined with hair, feathers and wool. Eggs: C/6 (1). Young: B/4 (1), B/6 (1). In **Inner Anatolia**, nest building was seen near Karapınar on 20 April 2004 (PC); a nest in a garden at Ankara (above) had five eggs on 21 May 1946, and subsequently a full clutch of six (Wadley 1951); an adult was carrying food north of Çankırı on 7 June 1975 (R. F. Porter); fledged young at Tuzla Gölü on 23 May 1970 indicate egg laying about 23 April; and fledged young were at Kumbet on 7 July 1907 (Ramsay 1914). In **Aegean**, there was a nest (no details) at Bafa Gölü on 12 May 1980 (Kasparek 1988a), and an adult was carrying food and alarm-calling at Karine Gölü on 9 May 2003 (PC). In Thrace (**Marmara**), a nest had young at Bebek on 29 June 1973, and fledged young were at Terkos on 20 June 1973; elsewhere in Marmara, fledged young were on the shore of the Sea of Marmara on 2 June 1966, and recently fledged young, next to their nesting crevice, were in Kocaçay Delta on 21 May 1967 (R. F. Porter), indicating egg laying about 21 April. In **Mediterranean**, a nest had six newly hatched young, in an ornamental rockery beside a swimming pool, north of Pozantı, on 12 May 2004 (N. Williams), indicating egg laying in late April. In **Black Sea Coastlands**, a nest had young at Abant Gölü on 2 June 1946 (Wadley 1951); a nest had three fully feathered young, with another young close by, on a bridge, north of Yoncalık, on 11 June 2004, indicating egg laying on 11 May (S. Williams); adults were feeding young at Yeniçağa Gölü on 12 June 1984 (Kiliç & Kasparek 1987); at Kızılırmak Delta, three pairs, each feeding four young, were seen in July 1971 (Dijksen & Kasparek 1985), several families were there in 1992, and a nest had young on 21 May 1992 (Hustings & van Dijk 1994); a fledged young, unable to fly, was at Sivrikaya on 14 June 2004 (PC); recently fledged young were at Fatsa on 10 June 1975 and Ikizdere on 12 June 1975, and copulation was seen near Boyabat on 9 June 1975 (R. F. Porter). In **East Anatolia**, a fledged young was with an adult at Sarıcan on 25 June 2004 (PC); juveniles were near Doğubeyazıt on 30 June 1969; a juvenile Common Cuckoo *Cuculus canorus* apparently reared by White Wagtails was at Kağızman on 17 August 1972. The records indicate double-brooding, which is usual in other (more southerly) parts of the range.

YELLOW-VENTED (WHITE-SPECTACLED) BULBUL
Arapbülbülü *Pycnonotus xanthopygos*

(Du: Arabische Buulbuul; Fr: Bulbul d'Arabie; Ger: Gelbsteißbülbül; Swe: Levantbulbyl)

Subspecies. Monotypic.

Status and distribution. National conservation status: Least Concern (population apparently stable). A fairly common resident, ubiquitous in the Mediterranean coastal region and the adjacent foothills of the Taurus (to 740m), including Antakya province. Extends from near Kaş, west of Antalya (recent records at Köyceğiz Gölü, Dalyan, in September 1988, Pazar Dağı, above Bodrum, in May 1989 [Roselaar 1995], near Bafa Gölü in May 1995, and Derköy, near Bozkır, Konya province, indicate a recent extension in range, though see below) to Birecik (Southeast Anatolia), where

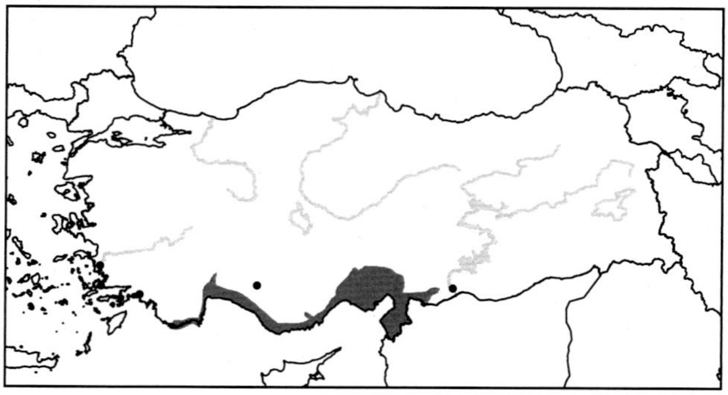

now known to be more frequent than originally supposed. Recorded as far north as Bucak, Burdur province; Kamışlı, north of Pozantı, Adana province and Haruniye, Adana province. Kumerloeve (1957b, 1961) mentioned a doubtful record from Izmir by Braun (1908b) in spring 1907. Found in a variety of scrub and woodland habitats, including olive and citrus groves, orchards, open pinewoods and *Ceratonia* scrub, to at least 1,000m (Kumerloeve 1957b), locally even reaching 1,260m, at least in the non-breeding season (Aslan & Erdoğan 2007). May favour gardens and other urban areas in winter, when occasionally recorded in larger concentrations, e.g. 100 were at Alanya in February 1972 (OST 1975).

Breeding. In bushes, thickets and trees, including olive and citrus groves, plantations, gardens and scrub. No data from Turkey but, elsewhere, the nest is a cup of fine twigs, grass, leaves and moss, lined with hair and rootlets, constructed in a bush or tree; usually three (2–4) eggs. All breeding records are from **Mediterranean.** In a hotel garden, near Side, on 17 May 2007, a nest contained four eggs, another three young *c.*2 days old, indicating egg laying in late April, and an adult was incubating on a nest in a climbing plant attached to a wall (J. Hollinrake). At Göksu Delta an agitated pair, one bird carrying food, was in a garden on 22 May 2004 (PC), and an adult was observed building a nest 1m up in a bush on 11 May 2005, with three nearby pairs calling, chasing and displaying, on 12 May 2005 (J. Hodson). Elsewhere in Mediterranean, a pair was displaying, but clearly not yet nesting, at Side on 25 April 2005 (T. Maunsell), and family parties were observed between Gazipaşa and Anamur on 27 June 2004 (K. P. Kendriks). In Israel, breeds late March–late August, often later, and hatching dates are 20 April–18 August, mainly in June (Shirihai 1996).

BOHEMIAN WAXWING — *Bombycilla garrulus*
İpekkuyruk

(Du: Pestvogel; Fr: Jaseur boréal; Ger: Seidenschwanz; Swe: Sidensvans)

Subspecies. Nominate and might be best considered monotypic. Certainly, the two Palearctic forms, *garrulus* and *centralasiae* are only weakly distinguished (see Vaurie 1958).

Status and distribution. A rare and highly irregular winter visitor, with one extremely unusual (and undocumented) record from late May. There is a Robson specimen held at the Natural History Museum, Tring (NHM 75.4.27.4), a male taken at the Forest of Belgrade, Istanbul, on 1 February 1875, and a total of four was taken by Alléon: a male in Constantinople in February 1892, a female at Makriköy [=Bakırköy, Istanbul] the same year and two separate females at San Stephano [=Yeşilköy, Istanbul] on 1 January and 9 January 1894 (National Museum of Natural History, Sofia). Mathey-Dupraz (1922) recorded the presence of three specimens in the Robert's College (Bebek, Istanbul) collection and that in winter 1890/91 the species was present at four localities on the European shore and two on the Asian side of the Bosphorus (Marmara). More recent records are as follows: one at Küçükcekmece, also near Istanbul, on 28 February 1962 (Kumerloeve 1970a); one ringed in Hungary was recorded in Thrace in late December 1965 or early January 1966 but no further details are available (Cornwallis & Townsend 1968); 30 were at Çankaya, Ankara (Inner Anatolia) in January 1966, whilst another was in the same garden in Ankara on 24 March 1967 (OST 1969); one at Ereğli (Black Sea Coastlands) on 25 March 1978 (Albrecht 1986 and pers. comm.) and one at Aphrodisas, Aydin province (Aegean) on 20 October 1982 (Martins 1989). There are only five subsequent records, despite much-increased levels of observer activity at the relevant season since the early 1990s, with four of these in winter 2005/06: single birds at Büyük Menderes Delta (Aegean) on 1 January 1994 (Kirwan & Martins 2000a), Toptepe Forestry Centre, near Samsun (Black Sea Coastlands) on 23 November 2005 (N. Yavuz *et al.*) and in Ankara on 27 January 2006 (M. Hughes *in litt.*

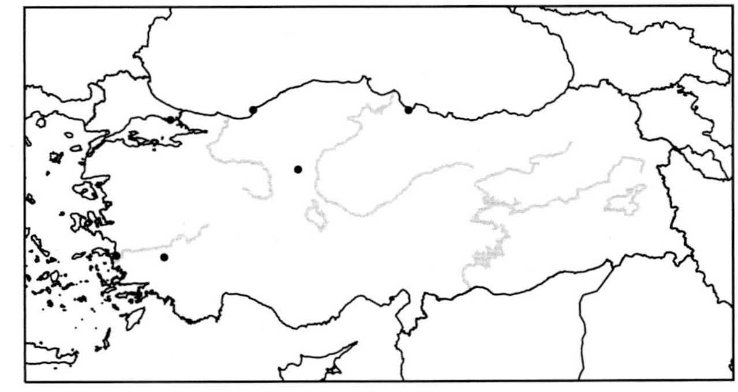

2006), with seven (photographed) at Asarcık, Samsun, on 2 and 7 April 2006 (S. Girişen) and three at Bandırma (Marmara) on 27 May 2006 (I. Sevim). The latter appears extraordinarly late. Kumerloeve (1957d) noted several records, including a flock of 25 birds, in adjacent Bulgaria, in winter 1941/42.

GREY HYPOCOLIUS *Hypocolius ampelinus*
Hurmayiyen

(Du: Zijdestaart; Fr: Hypocolius gris; Ger: Seidenwürger; Swe: Grå palmfågel)

Subspecies. Monotypic.

Status and distribution. Vagrant, with one authenticated record from Turkey, and a recent claim. One at Ağrı (East Anatolia) on 1 August 1986 was accepted by the Norwegian rarities committee (Kirwan & Martins 2000a). Additionally, one was briefly observed at Side, Antalya province (Mediterranean), on 19 November 2002 (A. Vittery *in litt.* 2002). No substantiative details are currently available for the latter claim.

Grey Hypocolius principally breeds in the Tigris/Euphrates valleys of Iraq, as far north as Mosul (Allouse 1953; *i.e.* within *c.*160km of the Turkish border) and in southern and eastern Iran. It winters to the south through much of eastern Arabia, but records further west are rare: there are several for Israel, all since 1987 (Shirihai 1996), but none elsewhere in the Levant and Near East.

WHITE-THROATED DIPPER *Cinclus cinclus*
Derekuşu

(Du: Waterspreeuw; Fr: Cincle plongeur; Ger: Wasseramsel; Swe: Strömstare)

Subspecies and taxonomy. Roselaar (1995) recognised a complex situation with *caucasicus* across much of northern Turkey and *aquaticus* in Thrace, *olympicus* in the rest of Marmara and throughout the Taurus (Mediterranean), and perhaps intergrades between *caucasicus* and *persicus* in the extreme southeast, though he admitted that some of his conclusions should be regarded as provisional. *C. c. rufiventris* must come close to Turkey in the extreme south, though Roselaar (*op. cit.*) seemed to reject that Taurus birds might represent intergrades between this form and *caucasicus*. Brewer & MacKay (2001, who almost certainly examined fewer specimens than Roselaar), considered all Turkish birds to be *caucasicus*, and that *olympicus* (confined by them to Cyprus and treated as extinct) might represent a synonym of *caucasicus*. Principally based on an examination of specimens in the Natural History Museum (Tring) and The Manchester Museum, we agree with Brewer & MacKay (*op. cit.*) that *caucasicus* probably occurs throughout Turkey; we have not attempted to review subspecies limits in this species, merely assign specimens to race. If those in the Taurus really are *olympicus*, as claimed by Roselaar, the differences between it and *caucasicus* can only be described as very minor and only appreciable in long series.

Although we maintain a traditional classification with wrens and dippers close to one another in the systematic list, recent genetic data suggest that these two genera cannot be viewed as closely related. Unfortunately no consensus concerning the correct placement of the Cinclidae is currently available: Barker (2004) found the family to be closest to either Old World flycatchers (Muscicapidae) or thrushes (Turdidae), whilst Voelker & Spellman's (2004) study raised the possibility that starlings (Sturnidae) or mockingbirds (Mimidae) were the dippers' closest relatives.

Status and distribution. National conservation status: Least Concern (population apparently decreasing). Widespread and locally not uncommon resident along fast-flowing upland streams and rivers, up to 2,200m in west and central Turkey, and up to 2,500m in the east. Apparently absent from Inner Anatolia and virtually entirely absent from Southeast Anatolia, and generally scarce in Marmara and Aegean. Breeding only confirmed at a small number of localities (see below).

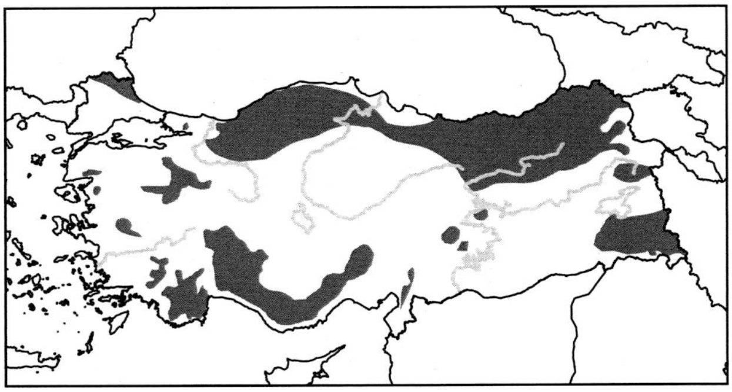

Post-breeding dispersal and severe weather-related movements are probable, as elsewhere in the species' range, e.g. one was found along a canal by Marmara Gölü (Aegean) on 6 January 1970, and one was at Paradeniz Gölü, Göksu Delta (Mediterranean), on 20 November 1991 (V. van den Berk *et al.*). The species is thus likely to occupy a wider geographic and altitudinal range during winter.

Breeding. On streams and rivers, mainly fast flowing, in montane or hill country. Nests in a raised site overlooking, and often overhanging, water; on a ledge or crevice in a bridge, rock face or bank. The nest is a domed structure of moss and some grass, with an internal cup of moss and grass, lined with dead leaves. No record of clutch size, but commonly five eggs elsewhere. At Sivrikaya (**Black Sea Coastlands**), on 14 June 2004, five nests, almost certainly belonging to one pair, were in adjacent holes on the central pillar of a stone bridge, *c.*4m above a fast-flowing river: two were new, with fresh excreta below the entrances, one slightly less fresh and evidently used for an earlier brood in the year, two were clearly old and used in earlier years but still intact, and the fifth was dilapidated and collapsed, on top of which a nest of Grey Wagtail *Motacilla cinerea* had young; an adult was also at a nearby nest, 5m above the water on a bridge, on 13 May 2004 (PC). In the Taurus (**Mediterranean**), Danford (1878) found an exposed nest on the face of a large boulder. Most records are from northeast Turkey, with nest building at Sivrikaya on 27 May 2002 (H. Dufourny), adults with food in June–July, and several records of fledged young in June–July; in the Çamlıhemşin–Ardeşen area, a pair was feeding two fledged young on 10 April 1993, but it was not until mid/late May that young were common along the rivers (Faldborg 1994); this extremely early record indicates laying of the first egg on *c.*26 February, with most eggs from early April. At Sarıcan (**East Anatolia**), a well-developed flying young (still with an adult) was seen on 27 June 2004 (PC). In **Mediterranean**, a pair feeding a juvenile at Ağla on 30 April 1989 indicates a first egg in mid March; an adult was carrying food at Akkopu on 15 May 1999. The wide range of dates implies double-brooding, as occurs elsewhere.

WINTER WREN — *Troglodytes troglodytes*

Çıtkuşu

(Du: Winterkoning; Fr: Troglodyte mignon; Ger: Zaunkönig; Swe: Gärdsmyg)

Subspecies. Considered problematic by Roselaar (1995), who favoured regarding Thracian birds as nominate *troglodytes* and those in the south and the rest of Marmara as *seilerni*, and noted that populations in the Black Sea Coastlands had variously been attributed to *hyrcanus, troglodytes* or *seilerni*, and that some represented intermediates between these. Brewer & MacKay (2001) failed to ascribe the Turkish birds to subspecies, omitting any mention of the country or a range that might encompass Turkey within their discussion of subspecific limits in the species. We have been unable to examine sufficient material to progress this issue but, without commenting on the validity of *cypriotes*, suggest that those in the Hatay (Mediterranean) might well prove to be this subspecies.

Status and distribution. National conservation status: Least Concern (population apparently stable). Widespread and fairly common resident in all types of woodland with dense undergrowth, occurring up to the treeline (2,500m) in the north. Breeding proved at relatively few localities (see below), but likely also in the Taurus (where recent surveys have proven it to be considerably more widespread and common than previously known), and the northern fringe of Inner Anatolia. Data for montane eastern Turkey are lacking and this area requires surveys for this and other similar species. A recent survey of Inner Anatolia revealed the species to be present in summer at the southern and eastern edges of the plateau, where principally recorded in montane areas, e.g. Aladağlar. There is only one breeding record from the east, in Erzurum province, according to Roselaar (1995), and reports at any season from Southeast and East Anatolia are rather few.

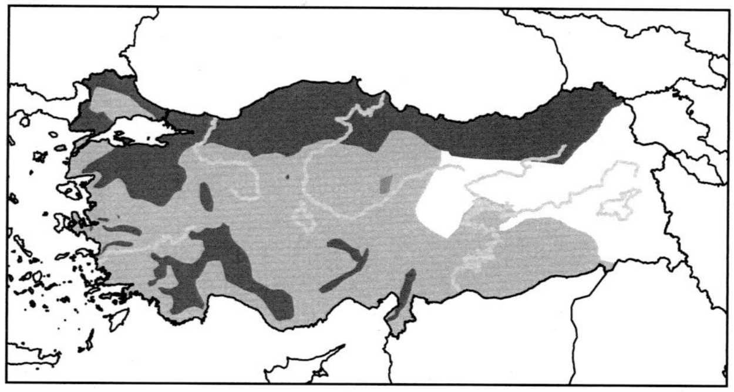

Some populations undertake long- or short-distance movements, either on a basically north–south axis or altitudinally. In winter, the species is present throughout much of the country, including the extreme southeast and especially near wetlands, probably as a result of both local dispersal and immigration from outside the country, and may perhaps be present in such areas until early May (2nd).

Breeding. Inhabits low vegetation in mature woodland, woodland edge and semi-open country, to the treeline on mountains. Two predator-destroyed nests at Sivrikaya (Black Sea Coastlands), 1m above ground in dense young conifers within mature woodland, are the only recorded nests in Turkey (PC); elsewhere, nests are commonly in hollows, usually on the side of a tree, bank or wall, or in low brambles, with many unusual sites reported. The nest is a domed structure of leaves, moss and grass, with a small entrance and lined with feathers; the lining of those at Sivrikaya also included hair and some fur; usual clutch 5–8. Song noted, mainly April–June, but five fledged young observed at Ikizdere (Black Sea Coastlands) on 31 August 1973 were obviously a second brood, which is usual elsewhere. In **Mediterranean**, five recently fledged young, barely able to fly, at Ağla on 11 May 2003 indicates egg laying in early April (PC); five fledged young were at Demirkazık on 10 June 2006 (M. Cowlard). In **Marmara**, a pair and four fledged young were observed at Kocaçay Delta on 19 May 1967 (R. F. Porter); a pair with two just fledged young were seen at Demirköy on 9 May 1967 (Rokitansky & Schifter 1971); and a pair reared three young on the Princes' Islands in 1975. In **Black Sea Coastlands**, family parties were observed at Abant Gölü on 2 June 1945 (Wadley 1951) and 6 July 1968 (Rokitansky & Schifter 1971), and a fledgling was at Kızılırmak Delta in early June 1992 (Hustings & van Dijk 1994).

DUNNOCK — *Prunella modularis*
Dağbülbülü

(Du: Heggenmus; Fr: Accenteur mouchet; Ger: Heckenbraunelle; Swe: Järnsparv)

Subspecies. Roselaar (1995) ascribed those in southern Marmara and, probably, the western Black Sea Coastlands to *euxina* (which is endemic to Turkey), and those in the northeast Pontics (Black Sea Coastlands) to *obscura*. Thracian birds he expected to be *meinertzhageni*, with nominate *modularis* perhaps also occurring in winter. A specimen at the Natural History Museum, Tring (88 April 1920.1244), taken by Robson in late February, might well be *modularis*. Other Turkish specimens we have examined are difficult to ascribe subspecifically, especially as some of the features ascribed by Roselaar (*op. cit.*) to various taxa, e.g. degree of saturation in underparts coloration, appear equally applicable to other named populations, which point was already made by Svensson (1992).

Status and distribution. National conservation status: Least Concern (population apparently stable). A fairly widespread and not uncommon resident throughout most of the western and eastern Black Sea Coastlands, in undergrowth of open mixed or coniferous forests, forest edge or alpine scrub at the border of the treeline. Only winter records are available from the central Black Sea Coastlands. Breeding restricted to montane forests in the north, from Bursa (at 1,500–2,500m on Uludağ) to Artvin (at 1,000–2,700m on Kaçkar), but may occasionally breed lower, although Kumerloeve (1961) found the species principally at 1,700–2,000m in Paphlagonia. There are a few probable breeding reports from Thrace and Kaz Dağı (Marmara), as well as the northern fringe of Inner Anatolia.

Fairly widespread in winter, mainly over the western two-thirds of the country, when descends from higher altitudes (e.g. Jetz 1995), but also reaches the extreme southeast at this season. Partial-migrant populations extend locally into adjacent parts of Aegean and Inner Anatolia, with regular reports from Marmara, Mediterranean and many fewer from Inner and East Anatolia (Suphan Dağı). Records prior to early May in the south perhaps refer to late migrants, but plausibly breeds in the Taurus in very small numbers and extremely locally, with late May and late August records from the Aladağ region (Mediterranean) being particularly suggestive of breeding. Extensive montane areas in eastern Turkey require investigation for this (and other) species, but there are no Southeast Anatolian records during the breeding season.

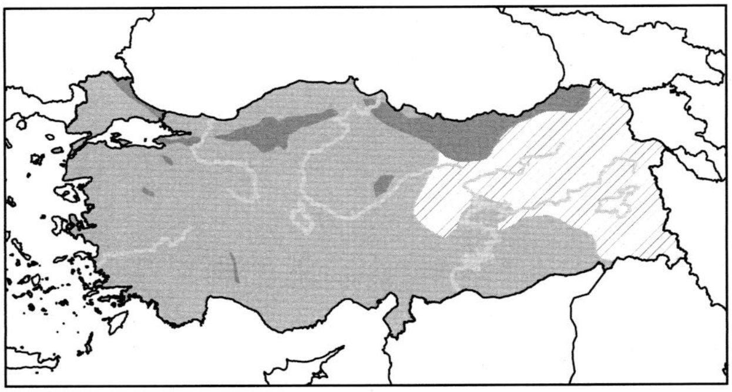

Breeding. Undergrowth and low cover, mainly in woodland and edge, but also in scrub above the treeline. Nests in a low shrub or bush; a cup of twigs, grass stems, leaves and moss, lined with hair and moss. Eggs: C/4 (1). Young: B/4 (1). Elsewhere, usual clutch 4–5. The only nest found in Turkey was in a low bramble at Abant Gölü (**Black Sea Coastlands**); it had four eggs on 29 May 1992, and four newly hatched young on 3 June 1992, indicating egg laying in mid May (PC). An adult carrying food at Çamlıhemşin (Black Sea Coastlands) on 8 May 1970 was presumably provisioning a nest, indicating egg laying in April. Song March–June at Kızılcahamam (**Inner Anatolia**) (Barış *et al.* 1984), apparently consistent in other regions of the country, but adults were still singing near the summit of Uludağ (**Marmara**) on 20 September 1966. At Çamlıhemşin–Ardeşen, local breeders were heard singing commonly in mid-May 1993 (Faldborg 1994). Elsewhere, usually double-brooded, occasionally treble-brooded.

SIBERIAN ACCENTOR — *Prunella montanella*
Sibirya Dağbülbülü

(Du: Bergheggenmus; Fr: Accenteur montanelle; Ger: Bergbraunelle; Swe: Sibirisk järnsparv)

Subspecies. Presumably nominate *montanella.*

Status and distribution. Vagrant: two records. A bird photographed at Rumelifeneri, Istanbul (Marmara), on 2 November 2006 (S. Bekir; *Birding World* 19: 454) and another, observed at Gelinkaya, near Erzurum (Black Sea Coastlands), on 12 October 2007 (S. Bekir, *Dutch Birding* 29: 390). Recorded as an accidental in eight European countries (Snow & Perrins 1998), with a single record from Lebanon (Porter *et al.* 1996).

RADDE'S ACCENTOR — *Prunella ocularis*
Sürmeli Dağbülbülü

(Du: Steenheggenmus; Fr: Accenteur de Radde; Ger: Steinbraunelle; Swe: Kaukasisk järnsparv)

Subspecies. Monotypic.

Status and distribution. National conservation status: Data Deficient (population trend unknown). A reasonably common resident and partial migrant but localised and breeds only above the treeline. Found on stony, scrub-covered slopes at 1,900–2,700m in the Anti-Taurus (Mediterranean/Inner Anatolia) and eastern Black Sea Coastlands, and at 2,100–3,000m (exceptionally 4,000m) in the east. Westernmost limits appear to be the Bolkar and Melendiz mountains (Eken & Magnin 1999), but apparently most abundant in the extreme southeast of the country, in the Hakkari range, *e.g.* around Başkale and the Karabel pass. At least some move lower in winter, when regularly recorded at Birecik (Southeast Anatolia), as well as in the Hatay, and Turkish birds are presumably also the source of the small numbers that winter in northern Israel, between October and March (Shirihai 1996). There is also a recent record from Lebanon as early as August (Barra 2003), suggesting that some may depart the Turkish breeding grounds very early. Probably more widespread, in both summer and winter, than available records suggest.

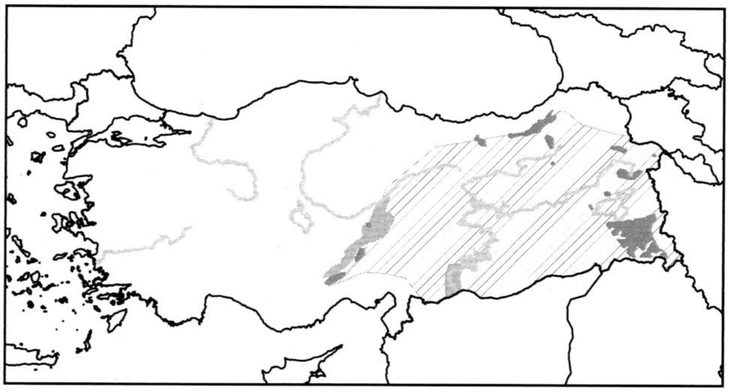

Breeding. Occurs in high mountains, above the treeline, usually above 2,000m, on stony or rocky slopes and in gullies, with scattered bushes, low scrub and/or tall herbs. The nest is usually less than 0.3m above ground, in a low bush, commonly juniper, or at 0.2–0.3m above ground in low 'heathery' plants, such as tall herbs (n=3); a cup of twigs, fine roots and plant down, lined with hair. Eggs: C/3 (2), C/4 (2). Young: B/3 (2), B/4 (3). At Demirkazık (**Mediterranean**), five nests had young 21–27 May, indicating earliest egg laying early–mid May; pairs were carrying food to two nests on 6 June 1991 and 10 June 1994; nest building was seen on 21 May 1993 and 28 May 1998 and, subsequently, the latter had four eggs on 8 June 1998 (T. Quittenden); a nest had three eggs on 12 May 2004, four eggs (full clutch) on 13 May 2004, and newly hatched young on 25 May 2004, confirming the incubation period as 12 days (H. Safak); pairs were seen with juveniles on 26 May 1985, 26 May 1986, 6 June 1993 and 16 June 1985; three young fledged from a nest on 28 June 1966. All these were probably first broods. Probable second broods are: a nest with two eggs on 27 June 1966 and three eggs (full clutch) on 29 June 1966; a nest had three young on 8 July 1976; adults were with juveniles on 24 July 1986, 12 August 1998 and 25 August 1986. At Nemrut Dağ (**East Anatolia**), nest building was seen on 5 June 2001 (T. Quittenden), a nest had one egg (incomplete clutch) in a low bush on 8 June 2004 (S. Williams), a nest in juniper had young on 9 June 1989 (D. Page), and a fledged young was seen on 8 July 1986; elsewhere in East Anatolia, a nest had three eggs near Yüksekova on 21 May 1989, and copulation was observed near Görentaş, Van, on 31 May 1969. In **Inner Anatolia**, a pair was feeding young at Erciyes Dağ on 1 August 1970.

ALPINE ACCENTOR *Prunella collaris*
Büyük Dağbülbülü

(Du: Alpenheggenmus; Fr: Accenteur alpin; Ger: Alpenbraunelle; Swe: Alpjärnsparv)

Subspecies. According to Roselaar (1995), those in eastern Turkey are *montana*, as perhaps are those in southern Marmara and the western Black Sea Coastlands (Kumerloeve 1961), although it is possible that like those in the Taurus they represent *subalpina*. Our own examination of material from the Taurus suggests that birds there are closest, on average, to *subalpina*, but there is potentially greater overlap between the latter and *montana* than admitted by Roselaar (*op. cit.*), e.g. in the intensity and extent of rufous streaking on the underparts and dark tips to the throat feathering, which is also subject to the effects of wear.

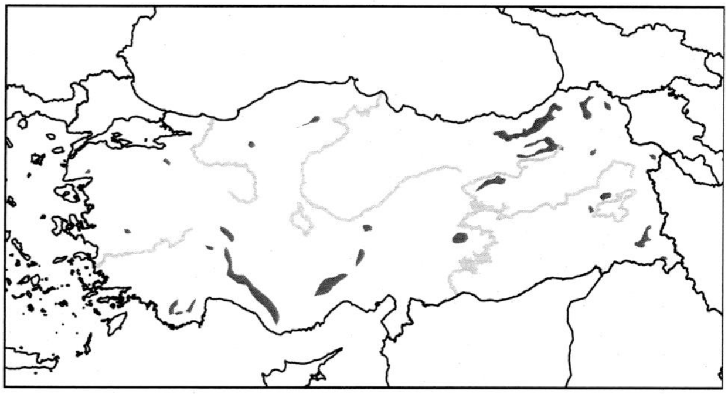

Status and distribution. National conservation status: Least Concern (population apparently stable). Apparently a local but not uncommon resident at high altitudes in eastern Black Sea Coastlands, Marmara, Aegean (perhaps very locally on Boz Dağ), Mediterranean and East Anatolia, but probably more widespread than available records suggest. Alpine grasslands with bare and/or rocky slopes between the treeline and snowline are preferred, and breeding altitude increases from west to east (above 1,900m in Marmara, to 3,500m in the Taurus and up to 4,200m on Suphan Dağı in East Anatolia). Perhaps occasionally found as low as 1,400m in the Taurus (Kumerloeve 1961). Particularly widespread in the Kaçkar range, the eastern Taurus, the western Taurus between Muğla and Antalya, but scarce on Uludağ (Marmara) and the Ilgaz Dağları (central Black Sea Coastlands). There are numerous breeding-season reports from the Anti-Taurus (Mediterranean), Işak Paşa Sarayı on the border with Iran, and from the mountains around Van Gölü (East Anatolia).

Winter status and distribution are still very poorly known, but apparently undertakes altitudinal and perhaps limited north–south movements at this season, when it has been recorded as low as 200m near the Mediterranean coast (Uhlig *et al.* 1996).

Breeding. Above the treeline, near the summit of high mountains, generally above 2,400m (1,900m), but up to 4,200m in the east; in rocky broken ground, typically near crags, boulders and scree in alpine meadows usually devoid of bushes. A nest at 2,900m near Sivrikaya (Black Sea Coastlands), was on steep rocky ground with boulders on 14 June 2004, when virtually the entire slope was snow-covered; it was 0.3m below a flat slab of rock, and mainly of grass stems and rootlets (no moss or lichen, which is usual, elsewhere), and lined with finer grass, some hair, and a tuft of animal fur; there were four eggs (S. Williams); on 24 June 2004, the mountainside was mainly clear of snow, and the nest had four young, 1–2 days old, indicating laying of the first egg on 3 June (PC); another nest at Sivrikaya was in a crevice 0.25m deep at 3m above ground in a roadside wall, and it contained three eggs (a complete clutch) on 15 June 2006 (M. Cowlard). Elsewhere, usual clutch 3–4 (5). At Uludağ (**Marmara**), song and display by three pairs was seen near the summit on 29–30 May 1966, and nest building, also near the summit, on 28 June 1976 (Jetz 1995). In **Mediterranean**, juveniles were above 3,000m on Karanfil Dağ in August 1966, and a juvenile was at 2,450m on Torosdağ on 17–23 July 1971. A juvenile at Nemrut Dağ (**East Anatolia**) on 9 June 1989 (D. Page), indicates egg laying in early May, about a month earlier than elsewhere; it was described as fairly numerous above 4,200m on Suphan Dağ in July 1966. In **Black Sea Coastlands**, juveniles were near Sivrikaya on 12 July 1986 (R. Webb), and two family parties were there on 9 August 1992 (J. Frost & J. Hornbuckle); a female and a juvenile were near Ispir on 14 August 1972; *c.*30 birds, including juveniles being fed, were at 3,000m between Ispir and Rize on 13 July 1975, indicating egg laying by 9 June; there are several records of large numbers of adults (up to 40) near Sivrikaya, mid June–early August, and described as common at Dilberduzu and Büyükdağ Duzu, in Kaçkar, in July. Elsewhere, breeding commences in late May, and two broods are reared in a year.

RUFOUS-TAILED SCRUB ROBIN (RUFOUS BUSH ROBIN) *Cercotrichas galactotes*

Çali Bülbülü

(Du: Rosse Waaierstraat; Fr: Agrobate roux; Ger: Heckensänger; Swe: Trädnäktergal)

Subspecies and taxonomy. Turkish birds relate to one or other of the grey-backed forms. As noted by Roselaar (1995), the differences between *syriacus* and *familiaris* are slight, subject to much intergradation, individual variation, wear and bleaching. Nevertheless, he treated those from about Gaziantep (Southeast Anatolia) westwards as *syriacus*, and those east of this line as *familiaris*, but was unable to label those in Inner and East Anatolia to subspecies, and admitted that some, even as far west as Burdur Gölü (Mediterranean) had upperparts as pale as *familiaris*. Rufous-backed and grey-backed birds may represent two species, based on their morphology: grey-backed forms have the crown, mantle and wings principally grey-brown (only occasionally showing any hint of rufous), contrasting with the rufous-brown rump and tail, which has less obvious white terminal spots but broader black subterminal spots and darker central tail feathers on the upper surface, more contrast in the head pattern, a different wing formula and a greyer tone to the breast and flanks (e.g. Svensson *et al.* 1999, Kok & van Duivendijk 2002). The two groups meet in the Levant, but very little information is available from the relevant regions and no qualitative study appears to have been conducted on their relationships there. Sometimes placed in the genus *Erythropygia* (e.g. Collar *in* del Hoyo *et al.* 2005).

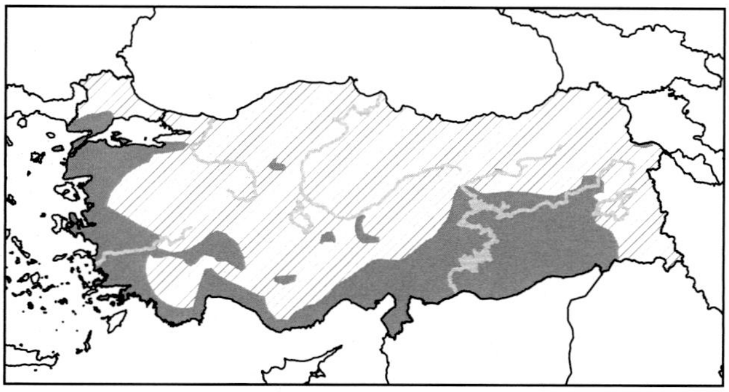

Status and distribution. National conservation status: Least Concern (population apparently decreasing). Widespread and fairly common summer visitor to lowland scrubby areas of Marmara, Aegean, Mediterranean and Southeast Anatolia. Regular around Ankara and elsewhere, locally, north of Tuz Gölü (Inner Anatolia), southern Thrace and the Anti-Taurus (Malatya) region to north of Erzincan (East Anatolia). Breeding altitude increases from 0–700m in Marmara/Aegean, to 900m on Aladağ, 1,100m in the Amanos Dağları and to 1,800m in eastern Turkey. Earliest records from regular breeding localities are 28 March in Mediterranean, 8 April in Aegean and 11 April in Southeast Anatolia. Arrival in breeding areas continues until early May, and departure commences towards late August, but is rarely present until mid September in Mediterranean (latest 24 September).

Breeding. Frequents scrub with open ground, cultivation with hedges, light woodland and groves, gardens, overgrown ditches, palm groves and coastal bushes on dunes. In Çukurova (Mediterranean) in 1987 especially common in plantations and forests with little undergrowth (van der Have *et al.* 1988). The nest is low, 0.4–1.0m above ground (n=12) in a bush, a recess in a bank or a low fork of a tree, usually well screened; or in unusual sites, in oil drums, tin cans and cavities in discarded breezeblocks, mainly at ground level (n=7, but many old or empty nests also seen) at Göksu Delta (Mediterranean) in 1992–93 (Castell 1996a); another unusual nest was at 2m in honeysuckle at Göksu Delta. The nest is an untidy cup of grass, plant stems and fibres, lined with roots, fine fibres, hair and wool. Eggs: C/3 (1), C/4 (4), C/5 (12); a clutch of six eggs near Akköy (Aegean) on 30 May 2004 (M. Demain) was exceptionally large. Those of 1–2 eggs (n=6) are considered incomplete clutches. Young: B/3 (1), B/5 (5). Adults assume territories in late April–early May. In Göksu Delta (**Mediterranean**), the fifth (and final) egg was laid in three nests on 24 May, 24 May and 28 May, and first eggs in these nests and one other can be calculated as 19 May, 20 May, 23 May and 24 May; two nests had newly hatched young on 8 June and 11 June, and two others had half-grown young on 10 June and 15 June; five young fledged from a nest on 17 June 1992, and young, barely able to fly, were seen on 16 June 1992, and probably fledged on 14 June 1992 (PC). At Birecik (**Southeast Anatolia**), a nest was almost complete on 19 May 1993 (PC), and two others also nearly complete on 10 May 2004 (J. Mason); two nests had one egg (incomplete clutch) on 11 May 2004 (R. Castell & M. Farmer); fledged young, being fed by adults, were observed on 26 June 2003 (C. Batty) and 6 July 1986 (R. Webb); at nearby Halfeti, an adult was carrying food on 11 June 1996 (GMK). In **Marmara**, a pair was collecting nest material near Manyas Gölü on 25 May 1967 (R. F. Porter). Around Izmir (**Aegean**), McNeile found 14 nests, all under construction or with eggs (none with young), on 19 May–3 June (1950–51 and 1954); the first egg in five of these can be calculated as 18 May, 22 May, 24 May, 27 May and 27 May, virtually identical to those in Göksu Delta; four of these were ready for eggs on 28 May (3) and 29 May (1); and one had four eggs of the robin and one of Common Cuckoo *Cuculus canorus*, all well incubated, on 31 May 1954. Fledged young were near Menemen (Aegean) on 22 June 1966 (R. F. Porter). More recently a nest at Bafa Gölü, found partially built on 19 May 1999, had five eggs on 29 May 1999, and two nests at Akköy had one egg on 24 May 1999 and four eggs on 26 May 1999 (J. Hollinrake); four nests had five eggs, five eggs, six eggs, and three eggs and two newly hatched young, at Akköy, 24 May–3 June 2004 (M. Demain); three juveniles and an adult at Bafa Gölü on 18 August 1996 were probably a second brood; elsewhere, the species is usually double-brooded, and presumably also in Turkey, but the single record is scant evidence.

EUROPEAN ROBIN *Erithacus rubecula*
Kızılgerdan

(Du: Roodborst; Fr: Rougegorge familier; Ger: Rotkehlchen; Swe: Rödhake)

Subspecies. Roselaar (1995) considered those in western Turkey to represent nominate *rubecula* (including *balcanicus*, which was described from Turkey) and that, as one moves east through the Black Sea Coastlands, the influence of *caucasicus* becomes increasingly prevalent. We agree with this interpretation, but consider recognition of *caucasicus* extremely doubtful, given such a broad zone of intergradation between these two forms. *E. r. hyrcanus* might be expected to occur in winter.

Status and distribution. National conservation status: Least Concern (population apparently stable). Fairly widespread and moderately common resident throughout the north of the country, in hilly or montane areas with all types of deciduous, mixed and coniferous forests and plantations. Breeding altitude increases from westernmost regions, where found at 350–600m, to 1,000–1,500m on Uludağ, 1,400–2,000m in central Black Sea Coastlands and up to 2,000m in eastern Black Sea Coastlands and East Anatolia. Breeds south along the Aegean coast to Izmir and even Fethiye (Mediterranean), with at least occasional breeding at Kızılcahamam and Akdağmadeni (Inner Anatolia), and breeds very locally in East Anatolia (in the Anti-Taurus and Erzurum province). Possibly more widespread as a breeder in the Taurus (Mediterranean) than mapped (see also The Gaps in our Knowledge).

More numerous and widespread on passage, which may be rather pronounced at some coastal localities, particularly in Thrace, Black Sea Coastlands and Inner Anatolia. Winters mainly in western and southern Turkey, occurring down to sea level at this season, although remains in reasonable numbers at some interior localities, east at least as far as Malatya. Apparently rather widespread in the Taurus (Mediterranean) in spring and, although some birds are probably late migrants, the possibility that the species breeds in this region should not be discounted, especially as spring passage in the south-coast deltas (where the species most certainly does not breed) is over by mid April.

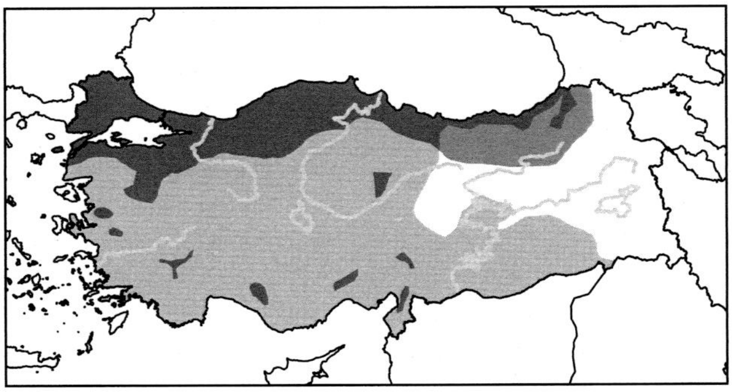

Breeding. In all types of woodland but more commonly in deciduous forest in some areas, e.g. Kocaçay Delta, Marmara (Ertan 1996), and breeds at all altitudes and forest types on Uludağ (Marmara), but apparently commonest in semi-open deciduous forest at 900m (Jetz 1995). Nest site, nest and clutch size undescribed in Turkey; elsewhere, nests in a cavity, low in a tree trunk, a bank or in undergrowth on the ground; builds a cup of dead leaves, grass and moss, lined with hair and roots; usual clutch 5–6. At Sivrikaya (**Black Sea Coastlands**), an adult fed a recently fledged young, yet unable to fly, on 14 June 2004, indicating laying of first egg in mid May (PC), and adults were carrying food on 5 June 1990. An adult was carrying food low on Dedegöl Dağ (**Mediterranean**), on 18 May 1998 (I. A. Green), and family parties were at Kızılcahamam (**Inner Anatolia**) in September 1967. Adults sing throughout the presumed breeding season (April–June) in some areas, e.g. Kızılcahamam, and the species is, therefore, likely to be double-brooded, as elsewhere.

THRUSH NIGHTINGALE *Luscinia luscinia*
Benekli Bülbül

(Du: Noordse Nachtegaal; Fr: Rossignol progné; Ger: Sprosser; Swe: Näktergal)

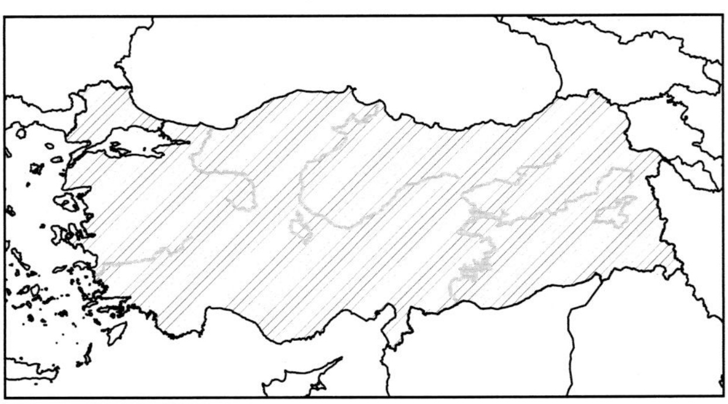

Subspecies and taxonomy. Monotypic. Genus occasionally merged within *Erithacus*.

Status and distribution. Passage migrant in small numbers across the western half of Turkey, which is probably considerably more abundant than records suggest. Most sightings are from western areas, particularly the Bosphorus and Inner Anatolia, and the species is regular on migration through parts of the southeast, e.g. at Birecik (Southeast Anatolia), but seemingly very rare in East Anatolia and the eastern Black Sea Coastlands, principally due to a lack of appropriate observer coverage. Spring migration continues apace from early April to mid May, with small numbers recorded until early June (3rd). Regularly heard singing on spring passage, and for this reason migrants are perhaps more evident then than in autumn. The earliest record is from Mogan Gölü (Inner Anatolia) on 10 March. Southbound passage is concentrated between mid August to mid September, when not uncommon in Black Sea Coastlands and at the Bosphorus, with the latest record from the latter (24 September).

COMMON NIGHTINGALE *Luscinia megarhynchos*
Bülbül

(Du: Nachtegaal; Fr: Rossignol philomèle; Ger: Nachtigall; Swe: Sydnäktergal)

Subspecies and taxonomy. Roselaar (1995) ascribed all those specimens he examined which were taken west of a line from Zonguldak (Black Sea Coastlands) south across Inner Anatolia to the Çukurova deltas (Mediterranean) to *baehrmanni*. Further east, he considered the situation to be decidedly less clear: those in the extreme eastern third might be *africana*, and given that this race occurs in neighbouring Iran and Armenia, this would be expected, but Roselaar (*op. cit.*) was unable to reach a certain diagnosis for those in the eastern half of Turkey. We have examined material only from western Turkey, but like Cramp (1988) and Collar *in* del Hoyo *et al.* (2005) we consider it very difficult to recognise *baehrmanni*, as among what can be assumed to be breeders from this region, several exhibit plumage characters of the nominate and there appears to be considerable overlap in both general upperparts' and underparts' ground colorations. Genus occasionally merged within *Erithacus*.

Status and distribution. National conservation status: Least Concern (population apparently decreasing). Widespread and common summer visitor to Black Sea Coastlands and Marmara, but more local elsewhere in the country with suitable habitat. Breeds particularly in dense deciduous growth in valleys, gullies and on slopes, or in dense undergrowth within open woodland. Breeding confirmed throughout the country, mainly at 0–1,000m (e.g. to 900m on Uludağ, Marmara), locally to 1,200m in central Black Sea Coastlands and to 1,700m in East Anatolia, except inland montane regions of Aegean, steppe-like areas around Tuz Gölü, arid parts of Southeast Anatolia (where principally restricted to the western half of the region) and most of the montane east (except orchards and woodlands of the Anti-Taurus and northeast of Van Gölü).

Singing birds on spring passage across the country may give the impression that the species is more widespread as a breeder than in truth it is. More widespread and numerous on spring passage, but apparently less common in autumn, at both seasons principally over the western half of the country, presumably reflecting observer activity. On spring passage, particularly abundant in May, with the earliest record on 13 March in the south and 6 April on Black Sea coast. There are insufficient data for autumn to draw many conclusions concerning the intensity, route and temporal nature of the passage, compared to spring, but latest record was on 2 October.

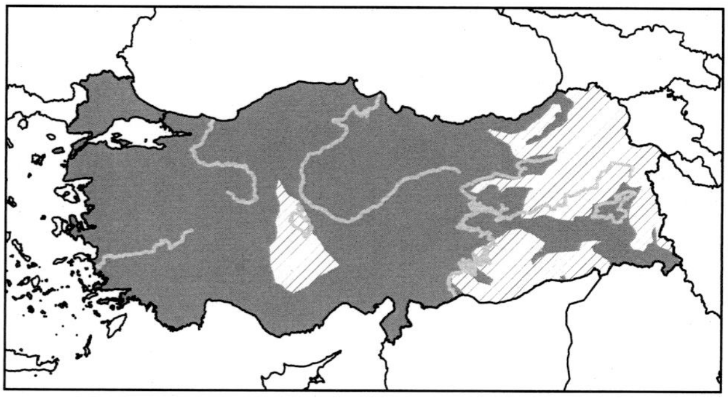

Breeding. In dense undergrowth in deciduous woodland and thickets, mainly near water or in damp areas; largest populations are at sea level and in lowlands, but occurs up to 1,700m, e.g. at Van Gölü (**East Anatolia**); also in *Liquidambar* woodland in the Dalyan area, Mediterranean (Kılıç & Kasparek 1989) and in the suburbs of Ankara, Inner Anatolia (Wadley 1951). At Kocaçay Delta (Marmara), population estimated at 3,000 pairs in 1993, with 4.13 singing males/ha in a 7.5ha plot and 3.16/ha in a 31ha plot in alluvial forest (Ertan 1996). In Çukurova (Mediterranean) in 1987 was in damp areas of plantations and forests, thickets and woods, and shrubs in coastal dunes, with 1.3/ha in an area of forestry on 12 May 1987 (van der Have *et al.* 1988). Thirteen singing males were in a 1km stretch along the River Aksu (Mediterranean) on 24 April 1967. At Kızılırmak Delta (Black Sea Coastlands) there were 1,200–1,400 breeding pairs, with 135 territories/100ha in Yürükler Forest, in 1992 (Hustings & van Dijk 1994). Nests in low dense undergrowth, usually well concealed under shrubby growth, or on the ground, in leaf litter and debris, screened by vegetation. A nest at Birecik was 0.7m above ground on a bushy tree stump in damp woodland; a loose, bulky cup of dead leaves and grass, lined with finer grass, strips of bark and some hair; it was ready, but empty, early on 11 May 2004, and had two eggs late morning on 12 May 2004 (D. Gregson). A nest in **Inner Anatolia**, which had five eggs on 15 May 1975 (Pforr & Limbrunner 1982), was similar to that above, but also had rootlets in the lining. At Kızılırmak Delta (**Black Sea Coastlands**), a nest contained four almost fully fledged young on 21 July 1972, indicating egg laying late June; a late date, and probably a replacement; an adult carrying food and fledged young were seen on 9 June 1975 (R. F. Porter); in 1992, most breeders arrived late April/early May, and adults were regularly observed carrying food and alarm-calling, from 27 May (Hustings & van Dijk 1994). Elsewhere in this region, an adult was observed carrying food at Ispir on 15 June 1987 (J. C. Eames), and an adult was alarm-calling at Gelinkaya on 23 June 2004 (PC). In **Marmara**, an adult was carrying food at Paşaköy on 26 June 1973, and at least one recently fledged young was near Bursa on 31 May 1966 (R. F. Porter), indicating egg laying in early May; a juvenile was at Çamlıca Hills in July 1966. An adult was feeding a young bird near Dursunbey (**Aegean**) in July 1966. In Lesbos (Greece), recently fledged young perched next to the empty nest on 27 May 1995, and an adult nest building on 30 May 1995 (T. Quittenden), indicate egg laying in early May and early June respectively. Males sing in Turkey mainly mid April–June and do so throughout the breeding cycle, even when feeding nestlings. A single brood per year.

BLUETHROAT *Luscinia svecica*
Mavigerdan

(Du: Blauwborst; Fr: Gorgebleue à miroir; Ger: Blaukehlchen; Swe: Blåhake)

Subspecies. Roselaar (1995) considered all those breeding in Turkey to be *magna*, a rather large subspecies that usually lacks any trace of a white central throat spot (though a faint indication of this may occasionally be evident). With the exception of one in the northeast Pontics (at Rizekent, north of Erzurum), which was apparently on territory and possessed a large, obvious white throat spot (Kirwan 1997c), all those we have observed breeding in the eastern third of the country are, indeed, *magna*. The identity of the Rizekent individual is unknown; several *L. s. magna* were found breeding in the same area. *L. s volgae* and, possibly, *L. s. cyanecula* (of which the latter is white-spotted and the former red-spotted) are known to occur as passage migrants in Turkey, and given that 'both forms were identified' at Mogan Gölü (Inner Anatolia) in 1967 (OST 1969) and Karakaya Reservoir, Malataya in April 2005 (MÖ *et al.*), and the status of the various taxa in Israel (Shirihai 1996, Morgan & Shirihai 1997), presumably at least one of the other red-spotted races, either *L. s. svecica* or *L. s. pallidogularis* might also have occurred (see also Kumerloeve 1961). However, as noted by Morgan & Shirihai (*op. cit.*), it is difficult to confidently ascribe many individuals to subspecies, given the presence of many intergrades and much individual variation.

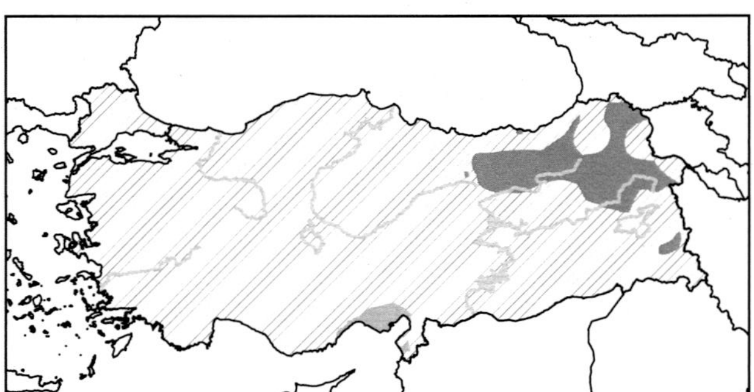

Status and distribution. National conservation status: Vulnerable (population apparently decreasing). Local and generally uncommon summer visitor to highlands in East Anatolia and adjacent parts of Black Sea Coastlands and Southeast Anatolia. Generally found in dense undergrowth in damp areas. Breeds in low tangled scrub and herbs, on stony slopes near water, as well as in bushes in damp subalpine meadows, above the treeline, at 1,700–2,600m. Breeds in montane areas of the east and northeast, and confirmed south to Hakkari and east to Van Gölü near the Iranian border, and plausibly even nests in the northeasternmost part of Inner Anatolia.

Uncommon on passage across Turkey at both seasons, when almost certainly more widespread than records suggest, but to some extent passage may obscure breeding status and distribution. Most regularly reported from Inner Anatolia, particularly at Mogan Gölü and Kulu Gölü; in Mediterranean, at Göksu Delta and in Hatay, and in East Anatolia at Karakaya Reservoir, Malatya, with the bulk passing in April and the earliest record 24 March (Kızılırmak Delta, Black Sea Coastlands). Autumn passage extends from mid September to mid October, with the latest record (27 October) from Erzurum (East Anatolia). Also winters scarcely in coastal areas of the western two-thirds of the country, particularly in the major southern deltas, where present until the end of the first week of April.

Breeding. In scrub near water, mainly above 1,800m, typically in damp, short (*c.*1m high) willow scrub and other bushes and herbs, in valleys along streams and small rivers, but also in scrub in damp meadows, and also in a small isolated patch of damp scrub adjacent to a black lava flow at Tendürek (East Anatolia). Nests on the ground, in a hollow in a bank or at the base of a shrub. In Black Sea Coastlands, a nest at Rizekent was in the muddy bank of a stream, where the roots of a large willow protruded; it had small young, a few days old, on 6 June 1997 (Kirwan 1997c). Nest and clutch size undescribed in Turkey; elsewhere, the nest is a cup of plant stems, grass and moss, lined with fine grass and hair, and usual clutch 5–7. Young: no data. In **East Anatolia**, a juvenile was in a gorge between Ağrı and Erzurum on 1 August 2000 (K. Haataja), and adults were alarm-calling, near Sarıcan on 16 June 2004 and 21 June 2004, but absent by 25 June 2004, and had probably moved away with fledged young (PC). In **Black Sea Coastlands**, adults were carrying food at Kopdağı pass on 10 June 1970, in the Sohun Valley on 15 June 1975 (R. F. Porter), and near Maden on 15 June 1975 (R. F. Porter); four pairs, and a total of four young were at Yoncalık on 12 July 1974, and a male and five young were near Bayburt, on 15 July 1975; two young birds were at two separate sites between Erzurum and Ispir on 28 June 1989. Elsewhere, usually single-brooded, but sometimes double-brooded in southerly regions; no data in Turkey.

WHITE-THROATED ROBIN *Irania gutturalis*
Taş Bülbülü

(Du: Perzische Roodborst; Fr: Iranie à gorge blanche; Ger: Weißkehlsänger; Swe: Vitstrupig näktergal)

Subspecies. Monotypic.

Status and distribution. National conservation status: Least Concern (population apparently increasing). Locally common summer visitor across much of the southern half of the country, but distinctly scarce and more local in the north. Occurs on scrub-covered rocky slopes or those with low-canopy deciduous or coniferous woodland, at 500–1,200m in west and central Turkey, to 2,300m in the Taurus and principally at 2,000–3,000m in the east. Largely absent from Marmara and Aegean. Fairly common in Mediterranean, more localised in Inner Anatolia (where principally restricted to the montane fringes), and very rare in Black Sea Coastlands. Widespread and reasonably common in Southeast Anatolia and in the western part of East Anatolia (a region still largely ignored by field ornithologists), and fairly common over the rest of this region, particularly the south and east. Most arrive on the breeding grounds in early May, with the earliest record of one in breeding habitat being on 10 April (Southeast Anatolia), usually departing by late August, with the latest record on 7 September (East Anatolia).

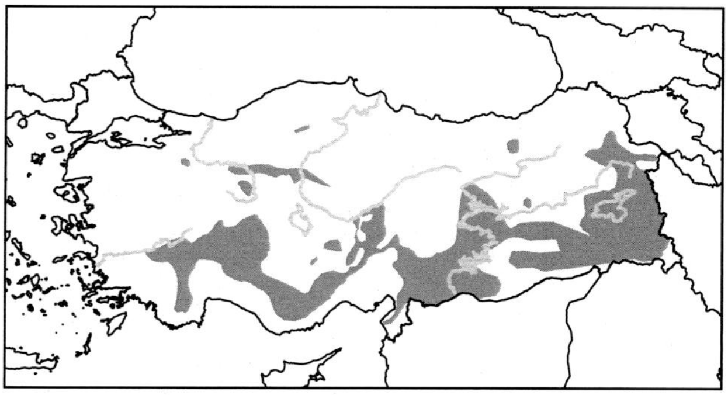

Breeding. Inhabits rocky slopes in dry hilly or montane areas, often in dense scrub, but also in open areas with only scattered bushes. Nests in a bush, 0.3–1.0m (n=15) above ground, with one at 1.6 m; builds a cup of small twigs, dead grass and bark fibres, lined with hair and plant down. Eggs: C/4 (14), C/5 (8). Young: B/4 (4), B/5 (2). Season fairly consistent in **Aegean**, **Inner Anatolia**, **Mediterranean** and **Southeast Anatolia**: display and nest building mainly early May, first eggs mainly from second week of May (earliest egg in Aegean 26 April), young in nest mainly from last week of May (earliest, two broods of newly hatched young, in Southeast Anatolia, on 17 May), and broods of fledged young mainly from 4 June (earliest, Inner Anatolia, 27 May). Perhaps also consistent in **Black Sea Coastlands**, where a pair was carrying food on 8 June 1975. In **East Anatolia**, earliest fledged young were on 16 June; an adult was carrying a faecal sac on 3 August 1974, a particularly late date, presumably a second attempt; records of family parties, on 23 July 1992 and on 9 August 1974, do not specify the development of the young. Nestlings were being fed by adults in late June 2005 in the Malatya area (MÖ, H. Kılınç).

[EVERSMANN'S REDSTART *Phoenicurus erythronotus*]
Kızıl Sırtlı Kızılkuyruk

(Du: Eversmanns Roodstaart; Fr: Rougequeue d'Eversmann; Ger: Sprosserrotschwanz; Swe: Altairödstjärt)

Subspecies. Monotypic.

Species not admitted to the Turkish list (Kirwan *et al.* 1999). Kumerloeve (1961) cast doubt on an observation by Taczanowski, mentioned without details by Hartert (1910–22), and Allouse (1953) recorded the species from Iraq in winter. Cramp (1988) erroneously listed the species as accidental in Turkey, apparently on this limited basis, but was corrected by Snow & Perrins (1998). The species is scarce and irregular in winter along the eastern fringe of Arabia (Gallagher & Woodcock 1980, Richardson 1990, Nightingale & Hill 1993, Hirschfeld 1995) and has been recorded in Iran at the same season (D. A. Scott), in Afghanistan and Pakistan, breeding in Central Asia, western China, northwest Mongolia and southwest Siberia (Sibley & Monroe 1990). It could wander to Turkey but there are presently no adequately substantiated records.

BLACK REDSTART *Phoenicurus ochruros*
Kara Kızılkuyruk

(Du: Zwarte Roodstaart; Fr: Rougequeue noir; Ger: Haurotschwanz; Swe: Svart rödstjärt)

Subspecies. Roselaar (1995) considered the vast majority of Turkish breeders to be nominate *ochruros*, perhaps with some influence of the Levantine form, *semirufus*, in the extreme south, and those in the westernmost quarter, in Marmara and much of Aegean, at least in winter, to be *gibraltariensis*, which form we can confirm from field observations to be widespread in Turkey at the latter season, at least in the western two-thirds of the country. We have examined insufficient specimen material from Turkey to comment further.

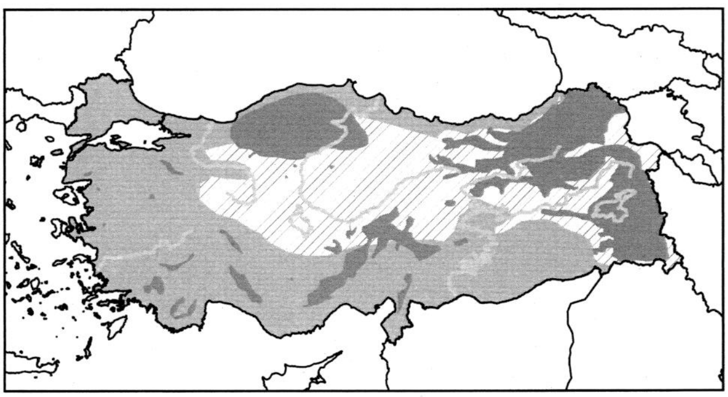

Status and distribution. National conservation status: Least Concern (population apparently increasing). Widespread and fairly common resident across most of Turkey. Partial migrant in high, montane areas, especially in the eastern third of the country. Breeds on rocky slopes and peaks, and in open forest bordering the treeline in East Anatolia, and reaches lower altitudes (1,200–1,500m) in western Black Sea Coastlands. In western and central Anatolia uses man-made structures such as houses and ruins for breeding, even penetrating central Ankara and Istanbul. Breeding reported in highlands of the east from 1,200m to at least 3,500m, and occasionally in the Taurus to 3,650m, but absent around Tuz Gölü (Inner Anatolia), the arid southeast (where appears surprisingly scarce, even in montane areas) and much of Thrace.

More abundant on passage throughout the country. Winters at lower altitudes, especially near wetlands, particularly in the south and west of the country: *e.g.* present around Bafa Gölü, Aegean, from late October to mid April, and at Fethiye, Mediterranean, from mid October to late April, as well as in Thrace.

Breeding. Mainly in high montane areas with cliffs and rock outcrops. Nests in a hole or crevice in a cliff, large boulder or inside a cave, or in a ruin or a building, serving as an artificial rock. The nest is a cup of dry grass, moss and plant fibres, lined with hair, wool and some feathers. Eggs: C/4 (2). Young: B/4 (1), B/5 (1); the limited data are consistent with elsewhere, where the usual clutch is 4–6. In **Inner Anatolia**, nest building was seen on 20 April 1993 and 22 May 1992 (PC), a nest had four eggs at Ürgüp on 18 April 1946 (Wadley 1951), adults were feeding young in the nest near Karapınar on 9 May 2006 (N. Burke), and fledged young were at Göreme on 25 May 1970 and Karadağ on 19 June 1996 (GMK). In **Black Sea Coastlands**, two nests, 20m apart, in a half-built house at Abant Gölü, had four and five young on 4 June 1945, and one of these broods fledged the same day (Wadley 1951); fledged young were at Sivrikaya in mid June and Ispir in late June. At Demirkazık (**Mediterranean**), nest building (built, but not fully lined) was noted on 13 May 2004 (N. Williams), and fledged young on 19/20 June 1990 (A. Forsten). At Uludağ (**Marmara**), breeds above 1,800m in rocky outcrops within alpine scrub, as well as at ski hotels, which serve as artificial habitats; fledged young were seen from early July to August (Jetz 1995). In **East Anatolia**, a nest had young at Işak Paşa Sarayı on 9 June 2001 (T. Quittenden); and another four eggs, high above Erzurum on 22 June 2004 (PC); there are several records of fledged young (earliest 25 May, latest August). The records indicate double-brooding, which is usual elsewhere.

COMMON REDSTART *Phoenicurus phoenicurus*
Kızılkuyruk

(Du: Gekraagde Roodstaart; Fr: Rougequeue à front blanc; Ger: Gartenrotschwanz; Swe: Rödstjärt)

Subspecies. Svensson (1992) remarked that those in Turkey appear to represent intergrades between nominate *phoenicurus* and the well-marked *samamiscus* race, which breeds in the Crimea, Caucasus and Central Asia, although Roselaar (1995) considered all Turkish populations to be pure *samamiscus*. In Turkey occasional examples of this species without wing patches are either very worn *samamiscus* or (more likely) nominate *phoenicurus* on migration (e.g. two from Erzurum in The Manchester Museum, BB06770 and BB06769). Both subspecies are described as common breeders and migrants in Armenia (Adamian & Klem 1999), and previous observers in Turkey have reported both. However, most, if not all, breeders have been noted as *samamiscus* (OST 1969, 1972, 1975, 1978). Most specimens from Turkey examined by us appeared to be nominate *phoenicurus* (but all of these could have been migrants), but the majority of specimens held in The Manchester Museum appear to be virtually pure *samamiscus*, although one of these is an *incognita* morph (BB12192). Our field observations support the intergrade option suggested by Svensson (*op. cit.*).

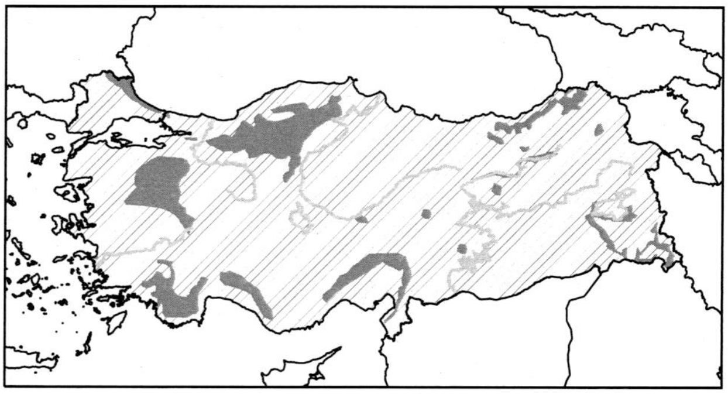

Status and distribution. National conservation status: Least Concern (population apparently decreasing). Fairly local but not uncommon summer visitor to upland areas in the eastern two-thirds of Turkey, reaching locally west in smaller numbers to Marmara, inland parts of western Black Sea Coastlands, the northern fringe of Inner Anatolia and the Istranca Mountains of Thrace. Recent (late 1990s) surveys have revealed the presence of a more significant population in the western Taurus than previously believed. Breeds at 1,000–2,100m in the east, but occasionally lower further west, e.g. down to 900m in western Black Sea Coastlands (Kumerloeve 1961), and plausibly even breeds at sea level in Kızılırmak Delta (Hustings & van Dijk 1994) and at the Bosphorus (Porter 1983).

Distinctly more widespread and numerous on passage, and even commoner in autumn than in spring. Spring migration typically commences in early April and peaks late April/early May, with the earliest record from 1 March. Autumn passage is especially marked between mid September and early October, with the latest record from 18 November, but substantial numbers move through northern areas from August and possibly even July. There is one recent winter report, from Samandağ, Antakya province (Mediterranean), on 23–31 December 2000 (A. Atahan).

Breeding. In open coniferous and deciduous woodland and woodland edges, parks, orchards, and occasionally in semi-open areas with trees, mainly in uplands, to 2,100m or higher, but also down almost to sea level in Aegean. Breeding greatly influenced by the availability of suitable nest holes, with natural sites being limited by forestry management practices, but in areas where nest boxes are provided, *e.g.* in *Pinus brutia* forestry near Antalya (Mediterranean), the species may nest quite commonly, although it normally shuns such woodland. Nests mainly in holes in trees, 1–9m above ground, including old woodpeckers' holes, but also in walls and nest boxes. Sole details of nest and clutch size are from Danford (1878), at Karanfil Dağ (**Mediterranean**), in 1876: he found a pair nest building on 24 April; a nest had six slightly incubated eggs, and another three fresh eggs, probably an incomplete clutch, on 25 April; two nests were in holes in cedars, and the other in a juniper; each was a loose cup of thin strips of juniper bark, with tufts of black goat's hair, and a little ibex wool, lined with goat's hair and a few feathers. More recent records in this region include a nest, which had at least four young, in a juniper at Pozantı on 16 May 1970; a pair entering a nest hole in a pine at Bolkar Dağları on 5 June 1999 (R. F. Porter); fledged young at Akseki on 11 May 2001 and Kovada National Park on 10 May 1995 (I. A. Green) indicate laying of first eggs about 8–9 April. In **Aegean**, near Marmaris, adults were taking food to a nest box on 5 June 1995 (J. Cooper), and two pairs were also using nest boxes there in May 1989 (P. J. Heath). In **Inner Anatolia**, two nests had young in holes in walls at Nevşehir on 20 May 1975, and adults were attending a nest box at Kızılcahamam on 12 June 1994 (P. Smith). Recently fledged young in an orchard at Ispir (**Black Sea Coastlands**) on 12 June 2004 indicate egg laying *c.*10 May (PC). Juveniles were seen in July in Aegean, Inner Anatolia, Mediterranean, Black Sea Coastlands and East Anatolia, but it is unclear whether these were older, well-developed young from nests in May or whether they might have been second broods.

[GÜLDENSTÄDT'S REDSTART *Phoenicurus erythrogastrus*]
Büyük Kızılkuyruk

(Du: Witkruinroodstaart; Fr: Rougequeue de Güldenstädt; Ger: Reisenrotschwanz; Swe: Bergrödstjärt)

Subspecies. Should the species be proven to occur in Turkey, the subspecies involved will presumably be the nominate (which occurs in the Caucasus). Note: the specific name is correctly spelled *erythrogastrus* not *erythrogaster* (David & Gosselin 2002a).

Species not admitted to the Turkish list (Kirwan *et al.* 1999). In the Western Palearctic this species breeds above *c.*3,500m, in small numbers in the Caucasus and Elburz, apparently locally through Georgia and Azerbaijan, and perhaps in northwestern Iran (Cramp 1988). It is considered casual in Armenia (Adamian & Klem 1997) and was unreliably reported (without details) in summer 1983 and 1984 from the Aladağ region, Niğde province (Mediterranean). Although the species could occur in Turkey (presumably undertaking movements during winter from the Caucasus) it seems unlikely that Güldenstädt's Redstart breeds. The species has been recorded in Bulgaria, Finland (Haukkovaara 1993), northern Saudi Arabia (Snow & Perrins 1994), Kuwait, and possibly the Red Sea Hills of northern Sudan (Cramp 1988).

[BLACKSTART — *Cercomela melanura*]
Karakuyruk

(Du: Zwartstaart; Fr: Traquet à queue noir; Ger: Schwarzschwanz; Sw: Svartstjärt)

Subspecies. Presumably the nominate, which breeds in the Near East.

Hypothetical. One recent claim, involving two individuals, from Suçeken, Batman province (Southeast Anatolia), on 3 October 2004, is supported by a somewhat basic description (Cofta *et al.* 2005), and one of the observers had previous experience of the species. Whilst it seems quite possible that the species has occurred in Turkey, and the record has been accepted by the Polish records committee, we prefer to await fuller details before admitting Blackstart to the country's avifauna, especially given that the species seems to be known only from extreme southern Syria, where there are records of dispersing birds from elsewhere (Baumgart *et al.* 1995, Snow & Perrins 1998), and even recent detailed and long-term studies of the central Syrian Desert have failed to reveal the species' presence there (Serra *et al.* 2005). Although Cofta *et al.* (2005) claim that records from northern Syria are available in the non-breeding period, no reference is given to support this assertion.

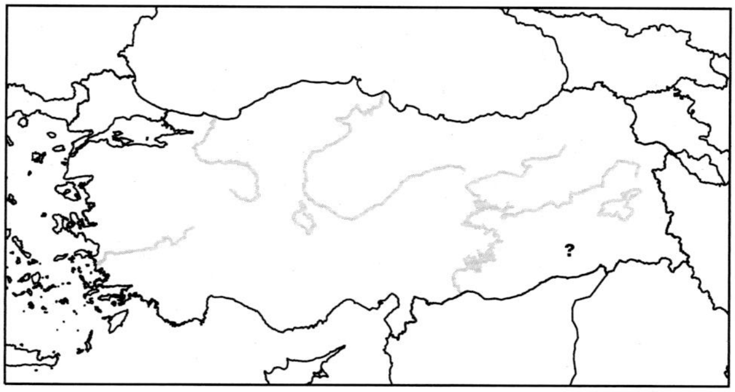

WHINCHAT — *Saxicola rubetra*
Çayır Taşkuşu

(Du: Paapje; Fr: Tarier des prés; Ger: Braunkehlchen; Swe: Buskskvätta)

Subspecies. Best considered monotypic (see Vaurie 1959a, Urquhart & Bowley 2002), although up to six subspecies have been described, including *senguni* from the region of Van Gölü (East Anatolia).

Status and distribution. National conservation status: Vulnerable (population apparently increasing). Rather local and fairly uncommon summer visitor to extensive damp meadows with scattered shrubs and herbs, at or above treeline, at *c.*1,700–2,300m. Breeding proven in eastern Black Sea Coastlands, in Ardahan province, and in valleys of extreme East Anatolia (in Kars and Ağrı provinces) as well as around Van Gölü. No certain breeding records for western Black Sea Coastlands or Thrace, but observed in late May and August in these regions, and plausibly breeds around Kastamonu, Bolu province, as well as at Akdağmadeni (Inner Anatolia), where discovered in 2004 (E. Yoğurtçuoğlu). Certainly does so in the Refahiye region, close to the border of Black Sea Coastlands/Inner Anatolia/East Anatolia, where observed above the treeline (BD). Observed singing in late May west as far as Bursa (Marmara), but no evidence of breeding.

A fairly common passage migrant across the country with largest numbers perhaps in the west. Earliest spring dates 10 March in Mediterranean, and 16 March in Black Sea Coastlands. Several hundred between Antalya and Side (Mediterranean) on 6 April were an unusually early concentration. Spring passage peaks between mid April and mid May, with records until

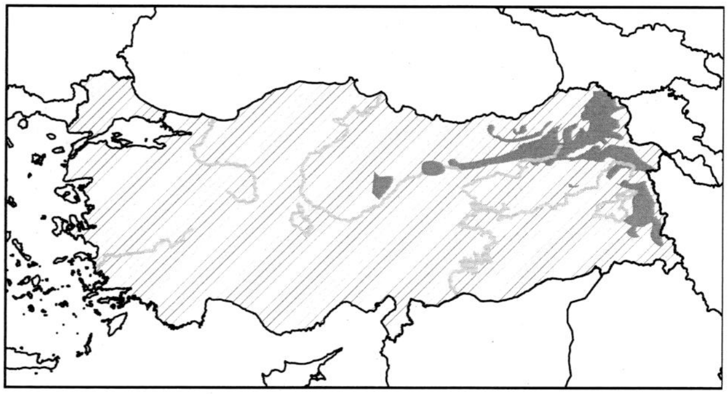

early June (3rd) on Black Sea coast. Exceptional numbers were present in the Eber Gölü basin on 19/20 April 1992, perhaps as many as 10,000 (Bradshaw & Kirwan 1992). Autumn passage commences mid August and peaks late August–late September, with the latest record 28 October. There are two reports of overwintering, from Bafa Gölü (Aegean) on 2 January 1987 (Kasparek 1988a), and from Göksu Delta (Mediterranean) on 15 December 1994 (N. Shelton).

Breeding. In rough pastures and meadows, with taller shrubs and sometimes thorn scrub, usually in uplands. Nests are on the ground and usually well concealed in dense vegetation; builds a cup of plant stems and dead grass, lined with finer grass and hair. Eggs: no data but, elsewhere, usual clutch is 5–7. Young: B/6 (1); records of fledged broods are mainly of 3–5. In **Black Sea Coastlands**, a nest had six young, about seven days old, near Eskipolat, on 11 June 2004, indicating egg laying on 17 May (PC); a nest had newly hatched young at Sivrikaya on 7 August 2001 (A. Comas), and may have been a second brood; two pairs were carrying food in the Sohun Valley on 15 June 1975 (R. F. Porter); the earliest record of fledged young, at Kopdağı pass on 15 June 1975, indicates egg laying in mid May; four fledged broods (five, four, three and three) on 15 July 1975. In **East Anatolia**, adults were carrying food at Erçek Gölü on 23 June 1968 (Kasparek & van der Ven 1983), there were fledged young on 21 June 1987 (J. C. Eames) and 12 July 1974, and adults were carrying food near Erzurum on 16 June 1991 (C. G. Bradshaw).

EUROPEAN STONECHAT *Saxicola rubicola*
Taşkuşu

(Du: Roodborsttapuit; Fr: Tarier pâtre; Ger: Schwarzkehlchen; Swe: Svarthakad buskvätta)

Subspecies and taxonomy. Nominate *rubicola* occurs throughout the Black Sea Coastlands and the western half of the country (Roselaar 1995). However, the precise limits between this taxon and the next are unclear (see below for full discussion). Note that, following a recent study of stonechat taxonomy (Wink *et al.* 2002), in common with Urquhart & Bowley (2002) we recognise three species of stonechats: African *S. torquatus*, European *S. rubicola* and Eastern (in preference to the commonly employed Siberian misnomer) *S. maurus*. Further refinements to this taxonomy, particularly with respect to the Asian taxa currently included within the *maurus* group, can be expected in the future. Any taxonomy that recognises multiple species within the *S. torquatus* complex *sensu lato* must employ the name *rubicola* for the European and west Asian forms, as the type locality for *torquatus* is the Cape of Good Hope, and it appears nonsensical to group *torquatus* with the European populations rather than with other African forms (see Urquhart & Bowley 2002).

Status and distribution. National conservation status: Vulnerable (population apparently increasing). Rather widespread but sparsely distributed breeder, being more abundant in East Anatolia (*armenicus*; see below). Populations in Black Sea Coastlands, Thrace, Marmara and Inner and East Anatolia are partially migratory, especially those at higher altitudes, and it has recently become apparent that at least some move southeast in winter, reaching across much of western and southern Iran at this season (Schweizer 2003, Kirwan & Bates in press). Occurs in varied scrubby habitats, principally in upland areas, but also down to sea level in the west and north. Breeding altitude increases from west to east: in westernmost areas at 100–900m, western Black Sea Coastlands at 1,000–1,500m, the central Taurus at 1,000–2,000m, and in the extreme east around Van Gölü (*armenicus*) and in Kars at 1,600–2,400m, occasionally to 3,000m. Seemingly absent as a breeder from arid areas of the southeast and west of Tuz Gölü (Inner Anatolia).

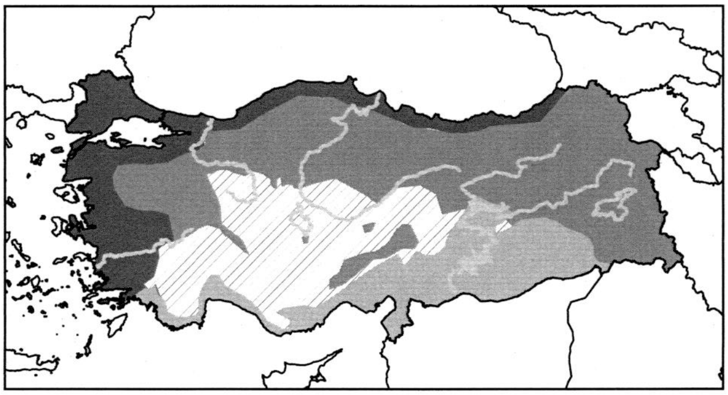

Somewhat more widespread on passage, with considerable evidence of immigration from further north, and passage at both seasons moderately noticeable in many areas, return movements commencing late February/early March in spring, although (surprisingly) seems not to move through the Bosphorus (Porter 1983). Winters most commonly in southern and southwest Turkey, often in considerable numbers locally, with birds generally departing wintering areas in the major deltas in early April, but some at least remain in the milder valleys of Inner Anatolia at this season.

Breeding. In open areas of grassland and heather, including drier edges of rough marshlands, and open grassy hillsides with scattered scrub, from sea level to 3,000m. Nests on the ground in herbage; builds a cup of grass and moss, lined with hair, wool and some feathers. Eggs: C/5 (1). Young: no data on nestlings, but up to four fledged young recorded. Elsewhere, usual clutch 5–6. In **Black Sea Coastlands**, a nest had five eggs near Bolu on 28 June 1977 (Pforr & Limbrunner 1982), a pair was carrying food north of Boyabat on 9 June 1975 (R. F. Porter), a juvenile was at Sivrikaya on 16 June 1984 (S. Harrap), and a pair was carrying food at Maden Koprubaşı, near Ispir, on 10 May 2004 (GMK). In **Inner Anatolia**, a pair was carrying food near Çankiri on 7 June 1975 (R. F. Porter), and a fledgling was on Karadağ on 19 June 1996 (GMK). In **Aegean**, there was a family party at Pamukkale on 10 May 1995, and another at Kuşadası on 17 May 1995 (A. J. Kane), a male was carrying food at Bolu Dağ on 12 May 1970 (R. F. Porter), and there were three fledged young at Bafa Gölü on 1 June 1984 (Kasparek 1988a). At Uludağ (**Marmara**), the season differs between the base and summit by about one month: at 800m, nest building was noted on 17 April 1992, and two nests had small young in mid-May 1993, when there was still much snow in the alpine region; first fledged young reported in late June in the alpine zone, and adults were feeding fledged young there on 12 August 1963 and 18 August 1977 (Jetz 1995). An adult with fledged young was near Terkos on 20 June 1973 (R. F. Porter).

EASTERN (SIBERIAN) STONECHAT *Saxicola (maurus) armenicus*
Sibirya Taşkuşu

(Du: Kaukasische Roodborsttapuit; Fr: Tarier pâtre; Ger: Schwarzkehlchen; Swe: Svarthakad buskvätta)

Subspecies and taxonomy. All relate to *armenicus*. The boundary between *armenicus* and *rubicola* in Turkey is unclear: the former certainly occurs around Van Gölü and south and east to the Iranian border, but neither we or Roselaar (1995) have seen specimen material from elsewhere in East and Southeast Anatolia to confirm whether *armenicus* occurs further west, if *armenicus* overlaps with *rubicola*, or if the two intergrade, as claimed by Beaman (1994), although there is no strong evidence of intergradation between *rubicola* and *armenicus*/*variegatus* in the eastern Caucasus, and the two might be separated by both habitat and altitudinal preferences. Complicating the situation even further, evidence has come to light recently that eastern *rubicola* might even breed, or at least oversummer, in western Iran (Kirwan & Bates in press). At least those birds in the eastern Black Sea Coastlands come rather close in the plumage of both sexes to *armenicus*, but nonetheless the latter still appears to represent a distinct taxon. GMK has recently gained in-depth experience with *armenicus* around Van Gölü and elsewhere. Those features that appear useful in its separation from both eastern *rubicola* and *variegatus* (e.g. tail pattern) seem constant, although in other respects birds are quite variable, e.g. in the amount of white on the wings in males and extent and intensity of colour saturation on the underparts in females).

The revised stonechat taxonomy proposed by Wink *et al.* (2002) recognised the following forms, additional to the nominate, within *S. maurus*: *stejnegeri* (treated as a synonym of *maurus* by some authorities), *variegatus*, *armenicus*, *indicus* and *przewalskii*. As suggested by Urquhart & Bowley (2002), *armenicus* and *variegatus*, which more or less share several plumage features distinctive within the *maurus* group as well as a disjunct breeding range from the rest of it, may represent a third Eurasian species within *S. torquatus sensu lato*. If this is the case (and we consider the possibility to require detailed evaluation) then *variegatus* would have priority for the species name (whilst a suitable English name would be Caucasian Stonechat). Roselaar (*op. cit.*) noted that there was no evidence of *variegatus* or *maurus* on migration through Turkey, but we have seen a recent published record, referring to a 'Siberian Stonechat', at Sivrikaya (Black Sea Coastlands), on 10 June 2000 (I. Hunt in *Birding World* 13: 273), which in our opinion is strongly suggestive of confusion with eastern *rubicola*. There is a possible record of *variegatus*, from Çukurova (Mediterranean), on 11 February 1999 (E. van der Burg & R. Hofland) and sight reports, by an experienced observer, of two *maurus* at Göksu Delta on 23 December 2004 with another or one of the same there on 4 January 2005 (V. Konrad). Note the correct spellings of *S. maurus variegatus*, *S. m. armenicus*, *S. m. indicus* and *S. maurus* (David & Gosselin 2002b).

Status and distribution. See previous species for discussion. It is presently unclear whether *armenicus* winters in Turkey or not.

Breeding. Basic details of nesting are apparently much as the previous species. Few data from Turkey. In **East Anatolia**, there were fledged young at Nemrut Dağ on 12 June 2001 (PC), Van on 13 June 1990 (A. Forsten), Erçek Gölü on 23 June 1968 (Kasparek & van der Ven 1983), Ağrı on 1 July 1969, Hamur on 19 May 2006 (GMK), Bulanık on 9 June 2006 (G. Pettersson) and near Van on 9 July 1986 (R. Webb). In the Van region in early May 2004, all birds seen were in pairs and there was much chasing (territorial defence) and displaying, but nesting did not appear to have started in earnest (GMK pers. obs.).

ISABELLINE WHEATEAR — *Oenanthe isabellina*
Boz Kuyrukkakan

(Du: Isabeltapuit; Fr: Traquet isabelle; Ger: Isabellsteinschmätzer; Swe: Isabellastenskvätta)

Subspecies. Monotypic.

Status and distribution. National conservation status: Least Concern (population apparently stable). Widespread and locally numerous summer visitor, but apparently absent as a breeder from coastal zones of Mediterranean and Black Sea Coastlands, and from high montane regions. Recorded in open areas, particularly steppe, to 1,850m in west and central Turkey, and principally at 700–2,300m in the east. Widespread records between early March (3rd) and late September, with a few exceptionally early records in February, the earliest being from Alanya (Mediterranean) on 25th, but collected as early as 29 February near Konya (Inner Anatolia), *i.e.* in potential breeding habitat, by Watson (1961b), with the latest autumn record on 5 November, in East Anatolia. In areas where the species does not breed, e.g. the north coast, spring passage may continue from late March to late May.

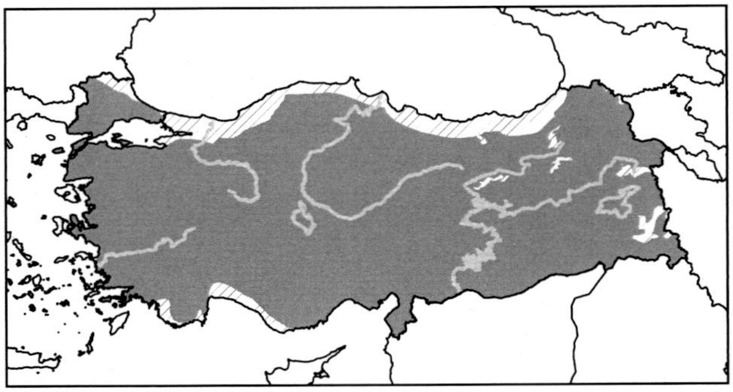

Breeding. On dry open steppe, plains and bare stony slopes. Largest density around Tuz Gölü (Inner Anatolia), where most commonly nests in ground holes of European Suslik *Spermophilus citellus*, but also in natural holes, rocks and piles of stones. The nest is a shallow cup of stems and grasses, lined with hair, feathers and wool. Very few nests examined. Eggs: C/6 (3); but broods of 3–5 flying young. Earliest fledged young in **Southeast Anatolia** (at Ceylanpınar) on 6 May 1970, indicates laying *c.*4 April. Most data from **Inner Anatolia**: display late March/April, first eggs 6/7 April, young in the nest in late April (earliest 20 April) and May, and fledged young from mid May; a male was feeding recently fledged young on 10 May 1993 while his mate was building a nest for a second brood (PC); there are records of females incubating in late May–June, feeding young in the nest in June, and fledged broods in June/July, almost certainly second broods. Records from other regions include: a nest with young on 7 May 2004 (GMK), and recently fledged young on 10 June 2001 (PC), in **East Anatolia**; an adult carrying food in **Black Sea Coastlands** on 15 June 1991 (C. G. Bradshaw); an adult incubating eggs, or brooding small young, in **Aegean**, on 28 May 1999 (J. Hollinrake); two pairs with fledged young in **Marmara** on 23 June 1973 (R. F. Porter); and several records of fledged young in **Mediterranean**, mainly in June.

NORTHERN WHEATEAR — *Oenanthe oenanthe*
Kuyrukkakan

(Du: Tapuit; Fr: Traquet motteux; Ger: Steinschmätzer; Swe: Stenskvätta)

Subspecies. All breeding birds were referred to *libanotica* by Roselaar (1995), but migrants appear to include at least some nominate *oenanthe* (e.g. NHM 98.9.1.2432, taken in the Taurus by Danford) and apparently *rostrata* (none seen by us). We have not examined the validity of these subspecific limits, but at least in more westerly parts of the range, including Turkey, the differences between *libanotica* and *oenanthe* do appear rather small.

Status and distribution. National conservation status: Least Concern (population apparently decreasing). Summer visitor that breeds throughout the country, apart from (with local exceptions) the Black Sea and Mediterranean littorals, and the open plains of Inner and Southeast Anatolia, where suitable habitat is scarce or lacking. Breeds mainly in montane areas, with altitude increasing from west to east: generally at 100–1,500m in the west (but to 2,300m on Uludağ), 1,100–2,700m in the Taurus and Pontics, and at 1,000–4,000m further east.

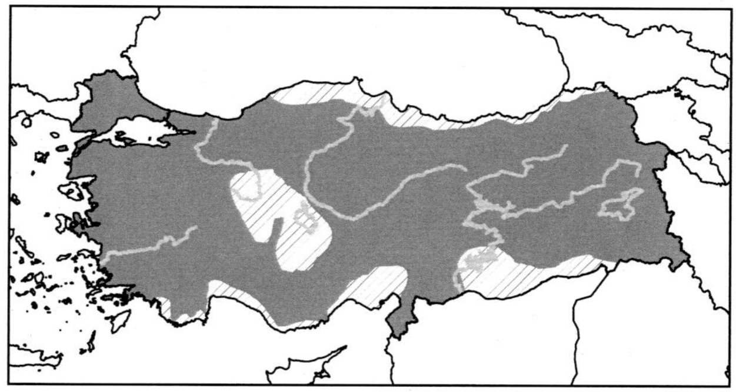

Migration commences early spring (earliest 28 February) and terminates late autumn (last 27 November), with peak numbers in early April–early May and mid September–mid October (16th), and passage more noticeable in spring than in autumn.

Breeding. On open stony slopes and pastures, with scattered bushes and boulders, often near rock outcrops and cliffs, and mainly in upland and montane areas. Nests in a hole or crevice in a rock, in piles of stones, under a rock or (quite commonly in Turkey) in ground burrows of mammals; base of plant stems, grass, moss and roots, with a shallow cup lined with fine grass, hair, wool and feathers. Eggs: C/4 (1). Young: B/4–5 (2). Earliest records, fledged young observed in the Aladağ (**Mediterranean**) on 12 May 1992 and 15 May 1970, indicate egg laying mid April. Also nest building at Sultandağlar on 19 April 1967 and copulation seen there on 7 May 1967 (R. F. Porter), nest building at Demirkazık on 13 May 2004 (N. Williams), two nests with half-grown and large young respectively in Aladağ on 12 June 1993 (PC), a nest with young at Karanfil Dağ on 1 July 1969 (Sutton & Gray 1972), and recently fledged young on Torosdağ on 23 July 1971. In **Inner Anatolia**, nesting generally by the first week of May (Wadley 1951); a nest had four eggs at Erciyes Dağ on 28 May 1972; and three broods of fledged young in June (earliest 4 June). In **East Anatolia**, nest building at Nemrut Dağ on 9 June 2004 (PC), and a nest had young there on 28 May 1990; two nests had young and two broods of recently fledged young near Doğubeyazıt on 8 June 2001, and there were three broods there on 16 June 2004 (PC); there were both eggs and young at Erzurum in the first week of June (McGregor 1917). In northeast **Black Sea Coastlands**, nest building was seen at Sivrikaya on 13 June 2004 (B. Pleasance) and fledged young were near Tosya on 7 June 1994 (Faldborg 1994). In **Marmara**, 'breeding seems to be late' at Uludağ, where there were three fledged young on 12 August 1963 (Jetz 1995).

PIED WHEATEAR *Oenanthe pleschanka*
Alaca Kuyrukkakan

(Du: Bonte Tapuit; Fr: Traquet pie; Ger: Nonnensteinschmätzer; Swe: Nunnestenskvätta)

Subspecies. Monotypic.

Status and distribution. National conservation status: Least Concern (population apparently stable). Breeding probable in East Anatolia and northeast Black Sea Coastlands, on barren montane slopes with scattered boulders or rocky outcrops, some fairly lush scrub and other low vegetation, principally at 800–2,300m. Presumably restricted as a breeder to areas around Van, the Kaçkar range (especially close to the Georgian and Armenian borders), at Doğubeyazıt, the Bingöl area and the Munzur range near Erzincan, which appears to represent the species' westernmost limit in the breeding season. Recently also reported in late June and July from the extreme southeast, but no evidence of breeding, and from the Diyarbakır area of Southeast Anatolia, but evidence of breeding is inconclusive and uncertainty persists as to the species identification. There are also reports in summer from further west (e.g. at Yozgat and Kayseri, Inner Anatolia), but breeding is uncertain and these sightings might reflect confusion with heavily worn O. *finschii*. The breeding range in Turkey, as in Iran (Haffer 1977), apparently overlaps with the closely related Eastern Black-eared Wheatear *O. melanoleuca*, and a possible hybrid has been reported from the extreme east (Ullman 1992). The two may favour different habitats in Turkey, but this requires clarification.

Otherwise an uncommon passage migrant, principally through the eastern two-thirds of Turkey, and very rare further west to coastal Aegean and Marmara. The earliest spring record is from Çukurova (Mediterranean) on 16 March, with passage continuing until at least 17 May, but with most between mid March and mid April, and the majority of records coming from the large (and comparatively well-watched) Mediterranean deltas. In recent years, *O. cypriaca* has proved equally if not more regular on the south coast of Turkey, suggesting that at least some former claims of *O. pleschanka* from this area (and perhaps even Inner Anatolia) actually involved *O. cypriaca*. Certainly there are comparatively few observations of the present species in autumn (from early August and perhaps even late July in the west), with the latest records from around Van Gölü (East Anatolia), on 1 October, and in Çukurova, on 29 October.

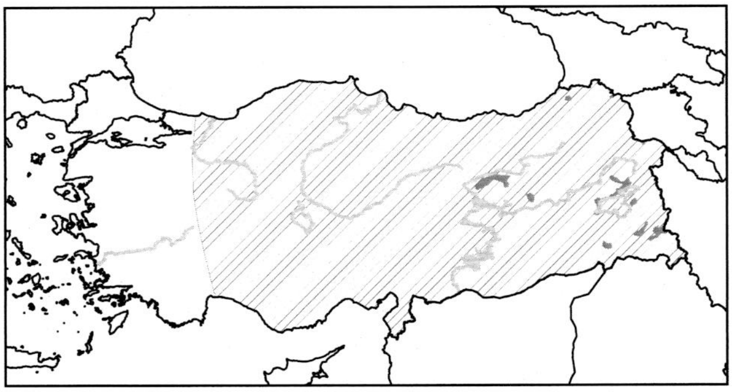

Breeding. Warncke (1964–65) found an empty nest, constructed of grass and roots, west of Gaziantep (**Southeast Anatolia**) on 2/3 May 1964, and a nest (no details) was found just west of Bingöl (**East Anatolia**) in July 1966, but these records require confirmation, given confusion between Pied and other wheatears, especially some Esatern Black-eared Wheatears *O. melanoleuca* (Beaman 1986). A pair was in suitable breeding habitat, above Maden Koprubaşı, south of Ispir (**Black Sea Coastlands**) on 22 June 1990 (Kirwan & Martins 1994). At Van (East Anatolia) there was a singing male on 1 May 1983 (Kirwan *et al.* 2003) and a pair with three juveniles on 29 July 1990 (Kirwan & Martins 1994), and a male and 3+ females/juveniles near the Armenian border, north-west of Iğdır (East Anatolia), on 6 August 1992 (Kirwan & Martins 2000a). At Kralkızı Reservoir, near Diyarbakır (Southeast Anatolia), on 28 April 2000, a female emerged from a cavity in a rock atop which a male was singing and the two flew higher up the hillside, and on 9 June 2000 a female was carrying possible nest material (Karakaş & Kılıç 2005b). Possibly a rare breeder in Armenia, but there are no nesting records (Adamian & Klem 1999). Elsewhere breeds on stony hillsides with rock outcrops, gullies and ravines, the nest placed in a hole or crevice in a bank or a rock, or under a stone; usual clutch 4–6, with two broods per year, and first eggs laid in late April/early May.

CYPRUS WHEATEAR *Oenanthe cypriaca*

Kıbrıs Kuyrukkakanı

(Du: Cyprustapuit; Fr: Traquet de Chypre; Ger: Zypernsteinschmätzer; Swe: Cypernstenskvätta)

Subspecies. Monotypic.

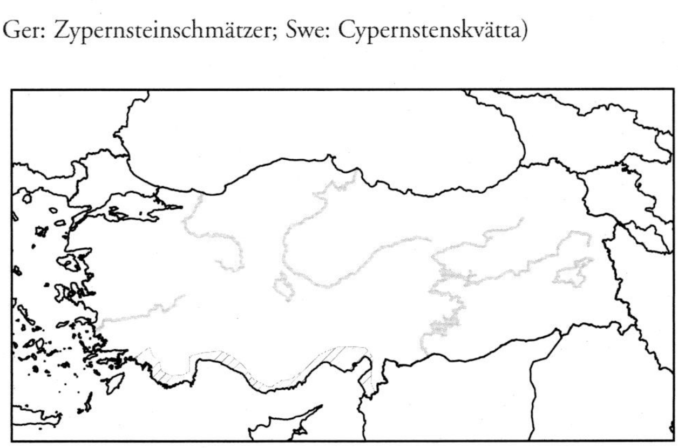

Status and distribution. Regular spring overshoot to south-coast localities, principally the well-watched Göksu and Çukurova deltas (Mediterranean), with the westernmost record from Köyceğiz Gölü (Mediterranean) on 16 April 1997 (Kirwan *et al.* 2003), but the species is considered regular at this season at nearby Fethiye (P. Hope) and has recently been recorded on several occasions in the environs of Antakya (A. Atahan). As mentioned under Pied Wheatear *O. pleschanka*, at least some (and possibly many) of the records previously attributed to the latter species from Mediterranean (and perhaps southernmost Inner Anatolia) in spring are now believed to have involved Cyprus Wheatear (see Martins 1989), given that the present species has now been demonstrated to be rather the commoner of the two in early spring in this region. Almost all records are between 11 March and 16 April, but with scattered reports until 17 May, and even one record as late as 4 June. No autumn records, but clearly should be searched for at this season, especially in the Hatay, as two of the four Lebanese records are dated October (Ramadam-Jaradi *et al.* 2008).

EASTERN BLACK-EARED WHEATEAR *Oenanthe melanoleuca*
Karakulaklı Kuyrukkakan

(Du: Oostelijke Blonde Tapuit; Fr: Traquet oreillard; Ger: Mittelmeer-Steinschmätzer; Swe: Medelhavsstenskvätta)

Subspecies and taxonomy. All are *melanoleuca* which until recently was usually treated as a subspecies of *O. hispanica* (Western Black-eared Wheatear). Since the publication of Sangster *et al.* (1999), the two are increasingly being recognised as species, despite broad intergradation between *hispanica* and *melanoleuca* in the Balkans; their field identification has recently been subject to a major review (Ullman 2003).

Status and distribution. National conservation status: Least Concern (population apparently stable). Widespread and fairly common summer visitor throughout the country, except the Black Sea littoral. Less confined to open rocky ground and boulders than other *Oenanthe*, instead requiring scattered shrubs and stunted trees for perches. Usually found at lower altitudes than Northern Wheatear *O. oenanthe*, but they overlap widely. Breeds mainly at 0–1,200m in the west and centre of the country, but up to 1,600m (–2,150m) in the Taurus, frequently to 2,300m in East Anatolia and reaches 3,000m on Nemrut Dağı.

Recorded regularly from across the whole country, including areas where the species does not breed, except coastal areas of Black Sea Coastlands and Thrace, between early April and late September, being a widespread migrant throughout the country especially in autumn. Spring passage continues through the south-coast deltas until at least mid May. The earliest spring record is from Antakya province (Mediterranean) on 4 March, and the latest in autumn from Kızılırmak Delta (Black Sea Coastlands), on 25 October.

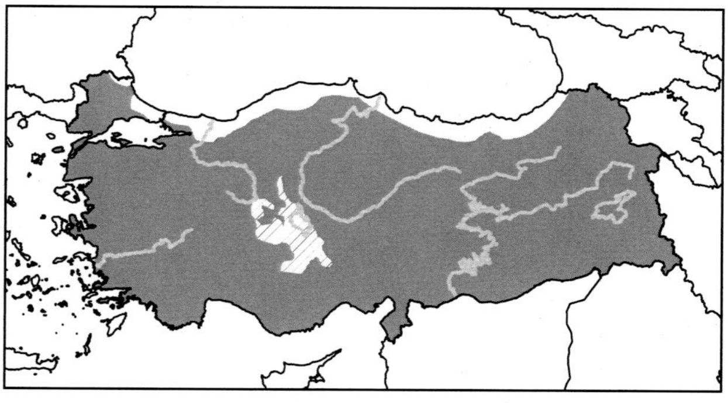

Breeding. Inhabits rocky ground, usually partially scrub-covered, preferring slopes; also cultivation, especially with stone walls. Nest usually in a hollow, often on the ground under a rock or stone, or in a crevice in rocks, a low cliff, or a stone wall; a base of plant stalks, grass and roots, with a shallow cup, lined with roots and hair. Eggs: C/4 (1), C/5 (2), C/6 (2). Young: B/4 (2), B/5 (2). In **Aegean**, nest building near Izmir on 5 May 1899 (Selous 1900) and near Muğla on 29 April 2001 (PC), three nests had 4–6 eggs, near Izmir, in the first half of May (McNeile), and one had six eggs, near Aydin, on 8 May 1899 (Selous 1900); a nest had five young, two days old, near Aliağa on 20 May 1970, another had four well-grown young, near Izmir, on 19 May 1951 (McNeile), and another small young, near Akköy on 25 May 2004 (M. Demain); fledged young were near Menemen on 21 June 1966 (R. F. Porter). At Uzuncaburç (**Mediterranean**), a female was nest building in a hole in the outer wall of a café, while the male displayed 1m away on a café table, on 7 May 2004; a nest had four young, four days old, on 10 June 1992, recently fledged young were seen the same day (PC), and several pairs with fledged young on 17 June 1996 (GMK); nest building was seen at Akseki (Mediterranean) on 19 April 2004 (PC); a nest had five young, 3–4 days old, near Mut on 13 May 2005 (J. Hodson). In **Southeast Anatolia,**

four nests had young on 1–2 June 1998 at Gaziantep (T. Quittenden), one nest there had large young on 6 June 2006 (S. Williams), and a nest had five eggs there on 18 May 2007 (PC). In **Inner Anatolia**, nest building was reported at Hasan Dağ, on 21 April 2004 (A. Dixon), on 28 April 1946 (Wadley 1951), and at Çubak on 7 May 1969. In **Marmara**, a pair with fledged young were at Uluabat Gölü on 15 June 1998 (Welch & Welch 1998b), a family party was at Vize, in Thrace, in June 1973 and a pair with four fledged young were at Kosedere on 26 June 1973 (R. F. Porter). The species is present early April–September, and egg laying commences mainly early May. Elsewhere, it is double-brooded, but the records above all clearly relate to first broods.

DESERT WHEATEAR *Oenanthe deserti*
Çöl Kuyrukkakanı

(Du: Aziatische Woestijntapuit; Fr: Traquet du désert; Ger: Wüstensteinschmätzer; Swe: Ökenstenskvätta)

Subspecies. Presumably the nominate form (Roselaar 1995), at least in the south, but the first Bulgarian record was considered referrable to Central Asian *atrogularis* (Nankindov 2003). However, *atrogularis* is rarely admitted by most authorities.

Status and distribution. National conservation status: Endangered (population apparently decreasing). Recently claimed breeding in Southeast Anatolia, and two additional records in recent years suggest that it is perhaps a very local resident in that region (and perhaps adjacent parts of East Anatolia). Records in Mediterranean (and elsewhere) presumably refer to vagrants. Peter (1995) reported a pair in courtship display and carrying nesting material to a small hole in the ground at Birecik (Southeast Anatolia) on 19 April–3 May 1985 and found a male in the same place on 18 April 1988. Brief plumage details are presented in support of the occurrence. Additional records are as follows: males at the former Amik Gölü, Antakya province (Mediterranean) on 3 September 1956 (Kumerloeve 1975b), and at 2,200m on Kuh Dağı, near Erçek Gölü (East Anatolia) on 17 August 1973 (Gallner 1976), a female at Fevzipaşa, Gaziantep province (Mediterranean) on 6 September 1988 (Kirwan & Martins 1994), with three males at Çınar-Göksu Reservoir, Diyarbakır (Southeast Anatolia), one at Cizre (Southeast Anatolia) on 13 June 1999 (Kirwan *et al.* 2003), one at Kuyucuk Gölü, Kars province (East Anatolia) on 25 September 2006 (Balmer & Betton 2007a), and another, photographed, at the Sakarya Delta (Black Sea Coastlands) on 14 November 2006 (S. Bekir). It is worth noting that substantiative details are unavailable for several of these records. In particular, Gallner (*op. cit.*), who reported a male, claimed a number of peculiarly unlikely birds (in some cases, patently incorrectly) from the Van region, casting a degree of doubt on all of his more remarkable observations. Also doubtfully reported from the Van area in April 1972 (Kumerloeve 1975b) and at Kralkızı Reservoir, Diyarbakır (Southeast Anatolia) on 12 May 2000 (Karakaş & Bırıcık 2005b).

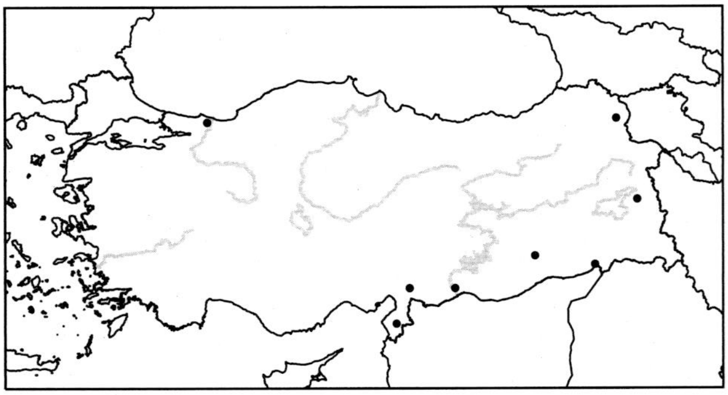

Breeding. For breeding evidence in Turkey see above and Peter (1994). The species does not breed in northern Syria, adjacent to the Birecik area, but does breed somewhat further south (Baumgart *et al.* 1995).

FINSCH'S WHEATEAR *Oenanthe finschii*
Ak Sırtlı Kuyrukkakan

(Du: Finsch' Tapuit; Fr: Traquet de Finsch; Ger: Felsensteinschmätzer; Swe: Finschstenskvätta)

Subspecies and taxonomy. Virtually all appear to be nominate *finschii*, whilst the extreme eastern quarter of Turkey, the Caucasus and western Iran represents a broad zone of intergradation between *finschii* and *barnesi*, where birds possess the body coloration of the former and tail pattern and overall size of the latter (Roselaar 1995). We have only examined one from the Turkish portion of the 'zone of intergradation', a female in the Natural History Museum, Tring (1907.12.21.131), but this seems identical in all respects to nominate *finschii*. In any case, *barnesi* appears to be a very weakly differentiated race, based on material in Tring and Chicago, although it was maintained by Collar *in* del Hoyo *et al.* (2005). Species has been considered conspecific with Mourning Wheatear *O. lugens*.

Status and distribution. National conservation status: Least Concern (population apparently stable). Widespread and locally common summer visitor, and passage migrant, with small numbers overwintering in the south and southwest, principally in Mediterranean, where it regularly reaches coastal areas down to sea level in small numbers, and the Anti-Taurus range. Entirely absent from Marmara and generally rare in Black Sea Coastlands, where largely confined to the extreme east, and in northern parts of Inner Anatolia. In the breeding season, occurs on open and sparsely vegetated slopes from at least 100–2,300m, although largely found above 500m and below 2,000m, and present in the breeding areas from about mid March to mid October, but may remain in the coastal wintering areas until at least the end of March. Nevertheless, small numbers winter well inland (e.g. across Inner Anatolia) and at moderate altitudes, whilst others move even further south, to the northern Levant, to winter.

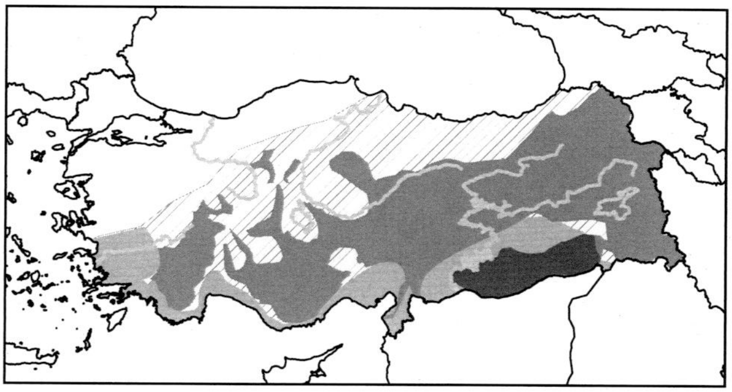

Breeding. On open rocky slopes and outcrops, and in steep-sided rocky valleys, generally bare or with only scattered scrub. Nests in a hole or crevice amongst rocks or in a small cliff; uses plant stems, grass and roots, lined with fine roots, hair and wool. Eggs: C/5–6 (2). Young: B/4 (1); broods of fledged young commonly only 2–3 (n=8), but also 4–6 (n=5). At Pamukkale (**Aegean**), nest building early April and fledged young on 15 May 1973, indicating egg laying in mid April. In **Mediterranean**, a nest had young, near Uzuncaburç, on 7 May 2004 (J. Mason), adults were carrying food in late May, and there were fledged young in June–July (earliest 1 June). In **Southeast Anatolia**, a nest had six eggs near Gaziantep on 2 May 1964 (Warncke 1964–65); a pair and four recently fledged young were near Ceylanpınar on 6 May 1970 (R. F. Porter), indicating egg laying in early April; many pairs with fledged broods, and some pairs already nest building again, were at Çınar-Göksu Reservoir, south of Diyarbakır, on 5 May 2004 (GMK). In **Inner Anatolia**, nest building was seen at Hasan Dağ on 21 April 2004 (A. Dixon), and a nest there had young, *c.*3 days old, on 24 April 2004 (PC); a nest had young at Şereflikoçhisar on 2 May 1973, and two nests had young at Hasan Dağ on 5 June 1992 and 18 June 1992 (PC); there are four records of fledged young May–August (earliest 8 May); evidence of double-brooding from Hasan Dağ on 27 May 1993, when a female was nest building (subsequently had five eggs), *c.*30m from the nest used for the first brood, which had young on 10 May 1993 (PC). In **East Anatolia**, four nests had young near Erçek Gölü in early June 2001 (T. Quittenden); there are four records of fledged young on 5 June, and six others in mid June, late July and early August.

[RED-RUMPED WHEATEAR *Oenanthe moesta*]
Büyük Kızılca Kuyrukkakan

(Du: Roodstuittapuit; Fr: Traquet à tête grise; Ger: Fahlbürzel-Steinschmätzer; Swe: Rödstjärtad stenskvätta)

Subspecies. The Middle Eastern form *brooksbanki* is the subspecies most likely to occur but differences are very slight and there is much overlap with the nominate form from North Africa, suggesting that variation within the species is best not recognised at nomenclatural level. Indeed, Collar *in* del Hoyo *et al.* (2005) treated the species as monotypic.

Species not admitted to the Turkish list (Kirwan *et al.* 1999). Three records: males between Gaziantep and Kilis, Southeast Anatolia and between Uzungeçit and Uludere, East Anatolia, in May 1981, and one north of Halfeti, Southeast Anatolia, in April 1983 (Glimmerveen & Hols 1986), are inadequately described, all records being from areas where Red-tailed Wheatear *O. xanthoprymna* is now known to occur. Following Roselaar (1995), we do not consider this species to have been reliably recorded in Turkey. It is interesting that there are no recent records from Syria, where the species formerly bred (Snow & Perrins 1998).

RED-TAILED WHEATEAR *Oenanthe xanthoprymna*
Kızılca Kuyrukkakan

(Du: Westelijke Roodstaarttapuit; Fr: Traquet oreillard; Ger: Mittelmeer-Steinschmätzer; Swe: Medelhavsstenskvätta)

Subspecies and taxonomy. Only nominate *xanthoprymna* is well known in the country although there is a recent published, but undocumented, claim of *chrysopygia*, at Nemrut Dağı, Adıyaman province (Ochiatto 2001) and MÖ has unpublished observations of birds belonging to this form from around Kemaliye, Erzincan province, and Arapgir, Malatya province, in May 2005. Roselaar (1995) had already speculated that *chrysopygia* might be found breeding in eastern Turkey, on the flanks of the Aras Valley and perhaps even west to Erzurum (East Anatolia), but there is a paucity of recent proof that *xanthoprymna sensu lato* even occurs in the regions he mentioned. However, *chrysopygia* has been discovered recently close to the easternmost frontier of Turkey with Iran and Armenia, in Ararat province, Armenia (Ananian & Ghasabian 2005). The relationships between *xanthoprymna* and *chrysopygia*, given occasional reports of intermediates demand further study, but there is only a limited zone where hybridisation might occur (in northern Malatya and southern Erzincan provinces), which is far less significant than that documented as occurring between Pied *O. pleschanka* and Eastern Black-eared Wheatears *O. melanoleuca* (Haffer 1977). Apparent hybrids between the two forms of red-tailed wheatear are sometimes given the name '*cummingi*', but the characters assigned to this form also appear to be shown by first-year *xanthoprymna* (Helbig 1984, Roselaar 1995). Bates (1935), Nikolaus (1987), Dickinson (2003), Rasmussen & Anderton (2004) and Collar *in* del Hoyo *et al.* (2005) all considered the two forms to be species, as had Ivanov (1941), but the newly invented English names Chestnut-rumped (*xanthoprymna*) and Rusty-tailed Wheatears (*chrysopygia*), whilst not entirely inappropriate, are decidedly forgettable, and we would advocate use of the geographic modifiers, Kurdish and Persian, for these two taxa. We agree that *xanthoprymna* and *chrysopygia* are better treated as two species.

Status and distribution. National conservation status: Vulnerable (population apparently decreasing). Status and distribution in Turkey earlier reviewed by Kumerloeve *et al.* (1984) and Kasparek (1986). A generally uncommon, sparsely distributed summer visitor to Southeast and East Anatolia, where proved breeding only in the foothills and mountains abutting steppic areas, e.g. around Erzincan, Malatya, the hills around Elazığ and is abundant in the Beydağları between Adıyaman and Malatya, and in the extreme east at the borders with Iraq and Iran. Breeding is also likely around Van Gölü, *e.g.* at Nemrut Dağı, Tatvan, and in Erzurum province. In Turkey, prefers dry rocky mountain slopes or valley bottoms, covered with patches of dwarf shrubs and herbs (e.g. *Ferula*), interspersed with bare rocks, scattered boulders or stunted trees. Breeds at 1,100–2,000m alongside the Euphrates, *c.*1,500m around Gaziantep and at 2,000–2,100m in Hakkari, with regular reports from Yesilce village, Halfeti, Nemrut Dağı near Adıyaman and the Hakkari region. Usually arrives in early May in the breeding areas and departs by early August, and has very rarely been recorded on passage in Mediterranean. There are exceptionally early records, from Halfeti (Southeast Anatolia), on 20 March and 12 April, and given the exceptionally mild climate in this region it is not impossible that the species might occasionally attempt to overwinter there.

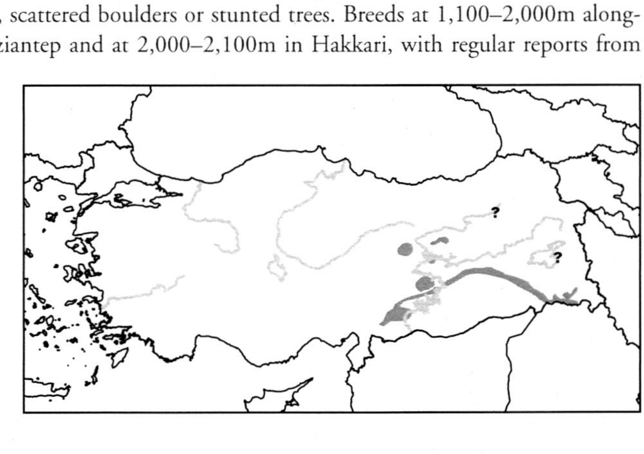

Breeding. On dry, rocky montane slopes, usually with scattered boulders and crags. Nests in a hole in a rock face, or in a crevice between fallen rocks; a base of dry grass and plant stems, with a shallow cup, lined with finer grass and hair. Eggs: no details but, elsewhere, usual clutch is 4–5. Young: B/4 (1); the number of fledged young most frequently observed is, however, only two (n=10). In **Southeast Anatolia**, most records are from Gaziantep area: young in nest mainly late May–early June (earliest 13 May), fledged young mainly first half of June (earliest 27 May) and the most common egg laying date can be back-dated to *c.*7 May; at Nemrut Dağı, a nest had young on 19 June, two broods of fledged young were seen on 15–16 June 1996 (GMK), and three pairs and fledged young on 12 July 1981 (J. Martens). In **East Anatolia**, adults were carrying food at Hakkari on 17 June 1978, and fledged young were there on 14 July 1990; a pair was feeding two newly fledged young, and another pair carried food to a rock crevice, near Uludere on 8 August 1981, which implies double-brooding.

[VARIABLE WHEATEAR *Oenanthe picata*]
Doğu Kuyrukkakanı

(Du: Picatatapuit; Fr: Traquet variable; Ger: Elstersteinschmätzer; Swe: Orientstenskvätta)

Subspecies. Unknown.

Species not admitted to the Turkish list (Kirwan *et al.* 1999). Four records all from the Mediterranean region: a male west of Silifke on 21 March 1967, a male north of Karataş on 23 March 1967 (Warncke 1968), three males in a young *Eucalyptus* plantation at the Tarsus Delta on 6 April 1971 (Warncke 1972) and a male observed in the Göksu Delta on 18 March 1972 (Eggers & Streese 1979), but all lack sufficiently detailed descriptions to warrant acceptance. Kasparek (1990a) suggested that confusion with Hume's Wheatear *O. albonigra* (note the correct spelling of the specific name; David & Gosselin 2002b) might have occurred in at least some instances. The latter would appear to be an even more unlikely vagrant in our opinion. It is possible that these records refer to Mourning Wheatears *O. lugens* of the dark morph breeding in northern Jordan and southern Syria, previously identified as Variable (Eastern Pied) Wheatear (Hollom *et al.* 1988, Andrews 1994, Tye 1994). The most plausible explanation, however, is that these records were of Cyprus Wheatear *O. cypriaca*, which is now known to be a regular overshoot to the Turkish south coast (Kirwan & Martins 1994, 2000a); dates of all these records fall within the passage period recorded for the latter species. There is also a recent claim of *O. picata* from Lebanon, in December 2000 (Ramadan-Jaradi *et al.* 2008).

MOURNING WHEATEAR *Oenanthe lugens*
Kara Sırtlı Kuyrukkakan

(Du: Westelijke Rouwtapuit; Fr: Traquet deuil; Ger: Schwarzrücken-Steinschmätzer; Swe: Sorgstenskvätta)

Subspecies and taxonomy. The subspecific identity of the first Turkish record was unrecorded, and the specimen, which was formerly housed in Munich, appears to have been lost (Kumerloeve 1975a), but the second individual was apparently of the form *halophila*, which breeds in North Africa from Morocco east as far as Egypt's western desert (Baha el Din & Baha el Din 2000). The Mourning Wheatear superspecies offers a fascinating challenge to taxonomists working with Western Palearctic birds; this complex may involve as many as five species. Fortunately, an extensive DNA and field study of this group is in progress (Shirihai *et al.* in prep.).

Status and distribution. Four records: a male was at Iskenderun (Mediterranean) on 28 February 1914 (Kasparek 1992a), a female *halophila* at the Göksu Delta (Mediterranean) on 15 April 1996 (Kirwan & Martins 2000a), a male south of Çatak (East Anatolia) on 23 May 2005 (T. Epple & U. Pittius *per* S. Pfützke), and, most recently, one apparently of the race *lugens* photographed at the Göksu Delta on 27 March 2008 (R. Wüst-Graf).

The species has also been recorded as a vagrant in Cyprus, in April 1993 (Snow & Perrins 1998) and March 2004 (*Birding World* 18: 32), but in neither case was the race concerned determined (P. F. Stewart *in litt.* 2003). The nominate form breeds quite widely in Syria, as far north as the Aleppo region (Baumgart *et al.* 1995), as well as in Lebanon (Ramadan-Jaradi *et al.* 2008).

WHITE-CROWNED WHEATEAR *Oenanthe leucopyga*
Ak Tepeli Kuyrukkakan

(Du: Witkruintapuit; Fr: Traquet à tête blanche; Ger: Saharasteinschmätzer; Swe: Vitkronad stenskvätta)

Subspecies and taxonomy. The form involved is not known, although Cyprus records have apparently involved the more westerly breeding nominate race, and Turkish occurrences have probably been from a similar vector. Collar *in* del Hoyo *et al.* (2005) suggested that race *ernesti* might be better considered a synonym of the nominate.

Status and distribution. Vagrant. Three records: a young bird at Pınarbaşı (Mediterranean) on 12 August 1993 has been accepted by the Italian rarities committee (Sorace 1996); an adult at Yumurtalık lagoon, Çukurova (Mediterranean) on 9 March 1996 was well described (Kirwan & Martins 2000a); and an adult photographed at Phaselis, south of Kemer, Antalya province (Mediterranean), on 2 March 2005 (T. Sacher & U. Lasch), has been submitted to the German rarities committee.

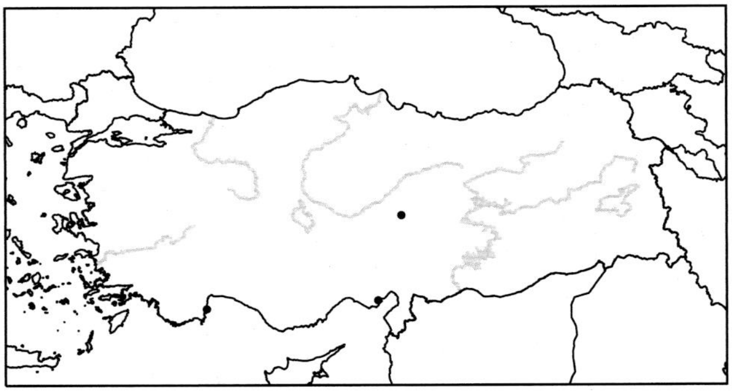

There are at least four records in Cyprus, all in spring between February and April (Flint & Stewart 1992, Kirwan 1999), with a single documented record from Greece, on Crete in April (Handrinos & Akriotis 1997).

RUFOUS-TAILED ROCK THRUSH *Monticola saxatilis*
Taşkızılı

(Du: Rode Rotslijster; Fr: Monticole de roche; Ger: Steinrötel; Swe: Stentrast)

Subspecies. All are nominate *saxatilis*, though Stepanyan (1964) described the subspecies *coloratus* from the little-explored Bingöl Dağları (East Anatolia) and Roselaar (1995), though sceptical of the validity of this form, ascribed it a potential range that covered much of the southeastern portion of East Anatolia, although he, presumably inadvertently, omitted the type locality from his map. Kirwan (2001a) did not admit the validity of *coloratus* and, in this, was followed by Collar *in* del Hoyo *et al.* (2005).

Status and distribution. National conservation status: Least Concern (population apparently stable). Fairly widespread and not uncommon summer visitor to high, montane areas, where seems rather ubiquitous above 1,500m and is widespread throughout the country, particularly in the eastern third but slightly more local westwards. In the breeding season, absent from Thrace, steppe-like areas of Inner Anatolia and much of Southeast Anatolia, except on isolated mountains. Breeds at 800–1,500m in the west and on the fringes of Inner Anatolia, at 1,500–3,000m on Uludağ and in the Taurus, at 2,000–3,000m around Van, and at 1,100–4,000m elsewhere in the east and the Pontics.

Spring migration may persist until early June, when somewhat more widespread, thus partially obscuring breeding distribution. Most are well established on the breeding grounds by early May and southbound passage is well under way by early September. The earliest record on the south coast is 28 March, and there are a few early April records from the south and Inner Anatolia. Return migration commences early August with the latest records on 16 September.

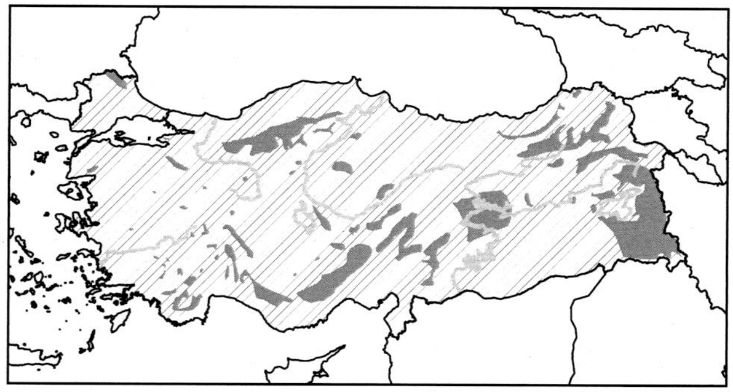

Breeding. On barren rocky slopes, usually in high mountains, in alpine scrub with boulders, and in gorges, cliffs and rocky areas, sometimes wooded, at lower elevations. Nests in a cavity in a rock or cliff, or on a ledge of a rock, partially screened by vegetation; builds a cup of plant stems, grass, roots and moss, lined with roots and fine grass. Eggs: C/5 (1). Young: no data. In **Marmara**, there were four pairs with fledged young at Uludağ on 8/9 July 1966 (R. F. Porter). In **Inner Anatolia**, nest building near Şereflikoçhisar on 22 May 1992 (PC); a nest concealed in a clump of *Salvia* on the slopes of Meketuzlazı had five fresh-looking eggs on 22 April 1970 (McNeile); at Karadağ, a pair was carrying food on 30 May 2007, there was a recently fledged young, just able to fly, on 10 June 1907 (Ramsay 1914), and several adults were carrying food on 19 June 1996 (GMK). In **Black Sea Coastlands**, a pair was at a nest at Sivrikaya on 15 June 1990 (A. Forsten), an immature was at Dilberduzu on 30 July 1996 (I. A. Green), and a female was carrying food near Ispir on 24 June 2004 (PC). In **Mediterranean**, nest building was seen in the Sultandağlar on 7 May 1967 (R. F. Porter); a recently fledged young, barely able to fly, at Demirkazık on 13 June 1993, indicates egg laying in mid May (PC); adults were carrying food near Çamardı on 15 May 1970 and Demirkazık on 9 June 1996 (GMK). Several adults were carrying food at Nemrut Dağ (**Southeast Anatolia**) on 16 June 1996 (GMK). At Nemrut Dağ (**East Anatolia**), a female flew to a nest in a hole, *c.*4m above ground, on 9 June 2004 (S. Williams), and juveniles were seen on 26 June 1989 and 8 July 1986 (R. Webb), and elsewhere adults were carrying food near Doğubeyazıt on 16 June 2004, at Erçek on 20 June 2004, and near Erzurum on 22 June 2004 (PC). Elsewhere, sometimes more than one brood per year.

BLUE ROCK THRUSH — *Monticola solitarius*
Gökardıç

(Du: Blauwe Rotslijster; Fr: Monticole bleu; Ger: Blaumerle; Swe: Blåtrast)

Subspecies. Those in the north appear to be all *solitarius*, with those in much of the south apparently intergrades between nominate and *longirostris*; specimens from the southeast quarter are unavailable, but the populations there probably relate to true *longirostris* (Roselaar 1995, Clement & Hathway 2001). Variation in this species throughout mainland Eurasia is clinal, and we have considerable difficulty in observing any diagnostic characters in specimens that warrant recognition of *longirostris*. One of the purported characters of *longirostris*, its smaller size, had already been questioned by Svensson (1992). As noted by Roselaar (*op. cit.*), the name *behnkei*, which has been applied to those in Crete and might be equally used for southern Turkey populations, must certainly be considered a synonym of either *longirostris* or nominate *solitarius*, as these birds merely represent intermediates between the other two forms (see also Cramp 1988, Collar *in* del Hoyo *et al.* 2005).

Status and distribution. National conservation status: Least Concern (population apparently stable). Fairly widespread and not uncommon resident, being locally common in Marmara, Aegean, Mediterranean, and in lower valleys of East Anatolia. Partial migrant in Mediterranean and southern Aegean, and more locally across much of the rest of the country. Apparently absent from Thrace, and scarce inland in Black Sea Coastlands (especially the west), and very local in Inner Anatolia, as well as being absent from much of Southeast

Anatolia, except the fringes and some isolated mountains. Appears commoner in the Taurus (Mediterranean) than any other part of the country, and probably prefers lower altitudes than *M. saxatilis*, although they frequently occur sympatrically. Mainly recorded at 0–1,700m in the west and in coastal areas, but to 2,500m on Uludağ, in the Taurus and montane East Anatolia.

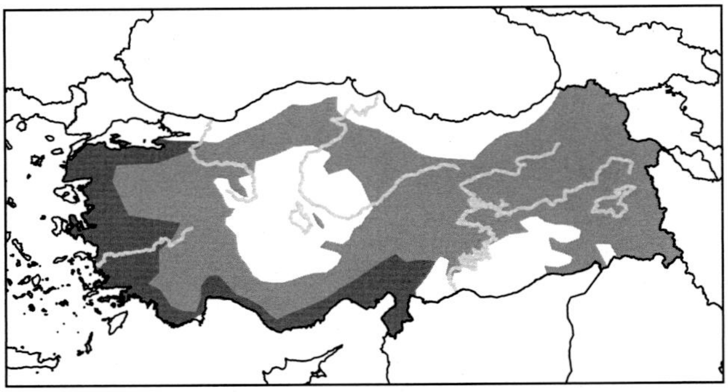

Winters (December–February) in moderate numbers throughout the Mediterranean littoral and at some localities in Aegean (e.g. Gediz and Büyük Menderes deltas). Departs breeding areas in early September, but migration continues until mid October, and in spring may not depart winter-only localities until as late as mid May.

Breeding. On rocky slopes with crags and rock outcrops, gorges, ruins and coastal cliffs, 0–2,500m, but generally below areas occupied by the Rufous-tailed Rock Thrush *M. saxatilis*. Nests in crevices in cliffs, in holes in caves, and once in a hole inside a disused school (PC); the nest is a cup of moss and grass, lined with roots and fine grass. Eggs: C/5 (1); elsewhere, the usual clutch is 4–5. Earliest record, a nest with young on 2 May 2003, which fledged on 9 May 2003, at Doğanbey (**Aegean**), indicates date of first egg as *c.*7 April. An adult and one juvenile were at Muğla on 15 July 1969. In **Mediterranean**, many males were on territory and singing, north of Silifke on 31 March 2000; a nest had three young, *c.*10 days old, south of Mut on 24 May 1993 (PC); a nest had young at Demirkazık on 7 June 1998 (T. Quittenden); recently fledged young were at Demirkazık on 13 June 1993 (PC); a pair was carrying food near Akseki, on 24–25 May 1999 (R. F. Porter); family parties were at various locations in June–July (some perhaps second broods). In **Inner Anatolia**, an adult female flew into a hole in a cliff (and remained there, presumably incubating) near Karapınar on 20 April 2004 (A. Dixon); a nest had young near Karapınar on 22 May 1998 (T. Quittenden), and adults carried food to a nest at Kızılcahamam on 8 June 1977 (Schubert 1979). In **Black Sea Coastlands**, a pair was carrying food south of Cimil Dağ on 12 June 1975 (R. F. Porter). In **Marmara**, an adult was carrying food at Babakale on 26 June 1973 (R. F. Porter). In **Southeast Anatolia**, a nest had five fresh eggs near Gaziantep on 2 May 1964 (Warncke 1964–65); at Işıklı a nest had young on 17 May 1993 (PC); adults were carrying food on 4 May 1996, 15 June 1996, and 13 May 2004 (GMK), and a female and two fledged young were there on 23 May 1994; a male was carrying food on 10 May 2004, and the pair subsequently had fledged young on 24 May 2004, within 100m of a displaying male (PC); a male carried food to a nest in a tall building near Durnalık on 23 May 2004 (PC). In **East Anatolia**, two pairs were carrying food near Görentaş on 1 June 1969 (R. F. Porter), an adult was carrying food near Doğubeyazıt on 16 June 2004 (B. Pleasance), and an adult was feeding nestlings at Nemrut Dağı in early June 2005 (MÖ, N. Özen).

[SIBERIAN THRUSH *Zoothera sibirica*]
Sibirya Ardıcı

(Du: Siberische Lijster; Fr: Grive de Sibérie; Ger: Schieferdrossel; Swe: Sibirisk trast)

Subspecies. The only mention of the species for Turkey refers to nominate *sibirica*.

Species not admitted to the Turkish list (Kirwan *et al.* 1999). Ergene (1945) stated that the nominate race had reached Turkey as a vagrant, but no details were presented. Porter *et al.* (1996) listed the species as a vagrant in Israel, but, as discussed by Shirihai (1996), this record has not been admitted to the Israeli list, being still under consideration by the records committee. There are no other records from the Middle East.

RING OUZEL *Turdus torquatus*
Boğmaklı Ardıç

(Du: Beflijster; Fr: Merle à plastron; Ger: Ringdrossel; Swe: Ringtrast)

Subspecies. Those in the northeast Pontics (Black Sea Coastlands) and further south in East Anatolia are *amicorum*, but those in the Taurus are intermediate between *alpestris* (which is probably the form in northwest Turkey) and *amicorum*. Variation between these two forms, and between nominate *torquatus* and *alpestris*, is clinal (Roselaar 1995; pers. obs.), although we disagree slightly with Roselaar (*op. cit.*) in considering those in southern Turkey to be overall much closer to *alpestris* than *amicorum*. Nonetheless, it is worth emphasising that pure *amicorum* represents a reasonably well-marked taxon that is clearly distinguishable from nominate *torquatus*.

Status and distribution. National conservation status: Least Concern (population apparently stable). A rather local but not uncommon resident, a partial migrant in uplands of East Anatolia and Black Sea Coastlands, and extremely locally and in very small numbers in all regions except Aegean and Southeast Anatolia. Principally breeds in the eastern two-thirds of the country, in the eastern Pontics (Kaçkar Dağları, Erzurum, Bayburt) and much of montane East Anatolia (e.g. around Van Gölü), as well as in the eastern Taurus, the northern and eastern fringes of Inner Anatolia, and Antakya (Mediterranean). Uludağ (Marmara) is the only certain breeding locality in the west, although there is a breeding report, at sea level, from southern Thrace (Kesan). Observations suggestive of breeding are also available from Bergama (Aegean), Burdur Gölü (Mediterranean) and Kızılcahamam (Inner Anatolia). Recorded at 1,300–1,500m on Uludağ and in the western Black Sea Coastlands, at 1,800–2,700m in the Taurus and 1,500–3,000m in the east.

More widespread on passage when reported from all regions, with large numbers recorded at the Bosphorus in autumn, and noted in moderate numbers in Inner Anatolia in spring. Autumn migration peaks in mid September to mid October at the Bosphorus, whilst spring peaks in Inner Anatolia are from early March to early April. Winters in small numbers at lower and middle altitudes in Mediterranean, and inland in southwest Aegean, more occasionally elsewhere.

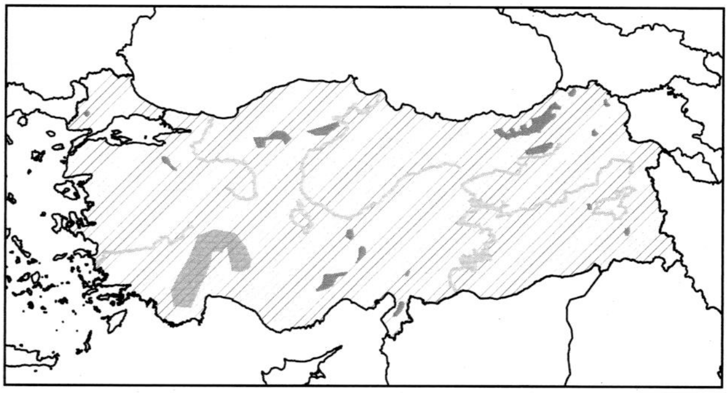

Breeding. In low shrubs above the treeline, on high rocky slopes, at up to 3,000m in the east. No details of nest and clutch size in Turkey but, elsewhere, the nest is usually on the ground, sometimes under a bush or shrub, or on a sheltered ledge of a bank or small rock outcrop, occasionally in a bush or a tree; a cup of grass, moss and leaves, with earth mixed in the base, lined with finer grass; usual clutch 4–5. In **Black Sea Coastlands**, a male with food was at a nest near Sivrikaya on 25 May 2002 (H. Dufourny). In **Mediterranean**, alarm-calling adults were on Karanfil Dağ in late April 1876 (Danford 1878) and at Torosdağ in mid-July 1971. A male was carrying food in the Aladağ on 29 June 2004 (K. P. Kendricks). In **East Anatolia**, fledged young were being fed in the crater at Nemrut Dağ on 7 June 1975, an adult was carrying food and alarm-calling there on 8 June 2004 (PC), and a pair was food-carrying at Yenimahalle, near Çatak, on 18 May 2006 (GMK *et al.*). In **Marmara**, a record of fledged young being fed at the edge of a pine forest at Kesan (a lowland area) on 21 May 1970, indicates date of first egg as about 21 April. In Armenia, nests in trees, bushes and on the ground; nests had four eggs on 23 April 1994, four newly hatched young on 11 May 1995 and three eggs on 13 June 1994; adults were carrying food on 14 June 1995, and a brood of three fledged young was observed on 2 June 1993 (Adamian & Klem 1999).

COMMON BLACKBIRD *Turdus merula*
Karatavuk

(Du: Merel; Fr: Merle noir; Ger: Amsel; Swe: Koltrast)

Subspecies. Clement & Hathway (2001) were content to consider all those in northern, western and much of southern Turkey as *aterrimus*, being replaced by *syriacus* in the southern coastal region. This was despite Roselaar's findings, which overlapped in respect of recognising the same range for true *syriacus*, but concluded that birds elsewhere in Turkey represented intergrades between *aterrimus* and *syriacus*. Roselaar's conclusion was based on the fact that their plumage matched one or the other, but mensural data were commensurate with the other taxon. Roselaar (*op. cit.*) also noted a male from Gaziantep (Southeast Anatolia) in winter, which was apparently nominate *merula*; it is certainly known to reach even further south, to the Levant, at this season. These findings demonstrate, rather classically, the fundamental difficulties in attempting to diagnose taxa that lack consistent, qualitative differences (see Sangster *et al.* 1998). A more far-reaching review of the taxonomy of this species than can be attempted here is clearly required, but we suggest, on the basis of the above and our review of specimen material in the Natural History Museum (Tring), that *aterrimus* at least be considered a synonym of *syriacus*.

Snow (1997) briefly discussed geographical variation in *Turdus merula*, as an example of the problems he considered inherent to application of the Phylogenetic Species Concept to very wide-ranging polytypic species, and speculated that *syriacus* might, along with *intermedius* from montane Central Asia, form a link between smaller forms of Western Europe and Macaronesia and very large forms in the eastern Palearctic.

Status and distribution. National conservation status: Least Concern (population apparently decreasing). A widespread and common resident, or partial migrant, over much of the country. Recorded year-round except over much of Inner Anatolia, where breeding is largely confined to isolated parkland areas and wooded montane isolates, and many parts of southeast Turkey, where principally a winter visitor. Breeds only very locally in the western and eastern mountains. Generally in wooded areas, from sea level to *c.*2,000m, higher in the east, but apparently not below 500m in much of the south, where seems largely absent from the Mediterranean littoral.

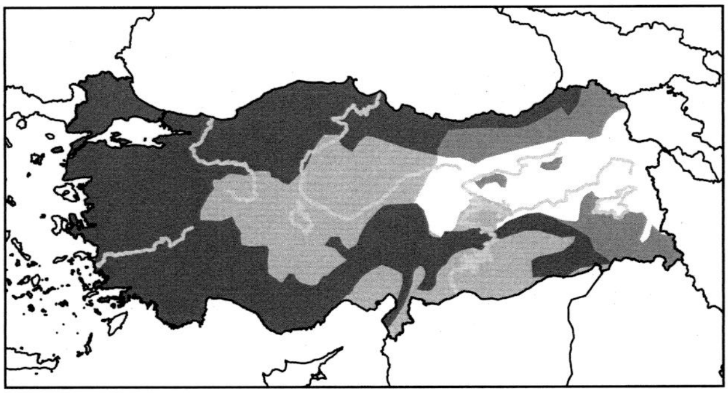

More widespread on passage, and there is considerable evidence of immigration into the country from further north, wintering mainly in the western two-thirds of Turkey. Much shyer of humans than in Western Europe, although occurs locally in parks in western regions of the country.

Breeding. Inhabits all types of forest and woodland, including plantations, gardens and agriculture with trees and bushes, and also scrub, where trees are few or absent, 0–2,000m; at Uludağ (Marmara) it is an abundant resident in all types of bushland and forests to the treeline, where it overlaps with Ring Ouzel *Turdus torquatus* (Jetz 1995). Nests in a bush or tree: eight records at 1.0–1.5m in bushes. A nest at Muğla (Aegean) was 0.6m up in a small oak, on a hillside of low scrub (PC); a nest at 2,100m on Karadağ (Inner Anatolia) was on the broken stump of a small oak (Ramsay 1914). The nest is a solid cup of grasses, moss, plant stems and roots, lined with mud, in turn lined with finer grasses. Eggs: C/3 (1), C/4 (5), C/5 (2). Young: B/2 (1), B/3 (1). In **Aegean**, around Izmir, two nests had five eggs on 10 May 1950 and 31 May 1954, and another young, almost ready to fledge, on 31 May 1954 (McNeile); a nest at Muğla had four eggs on 25 April 2002 (PC), and another there had four eggs on 27 May 2004 (M. Demain). In **Inner Anatolia**, a nest at Karadağ had two young, *c.*4 days old, on 26 May 1907 (Ramsay 1914); at Kızılcahamam, many were singing in mid-February 1999, and recently fledged young were observed in mid May. In **Mediterranean**, three nests, all in tall bushes, at Uzuncaburç on 7 May 2004 had two eggs (incomplete clutch), four eggs, and three young, *c.*7 days old (A. Rustell); a nest had three eggs at Akseki on 19 April 2004 (PC); fledged young were seen on the coast on

18 May 1992 (N. McMahon). In **Black Sea Coastlands**, a nest with an incomplete clutch of one egg at Yeniçağa Gölü, on 20 June 1992 was presumably a second brood, which is usual elsewhere (PC); in Kızılırmak Delta, fledged young were seen on 9–10 June 1975 (Dijksen & Kasparek 1985), and adults carrying food to nestlings were regularly seen in early June 1992 (Hustings & van Dijk 1994); an adult was carrying food north of Boyabat on 9 June 1975 (R. F. Porter). In **Marmara**, two nests had four eggs in Kocaçay Delta on 3/4 June 2006 (M. Farmer), adults carrying food to a nest near Eceabat on 25 June 1973, and four broods of recently fledged young on 26–28 June 1973 (R. F. Porter), indicate egg laying in late May, and were probably second broods.

BLACK-THROATED THRUSH *Turdus atrogularis*
Kara Gerdanli Ardiç

(Du: Zwartkeellijster; Fr: Grive à gorge noire; Ger: Bechsteindrossel; Swe: Taigatrast)

Subpsecies and taxonomy. Monotypic, if separated from *Turdus ruficollis* (Red-throated Thrush), as it is here. Taxonomy follows Dickinson (2003), Rasmussen & Anderton (2005) and Gill & Wright (2006), who separated *atrogularis* and *ruficollis* following Stepanyan (1990). The two forms were, for many years united under the senior name, *ruficollis*, and the vernacular Dark-throated Thrush, primarily on the basis of Vaurie (1959a), who reported that these taxa interbred extensively in the Altai, and elsewhere in Russia (Dement'ev & Gladkov 1954) and Mongolia, although in some parts of the Altai they are, in fact, segregated by habitat and mixed pairings are unknown. Nonetheless, the marked morphological differences between the two taxa and recent evidence that they appear to differ quite markedly in song (Arkhipov *et al.* 2003), seems sufficient reason to separate them specifically.

Status and distribution. Vagrant. Only recently recorded in Turkey: one, either adult female or perhaps first-winter, was at Altinpark, Ankara (Inner Anatolia), on 15 February 2006 (Yoğurtuoğlu 2006), together with many Fieldfares *T. pilaris* and Redwings *T. iliacus* during an unusually cold-weather period. This record was long overdue, as this species (*i.e. atrogularis*) winters widely in Iran, except the extreme northwest (Porter *et al.* 1996) and regularly reaches the opposite side of the Arabian Gulf, albeit in much smaller numbers, from southern Iraq to Oman and central Saudi Arabia, with vagrants having wandered west to Israel (at least 11 records; Shirihai 1996) and Lebanon (two records; Ramadan-Jaradi *et al.* 2008) in the Middle East. There are six records of *atrogularis* for Armenia (Adamian & Klem 1999) and one in Greece (Handrinos and Akriotis 1997).

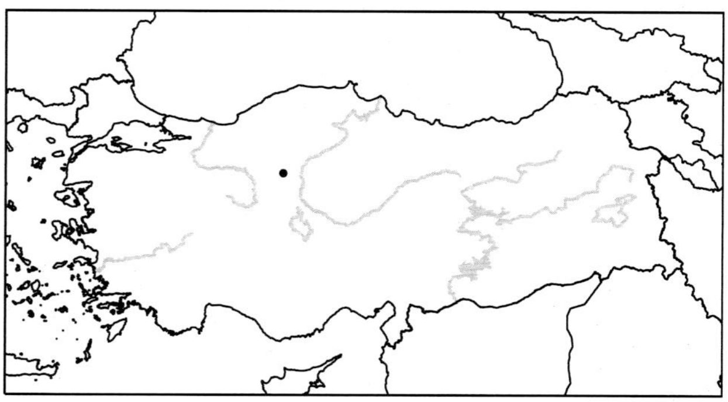

FIELDFARE *Turdus pilaris*
Tarla Ardıcı

(Du: Kramsvogel; Fr: Grive litorne; Ger: Wacholderdrossel; Swe: Björktrast)

Subspecies. Monotypic. We follow Clement & Hathway (2001) in electing not to recognise any geographical variation in Fieldfare.

Status and distribution. A widespread and fairly common winter visitor, particularly over the western two-thirds of Turkey, although perhaps scarcer in the southern littoral. Fairly common on passage, and less common further east. Generally found near wetlands and in areas with woods or plantations. Occasionally observed in city parks during cold winters. Regularly reported from late October to late March, when flocks of hundreds are usual, e.g. 400 in January 1969 between Manavgat and Beyşehir Gölü (Mediterranean) and, more exceptionally, 1,500 at Beyşehir Gölü on 22 January 2005 (BD *et al.*) and over 2,000 at Altinpark, Ankara (Inner Anatolia) on 5 February 2006 (KAB). Earliest records are from northeast Thrace on 17 September and Büyükçekmece (Marmara) on 25 October, and recorded until late March in Mediterranean and Inner Anatolia, with the latest records from Black Sea Coastlands on 29 April and Inner Anatolia on 17 May. The easternmost record involves a flock of over 100 individuals at Doğanşehir, Malatya, on 15 February 2006 (M. Erturhan).

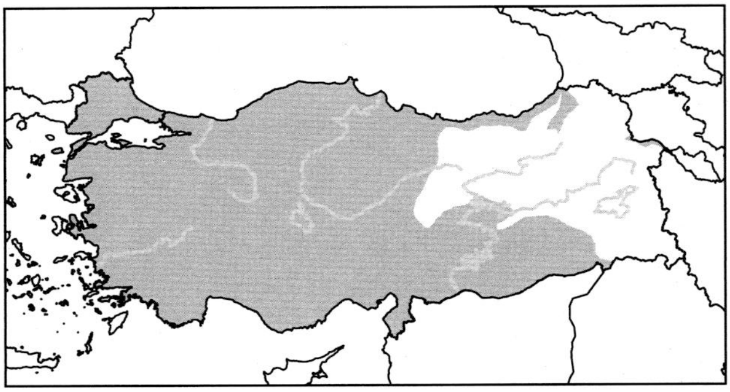

SONG THRUSH *Turdus philomelos*
Öter Ardıç

(Du: Zanglijster; Fr: Grive musicienne; Ger: Singdrossel; Swe: Taltrast)

Subspecies. Nominate.

Status and distribution. National conservation status: Least Concern (population apparently stable). Rather local and uncommon resident or partial migrant in Black Sea Coastlands, Marmara and Inner Anatolia, with a recent breeding record from Mediterranean, in which region it might prove reasonably widespread, albeit at low density, in the Taurus. Breeds in damp deciduous and mixed forests on hills and slopes at 500–1,900m, but locally occurs down to sea level in northwest Turkey. Singing birds in the southwest and in the Taurus, in May, could be late migrants, although the species at least occasionally breeds in these areas (see below).

More widespread on passage across the country (quite pronounced in the north and northeast), wintering in the west and south in considerable numbers, and in smaller numbers elsewhere in the western two-thirds of Turkey. Regularly uses forests, woodlands and orchards at this season, and may form mixed flocks with other thrushes. It is sometimes observed in cities and parks during the most severe winters. Immigrants arrive in the wintering areas from mid October and generally depart by mid April. The easternmost winter record appears to be from Malatya (East Anatolia) in December 2005.

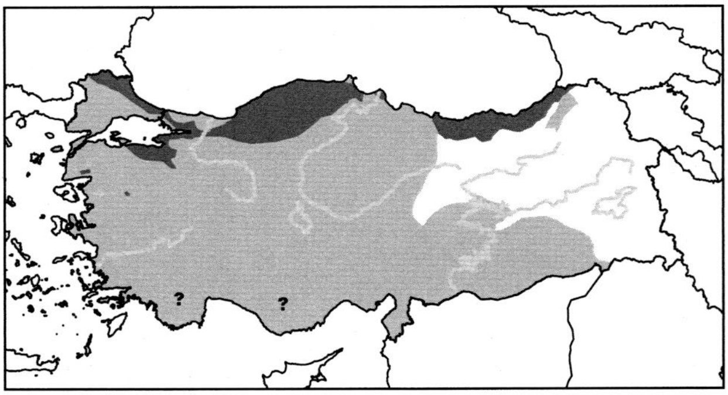

Breeding. Occurs in open coniferous, mixed and deciduous woodland, preferably with some younger bushy trees or undergrowth, from sea level in Thrace to *c.*2,000m in northeast Black Sea Coastlands. Nests in a tree or bush, but nest and clutch size undescribed in Turkey, though elsewhere the nest is a cup of thin twigs, grass, moss and lichen, with a smooth inner cup of mud; usual clutch 4–6. In **Marmara**, a nest with two eggs, in Forest of Belgrade, on 23 April 1970, is considered an incomplete clutch; at Kocaçay Delta, the population in 1993 was estimated at 300–350 breeding pairs, with 0.48 pairs/ha in a 31ha plot in alluvial forest, an adult was carrying food on 12 May 1967, and an adult was feeding recently fledged young on 27 May 1993 (Ertan 1996), indicating egg laying in late April; an adult feeding recently fledged young at Uludağ, on 6 June 1993 (Jetz 1995), indicates egg laying early May. In **Mediterranean**, an adult carrying food at Ağla, on 26 April 2002, indicates egg laying by 9 April (PC). At Kızılırmak Delta (**Black Sea Coastlands**), 25 males were singing on

24 March 1984 and one was alarm-calling on 12 May 1977 (Dijksen & Kasparek 1985). Song heard at Kızılcahamam (**Inner Anatolia**) March–July (Barış *et al.* 1984), and in mid-February 1999 (R. Hofland); in view of this, and the various dates of egg laying, probably double- or treble-brooded in Turkey, as elsewhere.

REDWING — *Turdus iliacus*
Kızıl Ardıç

(Du: Koperwiek; Fr: Grive mauvis; Ger: Rotdrossel; Swe: Rödvingetrast)

Subspecies. Nominate.

Status and distribution. Rather local but not uncommon winter visitor and passage migrant, mainly in coastal areas and in the western two-thirds of the country. Compared to Fieldfare *T. pilaris,* apparently generally less common with fewer records from Mediterranean, where it remains until the end of the first week of April, although recorded very irregularly as far south as Antakya province. Passage records most abundant from Black Sea Coastlands, including the eastern third, and Thrace. Generally reported from oak woodland, gardens and orchards. May form mixed flocks with other thrushes and is observed in city parks during very cold winters. Regularly reported from late October to late March, with the earliest record from Istanbul on 8 October. Recorded until late March in Black Sea Coastlands, Thrace, Mediterranean and, particularly, Inner Anatolia around Ankara. The easternmost winter record is from Yedikır Reservoir, Amasya province (Black Sea Coastlands) on 19 December, and the latest passage record is from 29 April (Albrecht 1986).

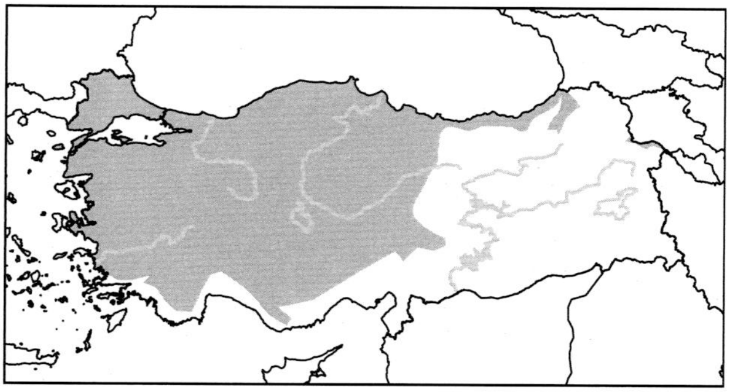

MISTLE THRUSH — *Turdus viscivorus*
Ökse Ardıcı

(Du: Grote Lijster; Fr: Grive draine; Ger: Misteldrossel; Swe: Dubbeltrast)

Subspecies. Nominate.

Status and distribution. National conservation status: Least Concern (population apparently stable). A fairly widespread and common resident in forested zones and woodlands over the western two-thirds of the country, as well as in the eastern Pontics and perhaps around Van Gölü (East Anatolia). Breeds mainly at 800–2,000m, lower in the extreme west, and up to 2,600m in the Taurus and the montane east. Breeds locally but widely in Marmara, parts of coastal Aegean, Black Sea Coastlands, the Taurus and Amanus mountains, and the Upper Aras Valley. Records in May and July in Van and Hakkari may also point to local breeding.

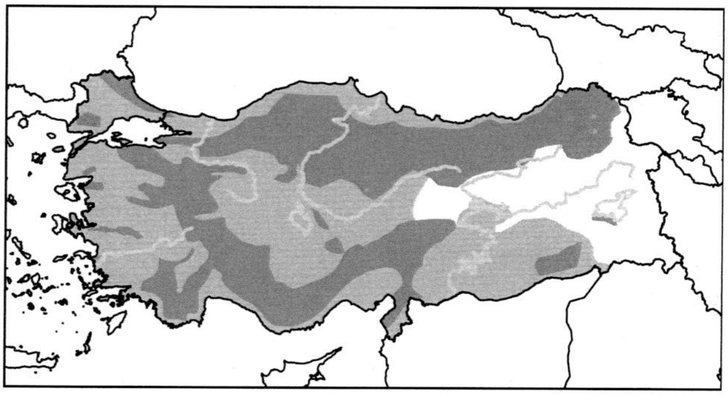

A partial migrant in the Black Sea Coastlands and in Thrace, with autumn movements observed at the Bosphorus and small-scale passage noted through Inner Anatolia, especially Ankara, at both seasons. In winter, reported extensively over the western two-thirds of

Turkey (both residents and immigrants), locally in orchards around Malatya (East Anatolia) and throughout the Black Sea Coastlands at lower altitudes. Compared to other thrushes, only scarcely reported from city parks, even in severe weather. Autumn passage peaks in October, and spring movement is most marked during April.

Breeding. In hilly or montane coniferous woodland; also in deciduous woodland, open agriculture and cultivation with trees, copses, lines of trees and even isolated trees in mountains. Nest usually 2.5m (or higher) above ground in a tree or bush; mainly of grass and moss, with mud mixed to form a solid cup, and lined with fine grass. Eggs: C/3 (1), C/4 (1). Young: B/4 (2). The earliest record of recently fledged young, at Akseki (**Mediterranean**) on 19 April 2004, indicates laying of the first egg *c.*19 March; elsewhere, early breeding often noted in February (and even January) and the relative lateness of this record may merely reflect the low number of Turkish records. In **Mediterranean**, a pair and two recently fledged young at Ağla on 7 May 1989, indicate egg laying about 7 April, an adult and two juveniles at Bolkar Dağları on 5 June 1999 (R. F. Porter), and a nest with four young, *c.*6 days old, at Akseki on 30 April 2005, indicating first egg *c.*7 April (PC). In **Inner Anatolia**, a nest had four eggs on 27 May 1993, and another four young, *c.*7 days old, at Hasan Dağ, on 10 May 1993 (PC), and an adult was incubating on a nest near Ankara on 17 May 1998 (T. Quittenden). In **Aegean**, a nest had three eggs at Muğla on 29 April 2001 (PC). In **Marmara**, there were two family parties at Uludağ on 9 July 1966, and a juvenile at Ayvacık on 26 June 1973. In **East Anatolia**, a family party was at Nemrut Dağ in early July. The later records above are probably second broods.

CETTI'S WARBLER *Cettia cetti*
Kamışbülbülü

(Du: Cetti's Zanger; Fr: Bouscarle de Cetti; Ger: Seidensänger; Swe: Cettisångare)

Subspecies. Roselaar (1995) ascribed those in Marmara and much of Aegean to *sericea*, with *orientalis* over the rest of the country. However, differences between the two forms are very slight, and they certainly intergrade (e.g. one trapped in northeast Turkey showed mixed characters between the two), rendering the recognition of such variation of low import in the understanding of real-world evolutionary dynamics. Baker (1997) chose not to recognise *sericea* and we suggest that even *orientalis* is only doubtfully diagnosable.

Status and distribution. National conservation status: Least Concern (population apparently stable). A widespread and usually fairly common resident breeder (or partial migrant, largely in Inner and East Anatolia), principally in the western half of the country, more locally in the east. The breeding population estimates from Black Sea Coastlands and Marmara (see Breeding), both made in recent years, provide some measure of the species' abundance in suitable areas of the country. Recorded from a variety of damp or wetland habitats, but also in dry areas with dense undergrowth, principally in the coastal lowlands and generally below 1,000m, but occasionally to 2,500m (especially in the northeast Pontics and the extreme southeast of the country), during the breeding season. Winters largely in coastal areas and inland wetlands in the west and south, and typically appears more common and obvious at this season.

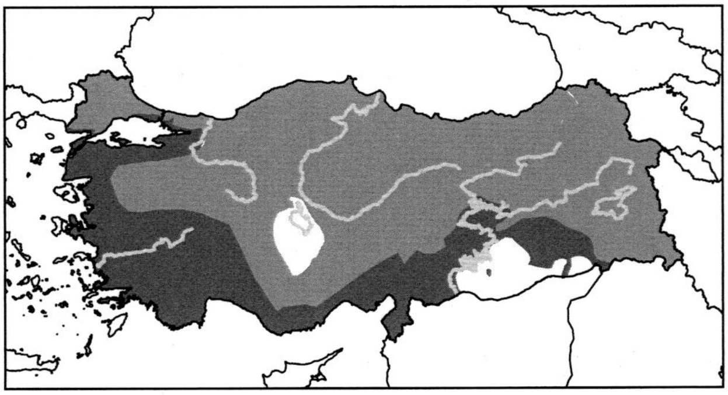

Breeding. Inhabits moist places with dense vegetation and bushes, especially thick bramble along ditches, streams and riverbanks, and where *Phragmites* or other dense aquatic vegetation grows up through dense bushes; also in clearings and edges of damp woodland and in *Tamarix* scrub near water, and occasionally in drier scrub and in olive groves with undergrowth. At Kocaçay Delta (**Marmara**), population estimated at 1,000–1,200 pairs, with 1.26 pairs/ha in a 31ha area in alluvial forest in 1993 (Ertan 1996). In Kızılırmak Delta (**Black Sea Coastlands**), population estimated at 1,400–1,600 pairs, with 90–150 territories/ha in Yürükler Forest

(Hustings & van Dijk 1994). A nest at Marmara Gölü (**Aegean**) was in dense brambles (McNeile), and two nests at Birecik (**Southeast Anatolia**) were *c.*1m above ground, in dense trailing brambles within damp poplars (D. Gregson). The nest is a deep cup of coarse grasses and dead leaves, including strips of reeds, lined with fine grasses and hair. Eggs: C/4 (1); elsewhere usual clutch four (3–5). At Birecik, on 11 May 2004, a nest was empty, but ready for eggs, and another had four eggs (D. Gregson). In **Aegean**, an adult was feeding fledged young at Büyük Menderes Delta on 19 May 1970, indicating egg laying in mid April; young had recently fledged at a nest at Marmara Gölü on 30 May 1950, indicating egg laying in late April; a brood of four recently fledged young, barely able to fly, at Bafa Gölü on 26 June 1999, indicates egg laying in late May and was, presumably, a second brood (PC). In Marmara, on 4 June 2006, there was a brood of at least two fledged young at Kocaçay Delta, and a brood of four fledged young was close to a used nest 1m up in dense brambles at Uluabat Gölü (PC). In Yürükler Forest, Kızılırmak Delta, song activity was very high, especially in late April, and an adult was seen carrying food in early June 1992 (Hustings & van Dijk 1994). These records suggest a peak in late April for commencement of first broods in Turkey. Song noted March–August, and the species is probably double-brooded, as elsewhere in the range.

ZITTING CISTICOLA *Cisticola juncidis*
Yelpazekuyruk

(Du: Graszanger; Fr: Cisticole des joncs; Ger: Cistensänger; Swe: Grässångare)

Subspecies. Roselaar (1995) was able only to examine specimen material from the west, around Izmir, which related to nominate *juncidis*, but speculated that those in southeast coastal Turkey probably involve *neuroticus* (see David & Gosselin 2000b for correct spelling of the latter form). One from Istanbul (Marmara) in the University Museum of Zoology, Cambridge (CUMZ27/Syl/23/t/7) is also *juncidis* (see below).

Status and distribution. National conservation status: Least Concern (population apparently stable). A rather local but not uncommon resident breeder principally in wetland areas and surrounding cultivation in the lowlands, from around Izmir (Aegean) south and eastwards, including the coastal littoral in the southern part of Aegean, Mediterranean (including the Hatay), western parts of Southeast Anatolia (where nesting is irregularly claimed at Birecik) and perhaps also very locally in Thrace. There is one recent summer record from southwest Marmara (Ayvalik). There are very few records from the latter region, but formerly it even occurred, at least erratically, in Istanbul, where Boisseneau took one in 1839 (see Subspecies). Less regularly recorded during the breeding season up to 900m at inland wetlands, e.g. Acıgöl (Mediterranean), in the same regions and also irregularly reported (but not recently) from several wetlands in Inner Anatolia, both in the breeding and post-breeding seasons, e.g. at Eber and Mogan lakes and the former marshes around Yarma, Konya province. Also recorded twice at Van Gölü (most recently in mid-September 2004) and once each at Erçek Gölü and Bulanık (East Anatolia).

Winters largely in its core breeding areas, but also rather more widespread at this season, having been recorded on a number of occasions in Marmara, and once in the Black Sea Coastlands, at Kızılırmak Delta. Apparently subject to dramatic fluctuations in population related to severe winter weather conditions, as elsewhere in its European range (Hagemeijer & Blair 1997), with a particularly obvious decline being noted in the latter half of the 1970s (Beaman 1986) and, locally, at the Göksu Delta (Mediterranean) in the early 1990s (van den Berk 1994).

Breeding. In open areas of dense low vegetation, e.g. meadows with long grass, and in *Juncus*, often damp, at the edge of marshes. Nest site, nest and clutch size unrecorded in Turkey but, elsewhere, the nest is well hidden in tall grass or rushes; a pear-shaped structure, with an entrance hole at the top, bound by spiders' webs to vertical grasses or rush stems, with grasses and plant down filling the cavity; usual clutch 4–6. In **Mediterranean**, *c*.5 recently fledged young were in Göksu Delta on 6 May 2004 (D. Marshall), indicating egg laying *c*.5 April, and adults were carrying food at Dalyan on 28 June 1999 (PC). There are many records of singing adults, mainly in Mediterranean, late March–early September, and at least two broods annually are likely, as elsewhere in range.

GRACEFUL PRINIA — *Prinia gracilis*
Dik Kuyruklu Ötleğen

(Du: Gestreepte Prinia; Fr: Prinia gracile; Ger: Streifenprinie; Swe: Streckad prinia)

Subspecies. The form *akyildizi* was described from, and is endemic to, southern Turkey, and Roselaar (1995) considered that all those populations in Mediterranean Turkey could be placed within it. No material from Birecik or elsewhere in Southeast Anatolia is available, precluding subspecific identification. From personal observations, they clearly belong to the *lepida*-group of subspecies (which includes *akyildizi*), and Roselaar (*op. cit.*) considered that they might refer to *irakensis*.

Status and distribution. National conservation status: Vulnerable (population apparently increasing). A local and generally uncommon resident breeder, in small to moderate numbers at several lowland wetlands in Mediterranean (principally in the eastern half) and Southeast Anatolia (only recorded at Birecik, Akçakale and localities on the Tigris), where rather more local and even less common. Recorded west as far as Antalya province and once north to Beyşehir Gölü (K. Haataja) and Burdur Gölü. Only common in the Göksu and Çukurova deltas and around the former Amik Gölü, Antakya province (Mediterranean), where generally found in areas of reed, *Scirpus* and *Juncus* beds, but also occasionally in cotton fields and other drier habitats. Some limited post-breeding dispersal may occur, although the evidence for this supposition is rather insubstantial.

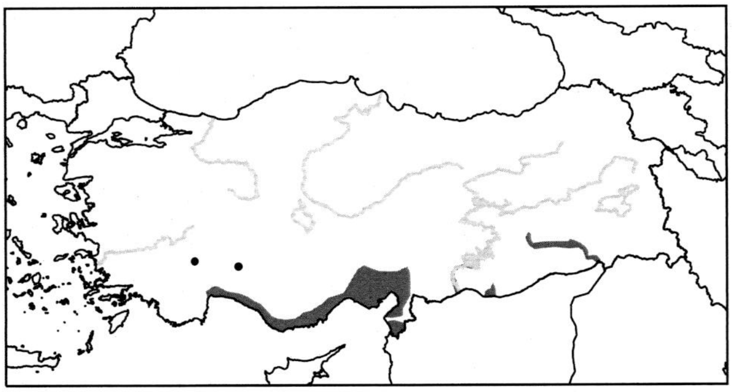

Breeding. Inhabits lowlands: beside the Euphrates at Birecik (**Southeast Anatolia**) but mainly in **Mediterranean**, where a common breeder in Göksu and Çukurova deltas. In Göksu Delta, population estimated at 400–600 pairs in 1991, favouring dense vegetation near the water's edge, along drainage canals, old river channels, the landward side of reedy wetlands, and *Tamarix* areas near the river mouth, as well as wheat fields (van den Berk 1994). In Çukurova, it occurs in dunes, along main rivers and meanders, in reeds, farmland with bushes and ditches, and plantations and forest clearings (van der Have *et al.* 1988). In Göksu, nine nests were in damp *Scirpus* and *Juncus* beside a wet area of *Phragmites*, and were in thick clumps of vegetation; three nests at Birecik were in similar situations, and all 12 nests were 0.3–1.0m above ground (PC). The nest is a domed structure, with a large side entrance; of fine grasses, mixed with plant debris, plant fibres and spider's webs, and lined with plant down. Eggs: C/4 (5), C/5 (7). Young: B/4 (4), B/5 (1). In Göksu Delta, fledged young reported early April–late August, indicating multiple broods (van den Berk 1994); an adult carried food to a nest in Çukurova on 11 September 1976, and a pair was feeding two newly fledged young near Alanya on 1 September 1977. At Birecik, four chicks hatched in a nest on 20 May 1993, another was empty, but ready for eggs, on 4 June 2001, and another had five eggs on 20 May 2007 (PC); a nest had five young, *c*.4 days old, on 10 May 2004 (D. Gregson). In Göksu, two nests had incomplete clutches of two and three (both subsequently had five eggs) on 1 June 1993 and 9 June 1993, two nests each had four young (4–6 days old) on 23 May 1993 and 1 June 1993, and nest building was observed on 23 May 1993 (PC); a nest had one egg (incomplete clutch) on 27 July 1971; in 2004, two pairs were observed building (half-built) nests, a nest had four young, 4–5 days old on 6 May, and a nest had five eggs on 22 May (PC); in 2005, a nest had small young on 6 May, another had four eggs on 11 May and nest building was observed on 12 May (J. Hodson). In Çukurova, a nest had five eggs on 10 May 1989 (J. Whitaker). A nest with four eggs and one newly hatched young, and several broods of fledged young, were near Side on 19 May 2007 (J. Hollinrake). The species is double- or treble-brooded elsewhere, but the number of broods per year in Turkey is unclear.

[SCRUB WARBLER *Scotocerca inquieta*]
Sürmeli Küçük Ötleğen

(Du: Maquiszanger; Fr: Dromoïque du désert; Ger: Wüstenprinie; Ökensnårsångare)

Subspecies. Unclear, see below.

Species not admitted to the Turkish list (Kirwan *et al.* 1999). A claim of one with Graceful Prinias *Prinia gracilis* at Akyatan Gölü, Çukurova (Mediterranean), has been published (van den Berk *et al.* 1986) but lacks additional documentation. Another, claimed at Side, Antalya province (Mediterranean) on 5 June 1999, was considered unproven by the *British Birds* Rarities Committee (M. J. Rogers *in litt.* July 2000); although the head pattern was well described, the lack of streaking/buff on the flanks is a feature exhibited only by the race *saharae* (from extreme southeast Morocco), an exceptionally unlikely vagrant to Turkey, whilst the bill and tail patterns and characteristic tail movements were either incorrectly described or not mentioned.

GRASSHOPPER WARBLER *Locustella naevia*
Çekirge Kamışçını

(Du: Sprinkhaanzanger; Fr: Locustelle tachetée; Ger: Feldschwirl; Swe: Gräshoppsångare)

Subspecies. The only specimen available (see below) relates to the form *obscurior*, a taxon that is very similar to nominate *naevia* (Kumerloeve 1967a, Roselaar 1995).

Status and distribution. National conservation status: Vulnerable (population apparently decreasing). One isolated breeding record, involving 8–10 singing birds and including a pair nest building, in the Murat Valley between Ağrı and Taşlıçay, Ağrı province (East Anatolia) in late May 1965. One specimen was obtained, a male with enlarged testes, further indicating breeding at the site (Kumerloeve 1967a). Roselaar (1995) mentioned that this record came from an area 'close to the breeding grounds in Armenia and Georgia', but only two records indicative of breeding exist for the former country, both in the north (Adamian & Klem 1999; *i.e.* a broadly equivalent situation to that in Turkey), with two singing birds discovered in late May and early June in more recent years (Ananian *et al.* 2002). In Georgia, it is a rare and patchy breeder, albeit locally common, principally in the south (on the Armenian border), with other populations including one in the Kura River valley close to the Turkish border (the Kura flows into Georgia from Turkey, rising in the hills south of Ardahan, Black Sea Coastlands), and a rare passage migrant whose numbers peak in late April and late September into the first ten days of October (A. Abuladze *in litt.* 2003).

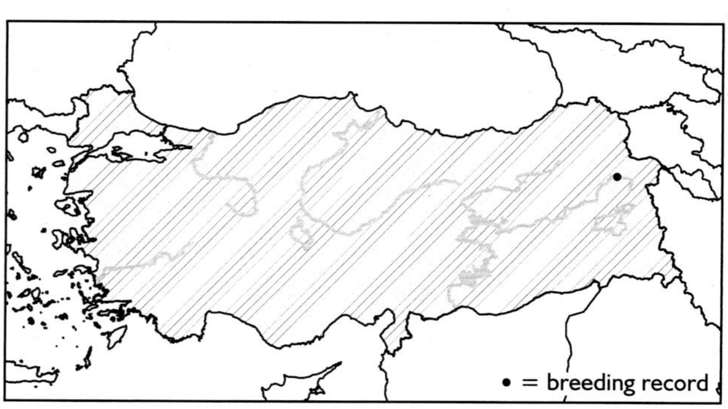

Principally known as a very rare, and less than annual passage migrant, largely through the western two-thirds of the country, although with the advent of a national ringing scheme the species has been perhaps detected with greater frequency (e.g. six in Kızılırmak Delta, Black Sea Coastlands, between 11 April and 11 May 2005). In spring there are just 26 published records, between 12 March and 19 May, with perhaps the majority in April, and only two in autumn; at Uluabat Gölü (Marmara) on 12 August 1986 and at the Göksu Delta (Mediterranean) on 30 October 1985 (Martins 1989), indicating a prolonged passage period. Mathey-Dupraz (1922) noted that the species occurred at the Bosphorus in spring. There are recent spring passage records from Southeast and East Anatolia (Kirwan *et al.* 2003), and two July reports, which perhaps also indicate breeding or at least unpaired males seeking a mate, although Roselaar (1995) considered them to reflect early post-breeding dispersal: Vielliard (1967) noted one singing at the former Amik Gölü, Antakya province (Mediterranean) on 15 July 1967 (Vielliard 1968) and another was at Sultan marshes (Inner Anatolia) on 5 July 2001 (Kirwan *et al.* 2003).

Breeding. See above. In addition, records of singing males in May, the presumed breeding season, but no evidence of nesting.

RIVER WARBLER — *Locustella fluviatilis*
Ağaç Kamışçını

(Du: Krekelzanger; Fr: Locustelle fluviatile; Ger: Schlagschwirl; Swe: Flodsångare)

Subspecies. Monotypic.

Status and distribution. An uncommon but secretive passage migrant in small numbers across Turkey. With the advent of a Turkish ringing scheme, it has proven to be more widespread and frequent than previous records suggested (e.g. previously only *c.*45 records in total including just four in Southeast and three in East Anatolia and, surprisingly, none specifically from Thrace, although the species has been recorded on the Asian side of Bosphorus and Mathey-Dupraz (1922) mentioned two specimens taken in the environs of Istanbul, but the collecting locality was not stated). Records of singing birds in summer (e.g. one at Kocaçay Delta, Marmara, on 7 June 1974, up to eight singing males at the Kızılırmak Delta, Black Sea Coastlands, in the first ten days of June 1992, three in Yeşilırmak Delta, Black Sea Coastlands, on 6–7 June 1993, and one at Manyas Gölü, Marmara, on 6 July 1986) raise the possibility that the species may at least occasionally have bred in Turkey; indeed, the species has been rapidly expanding its range southwards through Bulgaria (Nankinov 2000), suggesting that it should be specifically searched for in the breeding season in Marmara, especially Thrace, and elsewhere in the western two-thirds of Turkey during the breeding season. Roselaar (1995) remarked that the 'song. . .is frequently heard in spring and sometimes up to mid-June', but we have not seen evidence to indicate that the species is noted as late as the middle of the latter month (with the exception of the July record mentioned above).

In 2002, 21 individuals were trapped at three ringing stations in the western half of the country, with the vast majority of these in spring (KAD 2003), and in 2005 12 were ringed at Kars (East Anatolia) between 14 and 25 May. Spring passage commences in mid March (earliest 12th), but with few before late April, and continues until at least late May with the bulk in the first week of May (latest records, 7 June, which were perhaps breeding, although the species frequently sings on passage, which may continue as late as mid June in Kazakhstan; Cramp 1992). Autumn passage lasts from at least 10 August to 18 September (exceptionally 14 October), but habitats likely to hold the species, especially on passage, are not necessarily well covered by ornithologists in Turkey.

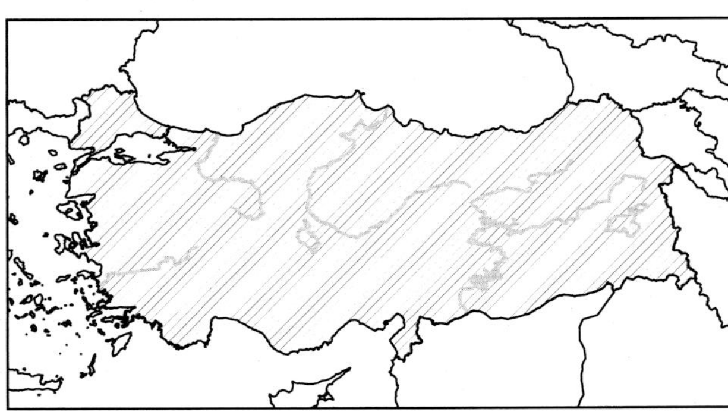

Breeding. In addition to the evidence cited above, there are records of singing males in May in Marmara, Aegean, Mediterranean and East Anatolia, although all of these could equally have still been on northward migration.

SAVI'S WARBLER *Locustella luscinioides*
Bataklık Kamışçını

(Du: Snor; Fr: Locustelle luscinioïde; Ger: Rohrschwirl; Swe: Vassångare)

Subspecies. All material examined from Turkey by Roselaar (1995) accorded with *fusca*, a generally greyer form above with whiter underparts, although given the evidence of Bauer *et al.* (1969) that nominate *luscinoides* occurs in the Evros Delta, on the Greek/Turkey border, it seems certain that this form extends into at least Thrace. In contrast, Kumerloeve (1969a) considered two male specimens taken at Van to be intermediate between nominate *luscinioides* and *fusca*, exemplifying the problems of recognising slight and largely clinal variation.

Status and distribution. National conservation status: Vulnerable (population apparently decreasing). A generally scarce but occasionally locally common summer visitor (e.g. about 100 singing males at Hotamış marshes, Inner Anatolia, in spring 1991; Kirwan 1993a), breeding in marshy wetlands across virtually the whole country. However, it is virtually absent as a breeder from Southeast Anatolia, where just two possible breeding localities have been identified during recent fieldwork. Most frequently recorded in Aegean, Black Sea Coastlands (500–750 pairs estimated at Kızılırmak Delta in spring 1992), and Inner and East Anatolia, distinctly less regular in Marmara (very rare in Thrace) and Mediterranean.

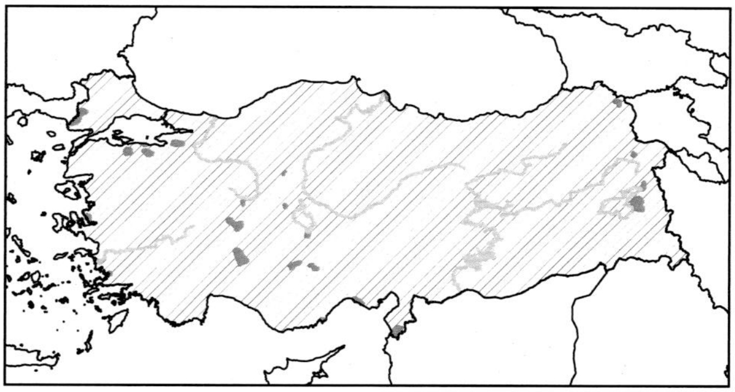

Rarely reported before mid April, though it is occasional from late March (earliest 10th) and possibly regular on the south coast during the latter month. Passage continues into May. In autumn most have departed by mid to late September and migration has certainly commenced by mid August. There are a few records available from early and even mid October (latest 17th) in southern areas of the country. Very rarely reported on passage (at non-breeding sites) across the country.

Breeding. In large reedbeds, especially dense *Phragmites*, in shallow water, but nest site and nest undescribed in Turkey. Elsewhere, nests low in reeds or sedges, usually well concealed in dense, matted vegetation above water; a loose cup of dead leaves of reeds, grasses and waterside plants, lined with finer leaves and fibres. At Kızılırmak Delta, (**Black Sea Coastlands**), there were 20–40 territories/100ha, with song activity intense in late April–early May (Hustings & van Dijk 1994). A nest had four eggs near Antakya (**Mediterranean**) on 19 May 1953 (Kumerloeve 1963a). An adult alarm-calling, with an insect in the bill, was at Gâvur (Southeast Anatolia) on 20 July 1967 (Vielliard 1968). Adults were alarm-calling and feeding young at Eşmekaya (**Inner Anatolia**) on 20–21 June 1977 (Schubert 1979). Adults sing at most sites mid April–mid July (earliest 27 March 1993 in Mediterranean), and the breeding records above indicate possible double-brooding (common elsewhere).

MOUSTACHED WARBLER *Acrocephalus melanopogon*
Bıyıklı Kamışçın

(Du: Zwartkoprietzanger; Fr: Lusciniole à moustaches; Ger: Mariskensänger; Swe: Kaveldunsångare)

Subspecies. The form *mimicus*, which is deceptively similar to Sedge Warbler *A. schoenobaenus*, at least for inexperienced observers, occurs throughout the vast majority of Anatolia (identification was well covered by Madge 1992). Roselaar's (1995) study of geographical variation was hampered by a chronic lack of specimens from Turkey, he being able to examine material only from Izmir, the Hatay and Van. Birds from both of the latter regions conformed to *mimicus*, but Roselaar (*op. cit.*) ascribed those from around Izmir (and by association populations elsewhere in Aegean and Marmara) to nominate *melanopogon*. We can confirm that *mimicus* is indeed widespread through Inner Anatolia and Mediterranean Turkey (as predicted by Roselaar), and it breeds in large numbers at the Kızılırmak Delta (Black Sea Coastlands), but we lack sufficient field experience with western birds to comment on their racial identity. The only specimen that we have examined from western Turkey, from Smyrna (=Izmir) closely accords with nominate *melanopogon* based on the almost solidly dark crown and extensively buffy underparts; though some *mimicus* from Mesopotamia may show near-solid caps, none that we have examined shows anything other than rather pale and very restricted buff on the underparts. Our field observations demonstrate that the songs of *mimicus* and *melanopogon* are very similar, although some eastern birds at least occasionally omit the 'start-up' phase of the song. The two forms show only very slight differences in mtDNA (Helbig & Siebold 1999, Parkin *et al.* 2004).

Status and distribution. National conservation status: Least Concern (population apparently decreasing). A relatively widespread and fairly common summer visitor, partial migrant and a resident at some south-coast localities. It breeds throughout Inner and East Anatolia, more locally in Black Sea Coastlands (where 1,000–1,500 pairs at Kızılırmak Delta) and Mediterranean, and probably parts of Aegean. Principally found in marshy wetlands, but also in adjacent wet grassland areas during the breeding season. Recorded in most of the inland and northern breeding areas (including those around Van Gölü) between early March and late September (e.g. recorded at Kızılırmak Delta between at least 15 March and 26 September).

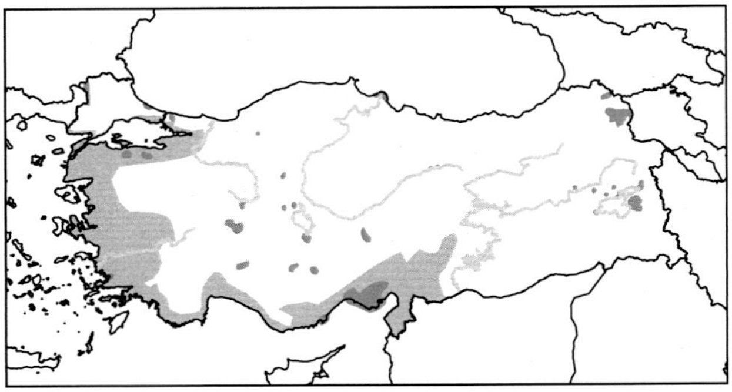

Apparently a rather uncommon passage migrant across the whole country. Winters, generally in relatively small numbers, in Marmara, Aegean, Mediterranean, and perhaps only formerly in western parts of Southeast Anatolia (e.g. recorded by Danford at Gâvur Gölü, Kahramamaraş province), generally vacating such areas in late March, more exceptionally mid April.

Breeding. In reedbeds, marshes and reed-fringed lakes, usually in wet sites. At Eşmekaya (Inner Anatolia), in 1998, a preference for wet areas, near open water, was noted, whereas Sedge Warblers *A. schoenobaenus* occupied drier sedge (T. Quittenden). At Kızılırmak Delta (**Black Sea Coastlands**), occurred in all types of marshland, but attained high densities of 30–80 territories/100ha, (sometimes 100–150) in dense old reedbeds, and mixed reed and reedmace; along Yürükler River, European Reed Warblers *A. scirpaceus* mainly occupied narrow reed strips beside the river, but Moustached Warblers preferred small islands with *Typha*, reeds, rough vegetation and scrub (Hustings & van Dijk 1994). Up to 100 pairs bred at Hotamış (**Inner Anatolia**) in 1991, with song April–July (Kirwan 1993a). A nest, at a small lake near Erçek Gölü (**East Anatolia**), was over water in reeds, in dense vegetation, but immediately adjacent to open water; a deep cup bound to vertical stems, of plant stems, grass and reed leaves, lined with reed flowers and feathers; it was empty, but ready, on 3 June 2001, and had four eggs on 11 June 2001 (T. Quittenden). Usual clutch, elsewhere, 3–4 (5–6). An adult was feeding fledged young at Kulu Gölü (Inner Anatolia), on 12 June 1998 (T. Quittenden). At Sultan marshes (Inner Anatolia), on 10 May 2006, along a drainage ditch, an adult was flushed from a new but empty nest, another had three eggs and a brood of recently fledged young was also observed, the latter indicating egg laying in mid April (N. Burke). No data in Turkey, but usually double-brooded.

AQUATIC WARBLER *Acrocephalus paludicola*
Sarı Kamışçın

(Du: Waterrietzanger; Fr: Phragmite aquatique; Ger: Segenrohrsänger; Swe: Vattensångare)

Subspecies. Monotypic.

Status and distribution. Conservation status: Vulnerable (BirdLife International 2000). Vagrant. There are several vague mentions in the historical literature and two modern records. Kumerloeve (1961) remarked that a small but unspecified number of observations were available from Izmir (Aegean) and that Braun was given the species during the autumn bird catches at Istanbul during the final years of the 19th century, whilst Mathey-Dupraz (1922) mentioned a specimen held in the Robert's College collection (at Bebek, Istanbul) that is no longer present (Kirwan 1997a). Another was reportedly held in Robson's private collection (Elwes & Buckley 1870). Smith (1960) mentions a rather improbable record (given that the species is known to have eggs from early May; Cramp 1992) of one, in June 1958, west of Trabzon (Black Sea Coastlands). Modern claims involve a first-winter at Anamur (Mediterranean) on 5 September 1991 (Kirwan 1992b), one at Kızılırmak Delta (Black Sea Coastlands) on 1 May 1992 (Hustings & van Dijk 1994), another at Fethiye (Mediterranean) on 5 April 1999 (P. Hope), and one trapped and ringed at Kızılırmak Delta on the rather early date of 28 March 2006 (Kirwan *et al.* in press). A detailed description is available only for the Anamur record, but brief notes were provided in support of the most recent report.

In neighbouring countries, records are available from Greece and Jordan (Snow & Perrins 1998). The lack of Middle Eastern records of this globally threatened species is unsurprising given its breeding range and migration route (Gorman 2002), although the recent discovery of a large breeding population in Belarus may mean that Aquatic Warbler is more likely to be recorded in spring in Turkey than was previously considered to be the case.

SEDGE WARBLER — *Acrocephalus schoenobaenus*
Kındıra Kamışçını

(Du: Rietzanger; Fr: Phragmite des joncs; Ger: Schilfrohrsänger; Swe: Sävsångare)

Subspecies. Monotypic.

Status and distribution. National conservation status: Near Threatened (population apparently decreasing). A relatively local but not uncommon summer visitor to marshy wetlands across the whole country, breeding in Aegean, Mediterranean, Black Sea Coastlands (e.g. 300–350 pairs in Kızılırmak Delta in spring 1992, or possibly as few as one-quarter the number of pairs of Moustached Warbler *A. melanopogon*), Inner and East Anatolia, and even more locally in Southeast Anatolia and probably Marmara.

Probably more common and widespread on passage throughout Turkey: in spring recorded in southern coastal regions regularly from late March (with the vanguard arriving early in the month, from 7th), but not usually noted in inland and northern areas until early/mid April, with passage continuing until late May (27th) and peak numbers usually occurring in late April. In autumn, passage is noted from early August to late September or late October in southern areas (latest 7 November).

There is some evidence of confusion among observers between this species and the race of Moustached Warbler that occurs across much of Turkey; *A. m. mimicus* is rather less strikingly marked than the nominate race. Such confusion is almost certainly responsible for two very late claims of the present species: by Strickland (1836) at Izmir, in December 1835, and near Edirne (Marmara) on 20 November 1970 (OST 1975). Both records are more indicative of Moustached Warbler in terms of locality and date.

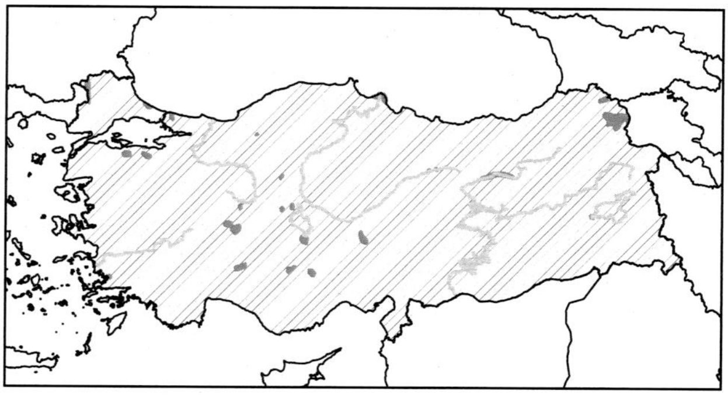

Breeding. Near water in rank herbage and shrubs. At Kızılırmak Delta (**Black Sea Coastlands**), the highest densities are in marshes or wet grasslands with dense *Euphorbia* (Hustings & van Dijk 1994). At Yeniçağa Gölü (Black Sea Coastlands), breeds in reedy and rushy fringes, but some singing males are also in rape and wheat fields (Kılıç & Kasparek 1987). At Kocaçay Delta (**Marmara**), estimated population in 1993 was 300–350 pairs, mainly along the river and canals, in areas with well-covered edges, preferring *Rubus* sp. rather than pure reed (Ertan 1996). At Eşmekaya (Inner Anatolia) in 1998 Sedge Warblers occupied drier areas of sedge, whereas Moustached Warblers *A. melanopogon* preferred wet areas beside open water (T. Quittenden). At Tatvan (**East Anatolia**) in 2004 noted breeding in patches of low dense shrubs (*c.*0.5m) in wet meadows (PC). No details of nest site, nest and clutch size in Turkey; elsewhere nest usually 30–60cm above ground (or water) in dense herbage; a deep cup, cylindrical or rounded, of plant stems, grass, moss and spider's webs, lined with leaves, grass flower heads, hair and plant down; usual clutch 5–6. In **Inner Anatolia**, a nest had four eggs at Sultan marshes on 30 April 1982 (Kasparek 1985) but, in view of the early date, was more likely a nest of Moustached Warbler; a recently fledged young, photographed at Eşmekaya on 21 June 1977 (Pforr & Limbrunner 1982), indicates laying of the first egg about 24 May; a nest was at Eber Gölü in early June 1966; and adults and juveniles were at Akşehir Gölü 5–9 August 1971.

PADDYFIELD WARBLER — *Acrocephalus agricola*
Doğu Kamışçını

(Du: Veldrietzanger; Fr: Rousserolle isabelle; Ger: Feldrohrsänger; Swe: Fältsångare)

Subspecies. Uncertain. On the basis of a photograph (see van den Berg & Bosman 1988), Roselaar (1995) ascribed Turkish birds to the form *capistratus*, rather than the more likely *septimus*. An additional photograph published since (van der Have & van den Berk 1995) provides contradictory evidence (although it could have been migrant from outside Turkey); the bird (a first-winter) shows rather pale (described as greyish-white) underparts, not the typical rufous-orange coloration of *capistrata* in this plumage. In any case, it seems far preferable to regard *agricola* as a monotypic species separate from the eastern form *tangorum* (Manchurian Reed Warbler), which is genetically and morphologically strikingly distinct (Parkin *et al.* 2004, Collinson 2006).

Status and distribution. National conservation status: Vulnerable (population apparently decreasing). Recently discovered to be a rare, perhaps uncommon but apparently spreading summer visitor, being regularly recorded from at least two localities in East Anatolia, where it is presumed to breed (van den Berg & Bosman 1988, Martins 1989). Since its discovery at Van marshes, in May 1986, birds have been recorded there in small numbers every year since (except apparently 1989–91), usually in May and June and nearly always apparently holding territory, with a maximum of 11 singing birds (van den Berk *et al.* 1993, Kirwan *et al.* press). In 1992–93, a second population of up to ten pairs was found at Çenge Gölü, near Erçek (Kirwan 1993c, Kirwan & Martins 2000a). Recorded at the latter site until late August (29th). Other records from elsewhere in East Anatolia suggest the possibility of additional breeding sites: three, including two singing birds, were at Saz Gölü, Karabulak, on 3 July 1989 (Kasparek 1990a), two, possibly migrants, were at Bulanık on 29 May 1993 (Kirwan 1993c), at least five, including three singing, were at the Bendimahi Delta, on 30 May 2002, six were singing and an additional three to four nest building there on 26 May 2003 and one was still present on 21 September 1994 (H. Schmaljohann *et al.*), with one singing at Ahlat Marsh, Van Gölü, on 24–25 May 2003 (H. Dufourny *et al.*), and several there on 6 May 2004 (GMK *et al.*). Finally, two were at Vali Uğur Boran Kuşcenneti, Bitlis, on 16 July 2006 (E. Yoğurtcuoğlu *et al.*). In 1998 it was recorded at two sites in the Konya Basin (Inner Anatolia), in late May and late June (Kirwan *et al.* 2003), and several were defending territories on river islands by Karkamis (Southeast

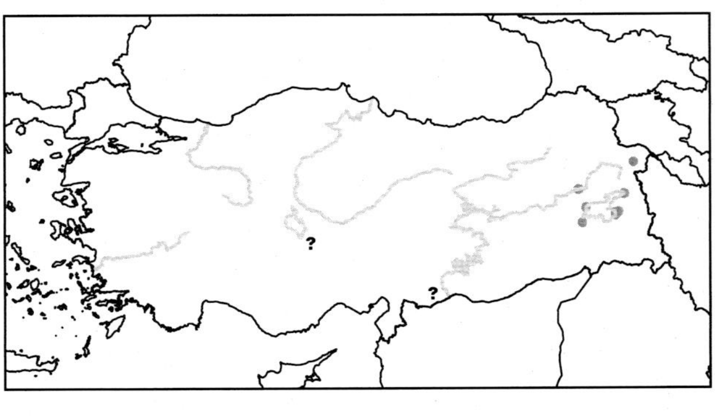

Anatolia), in early May 2004 and late May 2005 (GMK *et al.*), raising the possibility that the species breeds much further west in Turkey than supposed.

Elsewhere, migrants have been recorded three times in April (2004) and September (1988 and 1991) at the Göksu Delta (Mediterranean), as well as at Ereğli marshes (Inner Anatolia) on 15–17 August 1990. Latest record, 22 September (1991). The two original breeding-season records from Van and two of the migrant records have involved trapped birds, emphasising the need for such studies in Turkey and leaving open to question whether the species is genuinely a recent colonist or was simply overlooked in the past. In recent years (post-1995) Paddyfield Warbler has also been discovered breeding in Armenia, in the Araks (Aras) Valley, on the border with Turkey, these records also coinciding with increased observer activity in the country (Adamian & Klem 1999, Bussutil & Ananian 2001, Ananian *et al.* 2002).

Breeding. Singing birds observed in reeds at lake edges in **East Anatolia**, mainly at Van marshes and Erçek Gölü, in May–early July, the presumed breeding season; a fledged young was at Van on 23 June 1994. In Armenia, a nest had five recently laid eggs, in thickets of thin-stemmed reeds, in the Araks Valley, close to the border with Turkey, on 18 June 1995 (Adamian & Klem 1999).

BLYTH'S REED WARBLER *Acrocephalus dumetorum*
Kuzey Kamışçını

(Du: Struikrietzanger; Fr: Rousserolle des buissons; Ger: Buschrohrsänger; Swe: Busksångare)

Subspecies. Monotypic.

Status and distribution. Vagrant. Three or four records: one trapped at Ereğli marshes (Inner Anatolia) on 15–17 August 1990 (Kirwan & Martins 1994), one claimed at Tuzla Gölü, Çukurova (Mediterranean) on 19 May 2001 (Kirwan *et al.* 2003) and one ringed at Akyatan Gölü, Çukurova (Mediterranean) on 24 September 2003 (Balmer & Betton 2004a). We have not seen any substantiative details in support of the second report. Another record of a bird trapped and ringed in Kızılırmak Delta (Black Sea Coastlands), on 10 September 2004, is mentioned by Can (2004), but we have seen no details concerning this report. The species perhaps breeds in northeast Iran, where it occurs on passage in the east of the country. It is also known on passage through the United Arab Emirates, and has been listed as a vagrant to Bahrain, Cyprus, Israel, Oman and Syria (Porter *et al.* 1996).

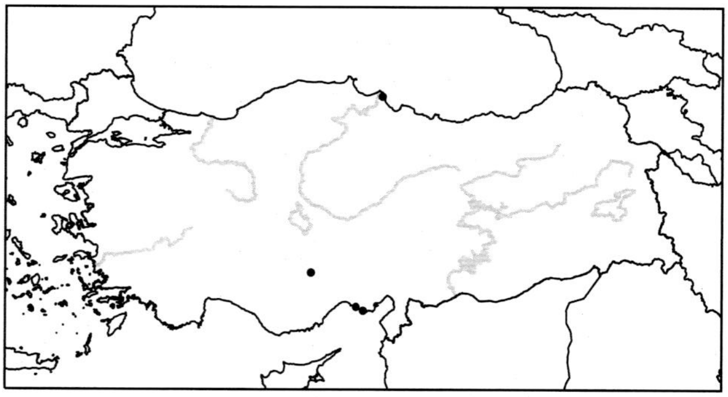

MARSH WARBLER *Acrocephalus palustris*
Çalı Kamışçını

(Du: Bosrietzanger; Fr: Rousserolle verderolle; Ger: Sumpfrohrsänger; Swe: Kärrsångare)

Subspecies. All material examined by Roselaar (1995) could be ascribed to nominate *palustris*.

Status and distribution. National conservation status: Least Concern (population apparently decreasing). A fairly common summer visitor, breeding in East Anatolia and the eastern Black Sea Coastlands, as well as perhaps immediately adjacent areas of other regions. Scattered records are available from Southeast Anatolia, in suitable breeding habitat, but at least some probably relate to late migrants. There are also isolated records during the breeding season from Thrace and elsewhere in Marmara, as well as in Aegean (e.g. at Büyük Menderes Delta). Also, recently recorded in summer just north of Beyşehir Gölü (Inner Anatolia) but not proven to breed. Principally breeds in low, dense scrub, from *c.*500m, more usually above 1,200m, to at least 3,000m.

A widespread passage migrant, usually in small numbers, perhaps exceptionally in very large numbers, e.g., 'probably thousands' at Çatalağzi, Zonguldak province (Black Sea Coastlands) on 23 August 1947 (Ogilvie 1954), although his description is rather vague and could equally involve European Reed Warbler *A. scirpaceus*. However, during a study at Manyas Gölü (Marmara), conducted between 27 March and 26 May 2003, 140 individuals were ringed, making it the third most numerous species to be trapped at the site that season (Didrickson *et al.* 2004). Spring passage is from early April (4th) until early June (3rd), with the peak occurring in the first two weeks of May and most breeding areas unoccupied until the second half of the month. In autumn, migration is recorded from mid August (10th) to mid October, although the majority passes through in September. Exceptionally recorded in late November (20th) at the Göksu Delta (Mediterranean).

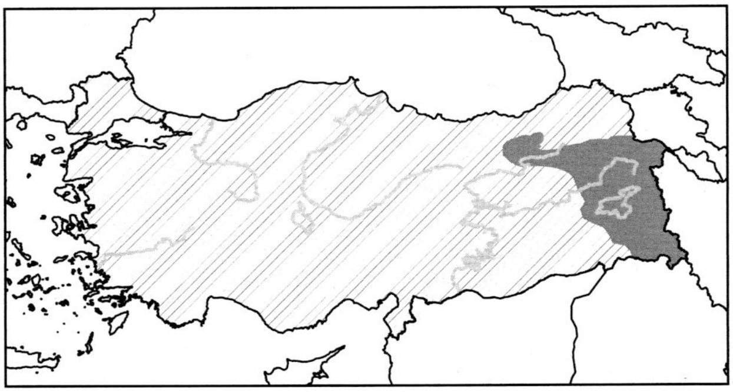

Breeding. In luxuriant vegetation, usually in flat areas, often damp and mainly near water, especially low willow scrub with dense mixed vegetation, including nettles, sometimes seasonally flooded, alongside rivers; also in dense weedy edges of a cornfield (PC); at least four were in unusual habitat within *Rhododendron caucasicum* scrub, high on a mountainside, at Sivrikaya (Black Sea Coastlands) on 12 July 1993 (A. Hogg). Usually in small colonies, typically 2–10 pairs, but 22 singing males and several other adults were in a broad floodplain near Aşkale (**East Anatolia**) on 15 June 1975. In a colony of *c.*10 pairs near Sarıcan (East Anatolia), three of four nests found were in nettles, a distinct preference, because this plant was not common in scrub dominated by low willow; the fourth nest was in slender vertical branches of a willow; all four nests were 0.5m above ground. The nest is a loose cup, attached to, and bound around, upright stems, forming distinctive 'basket handles'; of dead plant stems and grass, lined mainly with rootlets and fine grass. On 15 June 2004, two nests were partially constructed, and adults were incubating at two nests, which had full clutches of three and five; the latter had four newly hatched young and one egg on 25 June 2004, indicating first egg on 9 June 2004 (PC). In other parts of range, usual clutch 4–5. Elsewhere in East Anatolia, several family parties were at Bulanık on 31 July 1986 (J. McLoughlin), three adults were feeding a total of four fledged young, east of Erzurum, on 3 August 1974, an adult was feeding young at Van marshes on 4 August 1992 (J. Frost & J. Hornbuckle) and an adult was feeding young in reeds within an irrigation channel near Erzincan on 7 July 2004 (BD & O. Mülayim). Eight pairs, some carrying food, were at Yoncalık (**Black Sea Coastlands/East Anatolia**) on 12 July 1974.

EUROPEAN REED WARBLER *Acrocephalus scirpaceus*
Saz Kamışçını

(Du: Kleine Karekiet; Fr: Rousserolle effarvatte; Ger: Teichrohrsänger; Swe: Rörsångare)

Subspecies and taxonomy. Roselaar (1995) assumed that those breeding in Marmara and Aegean were *scirpaceus*, given that this form breeds in Greece, with *fuscus* elsewhere throughout Turkey. The same author was also able to point to a number of records of nominate *scirpaceus* from virtually throughout Turkey, all of which probably involved migrants. Pearson *et al.* (2002), who provide the most in-depth study of the identification of *fuscus*, also struggled with the distribution of this form in Turkey, but concluded, as had Roselaar (*op. cit.*), that it bred from central Turkey eastwards. Our experience of birds in central and eastern Turkey suggests that breeders can probably be assigned to *fuscus*, but we lack sufficiently detailed field experience of this issue from further west. This issue acquired much deeper significance due to the recent proposal to afford specific status to *fuscus* (Leisler *et al.* 1997). This molecular phylogenetic study suggested that *scirpaceus* and *fuscus* should not even be regarded as sister taxa, finding the former to be more closely related to African Reed Warbler *A. baeticatus* and

Mangrove Reed Warbler *A. avicenniae*. The latter taxon was also accorded specific status by Leisler *et al.* (*op. cit.*). Sangster (1997) proposed the vernacular name Caspian Reed Warbler for *fuscus*, in preference to Eastern Reed Warbler, a name that has also been employed for Oriental Reed Warbler *A. orientalis*. Presently, these various proposals lack widespread adoption or consensus in the ornithological literature, although they were initially followed by the Dutch committee for systematics, given their decision to abandon the Biological Species Concept in favour of the Phylogenetic Species Concept (Sangster *et al.* 1998, 1999). Subsequently, however, another molecular analysis, by Helbig & Seibold (1999), found no evidence for strong divergence between *fuscus* and *scirpaceus*. Indeed, there is some considerable evidence to suggest that *scirpaceus*, *baeticatus* and *avicenniae*, which was originally described as a subspecies of *baeticatus*, as well as, by implication, *fuscus*, represent a single, very widespread species (see Ash *et al.* 1989, Dawsett & Dowsett-Lemaire 1993, Collinson 2006). Pearson *et al.* (*op. cit.*) failed to find compelling evidence to even recognise *fuscus* as a taxon, based on its morphology. The evidence of our observations in central and eastern Turkey offers some support for this, as the purported characters defining *fuscus* are obviously variable and certainly not shown by all birds, whilst some purported *fuscus* characters may also be exhibited by *scirpaceus* (see also Parkin *et al.* 2004, Votier & Riddington 2005).

Status and distribution. National conservation status: Least Concern (population apparently decreasing). A widespread and fairly common summer visitor, breeding in marshy wetlands throughout the country, although rather localised in Southeast Anatolia, where only common as a breeder on the Euphrates south of Birecik. Generally rather less common than the next species, except on migration, across the whole country.

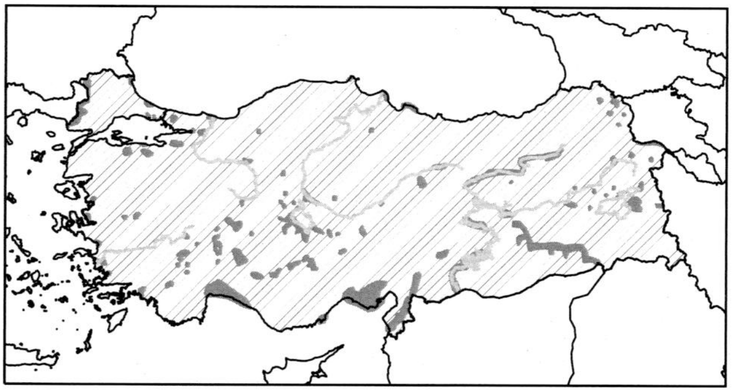

Fairly widespread and sometimes occurs in considerable numbers on passage: in spring from early March (2nd) to late May (28th), with the bulk from mid April to early May, and in autumn from early August (8th) to mid October (19th), more exceptionally to late October (30th), and even early November (9th) in Mediterranean. Most have departed by mid September. Recorded once in winter, at Iznik Gölü (Marmara) on 14 January 1972 (OST 1975).

Breeding. Mainly in areas of dense *Phragmites* in water, both extensive reedbeds or merely reed-fringed lakes, rivers and ditches. At Kızılırmak Delta (Black Sea Coastlands), the population was estimated at 1,500–2,000 pairs in 1992, and the highest densities (100–200 territories/100ha) were in areas with large and dense *Phragmites* (Hustings & van Dijk 1994). At Uluabat Gölü (Marmara) in 1998 often found in ditches, rather than in more extensive dense *Phragmites* (Welch & Welch 1998b). The nest is a deep cylindrical-walled cup, the sides woven around several upright reeds; of grass, flower heads and leaves of reed, lined with finer grasses. At Yeniçağa Gölü (Black Sea Coastlands), four nests in 1992 were 1.8–2.3m above ground in willows on damp ground (PC), and adults were also recorded in willow carr by the Murat River, East Anatolia (R. Webb). Nests are commonly built over water, but also in areas of dry reeds, where water has receded, e.g. six nests at Van marshes on 11 June 2001 (PC). Eggs: C/4 (11), C/5 (3). Young: B/4 (2). Whilst data are limited, records indicate that breeding commences earlier in Mediterranean, Southeast Anatolia, and Inner Anatolia than in other regions. At Göksu Delta (**Mediterranean**), adults were building two nests on 23 May 1993, four nests each had full clutches of four eggs on 23 May–10 June 1993, a nest had newly hatched young on 1 June 1993, and a recently fledged young was seen on 15 June 1992; the latter records indicate egg laying in mid May (PC); song heard mainly from mid April (earliest 22 March) in Çukurova in 1987 (van der Have *et al.* 1988). At Birecik (**Southeast Anatolia**), of six nests located on 11–12 May 2004, three were empty but complete, two had incomplete clutches of 1–2 eggs, and one had four eggs (L. Baxter). In **Inner Anatolia**, nest building was seen at Sultan marshes on 2 May 1982 (Kasparek 1985), 5–6 adults were already singing at Ereğli marshes on 30 March 1993 (J. van der Winden), and a juvenile was being fed at Mogan Gölü on 15 August 1969. In **East Anatolia**, two nests had eggs at Erçek Gölü on 3 June 2001 (T. Quittenden), and at Van marshes five nests had full clutches of 4–5 eggs and one nest had four newly hatched young on 11 June 2001, indicating egg laying late May (PC).

There were two nests with young in late June 2005 at Karakaya Reservoir, Malatya (MÖ *et al.*). In **Aegean**, a nest was empty, but ready for eggs, at Bafa Gölü on 29 May 1999 (J. Hollinrake). In **Marmara**, many adults were singing at Uluabat Gölü on 25 April 2003 (PC) and two nests there each contained four eggs on 4 June 2006 (R. Castell). A nest was located (contents unseen) at Manyas Gölü on 28 June 1972. In **Black Sea Coastlands**, there were several nests at Kızılırmak Delta in early June 1992 (Hustings & van Dijk 1994), and four nests in willows all had four eggs at Yeniçağa Gölü on 20–23 June 1992 (PC). Elsewhere, the species is sometimes double-brooded.

GREAT REED WARBLER *Acrocephalus arundinaceus*
Büyük Kamışçın

(Du: Grote Karekiet; Fr: Rousserolle turdoïde; Ger: Drosselrohrsänger; Swe: Trastsångare)

Subspecies. Nominate.

Status and distribution. National conservation status: Least Concern (population apparently decreasing). A reasonably widespread and fairly common summer visitor, breeding in marshy wetlands across the whole country, although rather scarcer in Southeast Anatolia due to a relative lack of suitable habitat: it is principally restricted to major rivers in this region. Rather local in Thrace and the Black Sea Coastlands. Recorded breeding up to at least 2,300m in East Anatolia.

More widespread and common on passage: recorded from early March on the south coast (7th), but not usually reported elsewhere, especially in inland and northern areas, until early to mid April, possibly even later in East Anatolia, with spring passage continuing until late May (27th). Autumn movements commence in early or mid August (8th), with most departing during September, but regularly recorded in Mediterranean and probably elsewhere until at least mid October (latest 18th), and once to early November (2nd).

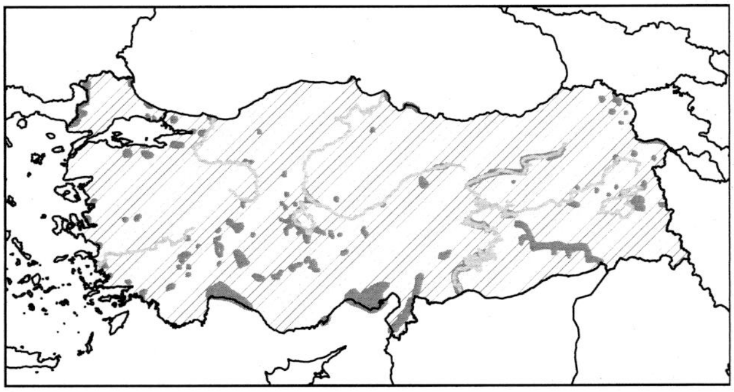

Breeding. In reedbeds with open water, reed-fringed lakes, rivers and canals, but singing birds also recorded in dense *Tamarix* and rice fields at Uluabat Gölü (Marmara), and in willow carr by the Murat River (East Anatolia). Nests mainly in tall dense reeds; builds a deep cylindrical cup, bound around three or more reeds, usually 1–2m above water level; of plant stems, fibres and grasses, lined with small leaves, reed flowers, down and hair. Eggs: C/3 (1), C/4 (3), C/5 (2). In **Inner Anatolia**, nest building was noted at Hotamış on 23 May 1993 (PC) and Uyuz Gölü on 18 May 1998 (T. Quittenden), nests had eggs at Sultan marshes on 23 May 1979 (Kasparek 1985) and on 29 May 1998 (T. Quittenden), and there were many fledged young at Yağlıören Pool on 19 June 1992 and at Uyuz Gölü on 21 July 1991 (Karauz Kiraç 1993). A nest had one egg (incomplete clutch) and 12 nests were empty, but complete, at Mogan Gölü on 15 May 2005 (J. Hodson). At Kocaçay Delta (**Marmara**) in 1993 the highest density was 21 singing males along 750m of river, with earliest singing on 14 April 1993 (Ertan 1996); elsewhere, there were two almost complete nests at Uluabat Gölü on 2/4 June 2006 (R. Castell). In Thrace, adults were carrying food at Küçükçekmece on 19 June 1973, and at Meriç Delta on 24 June 1973 (R. F. Porter). At Büyük Menderes Delta (**Aegean**), Selous (1900) saw many singing birds on 13 May 1899, but failed to find a nest, and considered that breeding had not yet commenced. At Kızılırmak Delta (**Black Sea Coastlands**) in 1992 regular song was heard from mid May, with densities of 10–30 territories/100ha, occasionally higher (Hustings & van Dijk 1994). In **Southeast Anatolia**, a nest had young at Birecik on 4 June 1998 (T. Quittenden). The breeding season in extreme **East Anatolia** is slightly later than in other regions; adults were nest building at Van Gölü on 1 June 2001 (T. Quittenden); near Erçek Gölü, several nests were under construction and five had eggs on 11 June 2001 (PC) and, on 19 June 2004, one nest was empty but complete, one had five eggs, and another two eggs and three newly hatched young, indicating laying of first egg on 1 June (B. Pleasance).

[BASRA REED WARBLER *Acrocephalus griseldis*]
Basra Kamışçını

(Du: Basrakarekiet; Fr: Rousserolle d'Irak; Ger: Basrarohrsänger; Swe: Basrasångare)

Subspecies and taxonomy. Monotypic. Previously was considered conspecific with Great Reed Warbler *A. arundinaceus*.

Species not admitted to the Turkish list (Kirwan *et al.* 1999). An individual reported as a 'possible', 10km north of Kilis, Gaziantep province (Southeast Anatolia) on 21 April 1987 (Bertault & Fremont 1993) is not sufficiently well described to eliminate similar species. A record from Cyprus, in June 1981, has recently been accepted (*Cyprus Orn. Soc. (1957) Ann. Rep.* 42), and there are also claims from spring 2005 (Richardson 2006) and in April 2006 (Yésou *et al.* 2007) in Syria. The recent discovery that the species apparently breeds in northern Israel (Perlman & Geffen 2007) clearly means that *A. griseldis* might reach southern Turkey, especially as a spring overshoot, in the future.

EASTERN OLIVACEOUS WARBLER *Hippolais pallida*
Ak Mukallit

(Du: Oostelijke Vale Spotvogel; Fr: Hypolaïs pâle; Ger: Blaßspötter; Swe: Eksångare)

Subspecies and taxonomy. Based on bill measurements, Roselaar (1995) considered those in the west, from Thrace east as far as Burdur Gölü (Mediterranean), to be *elaeica*, with *tamariceti* throughout the rest of the country. The latter is a very weakly marked taxon, which is almost certainly indistinguishable from *elaeica* in the field, and has frequently been considered a synonym of the latter. J. L. Copete (*in litt.* 2005) reports that birds trapped at Birecik are mensurally identical to *elaeica*. In common with Svensson (2001), we consider it of dubious efficacy to recognise geographical variation within the *elaeica* group at nomenclatural level, and consider all Turkish birds as *elaeica*. Furthermore, we have no hesitation in following the latter author's well-argued proposal to treat Western *H. opaca* and Eastern Olivaceous Warblers as two species, principally based on clear differences in morphology and vocalisations, and on the striking divergence in their mitochondrial DNA (Helbig & Seibold 1999). However, the problems posed by the three North African forms of Eastern Olivaceous Warbler; *reiseri*, the doubtfully diagnosable *laeneni*, and nominate *pallida* (which may intergrade with *elaeica*) have not yet been satisfactorily resolved (see also Parkin *et al.* 2004).

Hippolais pallida sensu lato and Booted Warbler *H. caligata* (*sensu lato*, see below) have recently been considered to be best placed within the genus *Acrocephalus* (Leisler *et al.* 1997). Their analysis of relationships among these genera presented a number of radical findings, but this particular hypothesis has yet to accrue wide support, and has indeed been rejected by Helbig & Seibold (1999), who preferred to place these species in the subgenus, *Iduna*, as well as by Helbig (2001). (Dickinson [2003] recognised *Iduna* at genus level.) Should this species be transferred to *Acrocephalus* then its specific name becomes *pallidus*.

Status and distribution. National conservation status: Least Concern (population apparently increasing). A widespread and common summer migrant (e.g. 2,000–2,500 pairs in Kızılırmak Delta, Black Sea Coastlands, in spring 1992), breeding throughout the country, principally in lowland areas to at least 2,000m in the south, west and north, but also at lower densities across the rest of the country, including Southeast Anatolia, where previously regarded as largely absent (e.g. OST 1975). Found in a variety of habitats, often favouring waterside vegetation, particularly willows and poplars, but also uses most types of scrubland, gardens, olive and citrus groves etc.

Also common and widespread on passage. In spring regular from mid March (16th) in south and southwest coastal localities, although more common and widespread from mid April to mid May (14th). In

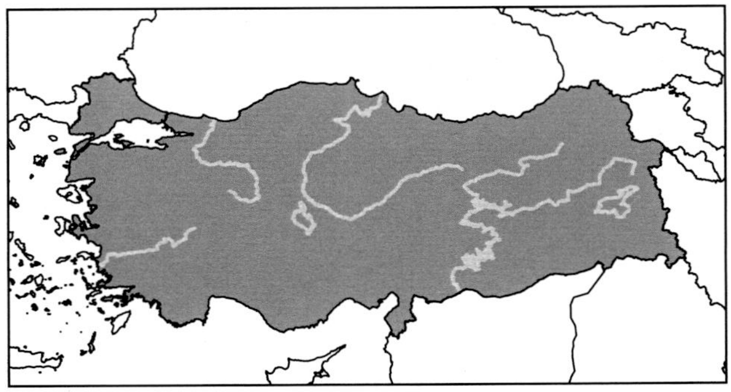

autumn migration is recorded from early August to mid September in most areas, and occasionally until the end of September or mid October (10th) on the south coast. Recorded once in winter, one at the Büyük Menderes Delta (Aegean) on 17 January 1969 (OST 1972). Two in Antalya (Mediterranean) on the exceptionally late date of 18 November 1990 (M. Bakker) may also have been potential winterers.

Breeding. In scrub, thickets, open woodland, orchards, olive groves and gardens; also near water, or in damp areas with willows, poplars and *Tamarix*. At Kızılırmak Delta (Black Sea Coastlands) in 1992 there were 160 territories/100ha in woodland (the most numerous breeder), but only 5 pairs/100ha in dunes and scrub, and most adults arrived in the last ten days of May (Hustings & van Dijk 1994). At Uluabat Gölü (Marmara), there was a population *c.*430 pairs in 1998, mainly in dense scrub, usually near water, and very high numbers in *Tamarix* (Welch & Welch 1998b). Nests in a bush or small tree, often in the outer branches, *c.*1.5m above ground (n=17), commonly *c.*3m and up to 5m; builds a neat, compact cup of plant stems, grasses, fibres and plant down, lined with finer fibres and hair. Eggs: C/3 (5), C/4 (7). Young: B/3 (2), B/4 (2). In **Aegean**, a nest had four eggs in a pomegranate at Aydin on 8 May 1899, and three nests each had four eggs near Izmir on 15 May 1899 (Selous 1900); a nest, at 5m in bamboo near Izmir had one egg on 12 May 1951 and a full clutch of four on 16 May 1951 (McNeile). In **Mediterranean**, recently fledged young were being fed by the adults at Dalyan on 29 June 1999 (PC); at Göksu Delta, recently fledged young on 9 June 1993 indicate egg laying from 12 May; a nest had three eggs on 21 May 1993, which hatched on 2 June; a nest had four eggs on 9 June 1992 which were still unhatched on 18 June 1992; a nest had large young on 17 June 1992; a new nest was empty, but ready for eggs, possibly for a second brood, on 15 June 1992 (PC), whilst an empty nest on 8 May 2004 had three eggs on 16 May 2004 (J. Mason). At Birecik (**Southeast Anatolia**), nest building was noted on 3 June 1993, three nests had eggs on 3–4 June 1993/98, and three nests had young on 4 June 1998 (PC & T. Quittenden); a nest had young on 7 June 1995 (I. A. Green), four nests were complete but still empty on 12 May 2004 (D. Gregson), an adult was nest building and a nest had three eggs on 23 May 2004 (PC); three fledged young at Işıklı on 14 June 1996 (GMK) indicates egg laying in mid May. An adult was carrying food at Boğazköy (**Inner Anatolia**) on 31 May 1975 (R. F. Porter). In **Marmara**, there was nest building at Gelibolu on 8 May 1970, a nest with one egg (incomplete clutch) at Uluabat Gölü on 4 June 2006 (M. Farmer), and an adult feeding a young bird at Çamlıca on 12 August 1966 (R. F. Porter). At Kızılırmak Delta (**Black Sea Coastlands**), there were eight pairs with young in July 1966, and ten pairs feeding young on 19 July 1975, and there was nest building at Gelinkaya on 11 June 2004 and 15 June 2004 (PC). In **East Anatolia**, young were being fed in a nest low down in riverside vegetation near Varto on 3 July 1969.

BOOTED WARBLER *Hippolais caligata*
Küçük Mukallit

(Du: Kleine Spotvogel; Fr: Hypolaïs bottée; Ger: Buschspötter; Swe: Gråsångare)

Subspecies and taxonomy. Two subspecies have traditionally been recognised within this species, *rama* (Sykes's Warbler) in Central Asia south to northwestern Pakistan and, locally, in Iran and eastern Arabia) and nominate *caligata* (including *annectens*) which breeds in the southern taiga from western Russia east and south to Kazakhstan and possibly Turkmenistan and Uzbekistan. Support for their treatment at the specific level mounted during the 1990s (e.g. Sibley & Monroe 1993) as did the available morphological, vocalisation and ecological evidence. These data were summarised and substantially added to by Svensson (2001), who provided the most detailed rationale available in the English language for such a treatment. Nonetheless, despite recent mtDNA evidence which strongly suggested that the Arabian breeders can be safely identified as *rama*, important anomalies persist, especially in their breeding biology (Castell & Kirwan 2005). Thus far, the available evidence (principally Kirwan 1994a and Dierschke *et al.* 1996), suggests that the form that has occurred in Turkey is *caligata*, based on general upperparts and underparts plumage tones, presence of a secondary panel on all birds for which we have seen descriptive details, bill length, overall proportions and head pattern. Given that it is the more westerly breeding of the two forms and that both winter in the Indian subcontinent, it might be assumed that *caligata* would also be the more likely to occur. Nonetheless, *rama* has reached northwest Europe and may even breed in northeast Africa (Svensson 2001, Pearson *et al.* 2004), so future observers of Booted Warbler in Turkey are urged to take detailed field notes and, preferably, photographs of any birds they encounter.

See additional taxonomic note concerning the correct generic placement of Booted Warbler *sensu lato* under the previous species and discussion in Sangster (1997b). If this species is transferred to *Acrocephalus* its specific name becomes *caligatus*.

Status and distribution. Apparently a vagrant, but possibly a near-regular passage migrant in very small numbers through the eastern two-thirds of the country. The following are the ten published records (all relate to single birds, unless otherwise stated): at Ardeşen (Black Sea Coastlands) on 9 September 1976 (Beaman 1986), at Rize (Black Sea Coastlands) on 20 August 1988 (Kirwan & Martins 1994), with further records of 2–3 on 15 August 1990 and single birds on 29 August 1991 and 5 August 1993 (the latter photographed; Dierschke *et al.* 1996), Birecik (Southeast Anatolia) on 24 May 1992, Mersin (Mediterranean) on 25–26 April 1993 (Kirwan 1995a, Kirwan & Martins 2000a), Erçek Gölü on 20 May 1993 (Kirwan 1994a), Beyşehir Gölü (Mediterranean) on 25 May 1998, between Muradiye and Çaldıran (East Anatolia) on 9 August 2001 (Kirwan *et al.* 2003) and at Balatos (East Anatolia) on 3 June 2002 (I. Weiss; Kirwan *et al.* in press). Descriptions are available for the majority of these reports, and the first Rize record was photographed and ratified by the German rarities committee.

UPCHER'S WARBLER — *Hippolais languida*
Dağ Mukallidi

(Du: Grote Vale Spotvogel; Fr: Hypolaïs d'Upcher; Ger: Dornspötter; Swe: Orientsångare)

Subspecies. Monotypic.

Status and distribution. National conservation status: Least Concern (population apparently increasing). Though described by Hemprich and Ehrenberg from specimens collected in the Lebanese mountains in summer 1824, and independently by Tristram from Palestine in 1864, the species was apparently not recorded in Turkey until its discovery around Gaziantep (Southeast Anatolia) by Bird (1937) in early June 1935. This area is a well-known one for the species until the present. Until comparatively recently, Upcher's Warbler was poorly known and rarely observed in Turkey away from a comparatively small number of areas; however, it is now well established to be a local and generally uncommon summer visitor, breeding in low scrub in rocky hills and montane country to *c.*2,000m, throughout Southeast Anatolia and immediately adjacent areas of western, southern and perhaps eastern East Anatolia (Kumerloeve 1969a), eastern parts of Mediterranean including Hatay and much of the eastern Taurus, and over the greater part of eastern and central Inner Anatolia (Eken & Magnin 1999). The true extent of the species' breeding range is unclear, as it may plausibly breed as far west as the Çukurova and even the Göksu deltas. Isolated breeding-season records from Ispir, Erzurum province (Black Sea Coastlands) and Van (East Anatolia) almost undoubtedly refer to displaced migrants and are not indicative of nesting, although breeding was recently proven in at least one locality in Kars province (East Anatolia) on the border with Armenia, and was suspected in mid-May 2005 at Yusufeli (Black Sea Coastlands). There is also a record of one singing (tape-recorded) in Istanbul, on 24 May 1998, which apparently represented the first record in Europe (Kirwan *et al.* 2003).

Slightly more widespread on passage through the same regions; e.g. in Mediterranean, only recorded as far west as Side, Antalya province (Deppe 1990). Arrives during the last week of April and the first week of May, with passage probably continuing until the end of the month. Departure is generally swift, with only relatively small numbers remaining into the second half of August, and apparently rare in September (latest reported occurrence, 29th; Mycock 1987).

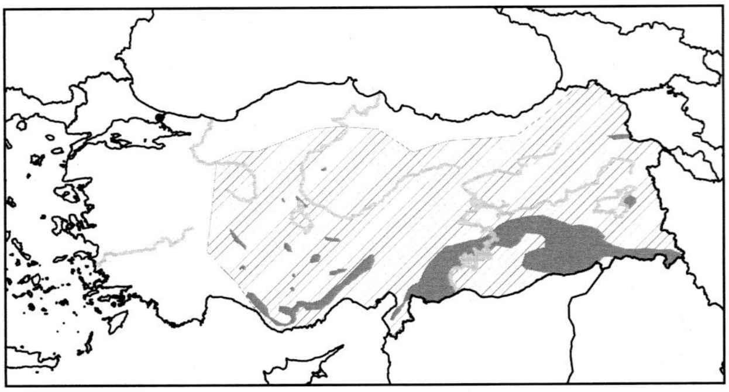

Breeding. In scattered scrub on arid rocky hills and in valleys, and in orchards: nests 1–2m (exceptionally 3m) above ground in a bush or tree (e.g. oak, pear, pistachio); the nest is a neat cup of grasses and thin stems, bound with spiders' webs, and lined with finer grasses, plant down and hair. Eggs: C/3 (4), C/4 (3). Young: B/3 (1), B/4 (3). At Gaziantep (**Southeast Anatolia**), a nest had four eggs on 2 June 2001, and four young, 7–8 days old, on 14 June 2001; another had four eggs on 3 June 2001, and four young, 11 days old, on 15 June 2001; another four young, ten days old, on 16 June 2001; and four recently fledged young, *c.*12 days old, were observed 15 June 2001; these three nests, and fledged young, can be back-dated to laying of first eggs on 18–21 May 2001; a female was nest building on 2 June 2001, and the nest had three eggs on 14 June 2001 (PC). Seven nests located at Gaziantep in 2006 were all 1.5–2.5m up in pistachio trees: on 7 June 2006 three of them contained respectively three eggs, four young two days old and four young five days old; the other four contained, on 12 June 2006, three eggs, three eggs, three young four days old and four young *c.*9 days old; most coincide with egg laying dates in 2001, but some were later and possibly replacements. At Hasan Dağ (**Inner Anatolia**), a nest had two eggs (an incomplete clutch) on 10 June 1998 (T. Quittenden). In **Mediterranean**, 4–5 pairs with fledged young were near Fevzipaşa on 6 July 1973. In **East Anatolia** (Tuzluca/Kars area), there were two nests on 3–4 June 1992, one of which was examined and had four eggs (G. Sandve). Probably double-brooded in some other parts of the range, but no evidence in Turkey.

OLIVE-TREE WARBLER — *Hippolais olivetorum*
Zeytin Mukallidi

(Du: Griekse Spotvogel; Fr: Hypolaïs des oliviers; Ger: Olivenspötter; Swe: Olivsångare)

Subspecies. Monotypic.

Status and distribution. National conservation status: Least Concern (population apparently stable). A generally local but not uncommon summer visitor, breeding in Marmara (including Thrace), Aegean and Mediterranean (including Hatay), more locally in Southeast Anatolia, as far east as the Mardin Mountains, and perhaps in the extreme western Black Sea Coastlands and southernmost parts of Inner Anatolia, from where there are several recent breeding-season records. Three recent records: near Bingöl (East Anatolia) on 20 May 1989, at Seyfe Gölü (Inner Anatolia) on 4 June 1981 and near Akdağmadeni, Yozgat province (Inner Anatolia) on 10 August 1992, surely refer to displaced migrants rather than potentially 'new' breeding areas. A record from Van

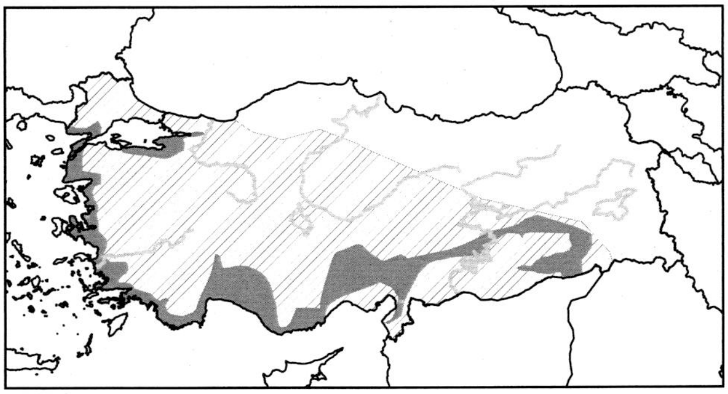

(East Anatolia), on 22 May 1968 (Kumerloeve 1969a) was surely also a migrant, if correctly identified. Found principally in lowland areas, to 800m, more occasionally to 1,400m in southern parts of Inner Anatolia (in the Taurus) during the breeding season, usually in olive or almond groves, orchards, open oak or pine woodland with patches of dense scrub, and less usually in other types of either dry or wet deciduous woodland.

Arrives from late March (31st), the majority in the first three weeks of April, and passage continues until mid May (17th). In autumn most have departed by the end of August, with very few records from September (most of those available were published by Beaman 1986), the latest ever being one north of Içel (Mediterranean) on 1 October 1974 (OST 1978). Slightly more widespread on passage, when also recorded somewhat further east.

Breeding. In olive groves, orchards, gardens; open dry woodland, often on slopes, with bushes and young trees (including pines); scrub with trees, and occasionally in damp woodland. Nests in forks of trees and bushes, 1.2–2.0m above ground, mean 1.5m (n=7), including three nests in wild pears in Aegean (McNeile), and one in an apple tree and another in an olive tree in Mediterranean (PC); builds a compact cup of grass, plant stems, fibres and down, bound with spiders' webs, lined with fine grass, hair, plant down, and occasionally feathers. Eggs: C/4 (7), C/3 (4). Young: B/2 (1), B/4 (1). In **Aegean**, McNeile found eight nests with eggs near Izmir in 1950–51 and 1954, and described the extent of incubation; back-dating from this, first eggs in most nests were laid from 21–29 May (earliest 11 May), and he also found an additional incomplete nest and a nest with young both on 3 June. More recently, there was a nest with three eggs at Intepe on 6 June 1974, a nest with four young at Erdek on 19 June 1987, and several records of fledged broods in July. At Uzuncaburç (**Mediterranean**), a nest had four eggs on 10 June 1992 (not hatched by 18 June 1992), and another two young, *c.*3–4 days old, on 14 June 1992, indicating egg laying on 31 May (PC); fledged young were seen on 10 June 1990. There was a nest with four eggs at 2.5m in an oak at Manavgat on 25 May 2007 (J. Hollinrake). In **Southeast Anatolia**, adults were feeding fledged young at Yesılce on 29 July 1992 (J. Frost & J. Hornbuckle). At Karapınar (**Inner Anatolia**), a nest had one egg (incomplete clutch), and another four eggs, on 11 June 1998 (T. Quittenden).

ICTERINE WARBLER — *Hippolais icterina*
Sarı Mukallit

(Du: Spotvogel; Fr: Hypolaïs ictérine; Ger: Gelbspötter; Swe: Härmsångare)

Subspecies. No specimen material from Turkey is apparently available (Roselaar 1995), but the species is probably best regarded as being monotypic.

Status and distribution. National conservation status: Near Threatened (population trend unknown). Rationale for inclusion in the Turkish Red Data Book (which only treats breeding species) appears highly suspect. Regular passage migrant, usually in rather small numbers over the entire country, although perhaps more common through the western two-thirds of Turkey. In spring, peak passage is recorded in the first half of May, with stragglers reported until early June (5th), and the earliest date being 9 April, other than two exceptionally early reports, from Marmaris, Muğla province (Aegean), on 25 March 1988 (R. Moore), and Fethiye (Mediterranean), on 16 March 1998 (P. Hope). There is one old breeding record of a pair feeding two almost fully fledged young at Izniköy, Bursa province (Marmara) on 6 July 1966, but no subsequent evidence of breeding. Occasional singing birds, usually in Thrace or adjacent areas of the Black Sea Coastlands, west as far as Kızılırmak Delta, and elsewhere in Marmara, are presumed to be on migration. However, at least one published record from July, at Bafa Golu (Aegean; Kasparek 1988a), suggests confusion with Eastern Olivaceous Warbler *Hippolais pallida*.

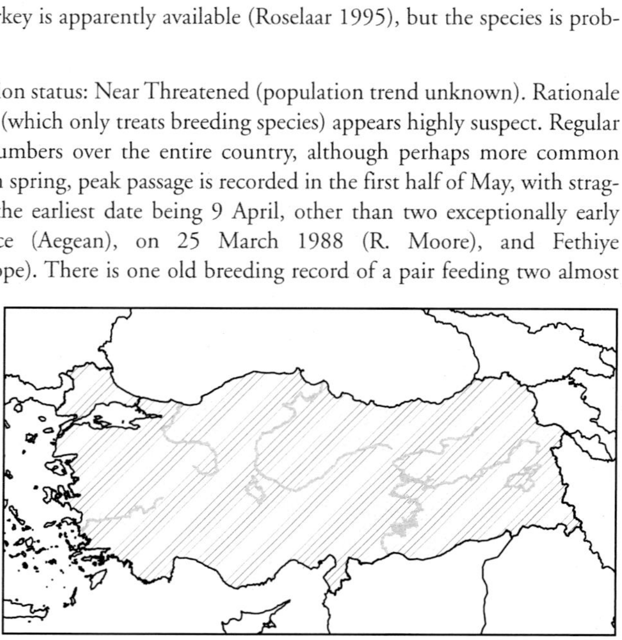

Extreme autumn passage dates are 28 July to 13 October, and peak numbers occur in late August and the first half of September. Recorded in all regions, although there are relatively few records from Southeast Anatolia at either season.

Breeding. See above for the only evidence of nesting in Turkey.

MELODIOUS WARBLER *Hippolais polyglotta*
Şarkıcı Mukallıt

(Du: Orpheusspotvogel; Fr: Hypolaïs polyglotta; Ger: Orpheusspötter; Swe: Polyglottsångare)

Subspecies. Monotypic.

Status and distribution. Vagrant. One record: a first-year trapped at Kavak, Çanakkale province, Thrace (Marmara), on 25 July 1973 (Eyckerman *et al.* 1992). Some details (biometrics) were presented. Roselaar (1995) speculated that a yellowish juvenile Eastern Olivaceous Warbler *H. pallida elaeica* might have been involved, but there is also an October 1982 record (also of a trapped bird) from Bulgaria (Nankinov 2003). The species is very rare in the eastern Mediterranean, with the only reports being a single verified and three non-ratified records from Greece (Handrinos & Akriotis 1997). The species is mentioned by Hollom *et al.* (1988) for Kuwait, but this is presumably in error.

BLACKCAP *Sylvia atricapilla*
Kara Başlı Ötleğen

(Du: Zwartkop; Fr: Fauvette à tête noire; Ger: Mönchsgrasmücke; Swe: Svarthätta)

Subspecies. Kasparek (1990c) considered that two subspecies breed in Turkey, with nominate *atricapilla* in the west and *dammholzi* in the east (see below). Vaurie (1954b) stated that all Turkish birds were *atricapilla*, but Shirihai *et al.* (2001) mapped both as breeding in northern Turkey. The presence of migrants, however, confuses the issue, and it is unclear whether Kasparek examined specimens from the eastern Black Sea Coastlands and Vaurie breeders (see Roselaar 1995). Our field observations confirm Roselaar's (*op. cit.*) observation that birds in the western Black Sea Coastlands (and migrants through central and southern Turkey) are nominate *atricapilla*. We lack sufficient experience of breeders in northeast Turkey to comment on the race that occurs there.

Status and distribution. National conservation status: Least Concern (population apparently stable). A fairly widespread and not uncommon summer visitor which breeds, in mixed or purely deciduous woodland, throughout much of northern Turkey, from Thrace to the extreme eastern Black Sea Coastlands, but also locally on the Aegean coast. Recently recorded in summer, without proof of breeding, in the southeast part of the Konya Basin (Inner Anatolia). Kasparek (1990c) proposed that the Black Sea population is disjunct, being divided into eastern and western areas, a pattern of distribution that reflects certain climatic parameters (namely the limits of 30mm of precipitation in the driest month). However, it would appear much more plausible that the species has a continuous distribution through the region, especially as there are several recent breeding-season records from the corridor of land (roughly from Sinop to Samsun) separating the two centres of distribution identified by the author. There are several more or less isolated records from other areas of the country during the breeding season, which may indicate a wider distribution; around Van Gölü (East Anatolia), in Ankara province, at Seyfe Gölü and near Eber Gölü (Inner Anatolia), near Pozantı, Adana province (Mediterranean) and at Bafa Gölü (Aegean). In addition, Schrader (1891) described the species as a 'summer bird around Mersin'; however, such a comment could equally refer to passage birds.

Common and widespread on passage throughout the country, though perhaps generally recorded in smaller numbers in the eastern third of Turkey. In spring migrants observed from early, more usually mid March (although due to the species' winter range, see below, it may be difficult to differentiate between immigrants and residents), with passage continuing until around the third week of May. Kasparek (*op. cit.*) identified the median spring date as 9 April, but this appears too early (in part because the author's observation sample was too small).

Peak passage numbers throughout southern and central areas, at least, occur from the third week of April to the first week of May: e.g., over 15,000 were estimated to have moved through Akşehir Gölü, Inner Anatolia, over several days in late April 1992 (Bradshaw & Kirwan 1992).

Perhaps less numerous on autumn migration, although there are fewer observations from this season. Recorded from mid August to mid October, although stragglers are reported in Inner Anatolia until mid November. The bulk would appear to move through during the latter half of September (Kasparek reported the autumn median as 16 September). Winters in fairly small numbers, principally at coastal localities in Aegean and Mediterranean, as well as perhaps Marmara, but also found at this season in the Taurus (Mediterranean) by Danford.

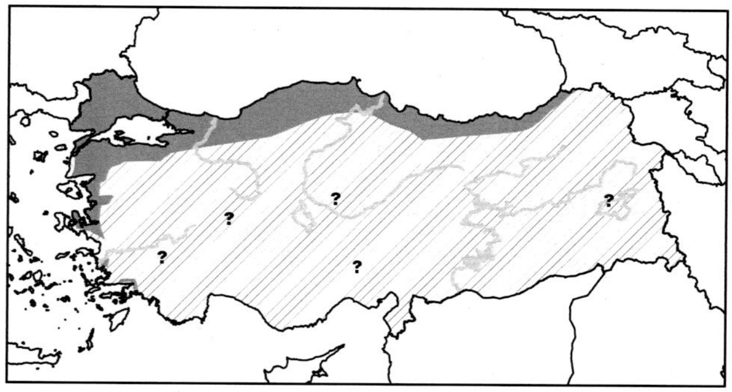

Breeding. In undergrowth of open woodland, both deciduous and coniferous, from sea level to 2,000m. In Kocaçay Delta (Marmara), a common breeder in alluvial forest, particularly at edges, with an estimated population of 300–400 pairs, and 0.65 singing males/ha in a 31ha plot in 1993 (Ertan 1996). Common, apparently nesting in May, at Çatalağzi, (Black Sea Coastlands) during 1946–48 (Ogilvie 1954), and one of the commonest species in semi-open land at 800m on Uludağ (Marmara) (Jetz 1995). At Kızılırmak Delta (Black Sea Coastlands) in 1992 the population was *c.*200–300 breeding pairs, mainly in Yürükler Forest, with 20 territories/100ha; first observed on 29 March and adults were still singing on 10 June, when the survey ended (Hustings & van Dijk 1994). In northeast Black Sea Coastlands, a common breeder from the coast to near the treeline (Faldborg 1994). Nest site, nest and clutch size unrecorded in Turkey; elsewhere nests in a low bush or low tree branch, usually in dense undergrowth, commonly *c.*0.5m above ground; builds a cup of dry grass and roots, lined with fine grass, roots and hair; usual clutch five (4–6). In **Marmara**, a pair was carrying nesting material in the Forest of Belgrade on 14 May 1970 (R. F. Porter), and there were young there 19–29 June 1973 (Kasparek 1990c). In **Black Sea Coastlands**, an adult was carrying food at Ispir on 12 May; in the Ereğli/Alapli area, there were three broods of flying young on 12–13/24 June 1978 (Kasparek 1990c).

GARDEN WARBLER — *Sylvia borin*
Boz Ötleğen

(Du: Tuinfluiter; Fr: Fauvette des jardins; Ger: Gartengrasmücke; Swe: Trädgårdssångare)

Subspecies. Apparently all *woodwardi* (Roselaar 1995 and pers. obs.) but given the broad degree of intergradation between this form and nominate *borin*, recognition of geographical variation within *S. borin* appears meaningless to an understanding of this species' evolutionary dynamics.

Status and distribution. National conservation status: Least Concern (population apparently decreasing). The species' precise status in Turkey is unclear. Kumerloeve (1961, 1967a, 1970a) was unable to point to any specific evidence of breeding, and only with the advent of systematic bird recording (under the auspices of the OST in 1966), and the consequent and coincident increase in records and research, was it proven to be a regular but perhaps uncommon and

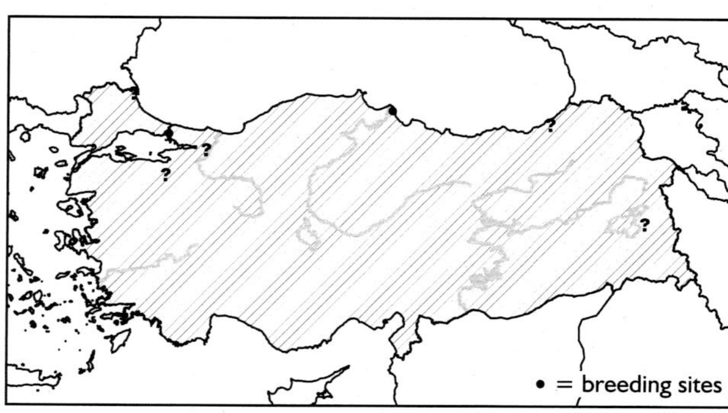

local breeder in parts of northern and eastern Turkey. Breeds in parts of Thrace, possibly elsewhere in Marmara, the Black Sea Coastlands, East Anatolia and plausibly in adjacent areas of Aegean and Inner Anatolia. During the breeding season, found in deciduous woodland with scrubby understorey, or in tall, well-developed secondary growth, occasionally in gardens, from sea level to 1,800–2,000m, but usually below 1,400m.

More common and widespread on passage, in moderate, occasionally large numbers, particularly through the eastern two-thirds of the country. In spring, reported from mid March (15th), but not usually widespread until mid April, and peak numbers move through during late April and early May, although significant numbers occur in eastern areas until the third week of May (latest 4 June was probably still on passage). Autumn migration lasts from early August (9th) to late September, with stragglers remaining into the third week of October (19th).

Breeding. There are many records of singing birds in late May–June, in deciduous and mixed woodland with undergrowth, mainly in coastal areas of Black Sea Coastlands; no records of nests and it is unclear whether some records relate to late migrants, rather than local breeders. A pair feeding five fledged young at Çamlıca (**Marmara**) on 11 August 1972 is the only definite breeding record. At Kızılırmak Delta (**Black Sea Coastlands**), in 1992, mainly migrants were seen, even in early June, but 50–100 breeding pairs were estimated, given the observation of 11 territories (ten in Yürükler Forest), a mean of five territories/100ha (Hustings & van Dijk 1994). Elsewhere, nests low in a shrub or young tree; builds a substantial, loose cup of dry grasses, lined with fine grass, hair and roots; usual clutch 4–5, laid in late May–June.

BARRED WARBLER *Sylvia nisoria*
Çizgili Ötleğen

(Du: Sperwergrasmus; Fr: Fauvette épervière; Ger: Sperbergrasmücke; Swe: Höksångare)

Subspecies. Nominate (Shirihai *et al.* 2001).

Status and distribution. National conservation status: Data Deficient (population apparently decreasing). Little evidence for the decline purported by the Turkish Red Data Book appears to be available. Apparently a local and generally uncommon summer migrant, breeding in all regions, although there are few records from Mediterranean; most reports come from northern Turkey, principally the Black Sea Coastlands. The first proven breeding records were from the Asian side of the Bosphorus (Alléon 1880, Mathey-Dupraz 1922). Confirmed instances of nesting remain relatively few, as passage frequently obscures breeding status and distribution (see below), but 125–175 pairs were estimated in Kızılırmak Delta (Black Sea Coastlands) in spring 1992, indicating the species' potential abundance in suitable areas. Breeds in open, cleared areas in deciduous woodland, and in upland and hilly areas with dense, often thorny scrub, from sea level to at least 1,000m, more locally to 1,700m in East Anatolia.

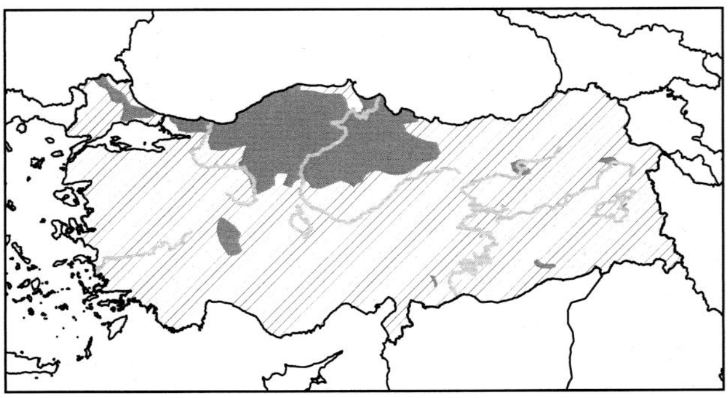

Moderately common and widespread on passage throughout the country. Recorded widely in southern regions from mid April (14th), with peak spring passage in late April to early May, and stragglers moving through until the fourth week of May (25th). In autumn, passage recorded from early August (9th) to late September, with the majority in the first three weeks of September, and late birds reported until mid October.

Breeding. In thorn thickets, mainly at edges of woodland, and in clearings, but also in more open country, in dense hedges and roadside verges with thick bushes. Nests in a dense bush; observations include one 1.2m up in a thorn bush 2.5m high and another 0.6m up in a low thorn bush (Hustings & van Dijk 1994). Nest and clutch size undescribed in Turkey; elsewhere, the nest is a deep cup of grass stems, lined with fine roots, hair and sometimes spiders' webs, and usual clutch five (4–6). Song heard, and adults seen, in most regions in May and June, but breeding status confused by northbound spring migration and subsequent post-breeding dispersal. Most records of breeding, or possible breeding, are from **Black Sea Coastlands**: at Kızılırmak Delta, nest building was reported on 3 June 1992 and 7 June 1992, and an adult with a brood patch was trapped on 3 June 1992 (Hustings & van Dijk 1994); and an adult was feeding young on 6 July 1990; at Ereğli, a male was carrying nesting material on 10 May 1978, with alarm-calling noted on 22 June 1978, and a pair was alarm-calling on 1 July 1977 (Albrecht 1986). In **Inner Anatolia**, a pair and a lone female were nest building in scrub south of Eber Gölü, on 6 May 1992; the male was alarm-calling and actively defending territory against another, unseen, singing male (Bradshaw & Kirwan 1994).

LESSER WHITETHROAT *Sylvia curruca*
Küçük Ak Gerdanlı Ötleğen

(Du: Braamsluiper; Fr: Fauvette babillarde; Ger: Klappergrasmücke; Swe: Ärtsångare)

Subspecies. Geographical variation and the relationships between the various forms included within this superspecies must surely rank among the most testing challenges for ornithologists and taxonomists studying Western Palearctic birds (Shirihai *et al.* 2001). Nonetheless, the latter authors provided one of the most detailed attempts to tackle the 'Lesser Whitethroat problem' to date. Roselaar (1995) had considered that three subspecies might occur: nominate *curruca* (regularly recorded on migration, and perhaps also as a breeder), *caucasica* (considered a synonym of *curruca* by Shirihai *op. cit.*) and *althaea* (Mountain Lesser Whitethroat, of which there is apparently a certain record, one trapped in Çukurova, Mediterranean, on 14 April 1987; van der Have *et al.* 1988). Based on our field experience of *S.* [*curruca*] in Turkey, we agree with Shirihai *et al.* (*op. cit.*) that the best treatment, on the available evidence, is to consider all Turkish breeding birds as *curruca*, albeit marginally greyer above and inclining towards *S.* [*c.*] *althaea*. The vast majority that we have studied vocally and in the hand give rattling songs and have plumage rather typical of, or close to, *curruca*, but some (principally those at the highest altitudes) have 'scratchy' songs, given in a song flight, closer to *S.* [*c.*] *althaea*. Given the recent recognition of four allospecies within the Lesser Whitethroat complex (Shirihai *et al. op. cit.*), the *althaea* record referred to above acquires renewed significance. Additionally, the Siberian form '*blythi*' has been reported on several occasions in autumn, but this name is considered a synonym of *curruca* by Shirihai *et al.* (*op. cit.*).

Status and distribution. National conservation status: Least Concern (population apparently stable). Local but generally not uncommon summer visitor, breeding in the interior in dry upland and montane areas with relatively sparse scrub, occasionally in open woodland, to at least 2,600m.

Common, occasionally abundant, and widespread on passage across the whole country. In spring generally recorded from late March/early April but occurs in small numbers from early March (5th), with passage peaking in the second half of April and continuing until mid May (20th). Autumn migration protracted and lasts from early August (3rd) until early November (7th), with an exceptionally late individual recorded at Birecik (Southeast Anatolia) on 24 November 1989. Peak numbers occur from mid September to mid October, perhaps even from early September in northern coastal localities in some years. Occasionally overwinters (as perhaps did the Birecik individual) in western areas, as evidenced by single birds at Büyük Menderes Delta (Aegean) on 18 January 1969 (OST 1972), Bafa Gölü (Aegean) on 3 December 1970 (OST 1975), Adana (Mediterranean) on 19 December 1996 (N. Shelton), Büyük Menderes Delta on 24 December 1996 (N. Shelton) and Karataş, Çukurova (Mediterranean) on 12 February 1999 (E. van der Burg & R. Hofland).

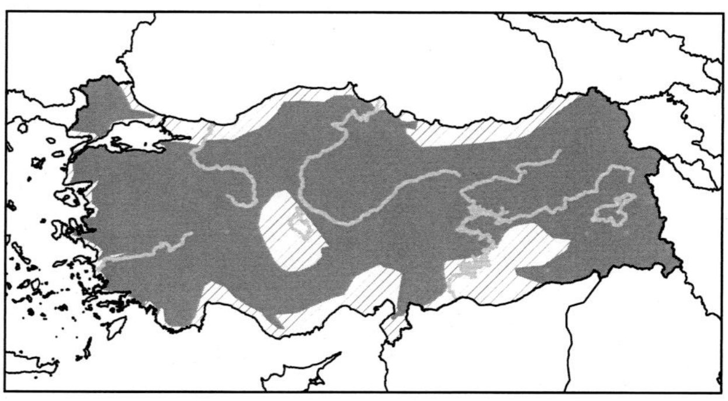

Breeding. Inhabits montane slopes with scattered bushes and stunted trees, and open woodland with undergrowth and scrub. Nests at Hasan Dağ (Inner Anatolia) and Gaziantep (Southeast Anatolia) were in scattered dog roses in open rocky areas, and in stunted conifers; a nest at Göreme (Inner Anatolia) was in a creeper growing over a wall. At Kocaçay Delta (Marmara), it breeds in forest clearings and agricultural areas (Ertan 1996). Most nests are low in bushes, typically 0.3–0.5m, but up to 1.5m; the nest is a small frail cup of dead twigs, grass and roots, lined with hair and fine roots. Eggs: C/2 (1), C/3(5), C/4 (2), C/5 (2); nests in 1907 had 4–6 eggs (Ramsay 1914). Young: B/3 (2), B/4 (5). In most regions, on territory early April–mid September. In **Southeast Anatolia**, nests had eggs near Gaziantep on 17 May 1993 and 2 June 1993 (PC), several broods of flying young were there on 14 June 1996 (GMK), six nests there on 8–9 May 2004 had three eggs (two nests) and newly hatched young (four nests), indicating egg laying *c.*24 April, and another nest had newly hatched young on 17 May 2004 (PC). In **Inner Anatolia**, two nests had eggs on 21 May 1998 and 31 May 1973; two nests had eggs at Hasan Dağ on 5 June 1992, from which young fledged by 18 June, indicating laying of first eggs *c.*21 May; a nest had young *c.*6 days old at Göreme, on 14 June 1993; a nest with young in Cappadocia on 26 July 1992 (J. Frost & J. Hornbuckle), and an adult feeding young at Kızılcahamam on 13 August 1967 were probably second broods; at Karadağ, nests had fresh eggs and others newly hatched young in the last week of May 1907 but, in another nest, eggs were not laid until after 8 June (Ramsay 1914). In **Mediterranean**, a nest had eggs at Demirkazık on 27 May 1992 (S. Williams), and a nest had small young at Torosdağ on 18 July 1971. A nest had three eggs at Gelinkaya (**Black Sea Coastlands**) on 15 June 2004 (PC).

EASTERN ORPHEAN WARBLER *Sylvia crassirostris*
Ak Gözlü Ötleğen

(Du: Oostelijke Orpheusgrasmus; Fr: Fauvette orphée; Ger: Orpheusgrasmücke; Swe: Mästersångare)

Subspecies and taxonomy. Few specimens from Turkey are available, all from the west and all of them relating to nominate *crassirostris* (Roselaar 1995). The latter author speculated that *crassirostris* probably occurs throughout much of Turkey, with the exception of the southeast north to at least the Aras Valley, where he postulated that *balchanica* might be expected to occur. In our view (and also apparently that of Shirihai *et al.* 2001), all Turkish birds are *crassirostris*, but we lack extensive field experience with this species in the easternmost part of the country. We follow Shirihai *et al.* (*op. cit.*) in considering Western *S. hortensis* and Eastern Orphean Warblers as allospecies.

Status and distribution. National conservation status: Least Concern (population apparently decreasing). Rather local but not uncommon summer migrant, breeding in small numbers in Marmara, Aegean, Mediterranean, and more locally in Inner Anatolia, western and also extreme eastern areas of the Black Sea Coastlands, as well as in parts of Southeast and East Anatolia, usually on scrub-covered slopes, and in orchards often in rocky or hilly areas. May also breed at lower altitudes in parts of Kars province (East Anatolia), e.g. in the Aras Valley, although nesting unconfirmed (Kumerloeve 1967a). Recorded to at least 1,800m in the eastern Taurus (Mediterranean) and to over 2,000m further east.

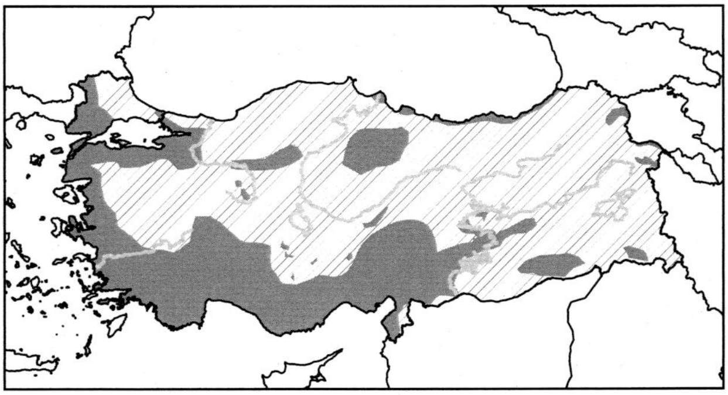

Distinctly more widespread on passage through all regions, although there are few published records from the eastern third of the country. Generally recorded from early April (late March on the south coast) to late May (27th) in spring, the earliest published record being 12 March. In autumn, passage reported from early August (7th) to late September, more rarely early October, the latest being one at Alanya, Antalya province (Mediterranean) on 15 October (1970).

Breeding. Nests are 1–2m above ground in bushes or low trees, especially oak, gorse, wild pear and olive scrub, preferring slightly taller bushes within low scrub. No records of high nests in Turkey (unlike in Western Orphean

Warbler *S. hortensis*, in Spain, where nests recorded at 5m). The nest is a cup of twigs, stems and grass, bound with spiders' webs, lined with grass, roots and hair. Eggs: C/3 (1), C/4 (8), C/5 (13), C/6 (5). In **Aegean**, earliest records, three nests with full clutches and a nest with an incomplete clutch on 16 April 1951, indicate nest building in early April; 11 further nests had full clutches and seven nests incomplete clutches on 18–26 April, all around Izmir in 1950–51 (McNeile); the earliest recent records are two nests with full clutches and two nests under construction during 23–30 April 2003 around Altınkum (PC); 33 nests had eggs in May, including several with well-incubated eggs in early May, and two with eggs of Common Cuckoo *Cuculus canorus* on 10–23 May; seven nests had young in May (including almost fledged ones on 11 May), and there were several broods of flying young from 8–17 May (earliest recently fledged on 8 May 2001), all in the Izmir/Altınkum/Muğla areas (McNeile, PC, M. Demain). In **Marmara**, three pairs were carrying food at Bandırma on 7–8 June 1966, and fledged young were at Gulpınar on 26 June 1973 (R. F. Porter). In **Inner Anatolia**, nests had fresh eggs C/4–6, and newly hatched young in the last week of May 1907, and there was another in which eggs were not laid until after 8 June (Ramsay 1914); a nest had five eggs at Hasan Dağ on 21 May 1998, and another two eggs at Karapınar on 11 June 1998 (T. Quittenden). In **Southeast Anatolia** (Gaziantep area), two nests each had five eggs on 2 June 2001 and 2 June 1993, and two nests six eggs and four eggs on 7–9 May 2004 and two nests had one egg (incomplete clutch) and three eggs on 12 June 2006 (PC); there are four records of juveniles in the Birecik/Halfeti area in early August. In **Mediterranean**, nest building was seen in Aladağ on 29 April 1970 and near Islahiye on 4 May 1970 (R. F. Porter); a nest had five eggs, and another one egg (incomplete clutch), at Uzuncaburç on 15 May 2004 (A. Rustell), and fledged young being fed were seen there on 17 June 1996 (GMK); at Uzuncaburç on 12 May 2005 a nest was empty, but complete, and two others had incomplete clutches of one and two eggs (J. Hodson); a male was carrying food at Dalyan on 5 May 1989. There is no proof of two broods, but nests in late May–June are possibly second broods.

ASIAN DESERT WARBLER — *Sylvia nana*
Çöl Ötleğeni

(Du: Woestijngrasmus; Fr: Fauvette naine; Ger: Wüstengrasmücke; Swe: Ökensångare)

Subspecies and taxonomy. Monotypic, given substantial evidence, recently presented, for viewing African *S. deserti* and Asian Desert Warblers as separate species (Shirihai *et al.* 2001).

Status and distribution. Vagrant. One record: a single bird at Karataş, Çukurova deltas (Mediterranean) on 27–28 November 1970 (OST 1975). All Middle Eastern and Western European records refer to this allospecies.

COMMON WHITETHROAT — *Sylvia communis*
Ak Gerdanlı Ötleğen

(Du: Grasmus; Fr: Fauvette grisette; Ger: Dorngrasmücke; Swe: Törnsångare)

Subspecies. Extensive intergradation exists between nominate *communis* and *icterops* (Roselaar 1995, Shirihai *et al.* 2001). The former provided the most detailed exposition of the problems facing any attempt to categorise Turkish populations racially: birds from the northwest, in Thrace, and apparently the entire Black Sea Coastlands are *communis*, with pure *icterops* occurring in the south and east, from the Iranian and Iraqi borders west almost as far as the western Taurus (Mediterranean), north to the Aras Valley (East Anatolia) and south of Ankara (Inner Anatolia). As noted by Roselaar, the form *traudeli*, described by Kumerloeve (1969a) from southeast Turkey, is a synonym of *icterops*. Populations that appear to represent intergrades between the two forms occur in the rest of western Turkey, from the Sea of Marmara south to the central Mediterranean littoral and east across a narrow swath of the country to the Armenian border.

Status and distribution. National conservation status: Least Concern (population apparently stable). A relatively common and widespread summer migrant, breeding in virtually all regions, although absent from much of northern Southeast Anatolia, rather local in Mediterranean and only present usually in smaller numbers over much of the northern, central and eastern parts of the country. Breeds principally in open, usually deciduous woodlands, typically oak *Quercus*, and in dense scrubby areas in both lowland and montane areas, exceptionally to 3,000m in East Anatolia and 2,100m in the Taurus Mountains (Mediterranean).

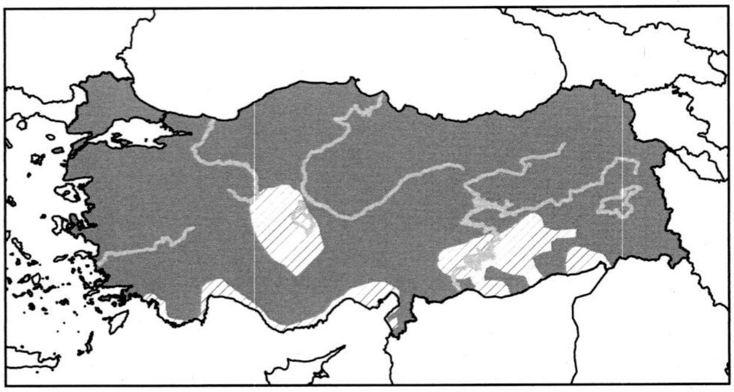

More common and widespread on passage across the whole of Turkey; in spring from mid March (12th), more atypically late February (27th), until late May (27th), with an obvious peak in the first half of April. In autumn from early August (3rd) to the third week of October (19th), although most have departed by the end of September. One winter record: one at Karataş, Çukurova (Mediterranean) on 12 February 1999 (E. van der Burg & R. Hofland).

Breeding. In scrub in open areas or in open woodland, especially clearings and edge, pastures with hedges and trees, and in bushy areas adjacent to arable land; especially oak scrub in Aegean. Nests in low bushes and brambles, usually below 0.5m: a nest near Izmir on 3 June 1950 was 'deep down in a prickly bush growing in an overgrown ditch at the edge of a cultivated field' (McNeile). The nest is a deep cup of dry grasses, lined with fine grass, hair and roots. Eggs: C/4 (2), C/5 (5). Young: B/5 (1). In **Black Sea Coastlands**, a nest at Abant Gölü ready for eggs on 29 May 1992, had an incomplete clutch of two on 3 June 1992, a complete clutch of five on 6 June 1992, and five young, 4–5 days old, on 22 June 1992 (PC); nest building was seen at Yeniçağa Gölü on 19 May 1978 (Kılıç & Kasparek 1987); at Kızılırmak Delta, a pair was feeding young on 9 June 1969 (Dijksen & Kasparek 1985), and an adult was carrying food on 4 June 1992 (Hustings & van Dijk 1994); and two nests each had five eggs at Gelinkaya on 10–11 June 2004 (PC). In **Marmara**, a pair and recently fledged young were at Uluabat Gölü in 18 May 1966, a pair was feeding four fledged young near Manyas Gölü on 25 May 1967, and adults were carrying food at Kilyos on 18 June 1973 and at Büyükçekmece on 19 June 1973 (R. F. Porter). In **Aegean**, a nest at Bornova, partially built on 22 May 1950, had four eggs on 3 June 1950, and a partially built nest at Izmir on 21 May 1954 was complete, but still empty, on 27 May 1954 (McNeile); an adult was carrying food at Denizkent in mid May. In **Inner Anatolia**, a nest had five eggs at Hasan Dağ on 21 May 1998 (T. Quittenden), and recently fledged young were at Göreme on 30 May 1999 (R. F. Porter), indicating egg laying in early May. In **East Anatolia**, a nest had eggs near Van on 13 June 2001 (T. Quittenden).

[DARTFORD WARBLER *Sylvia undata*]
Funda Ötleğeni

(Du: Provençaalse Grasmus; Fr: Fauvette pitchou; Ger: Provencegrasmücke; Swe: Provencesångare)

Subspecies. Unknown, although on distributional grounds the most likely to occur would be nominate *undata*, which reaches as far east as southern Italy (Shirihai *et al.* 2001).

Species not admitted to the Turkish list (Kirwan *et al.* 1999). One claimed in the environs of Istanbul (Marmara) on 17 August 1967 (Mathieu 1968) is insufficiently documented. There are two records from Greece: one collected in September 1934 and one observed in January 1967 (Handrinos & Akriotis 1997).

SPECTACLED WARBLER *Sylvia conspicillata*
Bozkır Ötleğeni

(Du: Brilgrasmus; Fr: Fauvette à lunettes; Ger: Brillengrasmücke; Swe: Glasögonsångare)

Subspecies. Nominate.

Status and distribution. National conservation status: Vulnerable (population trend unknown). Recently proven to breed in Southeast Anatolia, but prior to 2002 there were just six published records: a female at Side, Antalya province (Mediterranean) on 15 April 1969 (OST 1972), an immature east of Silifke (Mediterranean) on 19 October 1970 (OST 1975), an adult female/immature near Iskenderun (Mediterranean) on 4 October 1974 (OST 1978), one (no sex/age given) at the Belen Pass (Mediterranean) on 7 and 10 September 1981 (Beaman 1986), two at Burdur Gölü (Mediterranean) on 16 October 1982 and five at Küçük Kuyu, near Ayvacık (Marmara) on 11 November 1983 (Kasparek 1990a). However, at least some of these records, especially that from Marmara, suggest confusion may have occurred with either Common Whitethroat *S. communis* or Subalpine Warbler *S. cantillans* (the latter occasionally overwinters in Marmara and breeds at the locality mentioned above). The origin of the Turkish records was open to conjecture. The nearest population on Cyprus is resident or only partially migratory, the nearest wholly migratory population is in southern Italy and Sardinia. Subsequently, however, an apparently quite large breeding population was discovered in well-developed *Astragalus* scrub on the summit of Karacadağ, east of Siverek (Southeast Anatolia) in late May 2002 (Welch & Welch 2004). Thus far, this appears to be the only breeding location in Turkey.

Breeding. In a survey of Karacadağ, 45km east of Siverek (**Southeast Anatolia**), on 29–30 May 2002, 16 adults, some in pairs and one carrying food, were observed at 1,700m on sandy ground with low rocky volcanic outcrops and scattered low cushion-formation *Astragalus* scrub, which covered *c.*60% of the ground; the area surveyed was equivalent to *c.*200ha, of a total of *c.*8,900ha of similar habitat, thus indicating a potentially significant breeding population. A nest was found *c.*40cm up in a dense *Astragalus*; it was a cup of dried grasses, which had five young, close to fledging on 30 May 2002 (Welch & Welch 2004), indicating egg laying in early May. Photographs, included in the paper, show the habitat to be superficially similar to an area near Belchite, in Spain, where the species is also common, and the photograph of nestlings clearly shows the deep nest characteristic of Spectacled Warbler. Elsewhere, the outer part of the nest also includes roots, and sometimes leaves, and the lining of the inner cup includes not only grasses, but also plant down, hair, fine roots and occasionally wool; usual clutch 4–5 (3–6); two broods are reared annually; since snow may linger on Karacadağ until May or even June (Welch & Welch 2004), the record above is certain to be a first brood, with second broods likely in June–July.

SUBALPINE WARBLER *Sylvia cantillans*
Bıyıklı Ötleğen

(Du: Oostelijke Baardgrasmus; Fr: Fauvette passerinette; Ger: Weißbart-Grasmücke; Swe: Rödstrupig sångare)

Subspecies and taxonomy. *S. c. albistriata.* The races *S. c. moltonii* (on some West Mediterranean islands) and *S. c. albistriata* (from southeast Europe to western Turkey) are allopatric with nominate *cantillans* and comparatively well differentiated morphologically; they may represent incipient allospecies (Shirihai *et al.* 2001). North

African *inornata* is much closer to the nominate in its appearance. *S. c. moltonii* is probably best considered synonymous with *S. c. subalpina* (Baccetti *et al.* 2007).

Status and distribution. National conservation status: Least Concern (population apparently stable). A fairly widespread and not uncommon summer visitor, breeding in the western half of Aegean and adjacent areas of Marmara including southern Thrace, perhaps also in some areas of Mediterranean, but the southern and eastern extent of breeding range in Turkey is unclear. Found in areas of scrubland, to at least 1,100m (and higher on passage) and also in olive and almond groves and areas of open, deciduous or pine woodland during the breeding season.

More widespread on passage, usually in the western two-thirds of Turkey, being regular throughout Mediterranean and occasionally recorded in immediately adjacent areas of Southeast Anatolia (e.g. Bird 1937), as well as being moderately infrequently reported in Inner Anatolia. No records from the Black Sea Coastlands, but there is one recent published observation from East Anatolia; at Başkale, Van province, on 7 August 1986, which is one of the easternmost occurrences ever. Passage dates in spring are unclear, but the earliest was one at Karataş, Çukurova (Mediterranean) on 11 March (1972). In autumn, movements away from the breeding areas commence in early August, with peak numbers in late August and early September, and the latest being one at Büyük Çamlıca, near Istanbul, on 11 October (1981). Very infrequently recorded in winter from Aegean, e.g. two at Büyük Menderes Delta on 17 December 1969 (Kasparek 1988a) whilst the Ayvacık record (in November), cited under Spectacled Warbler *S. conspicillata*, appears very likely to refer to the present species.

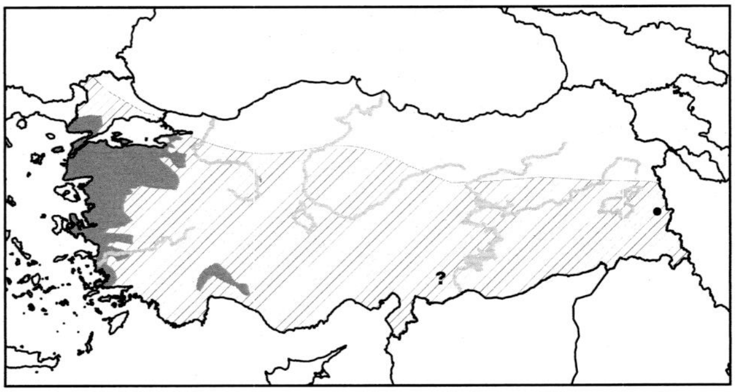

Breeding. In dense scrub, usually in open areas, but also in open woodland with undergrowth, mainly near the Sea of Marmara and coastal **Aegean**. Nest and clutch size undescribed in Turkey but elsewhere nests in a bush; builds a cup of grasses and plant down, lined with finer grasses, plant fibres and hair, and usual clutch 3–4; season commences mid April, and two broods are reared per year. In Turkey, fledged young and adults carrying food are reported mainly in the second half of May, with a late record, a pair with fledged young, in **Marmara**, on 26 June 1973 (R. F. Porter). In nearby Lesbos (Greece), two nests each had four eggs on 10 May 1995 and 18 May 1985, and three nests young on 6 May 1995, 12 May 1995 and 16 May 1995 (T. Quittenden); the earliest of these nests indicates laying of the first egg no later than 22 April.

SARDINIAN WARBLER *Sylvia melanocephala*
Maskeli Ötleğen

(Du: Kleine Zwartkop; Fr: Fauvette mélanocéphale; Ger: Samtkopf-Grasmücke; Swe: Sammetshätta)

Subspecies. Probably all nominate *melanocephala*, although the possibility that the Levantine form *momus* occurs in the extreme south, in the Hatay, as suggested by Roselaar (1995) requires confirmation through more detailed fieldwork in that region. Konrad (1985) reported *S. melanocephala* east to Halfeti, and presence, in small numbers, confirmed in recent years east as far as the Euphrates, but all those observed or trapped by GMK *et al.* (unpubl.) in Hatay, in spring 2004, were clearly nominate race birds. Should *momus* be confirmed to occur in southeast Turkey this would be interesting for the species' taxonomy. Although nominate *melanocephala* and *momus* are close in their DNA profiles, in most other respects, particularly morphology, they are distinct (and allopatric) and almost certainly merit allospecies status (Shirihai *et al.* 2001). Roselaar (*op. cit.*) recognised the indistinct form *pasiphae* and stated that birds in the southern Aegean region tend towards this subspecies, but like Shirihai *et al.* (*op. cit.*) we consider that recognition of such non-diagnosable variation at nomenclatural level is meaningless.

Status and distribution. National conservation status: Least Concern (population apparently stable). A reasonably widespread and fairly common resident breeder, or partial migrant in some areas, primarily in scrubland in coastal regions of Marmara, Aegean, western Mediterranean and the western two-thirds of the Black Sea Coastlands. Also recorded in adjacent areas of western and northern Inner Anatolia and apparently also breeds in western parts of Southeast Anatolia. Generally encountered in small or moderate numbers. The upper elevation of breeding unknown, but probably occurs at *c.*1,000m.

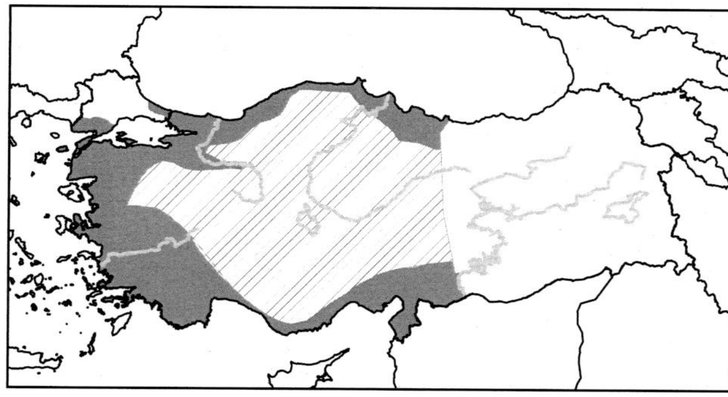

Slightly more widespread on passage across the western two-thirds of Turkey, although passage dates are difficult to discern among resident birds, wintering principally in coastal maquis and other scrublands in the west and south, less commonly elsewhere in the western two-thirds of the country.

Breeding. In dense scrub, mainly on dry rocky hillsides, but also in undergrowth in open woodland. Nest and clutch size undescribed in Turkey but, elsewhere, nests in a bush; builds a cup of grass, plant stems and spider's webs, lined with finer grasses, roots and hair; usual clutch 3–5, and two broods are reared per year, commencing in late March–April. A nest near Side (**Mediterranean**) was 0.6m up in a low thorn bush and contained five young, c.5 days old, on 17 May 2007, indicating first egg *c.*25 April (J. Hollinrake). A clutch of four, in Manchester Museum, was collected in Turkey on 17 May 1863. At Kızılırmak Delta (**Black Sea Coastlands**), two adults were feeding young on 27 May 1979, a male was carrying food on 9 June 1975, three pairs were with fledged young (2–4) in July 1971 (Dijksen & Kasparek 1985), and some adults trapped in June 1992 had brood patches (Hustings & van Dijk 1994). At Çatalağzi, on the Black Sea coast, it was the commonest warbler in 1946–48; it usually arrived in the second half of March, and immediately began to sing, often in flight, for the next six weeks, followed by shorter singing, only from cover (Ogilvie 1954). An agitated pair was observed near Izmir (**Aegean**) on 7 May 1950 (McNeile). In **Marmara**, a male was carrying nesting material at Uluabat Gölü on 22 May 1966, and an adult was carrying food at Kilyos on 18 June 1973 (R. F. Porter); there were a few nests on Marmara Island, at the west end of the Sea of Marmara in 1978 (Kaymas 1980a). Singing males are reported in most parts of the breeding range from late March/early April–June (the presumed breeding season), but there were several singing males near Adana (**Mediterranean**) and Gaziantep (**Southeast Anatolia**), in mid-February 1999 (R. Hofland & E. van der Burg), which may indicate earlier breeding in the south.

MÉNÉTRIÉS'S WARBLER *Sylvia mystacea*
Pembe Göğüslü Ötleğen

(Du: Ménétries' Zwartkop; Fr: Fauvette de Ménétries; Ger: Tamariskengrasmücke; Swe: Östlig sammetshätta)

Subspecies. Those in Southeast Anatolia are *rubescens*, but those in the Aras Valley and environs (East Anatolia) are nominate *mystacea*. These two subspecies' breeding ranges are apparently entirely allopatric. However, some of the differences between these races have been over-simplified in the literature (e.g. Shirihai *et al.* 2001). *Contra* these authors, spring males of *rubescens* (e.g. in the Hatay and along the Euphrates) quite commonly show a rather extensive and deep pinkish hue to the breast, although the extent and depth of the crown saturation appears to be much as described by Shirihai *et al.* (2001).

Status and distribution. National conservation status: Near Threatened (population apparently decreasing). A rather local but not uncommon summer migrant, breeding generally in small numbers in Southeast Anatolia and probably in adjacent (principally southwestern) areas of East Anatolia, although several such localities have been flooded in recent years as part of the GAP project. Recently also established to breed in the Aras Valley, around Iğdır, Kars province (East Anatolia), and also common in Hatay (Mediterranean), where Meinertzhagen

(1935) took the male of a pair breeding around the former Amik Gölü (specimen in the Natural History Museum, Tring). He reported the observation as referring to *S. melanocephala momus*, but the specimen is a *mystacea* (NHM 1965-M-13977). Breeds in dense scrub, especially *Tamarix*, and in olive and citrus groves, often in riverine vegetation, but more occasionally on rocky slopes, generally at 300–600m (900m in the Aras Valley) but perhaps considerably higher (1,600m) in some other areas of East Anatolia, and reported to at least 2,000m on passage in the latter region. There is one record from such higher altitudes in Black Sea Coastlands, at Gelinkaya, north of Erzurum, on 11 May 2004 (L. Svensson pers. obs.). Perhaps more commonly frequents drier-land habitats in Hatay and the Aras Valley than elsewhere in Turkey.

Although not mentioned by Danford during visits to the same areas, Weigold (1912) found the species around Urfa in April 1911, as did Bird (1937) around Gaziantep (Southeast Anatolia) in May and June 1935, although both these authors treated it as a subspecies of Sardinian Warbler *S. melanocephala*, as was the common perception until more recently. In addition, three females collected by Bird, held in the Natural History Museum (Tring), and attributed to Subalpine Warbler *S. cantillans*, were only correctly identified by Watson (Kumerloeve 1961). Indeed, the species' range in Turkey has only recently been elucidated, as fieldwork within appropriate parts of eastern and southeastern Turkey has become possible (albeit intermittently) in the last three to four decades.

Slightly more widespread and locally common on passage through the same regions, although there are three spring records from elsewhere, in addition to the Black Sea Coastlands record (mentioned above): one at Çankaya, Ankara (Inner Anatolia) on 3 May 1969, one northeast of Pozantı, Adana province (Mediterranean) on 5 May 1969 (which were erroneously described as the first Turkish records; OST 1972), and at Göksu Delta (Mediterranean) on 20 March 1996 (A. Royle). Excepting a vagrant in Portugal (in 1967), these constitute the westernmost records in Eurasia (Cramp 1992, Shirihai *et al.* 2001). Recorded in spring from 20 March (see above), although not recorded at Birecik prior to 2 April, as few observers visit the breeding areas before early May; passage is reported until mid or late May (Kumerloeve 1969a). In autumn migrants are reported from early August (8th), more usually from September, with no records beyond mid October. The bulk move through eastern areas between mid September and early October.

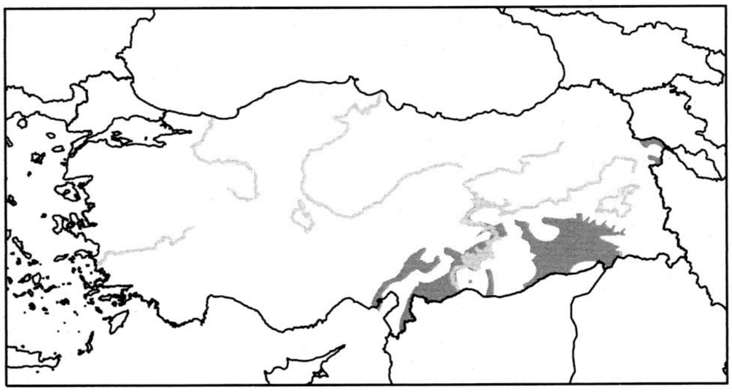

Breeding. Inhabits cultivation with brambles, hedges, vineyards, bushes in orchards, waterside vegetation, including *Phragmites*, and scrub on steppe and montane slopes. Nests in a bush or other dense vegetation, mainly 0.3–0.6m up, but sometimes up to 2.3m. The nest is a deep cup of thin twigs and plant stems, lined with fine grass and hair. Of ten nests at Birecik (**Southeast Anatolia**) on 18–19 May 1993, six were in bramble and four in *Phragmites* or other waterside vegetation: seven had four eggs, one had five eggs, one had one egg (incomplete clutch), and one had four newly hatched young (Castell 1996c); subsequent visits to some of the above nests, when they had young whose age could be determined, and several broods of recently fledged young seen on 3–4 June 1998 (T. Quittenden), indicate a peak period for date of first egg as 9 May, with first eggs at other nests only a few days later; this peak period was also confirmed during 10–12 May 2004, when five nests were under construction, two had one egg (incomplete clutches), and two had four and five eggs (PC). Later records at Birecik include nest building on 26 May 1991 (D. J. Bates), a nest with four young, *c.*3 days old, on 14 June 2001 (PC), a pair feeding fledged young on 7 July 1986 (R. Webb), a nest with young on 6 July 1987, and adults feeding young just out of the nest on 1 June 2003 (S. Harthill); records in July were possibly second broods, which is occasional elsewhere. In **East Anatolia**, nest building was noted near Iğdır on 2–3 June 1992 (G. Sandve).

RÜPPELL'S WARBLER *Sylvia rueppelli*
Kara Boğazlı Ötleğen

(Du: Rüppells Grasmus; Fr: Fauvette de Rüppell; Ger: Maskengrasmücke; Swe: Svarthakad sångare)

Subspecies. Monotypic.

Status and distribution. National conservation status: Least Concern (population apparently decreasing). A widespread and fairly common summer visitor, breeding on rocky, scrub-covered slopes, often in open, coniferous (typically *Pinus brutia*) woodland, less usually in deciduous woodland, in western Marmara (possibly including southern Thrace), Aegean, Mediterranean, immediately adjacent areas of Inner and perhaps Southeast Anatolia, and possibly very locally in the extreme west Black Sea Coastlands, where recorded in summer without proof of nesting. Recorded regularly up to 900–1,100m in the Taurus (Mediterranean) and more locally to at least 1,300m in the Karakuş and Sultan mountains, Isparta/Afyon provinces (Mediterranean/Inner Anatolia).

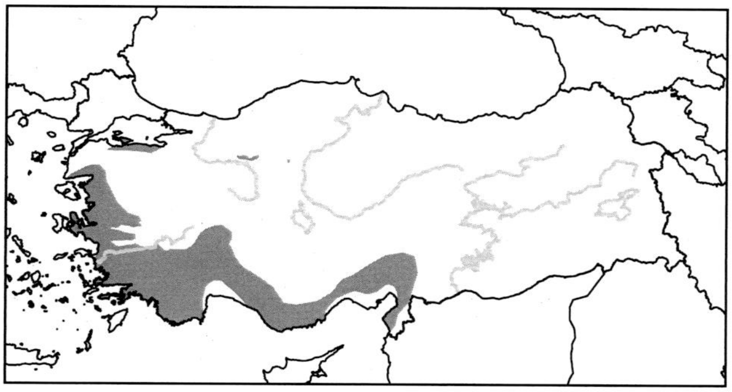

Slightly more widespread on passage through the same regions, arriving in mid (15th) to late March. An exceptionally early bird was seen on 6 March 1965 in the Hatay (Kumerloeve 1970a) and recorded from the first week of March 1990 in Çukurova (Kivit *et al.* 1994). The majority departs during September, there being only one published record from October (5th; OST 1972).

Breeding. Frequents rocky hillsides and valleys, mainly near the coast, or within *c*.50km of it, offering dense scrub of varied height, taller bushes and occasional trees, which provide higher perches for singing males. Inhabits a mix of bushes, including wild olive, prickly-oak and gorse, which last is commonly used for nesting. Found locally in open coniferous woodland. Nests in a bush, usually at 0.3–0.5m, or up to and occasionally higher than 1m. The nest is a neat cup of soft grey plant stalks (7.5–10cm in length) mixed with soft grey leaves and lined with rootlets (McNeile). Data below from **Aegean**, unless otherwise stated. Eggs: C/5 (15); a record of a full clutch of three eggs at Kuşadası on 2 June 1974 is considered to be a replacement. Young: B/4 (1), B/5 (7). Earliest records, of five recently fledged broods near Altınkum on 2 May 2001, indicate laying of first eggs on 2–3 April (PC). Most records are of nests with eggs in the second half of April–early May, and of nests with young in the first half of May; later records include nest building seen on 27 May 2004 and small young in a nest on 4 June 2004 (M. Demain); 26 records of fledged young in May–early June; and two later records, in late June and late July, which may refer to older, more developed, young. In **Mediterranean**, nest building was seen at Akseki on 30 April 1967 (R. F. Porter), adults were singing on territory on 30 March 1991 and 31 March 2000, with display on 12 April 1994 and 12 April 1999, and fledged young on 9 June 1993. Records indicate that only a single brood is reared in Turkey.

CYPRUS WARBLER *Sylvia melanothorax*
Kıbrıs Ötleğeni

(Du: Cyprusgrasmus; Fr: Fauvette de Chypre; Ger: Schuppengrasmücke; Swe: Cypernsångare)

Subspecies. Monotypic.

Status and distribution. Vagrant or perhaps a regular overshooting migrant in spring, although only six records are available as follows. A male was at Anamur (Mediterranean) on 4 April 1986 (Kasparek 1986d) and another was at Bodrum, Milas airport (Aegean) on 13 April 2003 (B. W. Stoneman). All other records are from the Göksu Delta (Mediterranean): a female on 10 April 1985, and single males in the same area on 10 April 1987 (Kasparek 1990a), 30 March 1999 and 29 March 2000 (Kirwan *et al.* 2003). The two latter records are well described. We have only seen (very scant) details for the first of the earlier two Göksu Delta reports (although both were accepted by the German rarities committee), and published details in support of the Anamur record are scant. The Bodrum record is accompanied by a detailed description and field sketches. The Turkish records are the northernmost for the species.

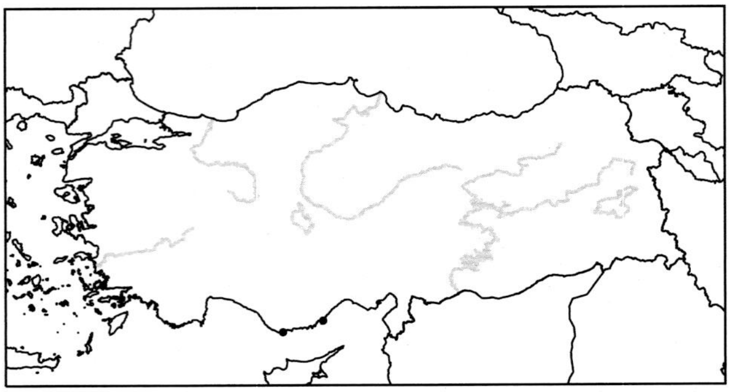

[GREENISH WARBLER *Phylloscopus trochiloides*]
Yeşilimsi Çıvgın

(Du: Grauwe Fitis; Fr: Pouillot verdâtre; Ger: Grünlaubsänger; Swe: Lundsångare)

Subspecies. Unknown, but presumably *viridanus*. For discussion of the relevant issues within the Greenish Warbler complex see the next species.

Potential vagrant. Five records: one or two in a garden in Çankaya, Ankara (Inner Anatolia) on 19 October 1967 (OST 1969), one singing in the grounds of Topkapi Palace, Istanbul (Marmara) on 19 June 1987, one singing at Denizkent, Göksu Delta (Mediterranean) on 6 June 1991 (Kirwan & Martins 1994), one singing in Yürükler Forest, Kızılırmak Delta (Black Sea Coastlands) on 29 April 1992 (Hustings & van Dijk 1994), and one again at the Göksu Delta on 29 April 1995 (Balch & Maybank 1996, Kirwan & Martins 2000a). These records are problematic. The first could equally have involved *nitidus*, as these two were usually treated as conspecific at this time, and we have seen no details concerning either of the more recent claims. The 1987 and 1991 reports were both made by observers familiar with *nitidus* (in the latter case including GMK) and involved singing individuals. However, at present, we prefer to consider these records as unproven, given that no sound recordings were made and the plumage descriptions do not unequivocally eliminate *nitidus* (see also Kirwan 1995a), based on the increased knowledge of the identification of *P. t. viridanus* and *nitidus* gained in recent years (see van der Vliet *et al.* 2001). Nonetheless, we consider it likely that *P. trochiloides* does occasionally occur on passage through Turkey, and urge any observer confronted with a potential individual of this species to document the record thoroughly. There is a record of a trapped bird from Bulgaria, in late August 1989 (Nankinov 2003). As noted by Roselaar (1995) *nitidus* in Turkey can be rather close to *P. t. viridanus* in their plumage characteristics.

GREEN WARBLER *Phylloscopus nitidus*
Yeşil Çıvgın

(Du: Groene Fitis; Fr: Pouillot du Caucase; Ger: Wacholderlaubsänger; Swe: Kaukasisk Lundsångare)

Subspecies and taxonomy. Monotypic, if considered a separate species from Greenish Warbler *P. trochiloides*. Following a period during the 1970s and 1980s when this taxon enjoyed general, popular acceptance as a species, Snow & Perrins (1998) treated *nitidus* as a subspecies of Greenish Warbler. L. Svensson (*in litt.* 1995) and K. Mild (*in litt.* 1995) also recommended this approach, but favoured use of the trinomial to spotlight the 'unique' status of this taxon within the Greenish Warbler complex. Subsequently, Svensson (*in litt.* 2006) reverted to favouring species status for this taxon. Mitochondrial DNA studies have revealed that genetic divergence between nominate *trochiloides* and *nitidus* is 'larger than between most subspecies studied so far at this locus' (Helbig *et al.* 1995). Thus, the latter authors elected to retain Green Warbler as an allospecies within the *trochiloides* superspecies.

Subsequently, however, the Greenish Warbler complex has come under renewed spotlight, particularly with respect to the undoubtedly problematic situation in central and eastern Asia (Irwin *et al.* 2001b, Collinson 2001b, Collinson *et al.* 2003), which provides a classic example of the occasional difficulties encountered when endeavouring to bring human order to the 'chaos' of speciation in progress. The result of such additional research has been to reaffirm subspecific status for *nitidus* (for now), based on the lack of a 100%-diagnosable combination of plumage and biometric characters, and the absence of clear and unambiguous differences in song between *nitidus* and *viridanus*.

Nonetheless, like Rasmussen & Anderton (2005) and Clement *in* del Hoyo *et al.* (2006), we prefer to accord *nitidus* 'higher' status, because relatively substantial and consistent differences between this and closely related taxa are long accepted, and the original grounds for no longer treating it specifically appeared to be that there is no overlap between *nitidus* and *P. trochiloides viridanus*, as had long been reported to be the case. That Greenish Warbler and Green Warbler are 'merely' allopatric appears insufficient rationale for their being treated as conspecific, especially given the degree of unanimity as to the 'uniqueness' of *nitidus*. Ticehurst (1938) and subsequent authors did not note any evidence for intergradation between *nitidus* and *trochiloides*. The field identification has been well covered, by van der Have (1987) and van der Vliet *et al.* (2001), and further summarised in Collinson *et al.* (2003). Much of the claimed perceived difficulty in identifying *nitidus* appears to stem from the paper of Albrecht (1984), wherein he reported considerable overlap between the appearance of *P. t. viridanus* and *nitidus*, both in the field and specimens. L. Svensson (*in litt* 2007) and GMK are, however, in agreement that whilst Albrecht wrote honestly about his experiences, his results have later been misinterpreted to constitute basis for the notion that *nitidus* cannot safely be separated from *viridanus* and the other forms of the Greenish ring. This is not so. The degree of morphological difference between *nitidus* and *viridanus* is certainly not large, but the majority of individuals of either taxon are probably identifiable, and the number of features known to be useful in their field or in-hand identification is as many as among several other 'difficult' congeners that are accorded specific status. That the song of *nitidus* does not differ substantially from *viridanus* is not necessarily surprising; *nitidus* is geographically comparatively well isolated within the Greenish Warbler complex and has not needed to develop the distinctive vocal repertoire of, for instance, *plumbeitarsus* in the zone of overlap with *viridanus*. Nevertheless, limited playback experiments in northeast Turkey (by GMK) suggest that Green Warbler responds rarely and only weakly to the song of Greenish Warbler, although this is not the consistent experience of K. Mild (*in litt.* 1995). On the wintering grounds, in the Indian subcontinent, the two have specific call notes, which readily differentiate them (Grimmett *et al.* 1998, Rasmussen & Anderton 2005; P. Willoughby & M. Newsome *per* A. Lassey), and they might also, to some extent, be segregated by habitat/altitude. Furthermore, as the mtDNA data of Helbig *et al.* (1995) demonstrated, *nitidus* is clearly following a different evolutionary trajectory to any other taxon within the Greenish Warbler complex, uncompromised by its lack of contact with other members of the ring of speciation linking *viridanus* to *plumbeitarsus*.

P. nitidus is the sister taxon of the more widespread and migratory *viridanus*, albeit occurring in temperate and boreal woodland at lower latitudes. Such patterns are artefacts of Pleistocene and Holocene vicissitudes. Modern revaluation of an analogous situation in the eastern Palearctic involving three, possibly four populations: *P. proregulus*, *chloronotus/simalaensis* and *kansuensis* – which were previously all treated as Pallas's Leaf Warbler *P. proregulus* – in the provinces of Qinghai, Gansu and Sichuan (China) (Alström & Olsson 1990, Alström *et al.* 1997) has demonstrated that *Phylloscopus* populations that are weakly morphologically differentiated may indeed comprise species-level taxa, based upon differences in vocalisations and habitat, once extensive fieldwork is undertaken. More detailed studies could yet reveal that *nitidus* is an analogous western example.

Status and distribution. National conservation status: Least Concern (population apparently stable). Generally a reasonably common summer visitor to upland deciduous and mixed forests, but also more occasionally in coniferous woodland and patchy *Rhododendron* scrub, throughout much of the Black Sea Coastlands, west at least as far as Düzce, Bolu province. Known from 800–1,500m in the west of this range and to at least 2,300m in the northeast. May occasionally breed closer to sea level in the latter region, but there are few data from the lower elevational range. Apparently most numerous in the eastern third of this region, with comparatively few records from the central third (which is poorly watched). There are no records from further west in Black Sea

Coastlands, but observed once recently at the comparatively frequently visited Uludağ (Marmara), where suitable habitat exists; 1–2 on 18 July 1998 (S. Aspinall & S. James), and has been recorded in summer, apparently holding territory, in the adjacent northern, montane fringe of Inner Anatolia, e.g. at Kızılcahamam National Park in mid-June 1996 (GMK).

On passage might be expected to occur across most of the eastern two-thirds of the country, but there are very few records at either season from any region: with the exception of the reports listed above for Greenish Warbler *P. trochiloides*, some or all of which may refer to the present species, passage records of *nitidus* are available from Aegean (see Kumerloeve 1958), Southeast and East Anatolia. Breeding areas are occupied from at least early May (5th) to early October, with passage birds recorded in late September and into mid May.

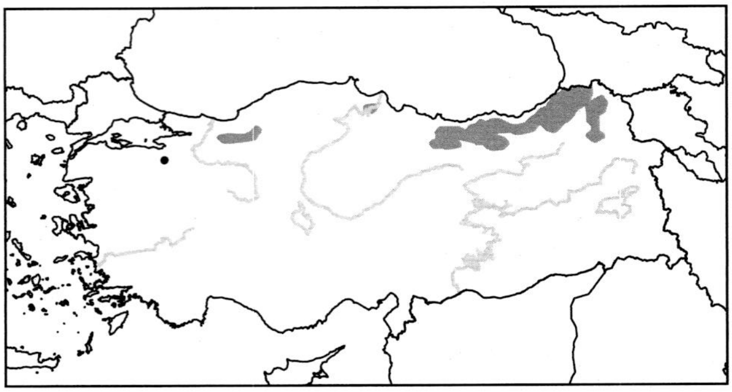

Breeding. Inhabits slopes, often steep and damp, in mature deciduous, coniferous or mixed forest; occurs at the edge of small glades, or in slightly open areas, where lower branches trail down almost to ground level, providing progressively lower perches for an adult returning to a nest on the ground (typical behaviour in several *Phylloscopus* spp.). Favours predominantly bare areas and those with low ground vegetation, tufts of grass or scant shrubs, or taller, scattered shrubs. All breeding records are from **Black Sea Coastlands**. Two nests were found near Abant Gölü in 1992 (Castell 1996f, and unpubl.): both were on the ground on near-vertical banks in deciduous woodland, predominantly beech; on 29 May, one nest was complete but empty and the other almost complete; on 3 June, they had three and two eggs, but incubation had not commenced; on 20 June, one nest had six deserted eggs and the other six newly hatched young; incubation had lasted a maximum of 14 days; each nest was in a hollow in the ground and slightly screened by a tuft of vegetation; the nests had a domed structure with a side entrance, of moss, grass and leaves, lined with finer grass and some hair. At Bafra, in the central Black Sea Coastlands, an adult was carrying food on 15 July 1986. All other breeding records are from eastern Black Sea Coastlands: an adult was carrying food to a nest on the ground in a clearing in spruce forest near Ikizdere on 13 June 1975; at Sivrikaya an adult was carrying a caterpillar on 12 July 1986 (R. Webb), and there were two family parties feeding in small pines with beech on 26 July 1986 (J. McLoughlin); at Sumela Monastery, adults were feeding young on 14 July 1986 (R. Webb) and 13 July 1990 (Eames 1991), and carrying food on 16 June 1975, and a family party was seen on 15 July 1992. There are many records of singing males, June–August (latest 8 September), suggesting possible double-brooding, but single broods are usual elsewhere.

ARCTIC WARBLER *Phylloscopus borealis*
Kuzey Çıvgını

(Du: Noordse Boszanger; Fr: Pouillot boréal; Ger: Wanderlaubsänger; Swe: Nordsångare)

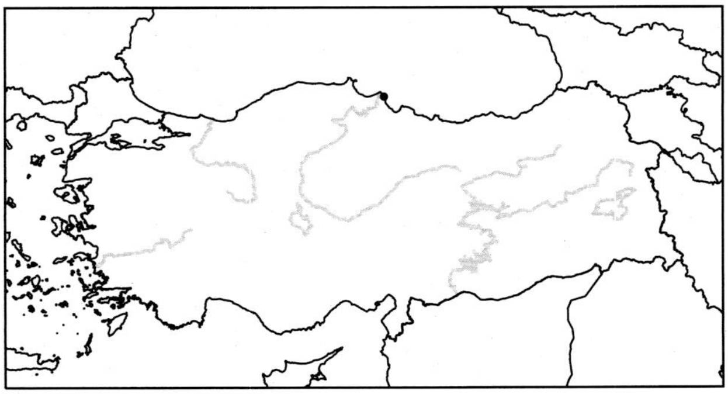

Subspecies. The only form likely to occur in Turkey is the nominate (e.g. Baker 1997), and the published photograph of the Turkish individual appears to confirm such identification.

Status and distribution. Vagrant. One record: an individual trapped and photographed in Kızılırmak Delta (Black Sea Coastlands) on 25 October 2002 (Erciyas & Özçam 2002). A detailed description and some biometric data were presented. Additionally, there is an unpublished claim 30km north of Manavgat, Antalya province (Mediterranean) in August 1988, but the supporting description does not preclude other similar *Phylloscopus* species, particularly Green Warbler *P. nitidus* (Kirwan *et al.* 1999). There are three records from Greece: sight records in April 1965, April 1967 and October 1993 (Handrinos & Akriotis 1997), and the species has also been recorded in eastern Saudi Arabia and Oman (Porter *et al.* 1996).

YELLOW-BROWED WARBLER *Phylloscopus inornatus*
Sarı Kaşlı Çıvgın

(Du: Bladkoning; Fr: Pouillot à grands sourcils; Ger: Gelbbrauen-Laubsänger; Swe: Taigasångare)

Subspecies and taxonomy. Monotypic, given the elevation of *humei* to species status (see below). The closest affinities of *P. h. mandellii,* however, are uncertain and this taxon may be best aligned with either the present species or the next, or treated as a species in its own right as recently proposed by Irwin *et al.* (2001a) and equivocally accepted by Rasmussen & Anderton (2005).

Status and distribution. Vagrant or perhaps a rare migrant. Five records: one at Çankaya, Ankara (Inner Anatolia) on 10 October 1967 (OST 1969), three at Karine in the Büyük Menderes Delta (Aegean) on 17–18 January 1969 (OST 1972), one at Ereğli, Zonguldak province (Black Sea Coastlands) on 16 December 1976 (Beaman 1986), a trapped individual at Manyas Gölü (Marmara) in autumn 2003 (Didrickson *et al.* 2004), and one at Ege University campus, Izmir (Aegean) on 29 October 2004 (Balmer & Betton 2005). We have seen no details for the Manyas bird, which was trapped on an unknown date between 26 August to 19 October 2003. Records from Gallipoli (Marmara) on 2 May 1915 and 21 October 1915 mentioned by Buxton (1916) were dismissed by Kumerloeve (1961) and certainly fail to meet modern standards for the documentation of rarity records. As noted by Hollom *et al.* (1988) the majority of winter records in the Middle East of *P. inornatus* probably refer to *P. humei*. The taxonomy and field identification of the latter have only recently been clarified (see e.g. Alström *et al.* 1991, Shirihai & Madge 1993). Whilst both species have been recorded in Israel (Shirihai 1996), it is clearly possible that at least some of the Turkish records currently ascribed to *inornatus* involved *humei*, given that three of the reports cited here pre-date strong interest in, and knowledge of, the features useful for separating the two forms in the field. There are also five late autumn records of Yellow-browed Warbler *sensu lato* from Cyprus (Flint & Stewart 1992) and three from Greece, in April, August and May (Handrinos & Akriotis 1997), and it is regarded as an irregular visitor to Bulgaria, where the first record was in 1987 (Nankinov 2003).

HUME'S (LEAF) WARBLER *Phylloscopus humei*
Yaprak Çıvgını

(Du: Humes Bladkoning; Fr: Pouillot de Hume; Ger: Tienschan-Laubsänger; Swe: Bergstaigasångare)

Subspecies and taxonomy. Presumably nominate *humei* (*mandellii* from China is an extremely unlikely vagrant to the Western Palearctic). Until recently *humei* was considered conspecific with Yellow-browed Warbler *P. inornatus*, but morphology and vocalisations point to the two being species-level taxa (Cramp 1992). The systematic position and status of *mandellii*, however, is less clear-cut, although Irwin *et al.* (2001a) proposed species status for this taxon.

Status and distribution. Vagrant. One record: one at Topkapi Palace, Istanbul, on 28 December 1994 was well described and has been accepted by the Dutch rarities committee (Ebels 1996). The species has recently proved to be a comparatively regular (annual) winter visitor to Israel (Shirihai 1996).

EASTERN BONELLI'S WARBLER *Phylloscopus orientalis*
Boz Çıvgın

(Du: Balkanbergfluiter; Fr: Pouillot oriental; Ger: Balkanlaubsänger; Swe: Balkansångare)

Subspecies and taxonomy. Monotypic, but was previously usually regarded as a subspecies of Western Bonelli's Warbler *P. bonelli*. Helbig et al. (1995) studied molecular phylogenetic relationships between western (*bonelli*) and eastern (*orientalis*) populations and concluded that the degree of genetic differentiation was substantially greater than that typical among subspecies of *Phylloscopus*. The proposal to treat the two as species has accrued universal support (e.g. Baker 1997, Beaman & Madge 1998, Snow & Perrins 1998). Eastern Bonelli's Warbler has become the English name in most common usage for this taxon, although Beaman & Madge (*op. cit.*) also suggested 'Balkan Warbler'.

Status and distribution. National conservation status: Least Concern (population apparently decreasing). A rather local and uncommon summer visitor, breeding in small numbers in hilly or montane areas in the western Black Sea Coastlands (including Thrace), Aegean and Mediterranean, and the northern and western fringes of Inner Anatolia. Perhaps more widespread than records suggest, e.g. breeding was proven at Yüksekova (East Anatolia) on 4 June 1969 and recently suspected at Şirnak (East Anatolia), and the species has been noted singing as late as 29 May at Sumela (eastern Black Sea Coastlands). Several areas of apparently suitable habitat have been identified but not surveyed in the Hatay (Mediterranean). Principally found in open, deciduous woodland (and typically in *Pinus brutia* in Mediterranean) at 300–2,000m, although usually occurs at the lower end of this elevational range in the Taurus (Mediterranean).

Distinctly more widespread on passage through the same regions, as well as Thrace, and locally in western parts of Southeast Anatolia. Scarce on passage through East Anatolia, and recorded once in eastern Black Sea Coastlands, at Trabzon, in May 2004 (G. Welch). Recorded from the third week of March (21st) in coastal areas of Mediterranean and Aegean, but not widespread in the interior and north until mid April. Passage continues until the third week of May, at least in eastern areas. In autumn passage reported from late July (24th) to mid October (14th), although most have departed by the end of September. At the Bosphorus in 1966 numbers peaked in the last week of July and swiftly declined thereafter (Porter 1983). Recorded twice in winter: one at Bafa Gölü (Aegean) on 17 December 1969 (Kasparek 1988b) and several in Köprülü National Park, Antalya province (Mediterranean) from November 1990 to at least February 1991 (M. Bakker).

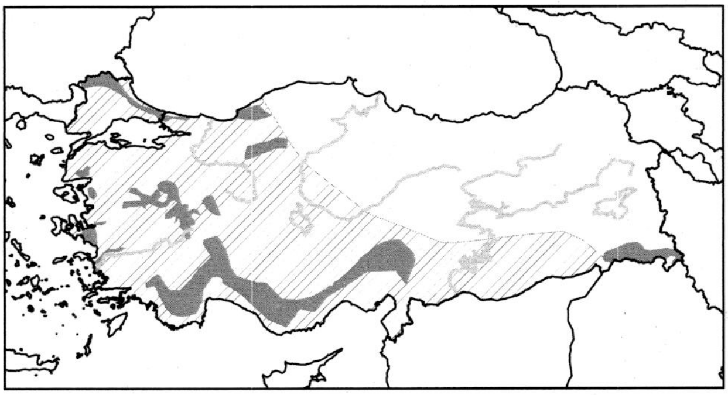

Breeding. Mainly in open woodland; coniferous, deciduous and mixed, usually on slopes, and also in scrub with scattered trees. Nests on the ground, in a hollow or in the side of a bank. The nest is a domed structure with a side entrance, of grass, leaves and moss, lined with finer grass, hair and rootlets. Eggs: C/5 (1). Young (in nest): no data. A nest had five eggs near Yüksekova (**East Anatolia**) on 4 June 1969; it was on the ground, and near a tree trunk, in woodland (R. F. Porter). In **Inner Anatolia**, a nest at Kızılcahamam was on the ground, beneath one of several small oaks, within mixed pines and oak, and had small young on 5 June 1996 (P. Doherty, GMK), indicating first egg on 19 May; also there, there were three fully fledged young on 20 June 1983, fledged young on 10 July 1973, and a family party of two adults and two young on 24 July 1974 (Barış *et al.* 1984). Many family parties were at Ayvacık (**Marmara**) on 16 June 1985. In **Mediterranean**, an adult was carrying food near Akseki on 25 May 1989 (A. Clarke), and a pair was at a nest in Kovada National Park on 15 June 1987 (I. A. Green). Males sing mainly late April–June, and the species is presumably single-brooded in Turkey, as elsewhere. A clutch of three eggs in the Natural History Museum, Tring, was collected by Krüper, in Greece, on 10 May 1880.

WOOD WARBLER — *Phylloscopus sibilatrix*
Orman Çıvgını

(Du: Fluiter; Fr: Pouillot siffleur; Ger: Waldlaubsänger; Swe: Grönsångare)

Subspecies. Monotypic.

Status and distribution. National conservation status: Data Deficient (population trend unknown). The rationale for inclusion in the Turkish Red Data Book is decidedly flimsy. Possibly has bred in Thrace, as five north of Dereköy, Kirkareli province, on 27 July 1976 (Beaman 1986) indicates, but the record from near Giresun (Black Sea Coastlands), on 21 July 1966 (OST 1969), was subsequently rejected (OST 1975: 272).

Otherwise a passage migrant through Turkey, being widespread and fairly common, or even locally abundant, e.g. there were 500 at Akşehir Gölü (Inner Anatolia) in late April 1992, although it is generally less common in the eastern third of the country. In spring recorded from mid March (11th) to early June (3rd), with the bulk occurring during the second half of April. On return migration observed from the end of July (see above) or early August (4th) until mid October, when recorded at Ankara (Inner Anatolia), Kastamonu (Black Sea Coastlands), Erzurum (East Anatolia; Kumerloeve 1961) and Izmir (Aegean; latest 16th; OST 1975), although most pass through in late August–mid September.

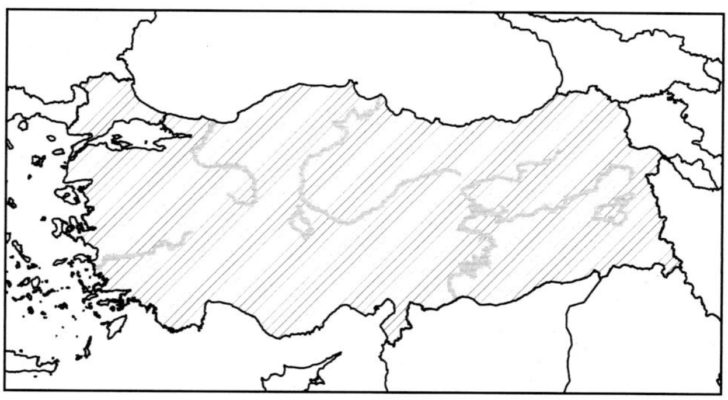

WILLOW WARBLER — *Phylloscopus trochilus*
Söğütbülbülü

(Du: Fitis; Fr: Pouillot fitis; Ger: Fitis; Swe: Lövsångare)

Subspecies. Three subspecies are generally acknowledged in the modern-day literature, namely nominate *trochilus* (which breeds from northwest Europe east to Poland and Romania), *acredula* (at higher latitudes from Scandinavia east to the Yenisey) and *yakutensis* (in central and eastern Siberia). All are known to occur on migration in Israel, with *yakutensis* by far the rarest of the three (Shirihai 1996), and the same is probably true in Turkey. In our experience and that of Schekkerman & van Roomen (1993) most of those passing through Inner and East Anatolia are *acredula*. However, there is evidence for intergradation between this form and nominate *trochilus* (e.g. Maas Geesteranus 1959), and whether *acredula*, at least, is meritorious of recognition appears moot. The difficulties in separating all three subspecies, even in the hand, are considerable, and were discussed by van Loon (1991). For *yakutensis*, it is interesting to note that well over 50% of the 256 birds handled in Kızılırmak Delta (Black Sea Coastlands) in spring 1998 had wing-lengths that accorded most closely with the range recorded for this subspecies. In the eastern Black Sea Coastlands, Faldborg (1994) noted *yakutensis*-type appeared later than those he ascribed to *acredula*, but their passage dates certainly overlap (GMK). Nominate *trochilus* is probably principally found in the west. Kumerloeve (1961) also mentioned records of two forms best regarded as synonyms, *eversmanni* (from Urfa, Southeast Anatolia, and Ankara, Inner Anatolia; best merged within *acredula*) and *fitis* (Amik Gölü, Mediterranean; considered synonymous with nominate *trochilus*).

Status and distribution. A widespread and common passage migrant throughout the country, occasionally recorded in very large numbers, e.g. there were *c.*3,000 at Büyük Çamlıca, Istanbul (Marmara) on 11 September 1970, and similar numbers were at Uludağ (Marmara) on 18 September 1968 and 28 August 1971. Recorded in spring from late March (22nd), or early April in the north and east. Passage occasionally continues as late as the first week of June (5th), although it generally peaks in the first half of May. In autumn, the species is reported from early August (4th), with peak numbers being recorded in September until mid October, and stragglers occurring until late October (29th), perhaps even later. Many observers fail to differentiate between this species and Common Chiffchaff *P. collybita* at passage localities, making it difficult to establish the extreme dates of migration. Recorded to at least 2,400m. Kasparek (1992a) mentioned occasional overwintering in Marmara and there are two published records from the Büyük Menderes Delta (Aegean), one singing (thus eliminating confusion with Common Chiffchaff, which commonly over-winters at the same locality) on 19 January 1969 and one or two on 24–29 January 1972 (OST 1972, 1975), with another claimed (on the basis of leg colour alone) at Dalyan (Aegean) on 31 December 1986 (Kılıç & Kasparek 1989).

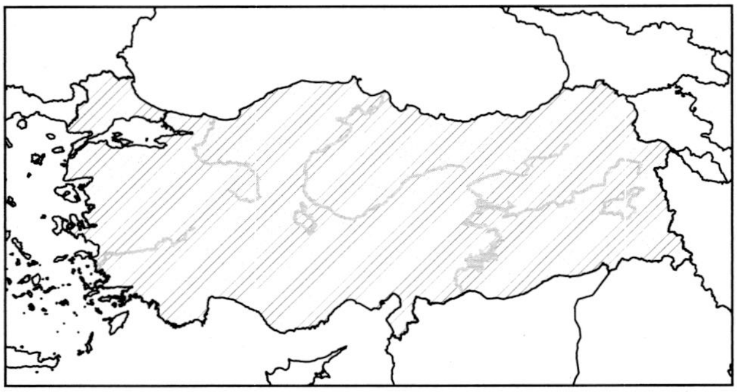

[PLAIN LEAF WARBLER *Phylloscopus neglectus*]
Küçük Çıvgın

(Du: Dwergtjiftjaf; Fr: Pouillot modeste; Ger: Eichenlaubsänger; Swe: Dvärggransångare)

Subspecies. Monotypic.

Status and distribution. There are several recent claims from the extreme east, all from the same area, but documentation of the species' presence in Turkey is currently rather weak. One was seen and heard on a sparsely vegetated hillside with scattered mature *Juniperus*, up to 3m in height, some *Quercus* and scattered shrubs of *Daphne mucronata*, beside the road to Çatak, south of Van Gölü (East Anatolia), on 24 June 2004 (I. A. Green), and on 25–26 June, 2–3 Plain Leaf Warblers were observed and sound-recorded in the same area (S. Pfützke *et al.*). Two birds, one of them singing, were seen in the same general area on 23 May 2005 (T. Epple & U. Pittius *per* S. Pfützke *in litt.* 2006) and one was apparently nesting (food-carrying to same area) on 13 June 2005 (D. Robel *et al. in litt.* 2005). GMK and C. G. Bradshaw *et al.* searched the same area, using playback of the species' calls (not song), without success on 18 May 2006. The sound-recording, although of rather poor quality, appears to be of this species but, as no detailed description of the birds is currently available, we treat the species as hypothetical for the present. However, given that the species breeds close to the Turkish border in Iran (see Porter *et al.* 1996), it is clear that it might be expected to breed, at least occasionally, in easternmost or southeasternmost Turkey.

Breeding. No certain records. In neighbouring Iran breeds in more southerly, drier mountains, mostly above the treeline, on open slopes with sparse shrubs, or in meadows with umbellifers and rose bushes, and in open degraded oak woods (Cramp 1992). The nest is in a bush, from almost ground level to 1m up; domed, of grasses, lined with feathers; 3–5 eggs from late April (Harrison & Castell 2002).

COMMON CHIFFCHAFF *Phylloscopus collybita*
Çıvgın

(Du: Tjiftjaf; Fr: Pouillot véloce; Ger: Zilpzalp; Swe: Gransångare)

Subspecies and taxonomy. Perhaps all breeding birds are *brevirostris* (certainly all those tape-recorded and observed by us appeared to belong to this form), although Roselaar (1995) speculated that it might not prove separable from nominate *collybita* (see also below). Whether *brevirostris* represents an intermediate form between *abietinus* and *P. lorenzii* (Caucasian Chiffchaff), which intergrades with the latter in northeast Turkey, as was postulated by Watson (1962b), is a question that has received insufficient attention, although neither Roselaar (*op. cit.*) nor GMK have found specimen evidence for such a hypothesis. Furthermore, our visits in the breeding season to the relevant area of the Pontics have not revealed any such intergradation, or, indeed, even the certain presence of chiffchaffs other than *lorenzii*, but one of several examined in the hand in May 2004 was perhaps closer to *brevirostris* than classic *lorenzii*. If *brevirostris* occurs in the northeast Pontics we expect that it will be proved to be in parapatry with *lorenzii*. Birds perhaps resembling *caucasicus* (which we consider to be synonymous with *brevirostris*) and with a song closer to European chiffchaffs were noted in this region of Turkey in 1992, but for now we regard these observations as uncertain because in at least one of the areas where they were conducted (around Sivrikaya) we have only ever found quite obvious *lorenzii* (at a range of altitudes).

P. c. brevirostris (endemic to Turkey; the holotype, taken by Strickland, is in the University Museum of Zoology, Cambridge, CUMZ27/Syl/63/e/1) was retained within *collybita* by Helbig *et al.* (1996), pending further investigation, because only small samples of this form's mitochondrial DNA were available to them, but the cytochrome *b*-gene sequence divergence between nominate *collybita* and *brevirostris*/*caucasicus*, and morphological differences, indicate that they have undergone comparatively long periods of separate evolution. Indeed, the available data suggest that *brevirostris*/*caucasicus* are more closely related to eastern *abietinus* than to *collybita*. However, what certainly requires renewed questioning is the extent to which *brevirostris* is truly diagnosable, notwithstanding Watson's (1962b) redescripton of this taxon, based on apparent breeders, rather than relying solely on Strickland's November specimen. What further appears certain is the dubious validity of recognising *caucasicus*.

On migration *abietinus* and *tristis* are also reported, although their frequency is generally unknown: Schekkerman & van Roomen (1993) caught none of the latter during their Inner Anatolian study; all of those they handled appertained to *abietinus*. Faldborg (1994) recorded several that he considered to be *tristis* in the northeast during mid to late April. The extremely weakly defined form *fulvescens* (which breeds between the Urals and the Yenisey, but is perhaps more closely related to *P. sindianus*) has been reported once in Turkey, at Uludağ (Marmara) on 21 September 1991 (Jetz 1995). The subspecific identity of all those wintering in Turkey is not certainly known. Some are presumably birds from farther north as, for instance, in Cyprus many appear to be nominate *collybita*, with perhaps some *abietinus* at this season (P. R. Flint *in litt.* 2004). The difficulties in certainly identifying *tristis* (especially without an in-hand examination) and of understanding the taxonomic position of *fulvescens* were ably enunciated by Dean & Svensson (2005), and it may well be the case that many claims of *tristis* are erroneous.

Status and distribution. National conservation status: Least Concern (population apparently stable). A widespread and fairly common, albeit relatively local, resident, or partial migrant in most regions, to much of Marmara and throughout much of the western two-thirds of the Black Sea Coastlands (but there are no certain records from within the range occupied by *P. lorenzii*), extending south into the northern fringes of Inner Anatolia and perhaps East Anatolia. There are also scattered records through the Aegean montane coastal fringe, the Taurus Mountains (Mediterranean) and even throughout the Hatay to the border with Lebanon. Perhaps also breeds in the extreme southeast part of East Anatolia. Though principally breeds in mature woodland at 500–2,000m, is reasonably frequent down to sea level, especially in the northwest and possibly ranges even higher in the east. Most winter in coastal regions of the west and south, but smaller numbers are scattered elsewhere throughout the western two-thirds of the country at this season.

Even more common and widespread on passage, when it is also frequent in the eastern third of the country including within the range of *P. lorenzii*. Given that large numbers are resident in Turkey, passage dates are difficult to identify, but arrivals in spring probably commence in early March with passage probably continuing in the east until the second or third week of May, although the majority have moved through southern and central areas by mid April and wintering areas in the south and west are entirely vacated by late April. Passage is also usually complete in the north by the end of April (see Albrecht 1986, Hustings & van Dijk 1994). In autumn migration almost certainly persists from early August until at least late October, and possibly later into November. Numbers on passage are typically considerably lower than for Willow Warbler *P. trochilus*.

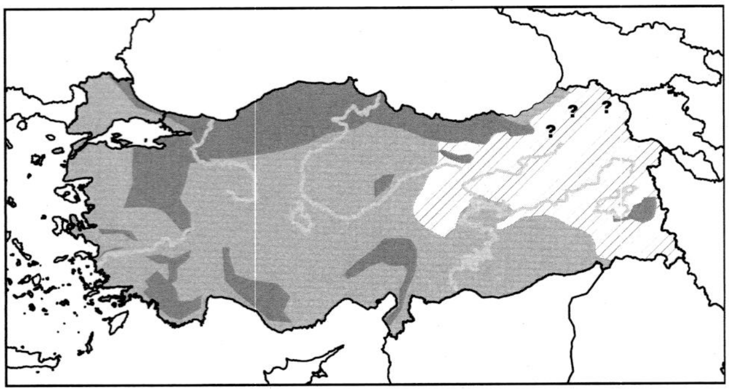

Breeding. In undergrowth in mature deciduous, coniferous and mixed woodland, on both level ground and on slopes, from sea level to 2,000m. A common breeder at sea level in alluvial forest at Kocaçay Delta (Marmara), with an estimated population of 350–380 pairs, and 0.52 singing males/ha in a 31ha plot in 1993 (Ertan 1996); two nests were found in April 1967 in low undergrowth and had large young on 19 May 1967, not April, as previously reported (R. F. Porter). An abundant breeder in all forest types and semi-open land at 800m on Uludağ (Marmara) (Jetz 1995). Nest usually built off the ground, low in a bush or in herbage. A nest at Kızılcahamam (**Inner Anatolia**), which had five eggs on 22 May 1996, was described as indistinguishable from nests in Europe (I. A. Green), *i.e.* domed with a side entrance, of plant stems, moss and leaves, and lined with feathers (Harrison & Castell 2002). Family parties were at Kızılcahamam in early June. An adult carrying food at Kopdağı pass (**Black Sea Coastlands**), on 15 June 1975, may have been *P. lorenzi* (R. F. Porter). In **Marmara**, an adult was carrying food in beech woods at 300m on Catal Dağ on 16 May 1966, and a pair with fledged young were near Demirköy on 22 June 1973 (R. F. Porter); five pairs bred in Çamlıca Hills in 1967. Singing males, on territory, are heard as late as August, perhaps indicating double-brooding.

CAUCASIAN CHIFFCHAFF *Phylloscopus lorenzii*

Kafkas Çıvgını

(Du: Kaukasische Tjiftjaf; Fr: Pouillot de Lorenz; Ger: Kaukasuslaubsänger; Swe: Kaukasisk gransångare)

Subspecies and taxonomy. Monotypic. We follow Stepanyan (1990) and Roselaar (1995), who treated *lorenzii* (from the northeast Pontics and the Caucasus) as distinct from *P. sindianus* (Mountain Chiffchaff, of central and southern Asia) with trepidation because it is unclear whether they possessed a common ancestor or were separate offshoots of *P. collybita tristis* (Siberian Chiffchaff) or, indeed, whether *tristis* might have been an offshoot of either of the two more southerly breeding taxa. Some vocal differences between *sindianus* and *lorenzii* have since been pointed to as additional evidence that the two may be less closely related than generally considered (Helbig *et al.* 2002), but no consensus to afford both taxa species status has been reached. The distinctiveness, as a group, of *sindianus*/*lorenzii* has been confirmed by mtDNA sequencing. Helbig *et al.* (1996) demonstrated that this grouping is genetically well differentiated from *P. c. brevirostris* (see above), which occurs elsewhere in Turkey, and from the recently described *P. c. caucasicus*, whose range is poorly known but occurs parapatrically with *lorenzii* in the Caucasus. Given recent, well-received proposals to recognise several species within Common Chiffchaff *P. collybita* (Helbig *et al. op. cit.*, Clement *et al.* 1998), the separation at species level of *sindianus* and *lorenzii* appears increasingly justifiable, although the results of DNA research do not confirm such a proposal. It has to be conceded that the issue presents considerable difficulties, as exemplified by the changing views of Sibley & Monroe, who initially regarded *sindianus* as a species, including *lorenzii* (1990), thereafter included *sindianus* within *P. collybita* and regarded *lorenzii* as a monotypic species (1993) and, finally, reverted to treating *lorenzii* as a subspecies of *P. sindianus* (Sibley 1996). Helbig *et al.* (*op. cit.*) continues to offer one of the most comprehensive discussions of relationships among West Palearctic chiffchaffs, including all of the forms treated here. In-the-hand identification of *lorenzii* and *brevirostris* was covered by Roselaar (*op. cit.*), and field separation of *lorenzii* was treated by Scott *et al.* (1999). See also Watson (1962b), whose call for research into the problems posed by the northeast Turkish chiffchaffs remains as urgent now as it was over 45 years ago.

Status and distribution. National conservation status: Data Deficient (population trend unknown). A fairly common, albeit relatively local summer visitor to much of the eastern third of the Black Sea Coastlands, extending south into the northeastern fringes of East Anatolia, east of the Zigana pass (Roselaar 1995). There are isolated records from the Bingöl Dağları (G. Magnin & L. Nauta). Very few records are available south of the Aras River valley. Dorèl *in* Roselaar (*op. cit.*) mentioned one showing characters of *lorenzii* at Kızılcahamam on 29 June 1991, but we have not seen details and the record may reflect confusion with *P. collybita brevirostris.* Principally breeds in mature pine and juniper woodland above 1,700m, but is also frequent in patchy, scrubby and light deciduous woodland, especially along streams and rivers.

Arrival and departure dates are very poorly understood: presumably arrives in April and perhaps into early May, though reported from mid March in the Caucasus. In autumn, migration almost certainly persists from early August (e.g. recorded near Çatak, south of Van Gölü, on 5 August 1981; Helbig 1984) until at least 11 October (latest available date from the breeding grounds), but there are almost no records in Turkey of this species away from the breeding areas, meaning that our knowledge of its movements, if any (many previous authors have speculated that *lorenzii* may, in fact, be little more than an altitudinal migrant, although there are two spring records of apparent *lorenzii* in Kuwait: Cleeve *et al.* 2004) is extremely limited. Reported once in winter, at Adana (Mediterranean) on 1 January 2005 (V. Konrad). In Armenia, birds may remain until November and large concentrations may occur in October (Scott *et al.* 1995).

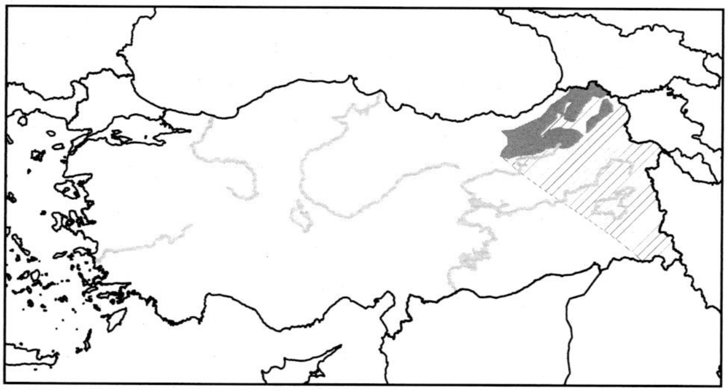

Breeding. In undergrowth in open deciduous and coniferous woodland, and in thickets and bushes, above 1,700m, and generally at higher altitudes than *P. collybita*. All data are from the best-known Turkish breeding locality, in riverside woodland and scrub between Gelinkaya and Yoncalık (**Black Sea Coastlands**). The area comprises small poplar plantations and copses of mature willows and other deciduous trees, mostly damp or partially waterlogged, in places with an understorey of scrub and dense vegetation, and with thickets of tall, dense willow bushes and other scrub, immediately alongside a river. Three pairs with young were observed on 11 July 1975. There were five nests in June 2004 (PC): two were virtually on the ground, not in hollows in the ground, as in some other *Phylloscopus*, but rather resting on ground vegetation; the other three nests were above ground, at 0.1–0.3m. The nest is domed, with a side entrance, like in *P. collybita,* but generally with fewer leaves in the outer structure; mainly plant stems and grasses, some leaves, thickly lined with feathers, and the cup lined mainly with hair and some fur. One nest was 120mm tall, 110mm wide at the base and 100mm wide at the top, with the entrance hole 30mm wide and 25mm deep. On 10 June 2004, one nest had four newly hatched young and one egg, and another two newly hatched young and one egg: both of the eggs had hatched by 11 June 2004, indicating egg laying on *c.*24–25 May 2004. Two nests had five eggs and six eggs on 11 June 2004, and the latter six young, 1–2 days old, on 21 June 2004, indicating egg laying on about 3 June. The fifth nest had four young on 23 June 2004, which left the nest on 25 June 2004. Both males and females regularly fed the young at two of these nests. Singing males are recorded as late as August, which may indicate double-brooding.

GOLDCREST *Regulus regulus*

Çalıkuşu

(Du: Goudhaan; Fr: Roitelet huppé; Ger: Wintergoldhähnchen; Swe: Kungfågel)

Subspecies. Roselaar (1995) referred all Turkish material to *buturlini*, a generally rather duller, paler and greyer subspecies than nominate *regulus*, although his map suggested the unconfirmed possibility that the latter taxon might reach Thrace.

Status and distribution. National conservation status: Least Concern (population apparently decreasing). A widespread and fairly common resident breeder in montane coniferous forest, from around 500m to at least 2,500m, in Thrace (where very local), elsewhere in Marmara, and Aegean, Mediterranean, Black Sea Coastlands and the northern fringes of Inner and East Anatolia. There is only one published record from Southeast Anatolia: one at Birecik on 17 March 1986 (Martins 1989).

Rather more widespread in most regions in winter, descending to sea level, and on passage, when there is some evidence from north-coast localities of immigration. Passage dates are unclear, but would appear to be in March and the second week of April, at least, in spring, and from early October (exceptionally in the first half of September) until at least late November in autumn, with large numbers occasionally moving through the northeast Pontics at the latter season. Much smaller numbers are recorded at the Bosphorus.

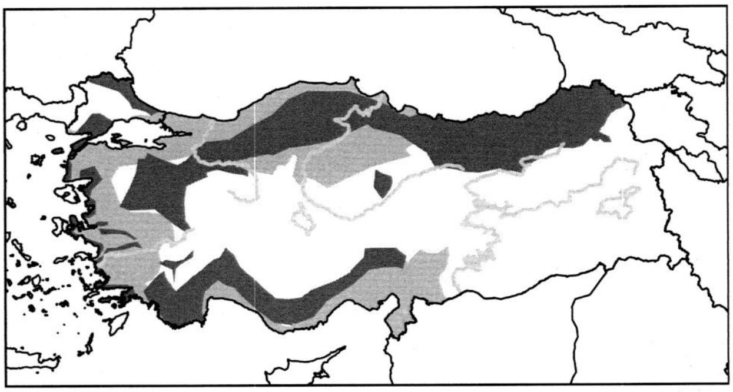

Breeding. In montane conifer forest, where singing birds are present April–July, the presumed breeding season, but there are no records of nests. In **Mediterranean**, fledged young were seen near Alanya on 28 May 2002 and an adult was observed food-carrying near Bozkır, Seydişehir, on 8 June 2002 (J. Scott).

FIRECREST *Regulus ignicapilla*

Sürmeli Çalıkuşu

(Du: Vuurgoudhaan; Fr: Roitelet à triple bandeau; Ger: Sommergoldhähnchen; Swe: Brandkronad kungsfågel)

Subspecies. No specimen material appears to be available from Turkey (see Roselaar 1995), but Vaurie (1959a) considered Turkish populations to represent the nominate form. In any case, as noted by Baker (1997), levels of geographic variation described for this species within the continental context appear so slight as to be unworthy of nomeclatural recognition. Subsequently, Red'kin (2001) published a description of a new subspecies, *tauricus*, from the southern Crimea, which apparently moves as far as the eastern Black Sea and Caucasus in winter (and thus has the potential to reach Turkey). It is described as having a very dark tone to the plumage. We have been unable to evaluate the validity of this form, but the available information suggests that it is insufficiently well differentiated to warrant recognition under the principles adopted here. Note the correct spelling of the specific name, *ignicapilla*, rather than *ignicapillus* (David & Gosselin 2002a).

Status and distribution. National conservation status: Data Deficient (population trend unknown). Generally a local and uncommon resident or partial migrant in some areas, in small numbers in either mixed or coniferous forests in montane areas of the Black Sea Coastlands, and (perhaps even more locally) in Marmara (excluding Thrace), Aegean, Mediterranean and at least one locality in Inner Anatolia. Apparently also present in the breeding season in immediately adjacent areas of Inner and East Anatolia, where presumably also breeds. Recorded from around 250 to 2,000m at least.

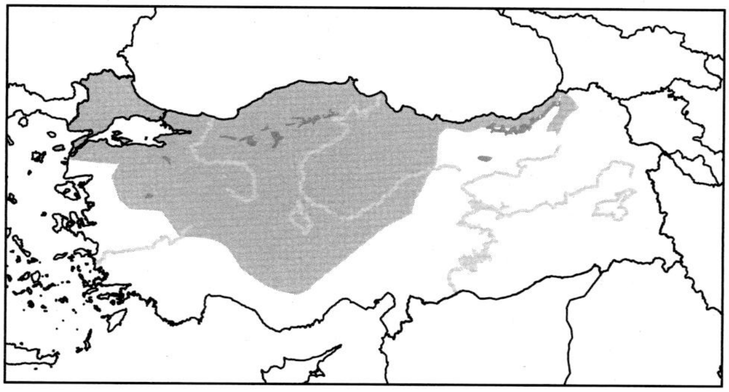

Slightly more widespread and common on passage and in winter (at which season Danford, almost certainly erroneously, described it as being more common in the Taurus, Mediterranean, than Goldcrest *R. regulus*), when also recorded in Thrace and to sea level at several south-coast localities. Some immigration is evident at this season, when recorded on passage at Black Sea and Thracian localities in early October–November and in the second half of March at least.

Breeding. In montane coniferous and mixed forest, where singing birds are present April–July, the presumed breeding season, but there are no records of nests. Elsewhere, the nest is suspended in a fork of twigs under foliage near the end of a conifer branch; it is a deep cup of moss, lichens and spiders' webs, lined with feathers; usual clutch 7–11. Several family parties were seen at *c.*1,800m near Dursunbey (**Aegean**) in July 1966 (OST 1969). Adults were observed feeding juveniles in spruce forest south of Trabzon (**Black Sea Coastlands**) on 6 July 1969 (OST 1972).

[ASIAN BROWN FLYCATCHER *Muscicapa dauurica*]
Esmer Sinekkapan

(Du: Bruine Vliegenvanger; Fr: Gobemouche brun; Ger: Braunschnäpper; Swe: Glasögonflugsnappare)

Subspecies. Unknown, but, given their relative distributions, only the nominate subspecies is likely to occur as a wild bird.

Species not admitted to the Turkish list (Kirwan *et al.* 1999). A record from Kulu Gölü (Inner Anatolia) on 21 May 1990 (*Birding World* 3: 196, Harvey 1992) was published without descriptive details. In Greece, there is an accepted sight record in September 1993 and two undocumented occurrences (Handrinos & Akriotis 1997).

SPOTTED FLYCATCHER *Muscicapa striata*
Benekli Sinekkapan

(Du: Grauwe Vliegenvanger; Fr: Gobemouche gris; Ger: Grauschnäpper; Swe: Grå flugsnappare)

Subspecies. Several previous authors (e.g. Svensson 1992) have considered the race in Turkey to be *neumanni*, but Roselaar (1995) found those in the extreme west to be identical to nominate *striata* and even further east, most if not all were the same race, albeit some showed influence of *neumanni*. Our own field and museum work confirms that *striata* is present throughout Turkey, and points to the unsurprising conclusion that *neumanni* is best treated as a synonym of the nominate. *M. s. inexpectata*, which is recognised by Svensson (*op. cit.*), breeds in the Crimea and most certainly occurs on passage through Turkey, but we consider this race, too, to be rather doubtfully distinct, although we have examined very few specimens.

Status and distribution. National conservation status: Least Concern (population trend unknown). A widespread and locally common summer visitor to Marmara, northern Aegean, Black Sea Coastlands and Inner Anatolia. Summer records from southern Turkey may well be indicative of quite widespread breeding, but there are no certain nesting records from the Taurus (Mediterranean).

Otherwise, widespread and fairly common on passage, although apparently scarcer in the east. Spring passage generally commences in early April (6th), with an exceptionally early migrant found exhausted at Alanya (Mediterranean) on 6 March. Numbers are generally small at this season, although occasionally recorded in low hundreds, primarily at coastal sites (Hustings & van Dijk 1994), but also reasonably common inland (Bradshaw & Kirwan 1992, Schekkerman & van Roomen 1993), with peak numbers probably passing between mid April and mid May, even in the south. Passage continues until at least the first week of June in the north, where breeders may not arrive until the end of April. Autumn migration generally commences early August (perhaps late or even mid July) in the north and continues until at least late October (latest 10 November in Mediterranean), with small numbers regular in all regions into the first half of October. Rather commoner at this season, with peak numbers in September, especially through western and central areas.

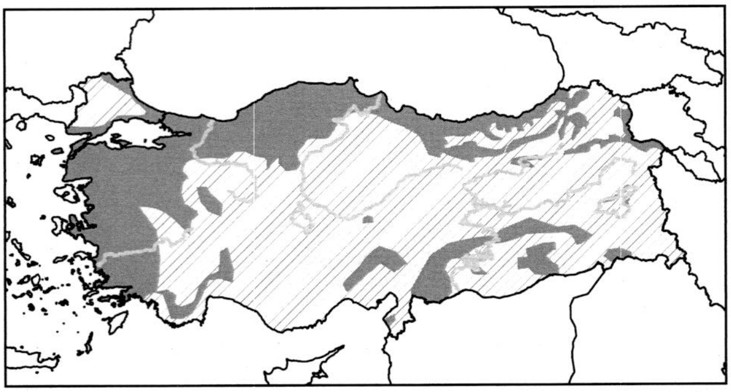

Breeding. In woodland, both deciduous and evergreen, including edges and scattered woodlots in farmland. Nest site, nest and clutch size unrecorded in Turkey; elsewhere, often nests in a crevice in a tree trunk, on a creeper or on a wall. The nest is a cup of roots, plant down and lichen, bound by spiders' webs, and lined with feathers and hair; usual clutch 4–5. At Kızılırmak Delta (**Black Sea Coastlands**), copulation was noted on 10 June 1975, there were fledged young on 19 July 1975, and two pairs each with three young in July 1971 (Dijksen & Kasparek 1985); in 1992 the estimated population was 150–250 pairs, 10–20 pairs/100ha, with first territories occupied mid May (Hustings & van Dijk 1994). At Kocaçay Delta (**Marmara**), in 1993, there were 0.29 pairs/ha in a 31ha area in alluvial forest, with an estimated population of 200–230 pairs (Ertan 1996). Fledged young were seen near Muğla (**Aegean**) on 8 June 2004 (H. Tholl-Schmitz). In **Southeast Anatolia**, adults were carrying nest material near Gerger on 25 May 2001 (H. Welch) and near Nusaybin on 8 May 2002 (S. Isfendiyaroğlu *et al.*).

RED-BREASTED FLYCATCHER *Ficedula parva*
Küçük Sinekkapan

(Du: Kleine Vliegenvanger; Fr: Gobemouche nain; Ger: Zwergschnäpper; Swe: Mindre flugsnappare)

Subspecies and taxonomy. Nominate *parva*. As discussed by Roselaar (1995) the name *colchica* is probably invalid. The eastern form *albicilla* is increasingly recognised at specific level (Taiga or Red-throated Flycatcher), given qualitative differences in vocalisations, morphology and ecology (see Svensson 1992, Cederroth *et al.* 1999).

Status and distribution. National conservation status: Near Threatened (population apparently stable). A widespread and often common passage migrant, especially in autumn, and a summer visitor, which breeds in very small numbers in Black Sea Coastlands, principally in the eastern half of this region, and probably at least in southern Marmara (Jetz 1995, Ertan 1996). Nests in deciduous and mixed forest, generally in montane regions to at least 1,400m, but perhaps locally to sea level. Seems very rare as a breeder, although its true breeding status is obscured by migrants passing through potential breeding areas during the early part of the nesting season. Breeding only recently proved (Albrecht 1981) and there are still very few instances of confirmed nesting (see below). One late July record from the northern fringe of Inner Anatolia might indicate nesting in this region (where suitable habitat is available) or may represent an early migrant.

A widespread and locally common migrant on the Black Sea coast, but only occurs in smaller numbers and extremely patchily elsewhere, principally in the western half of the country and the extreme northeast, with for example very few records from Southeast Anatolia (Martins 1989, Kirwan & Martins 1994). Spring passage commences in April, with the earliest noted on 5 April and the latest on 4 June (in Mediterranean), peaking during mid May, although numbers are nearly always small at this season. Autumn passage usually commences in August (earliest 24 July) and peaks in mid September/early October (Nisbet & Smout 1957, Porter 1983), with many hundreds probably regularly recorded at Black Sea coastal localities, e.g. Kızılırmak Delta, during this period. Passage declines rapidly, with latest dates generally 18 October (Marmara and Mediterranean) and 3 November (Inner Anatolia), although Kumerloeve (1961) noted birds in the latter region, at Ankara, as late as 26 November.

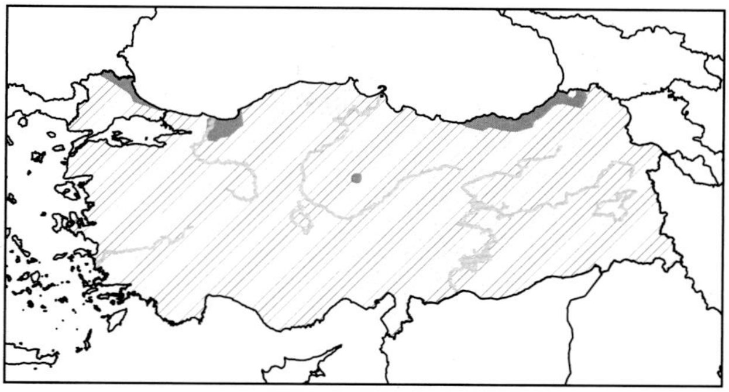

Breeding. In mature deciduous or mixed forest, mainly oak and beech. Nest site, nest and clutch size unrecorded in Turkey; elsewhere, nests 1–9m up, in a hollow or hole in a tree trunk, or on the side of a trunk, supported by small branches. The nest is a neat cup of moss, dead leaves and fibres, with lichen and thin twigs on the outside, lined with hair; usual clutch 5–6. All records are from **Black Sea Coastlands**. At Ereğli, in deciduous forest dominated by beech at *c.*1,000m, a pair was carrying food to a bush on 9 June 1978 and, on 12 June 1978, a different pair, *c.*1km away, was feeding a newly fledged young; it was calculated that eggs had been laid in mid May, hatching *c.*28 May (Albrecht 1981). One or two adults were carrying food at Sumela Monastery on 13 June 1985 (S. M. Andrews & A. J. Eardley), where a juvenile was seen on 7 July 1988 (M. Telfer). May also breed at sea level in Kızılırmak Delta where, in 1992, three singing males were on territories on 12 May 1992 and on 27 May 1992 (Hustings & van Dijk 1994).

SEMI-COLLARED FLYCATCHER *Ficedula semitorquata*
Alaca Sinekkapan

(Du: Balkanvliegenvanger; Fr: Gobemouche à demi-collier; Ger: Halbringschnäpper; Swe: Balkanflugsnappare)

Subspecies. Monotypic.

Status and distribution. Conservation status: Near Threatened (BirdLife International 2008); national conservation status: Vulnerable (population apparently decreasing). Historically, the status of this form in Turkey was unclear, as it was previously regarded as a subspecies of Collared Flycatcher *F. albicollis.* Breeds rather locally in northern Marmara, the northern fringe of Inner Anatolia (where probably even more local), across the Black Sea Coastlands and northern parts East Anatolia (where again very local), from sea level in Thrace to at least 2,000m in eastern Black Sea Coastlands; in woodland, groves, parkland and orchards. Nesting records are comparatively few and migrants passing through potential breeding areas, both early and late in the breeding season, obscure the species' true breeding status. Recorded at 2,150m in the Taurus (Mediterranean) in late July (Beaudoin 1976), where could plausibly breed, but other records from the southern half of the country, mapped by Roselaar (1995), almost certainly represent early migrants or post-breeding dispersal. Also noted in the Hakkari range (East Anatolia) in late June (Martins 1989), but it is unknown whether there is suitable breeding habitat in this region.

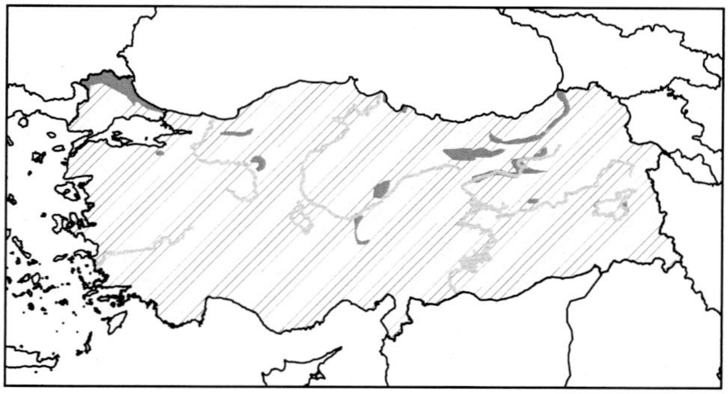

Rather more frequent, but still comparatively uncommon on passage across the country, but principally in the western half. Spring passage commences in late March (earliest 20 March in Southeast Anatolia) and continues through April (probably peaking in the second half of this month) and well into May (14th). There is a somewhat anomalous record from the treeless semi-steppe in Inner Anatolia as late as 16 June (GMK, TM), which was presumably still on passage. Autumn passage commences late July/early August with the latest date 22 September (Marmara), but problems inherent in identification of all black-and-white flycatchers at this season obscure passage dates. Nonetheless, numbers of this species are generally small at both seasons.

Breeding. In deciduous woodland, plantations, groves, riverine forest and orchards. In Forest of Belgrade (**Marmara**) at least three pairs bred in nest boxes in June 1991, and a pair carried food to a nest box on 4 June 1996 (GMK *et al.*). There are several records of adults with fledged young/juveniles, mainly in eastern **Black Sea Coastlands**, from 28 June. At Kızılırmak Delta (Black Sea Coastlands), in 1992, three males singing at the same location on 23 April and 3 June suggesting breeding (Hustings & van Dijk 1994). In Armenia (all details from Adamian & Klem 1999), breeds in deciduous or mixed forests and fruit orchards, with a preference for tall, older trees, with little undergrowth. Nests are in a natural tree hole or in a nest box; a base of dry leaves, with a cup of grass stems, lined with fine stems and bast fibres. Eggs: C/5–6 (3–7), mean 5 (31). The commencement of breeding varies according to weather, with earliest eggs 12–21 May in cold and wet years, and 19 April–23 May in warmer years. Nest building observed on 27 April and 27 June. Nestlings, *c.*5 days old, were seen on 4 July and fledglings on 24 May 1995 and 26 June 1998. Males sing, almost continuously, close to the nesting trees, whilst females are incubating.

COLLARED FLYCATCHER *Ficedula albicollis*
Halkalı Sinekkapan

(Du: Withalsvliegenvanger; Fr: Gobemouche à collier; Ger: Halsbandschnäpper; Swe: Halsbandflugsnappare)

Subspecies. Monotypic.

Status and distribution. Historically, the status of this form was uncertain as the identification features separating it and *F. semitorquata* (previously regarded as a subspecies of *F. albicollis*), particularly in female-type plumages, from Pied Flycatcher *F. hypoleuca*, were not adequately established. A widespread, seemingly locally fairly common, passage migrant, principally through the western half of the country. Occurs in woodland, parkland etc. Spring passage commences late March/early April, with the earliest 27 March (Inner Anatolia) and 30 March (Marmara), peaking in the second half of April (e.g. Bradshaw & Kirwan 1992, Hustings & van Dijk 1994) and continuing well into May (latest 30th). Autumn migration probably commences in late July with the earliest available date 19th. Passage continues into September, with the latest on 29th, perhaps 9 October in Inner Anatolia. Numbers are generally low at this season.

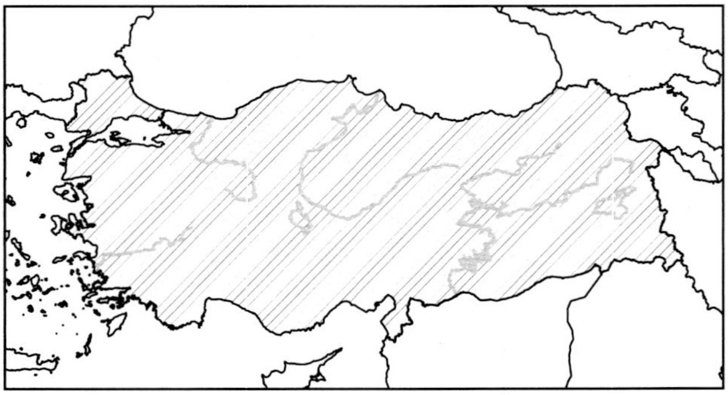

EUROPEAN PIED FLYCATCHER *Ficedula hypoleuca*
Kara Sinekkapan

(Du: Bonte Vliegenvanger; Fr: Gobemouche noir; Ger: Trauerschnäpper; Swe: Svartvit flugsnappare)

Subspecies. Nominate.

Status and distribution. Historically, the status of this species has been uncertain, due to many records, especially in autumn, being referred to unidentified *F. hypoleuca/albicollis*. Widespread and probably fairly common passage migrant, but principally recorded in the western half of the country. Spring passage is principally noted in April, with the earliest record being on 11 March (Mediterranean), and continues much reduced into May, with the latest on 5 June (Black Sea Coastlands). Numbers involved are generally small, although 302 moved through Kızılırmak Delta (Black Sea Coastlands) in spring 1992 (Hustings & van Dijk 1994). Autumn passage occurs from August, with the earliest specific date 9th, and continues until October, with the latest on 17th (Mediterranean). The majority apparently passes through in September, but numbers are never large.

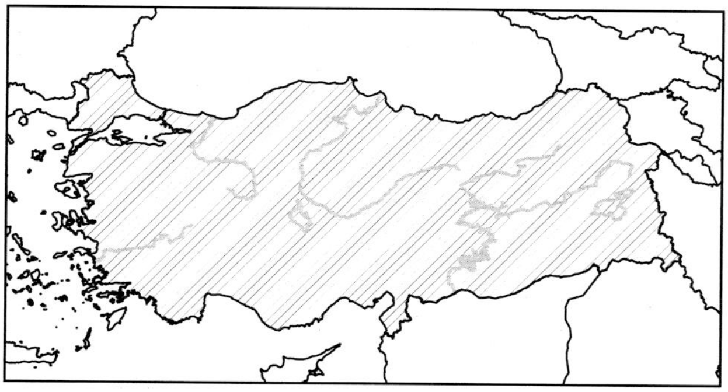

BEARDED REEDLING (BEARDED TIT) ***Panurus biarmicus***

Bıyıklı Baştankara

(Du: Baardman; Fr: Panure à moustaches; Ger: Bartmeise; Swe: Skäggmes)

Subspecies. In the west, in Marmara and Aegean, populations are probably representatives of nominate *biarmicus*, whilst in the Hatay the subspecies *kosswigi* was described from, and was apparently endemic to, the region of the former Amik Gölü. The fate of this population is unknown since drainage (completed in the mid 1970s), although Roselaar (1995) speculated that those in the south-coast deltas might also be of this form: we can confirm that the species breeds in the Göksu Delta (van den Berk & Guder 1992; pers. obs.) and it has recently been discovered breeding in Syria (Tavares *et al.* 2000, D. Murdoch *in litt.* 2003). Roselaar (*op. cit.*) ascribed the birds of Inner Anatolia and Black Sea Coastlands to *russicus*, and those around Van Gölü (East Anatolia) probably belong to the same form (pers. obs.). Those in Azerbaijan have been variously ascribed to *russicus* (Stepanyan 1990) and nominate *biarmicus* (Sluys 1983), whilst those in Armenia are considered *russicus* (Adamian & Klem 1999). Although we have not examined relevant material, it seems distinctly unlikely to us that Transcaucasian birds should really represent *biarmicus*, although Sluys (1983) apparently found specimens that were identical to Dutch birds. Such remarks, and our own limited experience of specimens from across the range, suggest the presence of much individual variation within the context of a poorly expressed cline from darker, more saturated birds in the south (and west), to paler birds in the north and east, with the species' irruptive behaviour, which permits it to colonise new areas and re-occupy former sites more easily, an important factor in this process. However, we have examined insufficient specimen material from Turkey to comment further on the issues, although it seems probable that neither *kosswigi* (certainly) nor *russicus* (quite possibly) would achieve recognition under the taxonomy employed in this book.

Status and distribution. National conservation status: Near Threatened (population apparently decreasing). Resident or a partial migrant, occurring at a relatively small number of wetlands, principally in Inner Anatolia but also in Marmara, Mediterranean and East Anatolia (principally around Van Gölü), and at one site (Kızılırmak Delta) in Black Sea Coastlands. Perhaps only formerly bred in Aegean (see below). Numbers generally lower in spring, but notably larger in August–September (post breeding) with a maximum of 100+ at a now-drained site in Inner Anatolia. There is some evidence for immigration at this season, but evidence of irruptive movements is poorly established, although it presumably occurs, as elsewhere within the species' range.

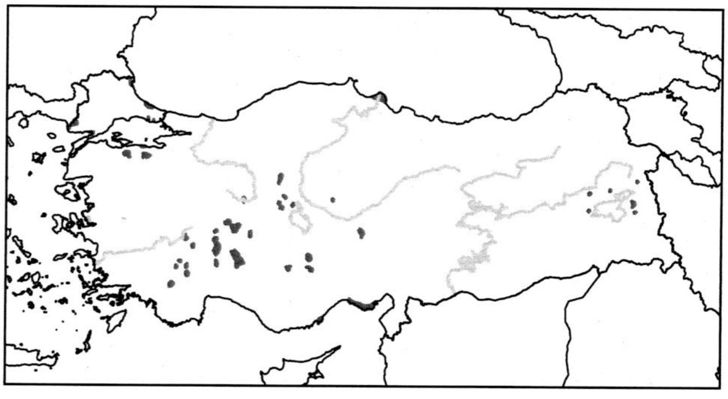

Certainly more widespread in winter, with records from coastal Aegean and more widely in Marmara, Mediterranean and Inner Anatolia at this season, although numbers are always small, and apparently never recorded in Southeast Anatolia.

Breeding. Inhabits reeds and sedges, in swampy ground, at lakes and marshes. Nests low in reeds and dense undergrowth; the nest is a deep cup of dead leaves of reeds and sedges, lined mainly with reed-flower heads. At Marmara Gölü (**Aegean**) on 15 May 1950 McNeile found seven nests, C/1 (incomplete clutch), C/4, C/4, C/5, C/5, C/5, C/6, all fairly close, in the base of isolated clumps of tall reeds; several clutches were well incubated, indicating laying of first eggs in late April; at the same place on 30 May 1951 he found only one nest with an incomplete clutch of one. In **Marmara**, there was an adult with food at Meriç Delta on 24 June 1973. At Ereğli marshes (**Inner Anatolia**), a pair was carrying nesting material on 9 May 1985, a nest had one chick, *c.*3 days old, three eggs and a further egg (on the ground) on 11 May 1987, and two family parties were observed on 22–23 June 1986 (M. Kasparek). Elsewhere in Inner Anatolia, a male was carrying nesting material at Eber Gölü on 18 April 1967, and several parties of juveniles were at Sultan marshes on 1 June 1999 (R. F. Porter); three nests were under construction in reeds at Kulu Gölü on 8 May 2005, each held one egg (incomplete clutches) by 14 May 2005, whilst another there had one egg on 8 May; three family parties (all recently fledged young) were there on 7 May 2006, indicating laying *c.*9 April, and there was a new empty nest on 12 May 2006 (N. Burke), the latter presumably for a second brood. One nest had three eggs (incomplete clutch) at Mogan Gölü on 15 May 2005 (J. Hodson). At Kızılırmak Delta (**Black Sea Coastlands**), the population was estimated at 100–150 pairs, and nest building was observed on 24 May 1992 (Hustings & van Dijk 1994). In **Mediterranean**, there was nest building in the Antakya area on 14 May 1953 (Kumerloeve 1963a); an adult and two juveniles were at Göksu Delta on 21 April 1996 (A. Royle), indicating date of first egg as *c.*21 March.Later records are probably of second broods.

IRAQ BABBLER *Turdoides altirostris*
Irak Yedikardeşi

(Du: Iraakse Babbelaar; Fr: Cratérope d'Irak; Ger: Rieddrossling; Swe: Irakskriktrast)

Subspecies. Monotypic.

Status and distribution. Previously rejected from the Turkish list (Kirwan *et al.* 1999) on the basis of an unsubstantiated and unpublished claim from the South Van marshes (East Anatolia) in July 1980, which was without precedent, as the species apparently shows no potential for vagrancy (Cramp 1993). However, more recently Iraq Babbler has been discovered to be resident outside the lower Euphrates and Tigris basin, in Iraq, along the entire length of the Euphrates in Syria (e.g. Balmer & Betton 2002a, Murdoch 2002), from the Iraq border to Jarabulus, on the Turkish border (I. Andrews *in litt.* 2003). Whether the Syrian population is a recent phenomenon, perhaps displaced by the drainage of the Mesopotamian marshes, or was previously overlooked, is unknown, but this discovery led to searches for the species around Karkamis (Southeast Anatolia) in early May 2004 and mid-May 2005. Despite use of tape playback these searches proved fruitless (GMK *et al.*). However, on 29–30 May 2006, four adults and two or three recently fledged but still dependent young were seen and photographed at Birecik, just to the north (S. Cale *et al.*; Donaghy 2006), and at least three birds were apparently still present on 4 July (*Dutch Birding* 28: 324), whilst at least four young were raised in late May 2007 (*Dutch Birding* 29:250–251). The species was again found (and photographed) at two sites near Birecik in May 2008 (J. L. Copate and H. Shirihai *in litt.* 2008).

Breeding. No data other than the observation of newly fledged young mentioned above. Elsewhere, breeds along river margins and in reedbeds; nests are less than 2m above ground in a tree or bush, or in reeds; builds a deep untidy cup of plant stems and grass, lined with grass, leaves and rootlets; usual clutch 3–4, and breeds March–July, with 2–3 broods per season.

LONG-TAILED TIT *Aegithalos caudatus*
Uzun Kuyruklu Baştankara

(Du: Staartmees; Fr: Mésange à longue queue; Ger: Schwanzmeise; Swe: Stjärtmes)

Subspecies and taxonomy. Across the species' broad range, at least 19 races are recognised, which are clumped into four groups by Harrap & Quinn (1996). Populations in Asian Turkey belong to the *alpinus* group (endemic to the Mediterranean and southwest Asia), taxa characterised (in comparison to the *europaeus* group) by their shorter wings and tail, darker heads, lacking virtually any pink and largely grey upperparts. However, the *europaeus* group does penetrate Turkey, in Thrace, where Roselaar examined specimens of *macedonicus* in the southwest and *europaeus* in the north, and speculated that intergrades might also be expected. Most of Anatolia is occupied by *tephronotus* (from the *alpinus* group), but *major* occurs in the northeast Pontics, and Iranian *passekii* (or intergrades) can be expected to represent the form in the extreme southeast (Roselaar 1995). We certainly support the delimitation of *A. caudatus* into four groups: those not mentioned above being nominate *caudatus*, over much of northern Eurasia, and *glaucogularis*, in China, which was formerly treated as a species (Silver-throated Tit) by some commentators. However, the continued recognition of the number of taxa afforded status by Harrap & Quinn (*op. cit.*) appears unnecessary to us; many of those currently recognised within the *europaeus* and *alpinus* groups are weakly defined, lack truly diagnosable and consistent characters, and some intergrade with as many as four other named forms. Continued nomenclatural recognition of such weak variation is not helpful in understanding real-world evolutionary processes.

Status and distribution. National conservation status: Least Concern (population apparently decreasing). A widespread and common resident in western Turkey, although largely absent from central Inner Anatolia. Scarce in the eastern Black Sea Coastlands and largely absent from East Anatolia, although some suitable areas have received very little attention from fieldworkers. The race *tephronotus* has very recently been discovered to be locally abundant in orchards around Malatya, in the westernmost part of the latter region. Recently discovered in Southeast Anatolia at a handful of sites and is obviously extremely local in this region. Perhaps slightly more widespread in autumn and winter, when recorded at a number of sites where unknown to breed, despite availability of apparently suitable habitat. Occurs in a variety of woodland and scrub, principally from sea level to 1,300m, but noted to 1,500m in the Taurus (Mediterranean) and 2,000m in the west-central Black Sea Coastlands.

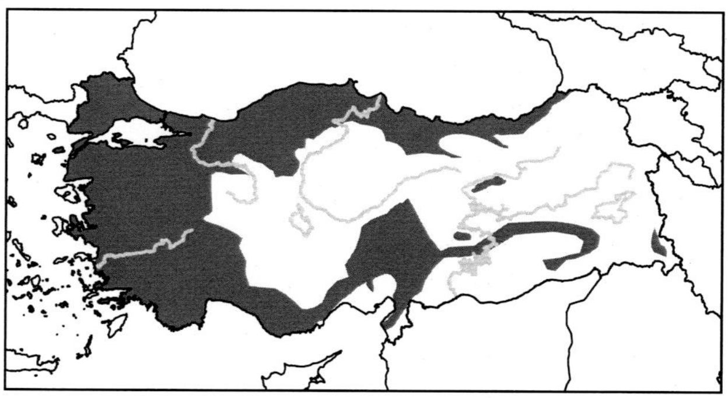

Breeding. Frequents open deciduous and coniferous woodland, in both lowland and higher areas, and tall scrub. No detailed information on nest site, nest or clutch size in Turkey. Elsewhere, usually nests in a bush, sometimes a tree; the nest is a vertically elongated and domed structure, with a small entrance hole near the top; of moss, bound with spiders' webs and hair, coated with lichen, and thickly lined with feathers; usual clutch 8–12 (5–16). At Kızılırmak Delta (**Black Sea Coastlands**), there were several nests in late April/early May 1992, and family parties in early June (Hustings & van Dijk 1994); three family parties were also there on 8–9 June 1969 (Dijksen & Kasparek 1985). In **Inner Anatolia**, there was a nest at Kızılcahamam on 8 May 1977 (Barış *et al.* 1984). In **Mediterranean**, there were family parties at Antalya on 9 June 1981 (H. Hols) and at Uzuncaburç on 14 June 1992 (PC) and on 18 June 1996 (GMK). Many fledged young were near Marmaris (**Aegean**) on 26 May 1995. Family parties were at Iğneada on 16 May 1967 (Rokitansky & Schifter 1971), at Uluabat Gölü (**Marmara**) on 3 June 1998 (Welch & Welch 1998b), and near Yüksekova (**East Anatolia**) on 4 June 1969.

MARSH TIT — *Poecile palustris*

Kayın Baştankarası

(Du: Glanskop; Fr: Mésange nonnette; Ger: Sumpfmeise; Swe: Entita)

Subspecies and taxonomy. Those in Marmara and east to the central Black Sea Coastlands appear to be *stagnatilis*, but no specimen material from the northeast Pontics was examined by Roselaar (1995) or by us. The population in this region might represent intergrades between *stagnatilis* and Caucasian *kabardensis* (Harrap & Quinn 1996). We follow Stepanyan (1990) in subsuming both these taxa within the nominate; variation within the *palustris* group (five taxa recognised by Harrap & Quinn *op. cit.*) is extremely slight and almost entirely clinal (Vaurie 1957c, 1959a). Formerly placed in the genus *Parus* (*Ibis* 147: 821–826; Gill *et al.* 2005).

Status and distribution. National conservation status: Least Concern (population apparently stable). A local and uncommon resident of deciduous or mixed woodlands, often in hilly or montane areas, in Black Sea Coastlands and Marmara, with isolated records further south in Aegean, e.g. around Izmir (see Roselaar 1995), perhaps solely reflective of winter dispersal. Breeds from sea level to 1,000m but restricted to northern Thrace (the Istranca Dağları), parts of the Black Sea Coastlands (especially the west), and montane southern Marmara (e.g., the Kaz Dağları and Uludağ), with isolated localities at Abant and Gerede near the border with Inner Anatolia. Regularly reported throughout the year from certain localities, e.g. Iğneada, Uludağ, Yedigoller and Zonguldak in the west, and the Kaçkar range in the east.

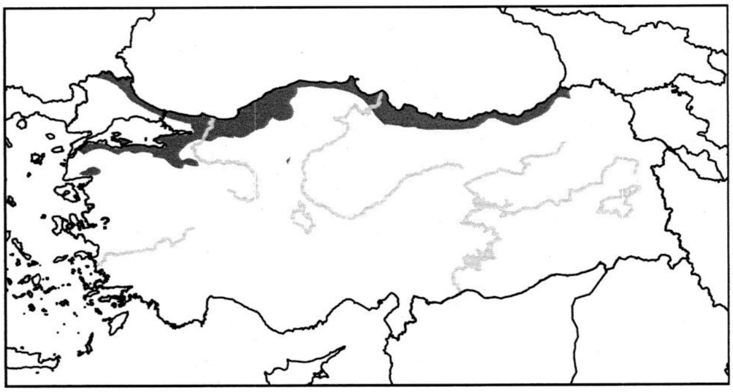

Breeding. In deciduous and mixed forests, often in hilly or montane areas. Nests in holes in trees and in nest boxes. Nest and clutch size undescribed in Turkey; elsewhere, nest is a cup of moss, lined with hair and some feathers, and usual clutch is 6–9. In **Black Sea Coastlands**, a nest had young near Alapli on 1 May 1978; a nest had young, in a nest box, near Abant Gölü on 29 May 1992 (PC). In **Marmara**, adults and fledged young were observed near Demirköy (Thrace) on 22 June 1973, and adults were carrying food in Forest of Belgrade on 29 June 1973.

SOMBRE TIT — *Poecile lugubris*

Ak Yanaklı Baştankara

(Du: Rouwmees; Fr: Mésange lugubre; Ger: Trauermeise; Swe: Balkanmes)

Subspecies and taxonomy. Roselaar (1995) considered those in Thrace to be *lugens* (treated as a synonym of nominate *lugubris* by Harrap & Quinn 1996) with *anatoliae* throughout the rest of the country, although those in the extreme southeast show evidence of intergradation with Iranian *dubius*. Harrap & Quinn (*op. cit.*), Roselaar (*op. cit.*) and Svensson (1992) agreed over the features separating *anatoliae* from *lugubris*/*lugens*, but we consider that geographical variation within at least the nominate group of this species is insufficiently marked to warrant

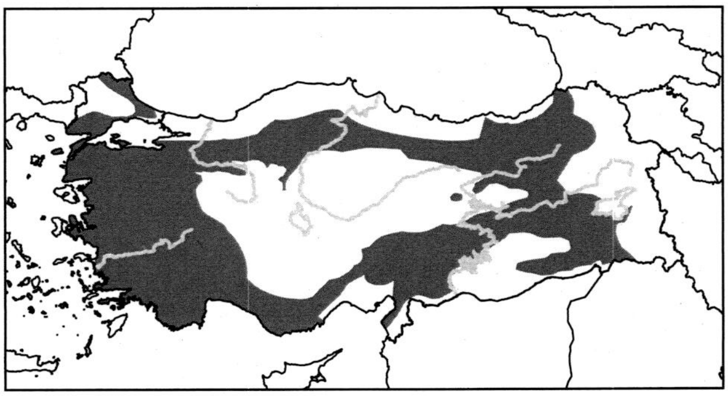

recognition of any other name than *lugubris* itself. Formerly placed in the genus *Parus* (*Ibis* 147: 821–826; Gill *et al.* 2005).

Status and distribution. National conservation status: Least Concern (population apparently stable). A widespread and fairly common resident in dry, open woodland and scrubby areas of Aegean, Marmara and Mediterranean, extending more locally across the rest of the country, and apparently absent from steppe-like areas of Inner and Southeast Anatolia, as well as parts of East Anatolia (e.g. around Malatya and Erzincan). Breeds at 0–800m in the west, 1,000–1,300m in the centre, to 2,200m in the Taurus and at 100–1,500m (locally to 2,200m) in the east. Breeds locally in Thrace, in both the northern and southern parts, and breeding is also rather rare in the Black Sea littoral (except east to Giresun), whereas it is a common breeder in coastal regions of Aegean and Mediterranean.

Breeding. Breeds in open deciduous and coniferous woodland with undergrowth, and on stony hillsides and in valleys with small trees and scrub, occasionally in gardens, from sea level to over 2,000m. Presumably nests mainly in holes in trees, but there are six records of nests in holes in rocks (and this practice is probably more common than records suggest); a nest was constructed in the top of a concrete umbrella stand at a petrol station (R. F. Porter). The nest is of dry grass, lined first with wool and then feathers (Danford 1878). In **Mediterranean**, a nest had seven eggs near Pozantı on 16 April 1876 (Danford 1878), a nest was occupied at Kovada National Park on 10 May 1995 (I. A. Green), a nest had several full-grown young in a nest box near Mut on 26 May 1998 (T. Quittenden); and a nest in a hole in a rock had five almost full-feathered young, near Uzuncaburç, on 7 May 2004 (M. Farmer), the last indicating egg laying in early April; family parties were observed north of Mut on 5 July 1986 (R. Webb) and 18 June 1996 (GMK), and at Uzuncaburç on 17 June 1996 (GMK). In **Southeast Anatolia**, near Gaziantep, nest building was observed on 11 April 1996 and 7 May 1992 (D. Griffin); a nest in a rock crevice had small young on 2 May 1964 (Warncke 1965); a nest in a hole in a rock had eggs on 20 May 1993 and young on 2 June 1993 (PC), and another had young on 28 July 1967 (Vielliard 1968), probably a second brood; on 8 May 2004, two broods of recently fledged young were in a rocky area of low scrub with no trees, and the nests must surely have been in holes in rocks, with egg laying in early April (PC); and records of four other family parties in late May–June (earliest 24 May); on 27 April 2005 an adult was carrying food to a nest deep in a hole in a rock, and remained there, presumably brooding small young (PC). A nest contained young at Uludere (**East Anatolia**) on 17 May 1975. In **Aegean**, an adult fed a recently fledged young on 13 May 1992 (M. Brown). In **Inner Anatolia**, a recently fledged young was observed in the Devres Valley on 8 June 1975 (R. F. Porter). In a study of species using nest boxes at 1,300–1,500m in Beynam Forest, near Ankara (Inner Anatolia) in 1978–80, mean dates were: start of nest building 18–23 April (19 nests), laying of first egg 1–6 May (25 nests), hatching of first young 15–20 May (21 nests), fledging 6–11 June (18 nests); eggs: of 24 first clutches, C/4 (1), C/5 (2), C/6 (5), C/7 (5), C/8 (7), C/9 (4), mean 7.1; mean size of 35 eggs (five clutches) 16.17 × 11.79mm; mean incubation period 13.3 days (17 nests) and mean fledging period 22.2 days (21 nests); 16.7% and 6.7% of pairs were proven (by ringing) to have second broods in 1979 and 1980 respectively (Kiziroğlu 1982); in the same study, 25 of 28 nests were successful; of 191 eggs laid, 164 hatched (85.9%); 161 young fledged (84.3% of eggs laid/98.2% of young hatched), with a mean 5.8 young per nest (6.4 per successful nest) (Kiziroğlu 1984). In **Marmara**, adults were feeding young in a nest near Balıkesir on 26 April 1970, a family party was at Manyas Gölü on 26 May 1966, and an adult with fledged young was at Alexandria on 25 June 1973 (R. F. Porter); a nest in a rock crevice, near Osmaneli, had seven eggs close to hatching on 22 April 1964 (Warncke 1964–65).

[CRESTED TIT *Lopophanes cristatus*]
Tepeli Baştankara

(Du: Kuifmees; Fr: Mésange huppé; Ger: Haubenmeise; Swe: Tofsmes)

Subspecies and taxonomy. See discussion below concerning the race most likely to occur in Turkey. Formerly placed in the genus *Parus* (*Ibis* 147: 821–826; Gill *et al.* 2005).

Species not admitted to the Turkish list (Kirwan *et al.* 1999). Kasparek (1990a) regarded the following evidence that the species occurs in Turkey as too circumstantial for inclusion in the main list. Rigler (1852) noted the species in a list of the birds of Istanbul, while Ganso & Spitzer's (1967) report of a bird at Abant Gölü, Bolu province (Black Sea Coastlands) in 1966, and five claimed in the Taurus above Pozantı (Mediterranean) on 31 August 1986 (Bijlsma & de Roder 1986), though not improbable, require more detailed documentation (R. G. Bijlsma *in litt.* 1999). Two recent claims in consecutive years from the same locality (Gümüşhane) in the central Black Sea Coastlands require substantiation, as does that of a bird apparently seen at Kapuşbasi Falls, in the Aladağlar between Niğde and Kayseri (Mediterranean) on 16 October 2007 (*Dutch Birding* 29:397). In addition, Ergene (1945) regarded Crested Tit as a Turkish bird, but provided no details other than stating that the subspecies involved were *P. c. baschkirkus* (from the Urals) and *P. c. derjugini* (no longer recognised), although, as noted by Roselaar (1995), *P. c. bureschi* from Albania, Greece and southern Bulgaria appears the race most likely to occur in Turkey, but the validity of *bureschi* was questioned by Vaurie (1957c).

COAL TIT — *Periparus ater*
Çam Baştankarası

(Du: Zwarte Mees; Fr: Mésange noire; Ger: Tannenmeise; Swe: Svartmes)

Subspecies and taxonomy. Harrap & Quinn (1996) and Roselaar (1995) recognised somewhat different levels of variation, though they agree over the essentials. Coal Tit was considered to comprise five groups by Harrap & Quinn (*op. cit.*), of which two occur in Turkey: *derjugini* in the northeast Pontics and nominate *ater* throughout Marmara, the western half of the Black Sea Coastlands and south through the Taurus (Mediterranean). In contrast, Roselaar (*op. cit.*), despite his ability to recognise minuscule gradients in geographical variation, was unable equivocally to assign those in the west and south to subspecies; although nearest to *abietum*, he also considered that the name *rufolateralis* might warrant resurrection for these populations (both taxa effectively considered synonyms of nominate *ater* by Harrap & Quinn *op. cit.*). In the northeast, Roselaar recognised *derjugini*, but slightly further south of the coastal range, to the east of Artvin and Sarikamis, he acknowledged a zone of intergradation between *abietum*/*rufolateralis* and *michalowskii*.

Given levels of individual variation in both the *ater* and Caucasus groups (the latter including both *derjugini* and *michalowskii*), intra- and inter-group intergradation, and that some authors (see Roselaar *op. cit.*) have recognised strong characters of *derjugini* as far west as the Ilgaz Dağları (based on specimens unavailable to us, Roselaar or Harrap & Quinn), the evidence for recognising even the entire Caucasus group as a separate evolutionary unit appears distinctly weak. Given the complex variation inherent to *Periparus ater*, we recommend a thorough review of the taxonomy of this species in order to establish which of the currently recognised taxa can truly be regarded as 'basal'. Snow (1997) also briefly discussed geographical variation in this species, but only briefly touched upon the situation in western Asia, although he admits that most variation is clinal.

Formerly placed in the genus *Parus* (*Ibis* 147: 821–826; Gill *et al.* 2005).

Status and distribution. National conservation status: Least Concern (population apparently stable). A widespread, common to abundant resident in coniferous forest in virtually all coastal regions, as well as more locally in relict forests at the northern and southern fringes of Inner Anatolia, e.g. Abant, Merzifon, Boyabat, Torul, Kızılcahamam, Beynam, Akdağmadeni and Ilgın. Absent elsewhere over much of Inner and Southeast Anatolia, as well as the montane east and southeast, most of which areas lack coniferous forests. Breeds in open or dense coniferous forest sometimes mixed with oak, but rarely in purely deciduous forest (in beech in Marmara) and only occasionally in juniper above the treeline (except in eastern Black Sea Coastlands). Altitudinal distribution reflects coniferous plantations: at 0–2,000m in the west, up to 2,200m in the centre (the Taurus and Ilgaz Dağları) and to at least 2,400m in the east. Slight dispersal occurs in the non-breeding season, when recorded more

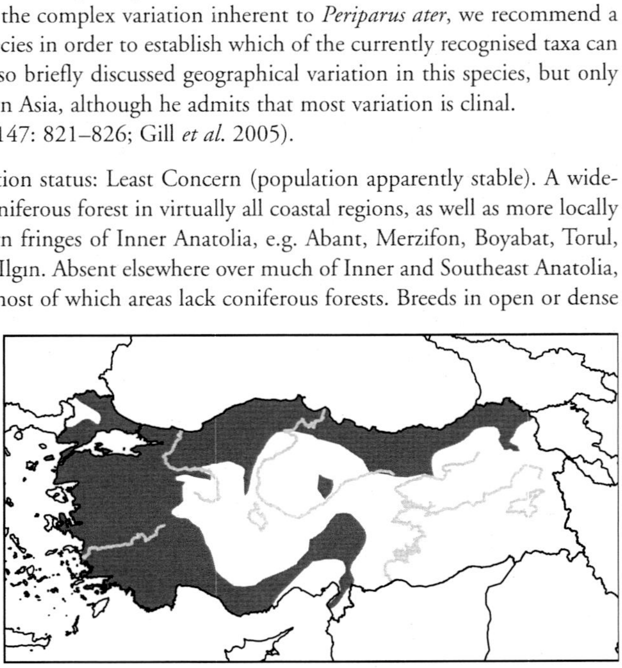

extensively, with winter records from lower altitudes in Inner Anatolia, e.g. at Mogan Gölü and Kayseri, but there are apparently no records from eastern areas even at this season.

Breeding. Breeds in various woodland types, with a distinct preference for conifers. Nests in holes; in trees, in the ground, in rocks and in walls. A nest in a hole in the ground under a rock, near Uzuncaburç (Mediterranean), was mainly of wool, with some moss, and lined with the same materials (N. Williams); a nearby nest, in a nest box, was mainly of wool and moss, with some grass and lined with finer grass, wool, hair and some feathers (PC). In a study using nest boxes at 1,300–1,500m in Beynam Forest near Ankara (**Inner Anatolia**) in 1978–80, mean dates were: start of nest building 20–27 April (12 nests), laying of first egg 2–6 May (19 nests), hatching of first young 19–22 May (18 nests), fledging 11–15 June (19 nests); eggs: of 20 first clutches C/4 (2), C/5 (3), C/6 (2), C/7 (3), C/8 (1), C/9 (4), C/10 (1), C/12 (3), C/13 (1), mean 7.95; mean size of 43 eggs (five clutches) 15.64 × 11.45mm; mean incubation period 13.2 days (11 nests) and mean fledging period 21.6 days (19 nests); 16.7% and 20% of pairs were proven (by ringing) to have second broods in 1978 and 1979 respectively (Kiziroğlu 1982); in the same study, 22 of 24 nests were successful; of 195 eggs laid, 162 hatched (83.1%); 153 young fledged (78.5% of eggs laid/94.4% of young hatched), a mean 6.4 per nest (7.0 per successful nest) (Kiziroğlu 1984). Also in Inner Anatolia, two nests had young in holes in conifers at Kızılcahamam on 29 May 1992 (PC), and a nest had young in a hole in the wall of a building at Daday on 9 May 1990. In **Mediterranean**, adults were building four nests in holes in the ground and under rocks, in pine forest at Akseki on 25 April 2004, and materials used included grass and moss (PC); an adult was incubating in a hole, 6m above ground in a pine at Ağla on 30 April 2001 (PC); at Uzuncaburç a nest had seven eggs on 26 April 2005 (PC) and another seven fully feathered young on 7 May 2004 (N. Williams), the latter indicating laying of first egg in late March/early April; several fledged young were at Uzuncaburç on 17 June 1996, and an adult fed a fledging at Delimahmutlu on 19 June 1996 (GMK); a juvenile was at Bolkar Dağları on 5 June 1999 (R. F. Porter). In **Black Sea Coastlands**, fledged young were near Boyabat on 9 June 1975 and a pair was feeding fledged young, near Ikizdere, on 12 June 1975 (R. F. Porter).

BLUE TIT — *Cyanistes caeruleus*
Mavi Baştankara

(Du: Pimpelmees; Fr: Mésange bleue; Ger: Blaumeise; Swe: Blåmes)

Subspecies and taxonomy. All Turkish populations belong to the nominate *caeruleus* group. Harrap & Quinn (1996) considered those in the west and north of the country to be *caeruleus*, with *satunini* in eastern and southern Turkey, and intergrading with the nominate race further west. The pattern of biogeographical variation recognised by Harrap & Quinn (*op. cit.*) would be distinctly unusual, if true. In contrast, Roselaar (1995) treated only Thracian birds as pure *caeruleus* with those elsewhere in the north showing intermediate characters between this race and *satunini*, and those in the south being apparently intermediate between *caeruleus* and *persicus*. Those elsewhere in the west he was unable to ascribe to taxon, due to a lack of material, though *calamensis* could perhaps also occur. We have been unable to access much specimen material from Turkey, but have examined material from virtually all of the above-mentioned forms and have found the variation to be largely clinal and slight. *P. c. persicus* is depicted as quite noticeably different from populations in Western Europe in Hollom *et al.* (1988) and Porter *et al.* (1996), but our examination of specimens suggests that the difference has been somewhat over-emphasised, perhaps due to vagaries in the colour reproduction in published works, and notwithstanding that Vaurie (1957b) considered *persicus* to be 'well differentiated'.

Salzburger *et al.* (2002) recently uncovered significant differences in the mitochondrial DNA profiles of the *caeruleus* (Europe/Middle Asia) and *teneriffae* groups (North Africa and the Canaries), and suggested that the two be recognised at the species level (a proposal that has since accrued widespread support). Unfortunately, no Middle Asian populations of *caeruleus* were included in their analysis. Formerly placed in the genus *Parus* (*Ibis* 147: 821–826; Gill *et al.* 2005).

Status and distribution. National conservation status: Least Concern (population apparently stable). A widespread and fairly common resident in all regions, but only very local in Southeast and East Anatolia, being absent or surprisingly rare in many areas which appear to possess suitable habitat, although not uncommon in many woodlands

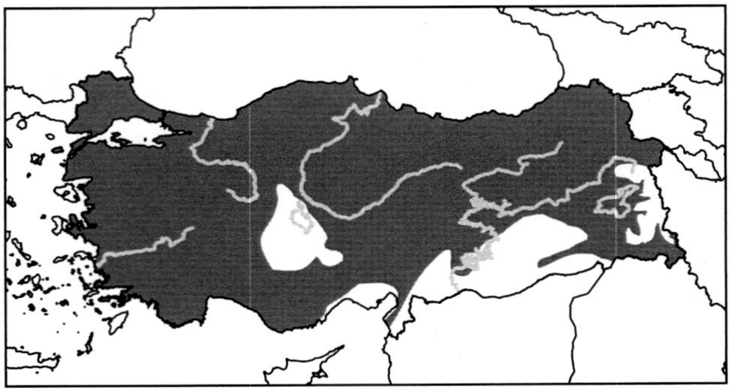

bordering the Euphrates. Breeds in open, mature deciduous and mixed forests, riverine woodland and scrub, plantations and orchards, at 0–750m in Thrace, 100–1,500m in Black Sea Coastlands, 500–1,700m in Mediterranean (but probably also lower) and up to 2,000m in East Anatolia. Disperses slightly in the non-breeding season, and autumn migration is clearly apparent at the Bosphorus, with winter immigration bringing records from various locations, e.g. parks in cities, where breeding is unknown.

Breeding. Breeds mainly in mature woodland, orchards and olive groves. At Kocaçay Delta (Marmara), highest densities were in humid dense alluvial forest, with 0.81 pairs/ha in a 31ha study plot in 1993 (Ertan 1996); and recorded only in wooded areas in Kızılırmak Delta (Black Sea Coastlands) in 1992 (Hustings & van Dijk 1994). Nests in holes in trees and also recorded nesting inside a metal pipe in an olive grove (Welch & Welch 1998b). Nest undescribed in Turkey; elsewhere, a cup of moss and grass, lined with hair and feathers. In a study using nest boxes at 1,300–1,500m at Beynam Forest near Ankara (**Inner Anatolia**), in 1978–80, mean dates were: start of nest building 21–29 April (18 nests), laying of first egg 4–9 May (27 nests), hatching of first young 20–29 May (21 nests), fledging 12–16 June (24 nests); eggs: of 30 first clutches C/4 (1), C/6 (2), C/7 (5), C/8 (7), C/9 (4), C/10 (7), C/11 (3), C/12 (1), mean 8.6; mean size of 41 eggs (5 clutches) 15.72 × 11.46mm; mean incubation period 13.6 days (27 nests) and mean fledging period 21.6 days (27 nests); 11% and 6% of pairs were proved (by ringing) to have second broods in 1979 and 1980 respectively (Kiziroğlu 1982); in the same study, 32 of 39 nests were successful; of 288 eggs laid, 241 hatched (83.7%); 223 young fledged (77.4% of eggs laid/92.5% of young hatched), a mean 5.7 young per nest (6.97 per successful nest) (Kiziroğlu 1984). In **Mediterranean**, adults entering and remaining in tree holes at Ağla on 6 May 2001 and 26 April 2002 were obviously incubating eggs; one of these nests was occupied by Krüper's Nuthatch *Sitta krueperi* the previous year (PC). In **Aegean**, a brood already able to fly was seen on 15 May 1951 (McNeile), indicating egg laying early April. In **Marmara**, the first family parties were observed at Erdek on 8 June 1966 (R. F. Porter). In **Black Sea Coastlands**, fledged young were observed near Yoncalık on 12 July 1974; at Kızılırmak Delta, there were family parties on 10 June 1975 and 15 June 1984 (Dijksen & Kasparek 1985); and several family parties were there in early June 1992 (Hustings & van Dijk 1994), indicating egg laying about mid April. Family parties were at various sites in Inner Anatolia and Black Sea Coastlands on 8–17 June 1975 (R. F. Porter).

GREAT TIT *Parus major*
Büyük Baştankara

(Du: Koolmees; Fr: Mésange charbonnière; Ger: Kohlmeise; Swe: Talgoxe)

Subspecies. This widespread and rather variable species is represented in Turkey by the *major* group. Both Svensson (1992) and Harrap & Quinn (1996) recognised only nominate *major* in Turkey, and Roselaar (1995) concluded that the vast majority of Turkish birds were inseparable from this subspecies. However, in the extreme southwest Aegean region, he suspected that the rather small race *aphrodite* might be present, based on mensural data for one specimen taken by Rokitansky & Schifter (1971), which Anita Gamauf (*in litt.* 2003) has confirmed as being *aphrodite*, through examination of the specimen (held in the Naturhistorisches Museum, Wien 72377) and comparative material of this and other subspecies held in the same institution. Roselaar (*op. cit.*) recognised some characters of the Levantine form *terrasanctae* in two specimens from the eastern Taurus (Mediterranean), but all those from the Taurus in the Natural History Museum (Tring) appear indistinguishable from *major* and, based on our review of specimen material held in the same institution, *terrasanctae* seems at best very weakly distinguishable from *major*, and probably only through examination of a very long series of each (see also Vaurie 1957c).

Status and distribution. National conservation status: Least Concern (population apparently stable). A widespread, common to abundant resident virtually throughout, but rather local in central parts of Inner Anatolia and parts of Southeast Anatolia, as well as in montane valleys of East Anatolia. Absent or surprisingly rare in some areas which appear to possess suitable habitat. Breeding from sea level to the treeline, but mainly in valleys and on lower slopes, to 1,700m in the Ilgaz Dağları (Black Sea Coastlands), to 2,000m on Uludağ (Marmara) and in the Taurus (Mediterranean), and to 2,300m in East Anatolia. Disperses slightly outside the breeding season, and autumn migration is clearly evident at the Bosphorus, with winter records more extensively distributed, e.g. near wetlands and parks, where breeding unknown.

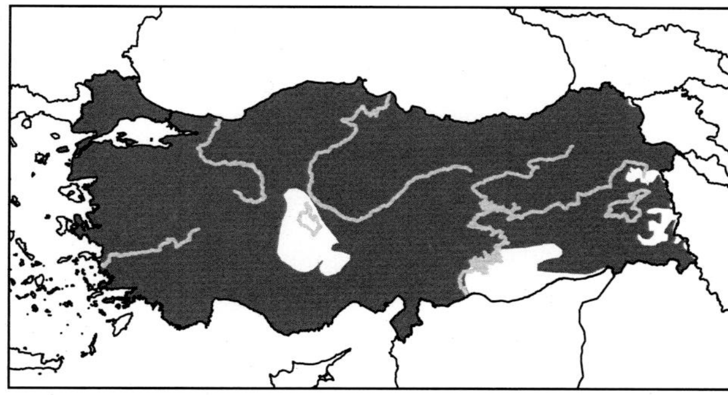

Breeding. Catholic but usually associated with trees, including all woodland types, copses, olive groves, gardens, parks, cultivation, even scrub. At Kocaçay Delta (Marmara), there were 1.16 pairs/ha in a 31ha study plot in alluvial forest in 1993 (Ertan 1996). Nests in holes in trees, buildings and walls. A nest in a hole in a wall at Sivrikaya (Black Sea Coastlands) was a cup of roots, grass, wool and some moss, lined mainly with hair (PC). In **Mediterranean**, a nest contained young at Gazipaşa on 15 April 1973, another had young at Uzuncaburç on 7 May 2004 (PC) and another contained young, *c.*12 days old, at Kargıhanı on 1 May 2005 (T. Maunsell), the latter indicating first egg in late March; newly fledged young were at Anamur on 15 April 1973 and at Dalyan on 14 June 1987 (Kılıç & Kasparek 1989), indicating egg laying in early March and early May, and possibly indicating double-brooding. In **Aegean**, an adult carried a faecal sac from a nest hole in a building near Altınkum on 22 April 2002, indicating egg laying late March (PC); several pairs were observed with fledged young at Büyük Menderes Delta on 19 May 1970, indicating egg laying by 10 April; and a juvenile on 3 May indicates egg laying in late March. In **Marmara**, fledged young were at Bebek on 17 June 1975 and Uluabat Gölü on 20 May 1966 (R. F. Porter), the latter indicating egg laying in mid April. In **Inner Anatolia**, breeding commences mid April (Wadley 1951) and fledged young are observed in June (earliest 10 June). In a study using nest boxes at 1,300–1,500m in Beynam Forest near Ankara, in 1978–80, mean dates were: start of nest building 23–27 April (125 nests), laying of first egg 7–8 May (181 nests), hatching of first young 22–24 May (135 nests), fledging 13–17 June (134 nests); eggs: of 193 first clutches, C/4 (3), C/5 (11), C/6 (34), C/7 (53), C/8 (47), C/9 (28), C/10 (13), C/12 (3), C/13 (1), mean 7.5; mean size of 343 eggs (45 clutches) 17.99 × 13.35mm; mean incubation period 13.6 days (105 nests) and mean fledging period 21.9 days (133 nests); 24.1%, 12.5% and 19.4% of pairs were proven (by ringing) to have second broods in the three years, virtually all in the immediate vicinity of boxes used for first broods (Kiziroğlu 1982); in the same study, 226 of 273 nests were successful; of 1,922 eggs laid, 1,665 hatched (86.6%); 1,607 young fledged (83.6% of eggs laid/96.5% of young hatched), a mean 5.9 young per nest (7.1 per successful nest) (Kiziroğlu 1984). In **Southeast Anatolia**, juveniles were observed on 4 May and 17 May, and others in June. At Kızılırmak Delta (**Black Sea Coastlands**), nests had young on 9 June 1969 and 27 May 1979, and fledged young were seen 10 June 1975 (Dijksen & Kasparek 1985); several nests were found on 2 May 1992, and families were seen during the first ten days of June (Hustings & van Dijk 1994). Elsewhere in this region, a nest had five young, *c.*3 days old, at Sivrikaya on 14 June 2004, indicating egg laying on 25 May (PC); two nests probably contained eggs at Ispir on 10 May 2004 (GMK), and two family parties were there on 12 June 2004 (B. Pleasance); a family party was at Sivrikaya on 13 June 2004 (S. Williams).

KRÜPER'S NUTHATCH *Sitta krueperi*

Küçük Sıvacıkuşu

(Du: Turkse Boomklever; Fr: Sittelle de Krüper; Ger: Türkenkleiber; Swe: Krüpers nötväcka)

Subspecies. Monotypic.

Status and distribution. Conservation status: Near Threatened (BirdLife International 2008); national conservation status: Least Concern (population apparently decreasing). Widespread and fairly common resident, occurring in coniferous and mixed forest, particularly favouring stands of *Pinus brutia.* Principally found at 500–1,000m, although occasionally at sea level and up to 2,500m in the west, and up to 2,000m in the east. Occurs in Marmara, Aegean, Mediterranean and Black Sea Coastlands, but largely absent from Inner Anatolia (except the northern fringe) and almost entirely absent from East and Southeast Anatolia (except the westernmost fringe of the latter, where only recently discovered). May undergo some post-breeding dispersal.

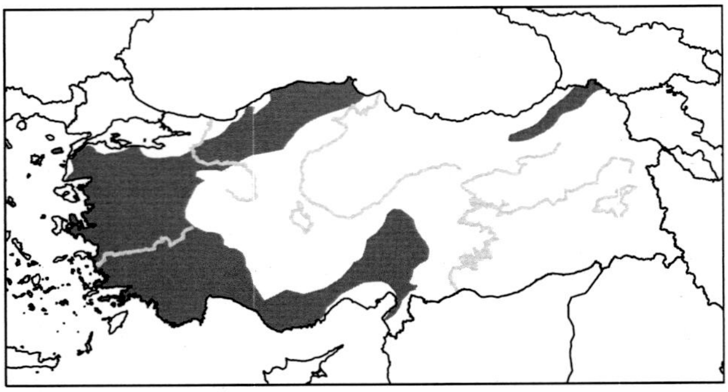

Breeding. In coniferous forest, almost exclusively in pines, especially *Pinus brutia,* but also in other conifers, sometimes mixed with deciduous trees, from sea level to 1,700–2,000m, but mostly above 1,200m (Frankis 1991). Usually nests in a natural pine hole, 0.3–4.0m to 7.0m above ground; nests also recorded in oak, aspen and apple trees, in areas of mixed woodland, and there are 25 records of nest boxes being used. At Pozantı (**Mediterranean**), on 17–18 April 1876, Danford (1878) found *c.*20 nests in lower fir woods, mostly with eggs, *c.*50% of which were well incubated. Clutch sizes mainly five, some six, and one of seven eggs. Nests were usually behind the bark of rotten tree stumps, in cavities partially excavated by the birds or in old woodpecker holes, and constructed mainly of juniper bark, lined with wood fibres and goat hair, but one nest had a thick pad of hare's fur and the feather of a Caspian Snowcock *Tetraogallus caspius.* Nests examined recently are mainly of long, narrow strips of bark, wood fibre and grass, lined with finer grasses, hair, fur and some feathers (PC). Recent records include nest building at Akseki on 13 April 1993, a nest with four young, aged 7–8 days, at Ağla on 30 April 2001, a nest with full-feathered young at Ağla on 6 May 2001 (PC), and other nests with young on 11 May 1987, 20 May 1996, 17 June 1996, 6 June 1999 and 8 June 2003; the young at this last nest were fed every four minutes over a period of 50 minutes (S. Harthill); fledged young were observed from 6 May; of six nests at Akseki on 25 April 2004, three were under construction (adults excavating holes), one had five eggs, one probably had eggs (adult remained in nest hole), and one had young, *c.*9 days old (PC); a nest had five newly hatched young at Akseki on 30 April 2005, and hatching is, therefore, synchronous, indicating that incubation commenced on completion of clutch (PC); three nest boxes at Uzuncaburç on 7 May 2004 had: young (age uknown); six young *c.*11 days old; and three unhatched eggs, in a nest from which young had recently fledged (R. Thomas) and, in nest boxes in the same area on 26 April 2005, two nests were under construction and another had six eggs (PC) whilst another, nearby, had five eggs on 12 May 2005 (J. Hodson). In a study in the Antalya area, 2000–03, the breeding period was from late March to late June; of 18 nests, nine were in *Pinus brutia,* five in *P. nigra,* three in cedar and one in a wooden pole; mean height of nest hole 11.84m; in 15 nests in nest boxes, there were 83 eggs (mean 5.5 per nest), of which 70 hatched (84%), and 54 of these fledged successfully, *c.*3.6 per pair (Albayrak & Erdoğan 2005). In **Aegean**, there were two nests at Bafa Gölü on 11 April 1981 and flying young were seen there on 21 May 1980 (Kasparek 1988a) In **Marmara**, nests with young, and fledged young, were found on Uludağ in late May (Jetz 1995; R. F. Porter). In **Inner Anatolia,** a nest had young at Kızılcahamam in mid May.

EURASIAN NUTHATCH *Sitta europaea*

Sıvacıkuşu

(Du: Boomklever; Fr: Sittelle torchepot; Ger: Kleiber; Swe: Nötväcka)

Subspecies. Members of the buff-breasted *caesia* group occur in Turkey (Harrap & Quinn 1996); though Roselaar (1995) examined two specimens from northern Thrace that were perhaps intergrades between nominate *europaea* (a white-breasted form) and *caesia*. However, the possibility exists that these birds represent intergrades between *caesia* and *levantina* (which if valid is endemic to Turkey). Elsewhere, those in Marmara and northern Aegean appear to be *caesia*, those in southern Turkey *levantina* and the populations in the Black Sea Coastlands *caucasica* (though some, in the west of this region, show evidence of being intergrades with *caesia* or *levantina*). In the extreme southeast, Roselaar (*op. cit.*) suspected the form involved would prove to be *persica*. Our own review of specimen material at the Natural History Museum (Tring) suggests the need to recognise three groups (two buff-breasted and one white-breasted) within *S. europaea*, as noted by Vaurie (1959a) and Harrap & Quinn (*op. cit.*), but variation within the *caesia* group is very poorly expressed and all of the six additional named forms could comfortably be regarded as synonyms of *caesia*.

Status and distribution. National conservation status: Least Concern (population apparently stable). Widespread but local in most forest and wooded habitats, as well as parks and gardens, at 300–900m in west and central Turkey and below *c.*1,800m in the east. Occurs in Marmara, Aegean, Mediterranean and Black Sea Coastlands. Largely absent from Inner Anatolia, except the northern fringe, and from most of East Anatolia, except locally in the west and extreme southeast of this region. In Southeast Anatolia is restricted to a few forested mountains and is very local. Generally recorded in small numbers, but is locally common in several areas.

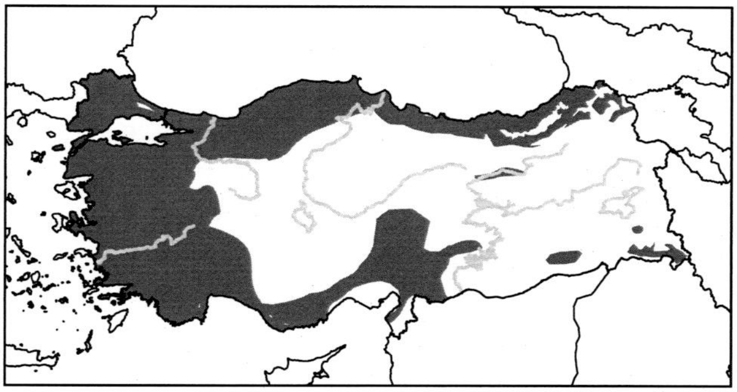

Breeding. In deciduous woodland, parkland, gardens, fruit orchards, olive and walnut groves, and even coniferous woodland. Nests up to 6m above ground, in natural tree holes, old woodpecker holes and a nest box; larger holes are reduced by mud plastered at the entrance. In Aladağ (Mediterranean), in 1876, common in oak woods and walnut groves, but not in conifers, and nested about mid April; a nest at Pozantı was constructed solely of thin flakes of the outside bark of a cherry tree (Danford 1878). No recent details of nest material or clutch sizes; elsewhere, nests are solely of dead leaves or of bark, and usual clutch is 6–9. In **Mediterranean**, a nest with large young coming to the nest entrance to be fed at Ağla on 6 May 2001, indicates laying of the first egg in late March (PC); two adults were repeatedly visiting a nest hole at Ağla on 5 May 1996, and a nest had young at Akseki on 22 May 2000 (P. & M. McGough). In **East Anatolia**, two pairs were feeding young west of Uludere, on 16–17 May 1975. At Kocaçay Delta (**Marmara**), there were fledged young on 19 June 1999 (PC). In **Black Sea Coastlands**, an adult was at a nest hole near Ispir on 26 May 1993 (K. Fahy), and large young were being fed at Abant Gölü on 29 May 1992, indicating egg laying in mid April (PC); family parties were seen at Abant Gölü on 13 June 1987 (E. Hirschfeld) and at Ispir on 15 June 1984 (S. Harrap) and 9 June 1994 (P. Smith).

EASTERN ROCK NUTHATCH *Sitta tephronota*

Büyük Kaya Sıvacıkuşu

(Du: Grote Rotsklever; Fr: Sittelle des rochers; Ger: Klippenkleiber; Swe: Östlig klippnötväcka)

Subspecies. Both Roselaar (1995) and Harrap & Quinn (1996) considered those throughout much of southeast and eastern Turkey to be *dresseri*, with the smaller race, *obscura*, likely to be present in the northeast, in the

Aras Valley and perhaps further west. Personal observations by GMK confirm that *obscura* is indeed the form concerned in the Aras (albeit, these are extraordinarily pale, like Armenian birds), but we are presently unconvinced as to the validity of recognising more than two of the seven described (3–4 generally accepted) races in this species. Nonetheless, it should be remarked that plumage characteristics within *obscura* are remarkably fluid and additional fieldwork should focus on elucidating the characteristics of the various populations of *S. tephronota sensu lato* in Iran.

In Iranian material held at the Field Museum of Natural History, Chicago, it is generally the case that subspecies *dresseri* and *obscura* are extremely close (*obscura* has priority), although *dresseri* does incline slightly larger. However, colour differences are not reliable, for some *dresseri* from the Zagros in Luristan are as dark as *obscura*, whilst others from further northwest, or southeast, are paler; the depth and length of the eyestripe is variable, with much overlap, and the same is true for underparts coloration. In contrast, the more northerly and easterly-ranging *tephronota* seems more easily diagnosable, having a generally pinkish-buff ventral region, much less prominent and narrower eyestripe, especially at the rear; the upperparts coloration might be generally considered intermediate between the other two races, but there is some variation, making it an unreliable feature. *S.t. iranica* was considered a synonym of *tephronota* by Greenway (1967), but defended by Harrap & Quinn (1996) and readmitted by Dickinson (2003); the 13 specimens in Chicago either from the range ascribed to this taxon, in northeast Iran, or from nearly arcas, can be considered easily within the nominate using morphology or measurements. This was apparently also the view of Charles Vaurie, as the labels of all 13 are annotated as being *tephronata* in his hand.

Status and distribution. National conservation status: Least Concern (population apparently decreasing). Until recently considered to be rather uncommon and local in Turkey, but increased fieldwork in its range in recent years has demonstrated the species to be rather common and widespread in suitable habitat in Southeast Anatolia, and also reasonably common in many parts of East Anatolia, where it inhabits cliffs and steep rocky slopes with stones and boulders. Found as far north as Erzincan and in the Upper Aras Valley (East Anatolia). It is not confined to open slopes, feeding in and occasionally breeding close to trees. Generally recorded above 1,000m, but its upper altitudinal limit is unknown, although noted to at least 2,700m.

Breeding. For habitat and nest site see Western Rock Nuthatch *S. neumayer*. Nests also similar, but those of the present species have a visibly broader spout and larger entrance hole. In Armenia, plant resins are used to coat the entire mud surface of the nest, and the lining is mainly of hair and feathers (Adamian & Klem 1999). Nests in Turkey are also decorated with feathers and insect casings, embedded in the surface of the mud. Feathers and polythene from a carrier bag were also stuffed into rock crevices, within 1m of a nest at Durnalık (Southeast Anatolia); this nest, examined after the chicks had fledged, was lined with a mass of wool and some hair. A nest at Işıklı was decorated with black, red and brown string, with tissue paper and several feathers wedged into the edges (PC). Nests may be repaired and used in subsequent years; two nests at Durnalık were used (by the same pairs?) in 2004–05 (PC). Eggs: C/5 (1); elsewhere, usual clutch 4–7 eggs. Young: B/4 (2), B/5 (1), B/7 (1). In **Southeast Anatolia**, breeding season apparently fairly consistent between years, with many family parties and recently fledged young in late May/early June, indicating laying of first eggs in mid April; adults were carrying food to nests from 5 May–3 June; at Durnalık in 2004 three nests had: four young, *c.*6 days old, on 9 May 2004 (D. Gregson), five young, *c.*16 days old, on 16 May 2004, and four young, *c.*16 days old, on 23 May 2004 (PC); at Durnalık, on 27 April 2005, an empty nest was complete and ready for eggs; another contained seven newly hatched young and hatching had been synchronous, indicating that incubation had commenced on completion of the clutch, with first egg laid *c.*5 April; a nest at Işıklı had five eggs on 28 April 2005 (PC); occupied nests of both Eastern and Western Rock Nuthatches were just 100m apart, near Gaziantep, on 13 May 2004 (GMK). In **East Anatolia**, a pair was food-carrying at Van on 17 May 2006 (GMK, C. G. Bradshaw).

WESTERN ROCK NUTHATCH *Sitta neumayer*

Kaya Sıvacıkuşu

(Du: Rotsklever; Fr: Sittelle de Neumayer; Ger: Felsenkleiber; Swe: Klippnötväcka)

Subspecies. Harrap & Quinn (1996) considered western and central Turkey to be occupied by *S. n. syriaca*, intergrading with *rupicola* north and east of Van Gölü (East Anatolia), but admitted that the differences between either of these taxa and nominate *neumayer* to be very slight and entirely clinal. Roselaar (1995), however, took a different view, based on more extensive museum work. In the southeast, he found examples of *syriaca.* For those in the extreme west Roselaar chose to recognise *zarudnyi* (a race endemic to the country, but considered a synonym of *syriaca* by Harrap & Quinn) and in the Mediterranean region a zone of intergradation between *zarudnyi* and *syriaca.* Those in the northeast are probably *rupicola* according to Roselaar, who was unable to locate specimens from this region of the country, whilst populations in the northwest Pontics appear to be *neumayer* on wing length. Mensural (wing and culmen) data for male specimens taken in the extreme northeast by Nesterov (1911), a source overlooked by Roselaar (*op. cit.*), accord with those presented by Harrap & Quinn (*op. cit.*) for *rupicola*, but even more closely match those for *zarudnyi* in Roselaar (*op. cit.*). Indeed, Nesterov (*op. cit.*) ascribed his specimens to *syriaca*! There is much variation in coloration of the under- and upperparts in *rupicola*, and although it is generally true that Azerbaijani birds have paler upperparts than *syriaca/neumayer*, there is too much variation to be resolute on this point – some are as dark on the upperparts and possess deeper and more extensively saturated underparts than birds from western Turkey in our experience. In our opinion, these discrepancies in reporting such miniscule variation merely prove the futility of such an exercise, and we regard *rupicola*, *syriaca* and *zarudnyi* as all being synonyms of *neumayer*.

Status and distribution. National conservation status: Least Concern (population apparently stable). Widespread and common across much of Turkey, although absent from Thrace, many parts of Black Sea Coastlands and the high Taurus (Mediterranean), but more widespread in Southeast Anatolia than previously realised. Found on cliffs and steep rocky slopes with stones and boulders, the species tolerates scrub and some trees but prefers more open areas than Eastern Rock Nuthatch *S. tephronota*. Occurs at 0–1,500m in west and central Turkey and generally up to 2,500m in the east, but reaches 2,700m in the Taurus (Mediterranean) and at least *c.*3,000m on the Nemrut and Suphan mountains, near Van (East Anatolia).

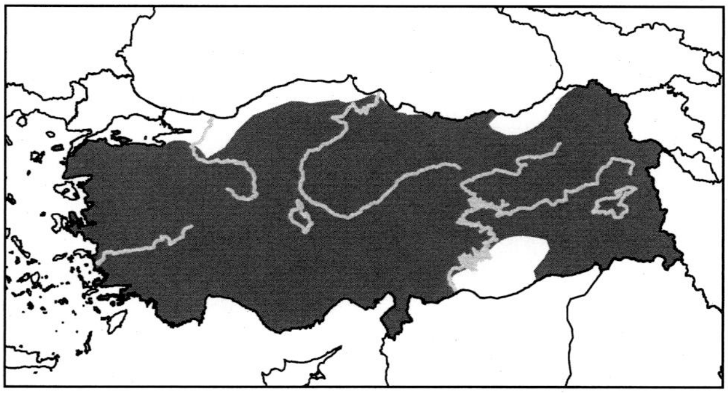

Breeding. Inhabits hillsides or gorges with cliffs, rocky outcrops and large boulders, and even open woodland or scrub with scattered boulders; prefers heavily eroded cliffs, typically limestone, with overhangs and niches affording some shelter for the bulbous nest of mud, which is attached to the rock surface. Nests may be 8m or more above ground on high cliffs, but sometimes just 1–2m up on low cliffs or boulders; nests also outside or within ruins and stone animal shelters, on walls, and even occasionally under the eaves of village houses. The nest is a rounded structure, mainly of hardened mud, up to 3cm thick, with some hair, dung, feathers and insect bodies used as binding, and a projecting entrance tube or spout. A nest examined by McNeile in Aegean was lined with a large pad of greyish wool, which looked almost like hemp, a bit of old cloth and some feathers mixed in. An unusual nest at Hasan Dağ (Mediterranean) was above the open doorway of an animal shelter, 2m above ground; it was built on to the wooden beam of the doorway, with a wood panel as the back, and with an old carpet as the roof (PC). Old nests can remain in good condition for several years, and are used by other species, such as House Sparrows *Passer domesticus*, Tree Sparrows *P. montanus* and Rock Sparrows *Petronia petronia*. Eggs: C/7 (2), C/9 (1), C/10 (1), but few nests examined, and three nests with young or from which young have been seen to fledge, have had broods of only 5–6 (PC). Nest building was observed in **Southeast Anatolia** on 22 March 1972, and in **Mediterranean** from 1 April; an early record in **Aegean**, on 10 January 1970, is considered exceptional. In most regions, nests have eggs and small young in late April–May, and broods of fledged young appear from May: the earliest fledged brood was on 4 May 2001 in Aegean. Nest building takes 10–18 days, and dates can be confusing, because adults will add to, and repair, nests at any stage of the breeding cycle. No evidence of double-brooding in Turkey; the complete breeding cycle, from nest building to fledged young, takes almost nine weeks, and fledged broods, in late June–July, will be first broods.

WALLCREEPER *Tichodroma muraria*
Duvar Tırmaşıkkuşu

(Du: Rotskruiper; Fr: Tichodrome échelette; Ger: Mauerläufer; Swe: Murkrypare)

Subspecies. Nominate.

Status and distribution. National conservation status: Least Concern (population apparently stable). Resident at low densities and a partial migrant. A rather local breeder in Mediterranean/Inner Anatolia and eastern Black Sea Coastlands, and rare in East Anatolia. Generally occurs on high cliffs and in gorges in the breeding season, usually between 2,000m and 3,200m but recorded at 2,300–3,900m in the Taurus (Mediterranean), and noted as low as *c.*1,000m in the breeding season in northeast Turkey. Disperses and moves lower in winter, even close to sea level, when known from various sites in Mediterranean, including Hatay (as late as May), westernmost East Anatolia, and at several sites in Southeast Anatolia, where generally noted from mid November to early April at 500–700m (Karakaş & Bırıcık 2005a), but noted at Birecik (where regular at this season) as early as July, on post-breeding dispersal. Recorded once in Thrace at this season (Balmer & Betton 2004b).

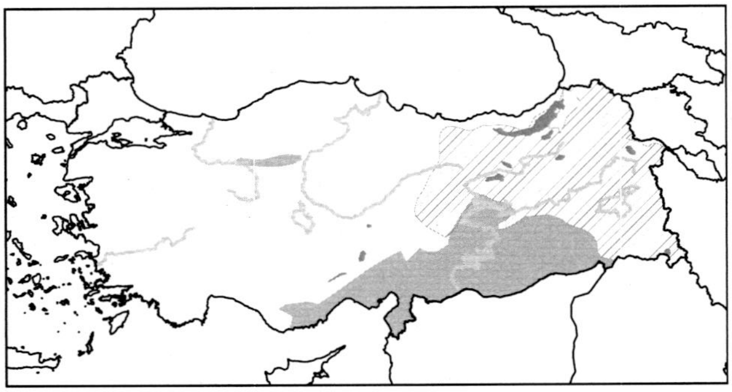

Breeding. Inhabits high mountainous areas with cliffs, gorges and precipices, usually above 2,400m and up to 3,900m. Nests in a hole or crevice in a cliff; elsewhere, also recorded nesting in walls of buildings, but there is no such record in Turkey. Nest and clutch size undescribed in Turkey; elsewhere, the nest is a cup of moss, grass, roots and wool, lined with hair and feathers, and usual clutch four (3–5). In **Black Sea Coastlands**, a pair was building a nest in a crevice, 12m above ground, at Sivrikaya on 9 June 1994 (A. Rosier), and a pair, presumed to be feeding young, was seen there on 12 June 1989 (D. Page); an adult was collecting food at Sumela Monastery in June 1975 (M. Rowntree); a pair was feeding young, near Yaylalar on Kaçkar Daği, on 9 July 1990 (Eames 1991), and a pair was carrying food to young in a nest on Kaçkar Daği on 27 July 1994 (I. A. Green). In **Mediterranean**, common on summit crags, above 2,900m, on Karanfil Dağ in July/August 1996, and all birds seen close to were juveniles (Sutton & Gray 1972); at Demirkazık, a pair was displaying at the base of crags, and another pair was at a possible nest site on 23 May 1988; and a juvenile was seen on 26 August 1986 (J. McLoughlin).

EURASIAN TREECREEPER *Certhia familiaris*
Orman Tırmaşıkkuşu

(Du: Taigaboomkruiper; Fr: Grimpereau des bois; Ger: Waldbaumläufer; Swe: Trädkrypare)

Subspecies. Roselaar (1995) recognised nominate *familiaris* in Thrace, but the remainder of the species' Turkish range he considered to be occupied by *caucasica*. In contrast, Harrap & Quinn (1996) stated that *persica* occurs throughout northern Turkey (presumably because they regard *caucasica* as a junior synonym of *persica*, a view with which we would concur, whereas Roselaar restricted the latter to the Caspian lowlands of northern Iran).

As noted by Harrap & Quinn (*op. cit.*), variation within northern Eurasian taxa of the *C. familiaris* complex is rather slight and clinal, principally based on decreasing rufous colour saturation from west to east, and it is probably best to regard most named variation within this group as synonyms of nominate *familiaris*. However, the relationship between the northern *familiaris* group (currently at least ten named forms) and the Sino-Himalayan group comprising four taxa, at present, does require further elucidation (see Collar & Pilgrim 2007).

Status and distribution. National conservation status: Least Concern (population apparently stable). Status uncertain owing to some confusion with *Certhia brachydactyla*, but apparently a local and uncommon resident in montane forests of Black Sea Coastlands and adjacent Marmara, Inner Anatolia and East Anatolia (in the Upper Aras Valley, with the single reports from Tunceli province and Kemaliye, Erzincan province, both requiring confirmation). Apparently overlaps with *C. brachydactyla* at moderate altitudes in the west (Kaz Dağı and Uludağ, Marmara, and the Istranca Dağları in Thrace) and north-central Turkey (Kızılcahamam and Bolu province). Reported in May 2008 from mountains south of Osmoniye (Mediterranean) at 700m (J. L. Copete *et al.*; not mapped), with an autumn record (in October 1969) from the Taurus (Karanfil Dağ) and winter records at sea level on the Black Sea coast and from Ankara suggest some dispersal in the non-breeding season, a wider breeding distribution or most probably misidentification of *C. brachydactyla*. Breeds mainly in coniferous or mixed forest, locally in deciduous forest, reaching to 1,500–2,000m in inland western parts of Black Sea Coastlands, at 125–550m in the littoral of the eastern Black Sea Coastlands, and 1,800–2,400m further inland in the northeast, but is found close to sea level in the Kızılırmak Delta. Sightings in the Taurus require careful documentation, given the record on Karanfil Dağ.

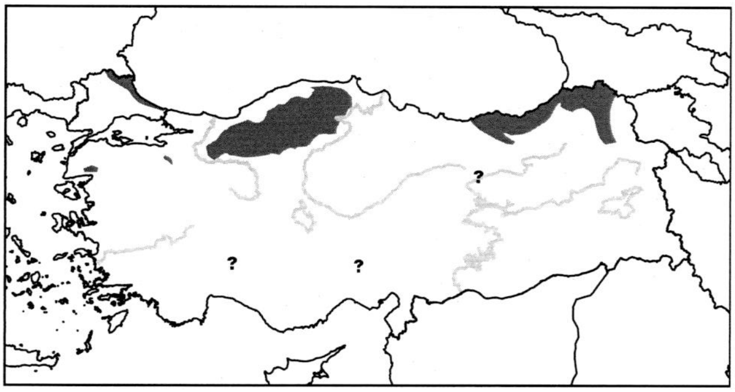

Breeding. Occurs in woodlands, generally at higher altitudes than Short-toed Treecreeper *C. brachydactyla*. Adults are recorded singing late March–July, the presumed breeding season, but there are no records of nests. A juvenile was trapped 50km north of Ankara on 1 August 1971.

SHORT-TOED TREECREEPER *Certhia brachydactyla*
Bahçe Tırmaşıkkuşu

(Du: Boomkruiper; Fr: Grimpereau des jardins; Ger: Gartenbaumläufer; Swe: Trädgårdsträdkrypare)

Subspecies. Whereas Roselaar (1995) endeavoured to tackle what he considered a potentially complex situation, made worse by a chronic shortage of specimen material, Harrap & Quinn (1996) considered Turkey to be inhabited solely by the race *harterti*. In contrast, firstly Roselaar (1995), probably correctly, given known biogeographic patterns for many other passerine taxa, considered that Thrace was probably populated by nominate *brachydactyla*. The situation in Anatolia is unclear: the names *harterti* and *stresemanni* have been applied to populations in the Pontics and Taurus respectively, but *stresemanni* has been considered a synonym of *harterti* and, alternatively, the latter a synonym of *brachydactyla*! Given such history, Roselaar (*op. cit.*) elected to uphold *stresemanni* (which is endemic to Turkey); in contrast, we consider such situations, involving morphologically

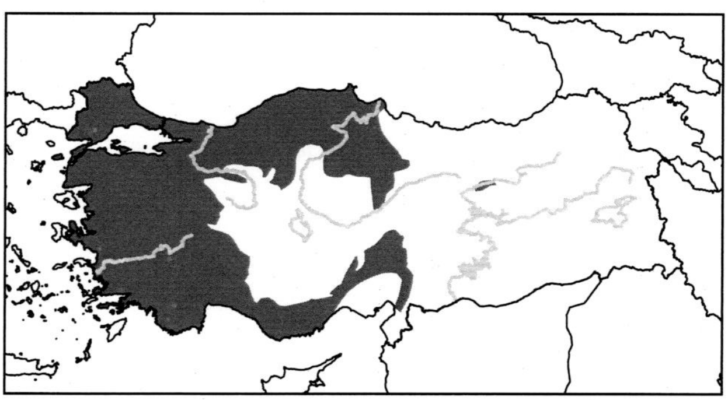

extremely similar and vocally identical taxa, subject to minimal degrees of clinal differentiation, to be unworthy of taxonomic recognition, and we consider both *stresemanni* and *harterti* to represent synonyms of *brachydactyla*.

Status and distribution. National conservation status: Least Concern (population apparently stable). Apparently a rather local but not uncommon resident of wooded areas in the western half of the country, but very local in Inner Anatolia, with most records from lower altitudes, although also occurs higher and overlaps with *Certhia familiaris* in the west and northwest, and perhaps elsewhere. Appears to disperse slightly in the non-breeding season, but apparently absent from eastern Black Sea Coastlands, much of Inner Anatolia, and the eastern half of the country. Breeds in deciduous, mixed or coniferous forests and in isolated stands of trees up to the treeline, as well as in copses, plantations, gardens and orchards, at 0–800m in Thrace and the extreme west, to 1,200m in Black Sea Coastlands, and at 800–2,000m in the Taurus. In winter, reported from parks in cities (e.g. Emirgan Park, Istanbul), where it could breed, and near wetlands in Inner Anatolia (e.g. Kulu Gölü).

Breeding. In deciduous and evergreen forest, orchards, gardens and lines of trees beside streams. Nest undescribed in Turkey but, elsewhere, usually behind loose tree bark at 0.3–4.0m; the nest is a narrow cup of twigs, roots, moss and grass, lined with feathers and wool. Eggs: C/5(1), elsewhere 5–7. Most breeding records are from Kızılırmak Delta (**Black Sea Coastlands**) where, in 1992, the population was estimated at 400–500 pairs, at a density of 20–55 breeding pairs/100ha, with an adult carrying food to a nest on 23 May and several family parties seen in early June (Hustings & van Dijk 1994); there were two pairs, each with four young, in July 1971 (Dijksen & Kasparek 1985), and an adult was carrying food on 11 May 1991 (R. Vos). An adult was at a nest at Akseki (**Mediterranean**) on 25 May 1989 (A. Clarke). In **Inner Anatolia**, an active nest with five eggs was discovered 20km west of Sultan marshes in 1991, set in a 1.5m-high *Pinus nigra* within a 15-year-old plantation (BD).

EURASIAN PENDULINE TIT — *Remiz pendulinus*
Çulhakuşu

(Du: Buidelmees; Fr: Rémiz penduline; Ger: Beutelmeise; Swe: Pungmes)

Subspecies and taxonomy. Nominate *pendulinus* appears to occur in western and perhaps southern coastal Turkey, although specimen evidence for this supposition is lacking (Roselaar 1995, Vaurie 1959a). Roselaar (*op. cit.*) ascribed all Inner and East Anatolian birds to *menzbieri*, and we suspect that this is also the race involved in the Black Sea Coastlands (pers. obs.). Some observers have reported nominate *pendulinus* in the Aras Valley (East Anatolia), on the border with Armenia, but this seems exceptionally unlikely. Differences between *pendulinus* and *menzbieri* (including *persimilis*) are very weakly defined and variable (though some *menzbieri* can be extremely pale and rather distinctive), and we would prefer to consider the latter as a synonym.

Taxonomy of the penduline tit complex is confused, being variously regarded as one, two, three or four species. However, few recent authorities, other than Dickinson (2003), have elected to consider the group as comprising any fewer than two species: Black-headed Penduline Tit (*R. macronyx*, including *neglectus*, *nigricans* and *ssaposhnikowi*) and bracketing the Eurasian/white-crowned/Chinese forms as a different species (Stepanyan 1990). Cramp & Perrins (1993) recommended that three species be recognised, with white-crowned and Chinese Penduline Tits being treated as one, but the Eurasian and black-headed groups both receiving specific status. In contrast, Gill & Wright (2006) recognised the *pendulinus*, *consobrinus* and *coronatus* groups specifically. Finally, Sibley & Monroe (1990) and Harrap & Quinn (1996) considered all four groups at the specific level, raising White-crowned Penduline Tit (*R. coronatus*, including *stolickzae*) and Chinese Penduline Tit (*R. consobrinus*) to the same status as the *pendulinus* and *macronyx* groups.

Status and distribution. National conservation status: Least Concern (population apparently decreasing). A fairly widespread and locally common resident in Marmara and Aegean, and slightly more locally also in Black Sea Coastlands, Mediterranean, Inner Anatolia and East Anatolia. A partial migrant, being more widespread in the non-breeding season, when dispersal (*cf.* van den Berk 1991a) and possibly some immigration, or even emigration, occurs. Almost invariably associated with willows beside waterways when breeding, principally in marshy areas. Mainly in the lowlands in the west, but regularly found up to 1,000m in Inner Anatolia and *c.*2,000m in East Anatolia. Absent or largely absent as a breeder from many parts of the montane east and Southeast Anatolia (where only two records). More surprisingly, breeding records are also lacking for the western (Zonguldak province) and eastern (in Kaçkar) extremes of Black Sea Coastlands. Recorded regularly throughout the year at breeding sites, but in winter records are available from many localities where breeding is unknown.

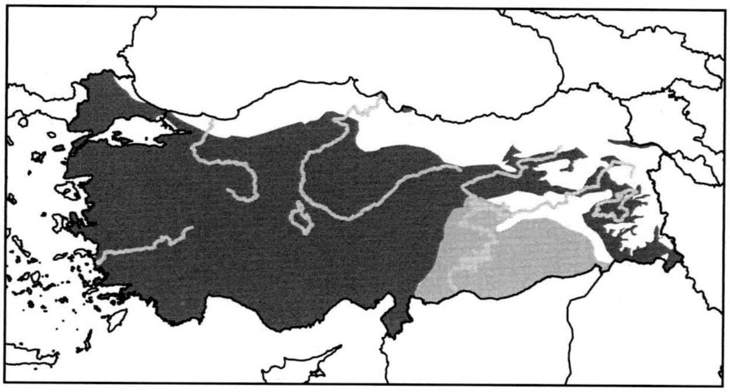

Breeding. In trees, near or alongside water, typically overhanging reed-fringed riverbanks, canals, pools and marshes, but occasionally in woodland away from water. Usually solitary, but several pairs may nest in the same area, e.g. 15 pairs (and nine nests) were along a 3–5km stretch at Bafa Gölü (Aegean) on 25/27 June 1966, and 21 nests were along the Berdan River between Tarsus and the Mediterranean Sea on 13–14 May 1987 (van der Have *et al.* 1988). Prefers willows, and poplars, *Eucalyptus* and *Tamarix*, but there are two records of nests in olive trees. The nest is suspended from thin twigs at the tip of a branch; at 5.0–9.2m (n=6), but in *Tamarix* must surely be much lower. A nest on a low thorn bush was only 0.6m above ground near Side (Mediterranean) on 17 May 2007 (J. Hollinrake), and a nest at Akköy (Mediterranean) was only 2.5m up in a lone tree, at the edge of a dry cornfield, with no water nearby, on 14 May 2007 (G. Jones). The nest is a domed structure with an entrance tube, whitish in colour, of closely felted plant down, commonly willow, mixed with fibres. Eggs: C/5–6 (5), smaller numbers considered incomplete clutches. Young: no data. In **Aegean**, earliest nest building (which takes *c.*2 weeks) noted on 5 April, with several records in April/first half of May (latest 1 June); a nest had six slightly incubated eggs, and another four eggs, at Ephesus on 5 May 1951 (McNeile); a nest had an incomplete clutch of one egg, and another was ready for eggs, at Miletos on 23 May 1999 (J. Hollinrake), and several nests had eggs, and some were ready, but still empty, on 14 May 1899 (Selous 1900); a nest had young on 14 May 1974; flying young were at Bafa Gölü on 28 May 1979 (Kasparek 1988a) and 26 June 1999 (PC). In **Mediterranean**, a nest had five eggs, and several other nests had fresh eggs, in Çukurova on 5 May 1964 (Warncke 1964–65); nest building was seen near Antalya on 24 April 1967 (R. F. Porter); there are many records of adults at nests, particularly in the Dalyan area, in May, including nest building as early as 8 April. In **Inner Anatolia**, nest building was noted at Sultan marshes from late April to late June (Kasparek 1985), with several records elsewhere, mainly in May; the latest a half-built nest near Ankara on 2 June 1974; recently fledged young at Seyfe Gölü on 2 August 1983 (Husband & Kasparek 1984) indicate egg laying late June. In **East Anatolia**, there are three records of nest building in the second half of May; a nest had an incomplete clutch of two on 20 May 1975; two nests were occupied at Haçlı Gölü on 23 May; a family party was at Van in early July; three juveniles (development not specified) were observed on 6 August 1986. In **Black Sea Coastlands**, a new nest, in a willow at Gelinkaya, was just 1m from a nest used the previous year, and an adult was incubating on 11 June 2004, and both adults were carrying food to it on 23 June 2004 (B. Pleasance). In **Marmara**, there were several nests in Kocaçay Delta on 23 April 1994 (Ertan 1996), nest building was observed at Manyas Gölü on 24–29 May 1967 (R. F. Porter), and nesting at Uluabat Gölü on 16 June 1998 (Welch & Welch 1998b). The breeding season is clearly protracted, possibly due to occasional second broods or, more likely, to replacements following earlier failure.

[PALESTINE SUNBIRD *Nectarinia osea*]
Filistin Nektarkuşu

(Du: Palestijnse Honingzuiger; Fr: Souimanga de Palestine; Ger: Jerichonektarvogel; Swe: Palestinasolfågel)

Subspecies. Should occurrence in Turkey be proven, the subspecies most likely to occur is the nominate, which is resident in the Near East and south through western Arabia (Cheke *et al.* 2001). Note: the latter authors placed this species in the genus *Cinnyris* following Fry *et al.* (2000), but incorrectly spelled the specific name *oseus* (David & Gosselin 2002b).

Species not admitted to the Turkish list (Kirwan *et al.* 1999). Listed as occurring around Antakya in the Hatay region of extreme south Turkey by Ergene (1945). Presumably because of a mistaken reading or interpretation of Kirwan *et al.* (1999), the species was again listed for Turkey by Cheke *et al.* (2001). No subsequent records or claims, although comparatively little ornithological fieldwork has been conducted in this part of Turkey, especially since the mid 1970s. In neighbouring Lebanon the species is restricted to the south of the country (Ramadan-Jaradi & Ramadan-Jaradi 1999), and the only occurrences in Syria have been in the extreme southwest (Baumgart *et al.* 1995).

EURASIAN GOLDEN ORIOLE *Oriolus oriolus*
Sarıasma

(Du: Wielewaal; Fr: Loriot d'Europe; Ger: Pirol; Swe: Sommargylling)

Subspecies. Nominate.

Status and distribution. National conservation status: Least Concern (population apparently decreasing). A widespread and locally common summer visitor to all regions, albeit perhaps commonest in Black Sea Coastlands and Thrace. Found in deciduous woodland, even in small groves of poplars. Breeding range not easily defined due to difficulties in distinguishing breeders from late migrants, but breeds in river valleys and plains with open forest, plantations or copses of tall deciduous trees, especially poplar and oak, occasionally sycamore and beech; at 0–1,200m in the west and centre, and up to 2,200m in the east. Absent as a breeder from treeless steppe and montane areas.

More widespread and more numerous on passage. Spring passage is heaviest in May, with the earliest records in Black Sea Coastlands on 14 April, the south coast from 16 April, Thrace on 19 April, and at Nallıhan (Inner Anatolia) on 21 April. Kumerloeve (1970a) mentioned an early record from 2 April. Passage may continue in earnest at this season until early June, at least in the north. Rather less common in autumn, but migration clearly peaks at the Bosphorus between late August and mid September. Passage is generally completed by late September, with a late record from Kızılırmak Delta (Black Sea Coastlands), on 15 October.

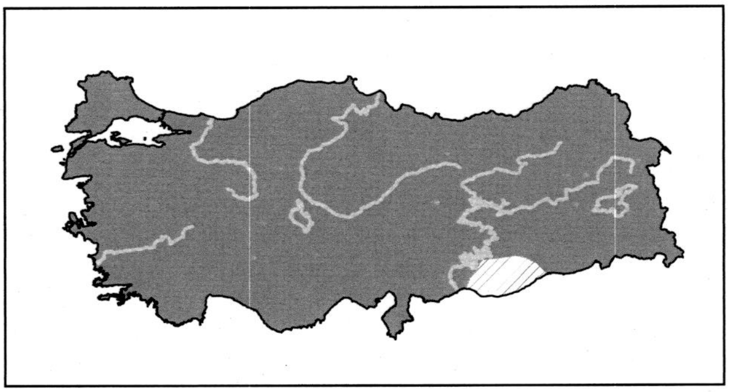

Breeding. In open, usually mature, deciduous woodland, with a preference, in some areas, for poplars and willows, especially along rivers and near lakes. Territories occupied mainly in May (some in late April). In Yürükler Forest, in Kızılırmak Delta (**Black Sea Coastlands**), seen from mid April in 1992, with large numbers arriving mid May, and new territories still being established early June; nest building was observed from 8 May, with a mean density of 25 territories/100ha, and highest density 55 territories/100ha (Hustings & van Dijk 1994). At Kocaçay Delta (**Marmara**), prefers old trees in open alluvial forest, with 0.3 singing males in a 31ha plot in 1993 (Ertan 1996). The nest is a cup, mainly of grass, between horizontal branches, usually high in a tree. Clutch size unrecorded in Turkey, but usually 3–4 (up to six) elsewhere. In Black Sea Coastlands, a newly fledged young being fed by adults at Kızılırmak Delta on 15 July 1971 (Dijksen & Kasparek 1985) indicates laying of the first egg on 12 June; a nest in roadside poplars at Gelinkaya was 15m above ground and the female was incubating 23–25 June 2004 (PC). In **Inner Anatolia**, nest building was seen near Ankara on 17 May 1998 (T. Quittenden); a nest found in 2004 at Hasan Dağ, was 5m above ground (A. Dixon). It appears that most pairs lay eggs from mid/late May.

ISABELLINE SHRIKE *Lanius isabellinus*
Kızıl Kuyruklu Örümcekkuşu

(Du: Daurische / Turkestaanse Klauwier; Fr: Pie-grièche isabelle; Ger: Isabellwürger; Swe: Isabellatörnskata)

Subspecies and taxonomy. Kryukov (1995) studied interactions between shrikes of the *cristatus* group, concluding that *L. phoenicuroides* should provisionally be regarded as a monotypic species, based upon plumage characters, biometrics and nesting periods. In south Turkmenistan these forms appear to be sympatric, although intermediates have been recorded in other parts of Turkmenistan, as well as in Iran and Dzungaria (Junggar Pendi, northwest China). Subsequent work has revealed that the original *cristatus* group may contain as many as six species-level taxa: three each of 'brown' (*cristatus, lucionensis* and *superciliosus*) and 'isabelline' (*phoenicuroides, isabellinus* and *arenarius*) shrikes. Their identification is best covered by Worfolk (2000) and future observers of one of these forms are urged to familiarise themselves with this paper, as several of these taxa are potential vagrants to Turkey. The following records in Turkey are referable to *phoenicuroides* (Turkestan Shrike): near Erzurum in August 1972, at Eceabat in July 1977, at Espiye in May 1991, at Kızılırmak Delta in May 1998 and at Samsun in November 2007 (for details see below). In addition, those at Burgaz in August 1972, at Mogan Gölü in April 2001, in the Göksu Delta in April 2004 and at Gediz Delta in January 2008 were referred to *isabellinus* (Daurian Shrike). Recent confusion concerning the type specimen of *isabellinus* and the rationale for considering *speculigerus* a synonym of *isabellinus* were discussed by Pearson (2000) and guardedly supported by Schodde & Bock (2002).

Status and distribution. Vagrant or perhaps a very scarce passage migrant. Fifteen records as follows (all pertain to single birds unless otherwise stated): a female near Silifke (Mediterranean) on 13 May 1969 (OST 1972), at Burgaz, Istanbul province (Marmara) on 31 August 1972, near Erzurum (East Anatolia) on 17 August 1972 (OST 1975), at Eceabat, Dardanelles (Marmara) on 25 July 1977 (Louette *et al.* 1977), south of Halfeti (Southeast Anatolia) on 9 July 1981 (an unusual date), at Uluabat Gölü (Marmara) on 15 September 1982, between Diyarbakır and Kahta (Southeast Anatolia) on 11 May 1984 (Kasparek 1990a), at Birecik (Southeast Anatolia) on 16 May 1987 (Kirwan & Martins 1994), at Manyas Gölü (Marmara) on 6 August 1987, between Doğubeyazıt and Adilcevaz (East Anatolia) on 28 August 1988 (Kasparek 1990a), a male at Espiye 12 km southwest of Tirebolu (Black Sea Coastlands) on 16 May 1991 (Kirwan & Martins 1994), a male trapped and photographed at Kızılırmak Delta (Black Sea Coastlands) on 3 May 1998, a male at Mogan Gölü (Inner Anatolia) on 15 April 2001 (Kirwan *et al.* 2003), a male photographed at Göksu Delta (Mediterranean) on 30 April 2004 (E. & S. Durand), one reported at Birecik (Southeast Anatolia) on 30 September 2006 (Balmer & Betton 2007a), a first-winter photographed at Gelemen, Samsun (Black Sea Coastlands), on 24 November 2007 (Balmer & Betton 2008), and a first-winter male photographed at Gediz Delta (Aegean) on 19 January 2008 (*per* E. Yoğurtcuoğlu).

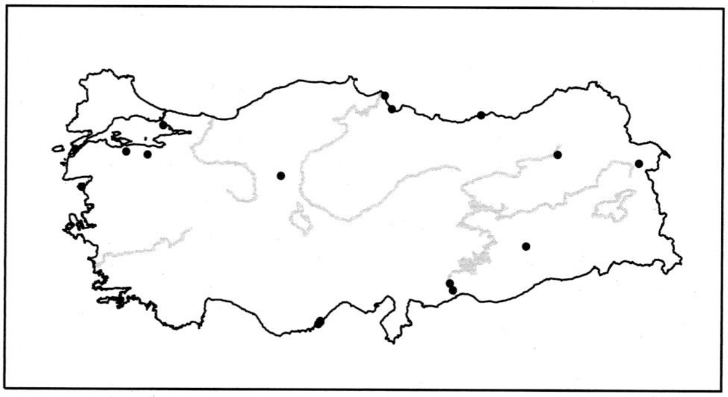

RED-BACKED SHRIKE *Lanius collurio*
Kızıl Sirtli Örümcekkuşu

(Du: Grauwe Klauwier; Fr: Pie-grièche écorcheur; Ger: Neuntöter; Swe: Törnskata)

Subspecies. Roselaar (1995) considered those in much of western and southern Turkey to be nominate *collurio*, with *kobylini* in the extreme eastern third and a broad area of intergradation in the central third of Turkey.

Several trapped in Çukurova (Mediterranean) in spring 1990 had plumage features suggestive of *pallidus*, which breeds in Siberia (Kivit *et al.* 1994). Given that Roselaar (*op. cit.*) stated that 'the difference between *collurio* and *kobylini* is slight and variation is strongly clinal', one can only speculate at the rationale for recognising *kobylini* at all. Given the additional evidence of Lefranc & Worfolk (1997) that 'even in western Europe (*collurio*), individual males may show characters of different races', and our own examination of specimens, we consider the suggestion of Stepanyan (1978, quoted in Lefranc & Worfolk *op. cit.*) to consider *L. collurio* as monotypic to be the only sensible treatment.

Status and distribution. National conservation status: Least Concern (population apparently decreasing). A widespread and generally common breeder throughout Turkey, being apparently largely absent only from the southern Aegean and Mediterranean littoral, at least parts of Southeast Anatolia and the treeless semi-steppes of Inner Anatolia. Principally occurs in open forest and scrubby areas with fields from sea level to at least 2,300m, especially in East Anatolia and the eastern Taurus (Mediterranean).

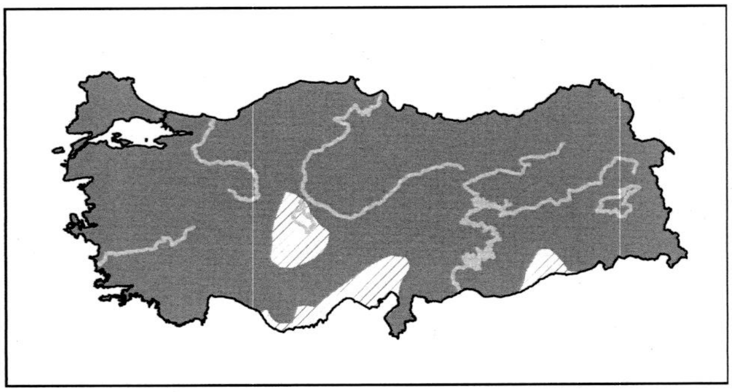

Even more common, occasionally even abundant, and widespread on passage across the entire country. Spring passage is from early/mid April (9th) to late May (30th), with most between late April and mid May. In autumn from mid August (29 July in the north) to late October, with a handful of November records (latest 17th) from the south, and the majority passing through in September.

Breeding. In scattered bushes, hedgerows, roadside undergrowth and, occasionally, isolated bushes; often at the edge of woodland, or in clearings, but also in open country, e.g. grassland or cultivation, provided bushes are present. At Kızılırmak Delta (Black Sea Coastlands), there were an estimated 650–700 pairs (25 territories/ha), with many at the edge of woodland or in very open woodland with thorny bushes or hedges, in pastures, also dunes with low bushes, even in small bushes in grassland or marshes (Hustings & van Dijk 1994). A dense breeding population inhabits the bushy dunes at the edge of the alluvial forest in Kocaçay Delta (Marmara). Nests 0.5–2.0m above ground (*n*=18) in a bush, preferably thorny and providing good cover; builds a cup of grass, plant stems and moss, lined with hair and some wool. Eggs: C/4 (4), C/5 (6), C/6 (2). Young: B/4 (5), B/5 (4). In **Mediterranean**, a nest at Eğirdir had eggs on 13 May 1997 (I. A. Green); three nests at Demirkazık full clutches on 17 May 1996, 27 May 1998 and 13 June 1993, and two nests had young on 7 June 1996, with newly hatched young on 7 June 1998, the latter indicating first egg *c.*18 May; a family party was at Akseki on 6 July 1989. In **Inner Anatolia**, fledged young at Kızılcahamam on 17 June 1983 (Barış *et al.* 1984), indicate first egg about mid May; nests had young, four days old, on 16 June 1993, and half-grown young on 27 June 1992, at Kızılcahamam, indicating first eggs in late May–early June (PC); two fully fledged young were in the suburbs of Ankara on 22 July 1969. At Kocaçay Delta (**Marmara**), nest building was observed on 21 May 1967 (R. F. Porter); several pairs were feeding fledged young on 24 June 1999, indicating egg laying by 21 May, two nests had young, about three-quarters grown, on 24 June 1999, and a late nest, possibly a replacement, had five eggs on 18 June 1999 (PC); three nests had eggs and another newly hatched young on 3–4 June 2006 (R. Castell), the latter indicating laying in mid May; elsewhere in Marmara, adults were observed carrying food at five localities in late June 1973, and a pair was feeding fledged young at Uluabat Gölü on 28 June 1973 (R. F. Porter). In **Black Sea Coastlands**, at Yeniçağa Gölü, nest building was observed on 20 June 1992; on 23 June 1992 a nest had one egg (incomplete clutch, and possibly a replacement), and another had four newly hatched young on 16 June 1993, a nest had four eggs, and another four young, *c.*3 days old, indicating egg laying about 25 May (PC); at Kızılırmak Delta, nest building was observed on 10 June 1975, the latest young fledged on 23 July 1971 (Dijksen & Kasparek 1985), and a male fed a female on the nest on 4 June 1992 (Hustings & van Dijk 1994), the latter presumably incubating. In **East Anatolia**, nests had four eggs at Tatvan on 7 June 1975, five eggs at Balık Gölü on 7 June 2001 (PC), and there were eggs at Lake Van on 13 June 2001 (T. Quittenden). At Işıklı (**Southeast Anatolia**), a nest had six eggs on 9 June 2006 (M. Cowlard). Young remain with the adults for some considerable time after fledging; family parties seen in August, for instance, may therefore have bred much earlier.

LONG-TAILED SHRIKE *Lanius schach*
Uzun Kuyruklu Örümcekkuşu

(Du; Langstaartklauwier; Fr: Pie-grièche schach; Ger: Schachwürger; Swe: Rostgumpad törnskata)

Subspecies. Uncertain, although the specimen generally approaches western *erythronotus* (see Shirihai & Golan 1994).

Status and distribution. Vagrant. One record: a first-winter male photographed, trapped and collected at Birecik (Southeast Anatolia) on 24 September 1987 (Shirihai & Golan 1994). The specimen is retained at the Natural History Museum, Tring (accession no. 1988.23.1). The only other records for the Near East involve an adult male (of the western race *erythronotus*) at Sede Boqer, Israel, from November 1982 until the third week of February 1983, which occurrence was described by the same authors; one photographed at Aqaba, Jordan, on 11–13 April 2004 (Dufourny 2006), and singles in Kuwait in October 2004 and between November 2006 and April 2007 (Balmer & Betton 2007a,b).

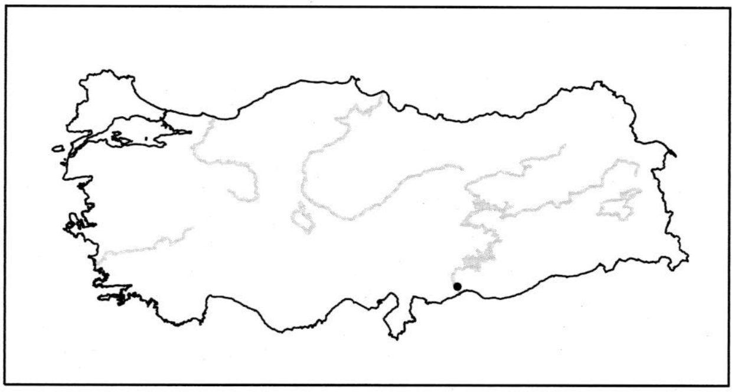

LESSER GREY SHRIKE *Lanius minor*
Kara Alınlı Örümcekkuşu

(Du: Kleine Klapekster; Fr: Pie-grièche a poitrine rose; Ger: Schwarzstirnwürger; Swe: Svartpannad törnskata)

Subspecies. Monotypic.

Status and distribution. National conservation status: Least Concern (population apparently decreasing). A fairly widespread but relatively uncommon summer visitor, breeding in all regions, albeit rather locally in Marmara (including Thrace), Aegean, Mediterranean and the Black Sea Coastlands (where extent of breeding range unclear; e.g. apparently no confirmed breeding records from well-watched Kızılırmak Delta, one of the few sites in this region likely to regularly support a breeding population). Commonest in Inner and East Anatolia, generally nesting in scattered trees amidst scrub in hilly or upland areas, up to at least 2,500m.

More widespread and common on passage throughout Turkey, especially in the eastern two-thirds. Overall, spring passage is from late March to at least late May, but in the Black Sea Coastlands, migrants rarely arrive before very late April (earliest 18th), with passage peaking in the middle two weeks of May and perhaps persisting into June (Albrecht 1986, Hustings & van Dijk 1994), and in Inner Anatolia passage possibly peaks in the first half of May (Schekkerman & van Roomen 1993). Autumn passage is from late July to mid October, more usually early October (latest 21st), peaking in late August to mid September.

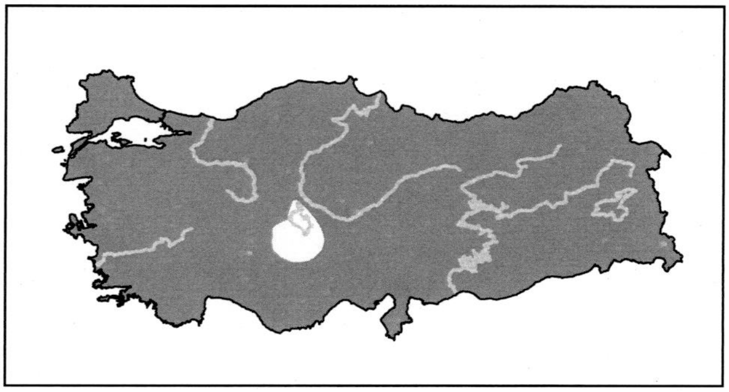

Breeding. Mainly in open areas with roadside trees and copses, and in scattered scrub and isolated trees, especially in uplands. Nests in trees, with considerable variation in height above ground, probably dependent on size of trees available, e.g. six nests on Karadağ (Inner Anatolia) were at 1.5–2.8m, in a type of thorny pear (Ramsay 1914), whereas three nests near Erçek Gölü (East Anatolia) were at 5m, 12m and 12m in poplars (PC). The nest is a cup of twigs, plant stems and rootlets, lined with feathers, roots, fine grass, hair and wool. Eggs: C/4 (1), C/5 (5), C/7 (1). Young: no details from nests, but fledged broods are mainly of 4–5. In **Inner Anatolia**, five nests on Karadağ each had five fresh eggs on 27–29 May 1907, and nest building was observed on 21 June 1907 (Ramsay 1914); a nest had hatching eggs near Karapınar on 10 June 1998 (T. Quittenden), indicating egg laying on 21 May; a nest had four eggs at 1,500m on Karadağ on 1 June 1972, and another occupied nest was in an olive tree near Ulukışla on 7 June 1971; two pairs with newly fledged young were seen on Karadağ on 24 July 1971 and a pair with newly fledged young were seen near Ankara on 22 June 1973; two adults were feeding 4–5 young, near Madenşehir, on 3 July 1989. In **Mediterranean**, nest building was observed at Akseki on 29 May 1991 (C. G. Bradshaw); a nest had seven eggs at Pınarbaşı on 12 June 1973, and several family parties were observed, including newly fledged young, 1–3 August 1971. In **East Anatolia**, adults were incubating at three nests near Erçek Golu, on 2 June 2001 and 11 June 2001 (PC).

GREAT GREY SHRIKE — *Lanius excubitor*
Büyük Örümcekkuşu

(Du: Klapekster / Zuidelijke Klapekster; Fr: Pie-grièche grise; Ger: Raubwürger; Swe: Varfågel)

Subspecies and taxonomy. Although *L. e. homeyeri*, which breeds as close to Turkey as Bulgaria, might be presumed to be the form most frequently involved (Lefranc & Worfolk 1997), the only specimen material from the country is of nominate *excubitor*: a female taken at Pendik, Marmara on 19 October 1871 (BB0792) and a male collected at Karanksi [locality untraced], Marmara, on 28 October 1872 (BB0791), held in the Manchester Museum, and three Robson specimens from the environs of Istanbul held at the Natural History Museum (Tring): a male (no location) collected on 12 November 1864 (BM 88.5.1930.391), a male taken at Pendik on 19 October 1871 (BM 72.11.4.19301) and a female from Haranköy on 22 October 1871 (BM 98.9.1920.572). Six Alléon specimens held in the National Museum of Natural History, in Sofia, are also all *excubitor*, including one from the unusually early date of 23 September (1898); all were taken in autumn in the environs of Istanbul. Only three observed in the field have been assigned to race, and these, from Ankara, Eskişehir and İstanbul, were also nominate *excubitor* (BD, E. Yoğurtcuoğlu). Mathey-Dupraz (1921) mentioned a specimen of Southern Grey Shrike *L. meridionalis* held in the Robert's College collection (Bebek, Istanbul) and described the species as rare, listing no other records from Turkey. No details of the specimen were provided and it no longer appears to be extant (Kirwan 1997a). Following the elevation to species status of the *meridionalis* complex (e.g. Clement 1995, Grant & Mackworth-Praed 1952, Isenmann & Bouchet 1993, Panov 1983, Panov 1995, Snow & Perrins 1998), this record acquires new significance. However, in the absence of further details it cannot be validated. Additionally, given its occurrence elsewhere in the Middle East, including Cyprus, observers in Turkey should also bear in mind the possibility of encountering the form *pallidirostris*, which is also increasingly recognised at species level (usually under the rather inappropriate English name, Steppe Grey Shrike). The suggestion that both *meridionalis*-complex and *pallidirostris* birds warrant specific status has recently acquired molecular support (Hernández *et al.* 2004). Future observers are encouraged to pay particular attention to specifically identifying grey shrikes encountered in Turkey, using modern identification literature (e.g. Lefranc & Worfolk 1997, Perttula & Tenovuo 2002).

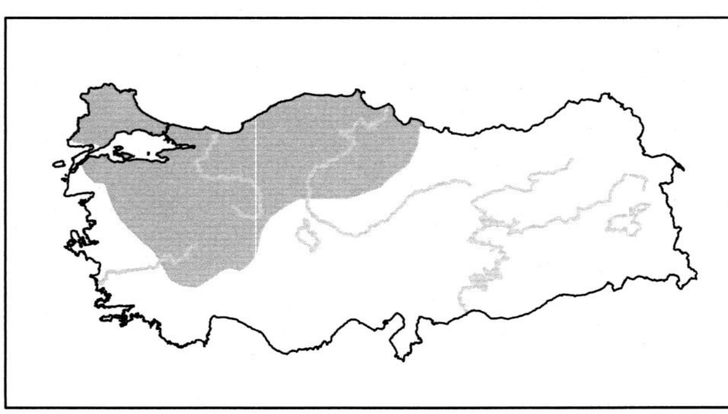

Status and distribution. An uncommon and generally fairly local passage migrant and winter visitor, principally in the western two-thirds of the country. Most appear from late October (15th), but occasionally recorded in the second half of September on passage in East Anatolia (earliest 20th) and Marmara (see above). Remains until mid/late March, or less usually, April (latest in East Anatolia, 20–21 April 1980). Most regular in Marmara (especially Thrace), Black Sea Coastlands and Inner Anatolia, with relatively few records from Aegean or Mediterranean. Confusion with Lesser Grey Shrike *L. minor*, especially immatures, is evident in autumn (e.g. Helbig 1984), and must also taint two May records from Samsun (Black Sea Coastlands) and Gelibolu, Çanakkale province (Marmara) mentioned by Kumerloeve (1961).

WOODCHAT SHRIKE *Lanius senator*
Kızıl Başli Örümcekkuşu

(Du: Roodkopklauwier; Fr: Pie-grièche a tête rousse; Ger: Rotkopfwürger; Swe: Rödhuvad törnskata)

Subspecies. Nominate *senator* (Thrace and elsewhere in Marmara south through Aegean to around Bodrum) and *niloticus* (east of a line from the Hatay north to Kars province). Situation in Mediterranean and extreme southwestern Aegean unclear and few data available, especially from the former area. Birds from Göksu Delta are likely to be *niloticus* and this subspecies may occur throughout the Mediterranean region, although *senator* has also been reliably recorded in this area, perhaps solely on migration. Some individuals in Marmara and Aegean show some white at the base of the central tail feathers, but this is usually much more restricted than on *niloticus*.

Status and distribution. National conservation status: Least Concern (population apparently stable). A fairly widespread and common summer visitor, breeding in Aegean, more locally in Marmara (including Thrace), but less commonly in Mediterranean, adjacent areas of Inner Anatolia and East Anatolia, and common over much of Southeast Anatolia. Perhaps a very local breeder elsewhere. Breeds, from early May, in a wide variety of usually lowland habitats; from scrub and semi-wooded areas in agricultural regions, especially citrus and olive groves. Recorded from sea level to 800m in Marmara and Aegean, to 1,200m in Mediterranean and 500–2,000m in East and Southeast Anatolia (Roselaar 1995). More common and widespread on passage, particularly in spring, from mid March (17th; the majority from mid April) to early May (8th). Less common in autumn from mid August (10th) to mid, exceptionally late October (31st), peaking in September.

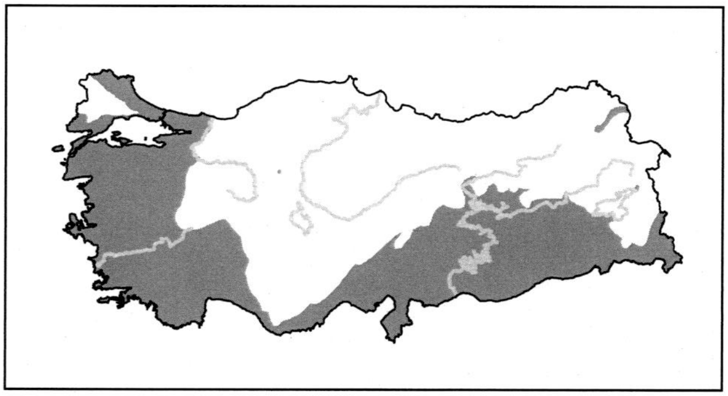

Breeding. In tall scrub and isolated bushes in open areas, in hedges, orchards, olive groves and open woodland. Nests in a bush or tree, 1.5–2.5m above ground (n=8), with no high nests recorded, possibly due to limited data (*c.f.* Spain, where nests up to 8m are frequent). The nest is cup of roots, grass and plant material, lined with wool, hair and feathers. Eggs: C/4–7, mainly 5–7 (n=12). Young: B/4 (1), B/6 (1). In **Aegean**, where more than 40 nests have been found, mainly by McNeile, around Izmir and Manisa, in 1950–51, nest building was observed in mid April; earliest nests had full clutches from 20 April through May (latest 3 June); mainly large young in late May, and fledged young were seen from 20 June (McNeile, PC, Selous 1900); fledged young were observed in the Menemen Hills on 22 June 1966, and at Marmara Gölü on 30 June 1966 (R. F. Porter). In **Southeast Anatolia** at Durnalık, three nests were empty, but ready for eggs, 9–17 May 2004, a nest there had newly hatched young on 2 June 2001 (PC) and another had six half-grown young on 7 June 2006 (R. Castell); a nest was empty, but ready for eggs, at Birecik on 11 May 2004 (J. Mason); recently fledged young have been seen, mainly in the Gaziantep area, from 9 June. In **Mediterranean**, a nest had five eggs at Karata, on 14 May (J. Whitaker), and fledged young were near Antalya on 5–6 July 1992 (N. Wheatley). In **East Anatolia**, a pair was observed incubating on 19 May 2005 at Malatya (H. Kılınç *et al.*). In **Marmara**, adults were feeding young at Vize on 21 June 1973, at Susurluk on 27 May 1967, and at Ezine on 27 June 1973 (R. F. Porter).

MASKED SHRIKE *Lanius nubicus*
Maskeli Örümcekkuşu

(Du: Maskerklauwier; Fr: Pie-grièche masquée; Ger: Maskenwürger; Swe: Masktörnskata)

Subspecies. Monotypic.

Status and distribution. National conservation status: Least Concern (population apparently decreasing by at least 10% in recent years, largely due to habitat loss: Perktaş 2004). A widespread and not uncommon summer migrant, breeding in Marmara (including Thrace where very local), Aegean and Mediterranean (where usually in the interior rather than coastal areas), probably rather locally in adjacent parts of Inner Anatolia, but widespread along the northern fringe of Southeast Anatolia (see e.g. Kirwan 1998a). Absent from most of Black Sea Coastlands and East Anatolia, where principally an uncommon passage migrant, although recent fieldwork has revealed the presence of a breeding population around Iliç and Kemaliye, Erzincan province (East Anatolia) and the species might also be expected to occur in the Kelkit Valley (Black Sea Coastlands) in summer, due to suitable climatic conditions there. Breeds from mid May in a range of semi-wooded habitats, from open oak *Quercus* or pine woodland to citrus and olive groves, to at least 2,000m but more usually below 1,100m.

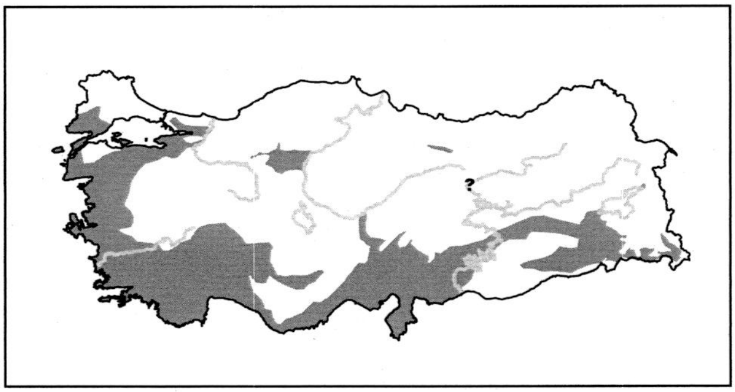

Normally light passage occurs from mid March (16th), peaking from mid April to early May in spring (with migrants recorded in the eastern two-thirds of the country and the Black Sea Coastlands into the second half of May, occasionally early June). Autumn passage is from early August (11th) to the third week of October (19th), most departing in September.

Breeding. In woodland edge and clearings, in deciduous and coniferous forests; lightly wooded areas, including olive groves and orchards; and more open agricultural areas with hedges and trees. At Kocaçay Delta (Marmara) there were 0.26 pairs/ha in a 31ha plot in the alluvial forest in 1993 (Ertan 1996). Nests in a tree or bush, including olive, apple, pear, fig, pine, usually 2–3m above ground, lowest 0.9m (McNeile), highest 8m (PC), and commonly built on a lateral branch; the nest is a cup of twigs and plant stems, lined with fine roots and plant fibres. Eggs: C/5 (5), C/6 (4), C/7 (1). Young: B/5 (1); broods of up to six fledged young recorded. In **Aegean**, Selous (1900) found several nests with eggs, at 2.5–3.0m above ground, all in olive trees, near Izmir, on 18 May 1899; McNeile also found nests in the Izmir area: one had six slightly incubated eggs on 10 May 1950, another seven eggs on 15 May 1951, and another an incomplete clutch of three on 8 May 1951, in which three further eggs had been laid by 12 May 1951; a nest at Söke had six eggs on 29 May 1999 (J. Hollinrake). In **Marmara**, a nest had five small young in Kocaçay Delta on 25 June 1999 (PC); there are three records of fledged young, the earliest at Ayvacık on 26 June 1973. In **Inner Anatolia**, nest building was observed at Kalecık on 7 June 1975; a nest had five eggs on 11 June 1998, another newly hatched young on 10 June 1998, both near Karapınar (T. Quittenden), and an adult was feeding two fledged young at Inözü Vadisi on 21 August 2005 (E Yoğurtcuoğlu). A nest was almost complete at Uludere (**East Anatolia**) on 17 May 1975 and two adults were attending a nest in a *Quercus* near Iliç on 7 July 2004 (BD & O. Mülayim). In **Mediterranean**, nest building was observed at Uzuncaburç on 7 May 2004 (A. Rustell), on 12 May 2005 (J. Hodson) and 22 May 1993 (PC), and at Taşağıl on 21 May 1996; two nests were found in Çukurova on 12–14 May 1987 (van der Have *et al.* 1988); a nest in a fig tree had five eggs at Intepe on 5 June 1974; at Uzuncaburç, a nest had one egg (incomplete clutch) on 14 June 1992, another five eggs on 10–18 June 1992 (PC) and another five eggs on 10 June

2006 (B. Pleasance); there was a nest with six eggs near Side on 25 May 2007 (J. Hollinrake). There are six records of fledged young in July–August, including a brood of five recently fledged at Aladağ on 30 July 1972.

EURASIAN JAY — *Garrulus glandarius*
Alakarga

(Du: Gaai; Fr: Geai des chênes; Ger: Eichelhäher; Swe: Nötskrika)

Subspecies. The *glandarius* group: *ferdinandi* (Istranca Mountains of Thrace), *hansguentheri* (endemic to the Bosphorus region, recorded from e.g. Belgrade Forest on the west side and Üsküdar in the east, with one record from the Istranca Mountains outside the breeding season; Keve 1973) and perhaps *graceus* (southwest Thrace, based on one record from the Kesan area). The *atricapillus* group: *krynicki* (northeastern Turkey south to Ağrı and west to about Samsun), *anatoliae* (from just east of Istanbul through western and southern Turkey east to Van Gölü, in northern Turkey east to Kastamonu), whilst the subspecies occurring in the central Black Sea Coastlands (between Kastamonu and Samsun) and east through parts of Inner and East Anatolia is unclear. Few specimens are available from these areas (Roselaar 1995).

Both *ferdinandi* and *graceus* are occasionally included within *glandarius* but are recognised by Roselaar (1995), whilst *hansguentheri* is closer to the black-capped group *atricapillus*, particularly to *anatoliae*. The subspecies *lendlii* (described from the Taurus Mountains) is a synonym of *anatoliae*, whilst *nigrifrons* described from Artvin (and Batumi in southwest Georgia) is a synonym of *krynicki* (Roselaar *op. cit.*), and we would also subsume *anatoliae* within *krynicki*. We can see some rationale for recognising the different groups of Eurasian Jays at nomenclatural level, but the naming of eight different races within the *glandarius* group and four in the *atricapillus* complex seems unnecessary for understanding real-world evolutionary relationships in the species.

Status and distribution. National conservation status: Least Concern (population apparently stable). A widespread, often common, resident through much of Turkey, although absent from much of Inner and Southeast Anatolia due to a lack of suitable habitat over much of these regions. Breeds in all types of wooded and semi-wooded country, principally in coniferous forest in the Taurus and throughout the Pontics although also fairly common in deciduous woodland, maquis scrub, olive and citrus groves, from sea level to at least 2,000m.

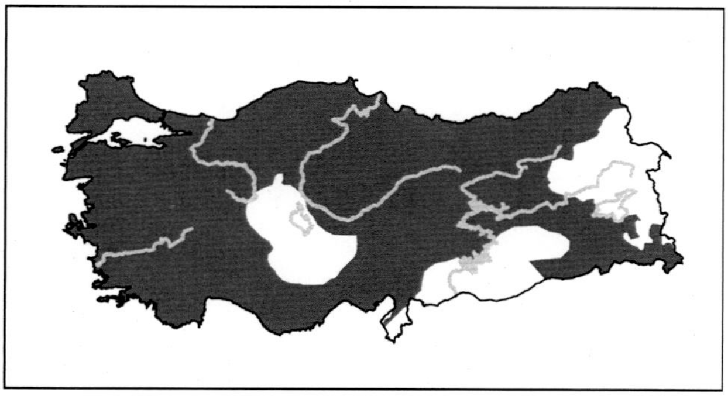

In some winters, generally relatively small numbers of immigrants may arrive in northern and perhaps western Turkey, e.g. up to 100+ per day were witnessed moving through Borçka (Black Sea Coastlands) during 24 September to 3 October 1980 (V. van den Berk *et al.*).

Breeding. In forest and woodland, both deciduous and coniferous; in olive groves and tall scrub. Nests in trees and tall bushes. A new, but empty, nest, at 3m up in an olive tree at Akköy (Aegean) on 1 June 2004, was exactly like nests elsewhere: a cup of twigs and stems, lined with fine roots and some hair (M. Demain). Two nests in orchards at Uzuncaburç at 3m in apple trees were also of twigs and stems, but both were lined only with narrow strips of nylon sacking; one was partially built but unlined on 7 May 2004, and had one broken egg on 15 May 2004, and the other was new but empty on 15 May 2004 (PC). Eggs: C/6 (1); elsewhere, usual clutch 5–7. In **Inner Anatolia**, there were family parties at Kızılcahamam in 1968–69, but no precise dates. In **Mediterranean**, a family party was at Uzuncaburç on 17 June 1996 (GMK), and several adults were near Osmaniye on 14 April 1935, 'when they just beginning to build their nests' (Bird 1937); at Uzuncaburç an adult was incubating or brooding small young on a nest at 4m in a thorn bush on 12 May 2005 (J. Hodson), a nest at 4m in a pear tree had six eggs on 10 June 2006, and there were three newly fledged young on the same day (M. Farmer). Three eggs, in Manchester Museum, were collected by Krüper, near Izmir (**Aegean**) on 1 June

1871, and probably represent an incomplete clutch; fledged young were seen near Izmir on 18 May 2004 (K. Çapaci) and an occupied nest was found near Muğla on 16 May 2004 (H. Tholl-Schmitt), and a nest in an olive tree at Akköy contained young, *c.*6 days old, on 14 May 2007, indicating first egg on 18 April (PC). In **Southeast Anatolia**, a nest contained young near Yaylacak, Mardin, on 7 May 2002 (B. Kurt), and an occupied nest was seen near Aran, also near Mardin, on 7 May 2002 (H. Welch). Fledged young were seen near Uludere (**East Anatolia**) on 8 June 2002 (T. Albayrak).

COMMON MAGPIE *Pica pica*
Saksağan

(Du: Ekster; Fr: Pie bavarde; Ger: Elster; Swe: Skata)

Subspecies and taxonomy. Nominate *pica* (throughout western Turkey east to Tokat in the north and Osmaniye in the south). No specimen material from northeastern Turkey has been examined but all are presumably *pica* (Roselaar 1995) whilst those in the southeast (from Gaziantep east to the Iranian border and south from Ağrı) are apparently intermediate between *pica* and *bactriana* (which occurs from southeast Russia and Iran eastwards) being slightly larger than the nominate race and showing more white in the primaries (Roselaar 1995). Ebels (2003) recently reviewed current knowledge and recent developments in our understanding of *Pica* taxonomy. In addition to the recent specific separation of Eurasian and North American black-billed magpies, which has already gained reasonably widespread adherence (e.g. AOU 2000), possibly five other Asian forms (*asirensis*, *bottanensis*, *camtschatica*, *mauritanica* and *sericea*) may also warrant species status, based on morphology, vocalisations and the results of preliminary DNA sequencing (see also Zink *et al.* 1995 and Choe 2002). Though none of the possible new arrangements would have any direct effect on the situation in Turkey, *bactriana* lying within the nominate group, it may prove to be the case that other named forms of *P. pica* are best regarded as synonyms of one or other of the taxa mentioned above, or of the nominate.

Status and distribution. National conservation status: Least Concern (population apparently increasing). A very common resident (and partial migrant in some regions) throughout most of Turkey, even in largely treeless steppe and semi-steppe areas of Inner and East Anatolia, where it frequently utilises telegraph poles and other man-made habitats for breeding. More typically occurs in open cultivated regions with scattered trees, bushes and scrub from sea level to 3,000m. Absent, or virtually so, from the Black Sea region, from Sinop east to Sarp (e.g. though described as a rare breeding bird in Kızılırmak Delta by Dijksen & Kasparek 1985, there was just one record in 1992 during three months intensive fieldwork at the site), and apparently absent from many parts of Southeast Anatolia, especially the southernmost portion. Apparently local in the southern coastal regions of Aegean and Mediterranean. Nests from early April. Some limited dispersal occurs in winter, particularly from central and eastern areas, when occasionally recorded in large communal roosts of up to 450 individuals.

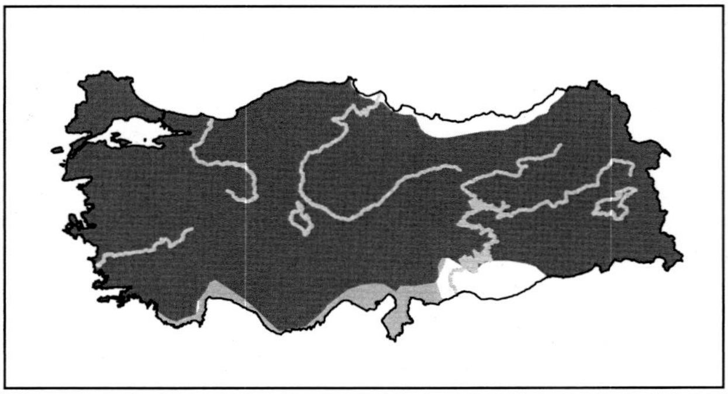

Breeding. In open cultivation with scattered trees or bushes, in scrub, orchards and lightly wooded areas. Nest usually 2–20m above ground, in a tree or bush, but also on telegraph poles and pylons in treeless areas of Inner Anatolia, and occasionally in reeds. In Inner Anatolia, nests in blackthorns on hills and in roadside acacias, with 25 nests counted within 1 mile (1.7km) (Wadley 1951). The nest is a bulky cup of sticks, thickly lined with mud, a finer lining of roots, and with a thick 'roof' or dome of twigs over the cup, leaving an opening at the cup edge. In some areas, the roof is absent (Harrison & Castell 2002), but this has not been not recorded in Turkey. In areas where trees absent, dead stalks of plants may be used in place of twigs; an unusual nest, in reeds at Marmara Gölü (**Aegean**) on 1 May 1951, was constructed of dead reeds, with a few sticks and thorns as a roof, and was only 15cm above water (McNeile). Eggs: C/5 (2), C/6 (1), C/7 (2), C/8 (2). Young: B/3–6 (6). The earliest record, a nest with five young, *c*.6 days old, in **Marmara** on 26 April 2003, indicates laying about 25 March (PC). Across Turkey nest building is mainly March–April, eggs mainly from mid April (latest record, a nest with five eggs, at Gelinkaya, **Black Sea Coastlands**, on 11 June 2004), young are in the nest May–early June, and broods of flying young are seen from late May. Nests with large young were in Kulu town (**Inner Anatolia**) on 12 June 1998, and recently fledged young were seen near Aksaray on 17 June 1998, although others were still in the nest (GMK). Later records, of adults carrying sticks to nests in May, are considered to be repair work, rather than initial nest building. Apparently the only species in Turkey parasitised by Great Spotted Cuckoo *Clamator glandarius*. Old nests are subsequently used by several species, including Long-eared Owl *Asio otus*, Common Kestrel *Falco tinnunculus* and (presumably) Eurasian Hobby *F. subbuteo*.

SPOTTED NUTCRACKER — *Nucifraga caryocatactes*
Göknar Kargası

(Du: Notenkraker; Fr: Cassenoix moucheté; Ger: Tannenhäher; Swe: Nötkråka)

Subspecies and taxonmy. Nominate. Though usually considered conspecific, Rasmussen & Anderton (2005) split nominate *caryocatactes* (Spotted Nutcracker) and *multipunctata* (Larger-spotted Nutcracker).

Status and distribution. Vagrant. Both Rigler (1852) and Mathey-Dupraz (1921) included the species in lists of birds recorded within the environs of Istanbul, the latter also reporting that Nutcrackers were regularly available for sale in autumn within the Istanbul markets during the period 1889–94. However, Kumerloeve (1961) questioned the provenance of such birds. One was in pines by Hirfanlı Reservoir, north of Tuz Gölü, Ankara/Kırşehir province (Inner Anatolia) on 6 November 1966 (OST 1969). Two (well described) were aboard the *Minerva II* during a large fall of migrants in the early morning of 4 October 2005 and eventually departed for the Turkish shore at Trabzon (Black Sea Coastlands) (D. R. Calder).

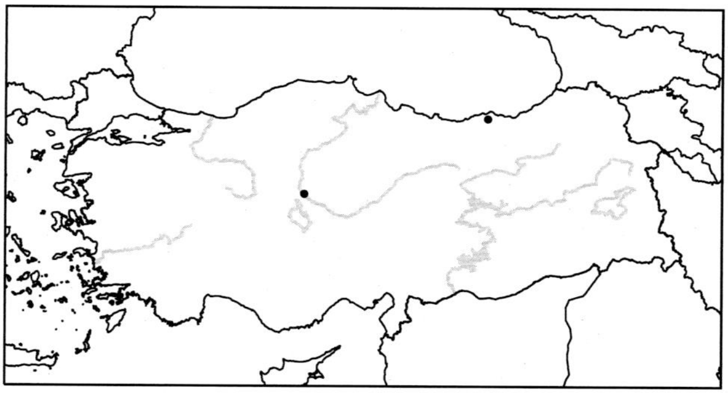

ALPINE CHOUGH — *Pyrrhocorax graculus*
Sarı Gagalı Dağ Kargası

(Du: Alpenkauw; Fr: Chocard à bec jaune; Ger: Alpendohle; Swe: Alpkaja)

Subspecies. Apparently *digitatus* in the Taurus and montane Aegean (e.g. from near Izmir) and this race also assumed to occur in the extreme southeast but confirmation is required. No specimens from Uludağ or the northeastern Pontics have been examined but these are assumed to be nominate *graculus*. Differences between the two are very slight, certainly clinal and apparently difficult to distinguish: Madge & Burn (1994) considered all Turkish birds to

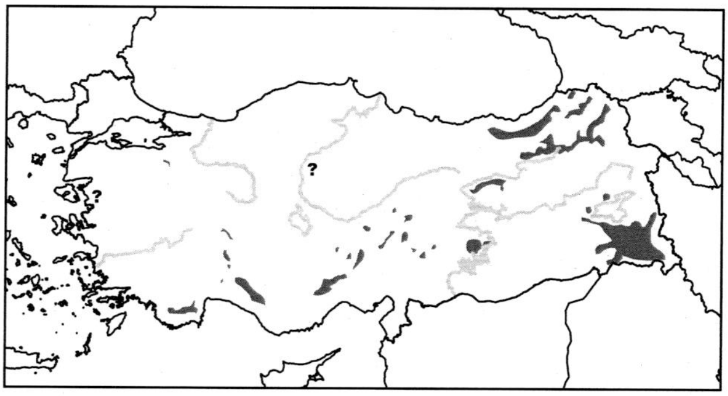

belong to the nominate form, *contra* Roselaar (1995). In the light of these facts, the species is almost certainly best considered monotypic.

Status and distribution. National conservation status: Least Concern (population apparently stable). A relatively local but not uncommon resident at high altitudes in montane areas in the eastern third of Turkey, extending westwards throughout the Taurus Mountains and more locally through the southern fringe of Inner Anatolia. Locally recorded in large numbers in all these areas. Also regularly recorded, usually in relatively small numbers, from Uludağ (Marmara). Status in Aegean is unknown, but is infrequently recorded in the mountains east of Izmir. Typically breeds above 2,500m, but recorded as low as 1,500m in Southeast Anatolia and at 1,400m in the Munzur range in East Anatolia. Undertakes limited dispersal to slightly lower altitudes in winter.

Breeding. Inhabits rocky outcrops and crags of high mountains. Six were carrying nesting material at up to 3,500m west of Görentaş (**East Anatolia**) on 31 May 1969, and there was an occupied nest in the Aladağlar (**Mediterranean**) on 14 August 2004 (A. Safak *et al.*); no other detailed data. Elsewhere, the nest is in a hole or crevice in a cliff; a cup of sticks and roots, lined with rootlets, grass and hair; four (3–6) eggs are laid, mainly in May. In the Natural History Museum, Tring, there are various clutches of eggs, collected by Krüper, at Parnassus, in Greece, in 1874–1900, including C/4 (20 May 1880), C/3 (26 May 1888), C/5 (8 May 1895), C/5 (22 May 1900).

RED-BILLED CHOUGH — *Pyrrhocorax pyrrhocorax*
Kırmızı Gagalı Dağ Kargası

(Du: Alpenkraai; Fr: Crave à bec rouge; Ger: Alpenkrähe; Swe: Alpkråka)

Subspecies. *P. p. docilis* occurs throughout Turkey. This forms apparently intergrades with *himalayanus*.

Status and distribution. National conservation status: Least Concern (population apparently decreasing). A widespread and fairly common resident in high mountainous areas, usually above 1,800m, occasionally as low as 700m (e.g. in Black Sea Coastlands). More frequent in foothills and steppe areas than Alpine Chough *P. graculus* especially in winter, and sometimes gathers into exceptionally large roosting flocks at this season, e.g. *c.*1,000 were at Demirkazık (Mediterranean) in mid-February 1999 (E. van der Burg & R. Hofland). Recorded to 3,000m on Suphan Dağı, Bitlis province (East Anatolia). Also present in other suitable habitats, e.g. cliffs on Akdamar Island, Van Gölü (East Anatolia). Occurs most extensively in the eastern third of Turkey but its range also extends westwards through the Taurus and the southern fringe of Inner Anatolia (where commoner than Alpine Chough), and on Uludağ (Marmara). Very local in Southeast Anatolia. Occasionally recorded at sea level on the south coast.

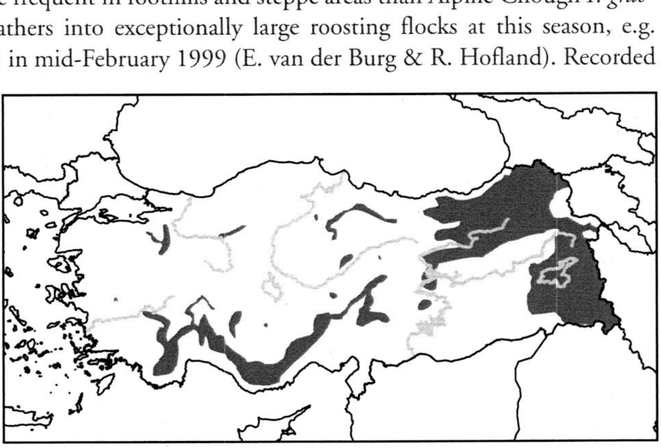

Breeding. Frequents rock outcrops, cliffs and gorges, generally high in mountains. A pair was at the castle in Ispir in April–May 1993 (Faldborg 1994), but there are no definite records of nesting in buildings, which is common in some countries, e.g. Spain. Usually in small colonies, comprising a few pairs, or solitary pairs, but up to 50 pairs recorded, e.g., on Akdamar Island (East Anatolia) on 8 June 1975. Nests in a crevice or cave in a cliff, occasionally on an open ledge; some nests are deep within crevices, but others are at, or near, the entrance, and clearly visible; builds a bulky cup of sticks and plant stems, thickly lined with hair and wool. Eggs: no data but, elsewhere, usual clutch is 3–4. Young: B/3 (1). At Demirkazık (**Mediterranean**), three nests had: young on

28 May 1998, noisy young (so probably large) on 7 June 1998, and three large young on 8 June 1998 (T. Quittenden), indicating egg laying in late April; a pair was observed copulating and nest building in the Aladağ on 29 April 1970 (R. F. Porter). In **Inner Anatolia**, an adult flew into a nesting crevice in a quarry at Ereğli on 21 April 2004 (A. Dixon). In **East Anatolia**, a nest had young at Anı (east of Kars) on 10 June 1975, and a nest at Nemrut Dağ had noisy young (probably large) on 12 June 2001, indicating egg laying in late April (PC); fledged young were observed near Van in late June–early July. In **Black Sea Coastlands**, nesting was recorded north of Erzurum on 13 May 1990 (E. Marsh), and 25–30 birds were seen at a colony near Çamlıhemşin on 10–11 May 1993 (Faldborg 1994); at Ispir, on 14 June 2004, a pair flew to a nest in a crevice and one flew out with a faecal sac, the other remaining at the nest, presumably brooding young (PC).

WESTERN JACKDAW — *Corvus monedula*
Küçük Karga

(Du: Kauw; Fr: Choucas des tours; Ger: Dohle; Swe: Kaja)

Subspecies. *C. m. soemmerringii* occurs throughout Turkey.

Status and distribution. National conservation status: Least Concern (population apparently decreasing). A widespread and common resident, or partial migrant in some areas. Largely absent from the coasts and their hinterland in the Mediterranean region and very local on the Black Sea coast during the breeding season. Usually breeds in small, loose colonies in or around habitation and ruins, alternatively in cultivated areas. Common in the lowlands below 1,500m but scarcer at higher altitudes.

More widespread and numerous in autumn and winter, although there is no direct evidence of immigration from counts at the Bosphorus or elsewhere in the north. Substantial movements do occasionally occur on the south coast in autumn. Large flocks congregate, generally in the western two-thirds of the country, including those coastal areas where the species does not breed, from late August (occasionally even late July) to early April, normally peaking from October to March. Roost concentrations frequently number several thousand; exceptionally there were 40,000 at Edirne (Marmara) in November 1970 (OST 1975).

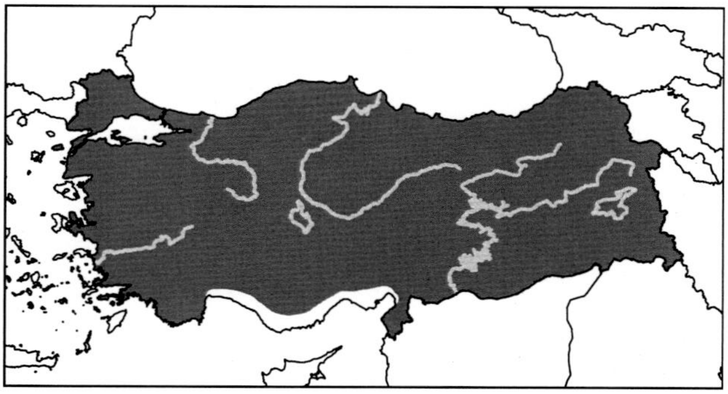

Breeding. Around habitation, nesting in holes in roofs, walls and chimneys; also in holes in cliffs (both inland and coastal) and in rock outcrops, mature trees, crevices in rocks on islands in lakes, sometimes quite remote, e.g. islands in Tuz Gölü (Inner Anatolia). Mainly colonial, colonies commonly numbering *c.*30 pairs, but isolated pairs sometimes occur. The nest is a base of twigs, variable in quantity according to cavity size, with a cup of wool, hair and plant material. Eggs: C/5–7 (2); most records are of adults flying into holes. At Manyas Gölü (**Marmara**), numbers breeding unknown but up to 6,000 were roosting in April 1967 (R. F. Porter). In **Inner Anatolia**, peak display is in April, with eggs in late April–early May, but nest building also observed on 8 May (Wadley 1951). In **Aegean**, nest building observed on 24 April 2003 (PC); many nests, some with eggs and some with young, were near Izmir on 14 May 1899 (Selous 1900), indicating earliest eggs from 20 April. In **Southeast Anatolia**, a recently fledged young bird at Birecik on 4 June 1993, indicates egg laying in mid April (PC). In **East Anatolia**, nest building was observed at Erzurum from mid-March 1910, but eggs were not laid until 30 April (McGregor 1917); a nest, examined on an island at Balık Gölü, had three eggs and two newly hatched young on 9 June 2001, indicating laying in mid May, somewhat later than elsewhere in Turkey (PC); several groups of fledged young, flying well and out of the nest for at least a week, were near Erçek Gölü on 26 June 2004, indicating egg laying in early May (PC).

ROOK
Ekin Kargası
Corvus frugilegus

(Du: Roek; Fr: Corbeau freux; Ger: Saatkrähe; Swe: Råka)

Subspecies. Nominate *frugilegus* occurs throughout Turkey.

Status and distribution. National conservation status: Least Concern (population apparently stable). Though first discovered as a breeding bird by Dickson & Ross (1839) near Erzurum in 1838 no further colonies were found until 1957, at Malazgirt, Muş province (East Anatolia) by Hellmich (1960). Since then the species has become established as a relatively common resident, sometimes in large numbers, principally in three disjunct areas: Thrace (Marmara), Inner Anatolia (principally in the south and west, but perhaps more widespread) and East Anatolia. The presence of breeding Rooks in Turkey itself, the most southerly ones in the Western Palearctic, raises interesting questions about the population's ecology, given that the species is excluded from almost all Mediterranean Europe by its dependence upon areas of moist grassland with relatively high winter temperatures and moderate to high annual rainfall.

Principally found in cultivated areas and in poplar stands around habitation, with tiny pockets of breeders in Southeast Anatolia. Kasparek (1989d) identified 60–65 colonies, totalling approximately 6,500 pairs; further research has revealed a population of well over double this figure, in more than 85 colonies (Kirwan 1995c and unpubl. data), and at least four other colonies in East Anatolia are mentioned by van den Berk *et al.* (1993). Given the species' extensive distribution in Turkey during the breeding season and the frequency of observations from the Aegean, Mediterranean and Black Sea Coastlands, its range may be wider in Turkey than these figures suggest. Many colonies go uncounted or are only partially counted by the majority of observers. However, East Anatolia is clearly a core area, with colonies being generally larger than elsewhere. The largest known colony is at Ağrı (East Anatolia), where over 1,200 active nests were counted in May 1993 (Kirwan 1995c).

More widespread on passage and in winter throughout Turkey, including Southeast Anatolia, although remaining most common in the breeding areas. Rare on the Mediterranean coast in winter. Post-breeding feeding flocks of up to 10,000 birds are regular, especially at roosts, with an exceptional total of 21,260 moving west at Karasu, Sakarya province (Marmara/Black Sea Coastlands) on 17–18 October 1983 (Kasparek 1992a). Substantial winter populations perhaps reflect significant immigration into Turkey at this season; counts at the Bosphorus in 1966 demonstrated that movement commenced in early October and continued until at least the end of the first week of November (Porter 1983). Immigration may continue until mid December. Return movements through northern Turkey continue through March into at least early April.

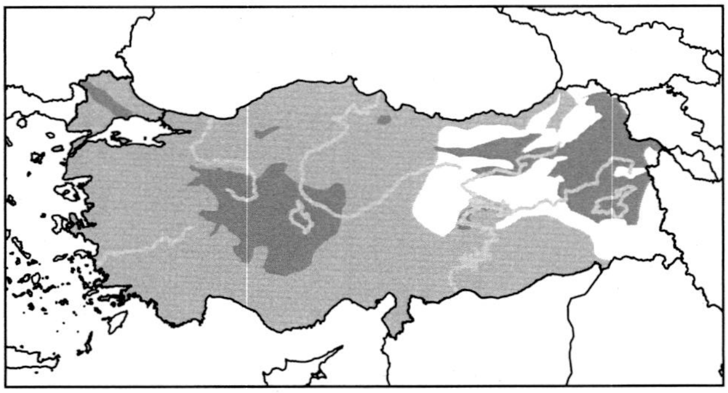

Breeding. Inhabits open country, nesting in small woods or in lines of trees, frequently poplars; commonly in, or on the outskirts of, villages and towns (and consequently near streams, rivers or lakes), adjacent to grasslands, steppe and cultivated areas; and in remote villages surrounded by extensive steppe and semi-desert; also breeds on an island in Beyşehir Gölü, Mediterranean (A. Kullberg). Nests in colonies number from seven to several hundred pairs, with larger colonies (and populations) in East Anatolia: e.g. there were at least 1,200 nests at Ağrı, and 820 nests along the roadside between Pasinler and Erzurum, perhaps not all occupied, in 1993 (Kirwan 1995c). Colonies are usually single-species, but trees near Balıkdamı (Inner Anatolia) are also used by Little Egrets *Egretta garzetta*, Black-crowned Night Herons *Nycticorax nycticorax* and Squacco Herons *Ardeola ralloides* (Magnin & Yarar 1997). Old nests are presumably used by Common Kestrel *Falco tinnunculus*, Eurasian Hobby *F. subbuteo* and Long-eared Owl *Asio otus*, but there are no records of this. Nests mainly high up, typically at *c.*5–20m; builds a cup of sticks, with earth mixed in, lined with grass, roots, wool and hair. Eggs: no data but the usual clutch elsewhere is 3–5. Young: B/2–3, observed from the ground, but a photograph in

Pforr & Limbrunner (1982) taken in Turkey, shows a nest with four nestlings. In **Inner Anatolia**, 11 nests were under construction at Eskişehir on 20 March 1968, and an adult was carrying nesting material at Polatlı on 1 April 1987 (Kasparek 1989d); in a rookery of *c.*200 pairs at Sultanhanı, adults were incubating eggs, brooding young, and feeding small young, not visible from the ground, on 23 April 2004 and, on 20 May 2004, most nests had 2–3 large feathered young, with some young on nearby branches, almost ready to fly, and others seen flying; a few adults were still on nests, presumably brooding small young; earliest eggs were laid late March (PC). Adults were incubating in a colony of *c.*30 pairs at Ilgın on 29 April 2005 (PC). It is noteworthy that these dates are in almost precise step with those for breeding Rooks in southern England, at least 12 degrees of latitude further north. In **Black Sea Coastlands**, young (some visible) were in most of 15 nests near Yeniçağa Gölü on 19 May 1992, indicating earliest egg laying in late March/early April. At Erzurum (**East Anatolia**) in 1910 adults returned from 20 March and were repairing nests by 30 March, but eggs were not laid until 30 April (McGregor 1917). Elsewhere in East Anatolia, most young had already fledged, but a few were standing close to nests, at Tutak on 10 June 2004; between Erzurum and Pasinler, a few fully feathered young were in nests, and others perching on nearby branches, on 10 June 2004, but virtually all nests were abandoned by 15 June 2004, indicating egg laying no later than about mid April; several feathered young were still in nests at Doğubeyazıt on 16 June 2004 (PC).

HOODED CROW — *Corvus cornix*
Leş Kargası

(Du: Bonte Kraai; Fr: Corneille noire; Ger: Aaskrähe; Swe: Kråka)

Subspecies and taxonomy. Unclear. Roselaar (1995) considered birds from throughout the northern half of Turkey and in the south as far west as Malatya to be intermediate between *pallescens* and *sharpii*, whilst birds from most of Southeast Anatolia are best referred to as *sharpii*. OST (1972, 1975) reported the possible occurrence of the all-black eastern subspecies *orientalis* (which is included within Carrion Crow, if Hooded and Carrion are split): one with a single *C. c. sardonius* north of Pozantı (Mediterranean) in October 1968, *c.*20 at Van (East Anatolia) in August 1971 and one at Kadınhanı (Inner Anatolia) in April 1972 but, as noted by Roselaar (*op. cit.*), this race is easily confused with immature Rook *C. frugilegus* and is largely non-migratory. (There is also a recent claim of *orientalis* in Lebanon, in October 2005: Prior 2007.) The latter author includes *sardonius* within *sharpii*. Whether Carrion (*corone*) and Hooded (*cornix*) Crows should be regarded as separate species appears almost to be a matter of personal choice, although the British Ornithologists' Union Records Committee has recently proposed to do so (Parkin *et al.* 2003), on the basis of their non-random mating preferences, differences in vocalisations and ecology, and the apparent reduced fitness of hybrid individuals, and we have elected to follow their lead.

Status and distribution. National conservation status: Least Concern (population apparently stable). A widespread and generally common resident throughout Turkey, although local and scarcer in certain limited areas, e.g. in some coastal areas in the Mediterranean region, between Ankara and Konya (Inner Anatolia), and in Southeast Anatolia. Generally absent from the highest montane areas (usually occurring to 2,500m) and from the arid, treeless areas of Inner Anatolia.

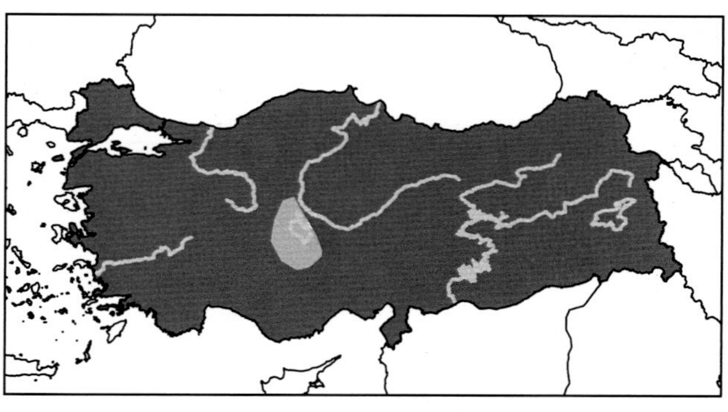

Rarely recorded in large flocks of up to 400 birds, usually in late autumn and winter. There is evidence to suggest some immigration into Turkey during the winter, with movements recorded at the Bosphorus from late September (26th) until mid December (16th), and return migration witnessed in spring from February (Kumerloeve 1957d).

Breeding. In woodland and woodland edges, copses, agricultural and urban areas, steppe with scattered or solitary trees, trees on islands in lakes, and in areas of bushes or taller scrub where trees are unavailable. Most nests are high in tall trees; but some nests in 1992 at Kızılırmak Delta (**Black Sea Coastlands**) were in low trees, and one nest was only 2.5m above ground (Hustings & van Dijk 1994); occasionally also breeds in bushes, e.g. a nest with three eggs was in a *Tamarix* beside Amik Gölü (**Mediterranean**) on 23 May 1933 (Meinertzhagen 1935). A nest south of Ispir (Black Sea Coastlands), in an area with few trees, was built on an electricity pylon and another was on a pylon at Halfeti (Southeast Anatolia) but, surprisingly, there are no such records from treeless areas of Inner Anatolia. The nest is a large cup of sticks and moss, bound with earth, and lined with wool, fine strips of bark and hair. Eggs: C/3 (1), C/5 (1); elsewhere, usual clutch 4–6. Young: B/2 (1), B/3 (1). Most records are of adults observed at (or on) nests in April (earliest 8 April) and May. In **Southeast Anatolia**, a nest had five eggs at Birecik on 11 May 2004 (L. Baxter). In Kızılırmak Delta in 1992 there were occupied nests from late April, the first nest with nestlings (one day old) was found on 8 May, and the first fledged young were seen on 1 June (Hustings & van Dijk 1994), the latter records indicating egg laying in mid April, and early–mid April, respectively; two nests at Gelinkaya and near Ispir had two and three fully feathered young on 11–12 June 2004 (PC). In **Marmara**, four recently fledged young and the adults were perched next to a nest in a roadside tree in Istanbul on 7 June 2004, indicating egg laying in mid April (PC).

BROWN-NECKED RAVEN *Corvus ruficollis*
Çöl Kuzgunu

(Du: Bruinnekraaf; Fr: Corbeau brun; Ger: Wüstenrabe; Swe: Ökenkorp)

Subspecies. Monotypic, as the form *edithae* from the Horn of Africa is almost certainly best considered at specific level (see Madge & Burn 1994).

Status and distribution. Vagrant, but perhaps best considered hypothetical. There is one recent documented record: seven feeding on garbage adjacent to the River Tigris at Cizre, Mardin province (Southeast Anatolia) on 9 July 1985, which appears to be one of the most northerly records ever (Jakobsen 1986). The description is reasonably detailed, by an observer with prior experience of the species, and offers some convincing information as to why the observation did not pertain to *C. corax*. However, other consistent and key features for the two species' separation, most importantly the wing/tail ratio, are not mentioned. It is doubtful whether the details presented in support of the observation would be considered sufficient to document a sighting in any Western European country for any species that is subject to similarly difficult identification problems, especially if involving a first country record. The species is also known from Syria, but only in the extreme south (Baumgart *et al.* 1995). Observers should be aware of the possibility of confusion with Common Raven *C. corax*, which may show brown on the nape in the Middle East. The two species apparently hybridise in some areas of Iran (Porter *et al.* 1996), but Roselaar (*in* Cramp & Perrins 1994) reported that there is little evidence of intergradation within areas of overlap in southern Iran, Turkmeniya and Israel.

COMMON RAVEN *Corvus corax*
Kuzgun

(Du: Raaf; Fr: Grand Corbeau; Ger: Kolkrabe; Swe: Korp)

Subspecies. Very few data available; Roselaar (1995) examined just three specimens, all from the northeast. The subspecies *laurencei* which occurs west and south of Turkey can be expected in the Aegean, Mediterranean and

Inner Anatolia regions, with nominate *corax* known from extreme northeast Turkey: at Rize and Verçenik and perhaps extending throughout the Black Sea Coastlands and Marmara, as well as south to Van (East Anatolia). Whether such variation is worthy of nomenclatural recognition appears extremely debatable. Neither Vaurie (1959a) nor Madge & Burn (1994) admitted *laurencei.*

Status and distribution. National conservation status: Least Concern (population apparently stable). A widespread and generally not uncommon resident (or altitudinal migrant in some areas) throughout much of the western two-thirds of Turkey, although apparently very local in Inner Anatolia and even absent from much of East and most of Southeast Anatolia. Breeds principally in upland or mountainous areas to at least 3,000m, often in forested regions. Locally common, e.g. there were over 100 on Uludağ (Marmara) in May 1969 (OST 1972).

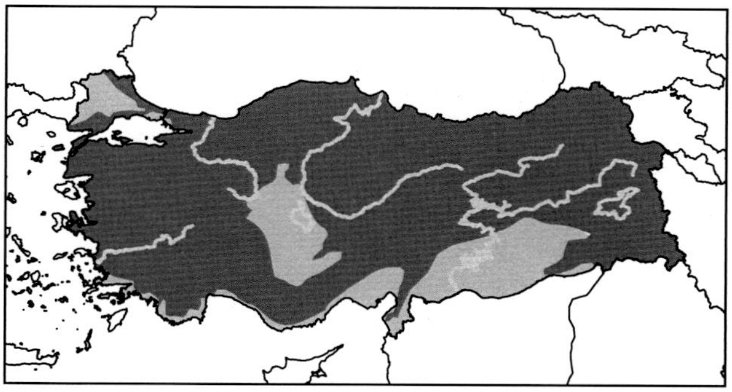

Descends in late autumn and winter, regularly dispersing, often down to sea level, into areas not normally occupied during the breeding season, when often reported in small flocks of up to 50 individuals.

Breeding. Mainly in hilly and montane areas, usually forested; in both coniferous and deciduous woodland; also in coastal lowlands, e.g. there were 7–10 pairs in Kocaçay Delta (Marmara) in 1993, breeding in flat alluvial forest and on nearby steep slopes (Ertan 1996). Nests are on cliffs, usually on a ledge, sheltered by an overhang, and high in trees; builds a large structure of sticks and twigs, bound with earth, grass, and moss, and thickly lined with wool and hair. Eggs: no data; usual clutch, elsewhere, 4–6. In **Mediterranean**, a pair was at a cliff nest north of Taşucu on 23 March 1993 (J. van der Winden) and another pair with a nest was at Ermenek on 13 April 1991 (M. Bakker); an adult mobbed a Long-legged Buzzard *Buteo rufinus* over mature pine forest near Uzuncaburç on 7 May 2004 (PC). In **Southeast Anatolia**, a pair was at an empty nest at Kilis on 8 April 1971, and a few pairs were nesting in caves near Gaziantep in March–April 1935 (Bird 1937). In **Black Sea Coastlands**, a pair nested at Ereğli Reservoir in 1976–78, and another pair probably bred (Albrecht 1986); the first fledglings were seen in late May in northeast Turkey (Pontic Alps) in 1993 (J. Faldborg), indicating egg laying in late March. In **Marmara**, three (a presumed pair and a fledged juvenile) were at Uluabat Gölü on 31 May 1998 (Welch & Welch 1998b). In **Inner Anatolia**, large feathered young in a nest on a crag at Karadağ on 26 April 2007, indicating egg laying in early March (A. Dixon).

FAN-TAILED RAVEN — *Corvus rhipidurus*
Kısa Kuyruklu Kuzgun

(Du: Waaierstaartraaf; Fr: Corbeau à queue courte; Ger: Borstenrabe; Swe: Kortstjärtad korp)

Subspecies. Unknown, but on geographical grounds *stanleyi* is clearly the most likely to wander to Turkey.

Status and distribution. Vagrant. One claimed at Belen Pass (Mediterranean), on 22 March 2005 (Welch & Welch 2006), was well described by observers with extensive previous experience of the species. Fan-tailed Raven breeds in Israel and Jordan, but apparently only exceptionally wanders far north of its breeding grounds, with no records known from Lebanon (Ramadan-Jaradi & Ramadan-Jaradi 1999) and perhaps just one in Syria, at Palmyra, in 1975 (Snow & Perrins 1998), although this record was mentioned neither by Baumgart *et al.* (1995) nor Serra *et al.* (2005).

COMMON STARLING *Sturnus vulgaris*
Sığırcık

(Du: Spreeuw; Fr: Etourneau sansonnet; Ger: Star; Swe: Stare)

Subspecies. The situation is complex and unclear, with *tauricus* in Inner Anatolia and much of Mediterranean, *purpurascens* in eastern Turkey from Baskale in the south and Erzurum in the north, and perhaps *oppenheimi* in Southeast Anatolia from Nizip and Birecik east to Cizre, Mardin province. Large areas of apparent intergradation exist. Birds in Thrace are apparently intermediate between nominate *vulgaris* and *tauricus* (such populations are occasionally treated as a separate subspecies *balcanicus*); *tauricus* and *purpurascens* apparently intergrade in the Upper Murat Valley (East Anatolia) and intergrades are also known other areas in east-central East Anatolia, e.g. at Eleşkirt and Elazığ. Roselaar (1995) suggested that birds in Aegean, Marmara and much of the Black Sea Coastlands are perhaps more likely to be intermediates between *vulgaris* and *tauricus*, than pure *tauricus*. He favoured the resurrection of *oppenheimi* (originally described from adjacent Syria) for populations in Southeast Anatolia, regarding them as being uniform in appearance rather than varied as with so-called *balcanicus* populations. In contrast, Vaurie (1959a) followed by Feare & Craig (1998) recognised just *tauricus* and *purpurascens* in Turkey. As suggested above much of the variation is strongly clinal and a re-evaluation of subspecific limits across the Common Starling's broad range might find good rationale for recognising many fewer races than are currently given currency.

Status and distribution. National conservation status: Least Concern (population apparently decreasing). A common and widespread resident over much of Turkey, breeding in all regions, although relatively scarce in a number of small areas, e.g. in the eastern Black Sea Coastlands, where only one pair was found in Kızılırmak Delta in spring 1992 (Hustings & van Dijk 1994). Apparently absent as a breeding bird from several others, notably much of southwest Aegean, the coastal regions of Mediterranean and eastern parts of Southeast Anatolia. Apparently a partial migrant in much of the eastern third of Turkey. Typically nests up to 3,000m.

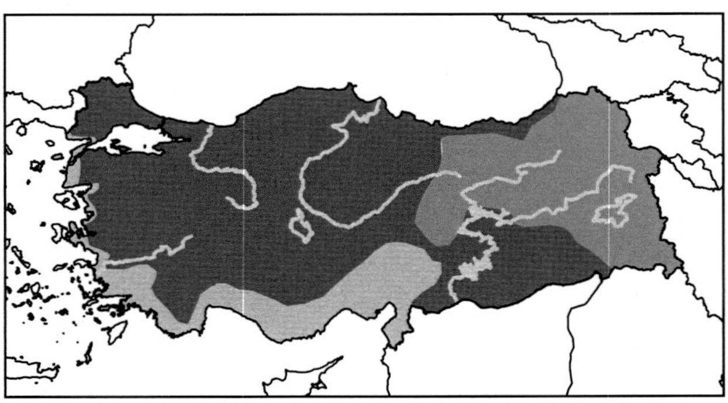

Even more widespread throughout the western two-thirds of the country, occurring in very large, locally enormous, numbers in autumn and winter. Winter roosting flocks are frequently estimated at 50,000–100,000 birds, exceptionally 500,000 at the Büyük Menderes Delta (Aegean) on 27 January 1972, 100,000 there on 7–10 January 1970 (OST 1975) and 500,000 at Balıkdamı, Eskişehir province (Inner Anatolia) on 3–6 January 1974 (OST 1978).

Breeding. Found mainly around habitation, in roofs and in holes in buildings, walls and ruins; also in open country with scattered trees, or in open woodland, where nests in holes in trees, including old woodpecker holes, and in nest boxes; also recorded nesting in rocks, in crevices in a ravine in Southeast Anatolia (R. F. Porter); nests from sea level to over 2,700m, e.g. a nest had young at 2,710m, in the roof of a roadside building at Günzeldere Geçidi, north of Başkale (East Anatolia), on 18 June 2004 (B. Pleasance). Generally solitary or a few pairs in loose association, but 100–150 nests were in a loose stone wall near Erçek Gölü (East Anatolia) on 27–29 May 1969 (Kasparek & van der Ven 1983). The nest is an untidy accumulation of grass, leaves and other plant material, the cup lined with finer grass and feathers. In a study using nest boxes, outside Ankara (Inner Anatolia), in 1976–85, breeding density was 4.2 pairs/ha (mean 41.6 pairs p.a.), the average of 155 clutches (mainly five eggs) was 4.4, and the mean number of fledged young per pair was 3.5 (Kiziroğlu *et al.* 1987). In

Southeast Anatolia, at Birecik, two nests in old Syrian Woodpecker *Dendrocopos syriacus* holes had large young on 10 May 2004, and several fledged young were there on 8 May 2001, indicating egg laying in early April; nest building was observed at Gaziantep on 4 May. In **Inner Anatolia**, nesting begins in late April, but mainly in early May (Wadley 1951); a nest had five eggs at Karapınar on 24 April 1988 (J. Whittaker), with several records of adults carrying food to nests, mainly in mid May; several broods of recently fledged young, barely able to fly, were at Eşmekaya on 20 May 2004, and at Konya on 21 May 2004, indicating egg laying in mid April (PC); fledged young were at Tüney on 13 June 1975 (R. F. Porter). In **East Anatolia**, young fledge about late May at Erzurum (McGregor 1917); two nests had young at Ahlat on 8 June 2001 and 9 June 2004, and recently fledged young, unable to fly, were there on 9 June 2004; a nest had young on 8 June 2001 near Doğubeyazıt, where fledged young were present on 4 June 2002; an adult and one fledged young were at Çatak on 1 June 1969 (R. F. Porter). In **Black Sea Coastlands**, young were being fed at Yeniçağa Gölü on 19 May 1978 (Kiliç & Kasparek 1987); a nest had young at Kızılırmak Delta on 27 May 1992 and 4 June 1992 (Hustings & van Dijk 1994); two nests had large young at Gelinkaya on 11 June 2004, indicating egg laying about 9 May, and an adult was observed carrying nesting material to the roof of a house at Çaldıran on 17 June 2004, presumably for the second brood of the year (PC). In **Aegean**, nest building was observed at Afyon on 22 April 1967 (R. F. Porter); an adult flew to a hole and remained there (presumably incubating eggs) at Manisa on 24 April 2003 (PC); juveniles were observed at Marmara Gölü on 1 July 1966. In **Marmara**, recently fledged young were at Bebek on 17 June 1973, and at Terkos Gölü on 20 June 1973 (R. F. Porter), and a pair was at a nest hole in Forest of Belgrade on 4 June 1996 (GMK). At Demirkazık (**Mediterranean**), a nest held young on 12 May 2004 (D. Gregson), indicating egg laying by 25 April. Elsewhere, the species is often double-brooded; some later records above are possibly second broods, but this is unclear.

[SPOTLESS STARLING *Sturnus unicolor*]
Kara Sığırcık

(Du: Zwarte Spreeuw; Fr: Etourneau unicolore; Ger: Einfarbstar; Swe: Svartstare)

Subspecies. Monotypic.

Species not admitted to the Turkish list (Kirwan *et al.* 1999). Mathey-Dupraz (1921) mentioned at least two undated claims of nesting birds near Istanbul in the late 1880s or 1890s, Aharoni (1930) that the species had bred in the region of Amik Gölü (Mediterranean) in summer 1910, and Wahby (1930) claimed the species at Sapanca Gölü, Kocaeli/Sakarya province (Black Sea Coastlands) on 7 July 1928. Both observations lack clear substantiation (Mathey-Dupraz mentioned that the birds had black plumage without any sheen or clear spotting) and appear improbable. Indeed, the last record was regarded as probably erroneous by Kumerloeve (1975a) and Kasparek (1990a), but both these authors apparently overlooked the earlier claim. Ergene (1945) considered the species hypothetical in Turkey, noting that Dresser (1902) mentioned its occurrence in Palestine (based on records by Tristram), but these are no longer considered acceptable (Shirihai 1996). The strongest evidence of the species' occurrence in Turkey lies in one that was photographed at Harran, near Urfa (Southeast Anatolia) on 5 May 2000 (Kirwan *et al.* 2003). A colour photograph has been published (Bradshaw & Kirwan 2000) which clearly shows a number of features indicative of *S. unicolor*, but the identification must be considered unconfirmed. Elwes & Buckley (1870) mentioned that the species was found by Lord Lilford in Corfu, and Handrinos & Akriotis (1997) noted two modern records: on Crete in May 1967 and on Paxoi in May 1992. Feare & Craig (1998) considered the species to be expanding its range eastwards.

ROSE-COLOURED STARLING *Sturnus roseus*
Pembe Sığırcık

(Du: Roze Spreeuw; Fr: Etourneau roselin; Ger: Rosenstar; Swe: Rosenstare)

Subspecies and taxonomy. Monotypic. Since the early 1940s, most researchers have placed the species in the genus *Sturnus*, principally based on its many undoubted similarities with other members of that genus and those taxa originally described within *Sturnia*, which are now also generally subsumed within *Sturnus*. However, other features (particularly the arched culmen and well-developed glossy crest) are shared with *Acridotheres*, while its reduced skull development is presumably due to its feeding specialisation, which (at least seasonally) involves much less prying than, e.g., *Sturnus vulgaris*. Additionally, its usual (but not exclusive) habit of breeding amongst rocks on the ground and its courtship display are most unlike those of other *Sturnus*, leading Feare & Craig (1998) to prefer conservation of the monotypic genus *Pastor* for the species.

Status and distribution. National conservation status: Least Concern (population apparently stable). Exact status remains unclear, reflecting the species' semi-nomadic movements. May breed comparatively widely throughout the eastern two-thirds of the country, principally in East Anatolia. However there are merely sporadic instances of confirmed nesting, perhaps in invasion years when it may be substantially commoner throughout all of Turkey. Recorded at to least 3,000m, but usually breeds in relatively inaccessible rocky, upland areas from at least 1,000–2,000m, often in large colonies, which can number several hundred or even several thousand pairs. These are rarely utilised in succeeding years. Breeding has been proven on at least four occasions in Inner Anatolia, three each in Mediterranean and Black Sea Coastlands, thrice in Southeast Anatolia and ten times in East Anatolia (most recently in 2001 and 2002), and is almost certainly more frequent in all these regions, especially the former and latter.

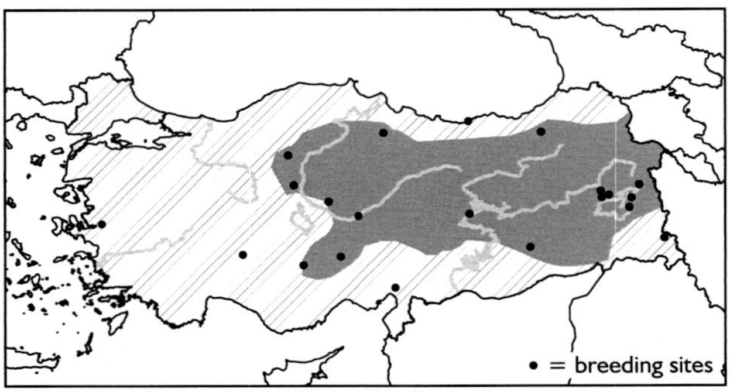

Otherwise recorded as a regular, occasionally very common, passage migrant in all regions (though usually comparatively rare as far west as Thrace, Marmara) during spring in May and early June, less regularly in autumn in August and September, occasionally late July, e.g. there were 5,000 at Tatvan, Van Gölü (East Anatolia) on 30 July 1990. Exceptionally, an estimated 27,000 were recorded around Van Gölü, on 24–25 June 2000 (*Birding World* 13: 373). Extreme dates in Turkey: 9 April–21 September.

Breeding. Found in extensive grasslands and arable fields, usually tree-less. Colonial-nesting in piles of boulders, on scree slopes, in crevices in low cliffs, and in walls. Highly erratic, depending on mass occurrence of grasshoppers (Roselaar 1995), but not dependent on villages for nest sites, as previously thought (OST 1969). Flocks wander widely, especially pre- and post-nesting, and may not indicate local breeding. A roadside colony of 4,000–5,000 nests was in a gently undulating area of large open cereal fields and virtually tree-less fallow grasslands at 840m, 20 km southwest of Diyarbakır (**Southeast Anatolia**), on 13 June 2001 (PC). The birds were nesting in two linear boulder piles, either side of a road, each c.1.3km in length, with *c.*2 nests per metre; nests were hidden in gaps between the boulders at soil level and at 0.3–1.0m from the outside; the nests were untidy shallow structures mainly of dry grass stems, some with seed heads, and usually with some feathers in the lining. Of 100–200 nests examined, only one had solely eggs, the rest held 3–5 (mainly four) small young, all less than four days old, or combinations of eggs and newly hatched young; incubation had commenced mainly from the third egg, which was most noticeable in broods of five, where three young were usually slightly larger, and two were newly hatched. Eggs much paler blue than those of Common Starling *S. vulgaris*; nests at either end of the colony were at the same stage as those nearer the centre, indicating laying of first eggs about 18 May; a very high degree of breeding synchrony, presumably based on a locally abundant food supply available for a limited duration. Adults were carrying mainly grasshoppers, but also caterpillars and other invertebrates. At least 400 adults were observed feeding juveniles, in an area that looked suitable for breeding, west of Diyarbakır (possibly the same site as above?), on 1 July 1978. Two adults and 12–14 juveniles were at Birecik (Southeast Anatolia) on 3 July 1970. In **Inner Anatolia**, 300, of which 70% were juveniles, were at Bögecık on 8 July 1977; five pairs were observed at 1,500m on a mountain near Ürgüp on 4 June 1971, when males were singing and females were inspecting piles of stones for nest sites. In **East Anatolia**, a colony of at least 200 pairs was nesting in black basalt stones at 1,300m, *c.*10 km south of Iğdır, on 20 July 1992 (G. Magnin & L. Nauta); hundreds of adults were seen flying towards a distant boulder-strewn ravine near Balık Gölü on 7 June 2001, and were probably breeding (PC); an estimated 6,000 were breeding under boulders on both sides of the road in a lava area north of Doğubeyazıt on 22 June 2002 (H. & J. Eriksen); considered numerous and breeding on the north side of Van Gölü in July–August 1966; juveniles were observed at Işak Paşa Sarayı on 29 July 2000 (MEBirdNet), at Bulanık there were 40, both adults and juveniles, on 10 July 1986 (R. Webb), and one juvenile was at Erçek Gölü on 29 August 2000 (A. & K. Gancz). Many juveniles were seen at four places in Turkey in August 1974, and hundreds of adults in three areas in June 1975. Three eggs, in the Manchester Museum, were collected by Krüper in the Izmir area (**Aegean**) on 18 June 1880. Elsewhere, the species is usually single-brooded, but occasionally double-brooded in years of exceptional food abundance.

COMMON MYNA *Acridotheres tristis*
Çiğdeci

(Du: Tristrams Spreeuw; Fr: Martin triste; Ger: Hirtenmaina; Swe: Brun myna)

Subspecies. Unknown. Some authors prefer to consider the species monotypic.

Status and distribution. Status uncertain, but probably an extremely local breeder (known only from Ankara and Istanbul) as a result of escapes from captivity (*A. tristis* is certainly imported into the country as a cagebird). First mentioned for Turkey by Sperl (1992), who remarked that the species occurred in the eastern Black Sea Coastlands, but failed to provide details. Snow & Perrins (1998) stated that it occurred west as far as Trabzon; thus far, there are no direct observations to support such a statement and we consider the assertion that Common Myna occurs in northeast Turkey to require proof. Subsequently, Common Myna was discovered at the Middle East Technical University campus, Ankara (Inner Anatolia) in May 1996 (Bilgin 1996), where the species bred annually until 2000 (Kirwan *et al.* 2003). The species has recently been discovered to be a breeding resident around Maltepe and Kartal, in Istanbul (A. I. Tireli, B. Bilgen). The only other published records for Turkey appear to be those noted by G. Magnin at Sirapınar, Istanbul (Marmara): two on 22 June 1997 and six on 1 July the same year (Kirwan *et al.* 2003), with one in Istanbul on 15 December 2003, seven on 29 May 2004, 22 on 16 April 2005, one on 12 August 2005 and nine on 10 September 2005 (Balmer & Betton 2005a,b, 2006).

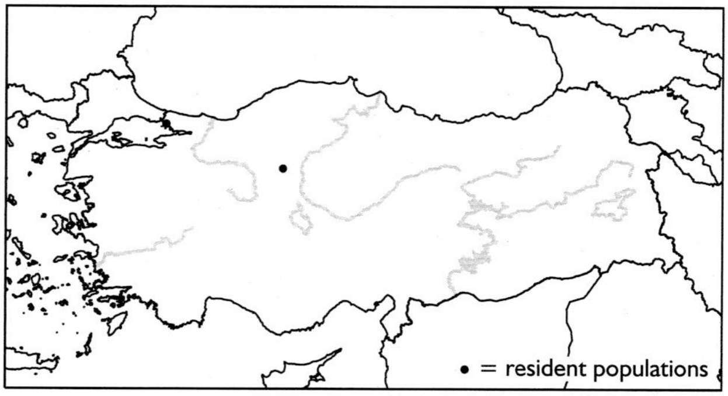

During the 1990s, Common Myna has been recently reported from several Middle Eastern countries, namely Egypt, Iraq and Lebanon (breeding) in addition to those populations shown by Porter *et al.* (1996). While these further records have generally been considered to relate to additional introductions or escapes from captivity, the possibility exists that they are part of the natural range expansion noted in Central Asia (see Feare & Craig 1998).

Breeding. A pair in Ankara (**Inner Anatolia**), first observed on 12 May 1996, was subsequently seen entering a cavity *c.*5m above ground in a nearby poplar tree; the pair was observed again with three fully fledged young on 21 July 1996 (Bilgin 1996). In the main breeding range, the species nests in any type of cavity – in buildings, walls, trees and banks, and among rocks; the nest is an untidy accumulation of debris, and natural material, such as grasses, lined with feathers, leaves and softer debris; usual clutch 4–5. An adult was observed carrying a lizard to a nest constructed on a pipe attached to a building in Kartal, Istanbul (**Marmara**) in summer 2004 (MÖ).

HOUSE SPARROW *Passer domesticus*
Serçe

(Du: Huismus; Fr: Moineau domestique; Ger: Haussperling; Swe: Gråsparv)

Subspecies. There is much variation and several populations are difficult to assign racially, but *balearoibericus* (throughout most of western and central Turkey; from southern Thrace and Marmara east to around Bolu and south throughout Aegean, much of Inner Anatolia, the Taurus and Mediterranean west of a line between Alanya and Niğde), *mayaudi* (endemic to montane eastern Turkey from around Erzurum east to Van and the Iranian border, and south to Yüksekova) and *domesticus* (northern and eastern Turkey; from northern Thrace west as far as Zonguldak, in the extreme northeast around Rize and Borçka and perhaps throughout the Black Sea Coastlands) have all been identified (Roselaar 1995).

Intermediates between *mayaudi* and *biblicus* (of the Levant, Syria and Iran) occur in Southeast Anatolia from Birecik east to Ceylanpınar, whilst elsewhere in the south and east those in Çukurova north and east to Osmaniye, Amik Gölü and Elazığ are intermediate between *balearoibericus*, *mayaudi* and *biblicus*. The subspecies occurring through much of the rest of northern and eastern Turkey is unclear due to a lack of museum material. *P. d. colchicus* Portenko, 1962, described from Artvin, northeast Turkey and western Georgia is regarded as invalid by Roselaar (1995). In contrast, Summers-Smith (1988) did not recognise *mayaudi* (described by Kumerloeve from Van) as a valid race, and treated all Turkish birds as belonging to *biblicus*. As can be seen from Roselaar's extensive work, the situation in Anatolia is rather complex and we have been unable to study this problem in any detail, although we suspect that much of the geographical variation recognised by others in this species would not be acknowledged under the criteria we have chosen to apply to the 'problem' posed by subspecies.

Status and distribution. National conservation status: Least Concern (population apparently decreasing). A widespread and common resident breeder throughout Turkey, e.g. there were an estimated 1,500–3,000 pairs in Kızılırmak Delta, Black Sea Coastlands in spring 1992 (Hustings and van Dijk 1993), breeding up to at least 2,700m (often in small, mixed colonies with Spanish Sparrow *P. hispaniolensis*). However, comparatively little is known about the species' precise status and distribution as it rarely attracts observers' attention. Apparent hybrids between this species and Spanish Sparrow are not uncommon and have been observed in most regions, while hybrids with Tree Sparrow *P. montanus* have been recorded at least twice. Frequently forms large mixed flocks in autumn and winter, most frequently with Spanish Sparrow, although the extent of movements if any, at this season is poorly known. In the Taurus the species is known to descend principally below 2,000m at this season. Some limited dispersal away from eastern areas is apparent.

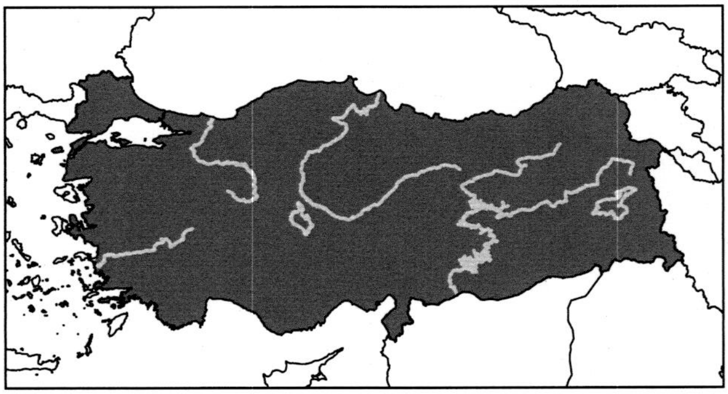

Breeding. Found mainly around habitation, in towns, villages and isolated buildings, usually in cultivated areas, but also in woodland. Generally nests in small colonies, e.g. typically 10 pairs nesting in crevices in a building, but occasionally as solitary pairs; larger colonies are usually in plantations (sometimes poplars and *Eucalyptus* near farms), where domed nests are built in the open, on branches, up to 20m above ground. In addition, nests in structures, such as motorway bridges and in holes in walls, trees and nest boxes, in old nests of Western Rock Nuthatch *Sitta neumayer* and Common House Martin *Delichon urbicum*, in burrows of European Bee-eater *Merops apiaster*, and (commonly) in the base of large stick nests, particularly those of White Stork *Ciconia ciconia*, with up to 15 pairs in a single stork nest. Nests in the open in trees, or in the bases of storks' nests, are rounded and domed, with a side entrance, enclosing an untidy cup of straw, grass, plant stems and debris, e.g. string, wool and paper, lined mainly with feathers and wool. Those in holes are usually loosely domed or are merely a cup. In a study outside Ankara (**Inner Anatolia**) in 1976–85, most laid five eggs (mean clutch 4.7), and mean number of fledged young was 4.5 (Kiziroğlu *et al.* 1987). In **Aegean**, two nests had four and five well-incubated eggs, and another almost fledged young, all at Izmir, on 21 May 1951 (McNeile), indicating egg laying in early May and late April; several nests had fairly large young in a building near Karine Gölü on 9 May 2003, indicating egg laying in mid April (PC); in a colony in an olive grove near Akköy on 1 June 2004, all nests were empty but ready for eggs (M. Demain), possibly a second brood for the year. In **Marmara**, an adult, feeding recently fledged young at Karacabey on 20 June 1999, indicates egg laying about 20 May (PC); recently fledged young were at Bebek on 17 June 1973 (R. F. Porter). In **Mediterranean**, a nest had five eggs in a half-built house at Göksu Delta on 5 May 2004, and recently fledged young were at nearby Taşucu on 6 May 2004, the latter indicating egg laying in early April (PC). In another building at Göksu, a nest had six eggs on 6 May 2005 (J. Whitaker) and, on 11 May 2005, most nests had young but one still held four eggs (J. Hodson). Adults were copulating near Adana on 12 June 2006, presumably for second broods (PC). In **Southeast Anatolia**, fledged young were observed near Ceylanpınar on 6 May 1970 (R. F. Porter), indicating egg laying early April. At Ispir (**Black Sea Coastlands**), a nest had young on 12 June 2004 (PC). There are no precise data from **East Anatolia**, but active colonies recorded in June. Elsewhere, breeding is very variable, but usually begins April–May, and there are 2–3 broods per season.

SPANISH SPARROW — *Passer hispaniolensis*
Söğüt Serçesi

(Du: Spaanse Mus; Fr: Moineau espagnol; Ger: Weidensperling; Swe: Spansk sparv)

Subspecies. Birds in Turkey are largely intermediate between nominate *hispaniolensis* and *transcaspicus*. Roselaar (1995) considered that smaller populations in western Turkey (Thrace and most of the rest of Marmara south through Aegean and west as far as Beyşehir (Mediterranean) and Akşehir Gölü (Inner Anatolia) were best included in nominate *hispaniolensis*, with larger populations in Göksu and Çukurova deltas north and east as far as the Upper Murat Valley and around Başkale assigned to *transcaspicus*. However, Kumerloeve (1970a) considered those as far east as Ceylanpınar (Southeast Anatolia) to be *hispaniolensis*, whilst prior to this, Vaurie (1959a) had considered all Near Eastern populations to be unassignable to subspecies.

Specimen material from western Mediterranean, the Black Sea Coastlands and most of Inner Anatolia was unavailable to Roselaar but personal observations suggest that nominate *hispaniolensis* may be that present in the Black Sea Coastlands with *transcaspicus* occurring through much of Inner Anatolia, at least to the south and east of Tuz Gölü, although additional specimen material is required for confirmation. However, we could consider such points to be, in any case, moot: as noted by Summers-Smith (1988) *transcaspicus* is only diagnosable in fresh plumage when the female, in particular, is paler and less chestnut than the nominate; in our opinion such levels of geographical variation do not require naming.

Status and distribution. National conservation status: Least Concern (population apparently decreasing). Mainly a widespread and common summer migrant (although locally resident in some areas), breeding throughout most of the country, though generally scarce, occasionally locally common in the eastern third of Turkey: e.g., there was a colony of 300 pairs at Siirt (East Anatolia) on 2 April 1964 (Kumerloeve 1967a) and colonies estimated as involving some thousands of pairs were at Ceylanpınar in recent years. Most common in Marmara (including Thrace), Aegean, western Black Sea Coastlands and Inner Anatolia. Slightly more local in Mediterranean and Southeast Anatolia, but there is a very large colony at Karkamis, on the Euphrates at the border with Syria. Colonies may hold several hundred pairs in one small area, and exceptionally a colony containing approximately 10,000 birds was south of Tarsus, Içel province (Mediterranean) in May 1965 (Kumerloeve 1970a), with

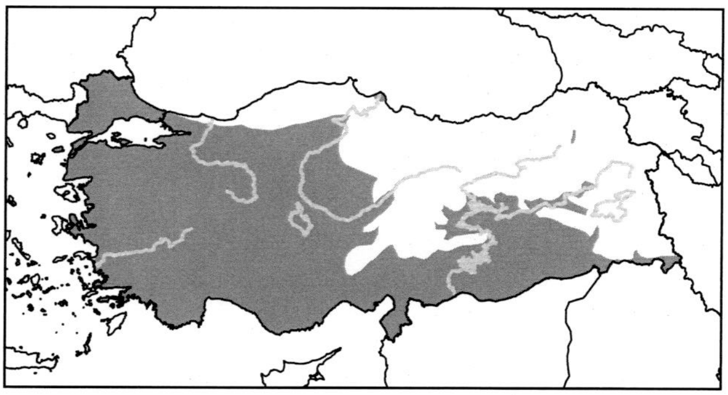

'many thousands' nesting in the same area in 1987 (van der Have *et al.* 1988). Generally in lowland and agricultural areas up to 1,500–2,000m, principally nesting in trees; usually shuns buildings as nest sites, unlike House Sparrow *P. domesticus*.

Slightly more widespread on passage, when highly gregarious: flocks may number several thousand, e.g. the majority of a roost of 50,000–100,000 at Uluabat Gölü (Marmara) in late September 1972 (OST 1975) were considered to be this species, and Robel & Bräuning (1992) noted a flock of at least 15,000 at the same site on 22 September 1990. Spring passage occurs during March and April, occasionally into May, and in autumn it lasts from August to early November. Small numbers irregularly overwinter usually in southern and western Turkey, but also more periodically in Inner Anatolia. Very occasionally recorded in larger numbers at this season.

Breeding. Cultivation, mainly in lowlands and valleys, often near habitation, but also in more remote areas. Nests are mainly in trees: in copses, plantations, lines of trees and solitary trees; closely linked with *Eucalyptus* in Çukurova (Mediterranean); other frequently recorded trees include poplar, willow, pine, *Acacia*, plane and cypress. Commonly nests in the base of stick nests of larger species, particularly White Stork *Ciconia ciconia*, but also Grey Heron *Ardea cinerea* and Long-legged Buzzard *Buteo rufinus*; also in tall bushes, and McNeile mentioned a colony in a strip of tall and 'murderously prickly' brambles alongside a road near Izmir (Aegean) in May 1950; reported nesting in solitary buildings in farmland (Hustings & van Dijk 1994) and in the nest of a Common House Martin *Delichon urbicum* in Thrace; and a cliff-nesting colony is reported at Silivri (Marmara). Colonies may comprise only a few pairs, most commonly 30–150 pairs, but 800 old nests were counted at a colony near Marmara Gölü (Aegean) on 6 January 1970, and there were colonies of many thousands in Çukurova (van der Have *et al.* 1988); 37 nests were built into the bases of 17 Grey Heron nests at Manyas Gölü (Marmara) on 5 June 1966; and up to 50 pairs were recorded breeding in the base of a single White Stork's nest 'so close together that they filled almost every interslice in the great pile of sticks' (Selous 1900). Mixed colonies of House *Passer domesticus* and Spanish Sparrows also nest in those of White Stork. Nests in trees are usually in forks and, in larger colonies, several may touch forming a contiguous mass. The nest is a bulky domed structure, often untidy, with a side entrance; of grass, often relatively long pieces, and often green, in the early stages of nesting, with a thick lining, mainly of feathers. Eggs: C/4–7 (see below). Young: no data. McNeile examined many nests near Izmir (**Aegean**): on 11 May 1950, most were half built, but a few had full clutches of 5–6 eggs, some considerably incubated (indicating earliest egg laying in late April); in another colony, near Izmir, nests had incomplete clutches, complete clutches and small young, on 17 May 1951 (earliest egg laying also late April), and most clutches were 3–7 eggs on 24 May 1951 (two clutches of four and five eggs were only slightly incubated); at this colony on 31 May 1954 most full clutches were five eggs but some nests were still empty. More recently, nests in a colony near Akköy had both eggs and young on 20 May 1999 (J. Hollinrake). Selous (1900) also examined many nests: they were lined, but still empty, near Acıgöl (**Mediterranean**) on 11 May 1899 but, in a colony of over 100 nests, near Izmir they contained mainly six eggs on 15 May 1899. At Amik Gölü, of eight nests examined on 1 May 1964, one was empty and seven contained eggs, some incomplete clutches (Warncke 1964–65). In **Marmara**, adults were observed carrying food at four sites, on 18–26 June 1973, and, of seven colonies, three were in poplars (R. F. Porter). In **Inner Anatolia**, a flock of *c.*50 adults was observed near Cihanbeyli on 23 April 2004, indicating that nesting had not yet commenced (PC); many nests were empty but new, and most only partially built, at Sarayonu on 9 May 1964 (Warncke 1964–65). Colonies in Turkey are active from early April, with nest building observed until June, and active colonies reported as late as August. The species is clearly double-brooded, as elsewhere in its range.

DEAD SEA SPARROW — *Passer moabiticus*
Küçük Serçe

(Du: Moabmus; Fr: Moineau de la Mer Morte; Ger: Moabsperling; Swe: Tamarisksparv)

Subspecies. Presumably *mesopotamicus* throughout breeding range in Turkey but only populations at Birecik and Ceylanpınar have been examined (Roselaar 1995).

Roselaar (*op. cit.*) chose to recognise the subspecies *mesopotamicus,* although usually only nominate *moabiticus* and the rather yellow race *yatii* of southeast Iran and southwest Afghanistan are acknowledged, because (as he claimed) *mesopotamicus* consistently exhibits a combination of the size of *yatii* and the plumage colour of *moabiticus*. Summers-Smith (1988) preferred not to acknowledge *mesopotamicus*, but Roselaar probably had the advantage of having examined more specimen material. *P. m. yatii* does, indeed, appear a rather distinctive form, but all those we have seen in the Natural History Museum (Tring) from within the purported range of *mesopotamicus*, in southwest Iran and Iraq, appear indistinguishable in morphology from nominate *moabiticus*, as do those we have observed in the field in southeast Turkey. However, C. S. Roselaar (*in litt.* 2004) reported that Zarudny collected some within the range of *mesopotamicus* showing limited yellow in the plumage; however, neither Roselaar nor ourselves have examined such birds. Indeed, Kumerloeve (1970c) considered a male taken at Ceylanpınar to be indistinguishable from nominate *moabiticus*. Variation in size appears insignificant in the context of meaningful geographic variation, although C. S. Roselaar (*in litt.* 2004) pointed out that it is not necessarily clinal. We agree with Ticehurst *et al.* (1923) that *mesopotamicus* should be considered a junior synonym of the nominate and, given this, the distinctive morphology and disjunct range of *yatii* (separated by over 1,200km from the nearest population of *moabiticus*) demands renewed taxonomic attention (Kirwan 2004b).

Status and distribution. National conservation status: Vulnerable (population apparently decreasing). A locally common summer visitor, or perhaps resident in small numbers (see discussion below) at a number of localities in the lowlands of Mediterranean and Southeast Anatolia. Also reported (formerly?) from one locality in East Anatolia. Typically breeds in scrub by water, usually by rivers or canals, between April and July in loose colonies of up to 100 birds in cultivated areas. The Turkish population appears to rely on live trees for nesting, in contrast to some other populations (Summers-Smith 1988).

A breeding colony, now expanded to a series of colonies along the River Euphrates, was found at Birecik in 1964 (Kumerloeve 1965) with *c.*100 nests located the following year (Kumerloeve 1966d). Another colony was found at Gâvur Gölü, south of Karamanmaraş, in 1967 (Kumerloeve 1969b). A further population was located at Ceylanpınar, on the banks of the Al Khabour, Şanlıurfa province (Southeast Anatolia) in 1968 (Kumerloeve 1969b). Further discoveries followed from the Mediterranean region in 1968 and 1972 when colonies were located on the River Ceyhan, just south of Adana, in Çukurova (Cramp 1971) and in the Göksu Delta, where birds were also present in 1970, before breeding was proven in 1972 (OST 1975). In 1974 a colony was discovered, at Kale, Malatya province (East Anatolia) on the Euphrates, the only report from this region (OST 1978). However only those colonies at Birecik and on the Ceyhan are currently definitely known to exist. The relatively small population in the Göksu Delta has been extirpated. Due to the difficulties of access to parts of Southeast Anatolia there are no reports from any other areas, although the colony at Ceylanpınar is known to have increased between 1968 and 1977 (Kumerloeve 1978).

Summers-Smith (1988) suggested that the Turkish population represented an expansion of the species' breeding range from central Iraq and southwest Iran, a view supported by Roselaar (1995). However, in a subsequent note (based on new evidence from Iraq), he surmised, probably correctly, that the lack of Turkish records before the 1960s was in part due more to an observational lacuna (Summers-Smith 1992). There was a lack of ornithological exploration in much of eastern Turkey, particularly the southeastern parts before the advent of modern recording (1966). However some recent westwards expansion does appear to have taken place, as illustrated by the colonisation of Cyprus (Flint & Stewart 1992). As mentioned by Roselaar (1995) it is unclear, without knowledge of the subspecific identity of populations in the Mediterranean coast deltas, whether the current Turkish range is a result of a natural expansion westwards from the Euphrates through crossing the watershed northwest of Gaziantep.

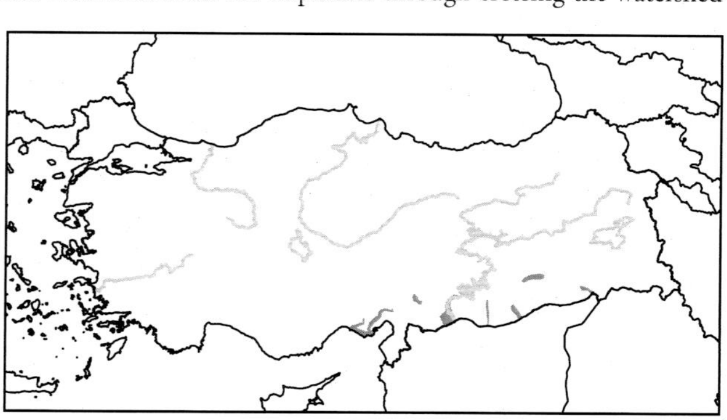

Summers-Smith (1988, 1990) also discussed the wintering areas of this species. The small Cyprus population is apparently migratory and that in Israel undertakes at least partial dispersal. In Turkey there is one record outside the breeding season, at Tarsus, Icel province (Mediterranean) (OST 1972) and there are several recent early winter records (late November and December) from the Birecik area, but none from later in the season; the first thereafter being on 6 April. Those in Çukurova appear to arrive in early March (Kivit *et al.* 1990) but in other years not until April, however winter departure is imperfectly known, although birds are present until at least late September. Recent records of large numbers wintering in eastern Arabia (Hirschfeld & Symens 1992) may indicate where the majority of the Turkish and/or Iraqi populations winter. The species' distribution and movements remain enigmatic. Further research in Turkey, and elsewhere throughout its disjunct range, is needed to clarify our knowledge.

Breeding. Mainly in riverine scrub and trees, occasionally flooded, typically nesting in willow, poplar, *Tamarix* and *Eucalyptus*, but also in pines and olives. Colonies usually have 10–20 pairs, but up to 100 birds recorded (Kirwan & Martins 2000b). Nests in a tree (or bush), mainly 4–5m above ground (n=52), but up to 12m and sometimes as low as 2m (n = 4); the nest is a rounded structure, almost the size of a football, with an entrance hole at the top; mainly of twigs, and thickly lined with feathers, hair, down and various soft material and debris. An old nest was used by a pair of Pallid Scops Owls *Otus brucei* near Halfeti (Southeast Anatolia) in 1987 (van den Berg *et al.* 1988). Eggs: C/5 (3), C/6 (4), C/7 (1). Young: B/5 (2), B/6 (1). All data is from Birecik (**Southeast Anatolia**) unless otherwise stated: adults had not yet arrived 8–19 April 1988 (C. & C. Bradshaw); there were several records of nest building early May, and one on 9–12 June 1996 (S. Barış), indicating double-brooding; a male took (or stole?) nesting material from a nest to his own on 23 May 2004; a nest had five eggs on 18 May 1993, and subsequently five large young on 6 June 1993; contents of eight nests examined on 3 June 1993 were five eggs (two nests), six eggs (three), seven eggs (one), five newly hatched young (one), and six newly hatched young (one), the last two records indicating egg laying in mid May; of 14 nests examined on 11 May 2004, 11 were empty, but freshly lined, and ready for eggs, two had incomplete clutches of one and two eggs, and one nest had six eggs (PC). Adults are present at Birecik until at least August, with 50 singing males noted on 3 July 1970, and there are records of fledged juveniles being fed in July until mid August, a clear indication of double-brooding, which is usual elsewhere. Many nests with attendant males were on islands in the Euphrates near Karkamis (Southeast Anatolia) on 3 May 2004 (GMK). In **Mediterranean** three nests were in flooded *Tamarix* at Göksu Delta on 29 April 1972 (no other details), and four pairs were on an island in the Ceyhan River on 2–4 July 1970.

EURASIAN TREE SPARROW *Passer montanus*
Ağaç Serçesi

(Du: Ringmus; Fr: Moineau friquet; Ger: Feldsperling; Swe: Pilfink)

Subspecies. *P. m. transcaucasicus* (northern and eastern Turkey west as far as Erzurum and Elazığ). Elsewhere in Turkey situation is unclear: birds from Ankara, Inner Anatolia and perhaps elsewhere through central and southern Turkey as well as parts of the Black Sea Coastlands most closely resemble nominate *montanus*. Populations in northwest Turkey from Istanbul (but presumably throughout Marmara and the western Black Sea Coastlands) were treated as *transcaucasicus* by Keve (1978). Given the probability that Inner Anatolian populations examined by Roselaar are apparently *montanus*, this seems unlikely and the latter author treats these birds as intermediates between *montanus* and *transcaucasicus*. In contrast, Summers-Smith (1988) chose to consider all Turkish populations as nominate *montanus*. As can be seen, variation between *montanus* and *transcaucasicus* is clinal, and further east the latter subspecies grades into *dilutus*. In cases such as these, we prefer not to recognise such geographical variation at nomenclatural level.

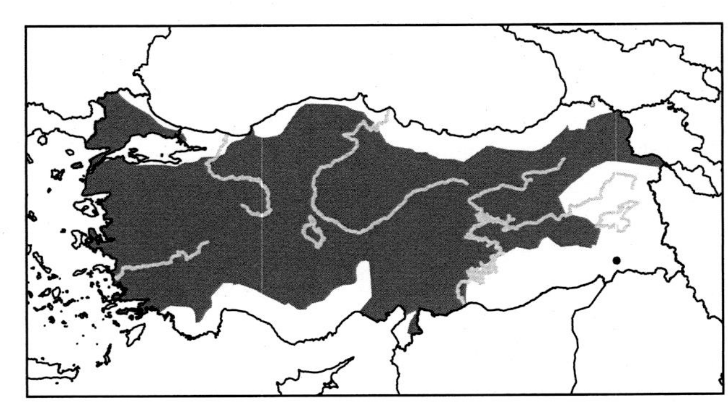

Status and distribution. National conservation status: Data Deficient (population trend unknown). A not uncommon resident or partial migrant, breeding in small or moderate numbers rather locally in Marmara (including Thrace), throughout much of the Black Sea Coastlands, northern and eastern parts of Aegean, much of Inner Anatolia and adjacent parts of Mediterranean (and once recently, in September, north of Kaş, Antalya province; M. P. Frankis) and East Anatolia. Recently recorded in Southeast Anatolia (e.g. around Birecik), and the southernmost (at Uludere in the Hakkari region) and easternmost (in the Aras Valley at Iğdır) parts of East Anatolia during the breeding season. Breeds in cultivated areas around habitation, including buildings, to at least 1,750m, more occasionally in montane areas, but still often near settlements. As elsewhere in Asia, will breed in urban areas, e.g. Ankara (Kiziroğlu *et al.* 1989).

More widespread in winter (though less prone to flocking than other sparrows), when recorded throughout the western two-thirds of Turkey.

Breeding. In cultivated areas, often near towns and villages, including parks and gardens, but also in woodland. Breeds solitarily or in small colonies of a few pairs, but up to 20 pairs were in a wall at Nevşehir (Inner Anatolia) on 20 May 1975. Nests in tree holes (including woodpecker holes), buildings and walls, and, occasionally, in old bee-eater tunnels. A pair was observed taking nesting material to an old nest of Western Rock Nuthatch *Sitta neumayer* at Göreme (Inner Anatolia), on 20 May 2004 (PC). Outside Ankara (Inner Anatolia), in 1976–85 the mean population was 11.4 pairs, 1.1 breeding pairs/ha, mean clutch 4.9, and a mean 4.6 young fledged per pair (Kiziroğlu *et al.* 1987). Builds a domed structure or a loose and untidy cup; of plant stalks and grass, lined mainly with feathers. In **Inner Anatolia**, pairs nesting around Ankara were established by April (Wadley 1951), and family parties were observed in early August 1968; an adult entered a hole in a willow near Eber Gölü on 21 April 1967 (R. F. Porter); and active colonies were at various locations in late May. In **Mediterranean**, copulation was observed in the Taurus on 3 June 1998 (I. A. Green), several adults were incubating in a roof, and copulation and nest building were noted at Demirkazık on 13–14 May 2004 (PC). In **Black Sea Coastlands**, small colonies were active at Ilgaz, Tosya and Ispir in early June; an adult was observed carrying food near Gümüşhane on 13 June 1975; and an adult flew out of a woodpecker's hole at Ispir on 12 June 2004 (PC). The records indicate double-brooding in Turkey, as elsewhere in the range.

PALE ROCKFINCH *Carpospiza brachydactyla*
Boz Serçe

(Du: Bleke Rotsmus; Fr: Moineau pâle; Ger: Fahlsperling; Swe: Blek stensparv)

Subspecies and taxonomy. Monotypic. The species' relationships are enigmatic and beg molecular research. As noted by Morony *et al.* (1975), Roselaar (1995, and *in* Cramp & Perrins 1994) and Fry & Keith (2004), its breeding biology and morphology suggest cardueline affinities (an appearance that is readily appreciated in life, although its skull structure is not cardueline), and the species' voice, behaviour and morphology do not suggest a particularly close relationship with other *Petronia* (with which it has been subsumed) or the recently resurrected *Gymnoris*. Indeed, this highly unusual species appears, on the basis of jaw morphology, tongue apparatus and skull structure, to represent a modern-day relict of an extremely early radiation of Passeridae, Emberizidae or Fringillidae. Aspects of breeding biology (colour and markings of eggs, and relatively narrow mouth and gape flanges of nestlings) more closely resemble Fringillidae (PC). Fry & Keith (2004) elected to transfer the species to the carduelines, but we prefer to await the results of DNA studies before following their course, especially given that Bock (2004) found skeletal evidence for its retention within the Passeridae.

Status and distribution. National conservation status: Data Deficient (population trend unknown). A local and usually rather uncommon summer migrant over much of Southeast and East Anatolia, as well as immediately adjacent parts of Mediterranean, although there are relatively few records of proven breeding. Sometimes considerably more widespread and common, occasionally recorded in large numbers, up to 60 at individual localities, in these regions. Exceptionally even recorded as far north as the southern part of the eastern Black Sea Coastlands, e.g. in 1986 and 2001. The years 2000 and 2001 were exceptional years during which the species was unusually widespread and numerous in the eastern third of Turkey (and in Armenia; Ananian *et al.* 2002). The dotted lines on the map depict the species' widest-known range, in years such as 2000 and 2001.

Usually a late migrant, mostly arriving in early/mid May (3rd), exceptionally as early as the first half of April (earliest 10th, at the Belen Pass), occasionally in the last week of April. Apparently departing by the end of

August (e.g. autumn migration through United Arab Emirates is in September and October; Richardson 1990) but information on autumn departure is based on few records and is very poorly understood. Latest record, 13 September (H. Schmaljohann *et al.*). Recorded as far west as Çukurova (Mediterranean) on spring migration. The breeding range apparently varies with seasonal conditions, and/or reflects apparently irruptive and semi-nomadic movements (as first noted by Kumerloeve 1975c). Found in both open grassland and semi-steppe areas, and on partially scrub-covered slopes, but more usually on rocky slopes, occasionally in vineyards and other crops. A number of core areas are occupied in most, if not all years, e.g. orchards and rocky hillsides around Yesilce, near Gaziantep (Southeast Anatolia) have been regularly occupied (albeit in variable numbers) during the last decade, and this area was first mentioned in the literature from observations in 1935 by Bird (1937). The status and distribution of this species in Turkey (and throughout its entire range) require clarification.

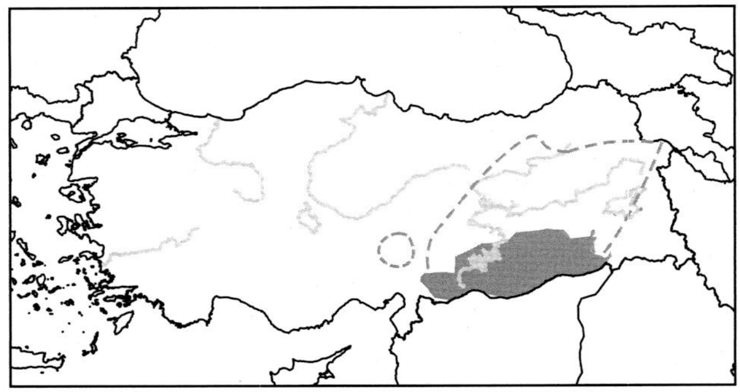

Breeding. Inhabits arid rocky slopes or flat stony plains in steppe or semi-desert, patchily covered with thorny scrub, isolated trees, or other scanty vegetation, mainly at 500–1,200m in the southeast, but to *c.*2,000m in the Van and Doğubayazıt area; in a dry stony valley with scattered grasses, herbaceous vegetation and *Astralegus* bushes near Erzurum, East Anatolia (P. Edwards); in open grassland and on partially scrub-covered rocky slopes or in vineyards and other crops (Beaman 1986), with 11 singing males in 1km^2 near Gaziantep (Southeast Anatolia) on 25 May 1971. Males sing persistently from prominent perches, commonly a tall plant stem or a rock. In **Southeast Anatolia**, territorial behaviour observed in the Idil area on 15 June 1983; many males were singing, and pair formation was beginning, at Çinar-Göksu Dam, on 5 May 2004 (GMK); adults were carrying food at Durnalık on 2 June 1998 (T. Quittenden) and at Işıklı on 14 June 1996 (GMK); at least two adults were feeding young at Işıklı on 28 June 2003 (C. Batty); two pairs were nest building (carrying grass stems) in undulating dry grassland with scattered low bushes, near Taluk, Sanlıurfa, on 1 May 2001, and another two pairs were nest building near Pınaryayla, Adıyaman, on 31 May 2001; an adult was carrying food at Cizre on 11 June 2001, and adults carried food to a nest, either on the ground or very low in a bush, on the low rocky slope of a narrow valley near Bozkır, Sanlıufra, on 10 June 2001 (G. Welch); fledged young (flying well) were seen at Cizre on 11 June 2001 (H. Welch). The species breeds in western Armenia, close to the Turkish border, and the information below (from Adamian & Klem 1999) should also be relevant: arrives on the breeding grounds mid–late May; pair formation by late May; copulation 4 June and 6 June; a female nest building on 6 June; nests low (mean 15.4cm above ground [14]) in low shrubs, particularly in thorny and prickly bushes, and in crevices and depressions in rocks and ruins; of 14 nests, ten were in *Atraphaxis spinosa*; on 2 June 1962 there were 20 singing males and 14 nests (in various stages of construction) in 2ha of suitable habitat; single-brooded; first eggs were laid in three nests near Jirvaz on 5 June (two) and 6 June (one); the nest is a cup of coarse stems, lined with leaves, wool and hair. The season in East Anatolia and Armenia is likely to be similar, but is much earlier in Lebanon, where three nests were in construction on 4 May 2003, 25cm above ground in *Tragacanth* vegetation (Ramadan-Jaradi *et al.* 2004). For other data from Lebanon see Knight & Beale (2004).

CHESTNUT-SHOULDERED SPARROW *Gymnoris xanthocollis*
Sarı Boğazlı Serçe

(Du: Indische Rotsmus; Fr: Moineau à gorge jaune; Ger: Gelbkehlsperling; Swe: Gulstrupig sparv)

Subspecies and taxonomy. The subspecies occurring in Turkey is surely *transfuga*, which is known from adjacent Iran and Iraq, although no specimen material from the country appears available to confirm this (Roselaar 1995). We have not attempted to examine the diagnosability of *transfuga*. Traditionally placed in *Petronia*, this species, along with three Afrotropical congeners – Yellow-spotted Petronia *P. pyrgita*, Yellow throated Petronia *P. superciliaris* and Bush Petronia *P. dentata* – has recently been reassigned to the resurrected genus, *Gymnoris*, on the basis of marked differences in plumage, structure, habitat and behaviour from Rock Sparrow *P. petronia* (Cramp & Perrins 1994, Roselaar 1995, Beaman & Madge 1998, Gill & Wright 2006).

Status and distribution. National conservation status: Near Threatened (population apparently increasing). Generally a scarce and apparently local summer visitor to Southeast Anatolia and parts of East Anatolia, though common in several areas, with one recent record from Mediterranean. Recorded from late April (27th) to mid July only, but may be presumed to remain longer. Discovered as a breeding bird in Turkey around the Halfeti and Birecik areas in 1977 (Beaman 1986), where the species has apparently steadily increased. More recently located at Cizre, Mardin, Batman and Sason (Southeast Anatolia), near Van (East Anatolia) and, west of the Euphrates watershed, at Yesilce, in 1996 (P. Doherty videotaped). Also found just north of Kırıkhan, Antakya province (Mediterranean) in May 2004 (A. Grieve), as well as in suitable habitat near Iliç, Erzincan province (East Anatolia), in July 2004 (Demirci & Mülayim in prep.). Inhabits most suitable areas of habitat within this range. Usually breeds in pistachio orchards or groves of oak *Quercus*, but is frequently unobtrusive and its undoubted recent spread, the result of the landscape-level agricultural development of Southeast Anatolia, is only now being documented. Further exploration of suitable breeding areas in Southeast Anatolia is required to clarify its status and distribution more fully.

Breeding. Mainly in pistachio orchards, where it nests in holes in trees, mainly old Syrian Woodpecker *Dendrocopos syriacus* holes (n=15), less than 2m above ground; perhaps partially dependent on the latter for provision of nest sites. The nest is an untidy accumulation of grass, leaves and various debris, lined with hair, grass, plant down and debris, e.g. wool and string; it may be merely a loose cup, but is sometimes domed, or partially domed, with an untidy tube of nest material forming an entrance spout, leading down to a cup (only one nest examined at Birecik was domed). Eggs: C/3 (female still building, so considered incomplete clutch), C/4 (1), C/5 (4). Young: B/4 (1), B/5 (1). Nest building was observed on 23 May 1990; 13 nests had eggs 5–14 June; a nest on 14 June 2001 had small young, and another three eggs and a newly hatched chick, indicating egg laying in late May (PC). Two adults were feeding fledged young on 7 July 1986, and immatures have been seen mid July. Double-brooded in some parts of the breeding range, but no evidence in Turkey, although males recorded singing at Birecik as late as 10 August.

ROCK SPARROW — *Petronia petronia*
Kaya Serçesi

(Du: Rotsmus; Fr: Moineau soulcie; Ger: Steinsperling; Swe: Stensparv)

Subspecies. Roselaar (1995) regarded the situation as being somewhat complex, considering birds in Aegean and southern Marmara as nominate *petronia*, those throughout much of northern, central and eastern Turkey as *exigua* and populations in Southeast Anatolia as being referable to *puteicola*. The subspecific identity of any population in Thrace and elsewhere in Marmara is unclear. There appear to be no records for Thrace, although it must surely occur there (records are available from adjacent Greece). However, as obliquely noted by Clement *et al.* (1993), there is very little readily distinguishable variation between most forms of Rock Sparrow, and several available names, including those mentioned above, may be better regarded as synonyms.

Status and distribution. National conservation status: Least Concern (population apparently stable). Generally a widespread and fairly common resident breeder, or partial migrant, in rocky and upland areas (including semi-steppe) at moderate or high altitudes (to 2,500m, more locally, 3,000m) throughout the eastern two-thirds of the country. More local further west and apparently absent from Thrace (see above). Breeds from late April, often in loose colonies; pairs frequently have young from early June. Occasionally (especially in semi-steppe areas) nests in small towns and villages, utilising similar niches to that typically occupied by House Sparrow *Passer domesticus* in such circumstances.

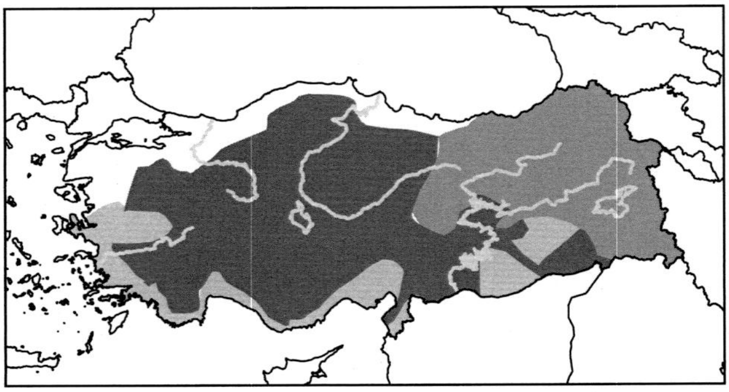

Descends to lower altitudes in winter (though still recorded in large groups above 1,000m at this season), often reaching the coast and observed in mixed flocks of up to 500 individuals with Crested Lark *Galerida cristata* (Uhlig & Uhlig 1999a). Typically more common and widespread in the western two-thirds of Turkey at this season, with most records from the west, the south and Inner Anatolia.

Breeding. In open montane areas with cliffs, gorges and boulders; in rocky areas within cultivation, and on plains with scattered trees. Forms loose colonies of up to 50 pairs, but more usually of only a few pairs, and sometimes solitary. Nests mainly in crevices and holes in cliffs, but also in holes in banks, riverbanks, road cuttings, trees, buildings, walls, wells and bridges; also uses nests of other species, including those of Western Rock Nuthatch *Sitta neumayer* and Common House Martin *Delichon urbicum*, tree holes of Syrian Woodpecker *Dendrocopus syriacus*, and breeding tunnels of European Bee-eater *Merops apiaster*. Builds a bulky nest of grass, lined with feathers (Selous 1900); in other regions, and probably in Turkey, the nest is domed but occasionally only a loose cup if the cavity is small; of plant stalks, grass and roots, lined with feathers, down and hair. Eggs: C/5–7 (3). Young: B/7 (1). In **Southeast Anatolia**, two nests had five and six eggs near Birecik on 2 May 1964 (Warncke 1964–65), and a pair fed fledged young on 8 May 1989, indicating egg laying early April; six other active nests noted May–June. In **Mediterranean**, four nests were empty, but ready for eggs, on 9–10 May 1899, and a nest had newly hatched young on 10 May 1899 (Selous 1900), indicating egg laying *c.*22 April; there were other active nests as late as 23 July 1971. In **Inner Anatolia**, a nest had seven eggs on 9 June 1977 (Pforr & Limbrunner 1982), three nests had young during 17 June–22 July; and one had feathered young on 24 May 1998 (T. Quittenden), indicating laying in late April; a pair was carrying nesting material near Inçesu on 1 June 1975, display was seen near Çankırı on 7 June 1975, and there were nests at various sites on 17–19 June 1975 (R. F. Porter). In **Black Sea Coastlands**, there were active nests at Gerede in mid-May 1967 and at Bayburt on 15 July 1975; an adult was carrying food and alarm-calling near Yeniçağa Gölü on 21 August 1985 (Kılıç & Kasparek 1987). In **East Anatolia**, nest building was seen near Erzurum on 22 June 2004 (PC); ten active nests noted from 1 June, including one near Bitlis as late as 18 July 1974; an adult carried food to a nest in an old building near Erçek Gölü on 26 June 2004 (PC), and there were two family parties near Van on 2 August 1986 (J. McLoughlin). Clearly double-brooded in Turkey, as elsewhere in the region.

WHITE-WINGED SNOWFINCH — *Montifringilla nivalis*
Kar Serçesi

(Du: Sneeuwvink; Fr: Niverolle alpine; Ger: Schneefink; Swe: Snöfink)

Subspecies. The form *leucura* occurs throughout and is endemic to Turkey (Roselaar 1995), though it is sometimes regarded as a synonym of *alpicola* (which is restricted to the Caucasus south to southwest Iran by Roselaar, but considered to include all other named forms east to Central Asia and Mongolia by other authors). The nominate reaches east from southwestern Europe to the Balkans. Geographical variation within White-winged Snowfinch demands further study, but *leucura* does appear to form a link between the nominate and *alpicola*, and the variation expressed between these forms is perhaps best viewed as clinal. We have been unable to examine specimens from Turkey and whilst acknowledging the clear differences between *alpicola* and nominate *nivalis*, we are unsure as how best to treat *leucura*, which is obviously closer to the former, and may (as often suggested) be really a synonym of *alpicola*. Watson's name *fahrettini*, applied to birds from the central Taurus, is certainly invalid, being a synonym of *leucura*.

Status and distribution. National conservation status: Least Concern (population apparently stable). A widespread and typically fairly common resident breeder in mountains in East Anatolia (e.g. throughout the Hakkari Dağları, Kockıran Dağları and Yüksekova province), the eastern part the Black Sea Coastlands (including much of the Doğu Karadeniz Dağları and Yalnızçam Dağları), the extreme northwest part of Southeast Anatolia and westwards through the Taurus (Mediterranean) and the southern fringe of Inner Anatolia to Sultan Dağları, Afyon province. Breeds as low as 950m in the west, though typically much higher, up to 4,500m, in the east, where principally recorded above 2,200m, but regularly to 1,500m. Post-breeding flocks may number up to at least 400 birds. Descends in winter, when widespread throughout extensive areas of Inner Anatolia, principally in steppe and semi-steppe areas, and occasionally in adjacent areas of other regions, e.g. there is a record of a flock in Bursa province in March 2006.

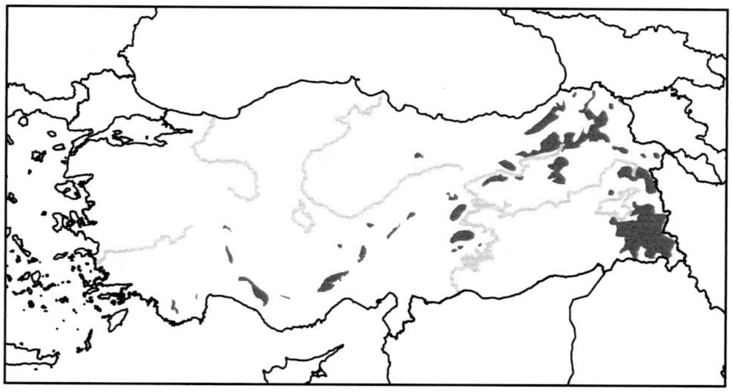

Breeding. In high mountainous areas: in pastures with rocky outcrops and boulders, and on scree, cliffs and in gorges, with short alpine vegetation, mainly at 2,000–2,700m. Nests in a hole or crevice in a cliff, up to 30m above ground, but usually much lower, or amongst rocks, under boulders, in mammal holes, and even in crevices in low roadside banks. Elsewhere, also nests in holes in buildings and man-made structures, but no records in Turkey. The nest is a substantial cup, mainly of plant stems and grass, with some fur and feathers, thickly lined, mainly with hair, but also fur and some feathers (PC). Eggs: C/3 (1); elsewhere, usual clutch 4–5. Young: B/2 (1), B/4 (1). In **East Anatolia**, nest building was seen at Erzurum on 13 May 1990; a nest contained three eggs, and another two large feathered young, in mammal burrows at 2,600m on Tendürek Dağı, north of Çaldıran, on 17 June 2004 (PC); several family parties of recently fledged young were at Balık Gölü on 5 June 2001, and four young were seen to fledge from a nest there on 9 June 2001, both records indicating laying in early May (PC); nests had young at Doğubayazıt on 4 June 2002, and at Erzurum on 24 June 1910 (McGregor 1917); adults were carrying food at Doğubayazıt on 21 July 1992 (G. Magnin), and at Erzurum on 22 June 2004 (PC). In **Black Sea Coastlands**, a nest had young near Sivrikaya in mid-June 1990 (A. Forsten). In **Inner Anatolia**, an adult was collecting nesting material near Sivas on 9 April 1969. A nest was partly constructed, and a female was apparently incubating at another nest, at Hodul Dağ on 11 May 2006 (N. Burke). At Demirkazık (**Mediterranean**), copulation was observed on 29 April 1970 (R. F. Porter), and display on 19 May 1996 (I. A. Green); two pairs were nest building, in crevices 6m and 18m up on cliffs, on 13 May 2004 (M. Farmer); there were two nests with large young being fed at the nest entrance on 12–13 June 1993, indicating egg laying in mid May (PC); four nests had young on 7 June 1998 (T. Quittenden), two nests had young on 20 June 1990 (A. Forsten), and there were several records of adults carrying food in the first half of June. Other records in Mediterranean include display at 2,200m on Sultandağlar on 7 May 1967 (R. F. Porter), young being fed on Karanfil Dağ in early July 1966 (Sutton & Gray 1972), adults feeding newly fledged young at Torosdağ on 23 July 1971, and a pair feeding young near Çamardı on 18 July 1986 (R. Webb). Elsewhere, sometimes double-brooded, but it is unclear whether any of the July records were second broods.

COMMON CHAFFINCH *Fringilla coelebs*
İspinoz

(Du: Vink; Fr: Pinson des arbres; Ger: Buchfink; Swe: Bofink)

Subspecies and taxonomy. Roselaar (1995) recognised a complex situation. Birds in Thrace were referred to nominate *coelebs*, with *caucasica* present throughout much of the rest of Marmara, the northern fringe of Inner and East Anatolia and east across the entire Black Sea Coastlands, possibly *schiebeli* in Aegean, and confusion surrounding which race to ascribe populations elsewhere in Turkey, partly due to a lack of good material.

Roselaar (*op. cit.*) tentatively suggested that those in southern parts of East Anatolia, Southeast Anatolia and the Hatay (Mediterranean) might be included in *syriaca*, but those in the Taurus and southern coastal regions of the Mediterranean region appeared intermediate between *schiebeli* and *syriaca*, and relevant specimen material was unavailable to categorise those present elsewhere in the western third of the country. Roselaar (*op. cit.*) conceded that the situation is problematic to resolve, given only slight differences between the named forms, and that these are strongly influenced by wear and bleaching. As noted by Cramp & Perrins (1994), the Common Chaffinch presents an extraordinary degree of variation across its Palearctic range, but three groups appear distinct: *coelebs* over continental Eurasia, *spodiogenys* in Africa and *canariensis* in the northeast Atlantic Islands. Their relationships and interactions, in southwest Europe, adjacent Africa and the Atlantic Islands, were the subjects of a review paper by Collinson (2001a) based on a genetic study by Marshall & Baker (1999). We consider that the study of geographical variation within Chaffinch is best and most meaningfully concentrated on the dynamics between these groups. Thus, like other authors (e.g. Clement *et al.* 1993), we prefer to include all those forms mentioned for Turkey by Roselaar within the nominate.

Status and distribution. National conservation status: Least Concern (population apparently stable). A widespread and common resident breeder or partial migrant over much of Turkey, although rather local in Inner, Southeast and East Anatolia, where it is typically restricted to cultivated areas around human settlements. Probably a partial or altitudinal migrant in all of these regions. Breeds in all well-vegetated habitats, though perhaps most common in coniferous and mixed woodland in upland regions at 1,100–2,000m, especially in the Black Sea Coastlands and Thrace.

More widespread on passage, large numbers normally moving through northern Turkey on a broad front during the first three weeks of October: e.g. 8,865 passed east over the Bosphorus between 24 September and 8 November 1966, and 727 passed Küçük Çamlıca, Istanbul, on 18 October 1980. However migration is recorded from late August (20th) to at least the end of the first week of November (8th) and probably even later in the south. Return movements in spring noted in March and April: e.g. 16 March–16 April at Erzurum, East Anatolia. Large numbers winter in Turkey, from early October to April, especially in Marmara, Aegean and Mediterranean, more occasionally in the Black Sea Coastlands, throughout all suitable habitat. Only occasionally recorded in very large concentrations at this season, e.g. there were many hundreds in Antalya (Mediterranean) during winter 1990/91.

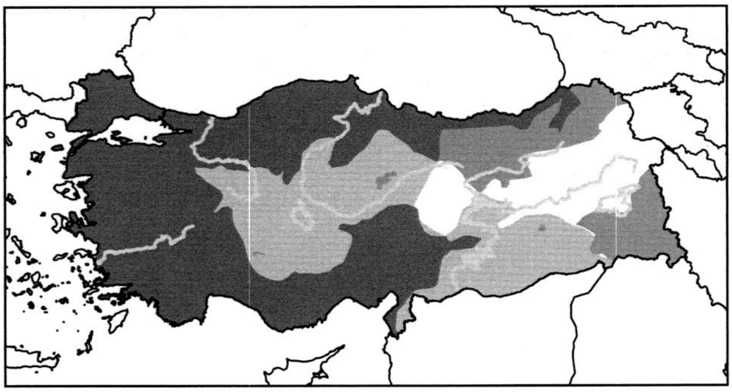

Breeding. In all types of woodland, including plantations, olive groves and orchards, and commonly in conifers in hilly country. In Marmara, the highest breeding density on Uludağ was in Black Pine *Pinus nigra* forest, at 1,100–1,400m (Jetz 1995); at Kocaçay Delta there were 1.3 pairs/ha in a 31ha plot in alluvial forest, with higher density in nearby deciduous forest (Ertan 1996). At Kızılırmak Delta (Black Sea Coastlands) there was a mean density of 15 territories/100ha in 1992 (Hustings & van Dijk 1994). The nest is a neat, deep cup; of moss mixed with lichen, grasses, roots and feathers, bound with spiders' webs, and lined with feathers, wool and plant down. Eggs: C/4 (2). Young: B/4 (1), B/5 (1). There are remarkably few breeding records. At Kızılcahamam (**Inner Anatolia**), a nest had four eggs on 26 April 1983 (Barış *et al.* 1984), and a recently fledged young, unable to fly properly, was observed on 25 May 1992 (PC). In **Black Sea Coastlands**, an adult was carrying food at Kızılırmak Delta on 1 June 1992 (Hustings & van Dijk 1994), and an adult was carrying food north of Boyabat on 9 June 1975 (R. F. Porter). In **Marmara**, an adult with fledged young was at Uvecık on 25 June 1973, and another was carrying nesting material near Paşaköy on 26 June 1973 (R. F. Porter). At Ağla (**Mediterranean**), a nest, 2m above ground, in a young pine, had four young, 3–4 days old, on 11 May 2003, indicating laying of the first egg on 23 April (PC). At Akseki, it was the commonest species in coniferous woodland in 1967, and a nest had four eggs on 28 April (R. F. Porter); and there were two half-built nests at 3m and 4m in spruce on 19 April 2004 (A. Dixon). A nest at Uzuncaburç had five young on 12 May 2005 and another near Mut had young on 13 May 2005 (J. Hodson). Also, a pair was feeding fledged young at Eğridir on 27 May 1999 (R. F. Porter).

BRAMBLING — *Fringilla montifringilla*
Dağ İspinozu

(Du: Keep; Fr: Pinson du Nord; Ger: Bergfink; Swe: Bergfink)

Subspecies. Monotypic.

Status and distribution. A generally uncommon, though occasionally locally numerous, winter visitor throughout the country, although there are generally rather few records from East Anatolia, mostly from around Malatya, where seemingly regular in winter. There are just three records from Southeast Anatolia: at Kabaklı Reservoir, Diyarbakır, in February 1993 (Biricik 1996), at Işıklı, near Gaziantep, in early April 1994 (C. Bradshaw), and 20 at Birecik, on 11 December 2004 (M. J. Grunwell), but we suspect that the species could occur in many parts of this region on an irregular basis. Most arrive between early October and late November in the Bosphorus, Thrace and the northeast, but recorded as early as 26 September in last-named region. Winters throughout the western two-thirds of Turkey, mostly in the north, west and Inner Anatolia. Irregularly recorded in large numbers, e.g. in the winters of 1966/67 and 1988/89 when flocks exceeding 1,000 individuals were seemingly not uncommon. Unusually large numbers were also observed in Inner Anatolia in early 1976. Recorded in all habitats up to 1,500m. Departure apparently commences in late March, by which time most have left central and southern areas, a few remaining until April (latest 21st). May rarely oversummer: there is one record in early June from Erzurum (East Anatolia), mentioned by McGregor (1917).

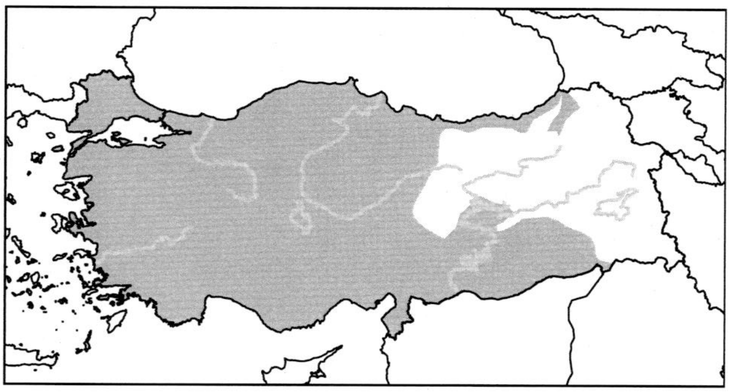

RED-FRONTED SERIN — *Serinus pusillus*
Kara İskete

(Du: Roodvoorhoofdkanarie; Fr: Serin à front d'or; Ger: Rotstirngirlitz; Swe: Rödpannad gulhämpling)

Subspecies: Monotypic.

Status and distribution. National conservation status: Least Concern (population apparently stable). A widespread, usually fairly common, resident breeder, and partial migrant, west to Söğüt Dağı, in the western Taurus. Perhaps scarcer in the eastern half of the Black Sea Coastlands, the northern fringe of Inner Anatolia, and throughout Southeast (where only noted in the extreme northwest and around Diyarbakir) and East Anatolia. An isolated population occurs on Uludağ (Marmara), which Jetz (1995) speculated was the result of a recent colonisation event. There are also at least two summer records in appropriate breeding habitat in the Ilgaz Dağları (central Black Sea Coastlands). Generally found at 1,700–2,800m, occasionally down to 1,000m, in low scrub, sometimes with scattered trees on rocky slopes, and has been recorded as high as 3,600m in summer. Breeding commences in late April, and flocks congregate from September.

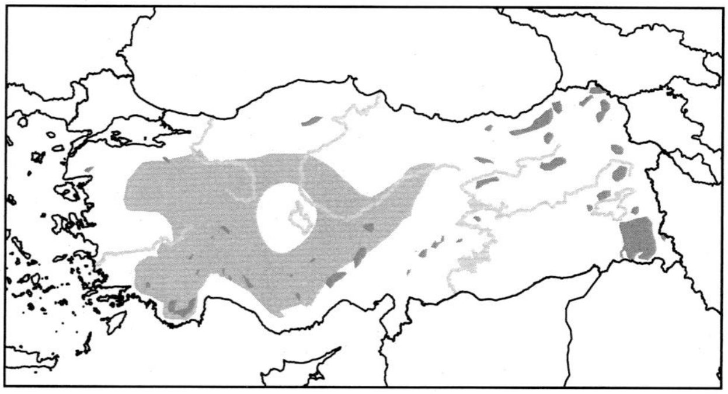

More widespread throughout its range, descending to lower altitudes (even to sea level in the Mediterranean region deltas) from late October to mid March (exceptionally the first week of April). Exceptionally reported from the Bosphorus, and the Bodrum and Izmir areas (Aegean) in winter, and at sea level in the central Black Sea Coastlands in early June (Hustings & van Dijk 1994). More extensively distributed, although still local in the steppe regions of Inner Anatolia at this season, e.g. very rare in the Ankara region.

Breeding. In high montane areas, on open rocky slopes with low scrub. Locally common in the subalpine zone of most Turkish mountains, mainly at 1,800–3,000m, where there is a mix of alpine meadows, sparse woodland and scrub, especially juniper (Green & Moorhouse 1995). Danford (1877–78) saw adults, both in flocks and in pairs, in firs interspersed with bushes and junipers, in the Aladağ (**Mediterranean**), during 19–28 April 1876; adults were observed nest building, and he considered that nesting was exclusively in junipers, which were numerous and large; eggs, almost ready for laying, were taken from birds shot; on 21 April 1876, a nest which had four eggs, taken from a juniper, 'tolerably high up', was brought to him; it was a cup of fine grasses, mixed with lichens and lined with dark feathers; one of the four eggs is held at Manchester Museum, but it is difficult to know how Danford could have possibly identified this egg (or the nest) confidently. Elsewhere in Mediterranean, it was the commonest bird on Karanfil Dağ from 2 July–16 August 1966, especially in ravines with *Berberis* and juniper; display flight by a male was observed on 14 July 1966, and ten juveniles were seen (or caught) on the summit on 12 August 1966, where none seen in July (Sutton & Gray 1972); most were in small roving groups, singing and displaying in areas of junipers at 2,150–2,300m on the northern slopes of Karanfil Dağ, on 9 May 1951 (Hollom 1955); a pair was observed nest building at Akseki on 28 April 1973; a flock of eight adults was feeding and displaying near Elmali on 11 April 1973 (B. Swann); nine adults and four juveniles were at Demirkazık on 25 August 1986 (J. McLoughlin). Three juveniles were in a gorge near Erkenek near Gölbaşı (**Southeast Anatolia**), on 26 July 1972. In **East Anatolia**, ten, obviously in pairs, were in a rocky valley, with relict juniper, 10km south of Görentaş on 1 June 1969; a pair, the female carrying nesting material, was in an area of stunted junipers at Nemrut Dağ on 5 June 2001 (T. Quittenden). In **Black Sea Coastlands**, a pair was seen with two fledged young at 2,600m at Yalnızçam Pass on 1 September 1973, and a male was singing at Maçka, south of Trabzon, on 16 June 1975. At Uludağ (**Marmara**), it was a common breeder, in 1993, around the treeline and in alpine scrub, gathering in flocks from July (Jetz 1995); a male and numerous probable females and juveniles were observed on 25 July 1967, two adults and a juvenile were there on 11 September 1990, and one was singing from a tree on 3 June 2002 (H. & J. Eriksen). Elsewhere, the nest is low in a dense bush or tree, sometimes high in a conifer, or in a crevice or hollow among rocks; a compact cup of grass, bark strips, stalks and moss, lined with plant down, hair, wool and feathers; usual clutch 3–5, and there are two broods per year.

EUROPEAN SERIN — *Serinus serinus*
Küçük İskete

(Du: Europese Kanarie; Fr: Serin cini; Ger: Girlitz; Swe: Gulhämpling)

Subspecies. Monotypic.

Status and distribution. National conservation status: Least Concern (population apparently stable). A generally widespread and fairly common resident breeder (in some upland areas probably a partial/altitudinal migrant) in much of Aegean, Marmara (including Thrace), Mediterranean (including Antakya province) and throughout the Black Sea Coastlands. Present locally on the northern, western and southern fringes of Inner Anatolia. Extremely local in western parts of Southeast Anatolia. Generally breeds in coniferous and mixed forests, and open woodland or tall scrub up to 2,000m.

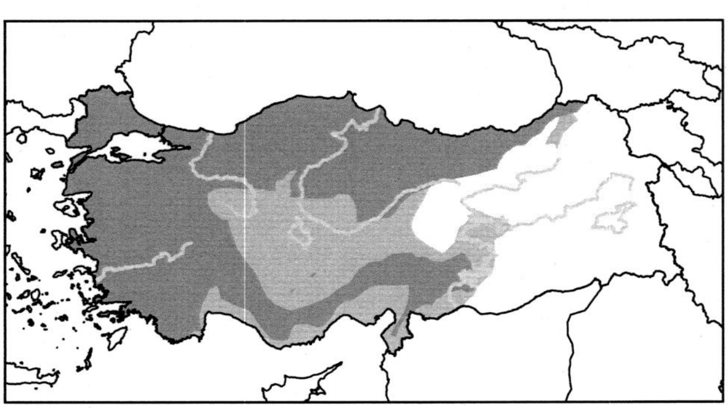

More widespread on passage: in autumn from mid August to early October, in spring from early March to late April (25th). Winters principally in the western third of Turkey and the south,perhaps as far east as western Southeast Anatolia; records from Birecik are available in November–April. Frequently recorded in coastal areas, including parts of the Black Sea Coastlands in winter, sometimes in large numbers (up to 500).

Breeding. In open woodland, orchards, parks, agriculture with trees and hedges, and scrub. Nests in a tree or bush; builds a neat cup of plant stems, grass and moss, lined with hair and plant down. Eggs: C/5 (1). Young: no data. In **Mediterranean**, in Çukurova 12 males were singing on 20 March 1987, and several nests were located (no details of contents) on 12–14 May 1987 (van der Have *et al.* 1988); a nest was built, but not yet lined, at 2.2m in a spruce at Akseki on 19 April 2004 (A. Dixon); a nest near Sögüt had five eggs on 7 May 2003, and was 1.6m above ground in a dense juniper on a hillside with scattered bushes (PC); fledged young were observed at Ağla on 30 April 1989, at Dalyan on 5 May 1996 (A. & S. Dijksen), near Nığde on 23 May 1970, and at Bolkar Dağları on 5 June 1999 (R. F. Porter), these records indicate egg laying in late March–early May. In **Aegean**, there were two adults and 2–3 juveniles at Ayvalık on 28 June 1984 (S. Harrap). In **Marmara**, at Küçük Çamlıca, there were fledged young in July 1966, and nest building was seen on 21 July 1966. The records above indicate double-brooding, as elsewhere in the range.

[SYRIAN SERIN *Serinus syriacus*]
Suriye İsketesi

(Du: Syrische Kanarie; Fr: Serin syriaque; Ger: Zederngirlitz; Swe: Grönhämpling)

Subspecies. Monotypic.

Conservation status: Near Threatened (BirdLife International 2000, 2004). Species not admitted to the Turkish list (Kirwan *et al.* 1999). Listed as a winter visitor to the east and the eastern Mediterranean region by Kiziroğlu (1989). No observations exist to substantiate this unlikely claim and the species is even rather poorly known in neighbouring Syria (Baumgart & Kasparek (1992).

[CITRIL FINCH *Serinus citrinella*]
Alp İsketesi

(Du: Citroenkanarie; Fr: Venturon montagnard; Ger: Zitronengirlitz; Swe: Citronsiska)

Subspecies and taxonomy. Should the species be proven to occur in Turkey, the nominate race will be involved. The subspecies *corsicanus*, which is endemic to Corsica, Sardinia, Capraia and Elba, in the western Mediterranean, is generally regarded as specifically distinct (initially by Thibault 1983), as it differs from typical *citrinella* in general coloration, size (wing and tail), habitat preference (lowlands) and, to some extent, genetic makeup. Others (e.g. Pasquet & Thibault 1997) have regarded the level of genetic variation within *citrinella sensu lato* as being only weak evidence to recognise more than one species.

Species not admitted to the Turkish list (Kirwan *et al.* 1999). Listed recently by Kasparek (1990a) as having been recorded reliably in Turkey. However, its inclusion on the Turkish list seems highly questionable (Kirwan *et al.* 1999); both Elwes & Buckley (1870) and Krüper & Hartlaub (1875) stated that Robson had observed the species in the Istanbul area but Kumerloeve (1961) either failed to observe such comments, or regarded them as being insufficiently precise to merit attention. Mathey-Dupraz (1922) reported that the 'serins were of the mountain species' and that they could be seen at the Bosphorus both on autumn passage and occasionally a flock would remain throughout the winter. The species was apparently relatively easy to obtain with lime-sticks. One was supposedly collected near Kadiköy (Istanbul) in September 1926 (Wahby 1930); although this record was subsequently questioned, no reasons for doing so were presented (Kumerloeve 1975). The specimen apparently no longer exists and its identity was apparently never verified. There is thus no record of this species from Turkey that may be adequately substantiated (the records of Mathey-Dupraz do not certainly eliminate *S. serinus*): the species cannot be placed on the Turkish list. Knox (1994), in stating the rationale for its removal from the

British and Irish list, commented on the very small number of records of vagrants of the species and advised the reassessment of other extralimital European records.

EUROPEAN GREENFINCH *Carduelis chloris*
Florya

(Du: Groenling; Fr: Verdier d'Europe; Ger: Grünling; Swe: Grönfink)

Subspecies and taxonomy. As with a number of carduelines, Roselaar (1995) was defeated in his efforts to categorise the forms occurring across much of the country. Only in Marmara and the northern Aegean, where populations are inseparable from *muehlei*, the Hatay (*chlorotica*) and the northeast (probably *bilkevitchi*) was Roselaar (*op. cit.*) able to make more or less certain diagnoses. Those in the Taurus and southern coastal Mediterranean appeared intermediate between *muehlei* and *chlorotica*, and the author speculated that *aurantiiventris* might occur in the southern Aegean region (no specimen material is available from this part of Turkey), but elsewhere in the country he was unable to speculate on the subspecies. Whether *muehlei* and *bilkevitchi* warrant recognition has been questioned in some literature, and we certainly regard them as invalid under the procedures adopted here.

Recent phylogenetic analyses of morphological characters and short mitochondrial DNA sequences suggest this species is not part of the *Carduelis* clade, but in fact is more closely related to *Pyrrhula*, leading to the proposal, already accepted by some (Sangster *et al.* 1999) to place this taxon in *Chloris* (Raikow 1978, Fehrer 1996), as it was in some 20th-century Russian literature, although some support for its preservation within *Carduelis* also exists (van den Elzen & Nemeschkal 1991).

Status and distribution. National conservation status: Least Concern (population apparently stable). A relatively widespread and fairly common resident breeder, locally a partial/altitudinal migrant, largely in the western half Turkey: in Marmara (including Thrace), Aegean, Mediterranean, and the entire Black Sea Coastlands. More local and largely occurs on the fringes of Inner Anatolia and the western half of Southeast Anatolia, although no records available from much of this region. Absent from most of East Anatolia, except from western areas, where local. Breeds in a wide variety of wooded habitats, up to *c.*2,000m in Marmara, the central Taurus (Mediterranean) and the eastern Pontics (Black Sea Coastlands).

More widespread on passage, particularly in autumn, when large numbers are recorded at the Bosphorus and on the Mediterranean coast between late September (25th) and early November, with most in the second half of October. Less obvious on spring migration in March and April. Winters largely in the western half of Turkey, especially in the Aegean and Mediterranean but also in Marmara (including Thrace), Black Sea Coastlands and Inner Anatolia in small numbers, usually forming mixed flocks with other Carduelines at this season.

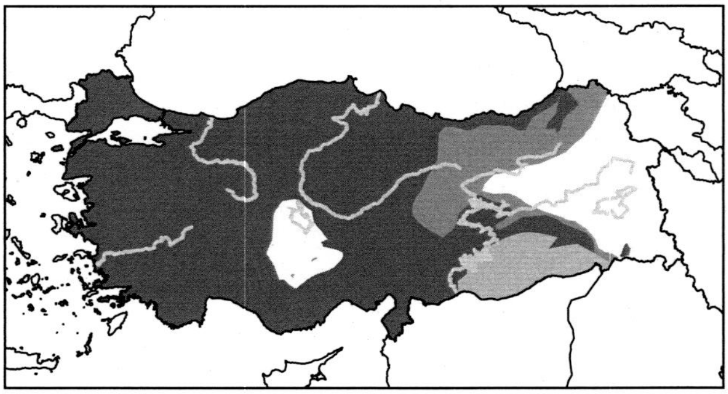

Breeding. Inhabits open areas with trees and bushes, including forest edge, orchards, olive groves, parks, gardens and rows of trees. Nests in a bush or tree, 2–4m above ground (n=3). Eggs: C/4 (2), C/6 (1). In **Mediterranean**, a nest had an incomplete clutch of one egg, another six eggs, and another three eggs and a newly hatched chick, all in a line of bushy conifers at Dalaman on 5 May 2003, the latter indicating egg laying about 19 April (PC). In **Aegean**, a nest had four eggs in bramble at Miletos on 23 May 1999 (J. Hollinrake). In **Marmara**, an adult was carrying food near Eceabat on 25 June 1973, and adults with recently fledged young were at Kücükkuyu on 26 June 1973 (R. F. Porter), the latter indicating egg laying in late May. Nest building was observed at Kızılcahamam (**Inner Anatolia**) on 8 April 1984 (Barış *et al.* 1984). Fledged young were at Işıklı (**Southeast Anatolia**) on 14 June 1996 (GMK). Fledged young were also seen at Kızılırmak Delta (**Black Sea Coastlands**) on 15 June

1984 (Dijksen & Kasparek 1985), and Yeniçağa Gölü on 8 July 1986 (Kılıç & Kasparek 1987). In **East Anatolia**, a juvenile was near Lake Van on 24 August 1972. These records indicate double-brooding, as elsewhere in the range.

EUROPEAN GOLDFINCH *Carduelis carduelis*
Saka

(Du: Putter; Fr: Chardonneret élégant; Ger: Stieglitz; Swe: Steglits)

Subspecies. Roselaar (1995) proposed to consider those in Thrace within *balcanica*, populations throughout much of the rest of the western two-thirds of Turkey within *niediecki* and those in the eastern quarter of the country as probably *loudoni*. Those in the intervening area, roughly from the central Black Sea Coastlands in the north and from the Euphrates to the Tigris in the south, were left unassigned by Roselaar (*op. cit.*). The latter admitted the difficulty of his task given the enormous influence of wear and bleaching. Most variation within black-headed races of Goldfinch appears clinal and it seems distinctly dubious, and unnecessary within the context of understanding real-world evolutionary processes, to recognise several, if not all, of the races mentioned here.

Status and distribution. National conservation status: Least Concern (population apparently decreasing). A widespread and common resident breeder in all regions. Commonest in Marmara, Aegean and Mediterranean, rather more local in East and Southeast Anatolia, although still reasonably common and widespread in both these regions. A partial migrant in some areas, especially in the eastern third of Turkey. Typically breeds in agricultural country to at least 2,300m, although principally occurs below 1,500m.

Widespread, occasionally in large numbers, on passage: in autumn from mid August (14th) to early November, with most in the latter half of October, when a mixed group, totalling at least 12,000 individuals of this species and *C. chloris*, has been recorded. Spring passage is from late February to early April, with no noticeable peak thus far detected. Winters principally in the western two-thirds of Turkey, often in large flocks of up to 500 birds, and frequently in mixed flocks with, e.g., Common Chaffinch *Fringilla coelebs*, European Greenfinch *C. chloris*, Common Linnet *C. cannabina* and even, occasionally, Great Tit *Parus major*.

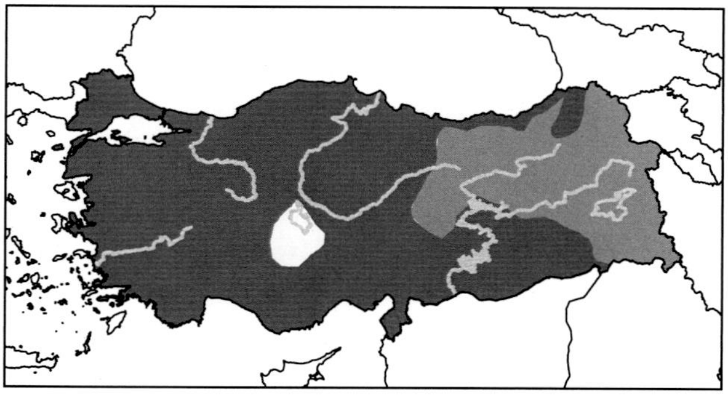

Breeding. Mainly in agricultural country, including orchards, gardens, olive and citrus groves, but also forest edge, in *Tamarix* and scrubby areas. Nests in a tall bush or tree; builds a neat cup of grasses and roots, lined with plant down, hair and wool. Eggs: C/4 (5), C/5 (2). Young (in nests): no data. In **Mediterranean**, nests were found April–May in Çukurova in 1987 (van der Have *et al.* 1988); and fledged young at Köyceğiz on 18 May 1992 (N. McMahon), and Belek on 24 May 1999 (R. F. Porter); in Göksu Delta, a female was observed nest building at 7m in a garden conifer on 22 May 2004, a nest had five eggs at 3m in a garden conifer on 6 May 2004, and recently fledged young were seen 22 May 2004 (PC); at Demirkazık, two nests were empty but ready for eggs, and two nests held incomplete clutches of one and two eggs on 12–13 May 2004 (D. Gregson). In **Inner Anatolia**, flocks disband in early March, with song and pairs in early April (Wadley 1951); records include a nest with eggs in early May (Warncke 1964–65), an empty but new nest on 8 June 1907, a nest with well-incubated eggs on 10 June 1907, and a pair with flying young on 29 June 1907 (Ramsay 1914). In **Marmara**, recently fledged young were at Gulpınar on 26 June 1973 (R. F. Porter), indicating egg laying in late May. In **Aegean**, there are seven records of nests with eggs from 12–22 May. In **Southeast Anatolia**, a nest at Birecik was almost complete on 10 May 2004, and a female was incubating on 23 May 2004 (PC); fledged young were at Işıklı on 14 June 1996 (GMK). Nest building was observed at Yeniçağa Gölü (**Black Sea Coastlands**) on 6 May 1978 (Kılıç & Kasparek 1987). Breeding records in June are probably second broods.

EURASIAN SISKIN *Carduelis spinus*
Kara Başlı İskete

(Du: Sijs; Fr: Tarin des aulnes; Ger: Erlenzeisig; Swe: Grönsiska)

Subspecies. Monotypic.

Status and distribution. National conservation status: Least Concern (population apparently stable). A relatively uncommon resident breeder, or partial/altitudinal migrant, in small numbers in montane coniferous (occasionally mixed) forests in Marmara (including Thrace) and Black Sea Coastlands, and perhaps in adjacent areas of Aegean, Mediterranean and Inner Anatolia, at 800–2,500m. Recently recorded in summer at Akseki, in the western Taurus (GMK).

More widespread between late September and early April, when there are at least three records from Southeast Anatolia (at Birecik, Diyarbakır and Halfeti) and the species reaches western parts of East Anatolia. Mainly winters in the west and south, but also in Thrace and the Black Sea Coastlands, usually in the lowlands to sea level. One record from the Kızılırmak Delta (Black Sea Coastlands) in early June is somewhat anomalous. Not uncommon on passage, largely through the western two-thirds of Turkey; relatively common at the Bosphorus in autumn between late September (26th) and early November (6th), with 2,810 passing Küçük Çamlıca in 1966, the majority in the middle two weeks of October. In some years, passage may continue until the end of November. Spring passage persists until at least late April.

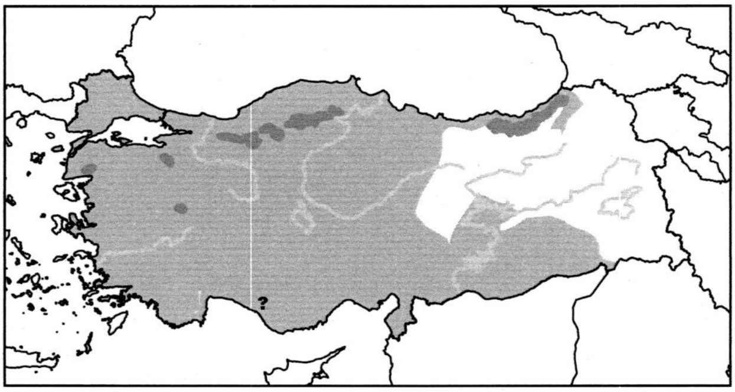

Breeding. In montane coniferous and mixed woodland in April–June, the presumed breeding season in Turkey; and in full song, with much display, at Abant Gölü (**Black Sea Coastlands**) in early June (Wadley 1951), and fledged young were observed in the Muğla area (**Aegean**) on 9 June 2004 (H. Tholl-Schmitt). Elsewhere, nests in conifers, usually high; a neat cup of twigs, grass and moss, lined mainly with hair and roots; 4–5 eggs and has two broods, the first usually in May.

COMMON LINNET *Carduelis cannabina*
Ketenkuşu

(Du: Kneu; Fr: Linotte mélodieuse; Ger: Bluthänfling; Swe: Hämpling)

Subspecies and taxonomy. All those breeding-season specimens examined by Roselaar (1995) were referable to *bella*, but nominate *cannabina* has been recorded in winter (e.g. at Erzurum, East Anatolia). He also speculated that those breeding in Thrace may pertain to *cannabina* or be intermediates between the latter and *bella*, but no material was available from this part of the country. Despite the possibility that the two intergrade, recognition of *bella* (which is sometimes afforded the name Turkestan Linnet, and whose variation is well treated by Vaurie 1949a) appears warranted, it being a larger, paler form than the nominate, with a virtually unstreaked white rump and uppertail-coverts, and a paler and less extensive rosy-red forehead than the nominate.

The genus *Acanthis* is available for this species, Twite *C. flavirostris* and the redpolls *C. flammea*, *C. cabaret* and *C. hornemanni*, but monophyly has not been established (Marten & Johnson 1986, van den Elzen & Nemeschkal 1991, Fehrer 1996, Arnaiz-Villena *et al.* 2001).

Status and distribution. National conservation status: Least Concern (population apparently stable). A generally widespread and fairly common resident breeder (although probably a partial/altitudinal migrant in more montane areas) almost throughout Turkey. Found principally in upland regions to 4,200m (more usually

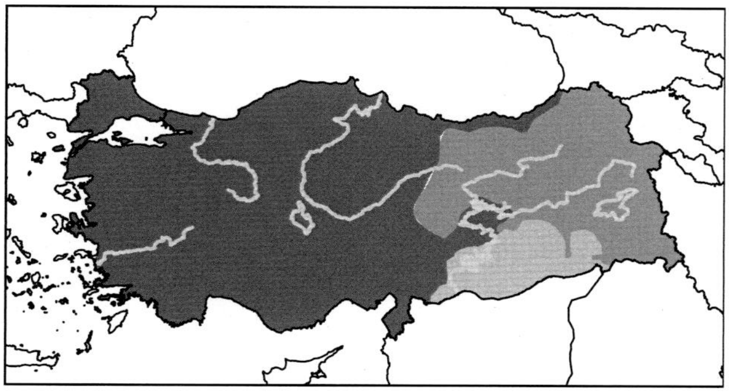

3,000m), more rarely in lowlands. Its range in Southeast Anatolia is largely restricted to western and northwestern areas. Usually breeds from April.

Much more widespread and common on passage: at the Bosphorus recorded between 11 August and 20 November, but even later on the south coast. Also in winter, when mainly recorded in the western two-thirds of Turkey; often in large flocks of up to 2,000 individuals and usually in lowland areas from October to March.

Breeding. In scrub, from sea level to above the treeline. Nests in a bush; builds a cup of grass, plant stems and moss, lined with hair and wool. Eggs: C/3 (3), C/4 (5), C/5 (6). Young: B/4 (2), B/5 (1). In **Inner Anatolia**, a nest at Karapınar was new, but still empty, on 28 April 1972, and another had four well-incubated eggs on 13 April 1970 (McNeile), indicating laying late March/early April; nests at Kızılcahamam had three eggs on 6 May 1993 and four eggs on 14 June 1993, and another at Hasan Dağ had three eggs on 27 May 1993 (PC); a nest had four young, two days old, near Karapınar on 20 April 2004 (S. Williams), indicating laying on 3 April. Adults were feeding fledged young near Aksaray on 18 June 1998 (GMK). In **Mediterranean**, nest building was noted at Aladağ on 28–29 April 1970, and a nest had five eggs near Islahiye on 8 May 1970; a pair was collecting nest material at Akseki on 30 April 1967 (R. F. Porter); many juveniles were at Aladağ in June 1966, and 5–6 juveniles were at Acıgöl on 5 June 1985 (Dijksen & Kasparek 1988); a nest had five eggs at Demirkazık on 13 May 2004 (M. Farmer); three nests near Mut had: four eggs (10 May 2005), four eggs (13 May 2005) and young (13 May 2005) (J. Hodson). In **Southeast Anatolia**, at Yesilce a nest held three eggs on 16 June 2001, and another an egg and four newly hatched young on 20 May 1993 (PC); four small young in a nest near Gaziantep on 2 May 1964 (Warncke 1964–65) indicate laying in mid April; several fledged young were at Işıklı on 15 June 1996 (GMK); a nest had five eggs near Osmaniye on 1 May 1970 (R. F. Porter); two nests had four and five eggs at Durnalık on 9 May 2004 (J. Mason). In **Black Sea Coastlands**, a male was with a juvenile at Abant Gölü on 3 June 1945 (Wadley 1951). In **Marmara**, a male was feeding two young at Uludağ on 21 July 1993 (Jetz 1995), and there was a nest with an incomplete clutch of one egg on 12 May 1964 (Warncke 1964–65); recently fledged young were observed at Alexandria and at Kosedere on 25 June 1973 (R. F. Porter). The late records above indicate double-brooding, as elsewhere in the range.

TWITE — *Carduelis flavirostris*
Sarı Gagalı Ketenkuşu

(Du: Turkse Frater; Fr: Linotte à bec jaune; Ger: Berghänfling; Swe: Vinterhämpling)

Subspecies. *C. f. brevirostris* occurs throughout the Turkish range, and thence east through the Caucasus, Transcaucasus, northern Iraq and northern Iran (Vaurie 1959a, Roselaar 1995). As noted by Roselaar (*op. cit.*) the ground colour on both surfaces of *brevirostris* is distinctly paler than nominate *flavirostris* from Scandinavia or *pipilans* from Britain, with duller and, particularly in males, more blotchy, compact and broader streaking on the underparts (*contra* Roselaar *op. cit.*, though noted by Sharpe 1888, Sushkin 1925, Kumerloeve 1967c). With the exception of the description and comparison of the underparts streaking, Roselaar's description of this race is accurate, but, in addition, the pinkish rump coloration is much deeper and redder in northwest European races. Five races (perhaps forming two groups) have been described from Central Asia and further east: these are even paler in their overall ground colours, with even less obvious, duller and more restricted rump patches. Thus, across the (sometimes broadly) disjunct range of this species there is a cline from west to east, put simply, from darker to paler birds, but geographical variation within Twite demands further investigation, and several of the described races are probably unworthy of recognition (Kirwan & Atkinson in prep.).

Status and distribution. National conservation status: Least Concern (population apparently stable). A fairly widespread and not uncommon resident breeder (and partial/altitudinal migrant in most areas), though distinctly local in the west of its range. Occurs in barren, rocky areas, usually at high altitude, in the eastern Black Sea Coastlands, throughout much of East Anatolia (above *c.*1,000m in the extreme south of this region) and extreme eastern Southeast Anatolia, extending west to the eastern and north-central fringe of Inner Anatolia and possibly the Mediterranean region (where recorded as low as 1,000m in the late breeding season). Typically breeds between 1,750m and 3,000m, but much higher (up to least 4,500m) on the highest mountains, e.g. Büyük Ağrı Dağı (East Anatolia). Typically nests around the snowline, and breeding commences from mid May. Post-breeding flocks may number as many as 400.

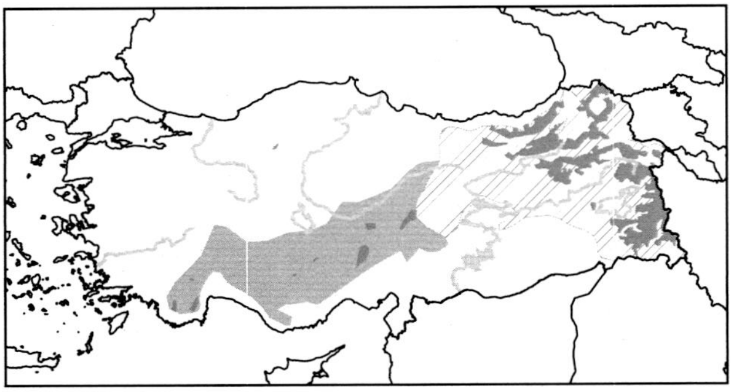

Descends in winter (usually between October and early April), being more extensively distributed in the west, down to 900m (e.g. very occasionally recorded in the Ankara region at this season), while also typically remaining common and widespread in flocks of up to 200 individuals throughout the eastern third of Turkey, up to 2000m.

Breeding. Inhabits barren, stony slopes with sparse vegetation, above the treeline. Nest and clutch size undescribed in Turkey but, elsewhere, the nest is on the ground in low cover, in hollows and holes of banks, and on rock ledges, occasionally in bushes; a cup of grasses and plant stems, lined with wool, hair and feathers; usual clutch 5–6. Young: no data. In **Black Sea Coastlands**, a fledgling, barely able to fly, was between Ispir and Rize on 13 July 1975, indicating egg laying in mid June; three adults and a juvenile were at Rize on 14 August 1972, and two family parties were near Bayburt on 11 August 1966. In **East Anatolia**, a pair copulated on the ground, and the female was observed nest building, on a steep hillside of short grass, herbs, and with rocky outcrops near Erçek on 7 May 2004; the nest was 2m above ground, on a narrow ledge of a low cliff, and partially screened by overhanging vegetation (GMK); a family party was in a lava field 45km north of Bendimahi on 5 August 1992 (J. Frost & J. Hornbuckle). In **Inner Anatolia**, a pair was behaving as if breeding at Tekke Dağ (east of Ürgüp) on 22 May 1972, and a fledgling was observed at Kızılcahamam on 19 August 2004 (E. Yoğurtçuoğlu).

COMMON REDPOLL *Carduelis flammea*
Huş İsketesi

(Du: Grote Barmsijs; Fr: Sizerin flammé; Ger: Birkenzeisig; Swe: Gråsiska)

Subspecies and taxonomy. All Turkish records may be assumed to be referable to nominate *flammea*. Two Alléon specimens (see below) are of *flammea* and Mathey-Dupraz (1922) noted a specimen (no longer extant) in the Robert's College (Bebek, Istanbul) collection that he specifically recorded as being of *flammea*; this form breeds from Scandinavia east to Kamchatka, and across Alaska and throughout northern Canada (Clement *et al.* 1993). Traditionally *C. flammea* was regarded as a polytypic species consisting of four subspecies. In addition to the nominate (Mealy Redpoll), these are *rostrata* (Greater Redpoll, on Baffin Island, Canada, and in coastal Greenland), *islandica* (Iceland) and *cabaret* (Lesser Redpoll, of northwestern Europe south and east to Switzerland and the Czech Republic). *C. f. cabaret* has been spreading in recent years, reaching southern Scandinavia in the 1950s, where it has subsequently been proven to breed sympatrically with *flammea*, without evidence for mixed pairings. Given that *cabaret* is diagnosably distinct from other redpolls within this complex, and apparently exhibits vocal and behavioural differences, it has been suggested to treat it at the specific level, with Mealy Redpoll comprising the other three taxa (Knox *et al.* 2001). Outside Europe, this proposal has already been followed by the AOU Check-list Committee (Banks *et al.* 2002), but it should be noted that the situation appears to be less clear-cut than was suspected originally and mtDNA work found no genetic differentiation between *flammea* and *cabaret* (Ottvall *et al.* 2002).

Status and distribution. Apparently a genuinely rare and irregular winter visitor, and perhaps also a passage migrant. There are a number of historical records from the Bosphorus and Istanbul area; Krüper & Hartlaub (1875) stated that the species occurred during the winter in this area, it having been reported in numbers by Robson, although no dates are given. A flock was observed in winter 1890/91 by Mathey-Dupraz (1922) at Üsküdar (on the eastern side of the Bosphorus), whilst Alléon took a male and a female at Büyükdere and Makriköy [=Bakırköy], on 30 January 1899 and 20 January 1899 respectively (National Museum of Natural History, Sofia). Braun (1903) also mentioned the species from Istanbul and Kasparek (1990a) noted (apparently based on Braun's observations) that a small invasion occurred in the winter of 1902/03. Ogilvie (1954) observed the species at Çatalağzi, Zonguldak province (Black Sea Coastlands) on 3 February 1947. Only four definite recent records have been published: up to ten at Burdur Gölü (Mediterranean) on 5 December 1987 (Magnin 1989), 25 migrating north at the Kızılırmak Delta (Black Sea Coastlands) on 26 March 1992 (date was previously incorrectly published as 25 March, in Kirwan 1992c) and ten at Uzungöl, Trabzon province (Black Sea Coastlands) in July 1995 (Kirwan & Martins 2000a). Most recently, one was photographed at Altınpark, Ankara (Inner Anatolia), on 15 November 2005 (E. Yoğurtçuoğlu).

Elsewhere in the Middle East doubtfully recorded from Israel (Porter *et al.* 1996). Five records from Cyprus, all in winter 1966 and 1974, (either trapped or collected; Flint & Stewart 1992), illustrate the potential for wandering birds to be recorded throughout Turkey. Kumerloeve (1957d) recorded the species twice in neighbouring Bulgaria in October and November 1941.

COMMON CROSSBILL *Loxia curvirostra*
Çaprazgaga

(Du: Kruisbek; Fr: Bec-croisé des sapins; Ger: Fichtenkreuzschnabel; Swe: Mindre korsnäbb)

Subspecies. Roselaar (1995) recommended that the names *vasvarii* (applied to birds from the western Black Sea Coastlands) and *caucasica* (for those in the Caucasus and Transcaucasia) are best regarded as synonyms of *guillemardi* (described from Cyprus), and that all Turkish populations be considered referable to this subspecies. We concur with his recommendation.

Status and distribution. National conservation status: Least Concern (population apparently stable). A very local but generally not uncommon resident breeder in coniferous mixed forests, principally in the eastern Black Sea Coastlands (apparently less numerous elsewhere in this region), southern Marmara (principally on Uludağ, irregular elsewhere), the northern fringe of Inner Anatolia (with perhaps a local population persisting in the Beynam Forest, near Ankara) and throughout the Taurus (Mediterranean). Recorded in the breeding season up to 2,500m, more typically at lower altitudes, e.g. around 800–900m in the Taurus. More widespread in autumn and winter, particularly during its rather irregular irruptive movements, which presumably also involve extralimital birds; spring records from other areas of Turkey may relate to transient populations.

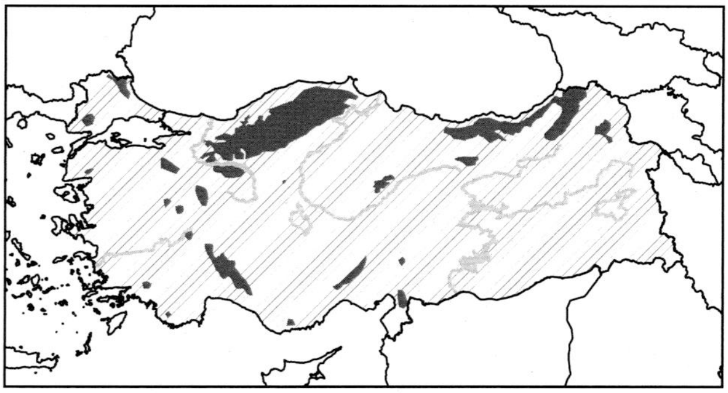

Breeding. In montane conifer forest. Other than observations of fledged young and juveniles, no breeding details from Turkey. Elsewhere, breeds mainly in fairly open conifer woodland, nesting 2–18m (usually at least 8m) above ground in a tree, mainly in the outer branches; nest base of twigs, built up with grasses, moss and lichen, with an inner cup of fine grass, hair and feathers; usual clutch 3–4 (2–5), and season variable, but often commences in March: sometimes double-brooded, given sufficient food. In **Mediterranean**, a female feeding two fledged young near Akseki on 19 May 1984 (S. Harrap), and a single juvenile at Akseki, 28 May 1996 (S. Mawby), indicate laying (if recently fledged) *c.*12 April and 21 April. In **Inner Anatolia**, a female and three juveniles at Daday on 9 May 1990 indicate laying in early April; adults and recently fledged young were in coniferous forest near Kızılcahamam on 29 May 1992 (S. Williams), indicating egg laying about 20 April. In **Marmara**, an adult was carrying food near Kesan on 21 May 1970. Juveniles were observed in **Black Sea Coastlands** in June–August, and in Inner Anatolia in August, the latter considered to be more mature juveniles, possibly fledged a few months earlier.

[PARROT CROSSBILL *Loxia pytyopsittacus*]
Koca Başlı Çaprazgaga

(Du: Grote Kruisbek; Fr: Bec-croisé perroquet; Ger: Kiefernkreuzschnabel; Swe: Större korsnäbb)

Subspecies and taxonomy. Monotypic, but relationships with other European crossbills apparently complex (see e.g. Summers 2002).

Species not admitted to the Turkish list (Kirwan *et al.* 1999). Kumerloeve (1961) recorded that Braun (1903) possibly noted one in the Istanbul bird market (no year given) without establishing its origin. There are no other Middle Eastern records (Porter *et al.* 1996) of this extremely unlikely vagrant.

(ASIAN) CRIMSON-WINGED FINCH *Rhodopechys sanguineus*
Alamecek

(Du: Aziatische Rode Woestijnvink; Fr: Roselin à ailes roses; Ger: Rotflügelgimpel; Swe: Bergsökenfink)

Subspecies and taxonomy. Nominate. See taxonomic remarks under Mongolian Finch *Bucanetes mongolicus*, and comments by Kirwan *et al.* (2006) on the probability that two allospecies are involved, given the distinctiveness of *R. s. alienus* (of the Atlas Mountains, in northwest Africa) in all plumages, as well as the evidence of morphometrics. Vocalisation data are currently insufficient to either support or negate this hypothesis, and molecular data for both taxa are not yet available. The type specimen of *R. sanguineus* is from Erzurum (East Anatolia). Note the correct spelling of the specific name (David & Gosselin 2002b).

Status and distribution. National conservation status: Least Concern (population apparently increasing). A fairly widespread and usually relatively common resident breeder (and altitudinal migrant in some areas) in high, typically rather arid upland areas principally in East, and parts of Southeast, Anatolia. Also in eastern and some southern parts of the Black Sea Coastlands and locally in the eastern Taurus and adjacent foothills (Inner Anatolia/Mediterranean) north through montane areas east of Tuz Gölü, where locally common. Two breeding season records, both in May (1967 and 1992) from the Eber Gölü area (Inner Anatolia) suggest breeding, perhaps indicating that a small outlying population may exist in this area, west of the main range. Additional breeding-season records are available even further south and west. Generally breeds on sparsely vegetated, rocky slopes to at least 4,200m (on Ağrı Dağı), regularly in agricultural areas in East Anatolia from 1,600m and

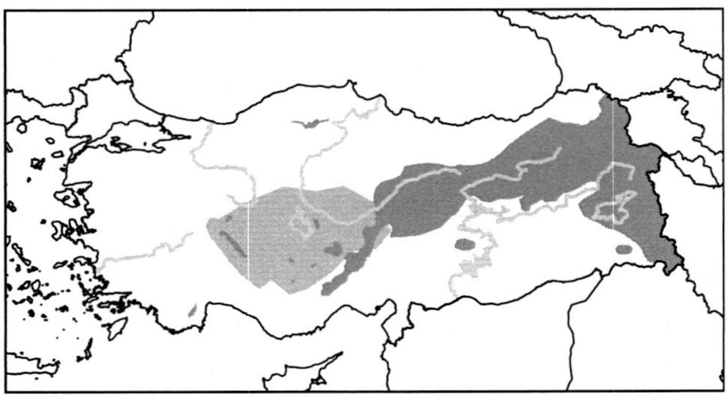

more regularly at lower altitudes to at least 900m in the west. Breeds from April, perhaps slightly later in eastern areas.

Descends in winter, when regular in lower steppe areas in Inner Anatolia. Occasionally recorded in Aegean, more exceptionally Marmara as far west as Biga, Çanakkale province (OST 1969). However, significant numbers remain in the eastern third of Turkey to at least 2,000m.

Breeding. On steep slopes with sparse vegetation, bare soil and stones, sometimes scattered scrub, and on lava slopes, almost devoid of vegetation; also on slopes with vineyards and scattered trees, east of Tuz Gölü (Inner Anatolia), and in rocky valleys and flat open country with low scrub (Roselaar 1995). In the crater of Meketuzlası (1,100m), near Karapınar (Inner Anatolia), Lehmann & Mertens (1969) found nests on 45° ashy slopes, under lava slabs 40 × 50 × 20cm thick, with the entrances facing east; subsequently nests were found in same place (McNeile), not only under slabs, but also under low clumps of *Salvia*; the nests were of coarse dry plant stems, and lined with finer grasses and plant fibres. Eggs: C/5 (3). Young: B/5 (1). At Meketuzlazı, a nest had four young, 2–3 days old, and an egg on 8 May 1967, a nest had eggs on 6 May 1968, and a third nest had young, almost ready to fledge, on 6 May 1968: laying of the first egg in each of these nests was estimated at 17 April, 20 April and *c.*1 April (Lehmann & Mertens 1969); four nests with eggs were located from 30 April–9 May 1970 (Lehmann 1971); one of two nests on 20 April 1970 had five fresh eggs and the other four eggs, in which a fifth was laid next day; and two nests on 22 April 1970 had five slightly incubated eggs and one egg (incomplete clutch), a further nest was empty but ready for eggs, and two nests had eggs on 21 April 1973 (McNeile); a nest had four eggs (incomplete clutch) under a flat rock on 4 May 2005 (J. Whitaker). Elsewhere in **Inner Anatolia**, copulation was seen in Sultan Dağları on 9 May 1967, a fledged juvenile was at Karadağ on 19 June 1996 (Kirwan 1998a), three newly fledged young were on Karadağ on 24 July 1971, and at least three pairs were carrying food (one pair feeding two fledged young) at Delice on 25 June 1977 (Schubert 1979). In **Mediterranean**, nest building and a pair displaying were noted at Demirkazık on 13 April 1994 (C. Bradshaw); adults were feeding recently fledged young at Torosdağ on 19 July 1971. In **East Anatolia**, an adult was carrying nesting material near Görentaş on 2 June 1969; a pair was at a nest near Van on 11 May 1992 (N. McMahon); several family parties were at Süphan Dağ, north of Van Gölü, in late July/early August 1966; and McGregor (1917) considered that the species reared two broods in a year around Erzurum, because he saw family parties, with young hardly able to fly, as late as 20 August. The species is considered possibly double-brooded elsewhere in its range.

DESERT FINCH *Rhodospiza obsoleta*
Boz Alamecek

(Du: Vale Woestijnvink; Fr: Roselin de Lichtenstein; Ger: Weißflügelgimpel; Swe: Ökenfink)

Subspecies. Monotypic. *Rhodospiza* was frequently subsumed within the genus *Rhodopechys* in the past, but strong, independently corrolated, genetic data supporting the distinctiveness of this species have been published recently (Groth 1998, Arnaiz-Villena *et al.* 2008). See also taxonomic remarks under the next species.

Status and distribution. National conservation status: Least Concern (population apparently increasing). First documented in Turkey by Weigold (1913) in mid-April 1911 at Maşık, north of Urfa (Southeast Anatolia) and Clarke discovered the species near Aleppo, in adjacent Syria in 1919 (Kumerloeve 1966c). There were no subsequent published records until 1964, when recorded around Gaziantep, and 1965, when discovered breeding at Birecik (Southeast Anatolia) (Kumerloeve 1967a). Presumed to be a resident breeder, but perhaps a partial migrant (as elsewhere in its range), although there are no winter records. Limited autumn dispersal is suggested by the available records. Previously recorded as a winter visitor to Syria and Iraq by Hartert & Steinbacher (1932/38), perhaps suggesting winter emigration from Turkey. Regularly recorded from early/mid April to late September, typically locally and in small numbers, usually in cultivated or arid, rocky country to at least 2,000m (although principally recorded below 1,000m). Recorded from the Gaziantep area east to Cizre and north to Adıyaman (Southeast Anatolia). Perhaps also breeds in adjacent parts of Mediterranean. Recently reported on several occasions from at least one locality around Van Gölü, as well as near Doğubeyazıt, and up to 2,100m on Kuh Dağ, Van (East Anatolia), suggesting a wider or expanding breeding range. Large flocks (up to several hundred) occasionally congregate especially at Birecik and Halfeti in August and early September. Three recent records from two separate localities in the eastern Taurus (Mediterranean) in late July and August perhaps reflect dispersal, although another record from this region in May 1967, near Kayseri (Inner Anatolia) suggested breeding, while later in the same month Kumerloeve found adults with young on the east side of Tuz Gölü (Kumerloeve 1970). Roselaar (1995) suggested that these records may have reflected confusion with (Asian) Crimson-winged Finch *Rhodopechys sanguineus* (which has been proven to be common in suitable areas of Inner Anatolia in recent years; Eken & Magnin 1999; pers. obs.), and Kirwan *et al.* (2000) postulated the idea that *Bucanetes mongolicus* was the species observed by Kumerloeve (see below).

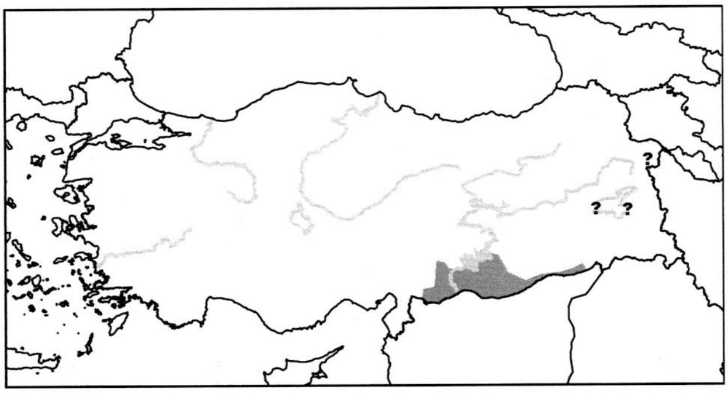

Breeding. In arid areas with olive groves, pistachio orchards, bushy areas and hedges, often close to cultivation or gardens. Nests in a tree or bush, 1.3–3.0m above ground (n=10), near the end of horizontal branches or in forks; builds a cup of twigs, grass and roots, lined with plant down (often white), finer grasses, hair and wool. Eggs: C/3 (2), C/4 (1), C/5 (1), C/6 (3). Young: B/2 (1); B/5 (2). Most records are from Gaziantep and Birecik (**Southeast Anatolia**), where it appears double-brooded: four nests had three eggs, three eggs, six eggs and one egg (incomplete clutch), some incubated for several days, on 23 April 1972 (McNeile), and another had eggs in early May; three nests had, respectively, young on 4 May 1964 (Warncke 1964–65), five young on 5 May 1970, and two young on 16 May 1985; two pairs, the females carrying nesting material, were seen on 10–11 May 2004 (R. Thomas), and a female was building a nest in a sapling on 23 May 2004 (PC); five eggs were in a nest on 17 May 2004, and a sixth egg was subsequently laid; five young, two days old, were in a nest on 7 June 2006 (PC); records of apparently second broods are: nest building on 4 June 1993, two nests with four and five eggs on 5 June 1993 (PC), a nest with six eggs on 12 June 2006 (A. Rustell), a male feeding at least three fledged young at Akçakale, 2 July 1978, and two family parties near Güreniz on 16 August 1974. A family party was near Doğubeyazıt (**East Anatolia**) on 5 August 1992.

MONGOLIAN FINCH — *Bucanetes (Eremopsaltria) mongolicus*
Doğu Alamaceği

(Du: Mongoolse Woestijnvink; Fr: Roselin de Mongolie; Ger: Mongolengimpel; Swe: Mongolfink)

Subspecies and taxonomy. Monotypic, although it was considered a subspecies of Trumpeter Finch *B. githagineus* by Cheng (1987), an arrangement that is clearly untenable given the suggestion of Panov & Bulatova (1972) that *mongolicus* is more closely related to Scarlet Rosefinch *Carpodacus erythrinus*, and, even more importantly, the fact that the two are broadly sympatric without interbreeding (e.g. Vaurie 1949a). Further east in Asia, *mongolicus* and *githagineus* are also apparently altitudinally segregated (Haffer 1989), but further west and in Turkey they have been, at least occasionally, recorded in total sympatry, and some evidence of character displacement has been observed (a phenomenon frequently observed among closely related congeners in areas of overlap). Nonetheless, in some respects, it was rather perceptive of Swinhoe to have described Mongolian Finch within *Carpodacus* (E. N. Panov *in litt.* 2003, reported that unpublished work on the vocalisations of *C. erythrinus* and *B. mongolicus* appears to support the idea of a close relationship between the two). However, to transfer the two 'trumpeter finches' to *Carpodacus* is not supported by the available evidence (although the caryotypes of *mongolicus* and *erythrinus* are rather similar and both differ more significantly from *githagineus*), and we do not favour the option of subsuming *mongolicus* alone within *Carpodacus* (as proposed by Panov & Bulatova *op. cit.*), because of the obvious and highly significant outward differences between the two, as well as the evidence of recently published genetic data (Arnaiz-Villena *et al.* 2008).

Both species of *Bucanetes* have been considered within an expanded *Rhodopechys*, along with Crimson-winged Finch *R. sanguineus* and Desert Finch *Rhodospiza obsoleta* (see e.g. Vaurie 1949a, Cheng 1987, Sibley & Monroe 1990, Clement *et al.* 1993; equally, some have viewed the four species within *Bucanetes*), but this arrangement, while holding some obvious advantages and clearly supported by a number of shared morphological, behavioural and other characteristics, also has significant negative results. For *Rhodopechys* cannot really be viewed as a natural or monophyletic grouping either: all four species are significantly different in many respects, and diverge more subtly in others (e.g. their breeding biologies; see Harrison & Castell 2002). Additionally, in some respects Crimson-winged Finch appears to be most closely related to Red-browed (Spectacled) Finch *Callacanthis burtoni*, a monotypic genus found in the Himalayas, which in turn is most closely allied to the rosefinches and the *Acanthis* group (subgenus) of *Carduelis* (C. S. Roselaar *in* Cramp & Perrins 1994), and these two may be congenerics, although mtDNA research suggests that this is not the case (Groth 1998). The eggs of *Rhodopechys sanguineus* and *Callacanthis* are remarkably similar, but differ from those of *Bucanetes* and *Rhodospiza*, which three species are generally rather similar (M. P. Walters unpubl. ms.). Further, Desfayes (1969) suggested that *Bucanetes* and *Rhodopechys* are probably not close relatives. Thus, an expanded *Rhodopechys* appears to be a 'catch-all' option for taxa whose relationships are not obvious, but are not necessarily congeneric. Genetic data, recently published, support a close relationship between *sanguineus* and *mongolicus*, which two species formed a clade in Groth's (1998) study. Arnaiz-Villena *et al.* (2008), whose molecular study unfortunately did not sample *sanguineus*, recovered high bootstrap support for a monophyletic clade containing *mongolicus*, *githagineus*, *Carpoclacus nipalensis* (Dark-breasted Rosefinch), *Leucosticte arctoa* (Asian Rosy Finch) and *L. tephrocotis* (Grey-crowned Rosy Finch). Kirwan *et al.* (2006) noted the likelihood that the rosy finches and sanguineus being close relatives. If such an arrangement becomes generally accepted, the genus *Leucosticte* would possess precedence.

For now, we reluctantly include *mongolicus* within *Bucanetes*, but given the species' vocalisations, we prefer to drop the 'trumpeter' modifier from the species' vernacular name. It is surely significant to note that no less a commentator than Voous (1977) struggled with the systematic position of many of the carduelines, he being forced to acknowledge the striking distinctiveness and the lack of obvious relationships of a number of taxa. Currently no other genus name exists to serve as a type for *mongolicus* (Sharpe 1888): although both 'trumpeter' finches were given the name *Erythrospiza* (Bonaparte; Bp. Faun. Ital., Ucc. Pl. 35 fig. 3 [1832–41]), Wolters (1975–82) considered this name to be a synonym of *Erythrina* (in its application as a subgenus) and *Carpodacus erythrina* (Pallas, 1770) is the type species of *Erythrospiza*, thereby invalidating its use for any of the 'desert or arid-zone finches'. See also Kirwan & Gregory (2005), who suggest *Eremopsaltria* as a new generic name should the species prove to be as distinctive as, in some respects, it appears to be. Note: if Mongolian Finch is included within *Rhodopechys* then its specific name should still be spelt *mongolicus*, whilst if treating Trumpeter Finch similarly its name must still be spelled *githagineus*.

Status and distribution. National conservation status: Near Threatened (population apparently increasing). Presumably a local resident or an altitudinal migrant (as elsewhere in range; Clement et *al.* 1993), generally in small numbers, in parts of extreme eastern East Anatolia (principally close to the borders with Iran and Armenia in Kars and Van provinces, but also south towards the Hakkari region and east to the Sultanbabadağı, in the Munzur range, Tunceli province). Perhaps spreading, as evidenced by a recent record from a well-watched area of the eastern Pontics (Sivrikaya, Black Sea Coastlands). Elsewhere, breeds through Iran and Afghanistan east to Mongolia. All modern records are from arid, or semi-arid, rocky upland areas at 800–2,700m (though it can be expected to occur considerably higher), usually with low scrub, herbs and grasses. About 60 published records are available from Turkey, all since 1977, involving groups of up to 25 individuals. The majority are listed in Kirwan & Martins (1994, 2000a) and Kirwan *et al.* (2003). It seems reasonable to assume that the lack of past records compared with the extensive recent series reflects a genuine and swift breeding range expansion, which perhaps persists. Records presumably reflect dispersal as well as local breeding. Given the species' presence far to the west of its originally supposed occurrence in Turkey, Mongolian Finch clearly should be searched for throughout the montane east, as far as the fringes of Inner Anatolia. Kirwan *et al.* (2000) drew attention to the fact that Kumerloeve (1970) reported Desert Finch *Rhodospiza obsoleta* twice in Inner Anatolia in the late 1960s (see that species) and, given that he misidentified the Tunceli birds as *R. obsoleta*, the possibility exists that these central Anatolian observations actually relate to *B. mongolicus*.

Poorly documented historical records exist from parts of what is now northeast Turkey. In May of 1911 or 1912, Bobrinskij (1916) observed this species near Bulgan (Naxcivan), near the border of Turkey, Armenia and

Azerbaijan. Subsequently Beme (Bohme) (1926) found a small flock near Tuzluca, Iğdır province (East Anatolia) in spring 1915. The species has recently been confirmed to occur and is almost certainly breeding, in southern Armenia, close to the Turkish border (Beddard *et al.* 2002, Ananian & Bussutil 2003). Kumerloeve (1961) mentions these records merely in passing but the extensive series of recent records lends weight to the significance of the historical records. The species' status in Turkey and the Western Palearctic were reviewed by Barthel *et al.* (1992), Kirwan & Konrad (1995) and Kirwan *et al.* (2000).

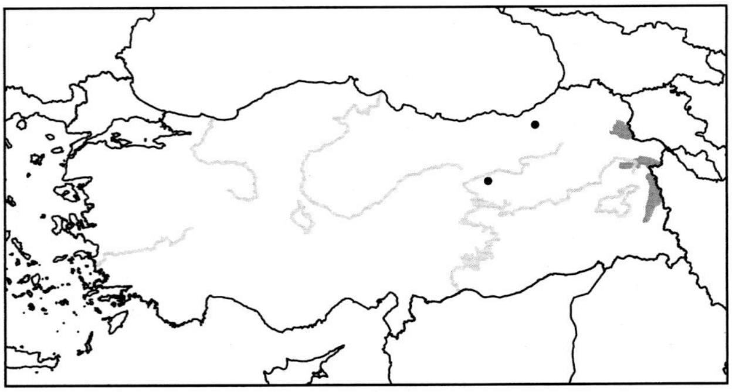

Breeding. In arid, rocky areas, on stony bare ground with sparse low shrubs, usually on higher steep slopes; on sparsely vegetated slopes, along the base of cliffs at Işak Paşa Sarayı, East Anatolia; in two dry valleys, with low herbs and scattered bushes 1m high and low herbs in the Tuzluca/Kars area, East Anatolia (G. Sandve), but probably also in extensive black lava fields with sparse vegetation (GMK). Nest site, nest and clutch size undescribed in Turkey; elsewhere, nests on the ground, in a small crevice or hollow, or sheltered by a low bush or overhanging rock; builds a loose, shallow cup of dry grass stems with strips of dead leaves, lined with hair and wool; usual clutch 3–5, laid in May, sometimes much later, and single-brooded. Young: B/3 (1). At Işak Paşa Sarayı, there were juveniles on 17 July 1988 and 21 June 1994, 20–42 adults, mainly in pairs in early June 1996, and a female and young bird on 9 August 2001 (A. Comas). Elsewhere in **East Anatolia**, a pair was feeding two young, near Aşağı Mutlu, north of Çaldıran, on 16 July 1990, and an immature was at Sultanbabadağıı on 23 August 1982; a nest had three young at Doğıubeyazıt on 23 June 2000, indicating egg laying no later than early June; a pair and three juveniles were near Serpmetaş on 28 July 2004 (Balmer & Betton 2005a).

TRUMPETER FINCH — *Bucanetes githagineus*
Küçük Alamacek

(Du: Woestijnvink; Fr: Roselin githagine; Ger: Wüstengimpel; Swe: Ökentrumpetare)

Subspecies. No specimen material from Turkey is available, but the subspecies involved can safely be assumed to be *crassirostris* (see Roselaar 1995). See taxonomic remarks under Mongolian Finch.

Status and distribution. National conservation status: Near Threatened (population trend unknown). Apparently a very local summer visitor in small numbers to parts of Southeast and East Anatolia (north as far as Doğubeyazıt), and perhaps adjacent areas of Mediterranean. Listed by Dementiev & Gladkov (1951/54) for Turkey, but this unsubstantiated comment was rejected by Kumerloeve (1961). Since 1974, as many as 22 records have been published, several with full descriptions. Nonetheless its status and distribution remain unclear. Probably a rare summer visitor (all records are during late April–mid August) to Mediterranean, Southeast and East Anatolia, with breeding proven in Southeast Anatolia (Martins 1989). Its occurrence perhaps reflects nomadic rather than regular seasonal movements.

The records are as follows: two males and two females near Tuzluca, Iğdır province (East Anatolia) with several more between there and Kağızman, on 24 May 1974; six in the same area, near Gaziler, between Tuzluca and Kağızman, on 6 June 1987; a probable at Nemrut Dağı, Tatvan province (East Anatolia), on 13 June 1977 (Krieger 1988); seven, including five juveniles, 5km south of Yesılce, west of Gaziantep (Southeast Anatolia), on 28 June 1977 (Martins 1989, with additional records from this area in June 1996, May 1997, May 1999 and May 2000; Kirwan & Martins 2000a, Kirwan *et al.* 2000); a pair between Alanya and Manavgat, Antalya province (Mediterranean), on 10 May 1979; a male at Nemrut Dağı, Tatvan province, on 17 August 1987 (Krieger 1988); one between Silifke and Mersin, Içel province (Mediterranean), on 24 April 1988; one between Birecik and Cizre, Urfa/Mardin province (Southeast Anatolia), on 17 May 1989; one at Birecik, on 16–17 June 1990 (Kirwan &

Martins 1994) and three there, on 17 May 1992 (Kirwan & Martins 2000a); one at Doğubeyazıt (East Anatolia) on 8 July 1997, with two there on 31 July–2 August 1998 (Kirwan *et al.* 2000) and two juveniles on 24 June 2004 (Balmer & Betton 2005a); three females or first-winters at Nemrut Dağı, Tatvan, on 10 August 1998 (J. L. Copete *in litt.* 2003); two at Çınar-Göksu Dam, Diyarbakır (Southeast Anatolia) on 29 April 1999 (Karakaş & Kılıç 2002); two males at the Göksu Delta (Mediterranean) on 13 June 1999 (Kirwan *et al.* 2000); a pair *c.*26km west of Erciş, on the north shore of Van Gölü (East Anatolia) on 30 May 2002 (H. Dufourny *et al.*); a singing male near Pervari, between Çatak and Siirt (East Anatolia) on 26 May 2005 (J. L. Copete *in litt.* 2005), and one at Milleyha, Samandağ (Mediterranean) on 1 April 2008 (A. Atahan). Whilst both Trumpeter and Mongolian Finches have been recorded from around Tuzluca and at Dogubeyazit, and confusion might be suspected, the occurrence of both species is substantiated by good descriptions (the two species have recently been confirmed to breed alongside one another in Azerbaijan; Clement *et al.* 1993).

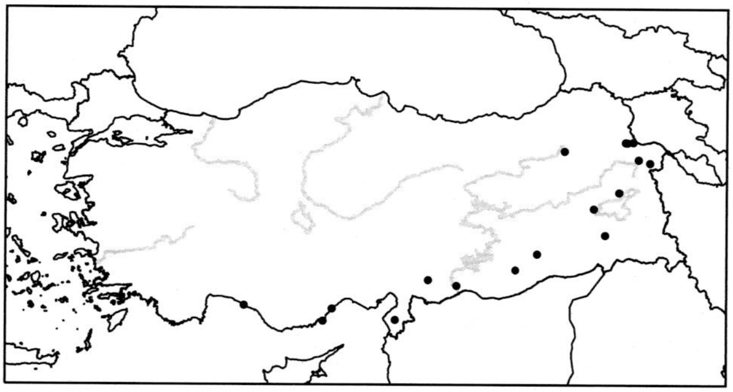

Breeding. Two fledged juveniles were at Doğubeyazıt (**East Anatolia**) on 24 June 2004 (Balmer & Betton 2005a) and five juveniles near Gaziantep (**Southeast Anatolia**) on 28 June 1977 (Martins 1989). There are no other specific details of breeding in Turkey, though a nest was apparently found at Doğubeyazıt in July 2006 (E. Yoğurtcuoğlu *in litt.* 2006). In Armenia, near Vedi, a breeding pair was present on 17 April 1961, pairs were observed on 7 July and 31 July 1995, a juvenile was seen on 31 July 1995, and an adult and four young (a family group) drank from a spring on 4 August 1995 (Adamian & Klem 1999). The species is at the northern edge of its range in Turkey (and Armenia). Elsewhere it breeds in rocky desert areas, and the nest is in a rock crevice or cavity in low cliffs (and in walls, terracing and old buildings), or sheltered under a rock or low shrub; a cup of plant stems, twigs and grass, lined with hair and wool; usual clutch 4–6. Breeding commences March–April, and there may be two broods.

COMMON ROSEFINCH — *Carpodacus erythrinus*
Çütre

(Du: Roodmus; Fr: Roselin cramoisi; Ger: Karmingimpel; Swe: Rosenfink)

Subspecies. *C. e. kubanensis* (northern and northeastern Turkey) and perhaps nominate *erythrinus* in extreme northwest Turkey (Roselaar 1995). Helb *et al.* (1995) considered *kubanensis* to represent a close link between central European populations and those further east.

Status and distribution. National conservation status: Least Concern (population apparently stable). A locally common summer visitor, principally from late April (earliest 2nd, though northeastern breeding areas are unoccupied apparently until at least three weeks later), and departs by late September/early October (12th). Breeds throughout northern Turkey, except perhaps in Thrace (one record of a singing bird close to the Bulgarian border), east from Uludağ, most common and widespread in Black Sea Coastlands; very localised in the north of Inner Anatolia (elsewhere in this region breeding has also been recorded, on several occasions, at Sultan marshes). Common where suitable habitat exists in East Anatolia as far south as the Hakkari Dağları, where apparently very local. Recorded once in northern Mediterranean in early June (Deppe 1990). Kumerloeve (1966c, 1975c) commented on the species' apparent expansion westwards through Turkey in the central third of the last century. May even breed in Southeast Anatolia, although there are no confirmed records, given availability of late spring records in suitable habitat. Breeds in mixed forests and areas of secondary growth or gardens, at moderate to high altitudes (e.g. from 400m on Uludağ to at least 2,700m).

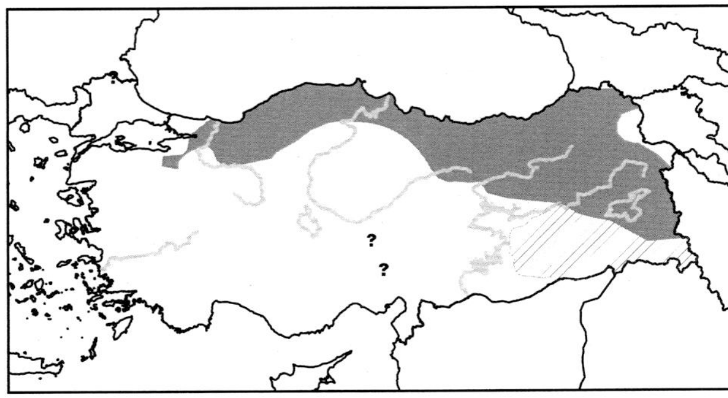

More widespread on passage, when recorded throughout the country, though very rare in the western third. Departure from breeding areas commences in the second half of August (migrants observed in East Anatolia from 23 August), continuing through September (latest 3 October). Exceptionally, reported once in winter, at Göksu Delta (Mediterranean) on 26 December 2004 (V. Konrad).

Breeding. In meadows with scattered bushes, hedgerows and scrub, including open woodland with undergrowth, copses, gardens and thickets in cultivation; often in damp areas and tangles, including willows near lakes and rivers. Nests in a bush or young tree, at 0.3–2.0m above ground (15). The nest is a loose cup of plant stems and grass, lined with roots and hair. Eggs: C/5 (6), C/4 (2). No records of young in nests in Turkey. In **Black Sea Coastlands**, at Yeniçağa Gölü there was nest building 20 May 1992 (but still not complete on 29 May), and nest building by two pairs on 20 June 1992, incomplete clutches of one egg on 22 June 1992 and two eggs on 20 June 1992, when three nests nearby had complete clutches of five eggs; three nests there contained four, four and five eggs on 16 June 1993 (PC); nest building was noted at Abant Gölü on 27 May 1992 (PC), and at Ismetpaşa on 30 June 1977 (Schubert 1979); fledged juveniles were seen at Kızılırmak Delta on 9 July 1970 (Dijksen & Kasparek 1985); at Yoncalık, a nest contained one egg on 11 June 2004 (and five eggs on 15 June 2004), and another had five eggs on 15 June 2004 (S. Williams). At Uludağ (**Marmara**), earliest song heard 11 May 1993 (Jetz 1995). At Sarıcan (**East Anatolia**), two nests were empty but ready for eggs, on 16 June 2004, and another contained two eggs (incomplete clutch) on 25 June 2004 (PC). Clearly a late breeder, and a record of a pair with three flying young at Sultan marshes (**Inner Anatolia**) on 9 June 1982 (Kasparek 1985), seems exceptionally early.

[GREAT ROSEFINCH *Carpodacus rubicilla*]
Büyük Çütre

(Du: Grote Roodmus; Fr: Roselin tacheté; Ger: Berggimpel; Swe: Större rosenfink)

Subspecies and taxonomy. Only the nominate form is likely to occur in Turkey (Roselaar 1995). Rasmussen & Anderton (2005) proposed that Central Asian and Himalayan *severtzovi* should be treated as a separate species (Spotted Great Rosefinch).

Species not admitted to the Turkish list (Kirwan *et al.* 1999). There are two claims from Turkey in the Erzurum area (East Anatolia). In June and July 1910 McGregor (1917) regularly noted three pairs, latterly two pairs, at 'Kyrk Deirmen', just southwest of Erzurum at around 1,800m, considering nesting probable. No plumage description was presented. It appears likely that this observation actually concerned Common Rosefinch *C. erythrinus*, given that the birds were apparently summer migrants (Great Rosefinch is a short-distance altitudinal migrant), while McGregor's description of the song closely recalls the more common species, although the two species' songs are rather similar. The habits and habitat of these birds also indicate Common Rosefinch. Perhaps most telling is that McGregor fails to mention Common Rosefinch in his list of birds recorded at Erzurum, although the species is now known to occur commonly in the environs of the city. Apart from Kosswig, who mentioned that he had seen Great Rosefinch near Erzurum to Kumerloeve (1966c, 1967a) in the 1940s, no other observers, several of whom spent considerable time in this area, have reported the species. Given the species' known range within the Caucasus it is conceivable that it could stray to Turkey. A 'possible' was perhaps briefly observed at Anı, Kars province (East Anatolia) in August 1992 (J. Hornbuckle *et al.*), but the description is unconvincing. Ergene (1945) considered occurrence in Turkey 'possible', while Kiziroğlu (1989) described the species as a summer visitor to the northeast, and Turan (1990) stated that it occurs in small numbers in the northeast part of eastern Anatolia, particularly in the Erzurum area, both presumably on the basis of McGregor's report. At present, therefore, there are no adequately documented records of the species and it cannot be admitted to the Turkish list.

EURASIAN BULLFINCH *Pyrrhula pyrrhula*
Şakrak

(Du: Goudvink; Fr: Bouvreuil pivoine; Ger: gimpel; Swe: Domherre)

Subspecies. *P. p. rossikowi* (northeast Turkey), with Roselaar (1995) erecting a new endemic subspecies, *paphlagoniae*, for those in the western Black Sea Coastlands, e.g. at Karadere, the Bolu Dağı, at Abant Gölü and in the Ilgaz Dağları, based on mensural data (it is generally smaller than *rossikowi*) and morphological characteristics, which are closer to European *germanica*. Further west in Turkey it is unclear whether *germanica* or *paphlagoniae* occurs. However, our examination of specimens in the Natural History Museum (Tring), the Manchester Museum, and the National Museum of Natural History (Sofia) taken throughout Europe and western Asia does not suggest the need for recognising the vast majority of the named geographic variation currently described for this species. Both *germanica* and *europoea* appear impossible to diagnose with certainty, even in longer series. Those ascribed to *rossikowi* closely approach western and central European birds; those taken in the Taurus (Mediterranean) in winter are seemingly close in their appearance to the nominate, though with some similarity to *rossikowi*; and, although (like Roselaar *op. cit.*) we have seen relatively few specimens from the breeding range of *paphlagoniae*, there seems to exist very little rationale to recognise such variation at nomenclatural level (for detailed discussion see Kirwan 2006). Whether nominate *pyrrhula* might wander to Turkey during the exceptional irruptions of presumed northern populations of this species that periodically occur is potentially unknown, but such birds were apparently recorded in Bulgaria during just such an influx to many parts of Europe in autumn 2004 (Pennington & Meek 2006).

Status and distribution. National conservation status: Data Deficient (population trend unknown). Generally a scarce and local resident breeder (or partial altitudinal migrant in some areas) usually in relatively small numbers in montane coniferous, more occasionally mixed, forest to at least 2,500m throughout the Black Sea Coastlands, more locally in Marmara (status in Thrace unclear) and northern Aegean regions. Probably a local resident throughout the northern fringe of Inner Anatolia, where it occurs regularly in winter, and there are several recent June and July records from the extreme northeastern Taurus (Mediterranean).

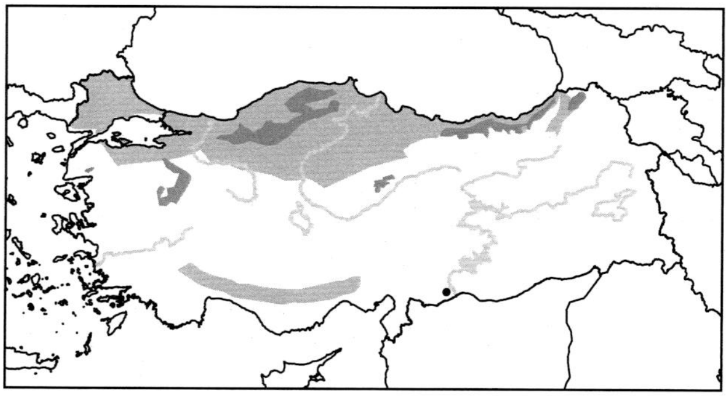

It is much more widespread, usually in small numbers, throughout the western two-thirds of the country between October and February (recorded once in Ankara, Inner Anatolia, as late as May), especially in coastal regions, and has exceptionally been recorded once at Birecik (Southeast Anatolia) at this season.

Breeding. In dense coniferous and mixed forests above 900m, mainly in Black Sea Coastlands. Site, nest and clutch size unrecorded in Turkey but, elsewhere the nest is 1–2m above ground in a bush or young tree; a shallow structure of fine twigs, moss and lichen, lined with roots and hair; usual clutch 4–5. A pair was observed feeding fledged young near Boyabat (**Black Sea Coastlands**) on 9 June 1975, indicating egg laying in early May; these birds were at *c.*1,300m on pine-covered limestone hills, interspersed with agriculture, some deciduous trees and scrub (R. F. Porter). Young were also observed at Sivrikaya on 25 August 2000 (in dense forest), and there were two at Ikizdere on 31 August 1973; these records are probably second broods, which is normal at this latitude elsewhere in the range.

HAWFINCH *Coccothraustes coccothraustes*
Kocabaş

(Du: Appelvink; Fr: Grosbec casse-noyaux; Ger: Kernbeißer; Swe: Stenknäck)

Subspecies. Probably all populations appertain to the nominate. Eastern and southern Balkan birds are apparently midway along a purported cline between nominate *coccothraustes* and Transcaucasian *nigricans* (see Roselaar 1995), and it appears safe to assume that at least western Turkish birds can be grouped with the nominate. Specimen material from eastern Turkey is apparently unavailable, and it is therefore impossible to comment on whether *nigricans* penetrates the northeasternmost part of the country. Roselaar (1995) suggests that the cline referred to above may not be valid because the species is not distributed throughout the Black Sea Coastlands of Turkey. We suggest that this absence of records may be the result of an observational lacuna and is not necessarily real.

Status and distribution. National conservation status: Least Concern (population apparently stable). A local but not uncommon resident in deciduous woodland in Marmara (locally very common in parts of Thrace), western Black Sea Coastlands and northern parts of Inner Anatolia. Present during the breeding season in the Aegean region but there is no evidence of breeding, although it almost certainly does so, at least in the north of the region. Danford (1878) reported the species breeding in the Mediterranean Taurus, but there are only two records indicative of this, an early July record in the east of this range (Roselaar 1995) and a mid-May report from just east of Mut, Içel province (P. Combridge *et al.*). Also reported from suitable breeding habitat in August in East Anatolia (Tunceli province; Helbig 1984) and July in the eastern Black Sea Coastlands (Trabzon province; Roselaar 1995). More widespread on passage across Turkey, principally from October (though there are relatively few records at major migration watchpoints such as Borçka and the Bosphorus), wintering throughout the western two-thirds of Turkey. Most winter records are concentrated along the Marmara, Aegean and Mediterranean coasts and their immediate hinterlands, where the species is locally common in a variety of wooded habitats. Also recorded from Inner and, more exceptionally, East Anatolia at this season, in which regions it probably also migrates altitudinally.

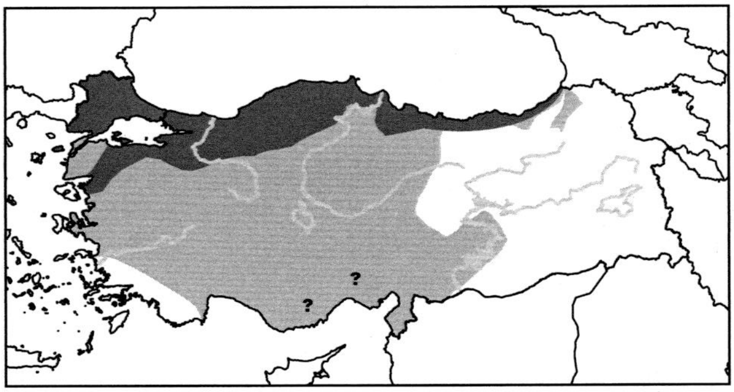

Breeding. Mainly in mature deciduous woodland, in Marmara, especially in Thrace, in western Black Sea Coastlands, and (rarely) in Inner Anatolia. Site, nest and clutch size undescribed in Turkey but, elsewhere the nest is usually higher than 4m above ground in a tree; a cup of twigs, roots and lichens, lined with finer roots, plant fibre and hair; usual clutch 4–5. In **Marmara**, a pair was carrying food in Forest of Belgrade on 14 May 1970; adults were carrying food there on 18–29 June 1973 and at Kavakköy on 21 June 1973, and an adult was feeding recently fledged young at Bebek on 17–19 June 1973 (R. F. Porter), the last record indicating egg laying about mid May. In **Inner Anatolia**, an adult was carrying food at Kızılcahamam on 5 June 1996 (GMK). In **Black Sea Coastlands**, an adult was carrying nesting material near Alapli on 9 June 1978; this record perhaps a replacement nest, or possibly a second brood, which is occasional in other parts of the range.

LAPLAND LONGSPUR *Calcarius lapponicus*
Mahmuzlu Çinte

(Du: Ijsgors; Fr: Bruant lapon; Ger: Spornammer; Swe: Lappsparv)

Subspecies. The subspecies involved is presumably the nominate, which breeds from northern Norway to eastern Siberia (Byers *et al.* 1995).

Status and distribution. Vagrant: one documented and another possible record. One was photographed at Rumelifeneri, Istanbul (Marmara), on 2 November 2006 (Bekir 2007). One briefly seen and heard at Kadılık, north of Çarsamba (Black Sea Coastlands) on 18 September 1987 (Kirwan & Martins 2000a) lacked sufficient details to be accepted as a first national record. However, the call note was convincingly described by two observers familiar with the species. Also mentioned by Braun (1901) for the Bosphorus without details. There are no other Middle Eastern records (Porter *et al.* 1996), although it has been recorded in Bulgaria and regularly occurs south to Italy in small numbers (Snow & Perrins 1998) and in larger numbers in Ukraine (Byers *et al.* 1995).

SNOW BUNTING — *Plectrophenax nivalis*
Alaca Çinte

(Du: Sneeuwgors; Fr: Bruant des neiges; Ger: Schneeammer; Swe: Snösparv)

Subspecies and taxonomy. No specimen material from Turkey is now available (see below) but the subspecies involved is almost certainly *vlasowae* which breeds from northeast European Russia to the Chukotski Peninsula, in Siberia, and winters south and west to Hungary and Romania (Byers *et al.* 1995). Klicka *et al.* (2003) recently suggested that the genus *Plectrophenax* should be subsumed within *Calcarius* on the basis that all taxa within the two currently recognised genera are from the same ancient clade. Other molecular data presented in recent years, e.g. Grapputo *et al.* (2001), also provide strong evidence of a very close relationship between the two genera, although other efforts to recreate phylogenies among the Emberizidae have not reached the same conclusion as Klicka *et al.* (*op. cit.*), namely that the taxonomic affinities of *Calcarius* appear to lie outside the sparrow assemblage.

Status and distribution. Listed by Braun (1901), Mathey-Dupraz (1922) and by subsequent authors (e.g. Kasparyan 1956, Kumerloeve 1961), based on records from the 19th century, as an irregular winter visitor and passage migrant at the Bosphorus, although only one specimen was taken (formerly lodged at the Robert's College, Istanbul; see Kirwan 1997a) to substantiate this statement. The species was apparently offered for sale in the markets in Istanbul in winter 1905/06 (Braun 1908). There are eight records from the modern period, of which four are from the Kızılırmak Delta (Black Sea Coastlands): eight, just north of Balık Gölü on 1 February 1992 (Dijksen & Klemann 1992), two flocks totalling 46 birds in the vicinity of Liman Gölü on 7 November 1992 (Kirwan 1993e), one in the delta on 1 November 2004 (O. Saglam & C. Özsemir) and three there on 28 November 2006 (Balmer & Betton 2007a). The other records involve single birds at Kulu Gölü (Inner Anatolia) on 13 February 2005 (I. Richardson), Vize (Marmara) on 21 November 2006, Iğneada (Marmara) on 27 November 2006 (Balmer & Betton 2007a) and Vize, again, on 22 October 2007 (Ö. Necipoğlu; photographed). Gallner (1976) claimed to have observed 30 individuals daily at Çaybağı, Van province (East Anatolia) between 30 August and 1 September 1973, at 2,100m, but the record undoubtedly reflects confusion with White-winged Snowfinch *Montifringilla nivalis*. In addition, a record of five or six observed by a roadside in mountains near Eleşkirt, west of Ağrı (East Anatolia) on 17 April 1981 (Goriup & Parr 1983), and a claim of one at Malatya (East Anatolia) on 19 September 2006 (Kirwan *et al.* in press) are also indicative of such confusion.

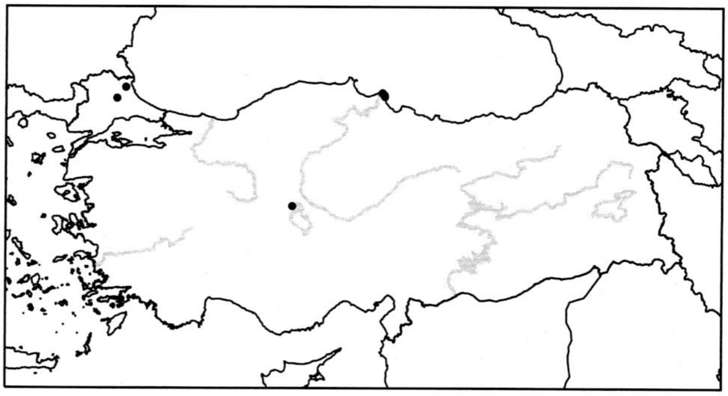

PINE BUNTING — *Emberiza leucocephalos*
Ak Başlı Çinte

(Du: Witkopgors; Fr: Bruant à calotte blanche; Ger: Fichtenammer; Swe: Tallsparv)

Subspecies. Material examined from Turkey (see below) unsurprisingly appertains to the nominate; the only other generally recognised subspecies, *fronto*, breeds in Qinghai and Gansu in western China (Byers *et al.* 1995).

Status and distribution. There are six specimens, all apparently collected during the 19th century (most by T. Robson, who was resident in Istanbul between 1861 and 1871). A male is lodged in the Liverpool Museum (National Museums & Galleries on Merseyside): it was taken at Istanbul on an unspecified date in winter 1858 (by H. Loftus) and formed part of the Canon Tristram collection (accession number T11583). A male collected at Ortaköy, now a suburb of Istanbul, on 31 October 1865 is held at the Natural History Museum, Tring (NHM 88.9.1912.524). Two further males, both adults, were taken by Robson in 1869 and sent to Dresser. They are now held in the Manchester Museum, and were taken at the 'Kagithane Hills, near Constantinople' on 21 October (BB03154) and simply 'Constantinople' on 8 November (BB03153). Two undated, one a female (presumably that noted by Mathey-Dupraz 1922), are lodged in the Robert's College collection, Istanbul (Kirwan 1997a). These were presumably obtained in the area, like most birds in this collection, but no details of locality or date are available. From the comments of Robson it appears that the species was a less than annual visitor on autumn passage, although other 19th-century authors cite it as being an uncommon winter visitor.

There are only four other documented records from Turkey: three males and three females at the Middle East Technical University campus, Ankara (Inner Anatolia) on 13 March 1974 (Beaman 1986, Kasparek 1986e), five including one male at the Middle East Technical University campus, Ankara on 2 February 1997 (Kirwan *et al.* 2003), a male photographed between Sandıklı and Kızılören, Afyon province (Aegean) on 30 November 2005 (O. Koçyiğit), and an adult female photographed at Büyükçekmece, Istanbul, on 10–16 January 2008 (F. Can; *Dutch Birding* 30:49). The Ankara area is a regular wintering locality for Yellowhammer *E. citrinella* and it is perhaps surprising that Pine Bunting has not been discovered there more frequently than has been the case.

Elsewhere in the Middle East, the species is a regular winter visitor in very small numbers to Iran (D. A. Scott), while it is a scarce but apparently regular winter visitor to northern and central Israel (Shirihai 1996), with two records in Jordan (Minshull 1996, Andrews *et al.* 1999). The species is listed as a vagrant to Iraq (Hollom *et al.* 1988). Three documented records, all of them recent, are known from Cyprus, in September, November and January/February (Flint & Stewart 1992), with a flock of 12 reported in the Troodos Mountains on 31 December 1993 (*Dutch Birding* 26: 68). In Saudi Arabia a female/immature was at Imhoff (Eastern Province) on 24 August 1979 (Bundy *et al.* 1989) and there are four recent records from Lebanon (Beale 2000, Ramadan-Jaradi *et al.* 2008).

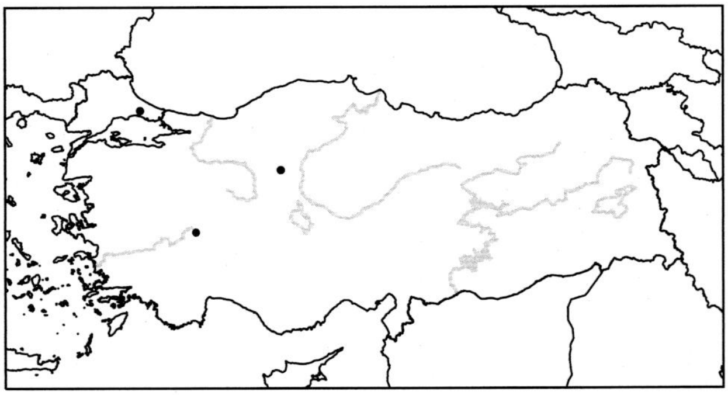

YELLOWHAMMER — *Emberiza citrinella*
Sarı Çinte

(Du: Geelgors; Fr: Bruant jaune; Ger: Goldammer; Swe: Gulsparv)

Subspecies. All apparently relate to *erythrogenys*, although breeding birds have not been examined (see Roselaar 1995). Recently, Panov *et al.* (2003) and, more cautiously, Byers *et al.* (1995), stated that a number of the traits appertaining to *erythrogenys* might, in fact, be a result of extensive hybridisation with *E. leucocephalos* and the resultant genetic overflow.

Status and distribution. National conservation status: Least Concern (population apparently increasing). Primarily a winter visitor: breeding status and distribution imprecisely known. Reported breeding in the northwest (Thrace) where probably resident in small numbers, though confirmation required. Presumably also resident locally in very small numbers in Mediterranean (in the central and western Taurus; see e.g. Deppe 1990, Kirwan & Martins 1994) and in southern Marmara (e.g. OST 1975, Martins 1989). Typically found during the breeding season in montane scrub and fragmented mixed woodland at moderate altitudes.

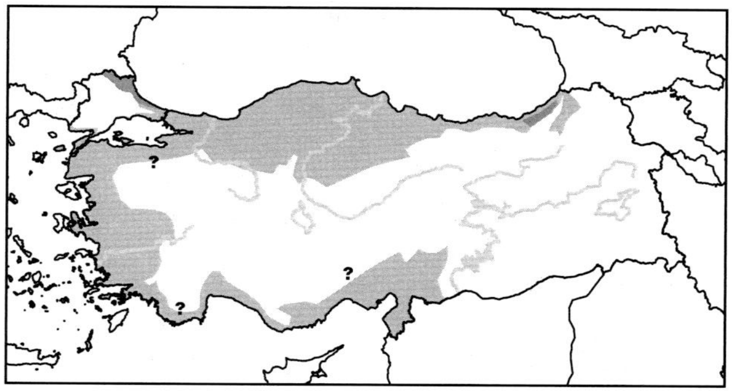

Rather less local, but generally uncommon, on passage across Turkey and as a winter visitor from early October (6th) to late March/mid April (18th) throughout the western two-thirds Turkey, with a few records of migrants from East Anatolia, where the species is known to be regular around Malatya and Elazığ in winter (MÖ, M. Ertuhan).

Breeding. Singing males have been observed, mainly in May–June (the presumed breeding season), usually in **Thrace** where breeding is probable, but there are no definite nesting records.

CIRL BUNTING — *Emberiza cirlus*
Bahçe Çintesi

(Du: Cirlgors; Fr: Bruant zizi; Ger: Zaunammer; Swe: Häcksparv)

Subspecies. Monotypic, following Svensson (1992). See also Collinson (2006).

Status and distribution. National conservation status: Least Concern (population apparently decreasing). A widespread, although probably no longer common (however, proof of any decline is unavailable) resident breeder or partial migrant in Marmara and northern, central and eastern Aegean (rare in the southern part). Also in western Black Sea Coastlands, becoming steadily more local throughout the rest of this region, and scarce in the northern fringe of Inner Anatolia and parts of the lower Taurus (Mediterranean). Breeding recorded in dense, lowland and Mediterranean scrub and deciduous woodland thickets (typically of oak *Quercus*), but more often in low hills with scrubby woodland. Such habitats rarely receive attention from observers in Turkey and thus the current status of the species requires clarification. Also occasionally recorded from the Mediterranean region in very small numbers in winter and perhaps on passage. Apparently, formerly more common in the latter region.

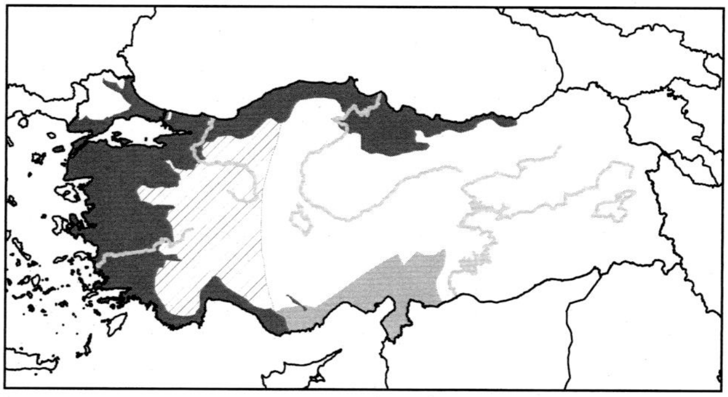

Breeding. In bushes, scrub and hedgerows, often in hilly areas, but also on flat ground and, where bushes are present, in woodland, forest edge, olive groves, and roadside verges. At Kocaçay Delta (Marmara) in 1993 it was a common breeder in open and dry parts of the alluvial forest, and in agricultural areas, with 0.39 pairs/ha in alluvial forest, but up to 1.2 pairs/ha in richly structured dry clearings (Ertan 1996). Nests were 0.7–1.6m up (n=5) in a variety of bushes, with one in a stunted pine (McNeile). The nest is a cup of grass, moss and roots, lined with hair and grass. Eggs: C/3 (1), C/4 (2), C/5 (2), C/6 (1); a record of two eggs is considered an incomplete clutch. Young: B/3 (1), B/4 (1). In **Aegean**, nest building was seen at Bafa Gölü on 3 April 1984 (Kasparek

1988a), and a nest had four almost full-grown young there on 30 April 2003, indicating egg laying in early April (PC); at Izmir, in 1950, 1951 and 1954, McNeile found seven nests: C/2 (incomplete clutch) on 22 May 1950, C/5 (slightly incubated) on 15 May 1951, C/6 (fresh) on 26 May 1951, C/4 and C/5 (fresh) on 28 May 1951, and C/4 (heavily incubated) on 28 May 1951 (some of these were possibly second broods); he also found a nest with feathered young on 27 May 1954. In **Inner Anatolia**, a nest had three newly hatched young at Kızılcahamam on 6 May 1993, indicating laying *c.*20 April (PC). In **Mediterranean**, fledged young were observed at Camiçi on 22 May 1989; a male was feeding a juvenile at Akseki on 15 July 2004 (K. P. Kendricks). In **Marmara**, a nest had three eggs at Kocaçay Delta on 3 June 2006 (A. Rustell), adults were carrying food at Çatal Dağ on 14 June 1966 and at Çamlıca on 11 June 1969 (R. F. Porter); several adults were carrying food at Kilyos on 18 June 1973, fledged young were near Vize on 21 June 1973, and recently fledged young were at Bebek on 17 June 1973 (R. F. Porter), the latter indicating laying about 21 May. In **Black Sea Coastlands**, 3–5 family parties were observed in July 1966, and adults were feeding a fledged young at Gerze on 9 June 1975.

ROCK BUNTING — *Emberiza cia*
Kaya Çintesi

(Du: Grijze Gors; Fr: Bruant fou; Ger: Zippammer; Swe: Klippsparv)

Subspecies. Complex and open to refinement. Roselaar (1995) considered the situation to be thus. Nominate *cia* occurs throughout northern Turkey, while those populations in southern Turkey, from at least the vicinity of Izmir (Aegean) to the Hatay (Mediterranean) are referred to *hordei* by Roselaar (*op. cit.*). Material examined by the latter author from extreme eastern Turkey appeared to relate to either *prageri* or intermediates between that form and *par*. Roselaar (*op. cit.*) was unable to ascribe a subspecies to the tiny populations elsewhere in Turkey. However, note that other authors (e.g. Vaurie 1956a, Byers *et al.* 1995) prefer to subsume *prageri* within *par* and *hordei* within the nominate, while neither of the two commentators mentioned acknowledged the validity of the name *barbata*, which is available for birds in southern Europe, the Balkans and Turkey.

Our own analysis of specimen material from Turkey held at the Natural History Museum (Tring) suggests that recognition of *hordei* is unwarranted, it being extremely similar to nominate *cia* in all of the characteristics listed by Roselaar for their separation. We have examined many examples of *par*, though none of *prageri*, but suggest that nomenclatural recognition of either form must be considered exceptionally dubious under the parameters we adopt here.

Status and distribution. National conservation status: Least Concern (population apparently stable). A fairly widespread and usually common resident breeder and partial migrant, though apparently highly localised and virtually absent from much of Thrace, Inner Anatolia (where probably restricted to isolated massifs, principally in the south of the region) and Southeast Anatolia (where known only from the extreme northwest) during the breeding season. Generally scarcer in the western third of the country. Nests in montane areas with rocky, scrub-covered slopes, occasionally with scattered trees. Also found in more wooded habitats, typically open *Pinus brutia* woodland, in the northeast Pontics (Black Sea Coastlands) especially. Principally recorded between 800 and 2,600m during the breeding season, more rarely at lower altitudes, e.g. to 500m in the Black Sea Coastlands and 600m in the Taurus.

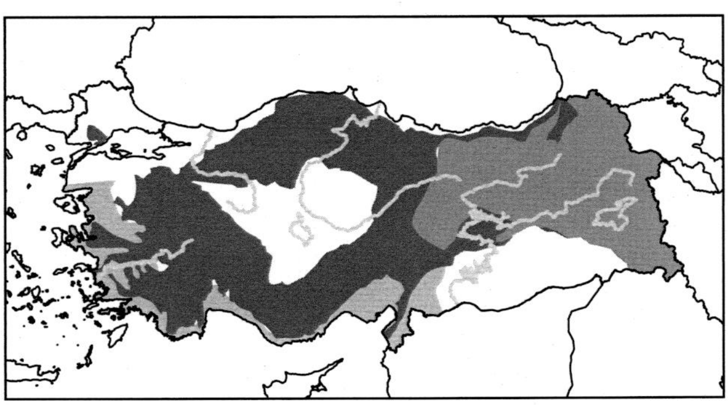

In winter the species is more widespread, descending to sea level, wintering largely in southern and western Turkey, although there are records from all regions at this season, when the species has even been recorded at the Bosphorus.

Breeding. On rocky slopes, especially limestone, with scrub and small trees; also in open woodland and gorges, and at Kızılcahamam (**Inner Anatolia**), in woodland clearings, in degraded oakwoods and on rocky slopes (Barış *et al.* 1984). Nests are on the ground partially sheltered by a shrub, low bush or rock, or in a slight cavity or recess in a low cliff or wall; a cup, mainly of grass, lined with finer grass, rootlets and some hair. Eggs: C/4 (2). A nest near Kızılcahamam had four eggs on 7 June 1977 (Pforr & Limbrunner 1982). A clutch of four in Manchester Museum was collected from a nest, under a stone, at Izmir (**Aegean**), on 28 April 1901. In **Mediterranean**, many pairs were feeding young at 1,700–2,050m on the north side of Torosdağ, on 17–23 July 1971; a pair with three fledged young at Sertavul Geçidi on 19 June 1996 (GMK), indicates laying about 22 May; an unusual nest in the same area was 1m above ground on a branch, close to the thin trunk of a small pine, and contained three recently hatched young and one egg on 13 May 2005 (J. Hodson), indicating egg laying in late April. At Nemrut Dağ (**East Anatolia**), a family party was seen 9 July 1985, and a pair was observed, carrying food and alarm-calling on 9 June 2004 (PC), and a pair was observed food-carrying at Yenimahalle, north of Çatak, on 18 May 2006 (GMK).

[HOUSE BUNTING *Emberiza striolata*]
Gri Başlı Çinte

(Du: Gestreepte Gors; Fr: Bruant striolé; Ger: Hausammer; Swe: Hussparv)

Subspecies and taxonomy. Should the species be proven to occur in Turkey, the form involved will be nominate *striolata*. However, the different forms of House Bunting appear, on the basis of vocal and morphological evidence, to form two species groups: *striolata* (Striped or Striated Bunting, including *jebelmarrae/saturiator*) in the Middle East and eastern Africa east to the Indian subcontinent, with *sahari* (House Bunting, including the weakly defined and almost certainly invalid *theresae* and *sanghae*) in northwest and western Africa (Kirwan & Shirihai 2007). This proposal has already been accepted by the Taxonomic Advisory Committee of the Association of European Records and Rarities Committees.

Species not admitted to the Turkish list (Kirwan *et al.* 1999). Kasparek (1990a, 1992a) included this species in his Turkish checklist, stating that Robson reportedly collected one in Istanbul during the second half of the 19th century, an incident that he deemed worthy of mentioning to Dresser (1871). However, the specimen was not lodged in the latter's collection, although Robson is also known to have supplied other European ornithologists with specimens, in addition to maintaining his own private collection. Indeed, Elwes & Buckley (1870) reported that the skin was retained by Robson but the specimen cannot be traced, and given the species' largely sedentary behaviour it appears appropriate to delete the species from the Turkish list. There is, however, a recent accepted record from Cyprus: one in May 1991 in the Armenokhori area (Flint & Stewart 1992), so future vagrancy is clearly possible.

CINEREOUS BUNTING *Emberiza cineracea*
Boz Çinte

(Du: Smyrnagors / Oostelijke Smyrnagors; Fre: Bruant cendré; Ger: Türkenammer; Swe: Gulgrå sparv)

Subspecies. *E. c. cineracea* (extreme southern Marmara to at least the eastern Taurus Mountains), and *semenowi* (from west of Gaziantep east to the Iranian and Iraqi borders). Roselaar (1995) placed the 'border' between the two forms further east, around the Euphrates, but study of photographs taken in the Gaziantep region and pers. obs. by GMK reveals that *semenowi* occurs there (see e.g. Martins 1989, Kirwan & Martins 2000a, Kirwan *et al.* 2003). Indeed, as long ago as 1974, field observers were noting that, at least some, birds in this area were *semenowi*. Roselaar (*op. cit.*) suggested that nominate *cineracea* occurs in this region but the Kumerloeve (1961) reference he put forward in support of this assertion is, in turn, based on Tristram (1882), who did not distinguish between the subspecies. Some observers, e.g. M. Schwalbe & K. M. Olsen, have separately claimed that both subspecies do occur, and perhaps inter-breed, in the region, but our own, more extensive experience with the species then has yet to yield evidence of either possibility. Nevertheless, the relationship between them demands study, as the two are perhaps allospecies, especially using vocalisations and molecular techniques, as none of those examined at the Natural History Museum, Tring, or the Manchester Museum appear to show reasonable trace of being intermediate, *contra* Roselaar (*op. cit.*). However, the potential overlap in the Gaziantep region, which requires confirmation, might account for the occasional reports of apparent intermediates on migration through the Levant.

Status and distribution. Conservation status: Near Threatened (BirdLife International 2000, 2004); national conservation status: Vulnerable (population apparently decreasing). First described from Turkey by Strickland in 1836 from specimens collected at Bornova, now part of Izmir (Aegean). The holotype is held in the Univ. Mus. Zool., Cambridge (CUMZ27Fri(E)/23/r/1). Its range has generally been considered disjunct in Turkey (but see discussion above). A rather local and usually comparatively scarce, though locally common, summer breeder on scrub-covered slopes in upland areas, and also at lower altitudes in relatively lush, sheltered valleys in southeastern regions. Recorded from Aegean and extreme southwestern Marmara, the southern and perhaps western and southeastern fringes of Inner Anatolia, central and eastern parts of Mediterranean (where particularly localised) and the entire Southeast and southern parts of East Anatolia north to Doğubeyazıt and Bitlis province (H. Dufourny). Both races, *cineracea* and *semenowi* breed, largely allopatrically (although see above) in Turkey; *cineracea* from Izmir and extreme southwest Marmara to the eastern Taurus Mountains, and *semenowi* from *c.*40 km west of Gaziantep east to the borders with Iran and Iraq (with an apparently isolated population breeding in the Zagros Mountains of southwest Iran). Elsewhere, the species breeds only on the Greek islands of Lesbos (Mytilene) just off the Turkish coast, Chios and Skyros, and perhaps Ikaria and Samos (BirdLife International 2000, Albayrak *et al.* 2002). In Turkey it breeds to *c.*2,000m, more usually at 800–1,200m. The winter range can only be inferred from the very few available records, all from Sudan, Eritrea, southwest Saudi Arabia and Yemen.

Recorded on spring passage, usually from the first week of April (exceptionally from mid March) on the south coast, with one recent record from western Inner Anatolia (Bradshaw & Kirwan 1994) and exceptionally on the Black Sea coast (in Kızılırmak Delta on 14 and 18 April 1992, *semenowi*; Hustings & van Dijk 1994). These records apparently constitute the third most-northerly records anywhere (the others are from Helgoland, Germany, in June 1877 [Gätke 1896] and Skagen, Denmark, in late May 2005 [*Dutch Birding* 27: 280]). Passage may persist as late as early June. Departure from the breeding areas is underway by early August, some birds dispersing northwards into areas not normally occupied (GMK, KAB, H. Schmaljohann *et al.*). A male *semenowi* in the Munzur Dağları, Tunceli province (East Anatolia), in late August was also probably a migrant. Departure from Turkey is usually complete by mid September, although stragglers have been claimed as late as 25 October (Kasparek 1985). There is a Meinertzhagen collection specimen (1965.M.16238), a male *semenowi*, in the Natural History Museum (Tring) dated 23 October and labelled as being taken at Damascus, Syria.

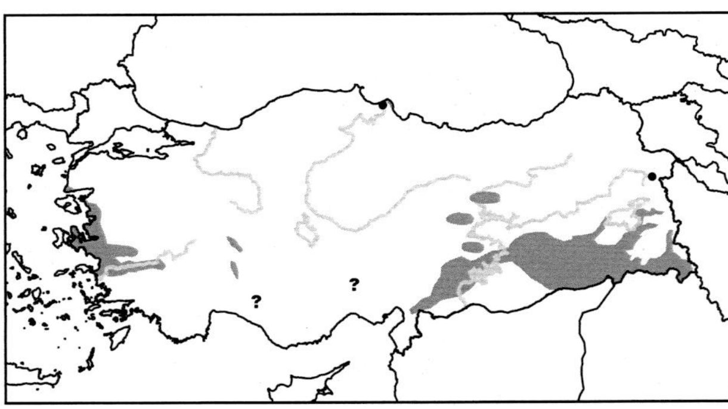

Breeding. On rock-strewn slopes with coarse short grass, scattered bushes and larger boulders, often with cliffs or large rock outcrops higher up. Nests on the ground, in sparse vegetation, sheltered on one side by a low rock or shrub. A new but empty nest at Durnalık (Southeast Anatolia), from which a female was flushed on 9 May 2004, was in an area of small, scattered trees on a rocky hillside; it was partially screened by a flat rock; a substantial cup of plant stems, grass and moss supported by a thick base of moss, and lined with rootlets (R. Castell). Eggs: C/3 (1), C/6 (1); elsewhere, usual clutch 4–6. A single egg, in Manchester Museum, was collected near Izmir (**Aegean**) on 10 May 1899. Young: B/4 (2). At Bornova, near Izmir, an agitated male with food on 23 May 1954 probably had young already out of the nest; and a fledged young, out of the nest and almost able to fly, was seen there on 20 May 1951 (McNeile), the latter indicating egg laying in late April; also near Izmir, a nest had eggs in the second half of April (Krüper 1875) and another a clutch of three eggs on 24 May (Hüe & Etchécopar 1970). Also in Aegean, 2–3 singing males were on territory at Pamukkale as early as 4 April 1996 (L. Vallotton), and there were two family parties between Menemen and Manisa on 1 July 1999 (PC). Near Baykan (**Southeast Anatolia**), on 23 May 1969, a nest had six fresh eggs and another held four young, 2–3 days old, indicating egg laying in early May (Chappuis *et al.* 1973). At Halfeti, an adult was feeding fledged young on 11 June 1996 (GMK). All other records are from near Gaziantep: copulation was observed on 8 June 1993 (PC); a female was carrying nesting material on 27 April 2005, and two nests were under construction on 18 May 2007 and 21 May 2007 (PC); there was a nest with young, 6–7 days old, on 2 June 1998 (T. Quittenden), indicating laying *c.*11 May; there was a nest with four young on 11 June 1990 (A. Forsten); adults were seen carrying food on 8 June, 14 June (twice) and from 23–26 June; several pairs were with recently fledged young on 6–7 June 2006 (S. Williams), indicating peak egg laying for first broods in early May; an adult was feeding a fledgling on 14 June 2001 (R. P. Martins); a female was feeding three fledged young 16 August 1974, probably a second brood. On Lesbos (Greece), a pair was observed was carrying food on 7 May 1995, and a nest (of a different pair) had large young on 14 May 1995 (T. Quittenden).

ORTOLAN BUNTING *Emberiza hortulana*
Kirazkuşu

(Du: Ortolaan; Fre: Bruant ortolan; Ger: Ortolan; Swe: Ortolansparv)

Subspecies. Monotypic.

Status and distribution. National conservation status: Least Concern (population apparently stable). A widespread and common summer visitor to dry, scrubby, cultivated and rocky areas, usually breeding in uplands at 750–2,600m, exceptionally to 3,300m in East Anatolia (Kumerloeve 1969a). Also breeds in wooded margins at lower altitudes in Aegean (80–300m) Marmara (above 300m) and Black Sea Coastlands (above 200m; Kumerloeve 1962b). Apparently absent as a breeder from much of Inner Anatolia and rather local in Thrace.

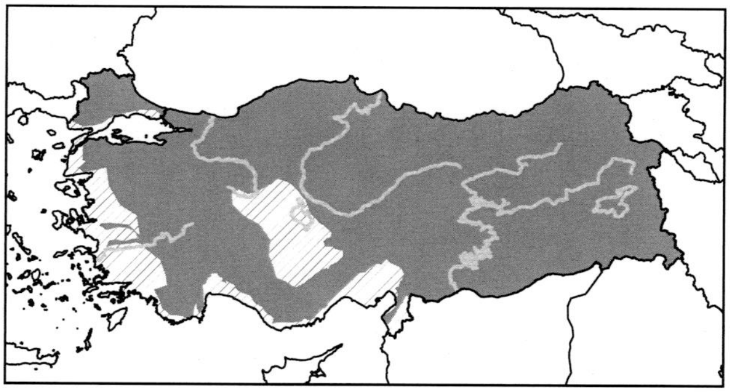

On both spring and autumn passage considerably more widespread and commonly recorded in all regions, occasionally in large concentrations exceeding 100 individuals e.g. in passage movements at Ardeşen, Black Sea Coastlands, in September 1973. Spring migrants arrive from the end of March (or, more exceptionally as early as 16 March) but are never widespread until at least mid April. Passage continues until at least mid May, even in the south. Autumn passage peaks in late August and the first half of September, a few remaining until the end of September (latest, 31 October).

Breeding. In open grassy areas, often hill slopes, with sparse scrub, sometimes with rocky outcrops or scattered trees; also in pastures and at edges of cornfields, at forest edges and in open woodland. Nests on the ground in grass, or on bare ground, sheltered by a tuft of vegetation, low shrub or rock; builds a cup of plant stems, grass and roots, lined with finer grass, roots and hair. Eggs: C/3 (1), C/4 (1). Young: no data. A nest had four eggs at Işıklı (**Southeast Anatolia**) on 8 May 1989 (I. A. Green). In **East Anatolia**, many juveniles were present in the Varto area in July 1969, and a family party was between Van and Erçek on 9 July 1986 (R. Webb). In **Black Sea Coastlands**, a nest had young, *c.*6 days old, near Bolu on 28 June 1977 (Pforr & Limbrunner 1982), indicating first egg on 7 June; adults were carrying food at Abant Gölü on 5 June 1996 (GMK), and at Kale on 11 July 1986 (R. Webb); a family party was at Ispir on 28 June 1969; two fledged young, fed by the adults, were near Eşkipolat on 11 June 2004 (PC). In **Inner Anatolia**, a nest had three eggs at Kızılcahamam on 28 June 1977, fledged young were seen there on 8 June 1977 (Barış *et al.* 1984), and an adult was carrying food on 5 June 1996 (GMK); fledged young were observed at Sultan marshes on 17 June 1977 (Kasparek 1985). In **Marmara**, a pair, with one fledged young, was at Çatal Dağ on 14 June 1966 (R. F. Porter), and a juvenile was near Ayvacık on 18 July 1988 (M. Telfer); recently fledged young at Kosedere on 26 June 1973 (R. F. Porter), indicate laying in late May. A single egg in Manchester Museum was collected by Krüper, near Izmir (**Aegean**), on 27 May 1876. Adults were feeding newly fledged young on Torosdağ (**Mediterranean**), 17–23 July 1971, indicating laying in late June. Song is heard mainly late April–July, and the records above indicate double-brooding, as elsewhere in the range.

GREY-NECKED BUNTING *Emberiza buchanani*
Doğu Kirazkuşu

(Du: Steenortolaan; Fre: Bruant à cou gris; Ger: Steinortolan; Swe: Bergortolan)

Subspecies. *E. b. cerrutii* but the level of geographical variation is slight (three races have been recognised) and the species is probably better regarded as monotypic.

Status and distribution. National conservation status: Least Concern (population apparently stable). A summer visitor, usually in relatively small numbers, principally to open upland areas in the eastern third of Turkey, in Southeast and East Anatolia, as far north as the Upper Aras Valley and as far west as the Munzur Dağları, Tunceli province (East Anatolia), and locally at Siverek (Southeast Anatolia), although absent further north-east. It arrives mainly in late April to early May. Breeds in dry uplands, generally above 1,800m (but occasionally below 1,000m) and to at least 3,000m. It overlaps with Ortolan Bunting *E. hortulana* at lower elevations in certain areas,: in others the two exclude one another (Martens 1979). Apparently nests at some density, at least locally (pers. obs.). There is a recent claim from the Ala Dağları (Mediterranean) in May 1990, well to the west of the species' known range (Kirwan 1990), which almost certainly reflects confusion with similar congeners. Certainly, there have been no other claims of this species from this comparatively well-watched area. However, there is a certain record, of a migrant, in a gorge south of Toprakkale (Mediterranean) on 9 May 2001 (Balmer & Betton 2002a) and another, involving a male, at the Göksu Delta (Mediterranean) on 15 April 1996 (J.-P. Paul). Autumn migration is little known, although birds certainly remain in the breeding areas until the latter half of September (latest available dated record 19th), though Kumerloeve (1969a) mentioned a small flock east of Doğubeyazıt (East Anatolia) in early November 1964. The species winters in west and central India (Sibley & Monroe 1990).

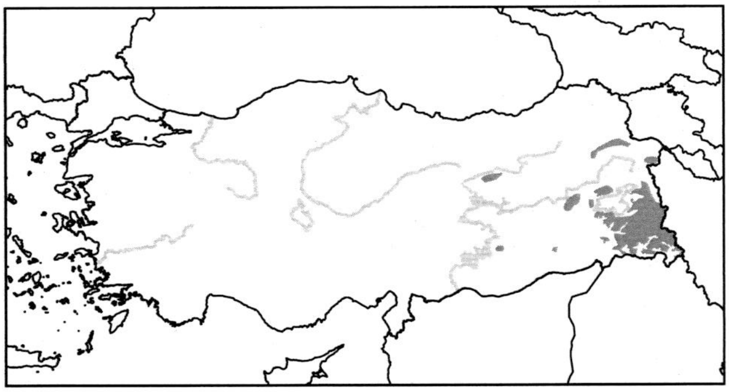

Breeding. On dry, high-altitude bare slopes (usually steep) and rocky areas with low, sparse vegetation, in extreme east and southeast Turkey. Nests are on the ground, sheltered by a low rock, shrub or tussock. A solitary breeder but eight singing males were evenly distributed along a 1.5km stretch of steep rocky hillside near Erçek, east of Van (**East Anatolia**), on 19 June 2004; most males were singing from a line of cliffs, and a nest (four eggs) was found just 4m from the base of the cliffs; it was on the ground and partially screened by a low shrub, 10cm in height; a cup of plant stems and grasses, higher on the exposed side, with a lining of fine grass and hair (S. Williams). The usual clutch, elsewhere, is five (3–5). Copulation was observed west of Görentaş (East Anatolia) on 2 June 1969 (R. F. Porter; erroneously cited as 31 May in the TBR). A nest had downy young near Van (East Anatolia) on 20 June 1990 (T. Bakker & W. Steenge), indicating egg laying in early June. Family parties/juveniles were observed in East Anatolia at Akdoğan Gölü on 23–24 July 1992 (G. Magnin), and at Işak Paşa Sarayı, near Doğubeyazıt, on 29 July 2000 (K. Haataja). In Armenia, nest building was observed on 21 May 1995 and 20 June 1972; a nest contained eggs on 29 May 1962, and another had five eggs on 15 June 1972, which hatched on 18 June 1972 (Adamian & Klem 1999). The species may be double-brooded in Turkey, as elsewhere.

CRETZSCHMAR'S BUNTING *Emberiza caesia*
Kızıl Kirazkuşu

(Du: Bruinkeelortolaan; Fre: Bruant cendrillard; Ger: Grauortolan; Swe: Rostsparv)

Subspecies. Monotypic.

Status and distribution. National conservation status: Least Concern (population apparently decreasing). A fairly widespread and locally common summer visitor, primarily to southern and western parts of Marmara, the Aegean coastlands, Mediterranean and Southeast Anatolia. Present locally in Thrace, the southern fringe of Inner Anatolia and even more locally in eastern parts of Southeast Anatolia and adjacent regions of East Anatolia. Has been recorded in summer in the western Black Sea Coastlands. Typically breeds on bare rocky hillsides with Mediterranean scrub or open pine forest, but occasionally in a variety of more vegetated habitats, from near sea level, typically between 200 and 800m and less usually to 1,200m.

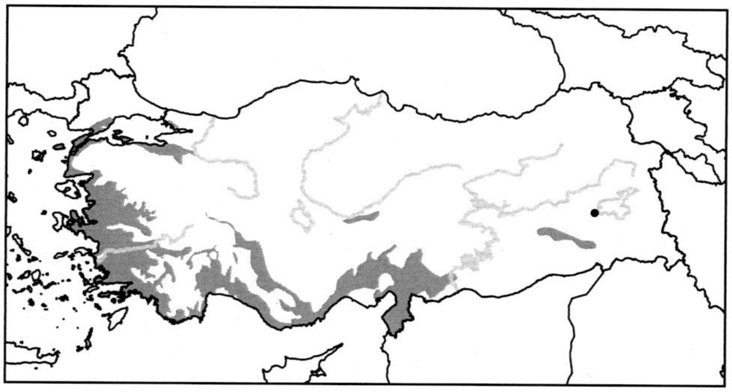

The main spring arrival is in early April, with the vanguard arriving from mid March (10th). Autumn migration generally commences in mid August. Rarely recorded after the third week of September (20th). Slightly more widespread on passage especially in autumn, when doubtfully recorded in extreme East Anatolia as late as early November (Kumerloeve 1967a).

Breeding. In rocky valleys and dry hill slopes with scattered scrub and bare ground with shorter vegetation, and high points, such as bushes, rocks and occasionally trees, for song posts. Nests are well concealed, on the ground, usually at the base of a shrub or rock, and partially screened by it; also nested in a field of barley in Mediterranean (Castell 1997). The nest is a cup of grasses and plant stems, lined with grass, roots and hair. Eggs: C/3 (1), C/5 (1). Young: B/4 (3), B/5 (2). In **Mediterranean**, adults were singing on territory, near Silifke, on 31 March 2000; a nest had five eggs near Acigöl on 10 May 1899 (Selous 1900); another had five half-grown young near Finike on 15 May 1974, another five young, *c.*2 days old, north of Silifke, on 22 May 1993 (PC), and another had young, north of Göksu Delta, on 30 May 1995 (I. A. Green); around Akseki, a pair was nest building on 24 May 1999 (R. F. Porter), a fledged young was seen on 22 June 1990 (A. Forsten), and recently fledged young were being fed by adults on 24 May 1999 (R. F. Porter), the latter indicating laying in late April. In **Aegean**, there were recently fledged young at Manisa on 4 May 2001, indicating egg laying in early April (PC); four young fledged from a nest near Izmir on 11 May 1951 (McNeile); McNeile also observed adults food-carrying at Manisa on 24 May 1950 and 8 May 1951, adults alarm-calling near Izmir on 2 May 1950 and 4 May 1951, and fledged young near Izmir on 20 May 1951; at Akköy, fledged young were observed on 22 May 1999 (J. Hollinrake), a nest had three eggs on 30 May 2004, and another four young, *c.*2 days old, on 1 June 2004 (M. Demain), the latter indicating laying in mid May. In **Inner Anatolia**, nest building was observed, and a nearby nest had young, at Hasan Dağ, on 10 June 1998 (T. Quittenden). At Işıklı (**Southeast Anatolia**), a nest had four young, *c.*6 days old, on 8 June 1993, two broods of recently fledged young were seen on 7 June 2006 (PC), and an adult was carrying food on 14 June 1996 (GMK). A male was feeding young near Tatvan (**East Anatolia**) on 10 July 1979 (van den Berg & Damhuis 1981). These records indicate double-brooding, as elsewhere in the range.

RUSTIC BUNTING *Emberiza rustica*
Ak Kaşlı Çinte

(Du: Bosgors; Fre: Bruant rustique; Ger: Waldammer; Swe: Videsparv)

Subspecies. Specimen material from Turkey, unsurprisingly, accords with the nominate race, which breeds from northern Scandinavia east to central Siberia (see Byers *et al.* 1995).

Status and distribution. There are five specimens from the latter part of the 19th century (probably all collected after 1860), including two from the Istanbul area, apparently collected by Robson, and two from the early 20th century. At least one, an immature female caught on lime twigs at Büyükdere on 24 October 1871, was sent to Dresser (Kasparek 1990a), and is currently held in The Manchester Museum (accession no. BB02963). The other specimen is not attributable to any year, but was taken to the north of Istanbul on 14 February. Further south, an immature male was collected by Schrader near Mersin, Içel province (Mediterranean) on 12 December, between 1882 and 1884, and deposited in the Laubmann collection (now held in the Zoologische Staatssammlung Munich, Germany; Roselaar 2003). Additionally, I. C. Parrot collected a Rustic Bunting in northwest Turkey at Pirgos (locality perhaps present-day Yuvacık, Çanakkale province) on 24 February 1904 (Kumerloeve 1966d, 1975a). Kirwan (1996b) mentioned an immature, possibly male, specimen held in the Robert's College, Bebek (Istanbul) collection, which apparently does not refer to any of the above-reported specimens, though it too was presumably taken in the environs of the city in the late 19th or early 20th century. More recently, two records documented photographically: one near Gaziantep (Southeast Anatolia) in October 2004 (details of photographer unavailable) and one, perhaps a male, at Karasu, Sakarya province (Marmara) on 18 January 2008 (S. Bekir).

Elsewhere in the Middle East, the species is listed as a rare migrant in Iran and Israel, and as a vagrant in Egypt, Kuwait, Oman, Saudi Arabia (Eastern Province), Syria, the United Arab Emirates (Hollom *et al.* 1988), Jordan (Tebb & Hamidan 2002) and Lebanon (Beale 2000, Ramadan-Jaradi *et al.* 2008).

LITTLE BUNTING — *Emberiza pusilla*
Küçük Çinte

(Du: Dwerggors; Fre: Bruant nain; Ger: Zwergammer; Swe: Dvärgsparv)

Subspecies. Monotypic.

Status and distribution. Vagrant, though perhaps best treated as hypothetical, given the lack of documentation for any of the following reports. Three claimed among a flock of Yellowhammer *E. citrinella* at or near Erzurum (East Anatolia) on 21 March 1910 are not supported by substantiative details (McGregor 1917). Mathey-Dupraz (1922) is vague concerning records from the Bosphorus (undated) though he saw two specimens held in the Robert's College (Bebek, Istanbul) collection and Elwes & Buckley (1870) considered it to be a 'rare winter visitor' in the region, while Krüper (1875) reported that von Gonzenbach collected the species, presumably around the middle of the 19th century, in the Izmir area (Aegean). More recently, G. Trommer (*in* Kumerloeve 1969a) observed seven at Tatvan (East Anatolia) on 10 November 1968, but subsequently withdrew the record. Two modern claims are known, both in the Kızılırmak Delta (Black Sea Coastlands): one seen and heard briefly on 15 April 1992 (GMK, R. Pahlplatz) and a male closely observed on 26 April 1992 (Hustings & van Dijk 1994), although full details have never been presented for either record.

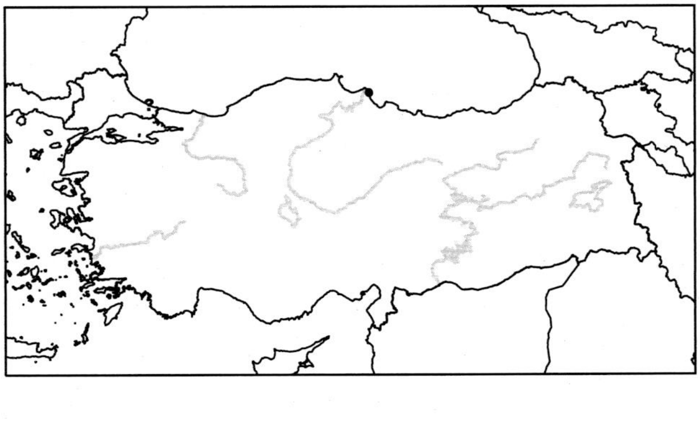

Little Bunting is a vagrant, or possibly very rare migrant, to the following Middle Eastern countries: Iran, Israel, Jordan, Kuwait, Lebanon, Oman and the United Arab Emirates (Hollom *et al.* 1988, Ullman 1991). A published record from Cyprus, in October 1957, was subsequently rejected by Flint & Stewart (1992).

[YELLOW-BREASTED BUNTING *Emberiza aureola*]
Söğüt Çintesi

(Du: Wilgengors; Fr: Bruant auréole; Ger: Weidenammer; Swe: Gyllensparv)

Subspecies. Should the species ever occur in Turkey, the subspecies involved would surely be the nominate which breeds from northern Scandinavia to Kamchatka (see Byers *et al.* 1995).

Species not admitted to the Turkish list (Kirwan *et al.* 1999). Though listed as a vagrant by Hollom (1971), Hollom *et al.* (1988) and Turan (1990), no evidence to support this has been traced. The error was corrected by Porter *et al.* (1996) and Snow & Perrins (1998). The species has been reliably reported from several other Middle Eastern countries, including Cyprus; one in May 1974 (Flint & Stewart 1992). There are four records from Greece: single birds in January 1956, May 1961, September 1974 and September 1976 (Handrinos & Akriotis 1997).

COMMON REED BUNTING *Emberiza schoeniclus*
Bataklık Çintesi

(Du: Rietgors; Fre: Bruant des roseaux; Ger: Rohrammer; Swe: Sävsparv)

Subspecies. Breeders belong to one, or perhaps more, of the intermediate or thicker billed races; *intermedia*, *reiseri* and *caspia* (which form a natural grouping, sometimes named Dark Reed Bunting, and apparently separate in their evolutionary lineage from Northern Reed Bunting *schoeniclus* or Pale Reed Bunting *E. pyrrhuloides*; see AOU 1998). The form *intermedia* was identified as breeding in the Black Sea Coastlands and East Anatolia by field observers (see OST 1975), but specimen evidence refutes this idea. Roselaar (1995) does not acknowledge the presence of *intermedia* in the Asia Minor region, and Byers *et al.* (1995) restricted its geographical range to Italy, Corsica and Adriatic Croatia. Roselaar (*op. cit.*) considered that those populations breeding in much of the western two-thirds of Turkey are ascribable to *reiseri*, with *caspia* occurring in the extreme east (these two appear to form a cline), and this view was seconded by Byers *et al.* (*op. cit.*). Birds taken at the former Amik Gölü (Mediterranean) were ascribed by their collector, Aharoni, to *pyrrhuloides*, but these and others breeding on the south coast of Turkey were considered plausibly to represent *korejewi* (Roselaar *op. cit.*), described from southeast Iran but considered to occur north into northeast Iran/Afghanistan and west to northern Syria by Roselaar (*op. cit.*). Byers *et al.* (*op. cit.*), however, limited *korejewi* to southeast Iran and questioned its validity. Winter immigrants belong largely to the nominate form or to intermediate-billed races: Kumerloeve (1966–67) considered those he collected at Amik in winter to be most similar to *E. s. ukrainae*, although Roselaar (*op. cit.*) considered *tschusii* to have been taken there. Our examination of specimen material from relevant related regions, e.g. the Balkans, suggests that winter immigrants, at least in western Turkey, are likely to involve many nominate/*stresemanni* and *tschusii*/*othmari*, but that *stresemanni* is insufficiently well marked to warrant separation from *schoeniclus* and that *tschusii*, *othmari* and, perhaps also *reiseri* are best treated under a single name, for which *othmari* would have priority.

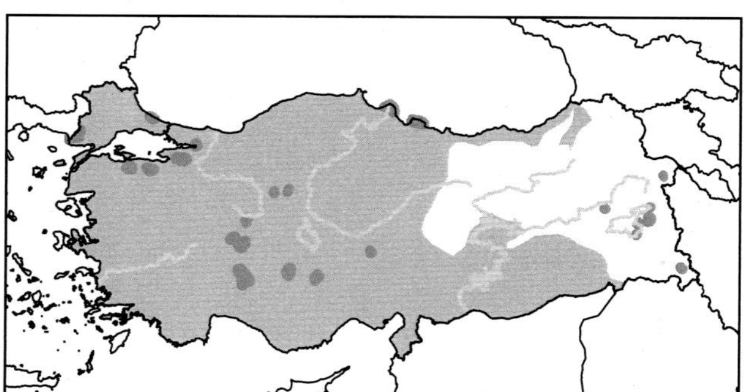

Status and distribution. National conservation status: Near Threatened (population apparently decreasing). A common resident breeder, largely in marshy wetlands of Black Sea Coastlands: where the 800–1,200 pairs in Kızılırmak Delta in spring 1992 almost certainly represents the largest national population. Also in Inner and East Anatolia (south as far as the Iraq/Iran borders), being less common and extremely local in similar habitat in the Marmara (including Thrace) and Aegean regions. Breeds to at least 2,000m in East Anatolia and at least 1,000m in central Turkey.

Widespread on passage, particularly in autumn, and usually common and widespread at most wetlands in the western two-thirds of the country during winter, from late October to May, sometimes forming flocks of up to 250 birds, especially to roost.

Breeding. In open marshes, reedbeds and in waterside vegetation at lakes. The nest is usually well concealed, close to the ground, in rank vegetation; of plant stems, grass and moss, lined with hair and fine grass. At Eşmekaya (**Inner Anatolia**), a nest held a complete clutch of three eggs on 25 May 1992, and another four newly hatched young on 23 May 1992, the latter indicating laying in early May (PC). Nest building was observed at Eber Gölü (Inner Anatolia) on 8 May 1967 (R. F. Porter). At Kızılırmak Delta (**Black Sea Coastlands**), a nest with eggs was found on 5 May 1992 (Hustings & van Dijk 1994); two pairs, each with four young, and a pair with three young, were seen in July 1972 (Dijksen & Kasparek 1985). At Van marshes (**East Anatolia**), an adult and juvenile were observed on 14 July 1986, and a family party on 4 August 1992. Whilst the records above are few, the number of eggs/young do not exceed four, whereas the usual clutch size, elsewhere, is 4–5. These records indicate double-brooding, as in other parts of the range.

[RED-HEADED BUNTING *Emberiza bruniceps*]
Kızıl Başlı Çinte

(Du: Bruinkopgors; Fre: Bruant à tête rousse; Ger: Braunkopfammer; Swe: Stäppsparv)

Subspecies. Monotypic.

Species not admitted to the Turkish list. Breeds in Central Asia, largely between the Caspian Sea and the Altai Mountains, and in northeastern Iran, northern Afghanistan and Pakistani Baluchistan (Roberts 1992). A record in spring 1901, of an individual in the bird market in Istanbul, which had 'obviously' been captured in the surroundings of the city (Kumerloeve 1966d, Kasparek 1990a), was perhaps not clearly a wild bird.

Records considered as referring to genuine vagrants are available from Israel, Kuwait and northwest Saudi Arabia (Hollom *et al.* 1988).

BLACK-HEADED BUNTING *Emberiza melanocephala*
Kara Başlı Çinte

(Du: Zwartkopgors; Fre: Bruant mélanocéphale; Ger: Kappenammer; Swe: Svarthuvad sparv)

Subspecies. Monotypic.

Status and distribution. National conservation status: Least Concern (population apparently increasing). A very common and widespread summer visitor to all regions, though apparently absent from much of Black Sea Coastlands, particularly the eastern part, during the breeding season. A characteristic breeding species in all types of agricultural and scrubby areas, usually below 2,400m, though occasionally nests as high as 2,900m, and recorded at even higher altitudes on migration.

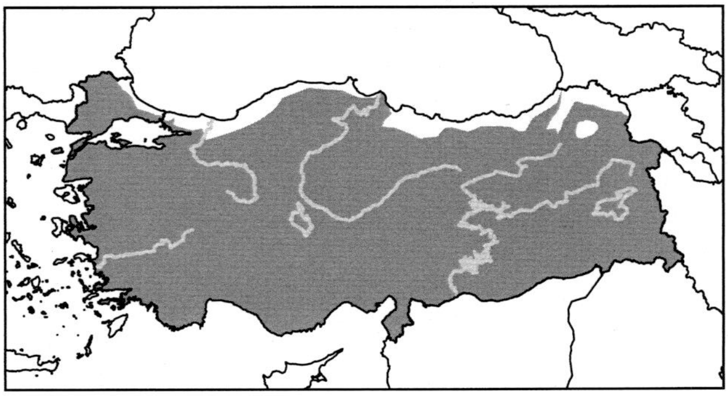

Spring arrival is from mid April on southern and western coasts and in Southeast Anatolia, reaching Inner Anatolia, Marmara (including Thrace) and Black Sea Coastlands from mid/late April or, more usually, early May, although passage through Turkey is clearly substantial until at least mid May. Earliest date is 28 March (twice). Typically occurs on migration in small parties, sometimes in flocks numbering hundreds of individuals, with males perhaps tending to arrive first, at least in some regions of the country. Breeding usually commences from early to mid May. Return migration commences in late August and is usually complete by late September (latest, 28th), with one October claim, from Van (East Anatolia), on 31 October (M. Grunwell).

Breeding. Mainly in fairly dense scrub, bushes and hedgerows adjacent to arable fields, but also in gardens, orchards, open woodland and at weedy edges of large open cornfields. Nests in a bush, small tree, or in weeds, thistles and tangled undergrowth, usually 0.5–1.0m above ground (n=31), but up to 2.5m; possibly also nests on the ground in open crops devoid of bushes, but no evidence. The nest is a cup of grass stems (often with seed casings still attached) and dead leaves, lined with finer grasses and hair. Eggs: C/3 (7), C/4 (28), C/5 (12); others considered incomplete clutches. Young: B/3 (1), B/4 (6), B/5 (5). Arrives on territory late April–early May in most regions, but from mid April in Mediterranean. At Altınkum (**Aegean**), there were very few birds on 23 April 2003 but by 3 May hundreds were present, with much chasing, singing and alarming, and a half-built nest on 4 May had two eggs on 8 May (PC). Elsewhere in Aegean there are records of 21 nests with eggs in May (earliest an incomplete clutch of two on 7 May); one had a cuckoo's egg on 19 May 1950 (McNeile); there were four nests with small young in late May. In **Southeast Anatolia**, nest building was observed on 5 May 1992 at Halfeti on 3 June 2001, on 10 May 2004 at Gaziantep, and on 11 May 2004 at Birecik (PC); a nest had four eggs on 7 May 1970; at Işıklı and Durnalık, two nests had eggs on 17 May 2004 (M. Farmer), and five nests with eggs were found in the first half of June (T. Quittenden); a nest had four small young at Yesılce on 3 June 2001 and there were four nests with young there on 6–12 June 2006 (PC); fledged young were observed on 12 June 1996, and several pairs with flying young on 14 June 1996 (GMK). In **Inner Anatolia**, five nests had eggs in the last week of May, and six nests eggs in the first half of June; nest building was observed on 10 June 1998, one nest had young on 10 June 1998 (T. Quittenden), and there was a pair with fledged young on 8 August 1998 (A. Barfull). In **Mediterranean**, nest building was observed at Amik Gölü on 2 May 1970 (R. F. Porter), two nests were under construction at Uzuncaburç on 15 May 2004, and one was empty, but ready for eggs, at Demirkazık on 13 May 2004 (N. Williams); eight nests held eggs mid May to mid June, a nest had newly hatched young in mid June, and four broods of fledged young were observed in June (earliest 10 June) and July. In **Black Sea Coastlands**, family parties were at Ispir on 26 July 1986. In **Marmara**, a nest had four eggs near Manyas Gölü on 6 June 1966, adults were carrying food at Büyükçekmece on 19 June 1973, and a pair with recently fledged young were at Kosedere on 25 June 1973 (R. F. Porter): the last record indicating laying in late May. In **East Anatolia**, four nests had eggs in the first half of June, the earliest an incomplete clutch of one on 5 June 2001 (T. Quittenden). The species is, presumably, double-brooded in Turkey, as elsewhere.

CORN BUNTING *Emberiza calandra*
Tarla Çintesi

(Du: Grauwe Gors; Fre: Bruant proyer; Ger: Grauammer; Swe: Kornsparv)

Subspecies and taxonomy. Roselaar (1995) and Byers *et al.* (1995) considered the forms occurring in Turkey to be nominate *calandra* (virtually throughout) and *buturlini* (from perhaps the Euphrates to Ceylanpınar, Southeast Anatolia). However, geographical variation within this species is generally rather slight, certainly clinal and may be best ignored taxonomically (see also Vaurie 1956a). Has been placed within *Emberiza* (e.g. Sibley 1996) and both Grapputo *et al.* (2001) and Lee *et al.* (2001) presented phylogenetic support for such a suggestion; Corn Bunting appears to be basal to a clade otherwise containing Yellowhammer *E. citrinella* and Cirl Bunting *E. cirlus*, which in turn is a sister group to the clade comprising Reed *E. schoeniclus*, Rustic *E. rustica* and Little Buntings *E. pusilla*. The removal of *calandra* from *Miliaria* to *Emberiza* now appears to be widely supported (see Collinson 2006) and is followed here.

Status and distribution. National conservation status: Least Concern (population apparently decreasing). A widespread and common breeding resident, and apparently partial migrant in Inner and East Anatolia (where it breeds in high densities in some areas), throughout most of the country, although more local in the Black Sea Coastlands and Southeast Anatolia (where suitable habitat probably increasing). In the former there were an estimated 1,500–2,000 pairs in Kızılırmak Delta, but it is obviously much rarer in many other parts of the region. Common in cultivation, scrubby semi-steppe and other open habitats, occurring to 2,300m (probably lower in the west and centre of the country).

Post-breeding flocks, numbering several hundreds, occasionally as many as 1,000, assemble between late August and October, and are apparently augmented by immigrants from outside the country, although there is little evidence of passage at the main migration watchpoints. Large movements of birds moving further south to winter elsewhere in the Middle East may occasionally be observed on the south coast in late autumn. In winter apparently recorded only from the western two-thirds of the country, sometimes in large flocks, e.g. there were 1,500 at Eşmekaya (Inner Anatolia) on 21 January 1992.

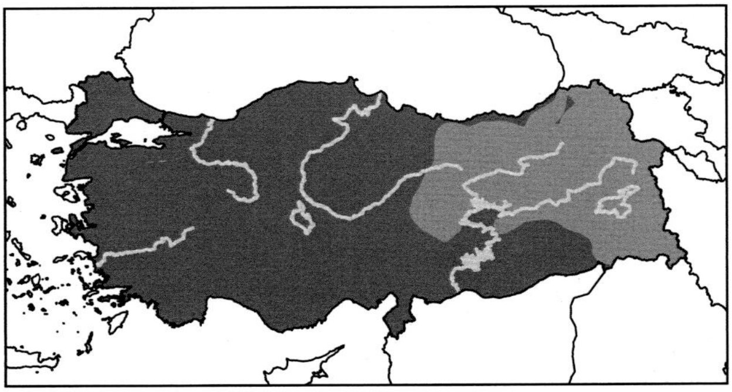

Breeding. Mainly in wheat and other crops, in grassland, and in open scrub, sometimes with hedgerows and scattered trees. Nests on the ground, screened by growing crops, or sheltered under a shrub (three), or above ground in a bush, or in tangled grass growing through a bush (seven); nests are more likely to be found in bushes, and it is probable that most nests are on the ground, despite the numbers above. The nest is a cup of grass, lined with finer grass, roots and hair. Eggs: C/4 (2), C/5 (1), C/6 (6). Young: B/3 (1), B/6 (1). In **Aegean**, two nests had six eggs on 20 April 2002, one nest had newly hatched young on 24 April 2002, indicating first egg on 6 April, and another young, *c.*5 days old, on 3 May 2001, all near Altınkum (PC); nearby, a nest had six eggs on 24 May 2004, and another six eggs on 29 May 2004, both nests at 0.8m in bushes (M. Demain); a nest had an incomplete clutch of three on 20 April 1951 (and a full clutch of six on 24 April 1951), and another had four heavily incubated eggs on 26 May 1951, both near Izmir (McNeile). At Kocaçay Delta (**Marmara**), in 1993, there were 0.76 pairs/ha in salt plains and 0.43 pairs/ha in fallow areas, with lower densities in agricultural areas and dunes (Ertan 1996). Near Bandırma, a nest, on the ground, had four eggs and two young on 8 June 1966 and another was three-quarters built on 25 April 1966 (R. F. Porter). At Göksu Delta (**Mediterranean**), a nest had two eggs on 5 May 2004, and subsequently a full clutch of five on 16 May 2004 (J. Mason). In **Southeast Anatolia**, a nest had four eggs in a bush at Birecik on 20 May 2007 (PC) and there were four fledglings at Halfeti on 11 June 1996 (GMK). In **Inner Anatolia**, arrives on high ground, singly or in pairs, in late February to mid March, with song March–July, occasionally October (Wadley 1951); at Erzurum (**East Anatolia**), in 1911–12, birds arrived in late April and left in early August (McGregor 1917). At Kızılırmak Delta (**Black Sea Coastlands**) in 1992 there were very high densities of 35–50 territories/100ha in drier grassland, on arable land and in very open dunes; the first nest with eggs was on 17 May, and many adults were carrying food from 1 June (Hustings & van Dijk 1994). Elsewhere in the range it usually rears two broods per season, and it is rather surprising that there are no observations of fledged young in July, or later, in Turkey.

BIBLIOGRAPHY

Acar, B., Beaman, M. and Porter, R. F. 1977. Status and migration of birds of prey in Turkey. *In*: Chancellor, R. D. (ed.) *World Conference on Birds of Prey, Vienna 1–3 October 1975*. International Council for Bird Preservation, Cambridge.

Adamian, M. S. and Klem, D. 1999. *Handbook of the Birds of Armenia*. American University of Armenia, Oakland.

Adızel, Ö. 1998. Van Gölü havzası ornitofaunası üzerine araştırmalar. Ph.D. thesis. Van 100 Yıl Üniversitesi.

Aharoni, J. 1930. Brutbiologisches aus dem Antiocha-See. *Beitr. Fortpfl. Biol. Vogel* 6: 145–151.

Aharoni, J. 1931. Brutbiologisches aus der Syrische Wüste und dem Libanon. *Beitr. Fortpfl. Biol. Vogel* 7: 161–166, 222–226.

Akçakaya, H. R. and Akçakaya, R. 1992. The extinction of the Northern Bald Ibis (*Geronticus eremita*) population in Birecik and evaluation of conservation efforts. *Turk. J. Zool.* 16: 1–12.

Akman, Y. 1995. *Türkiye Orman Vejetasyonu*. Ankara Üniversitesi Yayınları, Ankara.

Albayrak, T., Gürsoy, A. and Kirwan, G. M. 2002. International Species Action Plan for the Cinereous Bunting (*Emberiza cineracea*). BirdLife International, Wageningen.

Albayrak, T. and Erdoğan, A. 2005. Habitat requirement and the breeding ecology of Kruper's Nuthatch (*Sitta krueperi*) in Antalya, Turkey. Poster abstract. EOU Conf., Strasbourg, August 2005.

Albrecht, J. S. M. 1977. Winter birds in Karadeniz-Ereğli 1976–1977. *Orn. Soc. Turkey Bull.* 14: 5–7.

Albrecht, J. S. M. 1981. Red-breasted Flycatcher – a new breeding species for Turkey. *Sandgrouse* 3: 91–92.

Albrecht, J. S. M. 1984. Some notes on the identification, song and habitat of the Green Warbler in the western Black Sea Coastlands of Turkey. *Sandgrouse* 6: 69–75.

Albrecht, J. S. M. 1986. Notes on the birds of Eregli, Black Sea Coastlands, Turkey 1976–1978. *Sandgrouse* 8: 74–92.

Alléon, Comte A. 1880. Catalogue des oiseaux observés aux environs de Constantinople. *Bull. Zool. Soc. France* 5: 80–116.

Alléon, Comte A. 1886. Mémoire sur les oiseaux dans la Dobrodja et la Bulgarie. *Ornis* 2: 397–428.

Alléon, Comte A. and Vian, J. 1869–70. Des migrations des oiseaux de proie sue le Bosphore de Constantinople. *Rév. Mag. Zool.* (II) 21: 258–273, 305–315, 342–348, 369–374, 401–409; 22: 81–86, 129–138, 161–165.

Allouse, B. E. 1953. The avifauna of Iraq. *Iraq Nat. Hist. Mus. Publ.* 3.

Alström, P., Colston, P. and Lewington, I. 1991. *A Field Guide to the Rare Birds of Britain and Europe*. HarperCollins, London.

Alström, P., Kirwan, G. M., Grieve, A., Lassey, P. A., Mild, K. and Svensson, L. in prep. *Calandrella rufescens niethammeri* is a synonym of *C. r. aharonii*, and why it matters, with comments on other closely related Anatolian larks. *Bull. Brit. Orn. Club.*

Alström, P. and Mild, K. 2003. *Pipits and Wagtails of Europe, Asia and North America*. Christopher Helm, London.

Alström, P. and Olsson, U. 1988. Taxonomy of Yellow-browed Warblers. *Brit. Birds* 81: 656–657.

Alström, P. and Olsson, U. 1990. Taxonomy of the *Phylloscopus proregulus* complex. *Bull. Brit. Orn. Club* 110: 38–43.

Alström, P., Olsson, U. and Colston, P. R. 1997. Re-evaluation of the taxonomic status of *Phylloscopus proregulus kansuensis* Meise. *Bull. Brit. Orn. Club* 117: 177–193.

Alström, P. and Ranft, R. 2003. The use of sounds in avian systematics and the importance of bird sound archives. *In*: Collar, N. J., Fisher, C. and Feare, C. (eds.) *Why Museums Matter: Avian Archives in an Age of Extinction. Bull. Brit. Orn. Club* Suppl. 123A.

Amadon, D. and Short, L. L. 1992. Taxonomy of lower categories – suggested guidelines. *Bull. Brit. Orn. Club Centenary* Suppl. 112A: 11–38.

American Ornithologists' Union (AOU) 1998. *Check-list of North American Birds*. Seventh edn. American Ornithologists' Union, Washington DC.

American Ornithologists' Union (AOU) 2000. Forty-second supplement to the American Ornithologists' Union *Check-list of North American Birds. Auk* 117: 847–858.

Ananian, V. and Busuttil, S. 2003. Further observations and probable breeding of Mongolian Trumpeter Finch *Bucanetes mongolicus* in Armenia. *Sandgrouse* 25: 148–150.

Ananian, V., Busuttil, S. and Finn, M. 2002. Recent observations of some rare breeding birds in Armenia. *Sandgrouse* 24: 46–48.

Ananian, V. and Ghasabian, M. 2005. On the distribution of Red-tailed Wheatear *Oenanthe chrysopygia* (de Fillipi [*sic*], 1863) in Armenia. *Sandgrouse* 27: 157–159.

Ananian, V. and de Rouw, P. 2003. The first Long-tailed Duck *Clangula hyemalis* in Armenia. *Sandgrouse* 25: 65–66.

Andersen, G. S., Klann, M., Sandve, G. and Tveit, B. O. 1993. A large breeding colony of Citrine Wagtails in eastern Turkey. *Orn. Soc. Middle East Bull.* 30: 27–28.

Andersson, M. L. 1999. Phylogeny, behaviour, plumage evolution, and neoteny in skuas Stercorariidae. *J. Avian Biol.* 30: 205–215.

Andrews, I. J. 1994. Description and status of the black morph Mourning Wheatear *Oenanthe lugens* in Jordan. *Sandgrouse* 16: 32–35.

Andrews, I. J. 1995. *The Birds of the Hashemite Kingdom of Jordan*. Andrews, Musselburgh.

Andrews, I. J., Khoury, F. and Shirihai, H. 1999. Jordan Bird Report 1995–97. *Sandgrouse* 21: 10–35.

Andrews, J., Beaman, M., Fisher, P., Hereward, T., Heubeck, M., Morton, M., Porter, R. and Round, P. 1977. A 'new' raptor migration route through N. E. Turkey. *Orn. Soc. Turkey Bull.* 14: 2–5.

Anon. 1969. Summer migrants in south west Turkey. *Orn. Soc. Turkey Bull.* 3: 2–3.

Anstey, S. 1989. *The Status and Conservation of the White-headed Duck Oxyura leucocephala*. IWRB (Spec. Publ. No. 10), Slimbridge.

Arkhipov, V. Y., Wilson, M. G. and Svensson, L. 2003. Song of the Dark-throated Thrush. *Brit. Birds* 96: 79–83.

Arnaìz-Villena, A., Guillén, J., Ruiz-del-Valle, V., Lowy, E., Zamora, J., Varela, P., Stefanì, D. and Allende, L. M. 2001. Phylogeography of crossbills, bullfinches, grosbeaks and rosefinches. *Cell. & Mol. Life Sci.* 58: 1159–1166.

Arnaìz-Villena, A., Moscoso, J., Ruiz-del-Valle, V., Gonzalez, J., Reguera, R., Ferri, A., Wink, M. and Serrano-Vela, J. I. 2008. Mitochondrial DNA phylogenetic definition of a group of 'arid-zone' Carduelini finches. *Open Orn. J.* 1: 1–7.

Ash, J. S., Pearson, D. J., Nikolaus, G. and Colston, P. R. 1989. The mangrove reed warblers of the Red Sea and Gulf of Aden coasts, with description of a new subspecies of the African Reed Warbler *Acrocephalus baeticatus*. *Bull. Brit. Orn. Club* 109: 36–43.

Aslan, A. and Erdoğan, A. 2007. On the distribution of the White-spectacled Bulbul, *Pycnonotus xanthopygos* (Hemprich & Ehrenberg, 1833), in Turkey. *Zool. Middle East* 41: 31–34.

Aslan, A. and Kiziroğlu, I. 2003. A study on the ornithofauna of Sakaryabaşı/Eminekin pond and its vicinity. *Turk. J. Zool.* 27: 19–26.

Atkinson, P. W., Humpage, E. A., Jowitt, A. J. D., Oğurlu, I. and Rowcliffe, J. M. 1995. The distribution and status of Caucasian Black Grouse in north-eastern Turkey. *Proc. Intern. Symp. Grouse* 6: 131–133.

Austin, J. J. 1996. Molecular phylogenetics of *Puffinus* shearwaters: preliminary evidence from mitochondrial cytochrome *b* gene sequences. *Mol. Phyl. & Evol.* 6: 77–88.

Ayvaz, Y. 1990. The birds of the lake of Pınarbaşı, Malatya. *Turk. J. Zool.* 14: 139–143.

Ayvaz, Y. 1991. The birds of Çıldır Lake. *Turk. J. Zool.* 15: 53–58.

Baccetti, N., Massa, B. and Violani, C. 2007. Proposed synonymy of *Sylvia cantillans moltonii* Orlando, 1937, with *Sylvia cantillans subalpina* Temminck, 1820. *Bull. Brit. Orn. Club* 127: 107–110.

Baha el Din, M. and Baha el Din, S. 1996. The first Oriental Pratincole *Glareola maldivarum* in Egypt. *Sandgrouse* 18(1): 64–65.

Baha el Din, S. M. and Baha el Din, M. 2000. The occurrence of the North African subspecies of Mourning Wheatear *Oenanthe lugens halophila* in Egypt. *Sandgrouse* 22: 109–112.

Baker, K. 1997. *Warblers of Europe, Asia and North Africa*. Christopher Helm, London.

Balch, L. and Maybank, B. (eds.) 1996. The 1995 big day report and the 1995 list report. *Birding* Suppl. 28 (3).

Balkız, Ö., Béchet, A., Özesmi, O., Siki, M., Arnaud, A., Germain, C. & Eken, G. in prep. The Greater Flamingo (*Phoenicopterus roseus*) in Turkey: status and conservation issues.

Ballance, D. K. 1958. Summer observations on the birds of the Anatolian plateau and northwestern Cicilia. *Ibis* 100: 617–620.

Ballance, D. K. and Lee, S. L. B. 1961. Notes on autumn migration at the Bosphorus and in the Aegean. *Ibis* 103a: 195–204.

Balmer, D. and Betton, K. (compilers) 2001. Around the region. *Sandgrouse* 24: 156–160.

Balmer, D. and Betton, K. (compilers) 2002a. Around the region. *Sandgrouse* 24: 76–80.

Balmer, D. and Betton, K. (compilers) 2002b. Around the region. *Sandgrouse* 24: 156–160.

Balmer, D. and Betton, K. (compilers) 2003a. Around the region. *Sandgrouse* 25: 76–80.

Balmer, D. and Betton, K. (compilers) 2003b. Around the region. *Sandgrouse* 25: 154–159.

Balmer, D. and Betton, K. (compilers) 2004a. Around the region. *Sandgrouse* 26: 76–80.

Balmer, D. and Betton, K. (compilers) 2004b. Around the region. *Sandgrouse* 26: 160–168.

Balmer, D. and Betton, K. (compilers) 2005a. Around the region. *Sandgrouse* 27: 90–96.

Balmer, D. and Betton, K. (compilers) 2005b. Around the region. *Sandgrouse* 27: 170–176.

Balmer, D. and Betton, K. (compilers) 2006. Around the region. *Sandgrouse* 28: 90–96.

Balmer, D. and Betton, K. (compilers) 2007a. Around the region. *Sandgrouse* 29: 9–16.

Balmer, D. and Betton, K. (compilers) 2007b. Around the region. *Sandgrouse* 29: 122–128.

Balmer, D. and Betton, K. (compilers) 2008. Around the region. *Sandgrouse* 30: 14–21.

Balmer, D. and Kirwan, G. M. (compilers) 2003. News and information. *Sandgrouse* 25: 4–7.

Banks, R. C., Cicero, C., Dunn, J. L., Kratter, A. W., Rasmussen, P. C., Remsen, J. V., Rising, J. D. and Stotz, D. F. 2002. Forty-third supplement to the American Ornithologists' Union *Check-list of North American Birds*. *Auk* 119: 897–906.

Banks, R. C., Cicero, C. Dunn, J. L., Kratter, A. W., Rasmussen, P. C., Remsen, J. V., Rising, J. D. & Stotz, D. F. 2003. Forty-fourth supplement to the American Ornithologists' Union *Checklist of North American birds*. *Auk* 120: 923–931.

Banoğlu, N. A. & Burr, M. 1953. *Turkey: A Sportsman's Paradise.* Ankara.

Bara, T. 2002. Bird notes from Lebanon, including two new species. *Sandgrouse* 24: 44–45.

Bara, T. 2003. The first Radde's Accentor *Prunella ocularis* in Lebanon. *Sandgrouse* 25: 69.

Barış, S. 2004. Bird monitoring in Turkey. 16th Intern. Conf. EBCC, 6–11 September 2004, Kayseri, Turkey.

Barış, S., Akçakaya, R. and Bilgin, C. 1984. *The Birds of Kızılcahamam*. Birds of Turkey 3. Kasparek Verlag, Heidelberg.

Barış, S., Erciyas, K., Gürsoy, A., Özsemir, C. & Nowakowski, J. K. 2005. Cernek: a new bird ringing station in Turkey. *The Ring* 27: 113–120.

Barker, F. K. 2004. Monophyly and relationships of wrens (Aves: Troglodytidae): a congruence analysis of heterogeneous mitochondrial and nuclear DNA sequence data. *Mol. Phyl. & Evol.* 32: 486–504.

Barrowclough, G. F. 1982. Geographic variation, predictiveness, and subspecies. *Auk* 99: 601–603.

Barthel, P. H., Hanoldt, W., Hubatsch, K., Koch, H.-M., Konrad, V. and Lannert, R. 1992. Der Mongolengimpel, *Bucanetes mongolicus* in der Westpaläarktis. *Limicola* 6: 265–286.

Bashford, R. 1997. The first Cotton Teal *Nettapus coromandelianus* in Jordan. *Sandgrouse* 19: 142–143.

Başkaya, S. 2003. Distribution and principal threats to Caucasian Black Grouse *Tetrao mlokosiewiczi* in the eastern Karadeniz Mountains in Turkey. *Wildl. Biol.* 9: 377–383.

Bates, G. L. 1935. On *Oenanthe xanthoprymna* and *O. chrysopygia*. *Ibis* (13) 5: 198–201.

Bauer, W., von Helversen, O., Hodge, M. and Martens, J. 1969. *Catalogus Faunae Graeciae 2, Aves.* Thessaloniki.

Bauer, W. and Müller, G. 1969. Zur Avifauna des Ewros-Delta. *Beitr. Naturk. Forsch. Südw.-Dtl.* 28: 33–52.

Baumgart, W. and Kasparek, M. 1992. Notes on summer birds of Syria. *Zool. Middle East* 6: 11–19.

Baumgart, W., Kasparek, M. and Stephan, B. 1995. *Die Vögel Syriens: eine Übersicht.* Kasparek Verlag, Heidelberg.

Beale, C. 2000. Notes on the birds of Lebanon, autumn–winter 1999. *Sandgrouse* 22: 122–124.

Beaman, M. 1977. Further news on raptor migration in the north east. *Orn. Soc. Turkey Bull.* 15: 9.

Beaman, M. 1986. Turkey Bird Report 1976–81. *Sandgrouse* 8: 1–41.

Beaman, M. 1994. *Palearctic Birds. A Checklist of the Birds of Europe, North Africa and Asia North of the Foothills of the Himalayas.* Harrier Publications, Stonyhurst.

Beaman, M. and Madge, S. 1998. *The Handbook of Bird Identification for Europe and the Western Palearctic.* Christopher Helm, London.

Beaman, M. and Porter, R. F. 1977. A 'new' raptor migration route through NE Turkey. *Orn. Soc. Turkey Bull.* 14: 2–5.

Beaudoin, J.-C. 1967. La Marouette de Baillon *Porzana pusilla* nicheuse en Turquie. *Alauda* 35: 151–152.

Beaudoin, J.-C. 1976. Excursion ornithologique dans le Moyen-Taurus (Turquie). *Alauda* 44: 77–90.

Beddard, R., Ananian, V. and Finn, M. 2002. The first Mongolian Trumpeter Finch *Bucanetes mongolicus* in Armenia. *Sandgrouse* 24: 144–147.

Behrouzi-Rad, B. 1992. On the movements of the Greater Flamingo, *Phoenicopterus ruber*, in Iran. *Zool. Middle East* 6: 21–27.

Bekir, S. 2007. The first Siberian Accentor *Prunella ocularis* [*sic*] & Lapland Longspur *Calcarius lapponicus* in Turkey. *Sandgrouse* 29: 107–108.

Belik, V. P. and Lebedeva, E. A. 2002. International Species Action Plan for the Black-winged Pratincole (*Glareola nordmanni*). BirdLife International, Wageningen.

Beme, L. B. 1926. Liste der von P. J. Zhukov in den Jahren 1915–1917 in der nördlichen Türkei (asiatische Türkei) gesammelten Vögel. *Izv. Gorsk. Inst. Vladikavskas* 3: 97–102.

Beolens, B. and Watkins, M. 2003. *Whose Bird? Men and Women Commemorated in the Common Names of Birds.* Christopher Helm, London.

van den Berg, A. B. 1988. Grey-headed Woodpecker, *Picus canus*, in north-eastern Turkey. *Zool. Middle East* 2: 12–15.

van den Berg, A. B., Bison, P. and Kasparek, M. 1988. Striated Scops Owl in Turkey. *Dutch Birding* 10: 161–166.

van den Berg, A. B. and Bosman, C. A. W. 1998. Paddyfield Warbler, *Acrocephalus agricola*, at Van Gölü, eastern Turkey. *Zool. Middle East* 2: 16–18.

van den Berg, M. and Damhuis, R. 1981. Additional breeding records of Little Swift (*Apus affinis*) and Cretzschmar's Bunting (*Emberiza caesia*) in Turkey. *Orn. Soc. Middle East Bull.* 6: 9.

van den Berk, V. 1988. The Black Francolin in Turkey. *Sandgrouse* 10: 51–57.

van den Berk, V. 1990. The rapid movement of a Turkish-ringed Wryneck to Beirut, Lebanon. *Orn. Soc. Middle East Bull.* 24: 15–17.

van den Berk, V. 1991a. Visible migration of Sparrowhawk *Accipiter nisus* and Penduline Tit *Remiz pendulinus* in southern Turkey. *Sandgrouse* 13: 101–102.

van den Berk, V. 1991b. Impressions of autumn migration in mid-October along the Çukurova coast, near the Belen Pass, southern Turkey. *Orn. Soc. Middle East Bull.* 26: 16–19.

van den Berk, V. 1994. The severe 1991–1992 winter and its effect on Graceful Warbler (*Prinia gracilis*) and other resident breeding birds in the Göksu Delta, Turkey. *Sandgrouse* 16: 139–146.

van den Berk, V., Cronau, J. P. and van der Have, T. M. 1993. *Waterbirds in Van Province, Eastern Turkey, May 1989.* WIWO (Report No. 34), Zeist.

van den Berk, V., van Dorp, D., van Hoorn, O. and Vos, R. 1986. *Cranes and Waterfowl Counts of Some Turkish Wetlands.* WIWO (Report No. 10), Zeist.

van den Berk, V. and Guder, N. 1992. *Towards Integrated Management in the Göksu Delta, a Protected Special Area in Turkey.* Doğal Hayatı Koruma Derneği, Istanbul.

van den Berk, V. and van der Winden, J. 1992. Diurnal migration of Nightjars at the Göksu Delta, south Turkey. *Orn. Soc. Middle East Bull.* 28: 20–21.

van den Berk, V. and Kasparek, M. 1988. The White-breasted Kingfisher, *Halcyon smyrnensis*, in Turkey: on the occurrence of an endangered species. *Zool. Middle East* 2: 19–25.

Bertault, Y., Dubois, P. J. and Fremont, J.-Y. 1988. Some comments on the Armenian Gull in Turkey. *Orn. Soc. Middle East Bull.* 20: 20–21.

Bertault, Y. and Fremont, J.-Y. 1993. First observations of a possible Basra Reed Warbler in Turkey. *Orn. Soc. Middle East Bull.* 31: 18–20.

Bezzel, E. 1986. Lesser Sandplover, *Charadrius mongolus*, in Turkey. *Zool. Middle East* 1: 24–26.

Biber, J. P. 1990. *Action Plan for the Conservation of the Western Lesser Kestrel Falco naumanni Populations.* International Council for Bird Preservation (Study Report No. 40), Cambridge.

Bilgin, C. C. 1996. First record of the Common Myna (*Acridotheres tristis*) from Ankara, Turkey. *Zool. Middle East* 13: 25–26.

Bilgin, C. C. and Demirci, B. 2002. Kayıtlar (Mayıs–Aralık 2001). *Ibibik* 2: 30–31.

Birand, H. 1960. Erste Ergebnisse der Vegetations-Untersuchungen in der Zentralanatolischen Steppe. I. Halophytengesellschaften des Tusgölü. *Bot. Jahrb.* 79: 254–296.

Bird, C. G. 1937. The birds of southern Asia Minor from Mersin to the Euphrates. *Ibis* (XIV) 1: 65–85.

BirdLife International 2000. *Threatened Birds of the World.* BirdLife International, Cambridge & Lynx Edicions, Barcelona.

BirdLife International 2001. *Threatened Birds of Asia.* BirdLife International, Cambridge.

BirdLife International 2004. *Threatened Birds of the World 2004.* CD-ROM. BirdLife International, Cambridge.

BirdLife International 2008. *Threatened Birds of the World 2008.* BirdLife International, Cambridge.

Biricik, M. 1996. Birds of Kabaklı Reservoir, Diyarbakır. *Turk. J. Zool.* 20: 155–160.

Biricik, M. 1997. Winterbrut freilebender Palmtauben *Streptopelia senegalensis. J. Orn.* 138: 335–336.

Biricik, M., Kılıç, A. and Şahin, R. 1989. Fortpflanzungsverhalten der Palmtaube (*Streptopelia senegalensis*); Paarbildung bis Eiablage. *J. Orn.* 130: 217–228.

Bijlsma, R. G. 1987. *Bottleneck Areas for Migratory Birds in the Mediterranean Region. An Assessment of the Problems and Recommendations for Action.* International Council for Bird Preservation (Study Report No. 18), Cambridge.

Bjilsma, R. G. 1990. *Some Raptor Counts in the Vicinity of Istanbul, Autumn 1990.* SOVON, Beek-Ubbergen.

Bijlsma, R. and Roder, F. de 1986. *Notes on the Birds of Some Wetlands in Turkey, 17 August–10 September 1986.* WIWO (Report No. 14), Zeist.

Bobrinskij, N. A. 1916. Les résultats scientifiques des excursions ornithologiques dans les districts de Surmalin et d'Eczmiadzin du gouvernement d'Erivan pendant l 'été 1911 et 1912. *Izv. Kavkaz Muz.* 8: 171–250; 10: 1–148.

Bock, W. J. 2004. Affinities of *Carpospiza brachydactyla* (Passeriformes; Passeridae). *J. Orn.* 145: 223–226.

Boessreck, J. 1987. Frühbronzezeitliche Knochenfunde des Nonnenkranichs (*Grus leucogeranus*) aus der Südosttürkei. *Ecol. Birds* 9: 127–129.

Boev, Z. 2003. Specimens of extinct and threatened birds in the collections of the National Museum of Natural History in Sofia, Bulgaria. *In*: Collar, N. J., Fisher, C. and Feare, C. (eds.) *Why Museums Matter: Avian Archives in an Age of Extinction. Bull. Brit. Orn. Club* Suppl. 123A.

Bonaccorsi, G. 2003. Les Procellariiformes (Diomedeidae, Procellariidae et Hydrobatidae) non nicheurs en Méditerranée occidental: une synthèse. *Alauda* 71: 1–7.

van den Bossche, W. and Lens, L. 1994. Soaring bird migration at the Bosphorus (Turkey): the need for a multi-station survey. *Gerfaut* 84: 51–62.

Boulos, L., Miller, A. G. and Mill, R.R. 1994. South West Asia and the Middle East. *In*: Davis, S. D., Heywood, V. H. and Hamilton, A. C. (eds.) *Centers of Plant Diversity.* Information Press, Oxford.

Bourne, W. R. P. 1989. The relationship between the Armenian and Heuglin's Gulls. *Abstracts Second Mediterranean Seabird Symp.*, Mallorca, Spain, March 1989.

Bouwman, R. G. 1987. Crab Plover in Turkey in July 1986. *Dutch Birding* 9: 65–67.

Box, E. O., Fujiwara, K., Nakhutsrishvili, G., Zazanashvili, N., Liebermann, R. J. and Miyawaki, A. 2000. Vegetation and landscapes of Georgia (Caucasus) as a basis for landscape restoration. *Bull. Inst. Environ. Sci. & Tech., Yokomaha Natl. Univ., Japan* 26: 69–102.

Boyd, A. W. 1916. Birds seen during the Dardanelles campaign. *Zoologist* (IV) 20: 121–134.

Boyla, K. A., Aydemir, G. O. and Eken, G. 1998. The status and distribution of Ring-necked Parakeet *Psittacula krameri* in Turkey. *Turna* 1: 24–27.

Boyla, K. A. and Eken, G. 1998. Remarkable sightings. *Turna* 1: 35–40.

Bradshaw, C. G. and Kirwan, G. 1992. Impressions of spring passerine migration at Aksehir Gölü, Turkey. *Orn. Soc. Middle East Bull.* 29: 16–17.

Bradshaw, C. G. and Kirwan, G. M. 1994. The ornithological importance of the Aksehir Gölü wetlands, Central Plateau, Turkey. *Sandgrouse* 16: 118–138.

Bradshaw, C. G. and Kirwan, G. M. (compilers) 2000. Around the region. *Sandgrouse* 22: 156–160.

Braun, F. 1901. Einige Mitteilungen über die Alaudidae von den Ufern des Bosporus. *Orn. Monatsber.* 9: 65–67.

Braun, F. 1901. Einige Mitteilungen über die Alaudidae von den Ufern des Bosporus. *Orn. Monatsber.* 9: 65–67.

Braun, F. 1903. Ornithologisches aus Konstantinopel. *Orn. Monatsber.* 11: 20–21, 38–39, 65–69, 145–151.

Braun, F. 1908. Winterliches Vogelleben in Rumelien. *Natur. u. Haus* 16: 257–261.

Braun, M. J. and Brumfield, R. T. 1998. Enigmatic phylogeny of skuas: an alternative hypothesis. *Proc. Roy. Soc. Lond.* B 265: 995–999.

Bräuning, C. 1984. Wilson wassertreter (*Phalaropus tricolor*) in der Türkei. *Orn. Mitt.* 36: 139.

Brazil, M. A. 2003. *The Whooper Swan.* T. & A. D. Poyser, London.

Bretagnolle, V. and Lequette, B. 1990. Structural variation in the call of the Cory's Shearwater (*Calonectris diomedea*, Aves, Procellariidae). *Ethology* 85: 313–323.

Bretagnolle, V. and Zotier, R. 1998. Levels of endemism in Mediterranean Procellariiformes: evidence from morphometrics, behaviour and genetics. *In*: Walmsley, J., Goutner, V., El Hili, A. and Sultana, J. (eds.) Ecologie des oiseaux marins et gestion integrée du littoral en Méditerranée. *IV Symp. Méditerranéen des Oiseaux Marins.* Les Amis des Oiseaux & MEDMARAVIS, Tunis.

Brewer, D. and MacKay, B. K. 2001. *Wrens, Dippers and Thrashers.* Christopher Helm, London.

Brinkmann, R., Heins, J.-U., Köhler, B. and Rösler, S. 1990. *Menderes-Delta. Zustand und Gefährdungen eines OstMediterranen flussdeltas.* Hanover.

British Ornithologists' Union Records Committee (BOURC) 1997. Twenty-third report. *Ibis* 139: 197–201.

Broders, O., Osborne, T. and Wink, M. 2003. A mtDNA phylogeny of bustards (family Otididae) based on nucleotide sequences of the cytochrome b-gene. *J. Orn.* 144: 176–185.

Browne, S. 1997. The first White-rumped Sandpiper *Calidris fuscicollis* in Turkey and the Middle East. *Sandgrouse* 19: 146–147.

Brugière, D. 2002. Un nouveau cas de nidification du Vanneau à queue blanche *Vanellus leucurus* en Turquie. *Alauda* 70: 202.

de Bruin, S. 1989. Spotted Sandpiper in Turkey in August 1988. *Dutch Birding* 11: 121–122.

Brush, T. 2006. Common Sandpiper *Actitis hypoleucos* nesting on Mount Erciyes, Kayseri province, Turkey. *Sandgrouse* 28: 163–164.

Bundy, G., Conner, R. J. and Harrison, C. J. O. 1989. *The Birds of the Eastern Province of Saudi Arabia.* H. F. & G. Witherby, London.

Burgett, C., Rockmore, M. and Quinting, G. 1984. *Gazetteer of Turkey.* Vols. 1–2. Defense Mapping Agency, Washington DC.

Bussutil, S. and Ananian, V. 2001. Photospot: Armash Fishponds. *Sandgrouse* 23: 101–103.

Buxton, P. A. J. 1916. Notes on birds seen during the Gallipoli campaign. *Zool.* 4 (20): 413–416.

Byers, C., Olsson, U. and Curson, J. 1995. *Buntings and Sparrows: A Guide to the Buntings and North American Sparrows.* Pica Press, Robertsbridge.

Byfield, A. J. and Ozhatay, N. 1995. *Towards the Conservation of Turkey's Northern Sand Dunes.* Doğal Hayatı Koruma Derneği, Istanbul.

Çağlayan, E., Kılıç, D. T., Per, E. and Gem, E. 2005. *Türkiye Kış Ortası Sukuşu Sayımları 2005.* Doğa Derneği, Ankara.

Çakmak, M. and Işın, M. 2005. *Anadolu Kuş Adları Sözlüğü.* Kitap, Istanbul.

Can, O. K. 2004. Turkish national ringing scheme. *The Ring* 26: 79–80.

Castell, P. 1996a. Additional information on breeding biology of Rufous Bush Robin *Cercotrichas galactotes*. *Sandgrouse* 18(2): 67.

Castell, P. 1996b. Description of the nestlings of Olive-tree Warbler *Hippolais olivetorum*. *Sandgrouse* 18(2): 68.

Castell, P. 1996c. Additional information on Ménétries's Warbler *Sylvia mystacea* breeding biology in south-eastern Turkey. *Sandgrouse* 18(2): 68.

Castell, P. 1996d. Black-headed Bunting *Emberiza melanocephala* breeding behaviour in Turkey. *Sandgrouse* 18(2): 71.

Castell, P. 1996e. Nestlings of Bimaculated Lark. *Brit. Birds* 89: 363.

Castell, P. 1996f. Notes on the nesting of Green Warbler in northern Turkey. *Brit. Birds* 89: 363–365.

Castell, P. 1997. New information on the breeding biology of Cretzschmar's Bunting *Emberiza caesia*. *Sandgrouse* 19: 68–69.

Castell, P. & Kirwan, G. M. 2005. Will the real Sykes's Warbler please stand up? Breeding data support specific status for *Hippolais rama* and *H. caligata*, with comments on the Arabian population of 'booted warbler'. *Sandgrouse* 27: 30–36.

Cederroth, C., Johansson, C. and Svensson, L. 1999. Taiga Flycatcher *Ficedula albicilla* in Sweden: the first record in Western Europe. *Birding World* 12: 460–468.

Çetik, A. R. 1972. *Vegetation of Central Anatolia and its Ecology.* Cento Seminar on Agriculture Treaty Organization, Tehran.

Chantler, P. and Driessens, G. 1995. *Swifts: A Guide to the Swifts and Treeswifts of the World.* Pica Press, Robertsbridge.

Chantre, M. E. 1883. Rapport sur une mission scientifique dans l'Asie occidentale et specialement dans les regiones de L'Ararat et du Caucase. *Arch. Missions Sci. et Litt.* (3) 10: 199–263.

Chardine, J. W. 2002. Geographic variation in the wingtip patterns of Black-legged Kittiwakes. *Condor* 104: 687–693.

Chappuis, C., Heim de Balsac, H. and Vielliard, J. 1973. Distribution, reproduction, manifestations vocale et affinités du Bruant cendré *Emberiza cineracea. Bonn. Zool. Beitr.* 24: 302–316.

Cheke, R. A., Mann, C. F. and Allen, R. 2001. *Sunbirds: A Guide to the Sunbirds, Flowerpeckers, Spiderhunters and Sugarbirds of the World.* Christopher Helm, London.

Cheng Tso-hsin 1987. *A Synopsis of the Avifauna of China.* Science Press, Beijing & Paul Parey, Hamburg & Berlin.

Choe, J. C. 2002. A long-term phylogeographic study of magpies (genus *Pica*): behaviour, ecology, and evolution. http://klter.kookmin.ac.kr/EVENTS/Conference00/html/ choijaechun.htm

Chu, P. C. 1998. A phylogeny of the gulls (Aves: Larinae) inferred from osteological and integumentary characters. *Cladistics* 14: 1–43.

Chu, P. C. 2002. A morphological test of the monophyly of the Cardueline finches (Aves: Fringillidae, Carduelinae). *Cladistics* 18: 279–312.

Clancey, P. A. 1987. The Tree Pipit *Anthus trivialis* (Linnaeus) in southern Africa. *Durban Mus. Novit.* 14: 29–42.

Clark, W. S. 2005. Steppe Eagle *Aquila nipalensis* is monotypic. *Bull. Brit. Orn. Club* 125: 149–153.

Cleere, N., Kelly, D. and Pilcher, C. W. T. 2004. A mystery *Phylloscopus* warbler in Kuwait. *Sandgrouse* 26: 143–146.

Clement, P. 1995. Southern and eastern Great Grey Shrikes in northwest Europe. *Birding World* 8: 300–309.

Clement, P., Harris, A. and Davis, J. 1993. *Finches and Sparrows: An Identification Guide.* Christopher Helm, London.

Clement, P. and Hathway, R. 2001. *Thrushes.* Christopher Helm, London.

Clement, P., Helbig, A. J. & Small, B. 1998. Taxonomy and identification of chiffchaffs in the Western Palearctic. *Brit. Birds* 91: 361–376.

Clement, P. and Holman, D. 2001. Passage records of Amur Falcon *Falco amurensis* from SE Asia to southern Africa including first records from Ethiopia. *Bull. Brit. Orn. Club* 121: 222–230.

Cocker, M. and Temple Lang, J. 1991. A nest of the Caucasian Black Grouse *Tetrao mlokosiewiczi* in Turkey. *Sandgrouse* 13: 102–103.

Cofta, T., Lasecki, R. and Karakaş, R. 2005. The first record of Blackstart *Cercomela melanura* in Turkey. *Zool. Middle East.* 35: 107–108.

Cohen, B. L., Baker, A. J., Blechschmidt, K., Dittmann, D. L., Furness, R. W., Gerwin, J. A., Helbig, A. J., Korte, J. de, Marshall, H. D., Palma, R. L., Peter, H. U., Ramli, R., Seibold, I., Willcox, M. S., Wilson, R. H. & Zink, R. M. 1997. Enigmatic phylogeny of skuas. *Proc. Roy. Soc. Lond. Biol. Sci.* 264: 181–190.

Collar, N. J. 1996. Otididae (bustards). *In*: del Hoyo, J., Elliott, A. and Sargatal, J. (eds.) *Handbook of the Birds of the World.* Vol. 3. Lynx Edicions, Barcelona.

Collar, N. J. 2005. Turdidae (thrushes). *In*: del Hoyo, J., Elliott, A. and Christie, D. A. (eds.) *Handbook of the Birds of the World.* Vol. 10. Lynx Edicions, Barcelona.

Collar, N. J. and Pilgrim, J. D. 2007. Species-level changes proposed for Asian birds, 2005–2006. *BirdingASIA* 8: 14–30.

Collinson, M. 2001a. Evolution of the Atlantic-island Chaffinches. *Brit. Birds* 94: 121–124.

Collinson, M. 2001b. Greenish Warbler, 'Two-barred Greenish Warbler', and the speciation process. *Brit. Birds* 94: 278–283.

Collinson, M. 2004. Letters. Are there really two species of houbara? – a response from the TSC. *Brit. Birds* 97: 348.

Collinson, M. 2006. Splitting headaches? Recent taxonomic changes affecting the British and Western Palearctic lists. *Brit. Birds* 99: 306–323.

Collinson, M., Knox, A. G., Parkin, D. T. and Sangster, G. 2003. Specific status of taxa within the Greenish Warbler complex. *Brit. Birds* 96: 327–331.

Collinson, M., Parkin, D. T., Knox, A. G., Sangster, G. and Helbig, A. J. 2006. Species limits within the genus *Melanitta*, the scoters. *Brit. Birds* 99: 183–201.

Collman, J. R. and Croxall, J. P. 1967. Spring migration at the Bosphorus. *Ibis* 109: 359–372.

Cornwallis, R. K. and Townsend, A. D. 1968. Waxwings in Britain and Europe during 1965/66. *Brit. Birds* 61: 97–118.

Corso, A. and Catley, G. P. 2003. Separation of transitional second calendar-year Red-footed Falcon from Amur Falcon. *Dutch Birding* 25: 153–158.

Cowan, P. J. 2004. Letters. Are there really two species of houbara? *Brit. Birds* 97: 346–347.

Cramp, S. 1971. The Dead Sea Sparrow: further breeding places in Iran and Turkey. *Ibis* 113: 244–245.

Cramp, S. (ed.) 1985. *The Birds of the Western Palearctic*. Vol. 4. Oxford University Press.

Cramp, S. (ed.) 1988. *The Birds of the Western Palearctic*. Vol. 5. Oxford University Press.

Cramp, S. (ed.) 1992. *The Birds of the Western Palearctic*. Vol. 6. Oxford University Press.

Cramp, S. (ed.) 1993. *The Birds of the Western Palearctic*. Vol. 7. Oxford University Press.

Cramp, S. and Perrins, C. M. (eds.) 1994. *The Birds of the Western Palearctic*. Vol. 8. Oxford University Press.

Cramp, S. and Simmons, K. E. L. (eds.) 1977. *The Birds of the Western Palearctic*. Vol. 1. Oxford University Press.

Cramp, S. and Simmons, K. E. L. (eds.) 1980. *The Birds of the Western Palearctic*. Vol. 2. Oxford University Press.

Cramp, S. and Simmons, K. E. L. (eds.) 1983. *The Birds of the Western Palearctic*. Vol. 3. Oxford University Press.

Crivelli, A., Mitchev, T., Catsadorakis, G. and Pomakov, V. 1991. Preliminary results on the wintering of the Dalmatian Pelican, *Pelecanus crispus*, in Turkey. *Zool. Middle East* 5: 11–20.

Crochet, P.-A., Bonhomme, F. and Lebreton, J. D. 2000. Molecular phylogeny and plumage evolution in gulls (Larini). *J. Evol. Biol.* 13: 47–57.

Curzon, R. 1854. *A Year at Erzeroom and on the Frontiers of Russia, Turkey and Persia*. London.

Danford, C. G. 1877–78. A contribution to the ornithology of Asia Minor. *Ibis* (IV) 1: 261–274; 2: 1–35.

Danford, C. G. 1880. A further contribution to the ornithology of Asia Minor. *Ibis* (IV) 4: 81–99.

David, N. and Gosselin, M. 2002a. Gender agreement of avian species names. *Bull. Brit. Orn. Club* 122: 14–49.

David, N. and Gosselin, M. 2002b. The grammatical gender of avian genera. *Bull. Brit. Orn. Club* 122: 257–282.

Davidson, N. 1985. A possible hybrid Common Crane × Siberian White Crane in Turkey. *Orn. Soc. Middle East Bull.* 15: 1–3.

Davidson, P. & Kirwan, G. (compilers) 1995. Around the region. *Orn. Soc Middle East Bull.* 34: 32–44.

Davidson, P. and Kirwan, G. M. (compilers) 1997. Around the region. *Sandgrouse* 19: 76–80.

Davis, P. H. (ed.) 1965–85. *Flora of Turkey and the East Aegean Islands*. Vols. 1–9. Edinburgh University Press, Edinburgh.

Dean, A. R., Fortey, J. E. and Phillips, E. G. 1977. White-tailed Plover: new to Britain and Ireland. *Brit. Birds* 70: 465–471.

Dean, A. R. and Svensson, L. 2005. 'Siberian Chiffchaff' revisted. *Brit. Birds* 98: 396–410.

Dement'ev, G. P. and Gladkov, N. A. 1951/54. [*The Birds of the Soviet Union*.] Moscow. [In Russian.]

Dement'ev, G. P. & Gladkov, N. A. 1967. *Birds of the Soviet Union*. Vol. 4. Israel Program for Scientific Translations, Jerusalem.

Dement'ev, G. P. & Gladkov, N. A. 1968. *Birds of the Soviet Union*. Vol. 2. Israel Program for Scientific Translations, Jerusalem.

Demirci, B. 2002. Toygardan Kayıtlar (Ocak–Mart 2002). *Kuşçu Bulteni* 11: 6–7.

Demirci, B. 2003. Toygardan Kayıtlar (Nisan–Agustos 2002). *Kuşçu Bulteni* 12: 4–6.

Demirci, B. & Mülayim, O. in prep. A new prolific route for birdwatching in Turkey: Ankara to Erzincan.

Deppe, H.-J. 1990. Beobachtungsnotizen zur Vogelwelt der Türkei. *Orn. Mitt.* 42: 257–258.

Desfayes, M. 1969. Remarques sur les affinités des Fringillidés des genres *Rhodopechys* et *Callacanthis*. *Oiseau & Rev. Fr. Orn.* 39: 21–27.

Devillers, P. & Potvliege, R. 1981. Le Goéland leucophée *Larus cachinnans michahellis* en Belgique. *Gerfaut* 71: 659–666.

Dickinson, E. C. (ed.) 2003. *The Howard and Moore Complete Checklist of the Birds of the World*. Third edn. Christopher Helm, London.

Dickson, E. D. and Ross, J. J. 1839. A collection of bird-skins from the neighbourhood of Erzeroom. *Proc. Zool. Soc. Lond.* 7: 119–123, 130–135.

Didrickson, O. K., Didrickson, J. and Can, O. K. 2004. Ringing results in 2003 at Manyas Kuşcenneti station (Turkey). *The Ring* 26: 85.

van Diek, H., Felix, R., Hornman, M., Meininger, P. L., Willems, F. and Zekhuis, M. 2004. Bird counting in Iran in January 2004. *Dutch Birding* 26: 287–296.

Dierschke, V., Stühmer, F. and Stühmer, T. 1996. Records of Booted Warbler *Hippolais caligata* in north-eastern Turkey. *Sandgrouse* 18(2): 66–67.

Dijksen, L. J. and Kasparek, M. 1985. *The Birds of Kızılirmak Delta.* Birds of Turkey 4. Kasparek Verlag, Heidelberg.

Dijksen, L. J. and Kasparek, M. 1988. *The Birds of Acıgöl.* Birds of Turkey 7. Kasparek Verlag, Heidelberg.

Dijksen, L. J. and Klemann, M. 1992. The Snow Bunting in Turkey. *Orn. Soc. Middle East Bull.* 28: 21.

Dijksen, L. J. and van der Wolf, P. 1987. *Mid-winter Waterfowl Census in Turkey, January 1987.* WIWO (Report No. 18), Zeist.

Dimcheff, D. E., Drovetski, S. V. and Mindell, D. P. 2002. Phylogeny of Tetraoninae and other galliform birds using mitochondrial 12S and ND2 genes. *Mol. Phyl. & Evol.* 24: 203–215.

Doğal Hayatı Koruma Derneği (DHKD) 1992. *Results of the International Waterfowl Census Turkey 1992.* Bird Section Report No. 6. Doğal Hayatı Koruma Derneği, Istanbul.

Doğal Hayatı Koruma Derneği (DHKD) 1993. *Results of the International Waterfowl Census Turkey 1993.* Bird Section Report No. 7. Doğal Hayatı Koruma Derneği, Istanbul.

Doğal Hayatı Koruma Derneği (DHKD) 1996. *Results of the International Waterfowl Census Turkey 1996.* Bird & Wetland Section Report No. 9. Doğal Hayatı Koruma Derneği, Istanbul.

Doğal Hayatı Koruma Derneği (DHKD) 1999. *Results of the International Waterfowl Census Turkey 1999.* Biodiversity Programme Report No. 10. Doğal Hayatı Koruma Derneği, Istanbul.

Donaghy, N. 2006. Iraq Babbler – a new breeding bird for Turkey. *Birding World* 19: 283–284.

Donne-Gousse, C., Laudet, V. and Hanni, C. 2002. A molecular phylogeny of Anseriformes based on mitochondrial DNA analysis. *Mol. Phyl. & Evol.* 23: 339–356.

Dott, H. E. M. 1967. Le Traquet pie, *Oenanthe pleschanka* (Lepechin) en Turquie. *Alauda* 35: 151.

Dowsett, R. J. and Dowsett-Lemaire, F. (eds.) 1993. *A Contribution to the Distribution and Taxonomy of Afrotropical and Malagasy Birds.* Tauraco Press (Research Report No. 5), Liège.

Dresser, H. E. 1871. *A History of the Birds of Europe.* Vol. 1. London.

Dresser, H. E. 1878. *A History of the Birds of Europe.* Vol. 4. London.

Dresser, H. E. 1891. On a collection of birds from Erzeroom. *Ibis* (VI) 3: 364–370.

Dresser, H. E. 1902. *A Manual of Palearctic Birds.* London.

Driessens, G. and Svensson, L. 2005. Identification of Collared Pratincole and Oriental Pratincole – a critical review of characters. *Dutch Birding* 27: 1–35.

Dufourny, H. 2006. First record of Long-tailed Shrike *Lanius schach* for Jordan. *Sandgrouse* 28: 73–76.

Eames, J. 1989. Selected bird observations from Turkey: spring and summer 1987. *Orn. Soc. Middle East Bull.* 23: 6–13.

Eames, J. 1991. More selected bird observations from Turkey: spring and summer 1990. *Orn. Soc. Middle East Bull.* 27: 29–31.

Ebels, E. 1996. Hume's Yellow-browed Warbler in Istanbul, Turkey, in December 1994. *Dutch Birding* 18: 242–243.

Ebels, E. B. 2002. Brown Fish Owl in the Western Palearctic. *Dutch Birding* 24: 157–161.

Ebels, E. B. 2003. Trends in systematics. Speciation in *Pica* magpies. *Dutch Birding* 25: 103–116.

Ecsedi, Z. 1996. Great Black-headed Gulls in Europe. *Birding World* 9: 303–312.

Eggers, J. 1977. Weiteres zum Vorkommen des Klippenkleibers (*Sitta tephronota*) in der Südosttürkei. *Vogelwelt* 98: 25–27.

Eggers, J. and Lemke, W. 1964. Ornithologische Beobachtungen in der Türkei. *Orn. Mitt.* 16: 185–189.

Eggers, J. and Streese, U. P. 1979. Gesammelte Beobachtungen zur Avifauna der Türkei (II). *Mitteilungsblatt DBV Landesverband Hamburg, Sonderheft* 7: 79–84.

Eichhorn, G. and Khrokov, V. V. 2002. Decline in breeding Sociable Plover *Chettusia gregaria* in the steppes of Naurzum and Korgalzhyn, Kazakhstan. *Sandgrouse* 24: 22–27.

Eken, G. 1997a. The breeding populations of some species of waterbirds at Gediz Delta, western Turkey. *Zool. Middle East* 14: 53–68.

Eken, G. 1997b. Regular wintering by Scops Owl *Otus scops* in Turkey, at Havran Delta. *Sandgrouse* 19: 147.

Eken, G. 1997c. The importance of Turkish coastal islands for seabirds. *Türkiye Kiyilari '97 Konf. Bild. Kit.*: 453–466.

Eken, G. 1997d. The wintering populations of some waterbird species on the Mediterranean coastline of Turkey. *In*: Özhan, E. (ed.) *Proc. Third Intern. Conf. Mediterranean Coastal Env., MEDCOAST 97*: 61–69.

Eken, G. and Magnin, G. 1999. *A Preliminary Biodiversity Atlas of the Konya Basin, Central Turkey*. Biodiversity Programme Report 13. Doğal Hayatı Koruma Derneği, Istanbul.

Elliot, H. F. E. 1962. Migration on the Bosphorus and near the eastern Pyrenees. *Ibis* 104: 248.

Elwes, H. J. and Buckley, T. E. 1870. A list of the birds of Turkey. *Ibis* (II) 6: 59–77, 327–341.

van den Elzen, R. and Nemeschkal, H.-L. 1991. Radiation in African canaries (Carduelidae): a comparison of different classificatory approaches. *Acta Congr. Intern. Orn.* 20: 459–467.

Emirogullari, E. F., Sonmez, E., Özesmi, S. and Özesmi, U. in prep. Status of Ruddy Shelduck *Tadorna ferruginea* in Turkey. *Sandgrouse*.

Erard, C. and Etchécopar, D. 1968. Observations de printemps en Turquie. *L'Oiseau & R.F.O.* 38: 87–102.

Erciyas, K. & Özçam, P. 2002. Türkiye için yeni tür kaydı: Kuzey Çıvgını (*Phylloscopus borealis*). *Ibibik* 3: 32.

Erciyas, K., Özçam, P., Yavuz, N., Demirtaş, S. and Sağlam, Ö. 2008. First record of Lesser Yellowlegs *Tringa flavipes* for Turkey. *Sandgrouse* 30: 108–110.

Ergene, S. 1945. *Türkiye Kuşları*. Istanbul Üniversitesi Fen. Fak. Mon., Istanbul.

Erol, O. 1982. *Türkei: Naturräumliche Gliederung – Tübinger Atlas des Vorderen Orients*. Wiesbaden.

Ertan A., Kılıç, A. and Kasparek, M. 1989. *Türkiye'nin Önemli Kuş Alanları*. Doğal Hayatı Koruma Derneği, Istanbul.

Ertan, K. T. 1996. *The Birds of Kocaçay Delta*. Birds of Turkey 12. Kasparek Verlag, Heidelberg.

Eyckerman, R., Louette, M. and Becuwe, M. 1992. On the biometry of birds ringed in Turkey. *Zool. Middle East* 6: 29–37.

Faldborg, J. 1994. *Birds in Çamlıhemşın–Ardeşen area, North-east Turkey, April–May 1993*. Dansk Ornitologisk Forening, Copenhagen.

Farman, C. 1869. On some of the birds of prey of central Bulgaria. *Ibis* (2) 5: 199–204.

Feare, C. and Craig, A. 1998. *Starlings and Mynas*. Christopher Helm, London.

Fehrer, J. 1996. Conflicting character distribution within different data sets on Cardueline finches: atrifact or history? *Mol. Biol. Evol.* 13: 7–20.

Ferguson-Lees, J. and Christie, D. A. 2001. *Raptors of the World*. Christopher Helm, London.

Filchagov, A. V. 1993. The Armenian Gull in Armenia. *Brit. Birds* 86: 550–560.

Flint, P. R. and Stewart, P. F. 1992. *The Birds of Cyprus: An Annotated Check-list*. British Ornithologists' Union Check-list No. 6 (second edn.). British Ornithologists' Union, Tring.

Flint, P. R., Stewart, P. F. and Kirwan, G. M. 1997. A review of claimed records of Lesser Sand Plover *Charadrius mongolus* from Cyprus and Turkey. *Sandgrouse* 19: 133–139.

Forsman, D. 1993. Hybridising harriers. *Birding World* 6: 313.

Friesen, V. L., Anderson, D. J., Steeves, T. E., Jones, H. and Schreiber, E. A. 2002. Molecular support for species status of Nazca Booby (*Sula granti*). *Auk* 119: 820–826.

Frankis, M. P. 1991. Krüper's Nuthatch *Sitta krueperi* and Turkish pine *Pinus brutia*: an evolving association. *Sandgrouse* 13: 92–97.

Fry, C. H. and Keith, S. (eds.) 2004. *The Birds of Africa*. Vol. 7. Academic Press, London.

Fry, C. H., Keith, S. and Urban, E. K. (eds.) 2000. *The Birds of Africa*. Vol. 6. Academic Press, London.

Gallagher, M. and Woodcock, M. 1980. *The Birds of Oman*. Quartet, London.

Gallner, J.-C. 1976. Observations ornithologiques nouvelles dans la region de Van (Turquie). *Alauda* 24: 111–117.

Ganso, M. and Spitzer, G. 1962. Weitere Beiträge zur Avifauna Kleinasiens. *Egretta* 2: 9–25.

Gaston, A. J. 1968. The birds of the Ala Dagh Mountains, southern Turkey. *Ibis* 110: 17–26.

Gätke, H. 1896. *Die Vogelwarte Helgoland*. Bd. 1. Braunschweig.

Gaucher, P., Paillat, P., Chappuis, C., Saint Jalme, M., Lotfikhah, F. and Wink, M. 1996. Taxonomy of the Houbara Bustard *Chlamydotis undulata* subspecies considered on the basis of sexual display and genetic divergence. *Ibis* 138: 273–282.

Gavrilov, E. I. and Gavrilov, A. E. 2000. Possible ringing recoveries of Relict Gull in Bulgaria and Turkey. *Dutch Birding* 22: 219–221.

Gengler, J. 1920. *Balkanvögel.* Altenburg.

Gibbs, D., Barnes, E. and Cox, J. 2001. *Pigeons and Doves.* Christopher Helm, London.

Gill, F. B., Slikas, B. and Sheldon, F. H. 2005. Phylogeny of titmice (Paridae) II: species relationships based on sequences of the mitochondrial cytochrome-b gene. *Auk* 122: 121–143.

Gill, F. and Wright, M. 2006. *Birds of the World. Recommended English Names.* Christopher Helm, London.

Glimmerveen, U. and Hols, H. 1986. Red-rumped Wheatear, *Oenanthe moesta*, in Turkey. *Zool. Middle East* 1: 26–28.

Glutz von Blotzheim, U. N. and Bauer, K. M. 1982. *Handbuch der Vögel Mitteleuropas.* Bd. 8. Wiesbaden.

Glutz von Blotzheim, U. N., Bauer, K. M. and Bezzel, E. 1973. *Handbuch der Vögel Mitteleuropas.* Bd. 3. Frankfurt am Main.

Gokhelashvili, R., Reese, K. P. and Gavashelishvili, L. 2003. How much do we know about the Caucasian Black Grouse *Tetrao mlokosiewiczi*? *Sandgrouse* 25: 32–40.

von Gonzenbach, J. G. 1852. Einige ornithologische Notizen über Smyrna. *Naumannia* 2: 19–23.

Goodman, S. M. and Meininger, P. L. (eds.) 1989. *The Birds of Egypt.* Oxford University Press.

Goriup, P. D. and Parr, D. 1983. *Report on a Survey of Bustards in Turkey March 22 to May 10, 1981.* ICBP Study Report 1. International Council for Bird Preservation, Cambridge.

Goriup, P. D. and Parr, D. 1985. Results of the ICBP bustard survey in Turkey, 1981. *Bustard Studies* 2: 77–97.

Gorman, G. 2002. Aquatic Warbler. *Alula* 8: 62–65.

Gosney, D. 1993. *Finding Birds in Eastern Turkey.* Privately published, Sheffield.

Gosney, D. 1996. *Finding Birds in Turkey: Ankara to Birecik.* Privately published, Sheffield.

Gottschalk, T. K., Ekschmitt, K., Isfendiyaroğlu, S., Gem, E. and Wolters, W. 2007. Assessing the potential distribution of the Caucasian black grouse *Tetrao mlokosiewiczi* in Turkey through spatial modelling. *J. Orn.* 148: 427–434.

Granadeiro, J. P. 1993. Variation in measurements of Cory's Shearwater between populations and sexing by discriminant analysis. *Ring. & Migr.* 14: 103–112.

Grant, C. H. B. and Mackworth-Praed, C. W. 1952. On the relationship of the European and African Great Grey Shrikes. *Bull. Brit. Orn. Club* 72: 94.

Grapputo, A., Pilastro, A., Baker, A. J. and Marin, G. 2001. Molecular evidence for phylogenetic relationships among buntings and American sparrows (Emberizidae). *J. Avian Biol.* 32: 95–101.

Green, A. J. 1993. *The Status and Conservation of the Marbled Teal Marmaronetta angustirostris.* IWRB (Spec. Publ. No. 23), Slimbridge.

Green, A. J. 1996. The summer ecology of the Marbled Teal *Marmaronetta angustirostris*, Ferruginous Duck *Aythya nyroca* and other ducks in the Göksu Delta, Turkey in 1995. Doñana Biological Station, Seville.

Green, A. J. and Anstey, S. 1992. The status of the White-headed Duck *Oxyura leucocephala. Bird Conserv. Intern.* 2: 185–200.

Green, A., Hilton, G., Hughes, B., Fox, A. D. and Yarar, M. 1993. The White-headed Duck at Burdur Gölü, Turkey. *IWRB Threatened Waterfowl Res. Group Newsletter* 4: 7–8.

Green, I. A. 1997. Corncrake *Crex crex* breeding in north-east Turkey. *Sandgrouse* 19: 145–146.

Green, I. and Moorhouse, N. 1995. *A Birdwatchers' Guide to Turkey.* Prion, Perry.

Green, I. A., Moorhouse, C. N. and West, S. 1987. White-headed Duck in Turkey. A study of their breeding status and distribution. Unpubl. report.

Greenway, J. C. 1967. Family Sittidae. *In*: Paynter, R. A. (ed.) *Check-list of the Birds of the World.* Vol. 12. Mus. Comp. Zool., Cambridge, MA.

Gretton, A. 1991. *The Ecology and Conservation of the Slender-billed Curlew (Numenius tenuirostris).* International Council for Bird Preservation (Monograph No. 6), Cambridge.

Gretton, A., Yurlov, A. K. and Boere, G. C. 2002. Where does the Slender-billed Curlew nest, and what future does it have? *Brit. Birds* 95: 334–344.

Grieve, A., Hill, B. J. N., Lassey, P. A. and Wallace, D. I. M. 2005. The first American Golden Plover *Pluvialis dominica* in Oman and Arabia. *Sandgrouse* 27: 75–77.

Griffiths, C. S., Barrowclough, G. F., Groth, J. G. and Mertz, L. A. 2007. Phylogeny, diversity, and classification of the Accipitridae based on DNA sequences of the RAG-1 exon. *J. Avian Biol.* 38: 587–602.

Grimmett, R., Inskipp, C. and Inskipp, T. 1998. *Birds of the Indian Subcontinent*. Christopher Helm, London.

Grimmett, R. F. A. and Jones, T. A. 1989. *Important Bird Areas in Europe*. International Council for Bird Preservation, Cambridge.

Groh, G. 1968. Ornithologische Reiseeindrücke aus Griechenland und der Türkei. *Mitt. Pollichia* (III) 5: 163–170.

Groh, G. 1970. Jungfernkranich und Fahlsperling Brutvögel in der Türkei, mit Bemerkungen zu anderen Vogelarten. *Mitt. Pollichia* 18: 178–183.

Groh, G. 1986. Mass migration of the Red-necked Phalarope, *Phalaropus lobatus*, in eastern Turkey. *Zool. Middle East* 1: 28–29.

Groth, J. G. 1998. Poster abstracts 118: Molecular phylogeny of the cardueline finches and Hawaiian honeycreepers. *Ostrich* 69: 401.

Guidotti, G., Regato, P. and Jimenez-Caballero, S. 1986. *The Major Forest Types in the Mediterranean*. World Wildlife Fund, Rome.

Guillaument, A., Pons, J.-M., Godelle, B. and Crochet, P.-A. 2006. History of the Crested Lark in the Mediterranean region as revealed by DNA sequences and morphology. *Mol. Phyl. & Evol.* 39: 645–656.

Gürsoy, A., Erciyas, K., Özsemir, A. C. and Barış, Y. S. 2004. 2003 ringing summary of Cernek ringing station. *The Ring* 26: 85–86.

Gutiérrez, R. 1998. Flight identification of Cory's and Scopoli's Shearwaters. *Dutch Birding* 20: 216–225.

Gyllin, R. 1974. Spring migration of raptors over east Turkey and northwest Iran. *Vogelwelt* 27: 243–247.

Haaß, C. 1990. First record of the Storm Petrel, *Hydrobates pelagicus*, in Turkey. *Zool. Middle East* 4: 23–24.

Haffer, J. 1977. Secondary contact zones of birds in northern Iran. *Bonn. Zool. Monogr.* 10.

Haffer, J. 1989. Parapatrische Vögelarten der paläarktischen Region. *J. Orn.* 130: 475–512.

Haffer, J. 1992. The history of species concepts and species limits in ornithology. *Bull. Brit. Orn. Club Centenary* Suppl. 112A: 107–158.

Haffer, J. 1997. Species concepts and species limits in ornithology. *In*: del Hoyo, J., Elliott, A. and Sargatal, J. (eds.) *Handbook of the Birds of the World*. Vol. 4. Lynx Edicions, Barcelona.

Hagemeijer, W. J. M. and Blair, M. J. (eds.) 1997. *The EBCC Atlas of European Breeding Birds*. T. & A. D. Poyser, London.

Hales, S. 1996. Observations of seabirds on the Black Sea coast in Turkey 9–13 May 1995. *Sea Swallow* 45: 100–106.

Hamidan, N. A. 2003. The first Siberian White Cranes *Grus leucogeranus* in Jordan. *Sandgrouse* 25: 143.

Handrinos, G. and Akriotis, T. 1997. *The Birds of Greece*. Christopher Helm, London.

Haraldsson, T. 2008. The first Namaqua Dove *Oena capensis* for Lebanon? *Sandgrouse* 30: 90–91.

Harrap, S. 1987. Comments on historical records of the Siberian White Crane in Turkey. *Orn. Soc. Middle East Bull.* 19: 18–19.

Harrison, C. J. O. and Castell, P. 2002. *Birds Nests, Eggs and Nestlings of Britain and Europe with North Africa and the Middle East*. Revised edn. HarperCollins, London.

Hartert, E. 1910–22. *Die Vögel der paläarktischen Fauna*. Berlin.

Hartert, E. and Steinbacher, F. 1932/38. *Die Vögel der paläarktischen Fauna*. Berlin.

Harvey, P. 1992. The Brown Flycatcher on Fair Isle – a new British bird. *Birding World* 5: 252–255.

Haukkovaara, O. 1993. The Güldenstadt's Redstart in Finland. *Birding World* 6: 405.

van der Have, T. M. 1987. Mystery photographs 25: Green Warbler. *Dutch Birding* 9: 167–169.

van der Have, T. M. and van den Berk, V. M. 1988. Waders and waterfowl in southern Turkey, spring 1987. *Orn. Soc. Middle East Bull.* 20: 2–3.

van der Have T. M. and van den Berk, V. M. 1995. Paddyfield Warbler in Göksu delta, southern Turkey, in September 1991. *Dutch Birding* 17: 20–21.

van der Have, T. M., van den Berk, V., Cronau, J. P. and Langelveld, M. J. 1988. *South Turkey Project (A Survey of Waders and Waterfowl in the Çukurova Delta)*. WIWO (Report No. 22), Zeist.

Hazevoet, C. J. 1995. *The Birds of the Cape Verde Islands: An Annotated Check-list*. BOU Check-list No. 13. British Ornithologists' Union, Tring.

Heath, M. F. and Evans, M. I. 2000. *Important Bird Areas in Europe: Priority Sites for Conservation*. Vol. 2. BirdLife International (Conservation Series No. 8), Cambridge.

van Heezik, Y. and Seddon, P. J. 1999. Seasonal changes in habitat use by Houbara Bustards *Chlamydotis [undulata] macqueenii* in northern Saudi Arabia. *Ibis* 141: 208–215.

Heidrich, P., Ristow, D. and Wink, M. 1996. Differenzierung von Gelb- und Schwarzschnabelsturmtauchern (*Calonectris diomedea, Puffinus puffinus, P. yelkouan*) und Großmöwen des Silbermöwenkomplexes (*Larus argentatus, L. fuscus, L. cachinnans*). *J. Orn.* 137: 281–294.

Heidrich, P., Amengual, J. and Wink, M. 1998. Phylogenetic relationships in Mediterranean and North Atlantic shearwaters (Aves: Procellariidae) based on nucleotide sequences of mtDNA. *Biochem. Syst. Ecol.* 26: 145–170.

Heiser, F. 1984. Zum Vorkommen des Grauspechts (*Picus canus*) in der Türkei. *Orn. Mitt.* 36: 271.

Helb, H. W., Jetz, W., Helbig, A. and Wallschläger, D. 1995. Zur Herkunftsfrage der deutschen Einwanderungspopulation des Karmingimpels (*Carpodacus erythrinus*). *J. Orn.* 136: 351–352.

Helbig, A. 1984. Bemerkenswerte ornithologische Beobachtungen in der Türkei im Sommer 1981. *Bonn. Zool. Beitr.* 35: 57–69.

Helbig, A. J. 2001. Zur molekularen Identifizierung des Berliner Blaßspotters *Hippolais [pallida] elaeica*. *Limicola* 15: 155–156.

Helbig, A. J., Knox, A. G., Parkin, D. T., Sangster, G. and Collinson, M. 2002. Guidelines for assigning species rank. *Ibis* 144: 518–525.

Helbig, A. J., Kocum, A., Seibold, I. and Braun, M. J. 2005. A multi-gene phylogeny of aquiline eagles (Aves: Accipitriformes) reveals extensive paraphyly at the genus level. *Mol. Phyl. & Evol.* 35: 147–164.

Helbig, A. J., Martens, J., Seibold, I., Henning, F., Schottler, B. and Wink, M. 1996. Phylogeny and species limits in the Palaearctic Chiffchaff *Phylloscopus collybita* complex: mitochondrial genetic differentiation and bioacoustic evidence. *Ibis* 138: 650–666.

Helbig, A. and Seibold, I. 1999. Molecular phylogeny of Palearctic–African *Acrocephalus* and *Hippolais* warblers (Aves: Sylviidae). *Mol. Phyl. & Evol.* 11: 246–260.

Helbig, A. J., Seibold, I., Martens, J. and Wink, M. 1995. Genetic differentiation and phylogenetic relationships of Bonelli's Warbler *Phylloscopus bonelli* and Green Warbler *P. nitidus*. *J. Avian Biol.* 26: 139–153.

Hellmich, W. 1960. Vulkane und Seen in Ostanatolien. *Kosmos* 56: 205–211.

Hellyer, P. 2000. The first Yellow-billed Stork *Mycteria ibis* in Qatar and the Arabian Gulf. *Sandgrouse* 22: 125–126.

Heredia, B., Parr, S. J. and Yarar, M. 1997. A baseline survey of the Black Vulture *Aegypius monachus* in western Turkey. *Sandgrouse* 19: 126–132.

Hernández, M. A., Campos, F., Gutiérrez-Corchero, F. and Amecuza, A. 2004. Identification of *Lanius* species and subspecies using tandem repeats in the mitochondrial DNA control region. *Ibis* 146: 227–230.

Herrn, C.-P. 1966. Neue Sommerbeobachtungen in Anatolien, in Kilikien und im Hatay. *Vogelwarte* 23: 305–308.

Heunks, C., Heunks, E., Eken, G. and Kurt, B. 2001. Distribution and current status of Great Bustard *Otis tarda* in the Konya Basin, central Turkey. *Sandgrouse* 23: 106–111.

Heunks, C., Heunks, E., Eken, G. and Kurt, B. 2002. Distribution and status of Great Bustards, *Otis tarda*, in the Konya Basin, Turkey, spring 2000. WIWO (Report No. 76), Zeist.

Hirsch, U. 1979. Studies of West Palearctic birds 183: Bald Ibis. *Brit. Birds* 72: 313–325.

Hirschfeld, E. 1995. *Birds in Bahrain: A Study of their Migration Patterns 1990–1992.* Hobby Publications, Dubai.

Hirschfeld, E., Roselaar, C. S. and Shirihai, H. 2000. Identification, taxonomy and distribution of Greater and Lesser Sand Plovers. *Brit. Birds* 93: 162–189.

Hirschfeld, E. and Symens, P. 1992. First records of Dead Sea Sparrow *Passer moabiticus* in Arabia. *Sandgrouse* 14: 48–51.

Hollom, P. A. D. 1955. A fortnight in south Turkey. *Ibis* 97: 1–17.

Hollom, P. A. D. 1971. *Check List of the Birds of Turkey.* Ornithological Society of Turkey, London.

Hollom, P. A. D., Porter, R. F., Christensen, S. and Willis, I. 1988. *The Birds of the Middle East and North Africa.* T. & A. D. Poyser, Calton.

Howell, S. N. G. and Patteson, J. B. 2008. Variation in Cory's and Scopoli's Shearwaters. *Alula* 14: 12–21.

Howes, A. 1996. Autumn migration across the Bosphorus. *Leics. & Rutland Orn. Soc. Newsletter* 390: 5.

del Hoyo, J., Elliott, A. and Christie, D. A. (eds.) 2004. *Handbook of the Birds of the World.* Vol. 9. Lynx Edicions, Barcelona.

del Hoyo, J., Elliott, A. and Christie, D. A. (eds.) 2006. *Handbook of the Birds of the World.* Vol. 11. Lynx Edicions, Barcelona.

del Hoyo, J., Elliott, A. and Sargatal, J. (eds.) 1994. *Handbook of the Birds of the World.* Vol. 2. Lynx Edicions, Barcelona.

del Hoyo, J., Elliott, A. and Sargatal, J. (eds.) 1996. *Handbook of the Birds of the World.* Vol. 3. Lynx Edicions, Barcelona.

del Hoyo, J., Elliott, A. and Sargatal, J. (eds.) 1999. *Handbook of the Birds of the World.* Vol. 5. Lynx Edicions, Barcelona.

del Hoyo, J., Elliott, A. and Sargatal, J. (eds.) 2002. *Handbook of the Birds of the World.* Vol. 7. Lynx Edicions, Barcelona.

Huber, C. and Barbalat, A. 1990. Aperçus sur la migration automnale des rapaces en Turquie orientale. *Nos Oiseaux* 40: 289–293.

Hüe, F. & Etchécopar, R. D. 1970. *Les Oiseaux du Proche et du Moyen Orient.* Ed. Boubée, Paris.

Humphreys, P. N. and Kahrom, E. 1995. *The Lion and the Gazelle.* Comma International Biological Systems, Gwent.

Huş, S. 1974. Av Hayvanları ve Avcılık. *Istanbul Üniv. Orman Fak. Yay.* 202: 1–406.

Husband, C. and Kasparek, M. 1984. *The Birds of Seyfe Gölü.* Birds of Turkey 2. Kasparek Verlag, Heidelberg.

Hustings, F. and van Dijk, K. 1994. *Bird Census in the Kizilirmak Delta, Turkey, in Spring 1992.* WIWO (Report No. 45), Zeist.

Huxley, J. 1949. *From an Antique Land: Ancient and Modern in the Middle East.* London.

Idaghour, Y., Broderick, D., Korrida, A. & Chbel, F. 2004. Mitochondrial control region diversity of the Houbara Bustard *Chlamydotis undulata* complex and genetic structure along the Atlantic seaboard of North Africa. *Mol. Ecol.* 13: 43–54.

Irwin, D. E., Alström, P., Olsson, U. & Benowitz-Fredericks, Z. M. 2001a. Cryptic species in the genus *Phylloscopus* (Old World leaf warblers). *Ibis* 143: 233–247.

Irwin, D. E., Bensch, S. and Price, T. D. 2001b. Speciation in a ring. *Nature* 409: 333–337.

Isenmann, P. and Bouchet, M.-A. 1993. L'aire de distribution française et le statut taxinomique de la Pie-grièche méridionale *Lanius excubitor meridionalis. Alauda* 61: 223–231.

Isfendiyaroğlu, S., Welch, G. and Atad, M. 2007. The Caucasian black grouse *Tetrao mlokosiewiczi* in Turkey: recent survey results and conservation recommendations. *Wildl. Biol.* Suppl. 1: 13–20.

Ivanov, A. I. 1941. *Oenanthe chrysopygia* de Fil i *O. xanthoprymna* Hempr. & Ehrenb. *Izv. Akad. Nauk SSSR* 3: 381–384.

Jakobsen 1986. Occurrence of Brown-necked Raven, *Corvus ruficollis*, at Cizre in eastern Turkey. *Zool. Middle East* 1: 32–33.

Jetz, W. 1995. *The Birds of Uludağ.* Birds of Turkey 11. Kasparek Verlag, Heidelberg.

Johnsgard, P. A. 1978. *Ducks, Geese and Swans of the World.* University of Nebraska Press, Lincoln.

Johnson, A. R. 1989. Movements of Greater Flamingos (*Phoenicopterus ruber roseus*) in the Western Palearctic. *Rev. Ecol. (Terre Vie)* 44: 75–94.

de Juana, E. 2006. *Aves raras de España.* Lynx Edicions, Barcelona.

Juniper, T. and Parr, M. 1998. *Parrots. A Guide to the Parrots of the World.* Pica Press, Robertsbridge.

KAD (Turkish Bird Research Society) 2003. *Türkiye Ulusal Halkalama Programı ve 2002 Yılı Halkalama Raporu.* KAD, Ankara.

Kahl, M. P. 1975. Distribution and numbers – a summary. *In*: Kear, J. and Duplaix-Hall, N. (eds.) *Flamingos.* T. & A. D. Poyser, Berkhamsted.

Kanai, Y., Nagendran, M., Ueta, M., Markin, Y., Rinne, J., Sorokin, A. G., Higuchi, H. and Archibald, G. W. 2002. Discovery of breeding grounds of a Siberian Crane *Grus leucogeranus* flock that winters in Iran, via satellite telemetry. *Bird Conserv. Intern.* 12: 327–333.

Karakaş, R. and Kılıç, A. 2002. Birds of Göksu Dam (Diyarbakır) and new records in south-east Turkey. *Sandgrouse* 24: 38–43.

Karakaş, R. and Kılıç, A. 2005. The birds of Kralkızı Dam (Diyarbakır) southeast Turkey. *Sandgrouse* 27: 139–146.

Karakaş, R. and Bırıçık, M. 2005. Wintering Wallcreeper *Tichodroma muraria* in southeastern Anatolia. *Sandgrouse* 27: 150–151.

Karauz Kiraç, S. 1993. *The Birds of Cöl Lake, Uyuz Lake and Yağlıören Pool.* Birds of Turkey 10. Kasparek Verlag, Heidelberg.

Karauz Kiraç, S. and Kiraç, C. 1995. Population and distribution of breeding seabird colonies on Foça Islands, Aegean Sea, Turkey. MEDMARAVIS Conference, 11–16 April 1995, Tunisia.

Kasparek, M. 1985. *Die Sultanssümpfe: Naturgeschichte eines Vogelparadieses in Anatolien.* Kasparek Verlag, Heidelberg.

Kasparek, M. 1986a. On the occurrence of the Bittern, *Botaurus stellaris*, in Turkey. *Zool. Middle East* 1: 33–41.

Kasparek, M. 1986b. On the distribution and systematic position of the Barn Owl, *Tyto alba*, in Turkey and adjacent areas. *Zool. Middle East* 1: 44–51.

Kasparek, M. 1986c. New records of the Red-tailed Wheatear, *Oenanthe xanthoprymna*, in Turkey. *Zool. Middle East* 1: 51–54.

Kasparek, M. 1986d. The Cyprus Warbler, *Sylvia melanothorax*, recorded in Turkey. *Zool. Middle East* 1: 54–56.

Kasparek, M. 1986e. On records of the Pine Bunting, *Emberiza leucocephalos*, in Turkey from the last century. *Zool. Middle East* 1: 56–59.

Kasparek, M. 1987a. *The Birds of Kulu Gölü.* Birds of Turkey 5. Kasparek Verlag, Heidelberg.

Kasparek, M. 1987b. Historical records of the Siberian White Crane in Turkey. *Orn. Soc. Middle East Bull.* 18: 4–6.

Kasparek, M. 1988a. *Der Bafasee. Natur und Geschichte in der türkischen Ägäis.* Kasparek Verlag, Heidelberg.

Kasparek, M. 1988b. The Demoiselle Crane, *Anthropoides virgo*, in Turkey: distribution and population of a highly endangered species. *Zool. Middle East* 2: 31–38.

Kasparek, M. 1988c. Zum ursprünglichen Vorkommen und zur Wiedereinbürgerung des Fasans, *Phasianus colchicus*, in der Türkei. *Verh. Orn. Ges. Bayern* 24: 725–735.

Kasparek, M. 1989a. Status and distribution of the Great Bustard and Little Bustard in Turkey. *Bustard Studies* 4: 80–113.

Kasparek, M. 1989b. Brut und Zug des Fischadlers *Pandion haliaetus* in der Türkei. *Limicola* 3: 251–255.

Kasparek, M. 1989c. Zum Durchzug und zur Brutverbreitung des Wendehalses (*Jynx torquilla*) in der Türkei. *Ecol. Birds* 11: 251–256.

Kasparek, M. 1989d. Breeding distribution of the Rook *Corvus frugilegus* in Turkey. *Sandgrouse* 11: 89–95.

Kasparek, M. 1990a. Zum Vorkommen einiger in der Türkei seltener Vogelarten. *Bonn. Zool. Beitr.* 41: 181–202.

Kasparek, M. 1990b. On the migration of the Whimbrel, *Numenius phaeopus*, in Turkey. *Zool. Middle East* 4: 25–32.

Kasparek, M. 1990c. Zur Brutverbreitung und zum Zug der Mönchsgrasmücke (*Sylvia atricapilla*) im Nahen Osten. *Vogelwarte* 35: 169–176.

Kasparek, M. 1991. Zur Verbreitung der Palmtaube *Streptopelia senegalensis* im Vorderen Orient. *Orn. Verh.* 25: 51–64.

Kasparek, M. 1992a. *Die Vögel der Türkei: eine Übersicht.* Kasparek Verlag, Heidelberg.

Kasparek, M. 1992b. Status of Sociable Plover *Chettusia gregaria* and White-tailed Plover *C. leucura* in Turkey and the Middle East. *Sandgrouse* 14: 2–15.

Kasparek, M. 1992c. Zum Zug des Steinwälzers, *Arenaria interpres*, in der Türkei. *Beitr. Vogelkd.* 38: 200–208.

Kasparek, M. 1996a. Dismigration und Brutarealexpansion der Türkentaube *Streptopelia decaocto*. *J. Orn.* 137: 1–33.

Kasparek, M. 1996b. On the identity of *Ceryle rudis syriaca*. *J. Orn.* 137: 357–358.

Kasparek, M. 1998. Vorkommen und Ausbreitung der Türkentaube *Streptopelia decaocto* im Nahen und Mitteleren Osten. *Orn. Verh.* 25: 241–279.

Kasparek, M. and Bilgin, C. C. 1996. Kuşlar (Aves). *In*: Kence, A. and Bilgin, C. C. (eds.) *Türkiye Omurgalilar Tür Listesi.* TÜBITAK, Ankara.

Kasparek, M., Bilgin, C. C. and Akin, A. 1989. The Purple Gallinule, *Porphyrio porphyrio*, in the eastern Mediterranean. *Zool. Middle East* 3: 19–30.

Kasparek, M. and Ristow, D. 1986. On the occurrence of the Eleonora's Falcon, *Falco eleonorae*, in Turkey. *Zool. Middle East* 1: 60–69.

Kasparek, M. and van der Ven, J. 1983. *The Birds of Erçek Gölü*. Birds of Turkey 1. Kasparek Verlag, Heidelberg.

Kasparyan, A. 1956. A preliminary systematic list of the birds of Turkey. *Istanbul Fen. Fak. Mecm.* 21: 27–48.

Kaya, Z. and Raynal, D. J. 2001. Biodiversity and conservation of Turkish forests. *Biol. Conserv.* 97: 131–141.

Kaymas, C. 1980a. Marmara Island. *Orn. Soc. Middle East Bull.* 5: 2.

Kaymas, C. 1980b. Eleonora's Falcon – a new breeding species for Turkey. *Orn. Soc. Middle East Bull.* 5: 3.

Kear, J. (ed.) 2005. *Ducks, Geese and Swans*. Oxford University Press.

Keith, S., Urban, E. K. and Fry, C. H. (eds.) 1992. *The Birds of Africa*. Vol. 4. Academic Press, London.

Kennedy, M., Gray, R. D. and Spencer, H. G. 2000. The phylogenetic relationships of the shags and cormorants: can sequence data resolve a disagreement between behavior and morphology? *Mol. Phyl. & Evol.* 17: 345–359.

Keve, A. 1973. Über einige taxonomische Fragen der Eichelhäher des Nahen Ostens (Aves, Corvidae). *Zool. Abh. (Dresden)* 32: 175–198.

Keve, A. 1978. Revision der Unterarten des Feldsperlings (*Passer montanus* (Linné, 1758)). *Zool. Abh. (Dresden)* 34: 245–273.

Khaleghizadeh, A. 2004. Notes on three rare raptors in the Anzali wetlands, Iran. *Sandgrouse* 26: 155.

Kılıç, A. 1988. The Ereğli marshes: a new nesting site for the Greater Flamingo, *Phoenicopterus ruber*, in Turkey. *Zool. Middle East* 2: 39–42.

Kılıç, A. 1999. Karapınar (Konya) Yöresinin Kuşları. *Turk. J. Zool.* 23: 91–97.

Kılıç, A. and Karakaş, R. 2005. Recent observations on the Great Bustard, *Otis tarda*, in south-eastern Anatolia. *Zool. Middle East* 35: 99–101.

Kılıç, A., Karakaş, R. and Biricik, M. 2003 Observations on a newly detected breeding site of Bonelli's Eagle, *Hieraaetus fasciatus*, in south-eastern Anatolia. *Zool. Middle East* 30: 37–41.

Kılıç, A. and Kasparek, M. 1987. *The Birds of Yeniçağa Gölü*. Birds of Turkey 6. Kasparek Verlag, Heidelberg.

Kılıç, A. and Kasparek, M. 1989. *The Birds of the Köyceğiz-Dalyan Area*. Birds of Turkey 8. Kasparek Verlag, Heidelberg.

Kılıç, A. and Kasparek, M. 1990. The Ereğli Marshes: assessment of their biological importance and recommendations for conservation. Unpubl. report to International Council for Bird Preservation/World Wildlife Fund.

King, B. F. 1997. *Checklist of the Birds of Eurasia*. Ibis Publishing, Vista, California.

King, J. and Shirihai, H. 1996. Identification and ageing of Audouin's Gull. *Birding World* 9: 52–61.

Kinzelbach, R. 1985. Die Dreizehenmöwe (*Rissa tridactyla*) als Wintergast am Euphrat. *Vogelwarte* 33: 163–165.

Kirwan, G. (compiler) 1990. Around the region. *Orn. Soc. Middle East Bull.* 25: 42–46.

Kirwan, G. 1992a. The migration of the Great Snipe, *Gallinago media*, through Turkey. *Zool. Middle East* 7: 11–19.

Kirwan, G. 1992b. A record of Aquatic Warbler from Turkey. *Orn. Soc. Middle East Bull.* 28: 18–19.

Kirwan, G. (compiler) 1992c. Around the region. *Orn. Soc. Middle East Bull.* 29: 35–48.

Kirwan, G. 1992d. A freshwater breeding record of Greater Flamingo *Phoenicopterus ruber* in Turkey. *Sandgrouse* 14: 56–57.

Kirwan, G. M. 1993a. *The Birds of Hotamış*. Birds of Turkey 9. Kasparek Verlag, Heidelberg.

Kirwan, G. M. 1993b. Olive-backed Pipit, *Anthus hodgsoni*, new to Turkey. *Zool. Middle East* 9: 45–47.

Kirwan, G. 1993c. IBA survey work 1993. *Orn. Soc. Middle East Bull.* 31: 11–13.

Kirwan, G. M. (compiler) 1993d. Around the region. *Orn. Soc. Middle East Bull.* 30: 39–48.

Kirwan, G. M. (compiler) 1993e. Around the region. *Orn. Soc. Middle East Bull.* 30: 39–46.

Kirwan, G. 1994a. A record of Booted Warbler *Hippolais caligata* from Turkey. *Orn. Soc. Middle East Bull.* 33: 7–9.

Kirwan, G. M. 1994b. The breeding status and distribution of the White-headed Duck *Oxyura leucocephala* on the Central Plateau, Turkey. *Sandgrouse* 16: 66–75.

Kirwan, G. M. 1994c. American Golden Plover, *Pluvialis dominica*, in Turkey. *Zool. Middle East* 10: 23–26.

Kirwan, G. M. (compiler) 1994d. Around the region. *Orn. Soc. Middle East Bull.* 33: 32–44.

Kirwan, G. M. 1995a. Recent records of rare birds in Turkey. *Zool. Middle East* 11: 37–46.

Kirwan, G. M. 1995b. Breeding status of White-headed Duck in Turkey. *Threatened Waterfowl Res. Group Newsletter* 8: 14–15.

Kirwan, G. M. 1995c. Some comments on the distribution of the Rook *Corvus frugilegus* in Turkey. *Orn. Soc. Middle East Bull.* 35: 29–30.

Kirwan, G. M. 1996a. The status of the Red Knot, *Calidris canutus*, in the Middle East. *Zool. Middle East* 13: 17–23.

Kirwan, G. M. 1996b. A new specimen record of Rustic Bunting *Emberiza rustica* from Turkey. *Sandgrouse* 18 (2): 70–71.

Kirwan, G. M. 1997a. A list of bird specimens held in the Robert's College, Bebek (Istanbul, Turkey) collection, with some comments on Mathey-Dupraz (1920–24). *Sandgrouse* 19: 30–38.

Kirwan, G. M. 1997b. The status of the Ferruginous Duck *Aythya nyroca* in Turkey. *Bird Conserv. Intern.* 7: 345–356.

Kirwan, G. M. 1997c. White-spotted Bluethroats in Turkey. *Dutch Birding* 19: 237.

Kirwan, G. M. (compiler) 1997d. Around the region. *Sandgrouse* 19: 156–160.

Kirwan, G. M. 1998a. Ornithological observations on Karadağ, Konya province, Turkey. *Turk. J. Zool.* 22: 237–239.

Kirwan, G. M. (compiler) 1998b. Around the region. *Sandgrouse* 20: 157–160.

Kirwan, G. M. (compiler) 1999. Around the region. *Sandgrouse* 20: 188–192.

Kirwan, G. M. 2000. Spotted Sandgrouse at Birecik, Turkey, in June 1999. *Dutch Birding* 22: 17–18.

Kirwan, G. M. 2001a. The validity of *Monticola saxatilis coloratus* Stepanyan, 1964. *Sandgrouse* 23: 148–149.

Kirwan, G. M. 2001b. White-winged Lark status in Turkey. *Alula* 7: 124.

Kirwan, G. M. (compiler) 2001c. Around the region. *Sandgrouse* 23: 76–80.

Kirwan, G. M. 2004a. A Syrian record: the case of Aharoni's Thick-billed Larks *Ramphocoris clotbey*. *Bull. Brit. Orn. Club* 124: 227–232.

Kirwan, G. M. 2004b. The taxonomic position of the Afghan Scrub Sparrow *Passer* (*moabiticus*) *yatii*. *Sandgrouse* 26: 105–111.

Kirwan, G. M. 2005a. An early specimen of Richard's Pipit *Anthus richardi* from Syria? *Sandgrouse* 27: 148–149.

Kirwan, G. M. 2005b. Middle Spotted Woodpecker: range and subspecies in Turkey. *Dutch Birding* 27: 260–261.

Kirwan, G. M. 2006. Comments on two subspecies of passerine birds recently described from Turkey, *Eremophila alpestris kumerloevei* and *Pyrrhula pyrrhula paphlagoniae*, with remarks on geographical variation in related forms of Bullfinch from the Balkans and Caucasus. *Sandgrouse* 28: 10–21.

Kirwan, G. M. and Atkinson, P. W. in prep. Some remarks on subspecies of Twite *Carduelis flavirostris*.

Kirwan, G. M., Atkinson, P. W., van den Berg, A. B. and Shirihai, H. 2006. Taxonomy of the Crimson-winged Finch *Rhodopechys sanguineus*: a test case for defining species limits between disjunct taxa? *Bull. Afr. Bird Club* 13: 136–146.

Kirwan, G. M. and Bates, J. M. in press. The status of the European Stonechat *Saxicola torquatus* in the Islamic Republic of Iran. *Sandgrouse* 30.

Kirwan, G. M., Bechtolsheim, M. and Willig, S. 2000. Distribution of Mongolian Finch in Turkey. *Dutch Birding* 22: 149–150.

Kirwan, G. M. and Gregory, S. M. S. 2005. A new genus for the Mongolian Finch *Bucanetes mongolicus* (Swinhoe, 1870). *Bull. Brit. Orn. Club* 125: 68–80.

Kirwan, G. and Konrad, V. 1995. Little known Western Palearctic birds: Mongolian Trumpeter Finch. *Birding World* 8: 139–144.

Kirwan, G. M. and Martins, R. P. 1994. Turkey Bird Report 1987–91. *Sandgrouse* 16: 76–117.

Kirwan, G. M. and Martins, R. P. 2000a. Turkey Bird Report 1992–1996. *Sandgrouse* 22: 13–35.

Kirwan, G. M. & Martins, R. P. 2000b. Photospot: Dead Sea Sparrow in Turkey. *Sandgrouse* 22: 61–63.

Kirwan, G. M., Martins, R. P., Eken, G. and Davidson, P. 1999. A checklist of the birds of Turkey. *Sandgrouse* Suppl. 1: 1–32.

Kirwan, G. M., Özen, M. and Demirci, B. (compilers) in press. Turkey Bird Report 2002–06. *Sandgrouse* 30.

Kirwan, G. M., Özen, M., Kurt, B. and Martins, R. P. 2003. Turkey Bird Report 1997–2001. *Sandgrouse* 25: 8–31.

Kirwan, G. M. and Shirihai, H. 2007. Species limits in House Bunting. *Dutch Birding* 29: 1–19.

Kiss, J. B. and Szabó, L. 2000. First breeding record of White-tailed Lapwing in Romania (and Europe outside Russia). *Brit. Birds* 93: 400–401.

Kivit, H. A., Nijmeijer, H. and Ovaa, A. (eds.) 1994. *Wader and Waterfowl Migration in the Çukurova Deltas, South Turkey, Spring 1990*. WIWO (Report No. 48), Zeist.

Kiziroğlu, I. 1982. Brutbiologische Untersuchungen an vier Meisenarten (*Parus*) in der Umgebung von Ankara. *J. Orn.* 123: 409–423.

Kiziroğlu, I. 1984. Populationsdynamische Untersuchungen an 4 Meisenarten (*Parus* spp.) in der Umgebung von Ankara. *Beitr. Vogelkd., Jena* 30: 138–148.

Kiziroğlu, I. 1989. *Türkiye Kuşları*. Ankara.

Kiziroğlu, I. and Kiziroğlu, F. 1987. Die Vogelarten im Vogelparadies des "Kuş Cenneti/Bandirma"-Nationalparks und seiner Umgebung. *Verh. Orn. Ges. Bayern* 24: 515–532.

Kiziroğlu, I., Şişli, N. M. and Alp, U. 1987. Zur interspezifischen Konkurrenz in der Besiedlung von Nistkasten bei Ankara/Türkei mit Angaben zur Brutbiologie verschiedener Höhlenbrüter-Arten. *Vogelwelt* 108: 169–175.

Kiziroğlu, I., Turan, L. and Erdoğan, A. 1989. A bio-ornithological study on Sultansazlığı, one of the most marshy areas of Turkey and Europe: new bird species in the area and the current situation. Unpubl. manuscript.

Klein, R. and Buchheim, A. 1997. Die westliche Schwarzmeerkuste als Kontaktgebiet zweier Großmöwenformen der *Larus cachinnans*-Gruppe. *Vogelwelt* 118: 61–70.

Klicka, J., Zink, R. M. and Winker, K. 2003. Longspurs and snow buntings: phylogeny and biogeography of a high-latitude clade (*Calcarius*). *Mol. Phyl. & Evol.* 26: 165–175.

Knight, C. G. and Beale, C. M. 2004. Breeding habitat of the Pale Rock Sparrow *Carpospiza brachydactyla* in the Mount Lebanon range. *Sandgrouse* 26: 127–131.

Kock, D. 1990. Notes on mammals (Insectivora, Rodentia) taken by the Tawny Owl, *Strix aluco*, in NW Turkey. *Zool. Middle East* 4: 5–9.

Kok, D. & van Duivendijk, N. 2002. Masters of mystery. Solutions of sixth round 2001: Great Black-headed Gull and Rufous-tailed Scrub Robin. *Dutch Birding* 24: 32–38.

Kok, M. and Ongenae, J. P. 1995. Raptor migration in the north-east of Turkey, September 1990. *Orn. Soc. Middle East Bull.* 34: 8–11.

Knox, A. G. 1994. Removal of Citril Finch from the British & Irish List. *Brit. Birds* 87: 471–473.

Knox, A. G., Helbig, A. J., Parkin, D. T. and Sangster, G. 2001. The taxonomic status of Lesser Redpoll. *Brit. Birds* 94: 260–267.

Knox, A. G. and Walters, M. 1992. Under the skin: the bird collections of the Natural History Museum. *Bull. Brit. Orn. Club Centenary* Suppl. 112A: 169–190.

König, C., Weick, F. and Becking, J.-H. 1999. *Owls: A Guide to the Owls of the World*. Pica Press, Robertsbridge.

Konrad, V. 1985. Samtkopf-Grasmücke (*Sylvia melanocephala*) und Tamarisken-Grasmücke (*Sylvia mystacea*) doch zwei "gute" Arten? *Orn. Mitt.* 37: 81.

Konrad, V. 1994. Kennzeichen des Dunnschnabelbrachvogels (*Numenius tenuirostris*). *Orn. Mitt.* 46: 291–295.

Krajewski, C. 1989. Phylogenetic relationships among cranes (Gruiformes: Gruidae) based on DNA hybridization. *Auk* 106: 603–618.

Krajewski, C. and Fetzner, J. W. 1994. Phylogeny of cranes (Gruiformes: Gruidae) based on cytochrome-*b* DNA sequences. *Auk* 111: 351–365.

Krajewski, C. and King, D. G. 1996. Molecular divergence and phylogeny: rates and patterns of cytochrome *b* evolution in cranes. *Mol. Biol. Evol.* 13: 21–30.

Krieger, H. 1988. The Trumpeter Finch, *Bucanetes githagineus*, in Turkey. *Zool. Middle East* 2: 43–45.

Kruckenhauser, L., Haring, E., Pinsker, W., Riesing, M. J., Winkler, H., Wink, M. and Gamauf, A. 2004. Genetic vs. morphological differentiation of Old World buzzards (genus *Buteo*, Accipitridae). *Zool. Scripta* 33: 197–211.

Krüper, T. 1875. Beitrag zur Ornithologie Kleinasiens. *J. Orn.* 23: 258–285.

Krüper, T. and Hartlaub, G. 1875. Zeiten des Gehens und Kommens und des Brütens der Vögel in Griechenland und Jonien. *Griech. Jahreszeiten* 3: 153–330.

Kryukov, A. P. 1995. Systematics of small Palearctic shrikes of the "*cristatus* group". *Proc. Western Found. Vert. Zool.* 6: 22–25.

Kumerloeve, H. 1957a. Ökologie der Misteldrossel im Bosporusraum. *Vogelwelt* 78: 194–195.

Kumerloeve, H. 1957b. Zur Verbreitungsgrenze des Gelbsteißbülbüls in Kleinasien. *Anz. Orn. Ges. Bayern* 4: 575–576.

Kumerloeve, H. 1957c. Brutvogelbeobachtungen bei Savaştepe und Bergama (NW-Anatolien). *Anz. Orn. Ges. Bayern* 4: 712–720.

Kumerloeve, H. 1957d. Ornithologische Beobachtungen im "Zubringerraum" (Bulgarisch-rumänische Schwarzmeerküste) des Bosporuszuges. *Bonn. Zool. Beitr.* 3/4: 248–274.

Kumerloeve, H. 1958a. Grüne Laubsänger, *Phylloscopus trochiloides nitidus* Blyth, als Durchzügler in Westanatolien. *Vogelwarte* 19: 210–211.

Kumerloeve, H. 1958b. Vom Vorkommen des Gleitaars, *Elanus caeruleus* (Desf.), am Bosporus und in der Türkei. *Vogelwarte* 19: 210.

Kumerloeve, H. 1960. Zur Verbreitung des Rallenheihers *Ardeola ralloides* (Scop.) in Vorderasien. *Acta Orn.* 5: 301–305.

Kumerloeve, H. 1961. Zur Kenntnis der Avifauna Kleinasiens. *Bonn. Zool. Beitr.* 12: 1–318.

Kumerloeve, H. 1962a. Türkiye kuşları üzerine yeni araştırmalar (Bataklık ve su kuşları hariç). Weitere Untersuchungen über die türkische Vogelwelt (ausgenommen Sumpf- und Wasservögel). *Istanbul Üniv. Fen. Fak. Mecm.* B 27: 165–228.

Kumerloeve, H. 1962b. Zur Brutverbreitung der beiden Ortolan-Arten *Emberiza hortulana* L. und *Emberiza caesia* Cretzschmar in Kleinasien. *Bonn. Zool. Beitr.* 4: 327–332.

Kumerloeve, H. 1963a. L'Avifaune du lac d'Antioche (Amik Gölü-Gölbasi) et de ses alentours. *Alauda* 30: 110–136, 161–211.

Kumerloeve, H. 1963b. *Calandrella rufescens niethammeri*, eine neue Stummellerchenform aus Inneranatolien (Türkei). *Vogelwelt* 84: 146–148.

Kumerloeve, H. 1963c. Zur Brutverbreitung des Frankolins, *Francolinus francolinus* (L.) im Vorderen Orient. *Vogelwelt* 84: 129–137.

Kumerloeve, H. 1964a. Zur Sumpf- und Wasservogelfauna der Türkei. *J. Orn.* 105: 307–325.

Kumerloeve, H. 1964b. Akşehir Gölü und Burdur Gölü; zwei kleinasiatische Seen hoher Bedeutung fur Sumpf- und Wasservögel. *Natur u. Land.* 39: 162–163.

Kumerloeve, H. 1964c. Lachmöwen und Schwarzhalstaucher als Brutnachbarn auf dem Akşehir Gölü im südwestlichen Zentralanatolien. *Der Falke* 11: 208–209.

Kumerloeve, H. 1965. Le Moineau moabite, *Passer moabiticus* Tristram, pres Birecik sur l'Euphrate. *Alauda* 33: 257–264.

Kumerloeve, H. 1966a. Liste systématique revisée des espèces d'oiseaux de Turquie. *Alauda* 34: 165–182.

Kumerloeve, H. 1966b. Zu Brutverbreitung und Durchzug des Weißstorches *C. ciconia* (L.), in Kleinasien. *Vogelwarte* 23: 221–224.

Kumerloeve, H. 1966c. Tendances expansives chez des espèces de *Carpodacus*, *Rhodopechys* et *Serinus* en Asie mineure. *Nos Oiseaux* 28: 284–287.

Kumerloeve, H. 1966d. Ergänzungen zur Avifauna Kleinasiens. *Bonn. Zool. Beitr.* 17: 257–259.

Kumerloeve, H. 1966–67. Migration et hivernage sur le Lac d'Antioche (Amik Gölü, Hatay, Turquie); coup l'oeil sur son avifaune nidificatrice actuelle. *Alauda* 34: 299–308; 35: 1–19.

Kumerloeve, H. 1967a. Neue Beiträge zur Kenntnis der Avifauna Nordost- und Ost-Kleinasiens. *Istanbul Üniv. Fen. Fak. Mecm.* B 32: 79–213.

Kumerloeve, H. 1967b. Recherches sur l'avifaune de la République Arabe Syrienne. Essai d'un aperçu. *Alauda* 35: 243–266.

Kumerloeve, H. 1967c. Contribution a la connaissance de *Carduelis* (*Acanthis*) *flavirostris brevirostris* (Bonaparte) 1855. *Alauda* 35: 118–124.

Kumerloeve, H. 1967d. Starker Durchzug von Griefvögeln, Saatkrähen und Dohlen in NE-Kleinasien. *Beitr. Vogelkd.* 13: 29–32.

Kumerloeve, H. 1967e. Zum Brutvorkommen der Samtente, *Melanitta fusca* (L.), im transkaukasich-(armenisch)-nordostkleinasiatischen Hochland. *Anz. Orn. Ges. Bayern* 8: 63–65.

Kumerloeve, H. 1968. Zum Status des Würgfalken (*Falco cherrug* Gray, 1834) in Kleinasien. *Anz. Orn. Ges. Bayern* 8: 286–289.

Kumerloeve, H. 1969a. Zur Avifauna des Van Gölü- und Hakkârigebietes (E/SE Kleinasien). *Istanbul Fen. Fak. Mecm.* B 34: 245–312.

Kumerloeve, H. 1969b. The Dead Sea Sparrow: a second breeding-place on Turkish and the first-known breeding-place on Syrian territory. *Ibis* 111: 617–618.

Kumerloeve, H. 1969c. Recherches sur l'avifaune de la République Arabe Syrienne. Essai d'un aperçu. *Alauda* 37: 43–58.

Kumerloeve, H. 1969d. Remarques sur les collections d'oiseaux de Palestine. *Alauda* 37: 164–167.

Kumerloeve, H. 1969e. On the occurrence of the Pied Wheatear *Oenanthe leucomela* in Asia Minor and adjacent countries. *Ibis* 111: 238–239.

Kumerloeve, H. 1970a. Zur Kenntnis der Avifauna Kleinasiens und der europäischen Türkei (Ergänzungen – Hinweise – Fragestellungen). *Istanbul Fen. Fak. Mecm.* B 35: 85–160.

Kumerloeve, H. 1970b. Hinweise und Ratschläge für künftige ornithologische Studien in der asiatischen und europäischen Türkei. *Vogelwelt* 91: 138–149.

Kumerloeve, H. 1970c. Zur Vogelwelt im Raume Ceylânpinar [*sic*] (türkisch-syrisches Grenzgebiet). *Beitr. Vogelkd.* 16: 239–249.

Kumerloeve, H. 1971. Büyükçekmece, ein bedeutsames Brut- und Nahrungsgebiet für Wasser- und Sumpfvögel. *Turk. Biol. Der.* 21: 118–122.

Kumerloeve, H. 1975a. The history of ornithology in Turkey. *Orn. Soc. Turkey Bird Rep.* 1970–1973: 289–319.

Kumerloeve, H. 1975b. Zur Verbreitung der Steinschmätzer (*Oenanthe*)-Arten in der Türkei. *Bonn. Zool. Beitr.* 26: 183–198.

Kumerloeve, H. 1975c. Expansion du Moineau soulcie pâle *Petronia brachydactyla* et du Roselin cramoisi *Carpodacus erythrinus* en Turquie. *Alauda* 43: 324–325.

Kumerloeve, H. 1976. Zum Vorkommen und zur taxonomischen Beurteilung türkischer Populationen von *Phasianus colchicus* L. *Bonn. Zool. Beitr.* 27: 47–52.

Kumerloeve, H. 1978. Situation des Moineaux moabites nicheurs en Turquie. *Alauda* 45: 181–182.

Kumerloeve, H. 1984. A chronological review of birds first described from Turkey with their taxonomic status in 1984. *Sandgrouse* 6: 62–68.

Kumerloeve, H. 1986. Bibliographie der Säugetiere und Vögel der Türkei (rezente Fauna). *Bonn. Zool. Monogr.* 21.

Kumerloeve, H. 1989. Amik Gölü, southeast Turkey: a plea for national park status. *Orn. Soc. Middle East Bull.* 23: 5.

Kumerloeve, H. and Hollom, P. A. D. 1967. Zum Durchzug von *Phalaropus lobatus* (L.) in Kleinasien. *Vogelwarte* 24: 64–65.

Kumerloeve, H., Kasparek, M. and Nagel, K.-O. 1984. Der Rostbürzel-Steinschmätzer, *Oenanthe xanthoprymna* (Hemprich & Ehrenberg 1833) als neuer Brutvogel im östlichen Anatolien (Türkei). *Bonn. Zool. Beitr.* 35: 97–101.

Kumerloeve, H. and Niethammer, G. 1935. Beiträge zur Kenntnis der Avifauna Kleinasiens (Paphlagonien und Galatien). *J. Orn.* 83: 25–75.

Laine, L. J. 1996. The 'Borçka puzzle' – the first Western Palearctic Crested Honey Buzzard. *Birding World* 9: 324–325.

Lalanne, Y., Hémery, G., Cagnon, C., D'Amico, F., D'Elbée, J. and Mouchès, C. 2001. Discrimination morphologique des sous-espèces d'Océanite tempête: nouveaux résultats pour deux populations méditerranéennes. *Alauda* 69: 475–482.

Lamsdell, C. and Lamsdell, D. 2000. The first Namaqua Dove *Oena capensis* in Cyprus. *Sandgrouse* 22: 68–69.

Lane, B. A. and Rogers, D. I. 2000. The Australian Painted Snipe *Rostratula* (*benghalensis*) *australis*: an endangered species? *The Stilt* 36: 26–34.

Langeveld, M. J. and Ruiters, P. S. 1994. *Breeding Bird Census at Eber Gölü, Akşehir Gölü and Karamık Gölü.* Bird & Wetland Section Report No. 8, Doğal Hayatı Koruma Derneği, Istanbul.

Lassey, P. A. 1994. First record of Lesser Flamingo *Phoenicopterus minor* in Egypt. *Sandgrouse* 16: 52–53.

Lee, P. L. M., Richardson, L. J. and Bradbury, R. B. 2001. The phylogenetic status of the Corn Bunting *Miliaria calandra* based on mitochondrial control-region DNA sequences. *Ibis* 143: 299–303.

Lefranc, N. and Worfolk, T. 1997. *Shrikes: A Guide to the Shrikes of the World.* Pica Press, Robertsbridge.

Lehmann, H. 1969. The Greater Sand Plover (*Charadrius leschenaultii* Lesson) in Asia Minor. *Oologists' Record* 43: 30–54.

Lehmann, H. 1971. Der Rotflügelgimpel (*Rhodopechys sanguinea*) auf dem Hochplateau Zentral-Anatoliens. *Jahresberichte des Naturw. Vereins Wuppertal* 24: 89–100.

Lehmann, H. 1977. Steppenadler (*Aquila rapax orientalis* Cabanis) in Zentralanatolien. *Jahresberichte des Naturw. Vereins Wuppertal* 30: 125–129.

Lehmann, H. and Mertens, R. 1969. The Red-winged Bullfinch (*Rhodopechys sanguinea*) as a breeding bird in Central-Anatolia. *Oologists' Record* 43: 2–16.

Liebers, D. and Helbig, A. J. 1999. Phänotypische Charakterisierung und systematische Stellung der Armenienmöwe *Larus armenicus*. *Limicola* 13: 281–321.

Liebers, D. and Helbig, A. J. 2002. Phylogeography and colonization history of Lesser Black-backed Gulls (*Larus fuscus*) as revealed by mtDNA sequences. *J. Evol. Biol.* 15: 1021–1033.

Liebers, D., Helbig, A. J. and de Knijff, P. 2001. Genetic differentiation and phylogeography of gulls in the *Larus cachinnans–fuscus* group (Aves: Charadriiformes). *Mol. Ecol.* 10: 2447–2462.

Leisler, B. 1973. Die Jahresverbreitung des Mariskensängers (*Acrocephalus melanopogon*) nach Beobachtungen und Ringfunden. *Vogelwarte* 27: 24–39.

Leisler, B., Heidrich, P., Schulze-Hagen, K. and Wink, M. 1997. Taxonomy and phylogeny of reed warblers (genus *Acrocephalus*) based on mtDNA sequences and morphology. *J. Orn.* 138: 469–496.

Lerner, H. R. L. and Mindell, D. P. 2005. Phylogeny of eagles, Old World vultures, and other Accipitridae based on nuclear and mitochondrial DNA. *Mol. Phyl. & Evol.* 37: 327–346.

Lindroos, T. and Tenovuo, O. 2002. Black Lark – its identification in the field and distribution in Europe *Alula* 8: 22–28.

Livezey, B. C. 1991. A phylogenetic analysis and classification of recent dabbling ducks (tribe Anatini) based on comparative morphology. *Auk* 108: 471–507.

Livezey, B. C. 1995. Phylogeny and evolutionary ecology of modern seaducks (Anatidae: Mergini). *Condor* 97: 233–255.

van Loon, A. J. 1991. Mystery photographs [Willow Warbler]. *Dutch Birding* 13: 67–69.

Louette, M., Becuwe, M. and Eyckerman, R. 1977. Observations ornithologiques en Anatolie et en Thrace. *Gerfaut* 67: 427–436.

Maas Geesteranus, H. P. 1959. Ornithological report on a biological excursion to Asia Minor. *Ardea* 47: 111–157.

MacKinnon, J. and Phillipps, K. 2000. *A Field Guide to the Birds of China*. Oxford University Press.

Madge, S. C. 1984. White-headed Duck with black head. *Brit. Birds* 77: 154.

Madge, S. C. 1992. Identification of Moustached Warbler. *Birding World* 5: 299–303.

Madge, S. and Burn, H. 1988. *Wildfowl*. Christopher Helm, London.

Madge, S. and Burn, H. 1994. *Crows and Jays*. Christopher Helm, London.

Madge, S. and McGowan, P. 2002. *Pheasants, Partridges and Grouse*. Christopher Helm, London.

Magnin, G. 1989. *Falconry and Hunting in Turkey During 1987*. International Council for Bird Preservation (Study Report No. 34), Cambridge.

Magnin, G. 1991. A record of the Brown Fish Owl *Ketupa zeylonensis* from Turkey. *Sandgrouse* 13: 42–44.

Magnin, G. and Yarar, M. 1994. Some notes on the breeding of Greater Flamingo *Phoenicopterus ruber* and White Pelican *Pelecanus onocrotalus* in Turkey. *Orn. Soc. Middle East Bull.* 32: 28–30.

Magnin, G. and Yarar, M. 1997. *Important Bird Areas in Turkey*. Doğal Hayatı Koruma Derneği, Istanbul.

Makatsch, W. 1950. *Die Vogelwelt Macedoniens*. Leipzig.

Malling Olsen, K. and Larsson, L. 2003. *Gulls of Europe, Asia and North America*. Christopher Helm, London.

Marlow, T., Kirwan, G. M. and Güneş, B. 2001. Does Common Snipe *Gallinago gallinago* breed in Turkey? *Sandgrouse* 23: 147.

Marshall, H. D. and Baker, A. J. 1999. Colonization history of Atlantic Island Chaffinches (*Fringilla coelebs*) revealed by mitochondrial DNA. *Mol. Phyl. & Evol.* 11: 201–212.

Marten, J. A. and Johnson, N. K. 1986. Genetic relationships of North American Cardueline finches. *Condor* 88: 409–420.

Martens, J. 1979. Gesang und Verwandtschaft des Steinortolan (*Emberiza buchanani*). *Natur u. Mus.* 109: 337–343.

Martens, J. 1982. Ringförmige Arealüberschneidung und Artbildung beim Zilpzalp, *Phylloscopus collybita*. *Z. zool. Syst. Evolut.-forsch.* 20: 82–100.

Martins, R. P. 1989. Turkey Bird Report 1982–86. *Sandgrouse* 11: 1–41.

Martins, R. P. and Hirschfeld, E. 1998. Comments on the limits of the Western Palearctic in Iran and the Arabian Peninsula. *Sandgrouse* 20: 108–134.

Martins, R. and Robson, C. 1988. Selected bird observations from Turkey: spring and summer 1983. *Orn. Soc. Middle East Bull.* 20: 11–20.

Massa, B. and Fontana, P. 2004. New or little known birds for the Republic of Georgia (eastern Europe). *Riv. Ital. Orn., Milano* 73: 151–155.

Mathey-Dupraz, A. 1920–24. Notes ornithologiques de la région du Bosphore. *Orn. Beob.* 17: 25–29, 108–110; 18: 25–27, 38–41, 55–58, 101–104, 137–139, 157–158, 183–187; 19: 22–25, 41–43, 58–61, 116–119, 156–159; 20: 9–12, 24–27, 118–120, 135–137.

Mathieu, J. 1968. La Fauvette pitchou en Turquie. *Nos Oiseaux* 29: 202–203.

Mauve, L. 1938. Der Zug der Grossvögel über den Bosporus. *J. Orn.* 86: 261–301.

Mayer, H. and Aksöy, H. 1986. *Walder der Türkei.* Gustav Fischer Verlag, Stuttgart.

Mayr, E. and Greenway, J. C. (eds.) 1960. *Check-list of Birds of the World.* Vol. 9. Mus. Comp. Zool., Harvard University Press, Cambridge, MA.

McGregor, P. J. C. 1917. Notes on birds observed at Erzurum. *Ibis* (X) 5: 1–30.

McNeile, J. S. 1950, 1951, 1954, 1967, 1968, 1970, 1972, 1973. Personal diaries. Held in Edinburgh Museum.

Meinertzhagen, R. 1935. Ornithological results of a trip to Syria and adjacent countries in 1933. *Ibis* (13) 5: 110–151.

Mertens, R. 1974. Erstnachweis des Schieferfalken (*Falco concolor*) für die Türkei. *Jber. Naturw. Ver. Wuppertal* 27: 137–138.

Mertens, R. 1981. Erstnachweis des Rauhfusskauzes (*Aegolius funereus*) für die Türkei. *Vogelwelt* 102: 30–31.

Milchev, B. 1994. Breeding bird atlas of the Strandja mountains, south-east Bulgaria. *Sandgrouse* 16: 2–27.

Minshull, B. C. 1996. The first Pine Bunting *Emberiza leucocephalos* in Jordan. *Sandgrouse* 18 (2): 69.

Morgan, J. H. and Shirihai, H. 1997. *Passerines and Passerine Migration at Eilat.* International Birdwatching Centre in Eilat (Tech. Publ. 6), Eilat.

Morony, J. J., Bock, W. J. and Farrand, J. 1975. *Reference List of the Birds of the World.* New York.

Morozov, V. V. and Aarvak, T. 2004. Wintering of Lesser White-fronted Geese breeding in the Polar Urals. *Casarca* 10: 156–162.

Mrlík, V., Bobek, M., Pojer, F., Skopek, J. and Formánek, J. 1995. Raptor migration in north-east Turkey, autumn 1994. *Orn. Soc. Middle East Bull.* 35: 41–45.

Murdoch, D. 2002. Birding the Syrian Euphrates. *Birding World* 15: 297–298.

Murdoch, D. 2005. Observations from Syria with notes on 11 new breeding species. *Sandgrouse* 27: 37–45.

Murdoch, D., Andrews, I. J. and Hofland, R. 2004. The Syrian Wetland Expedition 2004: a summary. *Sandgrouse* 26: 94–104.

Murdoch, D. A. and Serra, G. 2006. The status of Sociable Lapwing *Vanellus gregarius* in Syria. *Sandgrouse* 28: 58–61.

Murphy, C. 1984. Recent trips to eastern Turkey. *Orn. Soc. Middle East Bull.* 13: 8–12.

Musil, P. and Musilová, Z. 2003. White-headed Ducks in eastern Turkey in late summer 2001. *Threatened Waterfowl Specialist Group News* 14: 23.

Mycock, J. 1987. Some autumn observations in Turkey. *Orn. Soc. Middle East Bull.* 18: 1–3.

Nakhutsrishvili, G. 1999. *The Vegetation of Georgia (Caucasus).* Centro Audovisivi e Stampa, Camerino.

Nankinov, D. N. 1996a. Coastal parks and reserves along the Black Sea and their importance for seabirds. *Marine Orn.* 24: 29–34.

Nankinov, D. 1996b. Status of Tengmalm's Owl, *Aegolius funereus,* in Bulgaria. *Riv. Ital. Orn., Milano* 66: 127–136.

Nankinov, D. N. 1998. First addition to the list of birds in Bulgaria. *Riv. Ital. Orn., Milano* 68: 183–187.

Nankinov, D. 1999a. Distribution of the Great Spotted Cuckoo, *Clamator glandarius,* in Bulgaria. *Zool. Middle East* 17: 9–14.

Nankinov, D. N. 1999b. Bar-tailed Godwit's [*sic*], *Limosa lapponica,* locations and migration times in Bulgaria. *Riv. Ital. Orn., Milano* 69: 97–104.

Nankinov, D. N. 2000. Range expansion and current breeding distribution of River Warbler *Locustella fluviatilis* in Bulgaria. *Sandgrouse* 22: 50–54.

Nankinov, D. N. 2003. Second addition to the list of birds in Bulgaria. *Riv. Ital. Orn., Milano* 73: 157–163.

Nankinov, D., Shurulinkov, P., Nikolov, I., Nikolov, B., Dalaktchieva, S., Hristov, I., Stankev, R., Rogev, A., Dutsov, A. and Sarov, M. 1998. Studies of the waders (Charadriiformes) on the wetlands around Sofia. *Riv. Ital. Orn., Milano* 68: 63–84.

Nesterov, P. V. 1911. Materialy po ornitofaune Yugo-Zapadnogo Zakavkaz'ya 1 severo-vostochnoi chasti Maloi Azii. *Ann. Proc. Zool. Mus. Acad. Sci.* 16: 311–408.

Niethammer, G. and Thiede, W. 1962. Der Fichtenammer, *Emberiza leucocephala* [*sic*], als Besucher Europas. *J. Orn.* 103: 289–293.

Nightingale, T. and Hill, M. 1993. *Birds of Bahrain.* Immel, London.

Nikolaus, G. 1987. *Distribution Atlas of Sudan's Birds with Notes on Habitat and Status.* Bonn. Zool. Monogr. 29.

Nikolov, I., Shurulinkov, P., Pankova, R. and Konaklieva, S. 2006. Observations of Barnacle Goose in Bulgaria. *Alula* 12: 92–93.

Nisbet, I. C. T. and Smout, T. C. 1957. Autumn observations on the Bosphorus and Dardanelles. *Ibis* 99: 483–499.

Ochiatto, D. 2001. Nemrut Dagi, Adiyaman. *Birding World* 14: 343–344.

Ogilvie, I. H. 1954. Bird notes from northern Asia Minor 1946–1948. *Ibis* 96: 81–90.

Ornithological Society of Turkey (OST) 1969. *Bird Report 1966–67.* OST, London.

Ornithological Society of Turkey (OST) 1972. *Bird Report 1968–69.* OST, London.

Ornithological Society of Turkey (OST) 1975. *Bird Report 1970–73.* OST, London.

Ornithological Society of Turkey (OST) 1978. *Bird Report 1974–75.* OST, London.

Ottens, G. 2006. African Darter in Israel in May 2004 and its WP occurrence. *Dutch Birding* 28: 214–218.

Ottvall, R., Bensch, S., Walinder, G. and Lifjeld, J. T. 2002. No evidence of genetic differentiation between Lesser Redpolls *Carduelis flammea cabaret* and Common Redpolls *Carduelis f. flammea. Avian Sci.* 2: 237–244.

Oustalet, E. 1882. [*Plotus chantrei.*] *Ann. Sci. Nat., Zool.*, Paris 6 (13): 7.

Ovaa, A. and Vos, R. 1990. A second breeding site of the Purple Gallinule in southern Turkey. *Orn. Soc. Middle East* 25: 25–26.

Özen, M. and Kurt, B. 2001. *Türkiye'nin Kuşları: 2000 yılı Gözlem Raporu.* Doğal Hayatı Koruma Derneği, Istanbul.

Panov, E. N. 1983. *Die Würger der Paläarktis.* Wittenberg, Lutherstadt.

Panov, E. N. 1995. Superspecies of shrikes in the former USSR. *Proc. Western Found. Vert. Zool.* 6: 26–33.

Panov, E. N. and Bulatova, N. S. 1972. [On the common habitats and interrelations of Trumpeter Finches (*Bucanetes githagineus* Licht. and *B. mongolicus* Swinh.) in Transcaucasia]. *Bull. Moscowskogo Obschestva Ispytatelei Prirody, Otdel biologii* 77 (4): 86–93. [In Russian.]

Panov, E. N., Roubtsov, A. S. and Monzikov, D. G. 2003. Hybridization between Yellowhammer and Pine Bunting in Russia. *Dutch Birding* 25: 17–31.

Parkes, K. C. 1982. Nomeclatural notes on the phalaropes. *Bull. Brit. Orn. Club* 102: 84–85.

Parkin, D. T., Collinson, M., Helbig, A. J., Knox, A. G. and Sangster, G. 2003. The taxonomic status of Carrion and Hooded Crows. *Brit. Birds* 96: 274–290.

Parkin, D. T., Collinson, M., Helbig, A. J., Knox, A. G., Sangster, G. and Svensson, L. 2004. Species limits in *Acrocephalus* and *Hippolais* warblers from the Western Palearctic. *Brit. Birds* 97: 276–299.

Parr, D. 1981. Notes on a journey through Turkey, spring 1981. *Orn. Soc. Turkey Bull.* 7: 4–6.

Parr, S. J., Collin, P., Silk, S., Wilbraham, J., Williams, N. P. and Yarar, M. 1995. A baseline survey of Lesser Kestrels *Falco naumanni* in central Turkey. *Biol. Conserv.* 72: 45–53.

Parr, S., J., Collin, P., Silk, S., Wilbraham, J., Williams, N. P. and Yarar, M. 1996. A baseline survey of White Storks *Ciconia ciconia* in central Turkey. *Sandgrouse* 18 (2): 46–51.

Parrot, I. C. 1905. Kritische Übersicht der palaearktischen Emberiziden. *Orn. Jahrb.* 16: 1–50.

Parry, S. J., Clark, W. S. and Prakash, V. 2002. On the taxonomic status of the Indian Spotted Eagle *Aquila hastata. Ibis* 144: 665–675.

Pasquali, R. 1991. *Degli Uccelli della Turchia (Europea ed Asiatica).* Lussemburgo.

Pasquet, E. and Thibault, J.-C. 1997. Genetical differences among mainland and insular forms in the Citril Finch *Serinus citrinella. Ibis* 139: 679–684.

Pavlova, A., Zink, R. M., Drovetski, S. V., Red'kin, Y. and Rohwer, S. 2003. Phylogeographic patterns in *Motacilla flava* and *Motacilla citreola*: species limits and population history. *Auk* 120: 744–758.

Pearson, D. J. 2000. The races of the Isabelline Shrike *Lanius isabellinus* and their nomenclature. *Bull. Brit. Orn. Club* 120: 22–27.

Pearson, D. J., Ash, J. S. and Bensch, S. 2004. The identity of some *Hippolais* specimens from Eritrea and the United Arab Emirates examined by mtDNA analysis: a record of Sykes's Warbler *H. rama* in Africa. *Ibis* 146: 683–684.

Pearson, D. J., Small, B. J. and Kennerley, P. R. 2002. Eurasian Reed Warbler: the characters and variation associated with the Asian form *fuscus*. *Brit. Birds* 95: 42–61.

Pegoraro, K., Föger, M. and Parson, W. 2001. First evidence of mtDNA sequence differences between Northern Bald Ibises (*Geronticus eremita*) of Moroccan and Turkish origin. *J. Orn.* 142: 425–428.

Pennington, M. G. and Meek, E. R. 2006. The 'Northern Bullfinch' invasion of autumn 2004. *Brit. Birds* 99: 2–24.

Perktaş, U. 2004. Breeding shrike populations in Turkey: status in 1998–2003. *Biol. Lett.* 41 (2): 71–75.

Perktaş, U. and Ayaş, Z. 2005. Birds of Nallıhan Bird Paradise (Central Anatolia, Turkey). *Turk. J. Zool.* 29: 45–59.

Perlman, Y. and Geffen, A. 2007. Basra Reed Warbler *Acrocephalus griseldis* in the Hula Valley, Israel, in 2006. *Sandgrouse* 29: 210–213.

Perttula, P. and Tenovuo, J. 2002. Great Grey Shrike subspecies *excubitor*, *sibiricus* and *borealis*. *Alula* 8: 54–60.

Peške, L. 1995. Via Pontica 95: census of migrating raptors in eastern Turkey. *Buteo* 7: 197–199.

Peter, H. 1990. *Waldrappdämmerung am Euphrat*. Kasparek Verlag, Heidelberg.

Peter, H. 1994. The first breeding record of the Desert Wheatear, *Oenanthe deserti*, in Turkey. *Zool. Middle East* 10: 27–29.

Peters, J. L. 1934. *Check-list of the Birds of the World*. Vol. 2. Mus. Comp. Zool., Cambridge, MA.

Pforr, M. and Limbrunner, A. 1982. *The Breeding Birds of Europe: A Photographic Handbook*. Croom Helm, Beckenham.

Pitra, C., Lieckfeldt, D. and Alonso, J. C. 2000. Population subdivision in Europe's Great Bustard inferred from mitochondrial and nuclear DNA sequence variation. *Mol. Phyl. & Evol.* 23: 63–74.

Pleasance, B. 1997. Pallid Harrier *Circus macrourus* breeding in central Turkey. *Sandgrouse* 19: 144–145.

Portenko, L. A. 1962. [New subspecies of passerine birds] (Aves, Passeriformes) II. *Trudy Zool. Inst. Akad. Nauk* 30: 385–394.

Porter, R. F. 1983. The autumn migration of passerines and near-passerines at the Bosphorus, Turkey. *Sandgrouse* 5: 45–74.

Porter, R. F. and Willis, I. R. 1968. The autumn migration of soaring birds at the Bosphorus. *Ibis* 110: 520–536.

Porter, R. F., Christensen, S. and Schiermacker-Hansen, P. 1996. *Field Guide to the Birds of the Middle East*. T. & A. D. Poyser, London.

Prior, R. 2007. First record of Carrion Crow *Corvus corone* (*C. c. corone* or *C. c. orientalis*) for Lebanon. *Sandgrouse* 29: 107.

Provençal, P. and Sørensen, U. G. 1998. Medieval record of the Siberian White Crane *Grus leucogeranus* in Egypt. *Ibis* 140: 333–335.

Quézel, P., Barbéro, M. and Akman, Y. 1980. Contribution à l'étude de la végétation forestière d'Anatolie septentrionale. *Phytocoenologia* 8: 365–519.

Quézel, P. and Pamukçuoğlu, A., 1973. Contribution a l'étude phytosociologique et bioclimatique de quelques groupements forestiers du Taurus. Feddes Rep. 8.

Radde, G. 1884. *Ornis Caucasica*. Kassel.

Ragyov, D. N., Popova-Wightman, L. G., Popov, K. S., Dalakchieva, S. Y., Nikolov, B. P. and Nikolov, I. P. 2003. The first Yellow-billed Stork *Mycteria ibis* in Bulgaria. *Sandgrouse* 25: 63–64.

Ramadan-Jaradi, G., Bara, T., Almécija, M. and Ramadan-Jaradi, M. 2004. Significant bird notes from Lebanon during 2002–2003. *Sandgrouse* 26: 29–34.

Ramadan-Jaradi, G., Bara, T. and Ramadan-Jaradi, M. 2008. Revised checklist of the birds of the Lebanon, 1999–2007. *Sandgrouse* 30: 22–69.

Ramadan-Jaradi, G. and Ramadan-Jaradi, M. 1999. An updated checklist of the birds of Lebanon. *Sandgrouse* 21: 132–170.

Ramsay, L. N. G. 1914. Observations on the bird-life of the Anatolian plateau during the summer of 1907. *Ibis* 10 (2): 365–387.

Rasmussen, P. C. and Anderton, J. C. 2005. *The Birds of South Asia: The Ripley Guide*. Vol. 2. Lynx Edicions, Barcelona.

Red'kin, Y. A. 2001. [A new subspecies of Firecrest *Regulus ignicapillus* (Temminck, 1820) (Regulidae, Passeriformes) from the mountains of Crimea]. *Orn.* 29: 98–104. [In Russian.]

Reiser, O. 1904. Zur Kenntnis der Vogelwelt von Konstantinopel. *Orn. Jahrb.* 15: 153–156.

Renkhoff, M. 1972. Bemerkenswerte Beobachtungen 1971 im Balik-Gölü Gebiet (Nordtürkei). *Orn. Mitt.* 24: 63–73.

Research Association for Rural Environment and Forest. 2000. *Türkiye'nin Tabiati Koruma Alanlari.* Donmez Ofset, Ankara.

Richardson, C. 1990. *The Birds of the United Arab Emirates.* Hobby Publications, Dubai & Warrington.

Richardson, C. 2006. Home and away: Birding Syria. *Birding World* 19: 76–77.

Richardson, C. and Aspinall, S. *The Shell Birdwatching Guide to the United Arab Emirates.* Hobby Publications, Liverpool & Dubai.

Richardson, I. M. 2003. A long-term bird survey of Kulu Gölü, Turkey (2001–2002). *Sandgrouse* 25: 110–121.

Ridgely, R. S. and Greenfield, P. J. 2001. *The Birds of Ecuador.* Vol. 1. Christopher Helm, London.

Riesing, M. J., Kruckenhauser, L., Gamauf, A. and Haring, E. 2003. Molecular phylogeny of the genus *Buteo* (Aves: Accipitridae) based on mitochondrial marker sequences. *Mol. Phyl. & Evol.* 27: 328–342.

Rigler, L. 1852. *Die Türkei und deren Bewohner in ihren naturhistorischen, physiologischen und pathologischen Verhältnissen vom Standpunkte Constantinopel's.* Wien.

Ristow, D. and Wink, M. 1995. Distribution of non-breeding Eleonora's Falcon *Falco eleonorae. Il Merill* 28: 1–10.

Robel, D. and Bräuning, C. 1992. Zum Greifvogelzug am Bosporus im September 1990. *Beitr. Vogelkd.* 38: 81–91.

Roberts, T. J. 1992. *The Birds of Pakistan.* Vol. 2. Oxford University Press.

Robinson, K., Holt, L., Buckley, Y., Pullen, T., Boyla, K. and Can, O. 1998. Breeding survey of White-headed Duck *Oxyura leucocephala* on the Central Plateau, Turkey. *Sandgrouse* 20: 47–48.

de Roeck, E. 1993. The status of Europe's rarer birds of prey: Imperial Eagle. *Birding World* 6: 239–242.

Rokitansky, G. and Schifter, H. 1971. Ornithologische Ergebnisse zweier Sammelreisen in die Türkei. *Ann. Naturhist. Mus. Wien* 75: 495–538.

Roselaar, C. S. 1995. *Songbirds of Turkey: An Atlas of Biodiversity in Turkish Passerine Birds.* Pica Press, Robertsbridge & GMB, Haarlem.

Roselaar, C. S. 2003. An inventory of major European bird collections. *In*: Collar, N. J., Fisher, C. and Feare, C. (eds.) *Why Museums Matter: Avian Archives in an Age of Extinction. Bull. Brit. Orn. Club* Suppl. 123A.

Rowlands, J. A. 1994. First record of Oriental Pratincole *Glareola maldivarum* in Cyprus. *Sandgrouse* 16: 56–57.

Rudebeck, G. 1974. Sträcket av rovfågel och stork över Bosporen-några kommentarer till höstresan 1973. *Anser* 13: 35–46.

Runemark, H., 1971. The phytogeographical foundations of the Middle East. *In*: Davis, P. H., Harper, P. C. and Hedge, I. C. (eds.) *Plant Life of South-West Asia.* Royal Botanic Garden Edinburgh, Edinburgh.

Safriel, U. N. 1980. Notes on the extinct population of the Bald Ibis *Geronticus eremita* in the Syrian Desert. *Ibis* 122: 82–88.

Şahin, R. 1996. Besetzung von Nestern in einer neu entstandenen Mechlschwalben-Kolonie (*Delichon urbica*) durch Haussperlinge (*Passer domesticus*) in Samsun/Türkei. *Ecol. Birds* 18: 45–54.

Salim, M. A. 2002. The first records, including breeding, of Black-winged Kite *Elanus caeruleus* in Iraq. *Sandgrouse* 24: 136–138.

Salim, M. A. 2008. The first Namaqua Dove *Oena capensis* in Iraq. *Sandgrouse* 30: 100–102.

Salzburger, W., Martens, J. and Sturmbauer, C. 2002. Paraphyly of the Blue Tit (*Parus caeruleus*) suggested from cytochrome *b* sequences. *Mol. Phyl. & Evol.* 24: 19–25.

Sandwith, H. 1856. *Geschichte der Belagerung von Kars.* Braunschweig.

Sangster, G. 1996. Trends in systematics. Taxonomy of Houbara and Macqueen's [*sic*] Bustards and neglect of intraspecific diversity. *Dutch Birding* 18: 248–254.

Sangster, G. 1997a. Trends in systematics. Species limits in flamingos, with comments on lack of consensus in taxonomy. *Dutch Birding* 19: 193–198.

Sangster, G. 1997b. Trends in systematics. *Acrocephalus* and *Hippolais*: shaking the tree. *Dutch Birding* 19: 294–300.

Sangster, G. 1998. Trends in systematics. Purple Swamp-hen is a complex of species. *Dutch Birding* 20: 13–22.

Sangster, G. 1999. Trends in systematics. Relationships among gulls: new approaches. *Dutch Birding* 21: 207–218.

Sangster, G., van den Berg, A. B., van Loon, A. J. and Roselaar, C. S. 2003. Dutch avifaunal list: taxonomic changes in 1999–2003. *Ardea* 91: 279–286.

Sangster, G., Collinson, M. J., Helbig, A. J., Knox, A. G. and Parkin, D. T. 2002. The specific status of Balearic and Yelkouan Shearwaters. *Brit. Birds* 95: 636–639.

Sangster, G., Collinson, M., Helbig, A. J., Knox, A. G. and Parkin, D. T. 2004. The taxonomic status of Macqueen's [*sic*] Bustard. *Brit. Birds* 97: 60–67.

Sangster, G., Collinson, M., Helbig, A. J., Knox, A. G., Parkin, D. T. and Prater, T. 2001. The taxonomic status of Green-winged Teal *Anas carolinensis*. *Brit. Birds* 94: 218–226.

Sangster, G., Collinson, M., Knox, A. G., Parkin, D. T. and Svensson, L. 2007. Taxonomic recommendations for British birds: fourth report. *Ibis* 149: 853–857.

Sangster, G., Hazevoet, C. J., van den Berg, A. B. and Roselaar, C. S. 1997. Dutch avifaunal list: taxonomic changes in 1977–97. *Dutch Birding* 19: 21–28.

Sangster, G. and Oreel, G. J. 1996. Trends in systematics. Progress in taxonomy of Taiga and Tundra Bean Geese. *Dutch Birding* 18: 310–316.

Sarıgül, G. 2001. Tepeli pelikanların ülkemizdeki durumu ve korunması. *İbibik* 1: 7–9.

Satunin, K. A. 1912. [Ornithological excursion to southern Transcaucasia.] *Mess. Orn.* 3: 1–15, 99–110. [In Russian.]

Savage, C. D. W. 1969. Wildfowl survey in south-west Asia: progress in 1968. *Wildfowl* 20: 144–147.

Schekkerman, H. and van Roomen, M. W. J. 1993. *The Migration of Waders and Other Waterbirds Through Inland Wetlands in Central Turkey, Spring 1988.* WIWO (Report No. 32), Zeist.

Schmidtke, K. and Utschik, H. 1980. Ornithologische Ergebnisse einer Türkeifahrt. *Anz. Orn. Ges. Bayern* 19: 57–74.

Schodde, D. and Bock, W. 2002. Does holotype priority always serve nomenclature? *Brit. Birds* 95: 593–596.

Schrader, G. 1891. Ornithologische Beobachtungen auf meinen Sammelreisen I. Kleinasien (Aidin und Mersina). III. Syrien. *Orn. Jber.* 2: 179–197, 215–223.

Schubert, W. 1979. Bemerkenswerte Brutnachweise und Brutzeitfeststellungen in Anatolien/Türkei. *Vogelwelt* 100: 151–155.

Schweiger, H. 1965. Ornithologische Beobachtungen in Anatolien während der Jahre 1959–1965. *Istanbul Univ. Fen. Fak. Mecm.* B 20: 177–189.

Schweizer, M. 2003. A European Stonechat *Saxicola rubicola* in northern Iran. *Sandgrouse* 25: 146–147.

Sclater, P. L. and Taylor, E. C. 1876. Ornithological notes from Constantinople. *Ibis* 6: 60–65.

Scott, D. A. 1994. The first Olive-backed Pipit *Anthus hodgsoni* in Iran. *Sandgrouse* 16: 59–61.

Scott, D. A. and Carp, E. 1982. A midwinter survey of wetlands in Mesopotamia, Iraq: 1979. *Sandgrouse* 4: 60–76.

Scott, M., Siddle, J. and Shirihai, H. 1999. Little known Western Palearctic birds: Mountain Chiffchaff. *Birding World* 12: 163–167.

Screiber, A., Stubbe, A. and Stubbe, M. 2001. Common Buzzard (*Buteo buteo*): a raptor with hyperpolymorphic plumage morphs but low allozyme heterozygosity. *J. Orn.* 142: 34–48.

Seçkin, S. and Coşkun, Y. 2005. Small mammals in the diet of the Long-eared Owl, *Asio otus*, from Diyarbakır, Turkey. *Zool. Middle East* 35: 102–103.

Şekercioğlu, C. H. 2006. A birder's guide to Turkey. *Living Bird* 25: 14–23.

Selous, F. C. 1900. A fortnight's egg-collecting in Asia Minor. *Ibis* (4) 4: 405–424.

Serez, M. 1992. Status, conservation and distribution of phasianid species in Turkey. *In*: Birkan, M., Potts, G. R., Aebischer, N. and Dowell, S. D. (eds.) First International Symposium on Partridges, Quails and Francolins, Fordingbridge, U.K., September 7–14 1991. *Gibier Faune Sauvage* 9.

Serra, G., Al Qaim, G., Abdullah, M. S., Kanani, A. and Assaed, A. K. 2005. A long-term bird survey in the central Syrian desert (2000–2004): Part 2 – a provisional annotated checklist. *Sandgrouse* 27: 104–125.

Sharpe, R. B. 1888. *Catalogue of the Birds in the British Museum*. Vol. 12. Trustees of the British Museum, London.

Sharpe, R. B. 1890. *Catalogue of the Birds in the British Museum*. Vol. 13. Trustees of the British Museum, London.

Sheldon, F. H. 1987. Phylogeny of herons estimated from DNA-DNA hybridisation studies. *Auk* 104: 97–108.

Sheldon, F. H., Whittingham, L. A. and Winkler, D. W. 1999. A comparison of cytochrome b and DNA hybridization data bearing on the phylogeny of swallows (Aves: Hirundinidae). *Mol. Phyl.& Evol.* 11: 320–331.

Shelton, N. 1997. The first Hoopoe Lark *Alaemon alaudipes* in Turkey. *Sandgrouse* 19: 149–150.

Shields, G. F. 1990 Analysis of mitochondrial DNA of Pacific Black Brant. *Auk* 107: 620–623.

Shirihai, H. 1996. *The Birds of Israel.* Academic Press, London.

Shirihai, H. 2002. *A Complete Guide to Antarctic Wildlife.* Alula Press Oy, Degerby.

Shirihai, H. and Christie, D. A. 1992. Raptor migration at Eilat. *Brit. Birds* 85: 141–185.

Shirihai, H., Gargallo, G. and Helbig, A. J. 2001. *Sylvia Warblers: Identification, Taxonomy and Phylogeny of the Genus Sylvia.* Christopher Helm, London.

Shirihai, H. and Golan, Y. 1994. First records of Long-tailed Shrike *Lanius schach* in Israel and Turkey. *Sandgrouse* 16: 36–40.

Shirihai, H. and Madge, S. 1993. Identification of Hume's Yellow-browed Warbler. *Birding World* 6: 439–443.

Shirihai, H., Yosef, R., Alon, D., Kirwan, G. M. and Spaar, R. 2000. *Raptor Migration in Israel and the Middle East: A Summary of 30 Years of Field Research.* International Birding and Research Center, Eilat.

Short, L. L. 1967. A review of the genera of grouse (Aves, Tetraoninae). *Amer. Mus. Novit.* 2289: 1–39.

Shurulinkov, P., Stoyanov, G., Tzvetkov, P., Vulchev, K., Kolchagov, R. and Ilieva, M. 2003. Distribution and abundance of Tengmalm's Owl *Aegolius funereus* on Mount Pirin, south-west Bulgaria. *Sandgrouse* 25: 103–109.

Sibley, C. G. 1996. *Birds of the World.* Version 2.0. Thayer Birding Software, Cincinnati.

Sibley, C. G. and Monroe, B. L. 1990. *Distribution and Taxonomy of Birds of the World.* Yale University Press, New Haven & London.

Sibley, C. G. and Monroe, B. L. 1993. *A Supplement to Distribution and Taxonomy of Birds of the World.* Yale University Press, New Haven & London.

Siegel-Causey, D. 1988. Phylogeny of the Phalacrocoracidae. *Condor* 90: 885–905.

Siki, M. 1988. [The bird species of Camaltı Saltpans–Homa fishery.] *Turk. J. Zool.* 3: 272–283. [In Turkish.]

Sluys, R. and van den Berg, M. 1982. On the specific status of the Cyprus Pied Wheatear *Oenanthe cypriaca. Ornis Scand.* 13: 123–128.

Sluys, R. 1983. Geographical variation and distribution of the Bearded Tit *Panurus biarmicus* (Linnaeus, 1758) (Aves). *Bijdr. Dierk.* 53: 13–32.

Smith, M. Q. 1960. Notes on the birds of the Trebizond area of Turkey. *Ibis* 102: 576–583.

Snow, D. W. 1997. Should the biological be superseded by the phylogenetic species concept? *Bull. Brit. Orn. Club* 117: 110–121.

Snow, D. W. and Perrins, C. M. (eds.) 1998. *The Birds of the Western Palearctic. Concise Edition.* Oxford University Press.

Somçağ, S. 1980. Winter in Istanbul. *Orn. Soc. Middle East Bull.* 5: 3–4.

Sorace, A. 1996. The first White-crowned Black Wheatear *Oenanthe leucopyga* in Turkey. *Sandgrouse* 18 (1): 68.

Sperl, J. 1992. Die Hirtenmaina, *Acridotheres tristis,* ein weiteres Ausbrietungsphanomen. *Falke* 39: 244–245.

Stegmann, B. 1932. *Calandrella minor pseudobaetica* subsp. nov. *Orn. Monatsber.* 40: 54.

Steiner, H. M. 1962. "Zilpzalp" und Grüner Laubsänger in NO-Kleinasien. *Egretta* 5: 57–60.

Steinfatt, O. 1932. Der Bosporus als Landbrüche für den Vogelzug zwischen Europa und Kleinasien. *J. Orn.* 80: 354–383.

Stepanyan, L. S. 1964. [Geographical variation of the Rock Thrush *Monticola saxatilis* L.] *Sb. Trudy Zool. Mus. MGU* 9: 228–231. [In Russian.]

Stepanyan, L. S. 1990. [*Conspectus of the Ornithological Fauna of the USSR.*] Nauka, Moscow. [In Russian.]

Strickland, H. 1836. List of birds, noticed or obtained by him in Asia Minor, in the winter of 1835 and spring of 1836. *Proc. Zool. Soc. Lond.* 4: 97–102.

Sultana, J., Gauci, C. and Beaman, M. 1975. *A Guide to the Birds of Malta.* Malta Ornithological Society, Valletta.

Summers, R. W. 2002. Parrot Crossbills breeding in Abernethy Forest, Highland. *Brit. Birds* 95: 4–11.

Summers-Smith, D. 1988. *The Sparrows.* T. & A. D. Poyser, Calton.

Summers-Smith, D. 1990. Letter: The winter quarters of migratory Dead Sea Sparrows. *Orn. Soc. Middle East Bull.* 24: 25–26.

Summers-Smith, D. 1992. Letter: The Dead Sea Sparrow in Iraq: a correction. *Orn. Soc. Middle East Bull.* 28: 25.

Sushkin, P. B. 1925. Notes on systematics and distribution of certain Palaearctic birds. *Proc. Boston Soc. Nat. Hist.* 38: 1–55.

Suter, W. 1990. Comments on the breeding range of Armenian Gull, *Larus armenicus. Orn. Soc. Middle East Bull.* 25: 12–15.

Sutherland, W. J. and Brooks, D. J. 1981a. Autumn migration of raptors, storks, pelicans and spoonbills at Belen Pass, southern Turkey. *Sandgrouse* 3: 1–21.

Sutherland, W. J. and Brooks, D. J. 1981b. The visible migration of passerines and near-passerines at the Belen Pass, southern Turkey, in autumn. *Sandgrouse* 3: 87–90.

Sutton, R. W. W. and Gray, J. R. 1972. Summer birds of Karanfil Dağ. *Orn. Soc. Turkey Bird Rep.* 1968–69: 186–205.

Svensson, L. 1992. *Identification Guide to European Passerines.* Fourth edn. Privately published, Stockholm.

Svensson, L. 2001. Identification of Western and Eastern Olivaceous, Booted and Sykes's Warblers. *Birding World* 14: 192–220.

Svensson, L., Grant, P. J., Mullarney, K. and Zetterström, D. 1999. *Collins Bird Guide.* HarperCollins, London.

van Swelm, N. D. 2001. The eye-colour of Eurasian Dabchick. *Alula* 7: 24–26.

van Swelm, N. D. 2002. Geographical differences in young Kentish Plover. *Alula* 8: 68–69.

Székely, T. 1998. The significance of Tuzla Gölü, Çukurova Delta, for shorebirds: a concise progress report. *Turna* 1: 28–30.

Szentirmai, I. and Székely, T. 2004. Diurnal variation in nest material use by the Kentish Plover *Charadrius alexandrinus. Ibis* 146: 535–537.

Tavares, J., Sá Pessoa, P. and Brito e Abreu, F. 2000. The first breeding record of Bearded Tit *Panurus biarmicus* in Syria. *Sandgrouse* 22: 145–146.

Tchihatchef, P. de 1864. *Le Bosphore et Constantinople.* Paris.

Tebb, G. and Hamidan, N. 2002. The first Rustic Bunting *Emberiza rustica* in Jordan. *Sandgrouse* 24: 53–54.

Temple-Lang, J. and Cocker, M. 1991. A nest of Caucasian Black Grouse *Tetrao mlokosiewiczi* in Turkey. *Sandgrouse* 13: 102–103.

van Tets, G. F., Meredith, C. W., Fullagar, P. J. and Davidson, P. M. 1988. Osteological differences between *Sula* and *Morus*, and a description of an extinct new species of *Sula* from Lord Howe and Norfolk Islands, Tasman Sea. *Notornis* 35: 35–57.

Thèvenot, M., Vernon, R. & Bergier, P. 2003. *The Birds of Morocco: An Annotated Checklist.* British Ornithologists' Union & British Ornithologists' Club, Tring.

Thibault, J.-C. 1983. *Les Oiseaux de la Corse.* Parc Naturel Regional de la Corse, Ajacco.

Thibault, J.-C., Bretagnolle, V. and Rabouam, C. 1997. *Calonectris diomedea* Cory's Shearwater. *BWP Update* 1: 75–98.

Thiede, W. 1988. A Swift over Istanbul in October. *Orn. Soc. Middle East Bull.* 20: 25.

Ticehurst, C. B. 1938. *A Systematic Review of the Genus Phylloscopus.* British Museum (Natural History), London.

Ticehurst, C. B., Buxton, P. A. and Cheesman, R. E. 1923. The birds of Mesopotamia. *J. Bombay Nat. Hist. Soc.* 28: 210–250.

Tomkovich, P. S. and Lebedeva, E. A. 2002. International Action Plan for the Sociable Lapwing (*Chettusia gregaria*). BirdLife International, Wageningen.

Tove, M. H. 1993. Field separation of Ring-billed, Mew, Common and Kamchatka Gulls. *Birding* 25: 386–401.

Trewick, S. A. 1997. Flightlessness and phylogeny amongst endemic rails (Aves: Rallidae) of the New Zealand region. *Philos. Trans. Roy. Soc. Lond.* B 352: 429–446.

Tristram, H. B. 1882. Ornithological notes of a journey through Syria, Mesopotamia, and southern Armenia in 1881. *Ibis* 4 (6): 402–419.

Tristram, H. B. 1889. *Catalogue of a Collection of Birds Belonging to H. B. Tristram.* Durham.

Tucker, G. M. and Evans, M. I. 1997. *Habitats for Birds in Europe: A Conservation Strategy for the Wider Environment.* BirdLife International, Cambridge.

Tucker, G. and Heath, M. F. 1994. *Birds in Europe: Their Conservation Status.* BirdLife International (Conservation Series No. 3), Cambridge.

Turan, L. 2006. Distribution and calendar of swift species in Turkey. *Sandgrouse* 28: 69–73.

Turan, N. 1990. *Kuşlar.* Ankara.

Tye, A. 1994. A description of the Middle Eastern black morph of Mourning Wheatear *Oenanthe lugens* from museum specimens. *Sandgrouse* 16: 28–31.

Ulfstrand, S. 1977. Plumage and size variation in Swedish Common Buzzards. *Zool. Scripta* 6: 69–75.

Uhlig, R. 1992. Gebäudebruten der Felsenschwalbe, *Ptyonoprogne rupestris* (Scopoli, 1769), in Bulgarien. *Beitr. Vogelkd.* 5–6: 304–318.

Uhlig, R., Uhlig, S., Kirwan, G. M. and Barış, Y. S. 1996. Winter status and distribution of Alpine Accentor *Prunella collaris* in Turkey. *Sandgrouse* 18 (2): 65.

Uhlig, S. and Uhlig, R. 1999a. Winterbeobachtungen an Steinsperlingen *Petronia petronia* in der Türkei. *Orn. Mitt.* 51: 245–246.

Uhlig, S. and Uhlig, R. 1999b. Winterbeobachtungen an Ohrenlerchen in der Türkei. *Orn. Mitt.* 51: 285–286.

Ullman, M. 1991. First record of Little Bunting *Emberiza pusilla* in Jordan. *Sandgrouse* 13: 53–54.

Ullman, M. 1992. Possible hybrid Pied × Black-eared Wheatear *Oenanthe pleschanka* × *O. hispanica* in eastern Turkey. *Sandgrouse* 14: 58–59.

Ullman, M. 2003. Separation of Western and Eastern Black-eared Wheatear. *Dutch Birding* 25: 77–97.

Urquhart, E. and Bowley, A. 2002. *Stonechats. A Guide to the Genus Saxicola.* Christopher Helm, London.

Vaassen, E. W. A. M. 1998. A February record of Demoiselle Crane *Anthropoides virgo* in Turkey. *Sandgrouse* 20: 144.

Vaassen, E. W. A. M. 2000. Some notes on urban raptors and their habitats in Turkey. *Intern. Hawkwatcher* 2: 34–38.

Vaassen, E. W. A. M. 2001. Status of the Griffon Vulture, *Gyps fulvus* in the central Taurus region, southern Turkey – a population estimate. *Vulture News* 44: 3–17.

Vader, W. J. M. 1965. Bird observations by the "Dutch Biological Expedition Turkey 1959". *Ardea* 53: 172–204.

Väli, Ü. 2006. Mitochondrial DNA sequences support species status for the Indian Spotted Eagle *Aquila hastata. Bull. Brit. Orn. Club* 126: 238–242.

Vauk, G. 1973. Ergebnisse einer ornithologischen Arbeitsreise an den Beyşehir-Gölü (SW-Anatolien) im April/Mai 1964. *Beitr. Vogelkd.* 19: 225–260.

Vaurie, C. 1949a. Notes on some Asiatic finches. *Amer. Mus. Novit.* 1424: 1–63.

Vaurie, C. 1949b. Notes on the bird genus *Oenanthe* in Persia, Afghanistan, and India. *Amer. Mus. Novit.* 1425: 1–47.

Vaurie, C. 1954a. Systematic notes on Palearctic birds. No. 10 Sylviinae: the genera *Cettia, Hippolais,* and *Locustella. Amer. Mus. Novit.* 1691: 1–9.

Vaurie, C. 1954b. Systematic notes on Palearctic birds. No. 11 Sylviinae: the genus *Sylvia. Amer. Mus. Novit.* 1691: 1–17.

Vaurie, C. 1955. Systematic notes on Palearctic birds. No. 14 Turdinae: the genera *Erithacus, Luscinia, Tarsiger, Phoenicurus, Monticola, Erythropygia,* and *Oenanthe. Amer. Mus. Novit.* 1731: 1–30.

Vaurie, C. 1956a. Systematic notes on Palearctic birds. No. 23 Fringillidae: the genera *Emberiza, Calcarius,* and *Plectrophenax. Amer. Mus. Novit.* 1805: 1–27.

Vaurie, C. 1956b. Systematic notes on Palearctic birds. No. 24 Ploceidae: the genera *Passer, Petronia,* and *Montifringilla. Amer. Mus. Novit.* 1814: 1–27.

Vaurie, C. 1957a. Systematic notes on Palearctic birds. No. 25 Motacillidae: the genus *Motacilla. Amer. Mus. Novit.* 1832: 1–16.

Vaurie, C. 1957b. Systematic notes on Palearctic birds. No. 26 Paridae: the *Parus caeruleus* complex. *Amer. Mus. Novit.* 1833: 1–15.

Vaurie, C. 1957c. Systematic notes on Palearctic birds. No. 27 Paridae: the genera *Parus* and *Sylviparus. Amer. Mus. Novit.* 1852: 1–43.

Vaurie, C. 1957d. Systematic notes on Palearctic birds. No. 30 The Certhiidae. *Amer. Mus. Novit.* 1855: 1–14.

Vaurie, C. 1958. Systematic notes on Palearctic birds. No. 32 Oriolidae, Dicruridae, Bombycillidae, Pycnonotidae, Nectariniidae, and Zosteropidae. *Amer. Mus. Novit.* 1869: 1–28.

Vaurie, C. 1959a. *The Birds of the Palearctic Fauna. Passeriformes.* H. F. & G. Witherby, London.
Vaurie, C. 1959b. Systematic notes on Palearctic birds. No. 35 Picidae: the genus *Dendrocopos* (Part 1). *Amer. Mus. Novit.* 1946: 1–29.
Vaurie, C. 1959c. Systematic notes on Palearctic birds. No. 37 Picidae: the subfamilies Jynginae and Picumninae. *Amer. Mus. Novit.* 1963: 1–16.
Vaurie, C. 1960a. Systematic notes on Palearctic birds. No. 40 Caprimulgidae. *Amer. Mus. Novit.* 1997: 1–14.
Vaurie, C. 1960b. Systematic notes on Palearctic birds. No. 41 Strigidae: the genus *Bubo*. *Amer. Mus. Novit.* 2000: 1–31.
Vaurie, C. 1960c. Systematic notes on Palearctic birds. No. 42 Strigidae: the genus *Athene*. *Amer. Mus. Novit.* 2015: 1–21.
Vaurie, C. 1960d. Systematic notes on Palearctic birds. No. 43 Strigidae: the genera *Otus*, *Aegolius*, *Ninox*, and *Tyto*. *Amer. Mus. Novit.* 2021: 1–21.
Vaurie, C. 1963. Systematic notes on Palearctic birds. No. 51 A review of *Burhinus oedicnemus*. *Amer. Mus. Novit.* 2000: 1–13.
Vaurie, C. 1965. *The Birds of the Palearctic Fauna. Non-Passeriformes.* H. F. & G. Witherby, London.
van der Ven, J. A. 1982. Birds in eastern Turkey 3. Unpubl. report.
van der Ven, J. A. and Gheyselinck, G. F. 1980. Birds in eastern Turkey 1. Unpubl. report.
van der Ven, J. A. and Gheyselinck, G. F. 1982. Birds in eastern Turkey 2. Unpubl. report.
Vielliard, J. 1968. Résultats ornithologiques d'une mission à travers la Turquie. *Istanbul Fen. Fak. Mecm.* 33: 67–170.
Vittery, A. 2000. Comments on the buzzards of the Central Plateau, Turkey. *Sandgrouse* 22: 135–136.
van der Vliet, R. E., Kennerley, P. R. and Small, B. J. 2001. Identification of Two-barred, Greenish, Bright-green and Arctic Warblers. *Dutch Birding* 23: 175–191.
Voous, K. H. 1977. *List of Recent Holarctic Bird Species.* British Ornithologists' Union, London.
Voous, K. H. 1992. Reflections on the genus in ornithology. *Bull. Brit. Orn. Club Centenary* Suppl. 112A: 261–266.
Voous, K. H., Hens, P. A. and van Marle, J. G. 1948. The distinguishing characters of the Steppe Buzzard. *Brit. Birds* 41: 77–82.
Votier, S. C. and Riddington, R. 2005. Tail pattern of Reed Warblers. *Brit. Birds* 98: 99–100.
Vural, M., Ekim, T., Duman, H., Aytaç, Z. and Adıgüzel, N. 1999. *Vegetation and Flora of the Konya Closed Basin.* Biodiversity Programme Report No. 12. Doğal Hayatı Koruma Derneği, Istanbul.
Wadley, N. J. P. 1951. Notes on the birds of central Anatolia. *Ibis* 93: 63–89.
Wahby, A. 1930. Les oiseaux de la région de Stamboul. *Bull. Orn. Romand* 1: 68–70.
Walters, M. 1988. Letter. *Orn. Soc. Middle East Bull.* 20: 44.
Warncke, K. 1964–65. Beitrag zur Vogelwelt der Türkei. *Vogelwelt* 85: 161–174; 86: 1–19.
Warncke, K. 1966. Ergänzung zu meinen vogelkundlichen Beobachtungen in der Türkei. *Vogelwelt* 87: 188–189.
Warncke, K. 1968. Nachwinterliche Beobachtungen an türkischen Brutvögeln. *Vogelwelt* 2: 89–102.
Warncke, K. 1970. Beitrag zur Vogelwelt des Zentralanatolischen Beckens. *Vogelwelt* 91: 176–184.
Warncke, K. 1971. The Flamingo (*Phoenicopterus ruber*) – a new breeding bird for Turkey. *Orn. Soc. Turkey Bull.* 7: 4–5.
Warncke, K. 1972. Beitrag zur Vogelwelt der Türkei im Beriech der Südgrenze. *Vogelwelt* 93: 23–27.
Watson, G. E. 1961a. Aegean bird notes 1. Descriptions of new subspecies from Turkey. *Postilla* 52: 1–15.
Watson, G. E. 1961b. Aegean bird notes including two breeding records new to Europe. *J. Orn.* 102: 301–307.
Watson, G. E. 1962a. Three sibling species of *Alectoris* partridge. *Ibis* 104: 353–367.
Watson, G. E. 1962b. A re-evaluation and redescription of a difficult Asia Minor *Phylloscopus*. *Ibis* 104: 347–352.
Watson, G. E. 1962c. A revision of Balkan, Aegean and Anatolian Crested Larks. *Bull. Brit. Orn. Club* 82: 9–18.
Welch, G. and Welch, H. 1998a. Results of a survey of wintering waterbirds along the Turkish Black Sea coast – 16 January to 7 February 1997. *Turna* 1: 16–23.
Welch, G. and Welch, H. 1998b. Breeding bird survey of Uluabat Gölü Ramsar site – 15 May to 20 June 1998. Unpubl. report.
Welch, G. R. and Welch, H. J. 2000. Important forest areas of Turkey: bird surveys, May and June 2000. Report to Doğal Hayatı Koruma Derneği, Istanbul.

Welch, G. and Welch, H. 2004. Spectacled Warbler *Sylvia conspicillata*: a new breeding species for Turkey. *Sandgrouse* 26: 55–58.

Welch, G. and Welch, H. 2006. Fan-tailed Raven *Corvus rhipidurus* – first record for Turkey. *Sandgrouse* 28: 68–69.

Welch, H. J. (ed.) 2004. *GAP Biodiversity Research Project 2001–2003 – Final Report.* Doğal Hayatı Koruma Derneği, Istanbul.

Weigold, H. 1912–1913. Ein Monat Ornithologie in den Wüsten und Kulturoasen Nordwest-Mesopotamiens und Innersyriens. *J. Orn.* 60: 249–297, 365–410; 61: 1–40.

Wilson, M. G. and Korovin, V. A. 2003. Oriental Turtle Dove breeding in the Western Palearctic. *Brit. Birds* 96: 234–241.

van Winden, A., Mostert, K., Ruiters, P., Siki, M. and de Waard, H. 1989. *Waders and Waterfowl in Spring 1988 at Eber Gölü, Turkey.* WIWO (Report No. 28), Zeist.

van der Winden, J. and van den Berk, V. 1997. Breeding birds of the Taurus foothills bordering the Göksu Delta, south Turkey. *Sandgrouse* 19: 45–50.

Wink, M., Clouet, M., Goar, J.-L. and Barrau, C. 2004. Sequence variation in the cytochrome b gene of subspecies of Golden Eagle. *Alauda* 72: 153–157.

Wink, M. and Sauer-Gürth, H. 2004. Phylogenetic relationships in diurnal raptors based on nucleotide sequences of mitochondrial and nuclear marker genes. *In*: Chancellor, R. and Meyburg, B.-U. (eds.) *Raptors Worldwide.* World Working Group on Birds of Prey & Owls, Berlin.

Wink, M., Sauer-Gürth, H. and Gwinner, E. 2002. Evolutionary relationships of stonechats and related species inferred from mitochondrial-DNA sequences and genomic fingerprinting. *Brit. Birds* 95: 349–354.

Winkel, E. and de Weerd, E. 2007. Barnacle Goose in Golestan, Iran, in January 2007. *Dutch Birding* 29: 91–92.

Wirth, H. 1996. Breeding record of the Imperial Eagle, *Aquila heliaca*, in Turkey. *Zool. Middle East* 12: 47–52.

Witt, H. H. 1976. Beobachtungen zum Vorkommen und zur Brut einiger Non-Passeres an der türkischen Südküste bei Silifke. *Vogelwelt* 97: 139–145.

Wolters, H. E. 1975–82. *Die Vogelarten der Erde.* Paul Parey, Berlin.

Worfolk, T. 2000. Identification of red-backed, isabelline and brown shrikes. *Dutch Birding* 22: 323–362.

Wynn, R. B. 2003. Further developments in 'Black Brant' identification, including the effects of body moult on the wintering grounds. *Brit. Birds* 96: 297–301.

Yarar, M. 1995. Aktaş Gölü: a new pelican breeding site on the Turkish–Georgian border. *Orn. Soc. Middle East Bull.* 35: 46–48.

Yarar, M. & Magnin, G. 1993. Bird sites in Turkey requiring research in 1993. *Orn. Soc. Middle East Bull.* 30: 10–17.

Yésou, P. 2002. Trends in systematics. Systematics of *Larus argentatus–cachinnans–fuscus* complex revisited. *Dutch Birding* 24: 271–298.

Yésou, P., Flohart, G. and Murdoch, D. 2007. First record of Basra Reed Warbler *Acrocephalus griseldis* for Syria. *Sandgrouse* 29: 214–215.

Yésou, P. & Paterson, A. M. 1999. Puffin yelkouan et Puffin des Baléares: une ou deux espèces? *Ornithos* 6: 20–31.

Yésou, P., Paterson, A. M., Mackrill, E. J. and Bourne, W. R. P. 1990. Plumage variation and identification of the 'Yelkouan Shearwater'. *Brit. Birds* 83: 299–319.

Yoğurtuoğlu, E. 2006. The first Black-throated Thrush *Turdus ruficollis atrogularis* in Turkey. *Sandgrouse* 28: 172–173.

Yöntem, O. 2007. An observation of Brown Fish Owl *Ketupa zeylonensis* in Turkey. *Sandgrouse* 29: 94–95.

Zalles, J. I. and Bildstein, K. L. (eds.) 2000. *Raptor Watch: A Global Directory of Raptor Migration Sites.* BirdLife International (Conservation Series No. 9), Cambridge, UK.

Zazanashvili, N. 1999. On the Colchic vegetation. *In*: Klotzli, F. and Walther, G. R (eds.) *Conference on Recent Shifts in Vegetation Boundaries of Deciduous Forests, Especially Due to General Global Warming.* Basel, Switzerland: 181–197.

Zeydanlı, U. Z., Welch, H. J., Welch, G. R., Altıntaş, M. and Domaç, A. 2005. *GAP Analysis and Priority Conservation Area Selection for Mediterranean Turkey: preliminary technical report.* Turkish Foundation for Nature Conservation (WWF-Turkey), Istanbul.

Zink, R. M., Rohwer, S., Andreev, A. V. and Dittmann, D. L. 1995. Trans-Beringia comparisons of mitochondrial DNA differentiation in birds. *Condor* 97: 639–649.

Zohary, M. 1971. Phytogeographical foundation of the Middle East. *In*: Davis, P. H., Harper, P. C. and Hedge, I. C. (eds.) *Plant Life of South-West Asia*. Royal Botanic Garden Edinburgh, Edinburgh.

Zohary, M. 1973. *Geobotanical Foundations of the Middle East*. Vols. 1–2. Gustav Fischer Verlag, Stuttgart.

Zusi, R. L. 1978. The interorbital septum in cardueline finches. *Bull. Brit. Orn. Club* 98: 5–10.

Index of scientific names

Index of Turkish common names

Index of English common names